TILINGS AND PATTERNS

Second Edition

Branko Grünbaum

University of Washington

G. C. Shephard

University of East Anglia

Dover Publications

Garden City, New York

To Zdenka and Helen

Bibliographical Note

This Dover edition, first published in 2016, is a slightly corrected, unabridged republication of the work originally published in 1987 by W. H. Freeman and Company, New York. For this Dover edition, the authors have provided a new Preface and a new Appendix.

International Standard Book Number

ISBN-13: 978-0-486-46981-2
ISBN-10: 0-486-46981-6

Manufactured in the United States of America
46981612
www.doverpublications.com

CONTENTS

PREFACE TO THE DOVER EDITION

Tilings and Patterns was originally published almost thirty years ago. It met with great acclaim in published reviews and in many personal communications by readers. According to the *Mathematical Reviews* database, the book was referenced in selected mathematical journals more than 200 times since the year 2000. Thus, the book remained popular despite the original publisher's decision to discontinue printing after the first run was exhausted in 1998. It is our hope that this edition will prove as popular as the original. In order to maintain consistency between the editions, the original has been left unchanged except for minor corrections of typographical errors. However, in view of the fact that much new material has appeared in the literature during the last two decades, we felt that the reader should have some help in locating these new developments. For that purpose we have added an Appendix that gives references for many of the additional results, keyed to appropriate locations in the book. By no means can these updates do justice to all the work that deserves to be mentioned—but it is the most we could do short of rewriting the whole book. We hope that the reader will understand our inability to engage in the enormous effort this would require.

Branko Grünbaum
University of Washington
Seattle, Washington

PREFACE

The increasing popularity of puzzles and games based on the interplay of shapes and positions illustrates the attraction that geometric forms and their relations hold for many people. Since space and motion are of great importance to most animals and all primitive peoples, this attraction probably has deep evolutionary roots. In addition, modern society increasingly demands that we cope with intricately shaped objects, follow the mazes of huge buildings and transportation systems, and orient ourselves on land and water and in the air (not to mention the problems of dealing with submarine or interplanetary space).

It is curious that almost all aspects of geometry relevant to the "man in the street" are ignored by our educational systems. Geometry has been almost squeezed out of school and university syllabuses, and what little remains is rarely of any use to people who wish to apply geometric ideas in their work—engineers, scientists, architects, artists, and the like. There are two causes of this state of affairs. At high-school level it has long been traditional to use geometry as a vehicle for teaching logical reasoning and the deductive method, without much regard for the geometric content. At the research level geometry has become no more than a specialized branch of algebra or analysis.

In each case the essence of the subject—its visual appeal—has been completely submerged in technicalities and abstractions. Other branches of mathematics have also suffered from this neglect of geometry, since many basic ideas in topology, analysis, measure theory and so on owe their origin to geometric intuition, and the workers in those fields would probably profit from some knowledge of geometry.

Each of us has long deplored this situation, and so, when we met at a conference in 1975, we decided to write a book on "visual geometry"—a rigorous book, but one that would also encourage geometric appreciation by the use of "pure" geometric reasoning. To cover many branches of the subject turned out to be too big a task. Eventually we decided on the study of tilings and patterns as a first step in carrying out the program we had in mind.

Perhaps our biggest surprise when we started collecting material for the present work was that so little about tilings and patterns is known. We thought, naïvely as it turned out, that the two millenia of development of plane geometry would leave little room for new ideas. Not only were we unable to find anywhere a meaningful definition of a pattern, but we also discovered that some of the most exciting developments is this area (such as the phenomenon of aperiodicity for tilings) are not more than twenty years old.

We have written this book with three main groups of readers in mind—students, professional mathematicians and non-mathematicians whose interests include patterns and shapes (such as artists, architects, crystallographers and others). Each of these groups will find some parts of the book more interesting than others, but we hope that we have struck a reasonable balance in both content and form. The whole book should be accessible to any reader attracted to geometry, regardless of (or even in spite of) his previous mathematical education. Our presentation is somewhat informal (but nevertheless precise) and we have used diagrams and illustrations in profusion. We have invested great effort to make these attractive since we feel that any visual delight the reader may experience will stimulate his interest and comprehension. In this respect we are rejecting the current fashion that geometry must be abstract if it is to be regarded as advanced mathematics, and that dispenses entirely with diagrams. To consider geometry without drawings as a worthy goal (as is frequently advocated by self-proclaimed "sophisticates") seems to us as silly as to extol the virtues of soundless music (suggesting, of course, that the sign of true musical maturity is to appreciate it by merely looking at the printed score!).

While assembling the material we realized that the field abounds with challenging but tractable problems, and that many previous publications contain serious errors. We have published a series of research papers on some of our results, but many more are being presented here for the first time. (The research was supported by grants from the National Science Foundation, and by a Fellowship from the John Simon Guggenheim Memorial Foundation.) Most of the chapters have a final section which contains historical and bibliographic references; we believe that the unattributed results are new. Though we cannot claim that these surveys of the literature are complete, we hope that we have omitted no important publications.

ORGANIZATION OF THE BOOK

The book falls naturally into two parts. The first, up to and including Chapter 7, can be used as the text for a geometry course at the undergraduate level—a course of a somewhat unconventional nature but nevertheless instructive and interesting. Some of the chapters are independent and there is no need to read them in sequence. The first few sections of Chapter 1 are fundamental, however. Chapter 2 deals mostly with tilings in which the tiles are regular polygons. It is an enlargement and updating of a paper that appeared in the *Mathematics Magazine*; we hope that it will appeal to students at high-school level and stimulate their interest in geometric ideas. The general theory of tilings is presented in Chapters 3 and 4; these chapters are rather more technical than the rest of the book, and so parts of them may be omitted if desired. In Chapter 5 we begin our discussion of the theory of patterns; this continues in Chapter 7. Chapter 6 deals with tilings which have sufficiently many symmetries so that all the tiles (or all the edges, or all the vertices) play the same role; such tilings are of special importance in certain applications.

The second part (Chapters 8 to 12) presents detailed surveys of various aspects of the subjects of patterns and tilings. These include colored patterns and groups of color symmetry, tilings by polygons, tilings in which the tiles are unusual in a topological sense, as well as a detailed and self-contained account of the intriguing topic of aperiodic tilings. Throughout, we have attempted to show the vitality of the topic by formulating open problems and stressing the gaps in present knowledge. In several sections the concepts under discussion have been illustrated by examples taken from nonmathematical disciplines and supplemented by references to nonmathematical literature.

To facilitate the use of the book as a textbook we have included many exercises. These are of three kinds. The unstarred ones are routine and do little more than test the understanding of the preceding text. The exercises marked in the margin by one star are more difficult, and those marked with two stars are—so far as we are aware—completely unsolved. Many of the latter could be taken as starting points for research projects.

ACKNOWLEDGMENTS

It would be impossible to thank individually all our colleagues and correspondents who have contributed to this book by their comments, criticisms and suggestions, as well as by sending us unpublished material. We are grateful to all of them, even if we cannot list all their names here. However, we must make special mention of Professor Marjorie Senechal, who used an early version of the text as the basis of a course at Smith College, and Professor Guy Valette, who ran a colloquium on tilings at the Free University of Brussels.

Almost all diagrams representing periodic tilings were drawn by computer, the majority at the University of Washington using a program specially written by Dr. Matthew Hackman, and the others at the University of East Anglia using GINO-F software. The rest of the diagrams were either drawn by hand or obtained from the sources acknowledged in the captions or in the text. We are also greatly indebted to Dr. Peter Renz, and to the staff at W. H. Freeman and Company for the great efforts they expended in getting this book into print.

We shall be glad to hear from readers who have information to add to what is presented here, or have experience in using the book in a college course. We should also be pleased if our attention is drawn to any errors in the text. For the next few years we shall endeavor to update the material presented here by preparing brief summaries of new results that appear in the literature. Readers interested in obtaining copies of these summaries should write to one of the authors.

Branko Grünbaum
University of Washington GN–50
Seattle, WA 98195, USA
grunbaum@math.washington.edu.

G. C. Shephard
University of East Anglia
Norwich NR4 7TJ, England

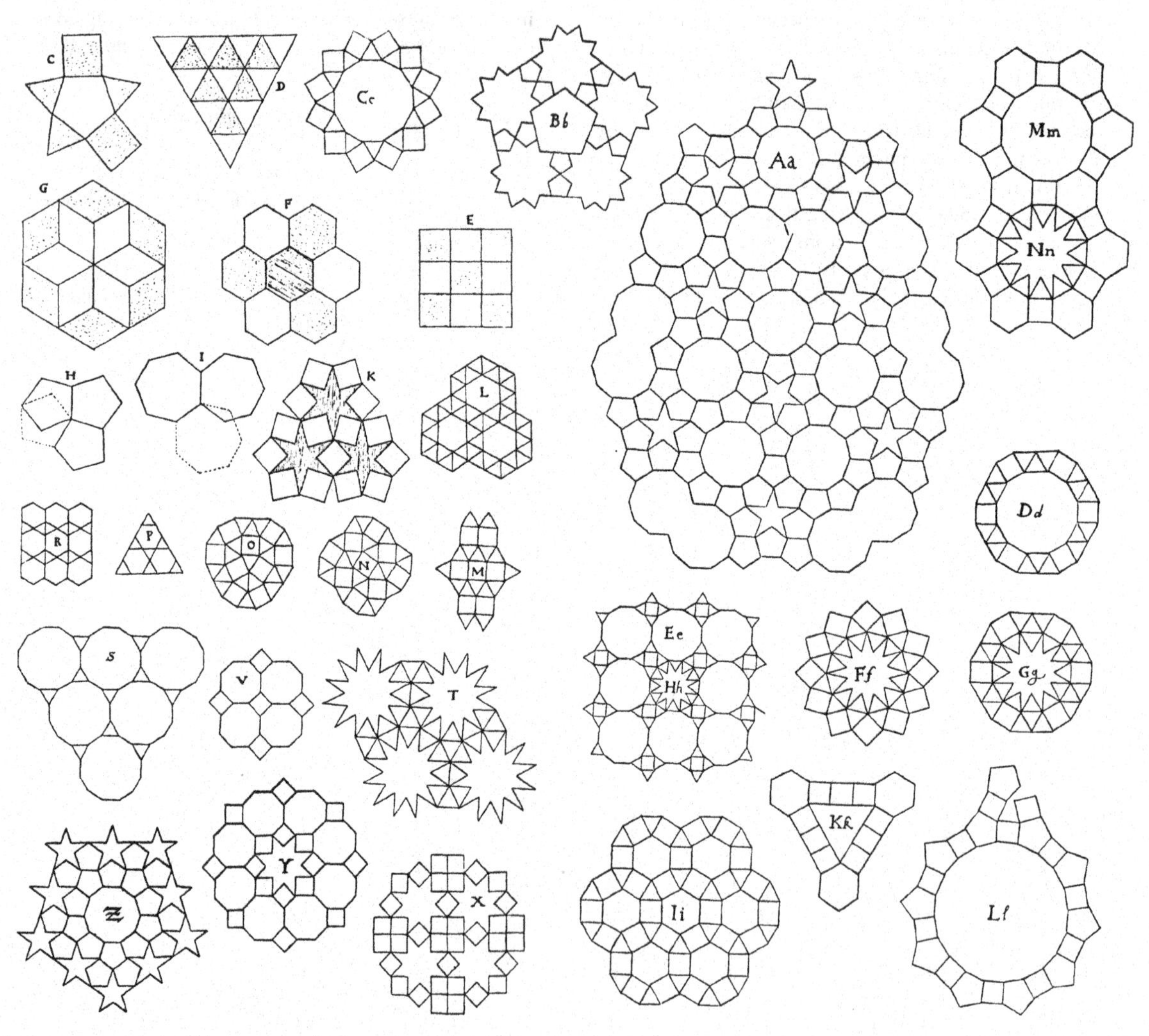

Johannes Kepler (1580–1630) is well known for his pioneering work in astronomy. He also made fundamental contributions to the theory of tilings, and some of his ideas have still not been fully investigated. We reproduce here drawings of tilings from his book *Harmonice Mundi* (volume 2) published in 1619. The tiling marked Aa can be extended over the whole plane as shown in Figure 2.0.1 and described in Section 2.5.

INTRODUCTION

The art of tiling must have originated very early in the history of civilization. As soon as man began to build he would use stones to cover the floors and the walls of his houses, and as soon as he started to select the shapes and colors of his stones to make a pleasing design, he could be said to have begun tiling in the sense that we shall use the word. Patterns—that is, designs repeating some motif in a more or less systematic manner—must have originated in a similar manner, and their history is probably as old as, if not older than, that of tilings.

Figure 1
A Roman mosaic (from Germany, see Parlasca [1959, plate 80]) showing busts of philosophers and stylized flowers in an intricate geometric setting.

Every known human society has made use of tilings and patterns in some form or another. In doing so, however, various cultures seem to have emphasized different aspects. For example, the Romans and some other Mediterranean peoples were largely concerned with portraying human beings and natural scenes in intricate mosaics; an example is shown in Figure 1. On the other hand, the artistic impulse of the Moors and Arabs frequently manifested itself in the use of a few shapes and colors of tiles to build up complex geometric designs. Famous examples are to be seen in the Alhambra at Granada in Spain (Figures 2 and 3), and especially impressive tilings can be found on many Moslem religious buildings (Figure 4). But throughout history, whatever kind of tiling was in favor, its art and technology always attracted skillful artisans, inventive practitioners and magnanimous patrons.

Similarly, artifacts of all cultures abound with patterns which are often surprising in their intricacy and complexity. In any situation in which some sort of decoration is required or desirable—on fabrics, carpets, baskets, utensils, weapons, wall covers—patterns of some sort are inevitably used. Moreover, the production of many objects inevitably leads to patterns; the prime examples here are such activities as weaving and basketry. Examples from various cultures are shown in

Figure 2
A view in the Alhambra showing the wealth of tilings used by its Moorish builders (Saladin [1926, plate 21]).

Figure 3
Details of several tilings from the Alhambra. Such tilings are widely known, in part due to sketches made in 1936 by the Dutch artist M. C. Escher (see Escher [1971, plates 83, 84]).

Figure 4
Persian tiling of the early 14th century, on the tomb of the prophet Daniel in Susa, Iran (Berendsen [1967, p. 25]).

Figures 5, 6 and 7. It is not surprising that similar motifs (representational, stylized, or abstract) often appear in patterns from widely separated times and places. Less expected by many people is the fact that the overall composition of patterns from different sources is also often very similar. We shall see later (especially in Chapter 5) that this is not accidental, but results from basic geometric considerations.

Over the ages, very many different shapes of tiles have been used, leading to some blurring of the distinction

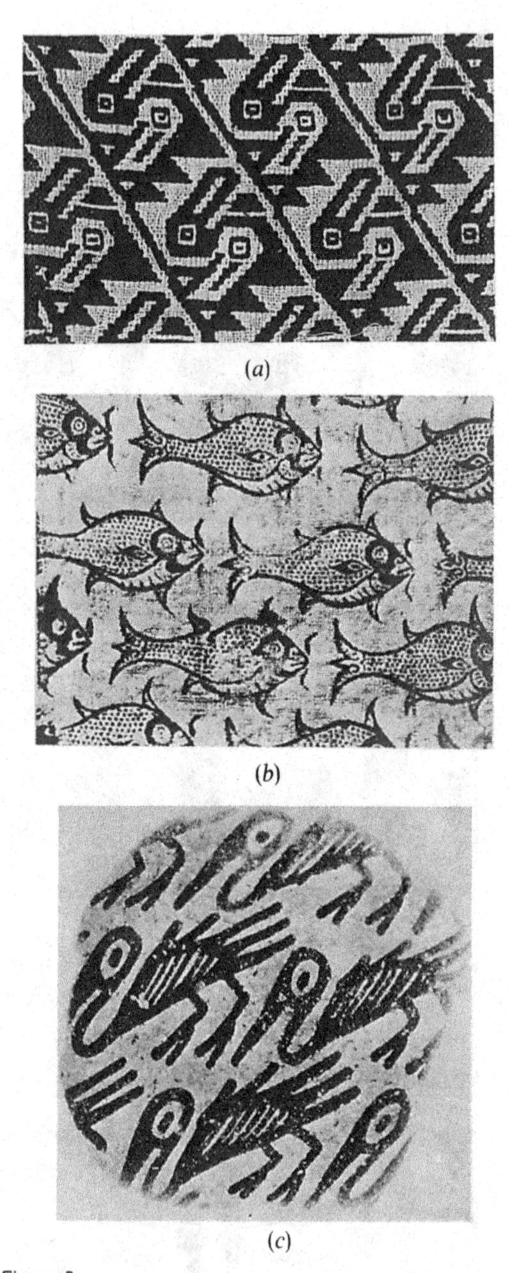

(*a*)

(*b*)

(*c*)

Figure 5
Patterns with animal motifs and very similar layouts.
(*a*) A pre-Inca fabric from Peru (Izumi [1964]). (*b*) A printed fabric from India (Pfister [1938]). (*c*) A pottery fragment from Arizona (Haury [1937]).

Figure 6
Stylized "snake design" patterns. (*a*) Wooden shield from the Bismarck Archipelago in Melanesia (Bodrogi [1959]). (*b*) A wall decoration from Ajanta, India (Smart [1971]).

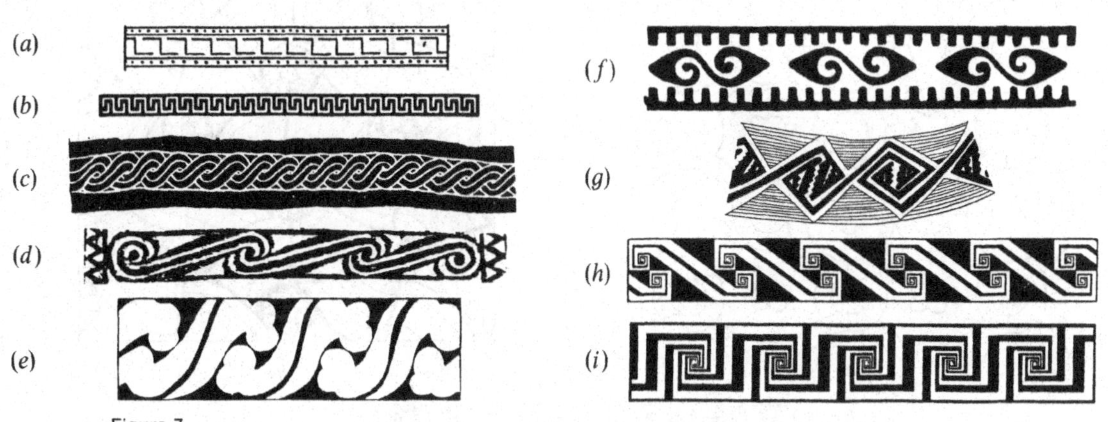

Figure 7
Similar strip decorations from widely separated sources: (*a*) Greece (Christie [1929]), (*b*) Tunis (Revault [1973]), (*c*) Egypt (Gerspach [1890]), (*d*) New Guinea (Lewis [1925]), (*e*) Japan (Menten [1975]), (*f*) Mexico (Enciso [1947]), (*g*) Arizona (Washburn [1977]), (*h*) and (*i*) relatively modern (Edwards [1932]).

between tilings and patterns. Some of the richness of possibilities is illustrated in Figures 8, 9 and 10, which show a number of tilings employed in antiquity and the Middle Ages for pavements, floors, walls and ceilings. Nowadays, by contrast, square and rectangular tiles seem to be almost universal; even so, there are many interesting possibilities (see the examples of brickwork patterns in Figure 11). Occasionally more imagination has been used in choosing the shape of tiles, and in Figure 12 we show a number of noteworthy modern street tilings. For an account of a street tiling with pentagonal tiles common in Cairo (Egypt), see Macmillan [1979].*

In addition to varying the shapes of the tiles, from the earliest times the aesthetic appeal of tilings has been increased by the use of colors or markings, as can be

* An author's name followed by a year in square brackets indicates a reference to the literature listed at the end of the book.

Figure 8
Some ancient tilings. (*a*) and (*b*) Central Asia before the Mongolian invasion in 1259. (*c*) Ely cathedral, England, c. 1325. (*d*) and (*e*) Breda castle, Netherlands, c. 1500. (Part (*c*) is from Wight [1975], the others from Berendsen [1967].)

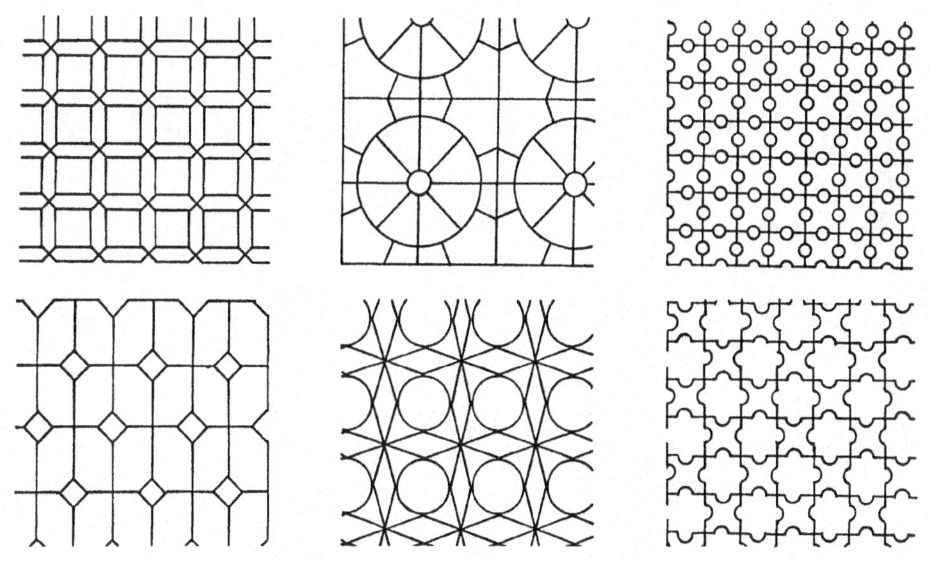

Figure 9
Ancient tilings from Portugal, 15th century (after Simões [1969]).

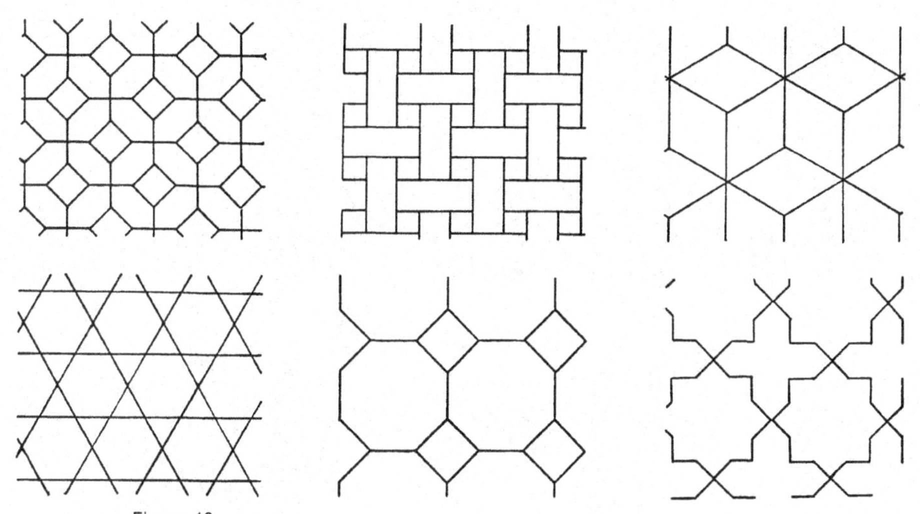

Figure 10
Various ancient tilings frequently encountered in Europe and the Middle East (see Berendsen [1967], Mars [1925], Parlasca [1959], Kiss [1973], and others).

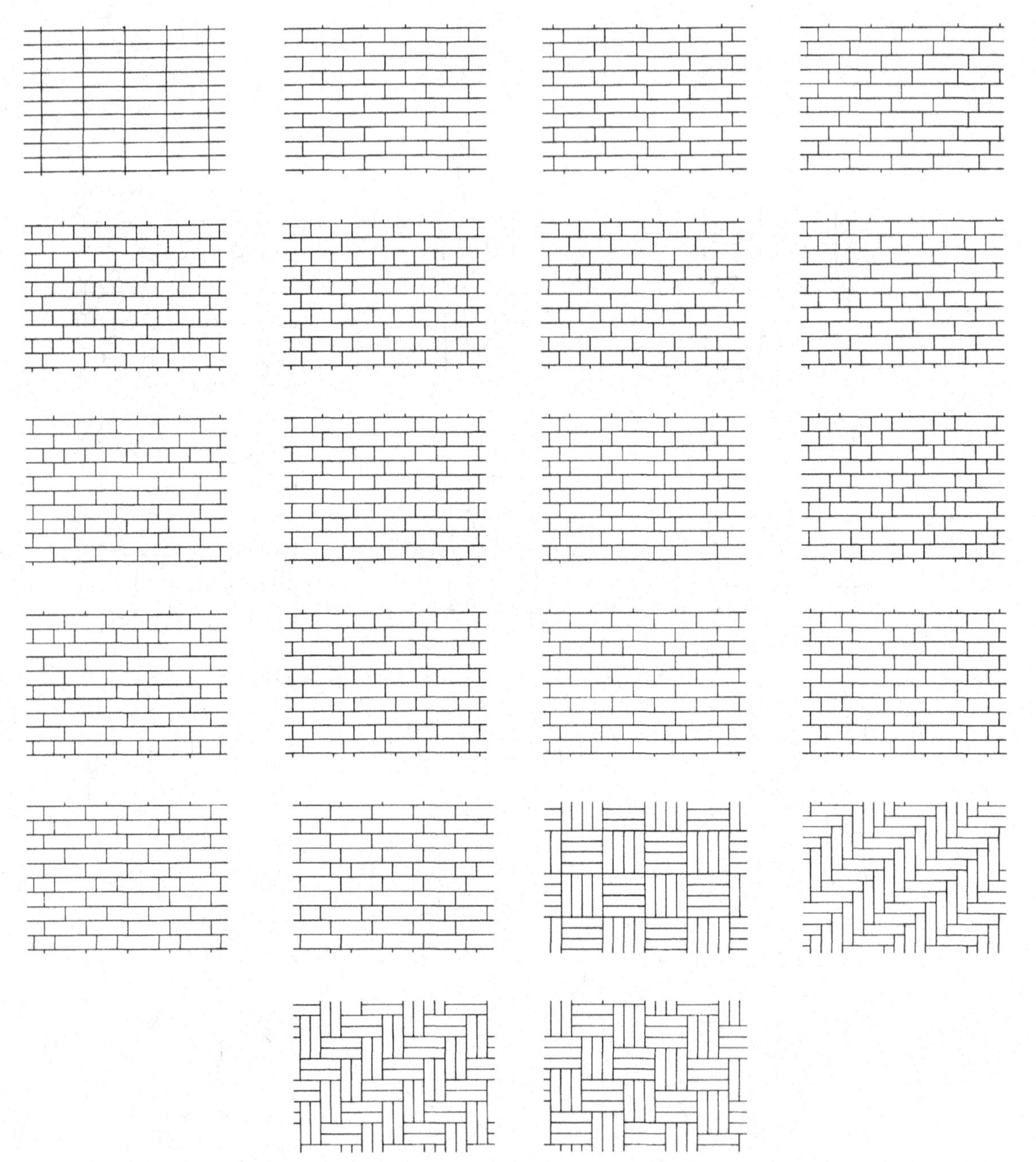

Figure 11
Diagrams illustrating the great variety of "tilings" possible with standard bricks. The number of possibilities can be vastly increased by using bricks of various colors and textures (see Plummer [1950]).

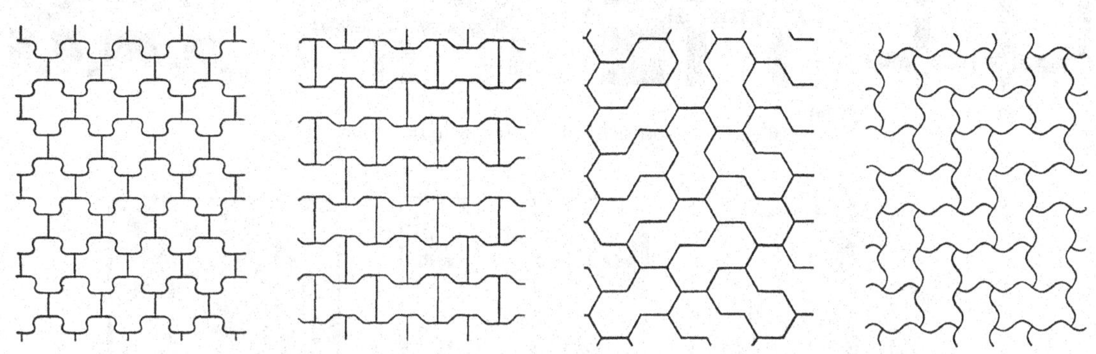

Figure 12
Some unusual street tilings seen by the authors in Europe and North America.

Figure 13
Some colored tilings executed in marble, from 13th century Roman churches (Christie [1929, plate XLV]).

seen by the examples in Figures 13, 14, 15 and 16. Such ideas, as well as other relations between patterns and tilings, turn out to be of mathematical importance as well, and will be mentioned frequently throughout the book.

So far, the tilings we have mentioned consist of tiles made of stone, ceramic or similar material, which fit together without appreciable gaps to cover the plane or some other surface. Clearly the same considerations apply to the fitting together of other physical objects such as items of wood, card, plastic or sheet metal. These have applications in modern engineering. For example, the manufacture of components by stamping them out of sheet metal is clearly most economical if the shapes can be fitted together without gaps—that is, if they form a tiling. This aspect of the subject is

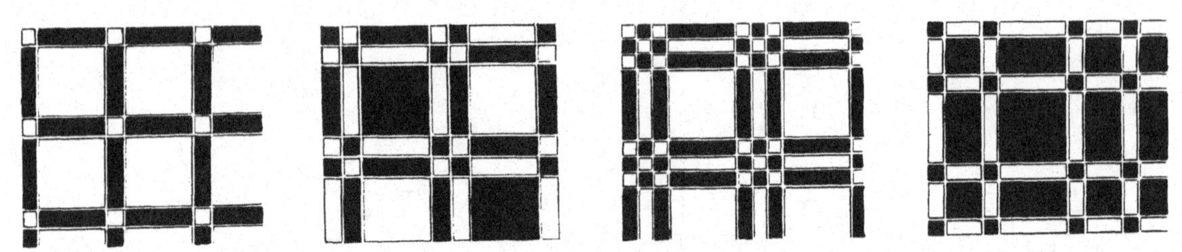

Figure 14
Colored tilings from Portugal, 15th and 16th centuries (Simões [1969]).

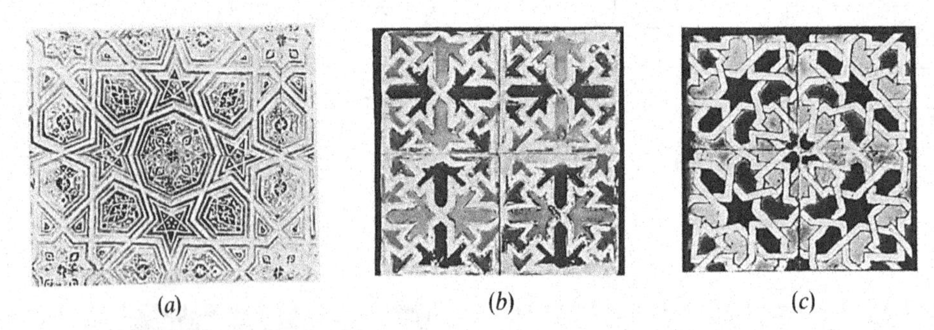

(*a*) (*b*) (*c*)

Figure 15
Marked tilings with intricate designs: (*a*) from Turkey (1391) (Öz [1954, plate 14]). (*b*) and (*c*) from Spain (15th and 16th centuries) (Berendsen [1967, p. 72]).

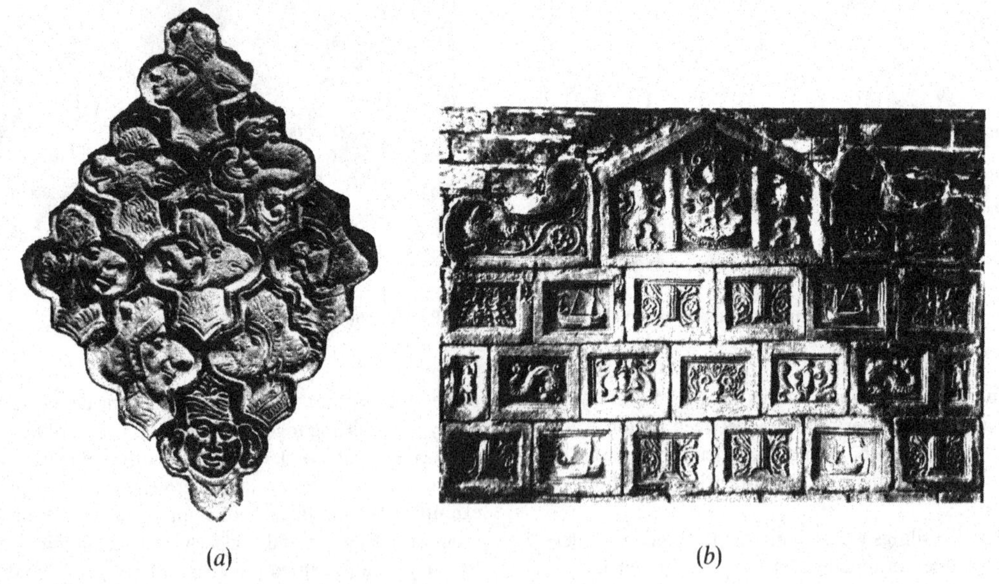

(*a*) (*b*)

Figure 16
Tilings marked in relief: (*a*) from France (12th century); (*b*) from the Netherlands (1660) (Berendsen [1967, pp. 62 and 71]).

discussed in the book by Heesch & Kienzle [1963], and we may mention that we possess a pamphlet on tiling printed in Germany during the Second World War and marked "Top Secret"! More recent work in this direction can be found in Stoyan [1975] and Fesenko [1981].

From many points of view, an extension of these ideas is both natural and useful. We shall consider a tiling to be any "partition" of the plane into "regions" (the tiles) regardless of whether or not this partition is realized (or can be realized) by physical objects. Thus a piece of engineering paper indicates a "tiling" with square "tiles". It is in this sense that the hundreds of drawings on the following pages are to be interpreted. This point of view is not new; since antiquity artists have drawn designs which may be considered as tilings in this sense. Examples from Mongolia are reproduced in Figure 17, and more recent examples can be found in the work of the Dutch artist M. C. Escher; related to it are the illustrations shown in Figure 18 and in Figure 6.0.1.* In this more general sense there are tilings in profusion all around us, not only man-made but also occurring in nature (the cells in the skin of an onion, the design of a spider's web, the honeycomb of a bee, patches of dried-up mud, and so on; see Figure 19).

Tilings of this kind, often referred to as "random tessellations" or "random networks", have been extensively studied because of their relevance to applications in science and engineering. We mention, as examples, metallurgy and geology (crystal structure in thin sheets of materials), biology (cell arrangements in skins and membranes of animals and plants) and communication theory (image enhancement and coding). Random tessellations can be constructed in various ways according to the proposed application, but a frequent method starts from a random distribution of points in the plane from which one constructs a Dirichlet tiling (see Section 5.4).

The *art* of designing tilings and patterns is clearly extremely old and well developed. By contrast, the *science* of tilings and patterns, by which we mean the study of their mathematical properties, is comparatively recent and many parts of the subject remain unexplored.

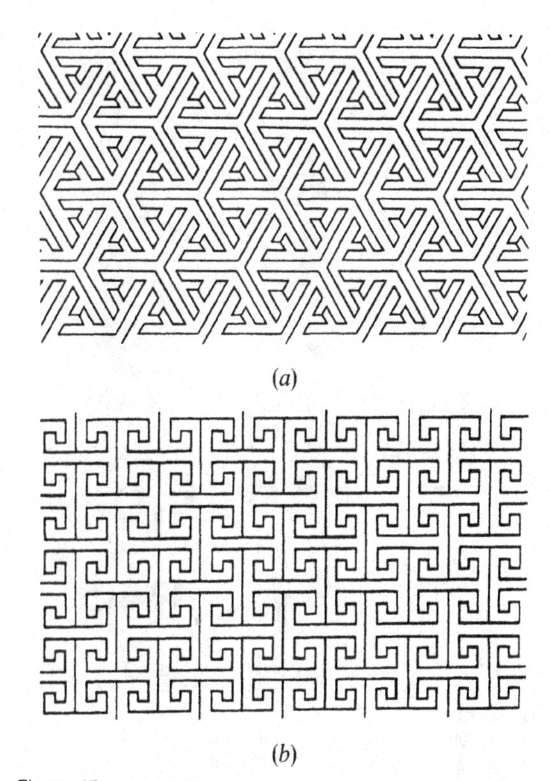

Figure 17
Mongolian designs which may be interpreted as tilings (Rinčen [1969, plates 50, 51]); similar designs are known from many other cultures.

Indeed, we have been unable to find in the literature a satisfactory mathematical treatment of the subjects of tilings and patterns. Many books deal with patterns, but they rarely do more than give numerous examples,

---

* Except for the figures in this Introduction, we refer to figures by a three-part symbol p.q.r, by which we indicate figure r in section q of chapter p. Completely analogous are the indications of statements, tables, and so on. In later chapters, figures in the Introduction will be prefixed with the letter I.

Figure 18
One of the less well known Escher sketches which may be interpreted as a tiling (Escher [1971, plate 92]).

accompanied by vague and often irrelevant commentary (see, for example, the "Publisher's Note" to the 1973 edition of Bourgoin [1879], or the "Vocabulary" of Proctor [1969, p. 9]), leading to completely useless definitions.

In the past, there have been many attempts to describe, systematize and devise notations for various types of tilings and patterns. However, without a mathematical basis such attempts could not succeed, despite the sometimes prodigious efforts devoted to them. The lack of success can be seen from such works as Bourgoin [1873], [1880], [1883], [1901], Day [1903], Dresser [1862], Edwards [1932], Meyer [1888], Schauermann [1892], Wersin [1953]. (Schauermann [1892] is noteworthy mainly for the extraordinary extent to which it plagiarizes Bourgoin [1883].) The books of Nobel prize-winning chemist Ostwald are also of interest. After developing a system for the classification of colors which is still in use today, Ostwald attempted to classify repeating patterns on the basis of their symmetries. And although he was aware of all the ways in which a plane

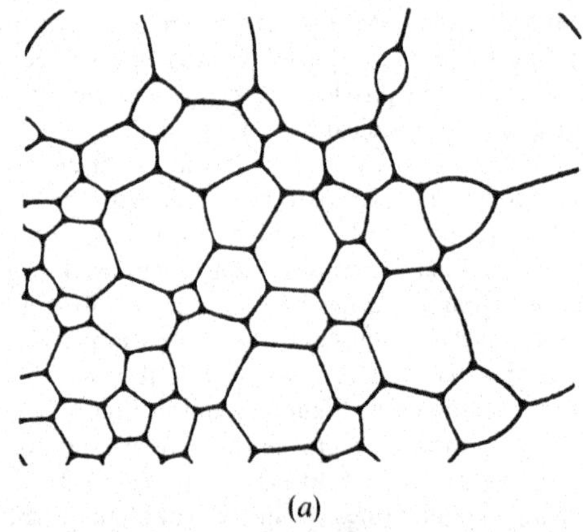

(*a*)

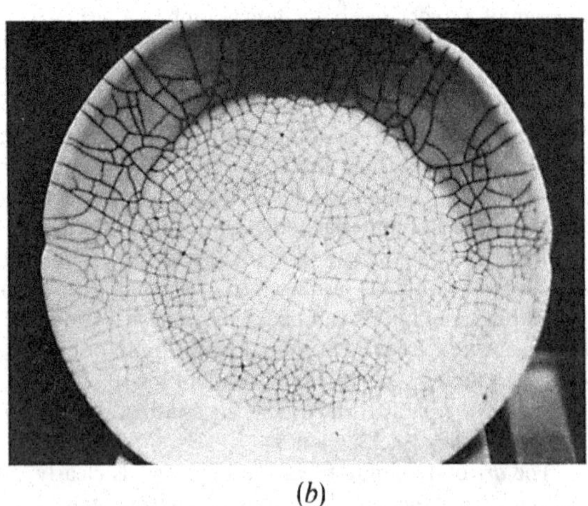

(*b*)

Figure 19
Two examples of tilings which occur in nature. (*a*) Froth of soap bubbles blown between parallel glass plates (Smith [1965, Figure 6.7]). (*b*) Craze lines on a glazed ceramic surface.

pattern can be symmetric (see Section 1.3 below), in the system expounded in Ostwald [1922a], [1922b], he inexplicably disregarded glide reflections and so missed the symmetry groups *pg* and *pgg* (see Section 1.4).

We shall attempt to remedy this situation, to a small extent, by developing a theory of one special type of pattern called a "discrete pattern" in Chapter 5, and for some other types in later parts of the book. Formal definitions and criteria for classification will be given in Chapters 5 and 7. We are under no misapprehension about the restricted nature of these investigations—far more remains to be done, and we hope that some of our readers will be tempted to continue research on this topic.

Some aspects of the geometry of tilings have been studied in the literature. Kepler wrote on tilings in his book *Harmonice Mundi* in 1619, and plates from this work appear as our frontispiece. But his investigations were largely forgotten till the early years of this century, as explained in detail in Section 2.10. In fact, one edition of Kepler's *Harmonice Mundi* omits the diagrams completely! A few papers on tilings were published towards the end of the nineteenth century, so for practical purposes the mathematical theory of tilings can be regarded as about a century old.

In studying the literature on tilings and patterns we were impressed by the fact that many contributions were made by engineers, architects, crystallographers and others with little or no formal mathematical training, as well as by distinguished mathematicians. But we were also impressed by the many errors which have been made, often through the use of badly formulated definitions and lack of rigor. In spite of great efforts at avoiding them, this book probably also contains errors, but we hope that they will be only superficial and not due to any serious mathematical deficiency.

We must remark on the fact that in our diagrams we are usually able to present only a small portion of a pattern or tiling, whereas we are almost always interested in one that extends over the whole plane. So our diagrams are to be understood as showing a randomly picked part of the tiling or pattern, and we implicitly assume that the reader will be able to understand from this how it could be continued indefinitely in all directions. As will be seen in Chapter 10, this assumption is much less natural or justified than one might be inclined to think. We must also appeal to the reader to forgive the occasional imperfection in our drawings. For example, in the illustrations taken from physical objects it is frequently necessary to transfer the pattern from a curved surface (bowl, shield and so on) to the plane, and this inevitably leads to distortion.

To us, tilings and patterns are fascinating subjects. In them visual appeal and ease of understanding combine with possibilities of applying both informally creative and systematic approaches. They concern topics and ideas equally useful in art, practical design, crystallographic investigation or mathematical research. This book will have served its purpose if it leads the reader to share some of our aesthetic and intellectual enjoyment of tilings and patterns. In addition, we hope that the book will be of help to many readers in their varied activities.

While preparing this Introduction, we have consulted many non-mathematical books which appeared to be relevant to the topic. We were surprised at their profusion, and frequently delighted by their contents. Probably the reader will derive a similar satisfaction if he should take the trouble to look them up in a library. He may even be inspired to pursue the artistic or the mathematical aspects of the subject. The following list contains the names of several such books, arranged alphabetically in groups dealing with related topics. These books are listed since they were accessible to us; there are many others of no less interest, and the reader is urged to peruse them as a source of information supplementing the mathematical treatment we shall give.

Tiles and brickwork: Aslanapa [1965], Berendsen [1967], Du Ry [1970], Foerster [1961], Godard [1962], Mars [1925], Öz [1954], Plummer [1950], Simões [1969], Wight [1975].

Mosaics and paintings: Budde [1969], Calvert [1904], Fisher [1971], Fořtová-Šámalová & Vilímková [1963], Germain [1969], Gerster [1968], Hoag [1977], Ipşiroglu [1971], Kiss [1973], Parlasca [1959], Saladin [1926], Sijelmassi [1974], Smart [1971], Speltz [1921].

Textiles and quilts: Clouzot [1931], Edmonds [1976],

Gerspach [1890], Harcourt [1975], Ickis [1959], Izumi [1964], Lantz [1976], Larsen & Gull [1977], Pfister [1938], Revault [1973], Safford & Bishop [1972], Trowell [1970].

Patterns and designs: Albarn et al. [1974], Audsley & Audsley [1882], Bain [1951], Bodrogi [1959], Bossert [1928], Bourgoin [1879], Christie [1929], Dye [1937], [1981], Edwards [1932], Enciso [1947], Ernst [1976], Escher [1971], [1982], [1983], Guiart [1963], Haury [1937], Hornung [1932], [1975], Jones [1856], Kelley & Mowll [1912], Lewis [1925], Menten [1975], Ōuchi [1977], Padwick & Walker [1977], Proctor [1969], Rinčen [1969], Shaffer [1979], Shepard [1948], [1956], Washburn [1977].

Patterns in nature and science: Bager [1966], Erdtman [1952], Kepes [1965], [1966], March & Steadman [1971], Prochnow [1934], Senechal & Fleck [1977], Šubnikov & Kopcik [1972].

1

BASIC NOTIONS

Figure 1.0.1

A monohedral tiling is one whose tiles are all of the same shape. Some of the most remarkable of such tilings are those that appear to be of spiral form. This figure shows one such tiling; its prototile admits many other forms of spiral, such as those shown in Figures 1.2.3(*b*) and (*c*). The first "spiral tiling" was discovered by Voderberg [1936] and is shown in Figure 9.5.1.

1

BASIC NOTIONS

Before we can apply mathematical techniques to tilings and patterns, it is necessary to define some of the terminology that will be required. We begin by making precise what we mean by a "tiling"* and also describing the restrictions that it is convenient to place on the sets that are eligible as "tiles". Later sections of the chapter deal with the important ideas of "symmetry" and "symmetry group".

The corresponding treatment of patterns will be postponed until Chapter 5. By then the theory of tilings will have been developed sufficiently for us to describe meaningful interrelations between the two subjects and to use each as a tool in the treatment of the other.

1.1 TILES, TILINGS AND PATCHES

A *plane tiling* $\mathscr{T}$ is a countable family of closed sets $\mathscr{T} = \{T_1, T_2, \ldots\}$ which cover the plane without gaps or overlaps. More explicitly, the union of the sets $T_1, T_2, \ldots$ (which are known as the *tiles* of $\mathscr{T}$) is to be the whole plane, and the interiors of the sets T_i are to be pairwise disjoint. By "the plane" we mean the familiar Euclidean plane of elementary geometry. It would be out of place to discuss its properties in detail here—for our purposes it will be sufficient that the concepts of straightness, length, angle, area and congruence are defined in it and are known to the reader.

Either of these two conditions can be imposed separately: A family of sets in the plane which has no overlaps is called a *packing*, and a family of sets which covers the plane with no gaps is called a *covering*. There is an extensive literature on both these topics. Most of it relates to estimates for the "density" of the closest possible packing, or the thinnest possible covering, by sets of some specified type. This question is, of course, trivial in the case of tilings which, being both packings *and* coverings, automatically have density 1. For additional information and references on packing and covering see L. Fejes Tóth [1953], G. Fejes Tóth [1983].

For most purposes the above definition of a tiling is too general. The countability condition excludes families in which every tile has zero area (such as points or line segments) but nevertheless the definition admits tilings in which some tiles have bizarre shapes and properties. A few examples, which do not by any means exhaust the possibilities, appear in Figure 1.1.1. This figure shows

- (*a*) a tile which is *not* connected, that is to say, it consists of two or more separate pieces;
- (*b*) a tile which is not *simply connected*, which means that it encloses at least one hole;
- (*c*), (*d*) tiles each of which becomes disconnected upon the deletion of a suitable finite set of points;
- (*e*), (*f*) tiles which are made up, in part, of line seg-

* In mathematical literature, the words *tessellation*, *paving*, *mosaic* and *parquetting* are used synonymously or with similar meanings. The German words for tiling are *Pflasterung*, *Felderung*, *Teilung*, *Parkettierung* and *Zerlegung*. The French words are *pavage*, *carrelage* and *dallage*. The Russian words are паркетаж, разбиение and замощение.

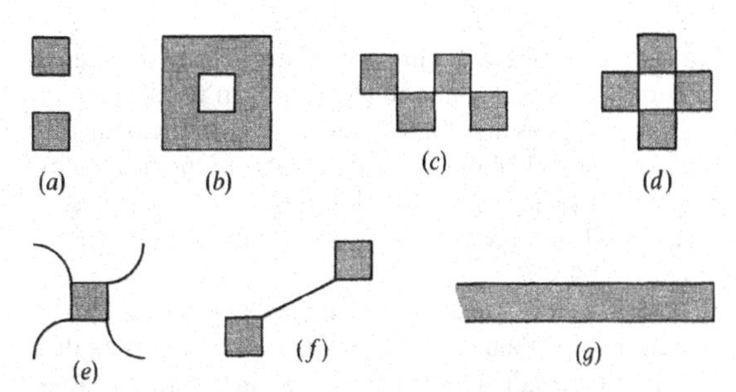

Figure 1.1.1
Some examples of sets that are not topological disks. In our treatment, we usually (but not always) exclude such sets as tiles.

ments, arcs of curves, or other figures of zero area;

(*g*) a tile which is *unbounded* in the sense that it cannot be enclosed by a finite circle, however large. In this particular example the tile consists of a part of an infinite strip. Note however, that it is a closed set since it contains all its boundary points.

Sometimes the study of tilings with such tiles is interesting and instructive, see Chapter 12, and we do not wish to remove them completely from our discussion. Nevertheless it is frequently convenient to restrict attention to "well-behaved" tiles and tilings. This question will be examined more carefully in Chapter 3. Here it suffices to explain the most natural restriction, which eliminates all the tiles of Figure 1.1.1: the requirement that each tile is a (closed) *topological disk*. The precise meaning will be given in Chapter 4 but for the present we may consider a topological disk to be any set whose boundary is a *single simple closed curve*. By this we mean a curve whose ends join up to form a "loop" and which has no crossings or branches. Thus in Figures 1.1.1(*a*) and (*b*) the boundary is not a *single* curve, in Figures 1.1.1(*c*), (*d*), (*e*) and (*f*) it is not a *simple* curve because it has branches or crossing points, and in Figure 1.1.1(*g*) it is not a *closed* curve.

Throughout this chapter and Chapter 2 we shall assume that each circular disk in the plane meets only a finite number of tiles. Tilings which do not satisfy this restriction will be considered in Chapter 3.

From the definition of a tiling we see that the intersection of any finite set of tiles of $\mathcal{T}$ (containing at least two distinct tiles) necessarily has zero area. For most of the tilings considered here, such an intersection may be empty or may consist of a set of isolated points and arcs. In these cases the points will be called *vertices* of the tiling and the arcs will be called *edges*, see Figure 1.1.2. (For examples of tilings in which other kinds of intersections occur see Chapter 3.) In particular, if the tiles are topological disks, then the simple closed curve which

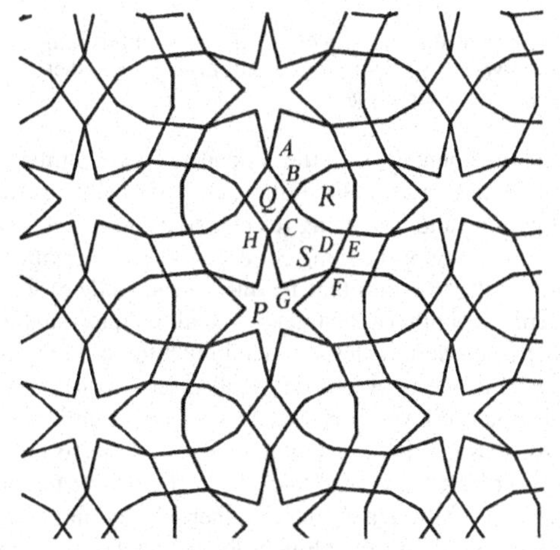

Figure 1.1.2
This tiling is adapted from one designed by Holiday [1978]. The tiles are polygons of four different shapes: a 12-gon (or star-hexagon) P, a parallelogram Q, a convex 9-gon R and a nonconvex 7-gon S. The line-segments BC, CD and DE are sides of the tile R (and also of S) but are not edges of the tiling. Their union, the polygonal arc $BCDE$, is an edge of the tiling and also of the tiles R and S, but is not a side of any tile. Similarly, FGH is an edge of the tiling and of the tiles P and S, while FG and GH are sides of these two tiles. The straight line-segment ABC is neither a side of a tile nor an edge of the tiling. The points A, B, E, F and H are vertices of the tiling, but C, D and G are not. Points B, C, D and E are four of the nine corners of the tile R. Point G is a corner of the tile S and also of the tile P but it is not a vertex of the tiling. Because every vertex is an end-point of four edges, the tiling is 4-valent.

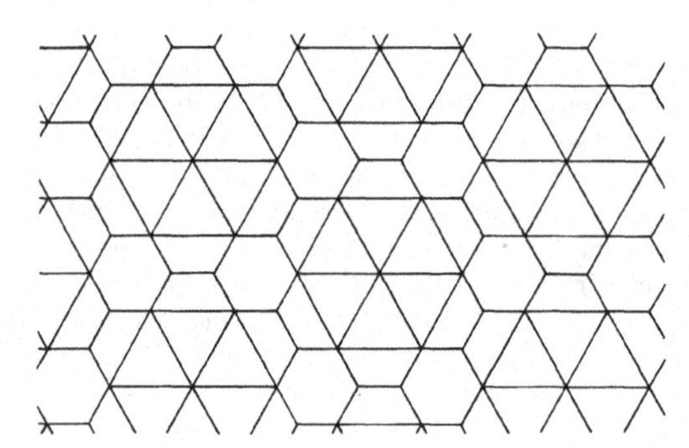

Figure 1.1.3
A tiling from Williams [1977]. Each vertex of this tiling is either 3-valent, 5-valent, or 6-valent. Each tile is either a 3-gon, a 4-gon, or a 6-gon.

forms the boundary of a tile is divided into a number of parts by the vertices of the tiling, each arc being an edge of the tiling and also being referred to as an *edge of the tile*. Each edge of the tiling coincides, of course, with the edges of the two tiles that lie one on each side of it. Except when otherwise stated, we shall be interested only in the case where each tile has a finite number of vertices. An edge connects two vertices (called the *endpoints* of the edge) and every vertex is the endpoint of a number of edges. This number is called the *valence* of the vertex, and is at least three, see Figure 1.1.3. If every vertex of a tiling $\mathscr{T}$ is of the same valence j, then we say that $\mathscr{T}$ is a *j-valent* tiling; thus the tiling in Figure 1.1.2 is 4-valent. The vertices, edges and tiles of a tiling are called its *elements*.

In the next chapter, and also later, we shall consider the special case of tilings in which each tile is a polygon. In the usual terminology a polygon has vertices and edges, but it is clear that it would lead to confusion if we used the same words for elements of the tiling to which the polygons belong. For this reason we shall refer instead to the *corners* and *sides* of a polygon. A polygon with k corners (and therefore k sides) will be called a k-gon. Of course, the corners and sides may coincide with the vertices and edges of the tiling—in which case we say the tiling by polygons is *edge-to-edge*—or this may not be so. It is clear that the tiling in Figure 1.1.3 is edge-to-edge, but the one in Figure 1.1.2 is not. The two tilings in Figure I.17 are also not edge-to-edge; in the second, the tiles are 28-gons and there are 6 vertices on the boundary of each tile, only 4 of which are corners of the tile. Each tile has 28 sides and 6 edges, but only two of the edges are sides. For a further explanation of these terms see Figure 1.1.4.

Two tiles are called *adjacent* if they have an edge in common, and then each is called an *adjacent* of the other. Two tiles are called *neighbors* if their intersection is non-empty. Thus, referring again to Figure 1.1.4, T has edges in common with T_2, T_3, T_4, T_5 and T_7, which are therefore its adjacents. In addition to these five tiles, T_1 and T_6 are also neighbors of T. Similarly, we shall say that two distinct edges are *adjacent* if they have a common endpoint. The word *incident* is used to denote the relation of a tile to each of its edges or vertices, and also of an edge to each of its endpoints. The relation of incidence is considered to be *symmetric*, by which we mean that if a tile is incident with a vertex then that vertex is also said to be incident with the tile; similarly in the other cases.

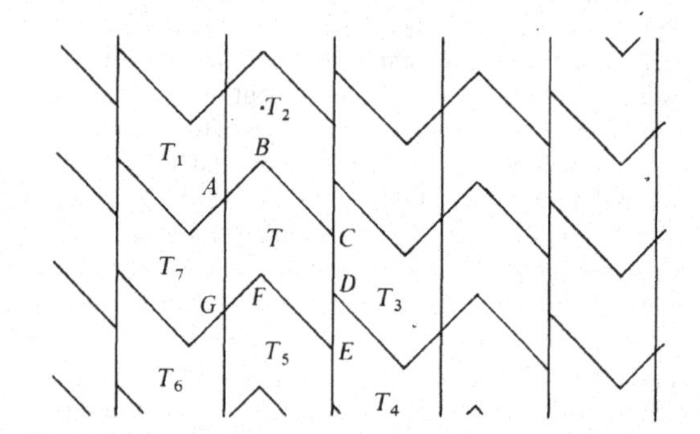

Figure 1.1.4
This diagram illustrates the differences between corners and vertices, sides and edges, neighbors and adjacents in the case of a polygonal tiling. The points A, B, C, E, F and G are corners of the tile T; but A, C, D, E and G are vertices of the tiling. The line-segments AB, BC, CE, EF, FG and GA are sides of T, while AC, CD, DE, EG and GA are edges of the tiling. The tiles T_2, T_3, T_4, T_5 and T_7 are adjacents (and neighbors) of T, whereas tiles T_1 and T_6 are neighbors (but not adjacents) of T.

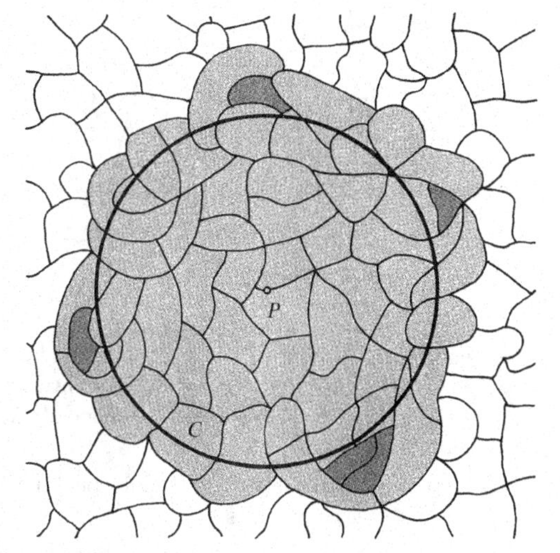

Figure 1.1.5
The construction of the patch $\mathscr{A}(D,\mathscr{T})$ generated by a set D. In this example, D is a closed circular disk with center P and with the heavily drawn circle C as boundary. We take all tiles that intersect D (*light shading*) and adjoin further tiles (*dark shading*) to make the union simply connected. The patch $\mathscr{A}(D,\mathscr{T})$ is the set of all these shaded tiles. The same patch is generated by the circle C; the point P alone generates a patch consisting of two tiles.

We shall say that two tilings $\mathscr{T}_1$ and $\mathscr{T}_2$ are *congruent* if $\mathscr{T}_1$ may be made to coincide with $\mathscr{T}_2$ by a rigid motion of the plane, possibly including reflection. More frequently we shall be interested in "equal tilings"; we shall say that two tilings are *equal* or *the same* if one of them can be changed in scale (magnified or contracted equally throughout the plane) so as to be congruent to the other. Equivalently, we can define this concept by saying that two tilings are equal if there is a similarity transformation of the plane that maps one of the tilings onto the other. For example, if $\mathscr{T}_1$ and $\mathscr{T}_2$ are two tilings, each by congruent regular hexagons (see Figure 1.2.1(*c*)), then $\mathscr{T}_1$ and $\mathscr{T}_2$ are necessarily equal; for this reason we are justified in referring to *the* tiling by regular hexagons. But $\mathscr{T}_1$ and $\mathscr{T}_2$ are congruent only if the hexagons in $\mathscr{T}_1$ are of the same size as those in $\mathscr{T}_2$. The two tilings labelled ($3^4.6$) in Figure 2.1.5 are congruent—but a reflection is needed to make one coincide with the other. In contrast, none of the tilings in Figure I.11 are equal, even though in several cases their rectangular tiles are all congruent.

In the following discussion we sometimes refer to a *patch* of tiles in a given tiling. By this we mean a finite number of tiles of the tiling with the property that their union is a topological disk—in other words, is connected and simply connected, and cannot be disconnected by the deletion of a single point.

There is a standard procedure for constructing patches of tiles, which we shall frequently employ. We take a connected set D—for example a point, or an arc, or a region of the plane. To obtain the *patch $\mathscr{A}(D,\mathscr{T})$ generated by* D, first consider the set $\mathscr{S}$ of all tiles of $\mathscr{T}$ that meet D. Assuming the tiles are topological disks, this set $\mathscr{S}$ may or may not form a patch (see Figure 1.1.5). The union of the tiles in $\mathscr{S}$ will clearly be connected but it may fail to be simply connected. Adjoining to $\mathscr{S}$ just enough tiles to fill up the "holes" (illustrated by the darker tiles in Figure 1.1.5), we obtain the required patch $\mathscr{A}(D,\mathscr{T})$. In Section 2.8 and in other places we shall discuss patches of tiles in some detail and draw attention to many interesting problems concerning them. Sometimes we shall abbreviate or modify the notation and write, for example, $\mathscr{A}(D)$ or $\mathscr{A}$.

EXERCISES 1.1

1. Find the number of adjacents, and the number of neighbors, of each tile in Figures 1.3.8, 1.3.9 and 4.1.3.
2. Find the number of vertices and number of corners of each type of tile in Figures 1.3.8 and 1.3.9. Indicate some corners of the tiles that are not vertices of the tiling, and some vertices of the tiling that lie on the boundary of a tile but are not corners of it. For each vertex determine its valence.

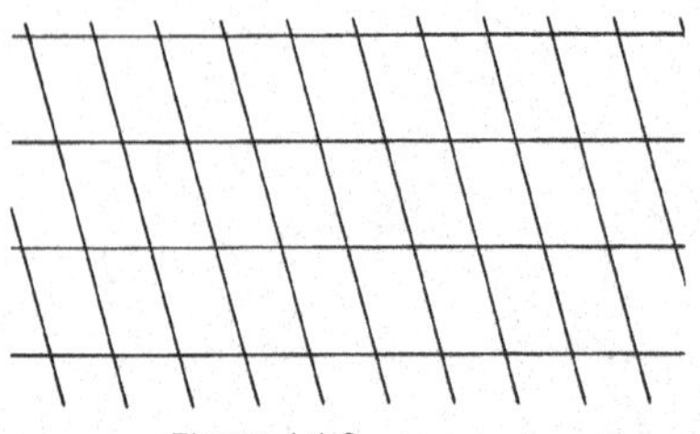

Figure 1.1.6
A parallelogram tiling.

3. For each of the sets of Figure 1.1.1, with one exception, construct a tiling in which each tile is congruent to the chosen set. Find the exceptional set for which such a construction is impossible.
4. For the tilings of Figures 1.2.7(*a*) and (*b*), determine the number of tiles, edges and vertices of the patches generated by the indicated square and rectangle (considered as closed sets). Verify that the numbers you obtain satisfy equation (3.1.4).
5. For a tiling $\mathscr{T}$ and a connected set D, let $\mathscr{A}_1(D) = \mathscr{A}(D)$ be the patch generated by D, and for $n \geq 1$ let $\mathscr{A}_n(D)$ be the patch generated by the union of the tiles in $\mathscr{A}_{n-1}(D)$. For each $n \geq 1$ and each of the tilings in Figure 1.2.1 determine the number of vertices, edges, and tiles of the tiling that belong to the patch $\mathscr{A}_n(D)$ when D is (*a*) a tile; (*b*) a vertex; (*c*) an edge.
6. Justify the assertion made at the beginning of Section 1.1 that the countability condition in the definition of a tiling precludes the possibility that every tile in a tiling can have zero area (for example, points or line-segments).
7. Find a tiling of the plane by (*a*) congruent triangles, (*b*) congruent quandrangles, (*c*) congruent pentagons, (*d*) congruent hexagons, (*e*) congruent heptagons and (*f*) congruent octagons. Can all the tiles be chosen to be convex polygons?

*8. Show that in the previous exercise any triangle or any quadrangle can be used in parts (*a*) and (*b*), but that in (*c*), (*d*), (*e*) and (*f*) the possibility of constructing the tiling depends essentially on the shape of the polygon that is chosen.

9. Show that every vertex in a tiling by polygonal tiles is a corner of at least two tiles.
10. Figure 1.1.6 shows an edge-to-edge tiling by congruent parallelograms with corresponding sides parallel. Such a tiling is possible for any parallelogram P and is called *the parallelogram tiling* for P.
    (*a*) Show that for every parallelogram P there exists an uncountable infinity of distinct tilings with all tiles congruent to P and corresponding sides parallel.
    *(*b*) Show that if P is not a rectangle there exists an uncountable infinity of edge-to-edge tilings with all tiles congruent to P.
    (*c*) For each $k \geq 5$ find a tiling by congruent parallelograms in which at least one vertex has valence k.
    (*d*) For some $k \geq 5$ show that there exist tilings by congruent parallelograms with infinitely many vertices of valence k.

1.2 TILINGS WITH TILES OF A FEW SHAPES

Most of the tiles we shall consider in this section will be topological disks and, as a further simplification, for the most part we shall be concerned with *monohedral* tilings. The word "monohedral" means that every tile in the tiling $\mathscr{T}$ is congruent (directly or reflectively) to one fixed set T—or, more simply, all the tiles are the same size and shape. The set T is called the *prototile* of $\mathscr{T}$, and we say that the prototile T *admits* the tiling $\mathscr{T}$. Familiar examples of tilings satisfying all our restrictions appear in Figure 1.2.1. These are the *regular* tilings we mentioned in the Introduction; they are monohedral and their prototiles are regular polygons. Monohedral tilings by parallelograms were mentioned in Exercise 1.1.10.

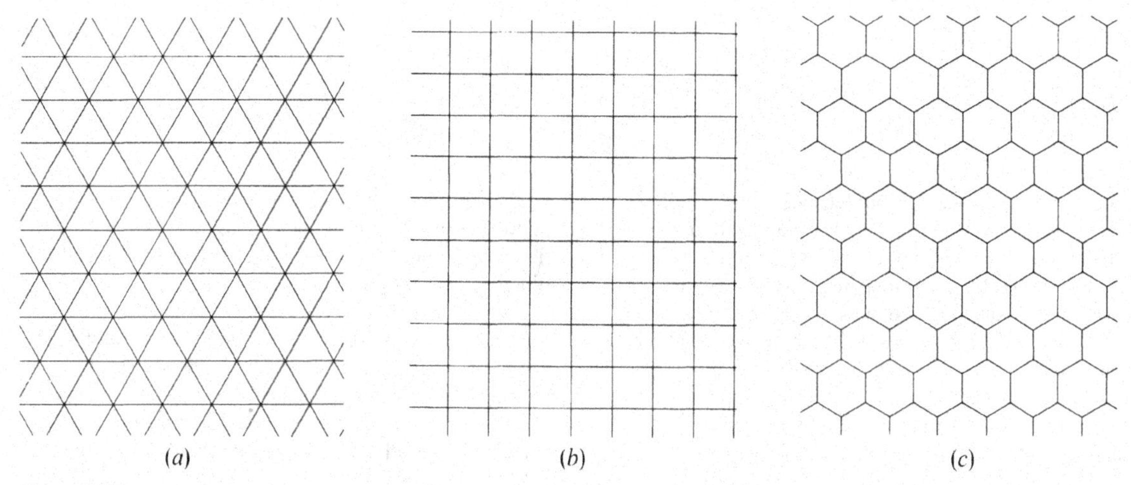

Figure 1.2.1
The three regular tilings (3^6), (4^4) and (6^3). The meaning of the word "regular" and the notation are explained in Sections 1.3 and 2.1

Figure 1.2.2
The twenty-four heptiamonds, each of which is a union of seven equilateral triangles. T. H. O'Beirne proposed the problem of finding which of these shapes is the prototile of a monohedral tiling. Gregory J. Bishop showed that only one of them is not. (For more details, see Gardner [1971, Chapter 24].)

At first glance it may appear that monohedral tilings (especially those whose prototiles are topological disks) are very simple, perhaps even trivial, from a mathematical point of view. At one time the great mathematician David Hilbert (1862–1943) seems to have been of that opinion, since in his famous collection of unsolved problems (Hilbert [1900]) he essentially ignored plane tilings and proposed questions concerning tilings in three or more dimensions. But no doubt he changed his mind later when one of his assistants, K. Reinhardt, began research into the subject, for it is evident from his thesis that he found the problems very far from trivial.

If further evidence is required on this point, the reader should ponder the fact that not only do we lack a method (algorithm) for determining whether a given set T is the prototile of a monohedral tiling, but there are good reasons to suppose that no such method (involving only a finite number of steps) can exist. This problem will be discussed in detail in Chapter 11, but to show how intricate is the question, we reproduce twenty-four shapes in Figure 1.2.2. These are the so-called "heptiamonds" formed by seven equilateral triangles (see Chapter 9) and exactly one of them is not the prototile of a monohedral tiling. It would be of interest to see how long it takes the reader to discover which is the exceptional one! In Section 9.4 we shall give examples of sets for which it is not yet known whether they are prototiles of a monohedral tiling or not. The extent of our ignorance is further illustrated by the fact that we do not know all the different shapes of convex pentagons which will tile the plane monohedrally. The interesting history of this problem and the available information will be given in Section 9.6.

In this book we give illustrations of many hundreds of monohedral tilings. Some of the more extraordinary ones are those which appear to be in the form of spirals with one or more arms. Examples are shown in Figures 1.2.3 and 1.0.1 of three monohedral tilings with the same prototile. As Stephen Eberhart pointed out to us, this prototile also admits many other unusual (non-spiral) tilings. In Section 9.5 we shall explain how other spiral tilings can be constructed.

Before concluding this introductory discussion of monohedral tilings, we produce a few examples of the additional possibilities that arise if we allow tiles which are not topological disks. Figure 1.2.4 shows an example of a prototile which is neither connected nor simply connected. In Figure 1.2.5 we show an example with a pro-

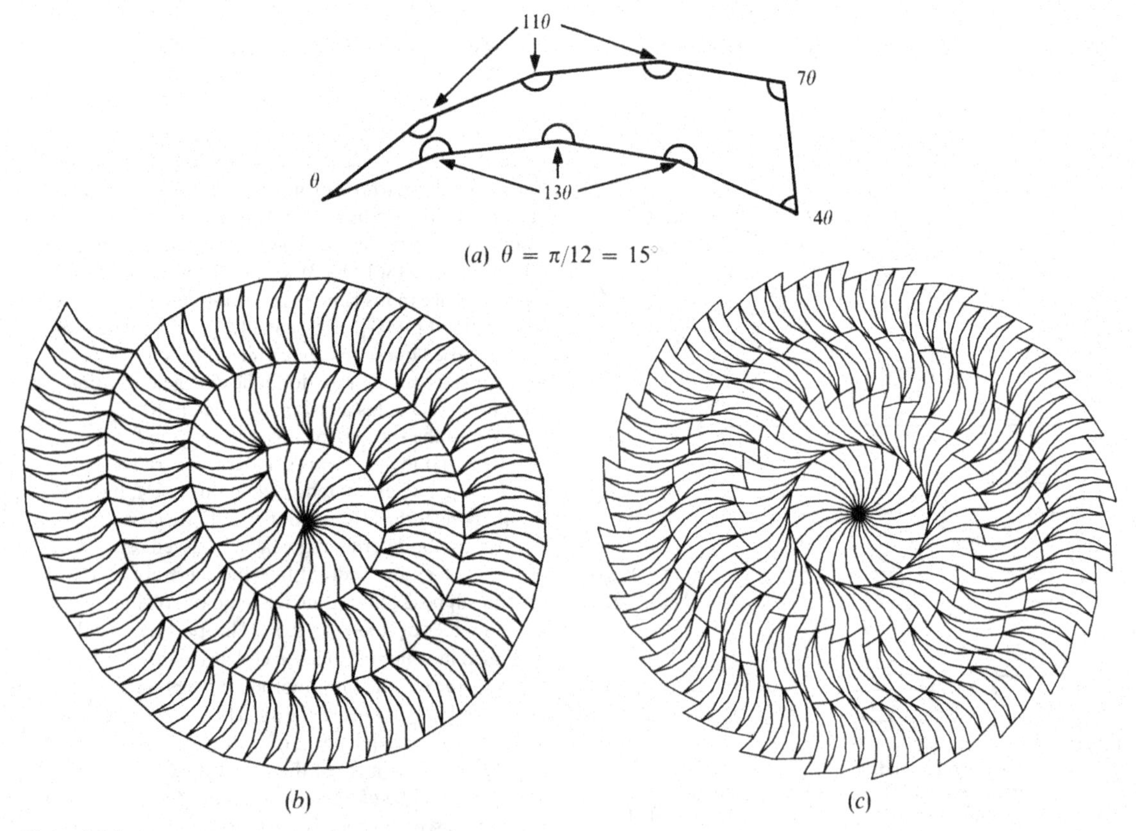

Figure 1.2.3
Two examples of monohedral tilings with the prototile shown in (*a*). Many other "spiral" tilings can be constructed with the same tile, see Section 9.5.

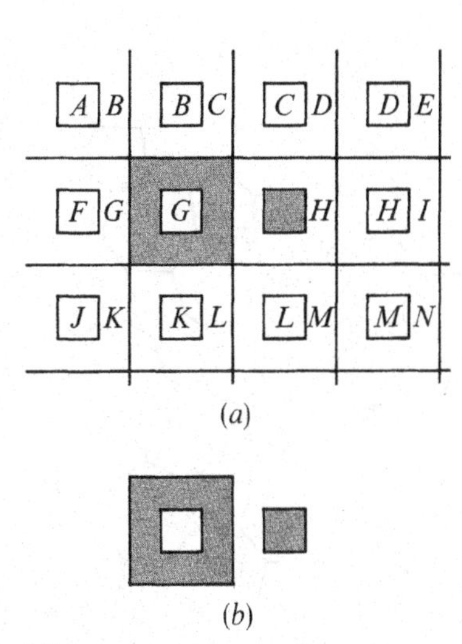

Figure 1.2.4
A monohedral tiling (*a*) with prototile (*b*) that is neither connected nor simply connected.

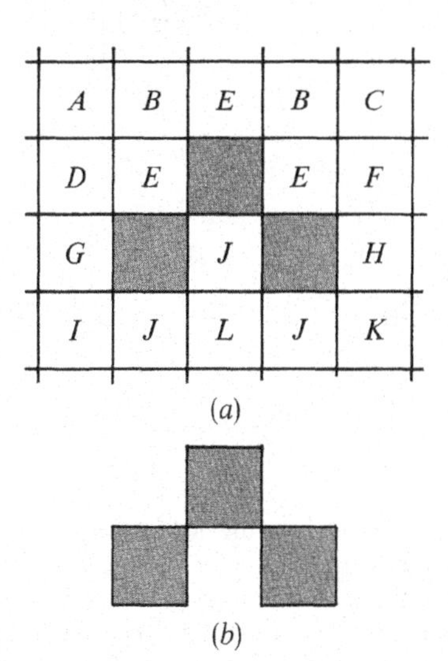

Figure 1.2.5
A monohedral tiling (*a*) with prototile (*b*) that is a connected set but is not a topological disk because it has "cutpoints".

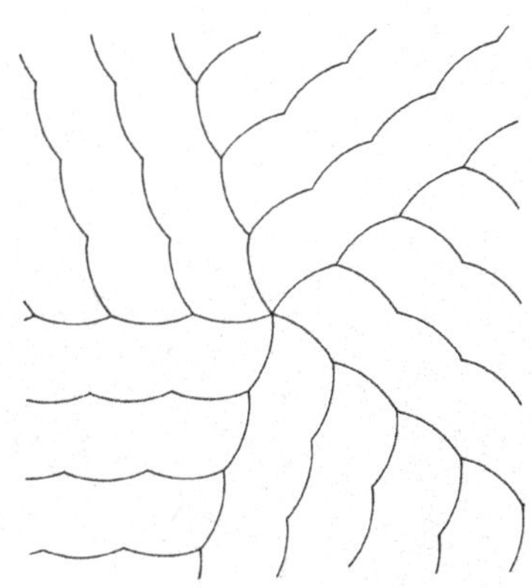

Figure 1.2.6
A monohedral tiling with unbounded prototiles (from Heesch [1968b]).

totile which is connected but is not a topological disk, and in Figure 1.2.6 a monohedral tiling with unbounded tiles. In Figures 1.2.4 and 1.2.5 it has been necessary to mark the different parts of each tile by suitable markings or shadings. This is, of course, to make the diagrams intelligible—we are interested here only in the shapes of the various tiles; they are not "marked tiles" or "colored tiles" as defined in the Introduction. In the case of a tile which is not connected, in the corresponding monohedral tilings the different parts must, of course, bear a fixed relation to one another.

The terminology we have introduced can be extended in the obvious way. By a *dihedral* tiling $\mathcal{T}$ we mean one in which every tile T_i is congruent to one or the other of two distinct prototiles T or T'. In a similar way we define *trihedral*, *4-hedral*, . . . , *n-hedral* tilings in which there are three, four, . . . , n distinct prototiles. If the tiling $\mathcal{T}$ uses a set $\mathcal{S}$ of prototiles we shall say that $\mathcal{S}$ *admits* the tiling $\mathcal{T}$. Two dihedral examples from crystallography are given in Figure 1.2.7; other examples, with $n = 2$, 3 and 4, appear in Figure I.9. For examples of n-hedral tilings by regular polygons, see Chapter 2.

We have remarked that there are many open problems concerning monohedral tilings. With n-hedral tilings ($n \geq 2$) the problems are much deeper and more challenging. See Chapter 10 for details of the interesting phenomenon of aperiodicity that occurs in some dihedral tilings but which seems to have no counterpart in the monohedral case.

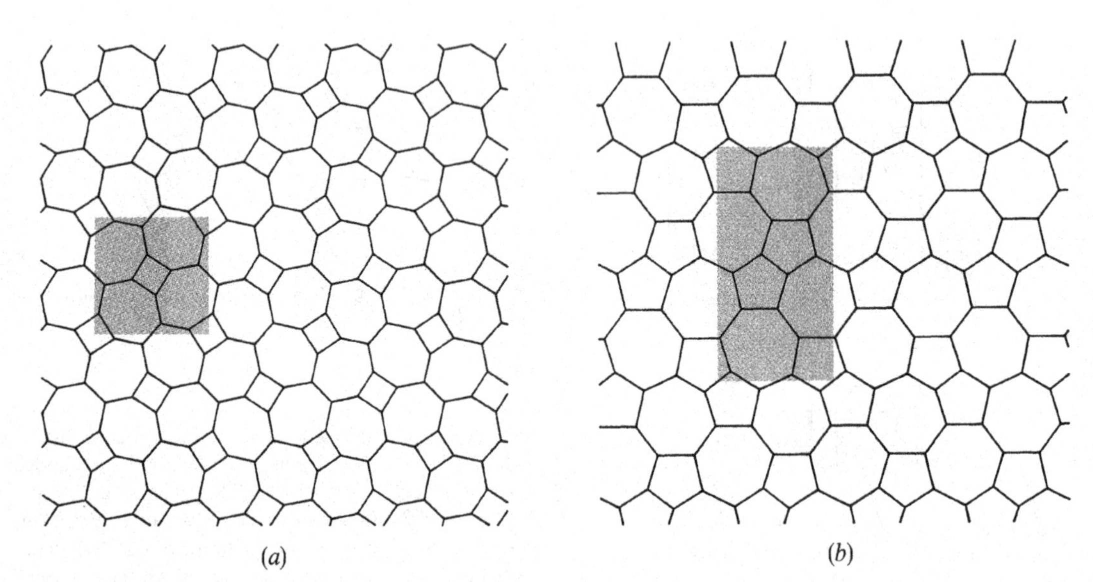

Figure 1.2.7
Two dihedral tilings whose prototiles are convex polygons. Both these tilings arise in crystallography. (*a*) is taken from Blum & Bertaut [1954] and (*b*) from Smith, Johnson & Nordine [1965]. The purpose of the rectangles will be explained later.

EXERCISES 1.2

1. Show that every pentagon with two parallel sides, and every hexagon with three pairs of equal parallel sides, is the prototile of a monohedral tiling.

2. Verify that in Figure 1.2.3 the tiles fit together exactly as shown.

3. (*a*) Let T be a closed topological disk and let the boundary of T be divided into six arcs by six points A, B, C, D, E and F in cyclic order. By describing a tiling method for each case, prove that T is the prototile of a monohedral tiling if at least one of the following conditions is fulfilled:
   (*i*) AB is a translate of ED (which means that there is a translation that maps point A into

point E, point B into point D, and the arc AB into the arc ED), BC is a translate of FE, and CD is a translate of AF (see Figure 1.2.8(*a*)); or

(*ii*) AB is a translate of ED and each of the four other arcs, BC, CD, EF and FA has a center of symmetry (see Figure 1.2.8(*b*)). This criterion is sometimes known as *Conway's criterion*, see Gardner [1975a], Göbel [1979], Schattschneider [1980].

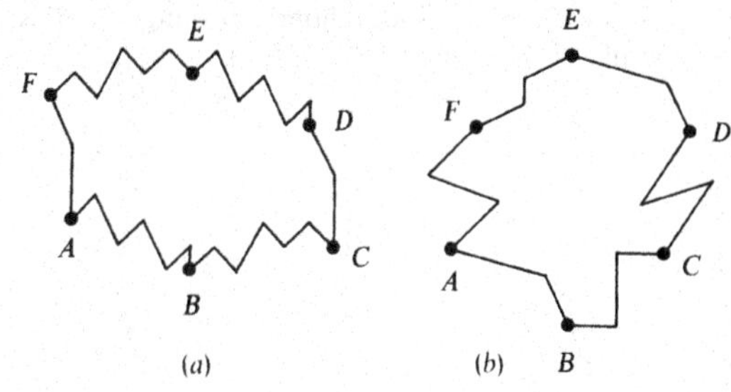

Figure 1.2.8
Prototiles of monohedral tilings that illustrate conditions (i) and (ii) of Exercise 1.2.3(*a*).

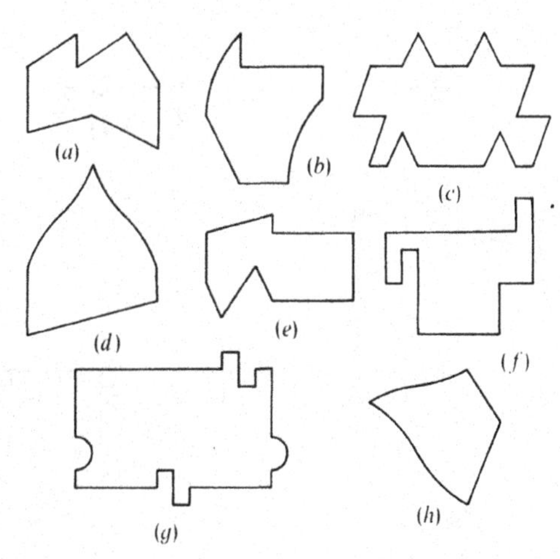

Figure 1.2.9
Some prototiles of monohedral tilings, see Exercise 1.2.4(*a*).

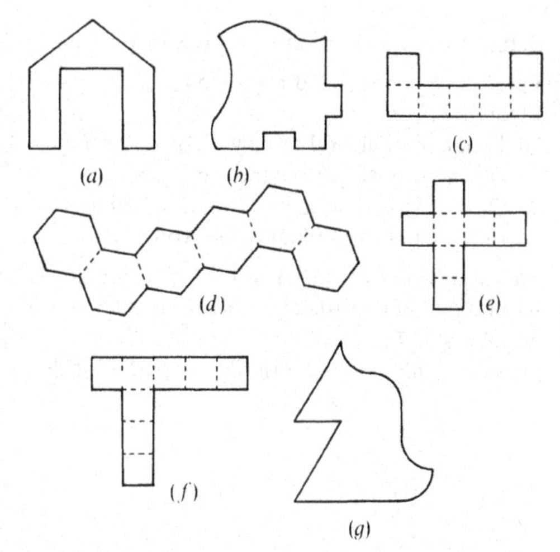

Figure 1.2.10
Are these sets prototiles of monohedral tilings? See Exercise 1.2.4(*b*).

Notice that in (*i*) and (*ii*) there is no need for the six points A, B, C, D, E and F to be distinct.

(*b*) Consider the tiles in the tilings of Figure I.12 and I.17. For each determine whether it satisfies condition (*i*), condition (*ii*), both, or neither, and also whether the tiling shown is equal to the one you obtained by the methods of part (*a*).

4. (*a*) Using Exercise 3 show that each of the tiles in Figure 1.2.9 is the prototile of a monohedral tiling, and sketch such a tiling.

   (*b*) Using either Exercise 3 or any other argument, determine whether the tiles of Figure 1.2.10 are the prototiles of monohedral tilings.

5. Find which one of the 24 heptiamonds shown in Figure 1.2.2 is not the prototile of a monohedral tiling. Find at least one tiling for each of the other 23.

6. Determine all the twelve pentominoes (that is, sets which are topological disks and are the union of five non-overlapping equal squares meeting along whole edges, see Chapter 9). Find which of the pentominoes are prototiles of a *unique* monohedral tiling. For those

that are not, determine at least three distinct tilings.

7. Let T be the unbounded prototile in the tiling of Figure 1.2.6.
   (*a*) Find a monohedral tiling with prototile T in which every vertex has the same valence.
   (*b*) Show that there is an uncountable infinity of distinct monohedral tilings with prototile T.

8. *The neighborhood* $\mathcal{N}(T)$ of a tile T in a tiling $\mathcal{T}$ is the set of tiles that consists of T and all the tiles that are neighbors of T.
   (*a*) Find a tiling $\mathcal{T}$ and a tile T in it such that the neighborhood $\mathcal{N}(T)$ of T does not coincide with the patch $\mathcal{A}(T)$ generated by T.
   (*b*) Give an example of a monohedral tiling $\mathcal{T}$ by triangles such that for each tile T there is a tile T' such that $\mathcal{N}(T) = \mathcal{N}(T')$.
   **(*c*) Show that there is no monohedral tiling in which the prototile is a polygonal disk, such that for each tile T there exist *two* other tiles T', T'' such that $\mathcal{N}(T) = \mathcal{N}(T') = \mathcal{N}(T'')$.
   **(*d*) Show that if $\mathcal{T}$ is a monohedral tiling and T is a tile of $\mathcal{T}$, then $\mathcal{N}(T) = \mathcal{A}(T)$.

1.3 SYMMETRY, TRANSITIVITY AND REGULARITY

Many important properties of tilings depend upon the idea of symmetry. In this section we explain what is meant by this term and give examples of tilings with various kinds of symmetry.

An *isometry* or *congruence transformation* is any mapping of the Euclidean plane E^2 onto itself which preserves all distances. Thus if the mapping is denoted by $\sigma: E^2 \to E^2$, and A, B are any two points, then the distance between A and B is equal to the distance between their images $\sigma(A)$ and $\sigma(B)$. It is not difficult to show (see, for example, Coxeter [1961, Chapter 3], Jeger [1966, Chapters 1 to 6], Guggenheimer [1967, Chapter 2], Yale [1968, Chapter 2], Gans [1969, Chapter 2], Ewald [1971, Chapter 1], Dodge [1972] Martin [1982]) that every isometry is of one of four types:

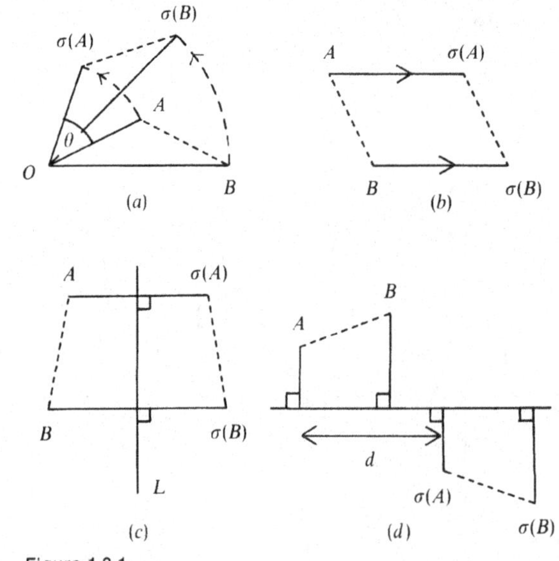

Figure 1.3.1
The four types of plane isometry: (a) rotation, (b) translation, (c) reflection and (d) glide-reflection.

1. Rotation about a point O through a given angle θ, see Figure 1.3.1(*a*). The point O is called the *center*

of rotation. In the particular case when $\theta = \pi$, the line joining A to $\sigma(A)$ will (for all A) be bisected by O, and in this case the mapping is sometimes called *halfturn, central reflection* or *reflection in the point O*.

2. *Translation* in a given direction through a given distance, see Figure 1.3.1(*b*).
3. *Reflection* in a given line L (the *mirror* or *line of reflection*), see Figure 1.3.1(*c*).
4. *Glide reflection* in which reflection in a line L is combined with a translation through a given distance d parallel to L, see Figure 1.3.1(*d*).

Isometries of types (1) and (2) are usually called *direct* because if points ABC form the vertices of a triangle named in a clockwise direction, then the same is true of their images under the isometry σ. If however σ is of type (3) or (4) then the images of the points ABC will form the vertices of a triangle named in a counterclockwise direction. These are called *indirect* or *reflective* isometries.

For any isometry σ and any set S we write σS for the image of S under σ. By a *symmetry* of a set S we mean an isometry σ which maps S onto itself, that is $\sigma S = S$. For example, any rotation about the center of a circular disk is a symmetry of the disk, and so also is the reflection in any line through the center of the disk. In the case of a square, the reflections in the lines L_1, L_2, L_3 and L_4 are symmetries (see Figure 1.3.2) as are rotations through angles $\pi/2$, π and $3\pi/2$ in a counterclockwise direction about its center O, which is called a *center of 4-fold rotational symmetry*. (More generally, if rotation through $2\pi/n$ about a point O is a symmetry of a given set, then O will be referred to as a *center of n-fold rotational symmetry*.)

There are two additional matters that must be mentioned. The first is that there is an isometry which maps every point onto itself. This is known as the *identity isometry* and is a symmetry of *every set*. Thus our list of symmetries of the square contains eight distinct members: four reflections, three rotations and the identity isometry. The second is that we do not distinguish between a counterclockwise rotation of θ and a clockwise rotation of $2\pi - \theta$, nor between a rotation of θ and a rotation of $\theta + 2\pi k$, for any integer k. As symmetries, these are regarded as identical. In other words, only the final result of the mapping is relevant, not the means of arriving at that result.

Clearly there are many other *movements* of the square that leave its position unchanged. For example, if we reflect the square in the line through the center which bisects the angle between L_2 and L_3, and follow this by a counterclockwise rotation through $\pi/4$, then the square coincides with its original position. But this is *not* a new symmetry. Because of the second remark made above it is regarded as coinciding with reflection in the line L_3. It will be seen that in fact the square has only eight symmetries—namely those listed above. (For a fuller discussion and further examples see Coxeter [1961, Chapters 3, 4] and Yale [1968].)

For any T we denote by $S(T)$ the set of symmetries of T. This set has algebraic properties—the symmetries can be combined by applying them consecutively and the result is another symmetry. Because of this algebraic structure $S(T)$ is known as a *group*, and the number of

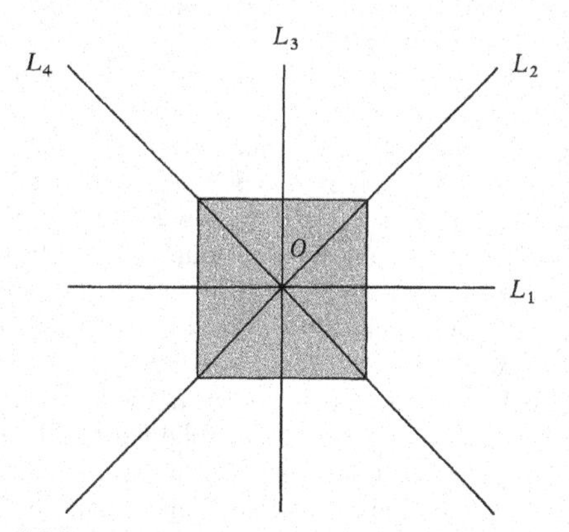

Figure 1.3.2
Reflections in the four lines L_1, L_2, L_3 and L_4 are symmetries of the square. The other symmetries of the square are the identity isometry, and counterclockwise rotations through angles $\pi/2$, π and $3\pi/2$ about the center O.

symmetries in $S(T)$ is called the *order* of the group. So the order of the symmetry group of the square is eight.

It is convenient to introduce some notation for groups that occur frequently. We shall use *c1* or *e* for the group consisting of one isometry only (the identity), and *cn* ($n \geq 2$) for the group consisting of rotations through angle $2\pi j/n$ ($j = 0, 1, \ldots, n-1$) about a fixed point. This is called the *cyclic group of order n* and is the symmetry group of the "*n*-armed swastika". Finally, we use *dn* ($n \geq 1$) for the group which includes all the isometries of *cn* together with reflections in *n* lines equally inclined to one another. This is called the *dihedral group of order* $2n$; for $n \geq 3$ it is the symmetry group of the regular *n*-gon (see Figure 1.3.2 for $n = 4$). When $n = 1$, the group *d1* (of order 2) consists of just the identity and reflection in a line; when $n = 2$, the group *d2* (of order 4) consists of the identity, reflections in two perpendicular lines, and rotation through angle π about the point in which the two lines of reflection meet. The *rotation group* $d\infty$ consists of all rotations about a given point and all reflections in lines through that point; it is the symmetry group of a circular disk. Note that the groups *cn* and *dn* ($n \geq 1$) each have the property of leaving at least one point of the plane fixed; in fact these are the only groups that can occur as symmetry groups $S(T)$ of *compact* (that is closed and bounded) sets T.

We extend the definition of symmetry in a natural way to structures more complicated than single sets. Thus in the case of a tiling $\mathcal{T}$ we say that an isometry σ is a *symmetry of* $\mathcal{T}$ if it maps every tile of $\mathcal{T}$ onto a tile of $\mathcal{T}$. For example, in Figure 1.3.3 we show a monohedral tiling which has symmetries of all four types. To begin with, there are rotations through $2\pi/3$ or $4\pi/3$ about each point in the figure marked by a small triangle (solid or open). Next, there are translations, which take any of the solid black triangles onto any other. In the third place there are reflections in each of the solid lines, and finally there are glide-reflections consisting of reflection in the dashed lines followed by translation along them. This translation is through half the distance between the solid black triangles and is marked in the diagram by half arrowheads. All these symmetries are indicated, except for the translations, and combinations of them form a complete enumeration of the symmetries of this particular tiling.

An easy and informal way to think of a symmetry of a tiling (though not without reservations as to its mathematical validity) is the following. Imagine we have drawn the tiling on an infinite piece of paper, and then traced it onto a transparent sheet. A symmetry corresponds to a motion of the latter (including the possibility of turning it over) such that, after the motion, the tracing fits exactly over the original drawing.

The idea of a symmetry can be extended to more general situations. Suppose, for example, we have a *marked tiling*—that is, one in which there is a *marking* or *motif* on each tile. Then a symmetry of the marked tiling is an isometry which not only maps the tiles of $\mathcal{T}$ onto tiles of $\mathcal{T}$, but also maps each marking on a tile of $\mathcal{T}$ onto a marking on the image tile. Thus, in the informal interpretation above, not only do we trace the tiles, we trace their markings also.

Similar considerations apply to colored tilings—that is, tilings in which to each tile is assigned one of a given set of colors. For these, a *color-preserving symmetry* is

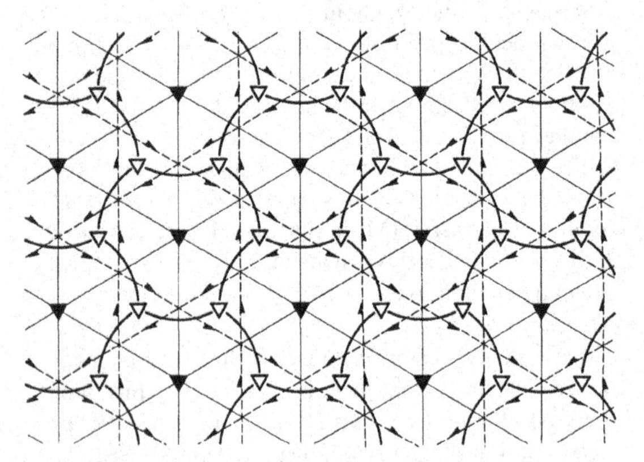

Figure 1.3.3
A monohedral tiling with its symmetries indicated. The triangles (solid and open) denote centers of 3-fold rotational symmetry, the solid lines are lines of reflection, and the dashed lines indicate glide-reflections. The half arrowheads indicate the translations associated with each glide-reflection.

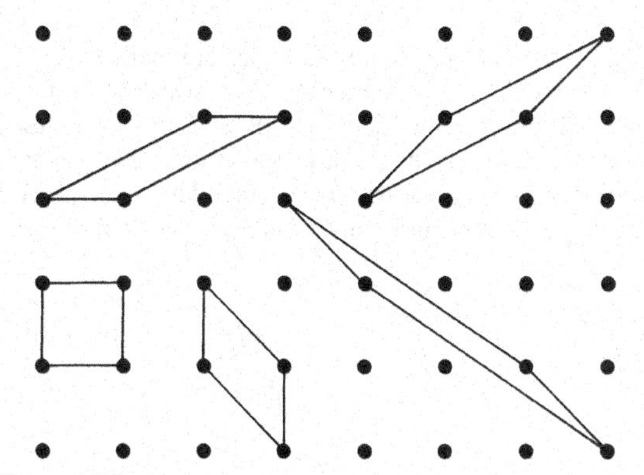

Figure 1.3.4
A lattice Λ of points in the plane and some parallelograms whose corners coincide with points of Λ. Each of these parallelograms is the prototile of a parallelogram tiling whose vertices coincide with Λ. In fact any parallelogram may be chosen so long as the only lattice points it contains are its vertices, and no lattice point lies in its interior or on its boundary. All such parallelograms have equal area.

defined as an isometry which maps each tile onto one of the same color. In Chapter 8 we shall introduce a modification of this idea called a "color symmetry" of a tiling, which—from some points of view—turns out to be a more interesting and fruitful concept.

For any tiling $\mathscr{T}$, we extend the notation introduced above and write $S(\mathscr{T})$ for the group of symmetries of $\mathscr{T}$. It is, of course, possible for $S(\mathscr{T})$ to consist of the identity alone—in which case we shall denote it by *c1* or *e*—or it may have many symmetries (as, for example, in the case of the regular tilings or that of Figure 1.3.3). It is clear that such facts can be used as a basis for classification. Since the groups which occur as symmetry groups of tilings are rather more complicated than those that arise in the case of a compact set, we shall postpone a discussion of them until Section 1.4.

If a tiling admits any symmetry in addition to the identity symmetry then it will be called *symmetric*. If its symmetry group contains at least two translations in non-parallel directions then the tiling will be called *periodic*. Many of the tilings we shall meet (such as the regular ones and most of the tilings illustrated in the next chapter) are periodic, and we note here that such tilings are very easily described. Let us represent the two non-parallel translations by vectors a, b. Then clearly $S(\mathscr{T})$ contains all the translations $na + mb$ where n and m are integers. All these translations arise by combining n of the translations a and m of the translations b in the manner described above. Starting from any fixed point O the set of images of O under the set of translations $na + mb$ forms a *lattice*. The most familiar example of a lattice is the set of points in the Euclidean plane with integer coordinates. This is known as the *unit square lattice*, see Figure 1.3.4; we have already met it as the set of vertices of the regular tiling (4^4). More generally, a lattice can be regarded as consisting of the vertices of the *parallelogram tiling* (see Exercise 1.1.10). For example, the solid black triangles (regarded as points) in Figure 1.3.3 or the points marked by black circles in Figure 1.3.5 are examples of lattices.

Thus with every periodic tiling $\mathscr{T}$ is associated a lattice, and the points of the lattice can be regarded (in many ways) as the vertices of a parallelogram tiling

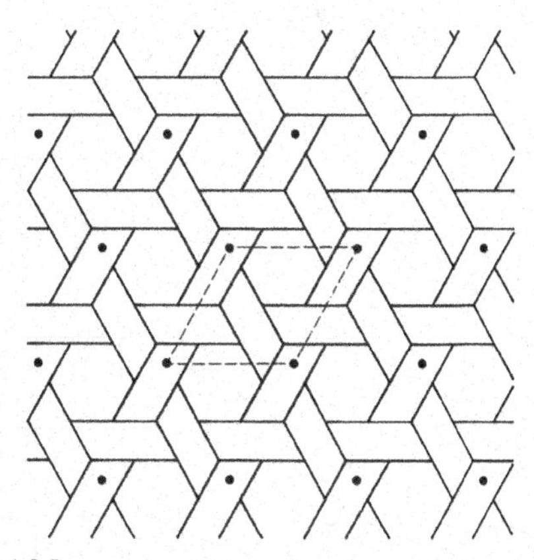

Figure 1.3.5
A periodic dihedral tiling with a corresponding lattice and one of the possible period parallelograms. This tiling is taken from a 16th century Persian drawing in the British Museum (see Christie [1929, Chapter 9]).

$\mathscr{P}$ (see Figure 1.3.4); the tiles of $\mathscr{P}$ are known as *period parallelograms.* If we know the configuration formed by the tiles, edges and vertices of $\mathscr{T}$ that are contained in one of the parallelograms of $\mathscr{P}$, then the rest of $\mathscr{T}$ can be constructed by repeating this configuration in every parallelogram of $\mathscr{P}$. A typical periodic tiling is shown in Figure 1.3.5. Here $\mathscr{P}$ and the lattice are marked.

Three examples of symmetric tilings which are not periodic are given in Figure 1.3.6. In the first two of these, the symmetries include rotations about one fixed point which may be called the *center* of the tiling (see Goldberg [1955]); the tiling in Figure 1.3.6(*b*) contains reflections

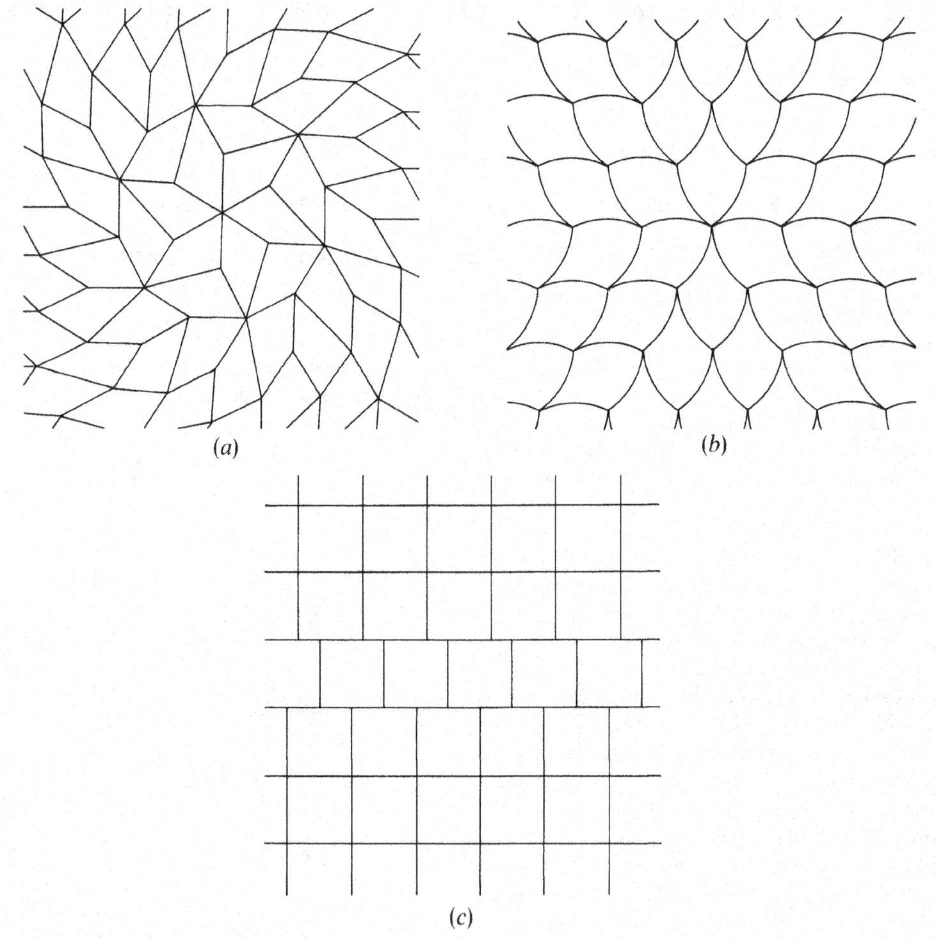

Figure 1.3.6
Some monohedral tilings that are symmetric but not periodic. Tiling (*a*) is from Goldberg [1955] and has symmetry group *c6*. Tiling (*b*) has symmetry group *d3*, and tiling (*c*) admits as symmetries only translations parallel to one direction.

as symmetries also. In the third example, the symmetry group is infinite and consists of translations parallel to one fixed direction.

Let T be a tile of any tiling $\mathscr{T}$. Then every symmetry of $\mathscr{T}$ which maps T onto itself is clearly a symmetry of T. But the converse is not, in general, true. For example, in Figure 1.3.6(*c*), the only symmetry of $\mathscr{T}$ which maps a tile onto itself is the identity—yet the tile itself, being a square, has seven other symmetries (as explained earlier). Hence we must carefully distinguish between $S(T)$, the group of symmetries of the tile T, and $S(\mathscr{T} \mid T)$, the group of symmetries of T which also are symmetries of the tiling $\mathscr{T}$. For brevity we shall often refer to $S(\mathscr{T} \mid T)$ as the *induced tile group* or as the *stabilizer* of T in $\mathscr{T}$. Figure 1.3.7 shows a more interesting example in which $S(\mathscr{T} \mid T)$ and $S(T)$ differ.

Two tiles T_1, T_2 of a tiling $\mathscr{T}$ are said to be *equivalent* if the symmetry group $S(\mathscr{T})$ contains a transformation that maps T_1 onto T_2; the collection of all tiles of $\mathscr{T}$ that are equivalent to T_1 is called the *transitivity class* of T_1. If all tiles of $\mathscr{T}$ form one transitivity class we say that $\mathscr{T}$ is *tile-transitive* or *isohedral*. An example of an isohedral tiling is given in Figure 1.3.7. Further examples are shown in Figure 1.3.3, as well as in Figures 1.1.4 and 1.1.6; the regular tilings in Figure 1.2.1 are also isohedral. The tilings of Figures 1.3.6 and 1.3.8 are monohedral but not isohedral. The distinction between isohedral tilings and monohedral tilings (in which each tile has the same shape) may seem slight, but it is very significant. This is illustrated by the fact that the problem of finding and classifying all monohedral tilings is unsolved (even when the tiles are convex polygons) whereas we shall explain in Chapter 6 how to describe and classify all the isohedral tilings. In Figure 6.2.4 we give illustrations of all the eighty-one different types possible according to that classification. If $\mathscr{T}$ is a tiling with precisely k transitivity classes then $\mathscr{T}$ is called *k-isohedral*. Figure 1.3.8 shows three monohedral tilings which are 2-isohedral. Figure 1.3.9 shows a 2-isohedral dihedral tiling. Generally, of course, if the tiles are of n different shapes then there will be at least n transitivity classes. In the case of a tiling which is not symmetric, every tile is a transitivity class on its own.

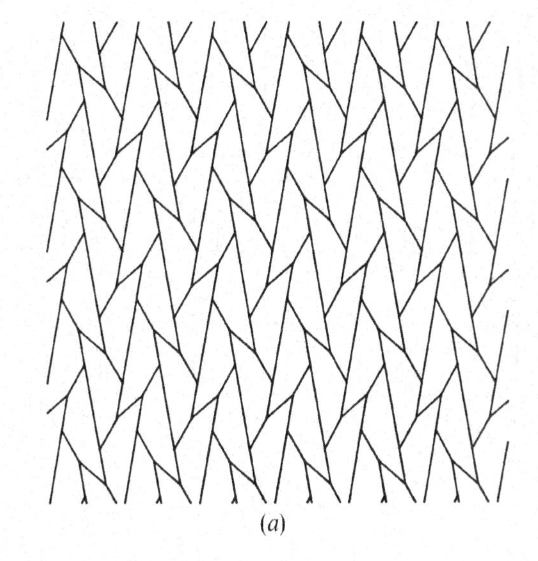

(*a*)

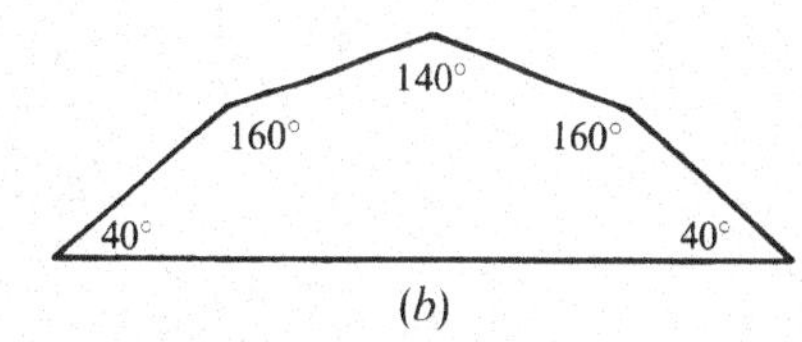

(*b*)

Figure 1.3.7
An isohedral tiling (*a*) with prototile (*b*) which has symmetry group of order 2, whereas the induced symmetry group consists of the identity only. It can be shown that there exists no other monohedral tiling with this same prototile (see Heesch [1968b, p. 46]).

The idea of transitivity and equivalence is applicable to other elements of a tiling also. If the symmetry group $S(\mathscr{T})$ of $\mathscr{T}$ contains operations that map every vertex of $\mathscr{T}$ onto any other vertex, then we say that the vertices form one transitivity class, or that the tiling is *isogonal*. For example, the three regular tilings and the tilings of Figures 1.3.3 and 1.3.9 are isogonal. So are four of the

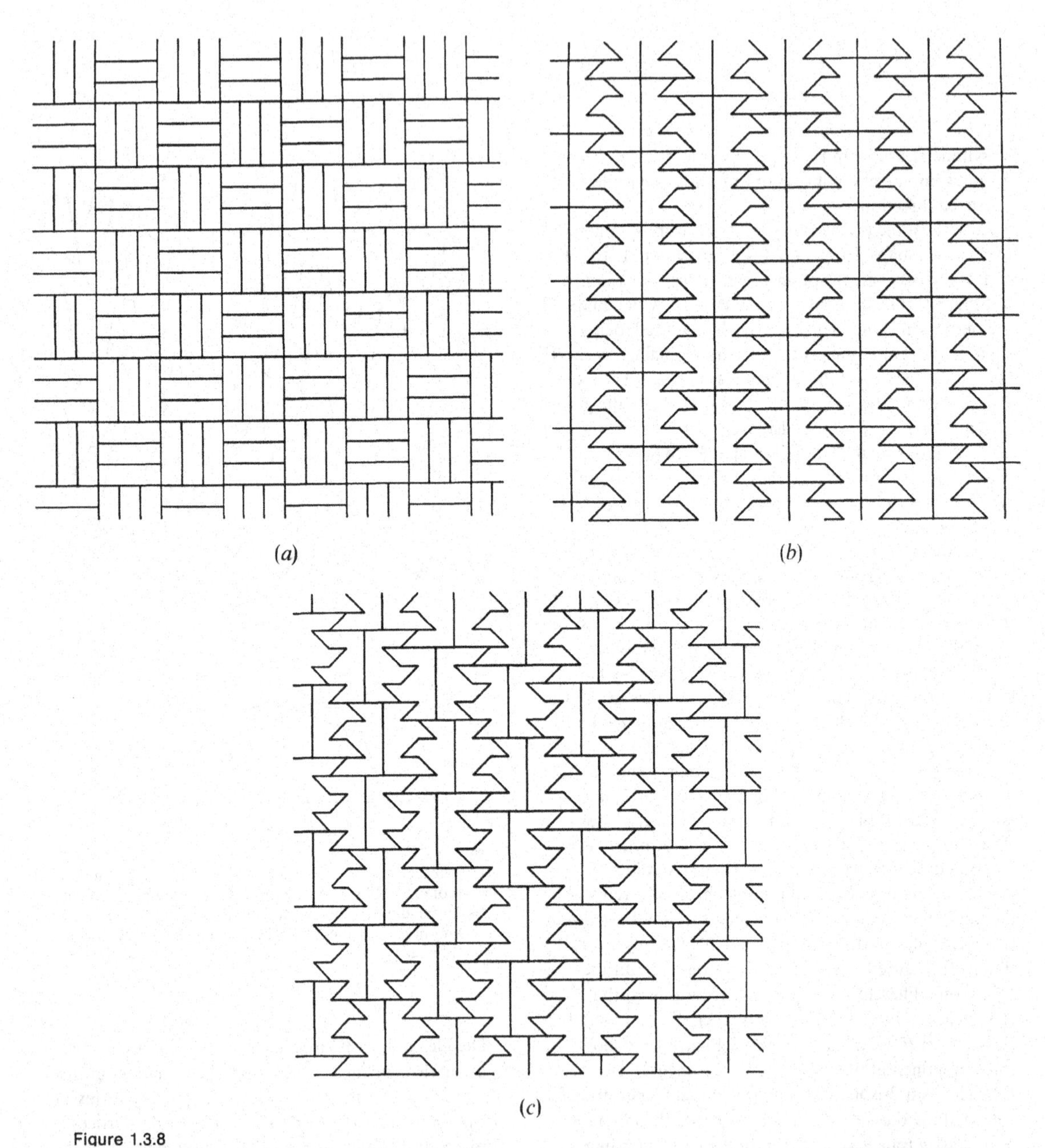

Figure 1.3.8
Three monohedral tilings that are 2-isohedral, that is, in which the tiles form two transitivity classes. Tiling (*b*) and (*c*) (reproduced from Goldberg [1955], using a prototile found by Heesch [1935]) are interesting because there is no way of tiling the plane using the same prototile in such a manner that the tiles belong to just one transitivity class.

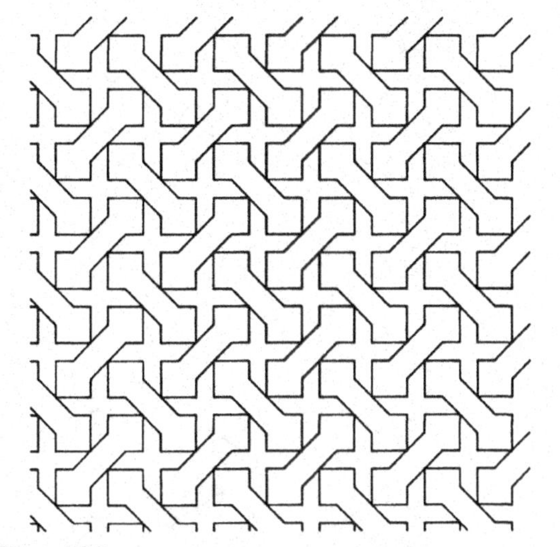

Figure 1.3.9
A dihedral and isogonal tiling in which the tiles belong to two transitivity classes. Designed by the Persian artist Mirza Akbar in the early 19th century (from Christie [1929, p. 239]).

tilings in Figure I.10. Section 6.3 will be devoted to a complete description of all isogonal tilings. In an analogous manner to that defined above, we may say that a tiling is *k-isogonal* if its vertices form k transitivity classes, where $k \geq 1$ is any integer. For example, the tilings of Figures 1.3.7 and 1.3.8(*a*) are 2-isogonal.

A *monogonal* tiling is one in which every vertex, together with its incident edges, forms a figure congruent to that of any other vertex and its incident edges. The distinction between isogonal and monogonal tilings is analogous to that between isohedral and monohedral tilings. An example of a tiling which is monogonal but not isogonal is shown in Figure 1.3.10.

Isotoxal tilings are tilings in which every edge can be mapped onto any other edge by a symmetry of the tiling. For example, the tilings of Figures 1.3.3 and 1.3.9 are isotoxal, the one in Figure 1.3.5 is not. In Chapter 6 we shall give a classification and complete description of all isotoxal tilings.

Figure 1.3.11 shows several examples of tilings which

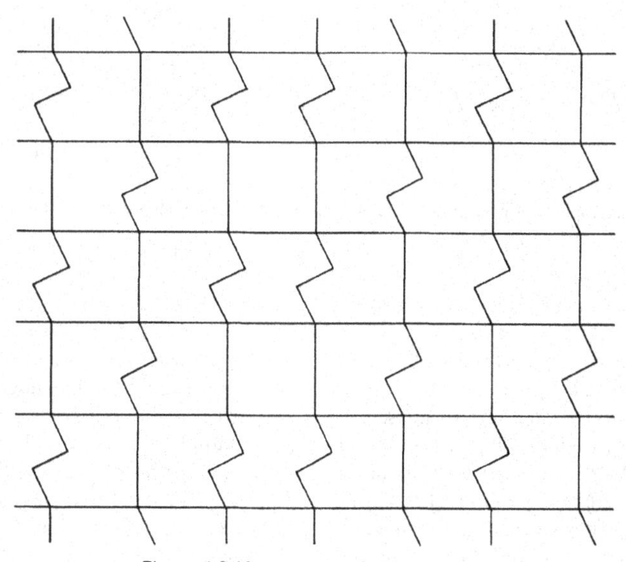

Figure 1.3.10
A monogonal tiling that is not isogonal.

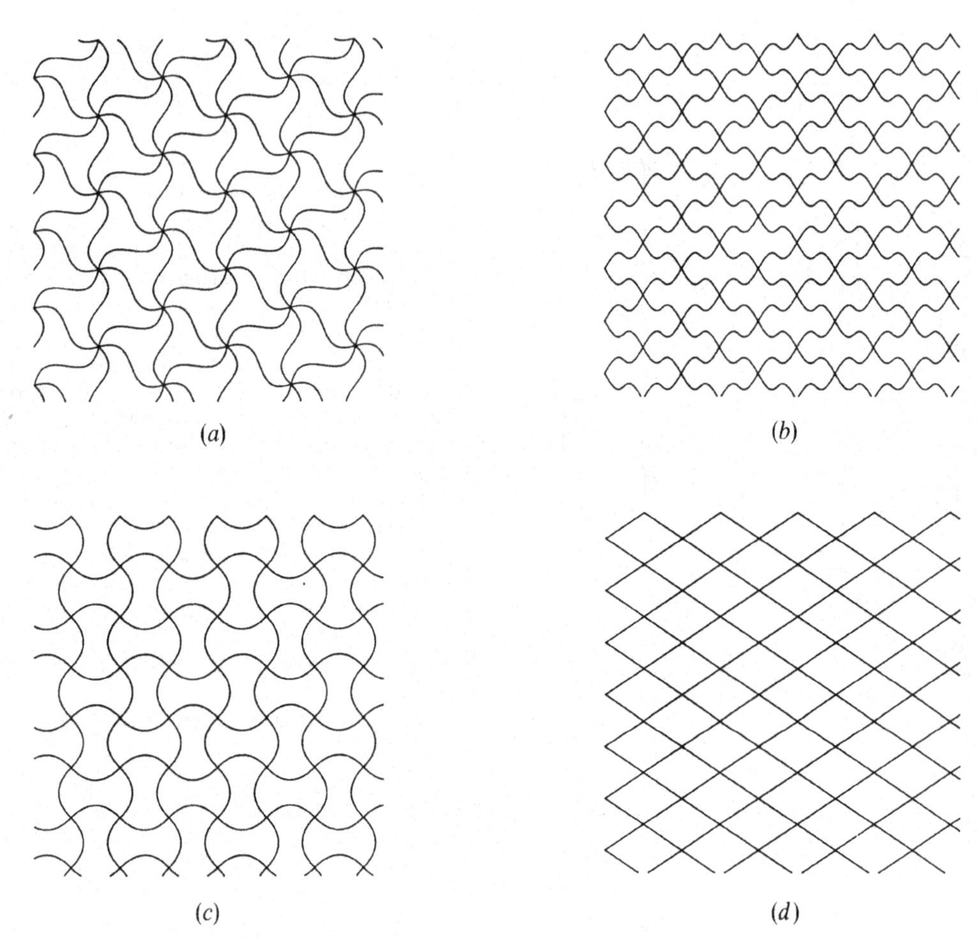

Figure 1.3.11
Some examples of tilings that are isohedral, isogonal and isotoxal but are not regular.

are isohedral, isogonal and isotoxal. They clearly display a considerable amount of "regularity", so we now consider what distinguishes these from the regular tilings of Figure 1.2.1. In order to define a regular tiling we again use the concept of transitivity, but in a very strong sense. By a *flag* in a tiling we mean a triple (V,E,T) consisting of a vertex V, an edge E and a tile T which are mutually incident. Examples of flags are indicated in Figure 1.3.12 by emphasizing the vertex and edge and shading the tile. We see that if T has n edges and n vertices then it belongs to precisely $2n$ flags—we may choose E in n different ways, and then V may be chosen to be one of the two endpoints of E. A tiling $\mathscr{T}$ is called *regular* if its symmetry group $S(\mathscr{T})$ is transitive on the flags of $\mathscr{T}$. Again referring to Figure 1.3.12 we see that in the first tiling (*a*) the two marked flags are equivalent under $S(\mathscr{T})$, and it is easy to check that this applies to all flags. Yet for the second tiling (*b*), there is no symmetry which maps one of the marked flags onto the other. It can be shown that the tiles of a regular tiling are necessarily regular polygons (see Exercise 1.3.7) but, as we shall see in Chapter 2, there are many tilings by regular polygons which are not regular. In fact there are only three regular tilings—namely those shown in Figure 1.2.1.

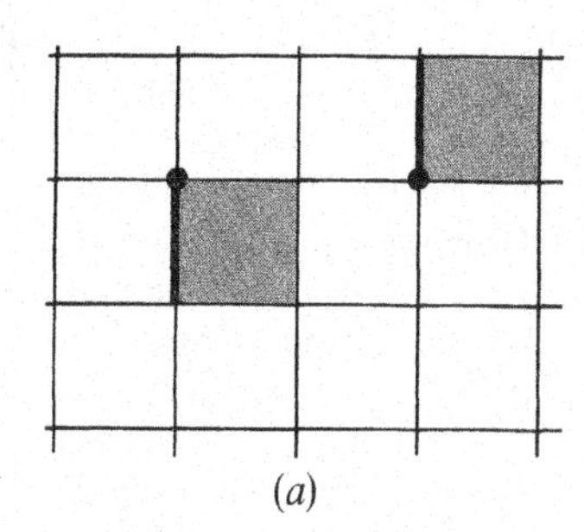

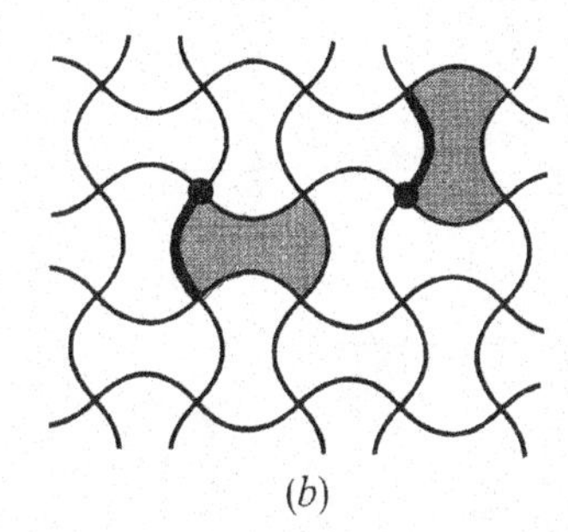

(a) (b)

Figure 1.3.12
Examples of tilings that are isohedral, isogonal and isotoxal. In each, two flags are marked. Tiling (*a*) is regular because all its flags are equivalent, but in (*b*) the flags belong to two distinct equivalence classes and so the tiling is not regular.

EXERCISES 1.3

1. Determine which of the tilings shown in Figure I.10 are (*a*) monohedral, (*b*) monogonal, (*c*) isohedral, (*d*) isogonal, (*e*) isotoxal.
2. For each $n \geq 2$ give an example of a monohedral tiling which is n-isohedral.
3. Show how each tile of Figure 1.3.3 can be marked so that the resultant marked tiling is still isohedral, isogonal and isotoxal, but there are no reflections or glide reflections in the symmetry group of the marked tiling. Can you modify the shape of the tile (leaving the vertices fixed) so as to achieve the same effect? Can you mark or modify the shapes of the tiles in the regular square tiling in an analogous manner?
4. For each of the tilings shown in Figures 1.1.2, 1.1.3, 1.2.7 and 1.3.8, determine the number of transitivity classes of (*a*) tiles, (*b*) edges, and (*c*) vertices.
5. Find a monogonal tiling which is not isogonal and each of whose vertices has valence (*a*) 3, (*b*) 5 and (*c*) 6.
6. Find an example of a monohedral tiling with Heesch's prototile (Figures 1.3.8(*b*) and (*c*)) in which the tiles belong to more than two transitivity classes.
7. Sketch a proof that the only regular tilings are by convex polygons and are, in fact, the three well-known regular tilings (Figure 1.2.1).

*8. For each integer $n \geq 1$ give an example of a tiling in which the flags belong to n transitivity classes.

9. Prove that the prototile of Figure 1.3.7 admits a unique monohedral tiling.

*10. For each of the prototiles in Figure 1.3.13, with three exceptions, find at least one isohedral tiling. (Note that where the tiles consist of several disconnected pieces, their mutual positions must be preserved in the tiling.)

*11. Consider the following three properties of a tiling $\mathscr{T}$:
   (*a*) The symmetries of $\mathscr{T}$ act transitively on pairs of tiles that share an edge.
   (*b*) The symmetries of $\mathscr{T}$ act transitively on triplets

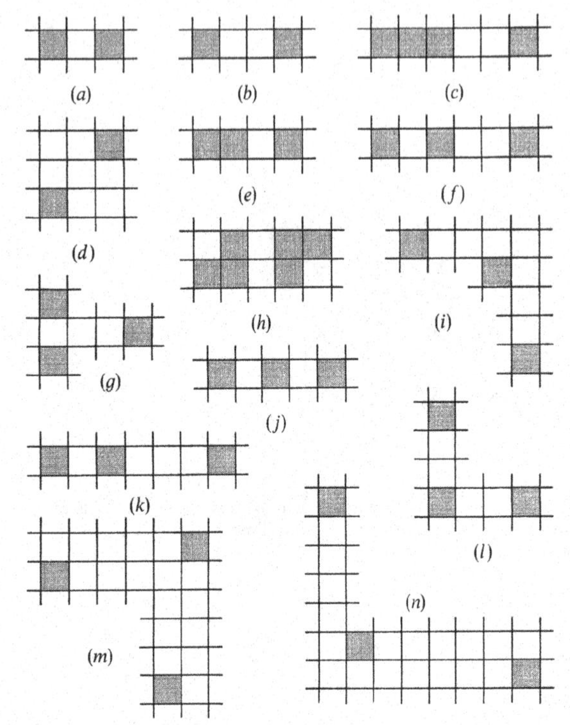

Figure 1.3.13
Are these prototiles of isohedral tilings? See Exercise 1.3.10.

of tiles that share a vertex.

(c) The symmetries of $\mathcal{T}$ act transitively on pairs of tiles that share a vertex but have no common edge.

Give examples to show that it is possible for a tiling to have property (a) alone, (c) alone, or (b) and (c) but not (a). (It seems that no tiling exists which has only property (b).)

*12. For the tiling of Figure 1.3.7 show how the edges may be modified (without changing the positions of the vertices) so that the tiling is still isohedral, but the induced tile group coincides with the symmetry group of the tile.

**13. In an isohedral tiling with $S(T) \neq S(\mathcal{T}|T)$ is it always possible to modify the edges of T and of the other tiles, without changing the positions of the vertices, in such a way that these two groups become equal? Can one insist that $S(\mathcal{T}|T)$ is preserved?

14. Show that there are an uncountable infinity of unequal monohedral tilings with the prototile shown in Figure 1.3.14 (the union of nine squares). How many of these tilings are isohedral?

15. Define *monotoxal tilings* by analogy with monohedral and isotoxal tilings; find examples of monotoxal tilings that are not isotoxal.

16. (a) Does there exist a tiling that is monohedral and monogonal, but is neither isohedral nor isogonal?
    (b) Does there exist a tiling that is monohedral, monogonal and monotoxal, but is neither isohedral, nor isogonal nor isotoxal?

17. Let $\mathcal{T}$ be a tiling which has m transitivity classes of tiles, n transitivity classes of edges and p transitivity classes of vertices. (See Chavey [1984a], [1984b] for results on problems of the kind considered here.)
    (a) Show that if $n = 1$ then $m \leq 2$ and $p \leq 2$.
    (b) Find a tiling with $p = 1$ and $m = n = 4$.
    (c) Find a tiling with $m = 1$ and $p \geq 5$.
    **(d) Find inequalities between m, n and p which are necessary and sufficient for $\mathcal{T}$ to exist.

**18. Suppose there exists a tiling $\mathcal{T}$ with prototile T such that every tile in $\mathcal{T}$ is a translate of T. Using only translates of T, is it always possible to tile the plane (a) periodically, (b) isohedrally, or (c) so that the centroids of the tiles form a lattice?

**19. Suppose there exists a tiling $\mathcal{T}$ with prototile T in which the tiles occur in r different *aspects* (that is to say, the set of tiles can be partitioned into r subsets

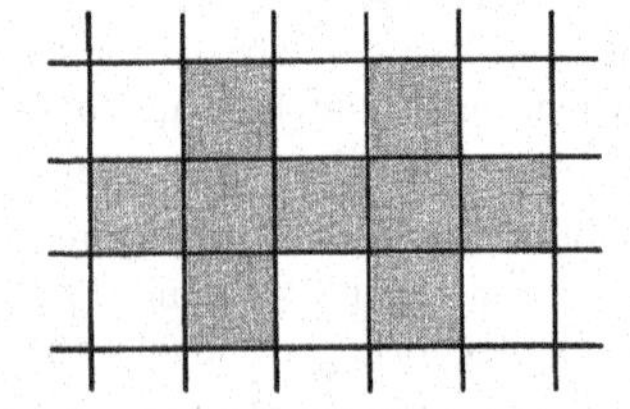

Figure 1.3.14
There exists an uncountable infinity of tilings with this prototile; see Exercise 1.3.14.

such that all the tiles in each subset are translates of one another). Is it always possible to tile the plane k-isohedrally with prototile T, for some $k \leq r$?

20. (*a*) Find monohedral periodic tilings with 1, 2, 3, 4, 5, 6, 7, 8 and 12 different aspects.
   **(*b*) Does there exist any monohedral periodic tiling whose tiles occur in 9 aspects? Or with any other number of aspects besides those mentioned in (*a*)?

21. Let $\mathcal{T}$ be a monohedral tiling in which the prototile T is a polygon.
   (*a*) Is there any upper bound on the number of aspects of the tiles in $\mathcal{T}$?
   **(*b*) Is there any prototile T such that the number of aspects in various monohedral tilings admitted by T is unbounded?
   **(*c*) Is there a monohedral tiling in which the tiles occurs in infinitely many aspects?
   **(*d*) What are the answers to (*b*), (*c*) if the prototile T is assumed to be convex?
   (*e*) Show that the answer to (*c*) is in the affirmative if dihedral tilings (whose tiles are not necessarily polygonal) are considered instead of monohedral ones.

22. Let G denote any one of the groups cn or dn, for $n \geq 1$.
   (*a*) For each G find a monohedral tiling $\mathcal{T}$ that has G as its symmetry group $S(\mathcal{T})$.
   **(*b*) Does there exist a monogonal tiling with symmetry group G?

23. Prove the assertion made early in Section 1.3 that every isometry is either a rotation, a translation, a reflection or a glide-reflection.

24. (*a*) Show that if two rotations, with different centers and through different angles, are combined, the result is another rotation. Give a geometric construction for the center and angle of the latter in terms of the centers and angles of the two given rotations.
   (*b*) Show that the result of combining a reflection and a glide-reflection is either a translation or a rotation. Show how, in the latter case, one can construct the center and angle of the rotation.

1.4 SYMMETRY GROUPS OF TILINGS*

In the previous section we explained how one can indicate, on a drawing of a given tiling $\mathcal{T}$, the elements of the symmetry group $S(\mathcal{T})$, see Figure 1.3.3. Here we use the word "elements" to mean the reflections, glide-reflections, rotations and translations in $S(\mathcal{T})$. Table 1.4.1 explains in detail the symbols we use to represent the elements diagrammatically; we shall usually adhere to this notation throughout the book.

A diagram in which the elements have been represented in this manner serves to define the group $S(\mathcal{T})$ precisely. We shall call it a *group-diagram* for $S(\mathcal{T})$. Examples of tilings and the corresponding group-diagrams appear in Figures 1.4.1 and 1.4.2.

The representation of the translations by arrows (as indicated in the last row of Table 1.4.1) differs from that of the other elements in two important respects.

1. Only the magnitude and direction of the arrow (vector) is important; unlike the lines of reflection and glide-reflection and centers of rotation, its actual position relative to the tiling is irrelevant. Thus it may be moved parallel to itself and it will still represent the same translation. Technically it

* The material in this section is not used (except in some exercises) until Chapter 5. The reader therefore may wish to postpone its study. However, a knowledge of the symmetry groups will enable the reader to appreciate better some of the material in Chapter 2.

Table 1.4.1 THE SYMBOLS USED IN REPRESENTATIONS OF THE SYMMETRY ELEMENTS IN DIAGRAMS

Symbol	Meaning
——	Line of reflection.
-⁄- - -⁀-	Line of glide-reflection. The half-arrowheads indicate the size of the translation (glide) associated with the reflection in the dashed line.
◊	Center of 2-fold rotation (reflection in a point).
△	Center of 3-fold rotation.
□	Center of 4-fold rotation.
⬡	Center of 6-fold rotation.
◆	Center of 2-fold rotation lying on a line of reflection; the elements of the group leaving this point fixed therefore form a dihedral group *d2*.
▲	The corresponding center for *d3*.
■	The corresponding center for *d4*.
⬢	The corresponding center for *d6*.
○⟶	Vector indicating a translation in the group (see notes (1) and (2) on pp. 37, 38).

is known as a *free vector*, that is, one that does not act at a particular fixed point or along a fixed line.

2. If $S(\mathscr{T})$ contains a translation, then it must contain an infinity of such. However, on the group diagram it is only necessary to indicate at most two. If the translations in $S(\mathscr{T})$ are all parallel then they can be represented by the set of vectors $\{na\}$ where a is a fixed vector and n runs through the integers, positive, negative and zero. Hence we need use only *one* arrow (representing a or $-a$) to specify all the translations in $S(\mathscr{T})$. If, on the other hand, $S(\mathscr{T})$ contains non-parallel translations, then the tiling is periodic and by the remark on page 29, the set of all translations may be written as $\{na + mb\}$ where a and b are fixed vectors and n, m run independently through the integers. Thus the translations in $S(\mathscr{T})$ can be specified by just two arrows, one corresponding to a and one to b. However, it should be observed that, as we have seen, the choice of a and b is not unique. (The vectors a, b corresponding to any two adjacent sides of any parallelogram in Figure 1.3.4 yield the same lattice and therefore the same set of translations.)

Because of these properties it is convenient to regard two group diagrams as the same if they can be made identical by movement (rigid motion) or by altering one's choice of vectors corresponding to translations as described in (1) and (2) above. In other words, we shall not distinguish between diagrams corresponding to the same group—or, as we shall sometimes say, between group diagrams that differ trivially.

In classifying symmetry groups, the most important concept is that of isomorphism. We say that two symmetry groups $S(\mathscr{T}_1)$ and $S(\mathscr{T}_2)$ are *isomorphic* if the group diagram of one can be made to coincide with the group diagram of the other by applying a suitable affinity.* An *affinity* (or *affine transformation*) is defined as any mapping of the plane onto itself representable by linear equations of the form

$$x' = px + qy + c$$

$$y' = rx + sy + d,$$

with $ps - qr \neq 0$. Geometrically, any given lattice (see page 29) can be brought into coincidence with any other lattice by applying a suitable affinity. Thus an affinity can be described as built up by successively applying a rigid

* Readers familiar with group theory will realize that we have defined "isomorphism" in a specialized (and to us much more useful) sense. We not only insist on the existence of a mapping between the groups that preserves their structure, but we also insist that elements be mapped onto like elements (reflections onto reflections, rotations onto rotations, etc.). The two concepts coincide except when the groups involved are *c2* and *d1*, or *p111* and *p1a1*, but it is far from trival to establish this fact.

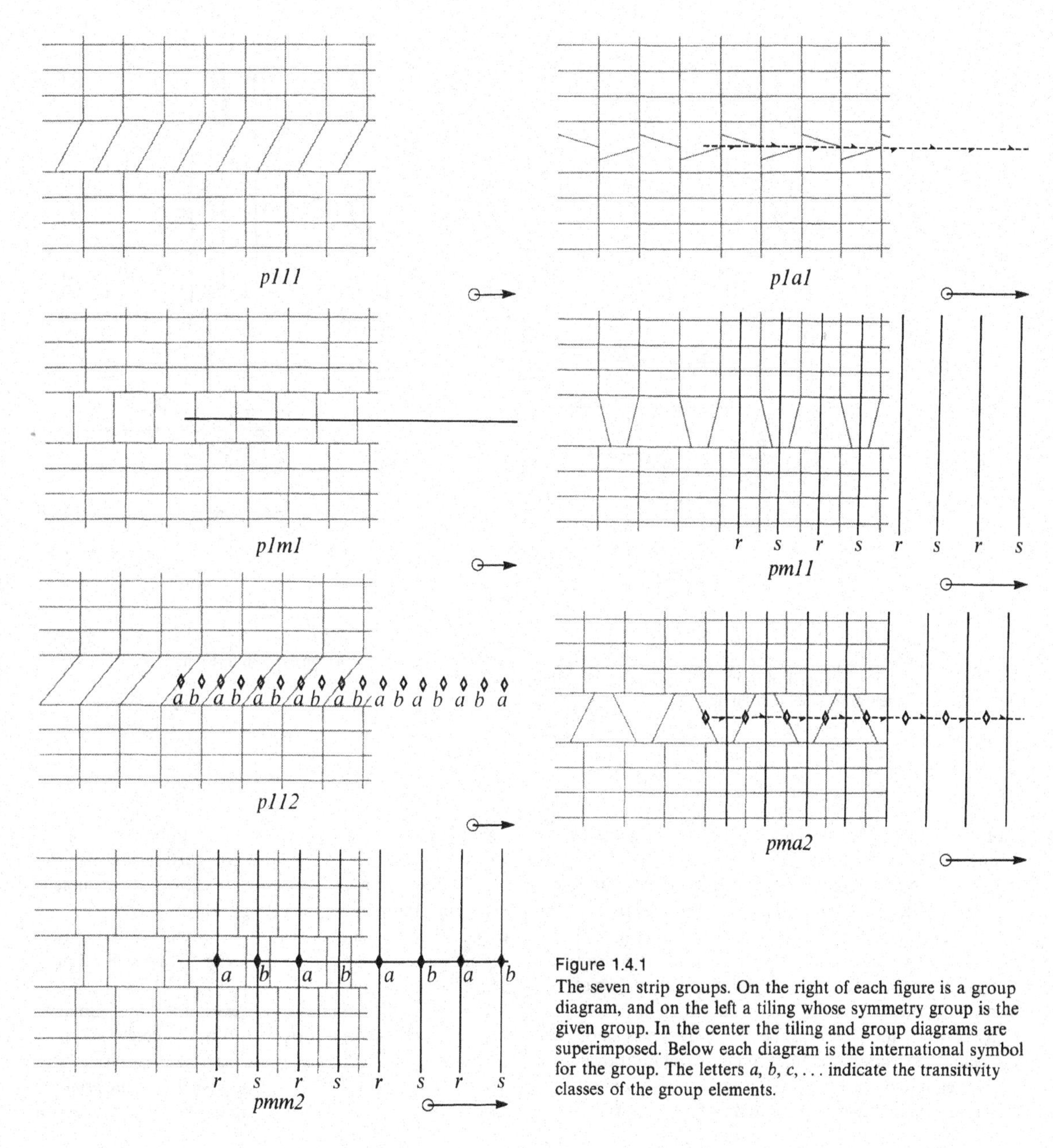

Figure 1.4.1
The seven strip groups. On the right of each figure is a group diagram, and on the left a tiling whose symmetry group is the given group. In the center the tiling and group diagrams are superimposed. Below each diagram is the international symbol for the group. The letters *a*, *b*, *c*, . . . indicate the transitivity classes of the group elements.

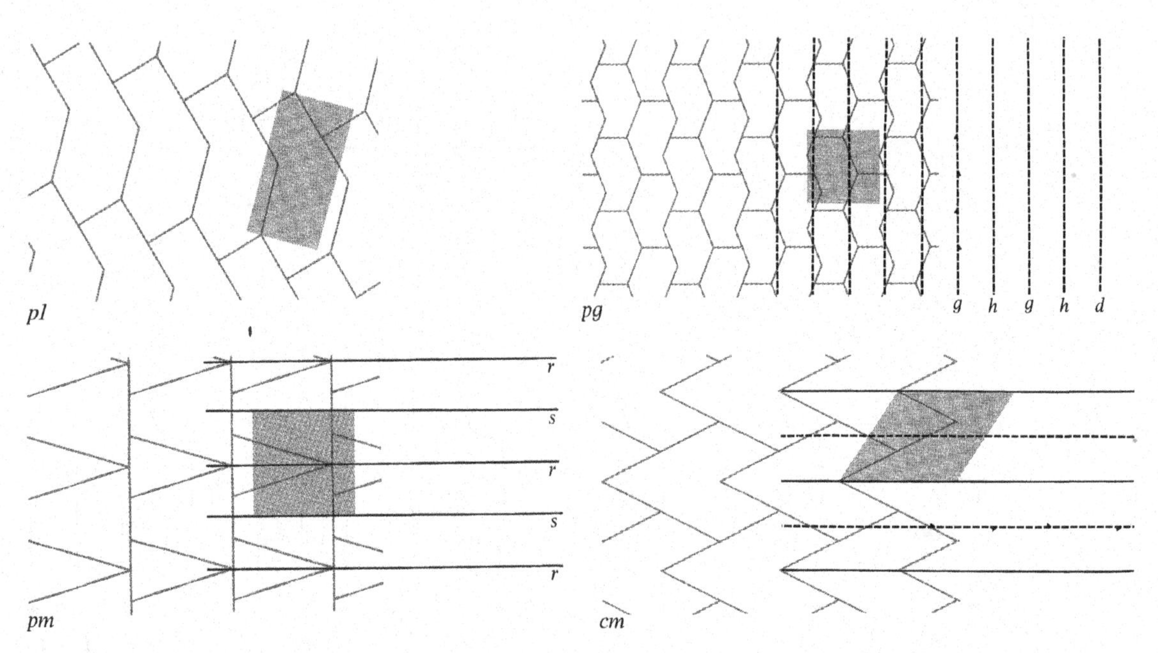

Figure 1.4.2
The seventeen crystallographic groups. Each diagram is arranged as explained in the caption to Figure 1.4.1. For clarity, instead of using two vectors to indicate translational symmetries, we show (shaded) the parallelogram they determine.

motion, a change of scale, and a shear (that is, a change of angle between the axes).

If $S(\mathscr{T}_1)$ and $S(\mathscr{T}_2)$ are isomorphic, then we shall say that $\mathscr{T}_1$ and $\mathscr{T}_2$ are of the same *symmetry type*. It is perhaps surprising that although the number of symmetry groups $S(\mathscr{T})$ is clearly infinite, if we restrict attention to tilings whose tiles are topological disks, the number of symmetry types is very limited. In fact when $S(\mathscr{T})$ contains no translations, it must be one of the types *cn* or *dn* defined in Section 1.3, and if it does contain translations then it must be of one of 24 types. In Figures 1.4.1 and 1.4.2 we display group diagrams of all these types, together with examples of appropriate tilings. We shall shortly give further details of these assertions, though we cannot here justify our statement that the list is complete. Later, in Chapter 6, we shall suggest how this can be done, or the reader may refer to the extensive literature on the subject (for example, Burckhardt [1966, Section 18], Guggenheimer [1967, Section III.3], Coxeter & Moser [1972, Section 4.5]). However, it is worth remarking here on a fact that turns out to be extremely useful if one is trying to enumerate symmetry types empirically: *Every symmetry in $S(\mathscr{T})$ is also a symmetry of the group diagram of $S(\mathscr{T})$.* Thus, for example, every group diagram

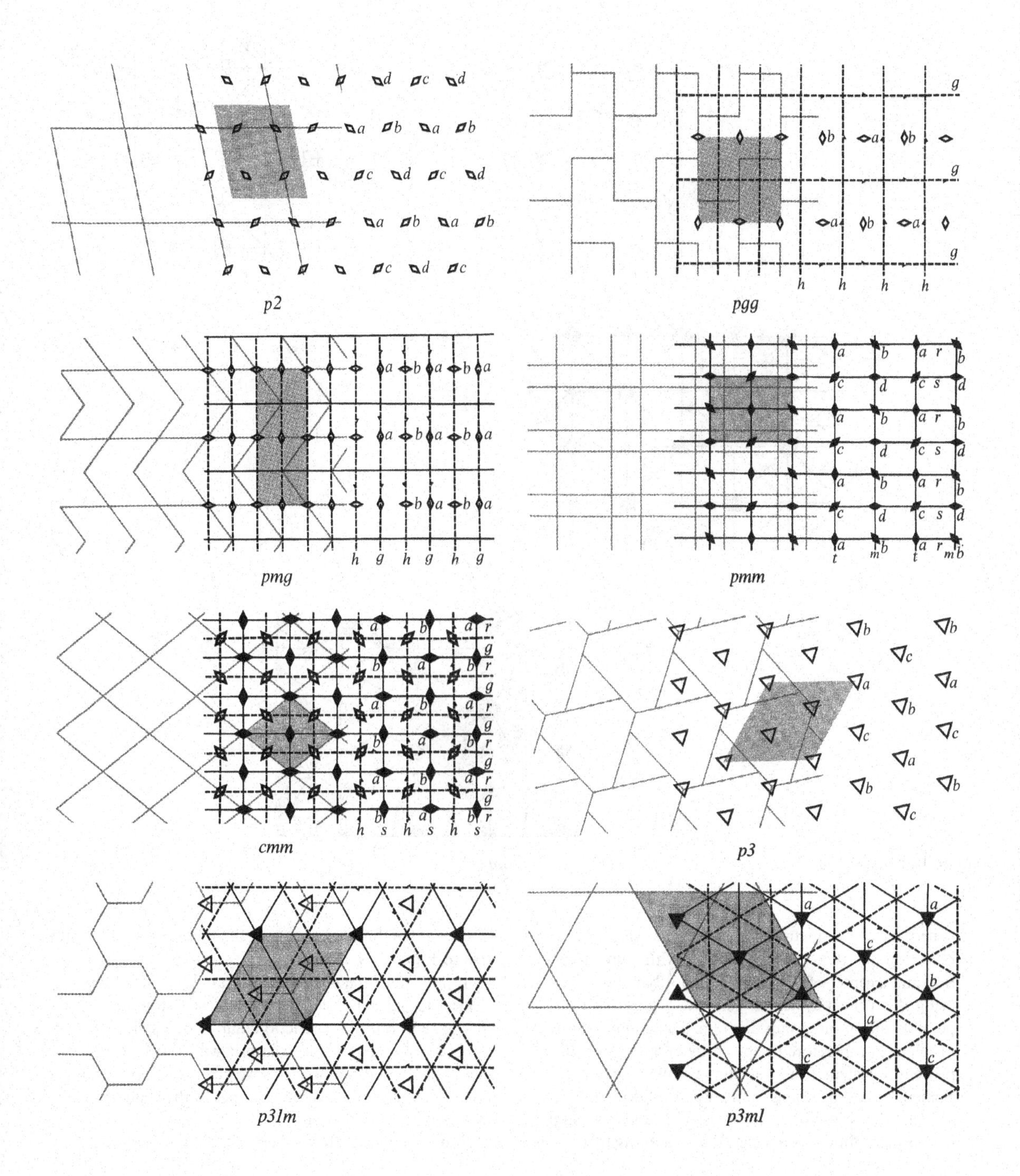

p2 *pgg*

pmg *pmm*

cmm *p3*

p31m *p3ml*

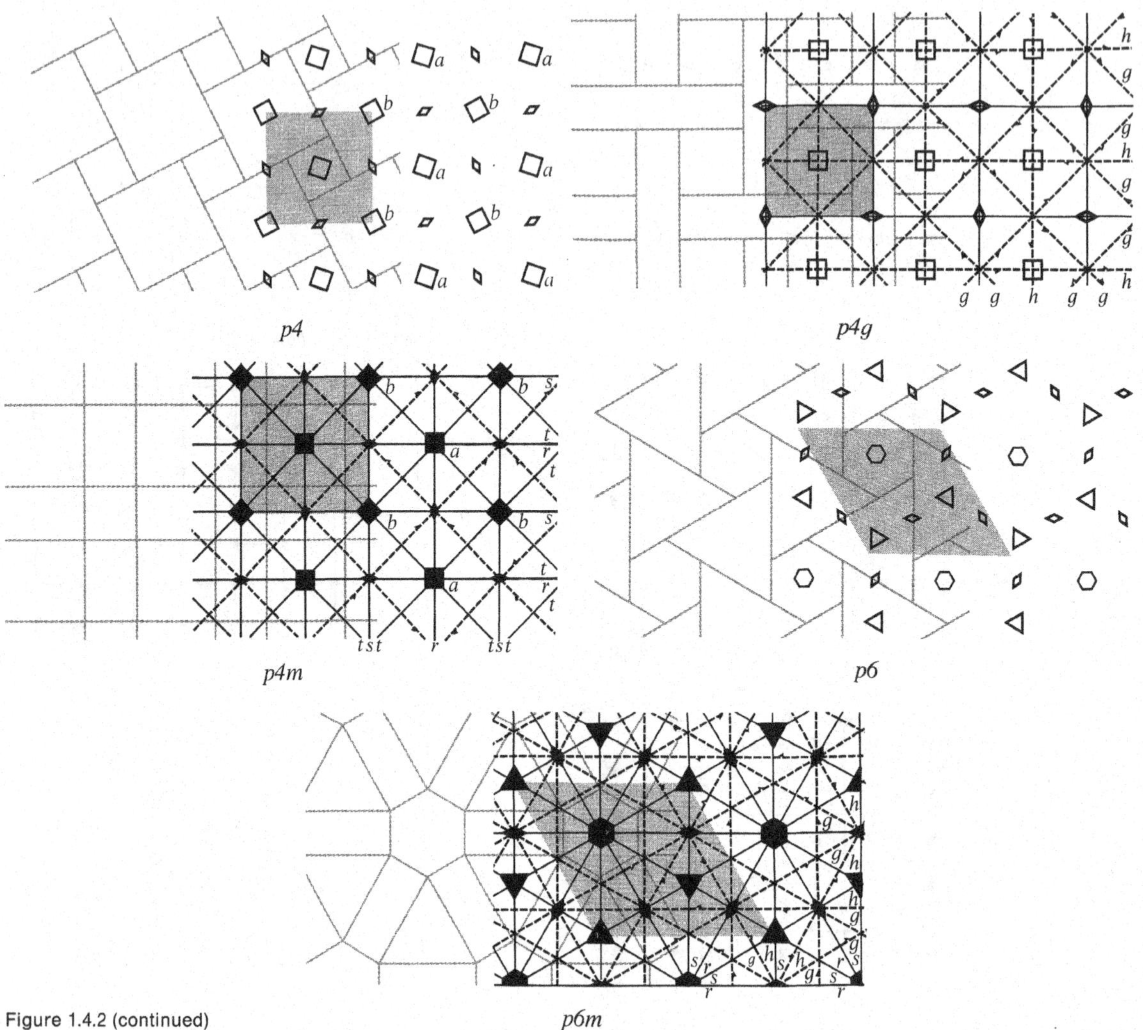

Figure 1.4.2 (continued)

is necessarily symmetric with respect to every line of reflection that it contains. If we start with a line of reflection and a center of n-fold rotation not lying on it, then using this principle we can generate a large part (or possibly all) of the group diagram. The enumeration problem can therefore be solved by carrying out this procedure systematically, starting with various group elements—and by doing so in a way that does not miss any possibilities! Notice, however, that a group diagram may possess more symmetries than the tiling from which it originated. See, for example, the diagram of the group *pm* in Figure 1.4.2.

It must be emphasized that this analysis and claim for completeness only hold when the tiles are topological disks—without this restriction many other symmetry types are possible. Some examples will be discussed in Sections 5.2 and 6.5. So far as we are aware, no systematic investigation of symmetry types in the general case has ever been carried out.

A detailed derivation of all 7 frieze groups and 17

crystallographic groups can be found in Martin [1982]. This book also contains a key for determining the type of symmetry group of a given tiling or pattern. Other keys for strip and periodic groups can be found in Rose & Stafford [1981]. Schattschneider [1978a] and Cundy [1979] point out that the crystallographic notations for the groups *p3m1* and *p31m* are frequently interchanged in the mathematical literature. The earliest example, and the source of the confusion, appears to be Speiser [1927]; the error is repeated in Budden [1972], Coxeter & Moser [1972] and many other works, but is corrected in Coxeter & Moser [1980].

Mackay [1969] mentions an ingenious method of displaying the crystallographic groups in the plane, suggested by N. M. Bashkirev in [1959]. He shows various jigsaw pieces that will only fit together in such a way as to display the 17 periodic groups.

Now let us consider the groups $S(\mathscr{T})$ in more detail. It is convenient to begin by considering three cases distinguished by the existence or otherwise of translations in $S(\mathscr{T})$.

1. *$S(\mathscr{T})$ contains no translations.* The non-trivial symmetries are necessarily rotations and reflections and not glide-reflections. One possibility is that $S(\mathscr{T})$ contains rotations only, and then it is of the type *cn*, the cyclic group of order n, for some value of n. If it contains more than one reflection then the corresponding lines of reflection cannot be parallel (for otherwise $S(\mathscr{T})$ would contain a translation) and they therefore meet in a point P. The product of the two reflections will be a rotation about P through angle $2v$, where v is the angle between the lines of reflection. The fact that the tiles are topological disks implies that v must be a rational multiple of π. Hence $S(\mathscr{T})$ is of finite order—it is of type *dn*, the dihedral group of order $2n$, for some value of n. Since we conventionally use *c1* and *d1* for the trivial group and the group containing only one reflection, we can assert that any symmetry group $S(\mathscr{T})$ without translations is of type *cn* or *dn* for an integer $n \geq 1$. In either case, it will be noted that there is at least one point of the plane (the *center* of the tiling) that is left fixed by every symmetry of $\mathscr{T}$. Examples of tilings of this kind appear in Figure 1.3.6(*a*) (type *c6*) and in Figure 1.3.6(*b*) (type *d3*).
2. *$S(\mathscr{T})$ contains translations, all of which are parallel to a given direction L.* Here $S(\mathscr{T})$ can contain reflections in lines perpendicular to L, and also at most one reflection in a line parallel to L. The only rotations that can occur are 2-fold, through angle π, and, if these exist, their centers must lie at equal distances on a line parallel to L. Using the principle explained on page 40 above, it is not difficult to see that only seven different types can arise. These are shown at right in the diagrams in Figure 1.4.1—in the left parts of the diagrams we show tilings illustrating these types. The seven groups are denoted by *p111*, *p1a1*, *p1m1*, *pm11*, *p112*, *pma2* and *pmm2*. They also occur as symmetry groups of strip patterns (see Chapter 5) and for this reason they are called *strip groups* or *frieze groups*. In fact, the tilings in Figure 1.4.1 are obtained from strip patterns of tiles by surrounding the strips with a "brick tiling". Any symmetry of the resulting plane tiling must superimpose the strip onto itself.
3. *$S(\mathscr{T})$ contains translations in non-parallel directions.* Here the tiling $\mathscr{T}$ is periodic and $S(\mathscr{T})$ can contain reflections, glide-reflections and rotations. The problem of enumerating the various possibilities is here much more laborious and we shall not attempt it since a more economical procedure will be described later. It turns out that the only rotations that can occur are of orders 2, 3, 4 or 6 and that there are only 17 distinct types of groups in all. These are indicated in Figure 1.4.2 along with representative tilings.

These 17 types are known as *wallpaper groups*, *periodic groups* or *(plane) crystallographic groups*. The latter name arose because they are analogous to the three-dimensional groups of crystallography. Perhaps the "wallpaper groups" terminology arose as a result of an early paper by Buerger & Lukesh [1937]. Associated with each group is a

Table 1.4.2 THE STRIP AND CRYSTALLOGRAPHIC SYMMETRY GROUPS

Column (1) gives the international symbol for the group.
Column (2) indicates whether it is a strip group (s) with all translations parallel or a crystallographic group (c) with translations in more than one direction.
Columns (3) give the number of transitivity classes of elements of each kind—glide-reflections, reflections and rotations of period 2, 3, 4 or 6.
All these groups are illustrated in Figures 1.4.1 and 1.4.2.

Symbol (1)	s or c (2)	Number of transitivity classes (3)					
				Rotations of period			
		Glide-reflections	Reflections	2	3	4	6
p111	s	0	0	0	0	0	0
p1a1	s	1	0	0	0	0	0
p1m1	s	0	1	0	0	0	0
pm11	s	0	2	0	0	0	0
p112	s	0	0	2	0	0	0
pma2	s	1	1	1	0	0	0
pmm2	s	0	3	2	0	0	0
p1	c	0	0	0	0	0	0
pg	c	2	0	0	0	0	0
pm	c	0	2	0	0	0	0
cm	c	1	1	0	0	0	0
p2	c	0	0	4	0	0	0
pgg	c	2	0	2	0	0	0
pmg	c	2	1	2	0	0	0
pmm	c	0	4	4	0	0	0
cmm	c	2	2	4	0	0	0
p3	c	0	0	0	3	0	0
p31m	c	1	1	0	2	0	0
p3m1	c	1	1	0	3	0	0
p4	c	0	0	1	0	2	0
p4g	c	2	1	1	0	1	0
p4m	c	1	3	1	0	2	0
p6	c	0	0	1	1	0	1
p6m	c	2	2	1	1	0	1

symbol, first introduced by crystallographers and now so universally adopted that it is known as the *international symbol*. An explanation of how these symbols are derived, and their relationship to the different groups, can be found in Schattschneider [1978a]. (Nowacki [1972], Coxeter & Moser [1972] and Schattschneider [1978a] present tables comparing the various notations for crystallographic groups that can be found in the literature.) We have also used analogous symbols from crystallography for the seven types of strip groups in Figure 1.4.1. These are not so generally adopted and, in fact, seem to us to be unsatisfactory in many respects. However, we do not introduce new symbols as this would tend to add to the confusion.*

In Table 1.4.2 we list the seven strip groups and 17 periodic groups together with information on the kind and number of transitivity classes of symmetry elements in each. A list of periodic subgroups of all the periodic plane groups, together with additional information, appears in Sayari, Billiet & Zarrouk [1978]. See also Billiet [1980]. Invariant one and two dimensional subgroups are listed in Senechal [1984].

EXERCISES 1.4

1. As Schattschneider [1978a] remarks, the identification of symmetry types of tilings can be fun! In order to practice their recognition, the reader is urged to identify the symmetry types of the many tilings shown in this book, and in particular of those in the Introduction and in Chapters 2 and 6.
2. Give an example of
   (*a*) a tiling whose symmetry type is unchanged by applying to it any affinity;
   (*b*) a tiling whose symmetry type can be changed to either of two other types by applying suitable affinities;
   (*c*) a tiling whose symmetry type can be changed to any of three other types by applying suitable affinities.
3. For each of the seven strip groups find a monohedral tiling with this group as its symmetry group. Determine for which strip groups it is possible to require that the prototile T has, as its symmetry group $S(T)$, (*a*) *c2*, (*b*) *d1*, (*c*) *d2*, (*d*) *c3*.
4. Which of the 17 crystallographic groups are symmetry groups of
   (*a*) monohedral tilings,
   (*b*) monohedral tilings with centrally symmetric prototile?
5. Determine which of the 17 crystallographic groups are symmetry groups of tilings in which all tiles are:
   (*a*) squares;
   (*b*) equilateral triangles;
   *(*c*) mutually similar (non-square) rectangles;
   *(*d*) mutually similar (non-square) rhombs.

* See the Appendix beginning on page 653.

1.5 MONOMORPHISM AND k-MORPHISM

In Section 1.2 we discussed the problem of deciding when a given tile is the prototile of a monohedral tiling. In this section we shall consider the possible numbers of distinct tilings admitted by a given prototile, or by a given set of prototiles.

A tile is called *monomorphic* if it is the prototile of a unique monohedral tiling of the plane. It is easy to find monomorphic tiles; a few examples appear in Figure 1.5.1. In (*a*) we show a hexagon with three pairs of equal parallel sides, each pair being of different length. The regular hexagon is also monomorphic, but a triangle is not; this is easily verified (see, for example, Figures 2.4.2(*a*) and (*b*)). We can however change a triangle into a monomorphic tile by replacing its sides by "zigzags" as shown in (*b*). If these zigzags are of sufficiently general shape (different for each side) then they ensure that the tiling is edge-to-edge. They also prevent reflections of the tile being used. The procedure just described, of replacing the edges of polygonal tiles by suitable curves in order to restrict the ways in which they can fit together, will often be used in the rest of the book. If we wish to focus attention on the fact that a polygonal tile P was altered by such replacements, then we shall say that the *basic shape* of the tile is P or that it is *basically* the polygon P with edges modified in the specified manner. Though this terminology may appear to be more psychological than mathematical—almost any shape of tile could be interpreted as basically any polygon!—it is very useful in understanding the structure of various tilings. This will be particularly evident in Chapters 6, 9 and 10.

In part (*c*) of Figure 1.5.1 we show a scalene quadrilateral with unequal sides and unequal angles, no two of which sum to π. Without these conditions a quadrilateral may not be monomorphic, as can be seen from Figure 9.1.2. In particular, a square admits an uncountable infinity of tilings by "sliding" the rows, see Figures 1.3.6(*c*) or 2.4.2(*c*), (*d*). However, it can be changed into a monomorphic tile by modifying its edges as shown in (*d*). The tile (*e*) is usually known as a *Greek cross*, and any two tilings using this prototile are either directly or

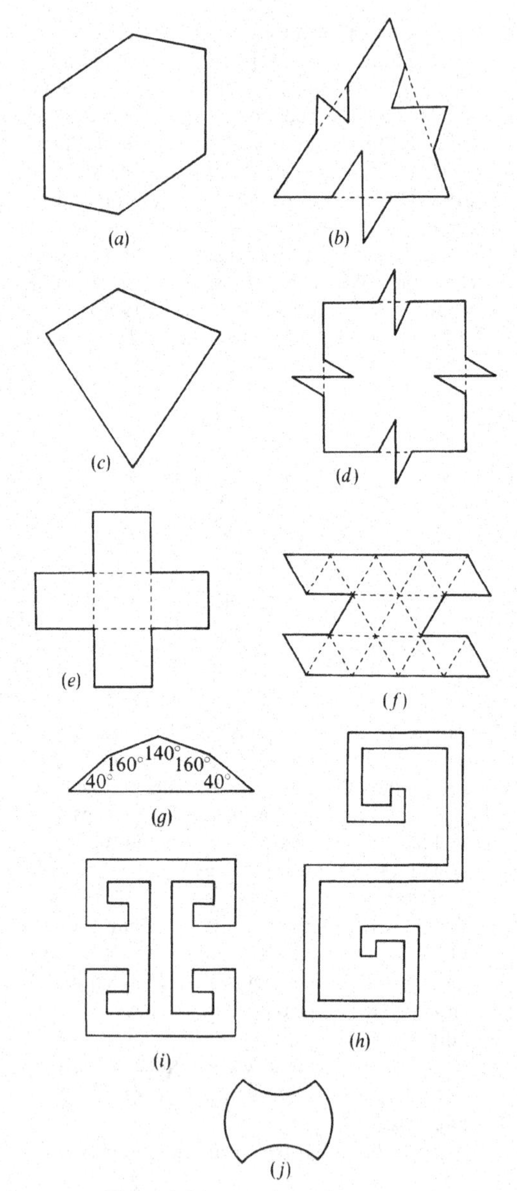

Figure 1.5.1
Examples of monomorphic prototiles.

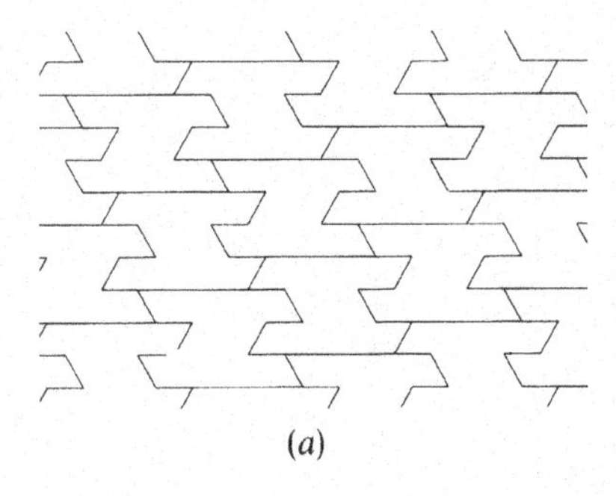

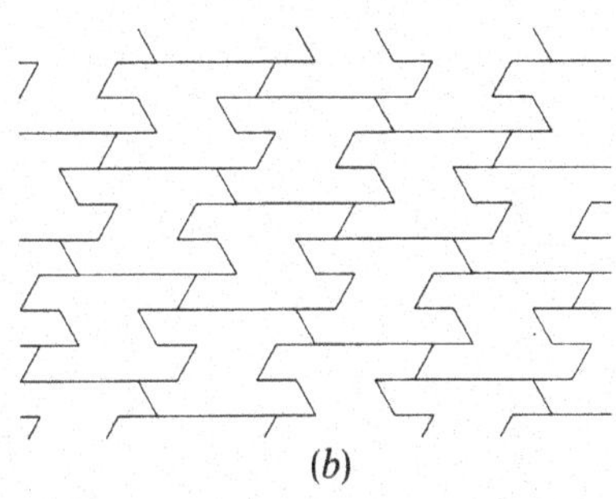

Figure 1.5.2
The two enantiomorphic tilings that are possible using the prototile of Figure 1.5.1(*f*).

reflectively congruent—and so are equal as defined in Section 1.1. The same applies to tile (*f*) and the corresponding tilings are shown in Figure 1.5.2. In this case the tile is the union of twenty equilateral triangles. The tile (*g*) leads to the unique tiling of Figure 1.3.7. Other examples are given in (*h*), (*i*) and (*j*); the tile (*i*) is the prototile of the tiling shown in Figure I.17(*b*). It is very easy to find many more examples of monomorphic tiles.

Puzzles in the form of interlocking lizards or clowns, which form monomorphic tilings (of type IH 17 in the notation of Section 6.2), are commercially available from Shmuzzles, Inc., of Chicago, IL 60680.

Now consider prototiles which are *dimorphic*, that is, admit *precisely* two distinct tilings. These are harder to find; a few examples are given in Figure 1.5.3. Tiles (*a*) and (*b*) are based on simple shapes, namely the union of three rhombs, and the union of seven equilateral triangles. Each of these shapes will tile the plane in an infinity of ways. To reduce the number of tilings to two we modify some of the sides as shown. The two tilings with tile (*a*) are shown in Figure 1.5.4 and the two tilings with tile (*b*) appear in Figure 1.5.5. Tiles (*c*) and (*d*) are based on the Greek cross of Figure 1.5.1(*e*). In (*c*) it has been sheared and then stretched; in (*d*) it has had its sides replaced by S-shaped curves. In each case the effect is to transform the two mirror-image tilings with the basic Greek cross into different tilings. In Figure 1.5.6 we show the two tilings that are possible with each of these tiles.

Tiles which are *trimorphic* are harder to find; one such tile is shown in Figure 1.5.7.

The examples we have given suggest the following open question: for every positive integer r, is it possible to find a tile which is r-morphic? For each $r \leq 10$, Fontaine & Martin [1983a], [1983b], [1984a], [1984b] give examples of r-morphic polyominoes. We believe that r-morphic tiles exist for every value of r, although we know of no method of generating an r-morphic tile for an arbitrarily prescribed value of r. Another open question is whether there exists a tile that admits a *countable* infinity of distinct tilings; we believe that the answer is negative.

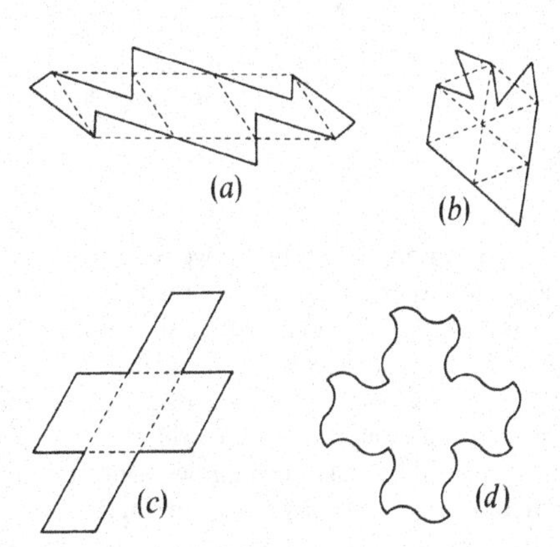

Figure 1.5.3
Some dimorphic prototiles.

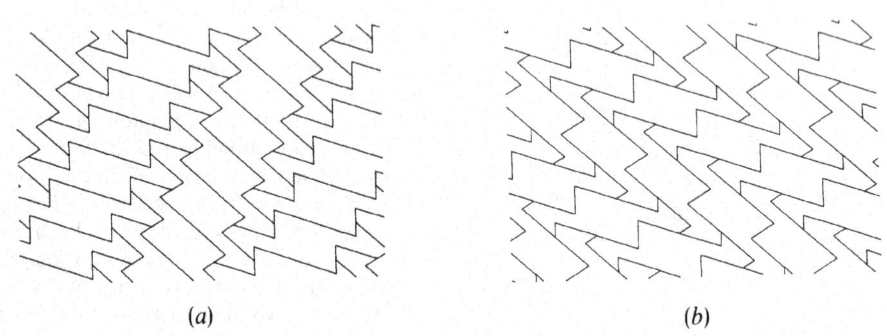

Figure 1.5.4
Two monohedral tilings with the same prototile. It can be shown that this prototile admits no other tilings.

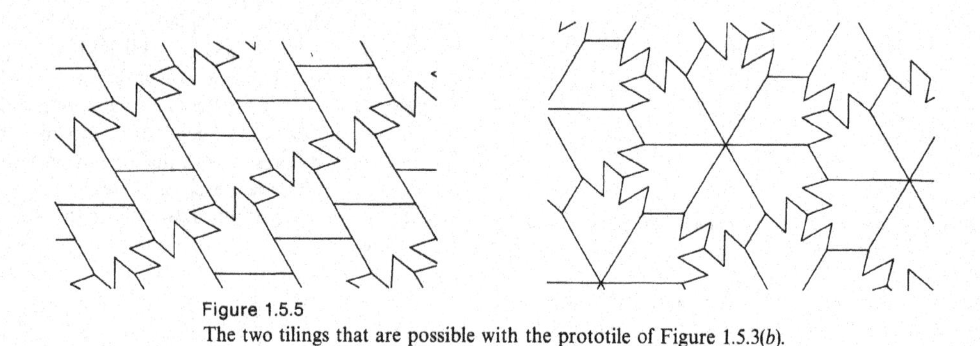

Figure 1.5.5
The two tilings that are possible with the prototile of Figure 1.5.3(*b*).

It should be pointed out that if we require the tilings to be *isohedral*, then for each $r \geq 1$ there exist tiles that admit precisely r such tilings. To see this consider the tile of Figure 1.5.8. Each side has r "teeth" as shown (in the figure $r = 7$). Then the tiles can be fitted together in rows, and each row can occupy r different positions relative to one of its neighboring rows. Once the relative position of two adjacent rows is given, then isohedrality determines the rest of the tiling uniquely. (In Figure 1.5.8 one such tiling is shown.)

We now ask the analogous question about sets of n prototiles ($n \geq 2$.) The following general result was recently proved by Harborth [1977a], [1977b]:

1.5.1 *For any $r \geq 1$ and $n \geq 2$ there exists a set of n prototiles which is r-morphic, that is, admits exactly r distinct n-hedral tilings.*

To begin the proof we note that this result is trivial for $r = 1$, 2 or 3. For we need only take any one of the r-morphic tiles mentioned above and split it into n parts. If these parts are of such shapes that they can only be

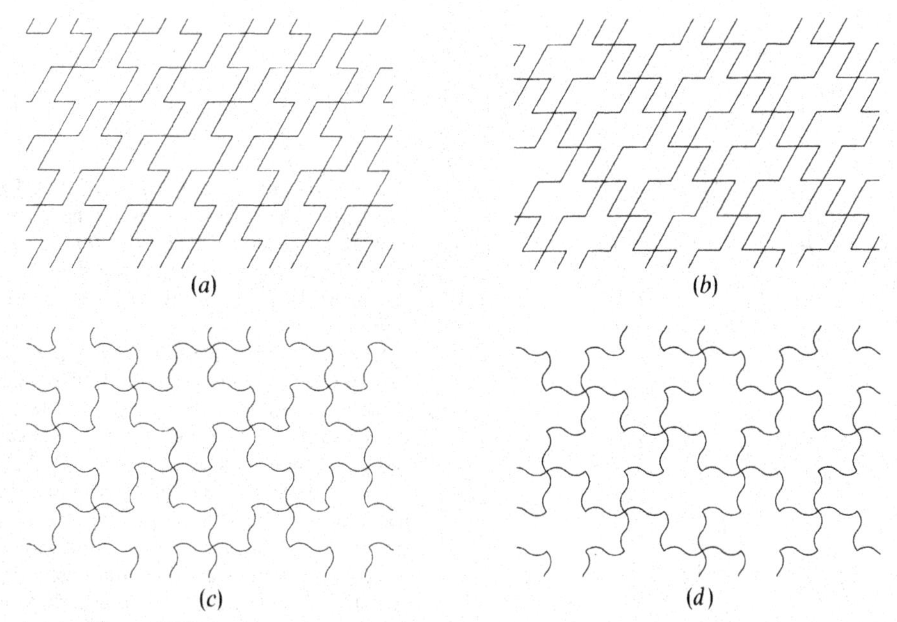

Figure 1.5.6
The two tilings that are possible with each of the prototiles of Figures 1.5.3(*c*) and (*d*).

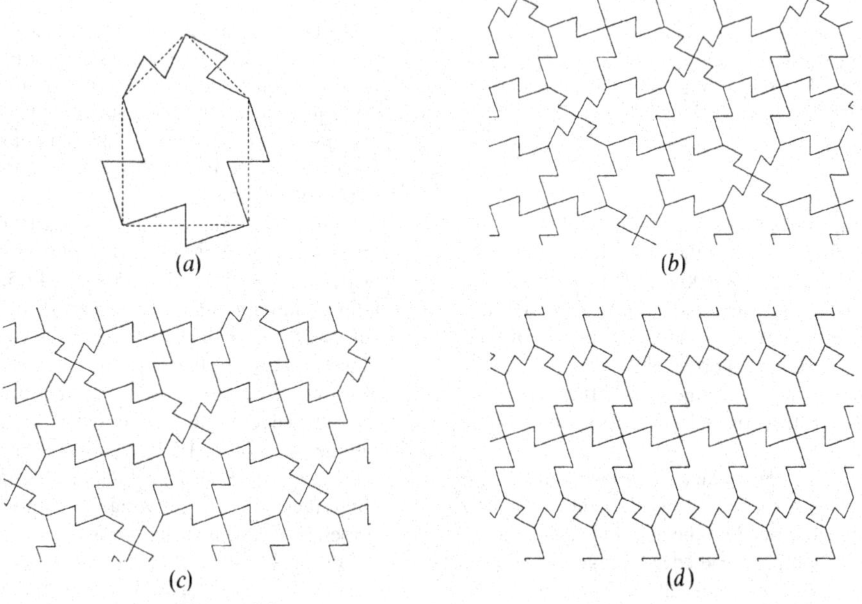

Figure 1.5.7
A trimorphic prototile and the three tilings which it admits.

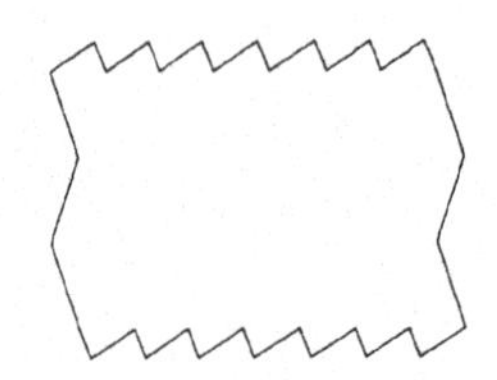

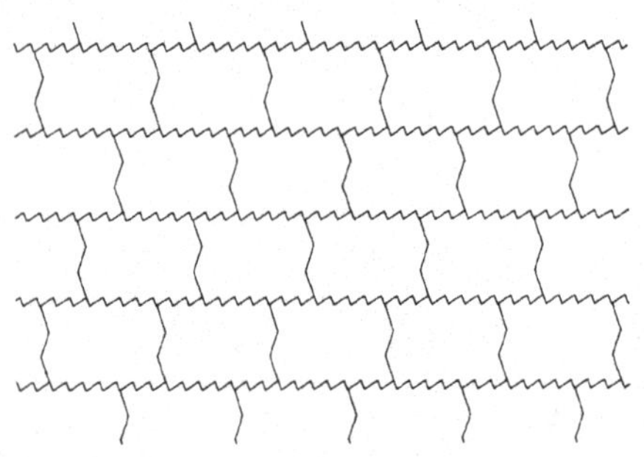

Figure 1.5.8
A prototile that admits exactly seven isohedral tilings, one of which is shown. In a similar way, one can construct prototiles that admit exactly n isohedral tilings, where n is any positive integer.

fitted together so as to recover the original tile, then clearly they form a set of n tiles with the required property. In a similar way, if we know of an r-morphic set of 2 tiles, then by splitting one or both of the tiles into parts, we can construct an r-morphic set with n tiles for any $n \geq 3$. Hence, to prove Statement 1.5.1 it will suffice to exhibit an r-morphic set of 2 tiles for $r \geq 4$. (In fact, the construction to be described works for any $r \geq 2$.)

Let $p = 6r - 7$, and take a rhomb with angles $2\pi/p$ and $(p - 2)\pi/p$ as the first tile T_1. For the other tile T_2 we take a block formed by the union of $(p - 2r + 3)/2 = 2r - 2$ such rhombs placed edge-to-edge with a common acute vertex, which we shall call the "apex" of T_2. Then it is easy to see that there is a unique tiling using just *one* copy of T_2—namely that in which the tiles "radiate" from the apex of T_2, as in Figure 1.5.9(*a*). There are also $r - 1$ tilings in which *two* tiles of type T_2 occur with coincident apexes, see Figures 1.5.9(*b*), (*c*) and (*d*). No other tilings are possible. One can show that these statements hold for all values of r, and so Statement 1.5.1 is proved.

Schmitt [1985a], [1985b], [1985c], has described several constructions for pairs of tiles which admit any prescribed number of tilings. Unlike Harborth's tiles, described above, here the tiles admit no monohedral tilings. (A dimorphic pair of tiles with this property is shown in Figure 3.2.3.) One of Schmitt's constructions leads only to periodic tilings, while another (in which one of the two tiles is a rhomb modified by making suitable projections and indentations on its sides, and the other is a "key-tile") leads to both periodic and non-periodic tilings. The latter construction is especially interesting because it can be modified slightly so as to yield a pair of tiles which admit a countable infinity of tilings.

Harborth's construction illustrates the fact that some pairs of prototiles have the following remarkable property. It may happen that there is an infinity of tilings using just one of the tiles, but the introduction of even a *single* tile of the second shape completely alters this situation. We conclude this section with some examples of this phenomenon.

Figure 1.5.10(*b*) shows how to construct an uncountable infinity of tilings using an equilateral hexagon with angles $3\pi/5$, $3\pi/5$, $4\pi/5$, $3\pi/5$, $3\pi/5$, $4\pi/5$. (The tiling contains an infinity of horizontal zigzags, one of which is emphasized in the figure. At any such zigzag the "slope" of the tiles can be changed from left to right or vice versa.) However, if we insist that a regular pentagonal tile of the same edge length is also used, then the tiling becomes unique (see Figure 1.5.10(*a*)). This tiling has been interpreted by Senechal [1980] as a model for a "hypothetical twin" growing from a nucleus in the presence of a pentagonal impurity.

For our second example we again use squares which, as we have remarked, admit an uncountable infinity of tilings. However, if we choose for our second tile a regular octagon of equal edge length, and insist on at least

Figure 1.5.9
Harborth's construction for two tiles with the property that there are exactly r tilings that use both tiles. In this diagram $r = 4$. One tile is a rhomb with angles $2\pi/p$ and $(p-2)\pi/p$ where $p = 6r - 7$, and the other tile is obtained by fusing together $2r - 2$ rhombs as shown.

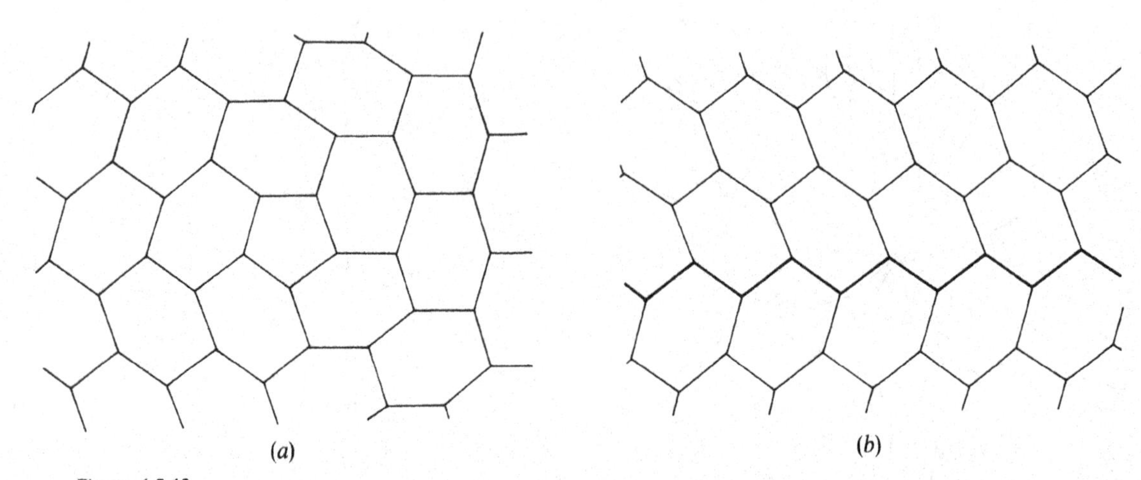

Figure 1.5.10
A dihedral tiling by hexagons and a regular pentagon is shown in (*a*). This tiling is unique. If, however the pentagon is omitted, an uncountable infinity of tilings becomes possible using hexagons alone. One of these is shown in (*b*).

one such tile occurring in the tiling, then the tiling becomes unique. It is the tiling denoted by (4.8^2) and shown in Figure 2.1.5. Notice that there is no tiling involving a finite (non-zero) number of octagons, whereas in the previous example there exists no tiling with more than *one* copy of the second prototile (pentagon).

For our final example we take as our tiles a rhomb with angles $\pi/5$, $4\pi/5$, $\pi/5$, $4\pi/5$, and a regular pentagon with the same edge length. This pair of prototiles has some unusual features. Note that there is an uncountable infinity of tilings using either rhombs alone, or rhombs and pentagons (see Figure 1.5.11(*a*)). (The famous artist and geometer Albrecht Dürer (1471–1528) produced many tilings, among them one using these same two prototiles and having pentagonal symmetry; see Dürer [1625], Pedoe [1976].) However, if we insist, in addition, *either* that only a finite non-zero number of pentagons occurs, *or* that pentagons occur and that two of the rhombs have an edge in common, then only three tilings are possible. These are shown in Figures 1.5.11(*b*), (*c*) and (*d*) and involve 1, 2 or 6 pentagons.

Some variants of this example are worth mentioning. If we either change the edge length of the pentagon so that it is not an integral multiple of that of the rhomb, or use a rhomb with angles $\pi/10$, $9\pi/10$, $\pi/10$, $9\pi/10$ in place of that described above, then only *one* tiling is possible using both tiles—namely a tiling similar in appearance to Figure 1.5.11(*b*). Of course, an uncountable infinity of tilings using the rhombs alone is also possible.

These examples show that the introduction of a few new tiles into a tiling may profoundly alter its nature. (An even more striking example of this phenomenon will be given in Section 10.5.) This effect may be regarded as a mathematical analogue of the well-known physical effects of introducing a foreign atom into a crystal. In fact, "fault lines", "planes of cleavage" and other crystallographic ideas all seem to have their analogues in the theory of tilings.

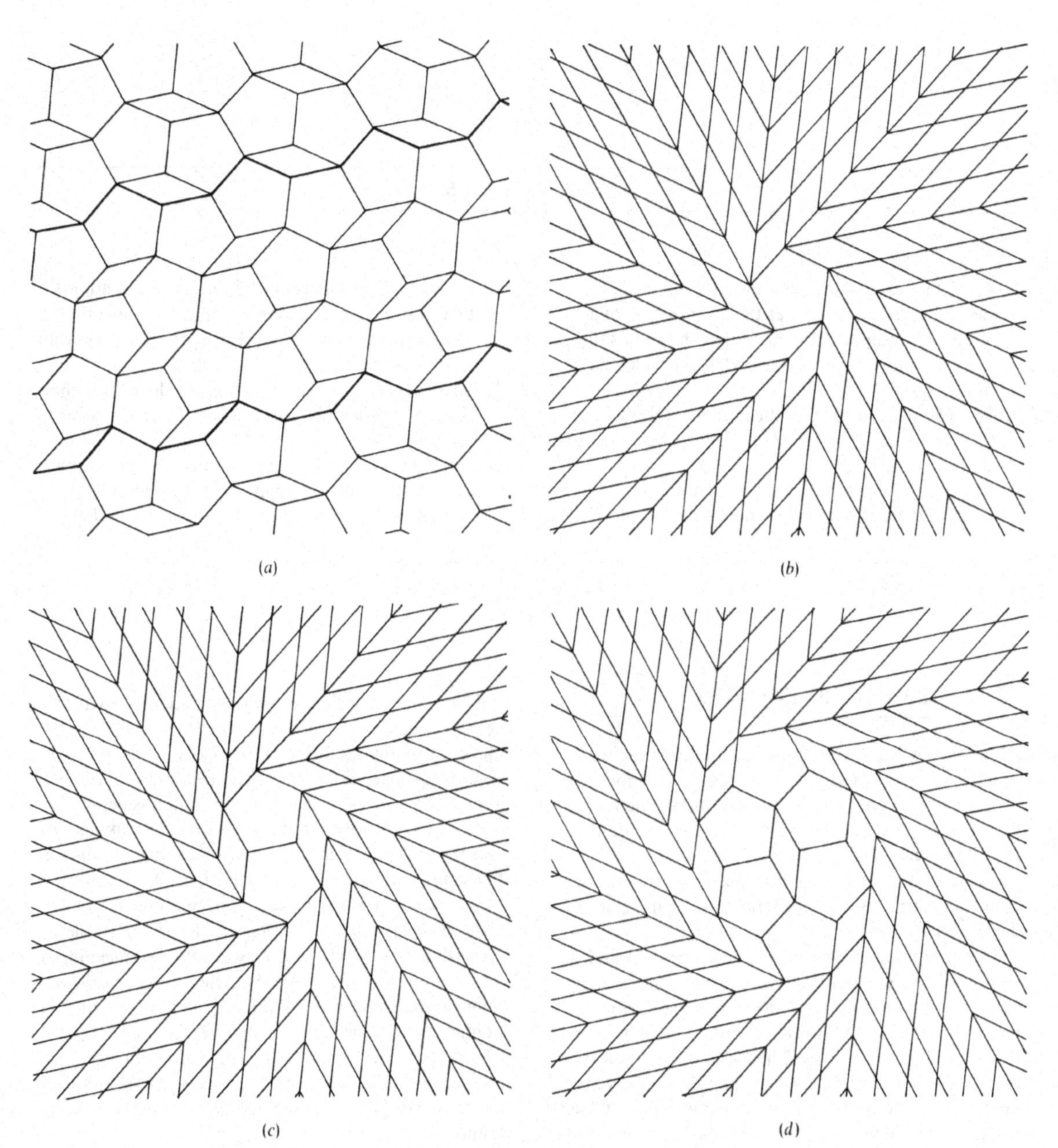

Figure 1.5.11
Dihedral tilings with regular pentagons and rhombs with angles $\pi/5$, $4\pi/5$, $\pi/5$ and $4\pi/5$. Changing the "tilt" of the tiling at any of the infinitely many polygonal lines (two of which are emphasized in (*a*)) leads to uncountably many different tilings.

EXERCISES 1.5

1. Verify that each of the prototiles in Figure 1.5.1 is monomorphic; this requires the construction of a tiling, and finding an argument that shows this tiling to be the only one possible with the prototile under consideration.
2. Consider the prototiles of the monohedral tilings in Figures I.12, I.17, and I.18. Which of these prototiles are monomorphic?
3. (*a*) Show by examples that for each $n = 1, 2, 3, 4, 6, 8, 12$ there exists a monomorphic prototile T_n that occurs in n aspects in the only tiling it admits.

   **(*b*) Does there exist a monomorphic prototile that occurs in its only tiling in n aspects for some value of n other than 1, 2, 3, 4, 6, 8, 12?

**4. Does there exist a dimorphic prototile T such that its symmetry group $S(T)$ is $c3$, $c6$ or dk for some $k = 1, 2, 3, 4, 6$?

**5. For which $r \geq 1$ and $n \geq 2$ do there exist r-morphic sets of n tiles with the property that no set of k of the tiles, with $k < n$, admits a tiling of the plane?

1.6 NOTES AND REFERENCES

Although most of the concepts introduced in the first two sections of this chapter have appeared in more or less explicit form in various publications, the literature contains no systematic account of the theory of tilings and there is no generally accepted terminology. Many of the ideas seem to have been introduced in the form of casual and imprecise remarks about tilings made in the discussion of other topics.

Some of our definitions may seem elementary and "obvious". But it is only by giving them explicitly that we can avoid the confusion arising from the various meanings assigned to the same words by different authors. Even the word "tiling" has been used in at least two senses. To most writers its meaning is that given here, but to others a tiling is a "decomposition" or "partition" of the plane—that is, every point of the plane belongs to one, and only one, tile. The difference between the definitions is illustrated by the fact that in most of our tilings the tiles are closed topological disks, whereas quite clearly there exists no decomposition of the plane into such sets.

Tilings in our sense can always be modified so as to become "decomposition tilings"—for example, if the tiles are topological disks we only need to "delete" an appropriate part of the boundary from each tile (though if the tiling is monohedral, then it is not in general possible to carry out this deletion in such a way that the tiles remain congruent to each other). Some examples of this construction are indicated in Figure 1.6.1. Conversely, a decomposition tiling can often be converted to a tiling as defined here by taking the closures of its tiles. Nevertheless, the difference between the two meanings is far-reaching, and the whole character of the investigations depends on the definition adopted. It is worth mentioning that the idea of a decomposition tiling is capable of wide generalizations, such as the "tiling" of a group (see, for example, Stein [1974], Conlan [1976] and Halsey & Hewitt [1978], where additional references can be found).

Decomposition tilings in which the "tiles" are arcs of smooth curves, or straight-line segments, have been investigated by Bothe [1964] and Conway & Croft [1964]. Among the results of the paper by Conway and Croft, we mention the following: there exists "monohedral" decomposition tilings of the plane in which the prototile is a closed line-segment, but none in which the prototile is an open line-segment.

Many of the early investigators of tilings (Fricke & Klein [1897], Hilbert [1900] and others) were preoccupied with "discrete groups" and "fundamental domains". A group G of isometries of the plane onto itself is called *discrete* if for each point P of the plane there is a positive distance $d = d(P)$ such that no image of P (distinct from P) under an element of G is at distance less than d from P. It can be shown that every discrete group of isometries of the plane is either *cn* or *dn* for a suitable n, or else is one of the 7 strip groups or one of the 17 plane crystallographic groups. A set F is a *fundamental domain* for a discrete group G if (*a*) F is a connected set with non-empty interior; (*b*) F contains no pair of points equivalent under G; and (*c*) F is maximal in the sense that it is not a proper subset of any set with properties (*a*) and (*b*). Fundamental domains are also known as fundamental regions, generating regions, or asymmetric units. If F is a fundamental region for G then clearly the family $GF = \{gF \mid g \in G\}$ is a covering of the plane. However, the geometric interest of these notions is rather limited. On the other hand, for most groups G no fundamental region F can be a closed set; hence GF is not a tiling in the sense used here. On the other hand, if G is not one of *c1*, *p111*, *p1a1*, *p1* or *pg*, then GF is not a decomposition tiling either (despite the statements in Hilbert & Cohn-Vossen [1932, pp. 58–59] and Santaló [1976, p. 128] that appear to claim otherwise).

The condition that every tile is a closed topological disk is much stronger than it appears at first sight; if we insist only that each tile is the closure of an open topological disk then many "pathological" examples would exist. For example, in this case the construction known as the "lakes of Wada" (see Yoneyama [1917, p. 61], Kuratowski [1924], Kerékjártó [1923, p. 118]) shows that it is possible for *every* point on the boundary of an open topological disk to be also on the boundary of *two* or more other open topological disks. The use of closed topological disks as tiles eliminates such disturbing possibilities.

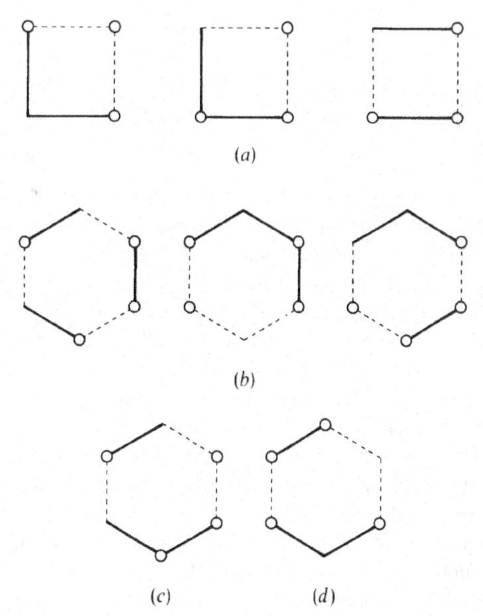

Figure 1.6.1
If any three vertices and the relative interiors of any two sides of a square are deleted, then we obtain the prototile of an isohedral "decomposition tiling" as defined in Section 1.6. In (*a*) we give three examples, the first two of which tile using translations only. (The deleted sides are dashed and the deleted vertices indicated by open circles.) In the case of regular hexagons, four vertices and three edges must be deleted, as in the examples in (*b*). However, in this case there are two essentially different hexagonal tiles (*c*) and (*d*) which do not tile isohedrally, though with (*c*) a monohedral tiling is possible. It is not possible to convert a tiling by equilateral triangles into a monohedral "decomposition tiling" by deleting edges and vertices in the manner just described (Conlan [1976]).

The "transparent sheet" method of checking the symmetries of a tiling (mentioned in Section 1.3) was first proposed by Fourrey [1907, p. 368] in connection with the enumeration of uniform tilings by regular polygons (see Section 2.1). Unfortunately he used it incorrectly and so was able to "prove" a false statement (see Section 2.10).

Crystallography provided the motivation for the first investigations of the periodic groups. This accounts for the surprising fact that the much more complicated enumeration of the three-dimensional crystallographic groups (there are 230 of them) was accomplished before the determination of the 17 groups in the plane. Both enumerations were first carried out by Fedorov (in [1885] and [1891]). The three-dimensional groups immediately attracted attention—independent enumerations were carried out by Schoenflies [1891] and in part by Barlow [1894], and several accounts of these groups were soon published in the mathematical and crystallographic literature (see, for example, Hilton [1903], Niggli [1919], Bogomolov [1932]). For a historical account of the discovery of the 230 space groups, see Burckhardt [1967a]. After many years of efforts it has been established that there are precisely 4783 classes of crystallographic space groups in 4-dimensional Euclidean space (see Brown, Bülow, Neubüser, Wondratschek & Zassenhaus [1978]). The number of classes in 5 dimensions is not known.

The study of the three-dimensional groups provided the motivation for the eighteenth of the famous problems of Hilbert [1900]. This asks, in part, whether the number of "essentially different" symmetry groups is always finite in a Euclidean space of finite dimension; for accounts of the solution of Hilbert's problem and for references to the original literature see Delone [1971] or Milnor [1976].

It is surprising that, apart from the independent investigations by Fricke & Klein [1897] undertaken in connection with the theory of functions of complex variables, Fedorov's determination of the 17 plane crystallographic groups was essentially forgotten for many years. Their rediscovery was due to Pólya [1924], Niggli [1924], Speiser [1927] and others. Discussions of these groups and their representation by group diagrams (as defined in Section 1.4) or by tilings or patterns that illustrate them can be found in many places. A particularly rich collection is Stevens [1980].

The thesis of E. Müller (Müller [1944]) is one of the first systematic applications to ornaments from a specific historic period of classification by symmetry groups. A short exposition of the results appears in Müller [1946]. Since then there have been many investigations in this direction; the following list can serve as a guide to further literature: Campbell [1983], Crowe [1971], [1975], [1981], [1982], Crowe & Washburn [1983], [1984], Donnay & Donnay [1985], Garrido [1952], Hanson [1985], Knight [1984a], [1984b], Shepard [1948], [1956], Washburn [1977], [1983], Zaslavsky [1973], Zaslow [1977], [1980], [1981], Zaslow & Dittert [1977a], [1977b]. According to Müller [1944], the widespread belief that all 17 classes of plane crystallographic groups can be found in the Moorish decorations of the Alhambra in Granada (Spain) is erroneous. This has been confirmed by more recent investigations, see Grünbaum, Grünbaum & Shephard [1985]. For a critique of this approach see Grünbaum [1984].*

The following list, which does not repeat the references already mentioned, contains the historically important sources for the 17 crystallographic groups, together with some more recent ones that are either easily available or visually attractive: Birkhoff [1933], Bradley [1933], Budden [1972], Buerger [1956], Burckhardt [1966], Cadwell [1966, Chapter 11], Coxeter & Moser [1972], Fejes Tóth [1965], Guggenheimer [1967, Section 3.3], Günzburg [1929], Heesch [1968b], Loeb [1971a], MacGillavry [1965], March & Steadman [1971], O'Daffer & Clemens [1976], Schattschneider [1978a], Schwarzenberger [1974], Šubnikov & Kopcik [1972, Chapter 7].

The first enumeration of the frieze groups seems to be due to Niggli [1926]. For other accounts of these groups see Budden [1972], Cadwell [1966], Coxeter [1961, Section 3.7], Crowe [1971], [1975], [1981], Fejes Tóth [1965], Speiser [1927], O'Daffer & Clemens [1976, pp. 228–230].

The analogues in three dimensions of the frieze groups are the "cylindrical (or rod) groups". These are the symmetry groups of patterns on the surface of a circular cylinder; they have been investigated by Alexander [1929]. See also Roman [1971], Koch & Fischer [1978]. Cylindrical patterns are important in the study of biological structures; see, for example, Caspar & Klug [1962], Iterson [1970], Erickson [1973], Salazar, Hutcheson, Tollin & Wilson [1978], Erickson, Tollin, Richardson, Burley & Bancroft [1982].

Section 1.5 is based on Grünbaum & Shephard [1977b] and Harborth [1977a], [1977b].*

* See the Appendix beginning on page 653.

2

TILINGS BY REGULAR POLYGONS AND STAR POLYGONS

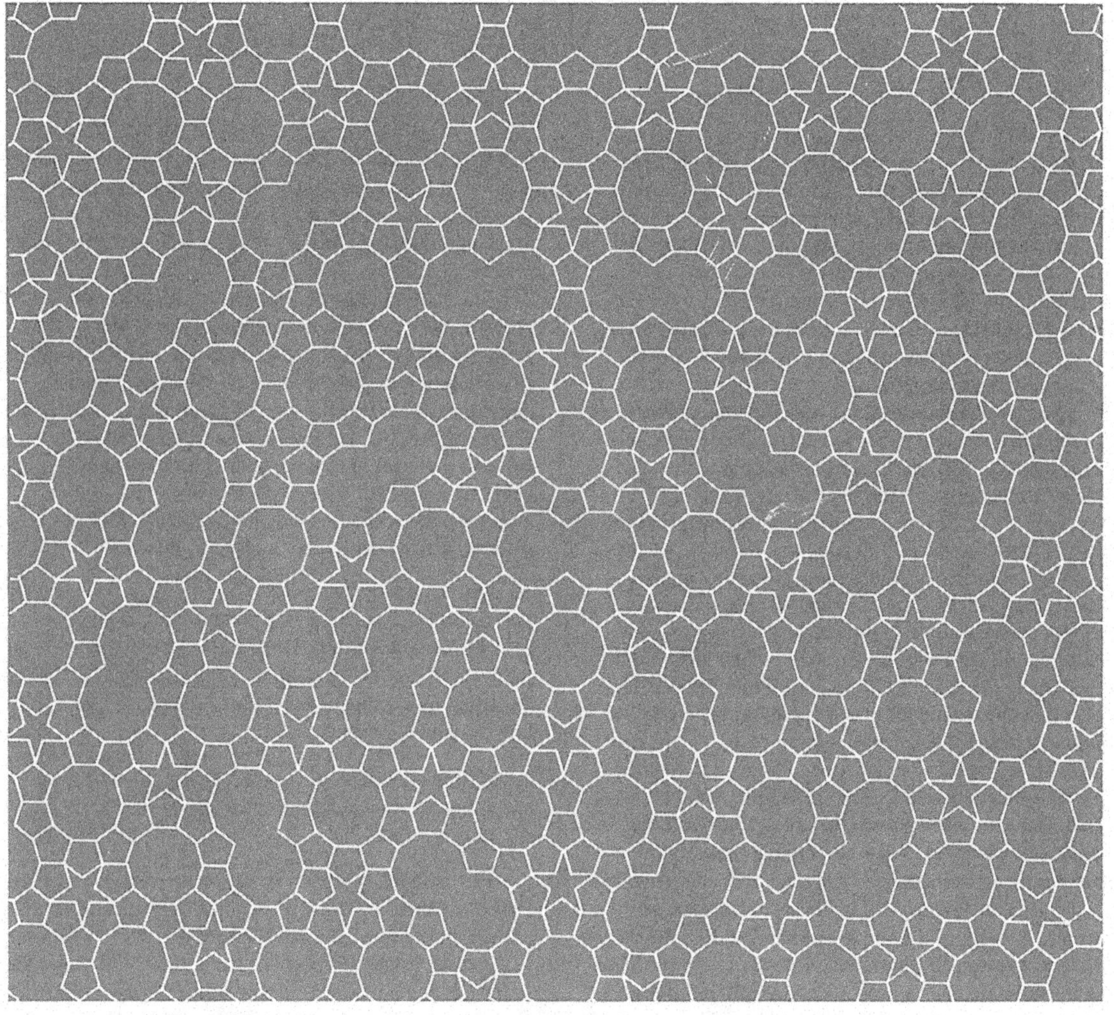

Figure 2.0.1
This is one of Kepler's more remarkable tilings. The tiles are regular pentagons, pentacles, regular decagons and "fused decagon pairs" or "monsters" as Kepler called them. Its construction is explained in Section 2.5. Other tilings involving pentagons, pentacles, decagons, star decagons and related polygons are shown in Figures 2.5.8 and 3.0.1 and in Chapter 10.

2

TILINGS BY REGULAR POLYGONS AND STAR POLYGONS

Many tilings by regular polygons—equilateral triangles, squares, regular hexagons, and so on—are well known. In Section 1.2 we mentioned the three regular tilings and other examples are shown in Figures I.10 and I.13. Recent school books that mention tilings seem to devote their attention almost exclusively to tilings of this type. Yet, as we shall see in this chapter, the subject has many facets which are rarely discussed and there remains a large number of open problems.

Tilings by regular polygons were the first kinds of tilings to be the subject of mathematical research. As we have already mentioned in the Introduction, they figure prominently in the pioneering work of Kepler [1619]. Through the centuries the concept of "more or less regular" tilings by regular polygons developed, and here we discuss this evolution. Quite apart from the simplicity and attractiveness of these tilings, interest in them is due to the fact that they sometimes occur as extremal solutions to various problems (see the detailed account in Fejes Tóth [1953]; see also Fejes Tóth [1960], [1963]), and that the treatment of more general problems can often be reduced to ones involving regular polygons (see, for example, Chapter 4).

In the first four sections of this chapter we shall use only regular convex polygons as tiles; if such a polygon has n sides and therefore n corners, we shall call it a *regular n-gon* and denote it by the symbol $\{n\}$. Thus $\{3\}$ is an equilateral triangle, $\{4\}$ is a square, and so on. Except in Sections 2.4 and 2.5 we shall restrict attention to tilings that are *edge-to-edge*. We recall that this means that every side of every tile is an edge of the tiling, and conversely; so each side of a tile is also a side of precisely one other tile.

Section 2.7 will be devoted to a related question—the existence of tilings whose *vertices* are regular. By this we mean that if, at any vertex, n edges meet (that is, the valence of the vertex is n), then the angle between neighboring pairs of edges is $2\pi/n$. Although such tilings have never been studied as extensively as those with regular polygonal tiles, they seem to display quite a number of interesting features.

While studying this chapter the reader will find it very useful to have available templates for drawing regular polygons with 3, 4, 6, 8 and 12 sides, all with the same edge-length. Templates for other regular polygons (5-gons and 10-gons) and for star polygons are also useful. These can be obtained from stores specializing in educational materials. For some of the exercises here and in later chapters it will also be convenient to prepare and have on hand several copies, in full-sheet size, of the three regular tilings and also of some of the Archimedean and Laves tilings described in Sections 2.1 and 2.7.

2.1 REGULAR AND UNIFORM TILINGS

We begin by stating formally the following simple, obvious and familiar fact about tilings:

2.1.1 *The only edge-to-edge monohedral tilings by regular polygons are the three regular tilings of Figure* 1.2.1. *They have, as prototiles, an equilateral triangle, a square and a regular hexagon, respectively.*

The term *regular tiling* is used here in the sense explained in Section 1.3, namely that the symmetry group of the tiling is transitive on the flags.

If we inquire about the possibility of edge-to-edge tilings of the plane that use as tiles regular polygons of several kinds, then the situation immediately becomes much more interesting. The interior angle at each corner of a regular n-gon $\{n\}$ is $(n-2)\pi/n$ radians (or $180(n-2)/n$ degrees) so that if an n_1-gon $\{n_1\}$, an

n_2-gon $\{n_2\}, \ldots,$ an n_r-gon $\{n_r\}$ meet at a vertex of a tiling then

$$\frac{n_1 - 2}{n_1} + \cdots + \frac{n_r - 2}{n_r} = 2. \qquad (2.1.2)$$

It is easy to check that only 17 choices of the positive integers $n_1, \ldots, n_r$ satisfy this equation and hence only 17 choices of polygons can be fitted round a vertex so as to cover a neighborhood of the vertex without gaps or overlaps. We call each such choice the *species* of the vertex, and we list the 17 possible species in Table 2.1.1. In four of the species there are two distinct ways in which the polygons in question may be arranged; the mere reversal of cyclic order is not counted as distinct. Hence there are 21 possible *types* of vertices, which are also listed in Table 2.1.1 and shown in Figure 2.1.1. A vertex around which, in cyclic order, we have an n_1-gon $\{n_1\}$, an n_2-gon $\{n_2\}$, etc., is said to be of type $n_1.n_2.\ldots$. Thus the three regular tilings have vertices of types 3.3.3.3.3.3, 4.4.4.4 and 6.6.6. For brevity we write these symbols as $3^6, 4^4$ and 6^3, and we use similar abbreviations in other cases. In order to obtain a unique symbol for each type of vertex we shall always choose that which is lexicographically first among all possible expressions.

Contrary to assertions frequently made (see Section 2.10), if we require of an edge-to-edge tiling only that it be composed of regular polygons and that all its vertices be of the same species, then there are infinitely many distinct tilings. This was noted already by Robin [1887]. For example, if at each vertex there are two triangles and two hexagons, it is possible to place each "horizontal strip" in two non-equivalent positions (see Figure 2.1.2); hence there is an uncountable infinity of distinct tilings. Similar is the situation (see Figure 2.1.3) if three triangles meet two squares at each vertex. Allowing three kinds of polygons, in case of vertices of species number 6 each "disk" of an infinite family can be put in two positions, again leading to an uncountable infinity of tilings (see Figure 2.1.4).

The preceding remarks make reasonable the restriction that only vertices of a *single type* be allowed, instead of the weaker requirement that all vertices be of a single species. In this case, since all edges have equal length, the tiling will be monogonal. If the type of each vertex is *a.b.c.*..., we shall denote the tiling by (*a.b.c.*...) and say that the tiling is of *type* (*a.b.c.*...), using superscripts to abbreviate when possible. This restriction indeed changes the situation completely and we have the following result, the history of which is discussed in Section 2.10.

2.1.3 *There exist precisely* 11 *distinct edge-to-edge tilings by regular polygons such that all vertices are of the same type. These are* (3^6), (3^4.6), ($3^3.4^2$), (3^2.4.3.4), (3.4.6.4), (3.6.3.6), (3.12^2), (4^4), (4.6.12), (4.8^2) *and* (6^3).

We recall from Section 1.2 that two tilings are equal if one may be made to coincide with the other by a rigid motion of the plane (possibly including reflection) followed by a change of scale. In fact, except for the tiling (3^4.6), reflections are never required to establish the equality of two tilings of the same type. But in the case of (3^4.6) reflections may be required and we describe this situation by saying that (3^4.6) occurs in two *enantiomorphic* (mirror image) forms.

The 11 tilings listed in Statement 2.1.3 are shown in Figure 2.1.5. They are usually called *Archimedean* tilings (although some authors call them homogeneous or semiregular), and they clearly include the three regular tilings.

Two not entirely trivial steps are required in order to prove Statement 2.1.3. First, it must be shown that for the ten types of vertices marked by one or two asterisks in Table 2.1.1 it is not possible to extend the tiling from the neighborhood of a starting vertex to an Archimedean tiling of the plane. To show this, one has only to examine the possible arrangements of polygons incident with the vertices of one of the n-gons with odd n in order to establish the impossibility. For the six species marked by two asterisks in Table 2.1.1 there is no edge-to-edge tiling by regular polygons that includes even a single vertex of the species. Second, it must be established that the remaining 11 types of vertices do actually lead to Archimedean tilings. This may be deemed obvious and trivial in view of Figure 2.1.5—but it is just this

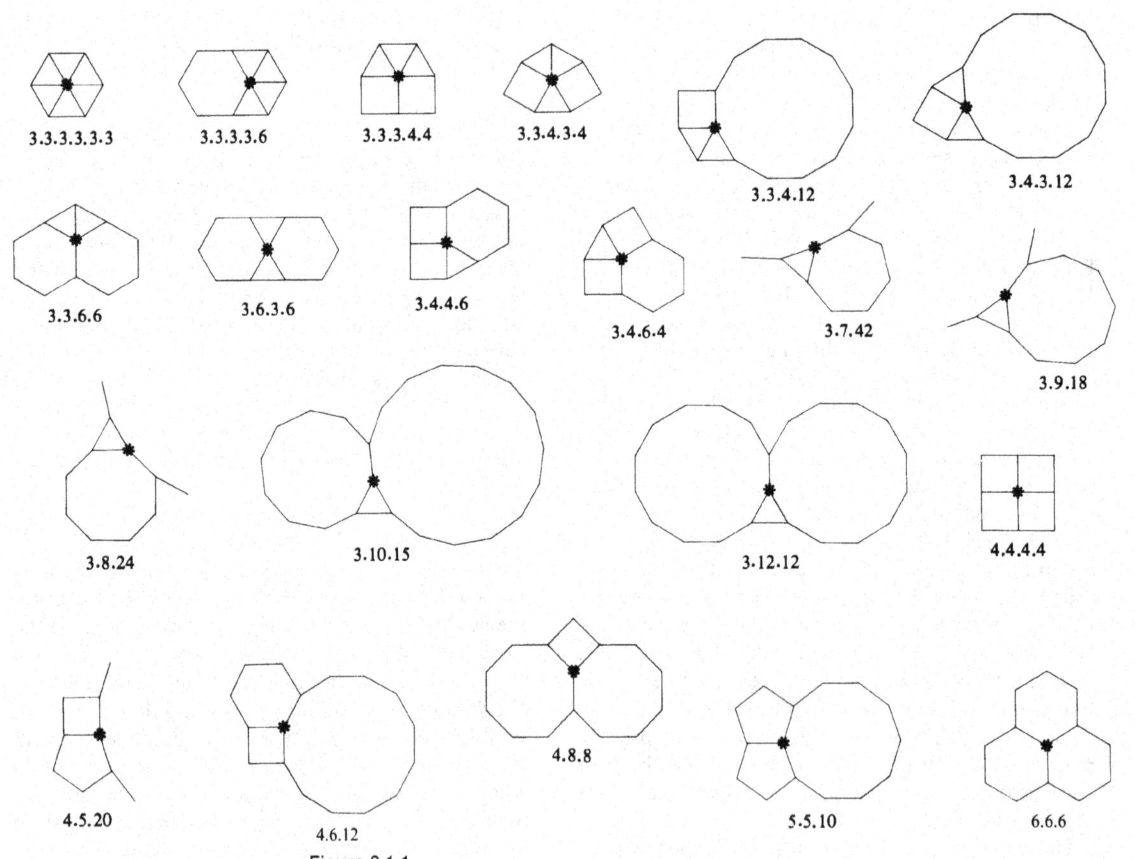

Figure 2.1.1
The 21 types of vertices possible with regular polygonal tiles. These correspond to the last column of Table 2.1.1.

TABLE 2.1.1 THE POSSIBLE SPECIES AND TYPES OF VERTICES FOR EDGE-TO-EDGE TILINGS BY REGULAR POLYGONS

The species of a vertex in a tiling is determined by the number of n-gons (for each value of n) that meet at the vertex. The various types of vertex arise from the different orders in which the n-gons meeting at a vertex can be arranged. This table gives details of the 17 species and 21 types of vertices that are possible. The meaning of the asterisks in the last column in explained in the text.

Species number	Number of n-gons meeting at a vertex														Type of vertex
	$n = 3$	4	5	6	7	8	9	10	12	15	18	20	24	42	
1	6														3.3.3.3.3.3
2	4			1											3.3.3.3.6
3	3	2													3.3.3.4.4 3.3.4.3.4
4	2	1							1						3.3.4.12 * 3.4.3.12 *
5	2			2											3.3.6.6 * 3.6.3.6
6	1	2		1											3.4.4.6 * 3.4.6.4
7	1				1									1	3.7.42 **
8	1					1							1		3.8.24 **
9	1						1				1				3.9.18 **
10	1							1		1					3.10.15 **
11	1								2						3.12.12
12		4													4.4.4.4
13		1	1									1			4.5.20 **
14		1		1					1						4.6.12
15		1				2									4.8.8
16			2					1							5.5.10 **
17				3											6.6.6

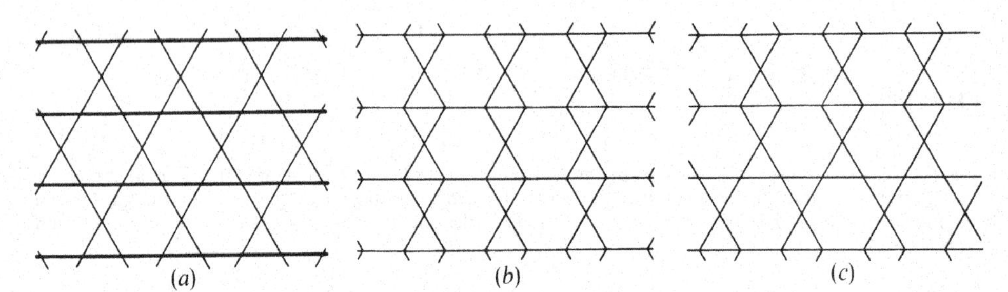

Figure 2.1.2
The uniform tiling (3.6.3.6) may be cut by parallel lines (emphasized on the left) into strips that may be slid independently of one another. In this way we obtain an uncountable infinity of tilings all of whose vertices are of species number 5.

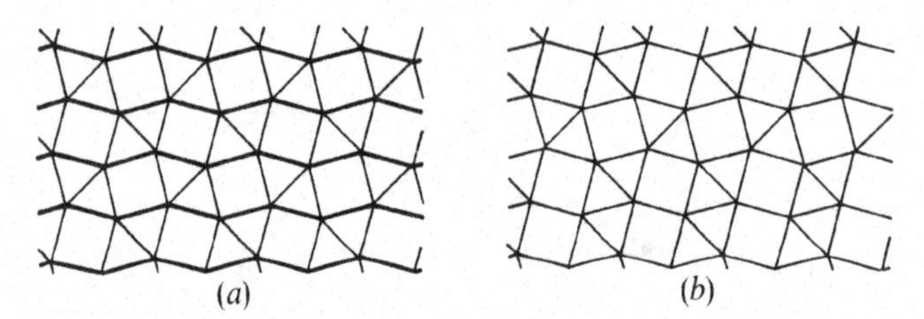

Figure 2.1.3
In the uniform tiling (3^2.4.3.4) there are zigzags of edges as shown on the left. The tiling may be cut along any such zigzag and then one half replaced by its mirror image. This process leads to an uncountable infinity of distinct tilings with all vertices of species number 3.

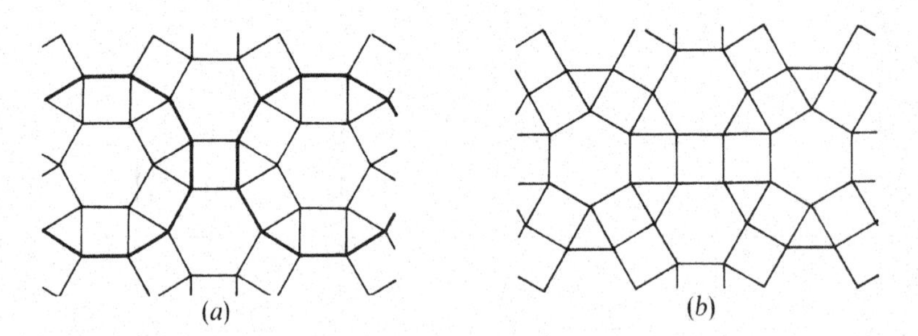

Figure 2.1.4
In the uniform tiling (3.4.6.4) there are "disks", each consisting of a hexagon and its neighbors, as indicated on the left. Any such disk may be rotated through the angle $\pi/6$ without altering the species of the vertices. Because pairwise disjoint disks can be independently rotated, this procedure yields an uncountable infinity of distinct tilings with all vertices of species number 6.

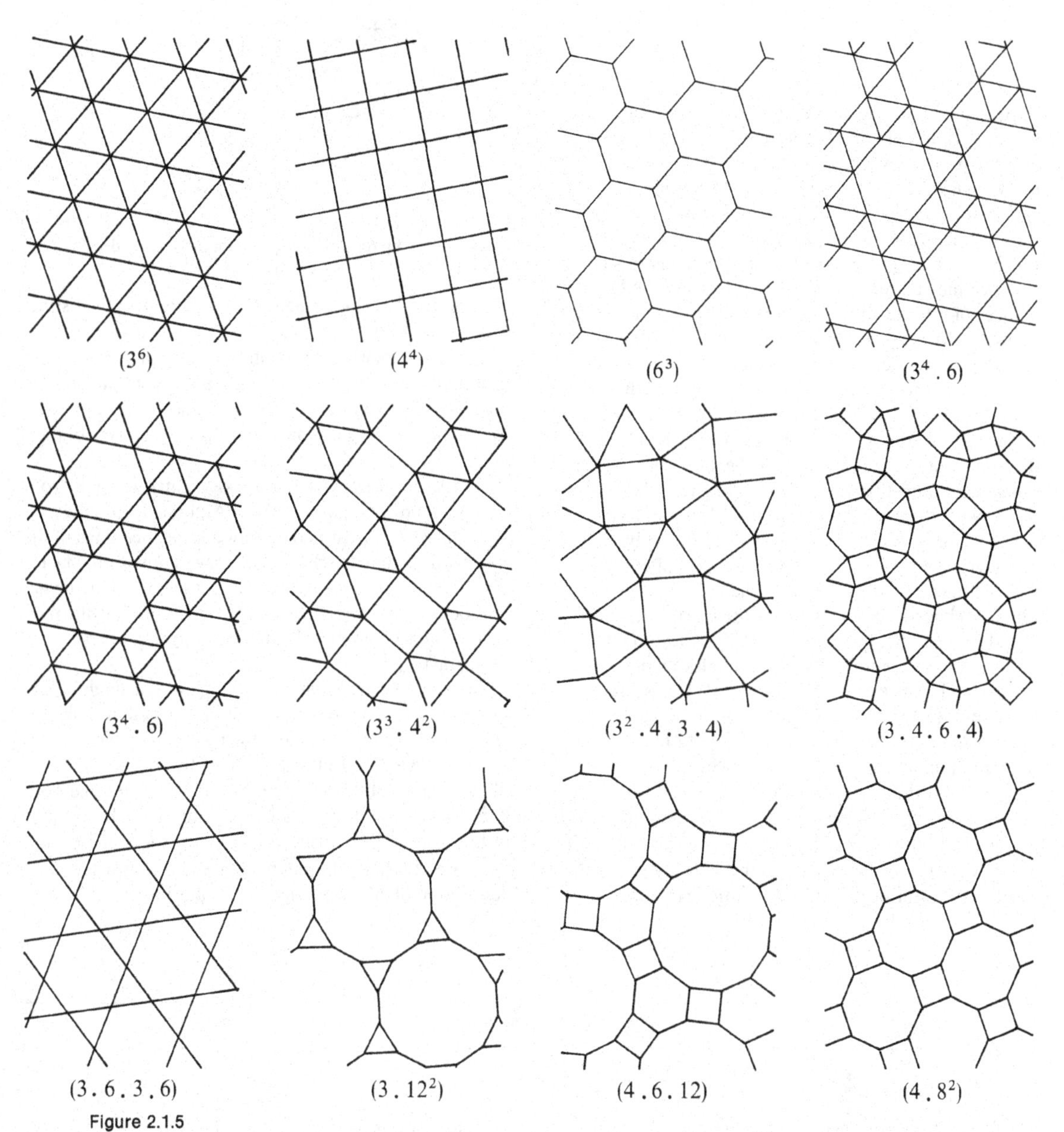

Figure 2.1.5
The eleven distinct types of Archimedean tilings. That of type $(3^4.6)$ occurs in two enantiomorphic forms, both of which are shown. These eleven tilings are also called *uniform tilings.*

"obviousness" that is dangerous. In Figure 2.1.6, adapted from a children's coloring book, we show a tiling that appears to consist of regular n-gons with $n = 4, 5, 6, 7, 8$ and to be (in the terminology we shall explain later) 4-uniform and equitransitive. Actually, this visual "proof" is a fraud since—using equation (2.1.2)—it is easy to check that the polygons in such a tiling cannot be regular. Similar to this are the circumstances of a tiling which appears—at first sight—to use equilateral triangles, squares and regular pentagons as tiles; this tiling occurs in the crystallographic representation of the structure of certain alloys; see Bhandary & Girgis [1977, Fig. 7]. On closer inspection it will be found that only the squares are exactly regular. Thus there is a real need to show that the 11 Archimedean tilings exist, and that the drawings in Figure 2.1.5 show tilings with "genuinely" regular tiles. It is easy to give direct proofs of existence for (4^4) and for (3^6) by considering two or three suitable families of equidistant parallel lines. Except for $(3^3.4^2)$, the existence of the other Archimedean tilings can be deduced from these two. For example, (6^3) may be obtained from (3^6) by taking suitable unions of six triangles as tiles; "cutting off" vertices of (6^3) and of (4^4) in a suitable way yields (3.12^2) and (4.8^2), and so on.

It should be noted that *a priori* it is not obvious that the limitation to a single type of vertices should lead to a unique tiling. It happens to turn out this way—but just barely so. On the one hand uniqueness does not always hold in analogous situations (see Section 2.2), and on the other hand it is a consequence of the fact that we allow reflections in our definition of equal tilings. Indeed, the

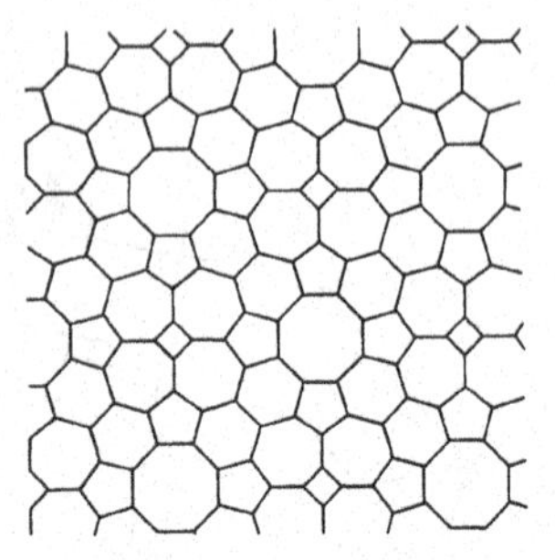

Figure 2.1.6
A fake "tiling by regular polygons" found in a children's coloring book, *Altair Design* (Holiday [1970]).

two enantiomorphic tilings $(3^4.6)$ are counted as distinct by some authors.

Another somewhat accidental but very important feature of the Archimedean tilings is the fact that each is *isogonal*. We recall from Section 1.4 that this means each two vertices are equivalent under some symmetry of the tiling.

In view of this remark about isogonality we shall, from now on, follow tradition and call Archimedean tilings also *uniform*. The distinction between the meanings of the two words is that "Archimedean" refers only to the fact that the tiling is monogonal, that is, the immediate neighborhoods of any two vertices "look the same", while the term "uniform" implies the much stronger property of isogonality.

Returning to the question of tilings with a single species of vertex we mention without proof that non-uniform tilings are possible only in case of species number 3, 5 and 6. In the last two of these cases *all* tilings can be obtained from the uniform ones (3.6.3.6) and (3.4.6.4) by the method explained at the beginning of this section. However, in case of species number 3 there are other possibilities as well and a complete description of all such tilings is still not known.*

EXERCISES 2.1

1. Prove that equation (2.1.2) has precisely the solutions given in Table 2.1.1.
2. Carry out the details of the proof of the assertion made in this section that the 11 Archimedean tilings exist.
3. Prove that each Archimedean tiling is isogonal.

* See the Appendix beginning on page 653.

4. Construct some tilings with all vertices of species number 3 that cannot be obtained by the procedure described on page 62.
5. Show that if a tiling by regular polygons contains an octagon, then it must be the tiling (4.8^2).
6. Find a tiling by regular polygons which contains 12-gons and squares, but no hexagons.
7. Find the number of distinct transitivity classes of flags in each of the uniform tilings.
8. Determine the symmetry group $S(\mathscr{T})$ of each of the 11 uniform tilings.
9. A polygon is called "angle-regular" if all its angles have the same angular measure as those of a regular polygon with the same number of sides. Clearly, for $n \geq 4$ there exist angle-regular n-gons which are not regular.
   *(*a*) Describe all monohedral edge-to-edge tilings by angle-regular quadrangles (that is, rectangles) and those by angle-regular hexagons.
   *(*b*) Devise a scheme for the description and classification of all edge-to-edge tilings by angle-regular quadrangles.
   **(*c*) Devise a scheme for the description and classification of all tilings by angle-regular hexagons.
10. (*a*) Is it possible for a tiling to have vertices of all the 15 types listed in Table 2.1.1 that are not marked by two asterisks?
    (*b*) Is it possible for a periodic tiling to have vertices of 14 of these types?
11. In a tiling $\mathscr{T}$ by equilateral triangles and squares, two triangular tiles T, T' are said to be related if there exists a finite sequence of triangular tiles

$$T = T_0, T_1, T_2, \ldots, T_k = T'$$

such that every two consecutive tiles in the sequence are adjacent, that is, have an edge in common. An exactly analogous definition of relatedness holds for square tiles. A *raft* is the union of all the tiles in an relatedness class. Show that the tiling $\mathscr{T}'$, whose tiles are the rafts of $\mathscr{T}$, is a 4-valent tiling by convex tiles (with the possibility that some tiles may be unbounded).

(Exercises 10 and 11 are based on results of F. Landuyt.)

2.2 *k*-UNIFORM TILINGS

The observation that the Archimedean tilings are uniform points the way to the following generalization. An edge-to-edge tiling by regular polygons is called *k-uniform* if its vertices form precisely k transitivity classes with respect to the group of symmetries of the tiling. In other words, the tiling is k-uniform if and only if it is *k-isogonal* (as defined in Section 1.3) and its tiles are regular polygons. In this terminology, uniform tilings are 1-uniform. If the types of vertices in the k classes are $a_1.b_1.c_1.\ldots$; $a_2.b_2.c_2.\ldots$; $\ldots$; $a_k.b_k.c_k.\ldots$ the tiling will be designated by the symbol $(a_1.b_1.c_1.\ldots; a_2.b_2.c_2.\ldots; \ldots; a_k.b_k.c_k.\ldots)$, with the obvious shortening through the use of superscripts, and with subscripts to distinguish distinct tilings in which the same types of vertices appear. The following result was obtained by Krötenheerdt [1969]:

2.2.1 *There exist* 20 *distinct types of* 2*-uniform edge-to-edge tilings by regular polygons, namely*: $(3^6; 3^4.6)_1$, $(3^6; 3^4.6)_2$, $(3^6; 3^3.4^2)_1$, $(3^6; 3^3.4^2)_2$, $(3^6; 3^2.4.3.4)$, $(3^6; 3^2.4.12)$, $(3^6; 3^2.6^2)$, $(3^4.6; 3^2.6^2)$, $(3^3.4^2; 3^2.4.3.4)_1$, $(3^3.4^2; 3^2.4.3.4)_2$, $(3^3.4^2; 3.4.6.4)$, $(3^3.4^2; 4^4)_1$, $(3^3.4^2; 4^4)_2$, $(3^2.4.3.4; 3.4.6.4)$, $(3^2.6^2; 3.6.3.6)$, $(3.4.3.12; 3.12^2)$, $(3.4^2.6; 3.4.6.4)$, $(3.4^2.6; 3.6.3.6)_1$, $(3.4^2.6; 3.6.3.6)_2$ *and* $(3.4.6.4; 4.6.12)$.

These tilings are shown in Figure 2.2.1; the solid dots indicate one vertex of each class.

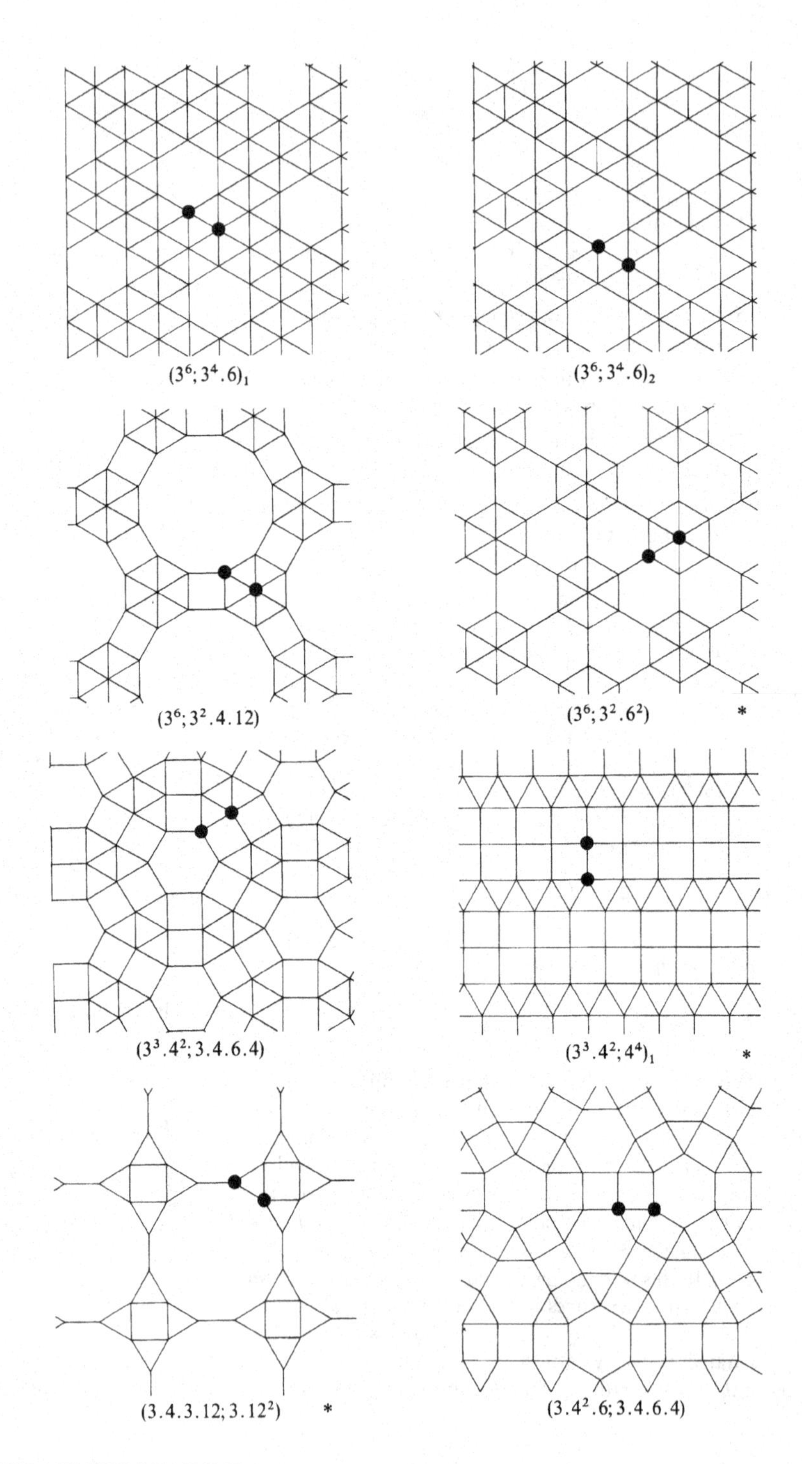

Figure 2.2.1
The 20 types of 2-uniform tilings (from Krötenheerdt [1969]). Only one of these, $(3^6;3^4.6)$, occurs in two enantiomorphic forms. The meaning of the asterisks is explained in Section 2.3.

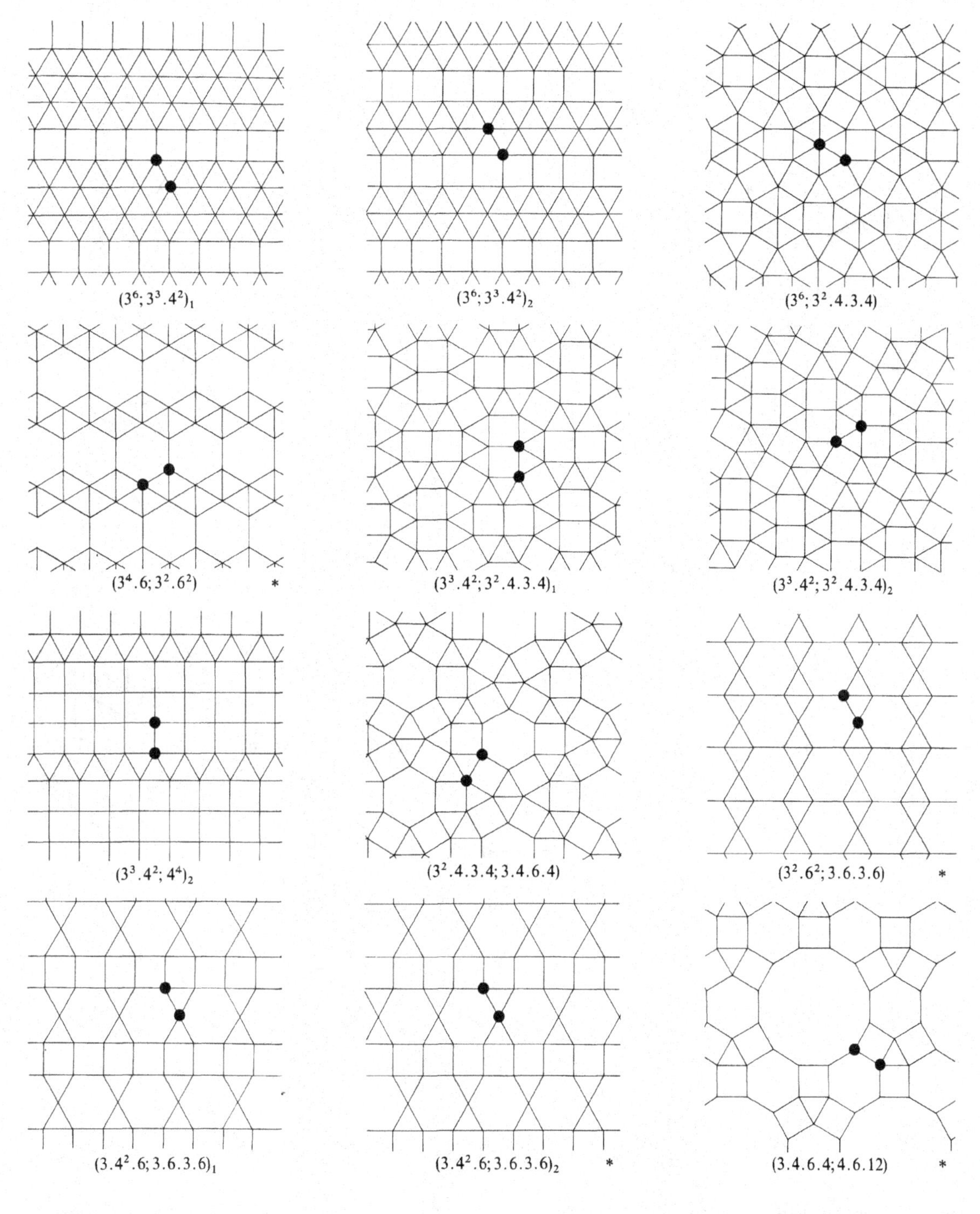
$(3^6; 3^3.4^2)_1$ $(3^6; 3^3.4^2)_2$ $(3^6; 3^2.4.3.4)$

$(3^4.6; 3^2.6^2)$ * $(3^3.4^2; 3^2.4.3.4)_1$ $(3^3.4^2; 3^2.4.3.4)_2$

$(3^3.4^2; 4^4)_2$ $(3^2.4.3.4; 3.4.6.4)$ $(3^2.6^2; 3.6.3.6)$ *

$(3.4^2.6; 3.6.3.6)_1$ $(3.4^2.6; 3.6.3.6)_2$ * $(3.4.6.4; 4.6.12)$ *

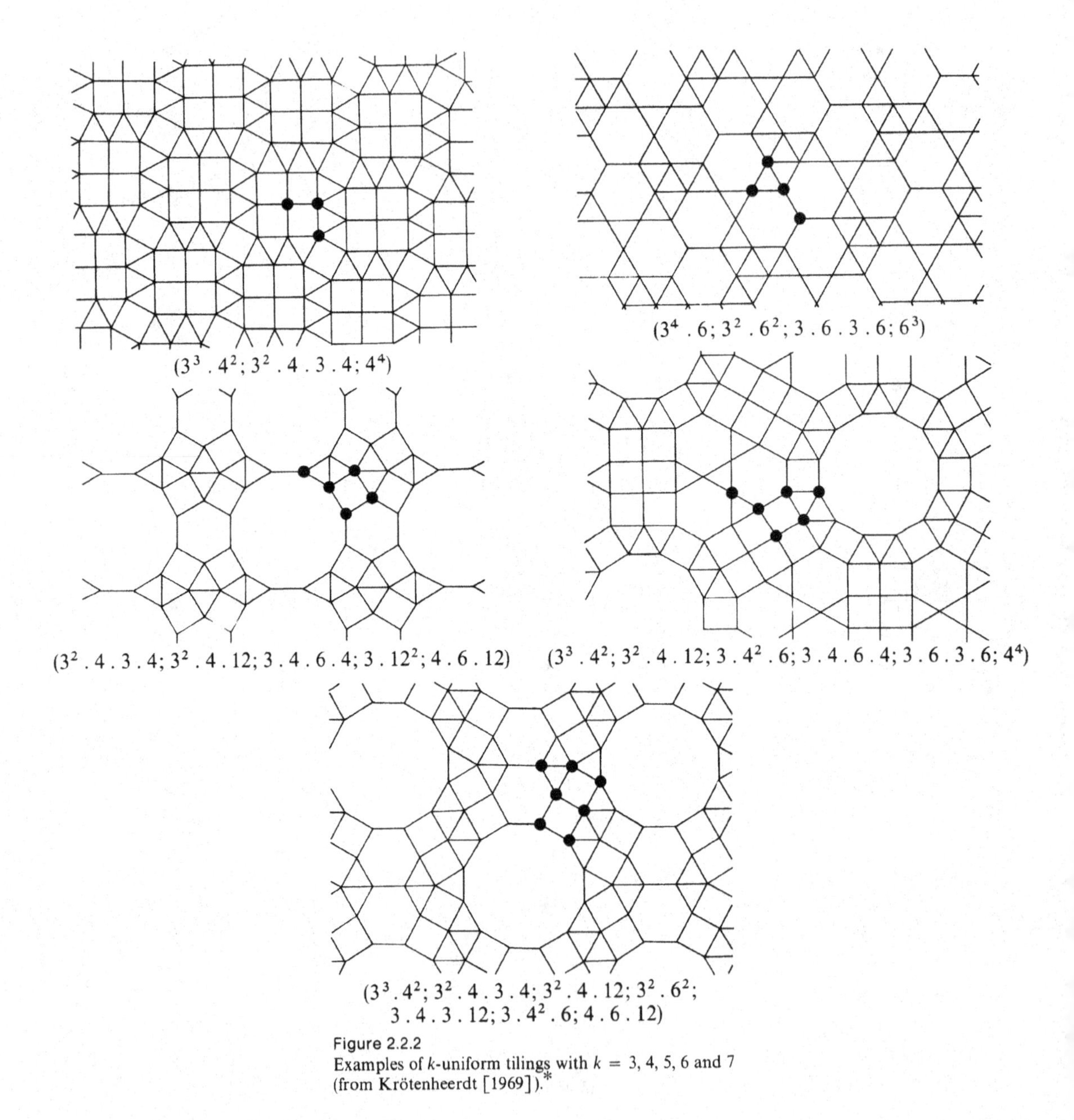

Figure 2.2.2
Examples of k-uniform tilings with $k = 3, 4, 5, 6$ and 7 (from Krötenheerdt [1969]).*

* See the Appendix beginning on page 653.

The proof of this statement may be carried out along lines analogous to those explained in connection with the proof of Statement 2.1.3 concerning the 11 uniform tilings. However, the details are much more intricate and we shall not give them here; a few of the steps are outlined in Exercises 1 and 2. It seems that the only published version of the complete proof is that of Krötenheerdt [1969].

It is easy to construct k-uniform tilings for each $k \geq 1$ (see Exercise 5). Some additional examples of a more complicated nature, which satisfy the Krötenheerdt condition discussed below, are shown in Figure 2.2.2. However, even for $k = 4$ it is not known how many k-uniform tilings exist, nor is any kind of asymptotic estimate available for the number of k-uniform tilings with large k. Using the fact that the symmetry group of each k-uniform tiling is one of the 17 crystallographic groups listed in Section 1.4, it can be shown that for each k the number of distinct k-uniform tilings is finite.

In a series of papers, Krötenheerdt [1969], [1970a], [1970b] examined a closely related question. He considered those k-uniform tilings in which the k transitivity classes of vertices consist of *k distinct types* of vertices. Although it is easily seen that for $k = 1$ and for $k = 2$ these coincide with the k-uniform ones, Krötenheerdt's condition is more restrictive for $k \geq 3$. If we denote by $K(k)$ the number of distinct Krötenheerdt tilings, his results are:

$$K(1) = 11,\ K(2) = 20,\ K(3) = 39,\ K(4) = 33,\ K(5) = 15,$$

$$K(6) = 10,\ K(7) = 7 \text{ and } K(k) = 0 \text{ for each } k \geq 8.$$

Krötenheerdt's method of proof is a natural extension of the one used in the determination of the uniform tilings.

EXERCISES 2.2

*1. Show that if each vertex of a k-uniform tiling is incident with a 12-gon, then either $k = 1$, and the tiling is (4.6.12) or (3.12²), or $k = 2$ and the tiling is (3.4.3.12; 3.12²), or $k = 3$ and the tiling is well-determined. Find its type in the latter case.

*2. Carry out the following parts of the proof of Krötenheerdt's theorem that there exist precisely twenty 2-uniform tilings by regular polygons.
   (*a*) Show that no vertex can be of a type which bears a double asterisk in Table 2.1.1.
   (*b*) Show that if the tiling contains a 12-gon then it must be of one of the following three types: (3^6; 3^2.4.12), (3.4.3.12; 3.12²) or (3.4.6.4; 4.6.12).

3. Determine whether the tilings in Figure 2.2.3 are k-uniform for some k. If so, find the appropriate symbols and check whether these satisfy Krötenheerdt's condition (described above).

*4. Find an example of a 3-uniform tiling which is *not* a Krötenheerdt tiling.

5. Prove that for each $k \geq 3$ there exist at least two different k-uniform tilings with all vertices of types:

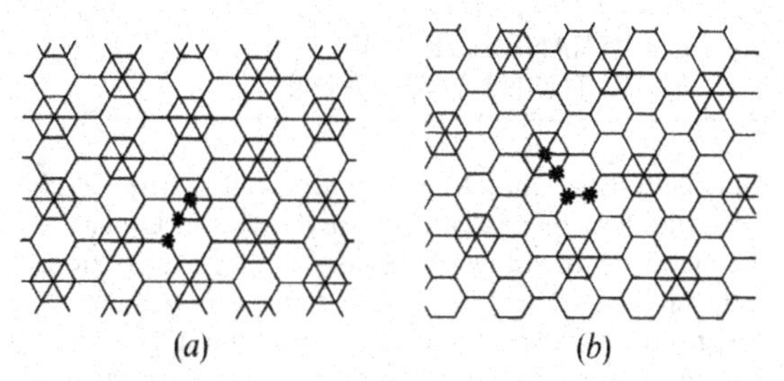

Figure 2.2.3
Are these tilings k-uniform for some value of k? See Exercise 2.2.3.

(a) 3^6 and $3^3.4^2$; (b) $3^3.4^2$ and 4^4; (c) $3.4^2.6$ and 3.6.3.6.

*6. Determine all the 3-uniform tilings. (Chavey [1984b] determined that there are 61 tilings of this kind.)

7. Determine the symmetry group of each of the tilings in Figures 2.2.1, 2.2.2 and 2.2.3.

8. A tiling by regular polygons is called *k-Archimedean* if its vertices belong to k distinct types, with at least one vertex from each of these types. As mentioned in Section 2.1, for some choices of types (such as $3^3.4^2$ and $3^2.4.3.4$, or $3^2.6^2$ and 3.6.3.6, or $3.4^2.6$ and 3.4.6.4) there exist infinitely many distinct 2-Archimedean tilings.
   (a) Show that there exist no 2-Archimedean tilings in which some vertices have type 4.8^2.
   (b) Show that if a tiling is 2-Archimedean and has some vertices of type 3.12^2, then the other vertices must be of type 3.4.3.12.
   (c) Show that there are uncountably many 2-Archimedean tilings with vertices of types 3.4.3.12 and 3.12^2.
   (d) Show that if a 2-Archimedean tiling has a vertex of type $3^2.4.12$, all its vertices are of types $3^2.4.12$ and 3^6, and it is actually a 2-uniform tiling.
   (e) Show that there exist infinitely many different 2-Archimedean tilings with vertices of types 3^6 and $3^2.4.3.4$.
   (f) Show that for all but one of the 17 periodic groups in the plane (see Section 1.4) there exists a 2-Archimedean tiling $\mathcal{T}$ which has this group as its symmetry group $S(\mathcal{T})$.
   (g) Find a 2-Archimedean tiling which has no translation as symmetry.
   (h) For each pair of distinct types of vertices determine whether or not there exist any 2-Archimedean tilings with these two types of vertices; if some such tilings exist, determine whether there are infinitely many distinct ones, or just a finite number.
   *(i) Show that if two given types of vertices occur in a 2-Archimedean tiling then there exists a 2-uniform tiling with the same two types of vertices.

2.3 EQUITRANSITIVE AND EDGE-TRANSITIVE TILINGS*

An edge-to-edge tiling by polygons is called *k-isohedral* if the tiles form precisely k transitivity classes under the symmetries of the tiling (see Section 1.3 for explanations and examples). We shall say that a tiling by regular polygons is *equitransitive* if each set of mutually congruent tiles forms one transitivity class. It is easily verified that all uniform tilings are equitransitive except $(3^4.6)$ which is dihedral but 3-isohedral. Other equitransitive tilings are the seven 2-uniform tilings with symbols marked by an asterisk in Figure 2.2.1.

It is rather surprising to find that there seems to be no consideration in the literature of equitransitive or k-isohedral tilings by regular polygons. A reasonable question to consider deals with the relations between k and h such that there exist k-uniform h-isohedral tilings. In view of the fact that every 1-uniform tiling is at most 3-isohedral, and every 2-uniform tiling is at most 5-isohedral (see Figure 2.2.1), it appears reasonable to expect that for each k there exists a least integer $h(k)$ such that every k-uniform tiling is also h-isohedral for some integer $h \leq h(k)$. Thus $h(1) = 3$ and $h(2) = 5$. The values of $h(k)$ for $k > 2$ are not known. Likewise there probably exists

* See the Appendix beginning on page 653.

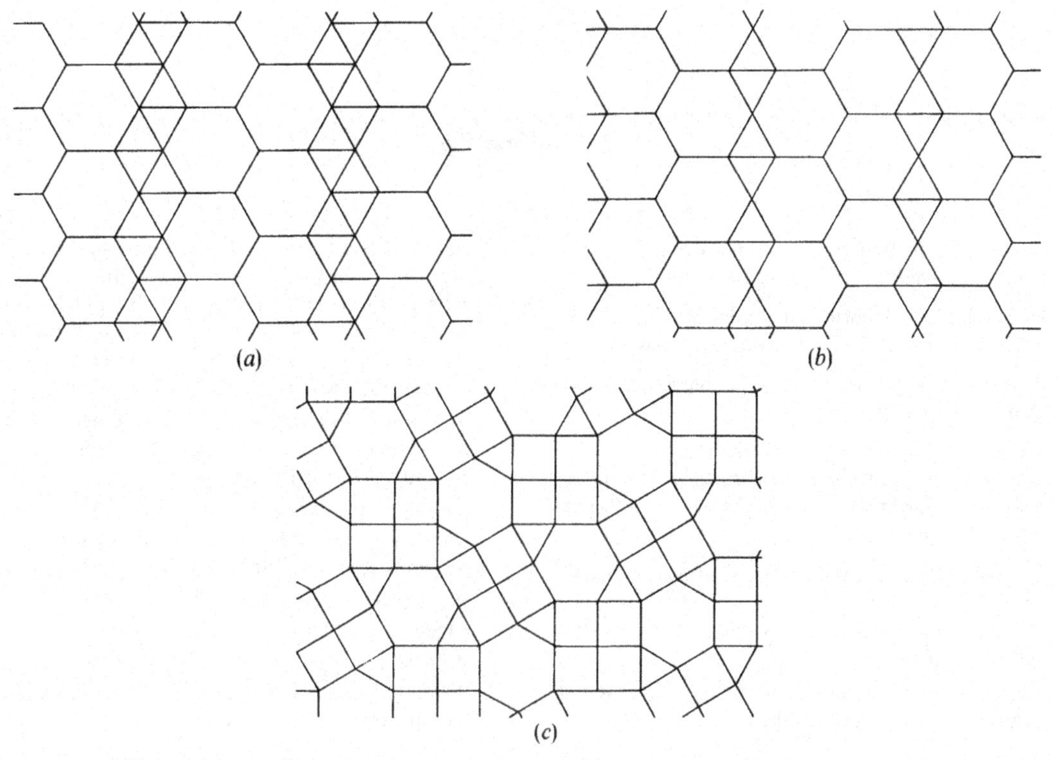

Figure 2.3.1
Three examples of equitransitive tilings. All are 3-uniform; the tilings (*a*) and (*b*) are 2-isohedral and tiling (*c*) is 3-isohedral (from Debroey & Landuyt [1979]).

a least integer $k(h)$ such that every h-isohedral tiling by regular polygons is also k-uniform for some $k \leq k(h)$. Clearly $k(1) = 1$, and the examples of Figure 2.2.3 show that $k(2) \geq 4$. (The existence of these bounds was established recently by Chavey [1984a], [1984b].)

Prompted by the account in Grünbaum & Shephard [1977a], I. Debroey & F. Landuyt [1979] and [1981] have recently established that there exist precisely 22 different equitransitive edge-to-edge tilings by regular polygons. In addition to the 19 such tilings shown in Figures 2.1.5, 2.2.1 and 2.2.3 they found the three tilings in Figure 2.3.1. This result corrects and solves a conjecture formulated in Grünbaum & Shephard [1977a, p. 236]. An examination of these 22 tilings shows that three of them are isohedral, thirteen are 2-isohedral, five are 3-isohedral and one is 4-isohedral; since 2-isohedral tilings by regular polygons are equitransitive, it follows that $k(2) = 4$. (These and related results were obtained independently by Chavey [1984b].)

Similar problems arise if we consider transitivity classes of edges. If there are j such classes in a tiling we call it a *j-isotoxal* tiling. We mention this idea only briefly because we believe that j-isotoxal edge-to-edge tilings by regular polygons have also not been considered in the literature. There are just four such 1-isotoxal (that is, isotoxal) tilings, namely (3^6), (4^4), (6^3) and $(3.6.3.6)$ (see Exercise 2.3.2 and Chapter 4), and there appear to exist only four 2-isotoxal such tilings (namely $(3^2.4.3.4)$, $(3.4.6.4)$, (3.12^2) and (4.8^2), a fact recently verified by F. Landuyt. For an account of isotoxal tilings with tiles of arbitrary shapes, see Chapter 6.

EXERCISES 2.3

1. For each of the twenty tilings in Figure 2.2.1 determine the number of transitivity classes of tiles.
2. For each of the uniform and 2-uniform tilings determine the number of transitivity classes of edges.

*3. Determine all 3-isotoxal tilings by regular polygons (Chavey [1984b]).

4. Consider tilings by equilateral triangles and hexagons in which the triangles form t transitivity classes and the hexagons h such classes. Show that all values of t and h are possible.
5. Consider tilings by equilateral triangles and squares in which the triangles form t transitivity classes and the squares s such classes. Show that all values of t and s are possible.

**6. Consider tilings by equilateral triangles, squares and regular hexagons, of which there are t, s and h transitivity classes respectively. Determine all the values of t, s and h for which tilings are possible. (Some results in this direction have recently been obtained by Delandtsheer & Vanden Cruyce [1980]. For example they have shown that all triples t, s, h are possible if $t \geq 92$, $s \geq 2$ and $h \geq 43$.)

7. Establish the result of Debroey & Landuyt [1981] that there exist precisely thirteen 2-isohedral edge-to-edge tilings by regular polygons.

*8. Prove that for each k there exists only a finite number of k-isohedral edge-to-edge tilings by regular polygons (Chavey [1984b]).

*9. Prove that for each k the number of distinct k-isotoxal edge-to-edge tilings by regular polygons is finite (Chavey [1984b]).

10. Prove that each equitransitive tiling contains at most four different kinds of tiles.

2.4 TILINGS THAT ARE NOT EDGE-TO-EDGE

We now consider tilings by regular polygons without the requirement that the tiling is edge-to-edge. Kepler briefly considered this possibility (see drawings Bb and Kk in the frontispiece), but again no further consideration seems to have been given to the mathematical possibilities for several centuries. In contrast, from the earliest times, ornamental designs have been based on tilings by regular polygons that are not edge-to-edge. See, for example, Figure I.13(a) and many examples in Dye [1937], [1981].

As will be immediately apparent if one attempts to construct such tilings, the number of possibilities is enormous, and so it is natural to impose various restrictions. To describe these we adapt the terminology introduced earlier and extend terms such as uniform, k-uniform and equitransitive to tilings by regular polygons that are not edge-to-edge. Thus *uniform* will mean that the symmetries of the tiling act transitively on the vertices, and equitransitive means that every tile is equivalent under some symmetry to every congruent tile.

First let us consider monohedral tilings (not necessarily uniform or equitransitive) by regular polygons. It is easy to see that the only such tilings by squares are those constructed (as in Figure 1.3.6(c)) by taking the regular tiling (4^4) and sliding the rows (or columns) of squares relative to one another. The monohedral tilings by equilateral

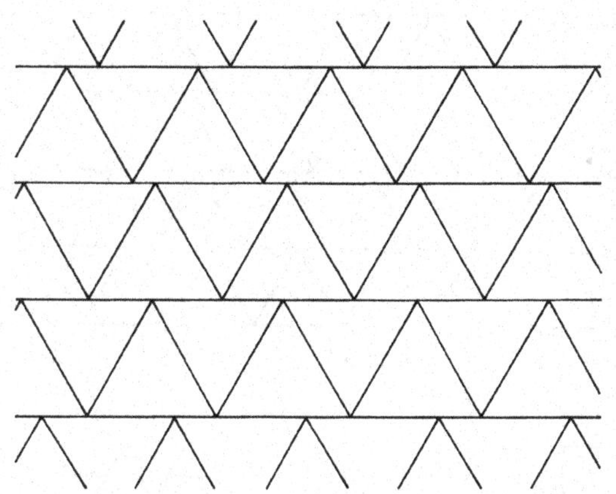

Figure 2.4.1
A monohedral tiling by equilateral triangles. All such tilings (even those that are not uniform) can be constructed from the regular tiling by equilateral triangles by sliding the rows relative to one another.

triangles can be similarly described—they are obtained by sliding the rows of (3^6) relative to one another, see Figure 2.4.1. The only monohedral tiling by regular hexagons is the regular tiling (6^3), and clearly no regular n-gons with $n \neq 3$, 4 or 6 can tile the plane monohedrally.

If we do not insist that the tiling be monohedral, the wealth of possibilities makes it reasonable to restrict attention to tilings which are uniform, equitransitive, or special in some other sense.

2.4.1 *The uniform tilings of the plane that are not edge-to-edge form eight families, each family depending on a real-valued parameter.*

These families are illustrated in Figure 2.4.2. The first four families are monohedral and the parameter indicates the fraction of overlap between sides of adjacent tiles. The tilings of the next three families are dihedral and the parameter indicates the ratio of the lengths of their sides. In the last family three different sizes of triangles appear and the parameter denotes the ratio of the side of the smallest triangle to that of the largest; in the exceptional case $\alpha = \frac{1}{2}$ only two sizes of triangles occur.

There exists a great variety of equitransitive tilings by regular polygons (see Figure 2.4.3 for several 2-uniform equitransitive tilings and it is easy to construct many other examples of a similar character). We shall therefore consider only tilings in which all tiles are similar, that is, all are n-gons for the same value of n. Several such tilings have been illustrated in Figure 2.4.2 and further examples are shown in Figures 2.4.4 and 2.4.5. The tilings of Figure 2.4.5 (unlike those of Figure 2.4.4) have the additional property that they are *unilateral*, by which we mean that each edge of the tiling is a side of *at most* one polygon. (In other words, in a unilateral tiling each edge of the tiling is a proper subset of a side of some tile.) It seems likely that the only equitransitive unilateral tiling by one kind of regular polygons other than squares is that of Figure 2.4.2(h), involving equilateral triangles of two or three sizes; but we have no proof of this. (Unilateral dissections of polygons have been considered by Kim [1976].)

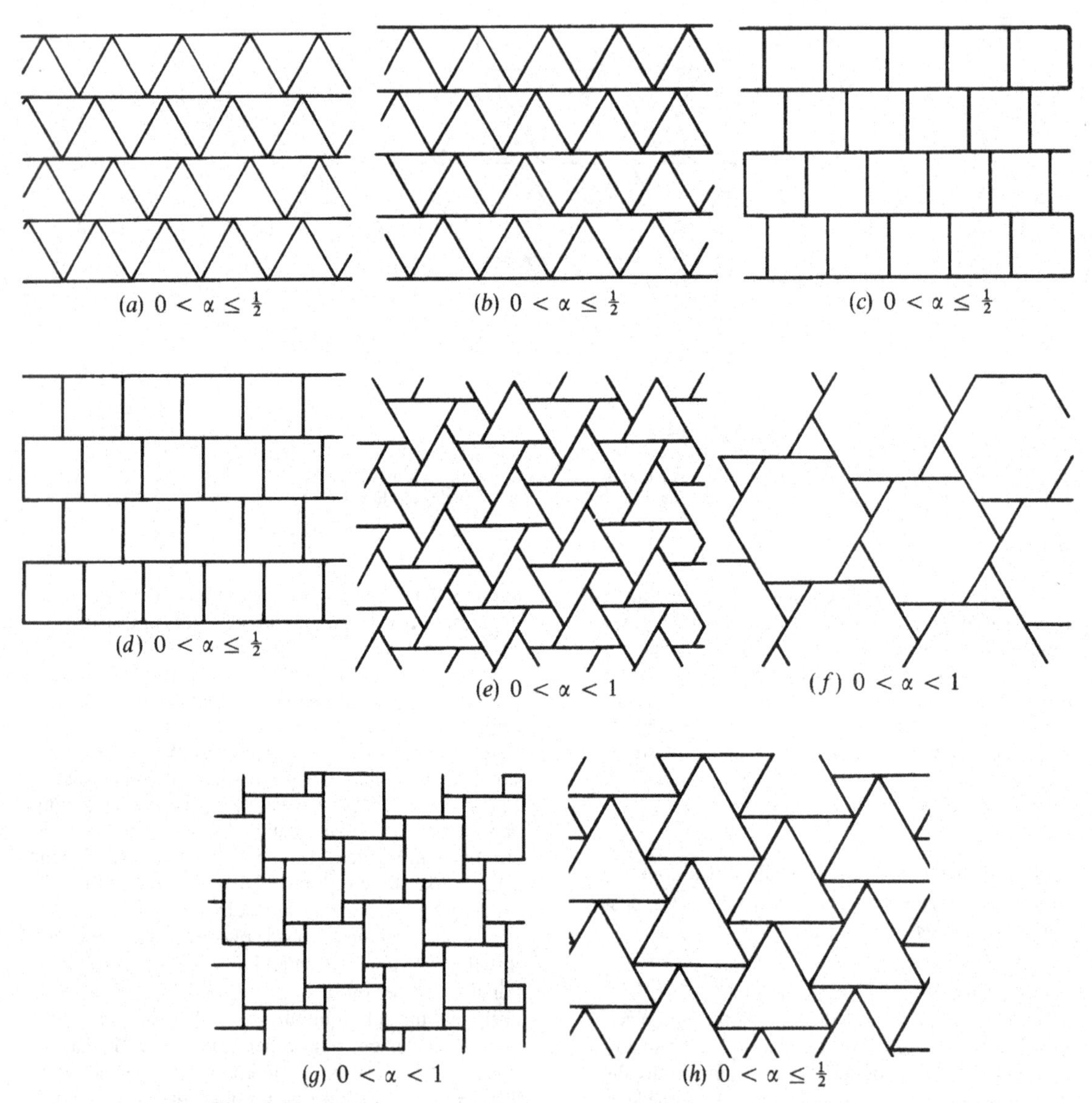

Figure 2.4.2
Representative examples of the eight families of uniform tilings that are not edge-to-edge. The parameters α are explained in the text.

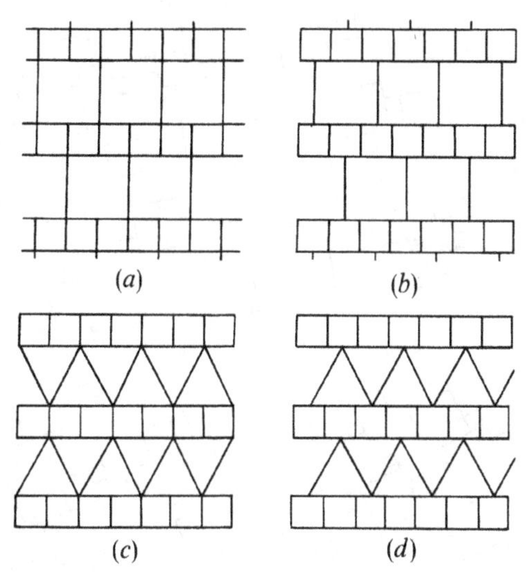

Figure 2.4.3
Examples of equitransitive 2-uniform tilings that are not edge-to-edge.

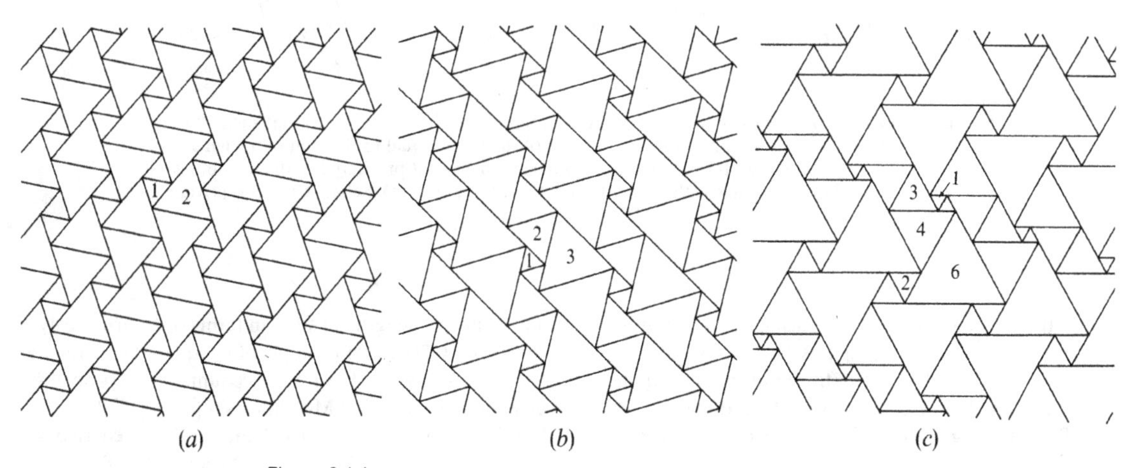

Figure 2.4.4
Equitransitive tilings with equilateral triangles of 2, 3 or 5 different sizes.

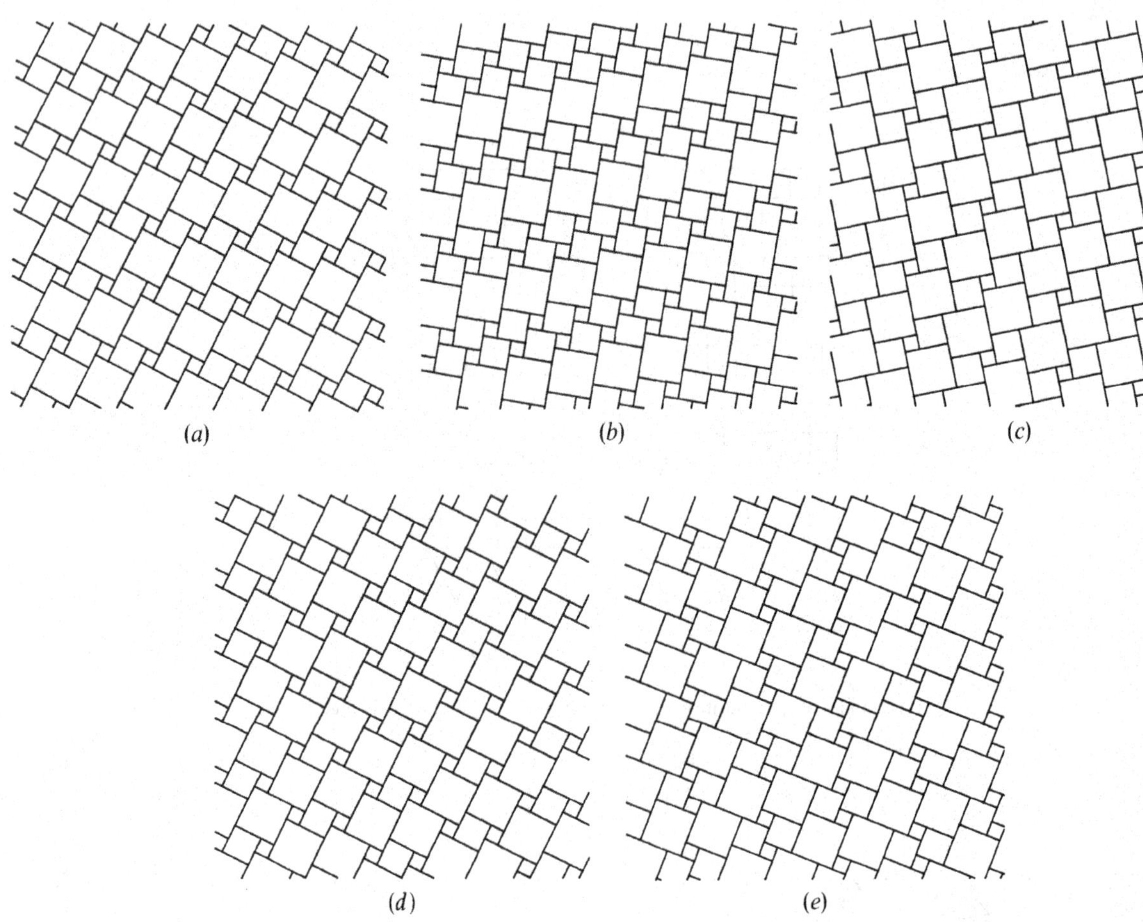

Figure 2.4.5
Five unilateral and equitransitive tilings by squares of three sizes. It can be shown that such a tiling is possible if and only if the length of sides of one size of squares is equal to the sum of the lengths of the sides of the other two sizes. If this condition is satisfied there exist precisely five distinct unilateral and equitransitive kinds of tilings with square tiles of these three sizes (D. Schattschneider, private communication).

Equitransitive unilateral tilings using squares of three* or more sizes are shown in Figures 2.4.5, 2.4.6 and 2.4.7. It is easy to construct tilings of this kind that use squares of any given number r of sizes. A method applicable for all sufficiently large values of r depends on the concept of a "squared rectangle" or "squared square" (see Figure 2.4.8). A square or rectangle is said to be "squared" into r squares if it is partitioned into r squares, all different and of integer lengths of sides in terms of an arbitrarily chosen unit. (The first example of a "squared square" was found by Sprague [1939]. For accounts of the problem of squared squares see Meschkowski [1960], Tutte [1965], Kazarinoff & Weitzenkamp [1973], Federico [1979], Duijvestijn, Federico & Leeuw [1982]). An equitransitive unilateral tiling is clearly produced by tiling the plane isohedrally by translates of a "squared"

* See the Appendix beginning on page 653.

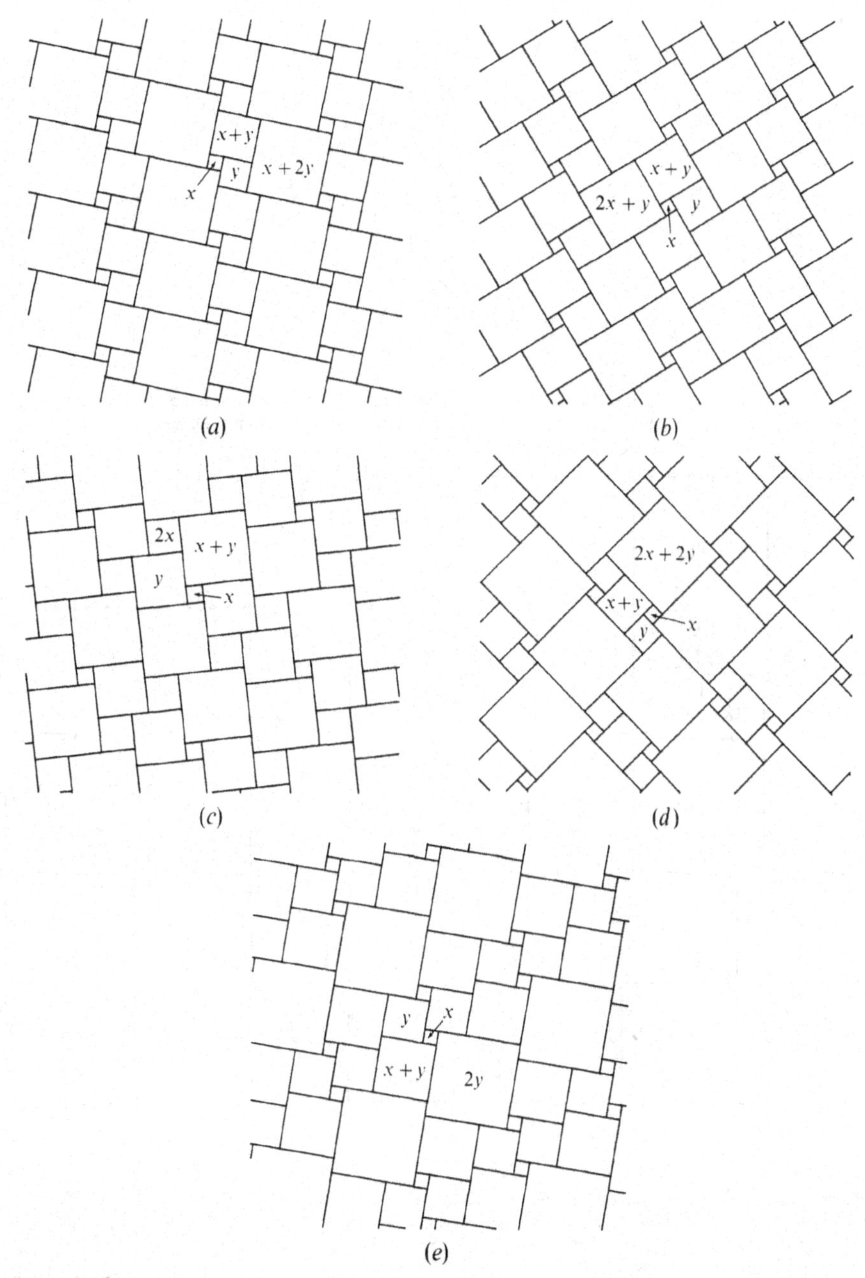

Figure 2.4.6
Five examples of unilateral and equitransitive tilings by squares of four sizes. In each case three of the sizes are x, y (with $x < y$) and $x + y$; the fourth size is $x + 2y$, $2x + y$, $2x$, $2x + 2y$, or $2y$. With each such quadruple of sizes, other tilings are possible (D. Schattschneider, private communication).

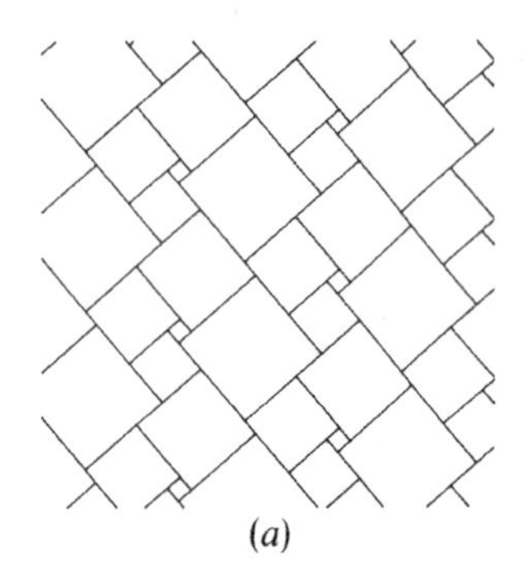

(a)

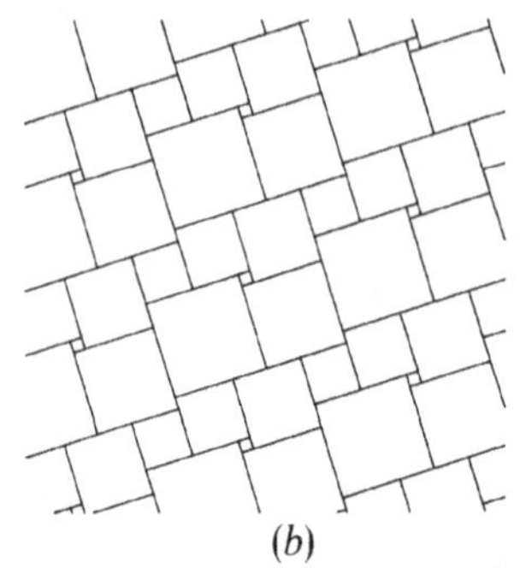

(b)

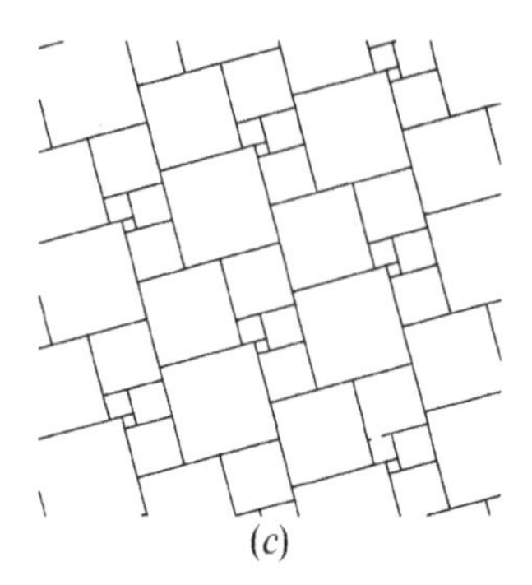

(c)

Figure 2.4.7
Examples of unilateral and equitransitive tilings by squares of five, six and seven sizes. In each of these tilings no tile has an equal tile as a neighbor.

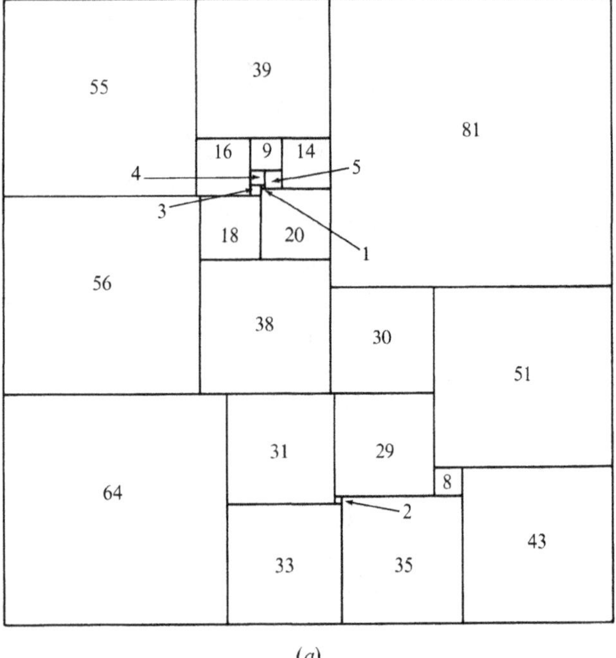

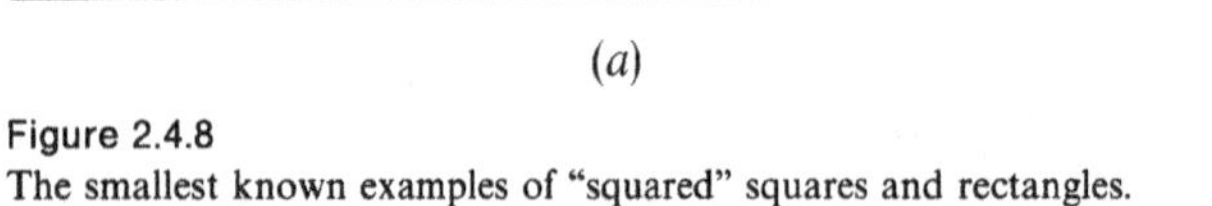

(a)

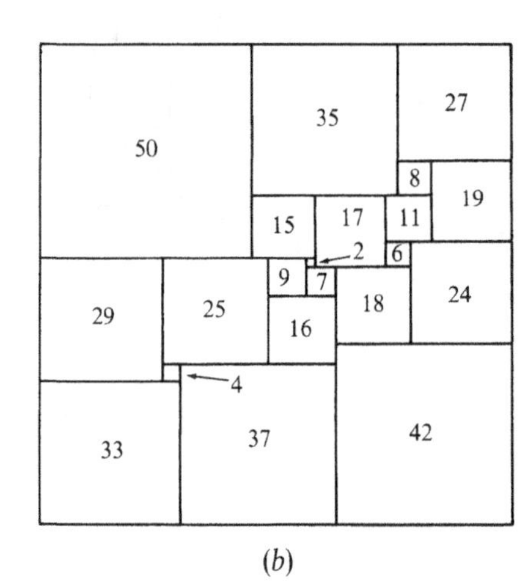

(b)

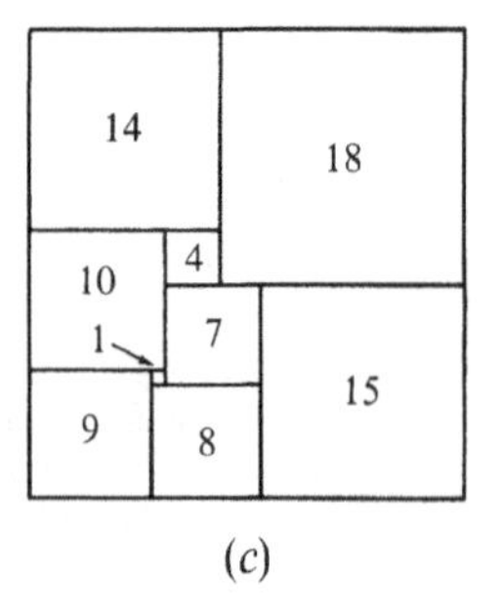

(c)

Figure 2.4.8
The smallest known examples of "squared" squares and rectangles. These and larger squared squares and rectangles can be used to construct equitransitive unilateral tilings by squares. The squared square with 24 squares shown in (a) is due to T. H. Willcocks (see Tutte [1966]); it was long believed to have the smallest possible number of squares. However, recently A. J. W. Duijvestijn found the squared square shown in (b) which has only 21 squares. It is known that there is no squared square with only 20 or fewer smaller squares. The smallest squared rectangle is shown in (c) (from Willcocks [1951]).*

* See the Appendix beginning on page 653.

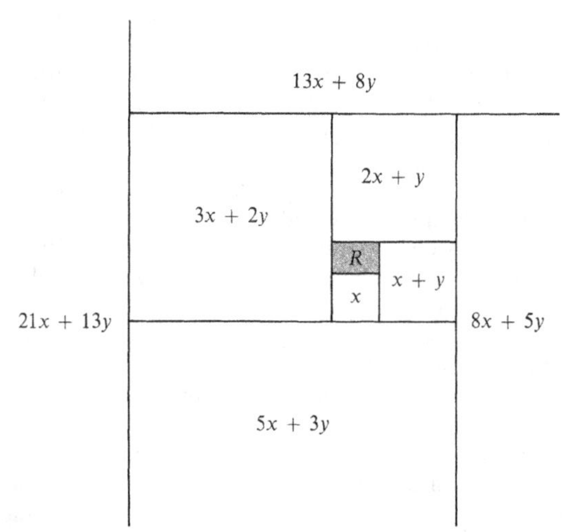

Figure 2.4.9
A construction for a tiling of the plane by squares of different sizes. We start with any "squared" rectangle R such as that of Figure 2.4.8(c) and adjoin successively bigger squares "spiralling" around it. If the original rectangle was of size x by y, then the squares are of sides x, $x + y$, $2x + y$, $3x + 2y$, $5x + 3y, \ldots$, each term being the sum of its two predecessors.

square (or of a rectangle, if its width is not the size of one of the squares used).

However, we do not know which sets of distinct numbers $p_1, \ldots, p_r$ can arise as the lengths of the sides of the squares in a unilateral equitransitive tiling, except when $r = 2$ or $r = 3$. In these case the possibilities are indicated in Figures 2.4.2(g) and 2.4.5. (The tilings with $r = 3$ and 4 have been investigated by Doris Schattschneider, Figures 2.4.5 and 2.4.6 are based on her unpublished results.)*

If $r \geq 5$ it is possible (as we see from Figure 2.4.7) to find tilings in which no two congruent squares are neighbors.

It seems a pity that, except for $r = 1$ or 2, the equitransitive tilings by squares appear to have been used in very few practical applications. They seem to us to be very attractive and have the advantage of using only tiles that are easily available commercially.

A problem of a rather different nature is that of tiling the plane by squares all of which are of different sizes. Two methods of doing this are as follows. The first depends upon the idea of "squared squares" mentioned above. Starting, for example, with Willcocks' "squared square" of Figure 2.4.8(a), we enlarge it in the ratio 175:1 and then replace the smallest square (which now has side 175) by the original "squared square". Proceeding inductively in this way, we arrive at a tiling with the required properties.

The second construction is shown in Figure 2.4.9. It can be based on any "squared" square or rectangle R, shaded in the figure. If we allow just two squares to be of the same size, then we may take R to be a pair of unit squares with common edge. The construction then leads to a tiling of the plane using squares with edge-lengths 1, 1, 2, 3, 5, 8, 13, 21, 34, ... which is the well known Fibonacci sequence. (For other properties of such "Fibonacci tilings" see Holden [1975], Hensley [1978].)

Herda [1981] also describes these two methods of tiling the plane by squares of different sizes. However, concerning the first method he adds a cryptic remark: "Here no tile remains fixed, outward growth occurs everywhere, and it is impossible to write down a sequence of side lengths of squares used in the tiling." This is clearly mistaken since the sizes that occur are easily determined and listed. Moreover, if the smallest tile at each step is kept in the same position, then even the position of each tile in the final tiling is well determined.

Both these constructions illustrate the fact (pointed out to us by Carl Pomerance) that for every known collection of different squares that tile the plane, the edge-lengths form a sequence that has an exponential rate of growth. No proof of this observation is known; in particular it is not known whether the plane can be tiled using *one* square of each edge-length 1, 2, 3, 4, ..., or any other prescribed (finite) number of tiles of each of these sizes. In Figure 2.4.10 we show a method of tiling a large part of the plane using one square of each integer size. For a further discussion of this problem, see Gardner [1979]. It is also unknown whether the plane can be tiled by equilateral triangles all of which are of

* See the Appendix beginning on page 653.

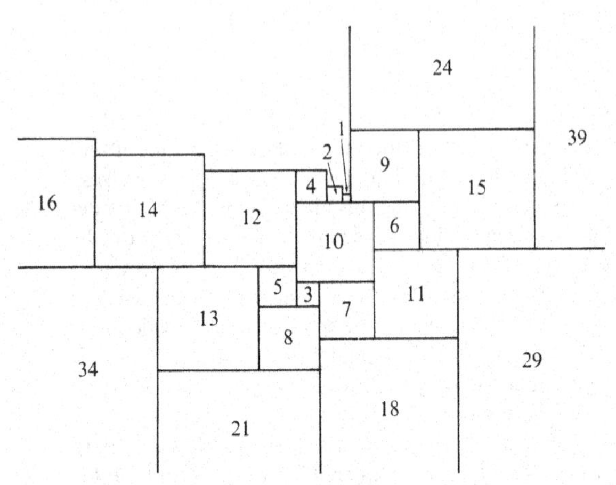

Figure 2.4.10
This figure shows how a large part of the plane can be covered by squares, one of each integer edge-length. The lower left part of the plane contains squares of edge-lengths 3, 5, 8, 13, 21, ..., the lower right part contains squares of edge-lengths 7, 11, 18, 29, 47, ..., and the upper right part contains squares of edge-lengths 6, 9, 15, 24, 39, ... In each of these sequences every term beyond the second is the sum of the two preceding terms, and it can be shown that no integer is contained in more than one sequence. All the squares omitted from the three sequences, namely 1, 2, 4, 10, 12, 14, ..., are placed in the upper left part of the plane. Here they are arranged so that the uncovered part is connected and simply connected—much more economical ways of arranging them can easily be devised. (This figure is adapted from one in Gardner [1979].)

different sizes. (For a partial result in this direction see Scherer [1983].) The constructions described above for squares cannot be adapted to this case for it is known that no equilateral triangle can be partitioned into unequal equilateral triangles (Tutte [1948], Honsberger [1970, pp. 59, 188]).

A recent account of the problem of "triangulating" by equilateral triangles, and references to previous work in this direction, can be found in Buchman [1981], Tutte [1981]. A variation of this idea, in which the plane is tiled by different triangular tiles each of which has an angle of 60° (and so may be thought of as that part of an equilateral triangle which is cut off by a line through its summit) is investigated by Mielke [1983].

Eggleton [1974], [1975] and Pomerance [1977] have shown how to tile the plane using different rationally-sided triangles (or various other special kinds of triangles) and moreover, how this can be done using precisely one representative of every congruence class of such triangles. This solves a problem originally raised by J. H. Conway and solved by D. C. Kay and others (see Conway [1965]). The results are related to interesting questions in algebraic geometry. The question whether the plane can be tiled using precisely one triangle of each congruence class of integral-sided triangles is still open.

EXERCISES 2.4

*1. Verify the statement made in this section that there exist precisely eight families of uniform non-edge-to-edge tilings by regular polygons.

2. Find five examples of equitransitive 2-uniform tilings that are non-edge-to-edge and are distinct from the ones in Figure 2.4.3.

3. (*a*) Find a 2-uniform equitransitive tiling with hexagons of two sizes and triangles.

**(*b*) Determine (at least for $n = 3, 4$) the least integer $k = k(n)$ such that there exists a k-uniform tiling with hexagons of n distinct sizes and triangles of suitable sizes.

4. Show that each unilateral tiling by squares of two sizes is equitransitive.

5. Find unilateral and equitransitive tilings by squares of sizes

(*a*) 1, 2, 3. (*b*) 1, 2, 3, 4. (*c*) 1, 2, 3, 4, 5.

**(*d*) 1, 2, 3, 4, 5, 6, or, more generally, any set of n consecutive integers, with $n \geq 6$.

6. (*a*) Prove the following assertion (made in the caption of Figure 2.4.5): a unilateral and equitransitive tiling by squares of three sizes $x < y < z$ is possible if and only if $z = x + y$.

*(*b*) Show that squares of given sizes $x < y < z = x + y$ admit precisely five distinct unilateral and equitransitive tilings.*

(*c*) Show that there are uncountably many unilateral but not equitransitive tilings by squares of sizes 1, 2, 3.

**(*d*) Show that for each unilateral (but not necessarily equitransitive) tiling by squares of three sizes $x < y < z$ the relation $z = x + y$ holds.

7. For each of the five quadruplets of sizes specified in Figure 2.4.6 construct a unilateral and equitransitive tiling different from the one shown in Figure 2.4.6.

8. (*a*) Find a unilateral and equitransitive tiling by squares of sizes 1, 2, 3, 4, 6 that has reflective symmetries.

*(*b*) Show that no unilateral and equitransitive tiling by squares of three sizes can admit any reflective symmetries.

**(*c*) Show that no unilateral and equitransitive tiling by squares of four sizes can admit any reflective symmetries.

9. (*a*) Show that for each $k \geq 2$ there exists a unilateral and equitransitive tiling by squares of k sizes.

(*b*) Prove that for each $k \geq 5$ there exist unilateral and equitransitive tilings by squares of k sizes, such that any two congruent squares are disjoint.

*(*c*) Show that for $k = 4$ there is no tiling with the properties mentioned in part (*b*).

**10. Show that there exists no tiling of the plane in which the tiles are equilateral triangles, each of a different size.

**11. Find all 2-uniform equitransitive tilings that are not edge-to-edge.

12. Show that there are just three equitransitive and 2-isotoxal tilings by squares.

*13. Determine all (not necessarily edge-to-edge) equitransitive and 2-isotoxal tilings by regular polygons.

14. Consider tilings in which exactly k square tiles have edge-length $L < 1$, and the remaining tiles are squares of edge-length 1.

(*a*) Find examples with $k = 1, 2, 3, \ldots$

*(*b*) Show that all values of k are possible and obtain a complete description of all tilings corresponding to a given value of k when L is irrational.

15. Consider tilings in which exactly k regular hexagonal tiles have edge-length $L < 1$ and the remaining tiles are equilateral triangles of edge-length 1. Show that $k = 0$ or 1.

16. Consider tilings in which exactly $k > 0$ tiles are equilateral triangles of edge-length $L < 1$ and the remaining tiles are regular hexagons of edge-length 1.

(*a*) Show that L must be of the form $L = 1/n$, where n is a positive integer.

*(*b*) Find all possible values of k and L.

17. (*a*) All the tiles of a tiling $\mathscr{T}$, with one exception, are equilateral triangles of edge-length 1. The exceptional tile T has diameter less than 1. Describe all possible shapes of T.

(*b*) Consider the analogous problem for square tiles.

---

* See the Appendix beginning on page 653.

2.5 TILINGS USING STAR POLYGONS

If we extend the definition of "regular polygon" to include regular star polygons, then several interesting possibilities emerge. Some are to be found in Kepler's drawings (reproduced as the frontispiece). It is clear from these that his approach to the subject was rather pragmatic and experimental, and that he was looking for various more or less "regular" tilings in some sense of the word. It is therefore to be expected that he would consider the possibility of tiles in the shape of star polygons, and it is surprising that no discussion along similar lines seems to have appeared prior to Grünbaum & Shephard [1977a].

Before we can proceed we must clarify the meaning of the term "regular star polygon". There are at least two possible interpretations, both of which occur in Kepler's work. In the first book of his *Harmonice Mundi* [1619], star polygons are obtained by extending the sides of regular (convex) polygons. In a modern spirit, Kepler treats only the endpoints of the extended sides as the corners of the star polygon, and not the corners of the original polygon. Thus the *pentagram*, usually denoted by $\{5/2\}$, has 5 corners and 5 sides (see Figure 2.5.1(*a*)). More generally, if n and d are coprime (that is, relatively prime) positive integers with $1 < d < n$, then the *star polygon*

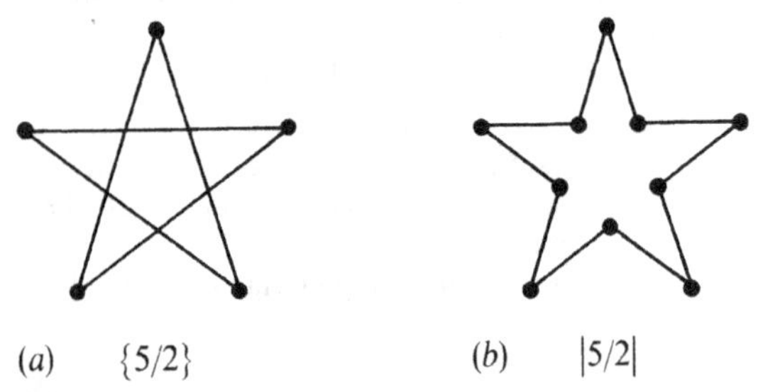

Figure 2.5.1
Two interpretations of the star pentagon: (*a*) is the *pentagram* $\{5/2\}$ with five corners and five equal sides, and (*b*) is a non-convex decagon with ten equal sides and angles of two sizes; this will be called a *pentacle* and denoted by $|5/2|$.

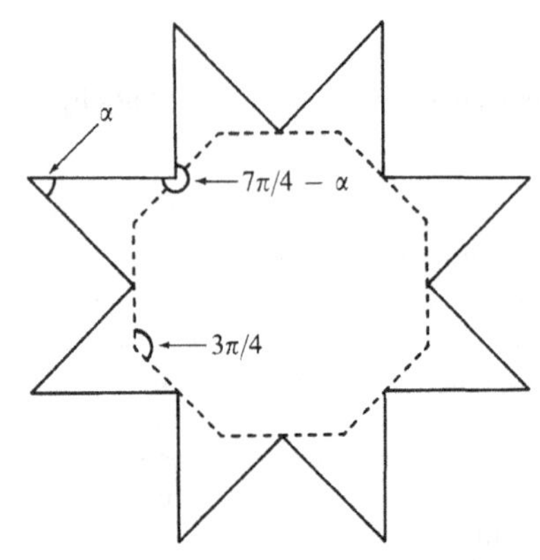

Figure 2.5.2
The star 8-gon (star octagon) $\{8_\alpha\}$. The restriction $\alpha < 3\pi/4$ is imposed in order to keep the polygon star-shaped (the dashed polygon is the regular octagon $\{8\}$). In this figure $\alpha = \pi/4$ so the polygon can also be denoted by $|8/3|$.

$\{n/d\}$ is obtained from the set of corners of a regular n-gon $\{n\}$ by joining each corner of $\{n\}$ to the dth one clockwise and counterclockwise on $\{n\}$. Although the interpretation of "tilings" with such "polygons" is not entirely obvious, they have been considered already by Badoureau [1878], [1881], and will be briefly described in Section 12.3; for a recent survey see Grünbaum, Miller & Shephard [1982].

On the other hand, in the second book of *Harmonice Mundi*, when dealing with tilings, Kepler treats the star pentagon as a non-convex decagon, with ten equal sides and five angles of each of two different sizes, see Figure 2.5.1(*b*). He considers other star polygons in a similar way. It is never made quite clear exactly what polygons may be used. At any rate he missed several possibilities and it is interesting to try to complete his enumeration of tilings under some definite set of rules.

One possibility is to allow as *regular star n-gons* all $(2n)$-gons with $2n$ equal sides, that have the same symmetries as $\{n\}$ (see Figure 2.5.2 for $n = 8$). Such a star, which will be denoted by $\{n_\alpha\}$, has n corners of angle α (where $0 < \alpha < (n - 2)\pi/n$) at the "points" of the star,

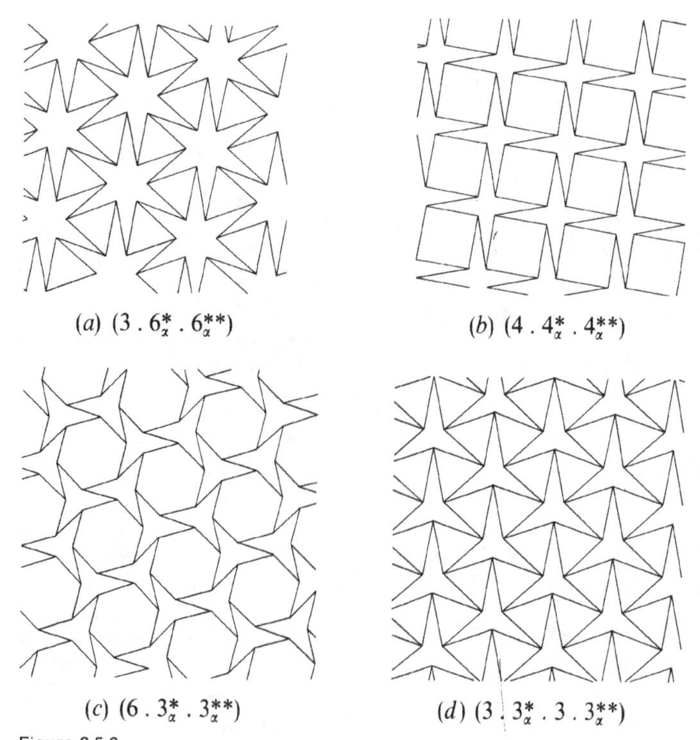

Figure 2.5.3
Uniform tilings by regular polygons and star polygons in which every corner of a tile is a vertex of the tiling.

and n corners of angle $2(n - 1)\pi/n - \alpha$ at the "dents" of the star. In particular, when $\alpha = (n - 2d)\pi/n$ we shall also use $|n/d|$ to mean the same as $\{n_\alpha\}$; if n and d are coprime then this α is the same as the angle at each corner of the star-polygon $\{n/d\}$. The star $|5/2|$ of Figure 2.5.1(*b*) will be called a *pentacle.*

The definitions of uniform, k-uniform and equitransitive may be extended in the obvious way to tilings in which star polygons are used. We stress, however, that *uniform* means that the symmetries act transitively on the *vertices* of the tiling, and not necessarily on the *corners* of the polygons.

It is easy to see that there are no monohedral tilings by regular star polygons, but without the monohedrality restriction we have:

2.5.1 *There exist four families (each depending on a real-valued parameter* α*) of uniform tilings by regular polygons and star polygons in which every corner is a vertex.*

The tilings in these families may be denoted by $(3.6_\alpha^*.6_\alpha^{**})$, $(4.4_\alpha^*.4_\alpha^{**})$, $(6.3_\alpha^*.3_\alpha^{**})$ and $(3.3_\alpha^*.3.3_\alpha^{**})$, where we use one or two asterisks to distinguish the two kinds of corners (points and dents) of the star polygons. The tilings of the first three of these families occur in two enantiomorphic forms. Examples of tilings of all four families are shown in Figure 2.5.3.

One proof of Statement 2.5.1 can be derived by first observing that at most one kind of star polygon can occur, and that each vertex of the tiling must be incident with one "point" and one "dent" of the star polygon; these two take up an angle of $2(n - 1)\pi/n$, so that all other tiles incident with the vertex have to contribute just $2\pi/n$. It is easily seen that then $n = 3, 4$ or 6, and the assertion follows.

If the requirement that all corners are vertices is removed, many additional uniform tilings are possible. In Figure 2.5.4 we show 17 uniform edge-to-edge tilings that include star polygons, in which not all corners are

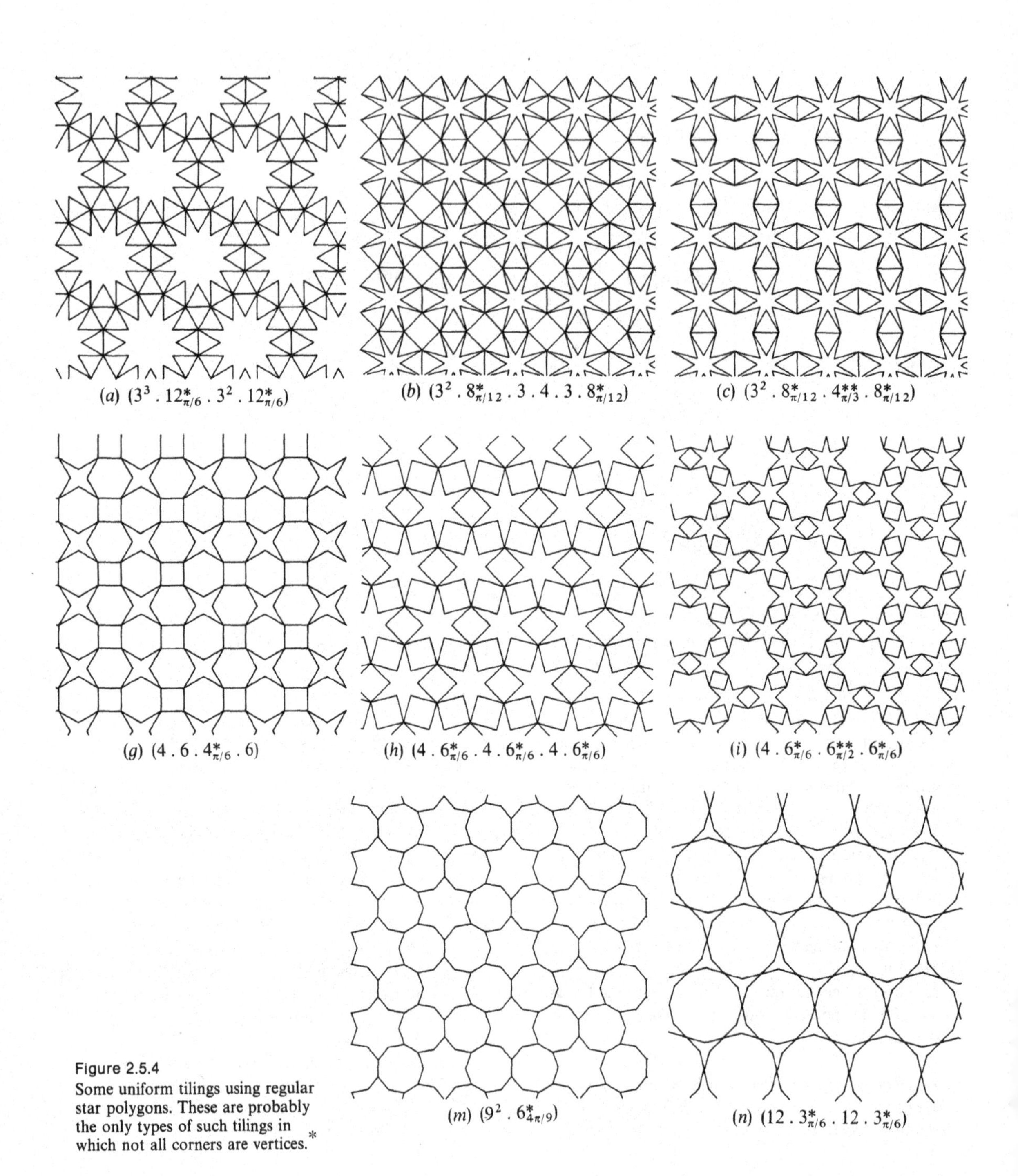

(*a*) $(3^3 . 12^*_{\pi/6} . 3^2 . 12^*_{\pi/6})$

(*b*) $(3^2 . 8^*_{\pi/12} . 3 . 4 . 3 . 8^*_{\pi/12})$

(*c*) $(3^2 . 8^*_{\pi/12} . 4^{**}_{\pi/3} . 8^*_{\pi/12})$

(*g*) $(4 . 6 . 4^*_{\pi/6} . 6)$

(*h*) $(4 . 6^*_{\pi/6} . 4 . 6^*_{\pi/6} . 4 . 6^*_{\pi/6})$

(*i*) $(4 . 6^*_{\pi/6} . 6^{**}_{\pi/2} . 6^*_{\pi/6})$

(*m*) $(9^2 . 6^*_{4\pi/9})$

(*n*) $(12 . 3^*_{\pi/6} . 12 . 3^*_{\pi/6})$

Figure 2.5.4
Some uniform tilings using regular star polygons. These are probably the only types of such tilings in which not all corners are vertices.*

* See the Appendix beginning on page 653.

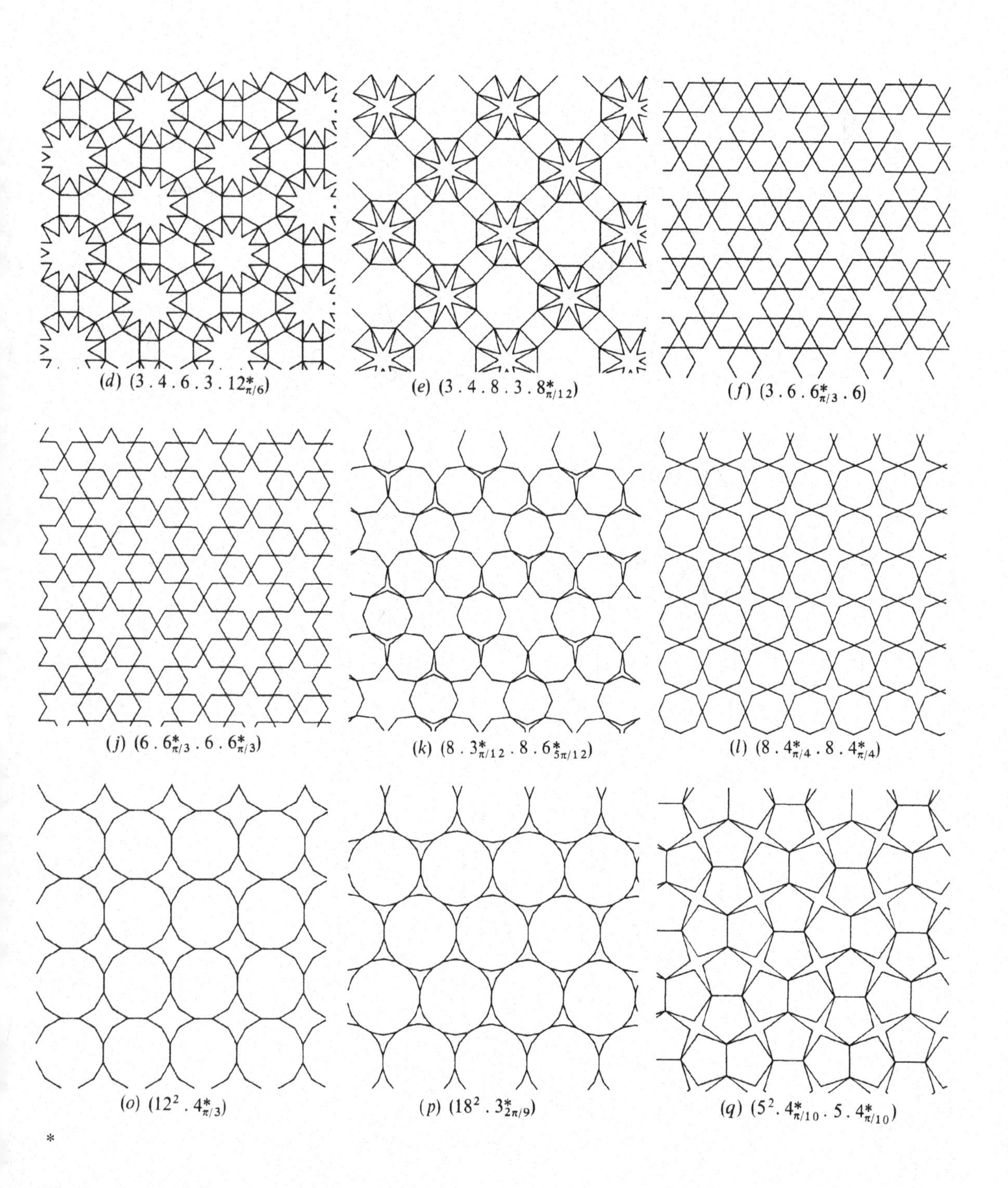

(*d*) $(3 . 4 . 6 . 3 . 12^*_{\pi/6})$

(*e*) $(3 . 4 . 8 . 3 . 8^*_{\pi/12})$

(*f*) $(3 . 6 . 6^*_{\pi/3} . 6)$

(*j*) $(6 . 6^*_{\pi/3} . 6 . 6^*_{\pi/3})$

(*k*) $(8 . 3^*_{\pi/12} . 8 . 6^*_{5\pi/12})$

(*l*) $(8 . 4^*_{\pi/4} . 8 . 4^*_{\pi/4})$

(*o*) $(12^2 . 4^*_{\pi/3})$

(*p*) $(18^2 . 3^*_{2\pi/9})$

(*q*) $(5^2 . 4^*_{\pi/10} . 5 . 4^*_{\pi/10})$

*

* See the Appendix beginning on page 653.

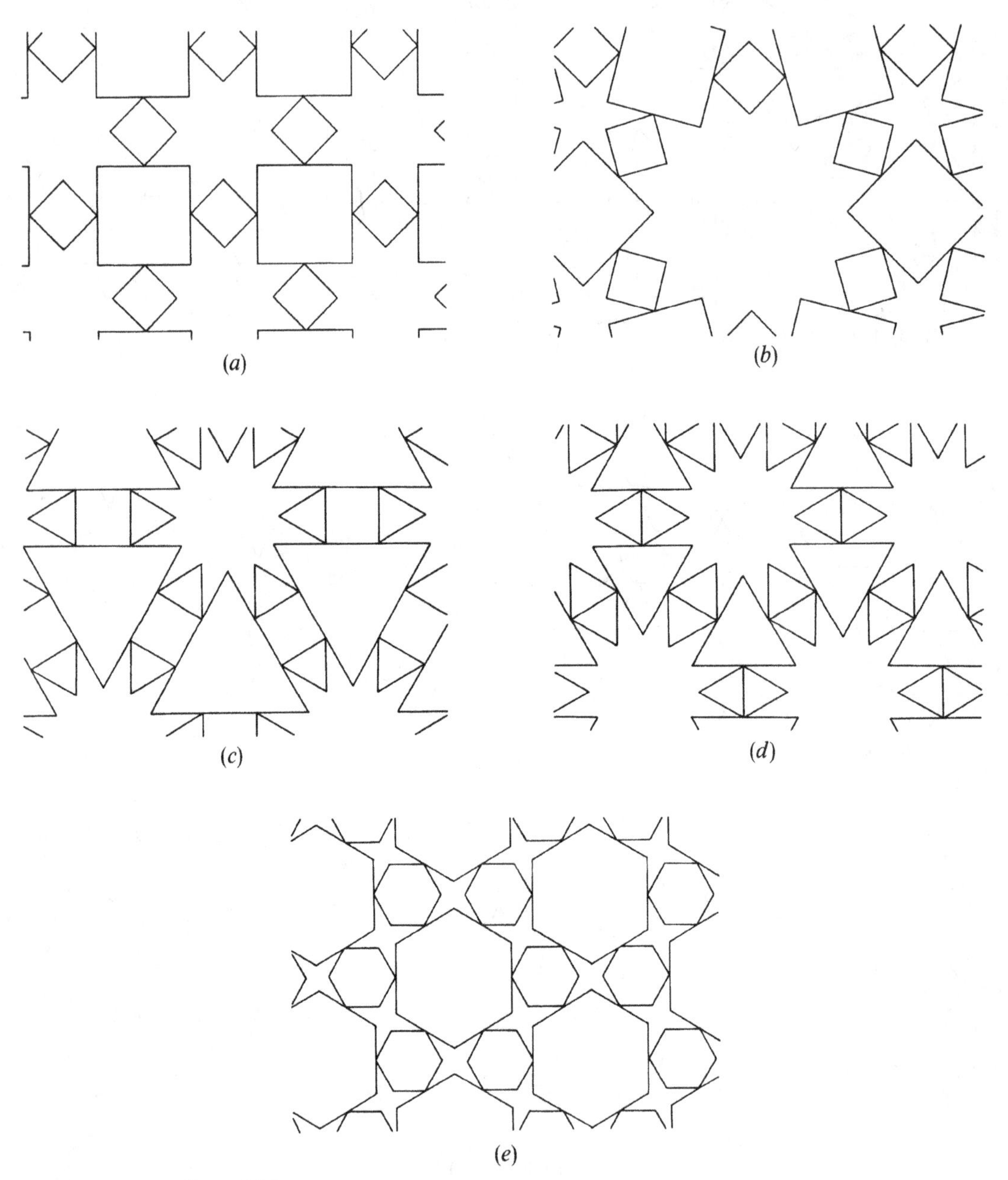

Figure 2.5.5
Five examples of uniform tilings which use star polygons and are not edge-to-edge.

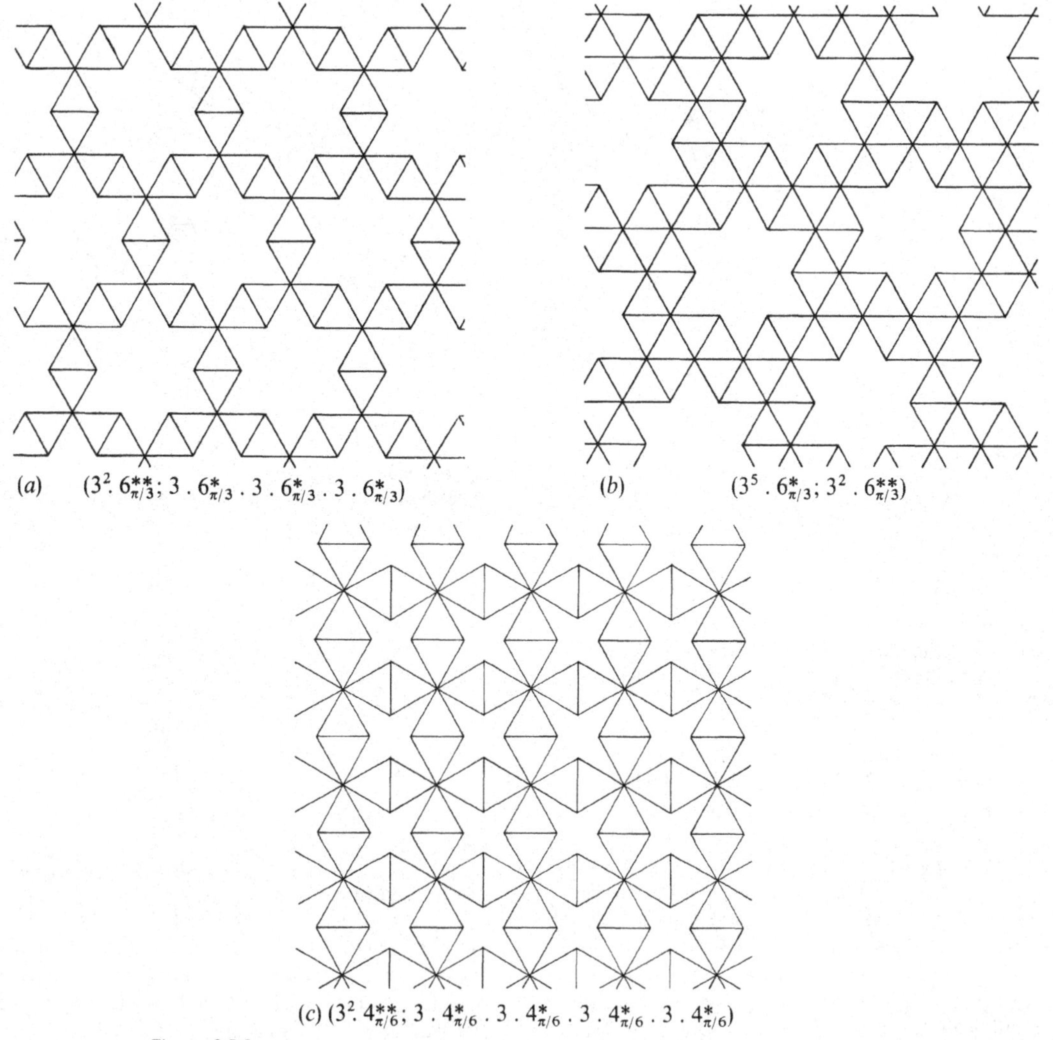

Figure 2.5.6
Three examples of edge-to-edge 2-uniform tilings by regular polygons and star polygons.

vertices. We believe that there are no other tilings with these properties.

Five examples of uniform tilings which are not edge-to-edge are shown in Figure 2.5.5, and three examples of edge-to-edge 2-uniform tilings are shown in Figure 2.5.6. Another example is shown in Kepler's drawing X, and if the tiling of Kepler's drawing Y is extended appropriately it will be seen to be 3-uniform. Most of these tilings are also equitransitive. With some patience it should be possible to determine all the 2-uniform and all the equitransitive tilings by regular polygons and star polygons.

It could be argued that the star polygons $\{n_\alpha\}$ should be called *alternating* (2*n*)-gons, and that in any case similar treatment should be given to the analogously defined convex polygons $\{n_\alpha\}$ with

$$(n-2)\pi/n < \alpha < (n-1)\pi/n \text{ and } n \geq 2$$

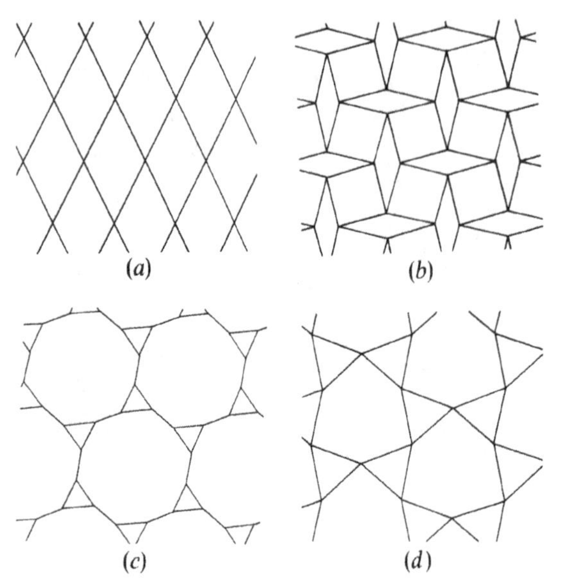

Figure 2.5.7
Some examples of uniform equitransitive tilings which involve only regular (convex) polygons and isosceles $(2n)$-gons $\{n_\alpha\}$ in which larger and smaller angles alternate.

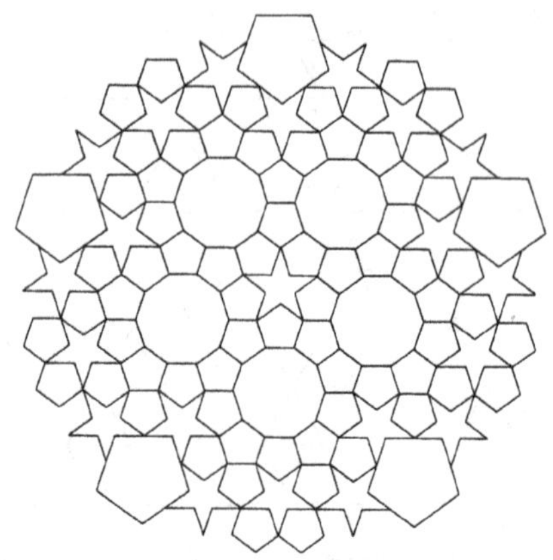

Figure 2.5.8
A patch of tiles in which each tile has symmetry group *d5*.

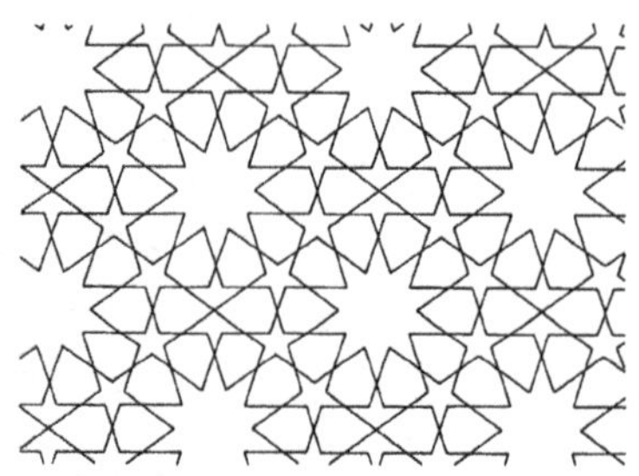

Figure 2.5.9
A tiling which contains pentacles $|5/2|$ and star decagons $|10/4|$, adapted from Islamic art.

in which larger and smaller angles alternate. Examples are given in Figure 2.5.7 of such tilings which are uniform and equitransitive. Many others exist.

Further extensions and generalizations are clearly possible, such as equitransitive tilings by arbitrary polygons. But this departs from the spirit of the present chapter. (All vertex-transitive tilings by arbitrary convex polygons will be determined in Chapter 6.)

Using arguments involving angles at the corners of regular polygons, in Section 2.1 we deduced that there is no tiling of the plane using regular pentagons, even in combination with other regular (convex) polygons. On the other hand all such arguments break down if we are allowed to introduce star polygons as well. For example, there is no *a priori* reason why a tiling by pentacles and other regular polygons should not be possible, but in fact it is not difficult to convince oneself that no tiling with these prototiles exists.

In Section 1.5 we have seen tilings which use regular pentagons $\{5\}$ and rhombs $\{2_{\pi/5}\}$, and which have some unusual features. Here, however, we shall report on tilings which use pentagons, pentacles, and a few other shapes of tiles as well, such as regular decagons. The problem of constructing such tilings challenged the ingenuity of medieval artists, but we found extremely few mathematical references to it in the literature. The paucity of available information is highlighted by the fact that we know of no tiling in which the symmetry group $S(T)$ of each tile T includes 5-fold rotational symmetries, but we are also unable to prove that no such tiling exists. In fact quite large patches of tiles, each with 5-fold symmetry can be constructed (see Figure 2.5.8; larger patches can be constructed using methods described in Danzer, Grünbaum & Shephard [1982]). The Arabic decorators reached a compromise by devising tilings in which many—but not all—tiles have 5-fold rotational symmetry. An example from Islamic art, which uses pentacles $|5/2|$ and ten-pointed stars $|10/4|$ is shown in Figure 2.5.9. It is adapted from a tiling used on a mosque in Tashkent, USSR (see Babakhanov [1962] and Bourgoin [1879, Plate 171]). Here the artist used

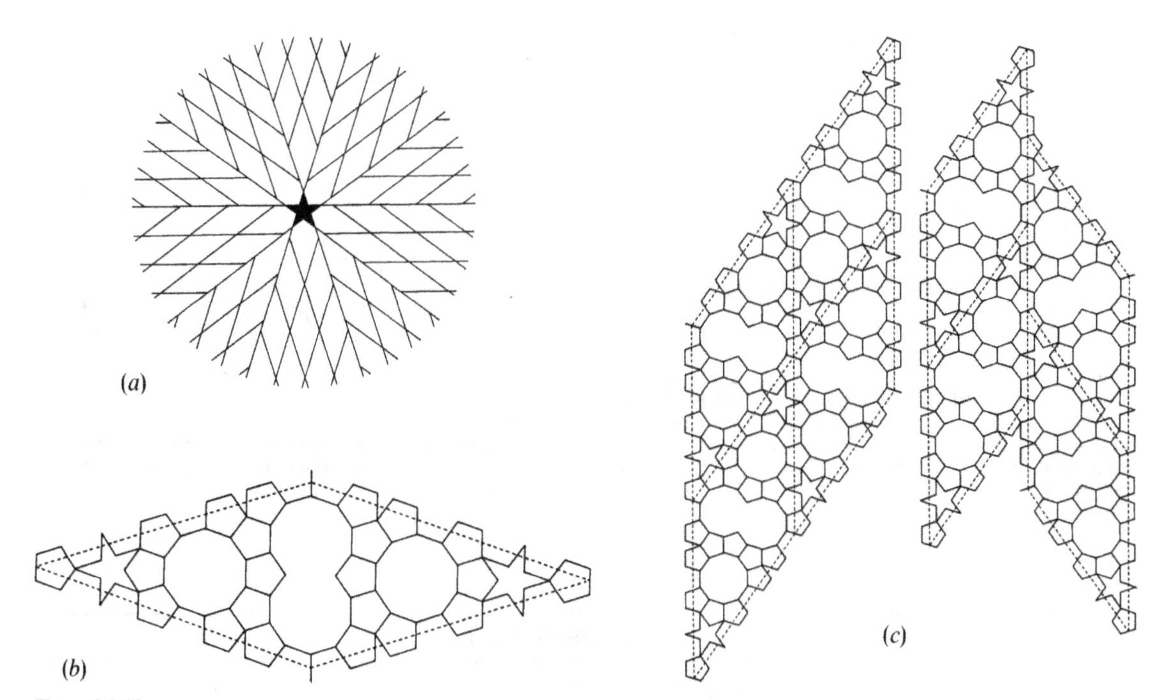

Figure 2.5.10
Dessecker's explanation of the construction of Kepler's tiling shown in Figure 2.0.1. The rhombs shown in (*b*) are arranged as indicated in (*a*). The construction is possible because of the different ways that the rhombs can fit against each other, see (*c*).

irregular hexagons and pentacles with two points cut off in order to complete the tiling which is nevertheless very decorative.

Kepler was also interested in the possibilities of such tilings. Extending the patch Aa shown in the frontispiece, we obtain the tiling of Figure 2.0.1. The method of construction was explained by Dessecker (see Bindel [1964], Eberhart [1975a]) and is indicated in Figure 2.5.10. The basic idea is to cover the plane with rhombs (having angles of 36° and 144°) as in Figure 2.5.10(*a*). These overlap in a pentacle (shown in black) at the center of the figure. Each rhomb is then covered with tiles as shown in Figure 2.5.10(*b*). The tiling is possible only because these subdivided rhombs fit together either edge-to-edge, or staggered (in 1/2 ratio), see Figure 2.5.10(*c*). This tiling is 4-hedral—the four kinds of tiles being pentagons, pentacles, decagons and "fused decagon pairs" or "monsters" as Kepler called them.

Kepler also indicated a patch that involves star-decagons $|10/3|$ in his Figure Bb. We do not know exactly what continuation he had in mind, but one possibility was hinted at by Caspar, who translated into German and annotated Kepler's *Harmonice Mundi* (Kepler [1619]; see Caspar [1939, p. 374]). (The question was considered also by Bindel [1964, p. 26], but the solution he proposes is unnecessarily complicated.) In fact, the same type of construction as above can be applied, and the rhombs of Figure 2.5.10(*a*) covered with tiles as in Figure 2.5.11. Again four kinds of tiles are required: pentagons of two sizes, star-decagons $|10/3|$ and "spiny monsters" consisting of two fused star-decagons. We suspect that this is what Caspar may have had in mind.

As observed by Eberhart [1975b], two related tilings may be obtained by the same technique, starting from the dotted rhomb in Figure 2.5.12. These tilings include regular pentagons of three sizes, pentacles and fused pairs of pentacles. With small changes in the interpretation of these fused pairs, the two tilings can be made 4-hedral.

Another kind of very remarkable tilings that use pentagons, pentacles, and tiles of other shapes, will be described in Chapter 10.

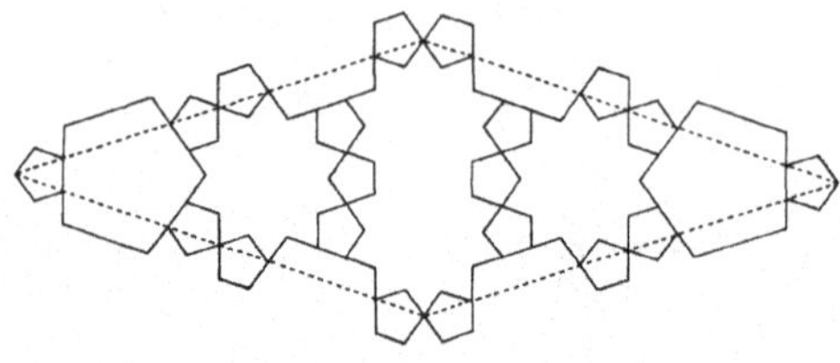

Figure 2.5.11
One possible method of extending Kepler's patch *Bb* shown in the frontispiece is to fit the dashed rhombs in the way indicated in Figure 2.5.10(*a*) (see Caspar [1939, p. 374]). Every vertex of the tiling obtained in this manner is the midpoint of an edge of Kepler's tiling of Figure 2.0.1.

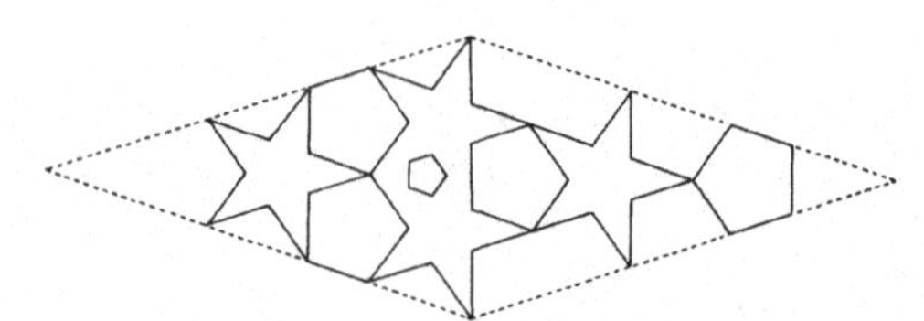

Figure 2.5.12
Eberhart's subdivision of the rhomb which leads to two interesting tilings using the construction indicated in Figure 2.5.10(*a*).

EXERCISES 2.5

1. Let n points be equally spaced on a circle. Label one of them A_0 and draw a line segment from A_0 to the dth point on the circle (which is labelled A_1) counterclockwise from A_0. Draw a line segment from A_1 to the dth point (labelled A_2) counterclockwise from A_1, and so on, until A_0 is reached again. Prove that a star polygon $\{n/d\}$ of n sides is obtained if and only if n and d are relatively prime.

2. Show that it is not possible to tile the plane using any collection of regular star polygons $|n/d|$.

3. Construct a 2-uniform and equitransitive tiling $(3^2.4^{**}_{\pi/6}; 3.4^{*}_{\pi/6}.4^{**}_{\pi/6}.4^{*}_{\pi/6})$.

4. Find at least five examples, different from those in Figure 2.5.6, of edge-to-edge 2-uniform tilings by regular polygons and star polygons.

**5. Show that the only uniform tilings that use star polygons and are not edge-to-edge are the five tilings shown in Figure 2.5.5.

**6. Do there exist any uniform tilings by regular polygons and star polygons other than those shown in Figures 2.5.3, 2.5.4 and 2.5.5?

7. Show that there exists a unique equitransitive dihedral tiling whose prototiles are a regular pentagon and a rhomb (alternating 4-gon) with angles $\pi/5$, $4\pi/5$.

8. The tilings in Figures 2.5.3(*d*) and 2.5.4(*h*), (*j*), (*l*), (*n*) are equitransitive and isotoxal. Show that these are the only isotoxal and equitransitive tilings with regular polygons and star polygons.

9. Several of the patches considered by Kepler (see the frontispiece) involve star polygons. Investigate the possibilities of extending them to tilings of the whole plane as follows:
   (*a*) Show that the patches Gg and Hh can be extended to 2-uniform edge-to-edge tilings, and that patch Y can be extended to a 4-uniform edge-to-edge tiling.
   *(*b*) Find a 6-uniform edge-to-edge tiling that extends the patch Ff and show that no extension to a k-uniform tiling is possible if $k \leq 5$.

10. (*a*) Find a 4-hedral periodic tiling $\mathscr{T}$ which uses all the tiles shown in Figure 2.5.13.
    (*b*) Find a tiling which uses the same tiles and has symmetry group $d5$.
    (*c*) Find a dihedral tiling which is periodic and uses only the first and fourth tiles of Figure 2.5.13.

11. Delete from Kepler's patch Ll (see the frontispiece)

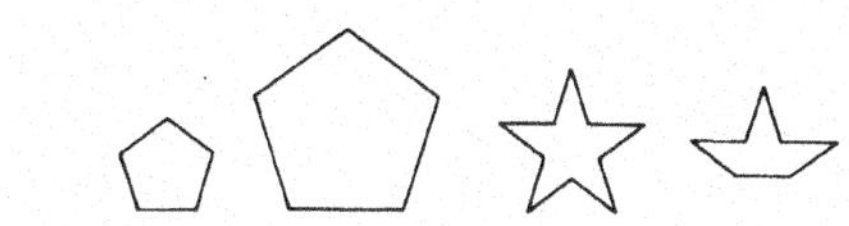

Figure 2.5.13
Prototiles for 4-hedral tilings. All sides have the same length except those of the large pentagon, which are twice as long.

one of the squares and consider the remaining regular 20-gon surrounded by alternating squares and pentagons. Show that this patch can be extended to a tiling in which, besides {4}, {5} and {20}, lens-shaped 10-sided tiles that have two angles of 72° and the other eight of 162° are used. (This construction is due to S. Eberhart.)

12. Early in this section we gave two different interpretations of the notion of star polygon. Consider the following related kinds of polygons, and in each case find some examples of tilings that are uniform and equitransitive:

(*a*) ($2n$)-gons with angles alternately α_1, α_2 and alternate edges of lengths e_1, e_2.

(*b*) ($3n$)-gons with angles $\alpha_1, \alpha_2, \alpha_3, \alpha_1, \alpha_2, \alpha_3, \ldots$ and all edges of the same length.

2.6 DISSECTION TILINGS

Every polygon P has the property that it can be cut into a finite number of pieces which can be rearranged to form the prototile of a monohedral tiling $\mathscr{T}$. This assertion, which applies to many other sets in addition to polygons, follows immediately from the well known fact that every polygon can be cut up and the pieces rearranged to form any other polygon of the same area. This is usually expressed by saying that every two polygons of the same area are *equidecomposable*. Any tiling $\mathscr{T}$ which arises in this way will be called a *dissection tiling* for P, and we shall refer to it as an *r-dissection* tiling if P is cut into r parts.

Our interest here is to minimize r for a given P. If r is minimal then $\mathscr{T}$ will be called a *minimal dissection tiling*.

In Figure 2.6.1 we exhibit a number of dissection tilings for regular n-gons which are obviously minimal for $n =$ 5, 8 and 10 since $r = 2$, and which are believed to be minimal for $n = 7$, 9 and 12. Figure 2.6.2 shows some other dissection tilings. These include an attractive (non-minimal) dissection tiling for the 10-gon and some dissection tilings for regular star polygons that seem to exemplify the extraordinary amount of ingenuity needed to construct tilings of this nature.

Goldberg & Stewart [1964] give examples of dissections of polygonal regions which could be used to construct dissection tilings. Most of our illustrations are taken or adapted from Lindgren [1972], who introduced dissection tilings as a means of solving geometric dissection problems. A discussion of the aesthetics of tilings, together with some examples of dissection tilings, appears in Senechal [1979b].*

* See the Appendix beginning on page 653.

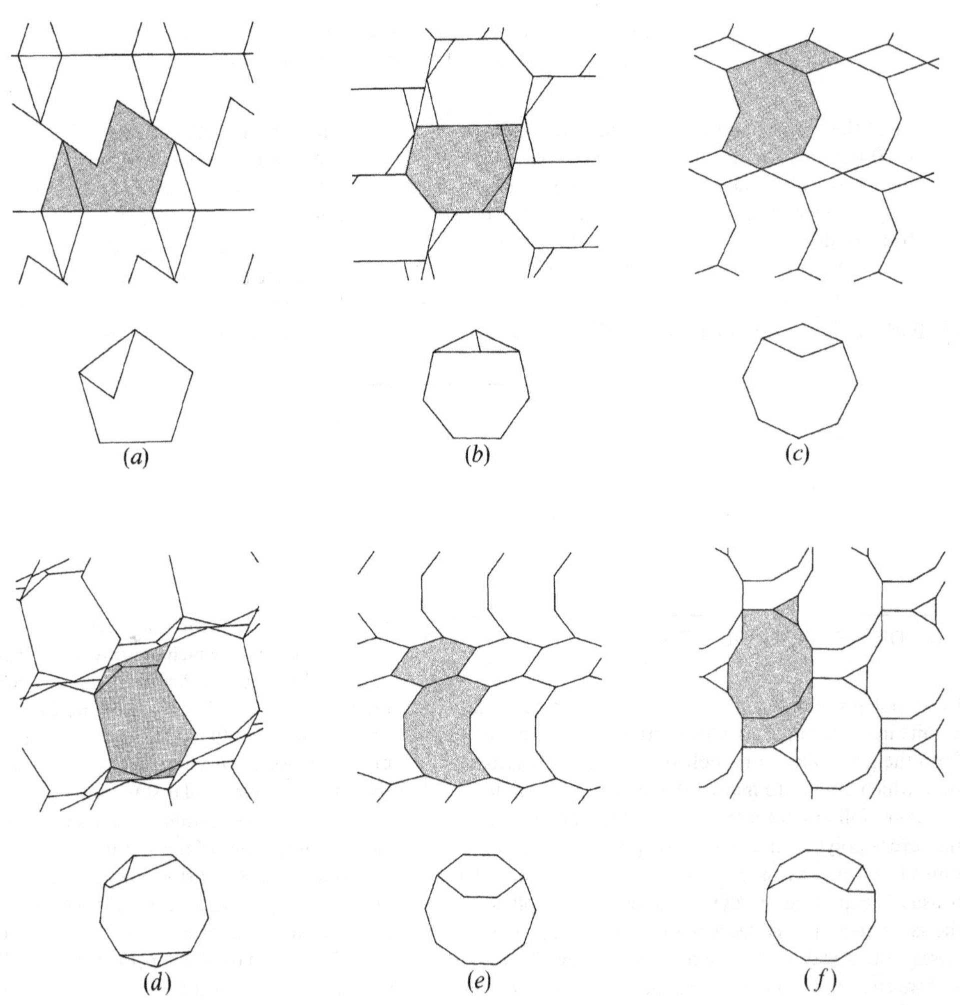

Figure 2.6.1
Some dissection tilings for regular n-gons: (*a*), (*c*) and (*e*) show minimal dissection tilings for $n = 5, 8$ and 10 respectively, and (*b*), (*d*) and (*f*) show dissection tilings for $n = 7, 9$ and 12 that are believed to be minimal (see Lindgren [1972]).*

* See the Appendix beginning on page 653.

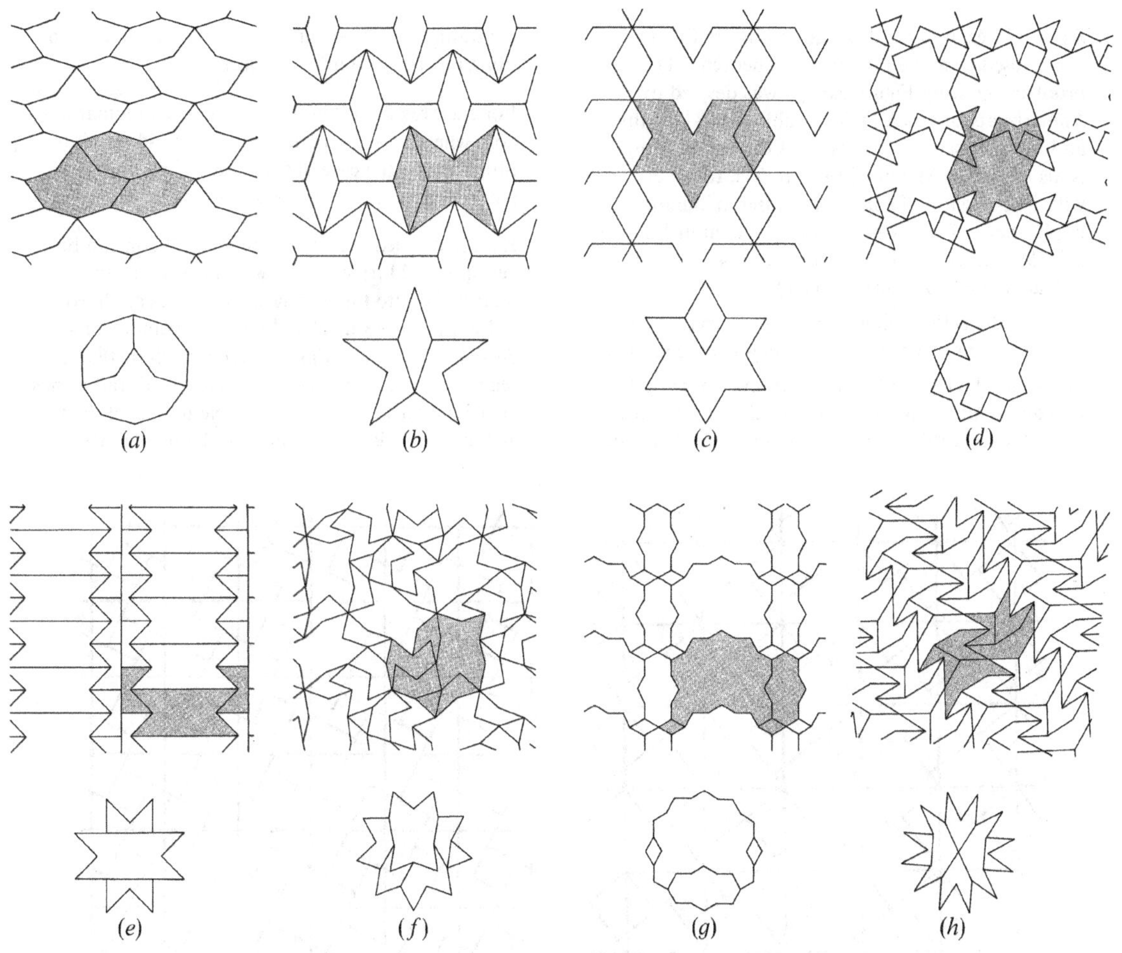

Figure 2.6.2
Additional examples of dissection tilings: (*a*) shows an attractive non-minimal 3-dissection tiling for the 10-gon, and (*b*) to (*h*) show dissection tilings for the regular star polygons |5/2|, |6/2|, |8/2|, |8/3|, |9/3|, |12/2| and |12/5|. These tilings were believed to be minimal (see Lindgren [1972]), but D. P. Chavey (private communication) has found a 2-dissection tiling for |8/3|.*

* See the Appendix beginning on page 653.

EXERCISES 2.6

1. Show how each of the dissection tilings in Figure 2.6.3 can be used to prove Pythagoras' Theorem. (The proof arising from Figure 2.6.3(*a*) was devised by Annairizi about 900 A.D. See Mahlo [1908] for this and other dissection proofs. This dissection tiling is also given in Magnus [1974, p. 53], together with a discussion of some interesting axiomatic implications. The dissection tiling shown in Figure 2.6.3(*b*) is given in Lindgren [1972, p. 43], where it is attributed to Henry Perigal.)
2. Find a 2-dissection tiling for the regular pentagon such that the two parts of the pentagon are congruent.
3. Show that the regular 9-gon can be cut up (by suitable diagonals) into four convex pieces that are the prototiles of a 4-hedral tiling. Explain why this does not contradict the conjecture that a minimal dissection tiling for the regular 9-gon has $r = 5$.
4. For each regular $(2k - 1)$-gon and each regular $2k$-gon (where $k \geq 2$ is an integer) find a k-dissection tiling, and in the case of the $2k$-gon show that the pieces can be chosen to be congruent.
5. Kürschak discovered that a regular 12-gon can be cut up into 24 triangles of two shapes that can be rearranged into three congruent squares as shown in Figure 2.6.4 (Kürschak [1898], Alexanderson & Seydel [1978]). Use this dissection to show that a regular 12-gon inscribed in a circle of unit radius has area 3, and also, by taking suitable unions of the triangles, find a 4-dissection tiling for the regular 12-gon.

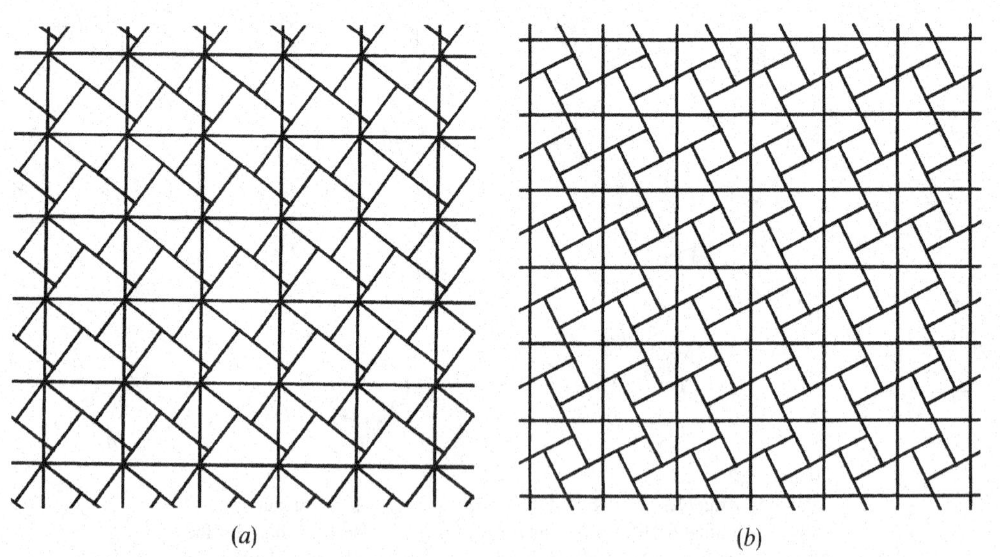

(*a*) (*b*)

Figure 2.6.3
Dissection tilings for the square which can be used to prove Pythagoras' Theorem, see Exercise 2.6.1.

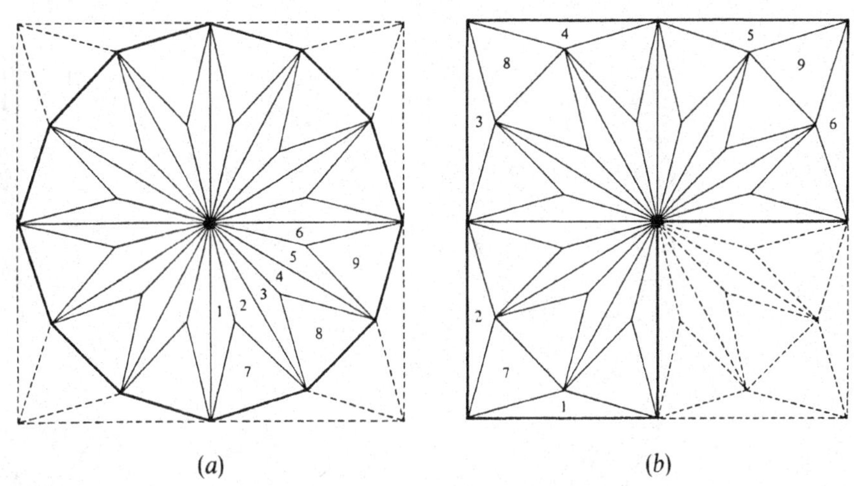

Figure 2.6.4
A dissection which shows that the area of a 12-gon inscribed in a circle of unit radius has area 3, see Exercise 2.6.5.

2.7 TILINGS WITH REGULAR VERTICES

If v edges meet at a vertex of a tiling (that is, if the valence of the vertex is v) then the vertex is called *regular* if the angle between each consecutive pair of edges is $2\pi/v$. Our object here is to enumerate the monohedral tilings with regular vertices.

Suppose that the prototile is an r-gon and that the valences at its vertices are $v_1, v_2, \ldots, v_r$. Then, since the sum of the angles at the corners of an r-gon is $(r - 2)\pi$, we see that

$$\frac{2\pi}{v_1} + \frac{2\pi}{v_2} + \cdots + \frac{2\pi}{v_r} = (r - 2)\pi$$

which, after a little rearrangement, becomes

$$\frac{v_1 - 2}{v_1} + \frac{v_2 - 2}{v_2} + \cdots + \frac{v_r - 2}{v_r} = 2. \qquad (2.7.1)$$

This is the same as equation (2.1.2) with v_i substituted for n_i. Hence the solutions are the same and are given in Table 2.1.1. From this we also deduce that there are 17 solutions which lead to 21 possibilities for the valences $v_1, \ldots, v_r$ taken cyclically around a tile. If we attempt to find monohedral tilings with these specifications we very simply find that we can do so only in 11 cases. These correspond to the solutions in the last column of the table which are not marked by asterisks.

Any monohedral tiling with tiles whose vertices have valences $v_1, \ldots, v_r$ will be denoted by $[v_1. \ldots .v_r]$, with superscripts used to shorten the symbols in the usual way. To certain symbols (namely $[3^4.6]$,

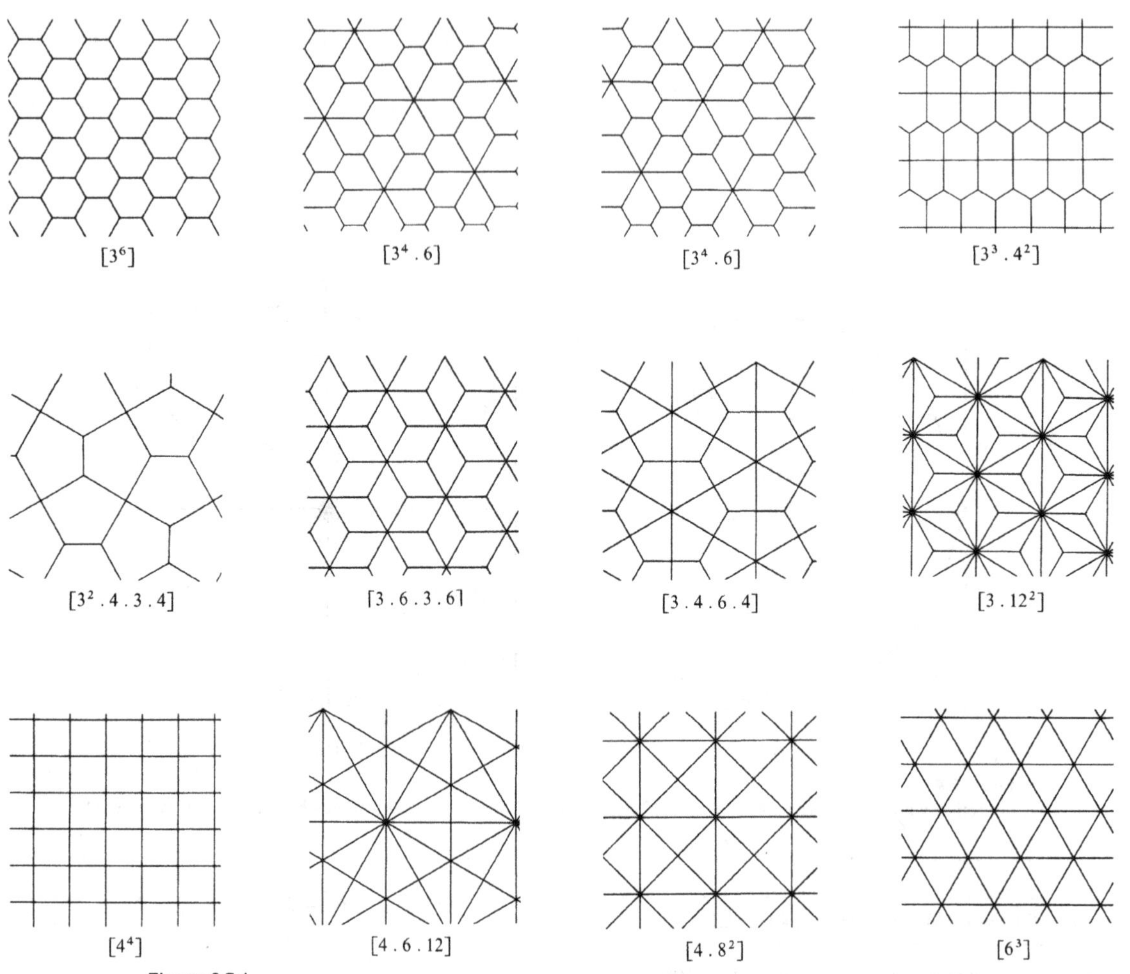

Figure 2.7.1
The eleven Laves tilings and their symbols. The tiling $[3^4.6]$ occurs in two enantiomorphic forms.

$[3^2.4.3.4]$, $[3.6.3.6]$, $[3.4.6.4]$, $[3.12^2]$, $[4.6.12]$, $[4.8^2]$, $[6^3]$) corresponds a unique tiling, but to others correspond tilings depending on one real-valued parameter ($[3^3.4^2]$, $[4^4]$) or two such parameters ($[3^6]$). In Figure 2.7.1 we show examples of all eleven tilings and in Figure 2.7.2 we show how the tilings of types $[3^3.4^2]$, $[4^4]$ and $[3^6]$ depend upon their parameters. In these last three cases, Figure 2.7.1 shows a very special example; for $[3^6]$ and $[4^4]$ we have chosen the regular tilings and for $[3^3.4^2]$ the prototile is the union of half a square and half a regular hexagon. The reason for these choices will be explained shortly. These eleven tilings are often called *Laves tilings*, after the famous crystallographer Fritz Laves whose work (which will be discussed in Chapter 4) led to the recognition of their importance in various classification problems (see Chapter 6).

It is worth stressing that each of the monohedral tilings with regular vertices turns out to be also isohedral. This fact is easily verified since the tilings

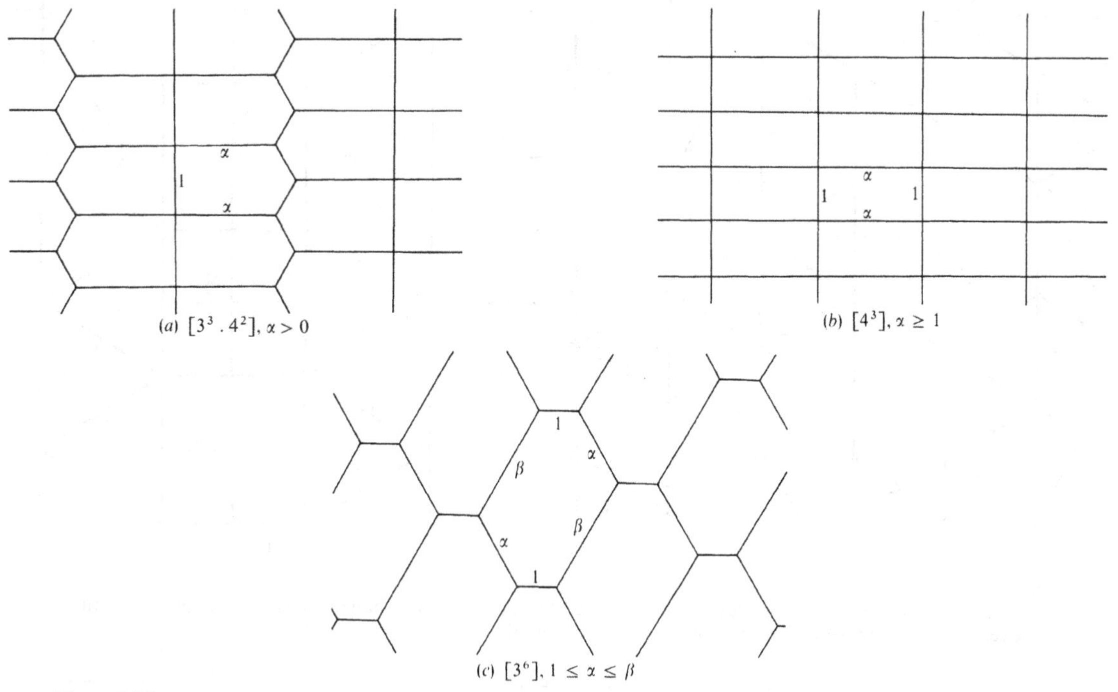

Figure 2.7.2
Explanation of the meaning of the parameters α, β in the tilings with regular vertices $[3^3.4^2]$, $[4^3]$ and $[3^6]$. The stated ranges of the parameters cover all possibilities.

in question have just been completely described. It is analogous to the observation made in Section 2.1 that Archimedean tilings are uniform, and it will be of importance in the determination of all isohedral tilings in Chapter 6.

The Laves tilings provide a good illustration of the principle of *duality*. We content ourselves here with a brief statement, postponing a more detailed discussion till Section 4.2. We say that $(v_1.v_2. \ldots .v_r)$ is *dual* to $[v_1.v_2. \ldots .v_r]$ because it is possible to set up a one-to-one correspondence between the tiles, edges and vertices of the first tiling and the vertices, edges and tiles of the second tiling, in such a way that inclusion is reversed. By this we mean that if, for example, in the first tiling a given tile contains a certain vertex, then in the second tiling the corresponding vertex is contained in the corresponding tile. In the case of the Laves tilings and the uniform tilings there are further geometric relationships between a tiling and its dual. These are illustrated in Figure 2.7.3, where two tilings and their duals are shown superimposed. It will be seen that this superposition can be done in such a way that the edges of the two tilings intersect at right angles. In fact, this is the reason for our special choice of parameters for the Laves tilings $[3^6]$, $[3^3.4^2]$ and $[4^4]$. For other parameter values such a superposition would not be possible.

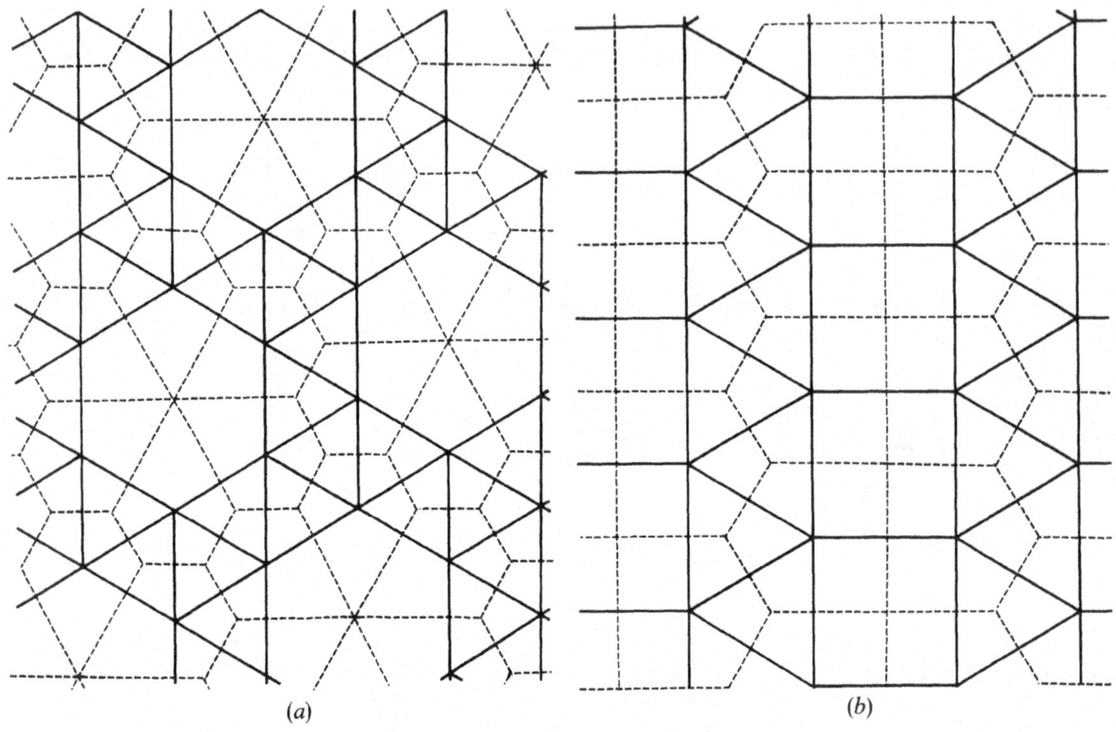

Figure 2.7.3
Two examples showing Laves tilings superimposed on the dual uniform tilings. Note that each vertex of a Laves tiling is the centroid of one of the tiles of the uniform tiling, and that the edges intersect orthogonally.

EXERCISES 2.7

**1. Determine all edge-to-edge monohedral tilings by convex tiles with all sides of same length. (For related material see Chapter 9, Schattschneider [1978b] and Hirschhorn & Hunt [1985].)

2. Verify that it is possible to superimpose a uniform tiling and its dual Laves tiling so that the edges intersect at right angles. Give an example of an isohedral tiling with regular vertices for which it is not possible to find a dual uniform tiling so that all edges intersect at right angles.

3. For the prototile of each Laves tiling determine whether it is monomorphic or admits more than one tiling. Among the non-monomorphic ones determine those that admit infinitely many edge-to-edge tilings.

4. Suggest a possible meaning for the phrase "a non-edge-to-edge tiling with regular vertices".
   (*a*) Find some examples.
   (*b*) Find all monohedral tilings that satisfy your definition, in which the prototile is (*i*) a square; (*ii*) an equilateral triangle.

*5. Find all isogonal tilings by vertex-regular polygons. (Here a polygon is called vertex-regular if it is an *n*-gon for which all angles are equal to those of the regular *n*-gon.)

2.8 EXTENSION OF PATCHES

In Section 1.1 we defined a patch of tiles in a tiling. Now (and in some later sections) we find it convenient to consider a slightly more general notion.

Let $\mathscr{S}$ be a given set of prototiles and

$$\mathscr{A} = \{T_1, \ldots, T_n\}$$

a finite set of tiles satisfying the following conditions:

(1) Each tile T_i is congruent to one of the prototiles in $\mathscr{S}$.

(2) If $i \neq j$ then $T_i \cap T_j$ has zero area; in other words, the tiles do not overlap.

(3) $\bigcup_{i=1}^{n} T_i$ is a topological disk.

Then $\mathscr{A}$ will be called a *patch* of tiles, with prototiles $\mathscr{S}$, and we shall say that $\mathscr{S}$ *admits* the given patch.

Every patch in the sense of Section 1.1 is clearly a patch in this new sense, but not conversely. The distinction is illustrated by the example shown in Figure 2.8.1. Here $\mathscr{S}$ consists of one square only (so the patch may be called *monohedral*). The set $\mathscr{A}$ of three squares shown in Figure 2.8.1 is a patch in the new sense, but not in the original sense for it is clearly not part of any monohedral tiling of the plane. Another way of expressing this is to say that the patch of Figure 2.8.1 cannot be *extended* to a tiling of the plane. More generally we shall use the word "extending", as applied to a given patch $\mathscr{A}$, to mean the adjoining to $\mathscr{A}$ of tiles, congruent to those in the given set $\mathscr{S}$ of prototiles, so as to form a larger patch.

In this section we shall consider only regular polygons as tiles and we insist that the tiles be arranged edge-to-edge. Moreover, we impose the additional condition that every vertex is of a given species, in the sense of Section 2.1, or at least *compatible* with it. By this we mean that the number of tiles of any kind meeting at a vertex on the boundary of a patch does not exceed the corresponding number for the given species. Interesting cases arise, of course, only when there is more than one type of vertex of a given species (numbers 3, 4, 5 and 6 in Table 2.1.1), for then a great variety of tilings and patches can arise by allowing the types to be mixed. The possibility

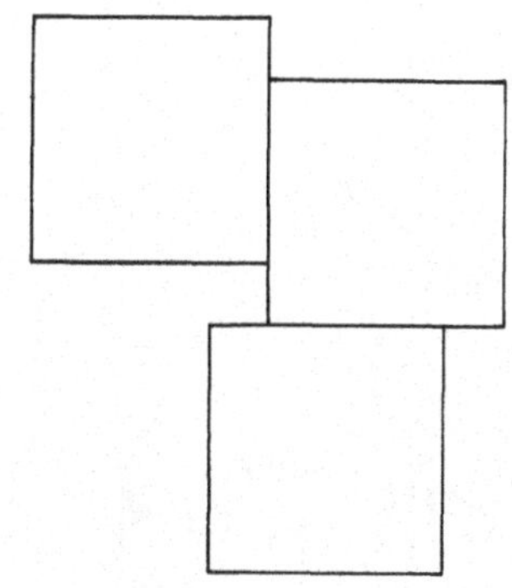

Figure 2.8.1
A monohedral patch which cannot be extended to a monohedral tiling of the plane.

of such tilings with mixed types of vertices was briefly discussed by Lévy [1891], [1894] and by Sommerville [1905], but they reached no discernible conclusions.

In the following subsections we give, for each of the relevant species, some results about the extendability of patches. In particular, we shall refer to convex patches, that is, patches for which the union of the tiles forms a convex region. Following each statement we give a brief justification of the assertions; detailed proofs and extensions are left as exercises.

Tilings with vertices of species number 3: all vertices of types $3^3.4^2$ or $3^2.4.3.4$.

2.8.1 *Arbitrarily large convex patches can be constructed with vertices of this species, and every convex patch can be extended to a convex patch enclosing it and so to a tiling of the whole plane.*

We shall say that a convex patch is *smooth* if every vertex of every tile in the patch belongs to at least one other tile in the patch. The method of extending a smooth convex patch is indicated in Figure 2.8.2. Starting from the patch $\mathscr{A}$ (shown shaded in the diagram), we extend it by adjoining a square to every boundary edge

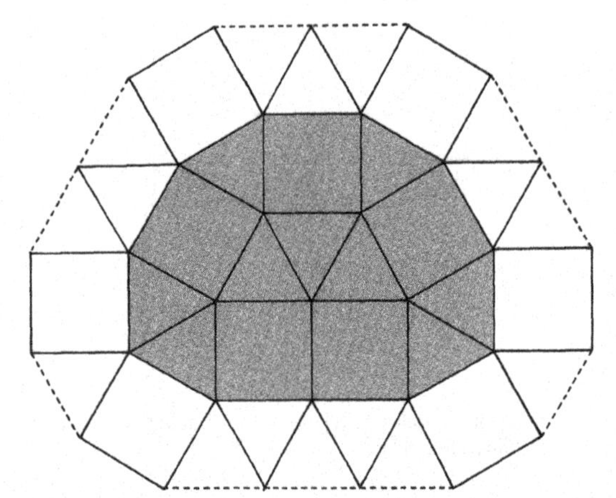

Figure 2.8.2
An example which shows how a smooth convex patch (shaded) with vertices compatible with species number 3 can be extended to a larger patch of the same kind.

which is a side of a triangle in $\mathscr{A}$, and adjoin a triangle to every edge which is a side of a square in $\mathscr{A}$. It is then a simple matter to adjoin further squares and triangles (as indicated by dashed lines in Figure 2.8.2) to form a larger smooth convex patch.

If a convex patch is not smooth, then the method of extension just described may fail (see Exercise 1(a)). On the other hand, it can be shown (see Exercise 1(c)) that every non-smooth convex patch can be extended to a smooth convex patch. Hence Statement 2.8.1 is proved.

Many non-convex patches are, of course, extendible, but not every such patch can be extended to a convex patch. The smallest examples of such patches contain 5 tiles; one is shown in Figure 2.8.3. This patch, moreover, cannot be extended in any way to a tiling of the plane (with vertices of the given species).

In general, tilings with vertices of species number 3 do not contain arbitrarily large convex patches. For example, the largest convex patch in the uniform tiling (3^2.4.3.4) contains 12 tiles and is shown in Figure 2.8.4. These statements follow from the fact that if the boundary of a convex patch is a polygon Q, then the interior angles of Q are at most $\pi/2 + \pi/3 = 5\pi/6$; hence Q cannot have more than 12 sides. In the tiling (3^2.4.3.4) no side of Q can be longer than the edge of a tile, and it is not hard to see that Q cannot have more than 10 sides in this case.

It is not known whether every patch $\mathscr{A}$ that can be extending to a tiling of the plane can also be extended to a convex patch.

Tilings with vertices of species number 4: all vertices of types 3^2.4.12 or 3.4.3.12.

No patch can be extended to a tiling of the plane, since no tiling has all vertices of species number 4 (see Section 2.1).

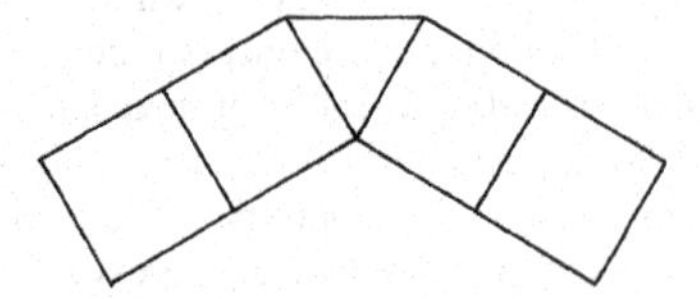

Figure 2.8.3
A patch of five tiles which cannot be extended to a tiling of the plane with all vertices of species number 3.

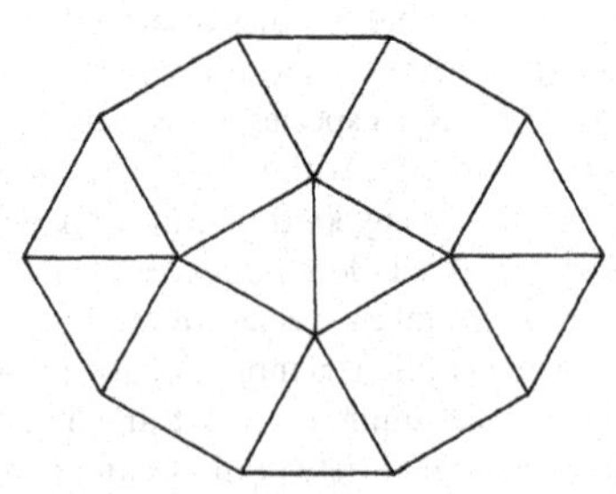

Figure 2.8.4
The largest convex patch in the uniform tiling (3^2.4.3.4).

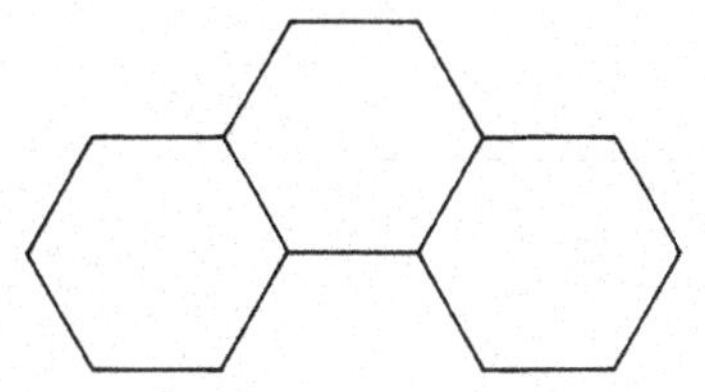

Figure 2.8.5
The smallest patch which cannot be extended to a tiling of the whole plane with all vertices of species number 5.

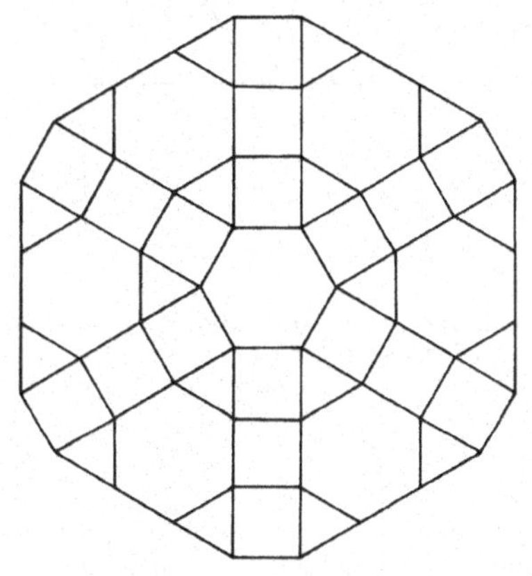

Figure 2.8.6
One of the two largest convex patches with vertices compatible with species number 6; it contains 43 tiles.

Tilings with all vertices of species number 5: all vertices of types $3^2.6^2$ or 3.6.3.6.

Because tilings with vertices of this species can be completely described (see Section 2.1) it is easy to deduce results analogous to those about tilings with all vertices of species number 3. For example, *not only do arbitrarily large convex patches exist, but every tiling contains such patches.* Notice that in the latter part of this statement "arbitrarily large patches" means patches containing arbitrarily large numbers of tiles—it does not mean patches that will cover arbitrarily large disks. For example, the 2-uniform tiling ($3^2.6^2$; 3.6.3.6) shown in Figure 2.2.1 contains no convex patch wider than twice the width of a hexagon.

2.8.2 *Every convex patch with vertices compatible with species number 5 can be extended to a convex patch enclosing it and therefore to a tiling of the plane with vertices of this species.*

The proof uses a method similar to that described above. Here we surround a given patch $\mathscr{A}$ by adjoining a triangle to each edge which is a side of a hexagon in $\mathscr{A}$, and adjoining a hexagon to each edge which is the side of a triangle in $\mathscr{A}$. One can then adjoin further triangles and hexagons to form the larger convex patch.

The smallest patch that cannot be extended to a convex patch, or to a tiling of the plane, contains three tiles; it is shown in Figure 2.8.5. It is not known whether every patch which can be extended to a tiling of the plane can also be extended to a convex patch.

Tilings with vertices of species number 6: all vertices of types $3.4^2.6$ or 3.4.6.4.

Here the largest possible convex patch contains 43 tiles. There are two such patches, one of which is shown in Figure 2.8.6. Every convex patch can be extended to a tiling of the plane. The smallest patches that cannot be extended to a tiling of the plane contain four tiles. An example of such a patch is shown in Figure 2.8.7.

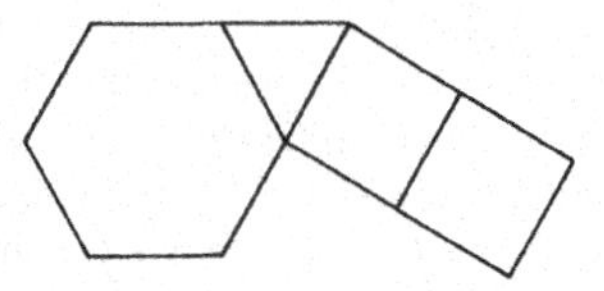

Figure 2.8.7
A patch which cannot be extended to a tiling of the whole plane with all vertices of species number 6.

EXERCISES 2.8

*1. (*a*) Show that the method of extending smooth convex patches with vertices of species number 3 described in the proof of Statement 2.8.1 may fail for convex patches that are not smooth. (Hint: Consider the patch in Figure 2.8.8.)

(*b*) Give a complete description of all finite convex patches with vertices of species number 3 that fail to be smooth.

(*c*) Show that every non-smooth convex patch can be extended to a smooth convex patch.

2. Show that there is just one convex patch with vertices of species number 6 and with 43 tiles, other than that shown in Figure 2.8.6.

3. For vertices of species number 3 find all patches of five tiles that cannot be extended to (*a*) convex patches, (*b*) tilings of the plane.

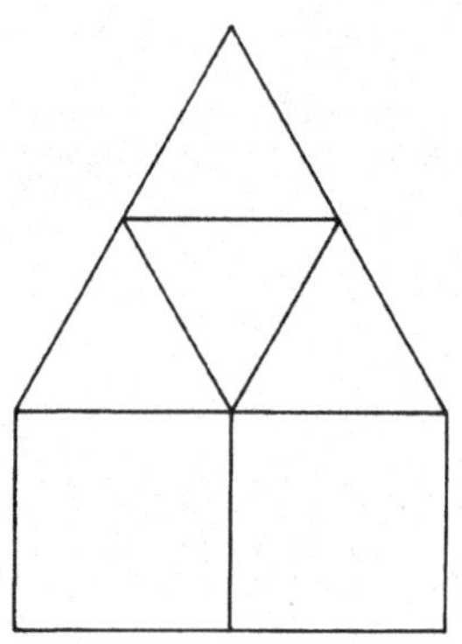

Figure 2.8.8
A patch which cannot be extended by the method described in the proof of Statement 2.8.1.

2.9 ARCHIMEDEAN AND UNIFORM COLORINGS*

In Section 1.3 we briefly introduced the idea of a *colored tiling*. We recall that this means that to each tile T of a given tiling $\mathcal{T}$ we assign one of a finite set of *colors*. The question we now consider is whether it is possible to color the tiles of a uniform tiling in such a way that it remains uniform—or just Archimedean—and if so, in how many ways this can be done. We can find no mention of this concept in the literature of tiling (though Heesch [1968a], [1968b], [1968c] considered the dual notion, as we shall explain at the end of this section). This is remarkable since it must have been frequently considered in connection with the design of actual, practical examples.

First let us say exactly what we mean by Archimedean and uniform colorings. In Figure 2.9.1 we show two colorings of the regular tiling (3^6) by two colors, indicated in the diagram by white and black. It will be seen that in each of these every vertex is surrounded by five white tiles and one black tile. By an obvious extension of the definition we gave earlier, these are *Archimedean colorings* because each vertex is surrounded by tiles of the same colors arranged in the same way. This means that the neighborhood of each vertex is the same as the neighborhood of every other vertex. On the other hand, the two tilings illustrated differ in one important respect. In Section 1.4 we defined a color-preserving symmetry of

* See the Appendix beginning on page 653.

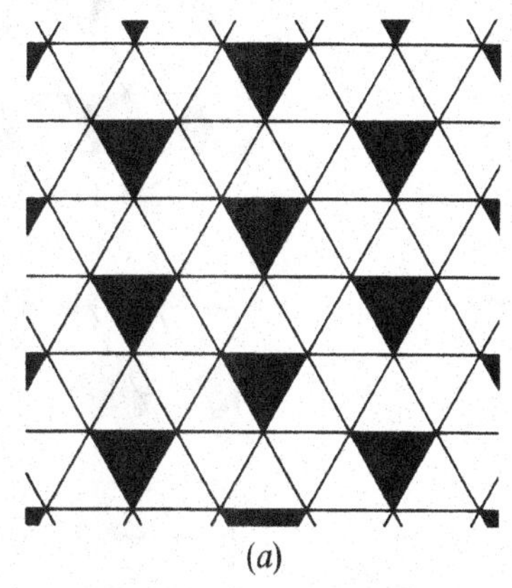

(*a*)

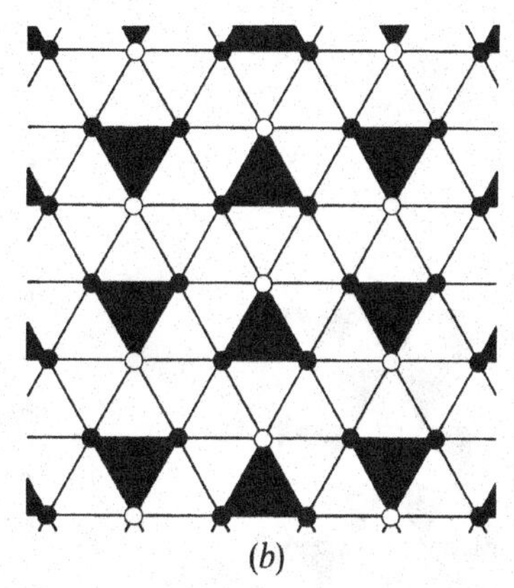

(*b*)

Figure 2.9.1
Two colorings of the regular tiling (3^6). In each case every vertex belongs to one black and five white tiles. Coloring (*a*) is uniform, while coloring (*b*) is Archimedean but not uniform.

a colored tiling as an isometry which maps each tile onto a tile of *the same color*. In the tiling of Figure 2.9.1(*a*), it is easy to verify that these symmetries act transitively on the vertices, so the colored tiling is *uniform*. On the other hand, in the tiling of Figure 2.9.1(*b*) the vertices belong to *two* transitivity classes (indicated by the small hollow and solid circles) with respect to the group of color-preserving symmetries, and so this tiling is not uniform.

The behavior of colored tilings differs in several ways from that of the "uncolored" tilings by regular polygons considered earlier in this chapter. In the first place there may be more than one coloring corresponding to a given vertex neighborhood, and in the second place not all Archimedean colorings are uniform. In fact, there is even a coloring of the tiling $(3^3.4^2)$ which is Archimedean, with a vertex neighborhood for which there is *no* uniform coloring.

To enumerate the uniform and the Archimedean colorings of the uniform tilings we agree not to consider as different the colorings obtained from each other by permutations or other changes in colors. In order not to introduce trivial extra cases we also insist that tiles in different transitivity classes (and hence, in particular, tiles of different shapes) have different colors. With these understandings we have the following result.

2.9.1 *The eleven uniform tilings admit precisely* 32 *uniform colorings, and four uncountably infinite families of non-uniform Archimedean colorings.*

In Figure 2.9.2 we illustrate all these colorings. For each coloring we show the symbol of the tiling and the arrangement of the colors round each vertex. The latter is indicated by the sequence of numbers that represent the various colors—1 for white, 2, 3 and 4 for grays and black—so that the colorings of Figure 2.9.1 correspond to the arrangement 111112 of the colors.

The method of enumerating these colorings is extremely elementary. We make use of a technique similar to that of Section 2.1, which will be applied many times in the sequel. First we list the tilings, and then list for each tiling *all possible* methods of coloring the tiles round a vertex. It is then necessary to test whether the proposed coloring leads to an Archimedean or uniform tiling (or several such tilings) by seeing whether it can be

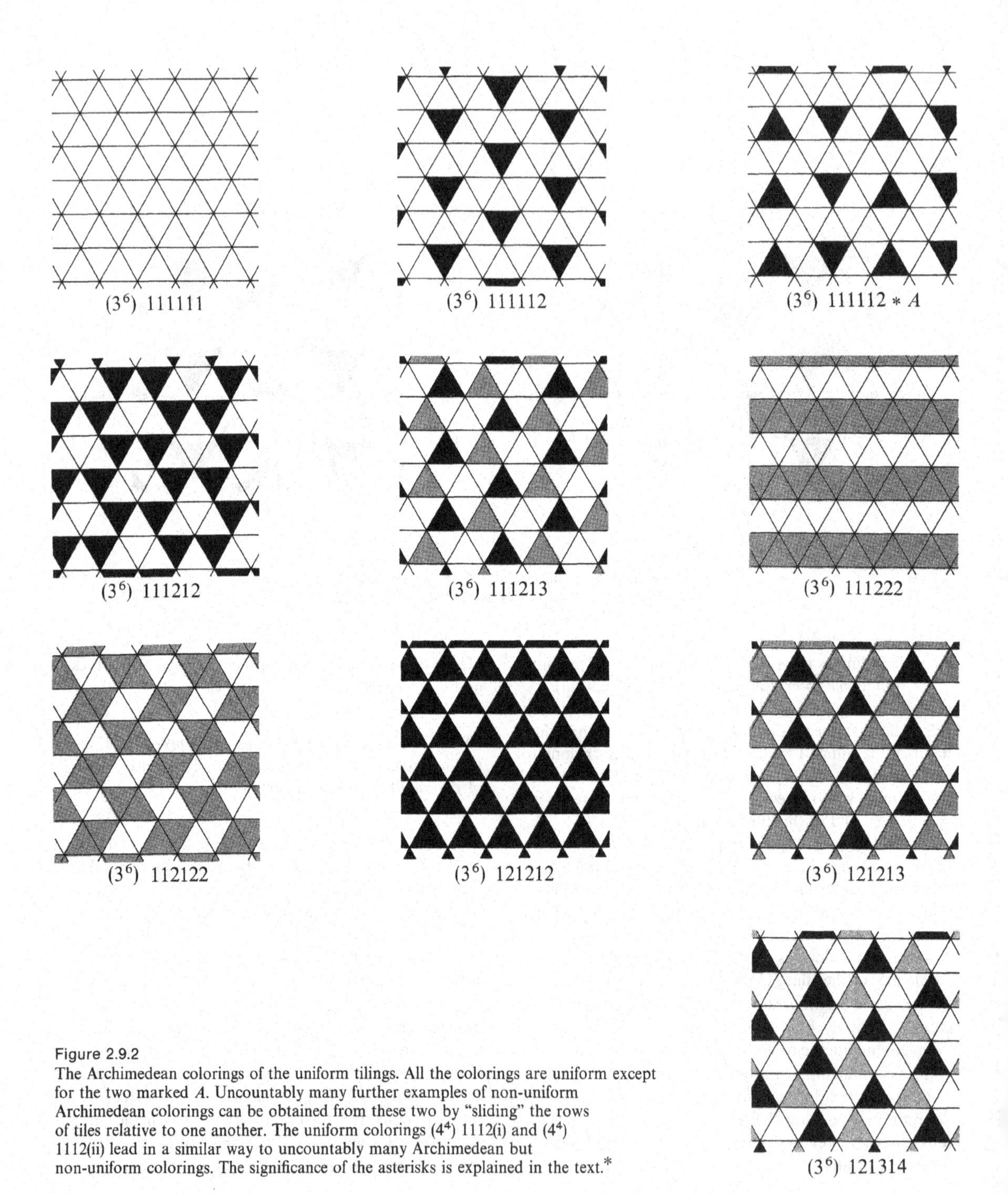

Figure 2.9.2
The Archimedean colorings of the uniform tilings. All the colorings are uniform except for the two marked A. Uncountably many further examples of non-uniform Archimedean colorings can be obtained from these two by "sliding" the rows of tiles relative to one another. The uniform colorings (4^4) 1112(i) and (4^4) 1112(ii) lead in a similar way to uncountably many Archimedean but non-uniform colorings. The significance of the asterisks is explained in the text.*

* See the Appendix beginning on page 653.

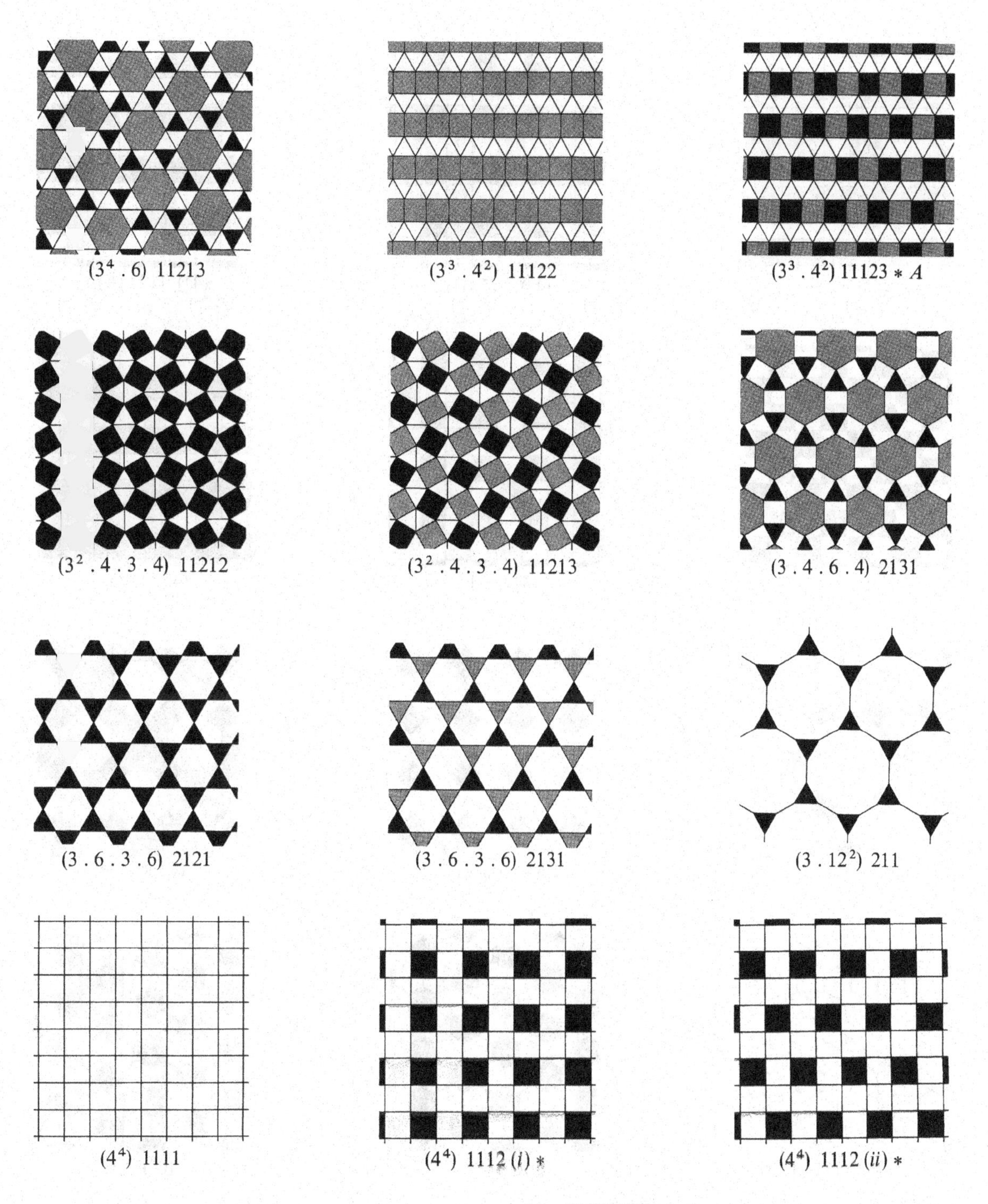

$(3^4 . 6)$ 11213

$(3^3 . 4^2)$ 11122

$(3^3 . 4^2)$ 11123 * *A*

$(3^2 . 4 . 3 . 4)$ 11212

$(3^2 . 4 . 3 . 4)$ 11213

$(3 . 4 . 6 . 4)$ 2131

$(3 . 6 . 3 . 6)$ 2121

$(3 . 6 . 3 . 6)$ 2131

$(3 . 12^2)$ 211

(4^4) 1111

(4^4) 1112 (*i*) *

(4^4) 1112 (*ii*) *

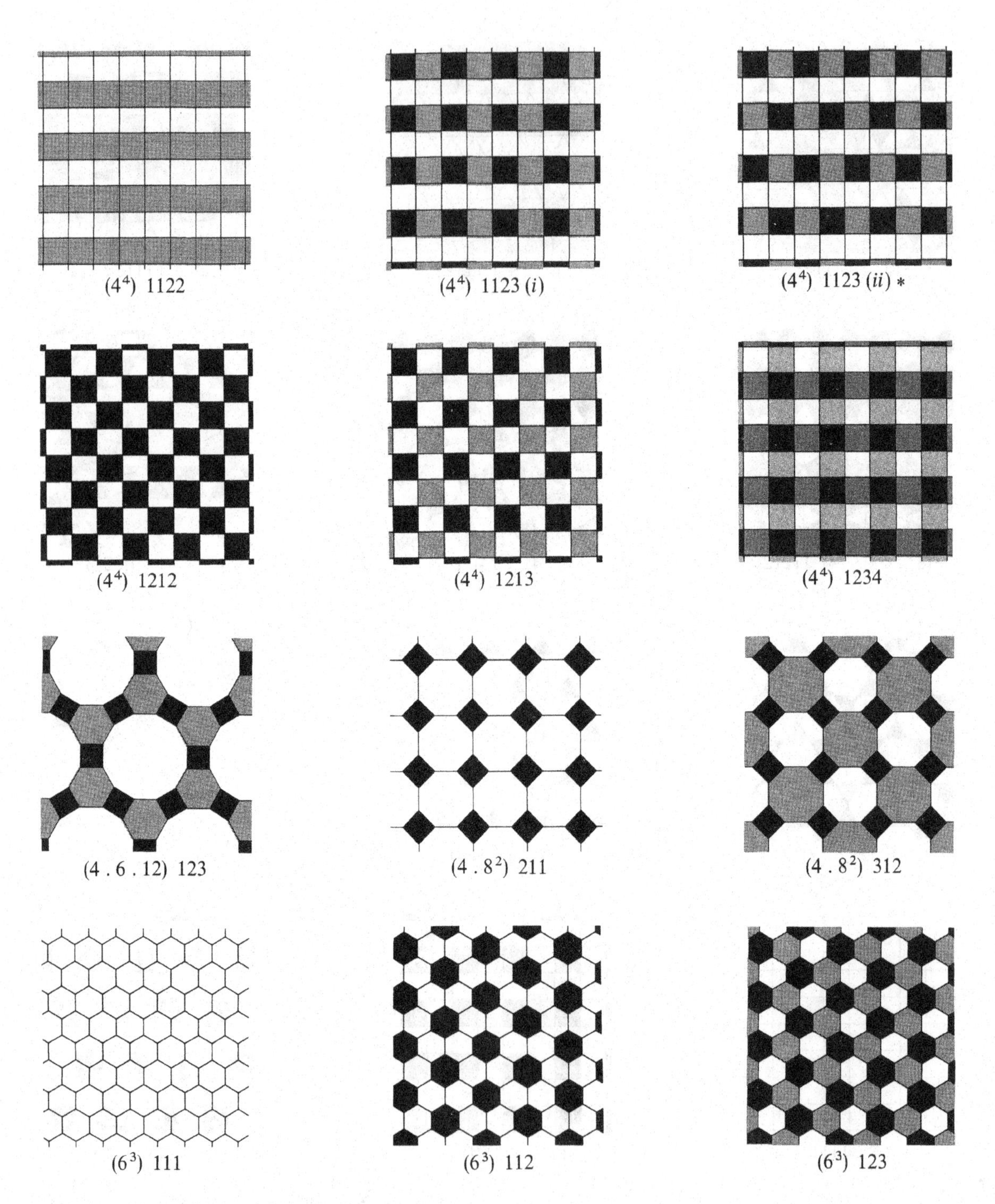

(4^4) 1122 (4^4) 1123 (*i*) (4^4) 1123 (*ii*) *

(4^4) 1212 (4^4) 1213 (4^4) 1234

$(4 . 6 . 12)$ 123 $(4 . 8^2)$ 211 $(4 . 8^2)$ 312

(6^3) 111 (6^3) 112 (6^3) 123

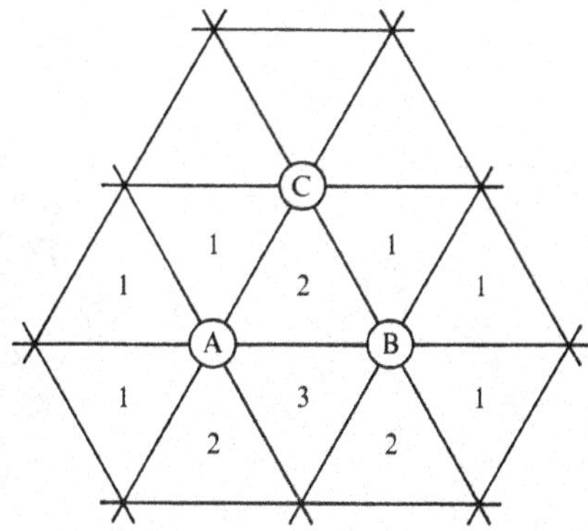

Figure 2.9.3
This figure illustrates the argument given in the text that the regular tiling (3^6) cannot be colored according to the scheme 111232.

extended consistently over the whole tiling. A typical situation is shown in Figure 2.9.3. Here we consider whether it is possible to color the regular tiling (3^6) with three colors according to the scheme 111232. Starting with the vertex labelled A, we color the surrounding six tiles as shown. Two of the tiles surrounding vertex B are already colored, and the remaining four can be colored in a unique way in accordance with the given scheme. But then vertex C is surrounded by tiles of which three consecutive ones are colored 121, which makes it impossible to color the tiles round C in the desired manner. We deduce that 111232 *is not* a possible coloring of (3^6).

We apply this method, in principle, to all the vertex neighborhoods and we are left with the thirty-four tilings of Figure 2.9.2. In fact the process is extremely easy except in the case of the tiling (3^6) where the number of possibilities is large. But there seems to be no obvious way in which to reduce the amount of work needed.

We observe that the above coloring problem can be equivalently formulated in terms of coloring the *vertices* of Laves tilings. In this form it was considered by Heesch [1968a], [1968b], [1968c], but his enumeration is not complete. The asterisks in Figure 2.9.2 correspond to the cases not listed by him.

EXERCISES 2.9

*1. Investigate the Archimedean (and the uniform) *oriented* tilings, that is tilings in which each tile is assigned one of the two orientations (clockwise, or counterclockwise) so that the neighborhoods of all vertices are the same (symmetries of the tiling that preserve or reverse the orientations of all the tiles are transitive on the vertices).

2. Let $\mathscr{T}$ be a uniform tiling, and let each edge of $\mathscr{T}$ be assigned one of a given set of c colors; then we shall say that $\mathscr{T}$ is *edge-c-colored*. An edge-c-coloring will be called *Archimedean* if the edges issuing from any two vertices can be mapped onto each other by a color-preserving isometry; it will be called *uniform* if for any two vertices there is a symmetry of $\mathscr{T}$ that preserves the colors of the edges and maps the first vertex onto the second. In Figure 2.9.4(a) we show a uniform edge-2-coloring of the regular tiling (3^6),

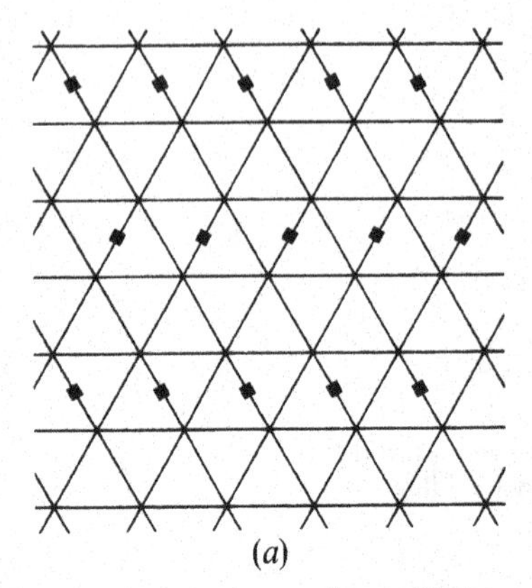

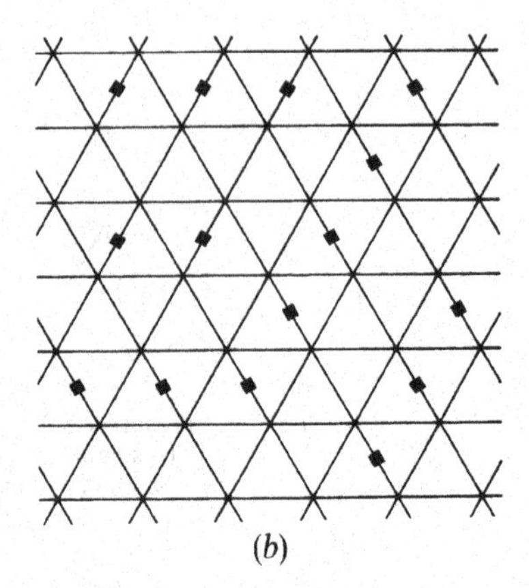

Figure 2.9.4
Examples of edge-2-colorings of the regular tiling (3^6). The two colors are distinguished by the presence or absence of square marks on the edges. The edge-2-coloring in (*a*) is uniform, that in (*b*) is Archimedean but not uniform.

while Figure 2.9.4(*b*) shows an Archimedean but not uniform edge-2-coloring of the same tiling.

(*a*) Show that the tiling (3^6) admits an uncountable infinity of Archimedean edge-2-colorings in which each vertex is incident with just one edge of the first color and with five edges of the second color.

*(*b*) Show that there exist precisely 10 different uniform edge-2-colorings of the tiling (3^6).

*(*c*) Show that in any uniform (and even in any Archimedean) edge-c-coloring of a uniform tiling we have $c \leq 5$.

**(*d*) Determine all the uniform edge-c-colorings of all the uniform tilings.

3. Verify that the enumeration of Archimedean and uniform colorings given in the text is complete. (This is a large task and the reader may wish to check part of the assertion only, for example, by restricting attention to the regular tilings.)

4. In this section we have restricted attention to colorings of tilings by regular polygons, but similar considerations apply to tilings by tiles of arbitrary shapes so long as the tiling is isogonal (which means that the symmetry group of the tiling is transitive on the vertices). In Figure 6.3.5 we show diagrams of 91 types of isogonal tilings, and it will be noticed that each of these corresponds to a uniform tiling in the sense that the numbers of edges of the tiles as we go round a vertex are the same in each case. (This is an example of topological equivalence discussed in Chapter 4.) Nevertheless, the "uniform" colorings of the isogonal tilings do not correspond exactly to colorings of uniform tilings. For example, the coloring of the tiling **IG69** shown in Figure 2.9.5, which "corresponds" to the uniform coloring (1234) of (4^4) is not even Archimedean in that the vertices are of four different types.

(*a*) Find other examples where this occurs.

(*b*) Find all possible uniform colorings of **IG69**.

(*c*) Find an example of a coloring of an isogonal tiling which corresponds to a uniform coloring of a uniform tiling, yet the vertices are of six different types.

5. Show that if it is not required that triangles in different transitivity classes must have different colors (but the color of the hexagons is distinct from the colors of triangles), then the uniform tiling $(3^4.6)$ admits precisely two uniform colorings not included in Statement 2.9.1 and in Figure 2.9.2. (This observation was made by G. Valette.)

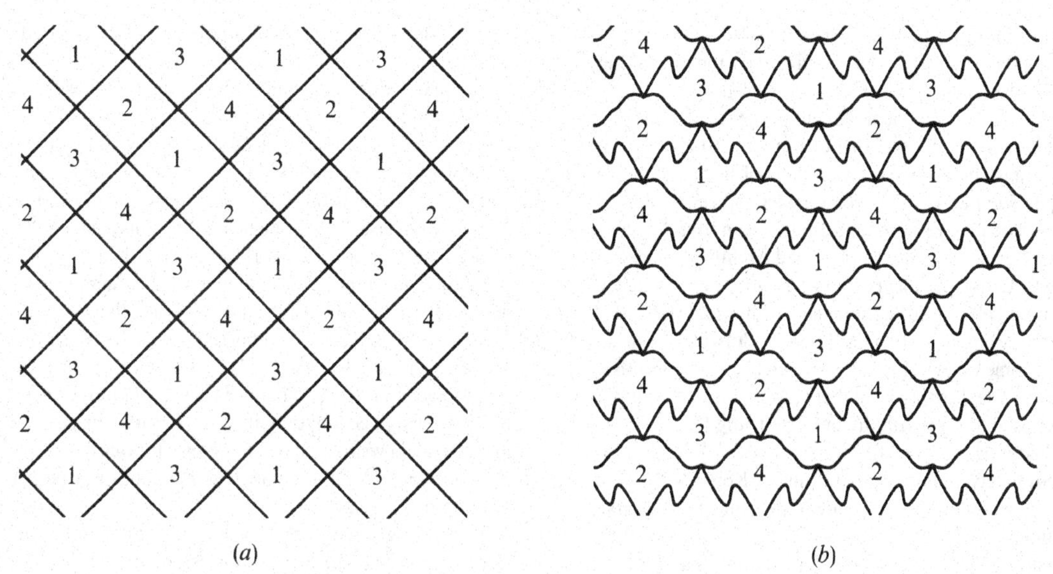

Figure 2.9.5
A uniform coloring 1234 of the tiling (4^4) and the corresponding coloring of the isogonal tiling IG69 . The latter is not colored uniformly because the vertices are of four different types.

2.10 NOTES AND REFERENCES

Tilings using only regular polygonal tiles have been well known since antiquity. Many examples can be found in photographs and drawings of ancient mosaics or medieval churches (see the references given in the Introduction, as well as those given below). However, it seems that there exists no systematic study which traces from a geometric-historical point of view the discovery, development and spread of these tilings in time and place.

The regular tilings are the starting point for several constructions, such as that of compound tessellations (see Coxeter [1948]; see also Section 12.4). These have been shown to have applications in crystallography (Takeda & Donnay [1965]), and also to have interpretations as (topologically) regular tilings on the torus. Applications to problems in theoretical geography and map-making can be found in Tarrant [1973], Lee [1973], Spilhaus [1983], [1984], [1985].

Crystallographers are interested in Archimedean tilings because the atoms in various crystals occur in layers which can be projected onto the vertices of such tilings. There is a very extensive literature devoted to such questions, starting with the pioneering work of Frank & Kasper [1958], [1959].

The first systematic treatment of the regular and Archimedean tilings is that of Kepler [1619]. (An English translation of the relevant parts of Kepler's book appears in Field [1979].) In a modern spirit, Kepler considered the regular and Archimedean tilings as analogues of the regular (Platonic) and Archimedean polyhedra. Using arguments similar to those in Section 2.1, he found all eleven Archimedean tilings and showed that they are the only ones. Kepler's drawings of these (and some other) tilings are reproduced in the frontispiece. (The drawing *M* in that figure is incorrect; it is supposed to represent the uniform tiling $(3^3.4^2)$ described in Kepler's text.)

Strangely—almost unbelievably—Kepler's work on tilings was completely forgotten for almost 300 years. Although Kepler's *Harmonice Mundi* [1619] was frequently quoted by authors interested in regular polyhedra (for he was the first to describe two of the regular star polyhedra) and although many of the same people were interested in tilings and wrote about them, the first reference to the fact that Kepler determined the eleven Archimedean tilings appears to be in a note added in proof to Sommerville [1905]. Similarly forgotten for a long time was the determination of the Archimedean tilings by a Rev. Mr. Jones in 1795; we have not seen his work and have learned about it from Ahrens [1921, p. 129] and Bradley [1933, p. 13].

Meanwhile, other workers were independently rediscovering these tilings, but the progress was unaccountably slow. Gergonne [1818] obtained several of them; his work was extended (and completeness claimed for the result obtained) by Badoureau [1878], [1881]. Although the latter paper is very interesting from several points of view (such as stressing uniformity as well as generalizing the kinds of tiles allowed), Badoureau's list of uniform tilings is incomplete since it does not contain $(3^4.6)$. This defective treatment was uncritically accepted by Lévy [1891], Brückner [1900], Fourrey [1907, pp. 363–371] and Bradley [1933, p. 17]. It is hard to understand how Thompson [1925] could miss $(3^3.4^2)$ in his list of Archimedean tilings, which is based on the work of Kepler [1619].

The first correct determinations in modern times of the eleven uniform tilings were carried out—independently of each other and blissfully unaware of any previous work—by Robin [1887], by Sommerville [1905] and by Andreini [1907]. The three papers actually enumerate and consider Archimedean tilings only and in this respect are inferior to the works of such earlier authors as Badoureau. (It is amusing to mention another determination of the Archimedean tilings, by Tachella [1931]. Even at that late date, the author as well as the writer of the review in *Jahrbuch über die Fortschritte der Mathematik* (Zacharias [1931]) believed that such tilings had not previously been investigated. Apparently, the attitude that considers writing papers more meritorious than reading them did not originate with the publish-or-perish syndrome of recent years!)

The proof given by Sommerville [1905] that there are just eleven Archimedean tilings is essentially the same as that indicated in Section 2.1. The same arguments are mentioned also in Ahrens [1901, pp. 66–71] but without a final list of possible tilings. Andreini [1907] uses the same method, but in a very cavalier way. He "finds" that there are just ten (!) possible species of vertices, and the impression is inevitable that he let his "knowledge" of the eleven uniform tilings influence his judgment concerning the possible solutions of Equation 2.1.2. Similarly inadequate is the treatment of that equation by Delone [1959], Williams [1972, p. 42], and Fedotov [1978], and of the analogous equation (2.7.1) by Šubnikov [1916]. (The second "proof" given in Andreini [1907] is even more seriously flawed, see Section 3.9). Related proofs and hints for the enumeration of the Archimedean tilings are given by Kraitchik [1942, p. 203], Bilinski [1948], Kravitz [1964], Critchlow [1970, p. 203], Watson [1973] and many others. A very nice treatment appears in O'Daffer & Clemens [1976]. Several other works present the eleven uniform tilings without proofs (Fejes Tóth [1953, Section 7], [1965, pp. 45–49], Steinhaus [1950, Chapter 4], Cundy & Rollett [1951, Section 2.9], Stevens [1974], Bezuszka, Kenney & Silvey [1977], Pearce [1978]. Very nicely executed versions of the uniform tilings are available as colorful posters (Cotleur [1977]). It is, perhaps, worth remarking that Wythoff's construction—so useful in constructing uniform polytopes in space of three or more dimensions—yields only eight of the uniform tilings (see Coxeter [1947]). The use of reflections also leads to the same eight uniform tilings (see Kingston [1957]).

Since so many authors have erred by leaving out some of the uniform tilings or solutions of equation (2.1.1), it is interesting to note that a modern text (Borrego [1968, pp. 132, 134]) on architectural design has one tiling too many: the author asserts (and "documents" by drawings!) that the tiling ($3^2.4.3.4$) exists—like ($3^4.6$)—in two enantiomorphic forms, not equivalent by motions without reflections. Field [1979, footnote 38] makes the same error.

Many authors have considered ways of generalizing Archimedean tilings by relaxing the requirement that all vertices are to be of the same type. In fact, as we have seen, even Kepler was interested in such tilings. For example, when discussing vertices of species number 3, Kepler remarks that they lead to two uniform tilings as well as to the tiling he denotes by O (see the frontispiece) that "may be continued non-uniformly". His tilings R, Dd and Ee are, in our notation, the 2-uniform ($3^2.6^2$; $3.6.3.6$), (3^6; $3^2.4.12$) and ($3.4.3.12$; 3.12^2), and his Cc may be extended to a 4-uniform ($3^2.4.3.4$; $3.4.3.12$; $3.4.6.4$; $3.4.6.4$). It is curious that Kepler states that his figure Kk cannot be extended without "mixing in" vertices of different species, while actually it appears to be part of the 2-uniform ($3.4^2.6$; $3.4.6.4$) all vertices of which are of species number 6. An imprecise statement of Kepler's can be misinterpreted as indicating that he considered the tiling R to be Archimedean. This led Bindel [1964, p. 17] to erroneously proclaim this 2-uniform tiling as uniform.

Kepler did not define precisely the kinds of tilings, more general than the Archimedean ones, in which he was interested. Several later authors were similarly vague, indicating (at best) only the desire to limit the species (or the types) of permitted vertices, or trying to obtain more or less symmetric tilings. Such discussions may be found in Lévy [1891], [1894] and especially in Sommerville [1905], while Ahrens [1901, Chapter 5], Fourrey [1907, pp. 370–371], Kraitchik [1942, pp. 205–207] and Steinhaus [1950, Chapter 4] present several examples. Ghyka [1946, Chapter 5] claims that there are just 14 non-uniform tilings by regular polygons. Similarly Critchlow [1970, p. 60] shows 14 non-uniform tilings with regular polygons and asserts that these are the only possible ones. This fallacious assertion is repeated by Williams [1972, p. 43], Critchlow [1976] and Bezuszka, Kenney & Silvey [1977].

Equitransitive and *k*-uniform tilings, *k*-isohedral tilings by regular polygons, and related topics have been investigated by Debroey & Landuyt [1979], [1981] and Chavey [1984a], [1984b].

Examples of non-edge-to-edge tilings are frequently found in the literature. Usually, consideration is restricted to square tiles; but even here, no systematic investigation seems to have been carried out. The observation that every tiling of the plane by congruent squares consists of "rows" of squares adjacent along whole edges was first made by Minkowski [1907, p. 74]. Its generalization to cubes in higher dimensions leads to remarkable connections with number theory and group theory; for a well written exposition of this material, with extensive references, see Stein [1974]. The conjecture of Keller [1930] that every tiling of Euclidean n-space ($n > 2$) by congruent and parallel cubes consists of "stacks" of cubes appears to be still open. For related material see Robinson [1979].

Williams [1972, p. 42] shows the three uniform tilings illustrated in Figures 2.4.2(*d*), (*f*) and (*g*). The enumeration of the uniform tilings given in Statement 2.4.1 was presented recently in Grünbaum & Shephard [1977a]. The unilateral tilings appear not to have been discussed in the literature at all.

Since antiquity, regular star polygons have frequently been associated with mystic or occult beliefs. (For a recent example of such association see Critchlow [1976].) Hence one can understand why mathematicians, in general, stayed away from them! Kepler, however, seems to have had a somewhat mystical attitude towards mathematics and astronomy; possibly this is the reason that he was not inhibited from considering star polygons as faces for regular polyhedra and as tiles for tilings. He produced several examples of tilings containing star polygons, but he did not investigate the question systematically. His attempts to find Archimedean tilings (or other tilings with some degree of regularity) that contain pentagons, pentacles, or decagons failed, though they led to remarkable tilings or patches such as those labelled Aa and Bb in the frontispiece. (See Section 2.5 for further details and references.) Other patches with such polygons were found by Bradley [1933, Chapter 6] and by Beard [1973]. Only one other related mathematical treatment of tilings with regular star polygons is known to us; it is Lévy's [1891] proof that no tiling of the plane can consist exclusively of regular star polygons. (The work of Badoureau [1881] involving regular star polygons is of a different character; see Grünbaum, Miller & Shephard [1982], as well as Chapter 12.) The tiling $(9^2.6^*_{4\pi/9})$ shown in Figure 2.5.4(*m*) appears also in Pearce [1978, p. 30], where several questions about tilings with regular polygons are considered; unfortunately, the statements made are imprecise and, in part, incorrect.

By contrast, in non-mathematical contexts the occurrences of star polygons in tilings and ornaments are very widespread. For example, Kepler's tiling K occurs in a marble panel of a 13th century Roman church (see Christie [1929, Figure 282]). Another example is shown in Figure I.13(*b*). Slightly distorted versions of $(4.4^*_\alpha.4^{**}_\alpha)$ appear frequently in Islamic art; see, for instance, Plates 104, 117 and 118 in Ipşiroglu [1971]. Plate 45 there shows the 2-uniform $(3.6.6^*_{\pi/3}.6;\ 6.6^{**}_{\pi/3})$. The tiling $(6.6^{**}_{\pi/3};\ 6.6^*_{\pi/3}.6.6^*_{\pi/3})$ is shown in Plate 1 of Bourgoin [1879], and $(3^2.4^{**}_{\pi/6};\ (3.4^*_{\pi/6})^4)$ occurs as the design of a patch-work quilt known as "windmill blades" (see Safford & Bishop [1972, Figure 173]).

The equidecomposability of polygons of same area is a well known classical result (see, for example, Boltyanskii [1956] for a detailed exposition of this and related facts). Problems concerning dissection tilings were first discussed in the literature of recreational mathematics (see, for example, Madachy [1966]). The contents of Lindgren [1972]—of which only a pale hint is given in Section 2.6—show the exquisite results that can be produced by ingenuity and patience.

The history of the tilings with regular vertices is noteworthy through its paucity. All such tilings except $[3^4.6]$ seem to have first appeared—as duals of the uniform tilings—in Brückner [1900, pp. 158–159]. But neither Brückner nor later mathematicians seem to have given them any serious thought—there is not even an accepted name for them in the literature. For the special tilings with regular vertices described in Section 2.7 we have adopted the name "Laves tilings" from the terminology of crystallographers who, however, mostly use this term in a less restrictive sense. It is convenient to give more details about these and related topics in Section 4.8, after a number of other properties of tilings have been established.

Similarly meager are the references that can be given concerning the topic of Section 2.9. Although some of the uniform colorings—especially of (4^4) and (6^3)—appear as decorations (and the 1212 coloring of (4^4) is the common checkerboard pattern) there seems to be no mathematical treatment of uniform colorings available in the literature. The closest we have been able to find is Heesch's [1968a], [1968b], [1968c] investigation of the dual problem of uniformly coloring the vertices of Laves tilings. As already mentioned, his results are incomplete.

Turner [1968] raised the somewhat similar question of determining all the *directed* uniform tilings, that is tilings obtainable from the uniform tilings, by assigning one of the two possible directions to each edge of the tiling in such a way that all the vertices are equivalent under symmetries preserving directions of the edges. He established that there are four distinct such directed tilings for which the underlying uniform tiling is (4^4).

In this section we have attempted to provide both a historic survey and also a list of references to the original books and papers. The latter is probably far from complete, but the literature is extremely scattered and no systematic exposition is available. In searching for references we have found frequent repetitions of the same facts and fallacies. It seems a pity that so much effort has been wasted duplicating previously known results, when there are still so many attractive and challenging problems to be investigated.

3

WELL-BEHAVED TILINGS

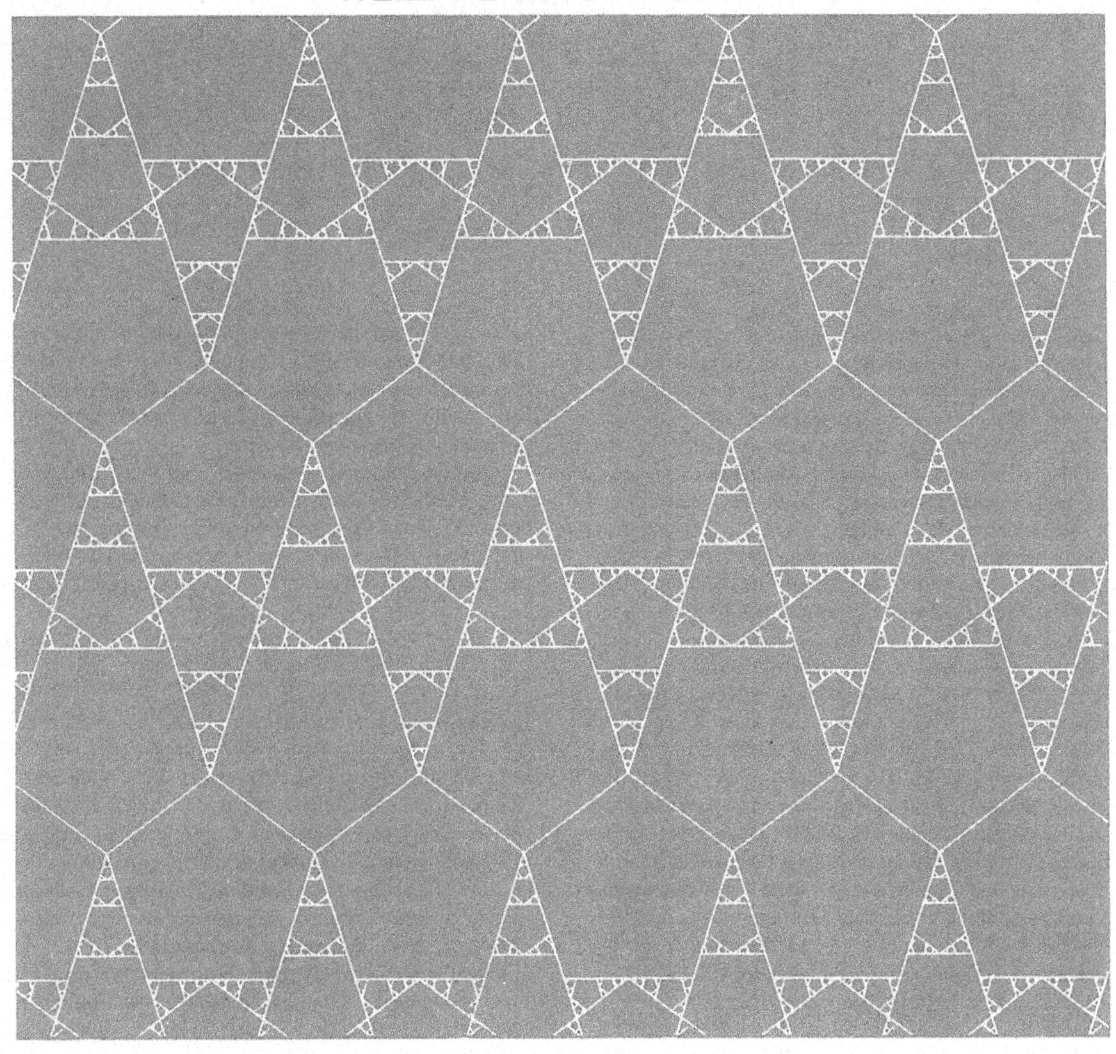

Figure 3.0.1
The tiles in this tiling are all regular pentagons or single points. The tiling possesses "singular points", that is, points of which every neighborhood meets an infinite number of tiles. Although such tilings are interesting and may be very decorative (as in this example) it usually is necessary to impose restrictions that eliminate them from our discussions. To do this, we introduce the idea of "normality" in Section 3.2.

3

WELL-BEHAVED TILINGS

In Chapter 2 we discussed tilings of a very special nature—either the tiles were regular polygons of a few sizes or the vertices were regular, and the tilings had many symmetries. Now, however, we widen the scope of our investigations by considering tilings in which neither the tiles nor the vertices are assumed to be regular. Somewhat unexpectedly, the classification and enumeration of these tilings, even of those which are quite symmetric, is rather unmanageable unless we impose severe restrictions both on the individual tiles and on the way they are arranged.

In Section 1.1 we illustrated a number of types of sets which are in general undesirable as tiles, and we showed how they can be eliminated from consideration by insisting that every tile is a topological disk. But even with this restriction it is still possible to construct tilings which are "pathological" in the sense that they have very strange properties and appear to lead to paradoxes. We begin by discussing a number of examples of this type; besides being interesting in their own right, they also serve as a a warning that it is very easy to make unjustified assumptions about tilings if one is familiar only with the rather "well-behaved" tilings that we have discussed so far. We feel that a warning of this kind is needed since the literature on tilings contains many errors caused by such assumptions.

In the following sections we shall formulate several conditions that exclude such "undesirable" tilings. With the increasing strength of the restrictions imposed we shall see how the tilings become "better behaved", and how a number of useful notions and techniques become available. In particular, we shall encounter the analogue for tilings of the Euler relation for polyhedra or planar maps, and see some of its consequences. In the final section we present results on the converse problem—the construction of tilings having certain prescribed properties and numerical characteristics.

The words "normal", "balanced", "strongly balanced" and "metrically balanced" for various kinds of "well-behaved" tilings are introduced in the hope that they will simplify the exposition as well as add precision to our treatment of the subject.

3.1 SINGULAR POINTS

In order to introduce the idea of singular points, consider the tiling indicated in Figure 3.1.1. This shows a tiling by similar triangles—each one has angles $\pi/2$, $\pi/4$ and $\pi/4$. The point P indicated in the diagram is called a *singular point* because every circular disk, however small, centered at P meets an infinite number of tiles. We sometimes say that the tiling is not *locally finite* at P. More generally, a tiling $\mathscr{T}$ is *locally finite* if every circular disk, centered at any point, meets only a finite number of tiles.

Tilings can have many, even an uncountable infinity of singular points. For example, in the tilings of Figure 3.1.2 every point on a line segment such as AB is a singular point.

Care is needed in the construction of such tilings if we wish to ensure that every tile is a topological disk. For example, although the tiling of Figure 3.1.2(b) appears very much like that of Figure 3.1.2(a), it will be seen to contains tiles which are single points (such as A). Similar

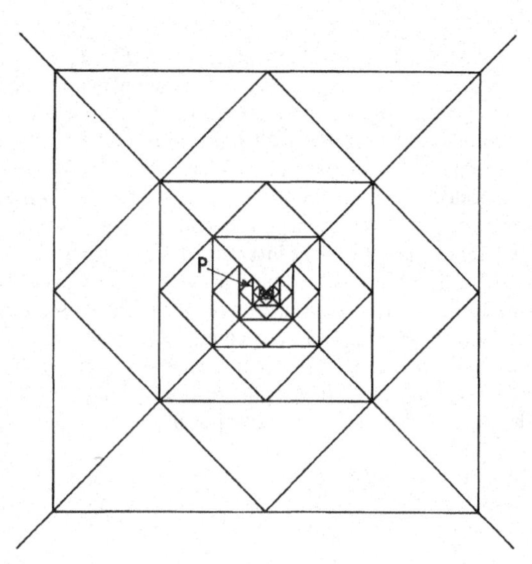

Figure 3.1.1
A tiling by right-angled isosceles triangles. This tiling has one singular point P.

considerations apply in the striking example of Figure 3.0.1. Here there are uncountably many tiles that consist each of a single point (see Exercise 3.1.8(c)). In order to eliminate examples of this nature, we shall require throughout the rest of this chapter that every tile be a topological disk.

It may seem surprising that singular points can occur even if we impose restrictions which imply that the tiles do not get too small, in some sense. For example, in the tiling of Figure 3.1.2(a) every tile T is a topological disk and has diameter $d(T) \geq 1$. (We recall that the diameter of a set is defined as the least upper bound of the distances between pairs of points in the set.) In a similar manner it is not difficult to find tilings with singular points in which every tile has a given area.

At first it may seem that singular points can only arise in an "abstract" sense and bear no relation to "practical" tilings. This is possibly so if we restrict the word "tiling" to mean that the plane is covered by a collection of physical objects. But with the wider interpretation that a tiling is a covering of the plane by closed sets, see Section 1.1, tilings which have singular points can certainly arise. For example, any photograph of a perspective view of a long brick wall, or of an extended tiled floor, conveys the feeling of a tiling with singular points. Similarly, the appearance of the shell in Figure 3.1.3 irresistibly suggests a tiling with a unique singular point.

In an unexpected way, tilings with one singular point have turned out to be relevant to the classification

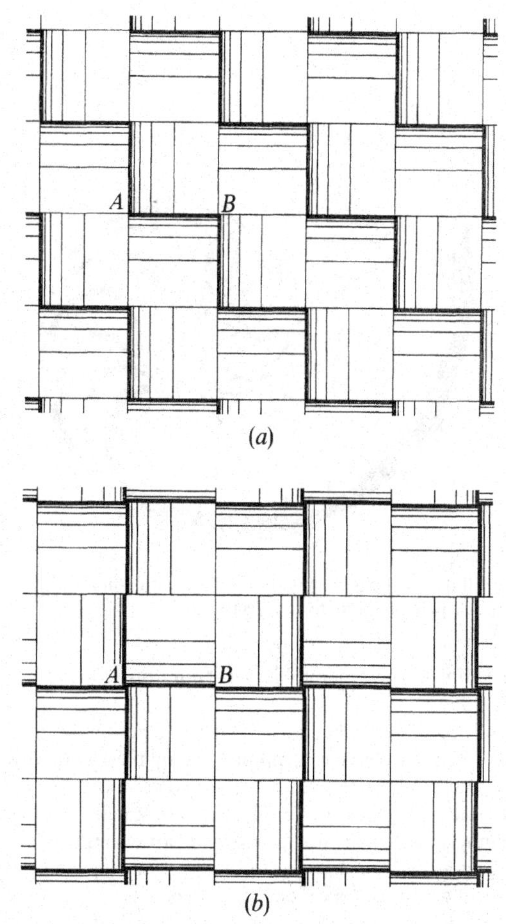

Figure 3.1.2
Two periodic tilings with an uncountable infinity of singular points occurring along segments such as AB. In (a) each tile is a topological disk (a rectangle of size 1×2^{-n} where $n = 1, 2, \ldots$), but in (b) some of the tiles are single points, such as the point A.

Figure 3.1.3
A shell that suggests a tiling of the plane with a unique singular point. (Compare also Section 10.1.)

of 3-dimensional manifolds; see Thurston [1982, pp. 362–363].

Although local finiteness excludes the possibility that some circular disk meets infinitely many tiles, or that two tiles meet in infinitely many connected components, it does not imply that a circular disk cannot meet one tile infinitely often. This situation occurs, for example, if an edge is a curve (such as a part of $y = x^2 \sin(1/x)$ near $x = 0$) which oscillates infinitely often.

Important consequences of local finiteness depend on the following result.

3.1.1 *If T is a tile of a locally finite tiling $\mathcal{T}$ then T has only a finite number of neighbors. For each neighbor T_i of T the intersection $T \cap T_i$ consists of a finite number of connected components.*

The proof of this result makes use of some basic facts from topology; readers not familiar with these may wish to skip this and similar proofs in the rest of the chapter.

If T had infinitely many neighbors then—by the compactness of the boundary of T—some point P of the boundary of T would have the property that every circular disk centered at P meets infinitely many tiles of $\mathcal{T}$, contrary to the assumption of local finiteness. Thus T has only a finite number of neighbors, say $T_1, T_2, \ldots, T_n$. But then, for each T_i, the intersection $T \cap T_i$ can consist of at most $n - 1$ connected components. To see this we recall that the intersection $A \cap B$ of two closed topological disks A, B has k connected components if and only if the complement of their union $A \cup B$ has k connected components (see Figure 3.1.4). This is a consequence of the Jordan Curve Theorem; for a proof of this theorem and of the result just quoted, see Newman [1951]. (For a very accessible discussion and proofs of the Jordan Curve Theorem and related results see Dostal & Tindell [1978].) Since each connected component of the complement of $T \cup T_i$ contains at least one of the $n - 1$ other neighbors of T, the number of these—and hence also of the connected components of $T \cap T_i$—is at most $n - 1$.

As a consequence of Statement 3.1.1 we have:

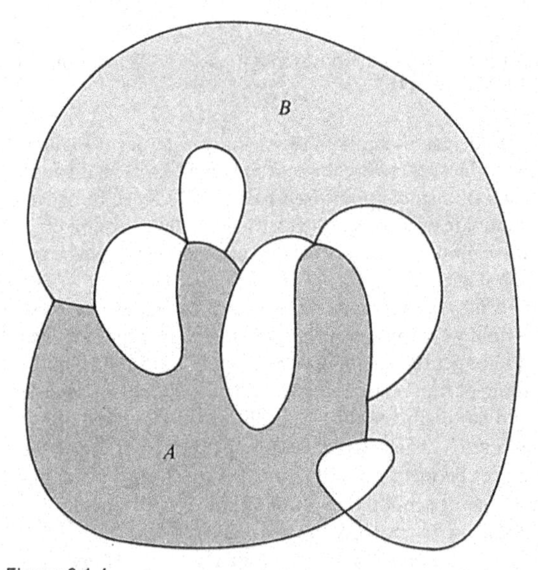

Figure 3.1.4
An illustration of the corollary of the Jordan Curve Theorem which asserts that the intersection $A \cap B$ of closed topological disks A and B has k connected components if and only if the complement of their union $A \cup B$ has k connected components. Here $k = 6$.

3.1.2 *In a locally finite tiling $\mathcal{T}$ each boundary point P of a tile T belongs to at least one other tile T' of $\mathcal{T}$.*

Indeed, by the local finiteness of $\mathcal{T}$ some circular disk $D = D(\varepsilon,P)$ centered at P meets only finitely many tiles of $\mathcal{T}$. Let $X_1, X_2, \ldots$ be a sequence of points of D that do not belong to T but converge to P (that is, $P = \lim_{k\to\infty} X_k$); such a sequence exists since P is a boundary point of T. Then each X_k belongs to at least one tile that meets D. Since there is only a finite number of such tiles, an infinite subsequence of points X_k belongs to the same tile $T' \neq T$. But T' is a closed set and hence the point P, which is the limit of that subsequence, also belongs to T'. This completes the proof of Statement 3.1.2.

Local finiteness is essential for the truth of Statement 3.1.2, as we can see from examples such as the tiling of Figure 3.1.2(*a*); any point on the (open) line segment AB lies on the boundary of tiles above this line, but does not belong to any tile below this line. So there are uncountably many points which belong to the boundary of only one tile.

In Section 1.1 we defined, for any tiling $\mathcal{T}$, a connected component C of the intersection of two or more distinct tiles of $\mathcal{T}$ to be either a *vertex* of $\mathcal{T}$ or an *edge* of $\mathcal{T}$ depending on whether C is a single point or not. Since a connected proper subset of the boundary of a closed topological disk is either a single point or else an arc, Statements 3.1.1 and 3.1.2 imply the following.

3.1.3 *In a locally finite tiling $\mathcal{T}$ the boundary of each tile T consists of a finite number of edges of $\mathcal{T}$; each edge E is a closed arc which is a connected component of the intersection of some two tiles T', T'' of $\mathcal{T}$; no point of E other than an endpoint belongs to any tile of $\mathcal{T}$ different from T' and T''. Each endpoint of an edge of $\mathcal{T}$ is a vertex of $\mathcal{T}$, and each vertex of $\mathcal{T}$ is a connected component of the intersection of some three tiles of $\mathcal{T}$.*

The reader should recall that we assumed that all tiles of $\mathcal{T}$ are closed topological disks. If this assumption is not made, strange phenomena may occur, as shown in Figures 1.2.5 and 1.2.6, in Exercise 3.1.8 and in Sections 3.2, 12.1 and 12.2. See also the remark about the "lakes of Wada" in Section 1.6.

Even locally finite tilings in which the tiles are closed topological disks can have unexpected properties if they possess a "singularity at infinity". A formal definition of this will be given in the next section, but here we can give examples of the sort of difficulties that can arise. Consider the tiling shown in Figure 3.1.5; it is locally finite, each tile is a closed topological disk (actually a convex quadrangle) and each vertex has valence 5. Nevertheless, as we shall soon see, this tiling is rather peculiar.

The numbers of tiles, edges and vertices in any tiling are, of course, infinite; but it seems reasonable to try to calculate their ratios. By this we mean to find constants v and e, which can be interpreted as the number of vertices and the number of edges that occur "on the average" per tile in the tiling. In many cases it is easy to determine these quantities. For example, in the regular tiling by squares each tile has 4 vertices and each vertex is shared by 4 tiles, so $v = \frac{4}{4} = 1$; similarly, each tile has 4 edges and each edge is shared by 2 tiles, so $e = \frac{4}{2} = 2$. For the regular tiling by hexagons we obtain $v = 2$ and $e = 3$ in the same way. In Figure 3.1.5 each tile has 4 vertices and each vertex is common to 5 tiles, so $v = \frac{4}{5}$; since each tile has 4 edges and each edge is common to 2 tiles, $e = \frac{4}{2} = 2$.

On the other hand, suppose that in a tiling $\mathcal{T}$ we construct a patch of tiles generated by a large disk D as described in Section 1.1. If we denote by $v(D)$, $e(D)$ and $t(D)$ the number of vertices, edges and tiles in the patch, then by the well known *Euler's Theorem for planar maps* (see, for example, Ore [1967, Chapter 4]),

$$v(D) - e(D) + t(D) = 1. \tag{3.1.4}$$

This implies that

$$\frac{v(D)}{t(D)} - \frac{e(D)}{t(D)} + 1 = \frac{1}{t(D)}.$$

Consider the limit as the disk gets larger and larger. The numbers $v(D)$, $e(D)$ and $t(D)$ tend to infinity, but

Figure 3.1.5
A locally finite tiling by convex quadrangles of unit diameter in which each vertex has valence 5. This tiling exhibits a "singularity at infinity".

intuitively at least, we expect $v(D)/t(D)$ to converge to the value v, $e(D)/t(D)$ to converge to the value e and $1/t(D)$ clearly converges to the value 0. Hence

$$v - e + 1 = 0. \tag{3.1.5}$$

This equation is satisfied by the values of v and e found earlier for the regular tilings by squares and by hexagons. However, equation (3.1.5) is not consistent with the values $v = \frac{4}{5}$, $e = 2$ calculated for the tiling in Figure 3.1.5 and so we arrive at a paradox.

In fact, the resolution of this paradox is rather more subtle than appears at first sight. It depends on the fact that the sort of limiting process we have indicated is not applicable to tilings in which the tiles get more and more "crowded" as we move outwards from a given point. It is similar to the sort of paradox that occurred in the early investigation of infinite series. At first, mathematicians were disconcerted by the fact that the sum of a conditionally convergent series could be altered (to *any* prescribed value!) by rearranging the terms. Yet we now know that such properties of series fit into a rigorous mathematical structure. The two situations are completely analogous. It only remains for us to identify those properties of tilings which lead to apparent paradoxes. This is especially important since in the sequel we shall often be required to calculate the ratios of various numbers of elements. Clearly we must know when such procedures are valid.

This question will be discussed in more detail in the next section, but we can say here that the difficulties are not avoided by imposing numerical conditions on the size of the tiles such as specifying their minimum area or diameter. (In Figure 3.1.5 we have drawn the tiling so that every tile has the same diameter, and it is easy to see how this can be extended to the whole plane.) On the other hand, in whatever way a tiling with 5-valent vertices and quadrangles as tiles is constructed, either the tiles become longer and longer, or they become thinner and thinner, and there is no way in which this can be avoided. The same phenomenon is also illustrated by the tiling of Figure 3.1.6 in which heptagonal tiles meet by threes at each vertex, and for which a similar paradox occurs. Again, the tiles get longer or thinner in any tiling satisfying these specifications. This observation gives us an indication as to how to proceed; we must, in some way, restrict the shape of the tiles.

One obvious way to avoid the difficulties due to a "singularity at infinity" is to restrict attention to periodic tilings for which they cannot occur. But some very interesting tilings are not periodic (see Chapters 10 and 11) and we must look for a wider class. What we require is some restriction on the shape of the tiles and how they fit together. The most convenient such restrictions are to insist that the tilings be *normal* or *balanced*. These two concepts form the topics of the next two sections.

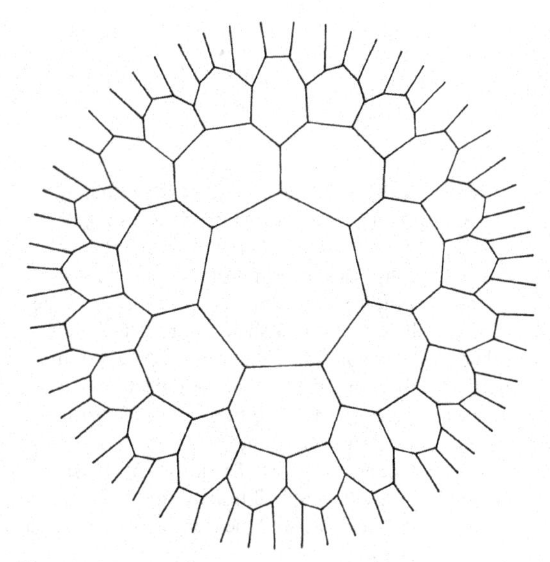

Figure 3.1.6
A locally finite tiling with convex heptagons, in which every vertex has valence 3. Here $v = \frac{7}{3}$ and $e = \frac{7}{2}$; note that $v - e + 1 = -\frac{1}{6}$. The tiling has a "singularity at infinity".

EXERCISES 3.1

1. Find a tiling in which each tile is a convex polygon of unit area, but the tiling has a singular point.

*2. Show that if a tiling $\mathcal{T}$, in which all tiles are polygons of either (*a*) the same area, or (*b*) the same diameter, has any singular points, then $\mathcal{T}$ has an uncountable infinity of singular points.

3. Describe a tiling in which each tile is a closed topological disk, but the intersection of some two tiles consists of infinitely many disjoint closed segments.

4. Let T be a tile in a locally finite tiling $\mathcal{T}$, in which each tile is a closed topological disk. Show that T has at least one neighbor T^* such that $T \cap T^*$ is a single edge of $\mathcal{T}$.

5. Let $f(n)$ denote the maximum of the numbers k with the following property: in some locally finite tiling $\mathcal{T}$ in which each tile is a closed topological disk, there is a tile T with n adjacents such that T has k edges.
   (*a*) Show that $f(2) = 2$, $f(3) = 4$.
   *(*b*) Show that $f(n) = 2n - 2$ for each $n \geq 2$. (See Reinhardt [1918, p. 7].)

6. (*a*) Show that there exist no locally finite tilings in which each tile is a closed topological disk with five adjacents and six neighbors.
   (*b*) Construct a locally finite tiling in which each tile is a closed topological disk with six adjacents and seven neighbors.

7. In tilings with singular points the definitions of edges and vertices given in Section 1.1 may not be satisfactory. For example, some of the points of the segment AB in the tiling of Figure 3.1.2(*a*) belong to just one tile only and some to two tiles—but it would appear

natural to call the latter *vertices* of the tiling, while the former constitute arcs whose closures might be called *edges* of the tiling. We refrain from giving precise definitions since their appropriateness (and the degree of complication necessary) depends on the generality of the tilings one wishes to consider. On the other hand, even in tilings that are not locally finite one can define the relations of neighborliness and adjacency among tiles. Two tiles are *neighbors* if their intersection is non-empty; they are *adjacent* if their intersection contains an arc. The following questions should illustrate the possibilities; all tiles are assumed to be topological disks, and the tilings are assumed to contain singular points.

(*a*) Construct a tiling in which all tiles have diameter $d \geq 1$ and each tile has at most six neighbors.

*(*b*) Find a tiling with tiles of diameter at least 1, in which each tile has the same number m of adjacents, where (*i*) $m = 3$, (*ii*) $m = 4$.

*(*c*) Solve parts (*a*) and (*b*) under the additional restriction that all tiles have area at least 1.

8. Consider *Cantor-type* tilings, that is, tilings which consist of tiles that are topological disks, and of single-point tiles, but a single-point tile occurs only if every open topological disk that contains that point meets some of the tiles that are topological disks. The Cantor-type tiling $\mathscr{C}$ in Figure 3.1.7 is periodic, with the unit square $S = \{(x,y) | 0 \leqslant x \leqslant 1, 0 \leqslant y \leqslant 1\}$ serving as the period parallelogram, and each tile in S that is a topological disk is a rectangle of height 1; the bases of these rectangles are the segments $[\frac{1}{3},\frac{2}{3}], [\frac{1}{9},\frac{2}{9}], [\frac{7}{9},\frac{8}{9}], [\frac{1}{27},\frac{2}{27}], [\frac{7}{27},\frac{8}{27}], [\frac{19}{27},\frac{20}{27}], [\frac{25}{27},\frac{26}{27}], \ldots$ of the x-axis. Each point not covered by these rectangles is a single-point tile.

(*a*) Prove that there are uncountably many single-point tiles in $\mathscr{C}$, and that each is a singular point of $\mathscr{C}$.

(*b*) Modify the construction of $\mathscr{C}$ to obtain a Cantor-type tiling in which the single-point tiles cover a set of positive Lebesgue measure.*

* For information about Lebesgue measure see, for example, Wheeden & Zygmund [1977].

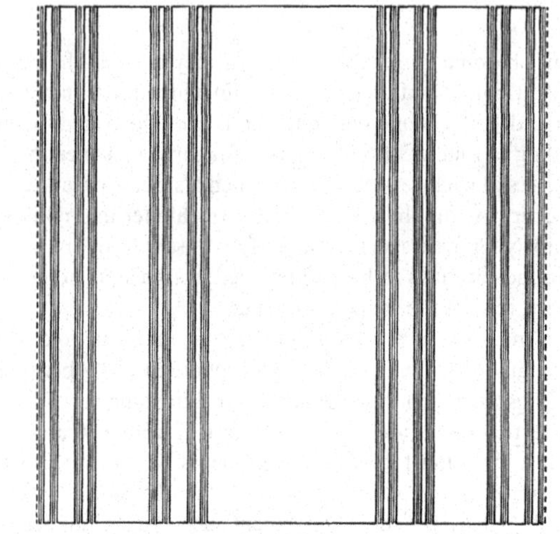

Figure 3.1.7
The period parallelogram of a Cantor-type tiling.

*(*c*) Prove that in the Cantor-type tiling by regular pentagons and single-point tiles shown in Figure 3.0.1 the sum of the areas of the pentagons contained in a period parallelogram equals the area of the parallelogram, but that there do exist uncountably many single-point tiles.

9. In Figure 3.1.8 we show a locally finite tiling with tiles of diameter at least 1, in which each tile has six adjacents and at least 8 neighbors.

(*a*) Construct a locally finite tiling in which each tile has area at least 1 and each tile has at least 9 neighbors.

*(*b*) Construct a locally finite tiling in which each tile has diameter at least 1, and each tile has precisely 5 adjacents and 10 neighbors.

*(*c*) Construct a locally finite tiling with tiles of diameter at least 1, in which each tile has at most 5 neighbors.

(*d*) Construct a locally finite tiling with tiles of diameter at least 1 in which each tile has at most 4 neighbors.

(*e*) Prove that there exists no locally finite tiling with tiles of diameter at least 1 in which each tile has at most 3 neighbors.

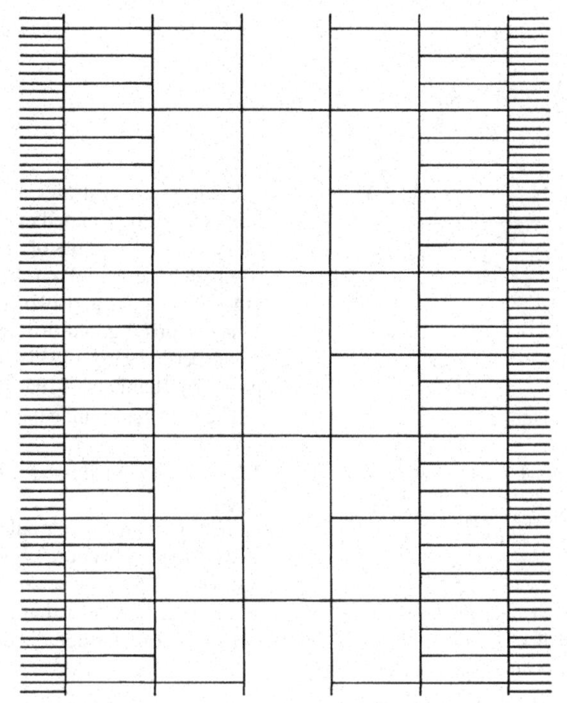

Figure 3.1.8
A locally finite tiling in which each tile has six adjacents and at least eight neighbors.

3.2 NORMAL TILINGS

In this section we shall discuss the consequences of three conditions (denoted here by N.1, N.2 and N.3) that can be imposed on a tiling $\mathscr{T}$. This will lead us to the consideration of normal tilings which play an important role in the subsequent development of the subject.

N.1 *Every tile of $\mathscr{T}$ is a topological disk.*

We recall from Section 1.1 that this condition implies that each tile is bounded by a simple closed curve. For an alternative definition see Section 4.1, and for a discussion of some tilings that do not satisfy condition N.1 see Chapter 12.

N.2 *The intersection of every two tiles of $\mathscr{T}$ is a connected set, that is, it does not consist of two (or more) distinct and disjoint parts.*

Thus in a tiling that satisfies conditions N.1 and N.2, the intersection of two tiles is either empty, or a single point, or an arc. However, in such a tiling not every boundary point of each tile necessarily belongs to another tile. For example, the tiling in Figure 3.1.2(*a*) has properties N.1 and N.2 but each of its tiles has boundary points that do not belong to any other tile.

N.3 *The tiles of $\mathscr{T}$ are uniformly bounded.*

By this we mean that there exist two positive numbers U and u, called the *parameters* of the tiling, such that

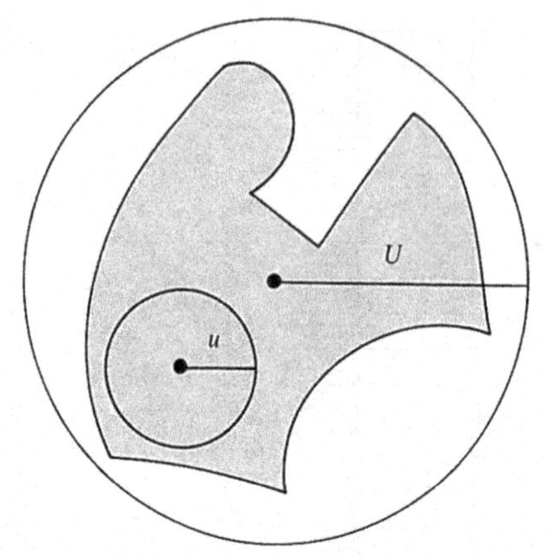

Figure 3.2.1
One of the requirements for a normal tiling is that there exist numbers U (the circumparameter) and u (the inparameter) such that every tile lies in some circular disk of radius U and contains some circular disk of radius u.

every tile contains some circular disk of radius u and is contained in some circular disk of radius U, see Figure 3.2.1. The numbers u (the *inparameter*) and U (the *circumparameter*) are to be fixed for the tiling in question, that is to say, the same two numbers apply for every tile in the tiling. This uniform boundedness property is the mathematical specification of the intuitive idea mentioned in the previous section that the tiles never get too long or too thin. For example, the tilings in Figures 3.1.5 and 3.1.6 do not satisfy condition N.3. As we shall see later, N.3 implies local finiteness.

A tiling $\mathcal{T}$ is called *normal* if it satisfies conditions N.1, N.2 and N.3.

Condition N.2 excludes tilings such as those in Figures 3.2.2, 3.2.3(a) and 3.2.4(a). In Figure 3.2.2 the intersection of tiles A and B has uncountably many components—in fact, it is the well known Cantor set on the segment CD. Each point of this Cantor set is a singular point. This tiling also fails to satisfy condition N.3. On the other hand, the non-normal tilings of Figures 3.2.3(a) and 3.2.4(a) satisfy both conditions N.1 and N.3.

The exclusion of tilings that do not satisfy N.2 from future considerations is mainly a matter of convenience. The tiling of Figure 3.2.3(a) is of interest for it shows that N.2 depends (unlike N.1 and N.3) on the way the tiles are arranged and not only on their shapes or sizes. Figure 3.2.3(b) shows another dihedral tiling with the same prototiles as the tiling of Figure 3.2.3(a), yet it satisfies N.2 and is normal.

Figure 3.2.4 illustrates the surprising fact that condition N.2 can be violated even in a monohedral periodic tiling. Here some pairs of tiles meet in such a way as to completely surround another tile (see Figure 3.2.4(b) and other pairs surround two other tiles (see Figure 3.2.4(c). This curious 9-gonal prototile was discovered by Voderberg [1936], [1937] in connection with a problem posed by Reinhardt [1934].

The uniform boundedness condition N.3 implies:

3.2.1 *Each normal tiling is locally finite.*

In other words, a normal tiling has no singular points as described in Section 3.1. To see this, we consider a disk $D(r,P)$ of radius r, centered at any point P of the plane. Then it is clear that any tile which meets $D(r,P)$ must lie entirely inside the disk $D(r + 2U,P)$, where U

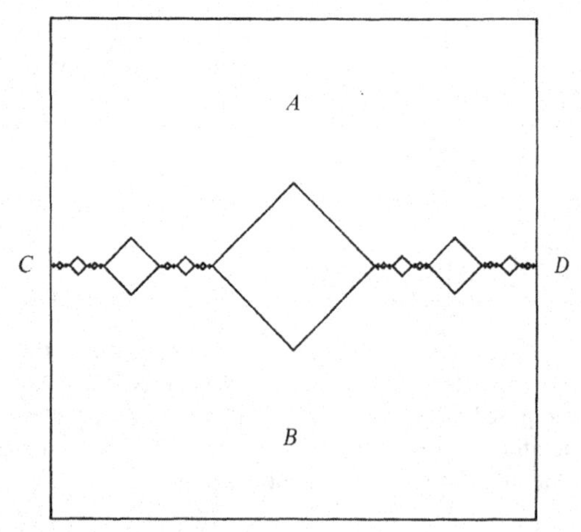

Figure 3.2.2
Starting from a tiling with rectangular tiles, we replace the middle third of the edge CD by a square tile set diagonally as shown. We then replace the middle third of each of the two remaining parts of CD by similar squares. Four parts of the edge CD then remain. The middle third of each of these is replaced by a square, and the construction continues in this manner. The tiles AB then intersect in uncountably many singular points (a Cantor set). The tiling can be seen to satisfy normality condition N.1, but not N.2 or N.3.

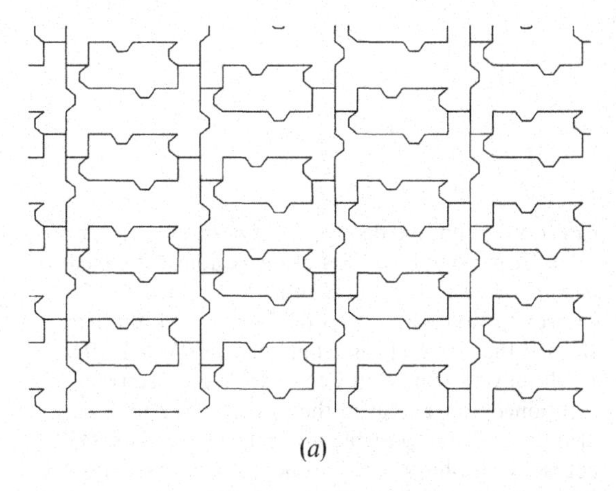

(a)

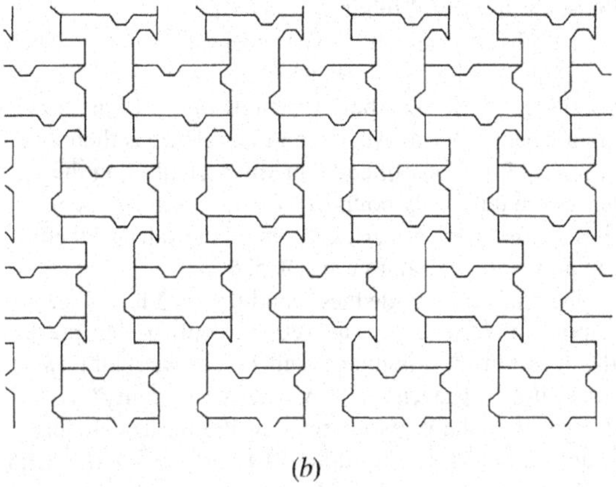

(b)

Figure 3.2.3
Two dihedral tilings with the same prototiles. Tiling (a) is not normal because the intersections of some pairs of tiles are not connected. Tiling (b) is normal.

is the circumparameter. Such a disk has area $\pi(r + 2U)^2$. But every tile contains a disk of radius u, and so it is of area at least πu^2. Thus not more than

$$M(r) = \pi(r + 2U)^2/\pi u^2$$

tiles can meet D. Since this is a fixed finite number we deduce that P is not a singular point, and since P is any point of the plane, we have demonstrated that the tiling has no singular points—it is therefore locally finite.

From Statement 3.2.1 and from the results of Section 3.1 it follows that the boundary of each tile consists of finitely many edges, that the endpoints of each edge are vertices of the tiling and that each vertex belongs to at least three tiles.

The reader will observe that we have, in fact proved slightly more than local finiteness. Taking $r = 1$ we have shown that there is a number $M(1)$ with the property that every unit disk in the plane meets at most $M(1)$ tiles. As the value of $M(1)$ is independent of which point P we chose to be the center of our unit disk, we express this property by saying that the tiling is *uniformly* locally finite.

Two other special values of r also lead to interesting

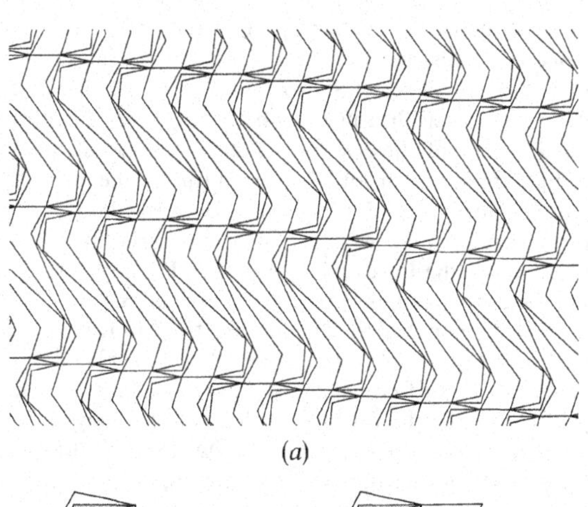

(a)

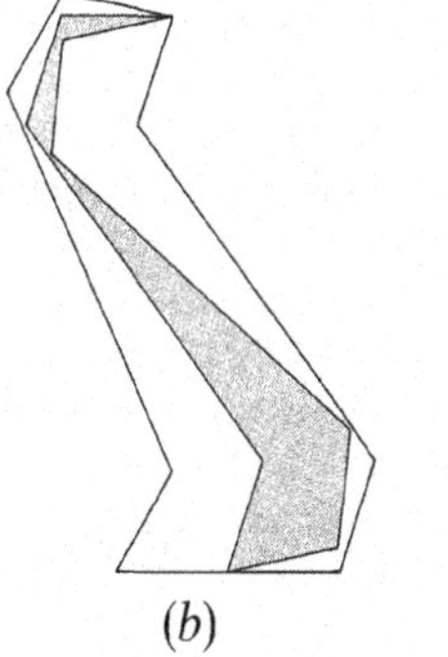

(b)

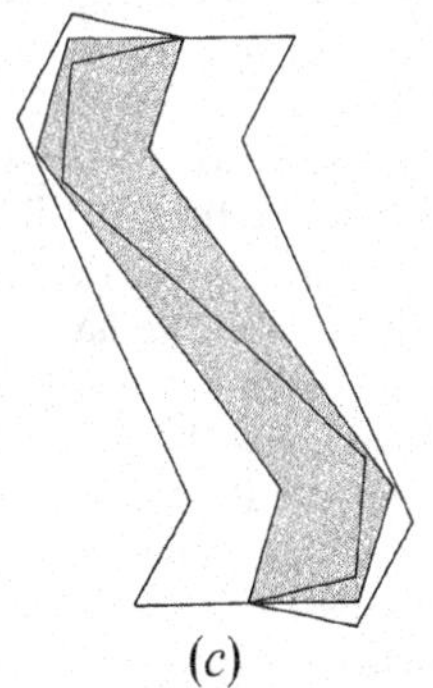

(c)

Figure 3.2.4
A monohedral periodic tiling (a) which is not normal because some pairs of tiles intersect in a nonconnected set. Voderberg [1936] discovered that these tiles have the remarkable property that two tiles can completely surround a third tile of the same shape (see (b)), or even two other tiles (see (c)).

properties of normal tilings. First we take $r = U$ and assume that the disk $D(U,P)$ contains a tile T. Then at most $M(U)$ tiles meet $D(U,P)$ and hence at most $M(U)$ tiles meet T. We deduce that T has at most $M(U)$ neighbors, where $M(U)$ is a fixed number depending only on the parameters of the tiling, $M(U) = 9U^2/u^2$. Of course, the number of neighbors can be made as large as we wish by taking U/u large (for example by making the tiles very long and thin).

The upper bound for the number of neighbors is also an upper bound for the number of adjacents of a tile. Concerning this, however, we have much more information. As we shall see in Statement 3.2.6, if every tile of a normal tiling has the same number of adjacents then this number can only be 3, 4, 5 or 6; this result can also be deduced from Statement 3.2.3. In contrast, if the number n of neighbors of each tile in a monohedral tiling is the same, a value as high as $n = 21$ is certainly possible (see the Laves tiling $[3.12^2]$ in Figure 2.7.1). On the other hand, in the case of a normal but not necessarily monohedral tiling, n can be arbitrarily large; see Fejes Tóth [1971]. Fejes Tóth [1975] conjectured that for convex tiles, $n = 21$ is the maximal possible number in any monohedral tiling; it may be conjectured that the same bound holds for monohedral tilings without the convexity assumption, but we must stress that *no* finite upper bound for n has ever been established. Bezdek [1977] asserts that in a normal tiling by convex tiles, in which each tile has the same number of vertices and the same number n of neighbors, n must have one of the values 6, 7, 8, 9, 10, 12, 14, 16 or 21; however, we have not seen the proof of this.

We note in passing that Wegner [1971] has constructed packings (which are not tilings) by congruent topological disks in which, for any $n \geqslant 3$, each disk has exactly n neighbors, as well as some rigidity properties. This extends earlier work by Fejes Tóth [1969b]. See also Gacs [1972].

A related question has been investigated by Linhart [1977], who gives references to earlier literature. Fejes Tóth [1969a] defined the *Newton number* N of an open bounded convex region R in the plane as the maximum number of regions, congruent to R, that can touch R and be disjoint from it and from each other. A packing by copies of the region R is called *maximal* if every copy of R touches N others. Improving an earlier result of Gacs [1972], Linhart showed that for every maximal packing $N \leqslant 21$. This result may be related to Fejes Tóth's conjecture mentioned above, that in the case of a tiling, 21 is also the maximal possible number of neighbors. Of course, such tilings need not be maximal packings in the sense used by Linhart.

The other special value of r we consider is $r = 0$. Then $D(0,P)$ is the single point P, which we take as a vertex of the tiling. We deduce that at most $M(0) = 4U^2/u^2$ tiles are incident with this vertex, and its valence is therefore at most $M(0)$. Consequently, in any normal tiling the valences are uniformly bounded. We shall show in Section 3.5 that if every vertex has the same valence, then the only possible values are 3, 4, 5, 6.

The uniform boundedness condition N.3 has other important consequences as well. Not only does it prevent the occurrence of singular points but, as we shall now show, it also prevents a "singularity at infinity" as described in the previous section. We recall that this situation occurs in the tilings of Figures 3.1.5 and 3.1.6.

Continuing to use the notation introduced earlier, we write $D(r,P)$ for the closed circular disk of radius r centered at P, and we construct the patch of tiles $\mathscr{A}(r,P)$ generated by this disk, as explained in Section 1.1 (see Figure 1.1.5). Modifying the notation of Section 3.1 we shall denote by $t(r,P)$, $e(r,P)$ and $v(r,P)$ the numbers of tiles, edges and vertices in $\mathscr{A}(r,P)$. As we can see from Figure 1.1.5, $\mathscr{A}(r,P)$ must lie inside the disk $D(r + 2U,P)$, where U is the circumparameter of the tiling.

Now let us perform the same construction starting with the disk $D(r + x,P)$ instead of $D(r,P)$, where x is any positive number. We arrive at a patch $\mathscr{A}(r + x,P)$ and use $t(r + x,P)$, $e(r + x,P)$, $v(r + x,P)$ to denote the numbers of tiles, edges and vertices in $\mathscr{A}(r + x,P)$. If a tiling $\mathscr{T}$ has the property that, for any value of $x > 0$ and any P,

$$\lim_{r \to \infty} \frac{t(r + x,P) - t(r,P)}{t(r,P)}$$

either does not exist, or does not have the value 0, then $\mathcal{T}$ is said to have a *singularity at infinity*. (The expression $(t(r+x,P) - t(r,P))/t(r,P)$ is a measure of how "crowded" the tiles are "round the outside" of the patch $\mathcal{A}(r,P)$. So the property that the limit is 0 corresponds to the intuitive idea that the tiles do not get too crowded.) For example, the above limit is infinite for both the tilings of Figures 3.1.5 and 3.1.6; an explicit calculation of the limit for the tiling of Figure 3.1.5 will be given in Section 4.7.

We can now prove the basic result which we shall need in the sequel.

3.2.2 Normality Lemma. *If $\mathcal{T}$ is a normal tiling then for every $x > 0$*

$$\lim_{r\to\infty} \frac{t(r+x,P)}{t(r,P)} = 1.$$

Equivalently,

$$\lim_{r\to\infty} \frac{t(r+x,P) - t(r,P)}{t(r,P)} = 0.$$

This implies that the tiling does not have a singularity at infinity.

To prove Statement 3.2.2 we make use of the fact that every tile has area at least πu^2 and at most πU^2. As the union of the tiles in $\mathcal{A}(r,P)$ contains the disk $D(r,P)$ of radius r we deduce that $t(r,P) \geq \pi r^2/\pi U^2 = (r/U)^2$. On the other hand, all the tiles that lie in $\mathcal{A}(r+x,P)$ but *not* in $\mathcal{A}(r,P)$ are contained in the annulus or "ring" between the boundaries of $D(r+x+2U,P)$ and $D(r,P)$. This ring has area

$$\pi(r+x+2U)^2 - \pi r^2 = \pi(x+2U)(2r+x+2U).$$

Thus

$$t(r+x,P) - t(r,P) \leq \pi(x+2U)(2r+x+2U)/\pi u^2.$$

From these two inequalities we deduce

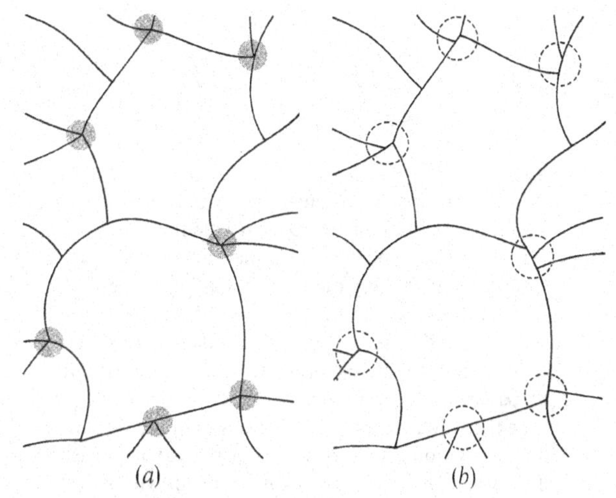

Figure 3.2.5
The tiling on the right is obtained by perturbing the vertices of the tiling on the left inside the regions indicated by shading. The new tiling has vertices all of valence 3. Note that this process cannot decrease the number of adjacents of any tile.

$$\frac{t(r+x,P) - t(r,P)}{t(r,P)} \leq \frac{(x+2U)(2r+x+2U)U^2}{u^2 r^2}$$

and clearly this tends to the value 0 as $r \to \infty$. This proves Statement 3.2.2.

As a final illustration of the implications of normality, we shall prove the following result:

3.2.3 *Every normal tiling $\mathcal{T}$ contains an infinite number of tiles each of which has at most six adjacents.**

This implies, for example, that since every tile in the tiling of Figure 3.1.6 has seven adjacents, this tiling is not normal. A very simple proof of Statement 3.2.3 is possible if we also assume that $\mathcal{T}$ is strongly balanced (see Section 3.5), but here we shall give a proof that depends on normality only.

To begin with, we remark that it suffices to prove the statement for the case in which every vertex is of valence 3. This follows because every vertex of valence $r > 3$ can be "pulled apart" and replaced by $r - 2$ vertices of valence 3, as indicated in Figure 3.2.5. Note that the "pulling apart" can be done in several different ways, and the effect of such a "perturbation" is to tend to increase the number of adjacents of each tile. Hence, if Statement 3.2.3 is true for the "perturbed" tiling, it is

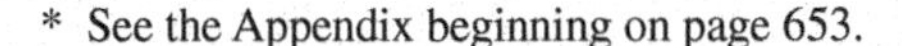
* See the Appendix beginning on page 653.

certainly true for the original one. From now on, we shall therefore assume that $\mathscr{T}$ is trivalent.

To prove Statement 3.2.3 we shall first show that the assumption that every tile in $\mathscr{T}$ has at least seven adjacents leads to a contradiction; this will imply that at least *one* tile has six or fewer adjacents. Let $\mathscr{T}$ be such a tiling, let T be any tile in $\mathscr{T}$, and let $\mathscr{A}_1 = \mathscr{A}(T)$ be the patch of tiles generated by T. For $n \geq 1$ define $\mathscr{A}_{n+1}$ to be the patch generated by the union of the tiles in $\mathscr{A}_n$, and conventionally we write $\mathscr{A}_0$ for the patch consisting of T alone. The following auxilliary result will be required.

3.2.4 *For every $n \geq 1$, each tile in the patch $\mathscr{A}_{n+1}$ which does not belong to $\mathscr{A}_n$ has either one or two edges in common with tiles of $\mathscr{A}_n$, and has exactly two edges in common with tiles of $\mathscr{A}_{n+1}$ that do not belong to $\mathscr{A}_n$.*

In effect, Statement 3.2.4 implies that the successive patches $\mathscr{A}_1, \mathscr{A}_2, \mathscr{A}_3, \ldots$ are each built up by adjoining a simple "border" of tiles round the previous patch—in particular, at no stage is it necessary (in constructing the patches as described in Section 1.1) to fill up "holes"; the initial part of the construction always yields simply connected sets of tiles.

We prove Statement 3.2.4 by induction. It is trivially true for $n = 1$, so assume it is true for a given value n—we shall show that this implies the truth of the statement for the value $n + 1$. Write T_0 for any tile of $\mathscr{A}_{n+1}$ that does not lie in $\mathscr{A}_n$. Clearly T_0 may have either a single edge in common with a tile of $\mathscr{A}_n$, or it may have two consecutive edges in common with two tiles of $\mathscr{A}_n$ (so that these two tiles and T_0 will meet at a vertex). We shall show that these are the only possibilities and that T_0 *cannot* have two non-consecutive edges in common with tiles of $\mathscr{A}_n$. Assume the contrary, namely that there is a tile T_0 with non-consecutive edges belonging to tiles T_1 and T_2 of $\mathscr{A}_n$ (see Figure 3.2.6; by condition N.2, T_1 is distinct from T_2). Then we can draw a simple closed curve C through T_0, T_1, T_2 and possibly other tiles of $\mathscr{A}_n$ not in $\mathscr{A}_{n-1}$ so that C passes through no vertex of $\mathscr{T}$, crosses no edge of $\mathscr{T}$ more than once, and does not contain $\mathscr{A}_{n-1}$ in its interior. Write D for the topological disk bounded by C, and let $\mathscr{M}$ be the (finite) planar map

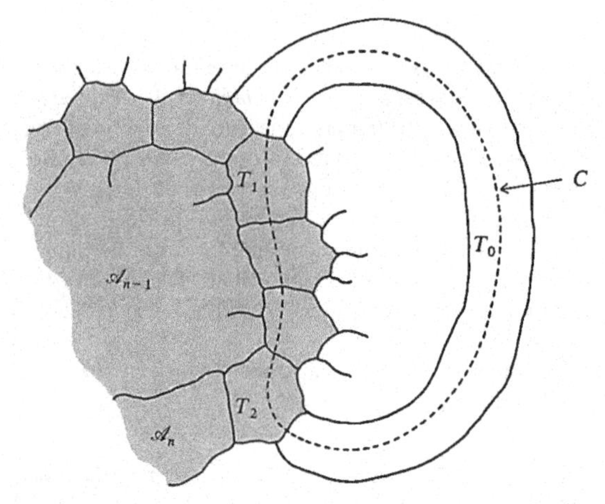

Figure 3.2.6
In the first part of the proof of Statement 3.2.4, we show that it is not possible for a tile T_0 to touch two tiles of $\mathscr{A}_n$ along non-consecutive edges, as shown in this diagram.

whose "countries" are the sets $T \cap D$ (as T runs through the tiles of $\mathscr{T}$) together with the exterior of D, regarded as one unbounded country W. If C passes through k tiles of T then clearly W has k adjacents in $\mathscr{M}$.

Since C passes through two non-consecutive edges of T_0, it is easy to see that $T_0 \cap D$ is a country with at least four adjacents (T_1, T_2, W and T_3, the country adjacent to T_0 and T_1 and contained in D). For similar reasons, the countries $T_1 \cap D$ and $T_2 \cap D$ each have at least four adjacents. Further, because of the inductive assumption, all the other countries $T \cap D$ (where T is a tile of $\mathscr{A}_n$ through which C passes) have at least six adjacents.

We now apply a well-known consequence of Euler's Theorem for Planar Maps (see, for example, Grünbaum [1975, p. 208]). It states that if exactly p_j countries in a trivalent planar map $\mathscr{M}$ have j adjacents each, then

$$3p_3 + 2p_4 + p_5 = 12 + \sum_{j \geq 7} (j - 6)p_j. \quad (3.2.5)$$

In the case under consideration, the unbounded country W can contribute at most 3 to the left side (this will happen if it has exactly three adjacents so $p_3 = 1$), the

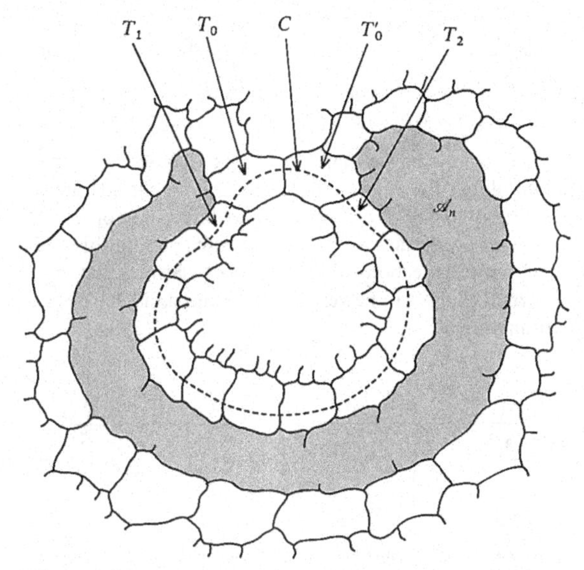

Figure 3.2.7
In the last part of the proof of Statement 3.2.4, we show that it is not possible for tiles T_0, T'_0 of $\mathscr{A}_{n+1}$ to touch each other as shown in this diagram.

countries $T_0 \cap D$, $T_1 \cap D$ and $T_2 \cap D$ can each contribute at most 2 to the left side (this will happen if they have four adjacents), and the other countries of $\mathscr{M}$, having at least six adjacents, contribute nothing. The left side of equation (3.2.5) is therefore at most $3 + 2 \cdot 3 = 9$ which is a contradiction because the right side is at least 12.

From this contradiction we deduce that our original assumption—that T_0 had non-consecutive edges in common with tiles $\mathscr{A}_n$—was incorrect. So the first stage of the proof is completed. In particular, no tile of $\mathscr{A}_{n+1}$ that does not belong to $\mathscr{A}_n$ can have more than two edges in common with $\mathscr{A}_n$.

To complete the induction it is necessary to show that no tile of $\mathscr{A}_{n+1}$ that is not in $\mathscr{A}_n$ can be adjacent to more than two other tiles of $\mathscr{A}_{n+1}$ that are not in $\mathscr{A}_n$. Intuitively this means that when we construct the patches $\mathscr{A}_1$, $\mathscr{A}_2, \ldots$ by adding "borders" of tiles, we never arrive at a situation such as that illustrated in Figure 3.2.7 in which tiles T_0, T'_0 of the ring each have three or more adjacents among the tiles in $\mathscr{A}_{n+1}$ not in $\mathscr{A}_n$. For suppose that such a situation occurs. In a manner similar to that described above we can draw a simple closed curve C through T_0, T'_0 and tiles of $\mathscr{A}_{n+1}$ not in $\mathscr{A}_n$, and such that C passes through no vertex of $\mathscr{T}$, crosses no edge of $\mathscr{T}$ more than once, and does not contain $\mathscr{A}_n$ in its interior. As before, we define a (finite) planar map $\mathscr{M}$ and write W for the country of $\mathscr{M}$ exterior to C. Then a similar calculation to that given above shows that the countries $T_0 \cap D$, $T'_0 \cap D$, $T_1 \cap D$ and $T_2 \cap D$ each have at least four adjacents and so can contribute at most 2 to the left side of equation (3.2.5) while, as before, W can contribute at most 3. The left side of equation (3.2.5) is therefore at most $3 + 4 \cdot 2 = 11$, which is a contradiction. This shows that no tile of $\mathscr{A}_{n+1}$ not in $\mathscr{A}_n$ can have three or more edges in common with tiles of $\mathscr{A}_{n+1}$ not in $\mathscr{A}_n$. It therefore completes the induction, and so establishes Statement 3.2.4.

It is now simple to complete the proof of Statement 3.2.3. We recall that we are assuming that $\mathscr{T}$ is trivalent, normal, and that every tile has at least 7 adjacents. Statement 3.2.4 shows that each tile T of $\mathscr{A}_{n+1}$ not in $\mathscr{A}_n$ has at most two edges in common with tiles of $\mathscr{A}_{n+1}$ not in $\mathscr{A}_n$, and at most two edges in common with tiles of $\mathscr{A}_n$. Hence it has at least three edges in common with tiles of $\mathscr{A}_{n+2}$ not in $\mathscr{A}_{n+1}$. From this it readily follows that each "border" of tiles must contain at least 3/2 times as many tiles as the previous ring. Since $\mathscr{A}_1$ has at least 7 tiles not in $\mathscr{A}_0$, we deduce that $\mathscr{A}_n$ has at least

$$\begin{aligned} 7(1 + \tfrac{3}{2} + (\tfrac{3}{2})^2 + \cdots + (\tfrac{3}{2})^{n-1}) &= 7((\tfrac{3}{2})^n - 1)/(\tfrac{3}{2} - 1) \\ &= 14((\tfrac{3}{2})^n - 1) \end{aligned}$$

tiles. Let the inparameter of $\mathscr{T}$ be u, so each tile has area at least πu^2. The total area of the tiles in $\mathscr{A}_n$ is therefore at least $14\pi u^2((\tfrac{3}{2})^n - 1)$. However, $\mathscr{A}_n$ lies in a circular disk of radius nU, where U is the circumparameter of $\mathscr{T}$. Its area is therefore at most $\pi(nU)^2$, from which we deduce that $14\pi u^2((\tfrac{3}{2})^n - 1) \leq \pi(nU)^2$. But clearly this cannot hold for arbitrarily large values of n and so we have a contradiction. This shows that our assumption that every tile in $\mathscr{A}_n$ has at least seven adjacents is false for some value of n, say $n = N$. In fact, the argument proves more—it shows that if we start from any tile T then the patch $\mathscr{A}_N$ we construct from it contains a tile with at most six adjacents. As N is a fixed number, determined by the parameters of $\mathscr{T}$, there are infinitely many disjoint patches of this type, and therefore infinitely many tiles with at most six adjacents. This completes the proof of Statement 3.2.3.

In a similar way we can now show that every normal tiling has an infinity of vertices of valence at most 6. (For an alternative proof see Exercise 4.2.3.)

If each tile in $\mathscr{T}$ has the same number of adjacents, Statement 3.2.3 clearly implies the following.

3.2.6 *If $\mathscr{T}$ is a normal tiling in which each tile has the same number k of adjacents, then $k = 3, 4, 5$ or 6.*

From now on we shall be chiefly concerned with normal tilings. But we shall not exclude from consideration any non-normal tilings that display interesting or important properties. Moreover, the discussion in this and the previous section should serve as a warning that non-normal tilings can be very puzzling and difficult objects to investigate.

EXERCISES 3.2

1. Find a tiling that satisfies conditions N.2 and N.3 and in which every tile consists of the union of two closed topological disks with (*a*) a single common point; (*b*) no common points.
2. Construct a tiling that satisfies conditions N.1 and N.3, in which every tile meets some other tile in exactly two points.

*3. Prove that there exists no tiling that satisfies conditions N.1 and N.3, in which every tile meets some other tile in exactly three points. (Valette [1981], Breen [1983b])

4. Show that there exists a (minimal) integer $c = c(u,U)$ such that for every tiling $\mathscr{T}$ that satisfies conditions N.1 and N.3 the intersection of any two tiles has at most c connected components. Prove that $c(1,2) = 2$.
5. Find a tiling $\mathscr{T}$ which has properties N.1 and N.2 and in which no point of a certain tile T belongs to any other tile of $\mathscr{T}$.
6. Give an example of a tiling $\mathscr{T}$ which is not normal although it satisfies conditions N.1 and N.3 of the definition of normality, and $\lim_{r\to\infty} t(r + x,P)/t(r,P) = 1$ for all $x > 0$ and P.
7. Give an example as in Exercise 6, but with condition N.2 substituted for N.3.
8. Show by example that the Normality Lemma (Statement 3.2.2) may fail if the tiling $\mathscr{T}$ is not normal.
9. Show that the Normality Lemma (Statement 3.2.2) is valid even for tilings that satisfy only conditions N.1 and N.3 of the definition of normality. In particular show that it holds for all monohedral tilings whose prototiles are topological disks.
10. Find a tiling $\mathscr{T}$ that satisfies conditions N.2 and N.3 in which each tile has 8 adjacents.
11. Show that Statements 3.2.1, 3.2.2 and 3.2.3 remain valid for tilings that satisfy N.1, N.2 and a weakened version of N.3 in which instead of the existence of incircles of radius u it is only required that each tile has area at least πu^2.
12. For $n = 1, 2, \ldots$ let D_n be a topological disk with diameter Δ_n and inradius δ_n, such that $\lim_{n\to\infty} \Delta_n = \infty$ and $\lim_{n\to\infty} \delta_n/\Delta_n = 1$. If a_n is the number of tiles of a normal tiling $\mathscr{T}$ that are contained in D_n and if b_n is the number of tiles of $\mathscr{T}$ that meet D_n but are

not contained in it, prove the following generalization of the Normality Lemma (Statement 3.2.2): $\lim_{n\to\infty} b_n/a_n = 0$.

13. Let $C(r,P)$ be the set of edges of a tiling $\mathscr{T}$ that are contained in a disk $D(r,P)$, and let $C^*(r,P)$ be the set of those edges of $\mathscr{T}$ that meet $D(r,P)$ but are not contained in it. Denote by $c(r,P)$ and $c^*(r,P)$ the numbers of edges in $C(r,P)$ and $C^*(r,P)$. Prove the following analogues of the Normality Lemma: If $\mathscr{T}$ is a normal tiling then

$$\lim_{r\to\infty} c^*(r,P)/c(r,P) = \lim_{r\to\infty} c^*(r,P)/t(r,P) = 0.$$

*14. Construct a normal tiling $\mathscr{T}$ such that the function $\phi(r) = c(r,P)/t(r,P)$ has no limit as $r \to \infty$ where $c(r,P)$ is defined as in Exercise 13. Determine bounds for $\phi(r)$ in terms of u and U.

15. Let T be a closed topological disk. We say that T has *property* $\mathscr{V}_n$ if there exist non-overlapping sets T' and T'', congruent to T, such that the complement of $T' \cup T''$ has a bounded component the closure of which is the union of n non-overlapping sets congruent to T. As shown in Figures 3.2.4(b) and (c), the prototile of the tiling in Figure 3.2.4(a) has properties $\mathscr{V}_1$ and $\mathscr{V}_2$.

*(a) Show that for each $n \geq 1$ there is a disk T with property $\mathscr{V}_n$.

**(b) Decide whether for some (or for each) $n \geq 3$ there exists a tile T with property $\mathscr{V}_n$ that is the prototile of a monohedral (or monohedral and periodic) tiling.

**16. A tile T which is a topological disk is said to have the property $\mathscr{V}_n^m$ if there exist non-overlapping sets T' and T'' congruent to T such that the complement of $T' \cup T''$ has m bounded components, the closure of which is the union of n non-overlapping sets congruent to T. Decide whether there are any sets T with property $\mathscr{V}_n^m$ in case $n \geq m > 1$.

*17. Let $\mathscr{T}$ be a normal tiling.

(a) Show that if each vertex of $\mathscr{T}$ is 4-valent, then $\mathscr{T}$ contains an infinite number of tiles each of which has at most four adjacents.

(b) Show that the conclusion of part (a) remains valid even if $\mathscr{T}$ is not required to be 4-valent, but only that each vertex is of valence at least 4.

(c) Show that if each vertex of $\mathscr{T}$ is at least 5-valent then $\mathscr{T}$ contains an infinite number of tiles with three adjacents each.

18. (a) Show that for each pair of natural numbers $n \geq 3$, $m \geq 1$ there exists a normal tiling $\mathscr{T}$ in which m tiles have n adjacents and the remaining tiles have 6 adjacents each.

**(b) Decide whether there exists a normal tiling $\mathscr{T}$ in which each tile has at least 6 adjacents, and infinitely many tiles have at least 7 adjacents each.

3.3 BALANCED TILINGS AND EULER'S THEOREM

When confronted with problems concerning the relative numbers of tiles, edges and vertices in a given tiling $\mathscr{T}$, it turns out that even normal tilings are too general. Hence we introduce the concept of a balanced tiling. Following the notation as used in the previous section, we denote by $t(r,P)$, $e(r,P)$ and $v(r,P)$ the numbers of tiles, edges and vertices in a patch $\mathscr{A}(r,P)$ of tiles generated by the circular disk $D(r,P)$. We shall say that $\mathscr{T}$ is *balanced* (*with respect to* P) if it is normal and satisfies the following condition.

B.1 *The limits*

$$\lim_{r\to\infty} \frac{v(r,P)}{t(r,P)} \quad \text{and} \quad \lim_{r\to\infty} \frac{e(r,P)}{t(r,P)} \tag{3.3.1}$$

exist and are finite.

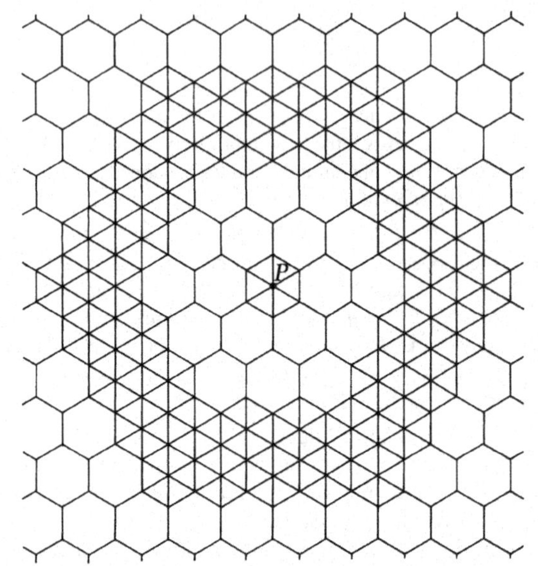

Figure 3.3.1
A method of constructing a dihedral tiling which is not balanced. Suitable "rings" of hexagons and of triangles centered at P can be chosen so that the ratios of the numbers of vertices and edges to tiles in the patches $\mathscr{A}(r,P)$ oscillate as $r \to \infty$.

We must emphasize the fact that not every normal tiling is balanced. This is illustrated by the tiling of Figure 3.3.1. We take the regular tiling by hexagons and split up some of the hexagons (lying in "rings" about P) each into six triangles. If *none* of the hexagons were split, then it is easy to show that $v(r,P)/t(r,P)$ would tend to the value 2, whereas if *every* hexagon were split, the value of $v(r,P)/t(r,P)$ would tend to $\frac{1}{2}$. Hence, by making the "rings" of split and unsplit hexagons larger and larger, in a suitable manner, it is clear that we can make the value of $v(r,P)/t(r,P)$ oscillate between any two numbers α, β satisfying $\frac{1}{2} < \alpha < \beta < 2$. In such a tiling the limit of $v(r,P)/t(r,P)$ does not exist, and the tiling is not balanced. On the other hand, there are non-normal tilings for which the limits in (3.3.1) exist, see for example Figure 3.3.2.

The procedure just described, of constructing a new tiling by combining two different tilings in "rings" about a point P, turns out to be useful in other contexts also. For example, we use it in the next section to illustrate the relation between balanced, strongly balanced and prototile balanced tilings. Various modifications and variations are also useful. Instead of combining two tilings in "rings", they can be combined in parallel "strips" of varying widths to yield examples of tilings with similar properties.

Our next result shows that the property of being balanced does not depend on the choice of P.

3.3.2 *If, for a given normal tiling $\mathscr{T}$ and a particular point P the limits $\lim_{r\to\infty} v(r,P)/t(r,P)$ and $\lim_{r\to\infty} e(r,P)/t(r,P)$ exist and are finite, then these limits exist and have the same value whenever P is replaced by any other point of the plane.*

It follows that if the limits do not exist for a particular choice of P (as in the tiling of Figure 3.3.1) and the tiling is normal, then the limits will not exist for any other point.

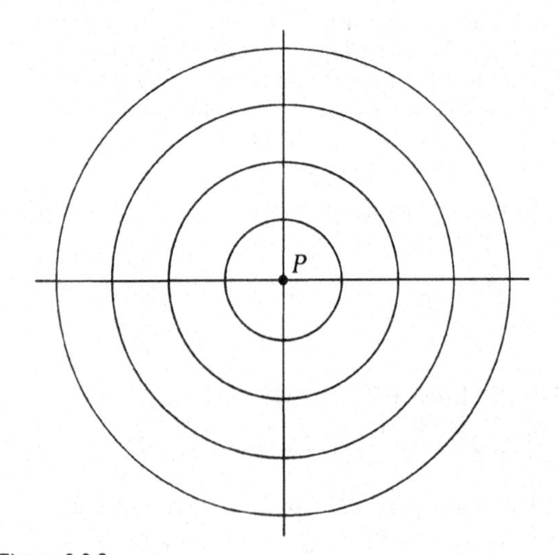

Figure 3.3.2
A tiling for which the limits of (3.3.1) exist although the tiling is not normal. The circles are of integer radius. As $r \to \infty$, $v(r,P)/t(r,P) = (4k + 1)/4k \to 1$ and $e(r,P)/t(r,P) = 8k/4k \to 2$, where $k = [r]$. It is easy to see that the existence and values of these two limits do not depend on the choice of P.

We shall now prove Statement 3.3.2. Let P and P' be any two points and let the distance from P to P' be $x/2$. Then clearly

$$D(r,P) \subset D(r + \tfrac{1}{2}x,P') \subset D(r + x,P)$$

(see Figure 3.3.3) and so

$$\mathscr{A}(r,P) \subset \mathscr{A}(r + \tfrac{1}{2}x,P') \subset \mathscr{A}(r + x,P).$$

We deduce that

$$t(r,P) \leq t(r + \tfrac{1}{2}x,P') \leq t(r + x,P)$$

and

$$v(r,P) \leq v(r + \tfrac{1}{2}x,P') \leq v(r + x,P).$$

Hence

$$\frac{v(r,P)}{t(r,P)} \cdot \frac{t(r,P)}{t(r + x,P)} \leq \frac{v(r + \frac{1}{2}x,P')}{t(r + \frac{1}{2}x,P')}$$

$$\leq \frac{v(r + x,P)}{t(r + x,P)} \cdot \frac{t(r + x,P)}{t(r,P)}.$$

Letting $r \to \infty$ and recalling from the Normality Lemma (Statement 3.2.2) that $\lim_{r\to\infty} t(r,P)/t(r + x,P)$ and $\lim_{r\to\infty} t(r + x,P)/t(r,P)$ both have the value 1, we see that the limits $\lim_{r\to\infty} v(r,P')/t(r,P')$ and $\lim_{r\to\infty} v(r,P)/t(r,P)$ are equal. This completes the proof for the assertion about v/t, and a precisely similar argument applies to e/t. Hence Statement 3.3.2 is established.

Although P may be chosen to be any point of the plane, *it must be kept fixed* as $r \to \infty$. In Section 3.6 we shall show that if one is allowed to translate the disk while it is increasing in size, then the limits in (3.3.1) may take different values, or may even fail to exist! In Section 3.6 we shall also show that the definition of "balance" is radically changed if we make use of patches based on topological disks other than circular ones. In fact our definition of "balance" is very weak; it suffices however for us to establish the following important result.

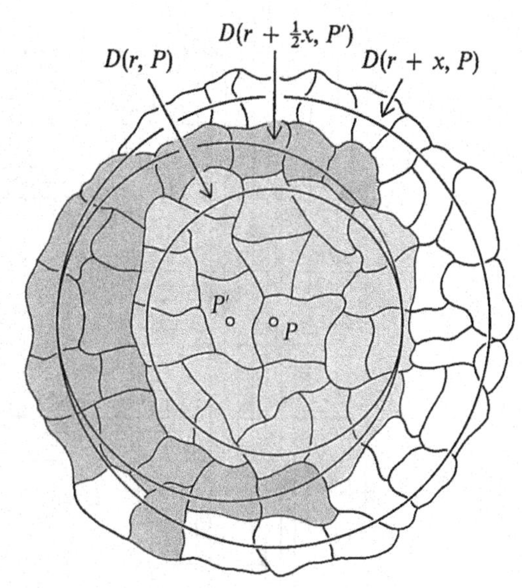

Figure 3.3.3
This illustrates the proof of Statement 3.3.2 that for a normal tiling the property of being balanced does not depend on the choice of the point P in the definition.

3.3.3 Euler's Theorem for Tilings. *For any normal tiling $\mathscr{T}$, if one of the limits $v(\mathscr{T}) = \lim_{r\to\infty} v(r,P)/t(r,P)$ or $e(\mathscr{T}) = \lim_{r\to\infty} e(r,P)/t(r,P)$ exists and is finite, then so does the other. Thus the tiling is balanced and, moreover,*

$$v(\mathscr{T}) = e(\mathscr{T}) - 1. \qquad (3.3.4)$$

To prove this we make use of Euler's Theorem for Planar Maps (equation (3.1.4)). Applied to the patch $\mathscr{A}(r,P)$ we obtain

$$v(r,P) - e(r,P) + t(r,P) = 1.$$

Divide this equation by $t(r,P)$ and consider the limit as $r \to \infty$. One obtains immediately both assertions of Statement 3.3.3.

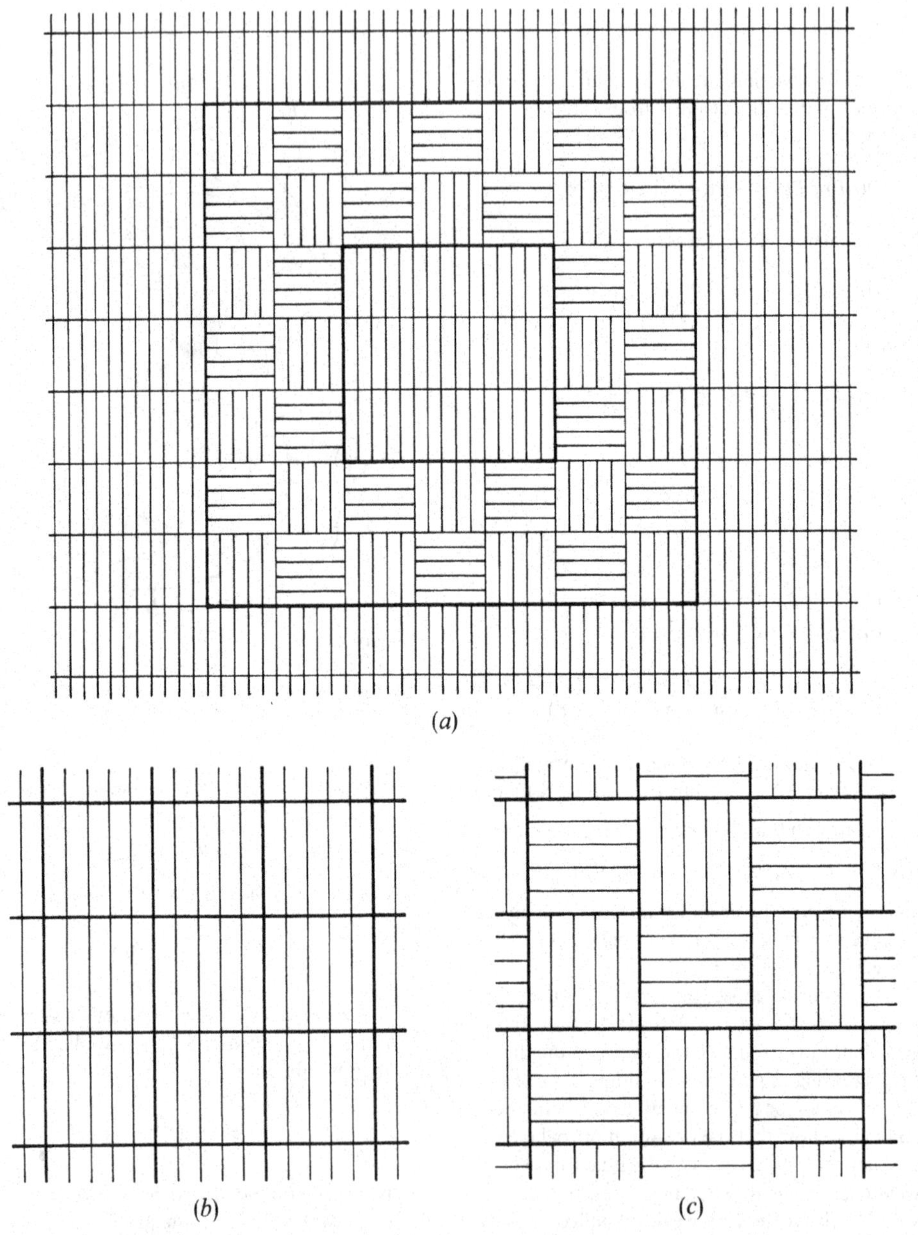

Figure 3.3.4
A monohedral tiling which is not balanced. The construction is explained in the text.

Euler's Theorem for Tilings is of fundamental importance, but it is a strong contender for one of the most frequently misquoted results in mathematics! Time and again it, or an equivalent statement, has been asserted as true for all tilings. This is manifestly false, as shown by the tilings of Figures 3.1.5 and 3.1.6. So, before Euler's Theorem can be used, it is essential that its applicability is established. From Statement 3.3.3 we see that in order to ensure this it is sufficient to show that the tiling is balanced.

Another typical error is to assume that every normal monohedral tiling is balanced. (We recall that a tiling is monohedral if all the tiles are congruent to a fixed prototile.) Consider the tiling of Figure 3.3.4(*a*). It consists of alternate "rings" of tiles arranged all parallel (as in Figure 3.3.4(*b*)) or in two orientations in the form of a checkerboard (as in Figure 3.3.4(*c*)). We now employ an argument similar to that used for the tiling of Figure 3.3.1. For the tiling in Figure 3.3.4(*b*) it is easy to show that $v(\mathscr{T}) = 1$, and for the tiling in Figure 3.3.4(*c*) that $v(\mathscr{T}) = \frac{9}{5}$. Hence, by arranging the "rings" of tiles in Figure 3.3.4(*a*) of suitable widths we can make $v(r,P)/t(r,P)$ oscillate between any two values α and β such that $1 < \alpha < \beta < \frac{9}{5}$. Thus we have a normal monohedral tiling which is not balanced. On the other hand, the following result clearly implies that every *isohedral* tiling is balanced:

3.3.5 *A normal tiling in which every tile has the same number of adjacents is balanced.*

To prove this we consider a patch $\mathscr{A}(r,P)$ of tiles as in Figure 3.3.5. If each of the $t(r,P)$ tiles has k adjacents, and $E(r,P)$ of the edges belong to one tile only (that is, they lie round the boundary of the patch), then a simple counting of edges yields

$$k \cdot t(r,P) = 2e(r,P) - E(r,P) \qquad (3.3.6)$$

or

$$\frac{e(r,P)}{t(r,P)} = \frac{k}{2} + \frac{E(r,P)}{2t(r,P)}. \qquad (3.3.7)$$

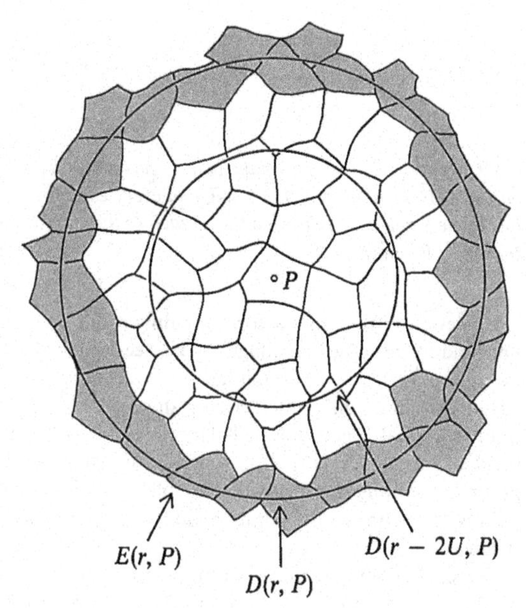

Figure 3.3.5
An illustration of the proof of Statement 3.3.5 that every normal tiling in which each tile has the same number of adjacents is balanced.

Now every one of the $E(r,P)$ edges on the boundary belongs to a tile of $\mathscr{A}(r,P)$ that does not meet the disk $D(r - 2U,P)$, nor is enclosed by tiles meeting the disk. These tiles are shaded in Figure 3.3.5 and there are at most $t(r,P) - t(r - 2U,P)$ of them. We deduce that $0 \leq E(r,P) \leq k(t(r,P) - t(r - 2U,P))$. Dividing this by $t(r,P)$ and letting $r \to \infty$, the Normality Lemma implies that the right side tends to zero. Hence

$$\lim_{r\to\infty} E(r,P)/t(r,P) = 0. \qquad (3.3.8)$$

Using this equation (3.3.7) becomes, as $r \to \infty$,

$$e(\mathscr{T}) = \lim_{r\to\infty} e(r,P)/t(r,P) = \tfrac{1}{2}k.$$

The fact that the tiling is balanced and that $\lim_{r\to\infty} v(r,P)/t(r,P)$ exists then follows from Euler's Theorem for Tilings (Statement 3.3.3). This completes the proof of Statement 3.3.5.

A slight modification of the above argument leads to the proof of the following:

3.3.9 *If every tile in a normal tiling $\mathscr{T}$ has either k_1 adjacents or k_2 adjacents ($k_1 \neq k_2$), and every edge in $\mathscr{T}$ lies in the boundary of just one tile of each of the two kinds, then $\mathscr{T}$ is balanced.*

However, we note that a corresponding result does not hold for all dihedral tilings. As we see from the example of Figure 3.3.1, even if the two kinds of tile each have a constant number of adjacents, then although the tiling is normal it need not be balanced. On the other hand, in Section 4.6 we shall show that if the symmetry group of the tiling is transitive on the tiles of each kind (that is, the tiling is 2-isohedral) then the tiling must be balanced.

We shall now prove the following result.

3.3.10 *A normal tiling in which every vertex has the same valence is balanced.*

Consider a patch $\mathscr{A}(r,P)$ of tiles, and suppose that $V(r,P)$ is the number of vertices in $\mathscr{A}(r,P)$ which belong to the boundary of the patch. Let j be the valence of every vertex in $\mathscr{T}$ (though, of course, in $\mathscr{A}(r,P)$ the valence of the $V(r,P)$ vertices on the boundary may be less than j). Since each edge has two endpoints, and each vertex in the interior of $\mathscr{A}(r,P)$ is the endpoint of j edges, we see that $2e(r,P)/j$ is approximately equal to $v(r,P)$. In fact the former is an underestimate due to the vertices on the boundary, and we have

$$0 \leq v(r,P) - 2e(r,P)/j \leq V(r,P). \qquad (3.3.11)$$

Since the union of the tiles in $\mathscr{A}(r,P)$ is a topological disk we may apply Euler's Theorem for Planar Maps, which yields

$$v(r,P) = e(r,P) - t(r,P) + 1,$$

and clearly

$$V(r,P) = E(r,P).$$

So (3.3.11) may be written in the form

$$0 \leq \left(1 - \frac{2}{j}\right)e(r,P) - t(r,P) + 1 \leq E(r,P).$$

Dividing through by $t(v,P)$, letting $r \to \infty$, and using (3.3.8) we obtain

$$e(\mathscr{T}) = \lim \frac{e(r,P)}{t(r,P)} = \frac{j}{j-2}$$

which, together with Statement 3.3.3, shows that $\mathscr{T}$ is balanced. Thus Statement 3.3.10 is proved.

A slight modification of the above argument enables us to prove the following.

3.3.12 *If every edge in a normal tiling $\mathscr{T}$ has a j_1-valent vertex at one end and a j_2-valent vertex at the other, then $\mathscr{T}$ is balanced.*

Examination of the proofs of Statements 3.3.5 and 3.3.10 shows that they depend upon an idea which will be used in the future. If we wish to establish a numerical relation for a given balanced tiling $\mathscr{T}$ then we first show that it holds approximately for patches of tiles of $\mathscr{T}$, any discrepancy being due to the tiles, edges and vertices that lie on the boundary of the patch. The essential part of the proof then consists in showing that this discrepancy becomes negligible as we take larger and larger patches. That we are able to do this is, of course, a direct consequence of the fact that the tiling is normal, and usually involves the use of the Normality Lemma.

Using this idea it is not difficult to prove the following useful result.

3.3.13 *Every normal periodic tiling is balanced.*

Thus, for example, all the uniform, k-uniform and Laves tilings are balanced, as well as the great majority of the others illustrated in this book. Periodic tilings will be investigated in more detail in Section 3.7 where a stronger property than balance is established.

EXERCISES 3.3

*1. Determine thicknesses of the "rings" which should be used in the tiling of Figure 3.3.1 so that the ratio $v(r,P)/t(r,P)$ oscillates indefinitely beyond $\frac{2}{3}$ and $\frac{4}{3}$.

*2. Show that if the tiling $\mathscr{T}$ is not normal then it is possible for just one of the limits $\lim_{r\to\infty} v(r,P)/t(r,P)$ and $\lim_{r\to\infty} v(r,P')/t(r,P')$ to exist and be finite.

3. Construct tilings that satisfy conditions N.1 and N.3 but not N.2 of the definition of normality (Section 3.2) and that (*a*) have, or (*b*) do not have, property B.1 used in the definition of balanced tilings.

4. Show that all the tilings in Figures 2.5.4 and 1.3.6 satisfy condition B.1. Find the values of $v(\mathscr{T})$ and $e(\mathscr{T})$ in each case. Determine which of these tilings are balanced.

5. Show that if $\mathscr{T}$ is a balanced tiling then the limit $\lim_{r\to\infty} e(r,P)/v(r,P)$ exists; express its value in terms of $v(\mathscr{T})$ and $e(\mathscr{T})$.

*6. Show that if $\mathscr{T}$ is a balanced tiling then $\frac{1}{2} \leq v(\mathscr{T}) \leq 2$ and $\frac{3}{2} \leq e(\mathscr{T}) \leq 3$.

7. For a given rational value of c with $\frac{1}{2} \leq c \leq 2$, construct a balanced tiling $\mathscr{T}$ with $v(\mathscr{T}) = c$.

8. For the uniform tilings and the Laves tilings discussed in Chapter 2 determine the values of $v(\mathscr{T})$ and $e(\mathscr{T})$.

9. For the tiling in Figure 2.4.9 determine whether it satisfies condition B.1 or not; if the limits exist, determine their values and investigate whether they depend on the choice of the point P. Is the tiling balanced?

3.4 STRONGLY BALANCED AND PROTOTILE BALANCED TILINGS

In calculations connected with the number of elements in a tiling we shall need the concepts of "strongly balanced tiling" and "prototile balanced tiling". For a given tiling $\mathscr{T}$ we write $v_j(r,P)$ for the number of vertices of valence j in a patch $\mathscr{A}(r,P)$, and $t_k(r,P)$ for the number of tiles with k adjacents in $\mathscr{A}(r,P)$. Then a tiling $\mathscr{T}$ is called *strongly balanced* if it is normal and satisfies the following condition.

B.2 *All the limits*

$$v_j(\mathscr{T}) = \lim_{r\to\infty} v_j(r,P)/t(r,P), \qquad t_k(\mathscr{T}) = \lim_{r\to\infty} t_k(r,P)/t(r,P) \tag{3.4.1}$$

exist.

Since $v(r,P) = \sum_{j\geq 3} v_j(r,P)$, the existence of the limits (3.4.1) implies that $\lim_{r\to\infty} v(r,P)/t(r,P)$ also exists and since the tiling is normal this limit must be finite. In fact, in a strongly balanced tiling, since

$$v(\mathscr{T}) = \sum_{j\geq 3} v_j(\mathscr{T}),$$

we deduce that each of the limits 3.4.1 is finite and also that every strongly balanced tiling is necessarily balanced. We note also, for future use, that

$$\sum_{k\geq 3} t_k(\mathscr{T}) = 1$$

in any strongly balanced tiling. Moreover, as in Statement 3.3.2, we can show that if the limits in (3.4.1) exist and are finite for a certain point P, the same holds for every choice of P.

However, a balanced tiling need not be strongly balanced. For example, consider tilings formed (like the one in Figure 3.3.1) by alternating suitable "rings" of hexagons (as in the regular tiling (6^3)) and "rings" taken from the tiling with pentagons and heptagons shown in Figure 3.4.1 (obtained from (6^3) by modifying quadruplets of hexagons into two pentagons and two heptagons). It is easily checked that such tilings $\mathcal{T}$ are balanced, with $v(\mathcal{T}) = 2$ and $e(\mathcal{T}) = 3$, but that for any α, β with $0 < \alpha < \beta < 1$ the ratios $t_6(r,P)/t(r,P)$ may be made to oscillate beyond α and β.

We shall now show that certain special kinds of tilings, which will be investigated in detail in Sections 4.3, 4.4 and 4.5, are strongly balanced.

3.4.2 *If every tile of a normal tiling $\mathcal{T}$ has k vertices, and these vertices have valences $j_1, j_2, \ldots, j_k$ (in some order), then $\mathcal{T}$ is strongly balanced.*

The j_i need not be distinct, but the same number of vertices of each valence must belong to each tile. In particular, each isohedral tiling satisfies the conditions of Statement 3.4.2; in this case, the values $j_1, \ldots, j_k$ lie in the same cyclic order round each tile.

To prove Statement 3.4.2 we use the procedure suggested at the end of the previous section. Suppose that $v_i(r,P)$ vertices of $\mathcal{A}(r,P)$ have valence i, and m_i of these belong to each tile (that is to say, exactly m_i of the numbers $j_1, \ldots, j_k$ have the value i). Then a simple counting argument shows that $m_i t(r,P)/i$ is approximately equal to $v_i(r,P)$. In fact it is an overestimate due to the $V(r,P)$ vertices on the boundary of $\mathcal{A}(r,P)$. Hence

$$0 \leq v_i(r,P) - m_i t(r,P)/i \leq V(r,P) \qquad (3.4.3)$$

(compare relation 3.3.11). Dividing by $t(r,P)$ and letting $r \to \infty$, we see that the right side tends to 0 and we obtain

$$v_i(\mathcal{T}) = \lim_{r \to \infty} v_i(r,P)/t(r,P) = m_i/i. \qquad (3.4.4)$$

Since this is true for each i, and since $t_k(r,P) = t(r,P)$ (so that the second condition of the definition is trivially

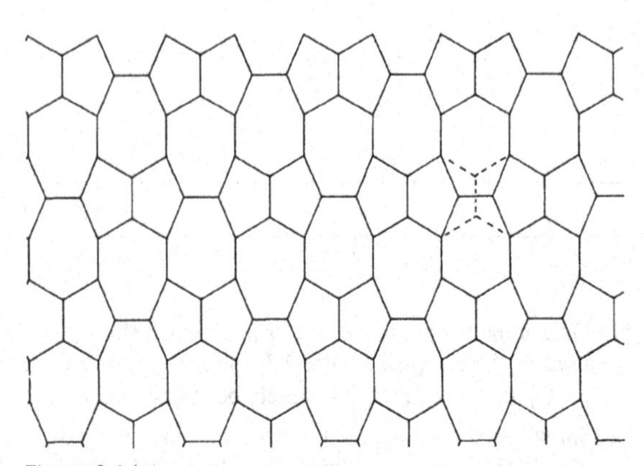

Figure 3.4.1
A normal and balanced tiling with pentagons and heptagons, used in the text to construct normal and balanced tilings which are not strongly balanced. It may be obtained by a modification of the regular tiling (6^3) (dashed lines).

fulfilled) we deduce that $\mathcal{T}$ is strongly balanced. Hence Statement 3.4.2 is proved.

We shall now consider a second kind of tilings.

3.4.5 *If every vertex of a normal tiling $\mathcal{T}$ has valence j, and is incident with tiles which have $k_1, \ldots, k_j$ adjacents (the order in which these are arranged round the vertices being of no consequence), then $\mathcal{T}$ is strongly balanced.*

The k_i need not be distinct, so long as there are the same number of tiles of each kind incident with each vertex. Clearly Statement 3.4.5 applies to isogonal tilings, and it is in this case that we shall use it in Section 4.4.

Suppose that exactly m_i tiles with i adjacents each are incident with each vertex. Then the number of vertices in a patch $\mathcal{A}(r,P)$ is approximately $it_i(r,P)/m_i$, and estimating the discrepancy as before we obtain the inequalities

$$0 \leq v(r,P) - it_i(r,P)/m_i \leq V(r,P) \qquad (3.4.6)$$

corresponding to the inequalities 3.4.3. Divide through by $t(r,P)$ and take the limit as $r \to \infty$. As before,

$$\lim_{r \to \infty} V(r,P)/t(r,P) = 0$$

since the tiling is normal, and $\lim_{r \to \infty} v(r,P)/t(r,P)$ tends to a finite limit $v(\mathcal{T})$ because the tiling is balanced (by Statement 3.3.10). Hence we obtain

$$\lim_{r \to \infty} t_i(r,P)/t(r,P) = m_i v(\mathcal{T})/i,$$

and we deduce that $\mathscr{T}$ is strongly balanced, as required.
Our next result is as follows.

3.4.7 *In a normal tiling $\mathscr{T}$, suppose that every tile has either k_1 or k_2 adjacents and every vertex is of valence j_1 or j_2. Then if every edge joins a vertex of one kind to a vertex of the other, and separates a tile of one kind from a tile of the other, the tiling is strongly balanced.*

In this case the inequalities corresponding to 3.4.3 and 3.4.6 are

$$0 \le e(r,P) - j_1 v_{j_1}(r,P) \le E(r,P)$$

$$0 \le e(r,P) - j_2 v_{j_2}(r,P) \le E(r,P)$$

$$0 \le e(r,P) - k_1 t_{k_1}(r,P) \le E(r,P)$$

$$0 \le e(r,P) - k_2 t_{k_2}(r,P) \le E(r,P).$$

They arise because each edge in $\mathscr{T}$ is incident with precisely one tile and one vertex of each kind. Divide the inequalities by $t(r,P)$ and then let $r \to \infty$. Since

$$\lim_{r\to\infty} E(r,P)/t(r,P) = 0$$

by relation (3.3.8), and

$$\lim_{r\to\infty} e(r,P)/t(r,P) = e(\mathscr{T})$$

by Statement 3.3.9, we immediately deduce that the tiling is strongly balanced.

In later sections we shall establish further results of a similar nature. In particular, in Sections 3.6 and 4.6 we shall show that periodic tilings and 2-tile-transitive tilings are strongly balanced.

The definition of a strongly balanced tiling $\mathscr{T}$ given at the beginning of this section depended upon the number of tiles of $\mathscr{T}$ that had various numbers of adjacents. It may, of course, happen that two or more prototiles of $\mathscr{T}$ have the same number of adjacents and in this case the definition tells us nothing about the frequency with which each prototile occurs in $\mathscr{T}$. For this reason we introduce a condition, stronger than B.1 which we shall call prototile balance.

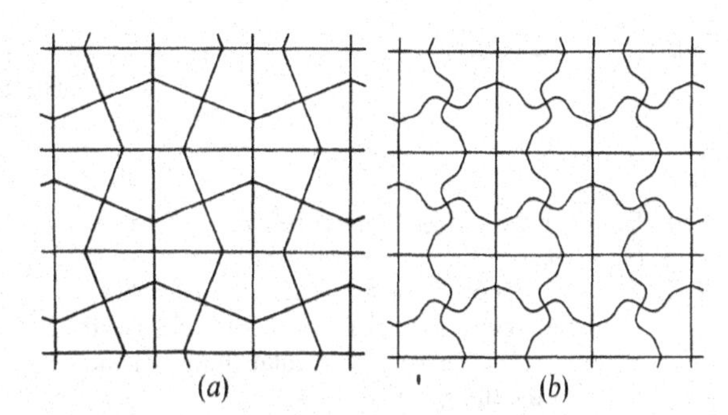

Figure 3.4.2
Using "alternating rings" obtained from the two tilings shown, it is possible to construct a strongly balanced tiling which is not prototile balanced.

Let $\mathscr{S} = \{T_1, T_2, \ldots, T_k\}$ be the set of prototiles for a tiling and write $t_{[T_j]}(r,P)$ for the number of copies of prototile T_j ($j = 1, \ldots, k$) that occur in a patch $\mathscr{A}(r,P)$. Then $\mathscr{T}$ is *prototile balanced* if the following condition is satisfied:

B.3 *The limits* $t_{[T_j]}(\mathscr{T}) = \lim_{r\to\infty} t_{[T_j]}(r,P)/t(r,P)$ *exist for each* $j = 1, \ldots, k$.

Clearly each of these limits, if it exists, must satisfy $0 \le t_{[T_j]}(\mathscr{T}) \le 1$ and, as in Statement 3.3.2, if the limits exist for a certain choice of P, then they necessarily exist for all choices of P.

Using the two tilings of Figure 3.4.2, the method of "alternating rings" illustrated in Figure 3.3.1 enables us to construct a tiling $\mathscr{T}$ which is strongly balanced (each tile has four adjacents) but not prototile balanced. Thus condition B.3 is not a consequence of B.2.

It is not difficult to see that Statement 3.3.13 can be strengthened as follows:

3.4.8 *Every normal periodic tiling is strongly balanced and prototile balanced.*

The numbers $v_i(\mathscr{T})$ and $t_k(\mathscr{T})$, as well as the values of $t_{[T_j]}(\mathscr{T})$ can be deduced by examining the configuration of tiles in a fundamental parallelogram of a periodic tiling $\mathscr{T}$, see Figure 3.7.4.

EXERCISES 3.4

1. Consider the tilings $\mathscr{T}$ described just before Statement 3.4.2.
   (*a*) Provide a proof for the assertion that such a tiling is balanced regardless of the widths of the "rings" chosen from the regular tiling (6^3) and from the tiling in Figure 3.4.1.
   (*b*) Give an estimate for the widths of the rings suitable to ensure that the ratio $t_6(r,P)/t(r,P)$ will oscillate indefinitely beyond $\frac{1}{4}$ and $\frac{3}{4}$.
   (*c*) Modify the construction described at the beginning of this section so as to obtain a strongly balanced tiling $\mathscr{T}$ with (*i*) $t_5(\mathscr{T}) = t_7(\mathscr{T}) = \frac{1}{4}$, $t_6(\mathscr{T}) = \frac{1}{2}$; (*ii*) $t_5(\mathscr{T}) = t_6(\mathscr{T}) = t_7(\mathscr{T}) = \frac{1}{3}$.
   (*d*) Find a strongly balanced tiling $\mathscr{T}$ which is periodic and has value of $t_j(\mathscr{T})$ as in (*i*), or (*ii*), of part (c).
2. Describe a 4-valent balanced tiling $\mathscr{T}$ which is not strongly balanced.
3. Describe the construction of a 3-valent balanced tiling $\mathscr{T}$ by convex quadrangles, hexagons and decagons which is not strongly balanced.
4. Construct a balanced tiling by triangles which is not strongly balanced.
5. Construct a balanced but not strongly balanced tiling in which all vertices are at most 5-valent, all tiles have at most 5 edges, but none of the limits

$$\lim_{r\to\infty} v_j(r,P)/t(r,P) \text{ and } \lim_{r\to\infty} t_j(r,P)/t(r,P) \quad (j = 3, 4, 5)$$

   exists.
6. Construct a strongly balanced periodic tiling $\mathscr{T}$ with $t_3(\mathscr{T}) = t_4(\mathscr{T}) = t_5(\mathscr{T}) = v_3(\mathscr{T}) = v_4(\mathscr{T}) = v_5(\mathscr{T}) = \frac{1}{3}$.
7. Construct a 3-valent strongly balanced tiling $\mathscr{T}$ with $t_6(\mathscr{T}) = 1$ such that $\mathscr{T}$ contains, besides hexagons, precisely: (*a*) one pentagon; (*b*) one heptagon; (*c*) two pentagons and two heptagons; (*d*) infinitely many pentagons and heptagons. (These tilings serve as counterexamples to the following assertion of M'Crea [1933]: "If in any system with three edges per vertex we know that every edge is either a side of a hexagon, or terminates at the vertex of a hexagon, it necessarily follows that *every* cell is a hexagon".)
8. Let $\mathscr{T}$ be a strongly balanced tiling. Prove that the limits $\lim_{r\to\infty} t_j(r,P)/v(r,P)$ and $\lim_{r\to\infty} v_j(r,P)/v(r,P)$ exist and calculate their values in terms of $t(\mathscr{T})$, $e(\mathscr{T})$, $v(\mathscr{T})$, $v_j(\mathscr{T})$, $t_j(\mathscr{T})$. Consider the analogous question with the terms $v(r,P)$ in the denominators replaced by $e(r,P)$.
9. Let $e_{i,j}(r,P)$ denote the number of edges in a patch $\mathscr{A}(r,P)$ with the property that one of the two tiles containing the edge has i adjacents and the other has j adjacents, and let $e_{i,j}(\mathscr{T}) = \lim_{r\to\infty} e_{i,j}(r,P)/t(r,P)$ if the limit exists.
   (*a*) For the tiling $\mathscr{T}$ shown in Figure 3.4.1 prove that the numbers $e_{i,j}(\mathscr{T})$ exist for all i, j, and determine them.
   (*b*) Construct a 4-valent strongly balanced tiling $\mathscr{T}$ with $e_{3,3}(\mathscr{T}) = \frac{1}{4}$, $e_{3,5}(\mathscr{T}) = 1$, $e_{5,5}(\mathscr{T}) = \frac{3}{4}$.
   (*c*) Construct a 4-valent strongly balanced tiling $\mathscr{T}$ for which $e_{3,4}(\mathscr{T})$ does not exist.
10. Construct a normal tiling $\mathscr{T}$ which satisfies condition B.3 but is not balanced.

*11. Construct a normal tiling $\mathscr{T}$ which satisfies conditions B.1 and B.3 but which is not strongly balanced.

12. Determine the numbers $t_i(\mathscr{T})$, $v_i(\mathscr{T})$ and $t_{[T_j]}(\mathscr{T})$ for the dihedral tilings in Figures 2.5.3(c), 2.5.4(*a*), (*h*), (*l*), 2.6.1(*a*), (*e*) and 2.6.2(*b*), and the trihedral tilings in Figures 2.5.4(*b*), 2.6.1(*f*) and 2.6.2(*d*). (Note that some of these tilings are not normal and have tiles with only two adjacents; in these cases $t_2(\mathscr{T}) > 0$.)

13. Compute the numbers $t_i(\mathcal{T})$, $v_i(\mathcal{T})$ and $t_{[T_j]}(\mathcal{T})$ for the 4-hedral prototile balanced tilings in Figures 2.0.1 and 2.5.9.

14. Determine the numbers $t_i(\mathcal{T})$, $v_i(\mathcal{T})$ and $t_{[T_j]}(\mathcal{T})$ for the tilings of Figures 2.4.4(*b*), 2.4.5(*b*), 2.4.6(*c*) and 2.4.7(*c*).

15. Let the figure formed by a vertex V of a tiling and the edges of $\mathcal{T}$ incident with V be called the *spider* of V. We say that $\mathcal{T}$ is k-gonal if each spider is congruent to one of k distinct ones.

    (*a*) Show that each tiling in Figures 2.4.4, 1.3.7, 1.3.8 and 2.4.5 is k-gonal for some value k, and determine k in each case.

    (*b*) If $S_1, \ldots, S_k$ are the k kinds of spiders in a k-gonal tiling $\mathcal{T}$, define the notion of *vertex-shape balanced* tilings and the numbers $V_{(S_j)}(\mathcal{T})$ in analogy to the definition of prototile balanced tilings and the numbers $t_{[T_j]}(\mathcal{T})$.

    (*c*) For each of the tilings listed in (a) decide whether the tiling is vertex-shape balanced and determine the numbers $V_{(S_j)}(\mathcal{T})$.

16. Let a tiling $\mathcal{T}$ be called *k-toxal* if among the edges of $\mathcal{T}$ there are precisely k noncongruent ones.

    (*a*) Show that each tiling in Figures 2.4.4, 2.5.6 and 2.7.1 is k-toxal for some k, and determine the value of k in each case.

    (*b*) If $E_1, E_2, \ldots, E_k$ are the k kinds of edges in a k-toxal tiling $\mathcal{T}$, define the notion of *edge-shape balanced* tilings and numbers $e_{\langle E_j\rangle}(\mathcal{T})$ in analogy to the definition of prototile balanced tilings and the numbers $t_{[T_j]}(\mathcal{T})$.

    (*c*) For each of the tilings mentioned in (*a*) decide whether the tiling is edge-shape balanced and determine the numbers $e_{\langle E_j\rangle}(\mathcal{T})$.

3.5 CONSEQUENCES OF EULER'S THEOREM

In this section we deduce a number of consequences of Euler's Theorem for Tilings.

3.5.1 *If every tile of a normal tiling $\mathcal{T}$ has k vertices, and these vertices have valences $j_1, \ldots, j_k$ in some order, then*

$$\sum_{i=1}^{k} \frac{j_i - 2}{j_i} = 2. \tag{3.5.2}$$

From Statement 3.4.2 we recall that under the assumed conditions the tiling $\mathcal{T}$ is strongly balanced. Denoting by m_i the number of i-valent vertices of each tile, $v_i(\mathcal{T}) = m_i/i$ by equation (3.4.4). From this we obtain

$$v(\mathcal{T}) = \sum_i v_i(\mathcal{T}) = \sum_i \frac{m_i}{i} = 1/j_1 + 1/j_2 + \ldots + 1/j_k.$$

Substituting in the Euler equation and recalling that $e(\mathcal{T}) = k/2$ since each tile has k adjacents, we obtain

$$1/j_1 + \cdots + 1/j_k = k/2 - 1. \tag{3.5.3}$$

A simple manipulation yields (3.5.2).

By a very similar argument we obtain the following statement (compare Statement 3.4.5):

3.5.4 *If every vertex of a normal tiling $\mathcal{T}$ has valence j, and is incident with tiles which have $k_1, \ldots, k_j$ adjacents, then*

$$\sum_{i=1}^{j} \frac{k_i - 2}{k_i} = 2. \tag{3.5.5}$$

It will be noticed that equations (3.5.2) and (3.5.5) are identical with equations (2.1.2). This fact will turn out to be of importance later. Note that whereas the derivation of equation (2.1.2) was *metrical* in that it depended on calculation of angles, equations (3.5.2) and (3.5.5) are of a *combinatorial* or topological nature in that their derivation depended only on valences and numbers of adjacents. (However, the assumption of normality can be regarded as a "hidden" metric condition.)

3.5.6 *In every strongly balanced tiling $\mathcal{T}$ we have*

$$2\sum_{j\geq 3}(j-3)v_j(\mathcal{T}) + \sum_{k\geq 3}(k-6)t_k(\mathcal{T}) = 0, \quad (3.5.7)$$

$$\sum_{j\geq 3}(j-4)v_j(\mathcal{T}) + \sum_{k\geq 3}(k-4)t_k(\mathcal{T}) = 0, \quad (3.5.8)$$

$$\sum_{j\geq 3}(j-6)v_j(\mathcal{T}) + 2\sum_{k\geq 3}(k-3)t_k(\mathcal{T}) = 0. \quad (3.5.9)$$

The derivation of all these relations is similar, so we shall prove only (3.5.7) in detail. In any patch $\mathcal{A}(r,P)$ a counting argument shows that $2e(r,P)$ is approximately equal to $\sum_{j\geq 3} jv_j(r,P)$, and also to $\sum_{k\geq 3} kt_k(r,P)$. Taking the limits as $r \to \infty$, and using the techniques of the previous section (which are applicable since $\mathcal{T}$ is strongly balanced and hence normal) we obtain

$$2e(\mathcal{T}) = \sum_{j\geq 3} jv_j(\mathcal{T}) \quad (3.5.10)$$

and

$$2e(\mathcal{T}) = \sum_{k\geq 3} kt_k(\mathcal{T}). \quad (3.5.11)$$

Adding twice equation (3.5.10) to (3.5.11) and using Euler's Theorem for Tilings on the left hand side, we obtain

$$6v(\mathcal{T}) + 6 = 2\sum_{j\geq 3} jv_j(\mathcal{T}) + \sum_{k\geq 3} kt_k(\mathcal{T}). \quad (3.5.12)$$

However, as we have already mentioned,

$$\sum_{j\geq 3} v_j(\mathcal{T}) = v(\mathcal{T}) \quad \text{and} \quad \sum_{k\geq 3} t_k(\mathcal{T}) = 1.$$

Substituting these values in the left side of equation (3.5.12) immediately yields equation (3.5.7). To obtain (3.5.9) we apply the same process except that we add (3.5.10) to twice (3.5.11); for (3.5.8) we add (3.5.10) and (3.5.11). (We may note that these computations are completely analogous to the more familiar ones dealing with convex polyhedra; see, for example, Grünbaum [1967, Section 13.3], [1975].)

Equations (3.5.7), (3.5.8) and (3.5.9) have many consequences. For example, since the valence of every vertex is at least three, (3.5.7) implies that $\sum_{k\geq 3}(k-6)t_k(\mathcal{T}) \leq 0$. Hence if every tile in a strongly balanced tiling has k adjacents, then $k \leq 6$. If, on the other hand, every vertex has valence 3, then $\sum_{k\geq 3}(k-6)t_k(\mathcal{T}) = 0$, and so if every tile has k adjacents then $k = 6$. Conversely, if every tile has 6 adjacents then 'nearly all' vertices have valence 3. By this we mean that $v_3(\mathcal{T}) = 2$ and $v_j(\mathcal{T}) = 0$ for all $j > 3$. This situation prevails if, for example, there are only a finite number of vertices of valence greater than 3.

Observing that equations (3.5.7) and (3.5.9) result from one another by interchanging the roles of v_j and t_k, similar remarks apply if "valences" and "number of adjacents" are interchanged. In particular, if every vertex in a strongly balanced tiling has the same valence j, then $j = 3, 4, 5$ or 6. Also from equation (3.5.8) we deduce that if the tiling is 4-valent and all tiles have k adjacents, then $k = 4$.

3.5.13 *For each strongly balanced tiling $\mathcal{T}$ we have*

$$\frac{1}{\sum_{j\geq 3} jw_j(\mathcal{T})} + \frac{1}{\sum_{k\geq 3} kt_k(\mathcal{T})} = \frac{1}{2} \quad (3.5.14)$$

where

$$w_j(\mathcal{T}) = v_j(\mathcal{T})/v(\mathcal{T}).$$

Thus $w_j(\mathcal{T})$ can be interpreted as that fraction of the total number of vertices in $\mathcal{T}$ which have valence j, and $\sum_{j\geq 3} jw_j(\mathcal{T})$ is the *average* valence taken over all the vertices. Since $\sum_{k\geq 3} t_k(\mathcal{T}) = 1$ there is a similar interpretation of $\sum_{k\geq 3} kt_k(\mathcal{T})$: it is the *average* number of adjacents of the tiles, taken over all the tiles in $\mathcal{T}$. The fact that the tiling is strongly balanced ensures, of course, that these averages exist and are uniquely defined.

To prove Statement 3.5.13, we use equations (3.5.10) and (3.5.11) to obtain

$$\frac{\sum_{k\geq 3} kt_k(\mathcal{T})}{\sum_{j\geq 3} jv_j(\mathcal{T})} = \frac{2e(\mathcal{T})}{2e(\mathcal{T})} = 1,$$

and hence, since $w_j(\mathcal{T}) = v_j(\mathcal{T})/v(\mathcal{T})$,

$$\frac{\sum_{k\geq 3} kt_k(\mathcal{T})}{\sum_{j\geq 3} jw_j(\mathcal{T})} = v(\mathcal{T}) = e(\mathcal{T}) - 1.$$

The last equality comes from Euler's Theorem for Tilings. Substituting again from (3.5.11), we get

$$\frac{\sum_{k\geq 3} kt_k(\mathcal{T})}{\sum_{j\geq 3} jw_j(\mathcal{T})} = \frac{1}{2}\sum kt_k(\mathcal{T}) - 1.$$

from which (3.5.14) follows immediately.

Equation (3.5.14) often proves useful when one wishes to construct a normal tiling which satisfies certain specifications. For example, suppose we ask for a 3-valent tiling in which every tile has 5 or 7 edges. Since then $w_3 = 1$, $w_j = 0$ for $j \geq 4$, and $t_k = 0$ for $k \neq 5, 7$, equation 3.5.14 implies $5t_5 + 7t_7 = 6$, which together with $t_5 + t_7 = 1$ yields $t_5 = t_7 = \frac{1}{2}$. It follows that the two kinds of tiles must occur in equal numbers, and so it is natural to attempt the construction by using a "hexagon" composed of a pentagon and a heptagon in juxtaposition (see Figure 3.5.1(*b*)). It is then easy to discover the tiling of Figure 3.5.1(*a*) which satisfies all our requirements. (A slight distortion of this tiling yields the one shown in Figure 3.4.1.)

A number of other tilings, with various values of $t_k(\mathcal{T})$ and $w_j(\mathcal{T})$ are shown in Figure 3.5.2. We note that

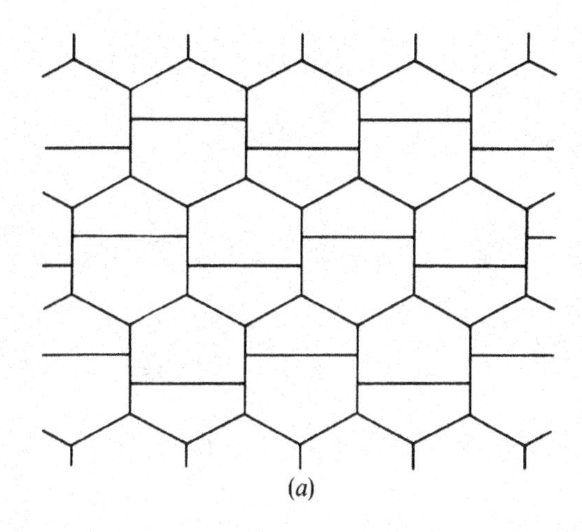

(*a*)

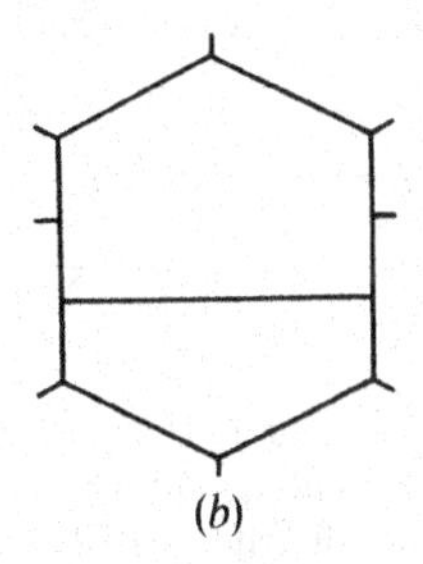

(*b*)

Figure 3.5.1
A 3-valent dihedral tiling using pentagons and heptagons in equal numbers. The construction is based on the hexagon shown in (*b*). This is the union of one pentagon and one heptagon.

(3.5.14) is a necessary but not sufficient condition for the existence of a strongly balanced tiling with specified w_i and t_i. For example, the values $w_3(\mathcal{T}) = 1$, $t_3(\mathcal{T}) = \frac{7}{10}$, $t_{13}(\mathcal{T}) = \frac{3}{10}$, with all other $w_j(\mathcal{T}) = t_k(\mathcal{T}) = 0$, satisfy equation (3.5.14). Moreover, since $v_3(\mathcal{T}) = 2$ and all $v_j(\mathcal{T}) = 0$ for $j > 3$, equations (3.5.7), (3.5.8) and (3.5.9) are also satisfied. But it is easy to see that no corresponding strongly balanced tiling exists.

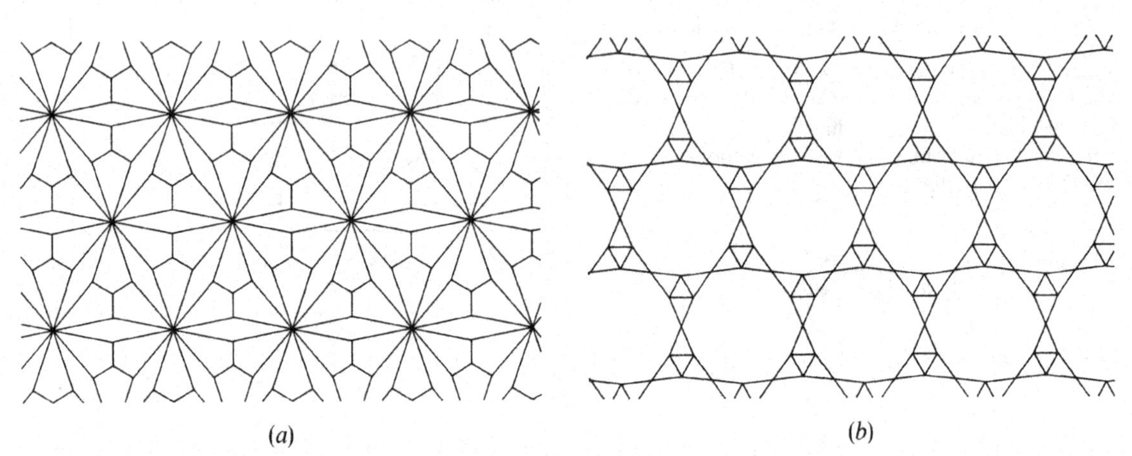

Figure 3.5.2
Two strongly balanced periodic tilings which are dual to each other. In (a) $t_4(\mathcal{T}) = 1$, $w_3(\mathcal{T}) = \frac{8}{9}$, $w_{12}(\mathcal{T}) = \frac{1}{9}$ and all other $t_i(\mathcal{T})$ and $w_i(\mathcal{T})$ are zero. In (b) $t_3(\mathcal{T}) = \frac{8}{9}$, $t_{12}(\mathcal{T}) = \frac{1}{9}$. $w_4(\mathcal{T}) = 1$ and all other $t_i(\mathcal{T})$ and $w_i(\mathcal{T})$ are zero.

EXERCISES 3.5

1. Verify the statement made at the end of the section that there exists no tiling with $w_3(\mathcal{T}) = 1$, $t_3(\mathcal{T}) = \frac{7}{10}$, $t_{13}(\mathcal{T}) = \frac{3}{10}$, and all other $w_j(\mathcal{T}) = t_k(\mathcal{T}) = 0$.
2. Find strongly balanced tilings with the following values of $w_j(\mathcal{T})$ and $t_k(\mathcal{T})$:
   (a) $w_3 = w_4 = \frac{1}{2}$, $t_3 = t_4 = t_7 = \frac{1}{3}$;
   (b) $w_3 = \frac{4}{13}$, $w_4 = \frac{7}{13}$, $w_6 = \frac{2}{13}$, $t_4 = 1$;
   and all other $w_j = t_k = 0$.

3.6 METRICALLY BALANCED TILINGS

The first four sections of this chapter were concerned with the restrictions that it is necessary to place on a tiling so that Euler's Theorem for Tilings can be stated and proved. As we saw in Section 3.5 this powerful theorem has many implications and it will prove an essential tool in the sequel. Even so, strongly balanced tilings can behave in unexpected ways; the present section is devoted to a discussion of some of these.

The first question we shall consider is this. Why were patches $\mathcal{A}(r,P)$ based on circular disks used in the definitions? We recall from Section 1.1 that any topological disk can be used to define a patch. What would happen if, for example, we used *squares* instead of circular disks? The answer is that not only could the limits corresponding to (3.3.1) take different values from those that arise using circular disks, but in certain instances the limits might not exist at all. Moreover, in the case of squares the values of the limits might depend on the particular orientations of squares that are chosen.

To illustrate these assertions, consider the tiling $\mathscr{T}$ shown schematically in Figure 3.6.1. The plane is divided into eight sectors radiating from a point P, as shown. For simplicity we shall take P as the center of all the circular disks and squares under consideration. Suppose that four of the sectors, shaded in the diagram, are tiled by equilateral triangles meeting six at each vertex, and the other four sectors are tiled by regular hexagons meeting three at each vertex. We choose the tiles to be of equal area; the way they "join up" along the boundaries of the sectors is unimportant. We recall from Section 3.3 that for the regular tiling $\mathscr{T}_1$ by equilateral triangles we have $v(\mathscr{T}_1) = \frac{1}{2}$, and for the regular tiling $\mathscr{T}_2$ by hexagons $v(\mathscr{T}_2) = 2$. Since any circular disk centered at P contains equal areas covered by each of these two regular tilings, it is reasonable to suppose that

$$v(\mathscr{T}) = \tfrac{1}{2}\cdot\tfrac{1}{2} + \tfrac{1}{2}\cdot 2 = \tfrac{5}{4} = 1.25.$$

This can be rigorously proved: to do so we must use the Normality Lemma (Statement 3.2.2) to show that the tiles round the boundary of the disk, as well as the tiles overlapping two or more sectors, can be ignored in calculating the limit.

Suppose on the other hand, we consider squares oriented like $S(r)$ of Figure 3.6.1. Then a simple calculation shows that $\sqrt{2} - 1 = 0.414\ldots$ of its area is shaded (and so is like the tiling $\mathscr{T}_1$) whereas $2 - \sqrt{2} = 0.586\ldots$ of its area is unshaded (and so is like the tiling $\mathscr{T}_2$). In this way the corresponding limit can be shown to be

$$(\sqrt{2} - 1)\cdot\tfrac{1}{2} + (2 - \sqrt{2})\cdot 2 = \tfrac{1}{2}(7 - 3\sqrt{2}) = 1.379\ldots.$$

The analogous calculation for squares oriented like $S'(r)$ yields

$$(2 - \sqrt{2})\cdot\tfrac{1}{2} + (\sqrt{2} - 1)\cdot 2 = \tfrac{1}{2}(3\sqrt{2} - 2) = 1.121\ldots.$$

In a similar way we can construct tilings with the property that for one shape of patch a limit exists, whereas for another shape or another orientation the limit does not exist.

Further problems can arise even if we restrict attention to patches based on circular disks. From Section 3.3 we

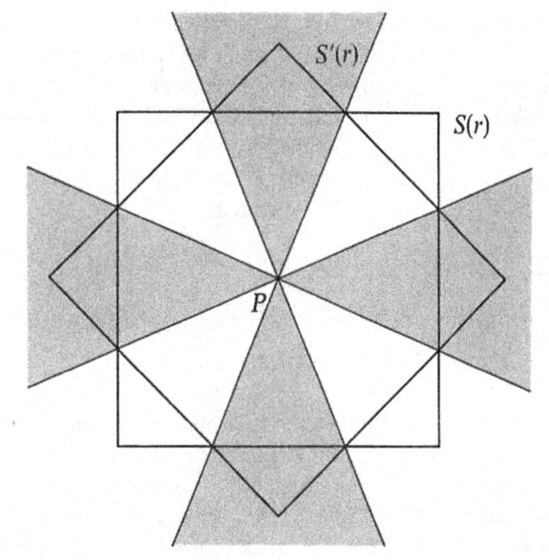

Figure 3.6.1
A schematic representation of a tiling for which the limits used in the definition of balance depend on the shapes of the disks used to define the patches.

recall that the value of the limits in (3.3.1) do not depend on the choice of the center P of the disk (Statement 3.3.2). On the other hand this assertion is not true if we allow *P to move while r is increasing*. For consider again the tiling indicated in Figure 3.6.1. It is possible to let a disk increase in size and, at the same time, move in such a way that it remains in one of the eight sectors. The values of the limits will then depend upon whether a shaded or unshaded sector is chosen.

It would be more satisfactory if we could frame our definitions in such a way that the various limits that arise depend *only* on the tiling and *not* on the shape of the patches, their orientation, or how they move as $r \to \infty$. To this end we introduce the much stronger condition of metrical balance. The intuitive idea of a metrically balanced tiling is that it is one for which it is sensible to calculate the numbers of tiles, edges and vertices *per unit area of the plane*.

We must introduce some notation. Let us choose axes in the plane and consider a square $S(r)$ of size $2r \times 2r$

with sides parallel to axes. We define a patch $\mathscr{A}(S(r),\mathscr{T})$ based on $S(r)$ as in Section 1.1 and we write $t_S(r)$, $e_S(r)$ and $v_S(r)$ for the numbers of tiles, edges and vertices in $\mathscr{A}(S(r),\mathscr{T})$. Then the fundamental definition is as follows:

A normal tiling $\mathscr{T}$ is called *metrically balanced* if, given any positive number ε, however small, there exist numbers R, $V(\mathscr{T})$, $E(\mathscr{T})$ and $T(\mathscr{T})$ such that

$$\left.\begin{aligned}(1-\varepsilon)T(\mathscr{T}) \leq t_S(r)/4r^2 \leq (1+\varepsilon)T(\mathscr{T})\\(1-\varepsilon)E(\mathscr{T}) \leq e_S(r)/4r^2 \leq (1+\varepsilon)E(\mathscr{T})\\(1-\varepsilon)V(\mathscr{T}) \leq v_S(r)/4r^2 \leq (1+\varepsilon)V(\mathscr{T})\end{aligned}\right\} \quad (3.6.1)$$

for all $r > R$. The number R depends only on the tiling $\mathscr{T}$ and the value of ε, but *not* on the position of the square $S(r)$. The definition states, in effect, that quantities like $t_S(r)/4r^2$ (the number of tiles per unit area) tend to a definite limit $T(\mathscr{T})$ as $r \to \infty$, and moreover this limit is independent of the position of the square in the tiling. As we shall show later, this is sufficient for a satisfactory solution of the question raised at the beginning of this section.

We can easily see that the tiling of Figure 3.6.1 is not metrically balanced. An arbitrarily large square can be taken in either the shaded or unshaded regions, and it is clear that then $e_S(r)/4r^2$ and $v_S(r)/4r^2$ cannot approach any fixed limits $E(\mathscr{T})$ and $V(\mathscr{T})$ as $r \to \infty$. On the other hand, in the next section we will show that all normal periodic tilings are metrically balanced. Other examples can be constructed from normal periodic tilings by introducing modifications such as splitting a tile up into smaller tiles, or removing edges so as to "weld" two or more tiles into a larger one. Then so long as we do not introduce too many modifications, the tiling will remain metrically balanced. In particular, this will be so if only a *finite* number of tiles in the original periodic tiling is affected. (A slightly stronger result can be proved, see Exercise 3.6.2.)

Constructions similar to that just described yield all the examples of tilings that we *know* to be metrically balanced, apart from the aperiodic tilings to be described in Chapter 10. It is an interesting problem whether other types of metrically balanced tilings exist.

A metrically balanced tiling is necessarily balanced. However, non-normal tilings such as that shown in Figure 3.3.2 also satisfy inequalities (3.6.1). It is convenient to eliminate such tilings from consideration by including the condition of normality in the definition of metrical balance.

For a set K and a positive real number r let rK denote the set consisting of all points rx with x in K (thus rK is similar to, and has the same orientation as K, but its position depends on the location of the origin); let $a(rK)$ be the area of rK and let $v(rK)$, $e(rK)$ and $t(rK)$ be the numbers of vertices, edges and tiles in the patch $\mathscr{A}(rK,\mathscr{T})$ generated in the tiling $\mathscr{T}$ by rK. Clearly $a(rK) = r^2a(K)$.

The main result of this section is the following statement, in which the numbers $T(\mathscr{T})$, $E(\mathscr{T})$ and $V(\mathscr{T})$ are those defined by relations (3.6.1).

3.6.2 *Let $\mathscr{T}$ be any metrically balanced tiling and let K be any plane convex region. Then*

$$\lim_{r\to\infty} t(rK)/a(rK) = T(\mathscr{T}),$$

$$\lim_{r\to\infty} e(rK)/a(rK) = E(\mathscr{T})$$

and

$$\lim_{r\to\infty} v(rK)/a(rK) = V(\mathscr{T}).$$

From Statement 3.6.2 we deduce immediately that although $T(\mathscr{T})$, $E(\mathscr{T})$ and $V(\mathscr{T})$ were defined in terms of squares with a given orientation, the same limits arise if squares of *any* orientation are used.

The proof of Statement 3.6.2 depends upon the following fact. For any convex set K it is possible to find unions of squares $S_i(K)$ and $S_o(K)$, such that

$$S_i(K) \subset K \subset S_o(K), \quad (3.6.3)$$

which approximate K arbitrarily closely (see Figure 3.6.2). By this we mean that if the small squares comprising $S_i(K)$ and $S_o(K)$ are of edge-length θ, then we can make $a(K) - a(S_i(K))$ and $a(S_o(K)) - a(K)$ as small as

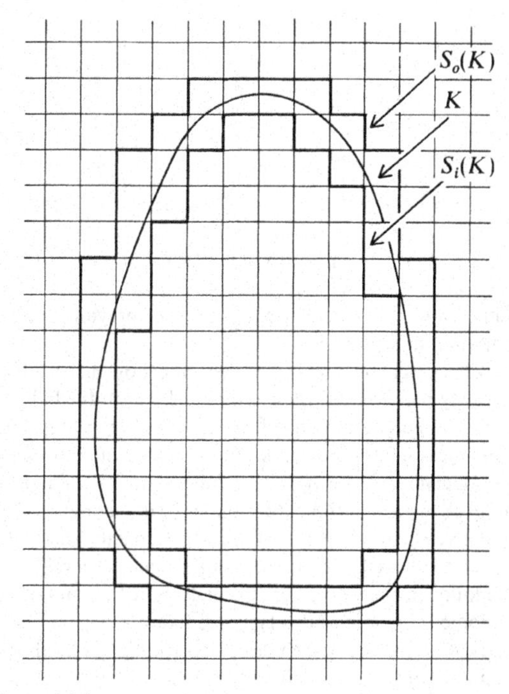

Figure 3.6.2
Approximation of a convex set K by unions of squares. $S_i(K)$ is the inner approximation and $S_o(K)$ is the outer one. The edge length of the squares is θ.

we wish by choosing θ sufficiently small. An alternative formulation of this property is as follows: for any positive value of η, however small, we can find a value of θ such that

$$(1 - \eta)a(S_o(K)) \leq a(K) \leq (1 + \eta)a(S_i(K)). \quad (3.6.4)$$

Here θ must be taken small relative to K, so the same effect is achieved by keeping the squares fixed in size, and replacing K by rK for a large value of r. Thus relation (3.6.4) is equivalent to saying that for any $\eta > 0$ the inequalities

$$(1 - \eta)a(S_o(rK)) \leq a(rK) \leq (1 + \eta)a(S_i(rK)) \quad (3.6.5)$$

hold for all sufficiently large values of r.

Now consider $t(rK)$, the number of tiles in the patch $\mathscr{A}(rK, \mathscr{T})$. To estimate its value we first note that every such tile must lie within a distance $2U$ of one of the squares of $S_i(rK)$, where U is the circumparameter of $\mathscr{T}$. (In fact every tile that meets rK necessarily meets one of the squares of $S_0(rK)$, but we recall that additional tiles may have to be adjoined to make $\mathscr{A}(rK,\mathscr{T})$ simply connected. Every such tile will be within a distance $2U$ of $S_o(rK)$.) Clearly then

$$t(rK) \leq n_o t_S(r\theta + 2U) \quad (3.6.6)$$

where n_o is any upper bound on the number of squares of side $r\theta$ in $S_o(rK)$. On the other hand, the tiles that belong to $\mathscr{A}(rK,\mathscr{T})$ certainly include the t tiles that meet $S_i(rK)$. To estimate t we remark that if $r\theta > 2U$ and if we replace each square $S(r\theta)$ in $S_i(rK)$ by a smaller concentric square $S(r\theta - 2U)$, then

$$n_i t_S(r\theta - 2U) \leq t \leq t(rK) \quad (3.6.7)$$

where n_i is the number of such squares. (They have to be decreased in size in this manner to avoid counting more than once the tiles that meet two or more of the original squares.) From relations (3.6.5), (3.6.6) and (3.6.7) we obtain

$$\frac{n_i t_S(r\theta - 2U)}{(1 + \eta)a(S_i(rK))} \leq \frac{t(rK)}{a(rK)} \leq \frac{n_o t_S(r\theta + 2U)}{(1 - \eta)a(S_o(rK))}. \quad (3.6.8)$$

The left term of (3.6.8) can be written

$$\frac{4n_i(r\theta)^2}{a(S_i(rK))} \cdot \left(\frac{r\theta - 2U}{r\theta}\right)^2 \cdot \frac{1}{1 - \eta} \cdot \frac{t_S(r\theta - 2U)}{4(r\theta - 2U)^2}.$$

Now $4n_i(r\theta)^2 = a(S_i(rK))$ and so the first factor is 1. As $r \to \infty$ the second factor tends to the value 1, and the third may be chosen arbitrarily near 1; by relations (3.6.1) the last factor tends to $T(\mathscr{T})$.

In a similar manner, the right term of (3.6.8) can be written

$$\frac{4n_o(r\theta)^2}{a(S_o(rK))} \cdot \left(\frac{r\theta + 2U}{r\theta}\right)^2 \cdot \frac{1}{1 + \eta} \cdot \frac{t_S(r\theta + 2U)}{4(r\theta - 2U)^2}.$$

Here the first factor is 1, the second tends to 1, the third may be chosen arbitrarily close to 1, and the last factor tends to $T(\mathscr{T})$ as $r \to \infty$. Hence relations (3.6.8) yield

$$\lim_{r\to\infty} t(rK)/a(rK) = T(\mathscr{T}).$$

In an exactly similar manner the corresponding results for $e(rK)$ and $v(rK)$ can be proved, and thus the proof of Statement 3.6.2 is completed.

In the above we have considered the case where K is convex. This restriction is introduced for convenience only; much more general sets can be used. The essential feature is that K should be approximable by unions of squares so that inequalities (3.6.4) hold. Sets with this property are called Jordan measurable. They include, for example, all sets which can be written as a finite union of convex sets.

The quantities $T(\mathscr{T})$, $E(\mathscr{T})$ and $V(\mathscr{T})$ are, for metrically balanced tilings, functions of $\mathscr{T}$ only. Thus we have achieved the goal stated at the beginning of the section. For a metrically balanced tiling, Euler's Theorem for Tilings becomes

$$T(\mathscr{T}) - E(\mathscr{T}) + V(\mathscr{T}) = 0. \tag{3.6.9}$$

This is an immediate consequence of the previous formulation in Section 3.3.

Without giving details it is clear now how metrically strongly balanced tilings and metrically prototile balanced tilings can be defined; in effect they are tilings for which it is sensible to speak of the number of vertices with a given valence and the number of tiles with a given number of adjacents, or the number of tiles congruent to a given prototile, per unit area of the plane. These are the best behaved of all tilings! For examples we refer to the following section where we shall show that all normal periodic tilings are of this type, see Exercise 3.7.4. Further, by adapting the idea described on page 144, other examples of strongly metrically balanced tilings can be constructed.

EXERCISES 3.6

1. Show that the tilings in Figures 1.1.3, 1.3.5, 1.3.8 and 2.4.5 are metrically balanced. Taking the length of the shortest edge as unit, compute $V(\mathscr{T})$, $E(\mathscr{T})$ and $T(\mathscr{T})$ for each. (In Figure 2.4.5 assume the squares have edge lengths 1, 2, 3.)

2. (*a*) Show that there exist tilings $\mathscr{T}'$, obtainable from a regular square tiling $\mathscr{T}$ by splitting infinitely many of the squares of $\mathscr{T}$ into four smaller squares each, and such that $V(\mathscr{T}) = V(\mathscr{T}')$, $E(\mathscr{T}) = E(\mathscr{T}')$, $T(\mathscr{T}) = T(\mathscr{T}')$.
   (*b*) Formulate and prove a result analogous to that of part (*a*) but applicable to every metrically balanced tiling $\mathscr{T}$.

3. (*a*) Define metrically strongly balanced tilings and the quantities $T_i(\mathscr{T})$ and $V_k(\mathscr{T})$.
   (*b*) Define metrically prototile balanced tilings and the quantities $T_{[T_j]}(\mathscr{T})$.
   (*c*) Give examples of metrically balanced tilings that are not metrically strongly balanced, and of metrically strongly balanced tilings that are not metrically prototile balanced.
   (*d*) Show that all the tilings mentioned in Exercise 3.6.1 are metrically strongly balanced, and metrically prototile balanced, and calculate the quantities $T_k(\mathscr{T})$ and $T_{[T_j]}(\mathscr{T})$ for each.
   (*e*) Formulate and prove the analogue of Statement 3.5.6 for metrically strongly balanced tilings.

3.7 PERIODIC TILINGS

We recall from Section 1.3 that a tiling $\mathcal{T}$ is called *periodic* if there exist, among the symmetries of $\mathcal{T}$, at least two translations in non-parallel directions. This implies that we may associate with $\mathcal{T}$, in infinitely many ways, a parallelogram tiling $\mathcal{P}$ (see Exercise 1.1.10) so that the parts of $\mathcal{T}$ in each tile of $\mathcal{P}$ are congruent and can be made to coincide by the translations that carry the tiles of $\mathcal{P}$ onto each other.

As in Section 1.4, to each translation τ which is a symmetry of $\mathcal{T}$, we associate a vector a. Then it is convenient to write $\lambda\tau$ for the translation corresponding to λa, and $\tau_1 + \tau_2$ for the translation $a_1 + a_2$, where τ_1 and τ_2 are given translational symmetries. If λ is an integer then both $\lambda\tau$ and $\tau_1 + \tau_2$ are symmetries of $\mathcal{T}$.

3.7.1 *Let $\mathcal{S}$ be a finite set of prototiles and suppose that each of these prototiles is a polygon. Suppose further that $\mathcal{S}$ admits an edge-to-edge tiling $\mathcal{T}_1$ which possesses a translation τ_1 as a symmetry. Then $\mathcal{S}$ admits a periodic tiling.*

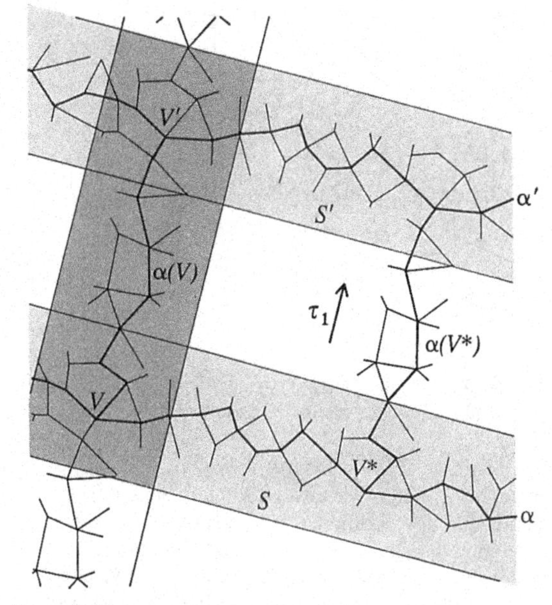

Figure 3.7.1
This illustrates the proof of the periodicity of the tilings considered in Statement 3.7.1.*

By a polygon we mean a topological disk bounded by a finite number of straight-line segments. Extending the terminology of Chapter 2, by an *edge-to-edge* tiling we mean that the vertices of the tiling lie only at the corners of the tiles and do not lie elsewhere on their sides.

We shall first prove Statement 3.7.1 and then discuss how it can be generalized.

We begin by observing that the tiling $\mathcal{T}_1$ necessarily satisfies conditions N.1 and N.3 for a normal tiling (see Section 3.2), the latter because $\mathcal{S}$ is finite. Hence we can define a circumparameter U. In a direction perpendicular to τ_1 construct a closed infinite strip S of width $2U$ as shown in Figure 3.7.1. It is easy to see that S contains an infinite edge-path α running along S, since no tile can cut completely across S. We now consider a second strip S' which is the image of S under a translation $\lambda\tau_1$. Here λ is any positive integer chosen so large that S and S' are disjoint. Then $\lambda\tau_1\alpha = \alpha'$ is an edge-path in S', congruent to and disjoint from α. For each vertex V on α construct a closed strip S'', perpendicular to S, of width $2U$ and with V on its midline. Then the image $V' = \lambda\tau_1 V$ of V will lie in $S' \cap S''$. Let $R(V)$ be the rectangle shown darkly shaded in the diagram. There will exist at least one edge-path from V to V' lying entirely in $R(V)$. Choose any such path and denote it by $\alpha(V)$. All the rectangles $R(V)$ are of equal size and each meets only a finite number of tiles, so it is clear that the number of distinct possible edge-paths that can join a vertex on α to its image on α' is finite. However, there are infinitely many vertices on α, from which we deduce that we can find two distinct vertices, say V and V^*, such that the edge-paths $\alpha(V)$ and and $\alpha(V^*)$ are congruent under some translation τ_2.

Now consider the patch of tiles bounded by $\alpha(V)$, $\alpha(V^*)$, the part of α lying between V and V^*, and the part of α' between V' and $V^{*\prime}$. It is obvious that this patch can be translated by $\lambda\tau_1$, by τ_2, and by linear combinations of these to yield a periodic tiling of the plane. This completes the proof of Statement 3.7.1.

* See the Appendix beginning on page 653.

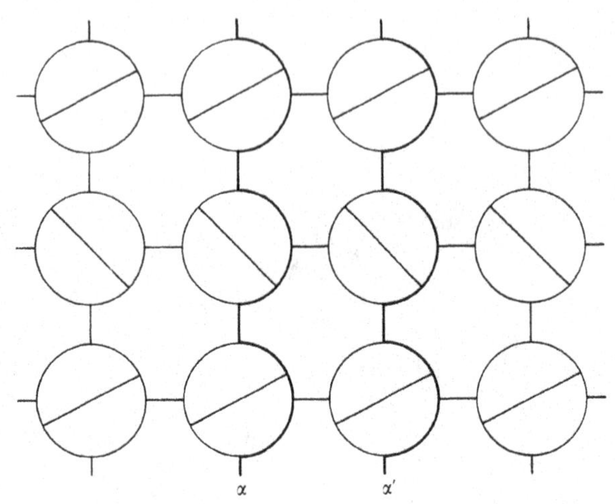

Figure 3.7.2
Modifications to the proof of Statement 3.7.1 are required in a tiling like this where "sliding" is permitted.

Statement 3.7.1 remains true if we allow the tiles to have curved edges. The only reason for using polygonal tiles is that the condition of being edge-to-edge seems natural in this case. The edge-to-edge condition eliminates the possibility of one tile "sliding" relative to an adjacent tile, that is, of assuming an infinite number of different relative positions. If this is allowed then we cannot assert that there exist only a finite number of incongruent edge-paths $\alpha(V)$, and the proof fails. In certain cases it may be possible to complete it using an auxiliary argument. For example, in Figure 3.7.2 we have a tiling where pairs of tiles fit together to form circular disks. These can be rotated to any position, and so the edge-paths $\alpha(V)$ may be all different. But it is clear how, in a case like this, a slight modification of the tiling $\mathcal{T}_1$ will enable the proof to be completed.

There are cases where no such modification appears to be possible. In Figure 3.7.3 we show a tiling using squares of two different sizes, the ratio of their sides being *irrational*. Then although the assertion of Statement 3.7.1 is true, the proof fails completely and it is not easy to see how it can be adapted to deal with this case—for here the tiles must be radically rearranged, using tiles of a single size, to obtain a periodic tiling.

Whether Statement 3.7.1 remains true if we allow as prototiles any topological disks and if we remove the edge-to-edge requirement is an open question. However, with minor modifications to the proof, Statement 3.7.1 can be strengthened as follows.

3.7.2 *Suppose that $\mathcal{S}$ is a set of prototiles of the type specified in Statement 3.7.1. Then the tiling $\mathcal{T}_1$ contains arbitrarily large patches of tiles each of which will generate a periodic tiling by translations in two non-parallel directions.*

This cannot be generalized if "sliding" is allowed, as the example of Figure 3.7.3 shows.

A periodic tiling need not be normal since it may possess singular points (see, for example, Figure 3.1.2), or the intersection of two tiles need not be connected (see Figure 3.2.2). For the remainder of this section, however, we shall implicitly assume that all the periodic tilings which we shall consider are normal. Further, in order to simplify the exposition, we shall assume that whenever a periodic tiling $\mathcal{T}$ is under consideration, the corresponding parallelogram tiling $\mathcal{P}$ has been chosen in such a way that no vertex of $\mathcal{T}$ lies on an edge of $\mathcal{P}$, and

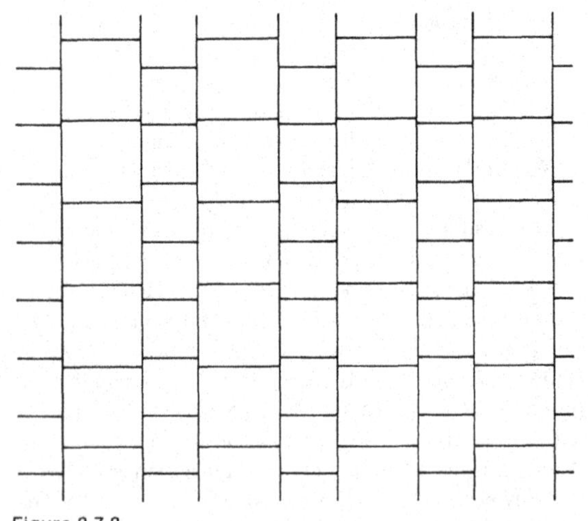

Figure 3.7.3
A tiling for which the proof of Statement 3.7.1 fails completely. The two squares have edge-lengths for which the ratio is irrational.

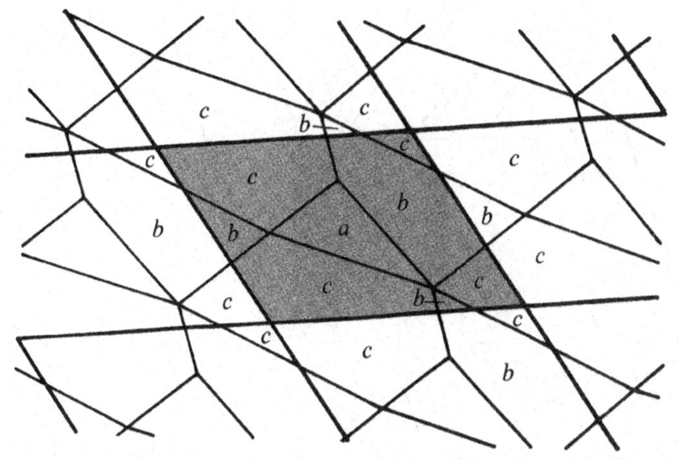

Figure 3.7.4
A periodic tiling with period parallelogram P. In this tiling $T = 3$ because the various fractions of the tiles inside P bearing the same letters can be fitted together to form three whole tiles. Similarly $V = 3$ and $E = 6$.

no vertex of $\mathscr{P}$ lies on an edge of $\mathscr{T}$. This assumption can clearly be fulfilled; for if either condition is violated then an arbitrarily small translation of the period parallelogram P (and correspondingly of $\mathscr{P}$) can be found which will satisfy the required conditions.

We now define three numbers V, E and T as follows. V is the number of vertices of $\mathscr{T}$ in P, E is the number of edges of $\mathscr{T}$ in P (which we can determine *either* as one-half of the sum of the valences of the vertices in P, *or* as the sum of the fractions of the various edges lying in P counted appropriately), and T is the number of tiles of $\mathscr{T}$ in P where, again, we count fractions appropriately, see Figure 3.7.4. The main result is the following:

3.7.3 *Every normal periodic tiling $\mathscr{T}$ is metrically balanced. Moreover*

$$V(\mathscr{T}) = V/A,\ E(\mathscr{T}) = E/A,\ T(\mathscr{T}) = T/A,$$

where A is the area of the parallelogram P, and $V(\mathscr{T})$, $E(\mathscr{T})$, $T(\mathscr{T})$ are the numbers defined in Statement 3.6.2.

It follows immediately from this statement that Euler's Theorem for Tilings takes the form

$$V - E + T = 0. \tag{3.7.4}$$

A proof of Statement 3.7.3 will be given below. But if we accept a result from elementary topology, a very simple proof of equation (3.7.4) is as follows.

We take the parallelogram P and identify each point on its boundary with the corresponding point on the opposite side. It is well known that we obtain, in this way, a torus (the surface of an American doughnut or a tire inner tube) and then the equation $V - E + T = 0$ coincides with Euler's Theorem for the torus (see, for example, Singer & Thorpe [1967, Section 6.1] or Coxeter [1961, Section 20.4]).

For a direct proof of Statement 3.7.3 we adapt the proof of Statement 3.6.2. It is clear that it is not necessary for the axes to be at right angles, and therefore we may use parallelograms instead of squares. This is sometimes expressed by saying that the concepts involved are invariant under affinities (see Section 1.4).

Let $P(r)$ be a parallelogram similar to one of the tiles P of $\mathscr{P}$ with edges parallel to those of P, and each edge multiplied by the number r. For each such parallelogram $P(r)$ we may define a patch as in Section 1.1 and write $v_P(r)$, $e_P(r)$ and $t_P(r)$ for the numbers of vertices, edges and tiles in this patch. Then our aim is to prove that

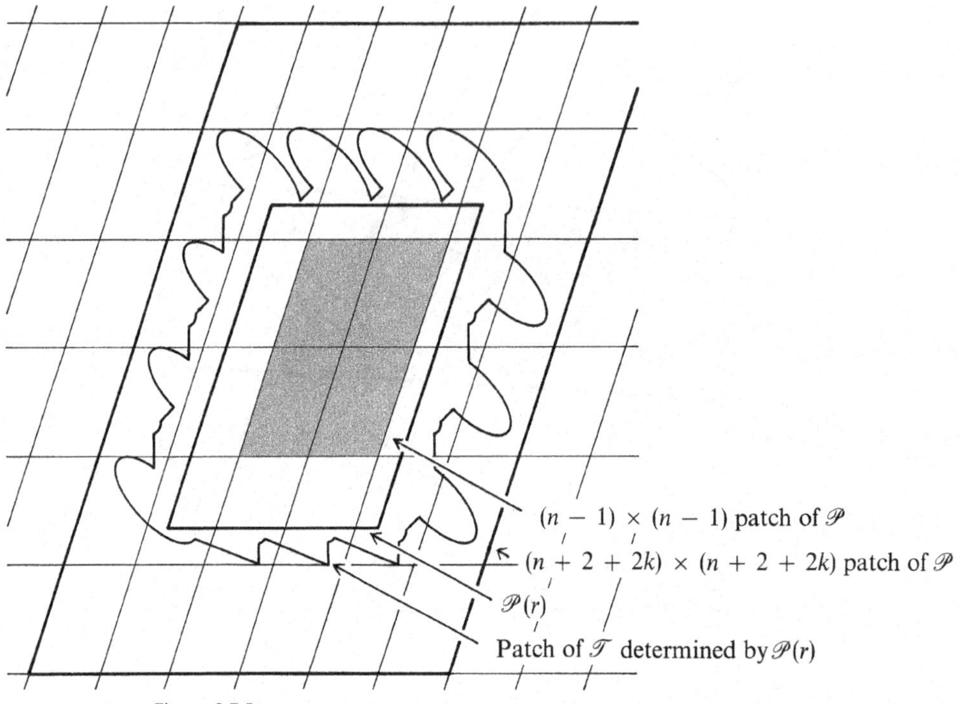

Figure 3.7.5
This illustrates the proof of the fact that a normal periodic tiling is metrically balanced. For clarity the individual tiles of $\mathscr{T}$ are not shown, but the parallelograms of $\mathscr{P}$ are indicated.

$$\left.\begin{aligned} \lim_{r\to\infty} v_P(r)/r^2A &= V/A \\ \lim_{r\to\infty} e_P(r)/r^2A &= E/A \\ \lim_{r\to\infty} t_P(r)/r^2A &= T/A. \end{aligned}\right\} \qquad (3.7.5)$$

By the above remarks, the suggested modification of Statement 3.7.2 will then show that the limits (3.7.5) exist and take the appropriate values; hence the tiling $\mathscr{T}$ will be metrically balanced.

Now to prove relations (3.7.5). From Figure 3.7.5 we see that if n is the largest integer that does not exceed r, then $P(r)$ necessarily contains $(n - 1)^2$ tiles of $\mathscr{P}$, whatever its position relative to the tiles of $\mathscr{P}$. On the other hand, let k be the smallest integer such that $kp_1 \geq 2U$ and $kp_2 \geq 2U$, where p_1 and p_2 are the widths of P in directions perpendicular to the sides, and U is the circumparameter of the tiling. Then the patch of tiles of $\mathscr{T}$ defined from $P(r)$ is necessarily contained in an $(n + 2 + 2k) \times (n + 2 + 2k)$ block of parallelogram tiles from $\mathscr{P}$, as shown in Figure 3.7.5. We deduce that

$$(n - 1)^2T \leq t_p(r) \leq (n + 2 + 2k)^2T$$

with similar inequalities for $v_P(r)$ and $e_P(r)$. Dividing through by r^2A and taking the limit as $r \to \infty$, we obtain

$$T/A \leq \lim_{r\to\infty} t_P(r)/r^2A \leq T/A.$$

This yields one of the identities of (3.7.5) and the other two follow in a similar manner. Hence the proof of Statement 3.7.3 is completed.

Slight modifications of this argument are needed to show that every periodic tiling $\mathscr{T}$ is metrically strongly balanced and metrically prototile balanced. However, the details are completely straightforward and will not be given here.

EXERCISES 3.7

1. Prove Statement 3.7.2
**2. Prove that if a finite set $\mathcal{S}$ of prototiles admits a tiling $\mathcal{T}_1$ which possesses a translation as a symmetry, then $\mathcal{S}$ admits a periodic tiling.
*3. Let $\mathcal{S}$ and $\mathcal{T}_1$ be as in Statement 3.7.1. Prove that any patch of $\mathcal{T}_1$ is also a patch of a periodic tiling admitted by $\mathcal{S}$.
4. Prove that every periodic tiling is metrically strongly balanced and metrically prototile balanced, and establish the appropriate analogues of the equations in Statement 3.7.3.
5. Use Statement 3.7.3 to derive alternative solutions to the questions in Exercise 3.6.1.

3.8 THE EXTENSION THEOREM

In the preceding sections of this chapter we were concerned with restrictions that can be placed on a tiling to ensure that it is "well-behaved". Now our outlook changes and we consider some conditions which guarantee the existence of tilings with prescribed properties.

The first of these results is the Extension Theorem (Statement 3.8.1). This gives a criterion for deciding whether a given set of prototiles admits a tiling. In order to state the theorem it is convenient to introduce the following terminology: a set of prototiles $\mathcal{S}$ is said to *tile over* a subset D of the plane if $\mathcal{S}$ admits a patch $\mathcal{A}$ such that the union of the tiles in $\mathcal{A}$ contains D. We shall also say that $\mathcal{A}$ *covers* D. In Figures 3.8.1 and 3.8.2 we show examples of patches that cover circular disks of various sizes; in each case the set $\mathcal{S}$ contains just one prototile.

3.8.1. The Extension Theorem *Let $\mathcal{S}$ be any finite set of prototiles, each of which is a closed topological disk. If $\mathcal{S}$ tiles over arbitrarily large circular disks D, then $\mathcal{S}$ admits a tiling of the plane.*

Before we prove this statement it is helpful to discuss a few examples. First consider that shown in Figure 3.8.1. From the given patches it is easy to see how arbitrarily large disks D can be covered and hence the Extension Theorem tells us (as is completely obvious in any case!) that the given prototile admits a tiling of the plane. This example is not typical for the following reason. Suppose the disks are translated so that they are concentric. Then each patch of tiles that covers a larger disk can be obtained by extending patches that cover smaller ones. In other words, in this case it happens exceptionally that any of the given patches can be extended arbitrarily far in all directions. The existence of a tiling $\mathcal{T}$ of the plane follows immediately, and the assertion of the theorem becomes trivial. (We shall sometimes describe this situation by saying that $\mathcal{T}$ is the *inclusion limit* of the given sequence of patches.)

However, consider the sequence of patches shown in Figure 3.8.2. Although the method of constructing larger patches in an analogous way is clear, none of these patches can be extended to cover a larger concentric disk D than that shown. According to the Extension Theorem the existence of the patches is enough to ensure

Figure 3.8.1
Patches of tiles covering circular disks. Each patch is obtained by extending the preceding patch, and the "inclusion limit" of these patches is a tiling of the plane.

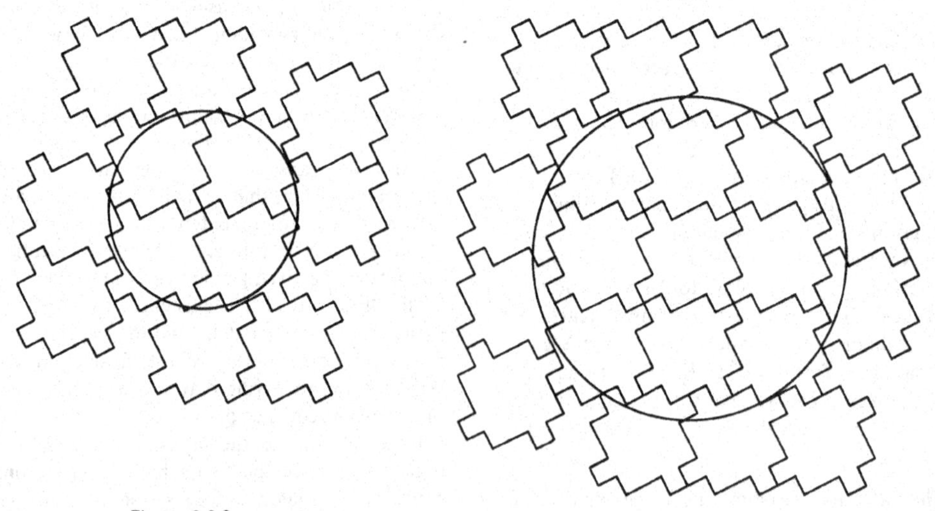

Figure 3.8.2
Here, in contrast to Figure 3.8.1, no patch is obtained by extending a preceding patch, yet the Extension Theorem implies that the given prototile admits a tiling of the plane.

that the given prototile admits a tiling $\mathscr{T}$ of the plane—*in spite of the fact that no such tiling $\mathscr{T}$ is an extension of any of the patches whose existence is postulated.* None of the larger patches contains one of the smaller patches—or, to put it another way, the Extension Theorem takes account of the fact that in order to obtain the tiling $\mathscr{T}$ (whose existence it asserts), it may be necessary to "re-arrange" the tiles in each of the given patches.

The conditions of the theorem cannot be significantly weakened. For example, if infinite sets of prototiles are allowed, then the result is false as we can see from the example of Figure 3.8.3. Here we take all the lattice points with integer coordinates in a quadrant Q of the plane, and join each nearest pair by a circular arc as shown, each arc being of a different radius. Then the infinite set of tiles which results from this construction will tile over arbitrarily large disks, but clearly it will not admit a tiling of the plane. On the other hand, the theorem implies that if $\mathscr{S}$ is finite and tiles over the quadrant Q, then $\mathscr{S}$ admits a tiling of the whole plane. Moreover this is true *even if the patch of tiles which covers Q cannot be extended in any way*! Here again the theorem takes account of the fact that the tiles in the patch will have to be "rearranged" to obtain the desired tiling.

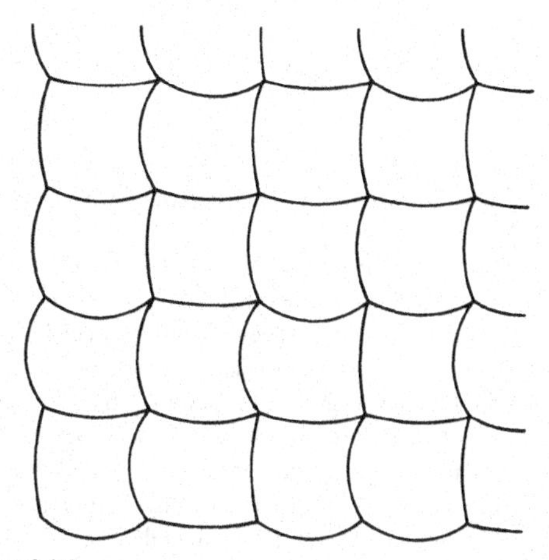

Figure 3.8.3
An infinite set of prototiles which admits a tiling of a quadrant of the plane but which does not admit a tiling of the plane. The edges are circular arcs, each of a different radius. This example shows that the Extension Theorem fails if infinite sets of prototiles are allowed.

In Figure 3.8.4 we give an example that shows the theorem to be false if unbounded tiles occur in $\mathscr{S}$. Here $\mathscr{S}$ consists of a single prototile—a semi-infinite strip with a semicircular end. The tiles may be "stacked" as shown so as to tile over a halfplane (and hence arbitrarily large circular disks), but clearly no tiling is possible. In fact, the requirement that the prototiles are topological disks was introduced specifically to ensure boundedness; easy generalizations of Statement 3.8.1 can be obtained which allow tiles that are not connected or simply connected.

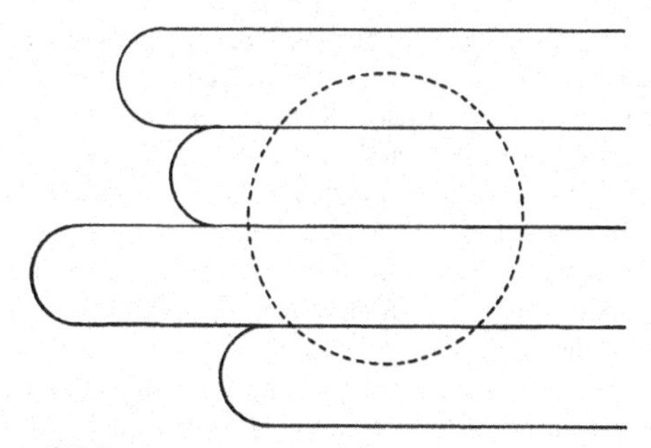

Figure 3.8.4
This figure indicates that the Extension Theorem fails if unbounded prototiles are allowed.

In Section 1.5 we saw how the introduction of a single new tile can profoundly alter the character of a tiling. One of the implications of the Extension Theorem is that if a set $\mathscr{S}$ of prototiles does not admit a tiling, then there exists no tiling in which all but a finite number of tiles are congruent to tiles in $\mathscr{S}$.

For the proof of Statement 3.8.1 we need to introduce some new ideas and terminology. The distance $\delta(T_1,T_2)$ between two tiles (or compact sets) T_1, T_2 is defined as the least number δ such that every point of T_1 lies at distance at most δ from some point of T_2, and every point of T_2 lies at distance at most δ from some point of T_1. Thus

$$\delta(T_1,T_2) = \max\left\{ \sup_{x_2 \in T_2} \inf_{x_1 \in T_1} \|x_1 - x_2\|, \sup_{x_1 \in T_1} \inf_{x_2 \in T_2} \|x_1 - x_2\| \right\}.$$

This is usually known as the *Hausdorff distance* between T_1 and T_2.

An infinite sequence $T_1, T_2, T_3, \ldots$ of tiles is said to *converge to the limit tile* T if $\delta(T_i,T) \to 0$ as $i \to \infty$. A special case of the Selection Theorem* tells us that if all the tiles T_i of an infinite sequence are congruent to a bounded tile T_0, and if there is some point P_0 common to all the tiles T_i, then it is possible to select a convergent subsequence from the given sequence. Moreover, in this case, the limit tile of the subsequence is also congruent to T_0. A proof of the Selection Theorem in this special case is easy to construct using the theorems of elementary analysis (see Exercise 3.8.2).

We now turn to the proof** of Statement 3.8.1. Since the set $\mathcal{T}$ of prototiles is finite, the tiles of $\mathcal{T}$ possess a common circumparameter U and a positive inparameter u. Consider the lattice Λ consisting of all points whose rectangular Cartesian coordinates are (nu,mu), where n and m are integers (positive, negative or zero). The lattice Λ has the property that every tile congruent to a prototile in $\mathcal{S}$ contains at least one point of Λ. We can arrange the points of Λ in an infinite sequence $L_0, L_1, L_2, L_3, \ldots$; for example, we can take L_0 to be the origin (0, 0), and determine the other points L_i by "spiralling" around L_0 as indicated in Figure 3.8.5.

For any positive number r let $D(L_0,r)$ be the closed circular disk of radius r centered at L_0, and let $\mathcal{A}(r)$ be any patch of tiles admitted by $\mathcal{S}$ that covers $D(L_0,r)$. When r is an integer sufficiently large for $D(L_0,r)$ to contain L_s we define T_{rs} to be that tile of $\mathcal{A}(r)$ which contains L_s. If L_s lies on an edge or at a vertex of $\mathcal{A}(r)$, then T_{rs} may be chosen to be any tile of $\mathcal{A}(r)$ incident with L_s. Since $\mathcal{S}$ is finite, the sequence S of tiles $T_{10}, T_{20}, T_{30}, \ldots$ must contain an infinite subsequence S'_0 consisting of tiles congruent to *one* tile T_0 in $\mathcal{S}$. Moreover, by the result quoted above, since all tiles of S'_0 contain L_0 there exists a convergent subsequence S_0 of S'_0 whose limit tile T'_0 will also contain L_0.

Now consider the sequence of tiles T_{r1} containing L_1, restricting attention to values of r that correspond to tiles in S_0. As above, we can show that there exists a subsequence S'_1 of this sequence, consisting of tiles all congruent to some tile T_1 of $\mathcal{S}$. Again a subsequence S_1 of

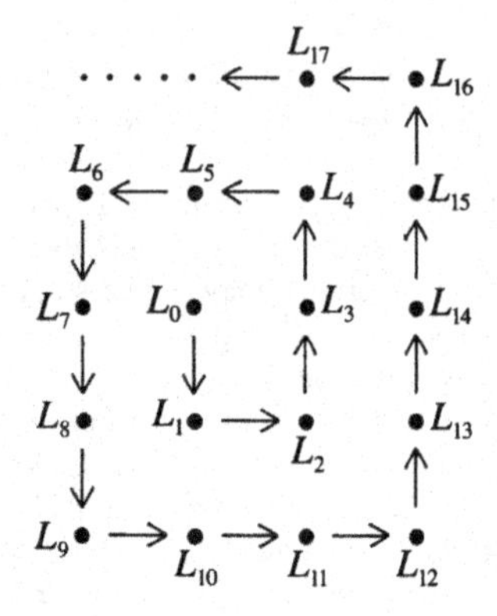

Figure 3.8.5
The points of the integer lattice Λ can be arranged as an infinite sequence by "spiralling" around the origin L_0 as shown.

S'_1 can be chosen which converges to a limit tile T'_1 containing L_1 and congruent to T_1.

We proceed in this way, at the mth stage picking a subsequence S_m of the preceding sequence S_{m-1} that ensures the convergence to a limit tile T'_m of congruent tiles containing the point L_m of the lattice Λ. Of course, for large m we must choose r large enough for $D(L_0,r)$ to contain L_m.

The rest of the proof depends upon establishing that the set of tiles $\mathcal{T} = \{T'_0, T'_1, T'_2, \ldots\}$ (from which we delete any duplicates that may be present) forms a tiling admitted by $\mathcal{S}$. To see this, let P be any point of the plane. Then we must show that P belongs to at least one of the tiles T'_i, but does not belong to the interior of any two distinct tiles T'_i. Continuing the notation introduced above we let $D(P,U)$ be the circular disk centered at P

* One version of the Selection Theorem states that, if an infinite family $\mathcal{T}$ of non-empty closed sets is *jointly bounded* (that is, a bounded set contains all members of $\mathcal{S}$), then it is possible to select from $\mathcal{S}$ a sequence of sets that converges to a limit set (see, for example, Nadler [1978, Theorem 0.8]).

** We are indebted to David Larman and Victor Klee for helpful comments that led to great simplifications of the original proof.

and of radius U (the circumparameter of the tiles in $\mathcal{S}$). Let L_m be the point of the lattice Λ in $D(P,U)$ with largest subscript, and restrict attention to the sequence of patches $\mathcal{A}(r)$ as r runs through values corresponding to the subsequence S_m. The sets of tiles

$$\mathcal{T}_r = \{T_{r0}, T_{r1}, T_{r2}, \ldots, T_{rm}\}$$

clearly converge, as $r \to \infty$ through these values, to the set $\mathcal{T}' = \{T'_0, T'_1, T'_2, \ldots, T'_m\}$. Since the tiles of each $\mathcal{T}_r$ have disjoint interiors and their union contains P, the same will be true of the tiles of their limit $\mathcal{T}'$. Thus P belongs to at least one tile of $\mathcal{T}'$ and cannot belong to the interiors of two such. We deduce that $\mathcal{T}$ is a tiling and so the Extension Theorem is proved.

There are several generalizations of the Extension Theorem. For example, the same proof goes through without difficulty if the tiles have colored edges which must be matched in color as well as fitted together. In this way an alternative proof of Wang's Theorem (see Statement 11.2.1) can be obtained.

A more significant generalization shows that, under certain circumstances, the theorem remains true even if $\mathcal{S}$ is not finite.

3.8.2 *Let $\mathcal{S}$ be a family of prototiles each of which is a closed topological disk. If $\mathcal{S}$ is compact and if it tiles over arbitrarily large circular disks, then $\mathcal{S}$ admits a tiling.*

Here, saying that $\mathcal{S}$ is *compact* means that $\mathcal{S}$ is jointly bounded (see the footnote on page 154) and that every infinite sequence of tiles from $\mathcal{S}$ possesses a subsequence which converges (in the sense of the Hausdorff distance) to some tile which also belongs to $\mathcal{S}$. The proof of Statement 3.8.2 is a straightforward modification of that of Statement 3.8.1.

Other related results are:

3.8.3 *Let $\mathcal{S}$ be a finite family of prototiles, each of which is a closed topological disk. If it is possible to tile over arbitrarily large circular disks so that the resulting patches all have a certain symmetry σ, then $\mathcal{S}$ admits a tiling of the plane with the same symmetry σ.*

3.8.4 *Let the finite family of prototiles $\mathcal{S}$ be minimal in the sense that it admits a tiling of the plane but no longer does so if any tile is removed from $\mathcal{S}$. Then there exists a number r such that, for every tiling $\mathcal{T}$ admitted by $\mathcal{S}$, every circular disk of radius r contains a copy of each of the prototiles in $\mathcal{S}$.*

We have already seen, in Section 2.8, that there are many interesting and challenging problems concerning the extension of patches of tiles. We conclude this section with some remarks on an old, and as yet unsolved, problem in this area. Some particular problems of a similar nature are suggested in the exercises.

Let $\mathcal{S}$ be a given set of prototiles, and $\mathcal{A}_0$ a given patch of tiles admitted by $\mathcal{S}$. Then we say that $\mathcal{A}_0$ can be *surrounded* (by tiles of $\mathcal{S}$) to form a patch $\mathcal{A}_1$ if the closure of that part of the plane not covered by $\mathcal{A}_1$ is disjoint from all the tiles of $\mathcal{A}_0$. The idea of surrounding a patch has already occurred implicitly several times. In Section 2.8 we used the process of surrounding $\mathcal{A}$ in order to construct extensions of $\mathcal{A}$, and in Section 3.2 the patch $\mathcal{A}_{n+1}$ (defined in connection with Figure 3.2.6) surrounds $\mathcal{A}_n$; it does so minimally.

Suppose $\mathcal{S}$ contains just one prototile T and $\mathcal{A}_0$ consists of just one copy of T. Then it may happen that $\mathcal{S}$ does not admit a tiling of the plane, yet T can be completely surrounded to form a patch $\mathcal{A}_1$, see Figure 3.8.6. (Another tile with the same property is the union of a unit circular disk with the curvilinear triangle determined by the original disk and two others congruent to it, tangent to it and to each other; see Lietzmann [1928, p. 242].) On the other hand, no prototile T is known which does not admit a tiling of the plane and yet can be surrounded *twice.* (By this we mean that patches $\mathcal{A}_1$ and $\mathcal{A}_2$ exist such that $\mathcal{A}_1$ surrounds T, and $\mathcal{A}_2$ surrounds $\mathcal{A}_1$.) From this fact Heesch [1968b, p. 23] was led to propose the following.

Heesch's Problem. *For which positive integers r does there exist a prototile T such that T can be surrounded r times, but not $r + 1$ times, by tiles congruent to T?* *

* See the Appendix beginning on page 653.

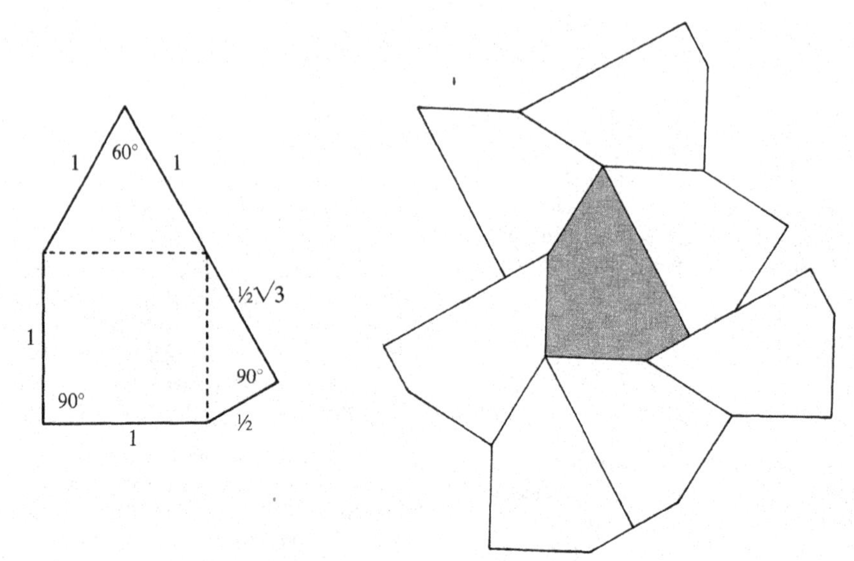

Figure 3.8.6
A prototile T which does not admit a tiling of the plane, yet T can be completely surrounded by tiles congruent to T as shown.

Whether or not $r = 1$ is the only such integer remains an open problem.

Similar questions arise if the initial patch contains more than one tile. It seems reasonable to conjecture the existence of a function $f(n,m)$ taking positive integer values, such that for *every* set $\mathcal{S}$ of n prototiles, if *every* patch containing m tiles can be surrounded at least $f(n,m)$ times, then $\mathcal{S}$ admits a tiling of the plane. About the value of $f(n,m)$ we have no information of any kind for any values of n and m, except to remark that by splitting the tile of Figure 3.8.6 into smaller tiles in a suitable manner one can show that $f(n,1) \to \infty$ as $n \to \infty$.

EXERCISES 3.8

1. Show that the Extension Theorem remains true if $\mathcal{S}$ is a *translative* set of prototiles (that is, only translates of the prototiles are allowed).

*2. Prove the special case of the Selection Theorem (as stated on page 154) using the following idea. A tile T, congruent to T_0, is specified by a point
$$P(T) = (x,y,\theta,\varepsilon)$$
in four-dimensioonal space R^4, where (x,y) are the coordinates of the centroid of T, $0 \le \theta \le 2\pi$ indicates the orientation of T, and $\varepsilon = \pm 1$ indicates whether or not it has been reflected. The set of points corresponding to tiles T that contain P_0 is a bounded set X in R^4. To every sequence of tiles of the kind under consideration corresponds a sequence of points in X. The required proof results from an application of the Weierstrass Theorem (see, for example, Rogers [1976, Theorem 1.6]) to such a sequence.

3. Give detailed proofs of Statements 3.8.3 and 3.8.4.
4. Let $\mathcal{S}$ consist of an equilateral triangle and a square of the same edge-length. Restrict attention in this exercise to patches and tilings all of whose vertices are of species number 3 (see Section 2.1).
   (*a*) For every positive integer r find a patch that can be surrounded r times but not $r + 1$ times.
   **(*b*) Show that if a patch $\mathcal{A}$ contains n tiles and can be surrounded $[n/3]$ times, then $\mathcal{A}$ can be extended to a tiling of the plane.
5. Let $\mathcal{S}$ consist of a suitably chosen centrally symmetric convex hexagon. For every integer r find a patch of tiles admitted by $\mathcal{S}$ that can be surrounded r times but not $r + 1$ times. Can one deduce anything about the values of the function $f(n,m)$ defined at the end of this section?

3.9 EBERHARD-TYPE THEOREMS

The numbers $v_j(\mathcal{T}) = \lim_{r\to\infty} v_j(r,P)/t(r,P)$ and $t_j(\mathcal{T}) = \lim_{r\to\infty} t_j(r,P)/t(r,P)$ convey information about the relative frequencies of vertices of various valences and tiles with various numbers of edges. In Statement 3.5.6 we have seen that in every strongly balanced tiling $\mathcal{T}$ these numbers satisfy several equations that are consequences of Euler's Theorem. One such equation is

$$\sum_{j\geq 3} (j-4)t_j(\mathcal{T}) + \sum_{k\geq 3} (4-k)v_k(\mathcal{T}) = 0. \quad (3.9.1)$$

We shall now investigate the converse question: given sequences of numbers t_j and v_k that satisfy equation (3.9.1), does there exist a tiling $\mathcal{T}$ with $t_j(\mathcal{T}) = t_j$ and $v_k(\mathcal{T}) = v_k$? By analogy with the results of a similar character known for convex polyhedra (Eberhard [1891], Grünbaum [1967, Section 13.3], Jendrol & Jucovič [1972], Jucovič [1973]) we shall call theorems of this nature *Eberhard-type* theorems. The first such result we shall prove is:

3.9.2 *Let t_j, $j \geq 3$, and v_k, $k \geq 3$, be non-negative rational numbers, of which only a finite number are non-zero, such that*

$$\sum_{j\geq 3} (j-4)t_j + \sum_{k\geq 3} (k-4)v_k = 0. \quad (3.9.3)$$

Then there exists a periodic tiling $\mathcal{T}$ and a positive rational number γ such that

$$t_j(\mathcal{T}) = \gamma t_j \qquad \text{for all} \qquad j \neq 4$$

and

$$v_k(\mathcal{T}) = \gamma v_k \qquad \text{for all} \qquad k \neq 4.$$

Moreover, if $\kappa = \sum_{j\neq 4} t_j + \sum_{k\neq 4} v_k \neq 0$, then we may choose $\gamma \geq 2/5\kappa$.

Clearly, Statement 3.9.2 is a reasonable converse to Statement 3.5.6. Moreover, it is a result that is best possible in several respects; for example, insisting that $\gamma = 1$, or that the conditions $j \neq 4$, $k \neq 4$ be removed, would render the result invalid.

The proof of Statement 3.9.2 is constructive. We begin by choosing positive integers ρ, τ_j and ν_k such that τ_j, ν_k are even and $t_j = \tau_j/\rho$, $v_k = \nu_k/\rho$ for $j, k = 3, 5, 6, 7, \ldots$. Starting with a rectangle S formed by squares, containing 5 rows and $\frac{1}{2}\sum_{j\geq 5}(\tau_j + \nu_j)$ columns, we shall show how it may be modified to obtain a rectangular patch $\mathcal{R}$ of tiles

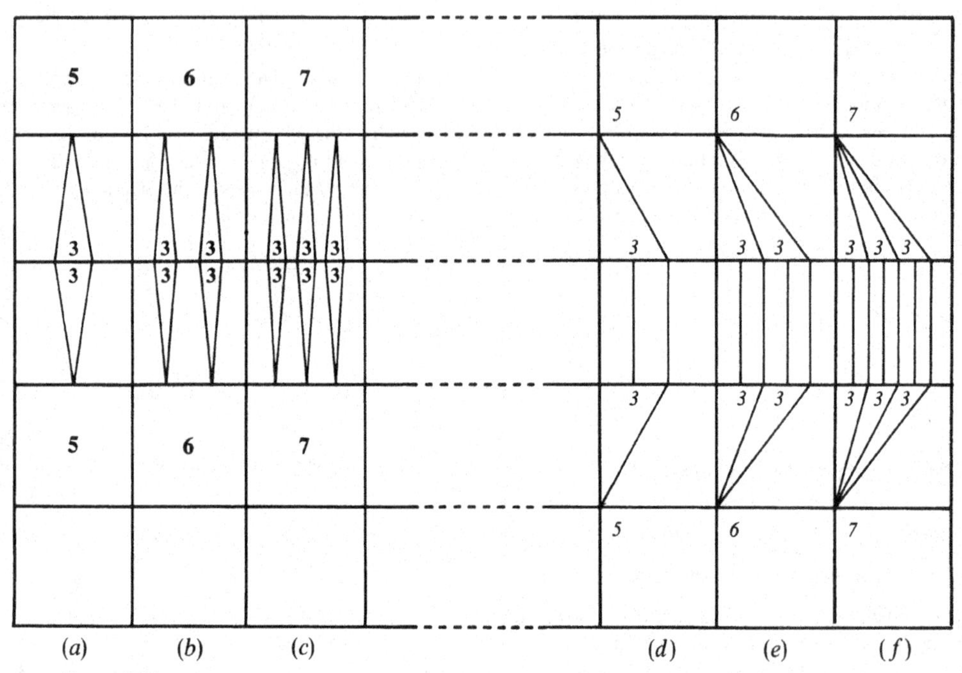

Figure 3.9.1
The construction of a period parallelogram of the tiling $\mathcal{T}$ described in the proof of Statement 3.9.2.

that contains τ_j j-gons and v_k vertices of valence k for all $j \neq 4$ and $k \neq 4$. In these modifications the number and valences of the vertices on the boundary of the rectangle are unchanged, so $\mathcal{R}$ may be used as the period parallelogram of a periodic tiling $\mathcal{T}$ which will be seen to satisfy all the requirements which are stated in Statement 3.9.2.

The *first stage* in the construction of $\mathcal{R}$ is illustrated in Figure 3.9.1. Here, except in the case of 4-sided tiles and 4-valent vertices, the number of edges of each tile is indicated by a bold-face numeral, and the valence of each vertex by an italic numeral. In the first column (*a*) of squares in S we show how, by inserting four new edges indicated by thinner lines, we can introduce two 5-gons and two triangles into the patch. If this construction is repeated on the first $\frac{1}{2}\tau_5$ columns of S we obtain the required number of pentagons. In columns (*b*) and (*c*) of Figure 3.9.1 we show the corresponding modifications for introducing 6-gons and 7-gons. It is clear how this technique, applied to the first $\frac{1}{2}\sum_{j\geq 5}\tau_j$ columns of S, produces a patch that contains τ_j j-gons for all $j \geq 5$. We note, for later use, that it also contains $\tau_3^* = \sum_{j\geq 5}(j-4)\tau_j$ triangles and, in the first $\frac{1}{2}\sum_{j\geq 5}\tau_j$ columns, $\mu_4^* = \sum_{j\geq 5}(j-\frac{5}{2})\tau_j$ quadrangles.

The right side of Figure 3.9.1 indicates the analogous modifications needed to introduce vertices of valence greater than 4. In column (*d*) we show how, by inserting four new edges, we can introduce two vertices of valence 5 (and two of valence 3). In columns (*e*) and (*f*) we show the corresponding constructions for vertices of valences 6 and 7. Generally, applying modifications of this kind to the last $\frac{1}{2}\sum_{k\geq 5} v_k$ columns of S, we obtain a patch with v_k vertices of valence k for all $k \geq 5$. These last columns also contain $v_3^* = \sum (k-4)v_k$ 3-valent vertices and $\mu_4^{**} = \sum_{k\geq 5} (2k - \frac{11}{2})v_k$ quadrangles.

If $\tau_3^* = \tau_3$ and $v_3^* = v_3$ then the construction of $\mathscr{R}$ is completed; otherwise we proceed to the *second stage* in which the numbers of triangles and 3-valent vertices are "adjusted" to the required values. If there are too many triangles, then we "shrink" pairs of them to edges, as indicated in Figure 3.9.2(*a*); if there are too few, then we delete edges as indicated in Figure 3.9.2(*b*). Since equation (3.9.3) implies $\tau_3^* + v_3^* = \tau_3 + v_3$, it follows that these two operations will always enable us to obtain τ_3 triangles and v_3 3-valent vertices, as required. This completes the construction of the patch $\mathscr{R}$.

It will be observed that if we write

$$\begin{aligned}\tau_4^* &= \mu_4^* + \mu_4^{**} \\ &= \tfrac{1}{2}\sum_{j\geq 5} (2j-5)\tau_j + \tfrac{1}{2}\sum_{k\geq 5} (4k-11)v_k\end{aligned} \qquad (3.9.4)$$

and write τ_4 for the number of quadrangles in the patch $\mathscr{R}$, then

$$\tau_3 + \tau_4 \leq \tau_3^* + \tau_4^*. \qquad (3.9.5)$$

In other words, the second stage tends to decrease the total number of triangles and quadrangles in the patch.

Finally, we need to establish the inequality claimed for γ. Assume $\sum_{j\neq 4} \tau_j \neq 0$ (since $\kappa \neq 0$ is assumed, and $\sum_{k\neq 4} v_k \neq 0$ can be dealt with in a similar manner). Then $t_j(\mathscr{T})$ is the "fraction" of j-gonal tiles in $\mathscr{R}$, so

$$t_j(\mathscr{T}) = \gamma t_j = \gamma\tau_j/\rho = \tau_j \Big/ \sum_{j\geq 3} \tau_j.$$

Hence, using relation (3.9.5),

$$\gamma = \rho \Big/ \sum_{j\geq 3} \tau_j \geq \rho \Big/ \Big(\tau_3^* + \tau_4^* + \sum_{j\geq 5} \tau_j\Big).$$

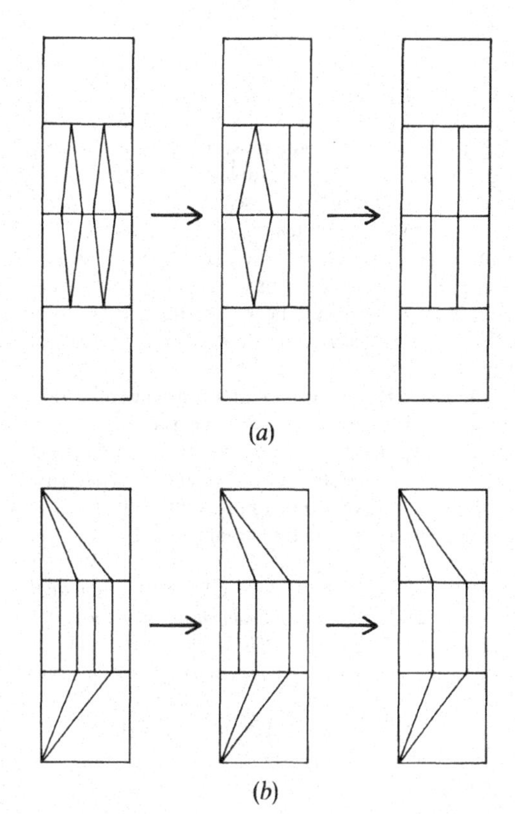

Figure 3.9.2
Illustrations for the "adjustments" of the number of triangles used in the proof of Statement 3.9.2.

Using the values of τ_4^*, τ_3^* and v_3^* obtained above and recalling that $\tau_3^* + v_3^* = \tau_3 + v_3$, we obtain:

$$\begin{aligned}\gamma &\geq \rho \Big/ \Big(\tau_3^* + \tfrac{1}{2}\sum_{j\geq 5}(2j-5)\tau_j + \tfrac{1}{2}\sum_{k\geq 5}(4k-11)v_k + \sum_{j\geq 5}\tau_j\Big) \\ &= \rho \Big/ \Big(2\tau_3^* + \tfrac{3}{2}\sum_{j\geq 5}\tau_j + 2v_3^* + \tfrac{5}{2}\sum_{k\geq 5}v_k + \sum_{j\geq 5}\tau_j\Big) \\ &= \rho \Big/ \Big(2\tau_3^* + 2v_3^* - \tfrac{5}{2}(\tau_3 + v_3) + \tfrac{5}{2}\Big(\sum_{j\neq 4}\tau_j + \sum_{k\neq 4} v_k\Big)\Big) \\ &> \rho \Big/ \Big(\tfrac{5}{2}\Big(\sum_{j\neq 4}\tau_j + \sum_{k\neq 4}v_k\Big)\Big) = \frac{2}{5\kappa},\end{aligned}$$

as required. It is clear that γ must be rational, and so the proof of Statement 3.9.2 is completed.

The method of proof just described leads to tilings with "large" period parallelograms $\mathscr{R}$, and so with many quadrangles and 4-valent vertices. It is not difficult to devise more "economical" constructions in which these numbers are reduced (and hence γ made correspondingly larger), but their description is not as simple as the one we have given.

If the prescribed numbers t_j and v_k are not all rational (and cannot be made so by multiplying them by a suitable constant), then they cannot correspond to a periodic tiling, and so the construction given above necessarily fails. However, it is possible to prove the following result, using suitable approximation arguments:

3.9.6 *Let $t_j, j \geq 3$, and $v_k, k \geq 3$, be non-negative real numbers of which only a finite number are non-zero, and which satisfy relation (3.9.3). Then there exists a strongly balanced tiling $\mathscr{T}$ and a real positive number γ such that*

$$t_j(\mathscr{T}) = \gamma t_j \qquad \textit{for all} \qquad j \neq 4$$

and

$$v_k(\mathscr{T}) = \gamma v_k \qquad \textit{for all} \qquad k \neq 4.$$

Moreover, if $\kappa = \sum_{j \neq 4} t_j + \sum_{k \neq 4} v_k \neq 0$, then we may choose $\gamma > 1/3\kappa$.

In order to prove Statement 3.9.6 we construct sequences $(t_j^{(n)})_{n=1}^{\infty}$, $(v_k^{(n)})_{n=1}^{\infty}$ of rational numbers such that, for all $j, k = 3, 5, 6, 7, \ldots$,

(1) equation (3.9.3) is satisfied for each n,
(2) $\lim_{n \to \infty} t_j^{(n)} = t_j$ and $\lim_{n \to \infty} v_k^{(n)} = v_k$,
(3) for all $n \geq 1$,

$$(t_j^{(n)} - t_j)(t_j^{(n+1)} - t_j) < 0$$

$$(v_k^{(n)} - v_k)(v_k^{(n+1)} - v_k) < 0,$$

(4) for each n only a finite number of $t_j^{(n)}$ and $v_k^{(n)}$ are non-zero.

Having chosen these sequences, for each n construct a rectangle $R^{(n)}$ using the numbers $t_j^{(n)}$ and $v_k^{(n)}$ and the method used in the proof of Statement 3.9.2. We can now construct the required strongly balanced tiling as follows.

First form a square $S^{(1)}$ using copies of $R^{(1)}$ in such a way that each side of $S^{(1)}$ is a multiple of each of the sides of $R^{(2)}$. Now construct a "border" round $S^{(1)}$ using the rectangles $R^{(2)}$, so as to form a larger square $S^{(2)}$ concentric with $S^{(1)}$ and such that each side of $S^{(2)}$ is a multiple of each of the sides of $R^{(3)}$. Generally, we border $S^{(n-1)}$ with copies of $R^{(n)}$ to form a larger concentric square $S^{(n)}$ whose sides are multiples of the sides of $R^{(n+1)}$. Moreover, because of condition (3), we may make the squares $S^{(n)}$ increase in size sufficiently rapidly, so that the ratios of the numbers of j-sided tiles and of k-valent vertices to the total number of tiles in $S^{(n)}$ closely approximate the corresponding values $t_j^{(n)}$ and $v_k^{(n)}$ for $R^{(n)}$ (to within, say, $\frac{1}{2}|t_j^{(n)} - t_j|$ and $\frac{1}{2}|v_k^{(n)} - v_k|$). This may be used to show that the tiling obtained as the union of all squares $S^{(n)}$ satisfies the requirements of Statement 3.9.6. The limit arguments needed are straightforward but somewhat lengthy, and the same applies to the estimation of γ. We omit these details.

It should be noted that Statements 3.9.2 and 3.9.6 both fail if an infinite number of the constants t_j and v_k are non-zero, since by the results of Section 3.2 the numbers of edges of tiles, and the valences of vertices, are bounded in every normal tiling.

Other Eberhard-type theorems result if one starts from equations (3.5.7) or (3.5.9). For example, we have:

3.9.7 *Let $t_j, j \geq 3$, and $v_k, k \geq 3$, be non-negative rational numbers of which only a finite number are non-zero, such that*

$$\sum_{j \geq 3} (6 - j)t_j + 2 \sum_{k \geq 3} (3 - k)v_k = 0. \tag{3.9.8}$$

Then there exists a periodic tiling $\mathscr{T}$ and a positive rational number γ such that

$$t_j(\mathcal{T}) = \gamma t_j \qquad \textit{for all} \qquad j \neq 6$$

and

$$v_k(\mathcal{T}) = \gamma v_k \qquad \textit{for all} \qquad k \neq 3.$$

The proof of Statement 3.9.7 may be constructed along lines parallel to those in the proof of Statement 3.9.2, but the amount of detail is much greater in this case. A description of the proof may be found in Grünbaum & Shephard [1982a].

A result analogous to Statement 3.9.6 but with condition (3.9.8) instead of (3.9.3) is also valid. Similarly, results analogous to those presented here hold for sequences t_j and v_k that satisfy equation (3.5.9).

EXERCISES 3.9

1. In each of the following cases construct a tiling $\mathcal{T}$ as described in the proof of Statement 3.9.2. Compare the numbers $t_4(\mathcal{T})$ and $v_4(\mathcal{T})$ with the corresponding numbers in the "most economical" tilings you can devise.
   (*a*) $t_3 = t_5 \neq 0$; $t_j = 0$ for $j \geq 6$, $v_k = 0$ for $k \neq 4$.
   (*b*) $t_3 = v_5 \neq 0$; $t_j = 0$ for $j \geq 5$, $v_k = 0$ for $k \neq 4, 5$.
   (*c*) $t_3 = 4t_8 \neq 0$; $t_j = 0$ for $j \neq 3, 4, 8$, $v_k = 0$ for $k \neq 4$.
   (*d*) $t_3 = 8t_{12} \neq 0$; $t_j = 0$ for $j \neq 3, 4, 12$, $v_k = 0$ for $k \neq 4$.
2. Find examples of sequences $(t_j)_{j=3}^{\infty}$ and $(v_k)_{k=3}^{\infty}$ that satisfy equation (3.9.3) and have the property that $t_4(\mathcal{T}) \neq v_4(\mathcal{T})$ for every tiling $\mathcal{T}$ with $t_j(\mathcal{T}) = t_j$ and $v_k(\mathcal{T}) = v_k$ for all $j \neq 4$ and $k \neq 4$.
3. For the following sets of values of t_i and v_k satisfying equation (3.9.8) construct the "most economical" tilings possible, that is, those with minimal values of $t_6(\mathcal{T})$ and $v_3(\mathcal{T})$. (Note that the proof of Statement 3.9.7 given in Grünbaum & Shephard [1982a] yields tilings which use far more 6-gons and 3-valent vertices than necessary.)
   (*a*) $t_4 = 1$, $t_7 = 2$; $t_i = 0$ for $i \neq 4, 6, 7$, and $v_k = 0$ for $k \neq 3$.
   (*b*) $v_7 = v_8 = 1$; $t_i = 0$ for $i \neq 6$, and $v_k = 0$ for $k \neq 3, 7, 8$.
   (*c*) $t_3 = 8$, $t_5 = 6$, $v_4 = 12$, $v_6 = 1$; $t_i = 0$ for $i \neq 3, 5, 6$, and $v_k = 0$ for $k \neq 3, 4, 6$.

3.10 NOTES AND REFERENCES

In view of the fundamental nature of the ideas and results of this chapter, it is surprising to report that they seem to have been completely ignored by many previous writers. The need for "normality" or a similar notion has occasionally been mentioned (see, for example, Laves [1931b]), but for the most part authors have just assumed that all tilings are "well-behaved". This uncritical approach has naturally led to many errors, some of which we shall mention later.

A notion related to (but less restrictive than) normality was first considered by Reinhardt [1918]. He dealt with tilings which admit a circumparameter and in which the areas of the tiles are bounded from below by a positive number. For such tilings Reinhardt established the analogue of our Normality Lemma (Statement 3.2.2) as well as results related to our Statements 3.3.2, 3.4.2 and 3.5.1. His work is important since it was the first one to indicate that stringent restrictions must be imposed on tilings if any meaningful general results are to be obtained. Unfortunately, Reinhardt's thesis is written so ponderously and organized so poorly that it was misunderstood by several later investigators, and ignored by many more.

Reinhardt [1927] and Niven [1978] proved that every tiling of this kind by convex polygons has at least one tile with at most six sides; this is the only precursor of Statement 3.2.3 we found in the literature which does not assume balance, or some similar condition on the tiling. A popular account of the result of Niven [1978] appears in Lavrič [1980]. Simplifications of parts of Niven's proof were given by Klamkin & Liu [1980a]. Under the assumption that the tiling is strongly balanced (Section 3.5), or that every tile has the same number of adjacents, such a result was proved by Reinhardt [1918].

For tilings in the hyperbolic plane, results giving bounds on the number of adjacents in terms of the circumparameter and inparameter were obtained by Reinhardt [1928b].*

The term "normal" seems to have been first used, in connection with tilings, by Fejes Tóth [1969b]. His definition is slightly less stringent than ours but is sufficient for his purposes—he defines a tiling to be normal if it satisfies condition N.3.

Tilings with singular points have not received much attention; a number of visually pleasing examples (mostly involving pentagons or pentacles) are shown in Beard [1973], others in some of the later tilings by Escher [1971]. The "similarity tilings" which are related to the tiling in Figure 3.1.3, and discussed in Section 10.1, lead naturally to singular points. The only direction in which considerable work involving singular points has been done was motivated by questions of the following type: in tilings of the plane by circular disks (or other mutually similar sets, as in Figure 3.1.3) and single-point tiles, what can be said about the "quantity" of single-point tiles that must be used? In the case of circular disks such tilings are usually called "Apollonian packings of circles". It turns out that the concept of "Hausdorff dimension" (see, for example, Hurewicz & Wallman [1941], Federer [1969], or Rogers [1970]) is the appropriate way to convey information about the set of single-point tiles. The reader interested in Apollonian packings is referred to the papers of Wilker [1967] and Boyd [1971], [1973], which contain some of the best results known, and provide references to the earlier literature. Eggleston [1953], [1957, pp. 160–165] investigated the case in which the tiles are equilateral triangles (all with the same orientation) and single points. For a stimulating discussion of these and more general questions, accompanied by beautiful illustrations, see Mandelbrot [1977].

For other approaches to questions about tilings with singular points, and for results on such tilings, see Breen [1983a], [1983b], [1985], Stone [1985], Valette [1980], [1981], Wollny [1969], [1974a], [1974b], [1983], [1984].

The problems posed by the phenomenon of "singularity at infinity" were first brought to attention by Laves [1931b]. The "paradox" we discussed on page 118 is attributed to H. Steinhaus and an account of it given by Gardner [1971, pp. 248–252]. Laves was led to these questions by his attempts to determine the topologically tile-transitive tilings which we discuss in Chapter 4.

Euler's Theorem for Tilings and its various corollaries are often quoted and used—usually without any indication of restrictions that must be imposed on a tiling to give meaning and validity to this procedure. In contrast to many other cases—in which a cavalier attitude towards mathematical rigor is an aesthetic shortcoming that does not affect the outcome—here many authors have claimed to have proved statements that are actually false. As recent examples we may mention Walsh [1972] and Loeb [1976, especially Chapter 9], whose assertions are disproved by tilings such as those in Figures 3.1.5 and 3.1.6. A typical "illegal" use of Euler's Theorem which nevertheless "yields" the correct result is the procedure of Andreini [1907]: he *asserts* that equation (3.5.2) is valid for *every* tiling by polygons, and then uses this to show that there exist precisely 11 types of uniform tilings! He had already made a similar error in an earlier paper

* See the Appendix beginning on page 653.

(Andreini [1902]; compare Exercise 4.7.2) and the fact that this procedure is invalid had been pointed out by Steinitz [1903]. Other types of reasoning containing similar flaws involve the (implicit) assumption that every tiling is balanced or strongly balanced, and the derivation of Euler's Theorem for Tilings as "the limiting form for infinite radius of the Euler Theorem on the sphere", etc. (see, for example Šubnikov [1916], Fedorov [1916], Haag [1932], [1933], Barrett [1939], Loeb [1976]).

Under very severe restrictions on the kind of tilings considered, equation (3.5.5) was derived by Turner [1968]; he also observed that it coincides with equation (2.1.2).

Whitworth [1884] produced two tilings in which each vertex is incident with three equilateral triangles and with two squares (that is, is of species number 3 in the terminology of Section 2.1). Although both clearly must have the same values $t_3 = 2/3$ and $t_4 = 1/3$ which he correctly assigns to one of them, for the second tiling—which happens to be the uniform tiling ($3^2.4.3.4$)—he asserts that $t_3 = 7/10$ and $t_4 = 3/10$!

The proof of Statement 3.7.3 on periodic tilings, using Euler's Theorem for the torus, seems to have been first proposed by Heesch [1933a]. But to establish that a tiling is periodic may be very hard. For example Fedorov [1885, Section 64] considers monohedral edge-to-edge tilings by polygons and claims to prove that the prototile of any such tiling admits a monohedral periodic tiling. While Fedorov's "proof" is clearly invalid, the truth of the assertion is undecided even today; we shall enlarge on this topic in Chapter 10.

Despite the fact that the question about the existence of tilings with prescribed frequencies of faces (or vertices) of various kinds is very natural, there seems to be no mention of it or of any relevant results in the literature. The material presented in Section 3.9 is clearly only the beginning of several possible directions of investigation.

Equation (3.5.14) implies that if all the vertices are of valence 3, then the "average" number of adjacents of each tile is 6. This fact was established, under very severe restrictions, by Graustein [1931]; an unconvincing proof of Graustein's result appears in M'Crea [1933]. Equation (3.5.14) appears in O'Keefe & Hyde [1980], but without any indication that it presupposes the tiling to be strongly balanced. (For another weak result in the same direction see Meretz & van der Waerden [1966]; however, this paper is very imprecisely formulated and the arguments presented are not sufficiently justified.) Graustein's work was motivated by the results of biological studies of epithelial tissues and artificial froths, in which similar conclusions were reached experimentally (see, for example, Lewis [1931]; for related investigations see Kikuchi [1956], Wheeler [1958], Lantuejoul [1978]). However, it seems that in many of these papers the theoretical discussions were based on a usually tacit but erroneous assumption. This assumption was clearly expressed by Smith [1954] who stated, speaking of patches in 3-valent tilings: "If one is dealing with large arrays of such cells, the average cell approaches a hexagon. Moreover, it is possible to subdivide the arrays into groups of cells, hexagons on the average, which would fit into a hole cut in a uniform hexagonal net with exact conformity to its boundary." Even if it is granted that the "real" tilings considered by Smith are normal, balanced and even strongly balanced, patches taken from the examples constructed in Exercises 3.4.7(*a*) and (*b*) clearly contradict Smith's statement. This erroneous statement is repeated, among several other incorrect assertions concerning tilings, in Smith [1978].

An interesting discussion of the geometric problems inherent in the division of cells in hexagonal arrangements (common in many epithelial and epidermal tissues) is presented in Abbott & Lindenmayer [1981].

Dress [1985b] has devised a very ingenious method (based on a homological idea) for the investigation of local perturbations possible in regular tilings.

A particularly intriguing open problem is to find variants of the results of Sections 3.5, 3.6 and 3.7 which are meaningful for non-normal tilings of certain kinds. A few relevant facts will be discussed in Section 4.6.

4 THE TOPOLOGY OF TILINGS*

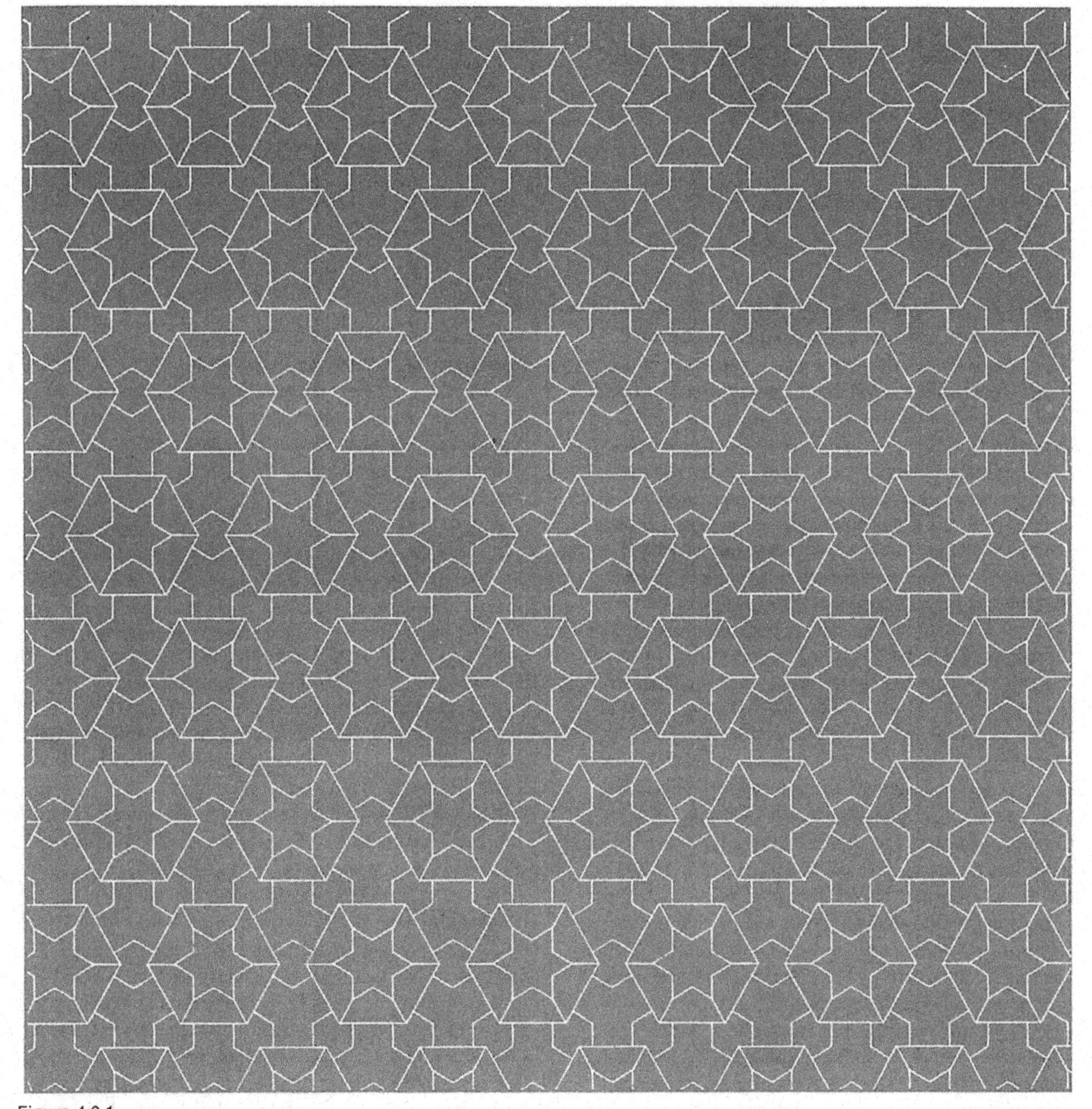

Figure 4.0.1
From a topological point of view, this tiling is equivalent to the regular tiling by regular hexagons in which three tiles meet at each vertex. The notion of "topological equivalence" plays a basic role in the study of tilings and provides us with one classification of tilings with certain transitivity properties.

* See the Appendix beginning on page 653.

4

THE TOPOLOGY OF TILINGS

The basic idea underlying this chapter is that of a "homeomorphism" or "topological transformation". These words will be formally defined in the first section; here it seems useful to try to explain them in intuitive terms.

In Section 1.3, in connection with our discussion of symmetries, we explained how an isometry could be interpreted using a tracing on a transparent sheet: two geometric figures F_1, F_2 are isometric (congruent) if we can trace F_1 onto a sheet which can then be moved so that the tracing "fits" exactly over F_2. In a similar way, two figures are homeomorphic if, in addition to sliding the tracing and possibly turning it over, we are allowed to stretch or compress (uniformly, or non-uniformly) the tracing paper, as long as we do not fold it or tear it. Thus an isometry is a very special case of a homeomorphism. Whereas an isometry is a transformation which preserves properties like straightness, angles between lines, and area, homeomorphisms need not preserve any of these. On the other hand, the application of a homeomorphism to a tiling preserves the valences of the vertices and the number of adjacents and neighbors of each tile. It is properties of this kind that we shall consider in this chapter.

A discussion of duality, already mentioned briefly in Section 2.7, fits naturally here, and will form the topic of Section 4.2. The later sections of the chapter are concerned with the classification, in topological terms, of tilings of certain special kinds—those in which the homeomorphisms that map the tiling onto itself are transitive on the tiles, or the vertices, or the edges. This is preparatory to the much fuller discussion of the classification of such tilings with respect to symmetries which will be presented in Chapter 6.

Section 4.7 is a digression: we shall mention some of the curious and unusual properties that arise when we extend our discussion to non-normal tilings.

4.1 HOMEOMORPHISMS AND TOPOLOGICAL EQUIVALENCE

A mapping $\phi: E^2 \to E^2$ of the plane onto itself is called a *homeomorphism* or a *topological transformation* if it is one-to-one and bicontinuous. *One-to-one* (or *bijective*) means that for any two points P, Q in the plane, $\phi(P) = \phi(Q)$ if and only if $P = Q$; this implies that there exists an inverse transformation ϕ^{-1} such that $\phi^{-1}(R) = P$ if and only if $\phi(P) = R$. The concept of continuity is familiar from elementary calculus—here it means simply that it is always possible to make $\phi(P)$ and $\phi(Q)$ as close together as we wish by taking P and Q sufficiently close. *Bicontinuity* means that both ϕ and ϕ^{-1} are continuous.

Using the idea of a homeomorphism we are now able to give a more useful and satisfactory definition of a *closed topological disk* than that proposed in Section 1.1. It is any plane set which is the image of a closed circular disk under a homeomorphism. Similarly, an *open topological disk* is the image of an open circular disk under a homeomorphism. Each topological disk, closed or open, is a bounded, connected and simply connected set.

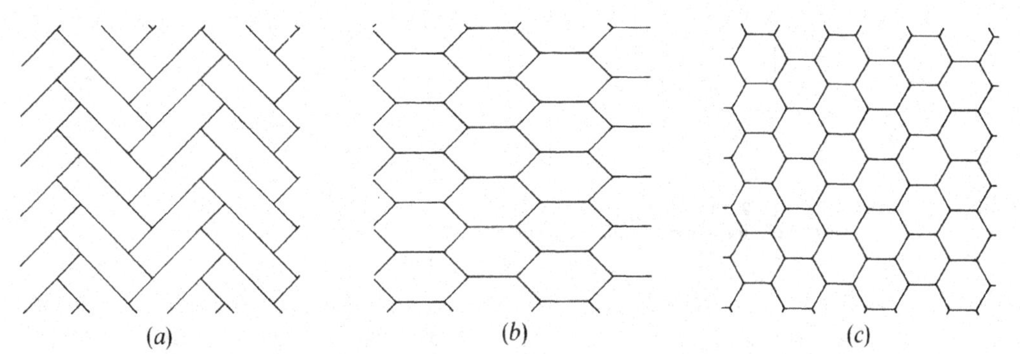

Figure 4.1.1
Three topologically equivalent monohedral tilings. The text describes the construction of homeomorphisms that map these tilings onto each other.

Two tilings are said to be of the same *topological type* (or to be *topologically equivalent*) if there is a homeomorphism which maps one onto the other. The fact that the composition of two homeomorphisms, and the inverse of a homeomorphism, are also homeomorphisms shows that topological equivalence is an equivalence relation (in that it is reflexive, symmetric and transitive). It therefore partitions the set of all tilings into "topological types".

Figures 4.1.1 to 4.1.4 show a number of tilings which will illustrate the definition of topological equivalence given above. In order to show that the "herringbone pattern" of rectangular tiles in Figure 4.1.1(*a*) is of the same topological type as the regular hexagonal tiling in Figure 4.1.1(*c*), it is convenient to use the "intermediate" tiling shown in Figure 4.1.1(*b*). Starting from the tiling in Figure 4.1.1(*a*) we first move the vertical "zigzags" up and down alternately to produce the tiling of Figure 4.1.1(*b*), then "shrink" the tiling horizontally to obtain that of Figure 4.1.1(*c*). Each of these operations corresponds to a homeomorphism and so we deduce that all three tilings are of the same topological type. It would be possible, if we wished, to represent the transformations just described using equations in the coordinates; but we do not do so for they would be complicated and would not help, in any way, to understand the necessary operations.

We could use similar arguments to establish the topological equivalence of the two tilings shown in each of Figures 4.1.2, 4.1.3 and 4.1.4. In these cases a description of the homeomorphisms would be quite complicated—we shall show shortly that there is a much simpler method of establishing topological equivalence without the necessity of producing explicit homeomorphisms.

The examples of Figures 4.1.1 to 4.1.4 show that, even in the monohedral case, topologically equivalent tilings may appear to be very different. All these tilings are normal, and it is natural to ask whether this situation is typical—if $\mathscr{T}_1$ is topologically equivalent to a normal tiling $\mathscr{T}_2$, is $\mathscr{T}_1$ necessarily normal? The answer, as we shall now show, is in the negative. We recall from Section 3.2 that a tiling is normal if the tiles are topological disks (condition N.1), if the intersection of any two tiles is either empty or a connected set (N.2), and if the tiles are uniformly bounded (N.3). Since the homeomorphic image of a closed topological disk is a closed topological disk, the fact that $\mathscr{T}_2$ satisfies N.1 shows that $\mathscr{T}_1$ does so also. In a similar way, we see that $\mathscr{T}_1$ satisfies N.2. However, it is possible that the tiles of $\mathscr{T}_1$ fail to be uniformly bounded, and so condition N.3 is violated.

For an example of this situation consider Figure 4.1.5. Here we show a tiling obtained from the regular tiling $[4^4]$ by applying a homeomorphism—we "shrink" it horizontally more and more as we move up or down

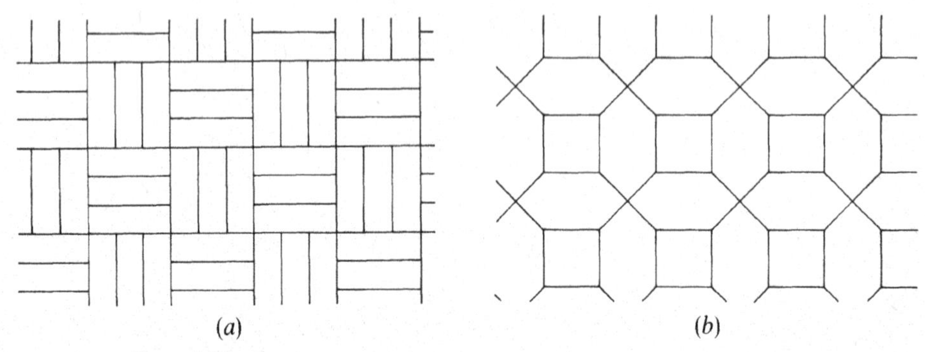

Figure 4.1.2
A monohedral tiling (*a*) which is topologically equivalent to a dihedral tiling (*b*).

away from a fixed horizontal line. It is easy to see that for every value of U, however large, there exists an infinite number of tiles each of which cannot be covered by any circular disk of radius U; and for every positive value of u, however small, there exists an infinite number of tiles each of which does not contain any circular disk of radius u. Hence the tiling does not satisfy the uniform boundedness condition N.3 although it is topologically equivalent to a regular tiling which is normal. (The reader to whom these ideas are new should be careful to convince himself that the tiling of Figure 4.1.5 is topologically equivalent to the regular tiling (4^4). To do this one can write down equations representing the homeomorphism. If (x,y) and (x',y') are coordinates then a transformation of the form $x = x'/(1 + y'^2)$, $y = y'$, or something similar, is required.)

Although homeomorphisms do not preserve normality it is worth noticing that if a tiling has no singular points then any topologically equivalent tiling also has no singular points. On the other hand, if a tiling $\mathscr{T}$ has a singularity at infinity, then it may be possible to find a topologically equivalent tiling without such a singularity.

We now introduce the notion of combinatorial isomorphism or combinatorial equivalence. Let $\mathscr{E}(\mathscr{T})$ denote the set of all *elements* of a tiling $\mathscr{T}$, that is, the set whose members are the vertices, edges and tiles of $\mathscr{T}$. A map Φ of $\mathscr{E}(\mathscr{T}_1)$ *onto* $\mathscr{E}(\mathscr{T}_2)$ is said to be *inclusion-preserving* if, whenever $e_1, e_2 \in \mathscr{E}(\mathscr{T}_1)$, then $\Phi(e_1)$ includes $\Phi(e_2)$ if and only if e_1 includes e_2. If there exists an inclusion-preserving map between $\mathscr{T}_1$ and $\mathscr{T}_2$, then $\mathscr{T}_1$ and $\mathscr{T}_2$ are said to be *combinatorially isomorphic* or *combinatorially equivalent*. If V is any n-valent vertex of $\mathscr{T}_1$, then $\Phi(V)$ will be an n-valent vertex of the combinatorially equivalent tiling

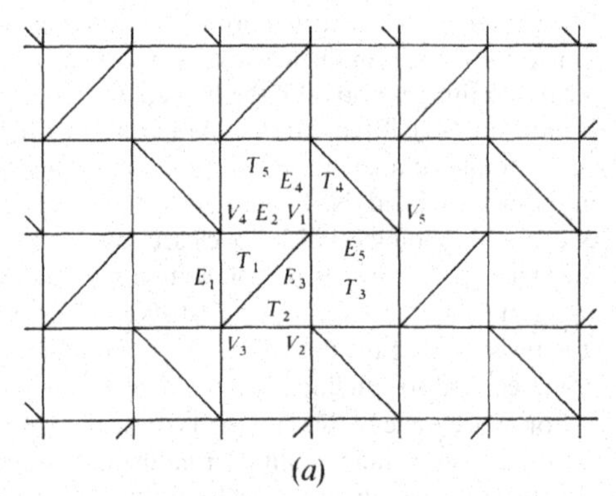

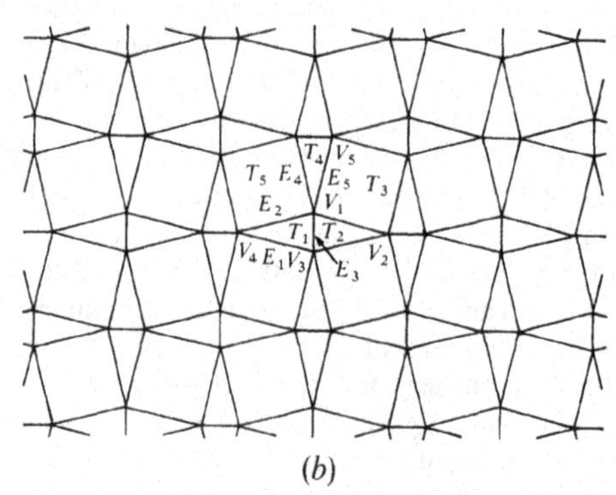

Figure 4.1.3
Two topologically equivalent dihedral tilings. Each is of the same topological type as the uniform tiling $(3^2.4.3.4)$.

Figure 4.1.4
Two topologically equivalent monohedral tilings. Each is of the same type as the regular tiling (4^4).

$\mathscr{T}_2$. Similarly, if a tile T of $\mathscr{T}_1$ has n adjacents, then so does the corresponding tile $\Phi(T)$ of $\mathscr{T}_2$.

The definition of combinatorial equivalence is illustrated in Figure 4.1.3. Here we have labelled some of the elements of the first tiling and used the same letters to indicate the elements of the second tiling which are the images of these under a mapping Φ. Then it is easy to verify that inclusion is preserved. For example, in both tilings the tile labelled T_1 includes vertices V_1, V_3, V_4 and edges E_1, E_2, E_3 and no others. Moreover it is clear how the labelling can be extended to all the elements of each tiling; we deduce that they are combinatorially equivalent.

The following result is fundamental.

4.1.1 *For normal tilings the concepts of topological equivalence and combinatorial equivalence coincide.*

The reason for introducing combinatorial equivalence is that it is conceptually simple. Moreover, it is useful in practice since an inclusion-preserving map can be indicated easily on a diagram. For example, in Figure 4.1.3 it may not be immediately clear that the two tilings are topologically equivalent but, as we have seen, marking corresponding elements of $\mathscr{T}_1$ and $\mathscr{T}_2$ with the same letters enables us to show with very little effort that they are combinatorially equivalent. In fact this idea can be further simplified: it is really necessary to label only the

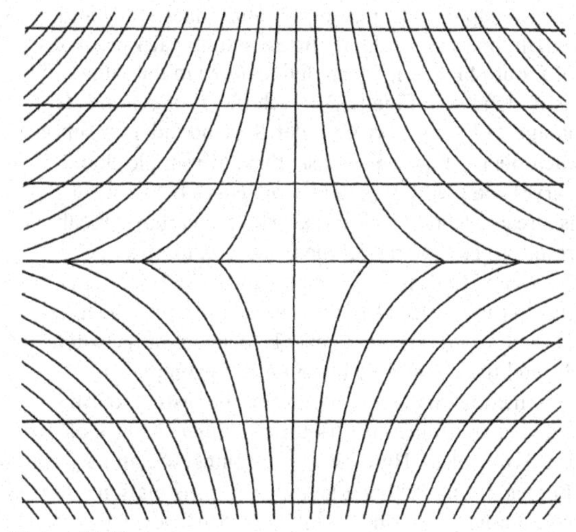
Figure 4.1.5
A non-normal tiling that is topologically equivalent to the normal, regular tiling by squares.

vertices in $\mathscr{T}_1$ and $\mathscr{T}_2$ since all the other elements are determined by these.

Using this procedure it is now simple to show that the pairs of tilings in Figures 4.1.1, 4.1.2 and 4.1.4 are combinatorially equivalent and so, by Statement 4.1.1, topologically equivalent.

To prove Statement 4.1.1 we first show that topologically equivalent tilings are combinatorially equivalent. Let $\phi: E^2 \to E^2$ be a homeomorphism between the planes, the first of which contains a tiling $\mathscr{T}_1$ and the second contains the image $\mathscr{T}_2$ of $\mathscr{T}_1$ under ϕ. Since ϕ maps each tile of $\mathscr{T}_1$ (pointwise) onto a tile of $\mathscr{T}_2$, and the same holds for edge and vertices, it induces a map $\Phi: \mathscr{E}(\mathscr{T}_1) \to \mathscr{E}(\mathscr{T}_2)$ which is obviously inclusion-preserving.

For the converse assertion let us suppose that Φ is an inclusion-preserving map of $\mathscr{E}(\mathscr{T}_1)$ onto $\mathscr{E}(\mathscr{T}_2)$. We are required to construct a homeomorphism $\phi: E^2 \to E^2$ that maps $\mathscr{T}_1$ onto $\mathscr{T}_2$; in fact we shall produce a homeomorphism ϕ that induces (as in the first part of the proof) the given map Φ. We define ϕ on the vertices of $\mathscr{T}_1$ by putting $\phi(V) = \Phi(V)$ for each vertex V. Next, for each edge E of $\mathscr{T}_1$ we define ϕ as any homeomorphism between E and $\Phi(E)$ that agrees with the already defined value of ϕ at the endpoints. For example, if the edges are of finite length, one way of doing this is to map each point X of E onto the point which divides $\Phi(E)$ in the same ratio as X divides E. Finally, for each tile T of $\mathscr{T}_1$ and its image $\Phi(T)$ in $\mathscr{T}_2$ we consider their boundaries. Since each tile is a topological disk, these are simple closed curves composed of finitely many edges between which ϕ is already defined. There is a well known (but difficult) result in the topology of the plane (see, for example, Newman [1951, p. 173], or Statement 7.1.2) which asserts that ϕ can be extended from the boundary to the interior of the tiles. In this way ϕ may be defined on the whole of E^2 and the proof of Statement 4.1.1 is completed.

Sometimes it is more natural or interesting to ask whether two tilings are "isotopic", rather than topologically equivalent. Here we say that tilings $\mathscr{T}_0$ and $\mathscr{T}_1$ are *isotopic* if there exists a family of transformations that continuously deforms $\mathscr{T}_0$ into $\mathscr{T}_1$. (More formally, $\mathscr{T}_0$ and $\mathscr{T}_1$ are isotopic provided there exists a family $\mathscr{T}(\lambda)$, with $0 \leq \lambda \leq 1$, of tilings that depend continuously on λ, and such that $\mathscr{T}_0 = \mathscr{T}(0)$ and $\mathscr{T}_1 = \mathscr{T}(1)$.) Isotopy appears to be more restrictive than topological equivalence but, since we know of no topologically equivalent tilings that are not isotopic, we conjecture that the two concepts actually coincide. The reader may appreciate the difficulty of establishing the validity of this conjecture by considering the two tilings in Figure 4.1.4.

To represent visually an isotopy between two tilings $\mathscr{T}_0$ and $\mathscr{T}_1$ it is convenient to draw the intermediate tilings $\mathscr{T}(\lambda)$ for sufficiently many values of λ so that the method of deformation is apparent; in effect we draw the "frames" in a cinematographic demonstration of the isotopy.

A concept akin to isotopy but distinct from it has been used by some artists; M. C. Escher utilized it in his famous woodcut "Metamorphosis III" and other works (see Escher [1971], [1982]). The tiling is represented on a long strip of paper and it "changes" gradually as one moves from one end of the strip to the other. For other examples of such tilings see Hofstadter [1983], Huff [1983]. It is quite challenging to design deformations of this kind in such a way that every tile can serve as the prototile of a monohedral tiling. Despite the claim to the contrary (see Hofstadter [1983, p. 14]), most of the tilings shown in Hofstadter's article include tiles which are not prototiles of any monohedral tilings.

Still other methods of continuous transformation of tilings, different from isotopy and the deformations just described, have been investigated by various authors. A classical work along these lines is Stott [1910]. More recently similar ideas have been proposed in a number of publications. For example, there is the "wandering vertex method" of Burt [1982] and the "transnets" of Lalvani [1977], [1981], [1982a], [1982b]. But as they lack the aesthetic appeal of the Escher-Huff approach and, more importantly, as their criteria and aims are not clearly formulated, they seem to be of very limited interest. In all these cases, the authors have been chiefly interested in polyhedra and higher-dimensional polytope, and the work concerning tilings appears to have been introduced only as an afterthought.

EXERCISES 4.1

1. Show that the tiling of Figure 3.2.3(*b*) is topologically equivalent to a uniform tiling. Determine which uniform tiling it is, and describe (as in Figure 4.1.1) a homeomorphism between these two tilings.
2. Construct a tiling, topologically equivalent to the tiling in Figure 3.7.4, in which all tiles with 5 adjacents are regular pentagons.
3. For each of the tilings in Figures 1.3.5 and 2.4.2 find a topologically equivalent edge-to-edge uniform tiling.
4. Extend the result of Statement 4.1.1 to tilings that satisfy N.1 and N.3, but for which the intersection of two tiles may have two connected components.

*5. Construct two tilings that satisfy N.2 and N.3 and have tiles that are either topological disks or topological rings, but for which the assertion of Statement 4.1.1 is not valid.

4.2 DUALITY

We are now able to explain in more detail the concept of duality briefly mentioned in Section 2.7 in connection with the uniform tilings and the Laves tilings. Except in the exercises and the last three paragraphs, we shall assume throughout this section that all tilings mentioned are normal.

Two tilings $\mathcal{T}$ and $\mathcal{T}^*$ are said to be *dual* to each other if there exists a one-to-one inclusion-reversing map Ψ from the set $\mathscr{E}(\mathcal{T})$ onto the set $\mathscr{E}(\mathcal{T}^*)$. (We recall that $\mathscr{E}(\mathcal{T})$ is the set of elements—vertices, edges and tiles—of the tiling $\mathcal{T}$.) By *inclusion-reversing* we mean that whenever e_1 and e_2 are elements of $\mathcal{T}$, then $\Psi(e_1)$ includes $\Psi(e_2)$ if and only if e_2 includes e_1. Naturally, such a map will send tiles, edges and vertices of $\mathcal{T}$ onto vertices, edges and tiles of $\mathcal{T}^*$, respectively. This is illustrated in Figure 4.2.1. Here some of the elements of $\mathcal{T}$ are labelled, and the images of these under Ψ are indicated by the same symbol with an asterisk. The reader should verify that it is possible, in a consistent way, to extend this labelling to all the elements of $\mathcal{T}$, and of $\mathcal{T}^*$.

The following results about duality are basic.

4.2.1 *If each of two tilings $\mathcal{T}_1$ and $\mathcal{T}_2$ is dual to the same tiling $\mathcal{T}^*$, then $\mathcal{T}_1$ and $\mathcal{T}_2$ are combinatorially equivalent and therefore topologically equivalent.*

This follows immediately from the definitions and Statement 4.1.1.

4.2.2 *For every normal tiling $\mathcal{T}$ there exists a normal tiling $\mathcal{T}^*$ which is dual to $\mathcal{T}$.*

Before starting the proof of Statement 4.2.2 it is convenient to introduce the following concept: two tilings $\mathcal{T}$ and $\mathcal{T}^*$ are said to be *dually situated* to each other if they lie in the same plane, every vertex of one is an interior point of a tile of the other, every tile of one contains precisely one vertex of the other, and every edge (with vertices V_1, V_2) of one tiling meets in a single point, and crosses, just one edge of the other. Moreover, the latter edge is the intersection of the tiles T_1, T_2 that contain the vertices V_1, V_2 of the other tiling in their interiors.

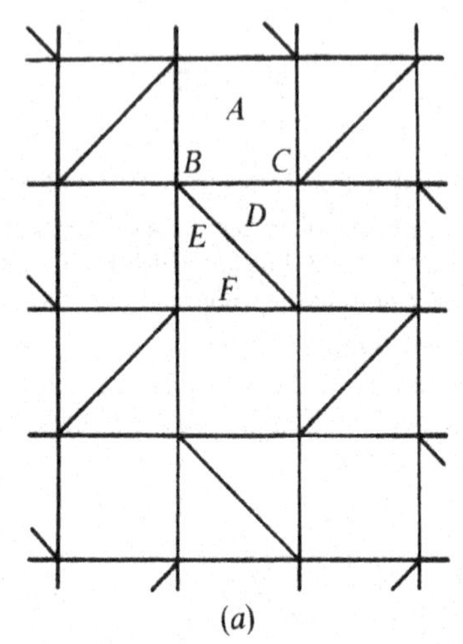

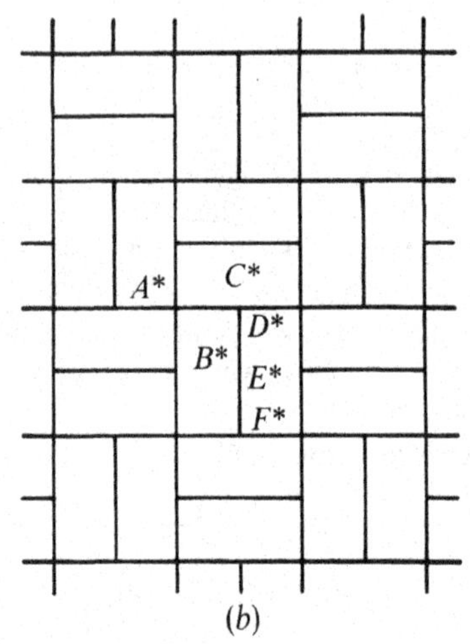

Figure 4.2.1
A pair of dual tilings. The labels indicate corresponding elements in the tilings.

Examples of dually situated tilings have already been given in Figure 2.7.3, and a further example appears in Figure 4.2.2. The edges of one tiling are indicated by solid lines, those of the other by dashed lines. From the example in Figure 4.2.1 we see that we cannot, in general, superimpose one tiling on a dual so as to ensure that they are dually situated, even if we allow any motion or change of scale of the one tiling relative to the other; this is true in spite of the fact that these tilings have straight edges and convex tiles, and each is homeomorphic to one of the two dually situated tilings of Figure 4.2.2.

To prove Statement 4.2.2 we first show how to construct, for a given normal tiling, a dually situated tiling whose tiles are topological disks. To do this, we choose an interior point of each tile T of $\mathscr{T}$ to be a vertex V^* of $\mathscr{T}^*$, and then choose one point $e(E)$ on each edge E of $\mathscr{T}$, distinct from the two vertices on that edge. Finally, we connect V^* to all the points $e(E)$ that we have chosen on the edges of T by arcs that are disjoint (except for containing V^*) and which lie (apart from the endpoints $e(E)$) in the interior of T. The two arcs that end at a point $e(E)$ together form an edge E^* of $\mathscr{T}^*$ that corresponds to the edge E of $\mathscr{T}$. The tile T^* of $\mathscr{T}^*$ that corresponds to a vertex V of $\mathscr{T}$ is the part of the plane bounded by the edges of $\mathscr{T}^*$ that correspond to those edges of $\mathscr{T}$ that meet at V.

Figure 4.2.3 gives an example of the construction. Here the points V^* are indicated by solid circles, and the points $e(E)$ by open circles. It will be seen that the tiles of $\mathscr{T}^*$ are topological disks. Moreover, $\mathscr{T}^*$ satisfies condition N.2 for normality, namely that the intersection of any set of tiles is empty or connected.

To complete the proof we need to show that, because of the normality of $\mathscr{T}$, it is possible to choose $\mathscr{T}^*$ to be normal, that is, in such a way that its tiles are uniformly bounded and so satisfy condition N.3. Let U and u be the parameters of the tiling $\mathscr{T}$, and let T^* be any tile of $\mathscr{T}^*$. Let V be the vertex of $\mathscr{T}$ that lies in T^*, see Figure 4.2.4. Then all tiles of $\mathscr{T}$ that contain V lie in a circular disk of radius $2U$ centered at V and, moreover, their union contains T^*. Hence $2U$ is a suitable choice for the circumparameter of $\mathscr{T}^*$.

The preceding argument is valid for any dually situated tiling, but in the rest of the proof it is necessary to select $\mathscr{T}^*$ with some care. For it is not difficult to see how certain unsuitable choices of the edges of $\mathscr{T}^*$ could

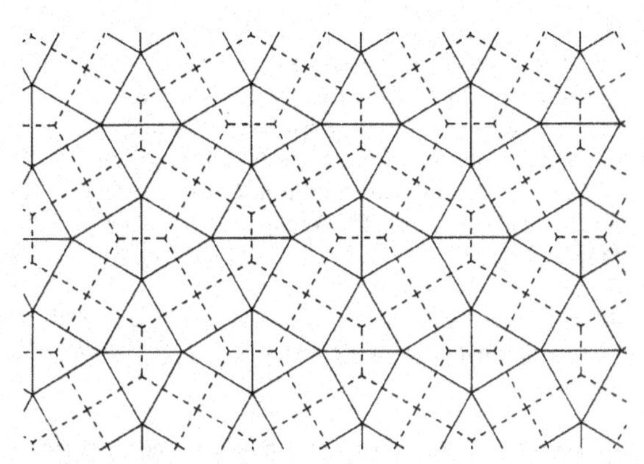

Figure 4.2.2
A pair of dually situated tilings; one is indicated by solid lines, the other by dashed lines. They are topologically equivalent to the pair of dual tilings in Figure 4.2.1.

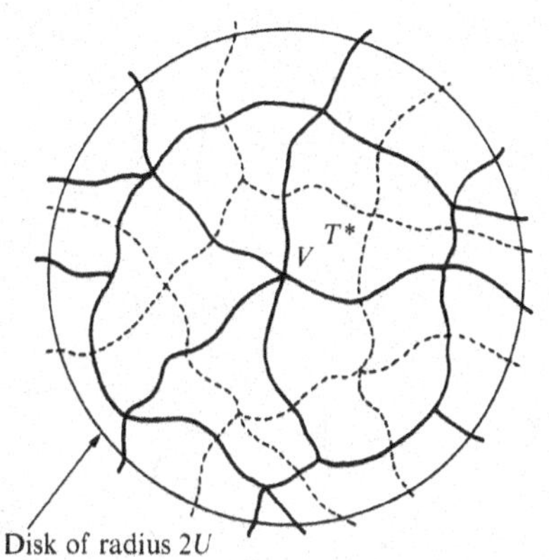

Figure 4.2.4
For pairs of dually situated tilings $\mathscr{T}$ and $\mathscr{T}^*$, each tile T^* of $\mathscr{T}^*$ is contained in the union of all the tiles of $\mathscr{T}$ which contain the vertex V corresponding to T^*. Hence $2U$ can serve as a circumparameter for $\mathscr{T}^*$, where U is a circumparameter for $\mathscr{T}$.

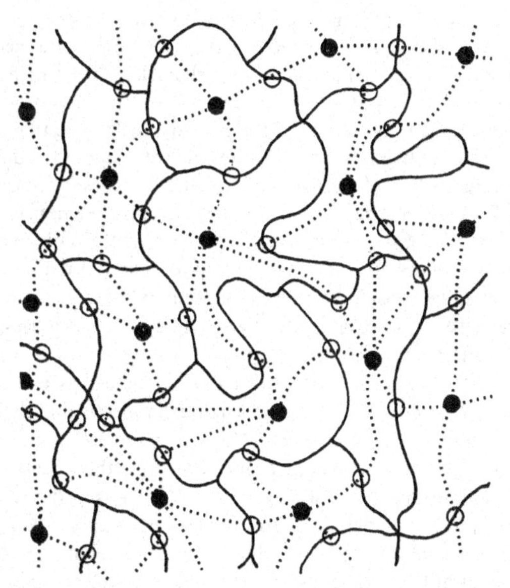

Figure 4.2.3
An example of the construction of a tiling (indicated by dotted lines) dually situated with respect to a given normal tiling (solid lines).

preclude the existence of a positive value of the inparameter.

The normality of $\mathscr{T}$ implies that every tile of $\mathscr{T}$ contains a circular disk of radius u. Let T_0 contain such a disk D_0 and choose the vertex V_0^* of $\mathscr{T}^*$ as the center of D_0. Let D_0' be another disk, concentric with D_0 but slightly smaller, of radius $(\sqrt{3} + 2)u/4$. In a similar way define points $V_1^*, V_2^*, \ldots, V_m^*$ and disks D_1' $D_2', \ldots$ in the other tiles $T_1, T_2, \ldots$ of $\mathscr{T}$ (see Figure 4.2.5). Then, instead of constructing, as edges of $\mathscr{T}^*$, *any* arcs connecting the points V_i^* in the tiles to the points $e(E)$ on their edges, we make sure that within the disks D_i' they are straight-line segments (radii) equally inclined to one another, that is, each segment makes an angle $2\pi/m$ with its neighbors. By elementary geometry it follows (see Figure 4.2.6) that since $m \geq 3$, each sector between two consecutive radii of a disk D_i' contains a disk of radius $\frac{1}{2}u \sin \pi/m$. Thus every tile of $\mathscr{T}^*$ contains a circular disk of that radius. Now, from Section 3.2 we recall that the number of adjacents of a tile in a normal tiling is uniformly bounded by $9U^2/u^2$. Since this applies to $\mathscr{T}$, it follows that each vertex of $\mathscr{T}^*$ has valence $m \leq 9U^2/u^2$. Each tile of $\mathcal{T}^*$ therefore contains a circular disk of radius $\frac{1}{2}u \sin(\pi u^2/9U^2)$. This quantity can be taken as an inparameter for $\mathcal{T}^*$. The existence of the circumparameter and inparameter shows that the tiles of $\mathcal{T}^*$ are uniformly bounded, condition N.3 is satisfied, and so $\mathcal{T}^*$ is normal. This completes the proof of Statement 4.2.2.

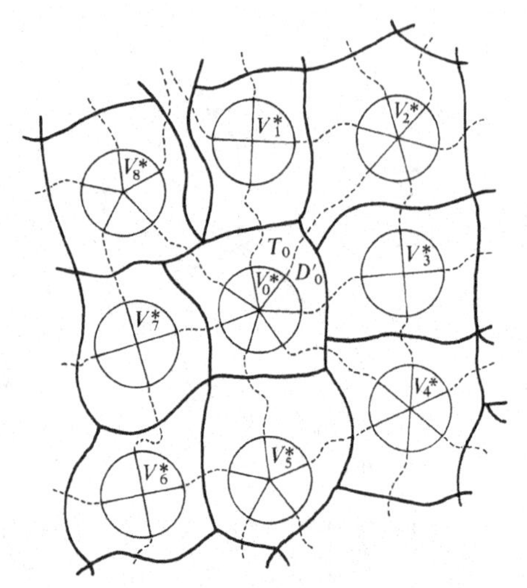

Figure 4.2.5
The construction, explained in the text, of a dually situated tiling in which tiles have incircles of radii bounded from below by a positive number.

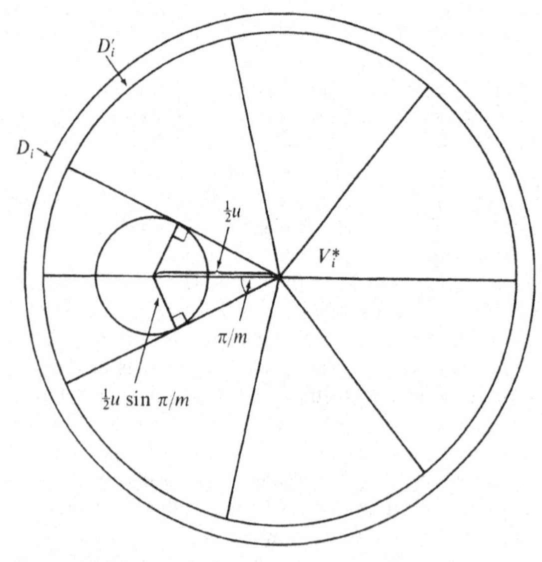

Figure 4.2.6
Computation of the bounds for the inparameter of a dually situated tiling, as explained in the text.

4.2.3 *For every normal (strongly) balanced tiling $\mathcal{T}$ there exists a normal (strongly) balanced tiling $\mathcal{T}^*$ which is dual to $\mathcal{T}$.*

The proof of this statement can be deduced from the construction of the normal tiling $\mathcal{T}^*$ described in the proof of Statement 4.2.2. It is not difficult to show that if $\mathcal{T}$ is (strongly) balanced then $\mathcal{T}^*$ is (strongly) balanced also.

The ideas of duality and dually situated tilings apply also to certain tilings that are not normal. For example, the construction indicated in Figure 4.2.3 can be carried out, with minor modifications, when the tiling satisfies conditions N.1 and N.3, but not N.2. As an illustration we show in Figure 4.2.7 a tiling $\mathcal{T}$ in which some tiles have only two adjacents. Such tiles give rise, in the dual tiling, to "pseudovertices" of "valence 2". These are not, of course, vertices in the sense that we have previously used the term, but may be regarded as fictitious objects invented just to extend the concept of duality. In a few cases, the consideration of such pseudovertices is useful, and we shall occasionally refer to them in the exercises of this and later sections.

We have now stated almost all that one can assert about duality of tilings. Particular examples of duality in interesting cases will be exhibited later, but essentially no other general results are known. We stress this point because it seems to be widely believed that—as in the case of convex polyhedra in three-dimensional space—there exists a *metrical theory* of duality for tilings. This is not so, or at least we know of no results in this direction. For example, it is not known whether a given tiling whose tiles are convex polygons possesses a dual whose tiles are also convex (though we know that no such dual can exist in every case if we also insist that it is dually situated).

Again, although the concepts of isohedrality and isogonality appear to be dual to each other, as we shall see in Chapter 6 it is impossible to deduce the types of isogonal tilings from those of isohedral tilings.

To summarize, we may say that duality is an important idea which is frequently useful. It must, however, be used with considerable care if errors are to be avoided.

A recent example of careless treatment occurs in Firby & Gardiner [1982, Section 8.4]. In describing the dual of the regular tiling (4^4), they claim to obtain the dual regular tiling (4^4) although they specify neither the position of the vertices of the dual tiling, nor the shape of its edges. Moreover, following their method of construction (which relies on the so-called Cayley graph of a group), *each* pair of the vertices of the dual tiling would have to be connected by an edge!

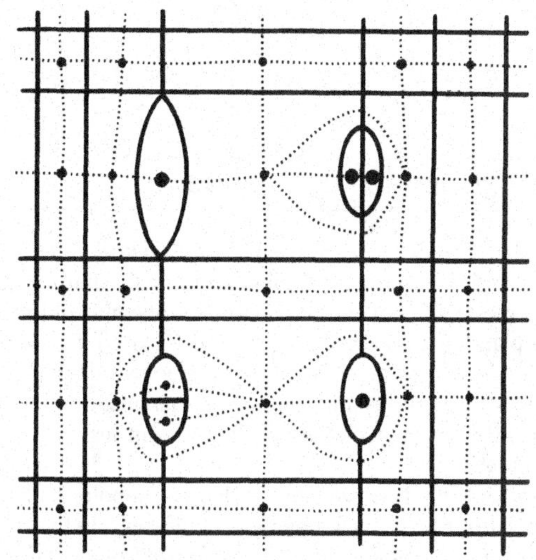

Figure 4.2.7
A tiling in which some tiles have only two adjacents. In the dual tiling (indicated by dotted lines), some of the vertices have valence 2 ("pseudovertices"); these are indicated by the large solid circles.

EXERCISES 4.2

1. Find an example of a normal tiling with convex tiles for which no dually situated tiling with convex tiles exists.
2. By considering the uniform and 2-uniform tilings illustrated in Sections 2.1 and 2.2 formulate a rule for constructing a tiling which is dually situated to a given tiling by regular polygonal tiles, and hence show that any such tiling has a dually situated tiling by convex tiles.
3. Use duality to deduce from Statement 3.2.3 that every normal tiling contains infinitely many vertices of valence at most 6.
4. For the isohedral tiling of Figure 1.3.7 construct a dually situated isogonal tiling by convex tiles.
5. Modify the proof of Statement 4.2.2 so as to obtain a tiling $\mathscr{T}^*$, dual to $\mathscr{T}$, whose circumparameter is smaller than $2U$ and whose inparameter is greater than $\frac{1}{2}u \sin(\pi u^2/9U^2)$.
6. Let $\mathscr{T}$ be any non-normal tiling that satisfies conditions N.1 and N.3. The tiling $\mathscr{T}^*$ is constructed as explained in the proof of Statement 4.2.2 and $\mathscr{T}_0^*$ is the same tiling as $\mathscr{T}^*$ except that each pseudovertex and the two edges incident with it are amalgamated to form an edge of $\mathscr{T}_0^*$. Let $\mathscr{T}_0$ be defined as the dual of $\mathscr{T}_0^*$. This process of constructing $\mathscr{T}_0$ from $\mathscr{T}$ is called *normalization*.

   (*a*) Show that if it is repeated a sufficiently large finite number of times it has the effect of converting $\mathscr{T}$ into a normal tiling.

   (*b*) For each n find a tiling $\mathscr{T}$ which requires the normalization process to be repeated at least n times before $\mathscr{T}$ is converted into a normal tiling.

4.3 HOMEOHEDRAL TILINGS

A tiling $\mathscr{T}$ is called *homeohedral* or *topologically tile-transitive* if it is a normal tiling and is such that for any two tiles T_1, T_2 of $\mathscr{T}$ there exists a homeomorphism of the plane that maps $\mathscr{T}$ onto $\mathscr{T}$ and T_1 onto T_2. Our object now is to describe all the homeohedral tilings. This is the first essential step in the classification of isohedral tilings, which we shall carry out in Chapter 6.

In Section 4.1 we pointed out that two topologically equivalent tilings might appear very different. In the same way, it may be quite difficult to recognize from a diagram that a tiling is homeohedral. For example, the tilings of Figures 1.3.9, 2.5.3(*c*), 2.5.4(*g*), (*l*), (*m*) and 2.5.7(*b*) are homeohedral, as is the equitransitive tiling of Figure 4.0.1.

It is necessary to introduce some notation. If T is a tile of the tiling $\mathscr{T}$ then we shall say that T is of *valence-type* $j_1.j_2.\ldots.j_k$ provided T has k vertices which, if considered in a suitable cyclic order, have valences $j_1, j_2, \ldots, j_k$. If all tiles of a normal tiling $\mathscr{T}$ have the same valence-type $j_1.\ldots.j_k$, we say that $\mathscr{T}$ is *homogeneous of type* $[j_1.\ldots.j_k]$. It is clear that every homeohedral tiling is homogeneous. Naturally, we do not distinguish between types if their symbols can be obtained from each other by a cyclic permutation or reversal of order. In order to standardize our notation we shall choose, among all possible symbols for a homogeneous tiling, the one that is lexicographically first. We shall also use superscripts to abbreviate the symbols when two or more consecutive numbers j_i take the same value.

The following is the main result.

4.3.1 *If $\mathscr{T}$ is a homogeneous tiling then it is of one of the eleven types* $[3^6]$, $[3^4.6]$, $[3^3.4^2]$, $[3^2.4.3.4]$, $[3.4.6.4]$, $[3.6.3.6]$, $[3.12^2]$, $[4^4]$, $[4.6.12]$, $[4.8^2]$, $[6^3]$. *Moreover, each homogeneous tiling is homeohedral, all homogeneous tilings of the same type are topologically equivalent to each other and each type is represented by one of the Laves tilings* (Figure 2.7.1).

To prove this we use Statement 3.5.1 which asserts that

$$\sum_{i=1}^{k} (j_i - 2)/j_i = 2. \tag{4.3.2}$$

As we have already pointed out, this relation coincides with equation (2.1.2). Hence there exist exactly the same 21 systems of integer solutions that are given in the last column of Table 2.1.1.

We must now pick out the integer solutions that correspond to actual tilings. Since the parity arguments used in Section 2.1 apply equally here, it follows, in precisely the same way, that at most eleven types of tilings exist. To illustrate this, consider just one case; all the other cases follow in an analogous manner. Suppose a tile T_1 has valence-type $3^2.4.12$ as marked in Figure 4.3.1. Then if the tile T_2 is to be of the same valence-type, the vertex A must have valence 3, whereas if tile T_3 is to be of the same valence-type vertex A must have valence 4 or 12. This is a contradiction from which we deduce that no homogeneous tiling of type $[3^2.4.12]$ can exist; this possibility can therefore be rejected. In the same way nine other possibilities can be eliminated, leaving us with the eleven types of tilings listed in Statement 4.3.1.

The fact that homogeneous tilings of these eleven types do indeed exist is now clear. We refer to Figure 2.7.1 which shows the eleven Laves tilings. These are not only homogeneous or homeohedral but are, in fact, even isohedral, since their symmetry groups act transitively on the tiles.

Finally it is necessary to check the uniqueness assertion, that is, we must show that two homogeneous tilings of the same type are topologically equivalent (and are also equivalent to the Laves tiling with the same symbol). We achieve this by constructing an inclusion-preserving map between the given tiling and the appropriate Laves tiling. The method of construction is similar in each case so we shall discuss one case only. Let us choose a tiling $\mathscr{T}$ of type $[3^2.4.3.4]$. Starting with a tile T_0 of $\mathscr{T}$ and an arbitrarily chosen tile T'_0 of the Laves tiling $\mathscr{L}$ of type $[3^2.4.3.4]$, we can establish a correspondence Φ between the vertices and edges of T_0 and those of T'_0 that preserves the valences of the vertices. (In this case Φ

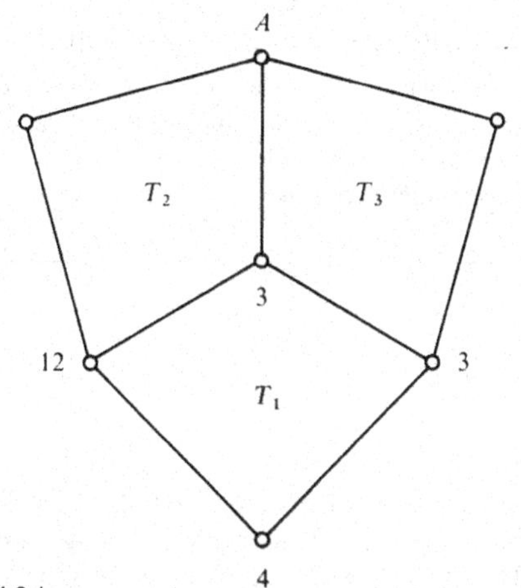

Figure 4.3.1
An illustration of the argument that establishes the non-existence of homogeneous tilings of type [3^2.4.12].

could be chosen in two ways—for other types the number of possibilities may range from 1 to 12.) Then we observe that Φ can be extended in a unique way to a map from the elements of the "ring" R_1 formed by the tiles that are neighbors of T_0 to the elements of the "ring" R'_1 formed by the neighbors of T'_0. We next consider the "rings" R_2 and R'_2 formed by the tiles neighboring those in R_1 and R'_1, and extend Φ accordingly. Proceeding in this way we extend Φ step by step to successive rings.

Because of the normality of both tilings the unions of these rings cover the planes; we deduce that there is an inclusion-preserving map between $\mathcal{T}$ and $\mathcal{L}$, and hence they are homeomorphic. This completes the proof of Statement 4.3.1.

Several remarks must be made in connection with the result we have just proved. They are, in part, analogous to the observations on Archimedean tilings made in Section 2.1.

First, there is an unexpected and very remarkable property of the eleven types of homeohedral tilings: each class contains at least one *very special* tiling, which has convex tiles and regular vertices and is *metrically* tile-transitive (that is, isohedral). The extent to which this is an accidental occurrence will be seen by considering non-normal topologically tile-transitive tilings (see Section 4.7), or the 2-homeohedral tilings (that is, tilings in which the tiles form two equivalence classes under topological equivalences of the tiling; see Section 4.6). Examples of 2-homeohedral tilings are shown in Figure 4.1.2; here the tiles form two classes under homeomorphisms of the tiling, and even under isometries, and the tiles are convex. However, in these two tilings the vertices are not regular, and it is easy to verify that no topologically equivalent tiling can have convex tiles and regular vertices.

The second remark is that it is quite fortuitous that all homeohedral tilings of the same type are topologically equivalent. In contrast to this we shall show, in Sections 4.6 and 4.7, that under very similar circumstances the situation can be quite different.

EXERCISES 4.3

1. Determine the valence-type(s) of the tiles in the following homogeneous tilings, and check whether the tilings are homeohedral.
   (*a*) The Escher tiling of Figure I.18.
   (*b*) The various brick walls in Figure I.11.
   (*c*) The tilings of Figures 1.3.5, 1.3.7, and 1.3.9.
   (*d*) The tilings of Figure 2.4.2.
   (*e*) The tilings of Figures 2.5.3 and 2.5.4.
2. Construct a normal tiling in which
   (*a*) one tile is a hexagon of valence-type 4^6, and all other tiles are quadrangles of valence-type 4^4;
   (*b*) one tile has valence-type 6^6, all other tiles 6^3;

(*c*) one tile has valence-type 3^{10}, all other tiles 3^6;
(*d*) one tile has valence-type 3.6.3.6.3.6, all other tiles 3.6.3.6.

3. Prove that there exists no normal tiling in which one tile has valence-type 3^4, four tiles have valence-type $3^2.4^2$ and all other tiles have valence-type 4^4.
4. Determine which of the eleven types of homeohedral tilings have representatives that are isohedral with
(*a*) a rectangle as prototile;
(*b*) a triangular prototile.

4.4 HOMEOGONAL TILINGS

A tiling $\mathcal{T}$ is called *homeogonal* or *topologically vertex-transitive* if it is a normal tiling and is such that for any two vertices V_1, V_2 there exists a homeomorphism of the plane that maps $\mathcal{T}$ onto $\mathcal{T}$ and V_1 onto V_2. The main purpose of this section is to classify such tilings and, in particular, to establish that there exist precisely eleven topologically inequivalent types.

Let V be a vertex of the tiling, and let the j tiles of $\mathcal{T}$ that contain V have, in a suitable cyclic order about V, k_1 vertices, k_2 vertices, ..., k_j vertices. Then we shall associate with the vertex V the symbol $k_1.k_2.\ldots.k_j$ and, if $\mathcal{T}$ is homeogonal (so that every vertex has the same symbol), we shall say that $\mathcal{T}$ *is of type* $(k_1.k_2.\ldots.k_j)$. We do not distinguish between symbols obtainable from each other by cyclic permutations or reversal of order, and in order to standardize the notation we shall always use the lexicographically first among the possible symbols. We shall also abbreviate the symbol by using superscripts whenever it is convenient to do so. Our main result is the following.

4.4.1 *If* $\mathcal{T}$ *is a (normal) homeogonal tiling then it is of one of the eleven types* (3^6), $(3^4.6)$, $(3^3.4^2)$, $(3^2.4.3.4)$, $(3.4.6.4)$, $(3.6.3.6)$, (3.12^2), (4^4), $(4.6.12)$, (4.8^2), (6^3). *Moreover, all tilings of the same type are mutually topologically equivalent, and each type is represented by one of the uniform tilings described in Section* 2.1.

The close analogy of this result with that established in Section 4.3 (concerning homeohedral tilings) is obvious. This analogy extends also to a proof which may be modelled on that given in Section 4.3 using Statement 3.5.4. However, while the carrying out of such a proof constitutes a good exercise, we prefer to give here a different proof, based on the notion of duality as defined and explained in Section 4.2. This approach is possible because of the following fact.

4.4.2 *Let* $\mathcal{T}$ *and* $\mathcal{T}^*$ *be normal, dual tilings. Then* $\mathcal{T}$ *is homeogonal if and only if* $\mathcal{T}^*$ *is homeohedral.*

In order to prove this assertion let us first assume that $\mathcal{T}^*$ is homeohedral. Let V_1, V_2 be any two vertices of $\mathcal{T}$ and T_1^*, T_2^* be the tiles of $\mathcal{T}^*$ that correspond to V_1, V_2 by duality. The topological tile-transitivity of $\mathcal{T}^*$ implies that there exists an inclusion-preserving map Φ^* of $\mathcal{T}^*$ onto itself which maps T_1^* onto T_2^*, and we show how this implies the existence of an inclusion-preserving map Φ of $\mathcal{T}$ onto itself which maps V_1 onto V_2. In fact, all we have to do is to take any element (vertex, edge or tile) e of $\mathcal{T}$, find the element (tile, edge or vertex) e^* of $\mathcal{T}^*$ that corresponds to e, and define $\Phi(e)$ to be the element of $\mathcal{T}$ that corresponds to $\Phi^*(e^*)$ by the duality. It is trivial that Φ, so defined, is inclusion-preserving. We deduce from the results of Section 4.1 that there exists a homeomorphism of $\mathcal{T}$ which maps V_1 onto V_2, and so $\mathcal{T}$ is homeogonal. A similar argument establishes the second half of the assertion.

In order to prove Statement 4.4.1 we now observe that if a vertex V of $\mathcal{T}$ corresponds to a tile T^* of a dual tiling $\mathcal{T}^*$, then the symbol of V and the valence type of T^* (as defined in the previous section) are the same. It follows from Statement 4.4.2 that $(n_1.n_2. \ldots .n_p)$ is the symbol of a homeogonal tiling if and only if $[n_1.n_2. \ldots .n_p]$ is the valence-type of a homeohedral tiling. Hence the validity of Statement 4.4.1 follows from Statement 4.3.1.

Moreover, comments analogous to those made about the homeohedral tilings apply here also. In particular we see that we may take the uniform tilings by regular polygons, described in Section 2.1, as representatives of the eleven types of homeogonal tilings.

EXERCISES 4.4

1. Consider the tilings in Figures I.9, 2.4.2, 2.5.4, 3.2.3, 3.2.4(*a*) and 3.4.1. Determine which are homeogonal, and for each of these determine its type.
2. Which of the tilings considered in Exercise 1 fail to be homeogonal, but would be classified as such if the normality condition N.2 were omitted from the definition of homeogonal tilings? For each of these tilings determine its type.
3. (*a*) Show that there exist infinitely many topological types of tilings in which homeomorphisms act transitively on the vertices, and which satisfy conditions N.1 and N.3 for normality but include digons.
   (*b*) Determine all the possible types of the tilings described in part (*a*).
   (*c*) Describe the relation of the result in (*b*) to the uniform edge-c-colorings discussed in Exercise 2.9.2.
4. Determine which of the eleven types of homeogonal tilings can be realized by tilings in which all tiles are: (*a*) squares; (*b*) congruent squares; (*c*) equilateral triangles; (*d*) congruent equilateral triangles.

4.5 HOMEOTOXAL TILINGS

A tiling $\mathcal{T}$ is called *homeotoxal* or *topologically edge-transitive* if it is a normal tiling and if for any two edges E_1, E_2 of $\mathcal{T}$ there exists a homeomorphism of the plane that maps $\mathcal{T}$ onto $\mathcal{T}$ and E_1 onto E_2.

With each edge E of a tiling $\mathcal{T}$ we may associate a symbol $n.m; p.q$ (see Figure 4.5.1) where n and m are the numbers of edges of the two tiles of $\mathcal{T}$ which contain the edge E, and p, q are the valences of the two vertices of E. If $\mathcal{T}$ is homeotoxal then every edge of $\mathcal{T}$ has the same symbol $n.m; p.q$ and then we shall say that $\mathcal{T}$ *is of type* $\langle n.m; p.q \rangle$. As before we do not distinguish between symbols that differ trivially, namely by interchange of n and m, or of p and q. The main result is the following analogue of the results of the previous two sections.

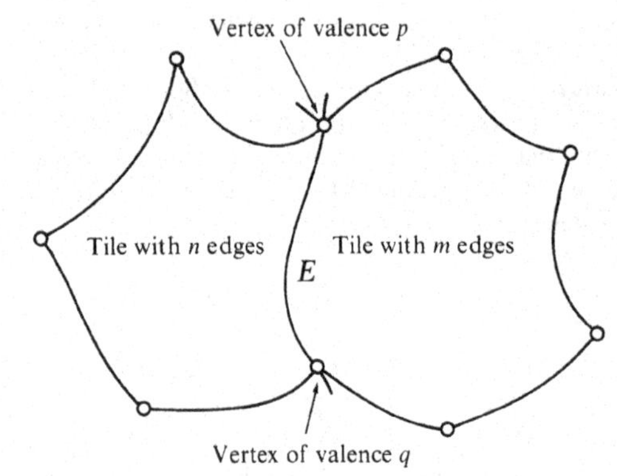

Figure 4.5.1
The explanation of the symbol $n.m; p.q$ for an edge E of a tiling.

4.5.1 *If $\mathscr{T}$ is a homeotoxal tiling then it is of one of the five types $\langle 3^2; 6^2\rangle$, $\langle 3.6; 4^2\rangle$, $\langle 4^2; 3.6\rangle$, $\langle 4^2; 4^2\rangle$, $\langle 6^2; 3^2\rangle$. Moreover, all tilings of the same type are topologically equivalent to each other.*

Types $\langle 3^2; 6^2\rangle$, $\langle 4^2; 4^2\rangle$, $\langle 6^2; 3^2\rangle$ are represented by the regular tilings by triangles, by squares and by hexagons, respectively. The type $\langle 3.6; 4^2\rangle$ is represented by the uniform tiling (3.6.3.6) and the type $\langle 4^2; 3.6\rangle$ is represented by the Laves tiling [3.6.3.6]. These especially symmetric representatives are shown in Figure 4.5.2.

Various proofs of Statement 4.5.1 are possible. We may use Statement 3.4.7 and Euler's Theorem for Tilings, which imply $1/n + 1/m + 1/p + 1/q = 1$. The various types of homeotoxal tilings can then be deduced from a

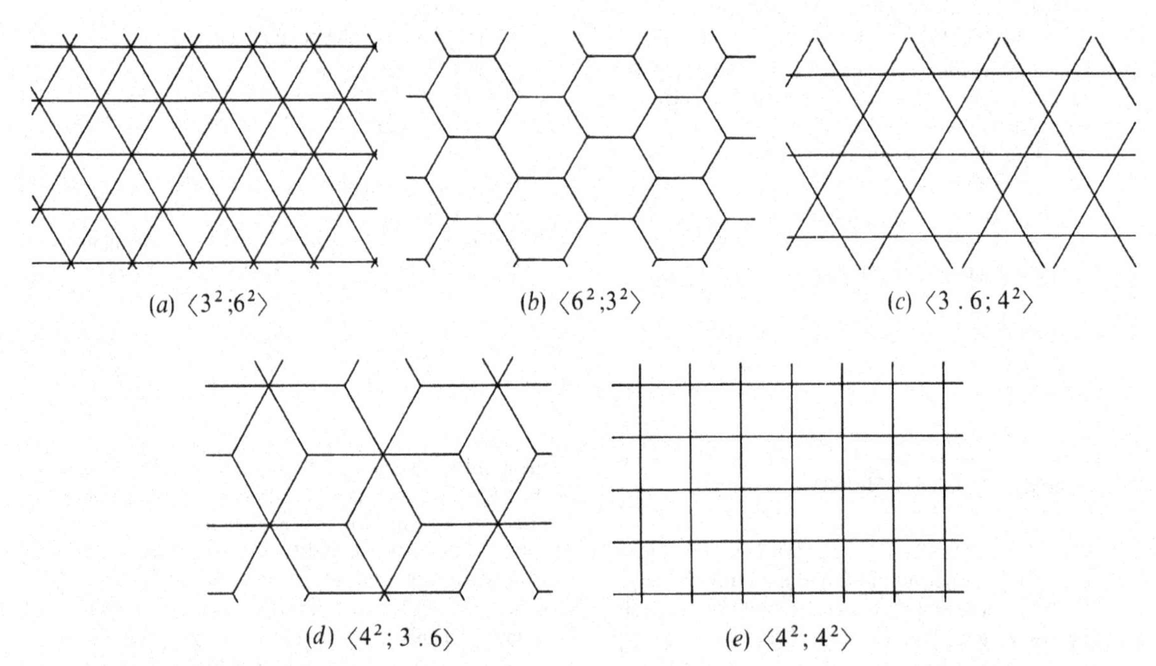

Figure 4.5.2
Examples of the five types of homeotoxal tilings. These particular tilings are actually edge-transitive (isotoxal).

study of the integer solutions of this equation (see Exercise 4.5.1).

Here, however, we shall present a proof based on a different idea, namely that of converting a homeotoxal tiling into one that is homeohedral. Statement 4.5.1 will then follow simply from the results of Section 4.3. The conversion depends upon the following construction, which also has other applications, as we shall see in Chapter 6.

Let $\mathcal{T}$ be a normal tiling. Define a tiling $\mathcal{T}_c$ as follows (see Figure 4.5.3). The *vertices* of $\mathcal{T}_c$ are the vertices of $\mathcal{T}$ together with one interior point in each tile of $\mathcal{T}$. Each *edge* of $\mathcal{T}_c$ is an arc from a vertex V of $\mathcal{T}$ to the chosen interior point of a tile T containing V. These arcs must be selected so that each lies in the tile containing its endpoints, and any two arcs are disjoint or meet at a common endpoint. The *tiles* of $\mathcal{T}_c$ are the domains of the plane into which it is divided by the arcs and vertices just constructed. It is clear that each edge of $\mathcal{T}$ is contained in precisely one tile of $\mathcal{T}_c$ and each tile of $\mathcal{T}_c$ contains precisely one edge of $\mathcal{T}$. The tiling $\mathcal{T}_c$ is said to arise from $\mathcal{T}$ by the process of *tile-centering*. Although $\mathcal{T}_c$ does not have to be normal (the choice of arcs may be made so as to preclude the existence of a positive inparameter u), it is easily seen that the arcs may be selected in such a way that $\mathcal{T}_c$ is normal.

If $\mathcal{T}$ is a homeotoxal tiling then it is easily seen (by arguments analogous to those used in the proof of Statement 4.4.2) that $\mathcal{T}_c$ is a homeohedral tiling. Moreover, if $\mathcal{T}$ is of type $\langle n.m;\ p.q\rangle$ then $\mathcal{T}_c$ must be of type

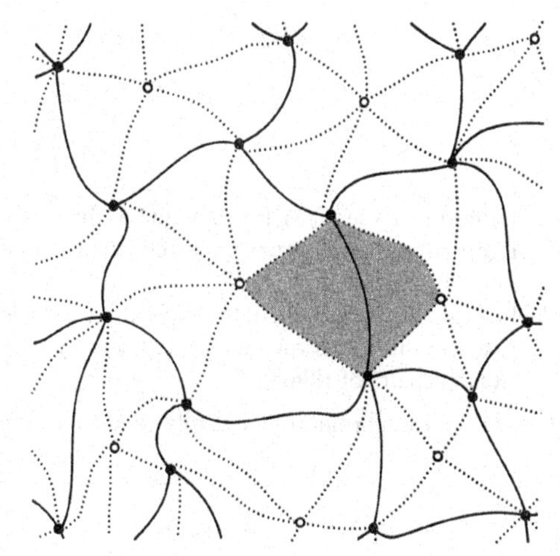

Figure 4.5.3
The construction of a tiling $\mathcal{T}_c$ from a tiling $\mathcal{T}$. The solid lines and circles denote the edges and vertices of $\mathcal{T}$. The dotted lines, open circles and solid circles denote the edges and vertices of $\mathcal{T}_c$. One of the tiles of $\mathcal{T}_c$ is shaded.

$[n.p.m.q]$. Since the only symbols for homeohedral tilings that involve four numbers (so the tiles have four edges each) are $[4^4]$, $[3.6.3.6]$ and $[3.4.6.4]$, tilings of these types are the only possible candidates for $\mathcal{T}_c$. Taking into account the various ways of rewriting these symbols, it is immediately obvious that the symbol of any homeotoxal tiling must be one of the following: $\langle 4^2;\ 4^2\rangle$, $\langle 3^2;\ 6^2\rangle$, $\langle 6^2;\ 3^2\rangle$, $\langle 3.6;\ 4^2\rangle$, or $\langle 4^2;\ 3.6\rangle$. But we know that each of these types exists. Hence to complete the proof of Statement 4.5.1 we need only establish uniqueness. This is easily done by showing that every two homeotoxal tilings with the same symbol must be combinatorially equivalent.

The proof of the assertion is thus completed.

EXERCISES 4.5

1. Carry out details of the alternative proof of Statement 4.5.1 that depends on solutions of the equation $1/n + 1/m + 1/p + 1/q = 1$ and does not use tile-centering.
2. Show that tilings of types $\langle n.m;\ p.q\rangle$ and $\langle p.q;\ n.m\rangle$ are dual to each other. Construct dually situated pairs of tilings by convex tiles with these symbols and with corresponding edges at right angles.
3. Show that if the requirement of normality is weakened by allowing digons or pseudovertices of valence 2, then there exist 14 additional types of topologically edge-transitive tilings.

(*a*) Determine these types, for each find a tiling which is topologically edge-transitive, and construct an isotoxal tiling of each type.

(*b*) Investigate which pairs of the types in (*a*) are dual to each other; construct in each case dually situated pairs of tilings.

*4. Describe a non-normal topologically edge-transitive tiling (with topological disks as tiles) of some type $\langle n.m;\ p.q\rangle$ for which the relation

$$1/n + 1/m + 1/p + 1/q = 1$$

does not hold. What weaker relation holds for all such tilings? (Such a relation can be established using the results of Section 4.7.)

4.6 2-HOMOGENEOUS AND 2-HOMEOHEDRAL TILINGS

As we have seen, the valence-type of a tile defined in Section 4.3 is useful in the description and classification of homeohedral and other kinds of tilings. We shall now introduce a more complex symbol which will be useful in the investigation of more general classes of tilings; we believe that this and related symbols can contribute greatly to the clarity of the subject.

Let T be a tile of a tiling $\mathcal{T}$, with vertices denoted by $V_1, V_2, \ldots, V_k = V_0$ in cyclic order and with adjacents $T_1, T_2, \ldots, T_k = T_0$. The notation is chosen so that V_j and V_{j+1} are the vertices on the edge $T \cap T_j$ of $\mathcal{T}$. The *neighborhood symbol* $H = H(T)$ is a sequence

$$H = \kappa_{1_{\sigma_1}} \kappa_{2_{\sigma_2}} \cdots \kappa_{k_{\sigma_k}}$$

of integers κ_j, σ_j written alternately on the line and below the line (as subscripts). Here κ_j is the valence of V_j and σ_j is the number of edges of T_j. For example, each tile in the tiling of Figure 4.6.1(*a*) has neighborhood symbol $3_4 3_4 3_{11} 3_{11}$ or $3_4 3_4 3_{11} 3_4 3_4 3_{11} 3_4 3_4 3_{11} 4_{11} 4_{11}$; those of Figure 4.6.1(*b*) have symbols $3_5 3_4 6_4$, $3_5 5_6 3_4 6_3$, $3_3 3_4 5_5 6_5 5_4$ or $3_4 5_4 3_4 5_4 3_4 5_4$. (Here, as usual, we do not distinguish between neighborhood symbols that can be obtained from each other by cyclic permutations or reversal of order.)

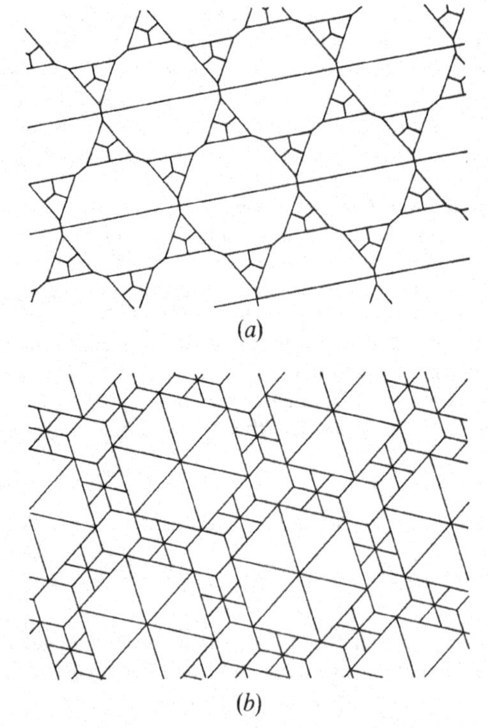

Figure 4.6.1
Illustration of the concepts of neighborhood symbols, *h*-homogeneous tilings and *h*-homeohedral tilings.

A tiling $\mathscr{T}$ is called *h-homogeneous of type* $[H_1][H_2]\ldots[H_h]$ if the neighborhood symbol of each tile is one of the H_i, and each H_i is the neighborhood symbol of some tile. We assume, moreover, that the symbols H_i are distinct. Thus the tiling of Figure 4.6.1 (*a*) is 2-homogeneous of type $[3_4 3_4 3_{11} 3_{11}]$ $[3_4 3_4 3_{11} 3_4 3_4 3_{11} 3_4 3_4 3_{11} 4_{11} 4_{11}]$, and that of Figure 4.6.1(*b*) is 4-homogeneous of type $[3_5 3_4 6_4][3_5 5_6 3_4 6_3][3_3 3_4 5_5 6_5 5_4][3_4 5_4 3_4 5_4 3_4 5_4]$. The monohedral tiling in Figure 1.3.6(*b*) is 2-homogeneous of type $[4_4 4_4 4_4 4_4][4_4 4_4 4_4 6_4]$ (though only six tiles have the second neighborhood symbol!); the one in Figure 1.3.8(*c*) is 2-homogeneous of type $[3_5 3_7 3_7 3_7 3_7][3_5 3_5 3_7 3_5 3_7 3_5 3_7]$.

A tiling $\mathscr{T}$ is called *h-homeohedral* if it is normal and if its tiles form h transitivity classes under homeomorphisms which map $\mathscr{T}$ onto itself. Since tiles equivalent under homeomorphisms of the tiling clearly have the same neighborhood symbols, it follows that each h-homeohedral tiling is h^*-homogeneous for some $h^* \leq h$. The two tilings in Figure 4.6.1 illustrate this: the tiling in (*a*) is 3-homeohedral (but 2-homogeneous), the one in (*b*) is 4-homeohedral (and 4-homogeneous).

We shall devote this section mainly to an investigation of 2-homeohedral tilings, aiming to provide another illustration of the use of topological tools in classification problems. This will occur in a setting that is technically more complicated than the examples considered earlier.

We begin by observing that a 2-homeohedral tiling is necessarily 2-homogeneous. To see this, we recall that for the reasons stated above a 2-homeohedral tiling is either 2-homogeneous or 1-homogeneous. But it cannot be 1-homogeneous, since we proved in Section 4.3 that such tilings are necessarily 1-homeohedral (that is, homeohedral) and so our assertion is proved. It follows that 2-homeohedral tilings can be designated by their types as 2-homogeneous tilings.

For 2-homeohedral tilings we can modify the neighborhood symbol in two ways, as follows.

Since all the tiles have one of the two symbols, instead of writing the subscripts we can place a circle if the adjacent tile is of the same transitivity class as the tile to which the symbol pertains, and an asterisk if it belongs to the other class. For example, the 2-homeohedral tiling in Figure 4.6.2(*a*) is of type $[3_4 8_3 12_3]$ $[3_4 8_3 3_3 8_4]$, which we may simplify to $[3_* 8_\circ 12_\circ][3_\circ 8_* 3_* 8_\circ]$; for the tiling in Figure 4.6.2(*b*) the corresponding symbols are $[4_4 6_3 6_3]$ $[4_3 6_3 4_4 6_4]$ and $[4_* 6_\circ 6_\circ][4_* 6_* 4_\circ 6_\circ]$.

For some purposes it is of more interest to have in the symbol a ready indication for the number of adjacent tiles of the other transitivity class, than it is to know precisely how they are distributed around the tile. In such cases we have found it convenient to replace all the subscripts by dots, and to prefix the neighborhood symbol modified in this way by an integer equal to the number of edges of the tile, with a subscript indicating the number of tiles of the other transitivity class adjacent to it. Applied to the tilings of Figure 4.6.2 this convention yields the symbols $3_1[3.8.12]4_2[3.8.3.8]$ and $3_1[4.6.6]4_2[4.6.4.6]$.*

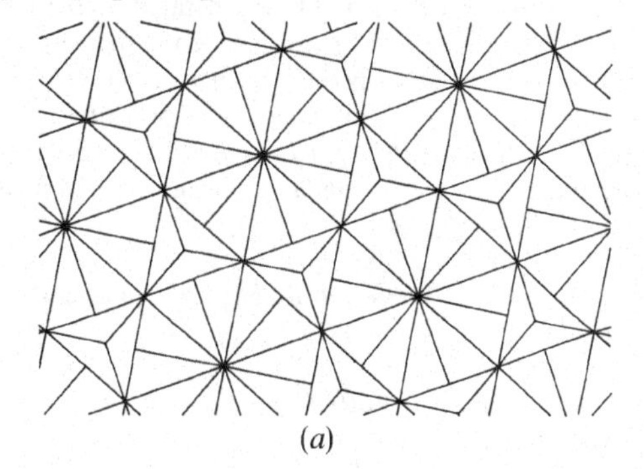

(*a*)

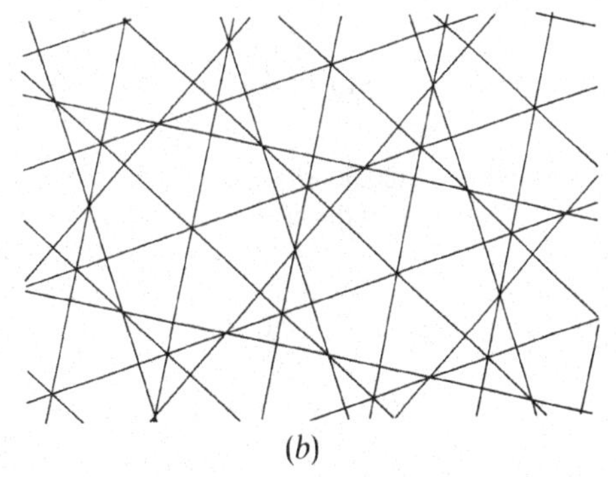

(*b*)

Figure 4.6.2
Examples of 2-homeohedral tilings.

* See the Appendix beginning on page 653.

Our next aim is to establish a necessary condition for the existence of a 2-homeohedral tiling of type $k_q[\kappa_1.\dots.\kappa_k]n_s[\nu_1.\dots.\nu_n]$, using the same general approach as in Section 4.3. There, in the enumeration of the homeohedral tilings, we first proved that such tilings are balanced. Here we follow a similar procedure, but we do slightly more, namely we estimate the values of $e(\mathcal{T})$ and $v(\mathcal{T})$ (as defined in Section 3.3) in terms of the parameters $k, q, \kappa_1, \dots, \kappa_k, n, s, \nu_1, \dots, \nu_n$ of the tiling $\mathcal{T}$. Application of Euler's Theorem (Statement 3.3.3) then yields the required Diophantine relation.

Let us assume, without loss of generality, that $k \le n$. Continuing with the notation introduced in Section 3.2, we denote by $D(r,P)$ a circular disk of radius r and center P, and by $\mathcal{A}(r,P)$ the patch of tiles generated by $D(r,P)$. We write $t(r,P)$, $e(r,P)$, and $v(r,P)$ for the numbers of tiles, edges and vertices in the patch $\mathcal{A}(r,P)$. Further we denote by $t_k(r,P)$ and $t_n(r,P)$ the numbers of tiles of the two types in $\mathcal{A}(r,P)$ so that $t_k(r,P) + t_n(r,P) = t(r,P)$. We begin by proving the following statement concerning the numbers of tiles of the two types.

4.6.1 *For every 2-homeohedral tiling the following equalities hold:*

$$\lim_{r\to\infty} t_k(r,P)/t(r,P) = s/(q + s)$$

and

$$\lim_{r\to\infty} t_n(r,P)/t(r,P) = q/(q + s).$$

To establish these relations, we consider a larger patch $\mathcal{A}(r + 2U,P)$ where U is the circumparameter of the tiling. Since every tile of the first kind in $\mathcal{A}(r,P)$ has q edges in common with tiles of the second kind, and the latter must clearly belong to $\mathcal{A}(r + 2U,P)$, we have

$$qt_k(r,P) \le st_n(r + 2U,P),$$

and similarly

$$st_n(r,P) \le qt_k(r + 2U,P).$$

From these we deduce

$$-q(t_k(r + 2U,P) - t_k(r,P)) \le qt_k(r,P) - st_n(r,P)$$
$$\le s(t_n(r + 2U,P) - t_n(r,P)).$$

Substituting

$$qt_k(r,P) - st_n(r,P) = (q + s)t_k(r,P) - st(r,P)$$

in place of the middle term, and dividing through by $(q + s)t(r,P)$ we obtain

$$\frac{-q}{q+s}\cdot\frac{t_k(r+2U,P)-t_k(r,P)}{t(r,P)} \le \frac{t_k(r,P)}{t(r,P)} - \frac{s}{q+s}$$
$$\le \frac{s}{q+s}\cdot\frac{t_n(r+2U,P)-t_n(r,P)}{t(r,P)}.$$

Letting $r \to \infty$ and noting that the extreme terms both tend to zero (by a simple application of the Normality Lemma) we obtain $\lim_{r\to\infty} t_k(r,P)/t(r,P) = s/(q + s)$ as required. The second limit can be obtained in a similar manner, or by noticing that

$$\lim_{r\to\infty} t_k(r,P)/t(r,P) + \lim_{r\to\infty} t_n(r,P)/t(r,P) = 1.$$

This completes the proof of Statement 4.6.1. Its meaning is, clearly, that the ratio of the numbers of tiles of the two kinds approaches $s:q$ as we take larger and larger patches. In view of this, the next statement concerning the edges is not surprising.

4.6.2 *In every 2-homeohedral tiling,*

$$e(\mathcal{T}) = \lim_{r\to\infty} e(r,P)/t(r,P) = \tfrac{1}{2}ks/(q + s) + \tfrac{1}{2}nq/(q + s).$$

To prove this we estimate $e(r,P)$ by a simple counting process. Most of the edges in $\mathcal{A}(r,P)$ will belong to two tiles, but some (those round the boundary of the patch) will belong to one tile only. Let us denote the latter by $E(r,P)$ (as in Section 3.2) and then we obtain

$$kt_k(r,P) + nt_n(r,P) = 2e(r,P) - E(r,P). \qquad (4.6.3)$$

Now every one of the $E(r,P)$ edges on the boundary belongs to a tile in $\mathscr{A}(r,P)$ which does not belong to $\mathscr{A}(r - 2U,P)$; hence, remembering that $k \leq n$ and so each tile has at most n edges, we obtain

$$0 < E(r,P) \leq n(t(r,P) - t(r - 2U,P)).$$

Dividing by $t(r,P)$ and letting $r \to \infty$, the Normality Lemma implies that

$$\lim_{r\to\infty} E(r,P)/t(r,P) = 0. \qquad (4.6.4)$$

Dividing equation (4.6.3) by $t(r,P)$, letting $r \to \infty$ and substituting from Statement 4.6.1 for the two limits on the left, we immediately obtain Statement 4.6.2 as required.

By similar reasoning we can show that

$$v(\mathscr{T}) = \lim_{r\to\infty} v(r,P)/t(r,P)$$

$$= \frac{s}{q+s}\sum_{i=1}^{k}\frac{1}{\kappa_i} + \frac{q}{q+s}\sum_{i=1}^{n}\frac{1}{\nu_i}. \qquad (4.6.5)$$

For the proof we apply another counting argument, this time to the vertices in $\mathscr{A}(r,P)$. We notice that, in general, a vertex of $\mathscr{A}(r,P)$ will belong to precisely κ_i (or ν_i) tiles if it is of valence κ_i (or ν_i) in the tiling, but that this will not be the case if the vertex in question is one of the vertices lying on the boundary of $\mathscr{A}(r,P)$. We denote the number of such vertices by $V(r,P)$. Ignoring these boundary vertices for the moment, we can estimate the number of vertices in $\mathscr{A}(r,P)$ by counting each vertex of valence κ_i (or ν_i) exactly $1/\kappa_i$ (or $1/\nu_i$) times for each of the tiles incident with it. This yields

$$t_k(r,P)\sum_{i=1}^{k}\frac{1}{\kappa_i} + t_n(r,P)\sum_{i=1}^{k}\frac{1}{\nu_i}.$$

However, because of the vertices on the boundary of $A(r,P)$ this is an underestimate. As the discrepancy must be numerically between 0 and 1 for each of the $V(r,P)$ boundary vertices we obtain

$$0 \leq v(r,P) - \left(t_k(r,P)\sum_{i=1}^{k}\frac{1}{\kappa_i} + t_n(r,P)\sum_{i=1}^{n}\frac{1}{\nu_i}\right) \leq V(r,P). \qquad (4.6.6)$$

We divide through by $t(r,P)$ and consider the limit as $r \to \infty$. By an argument similar to that which we used to prove relation (4.6.4) we can show

$$\lim_{r\to\infty} V(r,P)/t(r,P) = 0 \qquad (4.6.7)$$

and hence the estimate (4.6.6) with Statement 4.6.1 yields relation (4.6.5) as required. Incidentally we notice that (4.6.4) and (4.6.7) prove that 2-homeohedral tilings are balanced.

We are now in a position to establish the basic result of this section.

4.6.8 *If there exists a 2-homeohedral tiling of type* $k_q[\kappa_1 \ldots\ldots \kappa_k]n_s[\nu_1 \ldots\ldots \nu_n]$ *then*

$$\frac{1}{s}\left(\sum_{i=1}^{n}\frac{\nu_i - 2}{\nu_i} - 2\right) + \frac{1}{q}\left(\sum_{i=1}^{k}\frac{\kappa_i - 2}{\kappa_i} - 2\right) = 0. \qquad (4.6.9)$$

To prove this we substitute the values of $e(\mathscr{T})$ and $v(\mathscr{T})$ found in Statement 4.6.2 and equation (4.6.5) into the Euler relation (Statement 3.3.3) and simplify the result.

The Diophantine equation (4.6.9), which is a necessary condition for the existence of a 2-homeohedral tiling with given parameters, clearly has infinitely many solutions. This is still the case if we impose the various "obvious" constraints. (For example, we must have $1 \leq q \leq k$, $1 \leq s \leq n$ and, under our assumption that $k \leq n$, $3 \leq k \leq 6$; moreover, if $q \geq k - 1$ then all the values of the numbers κ_i must occur in the sequence of the numbers ν_j, etc.) In fact, even in very special cases we shall have to introduce many geometric arguments to reduce the number of potential candidates for 2-homeohedral tilings to manageable proportions.

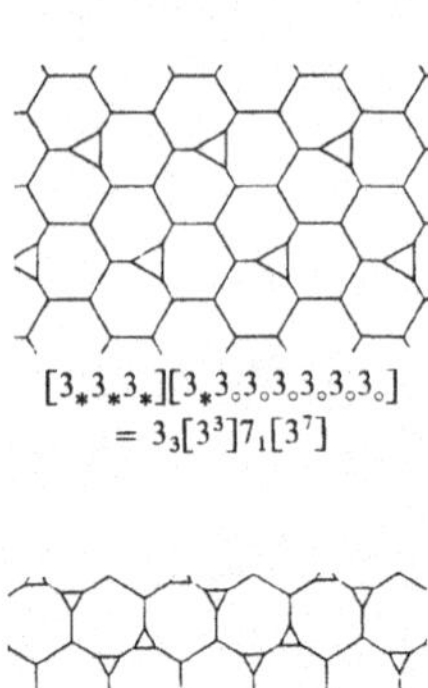

$[3_*3_*3_*][3_*3_\circ3_\circ3_\circ3_\circ3_\circ3_\circ]$
$= 3_3[3^3]7_1[3^7]$

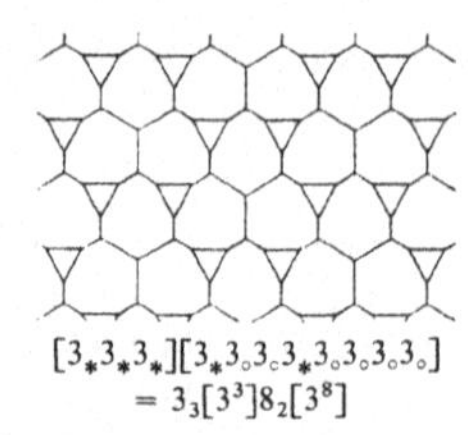

$[3_*3_*3_*][3_*3_\circ3_\circ3_*3_\circ3_\circ3_\circ3_\circ]$
$= 3_3[3^3]8_2[3^8]$

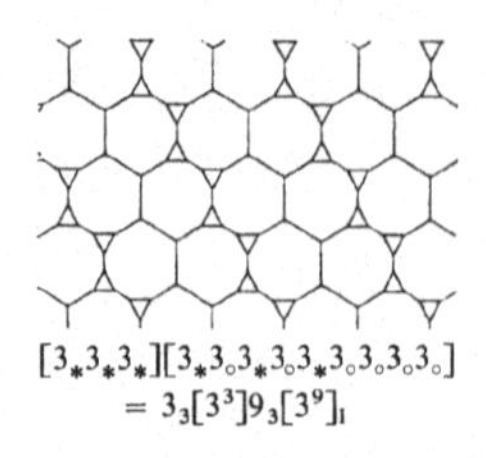

$[3_*3_*3_*][3_*3_\circ3_*3_\circ3_*3_\circ3_\circ3_\circ3_\circ]$
$= 3_3[3^3]9_3[3^9]_{\text{I}}$

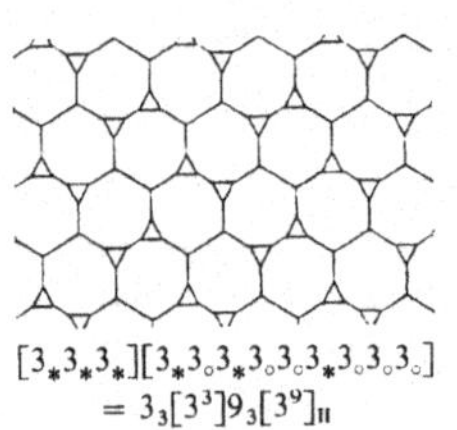

$[3_*3_*3_*][3_*3_\circ3_*3_\circ3_\circ3_*3_\circ3_\circ3_\circ]$
$= 3_3[3^3]9_3[3^9]_{\text{II}}$

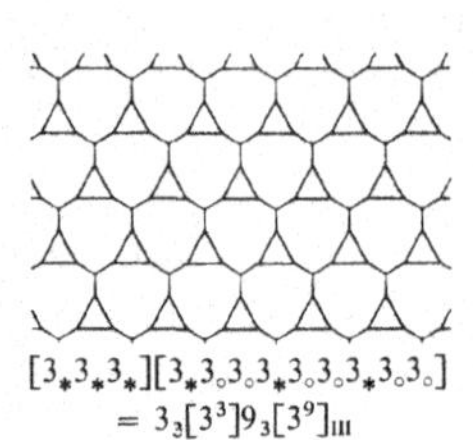

$[3_*3_*3_*][3_*3_\circ3_\circ3_*3_\circ3_\circ3_*3_\circ3_\circ]$
$= 3_3[3^3]9_3[3^9]_{\text{III}}$

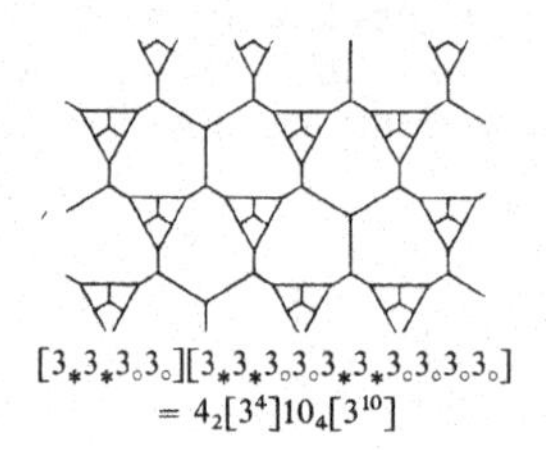

$[3_*3_*3_\circ3_\circ][3_*3_*3_\circ3_\circ3_*3_*3_\circ3_\circ3_\circ3_\circ]$
$= 4_2[3^4]10_4[3^{10}]$

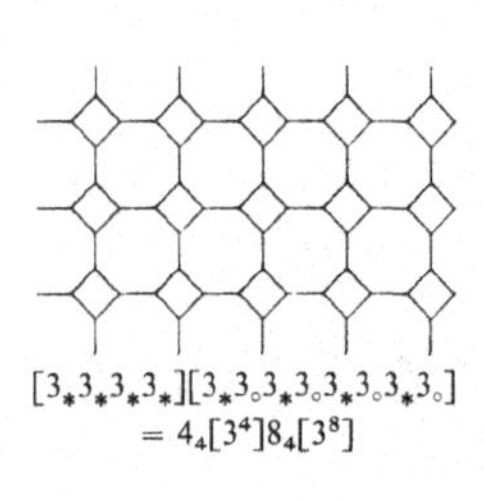

$[3_*3_*3_*3_*][3_*3_\circ3_*3_\circ3_*3_\circ3_*3_\circ]$
$= 4_4[3^4]8_4[3^8]$

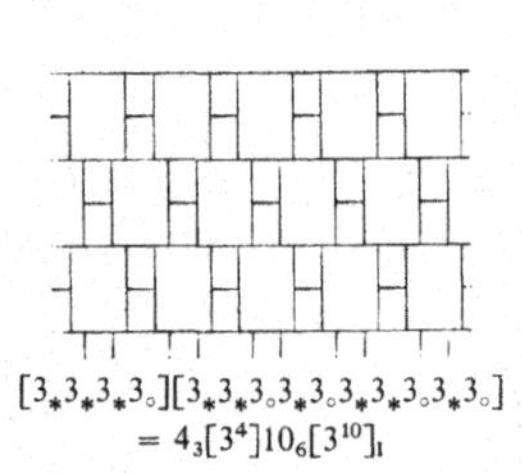

$[3_*3_*3_*3_\circ][3_*3_*3_\circ3_*3_\circ3_*3_*3_\circ3_*3_\circ]$
$= 4_3[3^4]10_6[3^{10}]_{\text{I}}$

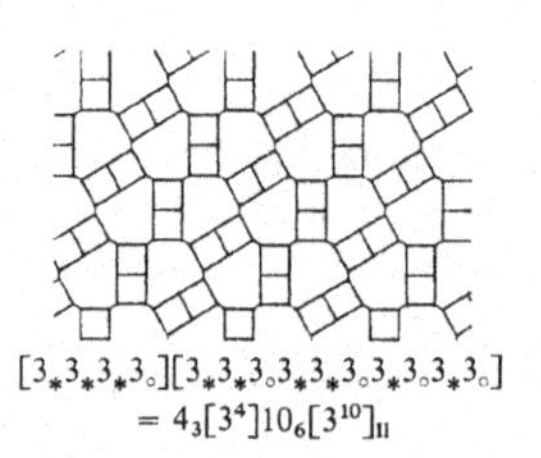

$[3_*3_*3_*3_\circ][3_*3_*3_\circ3_*3_*3_\circ3_*3_\circ3_*3_\circ]$
$= 4_3[3^4]10_6[3^{10}]_{\text{II}}$

Figure 4.6.3
A 2-isohedral representative of each of the 39 types of 2-homeohedral tilings in which all vertices have the same valence. The 26 types of 3-valent tilings, the 10 types of 4-valent tilings and the 3 types of 5-valent tilings are shown. For each tiling, we indicate the two modified type symbols by which it can be described. Roman numbers are used as subscripts to distinguish between equal symbols that correspond to distinct tilings.

As an example we shall first consider the case in which all the parameters κ_i, v_i are equal.

4.6.10 *There exist precisely 39 different types of 2-homeohedral tilings in which all vertices have the same valence v. Of these, 26 types correspond to $v = 3$, 10 types to $v = 4$, and 3 types to $v = 5$.*

Representatives of all these types are shown in Figure 4.6.3. We note, in particular, that each type is represented by a rather special tiling, namely by a 2-isohedral tiling. We conjecture that every 2-homeohedral type has a 2-isohedral representative, but have no proof of this.

In order to establish Statement 4.6.10 we first recall from Section 3.2 that if all vertices of a normal tiling have the same valence v then $3 \leq v \leq 6$. In the present case equation (4.6.9) simplifies to

$$2v(q + s) = (v - 2)(ks + nq). \tag{4.6.11}$$

Assuming, for definiteness, that $k \leq n$, we have $3 \leq k \leq 6$ and we are led to the consideration of the following eight possibilities, in each of which equation (4.6.11) reduces to the simpler form indicated.

$$\begin{array}{lll} v = 3 & k = 3 & 3s = (n - 6)q \\ & k = 4 & 2s = (n - 6)q \\ & k = 5 & s = (n - 6)q \\ & k = 6 & n = 6 \end{array}$$

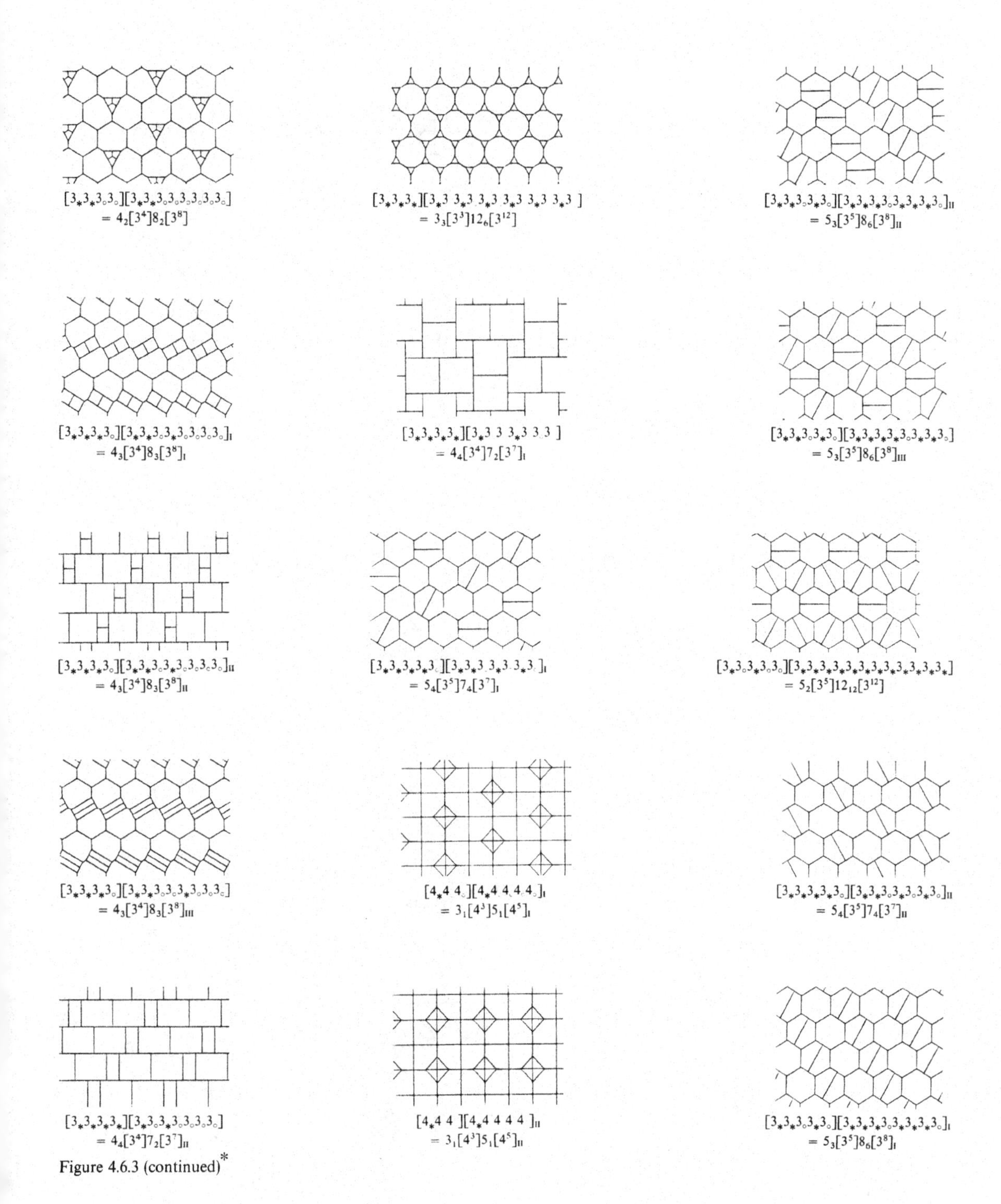

Figure 4.6.3 (continued)*

* See the Appendix beginning on page 653.

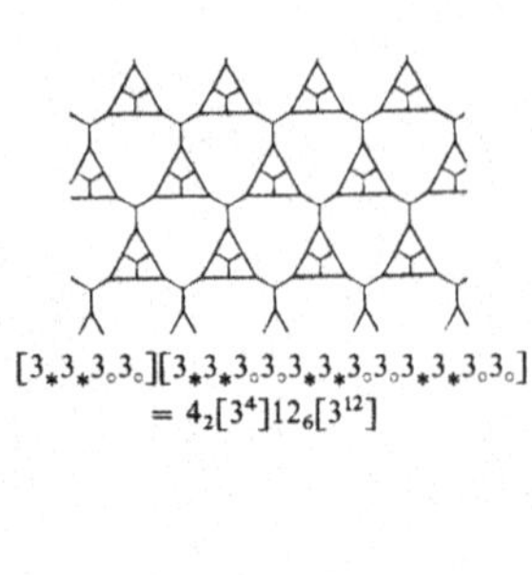

$[3_*3_*3_\circ3_\circ][3_*3_*3_\circ3_\circ3_*3_*3_\circ3_\circ3_*3_*3_\circ3_\circ]$
$= 4_2[3^4]12_6[3^{12}]$

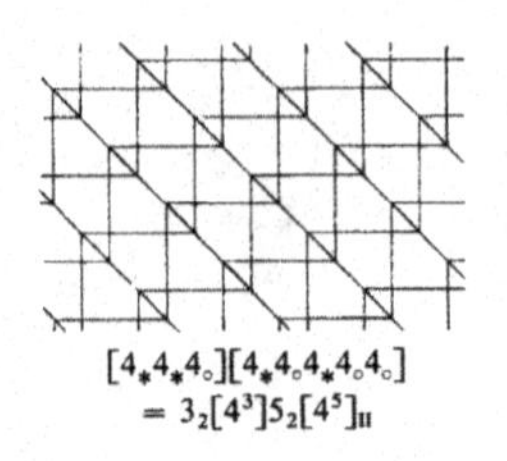

$[4_*4_*4_\circ][4_*4_\circ4_*4_\circ4_\circ]$
$= 3_2[4^3]5_2[4^5]_{\mathrm{II}}$

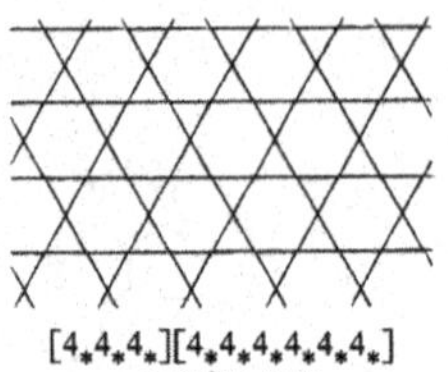

$[4_*4_*4_*][4_*4_*4_*4_*4_*4_*]$
$= 3_3[4^3]6_6[4^6]$

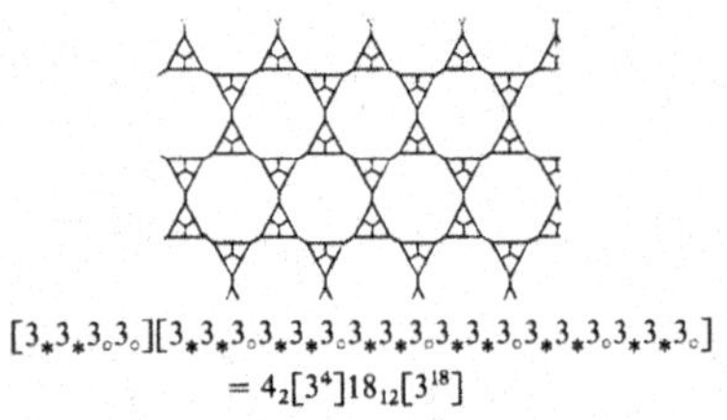

$[3_*3_*3_\circ3_\circ][3_*3_*3_\circ3_*3_*3_\circ3_*3_*3_\circ3_*3_*3_\circ3_*3_*3_\circ3_*3_*3_\circ]$
$= 4_2[3^4]18_{12}[3^{18}]$

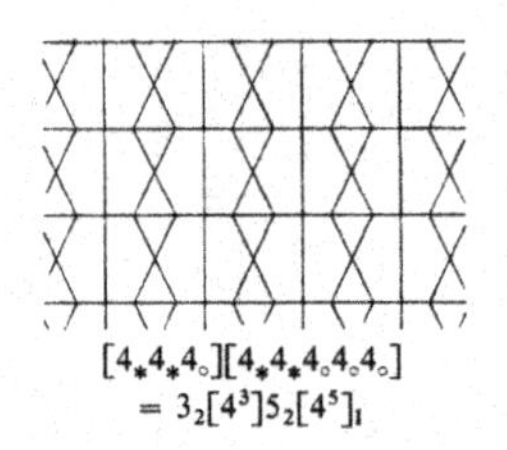

$[4_*4_*4_\circ][4_*4_*4_\circ4_\circ4_\circ]$
$= 3_2[4^3]5_2[4^5]_{\mathrm{I}}$

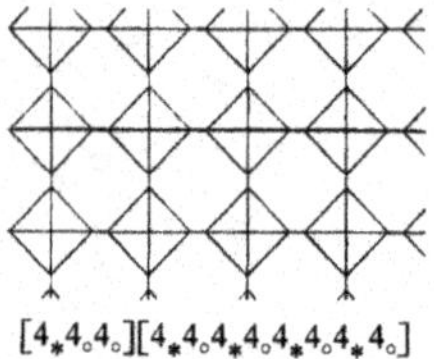

$[4_*4_\circ4_\circ][4_*4_\circ4_*4_\circ4_*4_\circ4_*4_\circ]$
$= 3_1[4^3]8_4[4^8]$

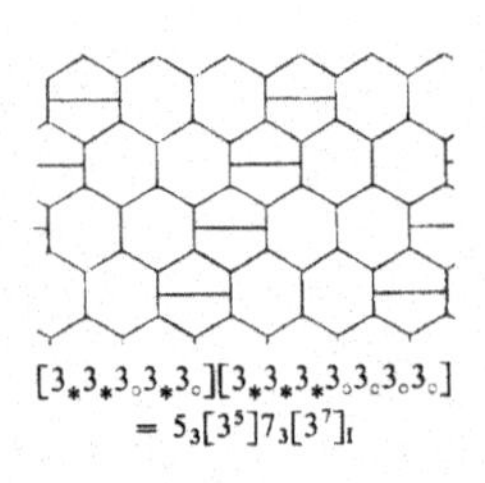

$[3_*3_*3_\circ3_*3_\circ][3_*3_*3_*3_\circ3_\circ3_\circ3_\circ]$
$= 5_3[3^5]7_3[3^7]_{\mathrm{I}}$

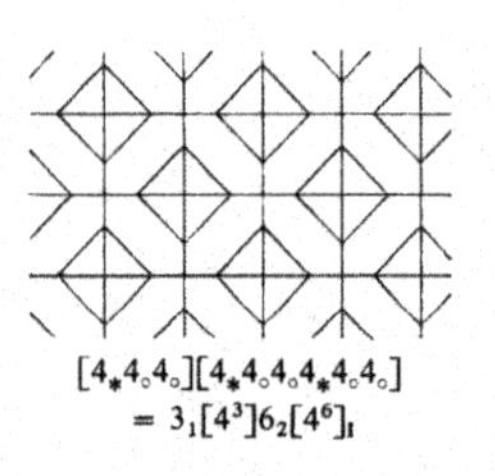

$[4_*4_\circ4_\circ][4_*4_\circ4_\circ4_*4_\circ4_\circ]$
$= 3_1[4^3]6_2[4^6]_{\mathrm{I}}$

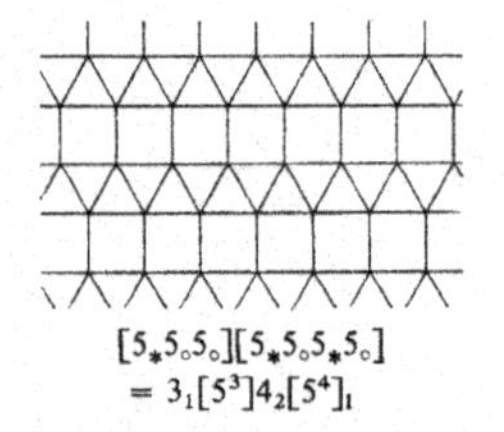

$[5_*5_\circ5_\circ][5_*5_\circ5_*5_\circ]$
$= 3_1[5^3]4_2[5^4]_{\mathrm{I}}$

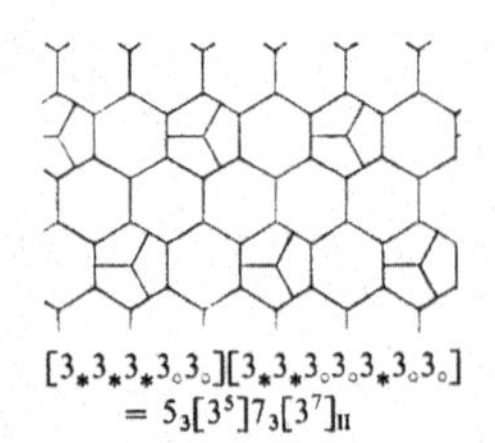

$[3_*3_*3_*3_\circ3_\circ][3_*3_*3_\circ3_\circ3_*3_\circ3_\circ]$
$= 5_3[3^5]7_3[3^7]_{\mathrm{II}}$

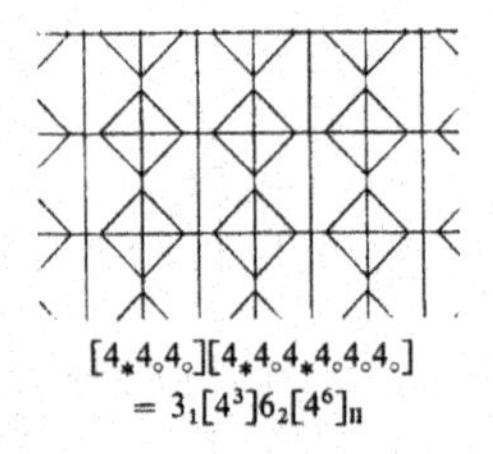

$[4_*4_\circ4_\circ][4_*4_\circ4_*4_\circ4_\circ4_\circ]$
$= 3_1[4^3]6_2[4^6]_{\mathrm{II}}$

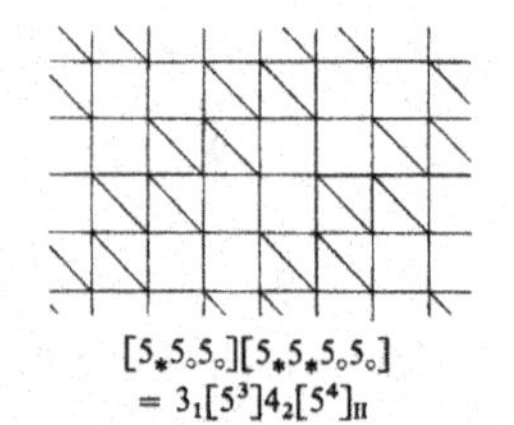

$[5_*5_\circ5_\circ][5_*5_*5_\circ5_\circ]$
$= 3_1[5^3]4_2[5^4]_{\mathrm{II}}$

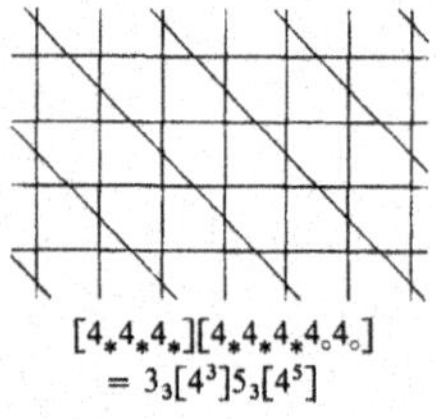

$[4_*4_*4_*][4_*4_*4_*4_\circ4_\circ]$
$= 3_3[4^3]5_3[4^5]$

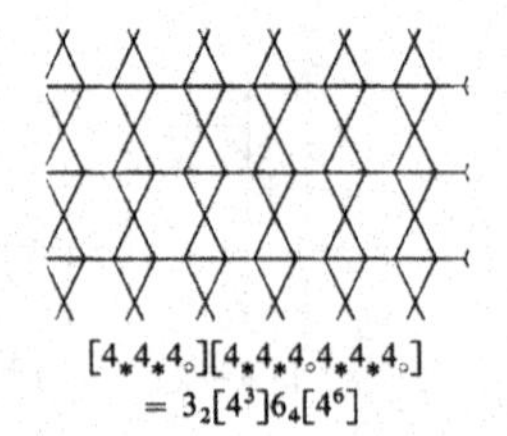

$[4_*4_*4_\circ][4_*4_*4_\circ4_*4_*4_\circ]$
$= 3_2[4^3]6_4[4^6]$

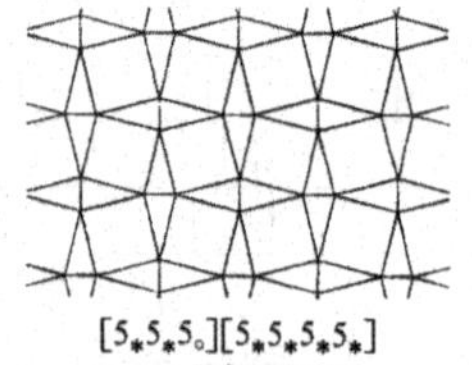

$[5_*5_*5_\circ][5_*5_*5_*5_*]$
$= 3_2[5^3]4_4[5^4]$

Figure 4.6.3 (continued)

$v = 4$	$k = 3$	$s = (n - 4)q$
	$k = 4$	$n = 4$
$v = 5$	$k = 3$	$s = (3n - 10)q$
$v = 6$	$k = 3$	$n = 3$

These equations still have infinitely many solutions, but all except a small finite number can be eliminated by arguments that vary from case to case. We shall not give all the details here, but only samples of the types of geometrical reasoning which need to be used.

In the case $v = 3, k = 3$ we must have $q = 3$. For if q were equal to 1, each triangle* would have a vertex shared only by triangles; but then there is no possibility of extending the tiling 3-valently beyond the three triangular tiles with a common vertex (see Figure 4.6.4(*a*)). If q were 2, then the triangles would share an edge (see Figure 4.6.4(*b*)) and hence the tiling would not be normal. But when $q = 3$, shrinking each triangle to a point (see Figure 4.6.5) leads from a tiling with n-gons and triangles to a tile-transitive 3-valent tiling by $(n - s)$-gons. By Statement 4.3.1 the only such tiling is $[3^6]$, that is, the tiling by hexagons. Hence each of the required tilings is obtained from $[3^6]$ by replacing some (or all) of the vertices of $[3^6]$ by triangles. As $s \leq 6$, the list of candidates is reduced to the following pairs (n,s): (7,1), (8,2), (9,3), (10,4), (11,5) and (12,6). The question whether a tiling with $s = n - 6$ exists is thus equivalent to the question of whether it is possible to replace by triangles $n - 6$ vertices of each tile of $[3^6]$ so that the resultant tiling is 2-homeohedral. Examination of the various possibilities leads to the first six tilings of Figure 4.6.3.

As a second example we consider the possibility $v = 5$ and $k = 3$. Then $s \leq n$ and $s = (3n - 10)q$ imply that for $q = 1$ we have $n = 4$ or 5, for $q = 2$ we have only $n = 4$, and for $q = 3$ there are no corresponding values of n. As candidates for 2-homeohedral tilings there are

* To avoid awkward expressions, throughout this section we use the words triangle, quadrangle, pentagon, . . . to mean tiles with 3, 4, 5, . . . edges. We do not imply by this terminology that the tiles are polygons.

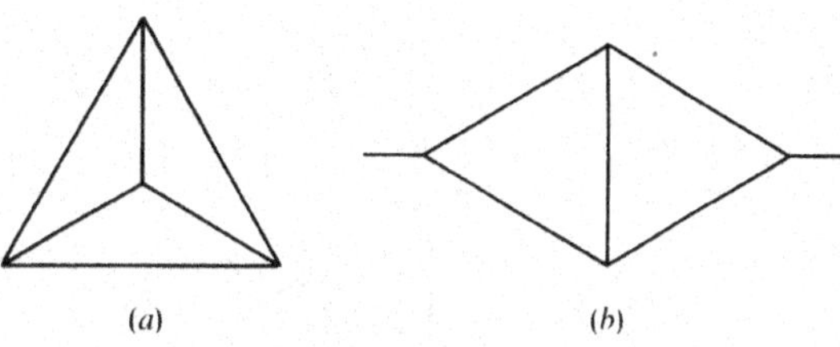

Figure 4.6.4
Illustrations for the argument that $q = 3$ when $v = k = 3$.

therefore only three possibilities: (*i*) $k = 3, q = 1, n = 4, s = 2$; (*ii*) $k = 3, q = 1, n = 5, s = 5$; (*iii*) $k = 3, q = 2, n = 4, s = 4$. The first case leads to two tilings shown in Figure 4.6.3, and the last case to a single type shown there. In order to see that in case (ii) no tiling exists we may argue as follows. If such a tiling were possible, the neighborhood of each pentagonal tile would have to look like that indicated in Figure 4.6.6. Considering the triangles C and D we see that the tiles A and B must both be pentagons; but this is impossible since no two pentagons can have a common edge in a tiling of this type. We deduce that no such tiling exists.

As a last example of the kind of reasoning used in establishing Statement 4.6.10 we consider the case $v = 6$, so that $k = n = 3$. Disregarding transitivity this means that we are interested in tilings of the plane in which *each* tile is a triangle and each vertex has valence 6. As we know from Statement 4.3.1 the only topological type of such tilings is $[6^3]$, in which tiles form *one*, rather than two, transitivity classes. Thus no 2-homeohedral tilings exist. Similar is the situation for $v = 4, k = n = 4$, and for $v = 3, k = n = 6$.

Analogous methods can be used to establish the following two results.

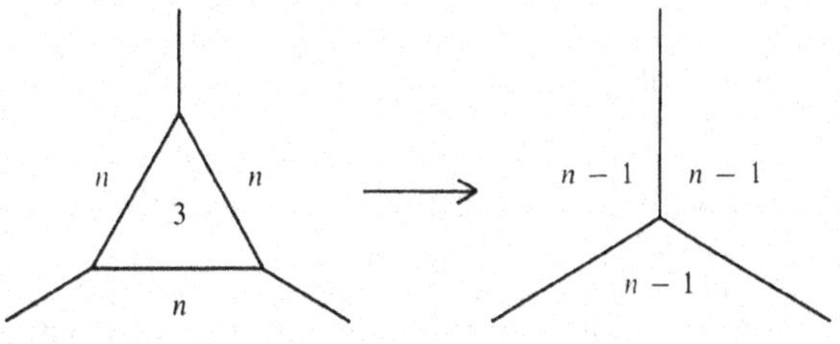

Figure 4.6.5
The method of transforming a 2-homeohedral tiling with n-gons and triangles into a homeohedral tiling by $(n - r)$-gons.

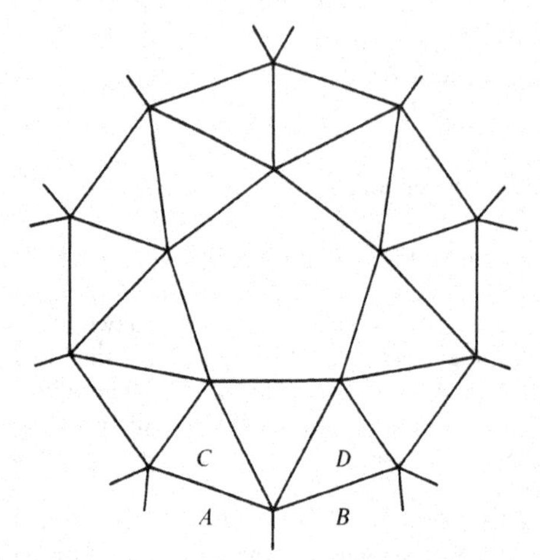

Figure 4.6.6
Illustration for the argument in the text that no 2-homeohedral tiling can exist for $k = 3$, $q = 1$, $n = r = 5$.

4.6.12 *There exist precisely* 18 *types of* 2-*homeohedral tilings in which the tiles of one transitivity class have three adjacents, those of the other class have four adjacents, and each tile is adjacent to precisely one tile from the other class.*

Representative examples of each type are shown in Figure 4.6.7. Again, all representatives shown are also 2-isohedral.

4.6.13 *There exist precisely seven types of* 2-*homeohedral tilings in which all tiles have three adjacents and each tile is adjacent to just one tile of the other transitivity class. There exist no* 2-*homeohedral tilings in which all tiles have four* (*or all tiles have five, or all tiles have six*) *adjacents, and each tile is adjacent to just one tile of the other transitivity class.*

Representative examples of each type are shown in Figure 4.6.8.

With enough time and patience it should be possible to determine all types of 2-homeohedral tilings without any difficulties of principle. We already know of 453 types of such tilings (see Section 4.8), and we believe that there are very few, if any, additional types.*

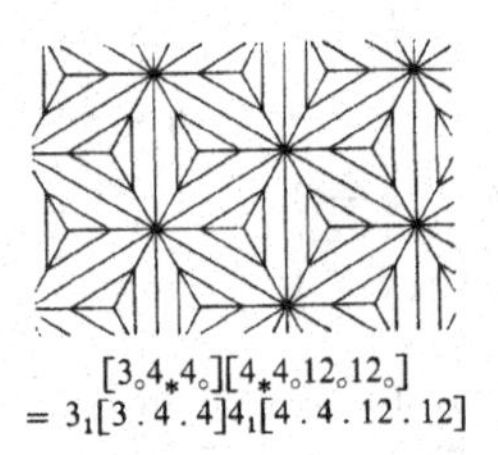

$[3_\circ 4_* 4_\circ][4_* 4_\circ 12_\circ 12_\circ]$
$= 3_1[3.4.4]4_1[4.4.12.12]$

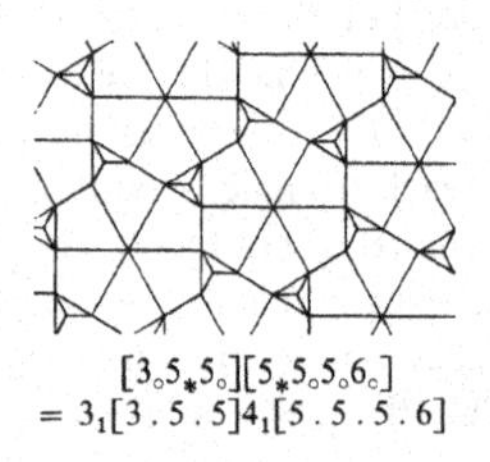

$[3_\circ 5_* 5_\circ][5_* 5_\circ 5_\circ 6_\circ]$
$= 3_1[3.5.5]4_1[5.5.5.6]$

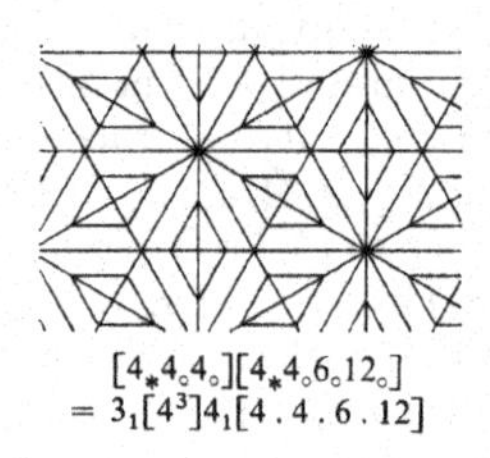

$[4_* 4_\circ 4_\circ][4_* 4_\circ 6_\circ 12_\circ]$
$= 3_1[4^3]4_1[4.4.6.12]$

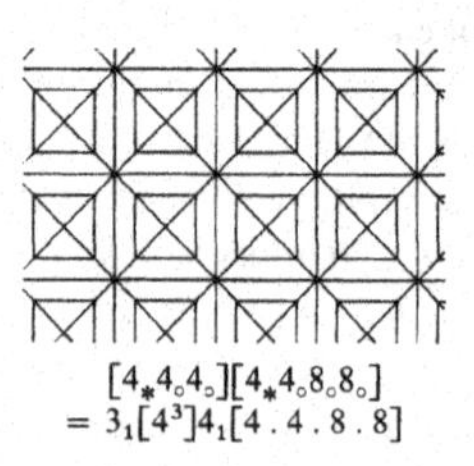

$[4_* 4_\circ 4_\circ][4_* 4_\circ 8_\circ 8_\circ]$
$= 3_1[4^3]4_1[4.4.8.8]$

Figure 4.6.7
A 2-isohedral representative of each of the 18 types of 2-homeohedral tilings in which the tiles of one transitivity class have three adjacents, those of the other class have four adjacents, and each tile is adjacent to a single tile of the other class.

* See the Appendix beginning on page 653.

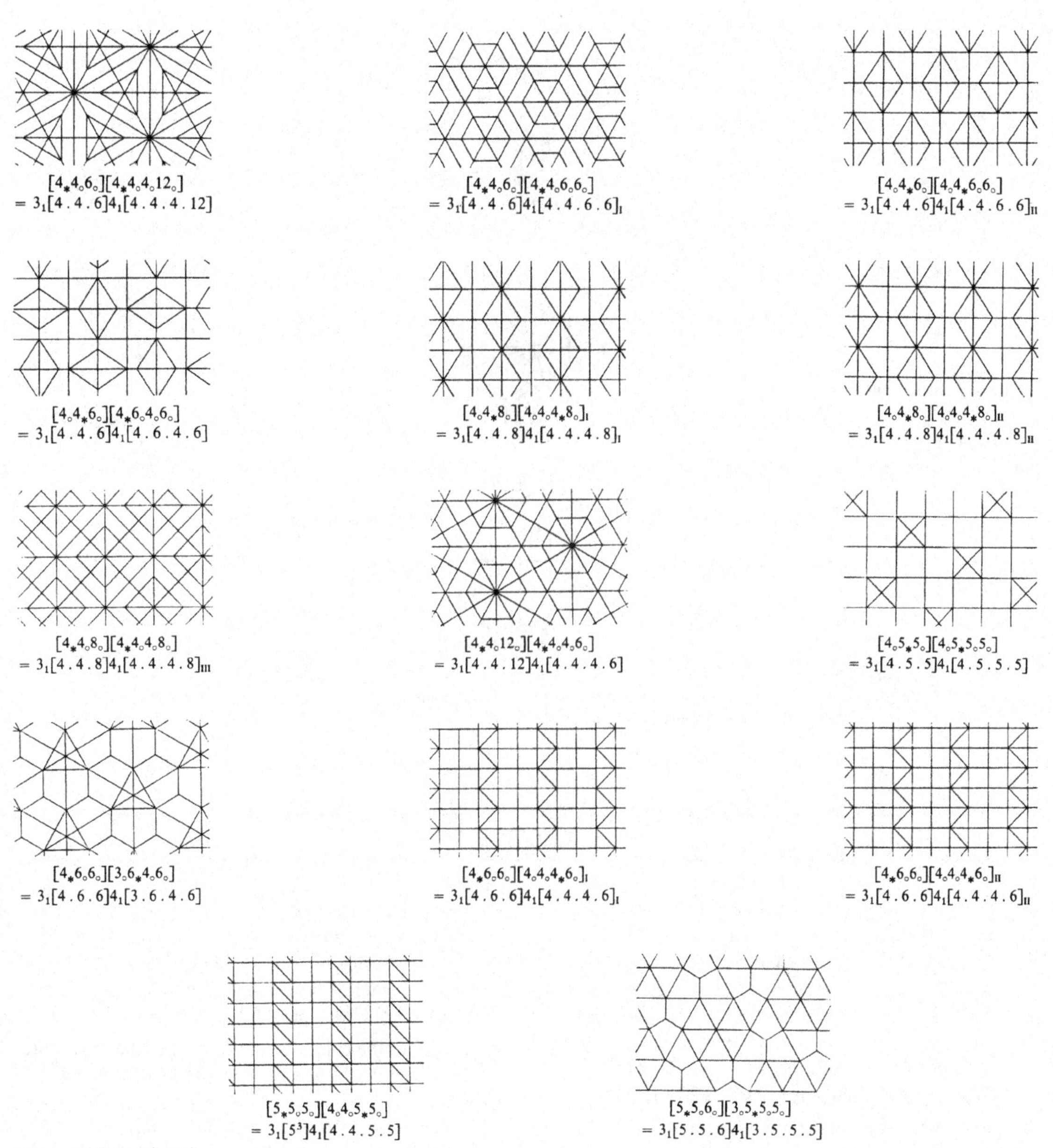

Figure 4.6.7 (continued)

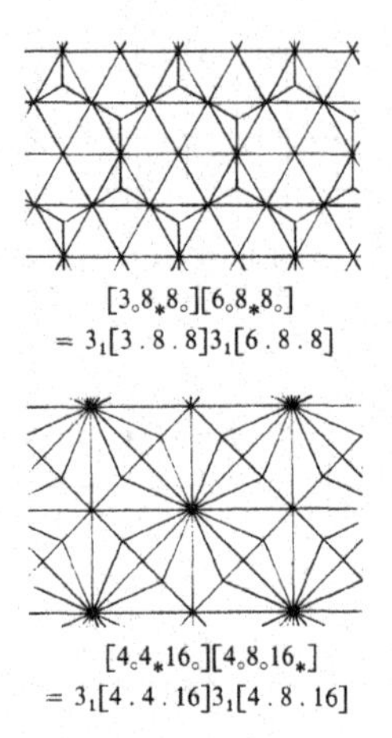

$[3_\circ 8_* 8_\circ][6_\circ 8_* 8_\circ]$
$= 3_1[3 . 8 . 8]3_1[6 . 8 . 8]$

$[4_\circ 4_* 16_\circ][4_\circ 8_\circ 16_*]$
$= 3_1[4 . 4 . 16]3_1[4 . 8 . 16]$

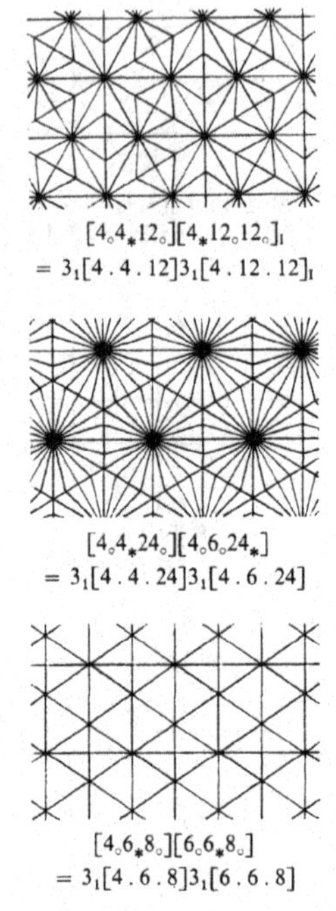

$[4_\circ 4_* 12_\circ][4_* 12_\circ 12_\circ]_{\mathrm{I}}$
$= 3_1[4 . 4 . 12]3_1[4 . 12 . 12]_{\mathrm{I}}$

$[4_\circ 4_* 24_\circ][4_\circ 6_\circ 24_*]$
$= 3_1[4 . 4 . 24]3_1[4 . 6 . 24]$

$[4_\circ 6_* 8_\circ][6_\circ 6_* 8_\circ]$
$= 3_1[4 . 6 . 8]3_1[6 . 6 . 8]$

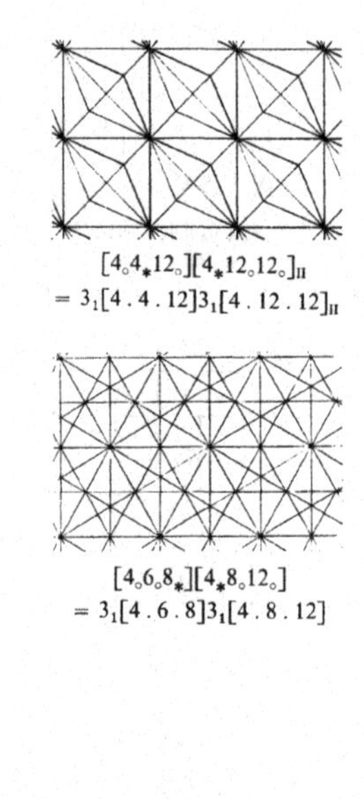

$[4_\circ 4_* 12_\circ][4_* 12_\circ 12_\circ]_{\mathrm{II}}$
$= 3_1[4 . 4 . 12]3_1[4 . 12 . 12]_{\mathrm{II}}$

$[4_\circ 6_\circ 8_*][4_* 8_\circ 12_\circ]$
$= 3_1[4 . 6 . 8]3_1[4 . 8 . 12]$

Figure 4.6.8
A 2-isohedral representative of each of the seven types of 2-homeohedral tilings in which each tile has three adjacents, one of which is in the other transitivity class.

EXERCISES 4.6

1. Check whether the tilings in Figures I.9, I.10, 2.1.5, 2.2.1, 2.2.3, 2.4.2, 2.5.3, 2.5.4 and 2.5.6 are 2-homogeneous or 2-homeohedral, and determine the types of those that are. Which would be 2-homeohedral if the normality requirement were weakened by not requiring property N.2? Find the type symbols of the 2-homeohedral tilings in Figures 2.2.1 and 2.5.4.
2. (*a*) Show that there exist infinitely many types of 2-homogeneous normal tilings.

   (*b*) Show by examples that Statement 4.6.1 is not valid for 2-homogeneous tilings.

   (*c*) Find where 2-homeohedrality was used in the proof of Statement 4.6.1.
3. Carry out the detailed examination of the possibilities and so establish the following parts of Statement 4.6.10.

   (*a*) There exist precisely six different types of 2-homeohedral tilings in which $v = 3$, $k = 3$.

(b) There exist precisely 3 types in which $v = 5$.
(c) There exist four types in which $v = 4$, $n = 6$.

*4. Establish that part of Statement 4.6.12 which deals with the situation in which each tile with 3 adjacents has one 3-valent vertex.

5.*(a) Show that there exist no 2-homeohedral tilings in which all tiles are triangles and each triangle is adjacent to precisely two triangles of the other transitivity class.
*(b) Find a 2-homeohedral tiling in which all tiles are quadrangles and each is adjacent to precisely two quadrangles of the other transitivity class.
(c) Determine all 2-homeohedral tilings by pentagons in which each tile is adjacent to precisely three tiles of the other transitivity class.

6.*(a) Show that there are only finitely many types of 2-homeohedral tilings.
**(b) Show that for each $n \geq 3$ there exist only finitely many types of n-homeohedral tilings.

**7. Show that there exist only finitely many types of 2-homogeneous tilings in which all vertices are 3-valent.

*8. Let a tiling $\mathscr{T}$ be called *2-homeotoxal* if it is normal and if the edges of $\mathscr{T}$ form two transitivity classes under the homeomorphisms that map $\mathscr{T}$ onto itself. Devise suitable symbols for 2-homeotoxal tilings, show that such tilings are balanced, find the analogue of relation (4.6.9) and determine all the types of 2-homeotoxal tilings.

4.7 NON-NORMAL HOMOGENEOUS AND HOMEOHEDRAL TILINGS

The first six sections of this chapter were concerned entirely with normal tilings. Here we make a digression and consider the additional possibilities that arise if we admit tilings which are not normal. Although this material will not be used in the sequel (and so may be omitted if the reader wishes) we feel that the non-normal topologically tile-transitive tilings have properties which are sufficiently important and unexpected to merit discussion here.

Of course, normal and balanced tilings can be investigated much more easily. In the important case of isohedral tilings (see Chapter 6) the fact that the normality condition is automatically satisfied explains why attention is usually restricted to this case.

In order not to introduce new terminology, *throughout this section* (*and this section only*) we shall use the terms *homogeneous, homeohedral* and *2-homeohedral* as defined earlier, but *without* the assumption of normality. We shall also use the term *topologically regular* to mean that the homeomorphisms which map a tiling onto itself are transitive on the *flags* of the tiling (compare Section 1.3). Except for Figure 4.7.1, however, we shall restrict ourselves to tilings that satisfy conditions N.1 and N.2, that is, whose tiles are topological disks, and where the intersection of any two tiles is either empty or a connected set.

To clarify the concepts we begin by presenting some examples. In Figure 4.7.1 we show a homeohedral tiling with an uncountable infinity of singular points. Clearly other tilings can be constructed with similar features, but we know of no homeohedral tiling with a non-zero finite number, or a countable infinity, of singular points. Also, we have not been able to construct a *periodic* homeohedral tiling with singular points, although there exist periodic 2-homeohedral tilings that have singular points (see Figure 4.7.2). We must emphasize that in these questions we are considering only tilings in which *every* tile is a topological disk. If this condition is also relaxed then it is easy to see that there are many more possibilities.

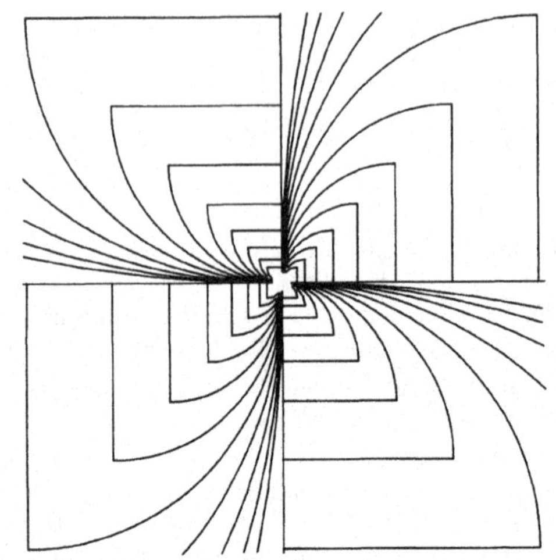

Figure 4.7.1
A non-normal homeohedral tiling with an uncountable infinity of singular points. These lie on the four "rays" radiating from the center of symmetry. This tiling is also a "similarity tiling" in the sense of Section 10.1.

Now consider tilings without singular points that satisfy N.1 and N.2 (but which may have a singularity at infinity). Extending the notation of Section 4.3 we call a tiling *homogeneous of type* $[j_1.j_2.\ldots.j_k]$ if each tile is of valence type $j_1.j_2.\ldots.j_k$. Clearly each homeohedral tiling is homogeneous of some type, and we inquire which symbols correspond to homogeneous tilings and which to homeohedral ones. The complete answer to this question is not known. However, in the special case where all the j_i are equal, the following surprising result holds:

4.7.1 *For every pair of positive integers j, k with*

$$1/j + 1/k \leq 1/2 \qquad (4.7.2)$$

there exists a (not necessarily normal) homeohedral tiling, without singular points, of the type $[j^k]$. *Such a tiling can be normal only if equality holds in* (4.7.2)*—that is, if* $[j^k]$ *is* $[3^6]$, $[4^4]$ *or* $[6^3]$*—but is topologically regular in all cases. If j and k do not satisfy inequality* (4.7.2) *then no homeohedral tiling of type* $[j^k]$ *exists.*

In order to prove that a tiling of a given type exists it is necessary to describe an explicit construction for it—and to show that a tiling of a given type does not exist, we must show that any attempt to construct such a tiling necessarily fails. Since the tilings are not assumed to be normal we cannot, of course, use Euler's Theorem for Tilings.

To prove the first assertion of Statement 4.7.1 we may construct a tiling of type $[j^k]$, with $k \geq 4$, in the following manner. (Figure 4.7.3 illustrates the case $[5^4]$; the construction for $k = 3$ is asked for in Exercise 4.7.9.) With a point V as center, draw circles $C_1, C_2, C_3, \ldots$ of radii $1, 2, 3, \ldots$, and join, by straight-line segments, V to j equidistant points on C_1. On each of the j arcs thus obtained we choose $k - 3$ equidistant points. The point V together with the $j(k - 2)$ points on C_1 will be vertices of the tiling, the $j(k - 2)$ arcs of C_1 together with the j line-segments meeting at V will be edges of the tiling; clearly each tile so determined has k vertices. Further vertices will lie on the circles $C_2, C_3, \ldots$ and further edges will be arcs of those circles together with line-segments between neighboring circles, determined as follows.

Suppose the construction has proceeded as far as determining all the vertices and edges in the (closed) disk determined by a circle C_n. Let x_n be the number of vertices on C_n that are joined by line-segments to vertices on C_{n-1}, and y_n be the number of vertices on C_n not so joined. Then first construct

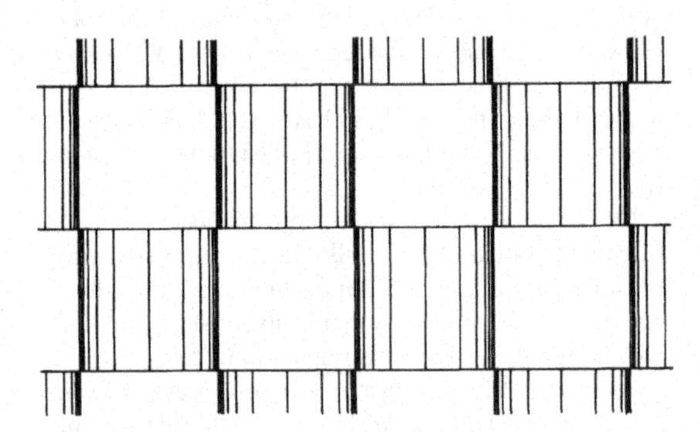

Figure 4.7.2
A non-normal 2-homeohedral periodic tiling which has an uncountable infinity of singular points.

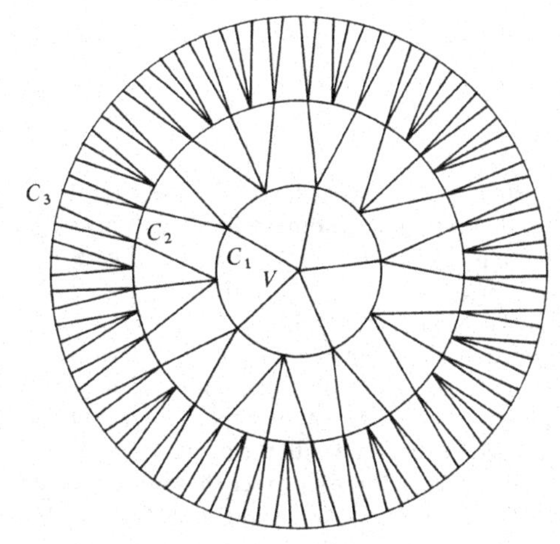

Figure 4.7.3
The construction of the topologically regular tiling $[5^4] = (4^5)$, which is typical for the construction of all tilings $[j^k]$. Note that this tiling is topologically equivalent to the tiling in Figure 3.1.5.

$$x_{n+1} = x_n(j-3) + y_n(j-2) \qquad (4.7.3)$$

points on C_{n+1}, and join $j-3$ by line-segments to each of the x_n vertices on C_n, and join $j-2$ by line-segments to each of the y_n vertices on C_n. Secondly we choose either $k-4$ or $k-3$ further points on C_{n+1} between each two adjacent points of the x_{n+1} so far constructed. We must do this in such a way that each of the x_{n+1} regions lying between C_n and C_{n+1} has k vertices. A simple calculation shows that the number of points introduced at this second stage is

$$y_{n+1} = (k-3)x_{n+1} - (x_n + y_n). \qquad (4.7.4)$$

Hence the total number of points (vertices of the tiling) on C_{n+1} is $x_{n+1} + y_{n+1}$. For definiteness we may assume that these points are equidistant on C_{n+1}. Each of the line-segments and arcs of C_{n+1} determined by the points we have constructed will be edges of the tiling.

We proceed in this way, constructing larger and larger disks. There are two possibilities: either the process goes on indefinitely, in which case we have constructed a tiling of type $[j^k]$, or else it terminates. (This happens, for example, if $j = 3$ and $k = 4$, as the reader will see if he attempts the construction.) To distinguish these cases, we need to solve the recurrence relations (4.7.3) and (4.7.4) with suitable initial values ($x_1 = j$, $y_1 = j(k-3)$). We can eliminate the numbers y_n to obtain

$$x_{n+2} - (jk - 2j - 2k + 2)x_{n+1} + x_n = 0. \qquad (4.7.5)$$

It is known that the positivity of x_n for all n depends upon the "indicial equation"

$$\lambda^2 - (jk - 2j - 2k + 2)\lambda + 1 = 0 \qquad (4.7.6)$$

(see, for example, Riordan [1958, p. 26] or Liu [1968, p. 60] for details and proofs of these assertions). If (4.7.2) holds then equation (4.7.6) has two real roots so the value of x_n increases with n. Then y_n also increases and the construction goes on indefinitely. However, if (4.7.2) does not hold then (4.7.6) has non-real roots, and this implies that x_n eventually becomes negative. This corresponds to the case where the construction terminates.

Although we have described the construction in metrically precise terms—using circles and equidistant points—it is, in a topological sense, universal. More precisely, to show that any tiling $\mathcal{T}$ of type $[j^k]$ is topologically equivalent to that just constructed, we may start from any vertex V' of the tiling $\mathcal{T}$. Let $\mathcal{A}_1$ be the patch consisting of the tiles containing V', $\mathcal{A}_2$ be the patch generated by the union of the tiles in $\mathcal{A}_1$, and, for $n \geq 2$, let $\mathcal{A}_n$ be the patch generated by the union of the tiles in $\mathcal{A}_{n-1}$. Exactly as in the proof of Statement 3.2.4 we can show that each successive patch is built up by adjoining a simple "ring" of tiles to the previous patch, and at no stage is it necessary to fill up "holes" to make the patch simply connected. Then it is clear how a homeomorphism can be set up which maps each of these rings of tiles onto the set of tiles between consecutive circles C_i, C_{i+1} in the tiling of type $[j^k]$ constructed above. This tiling is therefore topologically equivalent to $\mathcal{T}$, and the fact that our construction terminates when inequality (4.7.2) does not hold implies that no tiling $[j^k]$ with $1/j + 1/k > 1/2$ exists. However, it is worth noticing that the five solutions of this inequality correspond to five "tilings" of the sphere, and so to the five regular (Platonic) polyhedra. Also, since V' is any vertex of $\mathcal{T}$, one can deduce from the above argument that $\mathcal{T}$ is topologically regular.

The assertion in Statement 4.7.1 about normal tilings follows from Statement 4.3.1.

In the case $1/j + 1/k < 1/2$ the term *regular* may be justified by the fact that in the non-Euclidean hyperbolic (or Lobačevski) plane, such tilings can be realized using regular polygonal tiles—see, for example, Coxeter & Moser [1972, Chapter 5], Fejes Tóth [1965, p. 94].

Statement 4.7.1 resolves very satisfactorily the question of the existence of homeohedral tilings of types $[j^k]$, but no analogous result is available for types $[j_1.\cdots.j_k]$ where the j_i are not all equal. However, the following partial result can be proved.

4.7.7 *If $j_1, j_2, \ldots, j_k$ are positive integers such that*

$$\sum_{i=1}^{k} (j_i - 2)/j_i < 2 \tag{4.7.8}$$

then no homogeneous tiling of type $[j_1.j_2.\cdots.j_k]$ exists.

It can be shown that each solution of (4.7.8) is either eliminated on combinatorial grounds, or else corresponds to a "tiling" of the sphere. Each of the latter can be realized by the dual of an Archimedean polyhedron.

The case where

$$\sum_{i=1}^{k} (j_i - 2)/j_i = 2$$

has already been investigated in Section 2.7 and Section 4.4. However, when

$$\sum_{i=1}^{k} (j_i - 2)/j_i > 2, \tag{4.7.9}$$

the basic difficulty in any investigation is that the arithmetic implications of Euler's Theorem for Tilings no longer hold, and we are restricted to using arguments of a purely combinatorial nature. For example, it can be shown by simple parity considerations that homogeneous tilings of some types (such as $[3.4.5^2]$, $[3.5.4.5]$, $[4^2.5^2]$) do not exist. In other cases (such as the types $[3.5^3]$ or $[4.5^3]$) more detailed arguments are necessary to establish the non-existence of homeohedral tilings of that type. On the other hand, homeohedral tilings of type $[4^3.n]$ exist for each $n > 4$, of type $[4.n.4.m]$ for all $n > 4$, $m > 4$, and of types $[3^5.n]$ and $[3.n.3.n.3.n]$ for all $n > 3$.

We illustrate the proof of one of these assertions, namely the existence of a homeohedral tiling of type $[3^5.4]$. Further methods and examples are included in the exercises. Let us begin with the regular tiling $[8^3]$ (see Figure 4.7.4(*a*)). It is easy to see that we can "color" the tiles alternatively gray and white in such a way that no two adjacent tiles are of the same color (Figure 4.7.4(*b*)). Now proceed to split each of the white tiles into three triangles (as shown by the added lines in Figure 4.7.4(*b*)) and adjoin each of these to the adjacent gray tile. In this way we arrive at a tiling by hexagons with vertices alternately 3-valent and 4-valent, see Figure 4.7.4(*c*)). This is of type [3.4.3.4.3.4]. Now split each of the hexagons of the tiling into three triangles and three pentagons, as shown by the dotted lines in Figure 4.7.4(*c*). We amalgamate each triangle with one of the pentagons of the tile adjacent to it to yield a tiling by hexagons as in Figure 4.7.4(*d*). This is the required tiling of type $[3^5.4]$.

We mention briefly some other results on homeohedral tilings.

4.7.10 *If all j_i are even and satisfy relation* (4.7.9) *then there exists a homeohedral tiling of type $[j_1.\cdots.j_k]$.*

4.7.11 *If j_1, j_2, j_3 satisfy relation* (4.7.9) (*that is, if $1/j_1 + 1/j_2 + 1/j_3 < 1/2$*), *a homogeneous tiling of type $[j_1.j_2.j_3]$ exists if and only if either* (1) *all j_i are equal; or* (2) *all j_i are even; or* (3) *one of j_1, j_2, j_3 is odd, the other two are even and equal to each other. In each of these cases there exist homeohedral tilings of that type.*

4.7.12 *If j_1, j_2, j_3, j_4 satisfy* (4.7.9) (*that is, if $1/j_1 + 1/j_2 + 1/j_3 + 1/j_4 < 1$*) *a homogeneous tiling of type $[j_1.j_2.j_3.j_4]$ exists if and only if (possibly after a suitable change in notation) either* (1) *all j_i are even; or* (2) *j_1 is odd, j_2 is even and $j_3 \neq j_2 = j_4$; or* (3) *j_1 and j_2*

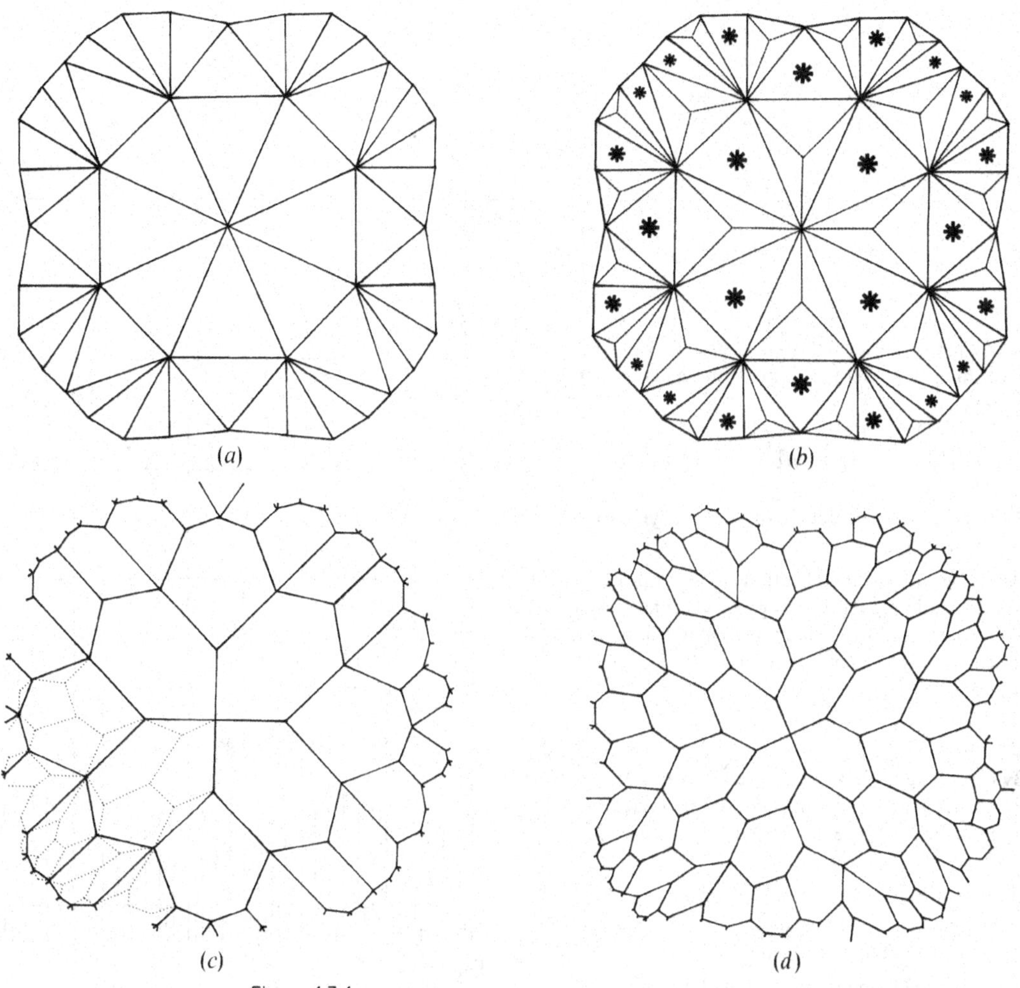

Figure 4.7.4
The construction of the homeohedral non-normal tiling $[3^5.4]$.

are odd, $j_3 = j_1$ *and* $j_4 = j_2$*; or* (4) $j_1 \neq j_2 = j_3 = j_4$ *and at least one of* j_1 *and* j_2 *is odd. A homeohedral tiling of type* $[j_1.j_2.j_3.j_4]$ *exists if and only if either one of the conditions* (1), (2), (3) *is fulfilled, or if* (4) *holds and* $j_2 \not\equiv \pm 1$ (mod 6).

Statement 4.7.10 can be found in Coxeter [1964]; Statements 4.7.11 and 4.7.12 are established in Grünbaum & Shephard [1979a]. From the last part of Statement 4.7.12 it follows, in particular, that there exist homogeneous tilings of types $[3.5^3]$, $[4.5^3]$, $[6.5^3]$, ... , $[3.7^3]$, $[4.7^3]$, $[5.7^3]$, . . . but that no such tiling is homeohedral. This is in marked contrast to the situation in case of normal tilings discussed in Section 4.3. A similar difference from the case of normal tilings is shown by the fact that two homeohedral tilings of the same type are not necessarily topologically equivalent (see Exercise 4.7.8). G. Valette (private communication) has constructed five homogeneous tilings of type $[3.5^3]$ which appear to be distinct.*

* See the Appendix beginning on page 653.

EXERCISES 4.7

*1. Prove parts of Statements 4.7.11 and 4.7.12 by establishing that there exist no homogeneous tilings of types
(a) $[k.n.m^2]$, (b) $[k.n.m.p]$,
(c) $[k^2.n^2]$, (d) $[k^2.n]$,
(e) $[k.m.n]$.
Here distinct letters denote distinct integers not less than 3, and k is odd.

*2. Applying suitable modifications to appropriate topologically regular tilings prove that there exist homeohedral tilings of the types listed below, where n denotes any integer not less than 7:
(a) $[3.n.3.n]$, (b) $[3.2n.2n]$,
(c) $[6.n.n]$ (n even), (d) $[3.4.n.4]$,
(e) $[4.6.4.n]$.

3. By modifying the argument given in connection with Figure 4.7.4, prove the existence of homeohedral tilings of types $[3.n.3.n.3.n]$ and $[3^5.n]$ for each $n > 3$.

4. Show that there exists a homeohedral tiling of type $[4^3.6]$, and that there exists a homogeneous but not homeohedral tiling of the same type.

5. Show that the homogeneous tiling of type $[3.4.3.4^2]$ shown in Figure 4.7.5 is not homeohedral. [Hint: Consider the possible ways in which one of the marked tiles could be mapped onto the other, and in each case try to extend the correspondence to all tiles that contain the marked vertices.]

6. Show that there exist several homogeneous tilings of type $[3.4.3.4^2]$ which are not topologically equivalent to each other.
**Do there exist uncountably many such tilings?

*7. Show that there exist homogeneous tilings of type $[3.5^3]$ but that no such tiling is homeohedral.

8. Show that there exist two topologically non-equivalent homeohedral tilings of type $[3.6^3]$.

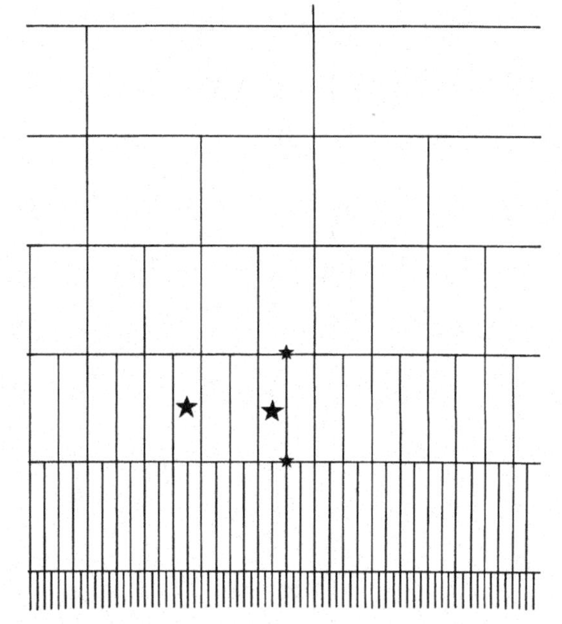

Figure 4.7.5
A homogeneous tiling of type [3.4.3.4.4] which admits no homeohedral tilings.

9. Describe in detail how the construction of a topologically regular tiling of type $[j^k]$ explained in connection with Figure 4.7.3 has to be modified in case $k = 3$. What is the recurrence relation which replaces equation (4.7.5) in this case?

**10. Show that if $\mathscr{T}$ is a homeohedral tiling which satisfies conditions N.1 and N.2 then $\mathscr{T}$ has no singular points.

11. (a) From the theory of recurrence relations it is known that the numbers x_n defined on page 194 are given by the expression

$$x_n = c_1\lambda_1^n + c_2\lambda_2^n,$$

where λ_1 and λ_2 are the two solutions of equation 4.7.6, and c_1, c_2 are constants which are determined by solving the equations

$$x_1 = c_1\lambda_1 + c_2\lambda_2 \qquad \text{and} \qquad x_2 = c_1\lambda_1^2 + c_2\lambda_2^2.$$

If (4.7.2) holds so the solutions of equation (4.7.6) are real we may assume $\lambda_1 > 1$ and $\lambda_1 > \lambda_2 > 0$. Find λ_1, λ_2, c_1, c_2 and hence obtain expressions for x_n and y_n in terms of j, k and n.

(*b*) Use the results of (*a*) to express t_n, the number of tiles inside the patch bounded by C_n, in terms of j, k and n. Show that

$$\lim (t_n - t_{n-1})/t_n = (\lambda_1 - 1)/\lambda_1 > 0,$$

thus explaining the "paradox" discussed in connection with Figure 3.1.5. (Compare Kárteszi [1957], Horváth [1964], Zeitler [1967].)

(*c*) Determine the limits as $r \to \infty$ of the ratio of vertices to tiles and also of the ratio of edges to tiles, in the patch bounded by C_n.

(*d*) Modify the construction of $[j^k]$ described in connection with Figure 4.7.3 by starting with a k-gon instead of a single vertex; compute the number t_n of tiles in the nth patch and find the limit $\lim_{r\to\infty} (t_n - t_{n-1})/t_n$ in this case.

(*e*) Determine, as above, the number t_n of tiles in the increasing sequence of patches in the homeohedral tiling of type $[3^5.4]$ shown in Figure 4.7.4(*d*)

4.8 NOTES AND REFERENCES

It is somewhat surprising that the idea of combinatorial equivalence of tilings seems to have originated with crystallographers, and that mathematicians paid no heed to it for a long time. The first explicit reference to what we call combinatorially equivalent tilings occurs in Laves [1931b]; similar ideas underlie Heesch [1932], [1933a], [1933b] and Nowacki [1935]. Delone [1959] and Heesch [1968b] also consider combinatorial equivalence. A number of other authors (Bilinski [1948], Loeb [1976]) appear to have in mind such a relation among tilings, but do not express their intentions explicitly. Strangely, we were unable to locate in the literature any mention of topological equivalence, or of the fact (Statement 4.1.1) that topological equivalence coincides with combinatorial equivalence in the case of normal tilings.

Duality of tilings, dually situated tilings, or notions close to these have been mentioned and considered by many authors, usually without any precise definitions, and mostly with just the uniform and Laves tilings in mind (see, for example, Williams [1972, pp. 37–42], Wells [1977, p. 185], Pearce & Pearce [1978, pp. 37–40]). The explanations and definitions are often illustrated by examples such as those in Figures 2.7.3 or 4.2.2, and no hint is given of the fact that very basic problems involving duality (as mentioned at the end of Section 4.2) are still unsolved. As a recent example of blatantly false statements concerning dual tilings we may mention the assertion of Šubnikov & Kopcik [1972] (page 179 of the English translation): "In order to construct [all isohedral tilings with convex tiles] we take [all isogonal tilings with convex tiles] and through the midpoints of their edges draw straight lines, which after mutual intersection form convex polygons of [isohedral tilings]."

The determination of all types of homeohedral tilings has been carried out by many authors. In every case, they followed the same procedure as that we described in Section 4.3, namely they enumerated homogeneous tilings. In many cases they implicitly assumed that this was sufficient to ensure tile-transitivity without realizing that such an assumption must be justified, and in any case is false without normality or some equivalent condition (see Section 4.7). The first enumeration of these tilings was by Laves [1931b]; although he did not state precisely the conditions under which his enumeration is valid, he realized that there are many additional possibilities if non-normal tilings are considered. He discussed in detail some of these and described them very adequately. Bilinski [1948] also obtained a correct enumeration on the assumption that the tiles are uniformly bounded (see page 121). Other authors escaped the need to formulate the condition of normality by assuming that the tilings in question are monohedral edge-to-edge tilings by convex polygonal tiles; see, for example, Delone [1959]. (However, this restrictive approach yields a much weaker result. Not only does it leave open the possibility that if curvilinearly bounded tiles, or tiles that are polygonal but not convex, are used there may exist other types of homeohedral tilings, but it even excludes from consideration tilings such as the one in Figure 4.2.1(*b*).) On the other hand, some recent treatments (such as Walsh [1972] or Loeb [1976]) are deficient even in comparison to Laves [1931b], since they ignore the existence of non-normal tilings.

The determination of all types of homeogonal tilings is frequently credited to Šubnikov [1916]. However, this paper has several serious shortcomings; it considers only polygonal tiles (and, in fact, only edge-to-edge tilings by convex tiles), tacitly assumes that each such tiling is balanced and strongly balanced, and is incomplete in the analysis of the solutions of equation (2.7.1) which Šubnikov uses to derive the possible types of such tilings. Moreover, the paper considers only isogonal tilings, and does not even mention homeogonal ones; the fact that each normal homeogonal tiling is topologically equivalent to an isogonal tiling by convex tiles is certainly not self-evident (and in case of tilings of the hyperbolic plane is not settled to this day; see Section 4.7). We shall present more details concerning Šubnikov's paper and some related later publications in Section 6.6 when we discuss isogonal tilings.

Homeotoxal tilings were introduced and studied by Heesch [1933a]. He correctly determined the five types possible for normal tilings, and the additional 14 types that arise if digons and 2-valent vertices are permitted (see Exercise 4.5.3).

Various examples of 2-homeohedral or 2-homogeneous tilings appear in crystallography (see, for example, Figure 1.2.7) and art (see, for example, Figures 1.3.5 and 1.3.9, many of the illustrations in the Introduction, or several of the tilings in Escher [1983]). A large number of examples which are of interest in chemistry appears in O'Keeffe & Hyde [1980]; they also show examples of k-homogeneous and k-homeohedral tilings for various values of $k > 2$. Collatz [1975], [1977] deals with questions remotely related to such tilings. Other examples of 2-homeohedral tilings appear in Heesch [1935], Penrose & Penrose [1958], and Wollny [1984], as well as in the works on monohedral tilings by convex polygons which we shall discuss in Chapter 9. Systematic (but incomplete) investigations of 2-homeohedral tilings can be found in Löckenhoff [1968] (see Figure 4.8.1) and Temesvári [1979], [1980]. A detailed account of the known results on the enumeration of 2-homeohedral tilings appears in Grünbaum, Löckenhoff, Shephard & Temesvári [1985]; however, it is not known whether the 453 types found there constitute a complete list. (The listing contains also 55 types of colored 2-homeohedral tilings, that is 2-colored homeohedral tilings in which the color-preserving homeomorphic maps of the tiling onto itself have two orbits of tiles.) Holladay [1983] determined the 38 types of 2-isohedral edge-to-edge tilings by triangles, in a classification by "incidence symbols" analogous to the classification of isohedral tilings discussed in Chapter 6; this coincides with the homeomeric classification (see Chapter 7) of such tilings.

In Grünbaum & Shephard [1983a] it was established that there exist precisely 51 topological types of normal 2-homeotoxal tilings. Additional results on 2-isohedral and related kinds of tilings can be found in Dress & Scharlau [1984], [1985].*

Homogeneous non-normal tilings appear to have been first considered by Schlegel [1883]; he proved (p. 360) that a (non-normal) tiling of type $[j^k]$ exists whenever $1/j + 1/k < 1/2$, and illustrated (in his Figure 9) the case

* See the Appendix beginning on page 653.

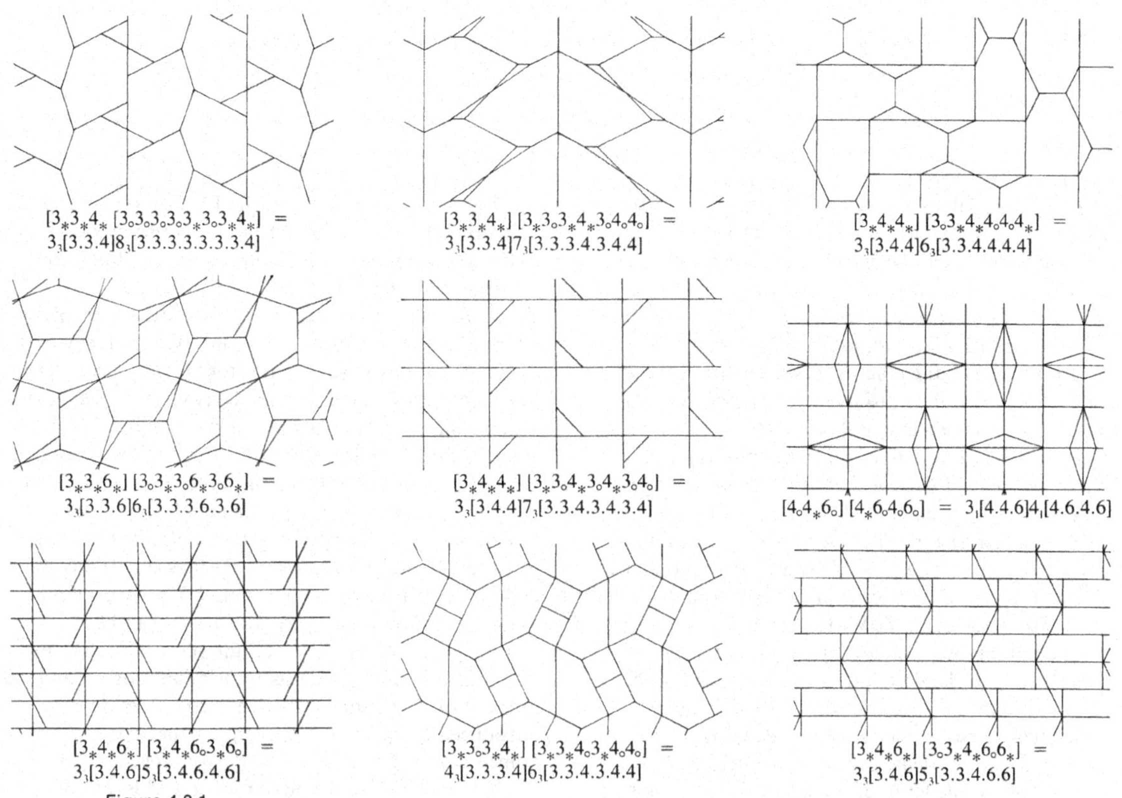

Figure 4.8.1
Nine 2-homeohedral tilings which are not included among the 435 types described by Löckenhoff [1968].

$[3^7]$. A much more detailed (but unfortunately rather inaccessible) study of tilings with all of same valence-type may be found in Bilinski [1948], [1949]; he determines all normal tilings of this kind, derives some necessary conditions for non-normal ones, and establishes the existence of some types of such tilings (for example, homogeneous tilings of types [4.4.4.5], [3.4.3.6.3.8], or [3.3.p.3.q] with $1/p + 1/q < 1/2$). But, as mentioned in Section 4.7, no complete characterization of types of non-normal homogeneous or homeohedral tilings is known. For derivation of equation (4.7.5) see Kárteszi [1957], Horváth [1964] or Zeitler [1967]. The corresponding calculations for the regular tilings of the Euclidean plane can be found in Ranucci [1971].

Tiling (and patterns) have been studied in the context of non-Euclidean geometry. For the geometry on the 2-sphere, all isohedral, isotoxal and isogonal tilings are listed in Grünbaum & Shephard [1982b] and the *dot patterns* have been classified *henomerically* and *homeomerically* in Grünbaum & Shephard [1981c] (for an explanation of the words in italics see Chapters 5 and 7). From these results it is easy to derive the corresponding classifications for elliptic geometry.

For a convenient survey of hyperbolic geometry see Fenn [1983]). The theory exhibits a number of interesting differences from the Euclidean case. As already pointed out in Section 4.7, in the hyperbolic plane there are infinitely many regular tilings (this was first established by Fricke & Klein [1897]; for more modern treatments of this material see, for example, Fejes Tóth [1965, p. 94] or Coxeter & Moser [1972, Chapter 5]), and a homogeneous tiling is not uniquely determined by the type of its vertices. Moreover, tiles may have sides which are "limiting parallel rays"; see Lehner [1964], Vermes [1971], [1974], [1979]. For a curious construction in this context see Richmond

[1937]. A partial solution of an interesting problem concerning regular tilings and Newton numbers in the hyperbolic plane can be found in Florian & Florian [1975].

Most authors represent the hyperbolic plane by the Poincaré model which, although very elegant, has the practical disadvantage that tilings and patterns in this model are very difficult to draw. The comments of M. C. Escher concerning his woodcuts "Circle Limit I, II and III" are of interest in this connection; they are reproduced in Coxeter [1979]. This latter paper also contains explanations of the remarkable geometry hidden in "Circle Limit III". A computer program that carries out the task of drawing hyperbolic tilings and patterns in the Poincaré model has been described by Dunham, Lindgren & Witte [1981].

Much of the literature on hyperbolic geometry is concerned with the description and classification of discrete groups of isometries of the hyperbolic plane; tilings then appear either as fundamental regions for the group or as *Dirichlet regions* (see Section 5.4 for a definition of this term) of the orbit of some point. If the group consists of orientation-preserving isometries only, it is known as a Fuchsian group. For a description of hyperbolic isometries see Wilker [1981], who also discusses the homeomeric classification of dot patterns in the hyperbolic plane. For more general descriptions and discussion of the groups see, for example, Baskan-Macbeath [1982], Macbeath [1967], Magnus [1974], Wilkie [1966], Zieschang, Vogt & Coldeway [1970], and the references given there.

For groups of isometries in some other plane geometries see, for example, Baltag & Garit [1974], Klotzek [1983].

Tilings on other 2-manifolds have also received some attention; see, for example, Huneke [1977], Senechal [1985a], [1985b]. Again there are complications that do not occur in the Euclidean case. For example, though the interior of a polygonal tile must be a topological disk, in discussing uniform tilings on manifolds it becomes necessary to allow for the fact that the closure of this disk need not even be simply connected, see Edmonds, Ewing and Kulkarni [1982].

Turning again to tilings of the Euclidean plane, we now briefly report on some results concerning questions of the following type: which tilings are topologically equivalent to tilings of geometrically special kinds. The most interesting case arises if the special property involved is the convexity of the tiles. We shall say that a tiling can be *convexified* if it is topologically equivalent to a tiling in which all tiles are convex. It is clear that a tiling which violates the normality condition N.2 from Section 3.2 cannot be convexified. As was pointed out in Grünbaum & Shephard [1981d] (using different terminology), it follows from results of Thomassen [1980] (see also Thomassen [1984]) that it is possible to convexify every tiling which satisfies conditions N.1 and N.2 and has no singular points. Moreover, if a tiling $\mathscr{T}$ is normal and periodic, then it can be convexified in such a way that the tiling $\mathscr{C}$ by convex tiles is also periodic, with its translations corresponding to those of $\mathscr{T}$. (Anticipating the terminology to be introduced in Chapter 7, we can say that the homeomorphism between $\mathscr{T}$ and $\mathscr{C}$ is *compatible* with all the translational symmetries of $\mathscr{T}$.) This was first established in a very complicated way by Mani-Levitska, Guigas & Klee [1979]; it follows more simply from the results of Thomassen [1980], [1984]. These results can be considered as analogous to the famous theorem of Steinitz that every *finite graph* which satisfies the obvious necessary conditions (planarity and 3-connectedness) is homeomorphic to the graph of a 3-dimensional convex polyhedron (see Steinitz [1922], Steinitz & Rademacher [1934], Grünbaum [1967, Section 13.1], Barnette & Grünbaum [1969]), and to the related results involving symmetries (Barnette [1970], Mani [1971]). However, the question whether every periodic tiling $\mathscr{T}$ can be convexified to a tiling $\mathscr{C}$ in such a way that the homeomorphism between $\mathscr{T}$ and $\mathscr{C}$ is compatible with *all* the symmetries of $\mathscr{T}$ seems to be still open. We conjecture that, more generally, every tiling that satisfies conditions N.1, N.2 and has no singular points can be convexified in such a way that the homeomorphism is compatible with all the symmetries of the original tiling.*

* See the Appendix beginning on page 653.

PATTERNS

Figure 5.0.1
A non-trivial discrete pattern whose design was inspired by a maze. The pattern is of type PP8 in the classification system explained in this chapter; see Figure 5.2.7.

5

PATTERNS

This chapter is concerned with patterns in the plane. The word "pattern" is used here in the same sense as it seems to have been used, without any precise definitions, in the writings of some crystallographers, designers and art historians. The patterns with which we shall be mostly concerned may be described as repetitions of a "motif" in the plane in a "regular" manner, subject to certain restrictions which we shall make precise in the first section; examples of these patterns are shown in Figures I.5 and 5.2.7. Most of our treatment is, we believe, new—a surprising fact considering the immense amount of effort that artists and architects have expended in designing and analysing patterns since time immemorial.

In this chapter we shall also show how the study of patterns yields a preliminary classification of certain kinds of tilings with transitivity properties. This is preparatory to the more detailed analysis of such tilings in the next chapter.

5.1 DISCRETE PATTERNS*

We begin by introducing some terminology and definitions.

By a *motif* we mean any non-empty plane set. A *monomotif pattern* $\mathscr{P}$ *with motif* M is a non-empty family $\{M_i \mid i \in I\}$ of sets in the plane, labelled by an index-set I, such that the following conditions hold:

P.1 *The sets* M_i *are pairwise disjoint.*

P.2 *Each* M_i *is congruent to* M *and called a* copy *of* M.

P.3 *For each pair* M_i, M_j *of copies of the motif there is an isometry of the plane that maps* $\mathscr{P}$ *onto itself and* M_i *onto* M_j.

By analogy with the definition given in Section 1.3, any isometry which maps $\mathscr{P}$ onto itself will be called a *symmetry* of the pattern $\mathscr{P}$. The set of all these symmetries forms a group under composition, called the *symmetry group* of $\mathscr{P}$ and denoted by $S(\mathscr{P})$. Condition P.3 clearly implies that $S(\mathscr{P})$ acts transitively on the copies of M in $\mathscr{P}$.

Except in some exercises, we shall deal only with monomotif patterns; for brevity, we shall call them simply "patterns".

Those symmetries of $\mathscr{P}$ which map a copy M_i of the motif M onto itself form a group denoted by $S(\mathscr{P} \mid M_i)$. It is easy to see that the groups $S(\mathscr{P} \mid M_i)$ are isomorphic for all copies M_i, and we define $S(\mathscr{P} \mid M)$ to be equal to each of these groups. We call $S(\mathscr{P} \mid M)$ the *induced* (*motif*) *group* of the pattern $\mathscr{P}$; sometimes $S(\mathscr{P} \mid M)$ is called the *stabilizer* of M in $S(\mathscr{P})$. If $S(\mathscr{P} \mid M)$ consists of the identity alone, the pattern is called *primitive*.

In Figure 5.1.1 we show some examples of patterns. As in the case of tilings we can usually represent in a diagram only a small finite portion of a pattern, and we tacitly assume that the reader can deduce from this how the pattern can be continued over the whole plane. In some cases (for example, in Figures 5.1.1(f) and (k))

* See the Appendix beginning on page 653.

the motif is unbounded, and so we cannot draw even one example of it completely. On the other hand, in patterns such as those of Figures 5.1.1(e), (m) and (o) the motif is repeated only a finite number of times, and so the complete pattern can be drawn.

In Figure 5.1.1(j) the motif is a line segment, and the pattern consists of an infinite number of these arranged radially in the form of an annulus. In such cases, and in Figure 5.1.1(l) where it is intended that adjacent line segments are separated by a single point, the diagram can do no more than give a general indication of what is intended. (See also Exercise 5.1.7.)

These examples show that the notion of a pattern is very general*. Just as in our discussions of tilings (see Section 1.1), where our original definition was too general to admit geometric analysis, so here it is convenient to restrict the kind of pattern to be considered.

We say that a pattern is *discrete* if the following conditions hold:

DP.1 *The motif M is a bounded and connected set.*

DP.2 *For some i there is an open set E_i which contains the copy M_i of the motif but does not meet any other copy of the motif; that is, $M_j \cap E_i = \varnothing$ for all $j \in I$ such that $j \neq i$.*

A discrete pattern is called *non-trivial* if it satisfies the following additional condition.

DP.3 *The pattern contains at least two copies of the motif.*

From now on we shall restrict attention to discrete patterns unless the contrary is explicitly stated. Condition DP.2 eliminates patterns such as that of Figure 5.1.1(j) as well as many other patterns which are difficult to draw! Of course, if condition DP.2 holds for some i, then it necessarily holds for every i—in other words, for every copy of the motif it will be possible to find an open set which contains it and is disjoint from every other copy of the motif. Somewhat surprising is the fact that it also implies that we can *simultaneously* find open sets $E'_i \supset M_i$ for all $i \in I$, with the E'_i pairwise disjoint. This is a consequence of Statement 5.1.1 given below.

Condition DP.1 eliminates the patterns of Figures 5.1.1(g), (h) and (l) because the motifs are not connected, and those of Figures 5.1.1(f), (k) and (n) because they are not bounded. Condition DP.2 eliminates the pattern of Figure 5.1.1(j) and condition DP.3 eliminates the pattern of Figure 5.1.1(i). Thus only the patterns of Figures 5.1.1(a), (b), (c), (d), (e), (m) and (o) are discrete and non-trivial, and of these all except the ones in Figures 5.1.1(d) and (m) are primitive.

If a pattern is not discrete because its motif is not connected, then it is sometimes possible to join the disconnected components with curves in such a way as to make the pattern discrete. In this way the pattern of Figure 5.1.1(g) can be made discrete. But this cannot be done for the patterns of Figures 5.1.1(h) and (l). For the former any attempt fails because the curves cannot be chosen in such a way that the copies of the motif are disjoint, and for the latter they can be chosen as disjoint but then they will necessarily violate condition DP.2.

The process just described is a particular case of what we shall call "engulfing". To be precise, we say that a pattern $\mathscr{Q}$ *engulfs* a pattern $\mathscr{P}$ if the copies N_i and M_i of the motifs of $\mathscr{Q}$ and $\mathscr{P}$ can be indexed by the same set I in such a way that the following two conditions hold.

E.1 *$N_i \supset M_i$ for each $i \in I$.*

E.2 *Every symmetry s of $\mathscr{P}$ is also a symmetry of $\mathscr{Q}$.*

These conditions clearly imply that if $s(M_i) = M_j$ then also $s(N_i) = N_j$. Note that condition E.2 does not require that every symmetry of $\mathscr{Q}$ is a symmetry of $\mathscr{P}$.

* Even greater generality can be obtained by admitting similarities, affine transformations, etc. as symmetries. Some of these possibilities will be explored in the exercises at the end of this section.

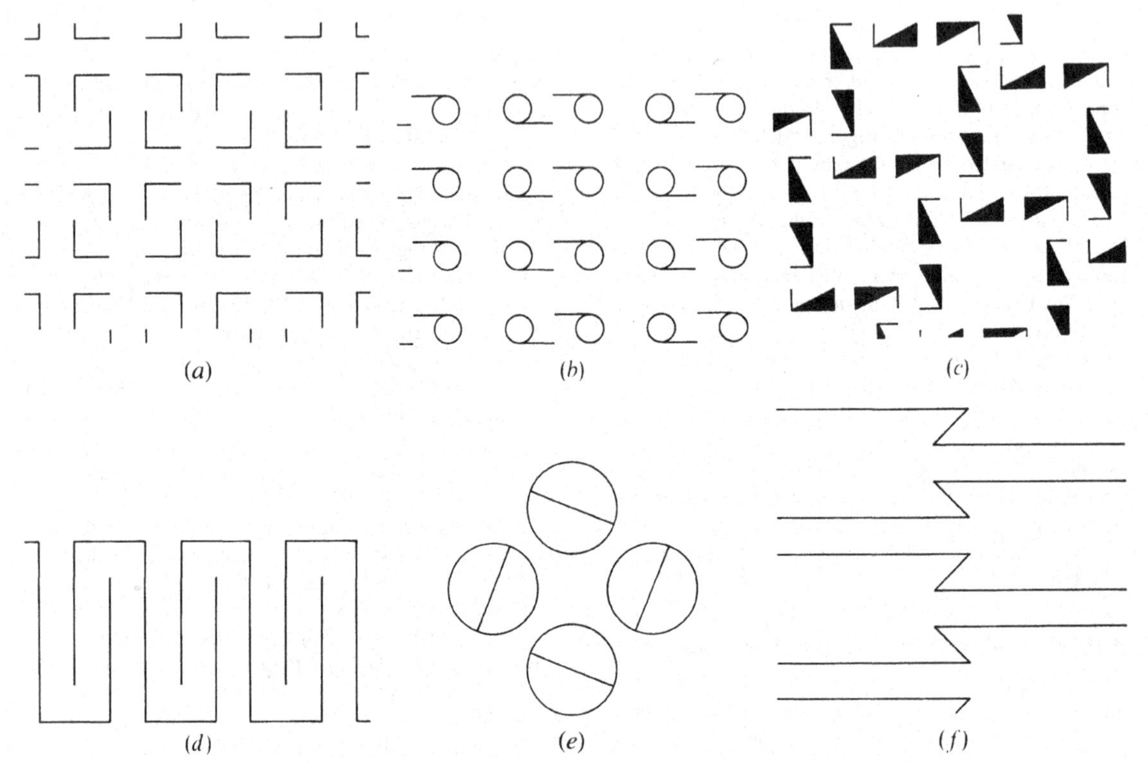

Figure 5.1.1
Some examples of patterns. In (*g*), (*h*) and (*l*) the motif is disconnected; the two pieces of each copy of the motif are labelled with the same numeral. In (*f*), (*k*) and (*n*) the motif is unbounded. The patterns in (*a*), (*b*), (*c*), (*e*), (*k*) and (*o*) are primitive. The induced motif group $S(\mathscr{P}|M)$ in (*d*), (*f*), (*g*) and (*h*) is *c2*, in (*j*), (*l*) and (*m*) it is *d1*, in (*i*) it is *d6* and in (*n*) it is *pma2*. In patterns (*e*), (*i*), (*k*), (*m*) and (*o*) the number of copies of the motif is finite; in (*i*) there is a single copy and the pattern is trivial.

Examples illustrating the definition of engulfing (and of subtending, to be defined shortly) appear in Figure 5.1.2.

Similarly, if a pattern is not discrete because its motif is unbounded, it may be possible to *truncate* the motif so that the groups $S(\mathscr{P})$ and $S(\mathscr{P}|M)$ are not altered and the pattern becomes discrete. Examples of patterns where this can be done are shown in Figures 5.1.1(*f*) and (*k*), but the technique is inapplicable to the pattern of Figure 5.1.1(*n*).

The process of truncation is related to that of "subtending"; this is, in a sense, the opposite of engulfing. We say that a pattern $\mathscr{P}$ with motif M *subtends* a pattern $\mathscr{Q}$ with motif N if the copies of the motifs can be indexed by the same set I in such a way that the following conditions hold.

S.1 *$M_i \subset N_i$ for each $i \in I$.*

S.2 *Every symmetry of $\mathscr{Q}$ is also a symmetry of $\mathscr{P}$.*

Note that condition S.2 does not require that every symmetry of $\mathscr{P}$ is a symmetry of $\mathscr{Q}$. Examples of sub-

(g) (h) (i)

(j) (k) (l)

(m) (n) (o)

tending, and illustrations of the relationship between subtending and engulfing, appear in Figure 5.1.2. Clearly, it is possible for $\mathscr{Q}$ to engulf $\mathscr{P}$ and for $\mathscr{P}$ to subtend $\mathscr{Q}$ only if the symmetry groups of $\mathscr{P}$ and $\mathscr{Q}$ are equal.

There are several special types of discrete patterns which we shall need to consider later. For these an obvious terminology is adopted: a *dot pattern* is one whose motif is a single point, a *closed disk pattern* is one whose motif is a closed topological disk, and a *closed circle pattern* is one whose motif is a closed circular disk. These are examples of *compact patterns*, so called because in each case the motif is a compact (that is, closed and bounded) set in the plane. Non-compact patterns will also be mentioned—in Section 7.3 we shall discuss *open circle patterns* (that is, discrete patterns whose motifs are open circular disks); of particular interest are also the more general *open disk patterns* whose motifs are open topological disks.

We shall now prove the result mentioned above.

5.1.1 *Every discrete pattern $\mathscr{P}$ can be engulfed by an open disk pattern $\mathscr{Q}$.*

Figure 5.1.2
Patterns that illustrate "engulfing" and "subtending". In each case $\mathscr{P}$ is drawn in heavy lines and $\mathscr{Q}$ in thin lines. The pattern $\mathscr{Q}$ engulfs $\mathscr{P}$ in (*a*) and (*b*) but not in (*c*) and (*d*). The pattern $\mathscr{P}$ subtends $\mathscr{Q}$ in (*a*) and (*d*) but not in (*b*) and (*c*). The symmetry groups of $\mathscr{P}$ and $\mathscr{Q}$ are equal in (*a*) and (*c*), but coincide only in (*a*).

The proof* of Statement 5.1.1 depends on the construction of the pattern $\mathscr{Q}$ in three steps, illustrated in Figures 5.1.3 and 5.1.4. As before, we let M be the motif of $\mathscr{P}$, and let E be an open set which contains the copy M_0 of the motif and is disjoint from all the other copies M_j.

The first step is to construct another open set E' with

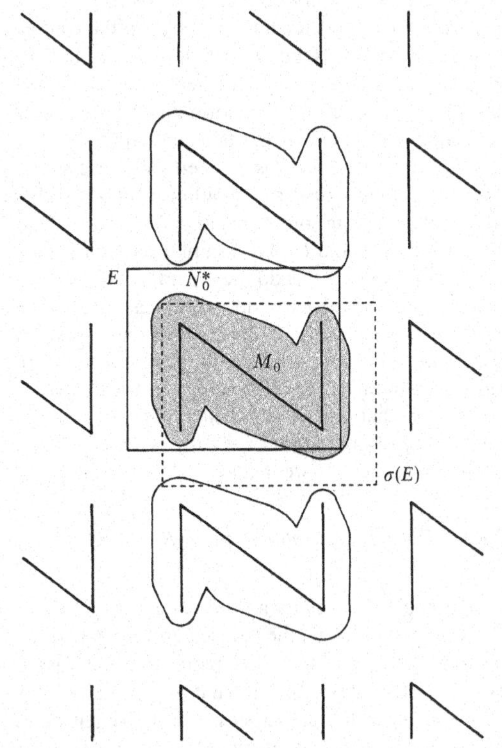

Figure 5.1.3
The construction of an open disk pattern that engulfs a discrete pattern (Statement 5.1.1). The motif M is a Z in heavy lines. The set E containing M_0 is indicated by a solid line, and $\sigma(E)$, where σ is reflection in the center of M_0, is indicated by a dashed line. The union $E \cup \sigma(E)$ is the set E'. The set N_0^* containing M_0 is indicated in gray, and the corresponding sets for two neighboring motifs are also shown. The essential part of the proof of Statement 5.1.1 is to show that these sets are disjoint. As the sets N_i^* are open topological disks, the final step in the proof is not needed here.

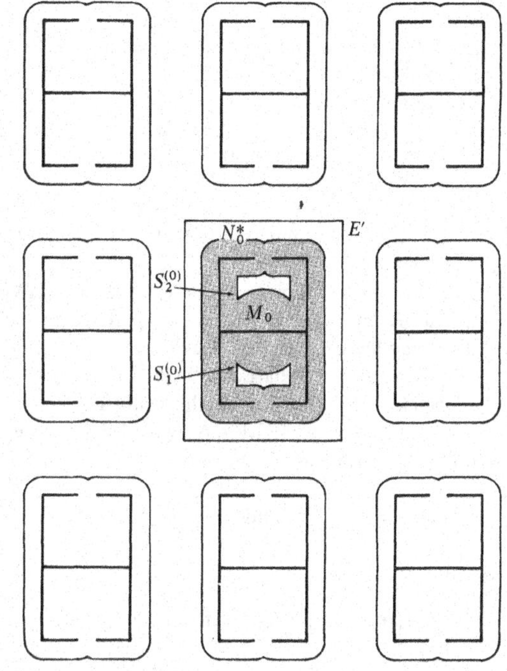

Figure 5.1.4
An illustration of the final step in the proof of Statement 5.1.1. The set N_0^*, shown in gray, is not simply connected and $S_1^{(0)}$, $S_2^{(0)}$ are the bounded components of its complement. The set N is the simply connected set $N_0^* \cup S_1^{(0)} \cup S_2^{(0)}$. Notice that even if M_0 (shown in heavy lines) and E' (outlined in a thin line) are simply connected, N_0^* may fail to be simply connected.

the same properties as E, which is (in general) more symmetric and larger than E. For each symmetry $s \in S(\mathscr{P}|M_0)$ it is obvious that $s(E) \supset M_0$, that $s(E)$ is open, and that $s(E) \cap M_j = \varnothing$ for all $j \neq 0$. Hence $E' = \bigcup_{s \in S(\mathscr{P}|M_0)} s(E)$ has the properties stated above.

In describing the second step we use the notation $D(r,P)$ for an open circular disk of radius r centered at P. For each point P of M_0 let r_P be the largest value of r

* The reader, if he wishes, may omit this proof and in place of condition DP.2 impose the stronger condition: *$\mathscr{P}$ can be engulfed by an open disk pattern.*

such that $D(r_P,P)$ is contained in E'. Then define N_0^* to be the union of all the disks concentric with $D(r_P,P)$ and of half the radius, that is,

$$N_0^* = \bigcup_{P \in M_0} D(\tfrac{1}{2}r_P,P).$$

For each $j \in I$ pick a symmetry $s_j \in S(\mathscr{P})$ which maps M_0 onto M_j, and define $N_j^* = s_j(N_0^*)$. Thus each copy N_j^* of N_0^* bears the same relation to M_j as N_0^* does to M_0. We claim that the family $\{N_j^* | j \in I\}$ forms a pattern $\mathscr{Q}$ which engulfs $\mathscr{P}$ and has an open set N_0^* as motif.

The only non-obvious part of this claim is that the sets N_j^* are pairwise disjoint. Suppose this is not so; then a point V satisfies $V \in N_i^*$ and $V \in N_j^*$ with $i \neq j$ and we can find points P', P'' in M_0 such that

$$V \in s_i(D(\tfrac{1}{2}r_{P'},P')) \quad \text{and} \quad V \in s_j(D(\tfrac{1}{2}r_{P''},P'')).$$

This implies that

$$\|V - s_i(P')\| < \tfrac{1}{2}r_{P'} \quad \text{and} \quad \|V - s_j(P'')\| < \tfrac{1}{2}r_{P''}.$$

(Here $\|A - B\|$ means the distance between the points A and B.) Suppose, without loss of generality, that $r_{P''} \leq r_{P'}$; then

$$\|s_i(P') - s_j(P'')\| \leq \|V - s_i(P')\| + \|V - s_j(P'')\|$$
$$< \tfrac{1}{2}r_{P'} + \tfrac{1}{2}r_{P''} \leq r_{P'},$$

or, since s_j is an isometry,

$$\|P' - s_i^{-1}s_j(P'')\| < r_{P'},$$

and

$$s_i^{-1}s_j(P'') \in D(r_{P'},P').$$

But this is a contradiction since $s_i \neq s_j$ implies that $s_i^{-1}s_j(P'')$ does not belong to M_0—yet $D(r_{P'},P') \subset E'$ does not meet any copy of the motif other than M_0. We deduce that our assumption about the existence of V must have been false and so $N_i^* \cap N_j^* = \varnothing$ as claimed.

The final step in the construction of $\mathscr{Q}$ is necessary to ensure that its motif is simply connected. If not so, consider the complement of N_0^* (see Figure 5.1.4). As the copies of the motif are congruent, it follows that the *bounded* components $S_i^{(0)}$ ($i = 1, 2, \ldots$) of the complement of N_0^* are disjoint from each N_j^* with $j \neq 0$, and also from the bounded components $S_i^{(j)}$ of the complements of these N_j^*. Hence we can define N_j to be the union of N_j^* with all the $S_i^{(j)}$. These N_j are open, disjoint and simply connected and so form an open disk pattern. This completes the proof of Statement 5.1.1.

Since the motif N of $\mathscr{Q}$ is an open set it necessarily contains a circular disk, say of radius u. Since N is bounded it is contained in some circular disk, say of radius U. The quantities u and U play a similar role in the theory of open disk patterns as do the inparameter and circumparameter in the theory of normal tilings, see Section 3.2. The argument on page 123 then shows that every disk $D(r,P)$ meets at most a finite number of copies of the motif N. Adapting the terminology in the obvious way, we can say that $\mathscr{Q}$ is *locally finite*. Since, by Statement 5.1.1, every discrete pattern $\mathscr{P}$ may be engulfed by an open disk pattern $\mathscr{Q}$ we deduce:

5.1.2. *Every discrete pattern is locally finite.*

In particular, for every non-trivial discrete pattern $\mathscr{P}$ based on the motif M, the induced group $S(\mathscr{P}|M)$ is necessarily finite. This fact will be required in the classification of patterns described in the next section.

Just as every discrete pattern $\mathscr{P}_1$ can be engulfed by an open disk pattern $\mathscr{Q}_1$ (Statement 5.1.1), so every compact discrete pattern $\mathscr{P}_2$ can be engulfed by a closed disk pattern $\mathscr{Q}_2$ (one whose motif is a closed topological disk). However, there is an essential difference between the two situations. Whereas $\mathscr{Q}_2$ may be chosen so that it is subtended by $\mathscr{P}_2$ (that is, so that every symmetry of $\mathscr{Q}_2$ is also a symmetry of $\mathscr{P}_2$), it will not, in general, be possible to choose $\mathscr{Q}_1$ so as to be subtended by $\mathscr{P}_1$ (Figure 5.1.5).

Also, from every discrete pattern $\mathscr{P}$ it is possible to derive a dot pattern $\mathscr{D}$ in the following way. First choose a point C_i which is mapped onto itself by every symmetry in $S(\mathscr{P}|M_i)$ for some M_i. (Thus C_i may be chosen to be the centroid of M_i or, if M_i is compact, as the center of the smallest circle containing M_i; many other choices are possible and may lead to several distinct dot patterns.) $\mathscr{D}$ is then defined as the set of images of C_i under the isometries of $S(\mathscr{P})$. We say that such a dot pattern $\mathscr{D}$ is *associated* with $\mathscr{P}$. It may happen that the point C_i does not lie in the motif M_i, and it is also possible that the points corresponding to distinct copies of the motif coincide. However, in the case of an open or closed disk pattern $\mathscr{P}$, $\mathscr{D}$ can be chosen so as to subtend $\mathscr{P}$, though in general $\mathscr{P}$ will not engulf $\mathscr{D}$.

The proofs of the above assertions are straightforward, and proofs or counterexamples are requested in the exercises.

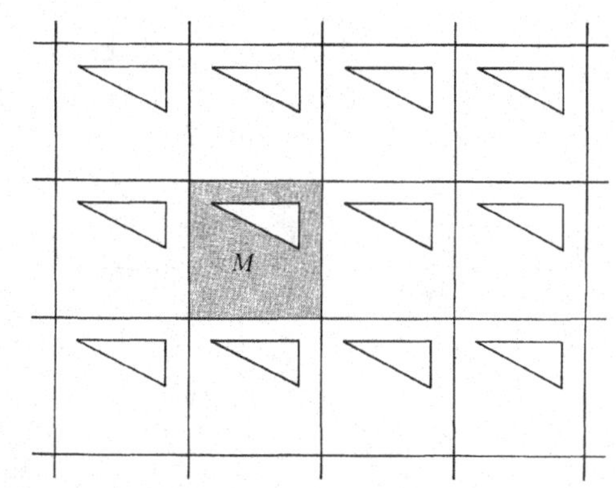

Figure 5.1.5
An example of a discrete pattern $\mathscr{P}$ which can be engulfed by an open disk pattern $\mathscr{Q}$, but $\mathscr{Q}$ cannot be chosen so as to be subtended by $\mathscr{P}$. The motif M of $\mathscr{P}$, shown in gray, is the open set consisting of points interior to a square and exterior to a triangle. There is only one open disk pattern that engulfs $\mathscr{P}$, namely that whose motif is an open square. This example hints at the close connection between patterns and isohedral tilings—a connection which will be explored more fully in Section 5.4.

EXERCISES 5.1

1. In Figure 5.1.6 we show several ways in which open, closed or partially open squares can be arranged in the plane. The method of representation is explained in the caption. For each arrangement (*b*) to (*g*) determine whether it is a pattern in the sense defined at the beginning of this section (noticing that in some cases the motif may be the union of two or more "squares"). When it is a pattern also determine whether it is:
   (*a*) discrete;
   (*b*) possible to engulf it by an open disk pattern;
   (*c*) possible to find a dot pattern that subtends it.

*2. Show that every compact pattern $\mathscr{P}$ can be engulfed by a disk pattern $\mathscr{Q}$ such that $\mathscr{P}$ subtends $\mathscr{Q}$.

3. (*a*) Give an example of a pattern for which, in an associated dot pattern, the same dot corresponds to two or more distinct copies of the motif.
   (*b*) Give an example of a pattern such that, in *every* associated dot pattern, each dot corresponds to two or more distinct copies of the motif.
   *(*c*) Can the situations described in (*a*) and (*b*) arise in a discrete pattern?

4. Give an example of an open disk pattern $\mathscr{P}$ which does not engulf any associated dot pattern $\mathscr{D}$.

5. State and prove an analogue of the Normality Lemma (Statement 3.2.2) for open disk patterns.

6. Extend each of the designs in Figure 5.1.7 to a discrete (monomotif) pattern with the same connected motif M, or prove that it is not possible to extend it to such a pattern. If the extension is possible, determine whether it is unique. In each case determine the symmetry groups of all possible extensions. Does the answer change if we insist that the induced motif group is $S(M)$?

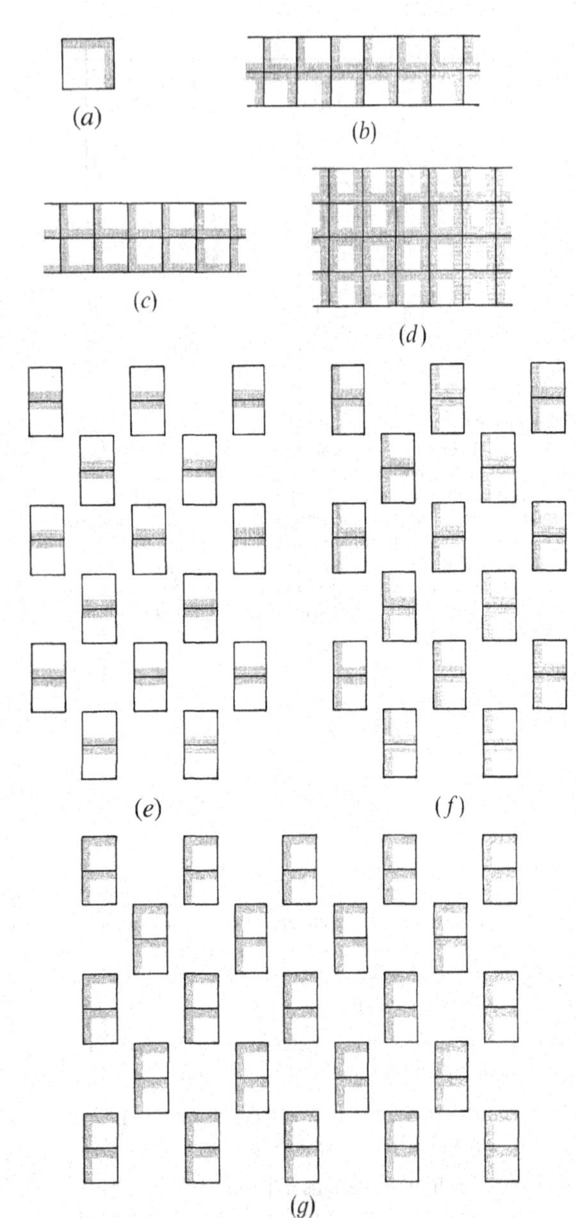

Figure 5.1.6
In (a) we show how a gray border may be added to the sides of a square to signify that they are open. Thus if the bottom left corner is taken as origin and the side of the square as 1, then the square shown is the intersection of the half-open strips $0 \leq x < 1$ and $0 \leq y < 1$. Parts (b) to (g) of this figure are referred to in Exercise 5.1.1.

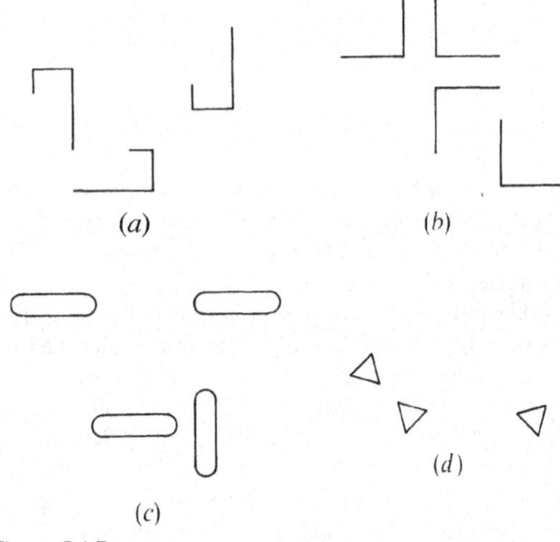

Figure 5.1.7
Can these arrangements of motifs be extended to discrete patterns? See Exercise 5.1.6.

7. Verify that the following family $\mathscr{P}$ of sets has properties P.1, P.2 and P.3, and hence is a (non-discrete) monomotif pattern. In the usual (x,y) coordinate system let M consists of the points $(x,0)$ with $0 < x < 2\pi$ together with the points (x,y) where x is rational, $0 < x < 2\pi$ and $0 < y < 1$. The family $\mathscr{P}$ consists of M, the set M^* obtained from M by a half-turn about the point $(\pi,\frac{1}{2})$ and all translates of M and M^* by vectors of the form $(2n\pi,2m)$ where n and m are integers.
8. In Figure 5.1.8 we reproduce diagrams of printed textiles from Proctor [1969]. The reader is invited, in this and the following exercise, to see how far the concepts of this section can be applied to practical examples such as these. Notice that none of the diagrams represents a discrete pattern since the motif is not connected, but in every case it can be made discrete by some simple modification of the motif. Each of the following parts may have several different answers depending upon which set is chosen as motif.
   (a) Find the symmetry group $S(\mathscr{P})$ and the induced motif group $S(\mathscr{P}|M)$.
   (b) Find an associated dot pattern.
   (c) Engulf the pattern by an open disk pattern.
   (d) Subtend the pattern by a closed disk pattern.
   (e) Determine whether, in parts (c) and (d), it is possible to choose the engulfing or subtending

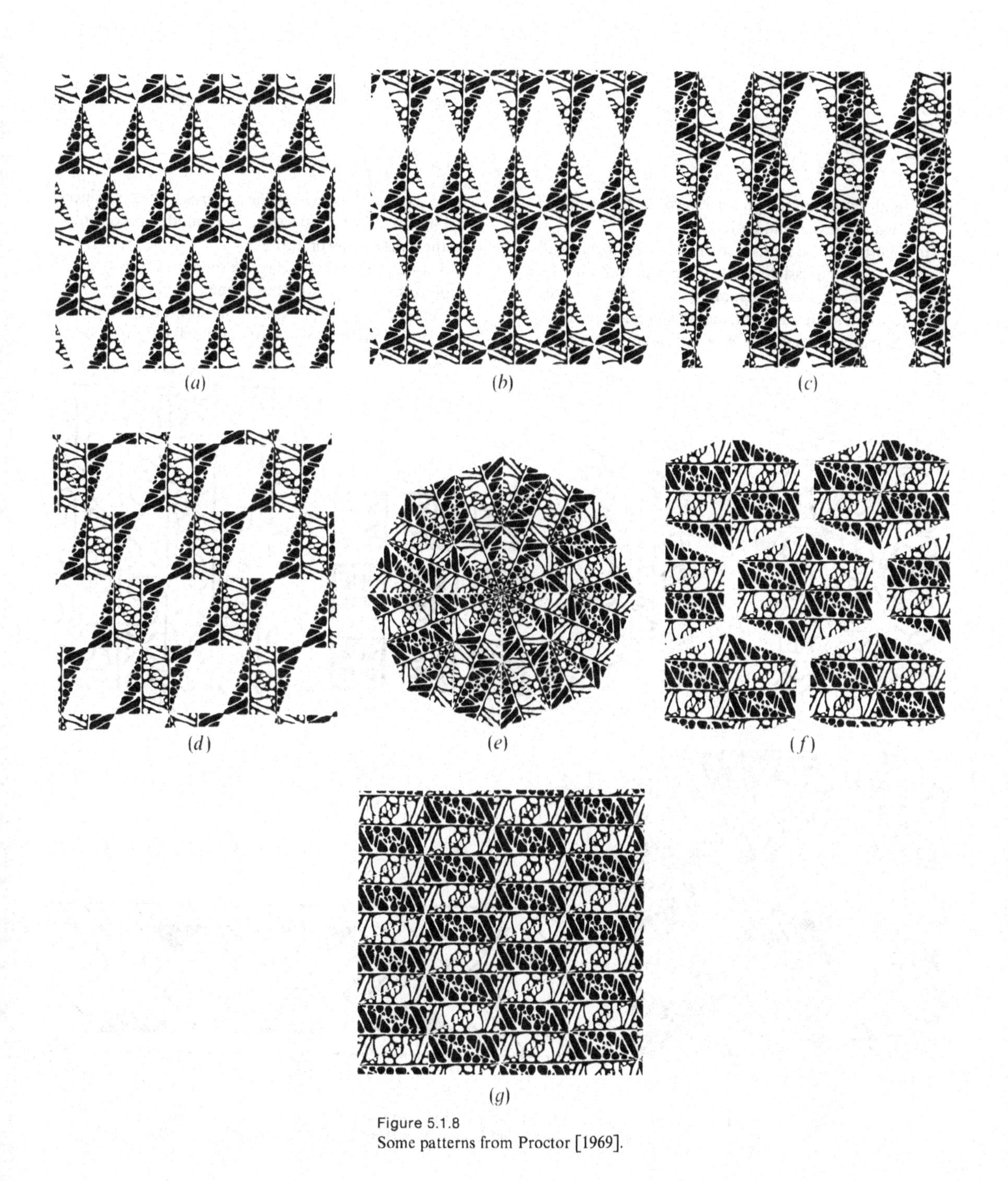

Figure 5.1.8
Some patterns from Proctor [1969].

pattern so that its symmetry group is the same as that of the original pattern.

9. In Figure 5.1.9 we reproduce designs from historical artifacts. It is possible (with a certain amount of goodwill) to interpret each of these as a (monomotif) pattern—in some cases the motif is a region printed black and in other cases it is a region which is outlined. Notice that none of the patterns is discrete since the motifs are disconnected or unbounded according to the chosen interpretation.

   Another interpretation of the diagrams is as the superposition of two patterns, one of which engulfs or subtends the other. Notice that in Figure 5.1.9(d) one should not "ignore" parts of the diagram—if one

Figure 5.1.9
Examples of patterns engulfed or subtended by other patterns. Patterns (*a*), (*b*), (*c*) and (*f*) are from Christie [1929], (*d*) is from Audsley & Audsley [1968], (*e*) is from Hornung [1975], and (*g*) is after Bain [1951].

does so there are many new possibilities which the reader is invited to explore.

A further matter for consideration is suggested by these diagrams. How can a "strip" pattern such as those of Figures 5.19(*b*), (*f*) and (*g*) be extended over the whole plane? This can always be done by simply laying the strips side by side, but in some cases there are other, more interesting possibilities. Modify the strip pattern in Figure 5.1.9(*f*) so that it can be extended to a pattern with symmetry group $p4g$.

10. (*a*) Adapt the definition of a (monomotif) pattern given at the beginning of this section to obtain a reasonable definition of a *dimotif pattern* $\mathscr{P}$ with motifs M and M'. Similarly, define *trimotif*, *tetramotif*, ..., *n-motif* patterns.
    (*b*) A dimotif pattern with motifs M and M' is called *restricted* if it is not possible to regard it as a monomotif pattern whose motif is the union of one copy of M and of one copy of M'. Give an example of a restricted dimotif pattern.
    *(*c*) Given any monomotif pattern, is it always possible to mark (or color) certain copies of the motif in such a way as to convert it into a dimotif pattern? And can both restricted and unrestricted dimotif patterns arise in this way?
    *(*d*) For trimotif patterns give a reasonable definition of the word "restricted" and answer the two questions in part (*c*) for such patterns.*

11. A monomotif pattern $\mathscr{Q}$ is said to engulf an n-motif pattern $\mathscr{P}$ if each copy of the motif N of $\mathscr{Q}$ contains exactly one copy of each motif of $\mathscr{P}$, and every symmetry of $\mathscr{P}$ is a symmetry of $\mathscr{Q}$. In Figure 5.1.10 several diagrams showing electron densities in crystals are shown. For various interpretations of parts of the diagrams as n-motif patterns $\mathscr{P}$ (whose motifs need not be connected) and other parts as monomotif patterns $\mathscr{Q}$, examine whether $\mathscr{Q}$ engulfs $\mathscr{P}$.

*12. Formulate and prove an analogue of Statement 5.1.1 for the concept of engulfing an n-motif pattern defined in the previous exercise.

13. Consider the images of the square M with vertices (1, 1), (2, 1), (2, 2) and (1, 2) under the transformation

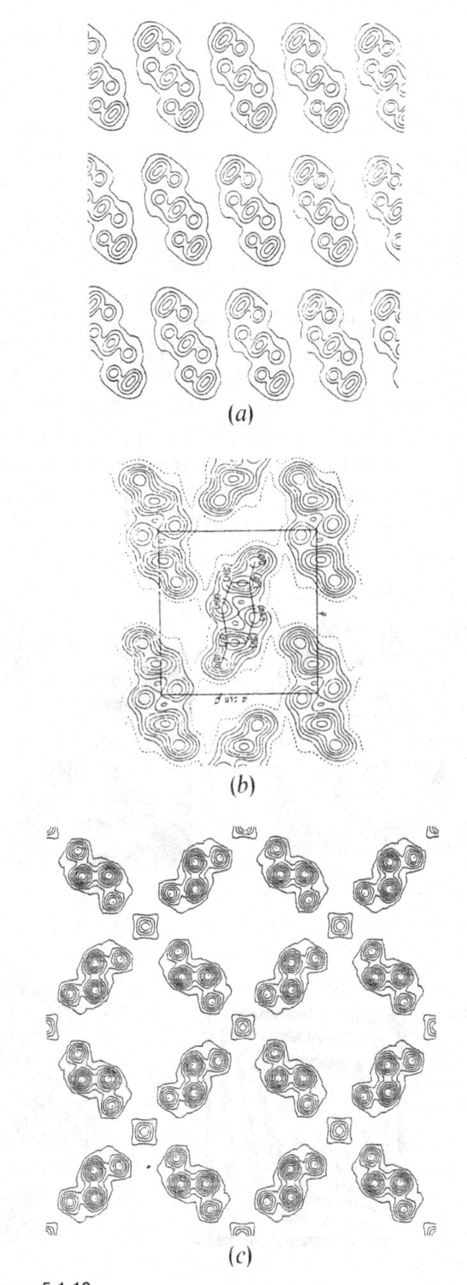

Figure 5.1.10
Patterns of electron densities of crystals of certain organic compounds. Parts (*a*) and (*c*) are from Luis & Amorós [1968]; (*b*) is from Kitaigorodskii [1955].

* See the Appendix beginning on page 653.

$T: (x,y) \to (\frac{1}{3}x, 3y)$ and the powers of T and its inverse. These sets form an *affine pattern* since it is the *affine symmetry group* $AS(\mathscr{P})$ (that is, the group of affine transformations which leave $\mathscr{P}$ fixed) which acts transitively on the copies of the motif. ($S(\mathscr{P})$ does not have this property.) Give at least two other examples of affine patterns which do not arise by minor modifications of the example given above.

14. *Similarity patterns* can be defined in complete analogy to the definition of affine patterns in the previous exercise except that we use the *similarity symmetry group* $SS(\mathscr{P})$ instead of $AS(\mathscr{P})$. Which of the diagrams in Figure 5.1.11 can be interpreted as similarity patterns? (In each case take a connected set as motif and imagine that the diagram extends outwards over the whole plane and also inwards to a central point using smaller and smaller copies of the motif.) If a diagram cannot be interpreted as a similarity pattern, decide whether a transformation of the plane (such as an affine transformation or a homeomorphism) can convert the given design into such a pattern.

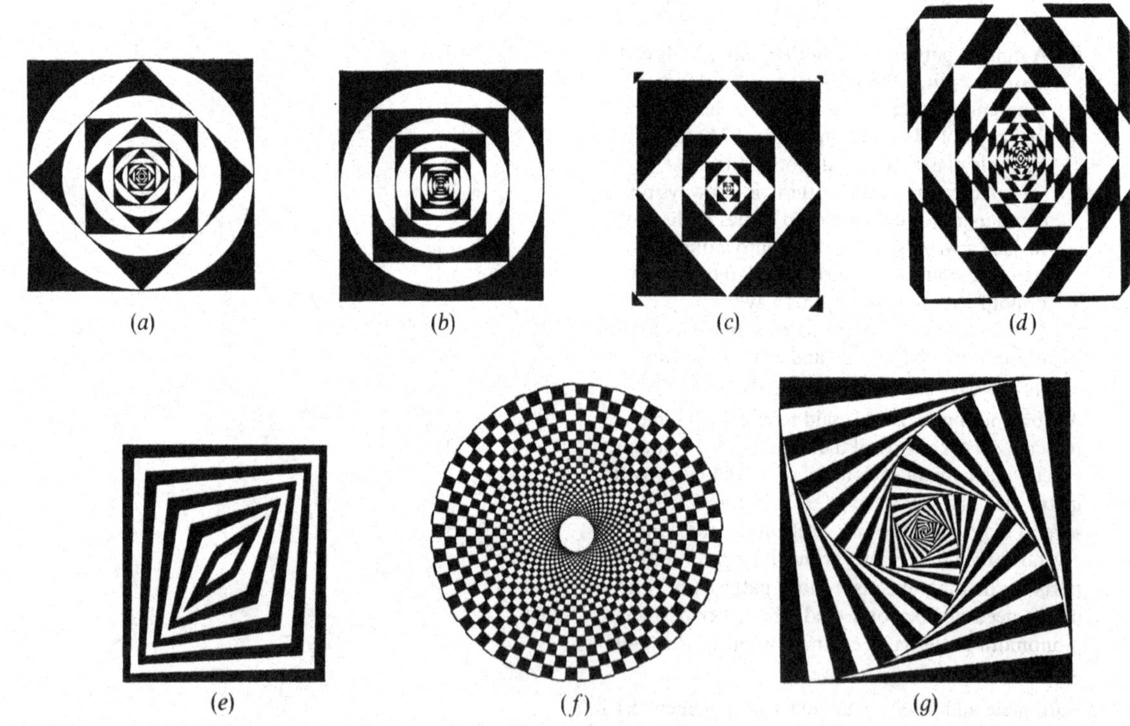

Figure 5.1.11
Examples of similarity and affine patterns and designs related to them. Parts (*a*) to (*i*) are adapted from Horemis [1970], (*k*) to (*p*) from Beard [1973] and (*q*) from Boyd [1949]. The curves in (*m*) are logarithmic spirals and the design in (*n*) is known as Baravelle spirals.

(h)
(i)
(j)
(k)
(l)
(m)
(n)
(o)
(p)
(q)

5.2 THE CLASSIFICATION OF PATTERNS

We begin by considering the possible symmetry group $S(\mathscr{P})$ of a pattern $\mathscr{P}$. As we shall see shortly, all the symmetry groups of tilings listed and illustrated in Section 1.4 can also occur as symmetry groups of patterns; however, without further restrictions, there are many other possibilities. In fact, at present, there is no known characterization of the possible groups; a few examples will illustrate the great variety of groups that can arise if the pattern is not discrete.

Let M consist of a single point, and let $\mathscr{P}$ be the pattern consisting of all points of the plane that have (in a Cartesian system) coordinates belonging to the field $Q(\sqrt{5})$ (that is, the set of all numbers expressible in the form $p + q\sqrt{5}$, with rational p and q). Then $\mathscr{P}$ is clearly a pattern (although not a discrete pattern) and it is easy to verify that $S(\mathscr{P})$ contains, among other symmetries, rotations by $2\pi/5$ about each point M_i of $\mathscr{P}$.

As a second example, let $\mathscr{P}$ consist of a family of parallel lines spaced equidistantly. Then in the direction of the lines *every* translation is a symmetry, whereas in every other direction the permissible translations are the integer multiples of a fixed translation. Later we shall denote this group by *smm*.

For a third example, let M be the union of the open circular disk of unit radius centered at the origin, with the set $\{(\cos 2\pi q, \sin 2\pi q) | q \text{ rational}\}$, and let $\mathscr{P}$ consist of just one copy of M, namely M itself. Then $S(\mathscr{P})$ obviously contains all rotations about the origin through rational multiples of π, and no rotations through irrational multiples of π. Clearly $S(\mathscr{P})$ also contains reflections, but it is not hard to modify the construction so that these are eliminated and the only symmetries left are the rotations through rational multiples of π. On the other hand, if M is the closed circular disk of unit radius centered at the origin and if $\mathscr{P}$ again consists of M alone, then $S(\mathscr{P})$ is $d\infty$ (that is, consists of all rotations about the origin, together with all reflections about lines through the origin). In this case, however, it is not possible to modify M so that the resulting pattern has all the rotations as symmetries but none of the reflections.

Further examples could be given, but the above will suffice to indicate some of the many interesting possibilities. In order to make the subject manageable with our present knowledge, it is convenient to consider only non-trivial discrete patterns. This restriction eliminates all the examples just given, and leads to the following result:

5.2.1 *The symmetry group of a non-trivial discrete pattern must be either cn for some integer $n \geq 2$ or dn for some integer $n \geq 1$, or one of the seven strip groups or seventeen crystallographic groups listed in Section 1.4. The symmetry group of a trivial discrete pattern can be cn or dn for some integer $n \geq 1$, or an infinite group.*

This statement illustrates the close analogy between the symmetry properties of non-trivial discrete patterns and those of normal tilings. The precise relationship will appear in Chapter 6, and it will lead to a proof of Statement 5.2.1.

The list of possible groups in Statement 5.2.1 enables us to distinguish between periodic patterns, strip patterns and finite patterns. A discrete pattern $\mathscr{P}$ is called *periodic* if its symmetry group $S(\mathscr{P})$ is one of the seventeen crystallographic groups (and hence contains translations in two independent directions) and $\mathscr{P}$ is called a *strip pattern* if $S(\mathscr{P})$ is one of the seven strip groups; clearly, $\mathscr{P}$ is non-trivial in all these cases. A discrete pattern $\mathscr{P}$ is called a *finite pattern* if it is non-trivial and hence its symmetry group $S(\mathscr{P})$ is either *cn* for some $n \geq 2$ or *dn* for some $n \geq 1$.

We shall now consider the possibility of a much finer classification of non-trivial discrete patterns. The basic question, of course, is to decide when we shall say that two patterns are of the "same type" or of "different types". It is clearly reasonable to distinguish patterns whose symmetry groups $S(\mathscr{P})$ differ, and likewise to distinguish between patterns with different induced groups $S(\mathscr{P}|M)$. However, on intuitive grounds as well as on mathematical ones, finer distinctions are often desirable. There exist patterns with the same motifs, the same symmetry groups and the same induced groups, which "appear" to be sufficiently different to be regarded as belonging to distinct types. An example is shown in Figure 5.2.1. Here the symmetry groups are indicated by group diagrams as explained in Section 1.4, and they are seen to be identical; the group is *p4m* in both cases. These patterns also have the same induced group *d1*. In order to explain how they can be formally distinguished, we need the following definition.

Let $T(\mathscr{P})$ be a subgroup of the symmetry group $S(\mathscr{P})$ of a given discrete pattern $\mathscr{P}$. Then $T(\mathscr{P})$ is called *motif-transitive* if it contains isometries that map any copy M_0 of the motif of $\mathscr{P}$ onto any other copy M_j. It is easy to see that if $\mathscr{P}$ is a primitive pattern (that is, if $S(\mathscr{P}|M)$ consists of the identity alone), it does not possess any motif-transitive groups other than $S(\mathscr{P})$ itself. On the other hand, if $S(\mathscr{P}|M)$ is non-trivial then there may (and, as it turns out, actually do) exist motif-transitive groups other than $S(\mathscr{P})$ (see, for example, Figures 5.2.1 and 5.2.2).

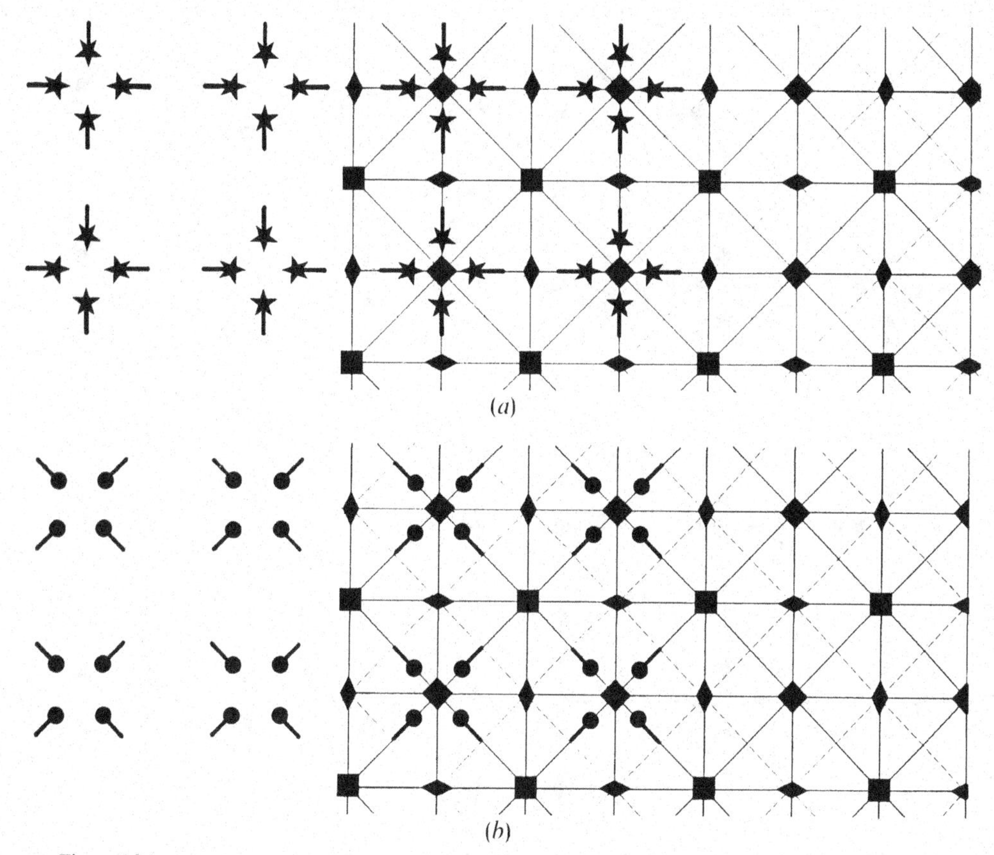

Figure 5.2.1
Two periodic patterns, with the same symmetry group and the same induced group, that are assigned to different pattern types by the definition given in the text. The elements of the symmetry group are indicated as in Section 1.4.

It is possible, of course, for $T(\mathscr{P})$ to be isomorphic to $S(\mathscr{P})$ and yet not contain all the elements of $S(\mathscr{P})$—in other words, to be a proper subgroup of $S(\mathscr{P})$ isomorphic to $S(\mathscr{P})$. (This happens, for example, in the case of the pattern in Figure 5.2.1(*a*).)

Using this idea, we are now able to formulate precisely what we mean by the "type" of a pattern. We say that two patterns are *of the same type*, or have the same *pattern type*, if and only if they have the same symmetry group, the same induced group, and the same set of motif-transitive subgroups. Patterns of the same pattern type are also said to be *henomeric* (from ἑνω = join, unite; μερος = part) or to have the same *henomeric type.*

We illustrate these definitions by reference to Figures 5.2.1 and 5.2.2. In Figure 5.2.2(*a*) we indicate the elements of a group which is motif-transitive for the pattern of Figure 5.2.1(*a*) but not for that of Figure 5.2.1(*b*), and in Figure 5.2.2(*b*) we indicate a group motif-transitive for

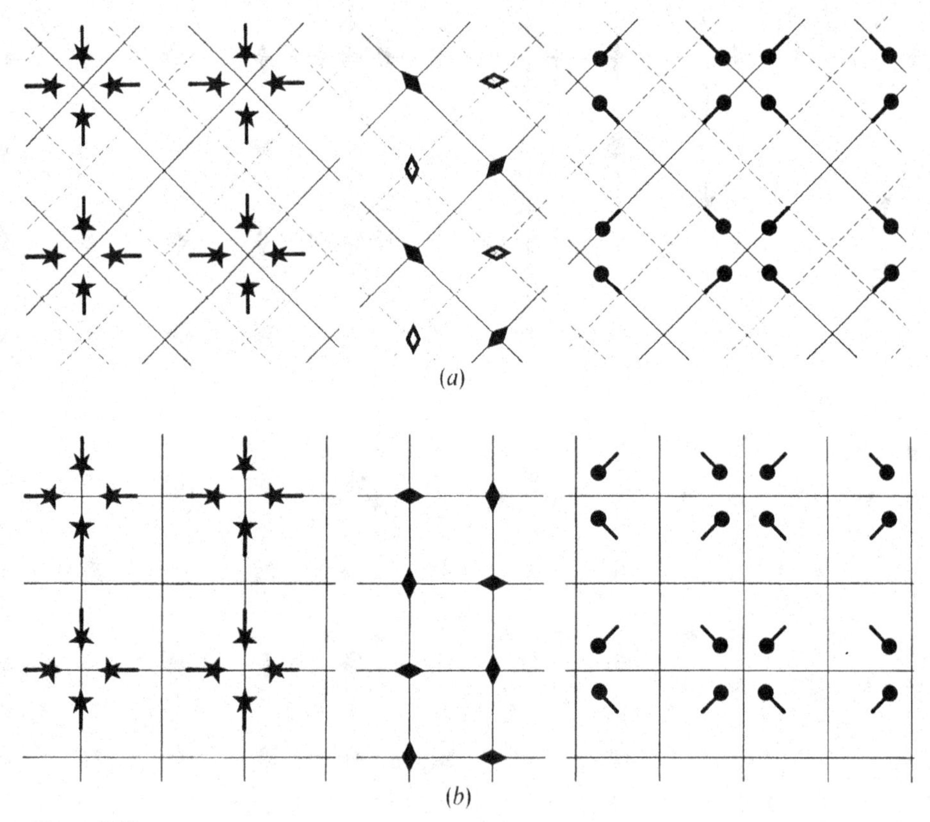

Figure 5.2.2
Motif-transitive subgroups of the symmetry groups of the patterns in Figure 5.2.1. The subgroup indicated in (*a*) is motif-transitive for the pattern in Figure 5.2.1(*a*), but not for the one in Figure 5.2.1(*b*), and the subgroup indicated in (*b*) is motif-transitive for the latter but not for the former. (For clarity, centers of 2-fold rotation are indicated only in the middle part.)

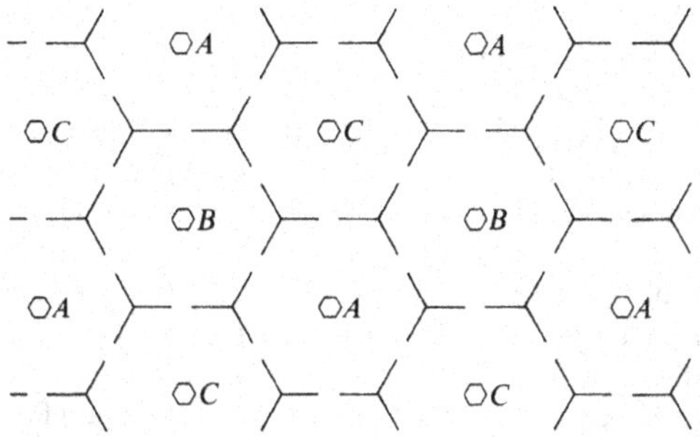

Figure 5.2.3
Motif transitive subgroups of a pattern of type PP50. Four groups of type *p6* are indicated: those with centers of six-fold rotational symmetry at *A*, or at *B*, or at *C*, or at all of *A*, *B* and *C*. The first three of these are not counted as distinct since each can be obtained from the other by an isometry σ (more specifically, by a translation that maps $\mathscr{P}$ onto itself). The last one is, of course, different from the other three.

the pattern of Figure 5.2.1(*b*) but not for that of Figure 5.2.1(*a*). Hence, as suggested above, these two patterns are of different types.

In order to apply our definition of "pattern type" it is clearly necessary to be precise as to when two motif-transitive subgroups are to be considered "the same" or not. In many cases, such as the above, there can be no doubt—if the groups are different or their elements are "situated differently" with respect to the motifs in the pattern then they must be counted as different. But in other cases some doubt can arise and the question becomes one of convention and definition. We find it convenient to consider two motif-transitive subgroups G_1 and G_2 of $S(\mathscr{P})$ as not different if there exists an isometry σ that maps the group diagram of G_1 onto that of G_2 in a manner that establishes a one-to-one correspondence between the copies of the motif of $\mathscr{P}$ in the planes of the two group diagrams. (Note that σ is not required to map $\mathscr{P}$ onto itself.) For example, consider the patterns of Figures 5.2.3 and 5.2.4, whose types we shall later designate by the symbols PP50 and PP29. In the caption of each diagram we state which of the indicated motif-transitive groups must be counted as different.

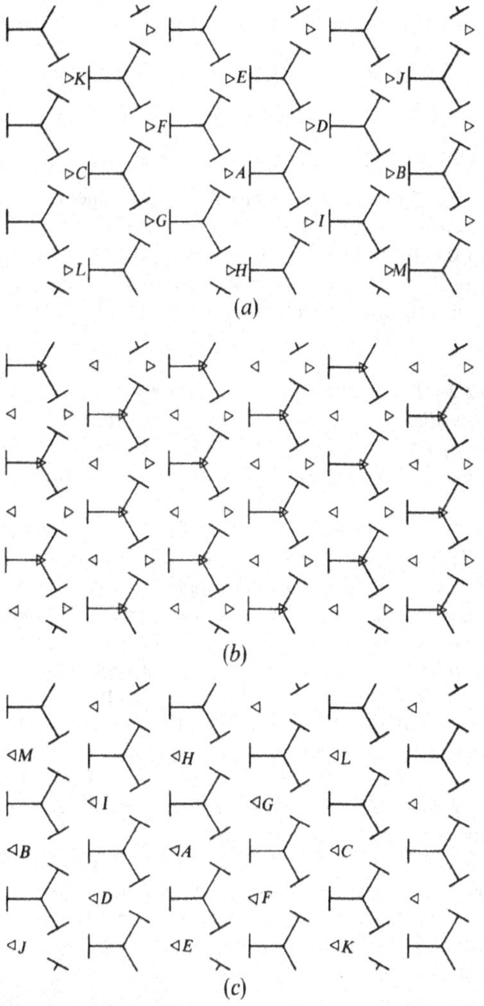

Figure 5.2.4
Three motif transitive subgroups *p3* for a pattern of type PP29. Those shown in (*a*) and (*b*) must clearly be counted as distinct since no isometry can map the centers of 3-fold rotational symmetry of one onto those of the other. In spite of the difference in appearance, the subgroups indicated in (*a*) and (*c*) must be considered to be the same. As the isometry σ mentioned in the text we can take a rotation by 180° which maps the equally labelled centers of 3-fold rotation onto each other. Note that this σ does not map the copies of the motif in (*a*) onto those in (*c*), but only establishes a one-to-one correspondence between them, as indicated by the labels. From such considerations we deduce that the group of symmetries of a pattern of type PP29 has precisely two different motif transitive subgroups *p3*.

The above definitions enable us to enumerate all possible types of patterns. The easiest way actually to do this is by the use of marked isohedral tilings, as described in the next chapter. Here we content ourselves with a statement of the main result.

5.2.2 *The non-trivial discrete patterns are classified into 3 families of types of finite patterns, each of which depends on a parameter n which takes positive integral values,* 15 *types of strip patterns, and* 51 *types of periodic patterns.*

These types are listed in Tables 5.2.1, 5.2.2 and 5.2.3, and patterns illustrating them appear in Figures 5.2.5, 5.2.6 and 5.2.7. In the diagrams it has been convenient to draw, as examples illustrating the different types, patterns in which the motifs are built up from a very simple L shape. When the induced groups are non-trivial, these Ls are partially superimposed to form Ts, swastikas, or the other shapes that appear in the diagrams.

The meaning of the first four columns in Tables 5.2.1, 5.2.2 and 5.2.3 is clear from their headings. The explanation of column (5) is as follows.

For each pattern $\mathscr{P}$, the induced motif group $S(\mathscr{P}|M)$ is a subgroup of the symmetry group $S(M)$ of the motif M of $\mathscr{P}$. Hence the classification by pattern type may be refined by specifying, for a given pattern $\mathscr{P}$ of a certain pattern type, which of the supergroups of $S(\mathscr{P}|M)$ is $S(M)$. As it turns out, not every supergroup is admissible in each case; it may happen that greater symmetry of the motif leads to additional symmetries of the whole pattern. For example, the pattern with rectangular motif shown in Figure 5.2.8(*a*) is of type PP40; if the rectangles are replaced by squares (or any other motif with symmetry group *d4*) so as to retain all the symmetries of pattern type PP40 (see Figure 5.2.8(*b*)), the glide-reflections indicated by dashed lines in Figure 5.2.8(*a*) become reflections, and the pattern type is PP41. In column (5) of Tables 5.2.1, 5.2.2 and 5.2.3 we have listed all the minimal supergroups of $S(\mathscr{P}|M)$ that may not occur as subgroups of $S(M)$ for patterns of that type. These groups are called *forbidden supergroups.*

Columns (7), (8) and (9) of Table 5.2.3 will be explained in Chapter 6.

From Figure 5.2.7 it can be seen that patterns of type PP48 can appear in two forms which exhibit a difference analogous to the difference between the two patterns in Figure 5.2.1. As we shall see in Chapter 7, there are also mathematical reasons for wishing to distinguish between such patterns. One of the ways in which this can be done is the following. Since the motif M of a discrete pattern is connected, if the stabilizer $S(\mathscr{P}|M)$ contains a reflection then the corresponding line of reflection L intersects M. For some pattern types such a line L is subdivided into inequivalent segments by points which are distinguished by the symmetries of the pattern (such as centers of rotation, or intersections of L with other lines of reflection). In these cases it is possible to refine the classification into pattern types by specifying which of the inequivalent segments of L are met by copies of M and which are not met. A consideration of all the pattern types shows that following this procedure a refinement of the classification is actually obtained only in the case of pattern type PP48. For that type the lines of reflective symmetry which pass through the centers of 3-fold rotations are partitioned into segments of two kinds: those with endpoints at 3-fold and 6-fold centers of rotation, and those with endpoints at 3-fold and 2-fold centers of rotation. If the copies of the motif intersect segments of the first kind we shall say that the pattern is of type PP48A, otherwise it is of type PP48B. When we wish to refer to the classification of patterns which takes into account these properties, we shall speak of *refined pattern types.*

For later use we record here the following fact which can be established by checking all possible cases with the help of Tables 5.2.1, 5.2.2 and 5.2.3. This solves in the affirmative a problem posed (in the special case of isohedral tilings) by Delone [1959, p. 372].

5.2.3 *Let G be a subgroup of $S(\mathscr{P})$ that acts transitively on the copies of the motif of a (finite, strip or periodic) discrete pattern $\mathscr{P}$. Then there exists a subgroup H of G such that H acts transitively on the copies of the motif of $\mathscr{P}$ and, moreover, the only element of H that maps a copy of the motif of $\mathscr{P}$ onto itself is the identity.*

Figure 5.2.5
Examples of the three families of discrete finite patterns $PF1_n$, $PF2_n$ and $PF3_n$. The value of n is 6 in each pattern.

Tables 5.2.1, 5.2.2 and 5.2.3 THE NON-TRIVIAL DISCRETE PATTERN TYPES

The three tables relate to finite, strip and periodic patterns, respectively. The meaning of the first four columns will be clear from their headings; **column (5)** is explained in the text, **column (6)** in Section 5.3, and the remaining columns in Table 5.2.3 will be explained in Chapter 6.

In **column (4)** of Tables 5.2.2 and 5.2.3, a number in parentheses following the designation of a group G indicates that the pattern admits this number of inequivalent motif-transitive proper subgroups of $S(\mathscr{P})$ isomorphic to G. An asterisk(*)indicates that $\mathscr{P}$ admits a motif-transitive proper subgroup of $S(\mathscr{P})$ that is isomorphic to $S(\mathscr{P})$.

Table 5.2.1 THE PATTERN TYPES OF FINITE PATTERNS

Pattern type (1)	Symmetry group $S(\mathscr{P})$ (2)	Induced group $S(\mathscr{P}\|M)$ (3)	Motif-transitive proper subgroups of $S(\mathscr{P})$ (4)	Forbidden supergroups (5)	Associated dot pattern types DPF (6)
$PF1_n$, $n \geq 2$	cn	$c1$	Primitive	$d\infty$ for $n \geq 2$	$\underset{=}{3}_n$ for $n \geq 2$
$PF2_n$, $n \geq 1$	dn	$c1$	Primitive	$d\infty$ for $n = 1$	3_2 for $n = 1$ $\underset{=}{2}_n$, 3_{2n} for $n \geq 2$
$PF3_n$, $n \geq 2$		$d1$	cn for all n $d\frac{n}{2}$ for even n		$\underset{=}{3}_n$ for $n \geq 2$

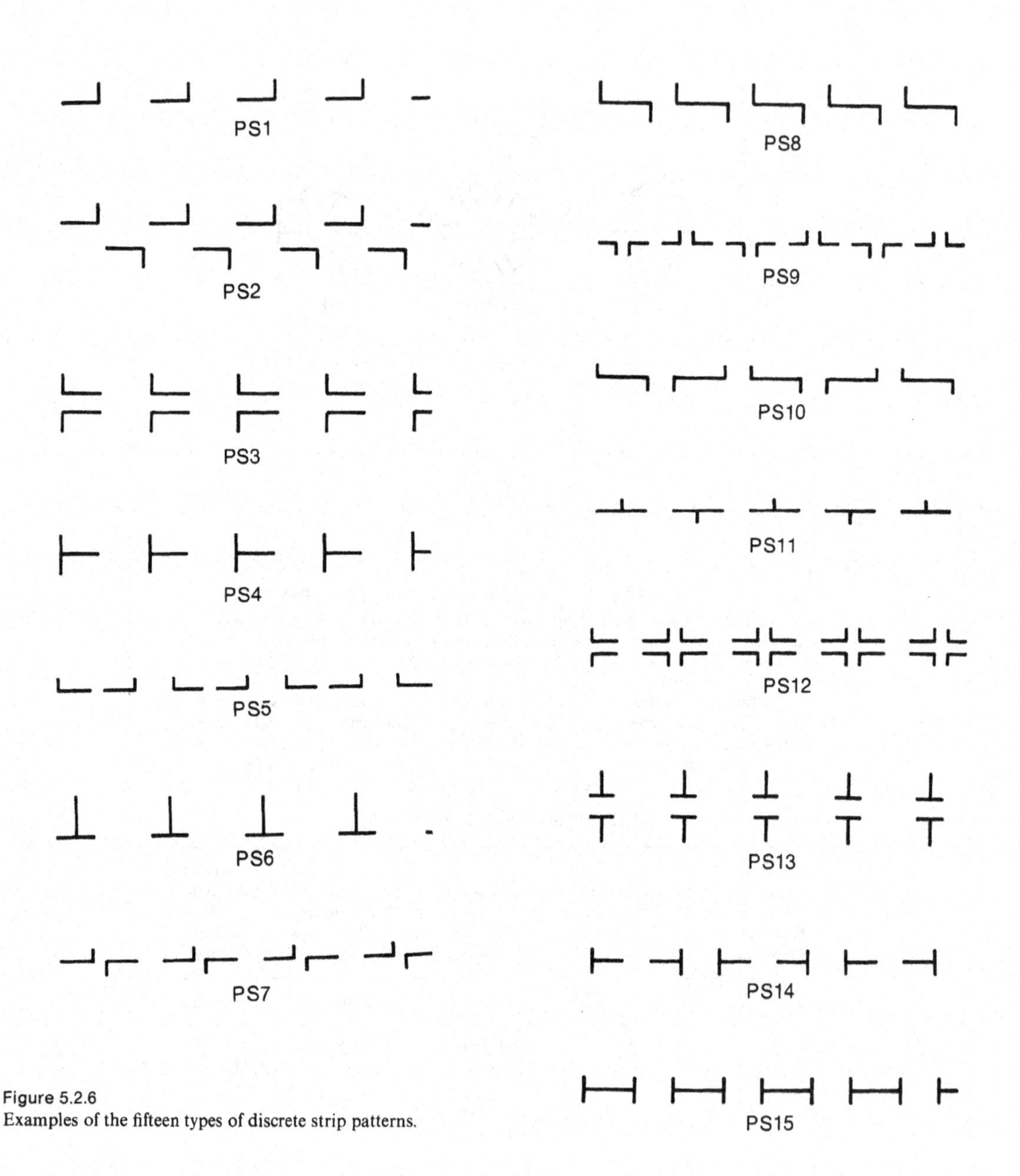

Figure 5.2.6
Examples of the fifteen types of discrete strip patterns.

Table 5.2.2 THE PATTERN TYPES OF STRIP PATTERNS

Pattern type (1)	Symmetry group $S(\mathscr{P})$ (2)	Induced group $S(\mathscr{P} \mid M)$ (3)	Motif-transitive proper subgroups of $S(\mathscr{P})$ (4)	Forbidden supergroups (5)	Associated dot pattern types DPS (6)
PS1	*p111*	*c1*	Primitive	*c2*	15
PS2	*p1a1*	*c1*	Primitive	$d\infty$	11, 15
PS3	*p1m1*	*c1*	Primitive	$d\infty$	13
PS4		*d1*	*p111*, *p1a1*	*d2*	15
PS5	*pm11*	*c1*	Primitive	$d\infty$	14, 15
PS6		*d1*	*p111*, *	*d2*	15
PS7	*p112*	*c1*	Primitive		7, 11, 13, 14, 15
PS8		*c2*	*p111*, *	$d\infty$	15
PS9	*pma2*	*c1*	Primitive		9, 13, 14, 15
PS10		*c2*	*pm11*	$d\infty$	15
PS11		*d1*	*p112*, *p1a1*		11, 15
PS12	*pmm2*	*c1*	Primitive		12, 13
PS13		*d1*	*p112*, *p1m1*, *pma2*, *		13
PS14		*d1*	*p112*, *pm11*, *pma2* (2)		14, 15
PS15		*d2*	*p111*, *p112* (2), *p1a1*, *p1m1*, *pm11* (2), *pma2* (3), *		15

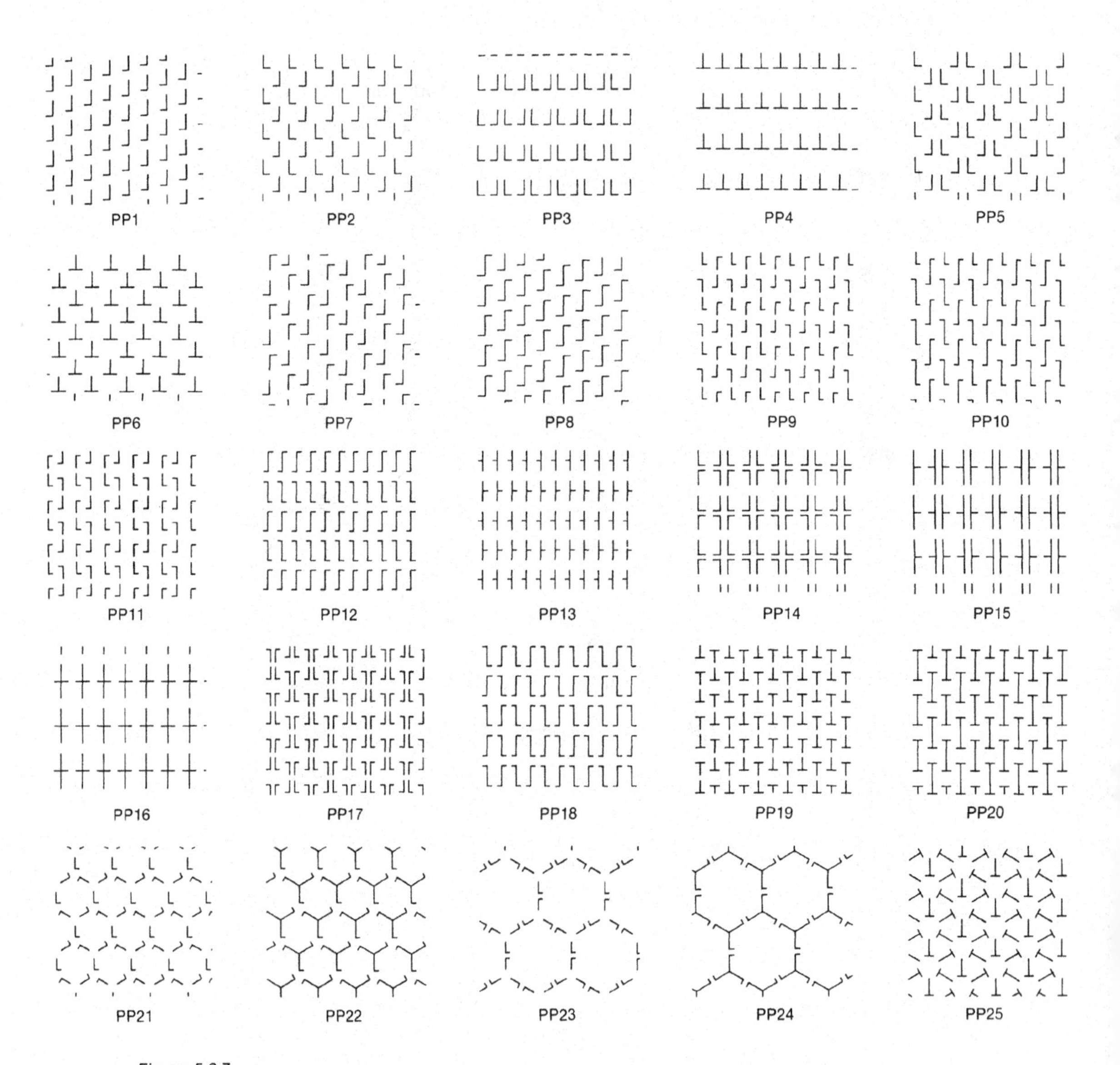

Figure 5.2.7
Examples of the 51 types of discrete periodic patterns. Two examples of type **PP48** are shown; they are of different refined pattern types.

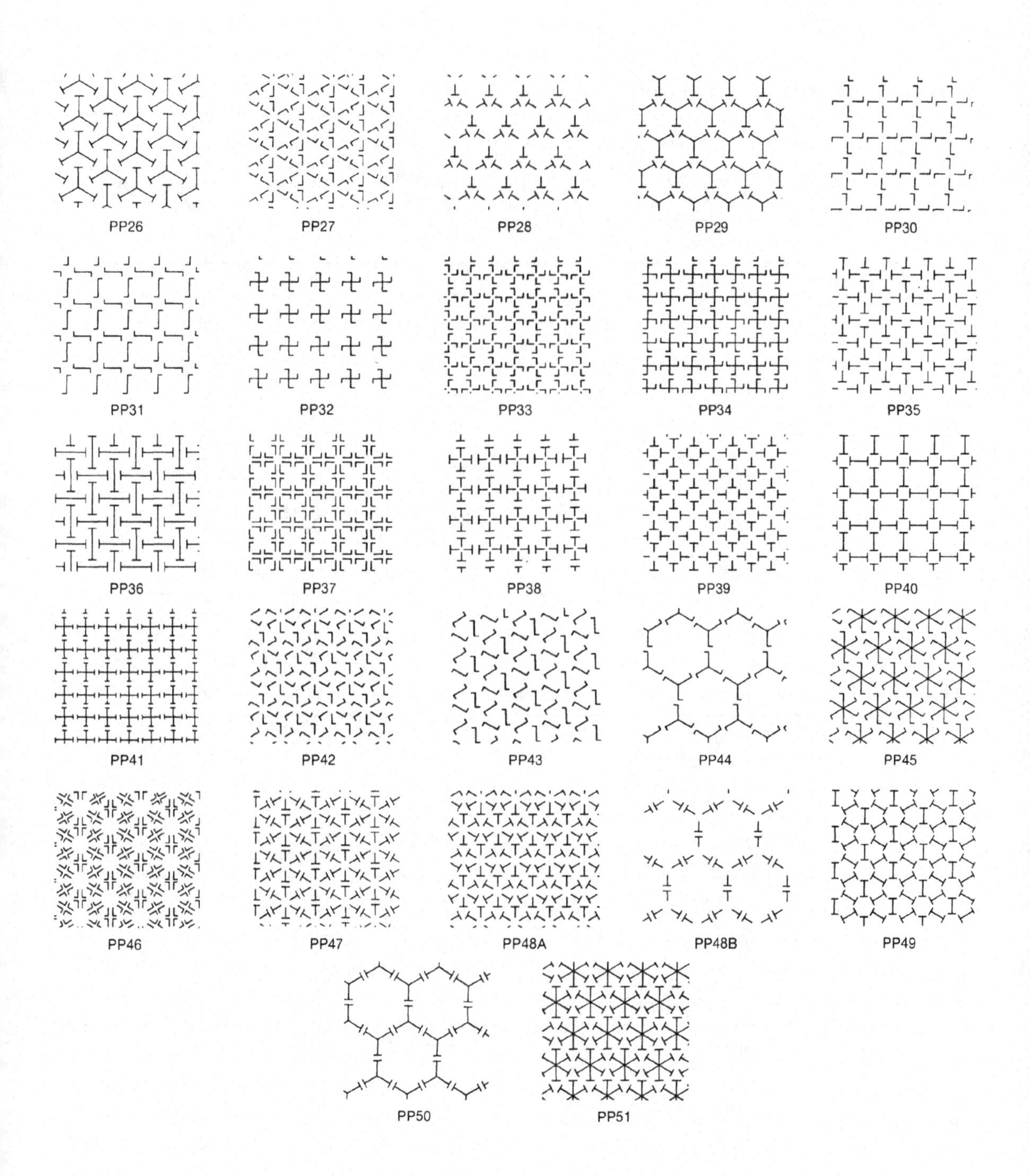
PP26
PP27
PP28
PP29
PP30
PP31
PP32
PP33
PP34
PP35
PP36
PP37
PP38
PP39
PP40
PP41
PP42
PP43
PP44
PP45
PP46
PP47
PP48A
PP48B
PP49
PP50
PP51

Table 5.2.3 THE PATTERN TYPES OF PERIODIC PATTERNS

Pattern type (1)	Symmetry group $S(\mathscr{P})$ (2)	Induced group $S(\mathscr{P}\|M)$ (3)	Motif-transitive proper subgroups of $S(\mathscr{P})$ (4)	Forbidden super-groups (5)	Associated dot patterns types DPP (6)	Isohedral tiling type IH (7)	Isogonal tiling type IG (8)	Isotoxal tiling type IT (9)
PP1	*p1*	*c1*	Primitive	*c2*	8, 16, 20, 41, 51	1, 41	1, 41	
PP2	*pg*	*c1*	Primitive	*d∞*	13, 16, 20, 41, 51	2, 3, 43, 44	2, 3, 43, 44	
PP3	*pm*	*c1*	Primitive	*d∞*	15, 16, 41	42	42	
PP4		*d1*	*p1, pg, cm*, *	*d2*	16, 41	64	64	
PP5	*cm*	*c1*	Primitive	*d∞*	16, 19, 41, 50	22, 45, 83	22, 45, 83	18
PP6		*d1*	*p1, pg*	*d2*	20, 41, 51	12, 14, 68	12, 14, 68	
PP7	*p2*	*c1*	Primitive		7, 8, 13, 15, 16, 19, 20, 41, 50	4, 23, 46, 47, 84	4, 23, 46, 47, 84	
PP8		*c2*	*p1*, *		8, 16, 20, 41, 51	8, 57	8, 57	
PP9	*pgg*	*c1*	Primitive		9, 13, 16, 19, 20, 41, 50, 51	5, 6, 25, 27, 51, 52, 53, 86	5, 6, 25, 27, 51, 52, 53, 86	17
PP10		*c2*	*pg*	*d∞*	20, 41, 51	9, 59	9, 59	
PP11	*pmg*	*c1*	Primitive		11, 15, 16, 19, 41, 50	24, 49, 50, 85	24, 49 50, 85	
PP12		*c2*	*pg, pm, pgg*, *	*d∞*	16, 41	58	58	
PP13		*d1*	*pg, p2, pgg*		13, 16, 20, 41, 51	13, 15, 66, 69	13, 15, 66, 69	
PP14	*pmm*	*c1*	Primitive		14, 15, 16, 39, 41	(48)	48	14
PP15		*d1*	*pm, p2, pmg*(2), *cmm*, *		15, 16, 41	(65)	65	
PP16		*d2*	*p1, pg, pm*(2), *cm, p2*(3), *pgg, pmg*(2), *cmm*(3), *(2)		16, 41	72	72	
PP17	*cmm*	*c1*	Primitive		15, 16, 17, 37, 38, 41	54, 78	54, 78	
PP18		*c2*	*cm, pgg, pmm*	*d∞*	16, 41	(60)	60	20
PP19		*d1*	*cm, p2, pgg, pmg*		16, 19, 41, 50	26, 67, 91	26, 67, 91	

Table 5.2.3 (continued)

Pattern type (1)	Symmetry group $S(\mathscr{P})$ (2)	Induced group $S(\mathscr{P}\|M)$ (3)	Motif-transitive proper subgroups of $S(\mathscr{P})$ (4)	Forbidden super-groups (5)	Associated dot patterns types DPP (6)	Isohedral tiling type IH (7)	Isogonal tiling type IG (8)	Isotoxal tiling type IT (9)
PP20		*d2*	*p1, pg, cm, p2*(2), *pgg*(2), *pmg*		20, 41, 51	17, 74	17, 74	
PP21	*p3*	*c1*	Primitive		21, 25, 28, 49	7, 33	7, 33	1, 26
PP22		*c3*	*p1*, ∗	*c6*	51	10	10	
PP23	*p31m*	*c1*	Primitive		23, 48	30, 38	30, 38	7, 10
PP24		*c3*	*cm, p3m1*	*d*∞	50	(89)	89	
PP25		*d1*	*p3*		25, 49, 51	16, 36	16, 36	(3), 29
PP26		*d3*	*p1, pg, cm, p3*(2)	*d6*	51	18	(18)	
PP27	*p3m1*	*c1*	Primitive		27, 47, 50	(87)	87	
PP28		*d1*	*p3*		28, 49	(35)	35	4, (28)
PP29		*d3*	*p1, pg, cm, p3*(2), *p31m*	*d6*	51	(19)	19	
PP30	*p4*	*c1*	Primitive		30, 35, 38, 39, 41	28, 55, 79	28, 55, 79	12, 15
PP31		*c2*	∗	*c4*	41	61	61	19
PP32		*c4*	*p1, p2*(3), ∗ (2)	*d*∞	41	62	62	
PP33	*p4g*	*c1*	Primitive		33, 38, 39, 41	56, 81	56, 81	13, 16
PP34		*c4*	*pg, cm, pgg*(2), *pmm, cmm*	*d*∞	41	(63)	63	
PP35		*d1*	*pgg, p4*		35, 41	29, 71	29, 71	(21), 24
PP36		*d2*	*pg, pgg, p4*(2)	*d4*	41	73	(73)	
PP37	*p4m*	*c1*	Primitive		37	(80)	80	
PP38		*d1*	*cmm, p4, p4g*, ∗		38	82	82	
PP39		*d1*	*pmm, p4, p4g*		39, 41	(70)	70	(22), 23
PP40		*d2*	*cm, pgg, pmm, cmm, p4*(2), *p4g*(2), ∗	*d4*	41	(75)	75	25

Table 5.2.3 (continued)

Pattern type (1)	Symmetry group $S(\mathscr{P})$ (2)	Induced group $S(\mathscr{P}\|M)$ (3)	Motif-transitive proper subgroups of $S(\mathscr{P})$ (4)	Forbidden super-groups (5)	Associated dot patterns types DPP (6)	Isohedral tiling type IH (7)	Isogonal tiling type IG (8)	Isotoxal tiling type IT (9)
PP41		*d4*	*p1*, *pg*(2), *pm*(2), *cm*(2), *p2*(3), *pgg*(3), *pmg*(3), *pmm*(3), *cmm*(4), *p4*(3), *p4g*(3), * (2)		41	76	76	
PP42	*p6*	*c1*	Primitive		42, 47, 48, 50	21, 31, 39, 88	21, 31, 39, 88	6, 9
PP43		*c2*	*p3*	$d\infty$	49	34	34	2, 27
PP44		*c3*	*p2*, *	$d\infty$	50	90	90	
PP45		*c6*	*p1*, *p2*(2), *p3*(2)	$d\infty$	51	11	11	
PP46	*p6m*	*c1*	Primitive		46	77	77	
PP47		*d1*	*p3m1*, *p6*		47, 50	(92)	92	
PP48		*d1*	*p31m*, *p6*		48	32, 40	32, 40	8, 11
PP49		*d2*	*p3*, *p31m*, *p3m1*, *p6*		49	37	37	5, 30
PP50		*d3*	*cm*, *pgg*, *pmg*, *cmm*, *p2*, *p31m*, *p3m1*, *p6*(2)		50	93	93	
PP51		*d6*	*p1*, *pg*(2), *cm*(2), *p2*(2), *pgg*(3), *pmg*(2), *cmm*, *p3*(2), *p31m*(2), *p3m1*, *p6*		51	20	20	

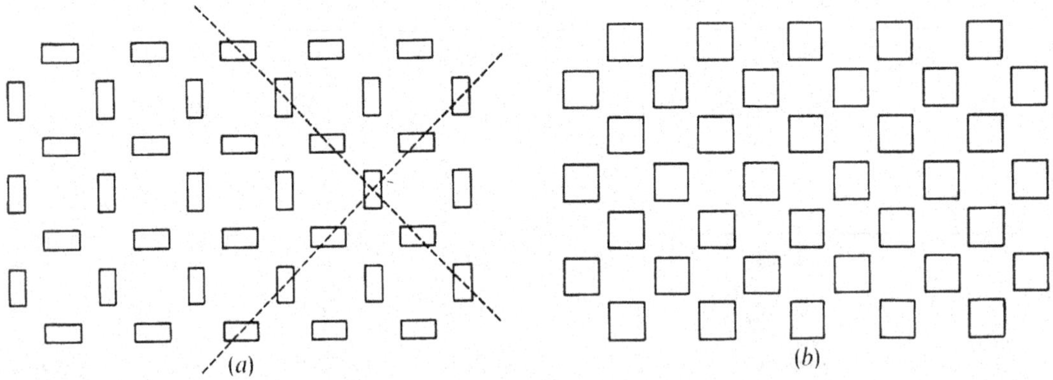

Figure 5.2.8
The replacement of a motif by a more symmetrical one can lead to a change of pattern type. For example the pattern of type PP40 shown in (*a*) becomes of type PP41 when the rectangular motif is replaced by a square.

EXERCISES 5.2

1. (*a*) Find a pattern the symmetry group of which consists of all rotations about a point through rational multiples of π.
   (*b*) Find a non-trivial pattern with the same symmetry group.

2. In Figure 5.2.9 we show eight periodic patterns with the same motif.
   (*a*) Determine the pattern type of each pattern.
   (*b*) With the same motif construct as many additional pattern types as you can.

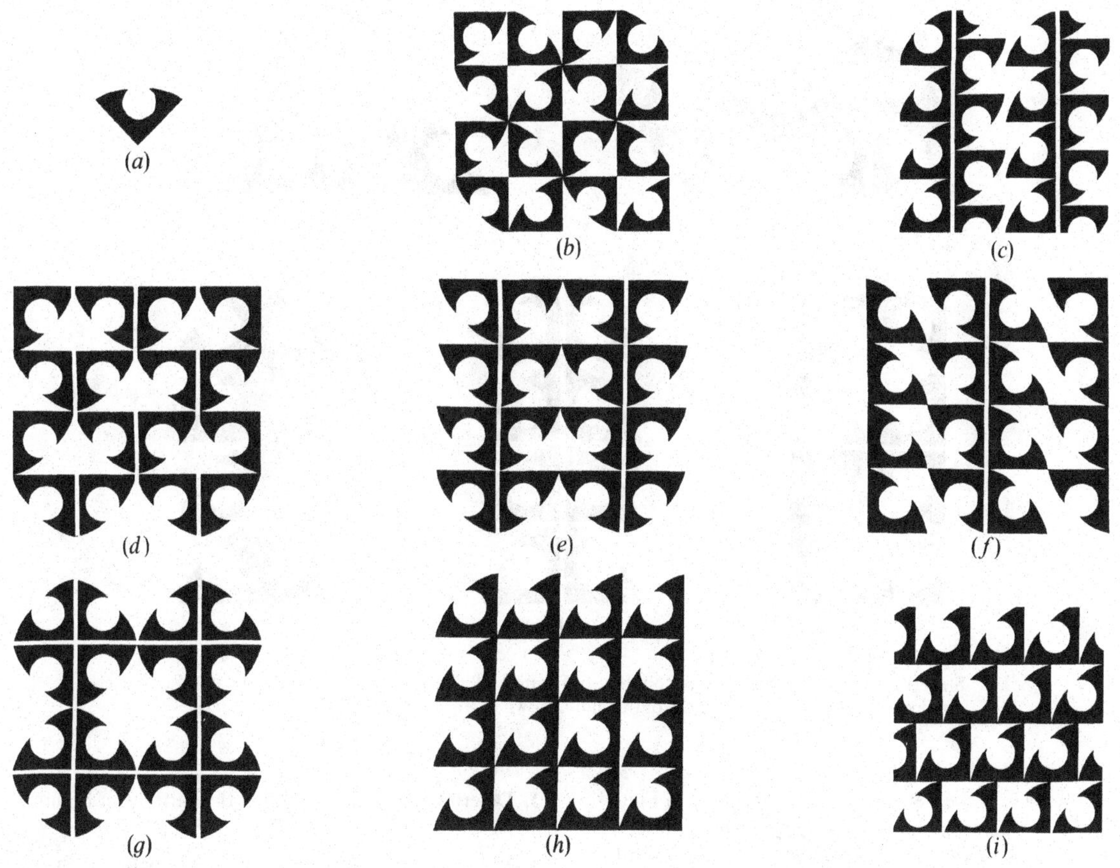

Figure 5.2.9
Periodic patterns of eight different types, all with the motif shown in (*a*). Adapted from Proctor [1969].)

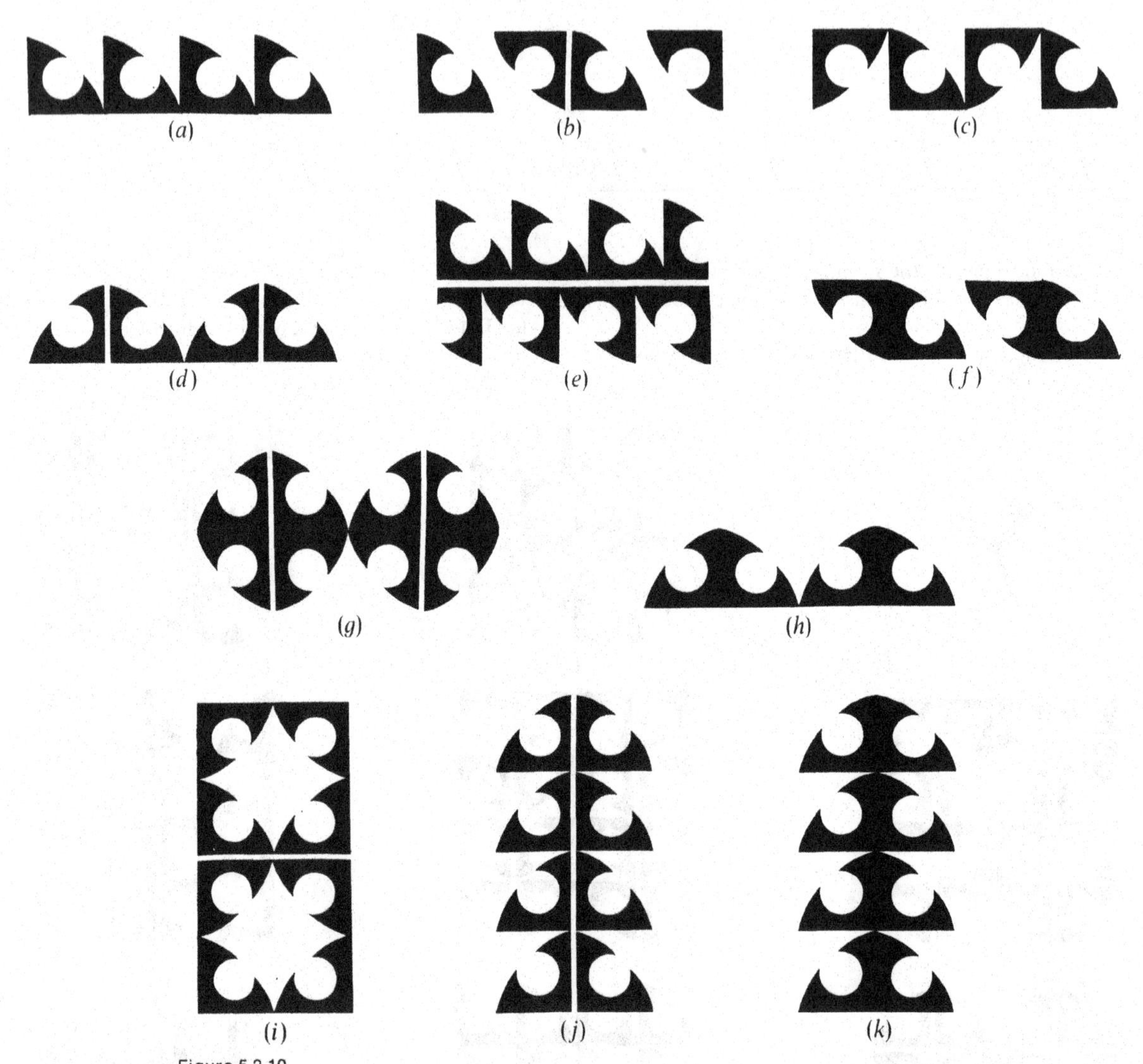

Figure 5.2.10
Strip patterns with motifs derived from that of Figure 5.2.9. Note that in some cases the motif is obtained by uniting several replicas of the motif shown in Figure 5.2.9(*a*).

3. For each of the eleven strip patterns shown in Figure 5.2.10 determine the pattern type. Construct at least two additional strip patterns with each of the motifs.
4. Carry out the tasks of Exercise 3 for the finite patterns shown in Figure 5.2.11.
5. Determine the type of each of the patterns in Figure 5.1.8.
6. In Figure 5.2.12 we show examples of patterns on various historical objects. The purpose of this exercise is for the reader to see to what extent the classifica-

tion of patterns introduced in this section can be applied to practical examples. In many cases it is necessary to disregard slight variations in copies of the motif, departures from strict regularity near the borders, and so on. Sometimes modifications are needed to convert the pattern into one that is monomotif and some markings may have to be ignored. With a certain amount of goodwill, however, each diagram or part of a diagram (as in Figures 5.2.12(*a*), (*g*) and (*m*)) is of a well-determined pattern type. Identify these.

7. In the 3-colored tilings of Figure 5.2.13 consider as copies of the motif the interiors of
   (*a*) only the black tiles,
   (*b*) the black and the stippled tiles,
   (*c*) all the tiles.

   In each case decide whether these copies form a pattern and if so determine the pattern type. (Ignore the colors of the tiles and consider only the shapes of the copies of the motif.)

8. (*a*) Give a full proof of Statement 5.2.3 for strip patterns of type **PS15**.
   (*b*) Show by an example that for suitable choices of the pattern $\mathscr{P}$ and group G it is possible to find several groups H that satisfy Statement 5.2.3.
   (*c*) Generalize Statement 5.2.3 by proving the following assertion. If G is a motif-transitive subgroup of $S(\mathscr{P})$, and if F is a subgroup of the *stabilizer* $G(\mathscr{P}|M)$ (that is, the subgroup of G that leaves a certain copy of M fixed), then there exists a subgroup H of G which acts transitively on the copies

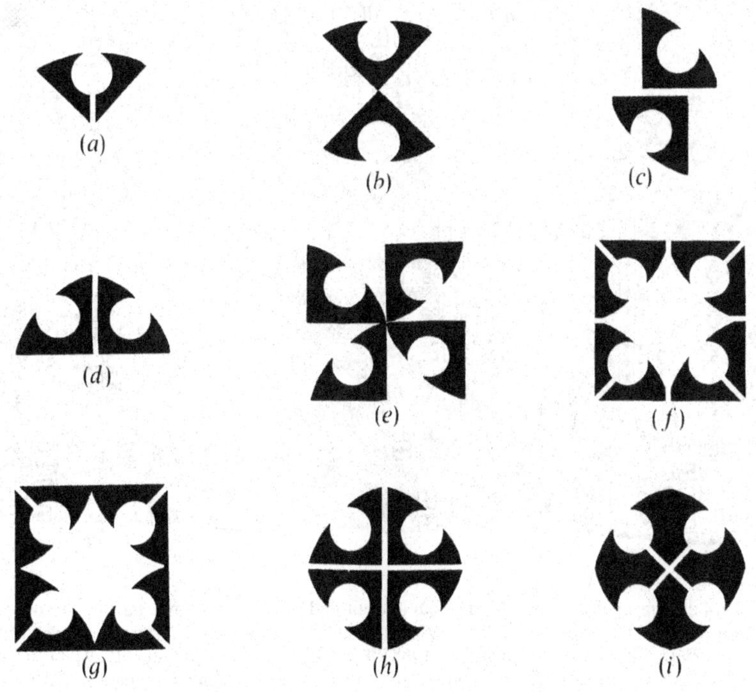

Figure 5.2.11
Finite patterns with motifs derived from that of Figure 5.2.9.

of the motif of $\mathscr{P}$ and is such that the stabilizer $H(\mathscr{P}|M)$ coincides with F.

9. To every motif-transitive subgroup of a pattern $\mathscr{P}$ corresponds a subgroup K of $S(\mathscr{P}|M)$, namely the group of those symmetries of the motif that are induced by the chosen motif-transitive subgroup.

Determine K in the following cases:

(*a*) Pattern type PP16 with its three motif-transitive subgroups of type *cmm*.

(*b*) Pattern type PP45 with two groups *p3*.

(*c*) Pattern type PP51 with two groups *pmg*, two groups *p3* and one group *p6*.

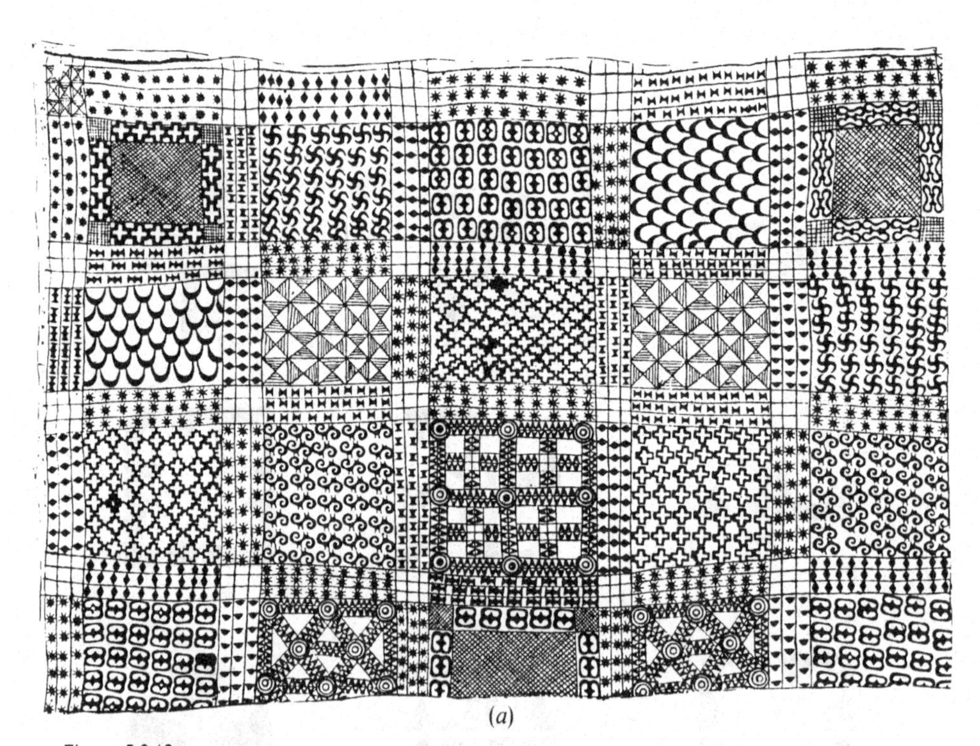

(*a*)

Figure 5.2.12
Examples of designs on decorated objects from ancient and primitive cultures. Part (*a*) is from Proctor [1969]; (*b*), (*c*), (*d*) and (*e*) are from Smart [1971]; (*h*) is from Guiart [1963]; (*i*) from Trowell [1960]; (*n*) and (*o*) from Bain [1951]; (*q*) from Justema [1968]; the others are from Christie [1929]. Observe that many of these designs are not "patterns" in the strict sense, but with some imagination and good will it is possible to modify them in such a way that they can be interpreted as patterns. The makers of these objects clearly had "regular repetitions" in mind and were very ingenious in their use of primitive techniques and materials in achieving this aim.

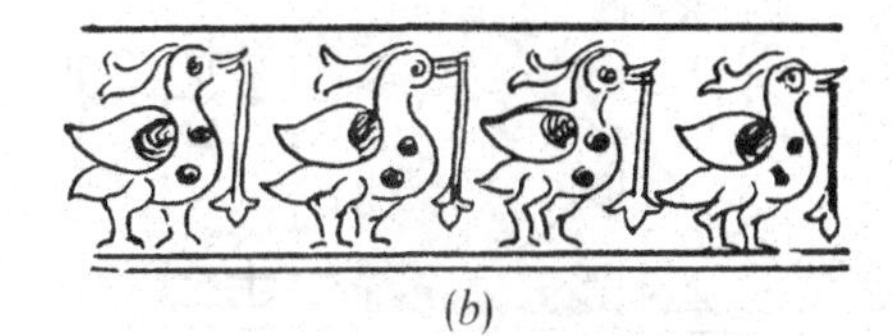

(*b*)

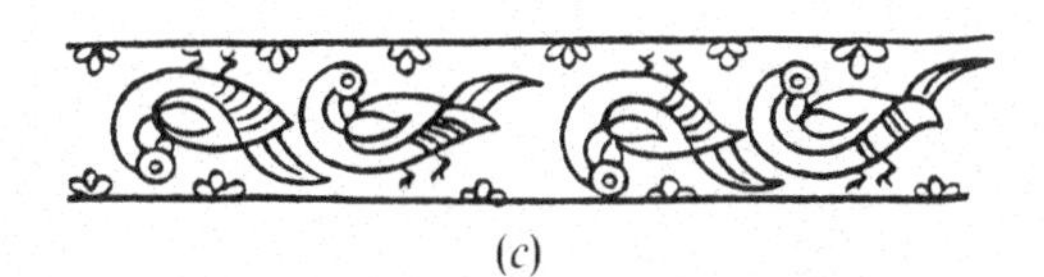

(*c*)

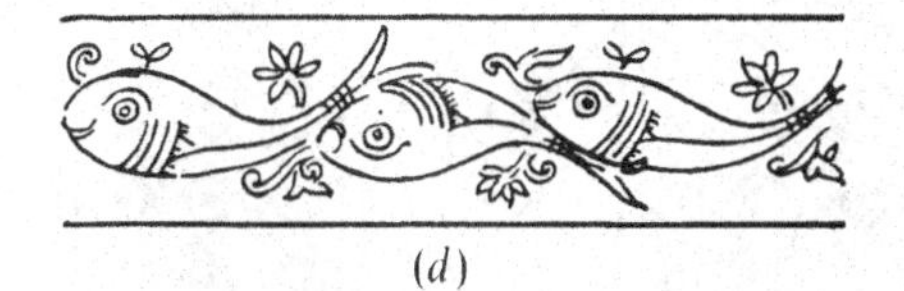

(*d*)

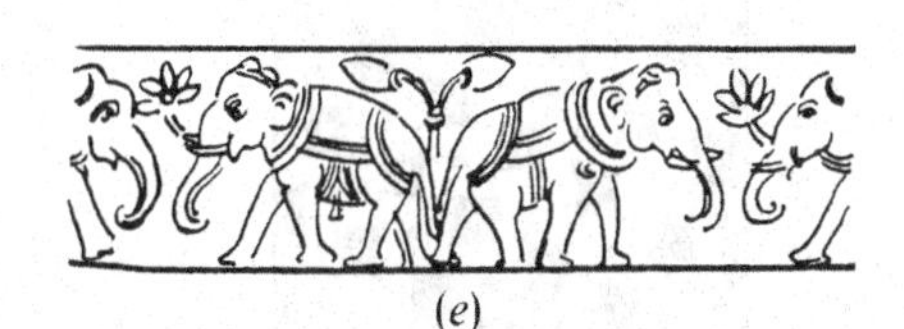

(*e*)

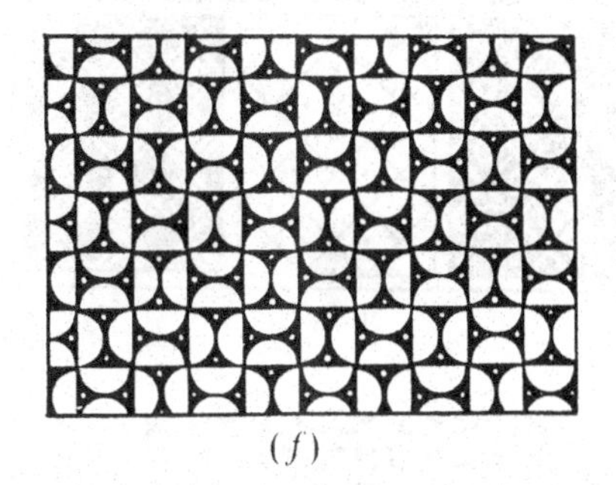

(*f*)

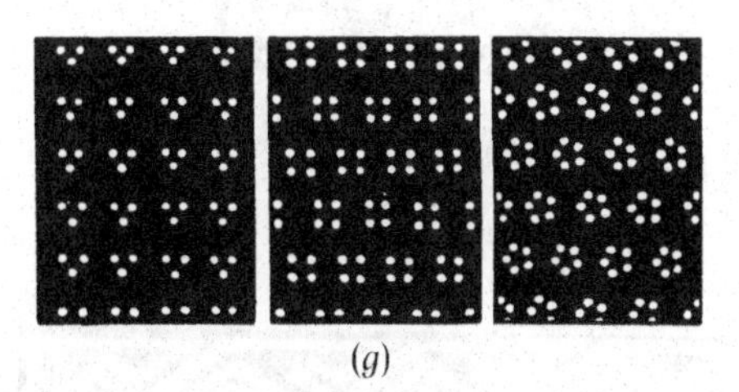

(*g*)

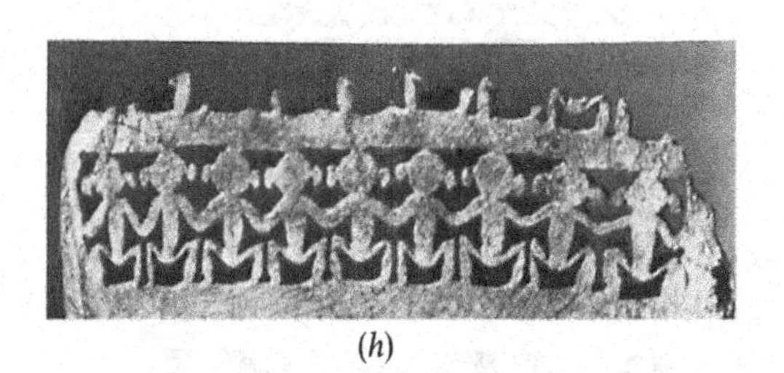

(*h*)

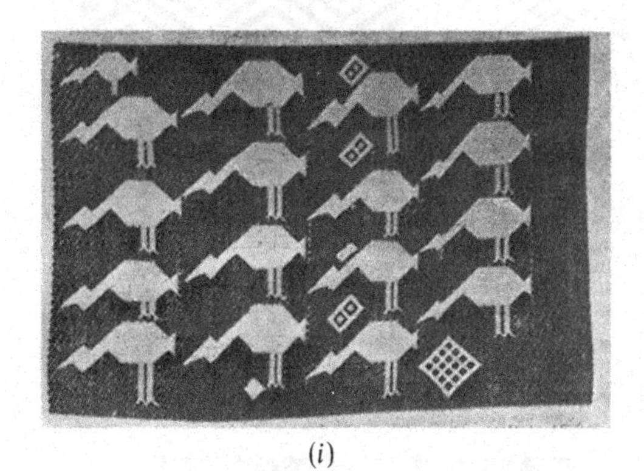

(*i*)

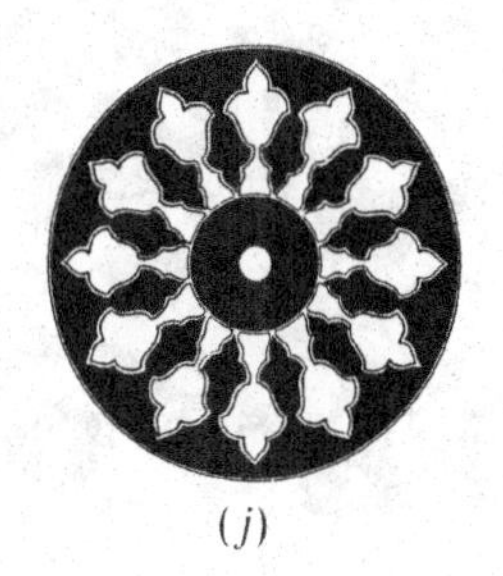

(*j*)

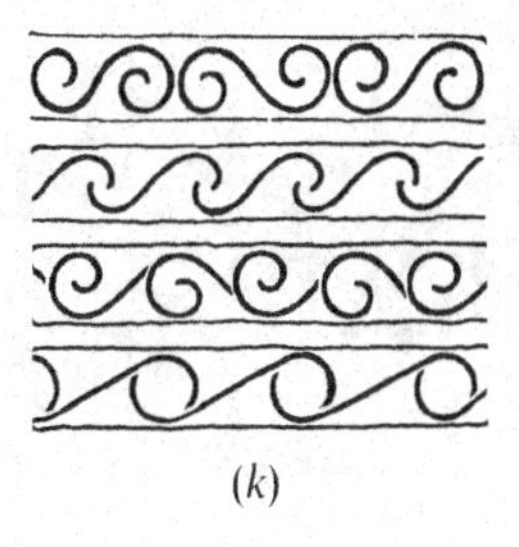

(*k*)

(*l*)

(*m*)

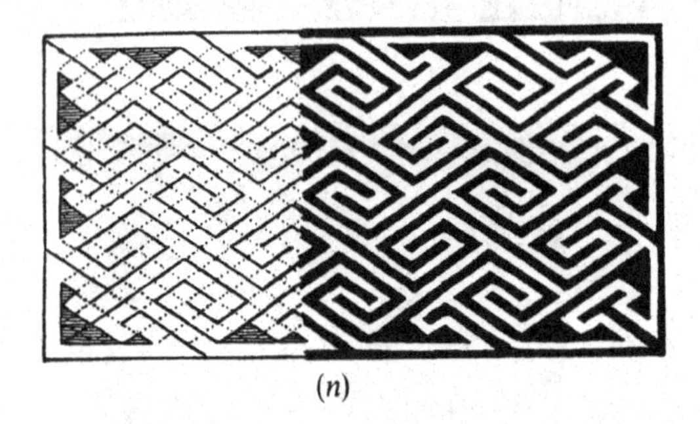

(*n*)

(*o*)

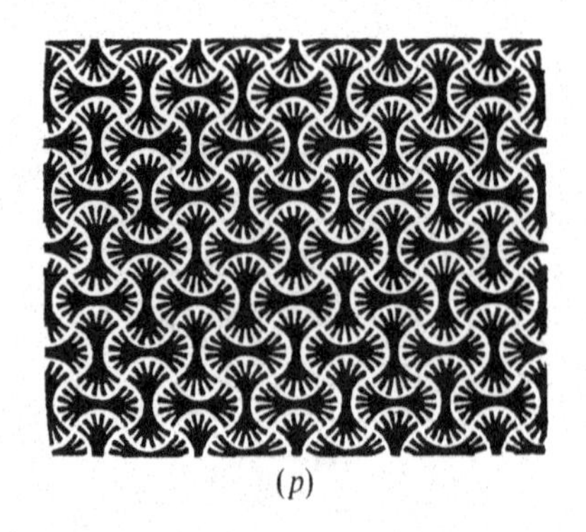

(*p*)

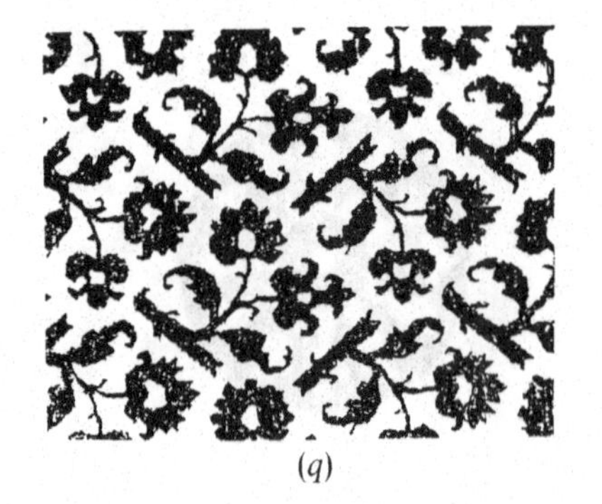

(*q*)

Figure 5.2.13
Some 3-colored tilings from Christie [1929].

5.3 DOT PATTERNS AND PATTERNS WITH PRESCRIBED MOTIFS

Let M be any given bounded and connected plane set. Then it is of interest to determine which types of discrete patterns exist with M as motif. In this section we shall show how to find a complete list of pattern types corresponding to *any* set M of the stated kind.

In order to do this, we begin by considering dot patterns, that is, patterns whose motif is a single point. It is easy to verify the following result:

5.3.1 *There exist two infinite families of types of nontrivial finite dot patterns, each depending on a positive integer n, seven types of strip dot patterns, and thirty types of periodic dot patterns.*

These are listed in column (6) of Tables 5.2.1, 5.2.2 and 5.2.3, and in column (1) of Table 5.3.1, and illustrated by examples in Figures 5.3.1, 5.3.2 and 5.3.3.

Statement 5.3.1 can be deduced from Statement 5.2.2. We examine, in turn, each type of discrete pattern to see whether it can be realized when the motif is a point. For

Table 5.3.1 THE TYPES OF FINITE, STRIP AND PERIODIC DOT PATTERNS

Column (2) gives the number of parameters needed to determine dot patterns of this type up to congruence; if patterns that differ only in scale are not distinguished, the number of parameters should be reduced by one.

Columns (3) to **(12)** list the pattern types, obtainable with motifs M having symmetry groups $S(M)$ indicated at the head of the columns, from each type of dot pattern by the procedure described in the text.

Column (13) shows the number of the diagram in Sohncke [1874] which illustrates this type of dot pattern.

Columns (14) and **(15)** indicate the symbols for the type used by Burzlaff, Fischer & Hellner [1968] and by Wollny [1969]. For the finite dot patterns, the subscript n can take integer values satisfying $n \geq 2$.

Dot pattern type (1)	Parameters (2)	*c1* (3)	*c2* (4)	*c3* (5)	*c4* (6)	*c6* (7)	*d1* (8)	*d2* (9)	*d3* (10)	*d4* (11)	(12)	(13)	(14)	(15)
DPF2_n	2	PF2_n												
DPF3_n	1	PF1_n					PF3_n							
(DPF3_2)		PF2_1												
(DPF3_{2n})		PF2_n												
DPS7	3	PS7												$\mathscr{P}'_5$
DPS9	3	PS9												$\mathscr{P}'_7$
DPS11	2	PS2, 7					PS11							$\mathscr{P}'_4$
DPS12	3	PS12												$\mathscr{P}'_6$
DPS13	2	PS3, 7, 9, 12					PS13							$\mathscr{P}'_3$
DPS14	2	PS5, 7, 9					PS14							$\mathscr{P}'_2$
DPS15	1	PS1, 2, 5, 7, 9	PS8, 10				PS4, 6, 11, 14	PS15						$\mathscr{P}'_1$

Table 5.3.1 (continued)

Dot pattern type (1)	Parameters (2)	*c1* (3)	*c2* (4)	*c3* (5)	*c4* (6)	*c6* (7)	*d1* (8)	*d2* (9)	*d3* (10)	*d4* (11)	(12)	(13)	(14)	(15)
DPP7	5	PP7										27	$mP2xy$	$\mathscr{P}''_2$
DPP8	3	PP1, 7	PP8									28	mP	$\mathscr{P}''_1$
DPP9	4	PP9										32	$og \cdot C2xy$	$\mathscr{P}''_8$
DPP11	4	PP11										31	$om \cdot P_a2xy$	$\mathscr{P}''_4$
DPP13	3	PP2, 7, 9					PP13					29	$o \cdot gP_a1y$	$\mathscr{P}''_3$
DPP14	4	PP14										20	$oP2x2y$	$\mathscr{P}''_7$
DPP15	3	PP3, 7, 11, 14, 17					PP15						$oP2x$	$\mathscr{P}''_6$
DPP16	2	PP1, 2, 3, 5, 7, 9, 11, 14, 17	PP8, 12, 18				PP4, 13, 15, 19	PP16					oP	$\mathscr{P}''_5$
DPP17	4	PP17										16	$oC2x2y$	$\mathscr{P}''_{11}$
DPP19	3	PP5, 7, 9, 11					PP19					26	$oC2x$	$\mathscr{P}''_{10}$
DPP20	2	PP1, 2, 7, 9	PP8, 10				PP6, 13	PP20					oC	$\mathscr{P}''_9$
DPP21	3	PP21										7	$hP3xy$	$\mathscr{P}''_{31}$
DPP23	3	PP23										2	$hP3x2y\bar{y}$	$\mathscr{P}''_{28}$
DPP25	2	PP21					PP25					9	$hP3x$	$\mathscr{P}''_{27}$
DPP27	3	PP27										19	$hP3x\bar{x}2y$	$\mathscr{P}''_{30}$
DPP28	2	PP21					PP28					8	$hP3x\bar{x}$	$\mathscr{P}''_{29}$
DPP30	3	PP30										10	$tP4xy$	$\mathscr{P}''_{18}$
DPP33	3	PP33										11	$t \cdot g \cdot C4xy$	$\mathscr{P}''_{17}$
DPP35	2	PP30					PP35					59	$t \cdot g \cdot C2xx$	$\mathscr{P}''_{16}$
DPP37	3	PP17, 37										17	$tP4x2y$	$\mathscr{P}''_{15}$
DPP38	2	PP17, 30, 37					PP38					13	$tP4x$	$\mathscr{P}''_{14}$
DPP39	2	PP14, 30, 33					PP39					12	$tP4xx$	$\mathscr{P}''_{13}$
DPP41	1	PP1, 2, 3, 5, 7, 9, 11, 14, 17, 30, 33	PP8, 10, 12, 18, 31		PP32, 34		PP4, 6, 13, 15, 19, 35, 39	PP16, 20, 36, 40		PP41			tP	$\mathscr{P}''_{12}$
DPP42	3	PP42										1	$hP6xy$	$\mathscr{P}''_{26}$
DPP46	3	PP46										18	$hP6x2y\bar{y}$	$\mathscr{P}''_{25}$
DPP47	2	PP27, 42					PP47					14	$hP6x$	$\mathscr{P}''_{22}$
DPP48 {A / B}	2	PP23, 42					PP48					{3 / 4}	$hP6x\bar{x}$	$\mathscr{P}''_{24}$ / $\mathscr{P}''_{23}$
DPP49	1	PP21	PP43				PP25, 28	PP49				61	hN	$\mathscr{P}''_{21}$
DPP50	1	PP5, 7, 9, 11, 27, 42		PP24, 44			PP19, 47		PP50		PP51	30	hG	$\mathscr{P}''_{20}$
DPP51	1	PP1, 2, 9	PP8, 10	PP22		PP45	PP6, 13, 25	PP20	PP26, 29				hP	$\mathscr{P}''_{19}$

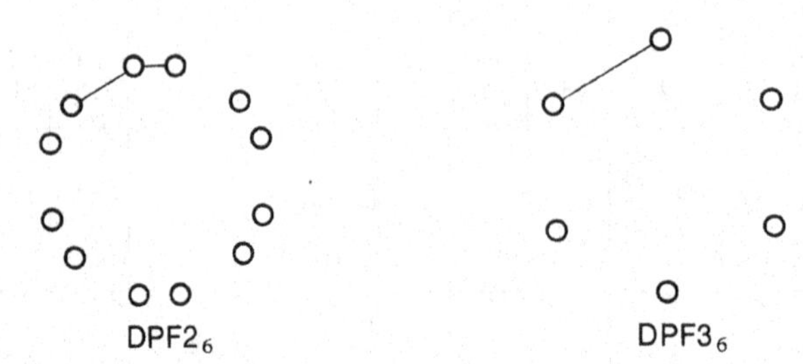

Figure 5.3.1
Examples of the two families of finite dot patterns, DPF2_n and DPF3_n. In both patterns the value of n is 6.

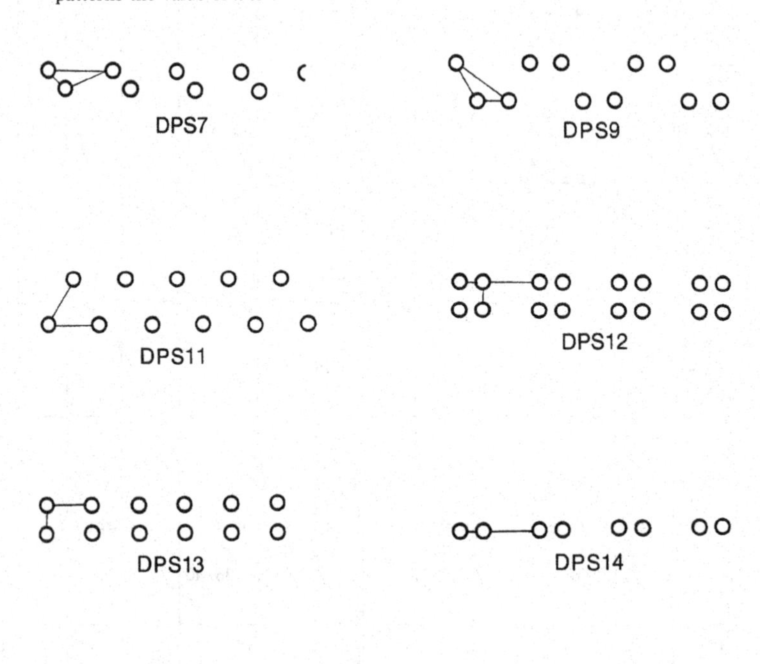

Figure 5.3.2
Examples of the seven types of strip dot patterns. In each, line segments indicate distances between points that can be used as the parameters mentioned in column (2) of Table 5.3.1.

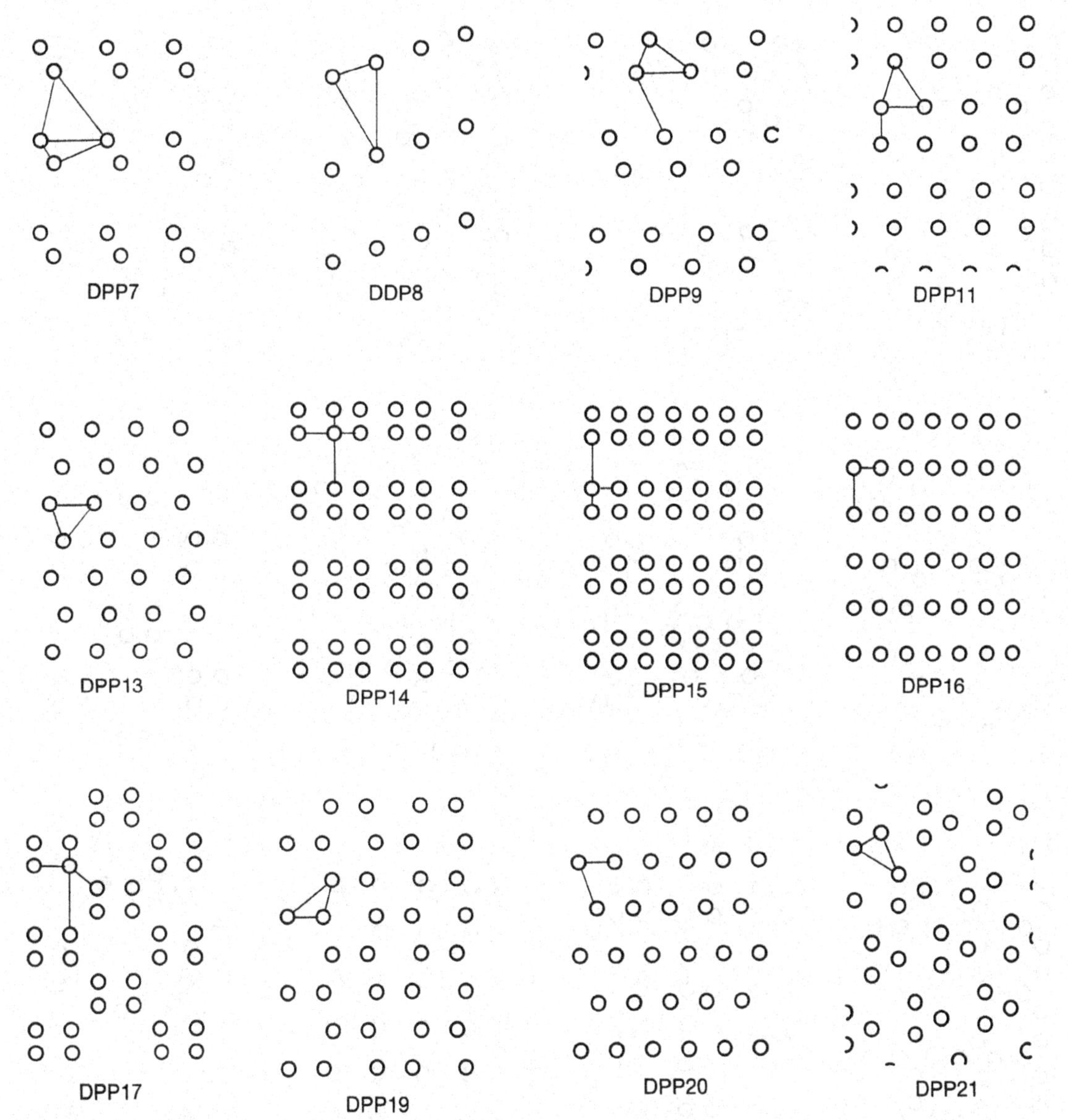

Figure 5.3.3
Examples of the 30 types of periodic dot patterns. Parameters are indicated as in Figure 5.3.2. Two examples of type DPP48 are shown, representing the two different refined pattern types DPP48A and DPP48B.

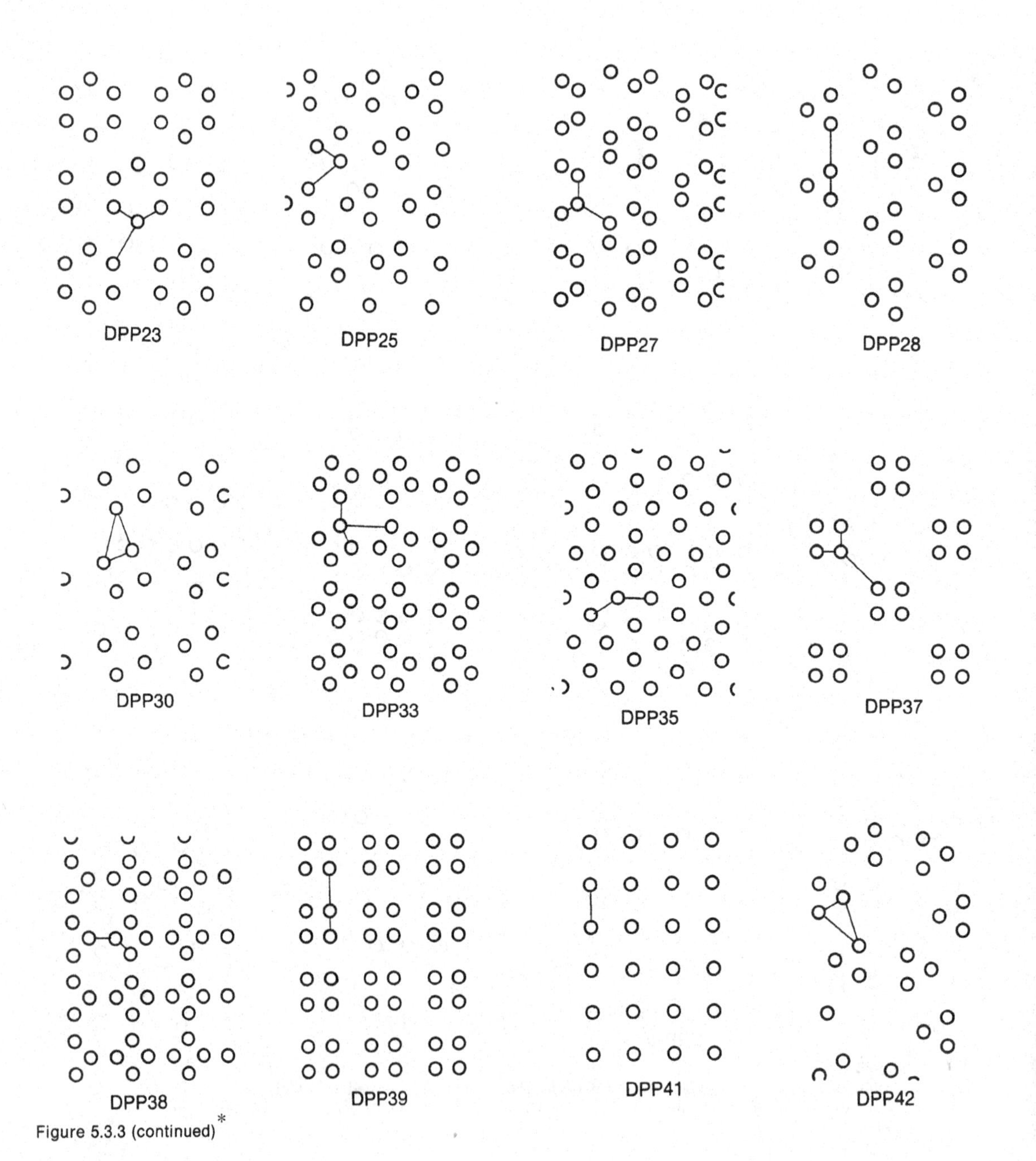

Figure 5.3.3 (continued)*

* See the Appendix beginning on page 653.

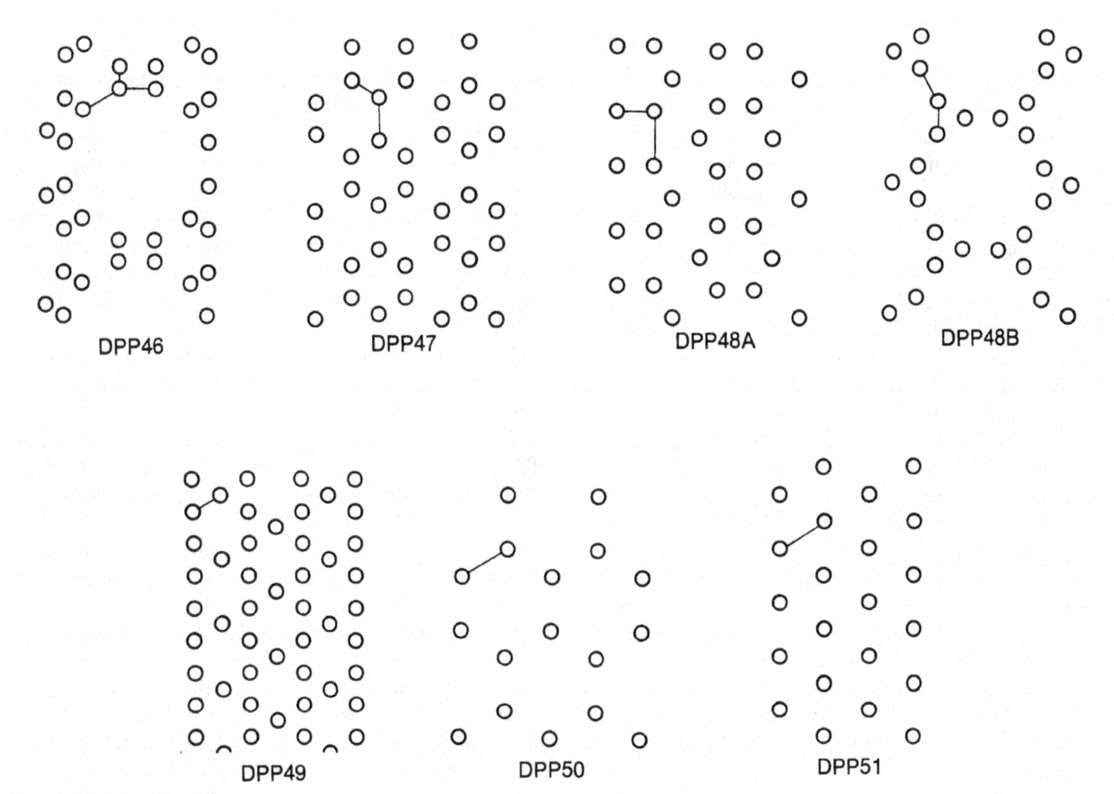

Figure 5.3.3 (continued)

21 of the periodic types (and for some strip and finite patterns) it turns out that such a choice of motif introduces additional symmetries into the pattern and hence leads to a change of type. For example, in the case of pattern type PP1 we obtain a lattice of points. But such a lattice necessarily has symmetry group at least *p2* (rather than *p1*) and possibly a much larger group; as a result the dot pattern is of one of the pattern types PP8, PP16, PP20, PP41 or PP51. Following an investigation of this kind for each pattern type, we eventually arrive at the list of dot pattern types. In column (6) of Tables 5.2.1, 5.2.2 and 5.2.3 we list, for each pattern type, all the dot patterns types to which it leads. (The possibility of additional translational symmetries, which may arise in dot patterns due to the choice of dots in special positions with respect to the symmetry elements, has been investigated by Wondratschek [1976] and Lawrenson & Wondratschek [1976]. Such considerations appear to be of considerable importance in crystallography.)

An equivalent explanation of these listings is the following. As mentioned in Section 5.1, for each pattern $\mathscr{P}$ we may form one or more associated dot patterns $\mathscr{D}$, by replacing copies M_i of the motif of $\mathscr{P}$ by suitably chosen points C_i. Assuming that the points C_i are distinct then each of the possible dot pattern types for $\mathscr{D}$, arising from patterns of a certain pattern type, is listed in column (6) of the tables in Section 5.2. In each case we have emphasized by underlining the *general* associated dot pattern, which results if the points C_i are chosen so as to minimize the symmetries of the resulting dot pattern ("in general position").

An alternative proof of Statement 5.3.1 can be obtained by considering the different patterns that arise when we apply the operations of each of the finite, strip and crystallographic groups to a single point. The pattern type obtained will depend upon where the initial point is chosen relative to the elements of the group. It is not hard to carry out this procedure systematically, and so to arrive at the required result.

The various dot pattern types allow a smaller or greater freedom in the choice of representatives. For example, it is obvious that any two dot patterns of type DPP41 differ (besides the position in the plane) only by the distance between neighboring points; hence we say that the dot patterns of type DPP41 *depend on one parameter*. Similarly, type DPP51 depends on one parameter, while types DPP16 and DPP48 depend on two parameters. In column (2) of Table 5.3.1 we list, for each of the types of dot patterns, the number of parameters on which it depends, and in Figures 5.3.1, 5.3.2 and 5.3.3 we indicate a suitable choice of parameters as the distances between dots joined by lines.

In column (13) of Table 5.3.1 we show the number of the diagram in Sohncke [1874] which illustrates this dot pattern type.

In columns (14) and (15) we give the designations of the dot pattern type (as "Gitterkomplex") in Burzlaff, Fischer & Hellner [1968] and (as "reguläres Punktsystem") in Wollny [1969].

Dot patterns may be used to obtain a complete answer to the following problem, the solution to which clearly extends the information given in column (5) of Tables 5.2.1, 5.2.2 and 5.2.3: *given a bounded motif M, determine all the pattern types that can be realized using the motif M, and give a method for their construction.*

We first observe that if the symmetry group of M is $d\infty$ then M can serve as motif in those and only those pattern types that can be represented by dot patterns. Therefore it suffices to discuss only the case in which $S(M)$ is finite. The procedure we shall employ is essentially a reversal of that described in Section 5.1 for finding a dot pattern associated with a given pattern.

Let M be given and choose any dot pattern $\mathscr{D}$. We now select any pattern type $\mathscr{P}$ which is compatible with both $\mathscr{D}$ and M. By "compatible with $\mathscr{D}$" we mean that $\mathscr{D}$ occurs in column (6) of Table 5.2.1, 5.2.2 or 5.2.3 on the row corresponding to $\mathscr{P}$. By "compatible with M" we mean that $S(M)$ contains the induced motif group of $\mathscr{P}$ as a subgroup, but does not contain a forbidden supergroup for $\mathscr{P}$ as a subgroup (see columns (3) and (5) of Tables 5.2.1, 5.2.2 and 5.2.3). We claim that any such pattern type $\mathscr{P}$ can be realized using the motif M. Hence by taking each dot pattern $\mathscr{D}$ in turn we are able to determine *all* the pattern types with M as motif.

In order to substantiate the above claim we proceed as follows. First choose any point C which is left invariant by all the symmetries of M. Observe that the choice of C is uniquely determined unless $S(M)$ is $c1$ or $d1$; in the former case C may be chosen anywhere in the plane, and in the latter case it is restricted to lie on the line of reflective symmetry of M. We now choose a representative of the dot pattern type $\mathscr{D}$ such that the following construction leads to a set of disjoint copies of the motif, which therefore form a pattern. To guarantee that this will happen it is enough to take as $\mathscr{D}$ a pattern in which the minimal distance between dots is sufficiently large (for example, larger than twice the diameter of the set $\{C\} \cup M$); but other choices of $\mathscr{D}$ may also be appropriate. We place a copy of M (say M_i) in the plane in such a position that C coincides with one of the points C_i of $\mathscr{D}$. We must do this in such a way that lines of reflective symmetry of M_i (if any) coincide with those of the induced motif group of $\mathscr{P}$. Then it is clear that the pattern, formed by applying to M_i all the symmetries of $\mathscr{D}$ that are also symmetries of $\mathscr{P}$, is a pattern of the required type with M as motif.

Further, it is easy to see that the above construction is universal in the sense that it will produce all such patterns. Except for the requirement that distinct copies of the motif must be disjoint, the possibilities depend not on M itself but only on its symmetry group $S(M)$. For this reason, if we are interested only in the pattern types that exist with M as motif and not in the construction of all such patterns, then it is possible to proceed in a more economical manner—see Statement 5.3.2 below. To present this we need to introduce the following terminology.

Let a motif M, with $S(M)$ finite, and a dot pattern $\mathscr{D}$ be given. We define the *maximal associated cyclic index* $n = n(\mathscr{D},M)$ as the largest value of k such that

(*a*) the cyclic group ck is a subgroup of the symmetry group $S(M)$ of the motif M, and

(*b*) there is at least one pattern type entered in Table 5.3.1 on the row corresponding to the dot pattern type of $\mathscr{D}$ and in the column corresponding to ck.

Obviously, $n(\mathscr{D},M)$ is one of the integers 1, 2, 3, 4 or 6.

For example, if $\mathscr{D}$ is of type DPP41 and $S(M)$ has $c2$ (or $c6$) as a subgroup but has no subgroup $c4$, then $n(\mathscr{D},M) = 2$.

We also determine the *maximal associated dihedral index* $m = m(\mathscr{D},M)$ which is the largest value of k such that

(*a′*) the dihedral group dk is a subgroup of $S(M)$;

(*b′*) there is at least one pattern type entered in Table 5.3.1 on the row corresponding to the dot pattern type of $\mathscr{D}$ and in the column corresponding to dk.

If M allows no reflections as symmetries, then there is no k as required in (*a′*) and hence no maximal associated dihedral index m. Notice that if m exists then necessarily $m = n$.

The entries in Table 5.3.1 referred to in (*b*) and (*b′*) (on the row corresponding to $\mathscr{D}$ and in columns cn and, possibly, dn) are called the *pattern types associated with* $\mathscr{D}$ *and* M.

We now have the following result.

5.3.2 *For any given bounded and connected set M with finite symmetry group $S(M)$, the only types of strip patterns and periodic patterns realizable with M as motif are those that are associated with M and each of the types of dot patterns as described above.*

This procedure is illustrated in Figure 5.3.4.

We do not give a formal proof of Statement 5.3.2, and remark only that the maximal groups cn and dm defined above are the only ones that can occur as induced motif groups $S(\mathscr{P}|M)$ with the given M. *Larger* groups are not possible because $S(\mathscr{P}|M)$ must be a subgroup of $S(M)$, and *smaller* groups cannot occur since they are precluded by the existence of additional relevant symmetries of M.

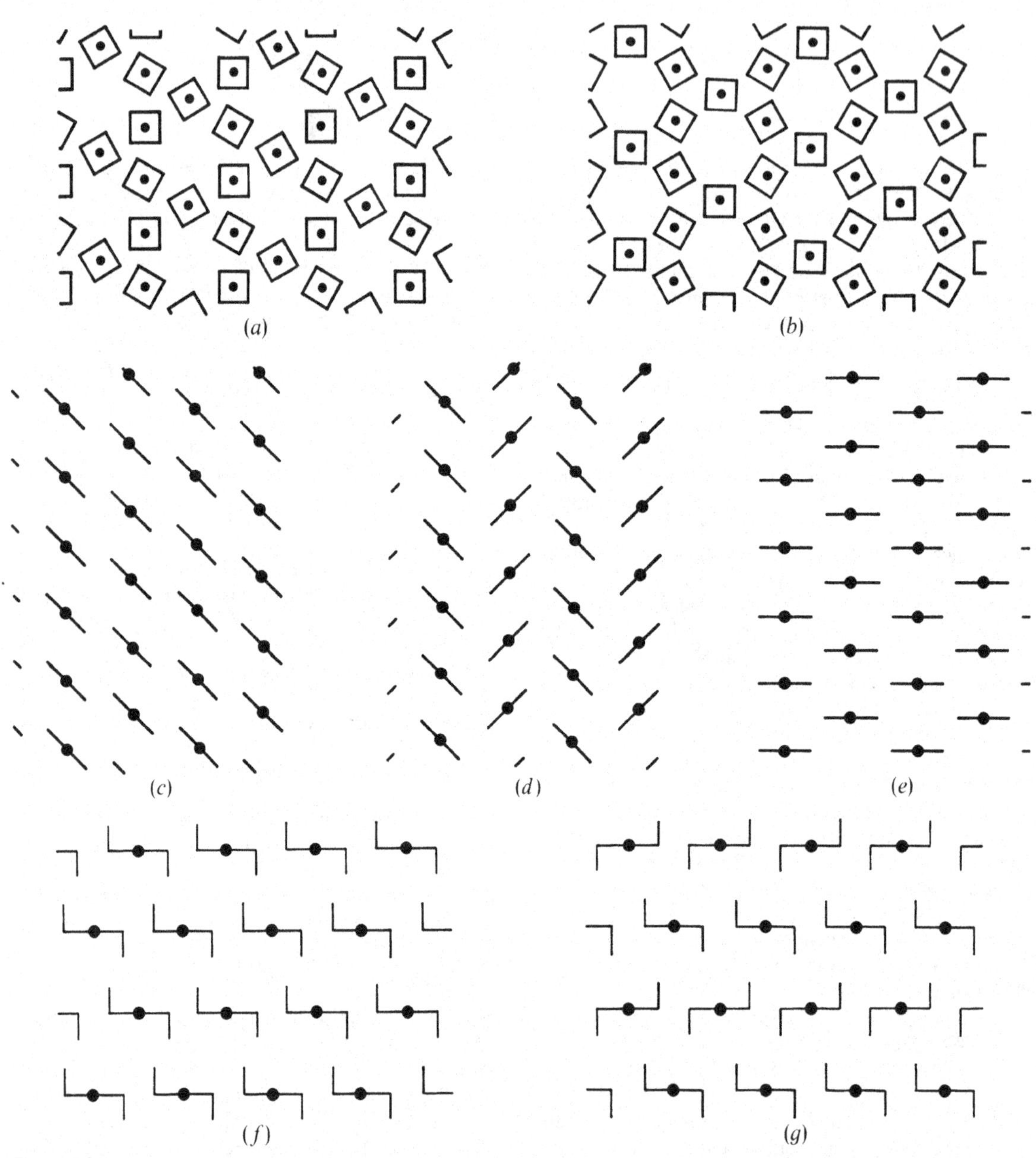

Figure 5.3.4
Examples showing the use of Table 5.3.2 in the determination of patterns. If the motif is a square and the associated dot pattern is DPP49, then $n = m = 2$ and the possible patterns are of types PP43 and PP49; these are shown in (*a*) and (*b*). If the motif is a straight-line segment (or a rectangle) and the associated dot pattern is DPP20, then again $n = m = 2$ and the possible types are PP8, PP10 and PP20, shown in (*c*), (*d*) and (*e*). If the motif has symmetry group *c2* and the associated dot pattern is DPP51, then $n = 2$, there is no maximal dihedral index *m*, and the possible pattern types are PP8 and PP10 shown in (*f*) and (*g*).

EXERCISES 5.3

1. By considering dot patterns associated with each of the fifteen types of strip patterns, show that only seven types of strip dot patterns exist (thus partially checking Statement 5.3.1). Verify this result by applying the operations of the strip groups to a single point, as suggested on page 244.

2. Carry out the details of proving Statement 5.3.2 in the case of strip patterns.

3. By applying Statement 5.3.2, determine all the types of strip and periodic patterns admitted by the motif in Figure 5.2.9(*a*). This will provide a check on the result of Exercise 5.2.2(*b*).

4. Show by examples that Statement 5.3.2 may fail if the set M does not satisfy the assumption that $S(M)$ is finite.

5. Considering the dark areas as motifs (and disregarding one-point contacts), for each of the patterns in Figure 5.3.5 determine all possible associated dot patterns and classify them according to their type.

6. In Figure 5.3.6 we show examples of finite *dimotif dot patterns*. These are dimotif patterns in which there are two "kinds" of dots (represented by "black" and "white" points, or by solid and hollow circles), with symmetries of the pattern acting transitively on dots of each kind. (Observe that the names of the "colors" are irrelevant, but that only color-preserving symmetries of the pattern are allowed. Dimotif dot patterns have attracted the attention of crystallographers in connection with crystals containing two sorts of atoms. An early paper on the subject is Laves [1930]; the subsequent literature is extensive.)

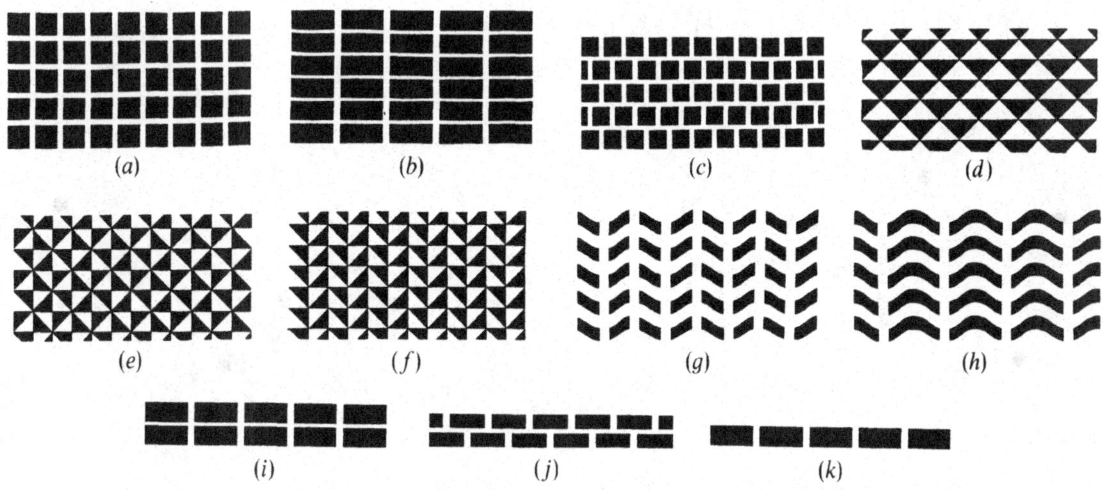

Figure 5.3.5
Examples of discrete patterns.

(*a*) Give a reasonable definition of "type" for dimotif dot patterns.
(*b*) Check whether all the examples in Figure 5.3.6 satisfy your definitions, and whether they all represent different types.
(*c*) Determine all types of strip dimotif dot patterns.

7. In Sections 5.1 and 5.3 we explained how a dot pattern $\mathscr{D}$ can be associated with a given pattern by replacing each copy M_i of the motif by a suitably chosen point C_i. Clearly dot pattern types other than those given in column (6) of Tables 5.2.1, 5.2.2 and 5.2.3 can arise if we allow the points to be chosen so that they are not distinct. Determine all the possibilities for each pattern type. (For example, find all dot patterns that can arise in this way from PP46.)
8. Show that every periodic dot pattern can be written as the union of at most twelve translates of one of the four patterns DPP8, DPP16, DPP41 or DPP51.
9. Explain why the pattern types of dot patterns are precisely the ones for which there is no entry in column (5) of the tables in Section 5.2.
10. Recall from Section 1.3 that a lattice Λ is a periodic dot pattern which admits a motif-transitive group of

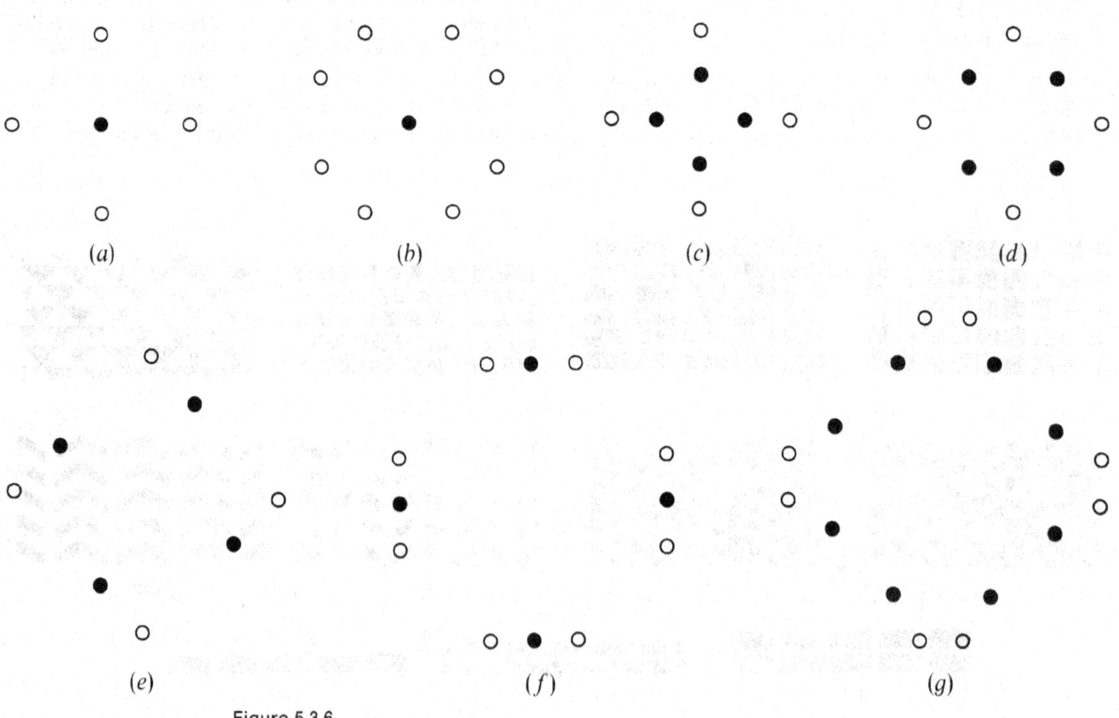

Figure 5.3.6
Examples of finite dimotif dot patterns from all families of types. Each family depends on an integer parameter n; the types with $n = 4$ are shown.

translational symmetries. Without using the results of this chapter, show that the symmetry group S(Λ) of any lattice Λ is one of *p2*, *pmm*, *cmm*, *p4m*, *p6m*, and that each of these groups is the symmetry group of some lattice.

11. A *restricted dot pattern* is a pair ($\mathscr{D}$,G) consisting of a subgroup G of $S(\mathscr{D})$ that acts transitively on the dots of $\mathscr{D}$; for convenience, G may be visualized by its group diagram (though care must be taken concerning translations). Examples of restricted dot patterns are shown in Figure 5.3.7. (The "marked dot patterns" described below coincide, except for the terminology and explanations, with the "sets of equivalent point positions" considered by crystallographers; see Section 5.6.)

    (*a*) Determine all possible restricted dot patterns obtainable from a dot pattern $\mathscr{D}$ of type DPP48, or one of type DPP49.

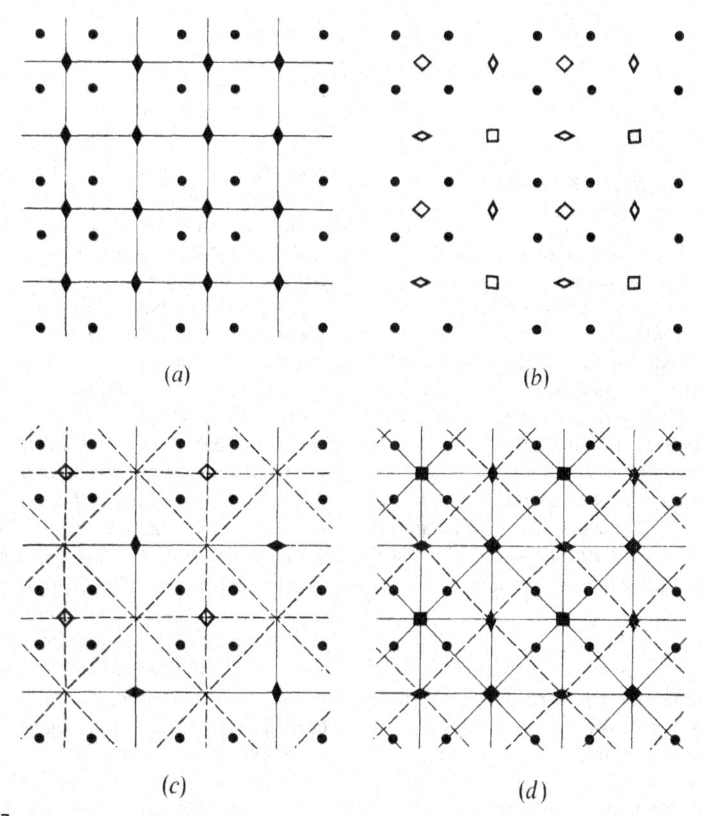

Figure 5.3.7
Examples of restricted dot patterns ($\mathscr{D}$,G) obtainable from a dot pattern $\mathscr{D}$ of type DPP 39. In each case G is represented by its group diagram. The types of the restricted dot patterns are: (*a*) PP14, (*b*) PP30, (*c*) PP33 and (*d*) PP39.

(*b*) In each of the restricted dot patterns ($\mathscr{D}$,G) obtained in part (*a*) "mark" each dot so that the symmetry group of the "marked dot" coincides with the stabilizer of the dot in G. (To visualize this marking it is convenient to replace each dot by a concentric closed circular disk.) Determine the pattern type of the resulting "marked dot pattern", and explain the relation of these pattern types to entries labelled **48** or **49** in column (6) of Table 5.2.3.

(*c*) Show that each pattern type is the pattern type of a suitable "marked dot pattern" obtained as in (*b*) from an appropriate restricted dot pattern ($\mathscr{D}$,G).

5.4 TILINGS AND TILING PATTERNS

We have already remarked on the close analogy that seems to exist between open disk patterns, and tilings by tiles which are topological disks. In this section we explore this relationship further.

We recall that an isohedral tiling $\mathscr{T}$ is a normal tiling whose symmetry group $S(\mathscr{T})$ acts transitively on the tiles. Illustrations of many isohedral tilings appear in Figure 6.2.4. The following basic fact is immediate:

5.4.1 *The interiors of the tiles of an isohedral tiling* $\mathscr{T}$ *form an open disk pattern* $\mathscr{P}_h(\mathscr{T})$ *whose motif is the interior of the prototile of* $\mathscr{T}$. *Moreover* $\mathscr{P}_h(\mathscr{T})$ *is maximal in the sense that it cannot be engulfed by any open disk pattern* $\mathscr{Q}$ *with* $\mathscr{Q} \neq \mathscr{P}_h(\mathscr{T})$.

An open disk pattern that is derived from an isohedral tiling in this way is called a *tiling pattern.* Such patterns are, of course, necessarily periodic.

We can classify isohedral tilings by the pattern type of $\mathscr{P}_h(\mathscr{T})$. However, not all the different pattern types actually occur. The numbers that appear in column (7) of Table 5.2.3 refer to the different types of isohedral tilings, as will be explained in Chapter 6, but for the present they may be interpreted as referring to the tilings shown in Figure 6.2.4. If the number in column (7) is in parentheses, then there is no tiling of the given pattern type, and the meaning of the number will be explained in Section 6.2.

Besides being able to associate a pattern with each isohedral tiling, it is possible to associate an isohedral tiling with each discrete pattern in the following way. We use a procedure that has several other applications in the theory of tilings and patterns.

Let $\mathscr{F} = \{F_i \mid i \in I\}$ be any non-empty family of pairwise disjoint sets in the plane; with each F_i we associate a tile $T(F_i)$ consisting of all the points P of the plane for which the distance from P to F_i is less than or equal to the distance from P to each F_j with $j \neq i$. Then $\{T(F_i) \mid i \in I\}$ is a tiling called the *Dirichlet tiling associated with* $\mathscr{F}$, which we denote by $\mathscr{D}(\mathscr{F})$ (see Figures 5.4.1 and 5.4.2). Note that the tiles of $\mathscr{D}(\mathscr{F})$ can be unbounded; in fact, if $\mathscr{F}$ is finite, $\mathscr{D}(\mathscr{F})$ necessarily contains unbounded tiles. If each set F_i is connected, then the tiles $T(F_i)$ are also connected. We are mainly interested in the case where $\mathscr{F}$ consists of all the copies of the motif in a discrete periodic pattern $\mathscr{P}$; then the corresponding Dirichlet tiling $\mathscr{D}(\mathscr{P})$ has bounded tiles. Examples of the construction of Dirichlet tilings in such cases appear in Figures 5.4.3 and 6.2.5.

Each tile of $\mathscr{D}(\mathscr{P})$ contains exactly one copy of the motif of $\mathscr{P}$ in its interior, and since the symmetries of $\mathscr{P}$ act transitively on the copies of the motif, the same isometries act transitively on the tiles of $\mathscr{D}(\mathscr{P})$. We deduce:

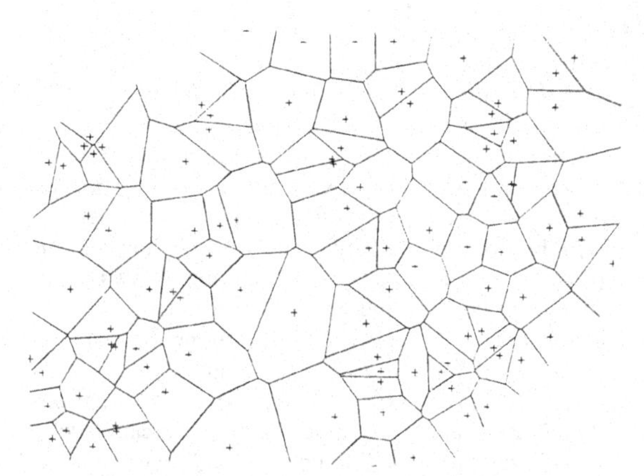

Figure 5.4.1
The Dirichlet tiling $\mathscr{D}(\mathscr{F})$ associated with a family $\mathscr{F}$ of points (From Green & Sibson [1978].)

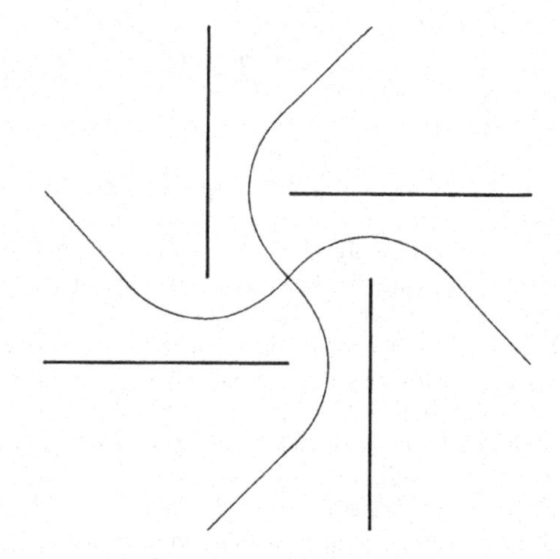

Figure 5.4.2
The Dirichlet tiling associated with a family of four line segments.

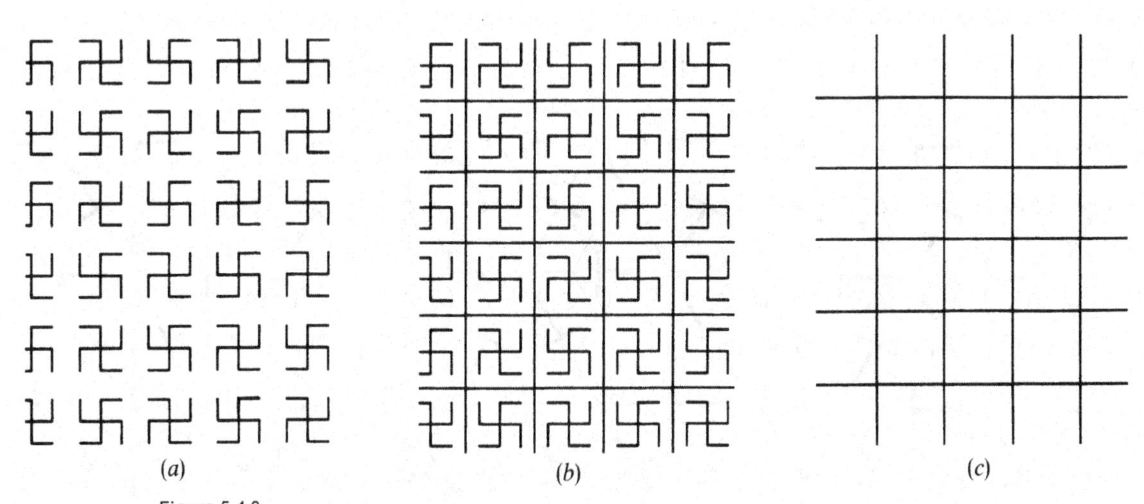

Figure 5.4.3
The construction of the Dirichlet tiling (*c*) from the pattern shown in (*a*). Notice that in this case the symmetry groups of the pattern and of the Dirichlet tiling are different.

5.4.2 *To every discrete periodic pattern $\mathscr{P}$ corresponds an isohedral tiling $\mathscr{D}(\mathscr{P})$, and the interiors of the tiles of $\mathscr{D}(\mathscr{P})$ form an open disk periodic pattern that engulfs $\mathscr{P}$.*

For brevity we shall write $\mathscr{P}_d$ instead of $\mathscr{P}_h(\mathscr{D}(\mathscr{P}))$ for this latter pattern. It is important to note that $\mathscr{P}$ and $\mathscr{P}_d$ are not necessarily of the same pattern type. For example, in Figure 5.4.3 we show (*a*) a pattern $\mathscr{P}$ of type PP34, (*b*) the pattern $\mathscr{P}$ and the corresponding Dirichlet tiling $\mathscr{D}(\mathscr{P})$ superimposed, and (*c*) the tiling $\mathscr{D}(\mathscr{P})$ alone. Here it is easily seen that the pattern $\mathscr{P}_d$, consisting of the interiors of the tiles of $\mathscr{D}(\mathscr{P})$, is of type PP41. In fact, since there is *no* tiling $\mathscr{T}$ for which $\mathscr{P}_h(\mathscr{T})$ is of type PP34, it is obvious that in cases such as this $\mathscr{P}_d$ cannot possibly be of the same type as $\mathscr{P}$.

On the other hand, if we wish to represent *every* type of periodic pattern by means of an isohedral tiling, we can do so by using the concept of a *marked* isohedral tiling. Marked tilings and their symmetries were defined in Section 1.3, and it is clear how the definitions of words such as transitivity, isohedrality etc. can be extended to such tilings. We shall write $\mathscr{D}'(\mathscr{P})$ for the *marked Dirichlet tiling* of $\mathscr{P}$ whose tiles are those of $\mathscr{D}(\mathscr{P})$ but each tile bears, as a mark, the copy of the motif it contains. It is easy to see that the marked interiors of the tiles of $\mathscr{D}'(\mathscr{P})$ form a pattern which is of the same type as $\mathscr{P}$.

The classification of isohedral tilings by pattern type is deficient in that it takes no account of one of the most important features of the tiling, namely its topological type. For example, in Figure 5.4.4 we show how a given pattern $\mathscr{P}$ can coincide with the markings on two distinct isohedral tilings $\mathscr{T}_1$ and $\mathscr{T}_2$. The latter must clearly be considered distinct since the marked tiling $\mathscr{T}_1$ is of topological type $[3^6]$, and $\mathscr{T}_2$ is of topological type $[3.6.3.6]$—yet $\mathscr{P}$, $\mathscr{P}_h(\mathscr{T}_1)$, and $\mathscr{P}_h(\mathscr{T}_2)$ are all of pattern type PP25. Exactly the same phenomenon can arise in the case of unmarked tilings, see Figure 5.4.5—in one case the tiles have four edges and in the other six, yet both lead to tiling patterns of the same type.

These comments provide some motivation for the treatment of isohedral tilings given in the next chapter—here it suffices to remark that many problems of classification are difficult to formulate and resolve in a completely satisfactory manner. This difficulty is borne out by the many attempts that have been made by authors in the past—with varying degrees of success.*

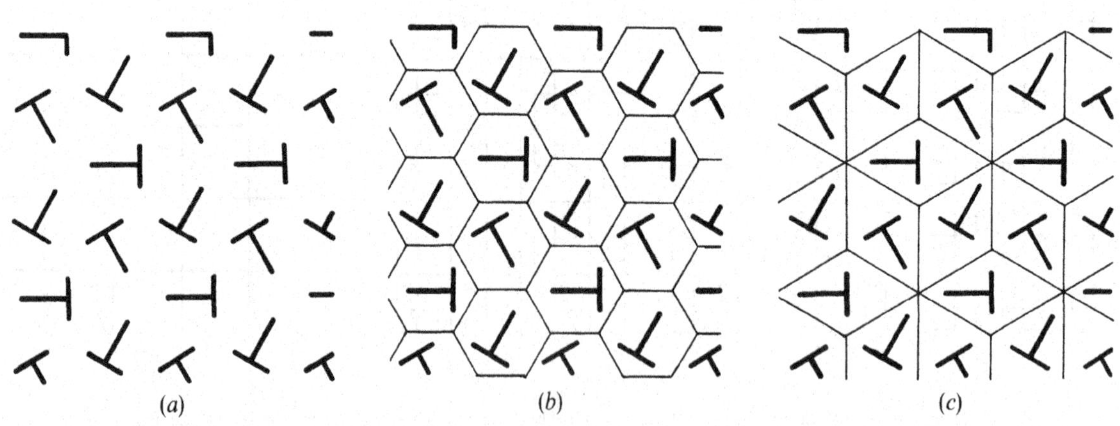

Figure 5.4.4
The pattern of type PP25 shown in (*a*) can be carried by tilings (*b*) and (*c*) of different topological types.

* See the Appendix beginning on page 653.

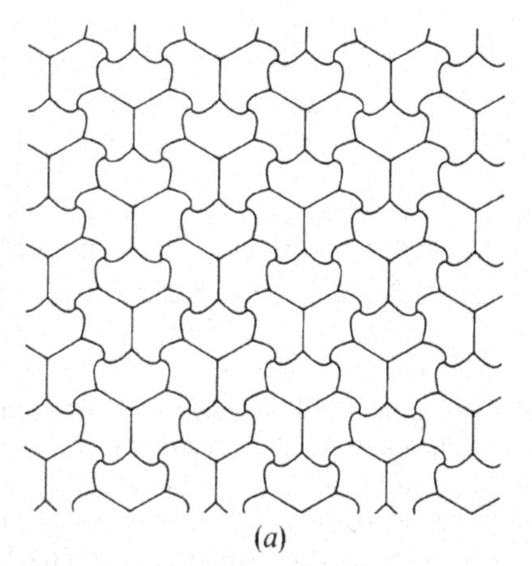
(a)

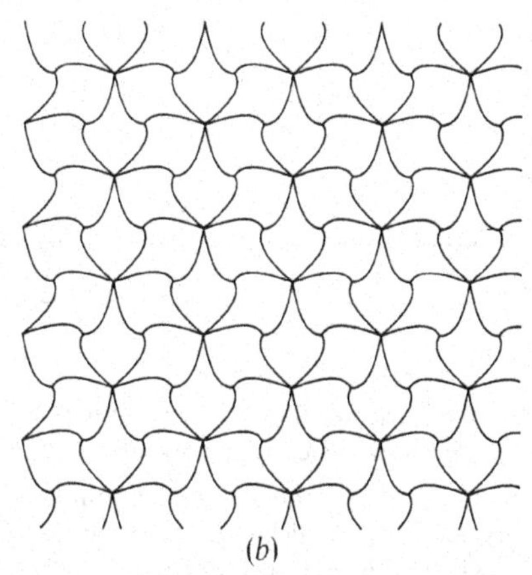
(b)

Figure 5.4.5
Two (unmarked) tilings which are of different topological types, yet the interiors of the tiles form patterns of the same type (**PP25**).

We conclude this section by briefly referring to the two other kinds of tilings defined by transitivity. In the first place, there are the isotoxal tilings, that is, tilings $\mathcal{T}$ whose symmetry groups $S(\mathcal{T})$ are transitive on the *edges* of $\mathcal{T}$. Diagrams of isotoxal tilings appear in Figure 6.4.2. Isotoxal tilings can be classified using the theory of patterns on account of the following property:

5.4.3 *If the vertices of an isotoxal tiling $\mathcal{T}$ are deleted, then the open arcs resulting from the edges are the copies of the motif in a discrete periodic pattern denoted by $\mathcal{P}_t(\mathcal{T})$.*

In column (9) of Table 5.2.3 we list, for each pattern type, the corresponding isotoxal tilings by type numbers that refer to the classification that will be explained in Section 6.4 and illustrated in Figure 6.4.2. Where no number appears, or the number is in parentheses, there is no isotoxal tiling of the given type. Hence only 18 types of periodic patterns can be derived from isotoxal tilings by the above method.

Finally, we consider isogonal tilings. We recall that a tiling is isogonal if its symmetry group is transitive on its vertices. Examples of isogonal tilings appear in Figure 6.3.5. Again, it is possible to associate a discrete pattern with each isogonal tiling $\mathcal{T}$, though here the procedure is slightly more complicated. First we must choose one point $P(e)$ on each edge e of the tiling $\mathcal{T}$ in such a way that the resulting set of points is invariant under every operation of $S(\mathcal{T})$. Thus if a symmetry of $\mathcal{T}$ maps an edge e_1 onto an edge e_2 then it must map $P(e_1)$ onto $P(e_2)$, and if a symmetry of $\mathcal{T}$ maps an edge e onto itself, interchanging the two endpoints, then $P(e)$ must be the "midpoint" of e, that is, the unique point of e that is mapped onto itself. Now delete all the points $P(e)$. Each vertex V of $\mathcal{T}$ belongs to a *star* (or *spider*) consisting of V and all the open arcs (parts of edges) incident with V. Then we have:

5.4.4 *The stars associated with the vertices of an isogonal tiling are copies of the motif in a discrete periodic pattern.*

We denote this pattern by $\mathcal{P}_g(\mathcal{T})$. In column (8) of Table 5.2.3 we list the various types of isogonal tilings, according to the classification of Section 6.3, and illustrated in Figure 6.3.5. The same convention concerning parentheses is used as for isohedral tilings. In this case it turns out that all but two of the pattern types (**PP26** and **PP36**) can be represented in this way.

The remarks made above about the classification of isohedral tilings by patterns being deficient, in that it ignores the topological type of the tiling, apply equally to the isotoxal and isogonal tilings. This will be discussed in detail in Chapter 6.

EXERCISES 5.4

1. Prove that the Dirichlet tiling associated with a periodic dot pattern has convex polygons as tiles.
2. The dot patterns of type **DPP13** depend on three parameters. Describe the dependence of the Dirichlet tiling associated with each such dot pattern on these parameters.
3. Find an isohedral tiling by convex polygons which is not the Dirichlet tiling of any dot pattern.
4. Show that each Laves tiling (see Section 2.7) is the Dirichlet tiling of the family of vertices of a suitable uniform tiling. (The particular case of the Laves tiling $[3 . 12^2]$ contradicts the assertion of Loeb [1976, pp. 119–120] to the effect that the prototile of an isohedral Dirichlet tiling by triangles cannot be an obtuse-angled triangle.)
5. Decide whether every isohedral tiling by polygons which has regular vertices (see Section 2.7) is the Dirichlet tiling associated with a suitable dot pattern.
6. Determine the Dirichlet tilings associated with the families of segments in Figure 5.4.6.
7. Show that the parallelogram tiling in Figure 5.4.7 (in which the prototile has angles of 30° and 150°) cannot be obtained as the Dirichlet tiling of any compact periodic pattern.

(*a*)

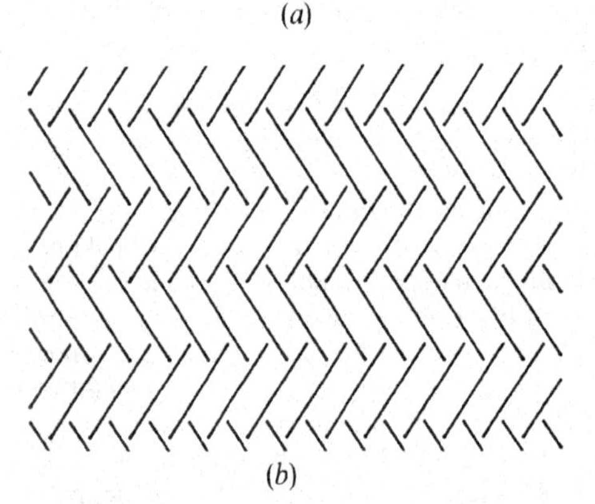
(*b*)

Figure 5.4.6
Two line-segment patterns. The associated Dirichlet tilings are asked for in Exercise 5.4.6.

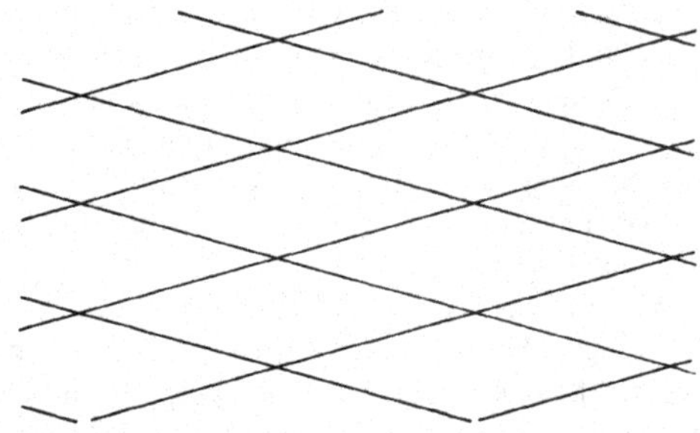

Figure 5.4.7
The parallelogram tiling considered in Exercise 5.4.7.

5.5 PATTERNS WITH UNBOUNDED MOTIFS

In this section we shall depart from the restriction to bounded motifs followed so far in this chapter, and present some preliminary remarks about patterns with *unbounded motifs*. As previously, we shall always assume that the symmetry group of the pattern is transitive on the copies of the motifs. In Sections 6.5 and 7.6 we shall discuss some of these patterns in detail; even under quite stringent conditions on the kinds of motifs allowed, the possible patterns are very plentiful and often quite attractive. Here our purpose is to show that the absence of boundedness leads to interesting complications and differences from the bounded case.

For definiteness, let us consider "curve patterns"; of course, many other kinds of motifs and patterns could be considered, and some of the possibilities will be mentioned later. By a *curve pattern* we mean one which satisfies conditions P.1, P.2 and P.3 as well as condition DP.2 of Section 5.1. Instead of condition DP.1 we shall impose the following conditon:

UP.1 *The motif M is a closed set obtained as the image of a straight line under a homeomorphism of the plane onto itself.*

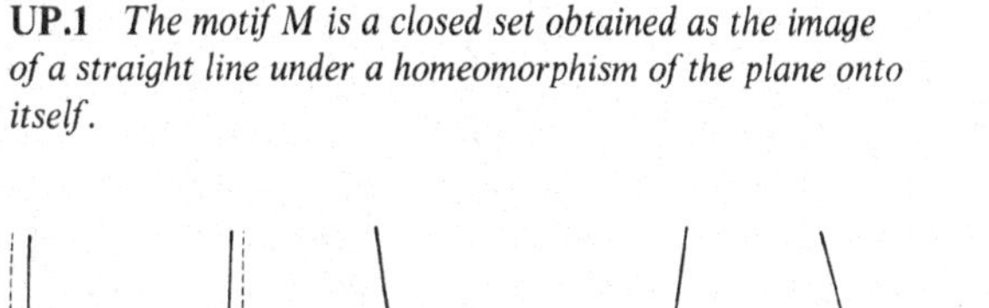

Informally, this condition means that M is a simple curve which stretches to infinity in both directions.

One way in which curve patterns differ from patterns with bounded motifs is the following. By the results of Section 5.3, if bounded motifs M_1 and M_2 have the same symmetry groups $S(M_1)$ and $S(M_2)$, then they give rise to the same types of discrete patterns. However, this is not the case for curve patterns. Consider, for example, the four motifs of Figure 5.5.1; here we take for M_1 the curve $\{(x,y) | y = (1 - x^2)^{-1}, -1 < x < 1\}$, for M_2 the parabola $\{(x,y) | y = x^2\}$, for M_3 the hyperbola $\{(x,y) | y^2 = k^2x^2 + 1, y > 0\}$, and for M_4 the curve $\{(x,y) | y = x^2 \cos 4\pi x\}$. (Naturally, in this and all other illustrations of this section only a finite portion of each motif can be shown, and it is in each case continued in the obvious manner or as indicated.) All four motifs have $d1$ as group of symmetries, but they admit different kinds of curve patterns. In the case of M_1 a pattern that has a truncation (see Section 5.1 for a definition of this term) of type PS6 is possible (see Figure 5.5.2), but no such pattern is admitted by M_2, M_3 or M_4. On the other

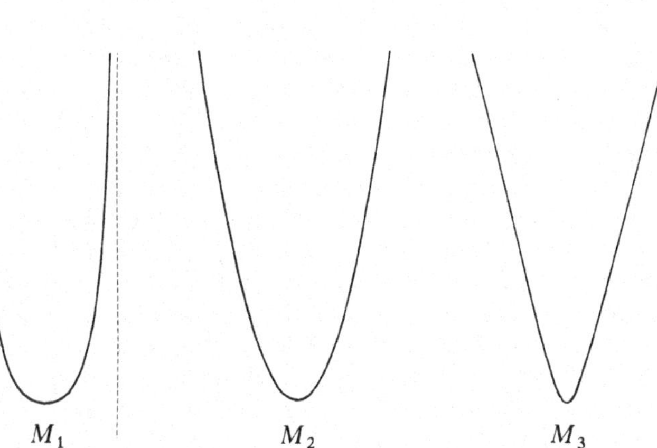

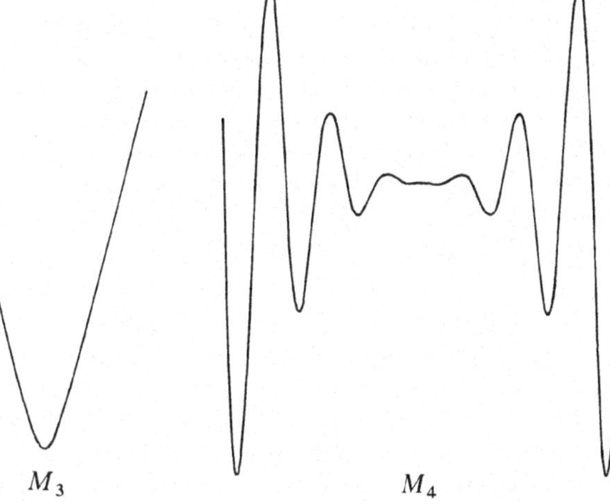

Figure 5.5.1
Four motifs for curve patterns.

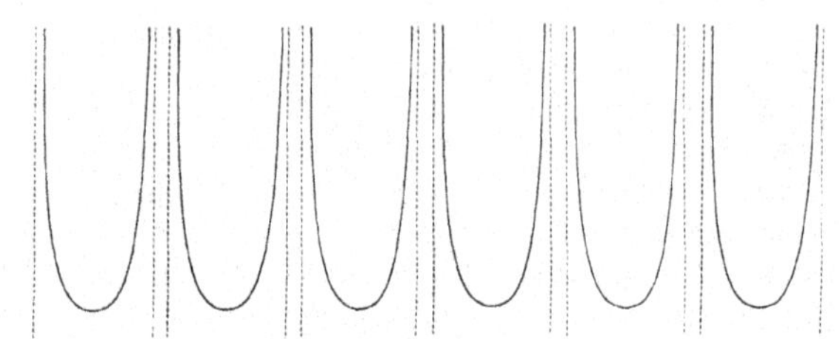

Figure 5.5.2
A curve pattern with motif M_1 which becomes of type PS6 on truncation. No such pattern is possible with the motifs M_2, M_3 or M_4 of Figure 5.5.1.

hand, since M_2 may be included in angular regions with arbitrarily small angles, it admits patterns with truncations of type $PF2_n$ for each $n \geq 1$ and types $PF1_n$ and $PF3_n$ for each $n \geq 2$; but each hyperbola M_3 is contained in a minimal angular region (of angle $\alpha = 2 \arctan 1/k$) and so admits patterns with truncations of type $PF3_n$ only for values of n such that $n\alpha \leq 2\pi$, while M_4 admits among finite patterns only the trivial one. It follows that any kind of enumeration of curve patterns possible with a given motif must take into account properties of the motif other than its symmetry group.

Another difference between curve patterns and discrete patterns is the possibility of non-crystallographic symmetry groups. Some of these "continuous groups" will be discussed in more detail below.

A special type of curve pattern which will be analysed in detail in Section 7.6 is called a *filamentary pattern.* This is a curve pattern $\mathscr{P}$ in which the motif M, the *filament* of $\mathscr{P}$, satisfies (besides UP.1) also the following condition:

UP.2 *M is contained in a (two-way) infinite strip between two parallel lines, but is not contained in any (one-way infinite) half-strip.*

In other words, a filament is the image of the midline of a strip under a homeomorphism of the strip onto itself.

In order to illustrate some of the possibilities we show, in Figure 5.5.3, a few examples of filamentary patterns. These indicate still another difference between such patterns and those with bounded motifs: filamentary patterns need not be locally finite. It will be seen, for example, that in Figures 5.5.3 (*c*) and (*d*) every circular disk centered at a point on one of the dashed lines necessarily intersects an infinite number of copies of the motif.

Among the simplest filamentary patterns are those in which the filament is a straight line. We call these *filamentary line patterns.* We can extend the notion of pattern type given in Section 5.2 to such patterns and it is easy to verify that there exist exactly four types, of which three are non-trivial (that is, contain more than one line). Examples of these are given in Figure 5.5.4.

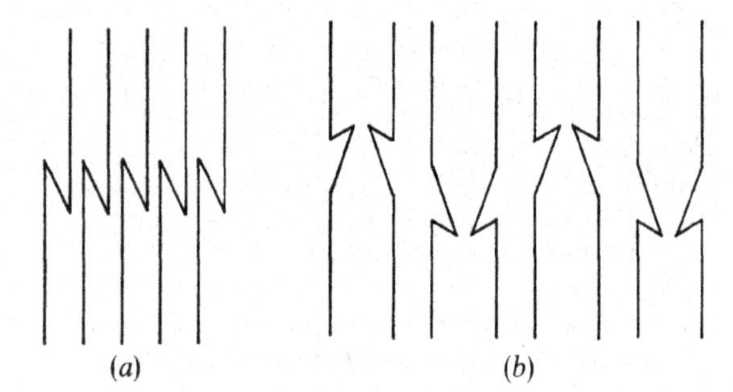

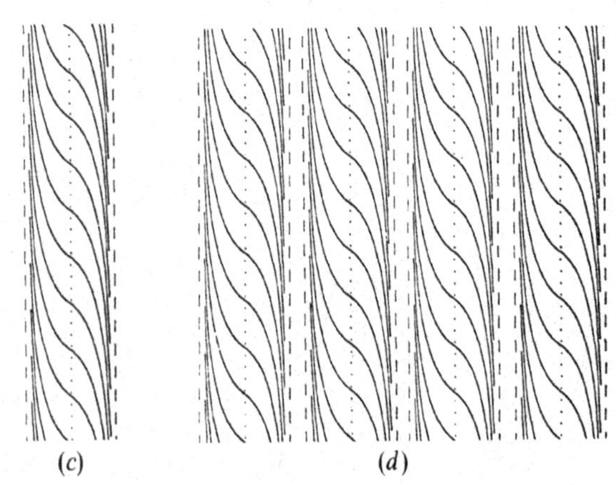

Figure 5.5.3
Examples of filamentary patterns.

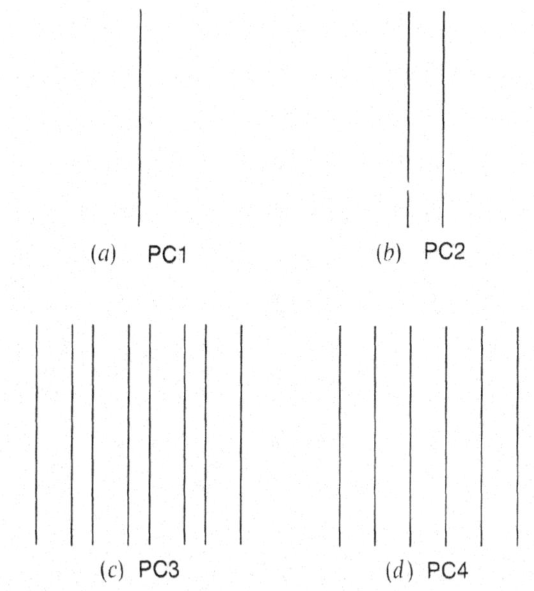

Figure 5.5.4
The four types of filamentary line patterns. The symmetry group $S(\mathscr{P})$ of the patterns PC1 and PC2 is *smm2*, and of the patterns PC3 and PC4 is *smm*. The induced motif group $S(\mathscr{P}|M)$ is *smm2* in PC1 and PC4 and is *sm11* in PC2 and PC3.

However, just as it is often more useful to consider marked tilings rather than unmarked tilings, so it is more useful to consider filamentary patterns with "sensed lines" rather than (ordinary) lines. By a *sensed line* we mean that the two sides of each line can be distinguished in some manner, and that either an orientation (along one of the directions of the line) is imparted to each side of the line, or both sides are unoriented. The five distinct ways in which a line can be sensed are indicated in Figure 5.5.5, and in Figure 5.5.6 we show examples of filamentary line patterns with sensed lines.

The symmetry groups of the patterns in Figure 5.5.6(*b*) (*c*) and 5.5.4(*c*) (*d*) are examples of *continuous periodic groups*. Here the word "continuous" means that all the translations along one line are symmetries, and "periodic" means that the symmetry group contains translations in at least two non-parallel directions. It can be verified that the patterns in Figure 5.5.6(*b*) have as their symmetry groups all possible continuous periodic groups except that consisting of *all* translations of the plane, and its supergroups. A symbol for the group, analogous to the international symbols for the crystallographic groups, is indicated near each pattern.

In the same vein, the groups of symmetries of the sensed lines in Figure 5.5.5 are representatives of all *the continuous strip groups*; these admit all translations along one line and possibly interchanges of sides and orientations.

For filamentary patterns with sensed lines, one can define pattern type in analogy with the definition given in Section 5.2. We shall not pursue this topic here, but remark that in Figures 5.5.5 and 5.5.6 we show all types of such patterns. Of these five contain only one line (Figure 5.5.5), and three contain two lines (Figure 5.5.6(*a*)); the symmetry groups of all these are the continuous strip groups. There are also twelve periodic patterns with sensed lines. These are shown in Figures 5.5.6(*b*) and (*c*). In Table 5.5.1 we list all pattern types of non-trivial filamentary patterns with sensed lines, together with information about them analogous to that contained in the tables of Section 5.2.

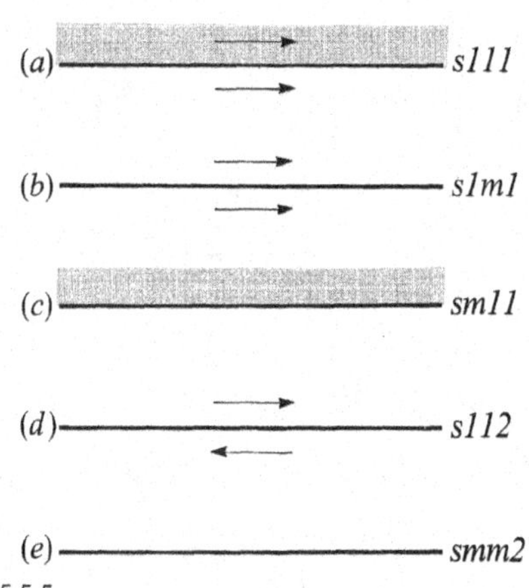

Figure 5.5.5
Examples of sensed lines that represent all continuous strip groups. If the two sides of the line are not equivalent, one of them is marked by a gray strip. If the sides are oriented, an arrow is used to indicate the orientation. The symmetries must preserve orientations (if present) and must not interchange the sides unless these are equivalent.

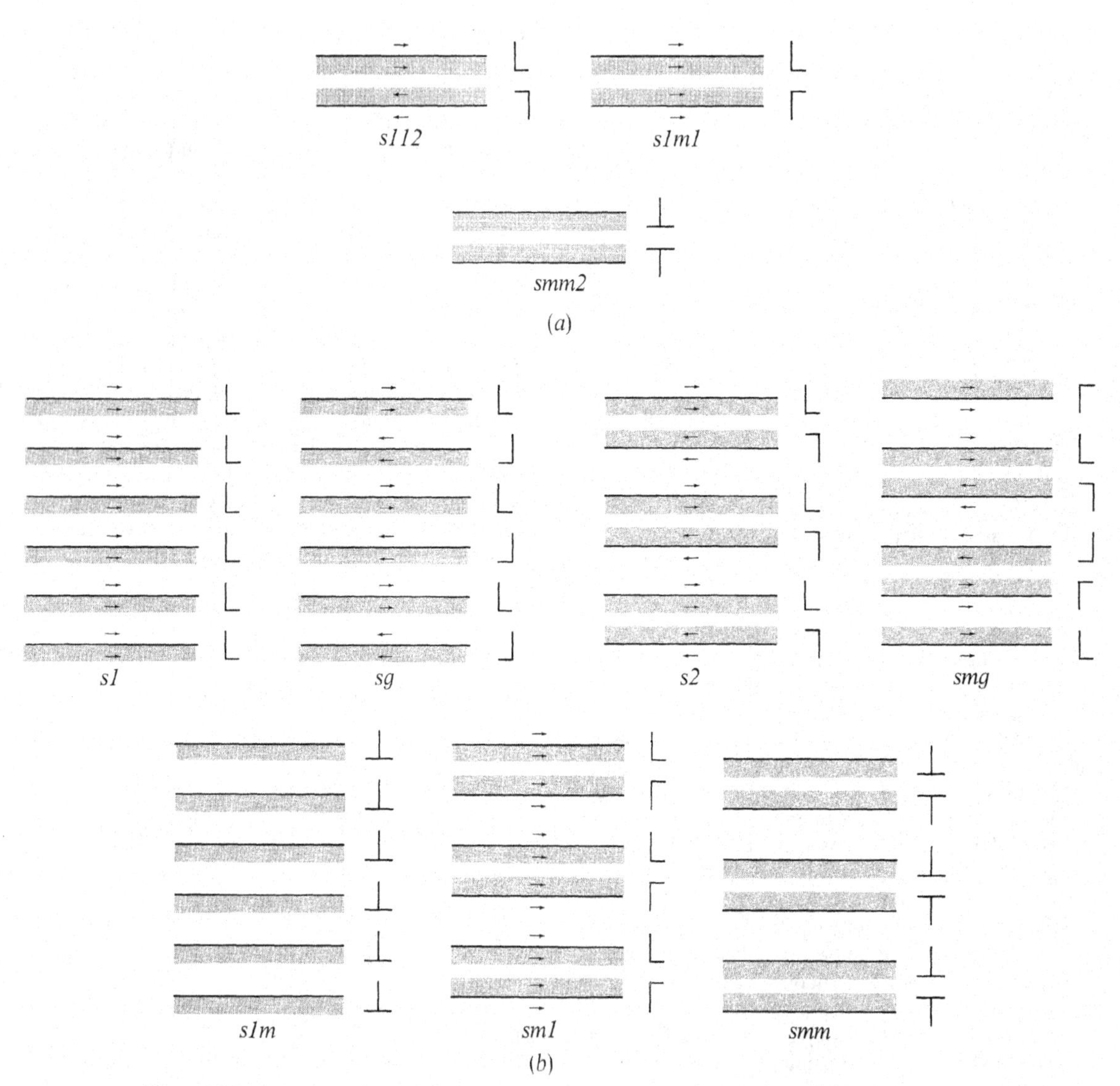

Figure 5.5.6
Examples of all the non-trivial filamentary patterns with sensed lines. In (*a*) each pattern contains just two lines; in (*b*) and (*c*) there are infinitely many lines and the patterns are periodic because their symmetry groups contain translations in non-parallel directions. By each pattern the symbol for its symmetry group is indicated. The patterns in (*b*) represent all seven continuous periodic groups.

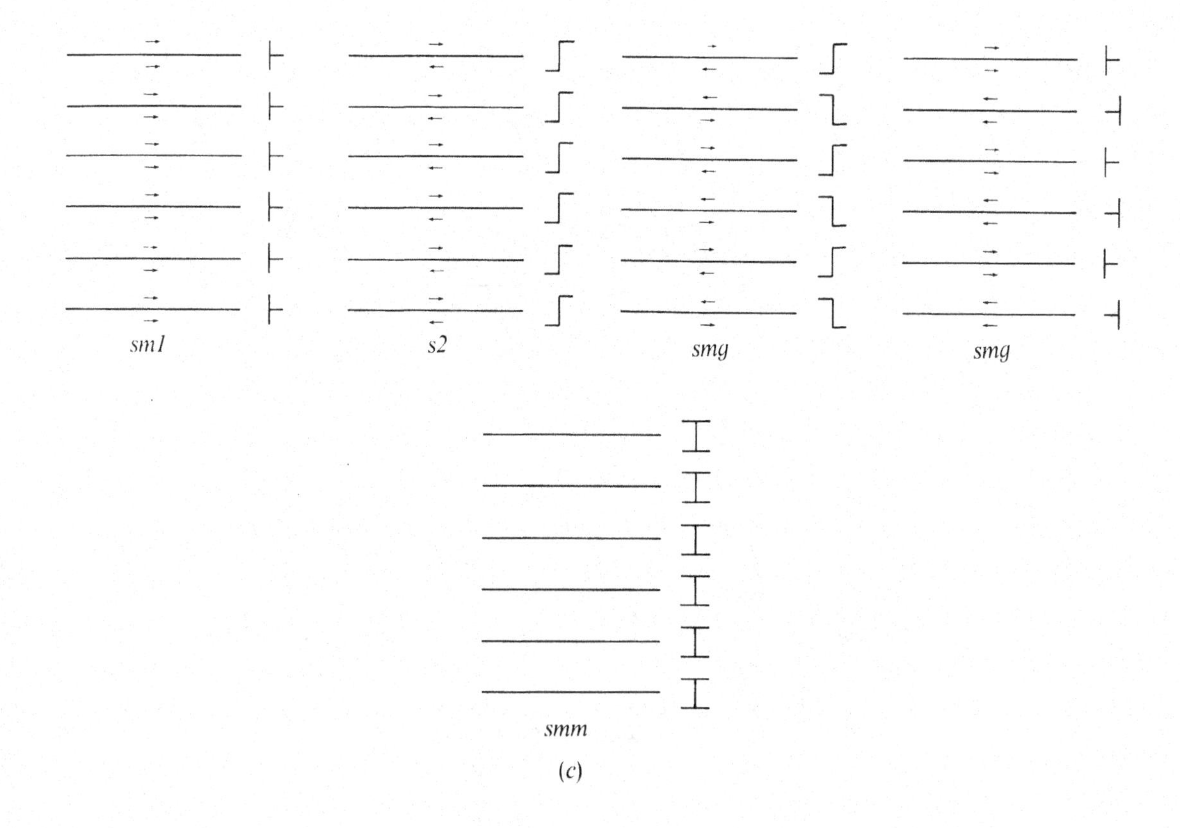

EXERCISES 5.5

1. Determine all types of curve patterns that have as motif:
   (a) the curve M_1 of Figure 5.5.1,
   (b) the parabola M_2 of Figure 5.5.1,
   (c) the hyperbola $\{(x,y) \mid 3y^2 = x^2 + 3, y > 0\}$.

*2. Verify the assertions made in the text that there exist precisely five continuous strip groups, seven continuous periodic groups, and four types of filamentary line patterns.

3. Show that there are just five kinds of sensed lines (that is, as far as symmetries are concerned, each sensed line is equivalent to one of those in Figure 5.5.5).

4. For each of the twelve pattern types of filamentary line patterns determine which of the five kinds of sensed lines can occur as motifs in patterns of that type.

5. (a) Can you find a connection between the twelve types of filamentary line patterns with sensed lines and the fifteen strip pattern types of Section 5.2?
   (b) Find an explanation for the fact that the number of continuous periodic groups is the same as the number of strip groups.

6. Let $T(a,b)$, $C(a,b)$, $H(a,b)$, $L(a,b)$ denote curves whose equations are, respectively, $y = a + \tan \pi(x - b)/2$ for $|x - b| < 1$, $y = a + 1/(\cos \pi(x - b)/2)$ for $|x - b| < 1$, $y = a + 1/(x - b)$ for $b < x < \infty$, and

Table 5.5.1 THE PATTERN TYPES OF FILAMENTARY LINE PATTERNS WITH SENSED LINES AS MOTIFS

Pattern type (1)	Symmetry group (2)	Induced group (3)	Motif transitive proper subgroups* (4)	Forbidden supergroups (5)
SLPF1	*s112*	*s111*	*c2*	*sm11*
SLPF2	*s1m1*	*s111*	*d1*	*sm11*
SLPF3	*smm2*	*sm11*	*d1, d2*	
SLPS1	*s1*	*s111*	*p111*	*s112, s1m1, sm11*
SLPS2	*sg*	*s111*	*p1a1*	*s112, s1m1, sm11*
SLPS4	*s1m*	*sm11*	*p111, p1a1, p1m1*	*smm2*
SLPS5	*sm1*	*s111*	*pm11*	*sm11*
SLPS6	*sm1*	*s1m1*	*p111, pm11*(2)	*smm2*
SLPS7	*s2*	*s111*	*p112*	*sm11*
SLPS8	*s2*	*s112*	*p111, p112*(2)	*smm2*
SLPS9	*smg*	*s111*	*pma2*	*sm11*
SLPS10	*smg*	*s112*	*pm11, pma2*	*smm2*
SLPS11	*smg*	*s1m1*	*p112, p1a1, pma2*	*smm2*
SLPS14	*smm*	*sm11*	*p112, pm11, pma2, pmm2*	
SLPS15	*smm*	*smm2*	*p111, p112*(2), *p1a1, p1m1, pm11*(2), *pma2*(3), *pmm2*(2)	

* Only finite and strip groups are listed.

$y = a + \ln(x - b)$ for $b < x < \infty$. Determine the symmetry group and the induced motif group for each of the following (non-discrete) patterns: $\{T(a,b) | \text{all real } a, \text{ and } b = 0\}$; $\{T(a,b) | \text{all real } a, \text{ and } b = 0, \pm 2, \pm 4, \pm 6, \ldots\}$; $\{C(a,b) | \text{all real } a, \text{ and } b = 0\}$; $\{C(a,b) | \text{all real } a, \text{ and } b = 0, \pm 2, \pm 4, \pm 6, \ldots\}$; $\{H(a,b) | \text{all real } a, \text{ and } b = 0\}$; $\{H(a,b) | \text{all real } a, \text{ and } b = a\}$; $\{L(a,b) | \text{all real } a, \text{ and } b = 0\}$; and $\{L(a,b) | a = 0, \text{ and all real } b\}$.

7. Determine all possible types of dimotif filamentary line patterns, that is, dimotif patterns with (unsensed) red and blue lines, such that color-preserving symmetries act transitively on the lines of each color.

*8. Interesting patterns arise if condition DP.2 is deleted from the definition of curve patterns; in order to avoid too great generality, we shall limit attention in this exercise to *continuous curve patterns* which admit among their symmetries a group *s111* which acts on different copies of the motif, and condition DP.2 is violated.
   (*a*) Show that there is one pattern type of continuous curve patterns with a straight line as motif, and that there are precisely five types of such patterns with sensed lines as motifs. Show that the latter pattern types determine four (doubly) continuous periodic symmetry groups, different from the groups we considered so far.
   (*b*) Explain why each of the patterns described in Exercise 6 is a continuous curve pattern, and show that each corresponds in a natural way either to one of the sensed lines, or to one of the pattern types with sensed lines as motifs, listed in Table 5.5.1.

(*c*) More generally, show that for each continuous curve pattern in which the motif is not a (sensed) straight line, the copies of the motif can be associated, in a natural way, with the points of either a single sensed line, or a pattern of sensed lines. This correspondence leads to a classification of such patterns.

(*d*) Show that no continuous curve pattern corresponds, in the classification described in part (*c*), to pattern types SLPF3, SLPS4, SLPS14 and SLPS15, but that all other types are obtainable for suitable choices of curves as motifs.

5.6 NOTES AND REFERENCES

The mathematical theory of patterns can be regarded as having started at the end of the nineteenth century with the enumeration of the crystallographic groups. It was followed by notable achievements in higher dimensions; see Section 1.6 for details and references. However, any exploration of other aspects of the general theory of patterns, even in the plane, seems to be completely lacking in the literature.

Many authors (see, for example, Schoenflies [1923, p. 216]) appear to have considered crystallographic groups to be so elegant and satisfactory that they abandoned (even if they ever started) any attempt to find other possible symmetry groups. Even if other groups (such as the group of symmetries of a circular disk, or of some straight-line patterns) are occasionally mentioned or investigated, they are either treated as unimportant and uninteresting exceptions which should be put aside as soon as possible, or else dealt with in such a superficial and imprecise manner that very few results of any value are obtained (Schubnikow [1929], Heesch [1930], Šubnikov [1951], Budden [1972, Chapter 26], Šubnikov & Kopcik [1972], Rosen [1975]). It seems to us that this is a very unsatisfactory approach to a genuine mathematical problem, and that a much more thorough investigation would be both justified and worthwhile.

Even if we agree to restrict attention to the crystallographic groups, the question immediately arises as to what restrictions we are going to impose on the patterns so as to ensure that they have the required sort of symmetry group. Again it is surprising that this never seems to have been done—and authors have been exceptionally vague about the sort of pattern which they are considering. It is for this reason that we introduce the concept of "discrete" patterns. The discrete patterns whose motifs are open disks seem to be analogous to normal tilings in many ways. The other kinds of patterns we consider also appear to be natural as well as mathematically interesting.

The vagueness just mentioned extends even to the very definition of "pattern" in the literature. In the works of some crystallographers, designers and art historians there are attempts at formulating a suitable definition, but always in a completely informal way that is irrelevant to subsequent developments (Buerger [1956, p. 3], [1971, p. 5], Loeb [1971a, p. 1], [1976, p. xvii], Wells [1956, Chapter 3], [1970, p. 170], Proctor [1969, p. 9], Christie [1929, p. 1]); many do not even try (Albarn et al. [1974], Horemis [1970], and many others). There seems to be not a single instance in the literature of a meaningful definition of "pattern" that is, in any sense, useful. Not only is this situation aesthetically objectionable, but it has led to many errors.

The classification of patterns given here arises in a very natural way, and it is surprising that no previous treatment along these lines seems to exist; brief accounts of the results of Sections 5.1 and 5.2 have appeared in Grünbaum & Shephard [1977e], [1979a]. Unfortunately, due to an error in the enumeration of motif-transitive subgroups, the refined pattern types PP48A and PP48B were presented as being of different pattern types, and it was asserted that there are 52 pattern types of discrete periodic patterns.

In Chapter 7 we shall show how the definition of pattern types described in this chapter fits into a general scheme of classification, which applies to many kinds of geometric objects.

The investigation of dot patterns started with Bravais [1850]; he proved that lattices can be classified into five types (see Exercise 5.3.10). To honor this achievement, crystallographers still use the name "Bravais lattices", in the plane as well as in three-dimensional space. Bravais' work was developed further by Sohncke [1874]. Although using quite different notions, Sohncke described the 31 refined pattern types of periodic dot patterns and illustrated all but the commonest ones (see Table 5.3.1). His main aim was the coarser division of periodic dot patterns into thirteen "systems", which correspond to the thirteen crystallographic groups that admit dot patterns. (Fedorov [1891] states that Sohncke "missed" the groups *p1*, *pg*, *pm* and *cm*. However, this is unjustified since Sohncke was listing dot patterns and not crystallographic groups.)

Crystallographers evinced an early interest in dot patterns, calling them "lattice complexes". At first, they mainly considered the analogous notions in the three-dimensional space (see, for example, Niggli [1919, p. 414], Hermann [1935, p. 419]). The 30 two-dimensional pattern types of dot patterns coincide with the 30 types of lattice complexes described by Smirnova & Potešnova [1966] and by Burzlaff, Fischer & Hellner [1968], and with the 30 "orbit types" of Matsumoto [1979] and Wondratschek [1979].

Using a definition of type of dot pattern equivalent to that of refined pattern type, Wollny [1969] determined the seven possible types of strip dot patterns and the 31 refined pattern types of periodic dot patterns. Dot patterns of a different nature have been defined by Meissner [1982]. These are based on Fibonacci sequences and lead to very attractive designs (which are not dot patterns in our sense).

Many kinds of "patterns" or designs could be considered besides those we have investigated so far. In Chapters 7 and 8 we shall deal with several such possibilities, but it is appropriate to mention here some kinds of designs (that are not patterns in the sense of Section 5.1) that have been used frequently in decorations and ornaments.

In Figures 5.6.1 to 5.6.4 we show examples of *patterns with overlapping motifs*; by this we mean designs that fail to be discrete patterns only because they do not satisfy condition P.1 which requires copies of the motif to be disjoint. Attractive examples of patterns with overlapping motifs, in which the motif is an equilateral triangle, appear in Loeb [1971b]. A classification of patterns $\mathscr{P}$ with overlapping motifs can be derived from that of discrete patterns in the following way. We first determine whether it is possible to find a dot pattern associated with $\mathscr{D}$ so that to different copies of the motif of $\mathscr{P}$ correspond different dots in $\mathscr{D}$. If $\mathscr{D}$ is such a dot pattern, we consider the pattern $\mathscr{P}_\lambda$ obtained by shrinking each copy M_i of the motif of $\mathscr{P}$ in the ratio $1:\lambda$ towards the point D_i of $\mathscr{D}$ associated with M_i. Then, for sufficiently small $\lambda > 0$, the pattern $\mathscr{P}_\lambda$ is discrete, and the pattern type of $\mathscr{P}_\lambda$ does not depend on the value of λ. It is therefore natural to assign to $\mathscr{P}$ the same pattern type as to all such $\mathscr{P}_\lambda$. If, on the other hand, no such dot pattern exists because several copies of the motif of $\mathscr{P}$ have coinciding centers of induced rotational symmetries, then, taking the union of all such copies of the motif of $\mathscr{P}$ as the motif of a new pattern $\mathscr{P}^*$, distinct dots correspond to distinct copies of the motif of $\mathscr{P}^*$ and we proceed as before.

As illustrations of these ideas consider the patterns in Figures 5.6.1 and 5.6.3. In each of these cases it is easily seen that the pattern is assigned to type PP51. On the other hand, in Figure 5.6.2 we have the second alternative, and to find the type of the pattern we consider as new motif the union of pairs of squares with the same center. The resulting pattern of eight-pointed star-shaped sets still has overlapping motifs, but by shrinking them, it can be seen to be of type PP41.

The classification just explained is not very satisfactory

for at least two reasons. First, although the patterns of Figures 5.6.1 and 5.6.3 are assigned to the same type, their appearance is sufficiently different to call for a finer classification. Second, in many cases it is possible to consider the same design in several different ways as a pattern with overlapping motifs. For example, Figure 5.6.2 can be interpreted as a pattern of eight-pointed stars (as described above) or of right-angled isosceles triangles (in which case it is of type PP37). In a similar manner, there are many possible motifs for the pattern of Figure 5.6.4; several are marked on the diagram. These lead to a number of different pattern types.

In view of the frequent occurrence of patterns of this kind in Islamic and Moorish art (see, among others, Bourgoin [1879], Critchlow [1976], El-Said & Parman [1976], Wade [1976]) as well as in oriental cultures (see, for example, Dye [1937]) a more satisfactory classification would appear very desirable.

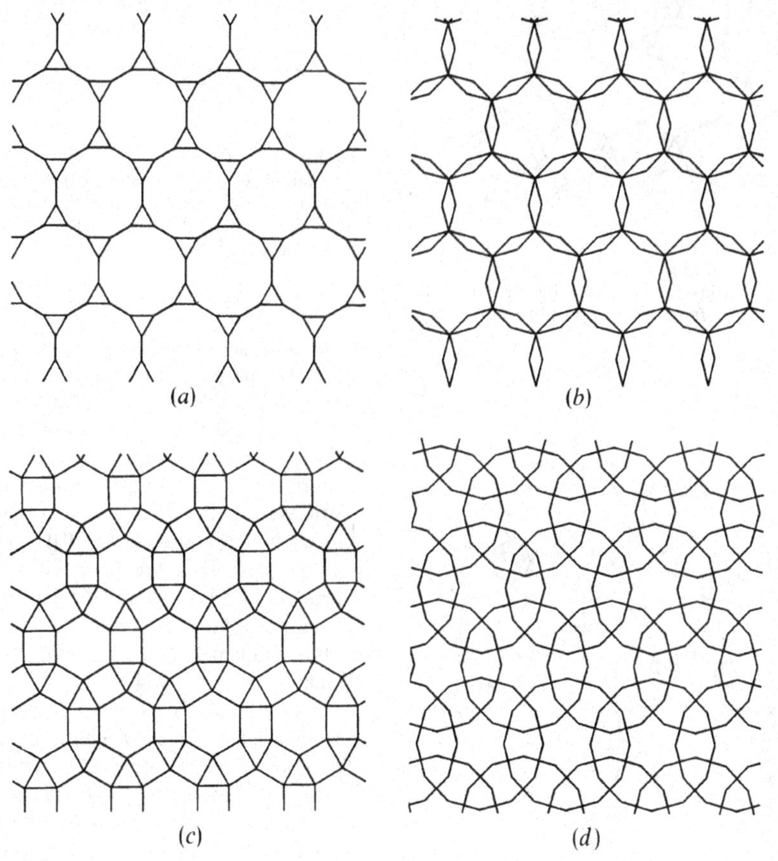

Figure 5.6.1
Examples of patterns with overlapping motifs, after Wade [1976]. In each case the motif is a regular dodecagon.

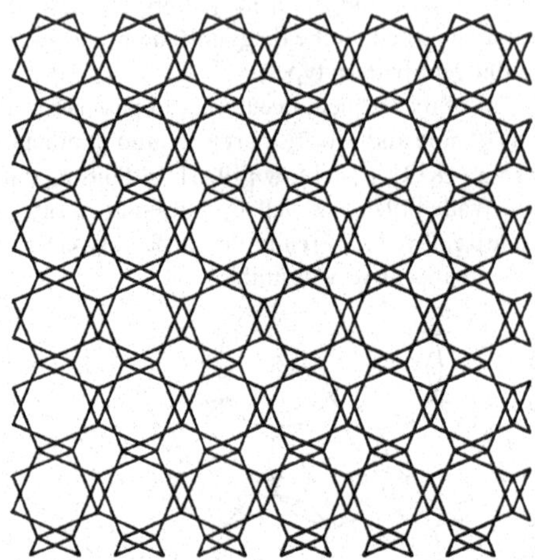

Figure 5.6.2
Another example of a pattern with overlapping motifs, after Wade [1976]. Here the motif is a square. For purposes of classification it is more convenient to interpret the pattern as having the union of two concentric squares as its motif.

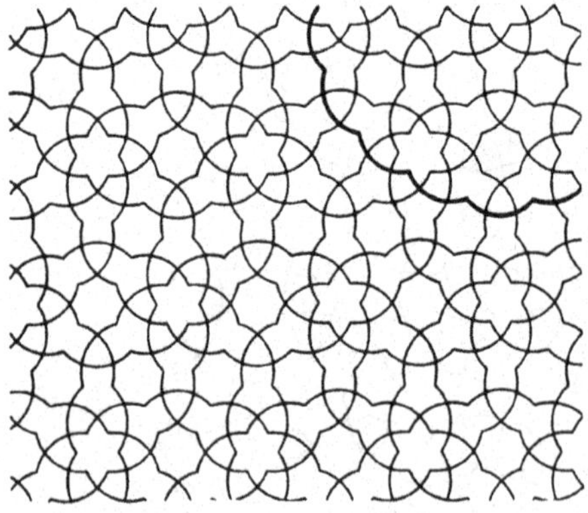

Figure 5.6.3
An example of a pattern with overlapping motifs from Bourgoin [1879]. The motif is a closed curve made up from 18 circular arcs of the same radius; part of one such curve is emphasized by heavy lines.

Figure 5.6.4
An example of a pattern which can be regarded as consisting of overlapping copies of a motif in many different ways. Seven possible motifs are indicated by heavy lines. This pattern is from Bourgoin [1879].

An analysis of the symmetry groups in Islamic geometric patterns, and the relative frequency of their occurrence can be found in Makovicky & Makovicky [1977]. Attractive illustrations of Islamic art are contained in the collection of slides and other teaching materials published under the auspices of the New York Metropolitan Musem of Art (Niman & Norman [1978], Norman & Stahl [1979]). Possible constructions for Islamic patterns are described by El-Said & Parman [1976], Wade [1976]. However, an interesting account of a method of construction actually used by Islamic craftsmen appears in Hankin [1925a]; see also Hankin [1925b], [1934].

Another kind of designs even more widely used are the *layered patterns*, in which it is convenient to think of the motif as cut out of paper or similar material of negligible thickness. The copies of the motif may "pass over and under" each other, so these patterns are not really subsets of the plane. As an illustration of this kind of pattern we show, in Figure 5.6.5, a layered pattern that is derived in an obvious way from that of Figure 5.6.4 interpreted as having a self-intersecting closed curve as motif. For further information on layered patterns see Grünbaum & Shephard [1983b].*

A special kind of layered pattern with unbounded strips as motifs is known as an *isonemal fabric*. These are mathematical models of woven fabrics; for details see Grünbaum & Shephard [1980a], [1985a], [1985b], [1985f].

In various publications, especially crystallographic ones, there is an appreciation of the idea that the choice of motif restricts the type of pattern that can be formed. Actually, the awareness of this dependence between motif and pattern appears in a somewhat vague form, since neither of the concepts is defined with any precision. For example, Šubnikov [1963b] correctly criticizes the (rather widespread) custom among crystallographers of using a point (or a small circular disk) as motif when illustrating the various groups. Šubnikov points out, in particular, that an attempt to illustrate the group *p1* in this manner shows, instead, *p2* or some other group; similarly, each representation of *pm* by a one-point motif has at least the symmetry *pmm*. This leads Šubnikov to insist on the use of an asymmetric triangle (or some other asymmetric motif) for such illustrations. This is, indeed, frequently done; examples may be found in Buerger [1956], Fejes Tóth [1965], Burckhardt [1966], Budden [1972] and many others; but it should also be pointed out that many authors show no recognizable consistency in the choice of their motifs (see, for example, Pólya [1924], Speiser [1927], Bradley [1933], Cadwell [1966], Loeb [1971a], O'Daffer & Clemens [1976]). The problem of deciding what types of patterns can be obtained with a given motif seems to have been treated only in the case of circles (Sinogowitz [1939]) and ellipses (Nowacki [1948]); however, since these authors understood "type" in a sense different from that used here, we delay the discussion of their results until Chapter 7.

Figure 5.6.5
An example of a layered pattern, after Wade [1976].

The "sets of equivalent point positions" with respect to some group of symmetries (which we have mentioned in Exercise 5.3.11) have been used by crystallographers in the solution of various problems concerning patterns. In fact, in the case of reasonably simple motifs (such as circular disks) their informal approach is essentially equivalent to the use of Statement 5.3.2 (see, for example, Niggli [1927]). With this and other applications in mind, lists of "sets of equivalent point positions" have been prepared (see Henry & Lonsdale [1965, pp. 55–72]; other authors (such as Sinogowitz [1939], Nowacki [1948]) appear to have used them as well, without publishing their classifications. The "sets of equivalent point positions" essentially coincide with the pattern types we introduced in Section 5.2; however, their utilization seems to have required a well-developed intuitive feeling for patterns, since no definitions of "type of pattern" were available.

Surveys of early work on Dirichlet tilings can be found in Laves [1931a] and Nowacki [1933], [1935]. Dirichlet tilings are used and discussed widely in scientific literature, where Dirichlet domains are also variously known as Voronoi polygons, Thiessen polygons, Wigner-Seitz cells, or *Wirkungsbereiche*. They have mathematical applications in packing and covering problems (Rogers [1964]), spatial data analysis (Sibson [1980]) and interpolation (Lawson [1977]). Applications to geography and meteorology have been described by Rhynsburger [1973] and Brassel & Rief

* See the Appendix beginning on page 653.

[1979]. Gilbert [1962] mentions them in connection with crystal growth; they have also been used by physicists, biochemists, material scientists, physical chemists, sociologists, and in the theory of communications and computer graphics. It would lead us too far to give references to all these applications; the interested reader can find them through the references in the literature already quoted, and that we shall mention below. Because of Dirichlet tilings' many applications, efforts have been made to find efficient algorithms for computing the Dirichlet tilings obtained from a given set of points, a family of sets, or even a set of "weighted points". See, for example, Ash & Bolker [1984], Avis & Bhattacharya [1982], Lee [1980], Lee & Drysdale [1981]. Loeb [1976, p. 119–120] asserts that a triangulation of the plane, which is the Dirichlet tiling of a dot pattern, has to be 6-valent and the angles of the triangles cannot exceed 90°. Similar claims have been made in Loeb [1978] and Gasson [1983, p. 227]. The examples constructed in Exercise 5.4.4 show that these assertions are erroneous.

Other investigations concern the determination of statistical data about the polygons that arise in Dirichlet tilings obtained from random distributions of points; see, for example, Crain [1972], Lantuejoul [1978], Santaló [1980]. The literature is too extensive to review here, and much of it relates to three-dimensional space or to systems which are not stationary but evolve as in some cell-division processes. The reader interested in these topics should refer to texts and expository papers on geometric probability, such as Moran [1966], [1969], Little [1974], and Baddeley [1977], or to the relevant section of *Mathematical Reviews.*

Associated with a discrete set of points in the plane is a "Delaunay triangulation". It is a dual of the Dirichlet tiling, with vertices at the given points and edges joining two points if and only if the corresponding Dirichlet domains are adjacent. (Clearly, this yields a triangulation only if no four Dirichlet domains meet at a point—if the latter occurs, then a special procedure is needed to obtain a triangulation; see Sibson [1978].) The computing algorithms mentioned above usually aim at identifying the edges of the Delaunay triangulations. This is the so-called "nearest neighbor problem" (Knuth [1973]). As interesting result of Sibson [1978] shows that, for a prescribed set of points as vertices, the Delaunay triangulation is the "nearest" to being equiangular in the sense that the angles of the triangles differ from 60° by the smallest possible amount.

The seven continuous periodic groups we discussed in Section 5.5 (or groups isomorphic to them but defined in slightly different ways) have been mentioned in the literature; see, for example, Heesch [1930], Šubnikov & Kopcik [1972, pp. 185–187 of the English translation]. However, most earlier authors seem not to have considered the possibility of such groups occurring as the symmetry groups of patterns (such as those discussed in Exercise 5.5.6). In fact, although Heesch was aware of the existence of these groups, some passages in his paper [1930, in particular p. 347] seem to be contradicted by such patterns.

The groups of the similarity patterns as defined in Exercise 5.1.14 have been considered by Šubnikov [1960], Galyarskii & Zamorzaev [1963] and Galyarskii [1967], [1974a], [1974b], but the various possibilities of affine and of similarity patterns clearly deserve a more detailed investigation.

CLASSIFICATIONS OF TILINGS WITH TRANSITIVITY PROPERTIES

Figure 6.0.1
An isohedral tiling in which the tiles resemble human figures! The drawing is based on a sketch by M. C. Escher (see Escher [1971, Number 98]). This tiling is of isohedral type IH16 in the classification described in Section 6.2—see Figure 6.2.4 for examples of all 81 types of isohedral tilings.

CLASSIFICATIONS OF TILINGS WITH TRANSITIVITY PROPERTIES

This chapter is devoted to a discussion of isohedral, isogonal and isotoxal tilings. We have already met two methods of classification of such tilings. In Chapter 4 we classified them according to their topological type, showed that the number of such types is quite small, and found that each type can be conveniently represented by a short numerical symbol. Using a completely different approach we explained (in Section 5.4) how each such tiling can be regarded as a pattern and so allocated to one of the 51 different pattern types.

At first sight it would appear, therefore, that it would be satisfactory to base a classification of these tilings upon the idea that two tilings are of different types if they differ in their topological types or in their pattern types. In fact this is not sufficient. For example, the three tilings of Figure 6.0.2 have topological type $[4^4]$ and pattern type PP9, yet—at least in an intuitive sense—they are "really different" and so should be distinguished in some way. (For example, it will be observed that tiles of the same orientation are arranged differently in each of the tilings.) A closer examination of the diagrams reveals that the "difference" depends in an essential way upon the manner that each tile is related to each of its four adjacents. In every case the isometry that maps a tile

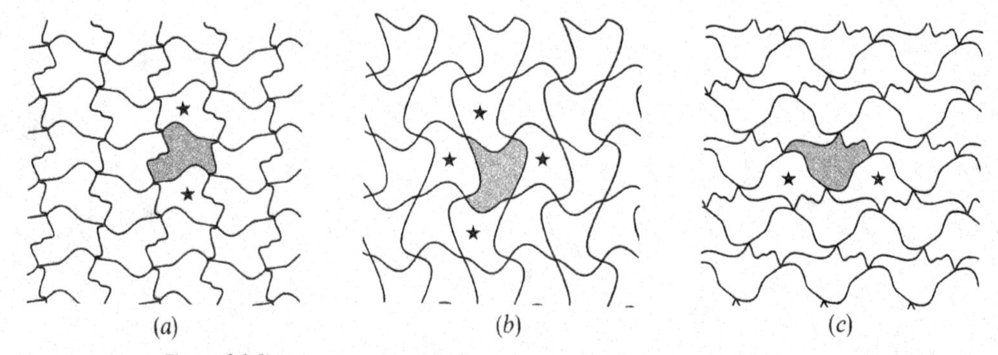

Figure 6.0.2
Three isohedral tilings of the same topological type $[4^4]$ and of the same pattern type PP9 that are, nevertheless, essentially different. One way to see this is to consider which adjacents of a tile are related to it by glide-reflections. (For the gray tile in each tiling these adjacents are indicated by asterisks.)

onto one of its adjacents is either a rotation through angle π about the midpoint of an edge (a central reflection) or a glide-reflection. In the tiling of Figure 6.0.2(*b*) the four adjacents of T are related to T by glide-reflections, whereas for the tilings of Figures 6.0.2(*a*) and (*c*) two glide-reflections and two central reflections are required. The latter two tilings are distinguished by the different arrangements of the two types of isometries.

Our first task, which will occupy Sections 6.1 and 6.2, will be to show how, in the case of isohedral tilings, we can encode and record this relationship between a tile and its adjacents in the form of a convenient symbol called an "incidence symbol". We shall say that two isohedral tilings are of the same type if their incidence symbols differ in a "trivial" way. The following two sections will extend this idea to a similar classification of isogonal and of isotoxal tilings.

Our classification, using incidence symbols, makes precise the classification system which various authors have previously attempted to formulate without full success. Moreover, it has the advantage that, using the techniques to be described, a brief examination of a given tiling enables one to identify its type quickly and without ambiguity.

One aspect of the classification deserves special emphasis here, since it deals not with technical details but with a basic point of view. It is the fact that we directly classify *marked tilings* only, that is, tilings in which each tile carries a copy of a *motif*. The classification of (unmarked) tilings is then obtained by the stratagem of considering the interiors of the tiles (or some other elements of the tiling) as markings on the tiles. The surprising consequences of this distinction between tilings and marked tilings will be given in detail for each of the classifications investigated in the first four sections; the pitfalls into which previous writers on these questions have stumbled by ignoring this distinction will be discussed in Section 6.6.

Two objections may be raised to our classification procedure. The first is that it may seem rather elaborate. But the basic idea is very simple (and fruitfully applicable in many other cases), and the complications are only of a superficial nature. In any case, considerable detail is to be expected in any system of classification which, like ours, simultaneously refines all classification schemes proposed earlier.

The second objection is somewhat more serious. It may appear from what we have said so far that every time we encounter a classification problem, we invent an *ad hoc* solution that applies only to the particular case under consideration, and is devised only as a matter of convenience. In Chapter 7 we shall meet this objection by showing that all our classification methods (of both tilings and patterns) depend on one fundamental and universal idea. We shall consider "objects" in the plane and state exactly what we mean when we say that two such objects are of the same *type* with respect to symmetries. It turns out that this idea has many applications and we illustrate some of them in the next three chapters. We shall classify patterns with circular disks, elliptical disks, or line-segments as motifs in Chapter 7, tilings with colored tiles in Chapter 8, and tilings by polygons in Chapter 9.

We do not proceed immediately to a discussion of this general approach since the use of incidence symbols, to be described in this chapter, has an extremely important advantage—in addition to providing us with a *classification* of the isohedral, isogonal and isotoxal tilings it also enables us to *enumerate* the different types that exist. Besides being of interest in its own right, this also leads to a geometric proof that the enumeration of the crystallographic groups given in Section 1.4 is complete.

6.1 INCIDENCE SYMBOLS AND ADJACENCY DIAGRAMS

In order to introduce the idea of incidence symbols we shall consider, in detail, a special case that exhibits most of the features one meets in a general treatment of isohedral tilings. The example we have chosen is that of marked isohedral tilings $\mathcal{T}'$ that can be obtained by adding markings (copies of a motif) to the regular tiling $\mathcal{T}$ by squares $[4^4]$. The symmetry group $S(\mathcal{T})$ of $\mathcal{T}$ is *p4m*, and its elements are shown in the group diagram in Figure 1.4.2. Both the induced tile group $S(\mathcal{T}|T)$ and the symmetry group $S(T)$ are $d4$, and the associated periodic pattern is of type PP41.

When we introduce markings on the tiles of $\mathcal{T}$ the groups $S(\mathcal{T})$, $S(\mathcal{T} \mid T)$ and $S(T)$ will tend to be decreased, since isometries that are symmetries of $\mathcal{T}$ may fail to be symmetries of the marked tiling $\mathcal{T}'$. In other words, the groups for the marked tilings are subgroups of the corresponding groups for the *underlying* (unmarked) tiling $\mathcal{T}$.

Since we are concerned with classification under symmetry, only the effect that the introduction of markings has on the groups is important, and not the actual form or shape of the markings themselves. We therefore choose to mark the tiles in a very simple manner, which has the desired effect. We use as motifs L-shaped markings or their combinations, with uprights and bases parallel to the sides of the square tiles; these motifs are similar to the ones used for patterns in Chapter 5. The eight possibilities corresponding to different tile groups $S(T)$ are indicated in Figure 6.1.1. It would clearly make no difference if, for example, in case of the group $c2$ the marking were replaced by its mirror image; both correspond to central reflection in the center of the tile. However, we must be careful to distinguish the two cases of $d2$ and the two cases of $d1$ that arise. These differ essentially because in one case the reflections are in lines parallel to the sides of the squares, whereas in the other the lines of reflection are parallel to the diagonals. The cases are notationally distinguished by adding to the group symbol either s for "short" or l for "long". It is easy to verify that Figure 6.1.1 illustrates all possibilities.

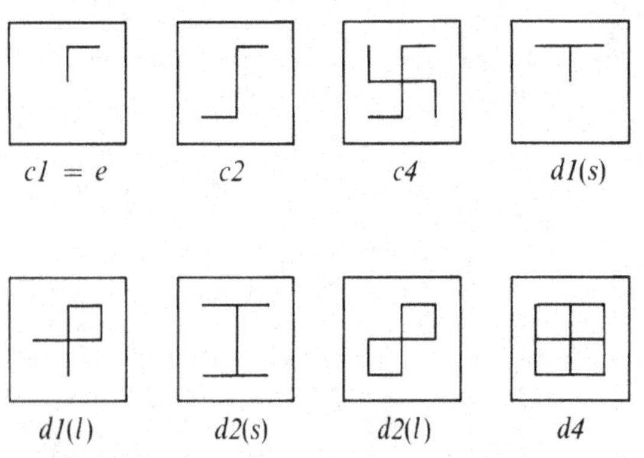

Figure 6.1.1
Examples of the various markings that may be added to tiles of the regular tiling $[4^4]$ to transform it into a marked isohedral tiling. Under each of these marked tiles T we have indicated its symmetry group $S(T)$.

Now let us assume that all the tiles of $\mathcal{T}$ are marked in such a way that the resulting marked tiling $\mathcal{T}'$ is isohedral, and that the induced group $S(\mathcal{T}' \mid T)$ of each marked tile T coincides with its symmetry group $S(T)$. In other words, we suppose that for any two tiles of $\mathcal{T}'$, there exists a symmetry of $\mathcal{T}'$ that maps the first of the marked tiles onto the second, and that for any marked tile T the symmetries of T are in a one-to-one correspondence with those symmetries of $\mathcal{T}'$ that map T onto itself. It would be reasonably easy to find all possible marked tilings $\mathcal{T}'$ by trial and error, but the process is greatly simplified by the introduction of symbols in the following manner (see Figures 6.1.2 and 6.1.3 for illustrative examples).

The first step is the labelling of all the edges of all the tiles of $\mathcal{T}'$. To do this, we assign a letter, say a, to any directed (oriented) edge of any one of the tiles. This is conveniently indicated by a small arrow, labelled with the letter a, placed near to and parallel to the chosen edge. (We recall that each edge of the tiling coincides with edges of two tiles, one lying on each side of it. Hence each edge of the tiling will ultimately carry two labels.) Now apply all the operations of the symmetry group $S(\mathcal{T}')$. The effect will be to carry the letter a onto the other tiles, and since $\mathcal{T}'$ is isohedral, the letter a will be associated with at least one oriented edge of every tile in $\mathcal{T}'$. It may happen that a is assigned to two or more oriented edges of the same (and therefore of each) tile. It may also happen that the same letter is assigned to the same edge of the same tile, but with reversed orientation; this occurs if the reflection in a line joining the midpoints of opposite sides of a square is a symmetry of $\mathcal{T}'$. In this case we shall say that a is assigned to an *unoriented* edge of the tile and in the figures we replace the small arrow by a doubled-headed arrow.

Possibly all the edges of all tiles will be labelled a; if this is the case then our labelling is finished. If not, we

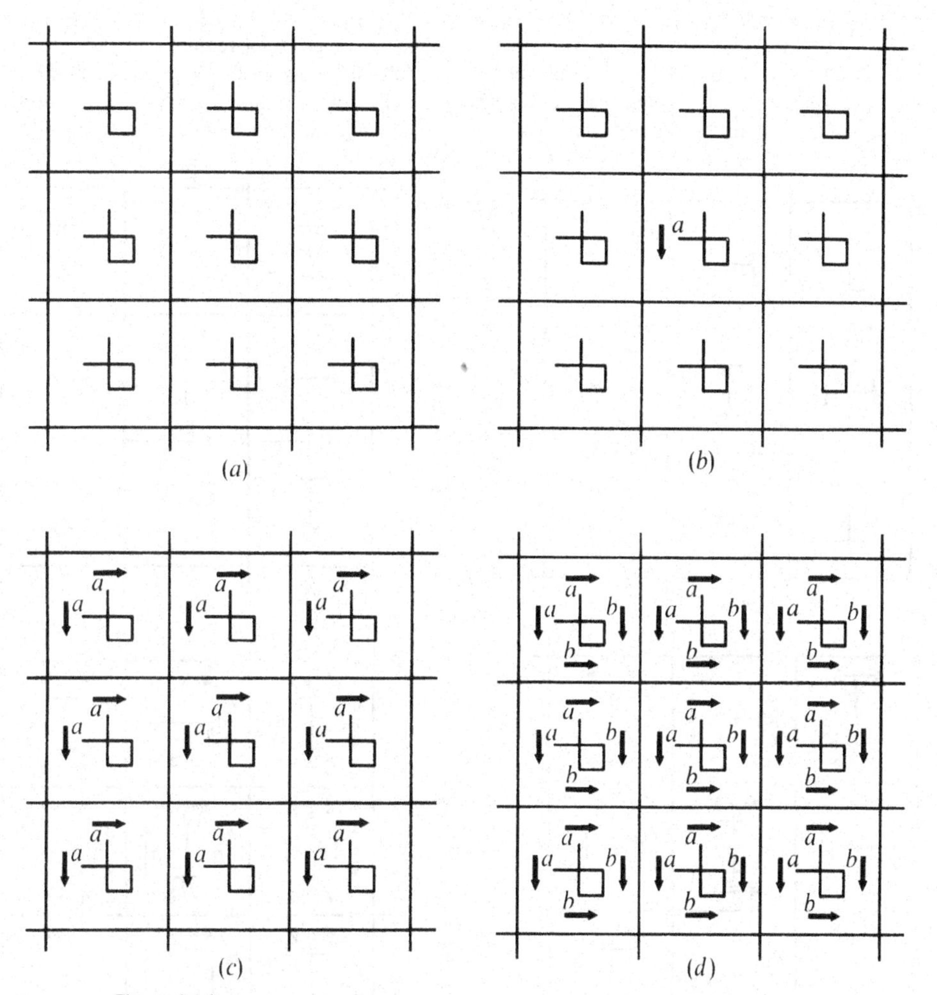

Figure 6.1.2
An illustration of the labelling process for the edges of tiles in a marked isohedral tiling $\mathscr{T}'$ by squares. The tiling is shown in (*a*) and it can be verified that $S(\mathscr{T}'|T) = S(T) = d1$ for each marked tile T. In (*b*) one arbitrarily chosen oriented edge of a tile is labelled with the letter *a*. In (*c*) we show how the symmetries of $\mathscr{T}'$ carry this label to all tiles of $\mathscr{T}'$. Repeating the procedure with a different label *b* assigned to one of the edges unlabelled in (*c*) leads to the labelling shown in (*d*). Since there are now no unlabelled edges, the process is complete.

choose an edge of a tile which is not already labelled, orient it and assign another letter, say b, to it. Again we apply the operations of $S(\mathscr{T}')$, and then at least one oriented edge of every tile will be labelled b. We continue in this way, and since squares have four sides it will never be necessary to use more than four letters; hence the labelling comes to an end after at most four steps. The examples in Figures 6.1.2 and 6.1.3 use two letters; those in Figure 6.1.4 require one and four letters.

After labelling all the edges in the manner just described we obtain a *tile symbol* by reading the letters on the edges as we go around a tile, and adding a super-

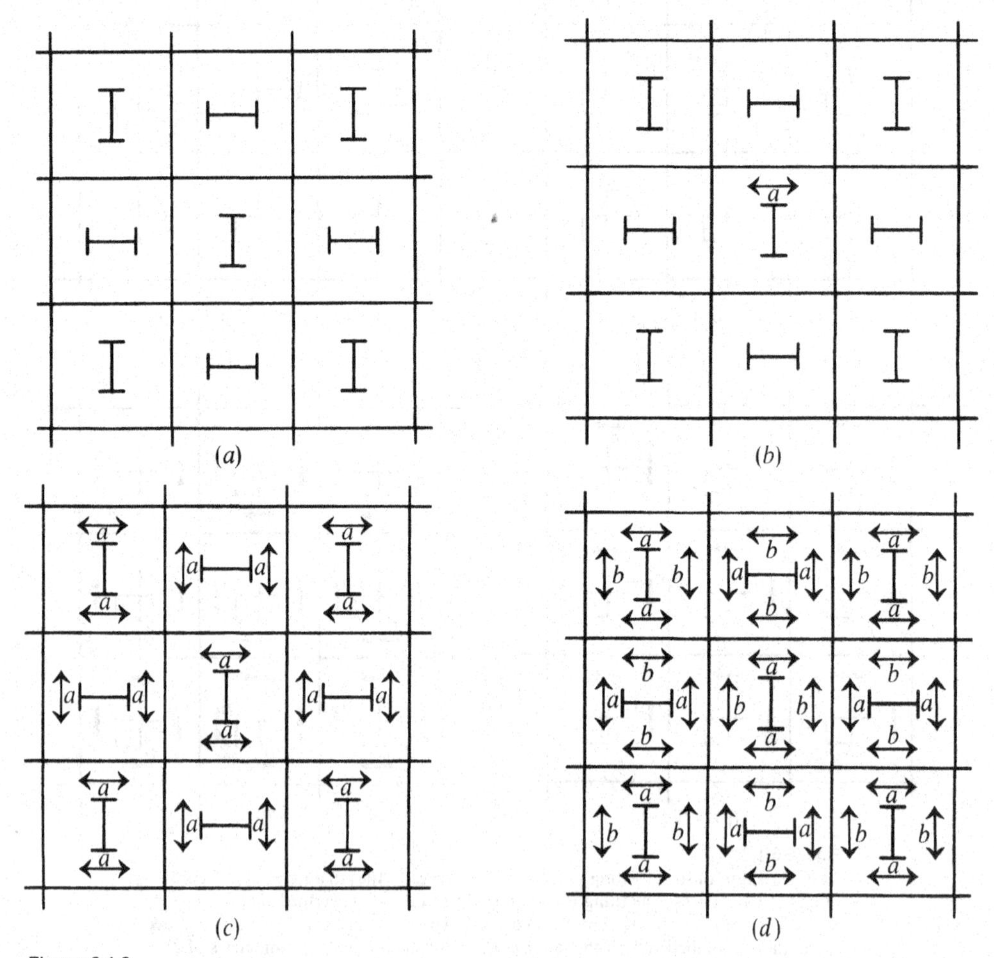

Figure 6.1.3
Another illustration of the labelling process. The steps are the same as in Figure 6.1.2 and again only two labels a and b are needed.

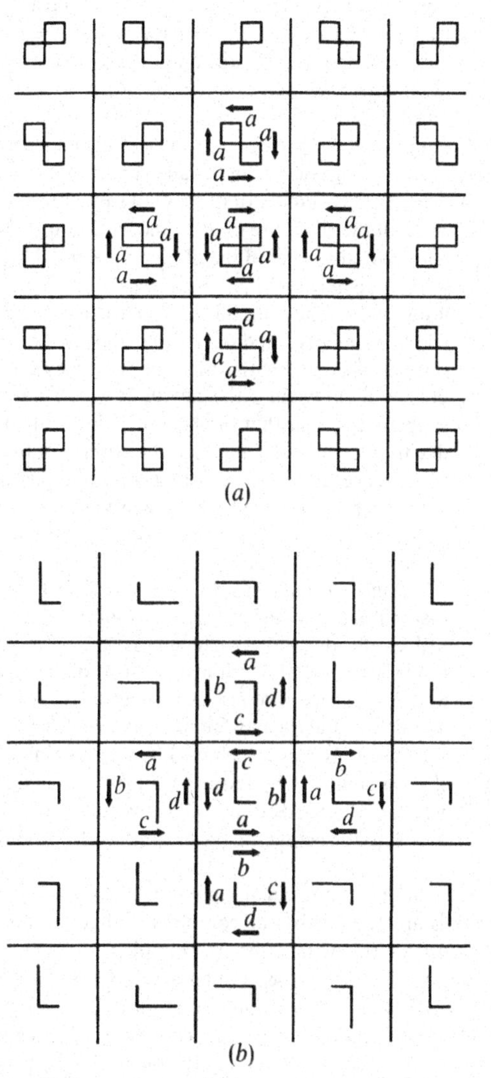

Figure 6.1.4
Two more examples showing the labelling of the edges in an isohedral tiling. In (*a*) only one label is used, but in (*b*) four are needed.

script + or − to indicate whether the corresponding edge is coherently or oppositely oriented. (That is to say, whether the arrow associated with the letter points in the direction in which we are going around the tile, or in the opposite direction.) If the edge is unoriented then no superscript is used. Thus a tile symbol of the tiling in Figure 6.1.2 is $a^+b^+b^-a^-$, for the tiling in Figure 6.1.3 it is *abab*, and for those in Figure 6.1.4 the symbols are $a^+a^-a^+a^-$ and $a^+b^+c^+d^+$. A tile symbol clearly indicates the induced tile group $S(\mathcal{T}\,|\,T)$. Thus in Figure 6.1.4(*a*), the fact that the tile symbol is $a^+a^-a^+a^-$ tells us that the induced tile group contains operations which map every edge of a tile onto every other edge, with the indicated orientations. Hence, in this case, the group must be $d2(l)$. For the tiling of Figure 6.1.4(*b*) where every edge bears a different letter, $S(\mathcal{T}\,|\,T) = e$. When two letters occur in a tile symbol then the induced tile group must be $c2$, $d1(l)$ or $d2(s)$.

The labelling of the edges of the tiles also provides a simple way of recording how a tile is related to its adjacents. For this purpose we introduce an *adjacency symbol*. Suppose that a tile symbol L has been chosen and that the first letter (edge symbol) that occurs in L is w (where w represents a, b, c or d). Then on an adjacent tile the corresponding edge (that is to say, the edge that abuts against an edge of T that is labelled w) will bear a letter, say x, where x is also one of a, b, c or d. The first component of the adjacency symbol is defined to be the letter x *either* with the superscript + or − to denote whether the edge x is oriented oppositely to or in the same direction as the edge w, *or* with no superscript if the edge w (and therefore also x) is unoriented. Now take the second letter, distinct from w, that occurs in the tile symbol L, say y, and if the corresponding edge of an adjacent tile is z, then the second component of the adjacency symbol is z, either with or without a superscript—determined in the same manner as described above for the first component x. The third component of the adjacency symbol is defined in the same way using the third distinct letter (if any) that occurs in L, and so on. We proceed until all the distinct letters in L have been used up. Thus an adjacency symbol is a sequence of letters, with or without superscripts, whose length is equal to the number of distinct letters which appear in the tile symbol.

For example, if we choose the tile symbol $a^+b^+b^-a^-$ for the tiling of Figure 6.1.2, then each edge of a tile which is labelled with the first letter a of the tile symbol abuts an edge on the edge labelled b on the adjacent tile; since the orientations of a and b are the same, the first component of the adjacency symbol will be b^-. Similarly, the second component turns out to be a^-, and so the adjacency symbol is b^-a^-. For the tiling in Figure 6.1.3 the adjacency symbol is ba, and for the tilings in Figure 6.1.4 the symbols are a^- and $b^-a^-c^+d^+$.

We have remarked how a tile symbol indicates the group $S(\mathscr{T}|T)$ and it is evident that the adjacency symbol tells us how each tile is related to its adjacents. Since the whole tiling is determined by this information, we put the symbols together in the form $[L; A]$, where L is a tile symbol and A is the corresponding adjacency symbol. This combination $[L; A]$ is called an *incidence symbol* of the tiling.

Of course, an incidence symbol $[L; A]$ depends on the original choice of letters and orientations of edges of the tiles, as well as on the choice of tile symbol. If two incidence symbols can be made identical by reallocation of letters and orientations, then we shall say that they differ *trivially*. This enables us to formalize the definition of a type of tiling: two isohedral tilings are of the *same type* if their incidence symbols $[L; A]$ differ trivially.

The use of incidence symbols provides us with an algorithmic procedure for determining all types of tilings in cases such as the isohedral tilings by marked squares. We first determine all possible induced tile groups $S(\mathscr{T}|T)$ (eight in this case) and write down corresponding tile symbols. We then write down all possible sequences of letters with or without superscripts, as potential adjacency symbols. Finally, we reject those sequences which, for one reason or another, are not adjacency symbols of tilings. The reasons for rejection are of three kinds:

(*a*) *The transposition condition.* If an edge labelled a abuts an edge labelled b, with $b \neq a$, then b must abut a, and the corresponding letters in the adjacency symbol will be b^+a^+ or b^-a^- or ba. We shall say, under these circumstances, that a and b are transposed. Clearly the adjacency symbol must consist of transposed pairs and of single letters, the latter corresponding to those edges that abut edges labelled with the same letter. For example, in Figure 6.1.4(b) the pair a, b is transposed while c and d occur as single letters in the adjacency symbol.

(*b*) *The fitting condition.* In the case of square tiles, since all the edges are of the same length, any edge may abut any other in the tiling. In general, however, there may be some restrictions. For example, if we were considering tilings by marked rectangles and the symbols a and b had been allocated to sides of different lengths, then the adjacency symbol could not begin with ba (with or without superscripts). It is clear that similar restrictions can arise due to the size of the angles at the corners of the tiles. Hence, although the fitting condition places no restriction on the adjacency symbols in the case of square tiles, in many other cases it is very important.

(*c*) *The combinatorial condition.* We have already remarked that if we arbitrarily pick four adjacents to a given tile, then it may happen that the tiling cannot be continued isohedrally. In terms of the symbols, a sequence of letters with superscripts that satisfies (a) and (b) may fail to be an adjacency symbol. The only way which we know to eliminate this possibility is by trying to construct a tiling which corresponds to the sequence under consideration, and rejecting it if no such tiling can be found.

For marked squares, many potential adjacency symbols are rejected by the combinatorial condition. We illustrate this with a simple example.

Suppose we ask whether a tiling of type $[a^+b^+c^+d^+; b^+a^+c^-d^+]$ exists. We begin (see Figure 6.1.5(a)) by labelling the edges of a tile T in accordance with the tile symbol $a^+b^+c^+d^+$, and marking the tile in some manner appropriate to the induced group $c1$ that is implied by this tile symbol. The four adjacents are then determined by the adjacency symbol $b^+a^+c^-d^+$ (see Figure 6.1.5(b)) and the next stage is to consider whether we can continue the tiling, in a manner consistent with the given symbols, to all the neighbors of T. In this particular case the process fails since it is clear that no tile can fit into the corner marked $\star$. We deduce that $b^+a^+c^-d^+$ is not an adjacency symbol corresponding to the tile symbol $a^+b^+c^+d^+$. In fact the situation we have just described is typical on account of the following result:

6.1.1 *If it is possible to surround completely a (marked) tile T_0 by copies of T_0 in a way that is consistent with a given incidence symbol, then this same incidence symbol corresponds to an isohedral tiling of the plane with T_0 as prototile.*

By "completely surround" T_0 we mean, of course, construct a "ring" R_1 of tiles around T_0 as described in Section 4.3. Since an incidence symbol specifies the adjacents (and therefore the neighbors) of each tile uniquely, it is clear that the ring R_1 can be surrounded by tiles (in a way consistent with the given incidence symbol) to form a ring R_2. Similarly, R_2 can be surrounded to form a ring R_3, and so on. Since the tiling is isohedral it is therefore normal and so the union of these rings covers the plane, thus leading to the required tiling.

Carrying out the procedure described above in a systematic manner we find:

6.1.2 *There exist precisely 36 different types of marked isohedral tilings with underlying tiling* $[4^4]$.

The incidence symbols of these types are displayed in Figure 6.1.6 together with their "adjacency diagrams", which will be described shortly.

The above description of the classification method seems to indicate that the number of potential incidence symbols that need to be examined in the proof of State-

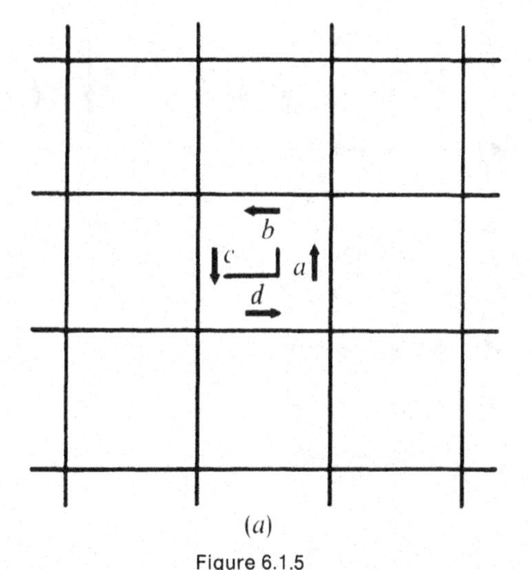

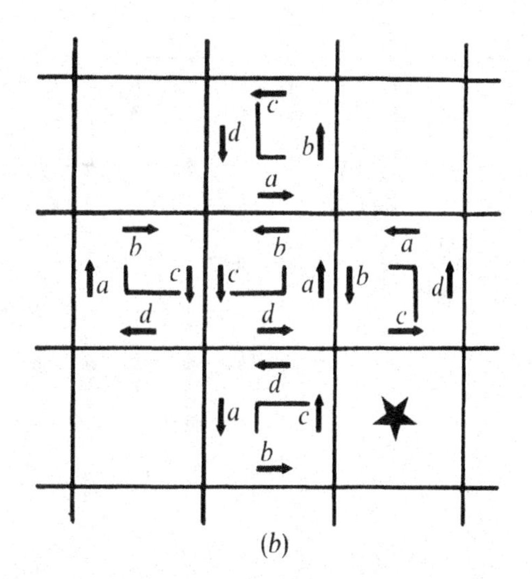

Figure 6.1.5
An illustration of the procedure for checking whether a proposed incidence symbol corresponds to a marked isohedral tiling by squares. The symbol investigated here is $[a^+b^+c^+d^+; b^+a^+c^-d^+]$; it does not correspond to a tiling.

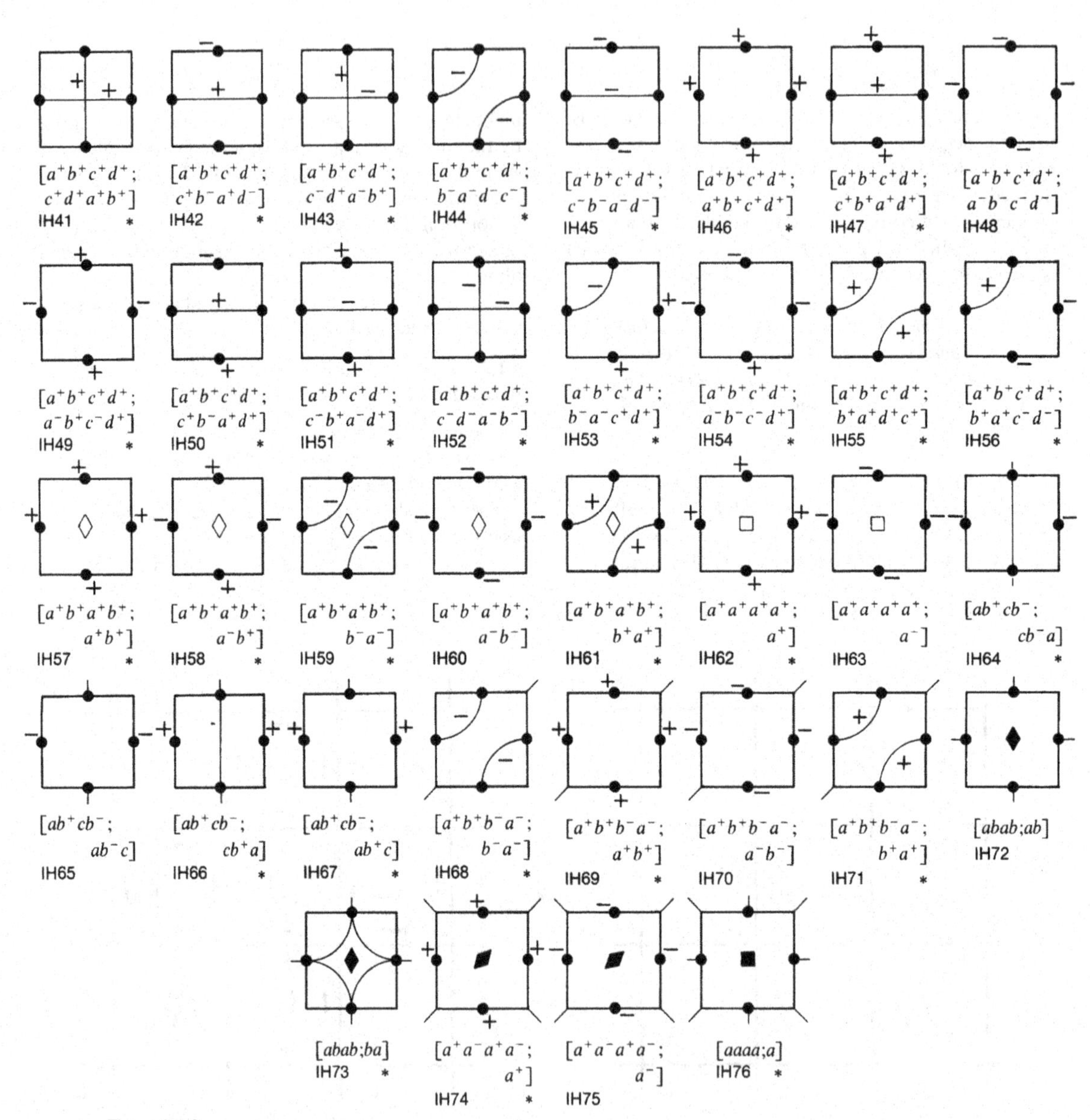

Figure 6.1.6
Adjacency diagrams and incidence symbols of the 36 types of isohedral tilings by marked squares. Near each adjacency diagram we give the list number in the complete enumeration of isohedral types; see Table 6.2.1. Asterisks indicate diagrams that correspond to unmarked topologically square tilings.

ment 6.1.2 is terrifyingly large. In fact, the three conditions stated above eliminate vast numbers with very little effort, and it is a simple matter to arrive at the enumeration in Figure 6.1.6.

In practice, while carrying out an enumeration of this nature, the main difficulty seems to be that of deciding when two different incidence symbols correspond to the same type of tiling (that is, differ trivially). In the case of marked squares it is not too hard to do this—but confusion is possible, and we have found it convenient and time-saving to use *adjacency diagrams*, which are based on an idea of Delone [1959].

The adjacency diagram consists of a drawing of a tile T with each edge marked by a central dot ●. If an oriented edge of T labelled w abuts an edge of an adjacent tile T' labelled with the same letter w then we mark the dot with + or − depending upon whether T' is the image of T under a half-turn about the dot, or under reflection in the corresponding edge. If w is an unoriented edge of T then the dot is left unmarked (see Figure 6.1.7(*a*)). If an oriented edge w of T abuts an edge of an adjacent tile T' bearing a different letter x, then the dots on the edges of T labelled w and x are joined by an arc. This arc is labelled + if T' is the image of T under a translation or rotation, and is marked − if T' is the image of T under a glide-reflection. If w is an unoriented edge, then x must also be unoriented and the arc is left unmarked (see Figure 6.1.7(*b*)).

Using the above rules it is easy to construct an adjacency diagram for a given tiling $\mathscr{T}$ by inspection of a drawing of $\mathscr{T}$. However, it is also easy to do this from an incidence symbol. We proceed as follows. First we draw the tile T with dots on its sides and label its edges in accordance with the given tile symbol. If the letter w in the adjacency symbol corresponds to the same letter w in the tile symbol (which means that two edges bearing the same letter abut) then the dot on the edge w is marked + or − or left unmarked according to the superscript attached to w in the adjacency symbol. If the letter x in the adjacency symbol corresponds to a different letter w in the tile symbol then the dots on the edges of T labelled w and x are joined by an arc marked + or − or left unmarked. The appropriate

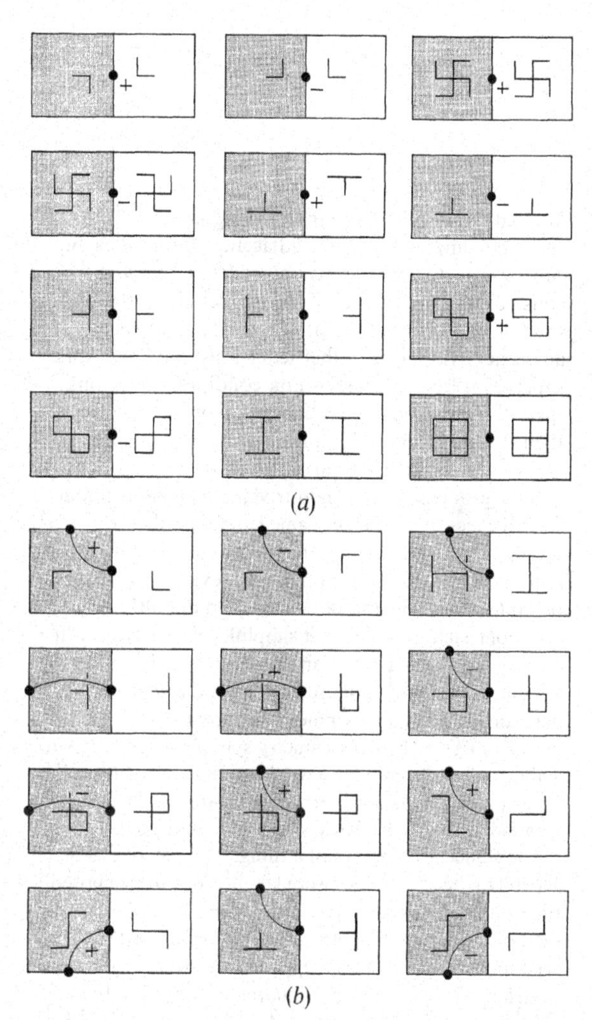

Figure 6.1.7
Some examples showing how the relationship between adjacent tiles is indicated on an adjacency diagram. In each case the left (gray) tile T is marked to record the position of the right tile T' relative to it. In (*a*) the mark on the adjacency diagram is a dot placed centrally on an edge; in (*b*) it consists of two dots on edges joined by an arc. A symbol + or − (or no symbol) is added to the arc as shown and as explained in the text. In the second and third rows of (*b*) it will be seen that where $S(\mathscr{T}|T)$ is not trivial there may be more than one way to indicate the relationship between T and T'. An adjacency diagram consists of a drawing of T with all the marks (including alternatives) that indicate the relationship between T and its adjacents. Sometimes it is permissible to omit some of the marks without losing essential information, and in such cases we shall usually simplify as much as possible.

alternative is decided in the following manner. If the corresponding letters in the adjacency symbol bear no superscripts, then the arc is unlabelled. Otherwise we consider the superscripts of the two letters in the tile symbol, and the superscript of one of them in the adjacency symbol (the other letter must have the same superscript by the transposition condition). If among these three signs there are either one or three signs $-$, then the arc is labelled $-$; if there are one or three signs $+$, then the arc is labelled $+$.

If the group $S(\mathscr{T} \mid T)$ is non-trivial, it can happen that several edges of T bear the same letter, and then we should, in accordance with the procedure just described, join the dot on one edge to those on several other edges. It is often convenient to omit some (but not all!) of the arcs from each dot. In fact, it simplifies the diagram without losing any essential information if we include only the minimum number of arcs such that the adjacency diagram has the same symmetry group as $S(\mathscr{T} \mid T)$, and indicate $S(\mathscr{T} \mid T)$ by the symmetry symbols introduced in Table 1.4.1. In Figure 6.1.6 we show all the possible adjacency diagrams for tilings by marked squares; these correspond to the 36 incidence symbols as indicated.

The adjacency diagram of a tiling does not, of course, depend on how the edges were labelled or on the chosen tile symbol. For this reason it provides a very easy method of deciding whether two tilings are of the same type. For example, let us suppose that we are given an isohedral tiling by marked squares from which, by assigning letters to the oriented edges of the tiles, we deduce the tile symbol $a^- d^+ b^- c^+$ and the adjacency symbol $a^+ b^+ d^+ c^+$. (Here, as an illustration, we have chosen our lettering and orientations in a bizarre manner!) Then the above rules lead to the adjacency diagram of Figure 6.1.8. Comparing it with the adjacency diagrams in Figure 6.1.6, we see that the tiling under discussion is of type IH53.

The process we have described above is not only relevant to the classification of marked tilings, but has implications for the classification of tilings by unmarked tiles. It leads to an answer to the question: what different types of (unmarked) isohedral tilings exist, topologically equivalent to the tiling $\mathscr{T}$ with vertices in the same positions in the plane as those of $\mathscr{T}$? Calling such tilings *topologically square isohedral tilings*, we observe first that each tiling of this kind may be assigned to one of the 36 types of Statement 6.1.2. Indeed, for any given topologically square isohedral tiling $\mathscr{T}$ we can use the same technique as described above, of assigning letters to oriented edges of tiles, to derive an incidence symbol $[L; A]$ for $\mathscr{T}$; this incidence symbol obviously represents one of

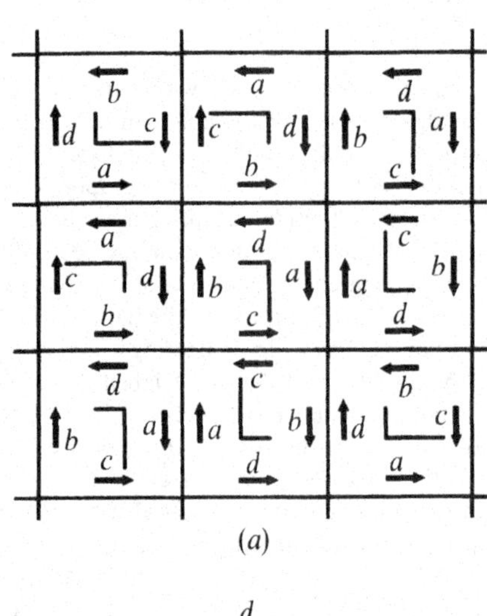

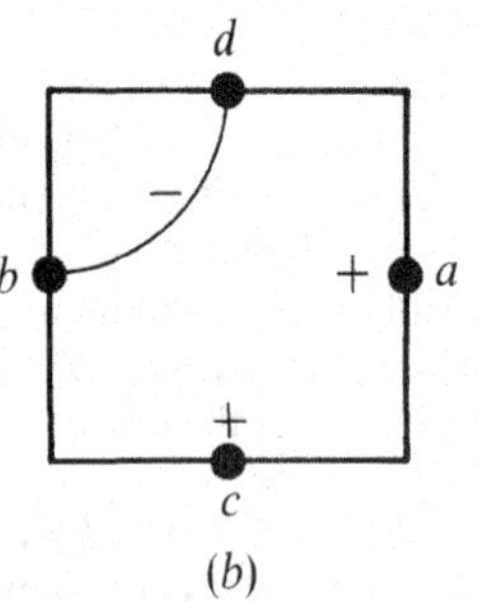

Figure 6.1.8
A tiling with incidence symbol $[a^- d^+ b^- c^+; a^+ b^+ d^+ c^+]$ and the corresponding adjacency diagram.

the types listed in Figure 6.1.6, and we shall say that $\mathscr{T}$ is of that same type.

Conversely, if we have a marked isohedral tiling $\mathscr{T}'$ with underlying tiling $[4^4]$ we may construct an (unmarked) topologically square isohedral tiling $\mathscr{T}$ by replacing each edge of $\mathscr{T}'$ with a suitable simple curve according to the following rules:

(*a*) All mutually equivalent edges of $\mathscr{T}'$ are replaced in the same manner by congruent curves.

(*b*) If two oriented edges of tiles, coinciding with an edge E of the tiling, bear the same labels and are oriented in opposite directions, then the edge E may be replaced by any S-*curve*, that is, any centrally symmetric arc whose center of symmetry is the midpoint of the edge E.

(*c*) If an unoriented edge abuts an edge bearing a different label (which must, of course, also be unoriented), then we may replace the edge by a C-*curve*, that is, by any arc for which the perpendicular bisector of the edge is a line of reflective symmetry.

(*d*) If an oriented edge abuts an edge bearing another label, then we may replace it by a J-*curve*, that is by any arc which does not have any non-trivial symmetry.

(*e*) In all other cases the edge is not replaced, but is left as a straight line-segment.

(*f*) Each curve introduced above misses all the vertices of $\mathscr{T}'$ except the two at its endpoints, and it meets no other curve except at a common endpoint.

In Figure 6.1.9 we illustrate the above procedure by showing how two marked tilings by square tiles can be converted into tilings with unmarked non-square tiles with vertices coinciding with those of $[4^4]$. Clearly this conversion preserves all the symmetries involved, and it is the most general procedure to do so. But it is well possible that the (unmarked) tiling $\mathscr{T}$ has additional symmetries, which are not present in the marked tiling $\mathscr{T}'$. To see how this can happen, consider the tiling $\mathscr{T}'$ of type $[abab; ab]$, see Figure 6.1.10(*a*). The above rules tell us that every edge must be left as a straight line-segment; the resulting tiling is necessarily $[4^4]$, with incidence symbol $[aaaa; a]$, which is not of the same type as $\mathscr{T}'$.

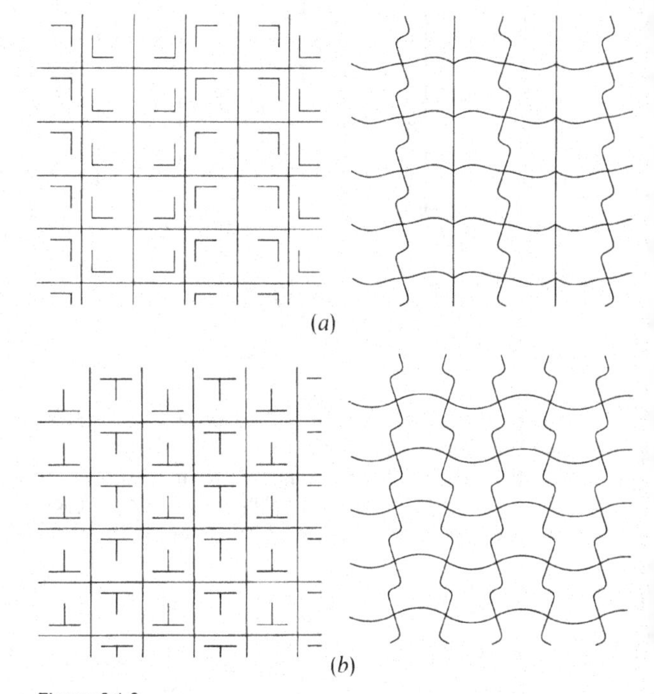

Figure 6.1.9
Two isohedral tilings by marked squares and the corresponding (unmarked) topologically square tilings. These are of types IH50 and IH66 in the enumeration given in Table 6.2.1.

A detailed investigation of the 36 types of marked tilings in Figure 6.1.6 shows:

6.1.3 *There exist precisely 29 different types of topologically square isohedral tilings.*

These are the types marked by an asterisk in Figure 6.1.6.

If we allow ourselves further freedom by specifying the

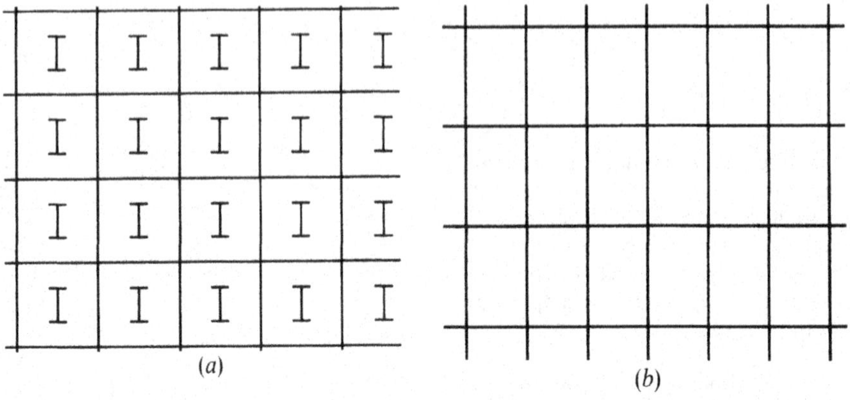

Figure 6.1.10
(*a*) An isohedral tiling by marked squares for which there is no corresponding (unmarked) topologically square tiling. There is, however, an unmarked tiling of this type by rectangles (see (*b*)). This tiling is of type IH72 and its incidence symbol is $[abab; ab]$.

topological type of the isohedral tiling, but not insisting on the positions of the vertices, there are additional possibilities. For example, by changing scale in one direction we can convert the square tiling to one by rectangles, see Figure 6.1.10(*b*). This is of the same isohedral type as the tiling $\mathscr{T}'$ of Figure 6.1.10(*a*), and so it is reasonable to assign to it the same symbol $[abab; ab]$.

With these considerations behind us, we can now carry out the enumerations of tilings described in the next three sections.

EXERCISES 6.1

1. For each of the 29 types with adjacency diagrams marked by an asterisk in Figure 6.1.6 construct a topologically square isohedral tiling of that type.
2. Show that there exist no topologically square tilings of each of the seven types with adjacency diagrams not marked by an asterisk in Figure 6.1.6.
3. Prove that six of the seven types mentioned in Exercise 2 cannot be realized by any (unmarked) tiling in which the vertices occupy the lattice-points in a rectangular lattice.
4. Find which of the 29 types of Exercise 1 cannot be realized by a topologically square isohedral tiling in which the tiles are polygons with at most (*a*) six sides, (*b*) eight sides?
5. Determine which of the tile groups $S(T)$ indicated in Figure 6.1.1 can be achieved by marked square tiles in which the marking consists of a single, suitably chosen (*a*) circle, (*b*) square, (*c*) rectangle, (*d*) regular hexagon.
6. Consider the following symbols, and for each decide whether it is the incidence symbol of a marked isohedral tiling by squares. If it is, construct its adjacency diagram and use it to identify the type with one of those in Figure 6.1.6. Find the change of orientations and labels which identifies the given symbol with that given in Figure 6.1.6.

   (a) $[a^+a^-b^+b^-; b^+a^+]$
   (b) $[c^-d^-a^+b^+; b^+a^-d^-c^+]$

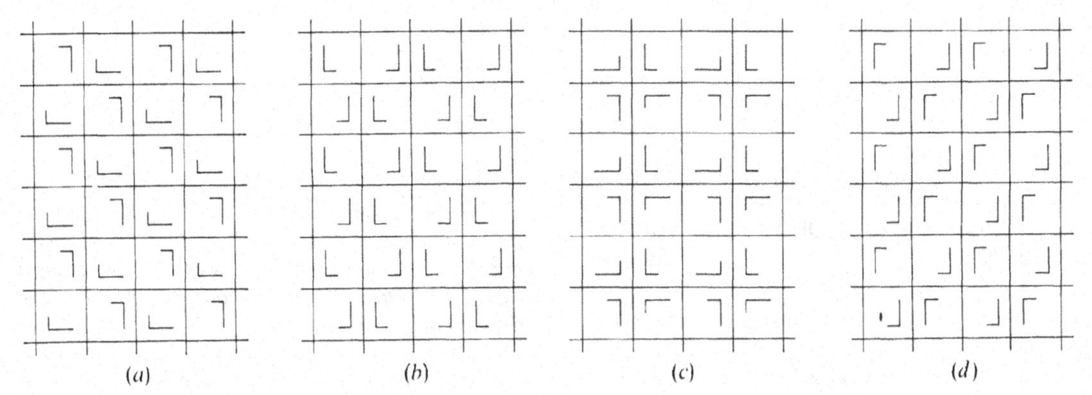

Figure 6.1.11
What types are these four isohedral tilings by marked squares? See Exercise 6.1.7.

(c) $[b^- d^+ c^+ a^-; a^+ c^+ d^+ b^+]$
(d) $[a^+ d^+ b^- c^+; b^- c^- a^- d^-]$
(e) $[c^+ d^- a^+ b^-; b^+ a^- c^+ d^+]$

7. For each of the marked isohedral tilings in Figure 6.1.11 determine the incidence symbol, the adjacency diagram and the type as indicated in Figure 6.1.6.
8. For each of the topologically square isohedral tilings in Figures 4.1.4 and 6.1.12 determine the incidence symbol, the adjacency diagram and the type as indicated in Figure 6.1.6.

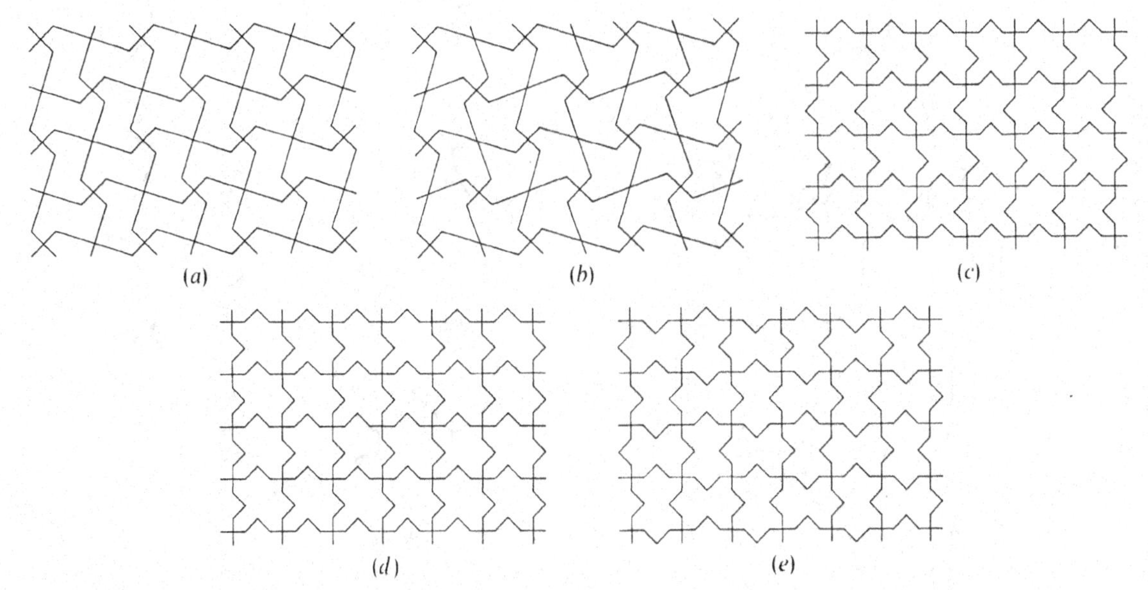

Figure 6.1.12
What types are these (unmarked) topologically square isohedral tilings? See Exercise 6.1.8.

6.2 THE CLASSIFICATION OF ISOHEDRAL TILINGS

The classification of isohedral tilings uses the method described in the previous section, and we shall continue with the terminology and notation introduced there.

Suppose that $\mathcal{T}$ is an isohedral tiling or a marked isohedral tiling. We allocate letters $a, b, \ldots$ to the oriented edges of the tiles exactly as described for marked squares. Two examples are shown in Figure 6.2.1. Here the tilings have incidence symbols $[a^+b^+c^+d^+e^+f^+; a^+e^+d^-c^-b^+f^+]$ and $[ab^+c^+dc^-b^-; dc^-b^-a]$. We distinguish different types of isohedral tilings as in the case of tilings with underlying tiling $[4^4]$, and say that *two tilings are of the same isohedral type if and only if they are of the same topological type and their incidence symbols* $[L; A]$ *differ trivially*. Whether or not a difference between incidence symbols is trivial can be checked either by manipulation of the symbols, or by using an adjacency diagram constructed as in Section 6.1.

Our first task is to explain how the different marked isohedral types can be enumerated. To do this we make use of the fact that the underlying tiling of a marked isohedral tiling is necessarily topologically tile-transitive. By the results of Section 4.3 such a tiling must be topologically equivalent to one of the eleven Laves tilings shown in Figure 2.7.1. We take each of these tilings one at a

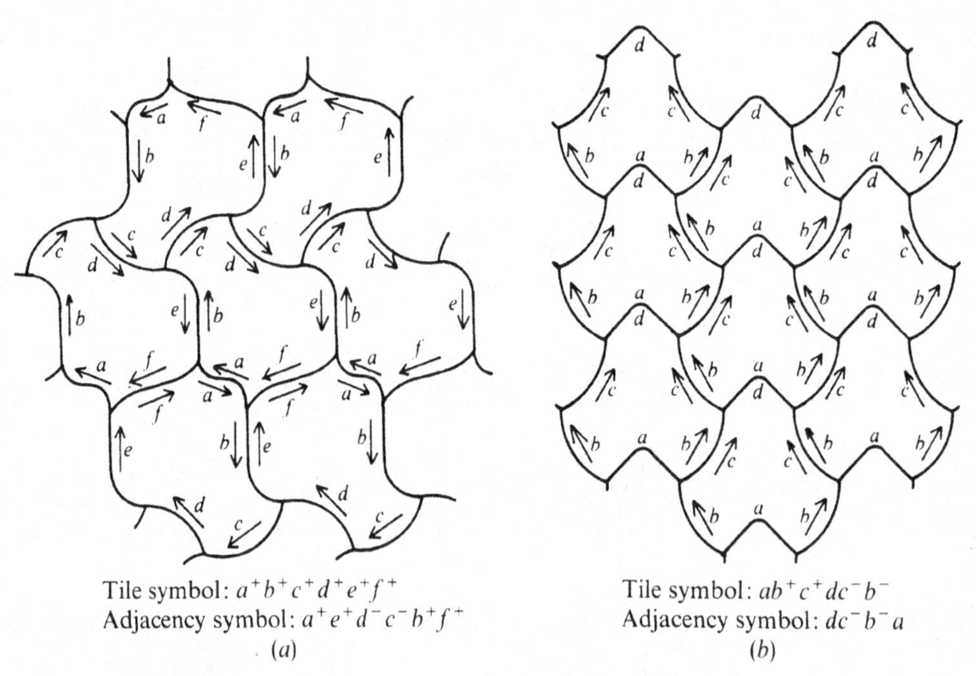

Figure 6.2.1
Two examples of isohedral tilings and the determination of their tile and adjacency symbols by the same procedure as that described in Section 6.1.

time and consider the possible induced tile groups. In many cases these are restricted by the topological type of the tiling.

For example, consider the tiling $[3.12^2]$ (see Figure 6.2.2). The symmetry group of a triangle is at most *d3* (realized in the case of an equilateral triangle), the six operations of which correspond to the permutations of the three vertices. But in the tiling $[3.12^2]$ one vertex of each triangle differs essentially from the other two by its valence. Hence the induced tile group $S(\mathcal{T} \mid T)$ can be either *c1*, which leaves each vertex fixed, or *d1*, which interchanges the two equivalent vertices.

For each group $S(\mathcal{T} \mid T)$ we can write down a corresponding tile symbol—for $[3.12^2]$ the tile symbols are $a^+b^+c^+$ and ab^+b^-. In Figure 6.2.3 we show the 39 cases that arise from the eleven Laves tilings.

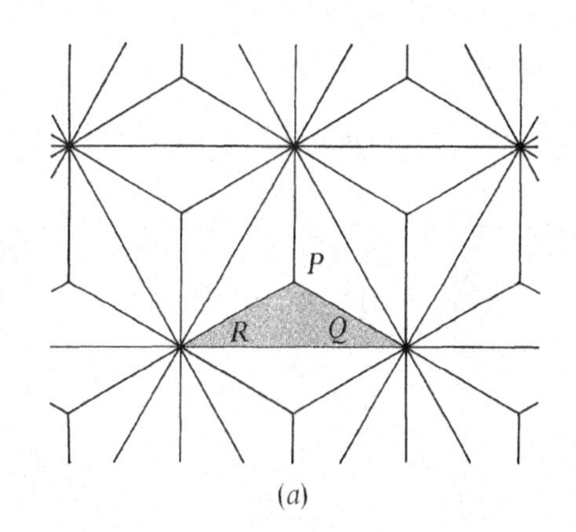

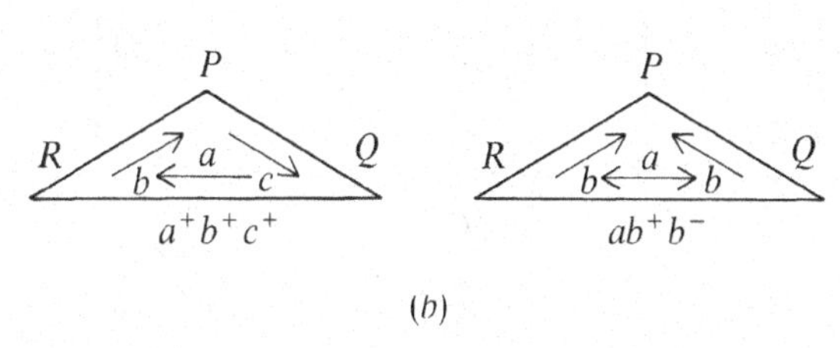

Figure 6.2.2
In tilings of topological type $[3.12^2]$ the induced tile group is either *c1* or *d1*, and the tile symbol can then be taken as $a^+b^+c^+$ or ab^+b^-, respectively.

After determining all possible tile symbols we take sequences of letters as potential adjacency symbols and test whether the adjacencies that any such sequence specifies can be continued isohedrally over the whole plane, as a tiling of the prescribed topological type. The procedure for testing is exactly the same as that described for isohedral marked tilings by squares described in the previous section. In this way we find:

6.2.1 *There exist precisely* 93 *distinct types of marked isohedral tilings.*

These types are listed in Table 6.2.1, together with additional information on each type. It should be observed that in some cases the incidence symbol $[L; A]$ alone is not sufficient to determine the type of the isohedral tiling (see, for example, the types IH39 and IH79 in Table 6.2.1); however, it is sufficient when the topological type of the tiling is known. Usually this can be inferred from the context, but if any confusion is possible then it must be stated explicitly.

The 93 entries in Table 6.2.1 do not, of course, all correspond to isohedral tilings with unmarked tiles. We have enumerated the isohedral types of *marked Laves tilings.* To see in which cases a corresponding *unmarked* tiling exists we employ the method, described in the previous section, of replacing edges by S-curves, C-curves and J-curves. The procedure is exactly the same as for square tilings except that now, in addition, we may distort the tiling by altering the lengths or positions of edges so long as the symmetries of the tiling are not affected.

It turns out that 12 of the types in Table 6.2.1 do not correspond to unmarked tilings; in each case the assumed symmetries and the relative positions of the adjacents force additional symmetries which change the type of the tiling. Thus we obtain:

6.2.2 *There exist precisely* 81 *distinct types of isohedral tilings.*

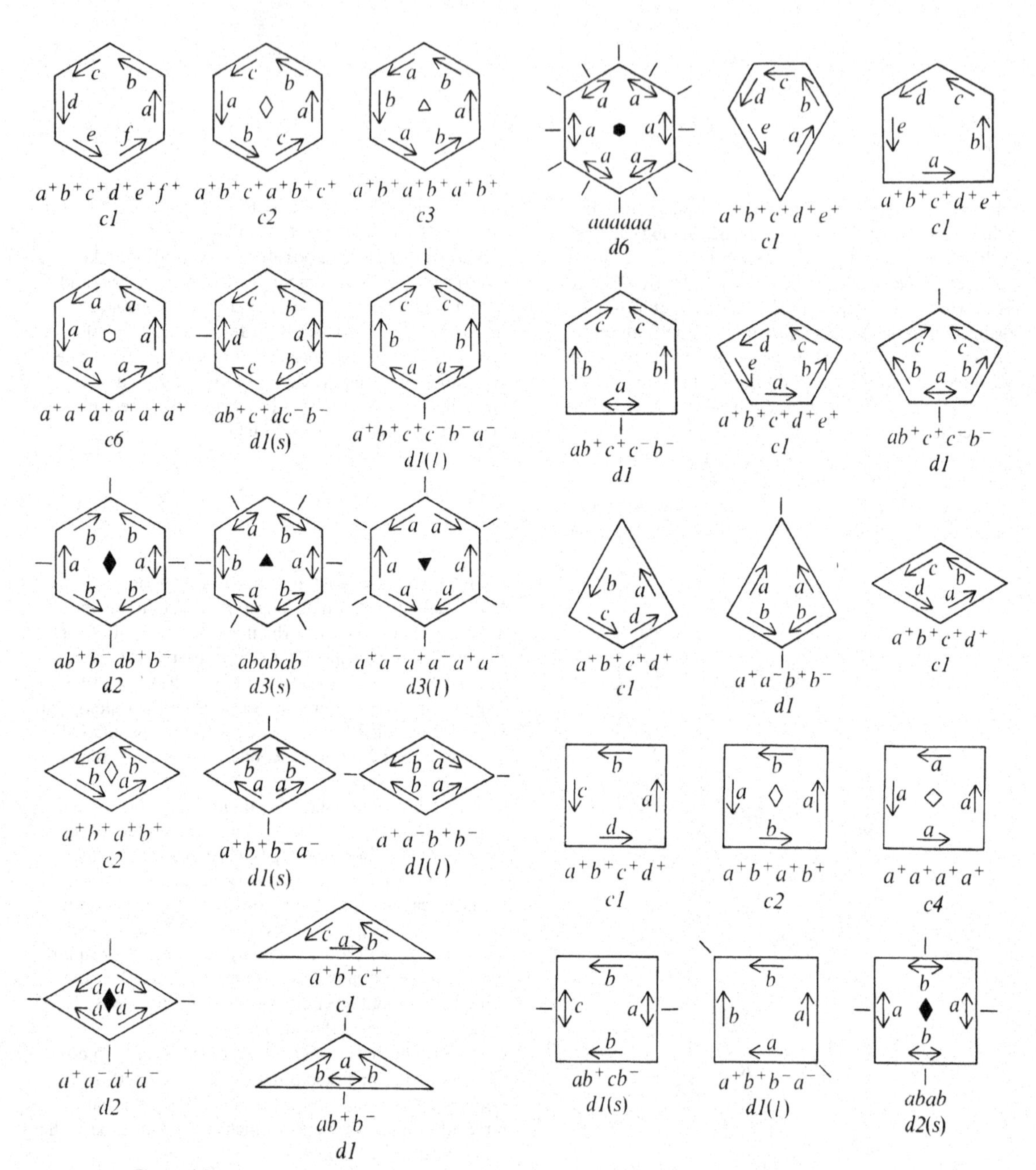

Figure 6.2.3
The 39 possible induced tile groups of isohedral tilings (or of marked isohedral tilings). Near each tile we indicate its tile symbol and its induced group (which is also indicated on the tile using the symbols of Table 1.4.1).

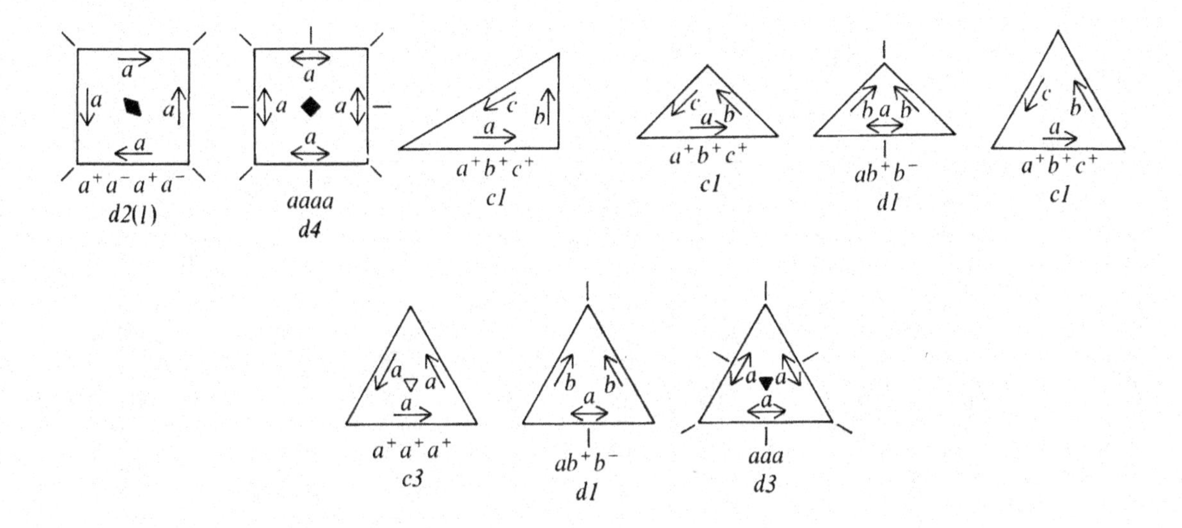

Table 6.2.1 THE 93 TYPES OF MARKED ISOHEDRAL TILINGS

Columns (1), (2) and (3) give the type, topological type and incidence symbol of each type of tiling.
Columns (4) and (5) indicate the symmetry group $S(\mathcal{T})$ and induced tile group $S(\mathcal{T} \mid T)$ as denoted in Figure 6.2.3.
Column (6) shows whether the type is realizable by a tiling with unmarked convex tiles (C), or with unmarked non-convex tiles (N). Every type can be realized by marked tiles; type which cannot be realized by unmarked tiles are indicated by M.
Columns (7) and (8) indicate the transitivity classes of the vertices and edges under symmetries of the tiling. The symbols α, β, . . . indicate the classes to which the vertices or edges, taken in cyclic order round the tile, belong. It will be observed that whereas the edges can fall into as many as five transitivity classes (type IH4), for no type do the vertices belong to more than four classes.
Column (9) gives the number of aspects of the tiles in a tiling of the stated type, that is to say the number of equivalence classes under translations. nD indicates that the tiles occur in n direct aspects, mR that they also occur in m reflected aspects, and an integer (without D or R) is used if the tile has a line of reflective symmetry so that direct and reflected aspects coincide.
Column (10) shows the pattern type of the associated tiling pattern in the notation of Section 5.2.

Type (1)	Topological type (2)	Incidence symbol (3)	Symmetry group (4)	Induced tile group (5)	Realizations (6)	Transitivity classes of vertices (7)	Transitivity classes of edges (8)	Aspects (9)	Pattern type (10)
IH1	$[3^6]$	$[a^+b^+c^+d^+e^+f^+; d^+e^+f^+a^+b^+c^+]$	$p1$	$c1$	N	$\alpha\beta\alpha\beta\alpha\beta$	$\alpha\beta\gamma\alpha\beta\gamma$	$1D$	PP1
IH2		$[a^+b^+c^+d^+e^+f^+; b^-a^-f^+e^-d^-c^+]$	pg	$c1$	C, N	$\alpha\alpha\alpha\beta\beta\beta$	$\alpha\alpha\beta\gamma\gamma\beta$	$1D, 1R$	PP2
IH3		$[a^+b^+c^+d^+e^+f^+; c^-e^+a^-f^-b^+d^-]$	pg	$c1$	C, N	$\alpha\beta\alpha\beta\alpha\beta$	$\alpha\beta\alpha\gamma\beta\gamma$	$1D, 1R$	PP2
IH4		$[a^+b^+c^+d^+e^+f^+; a^+e^+c^+d^+b^+f^+]$	$p2$	$c1$	C, N	$\alpha\alpha\beta\beta\beta\alpha$	$\alpha\beta\gamma\delta\beta\varepsilon$	$2D$	PP7
IH5		$[a^+b^+c^+d^+e^+f^+; a^+e^+d^-c^-b^+f^+]$	pgg	$c1$	C, N	$\alpha\alpha\beta\beta\beta\alpha$	$\alpha\beta\gamma\gamma\beta\delta$	$2D, 2R$	PP9
IH6		$[a^+b^+c^+d^+e^+f^+; a^+e^-c^+f^-b^-d^-]$	pgg	$c1$	C, N	$\alpha\alpha\beta\beta\alpha\beta$	$\alpha\beta\gamma\delta\beta\delta$	$2D, 2R$	PP9
IH7		$[a^+b^+c^+d^+e^+f^+; b^+a^+d^+c^+f^+e^+]$	$p3$	$c1$	C, N	$\alpha\beta\alpha\gamma\alpha\delta$	$\alpha\alpha\beta\beta\gamma\gamma$	$3D$	PP21
IH8		$[a^+b^+c^+a^+b^+c^+; a^+b^+c^+]$	$p2$	$c2$	C, N	$\alpha\alpha\alpha\alpha\alpha\alpha$	$\alpha\beta\gamma\alpha\beta\gamma$	$1D$	PP8
IH9		$[a^+b^+c^+a^+b^+c^+; a^+c^-b^-]$	pgg	$c2$	C, N	$\alpha\alpha\alpha\alpha\alpha\alpha$	$\alpha\beta\beta\alpha\beta\beta$	$1D, 1R$	PP10
IH10		$[a^+b^+a^+b^+a^+b^+; b^+a^+]$	$p3$	$c3$	N	$\alpha\beta\alpha\beta\alpha\beta$	$\alpha\alpha\alpha\alpha\alpha\alpha$	$1D$	PP22
IH11		$[a^+a^+a^+a^+a^+a^+; a^+]$	$p6$	$c6$	N	$\alpha\alpha\alpha\alpha\alpha\alpha$	$\alpha\alpha\alpha\alpha\alpha\alpha$	$1D$	PP45

Table 6.2.1 (continued)

Type (1)	Topological type (2)	Incidence symbol (3)	Symmetry group (4)	Induced tile group (5)	Realizations (6)	Transitivity classes of vertices (7)	Transitivity classes of edges (8)	Aspects (9)	Pattern type (10)
IH12	$[3^6]$	$[ab^+c^+dc^-b^-; dc^-b^-a]$	*cm*	*d1(s)*	*N*	αααααα	αββαββ	1	PP6
IH13		$[ab^+c^+dc^-b^-; db^+; db^+c^+a]$	*pmg*	*d1(s)*	*C, N*	αααααα	αβγαγβ	2	PP13
IH14		$[a^+b^+c^+c^-b^-a^-; c^-b^-a^-]$	*cm*	*d1(l)*	*C, N*	αβαβαβ	αβααβα	1	PP6
IH15		$[a^+b^+c^+c^-b^-a^-; a^+b^-c^+]$	*pmg*	*d1(l)*	*C, N*	ααβββα	αβγγβα	2	PP13
IH16		$[a^+b^+c^+c^-b^-a^-; a^-c^-b^+]$	*p31m*	*d1(l)*	*C, N*	αβγβγβ	αββββα	3	PP25
IH17		$[ab^+b^-ab^+b^-; ab^+]$	*cmm*	*d2*	*C, N*	αααααα	αββαββ	1	PP20
IH18		$[ababab; ba]$	*p31m*	*d3(s)*	*N*	αααααα	αααααα	1	PP26
IH19		$[a^+a^-a^+a^-a^+a^-; a^-]$	*p3m1*	*d3(l)*	*M*	αβαβαβ	αααααα	1	(PP29)
IH20		$[aaaaaa; a]$	*p6m*	*d6*	*C*	αααααα	αααααα	1	PP51
IH21	$[3^4.6]$	$[a^+b^+c^+d^+e^+; e^+c^+b^+d^+a^+]$	*p6*	*c1*	*C, N*	αβγββ	αββγα	6*D*	PP42
IH22	$[3^3.4^2]$	$[a^+b^+c^+d^+e^+; a^-e^+d^-c^-b^+]$	*cm*	*c1*	*C, N*	ααβββ	αβγγβ	1*D*, 1*R*	PP5
IH23		$[a^+b^+c^+d^+e^+; a^+e^+c^+d^+b^+]$	*p2*	*c1*	*C, N*	ααβββ	αβγδβ	2*D*	PP7
IH24		$[a^+b^+c^+d^+e^+; a^-e^+c^+d^+b^+]$	*pmg*	*c1*	*C, N*	ααβββ	αβγδβ	2*D*, 2*R*	PP11
IH25		$[a^+b^+c^+d^+e^+; a^+e^+d^-c^-b^+]$	*pgg*	*c1*	*C, N*	ααβββ	αβγγβ	2*D*, 2*R*	PP9
IH26		$[ab^+c^+c^-b^-; ab^-c^+]$	*cmm*	*d1*	*C, N*	ααβββ	αβγγβ	2	PP19
IH27	$[3^2.4.3.4]$	$[a^+b^+c^+d^+e^+; a^+d^-e^-b^-c^-]$	*pgg*	*c1*	*C, N*	ααβαβ	αβγβγ	2*D*, 2*R*	PP9
IH28		$[a^+b^+c^+d^+e^+; a^+c^+b^+e^+d^+]$	*p4*	*c1*	*C, N*	ααβαγ	αββγγ	4*D*	PP30
IH29		$[ab^+c^+c^-b^-; ac^+b^+]$	*p4g*	*d1*	*C, N*	ααβαβ	αββββ	4	PP35
IH30	$[3.4.6.4]$	$[a^+b^+c^+d^+; a^-b^-d^+c^+]$	*p31m*	*c1*	*C, N*	αβαγ	αβγγ	3*D*, 3*R*	PP23
IH31		$[a^+b^+c^+d^+; b^+a^+d^+c^+]$	*p6*	*c1*	*N*	αβαγ	ααββ	6*D*	PP42
IH32		$[a^+a^-b^+b^-; a^-b^-]$	*p6m*	*d1*	*C*	αβαγ	ααββ	6	PP48
IH33	$[3.6.3.6]$	$[a^+b^+c^+d^+; d^+c^+b^+a^+]$	*p3*	*c1*	*N*	αβγβ	αββα	3*D*	PP21
IH34		$[a^+b^+a^+b^+; b^+a^+]$	*p6*	*c2*	*N*	αβαβ	αααα	3*D*	PP43
IH35		$[a^+b^+b^-a^-; a^-b^-]$	*p3m1*	*d1(s)*	*M*	αβγβ	αββα	3	(PP28)
IH36		$[a^+a^-b^+b^-; b^-a^-]$	*p31m*	*d1(l)*	*N*	αβαβ	αααα	3	PP25
IH37		$[a^+a^-a^+a^-; a^-]$	*p6m*	*d2*	*C*	αβαβ	αααα	3	PP49
IH38	$[3.12^2]$	$[a^+b^+c^+; a^-c^-b^+]$	*p31m*	*c1*	*N*	ααβ	αββ	3*D*, 3*R*	PP23
IH39		$[a^+b^+c^+; a^+c^+b^+]$	*p6*	*c1*	*N*	αββ	αββ	6*D*	PP42
IH40		$[ab^+b^-; ab^-]$	*p6m*	*d1*	*C*	ααβ	αββ	6	PP48
IH41	$[4^4]$	$[a^+b^+c^+d^+; c^+d^+a^+b^+]$	*p1*	*c1*	*N*	αααα	αβαβ	1*D*	PP1
IH42		$[a^+b^+c^+d^+; c^+b^-a^+d^-]$	*pm*	*c1*	*N*	αββα	αβαγ	1*D*, 1*R*	PP3
IH43		$[a^+b^+c^+d^+; c^-d^+a^-b^+]$	*pg*	*c1*	*N*	αααα	αβαβ	1*D*, 1*R*	PP2
IH44		$[a^+b^+c^+d^+; b^-a^-d^-c^-]$	*pg*	*c1*	*N*	αααα	ααββ	1*D*, 1*R*	PP2
IH45		$[a^+b^+c^+d^+; c^-b^-a^-d^-]$	*cm*	*c1*	*N*	αβαβ	αβαγ	1*D*, 1*R*	PP5
IH46		$[a^+b^+c^+d^+; a^+b^+c^+d^+]$	*p2*	*c1*	*C, N*	αααα	αβγδ	2*D*	PP7
IH47		$[a^+b^+c^+d^+; c^+b^+a^+d^+]$	*p2*	*c1*	*N*	αββα	αβαγ	2*D*	PP7
IH48		$[a^+b^+c^+d^+; a^-b^-c^-d^-]$	*pmm*	*c1*	*M*	αβγδ	αβγδ	2*D*, 2*R*	(PP14)
IH49		$[a^+b^+c^+d^+; a^-b^+c^-d^+]$	*pmg*	*c1*	*C, N*	αββα	αβγδ	2*D*, 2*R*	PP11
IH50		$[a^+b^+c^+d^+; c^+b^-a^+d^+]$	*pmg*	*c1*	*C, N*	αββα	αβαγ	2*D*, 2*R*	PP11

Table 6.2.1 (continued)

Type (1)	Topological type (2)	Incidence symbol (3)	Symmetry group (4)	Induced tile group (5)	Realizations (6)	Transitivity classes of vertices (7)	Transitivity classes of edges (8)	Aspects (9)	Pattern type (10)
IH51	$[4^4]$	$[a^+b^+c^+d^+; c^-b^+a^-d^+]$	*pgg*	*c1*	*C, N*	αααα	αβαγ	*2D, 2R*	PP9
IH52		$[a^+b^+c^+d^+; c^-d^-a^-b^-]$	*pgg*	*c1*	*N*	αβαβ	αβαβ	*2D, 2R*	PP9
IH53		$[a^+b^+c^+d^+; b^-a^-c^+d^+]$	*pgg*	*c1*	*C, N*	αααα	ααβγ	*2D, 2R*	PP9
IH54		$[a^+b^+c^+d^+; a^-b^-c^-d^+]$	*cmm*	*c1*	*C, N*	αβγα	αβγδ	*2D, 2R*	PP17
IH55		$[a^+b^+c^+d^+; b^+a^+d^+c^+]$	*p4*	*c1*	*N*	αβαγ	ααββ	*4D*	PP30
IH56		$[a^+b^+c^+d^+; b^+a^+c^-d^-]$	*p4g*	*c1*	*C, N*	αβαγ	ααβγ	*4D, 4R*	PP33
IH57		$[a^+b^+a^+b^+; a^+b^+]$	*p2*	*c2*	*C, N*	αααα	αβαβ	*1D*	PP8
IH58		$[a^+b^+a^+b^+; a^-b^+]$	*pmg*	*c2*	*C, N*	αααα	αβαβ	*1D, 1R*	PP12
IH59		$[a^+b^+a^+b^+; b^-a^-]$	*pgg*	*c2*	*N*	αααα	αααα	*1D, 1R*	PP10
IH60		$[a^+b^+a^+b^+; a^-b^-]$	*cmm*	*c2*	*M*	αβαβ	αβαβ	*1D, 1R*	(PP18)
IH61		$[a^+b^+a^+b^+; b^+a^+]$	*p4*	*c2*	*N*	αβαβ	αααα	*2D*	PP31
IH62		$[a^+a^+a^+a^+; a^+]$	*p4*	*c4*	*N*	αααα	αααα	*1D*	PP32
IH63		$[a^+a^+a^+a^+; a^-]$	*p4g*	*c4*	*M*	αααα	αααα	*1D, 1R*	(PP34)
IH64		$[ab^+cb^-; cb^-a]$	*pm*	*d1(s)*	*N*	αααα	αβαβ	1	PP4
IH65		$[ab^+cb^-; ab^-c]$	*pmm*	*d1(s)*	*M*	ααββ	αβγβ	2	(PP15)
IH66		$[ab^+cb^-; cb^+a]$	*pmg*	*d1(s)*	*N*	αααα	αβαβ	2	PP13
IH67		$[ab^+cb^-; ab^+c]$	*cmm*	*d1(s)*	*C, N*	αααα	αβγβ	2	PP19
IH68		$[a^+b^+b^-a^-; b^-a^-]$	*cm*	*d1(l)*	*N*	αααα	αααα	1	PP6
IH69		$[a^+b^+b^-a^-; a^+b^+]$	*pmg*	*d1(l)*	*C, N*	αααα	αββα	2	PP13
IH70		$[a^+b^+b^-a^-; a^-b^-]$	*p4m*	*d1(l)*	*M*	αβγβ	αββα	4	(PP39)
IH71		$[a^+b^+b^-a^-; b^+a^+]$	*p4g*	*d1(l)*	*N*	αβαβ	αααα	4	PP35
IH72		$[abab; ab]$	*pmm*	*d2(s)*	*C*	αααα	αβαβ	1	PP16
IH73		$[abab; ba]$	*p4g*	*d2(s)*	*N*	αααα	αααα	2	PP36
IH74		$[a^+a^-a^+a^-; a^+]$	*cmm*	*d2(l)*	*C, N*	αααα	αααα	1	PP20
IH75		$[a^+a^-a^+a^-; a^-]$	*p4m*	*d2(l)*	*M*	αβαβ	αααα	2	(PP40)
IH76		$[aaaa; a]$	*p4m*	*d4*	*C*	αααα	αααα	1	PP41
IH77	$[4.6.12]$	$[a^+b^+c^+; a^-b^-c^-]$	*p6m*	*c1*	*C*	αβγ	αβγ	*6D, 6R*	PP46
IH78	$[4.8^2]$	$[a^+b^+c^+; a^+b^-c^-]$	*cmm*	*c1*	*C, N*	ααβ	αβγ	*2D, 2R*	PP17
IH79		$[a^+b^+c^+; a^+c^+b^+]$	*p4*	*c1*	*N*	ααβ	αββ	*4D*	PP30
IH80		$[a^+b^+c^+; a^-b^-c^-]$	*p4m*	*c1*	*M*	αβγ	αβγ	*4D, 4R*	(PP37)
IH81		$[a^+b^+c^+; a^-c^+b^+]$	*p4g*	*c1*	*N*	ααβ	αββ	*4D, 4R*	PP33
IH82		$[ab^+b^-; ab^-]$	*p4m*	*d1*	*C*	ααβ	αββ	4	PP38
IH83	$[6^3]$	$[a^+b^+c^+; b^-a^-c^-]$	*cm*	*c1*	*N*	ααα	ααβ	*1D, 1R*	PP5
IH84		$[a^+b^+c^+; a^+b^+c^+]$	*p2*	*c1*	*C, N*	ααα	αβγ	*2D*	PP7
IH85		$[a^+b^+c^+; a^-b^+c^+]$	*pmg*	*c1*	*C, N*	ααα	αβγ	*2D, 2R*	PP11
IH86		$[a^+b^+c^+; b^-a^-c^+]$	*pgg*	*c1*	*N*	ααα	ααβ	*2D, 2R*	PP9
IH87		$[a^+b^+c^+; a^-b^-c^-]$	*p3m1*	*c1*	*M*	αβγ	αβγ	*3D, 3R*	(PP27)
IH88		$[a^+b^+c^+; b^+a^+c^+]$	*p6*	*c1*	*N*	αβα	ααβ	*6D*	PP42
IH89		$[a^+a^+a^+; a^-]$	*p31m*	*c3*	*M*	ααα	ααα	*1D, 1R*	(PP24)
IH90		$[a^+a^+a^+; a^+]$	*p6*	*c3*	*N*	ααα	ααα	*2D*	PP44
IH91		$[ab^+b^-; ab^+]$	*cmm*	*d1*	*C, N*	ααα	αββ	2	PP19
IH92		$[ab^+b^-; ab^-]$	*p6m*	*d1*	*M*	ααβ	αββ	6	(PP47)
IH93		$[aaa; a]$	*p6m*	*d3*	*C*	ααα	ααα	2	PP50

These 81 types are indicated in Table 6.2.1 by symbols *C* or *N* in column (6). Diagrams showing representatives of each of the 81 types appear in Figure 6.2.4. The 12 types of marked isohedral tilings that are not realizable by unmarked tilings are indicated by the letter *M* in column (6) of Table 6.2.1; marked Laves tilings of these types are shown in Figure 6.2.5.

As in the preceding section, the practical identification of the type of a (marked or unmarked) isohedral tiling is simplified by the use of adjacency diagrams. These are defined in exactly the same way as in Section 6.1. The adjacency diagrams that correspond to the 57 types of isohedral tilings not based on $[4^4]$ are shown in Figure 6.2.6, which thus complements Figure 6.1.6.

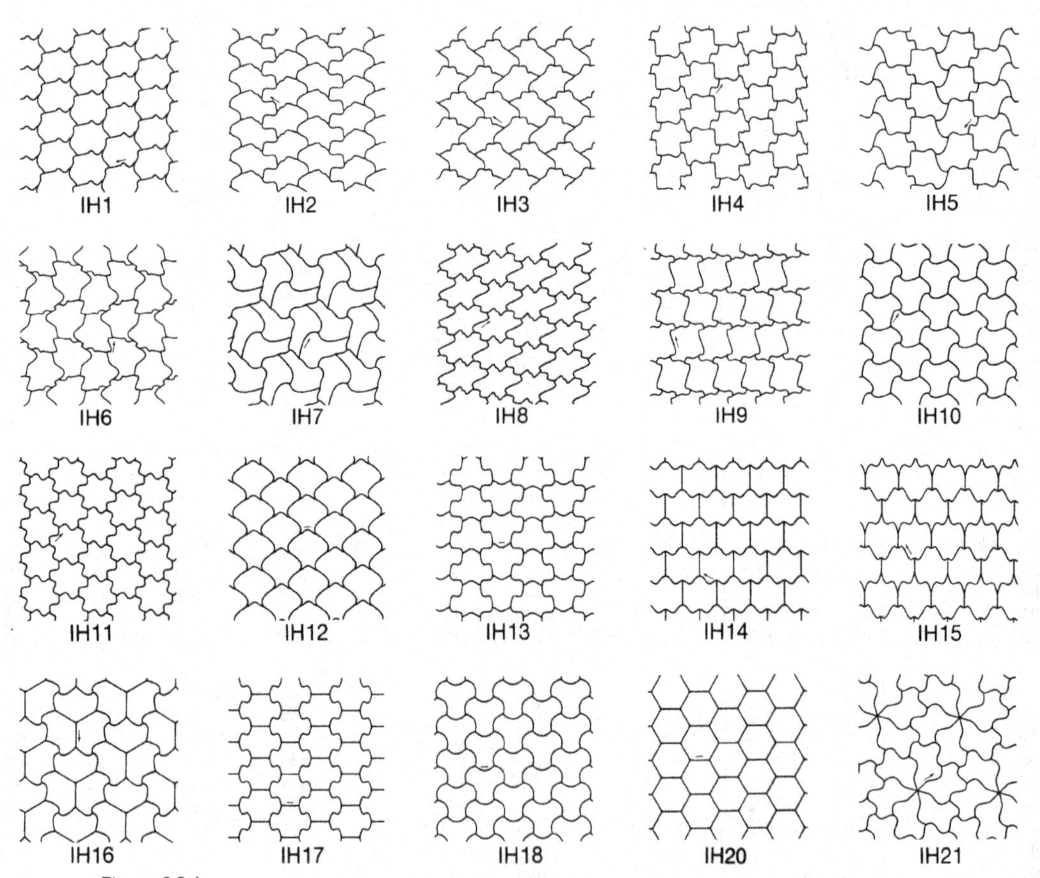

Figure 6.2.4

Examples of all 81 types of (unmarked) isohedral tilings. For help in identifying the corresponding incidence symbols in Table 6.2.1 we have indicated, in each tiling, an edge labelled a^+ by $\uparrow$ or an edge labelled a by $|$. For clarity, all other labels are omitted.

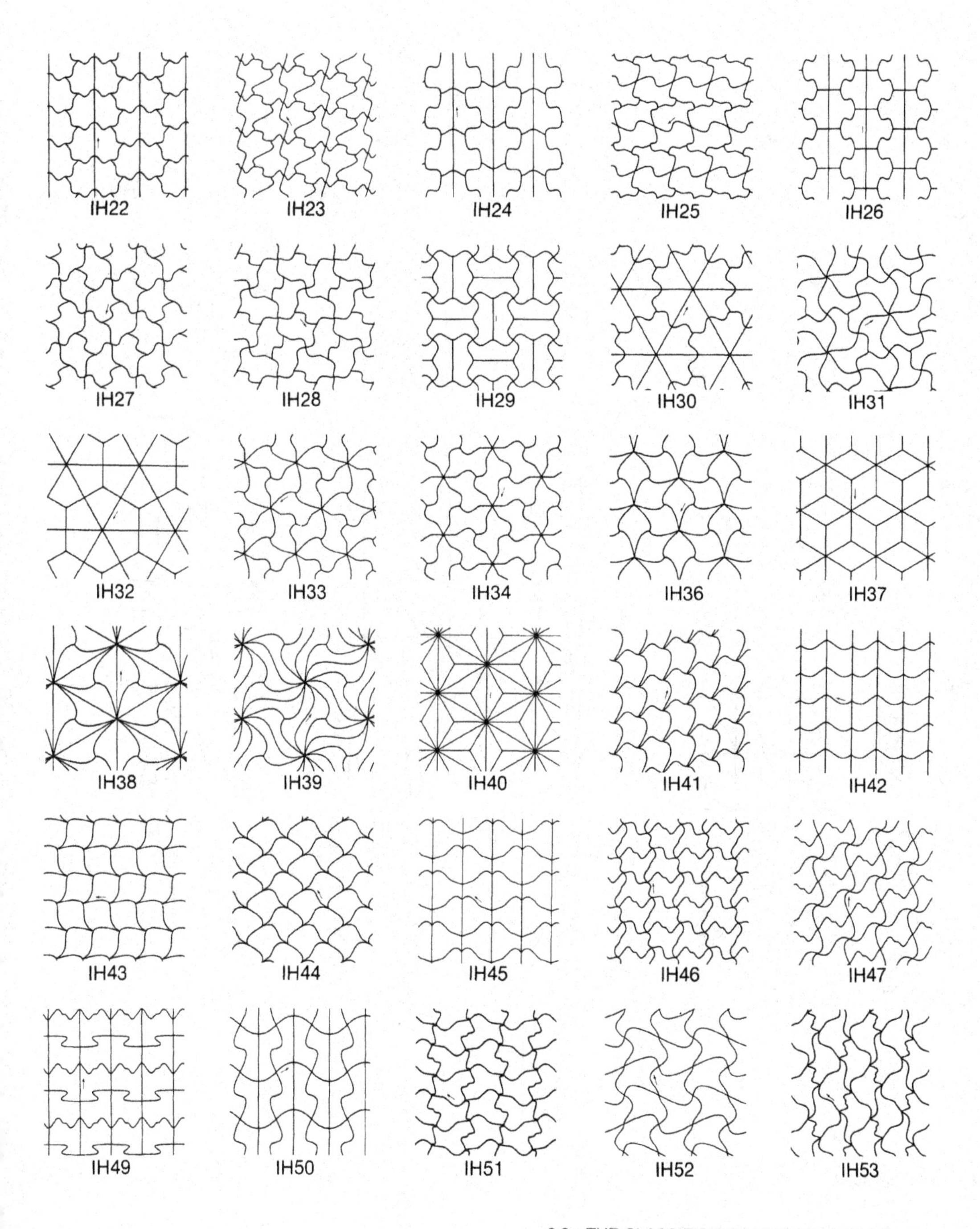
IH22
IH23
IH24
IH25
IH26
IH27
IH28
IH29
IH30
IH31
IH32
IH33
IH34
IH36
IH37
IH38
IH39
IH40
IH41
IH42
IH43
IH44
IH45
IH46
IH47
IH49
IH50
IH51
IH52
IH53

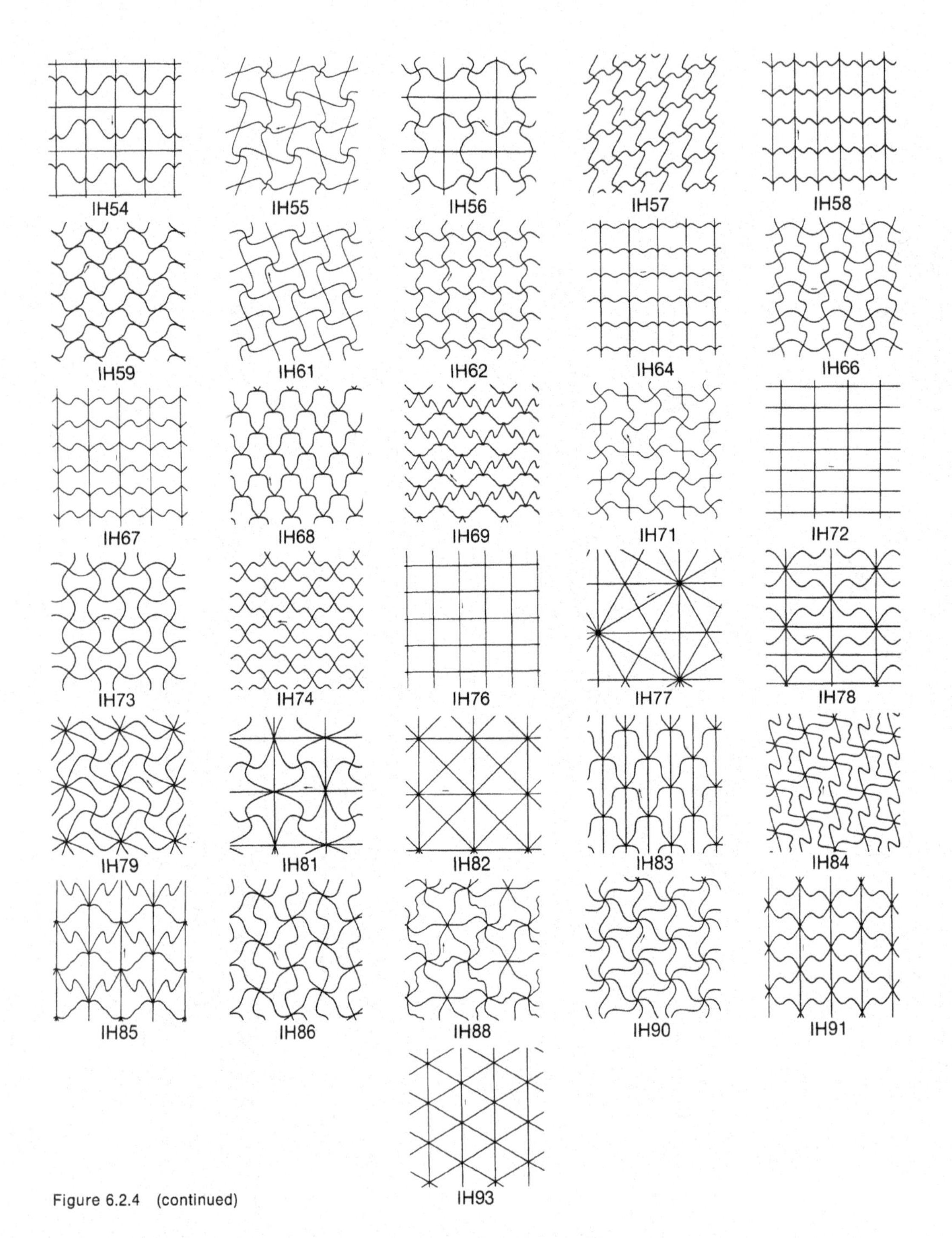

Figure 6.2.4 (continued)

The enumeration of the types of isohedral tilings carried out in this section provides us with a simple method of enumerating the types of symmetry groups of *all* periodic discrete patterns. Starting from any periodic pattern $\mathscr{P}$ (which can also be a tiling pattern derived from a given isohedral tiling by the method described in Section 5.4), let $\mathscr{D}(\mathscr{P})$ be the Dirichlet tiling obtained from $\mathscr{P}$ (see Section 5.4). Then $\mathscr{D}(\mathscr{P})$ is a marked isohedral tiling with the same symmetry group as $\mathscr{P}$. However, we now know *all* possible types of marked isohedral tilings, and inspection of Table 6.2.1 shows that only 17 types of groups can arise as symmetry groups of such tilings. These are the 17 crystallographic groups discussed in Section 1.4.*

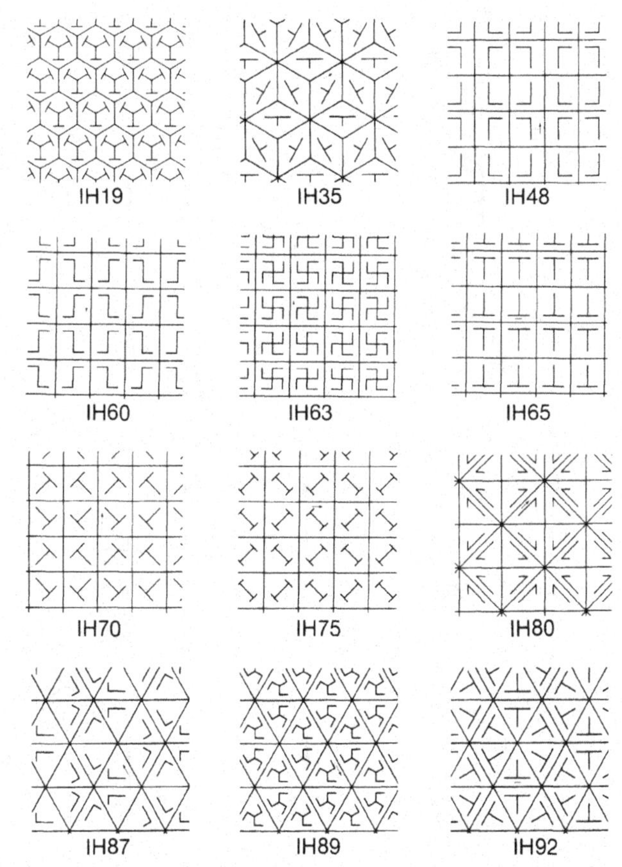

Figure 6.2.5
Examples of the twelve types of marked isohedral tilings that have no unmarked representatives.

* See the Appendix beginning on page 653.

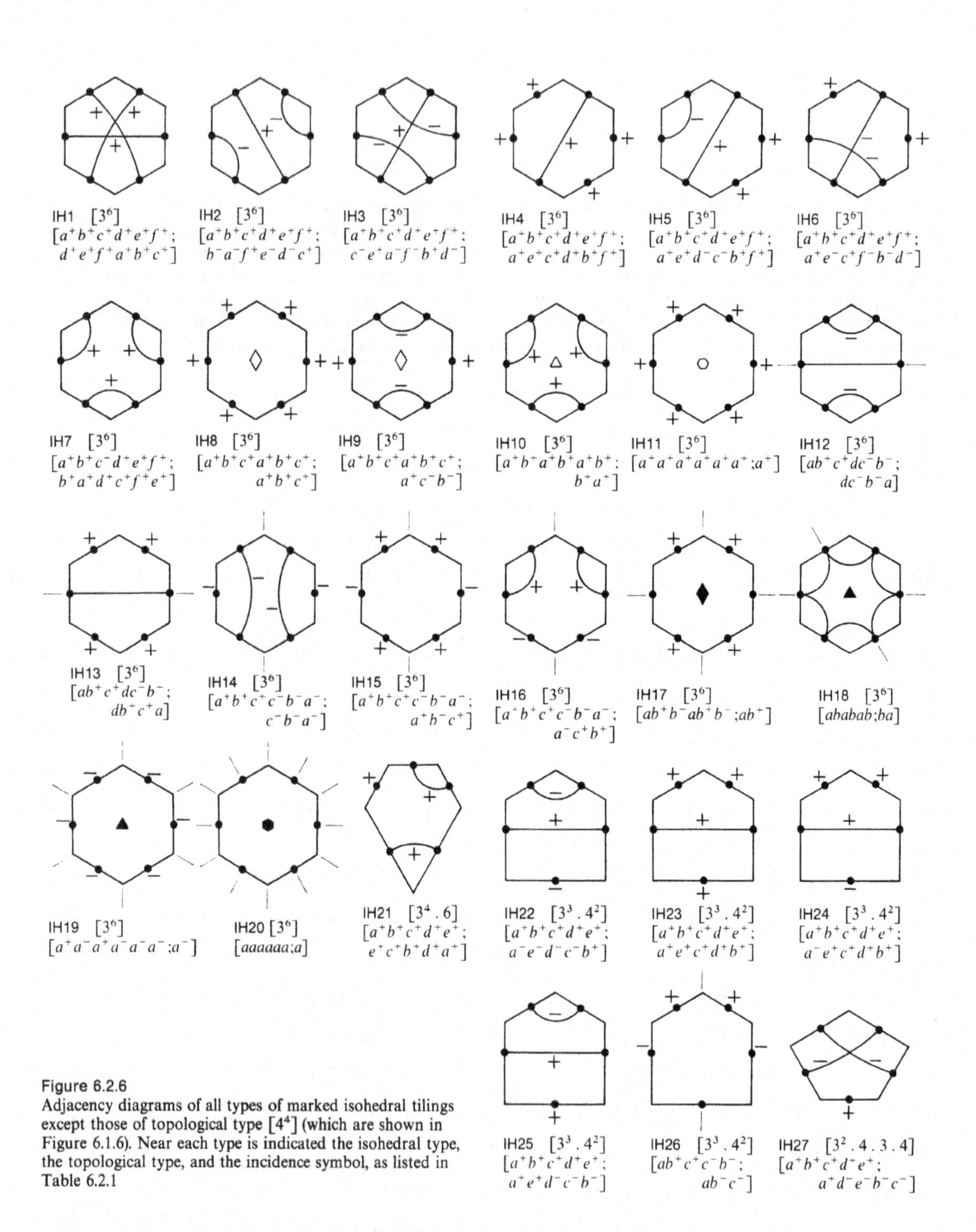

Figure 6.2.6
Adjacency diagrams of all types of marked isohedral tilings except those of topological type $[4^4]$ (which are shown in Figure 6.1.6). Near each type is indicated the isohedral type, the topological type, and the incidence symbol, as listed in Table 6.2.1

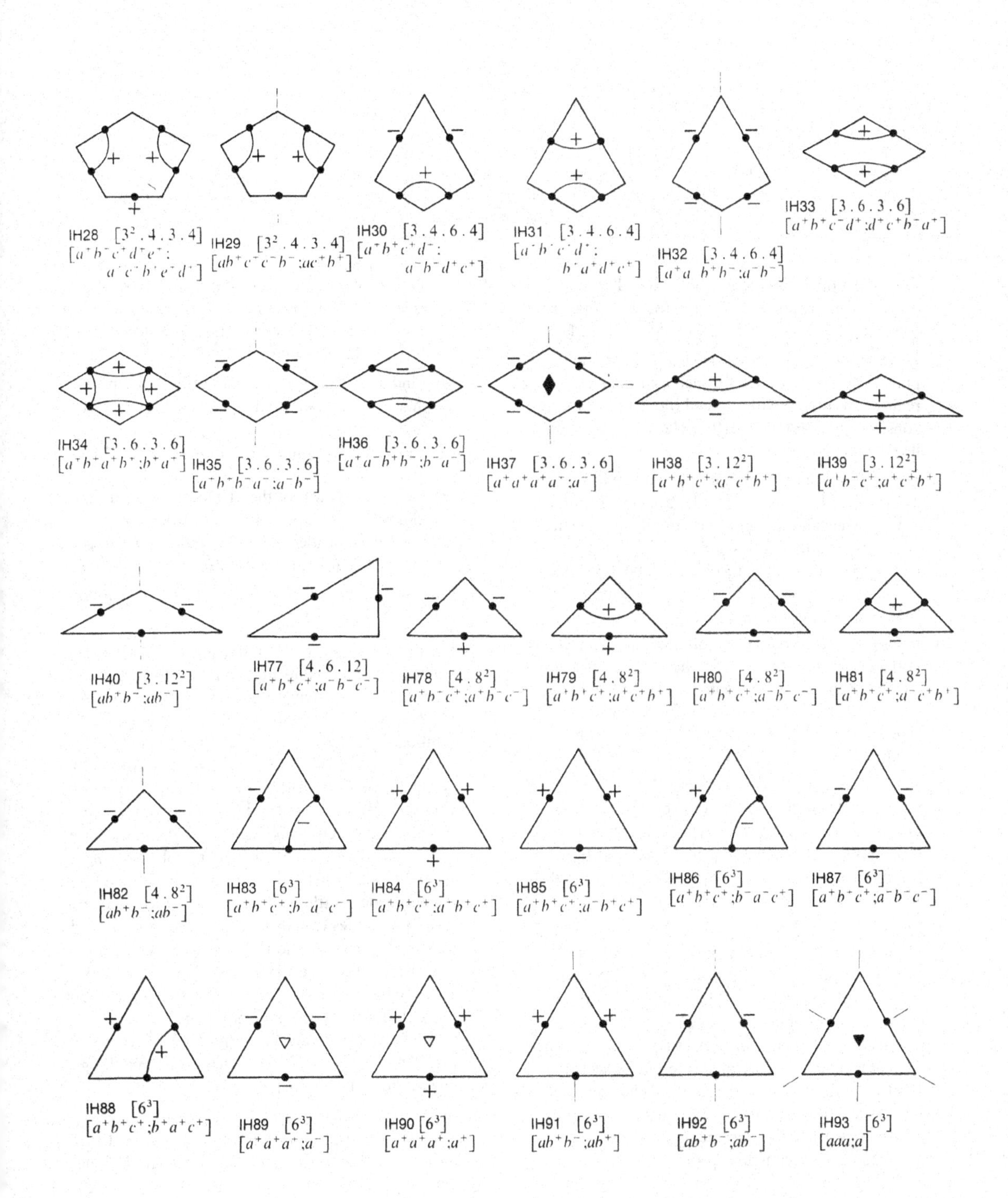
IH28 [3^2 . 4 . 3 . 4]
IH29 [3^2 . 4 . 3 . 4]
IH30 [3 . 4 . 6 . 4]
IH31 [3 . 4 . 6 . 4]
IH32 [3 . 4 . 6 . 4]
IH33 [3 . 6 . 3 . 6]
IH34 [3 . 6 . 3 . 6]
IH35 [3 . 6 . 3 . 6]
IH36 [3 . 6 . 3 . 6]
IH37 [3 . 6 . 3 . 6]
IH38 [3 . 12^2]
IH39 [3 . 12^2]
IH40 [3 . 12^2]
IH77 [4 . 6 . 12]
IH78 [4 . 8^2]
IH79 [4 . 8^2]
IH80 [4 . 8^2]
IH81 [4 . 8^2]
IH82 [4 . 8^2]
IH83 [6^3]
IH84 [6^3]
IH85 [6^3]
IH86 [6^3]
IH87 [6^3]
IH88 [6^3]
IH89 [6^3]
IH90 [6^3]
IH91 [6^3]
IH92 [6^3]
IH93 [6^3]
[aaa;a]

EXERCISES 6.2

1. Determine the incidence symbols, the adjacency diagrams and the isohedral types for the tilings in Figures 1.3.3, 1.3.7, 1.5.4, 1.5.5, 1.5.7 and 6.0.1.
2. For each of the 81 types of isohedral tilings determine the smallest value of n such that the type can be realized by a tiling with n-gonal tiles. (For example, consider types IH1, IH7, IH10, IH12, IH22, IH25, IH29, IH33, IH84 and IH90.)
3. For each of the types marked by N in column (6) of Table 6.2.1 verify that if the tiles are convex and if the tiling has all the symmetries implied by the incidence symbol, then it has additional symmetries and is therefore of a different isohedral type. (For example, consider types IH1, IH14, IH31, IH33, IH39, IH47, IH64, IH79 and IH86.)
4. For each type marked C in column (6) of Table 6.2.1 find a tiling of that type which has convex tiles. Try to find a general rule for doing this rather than considering every type individually. Note that some types are already represented by convex tiles in Figure 6.2.4.
5. Verify that if $\mathscr{T}$ is an (unmarked) tiling which has all the symmetries indicated by the incidence symbol and is of one of the twelve types marked M in column (6) of Table 6.2.1, then $\mathscr{T}$ has additional symmetries and is therefore of a different isohedral type.
6. (*a*) Show that *c2* is a *forbidden supergroup* for the induced tile group of the isohedral type IH1. By this we mean (as in Section 5.2) that if an unmarked tiling $\mathscr{T}$ has all the symmetries of the type IH1 and the prototile T has symmetry group $S(T) = c2$, then $\mathscr{T}$ possesses additional symmetries and is therefore of a different isohedral type.
   (*b*) Is *d1* a forbidden supergroup for the type IH1?
   (*c*) Show that *d2* is not a forbidden supergroup for the type IH29, by constructing a tiling of that type with a rectangular prototile.
   **(*d*) Determine the forbidden supergroups for all isohedral types of (unmarked) tilings. (For example, consider IH10, IH12, IH30, IH33, IH46, IH56, IH57, IH68, IH73 and IH86.)
7. (*a*) Find an (unmarked) isohedral tiling of type IH12 that is monotoxal (that is, all edges are congruent to each other).
   (*b*) Show that no (unmarked) tiling of type IH14 is monotoxal.
   (*c*) Determine which of the 81 types of isohedral tilings have monotoxal representatives. (For example, consider IH5, IH17, IH24, IH29, IH33, IH38, IH43, IH66, IH79 and IH85.)
8. Find an unmarked isohedral tiling which is monogonal but is not isogonal.
9. Recall from Section 4.3 that a tiling $\mathscr{T}$ is called *homogeneous* if it is normal and if all tiles of $\mathscr{T}$ have the same valence sequence. A *proper labelling* of a tile T in a homogeneous tiling is an assignment of symbols (labels $a, b, \ldots$) to oriented or non-oriented edges of T in such a way that each (combinatorial) symmetry of T implied by the labelling is compatible with the valences of the vertices. (For example, the assignments of letters in Figures 6.2.7(*a*) and (*b*) are proper labellings, the other two are not.) The *tile symbol* of a labelled tile is defined as in Section 6.1. A homogeneous tiling $\mathscr{T}$ admits an *incidence symbol* $[A; B]$ if its tiles can all be labelled so as to have the same tile symbol A, and if the labels on adjacent tiles conform to the adjacency symbol B.

   Prove the following theorem. Every homogeneous tiling that admits an incidence symbol $[A; B]$ is topologically equivalent to a marked Laves tiling with the same incidence symbol (that is, to a tiling of one of the 93 types listed in Table 6.2.1).
10. In Table 6.2.1 there are 46 *basic* types of (unmarked) isohedral tilings, that is, types in which the induced tile group is the trivial group *c1*. (Compare Delone

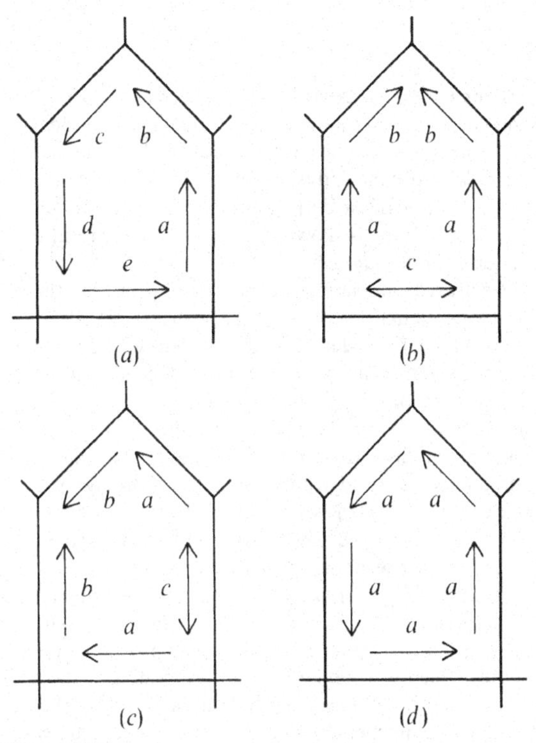

Figure 6.2.7
The labelling of a tile in a homogeneous tiling. Labellings shown in (a) and (b) are proper, those in (c) and (d) are not.

[1959], Bantegnie [1978].) All other types of isohedral tilings can be obtained by "uniting" several tiles of a suitable basic tiling. For each of the non-basic types find to which basic types is it related in this manner.

11. An edge E of a tiling $\mathcal{T}$ is *marked* if there is a set (the *marking*) associated with it. Here we shall use markings of edges only in connection with Laves tilings $\mathcal{T}$, and the only markings we shall be concerned with consist of arrowheads in different positions, and of different shapes if we need to distinguish several of them. Thus we are interested only in edge-markings similar to the ones shown in Figure 6.2.8.

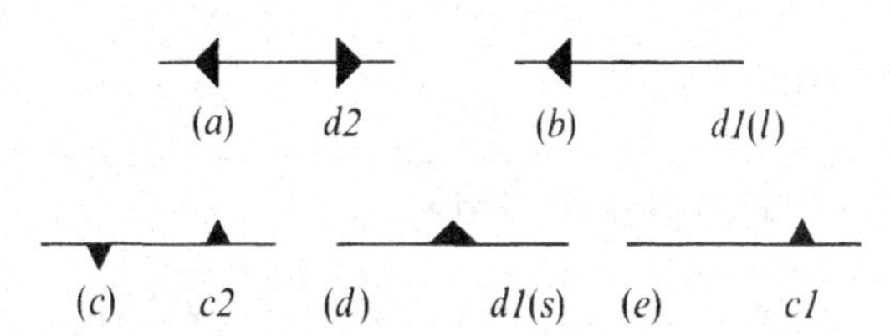

Figure 6.2.8
Edge markings used in Exercise 6.2.11. The symmetry groups and corresponding types of edge-curves are indicated.

If $\mathcal{T}$ is a tiling with marked edges, a *symmetry* of $\mathcal{T}$ is an isometry that maps $\mathcal{T}$ onto itself and preserves the edge-markings. Clearly it is possible to define isohedral tilings with marked edges and their incidence symbols in the same way as for isohedral tilings. For example, the edge-marked isohedral tiling in Figure 6.2.9 is easily seen to be of type IH13.

Show that all 93 isohedral types can be represented by edge-marked Laves tilings. Try to find a general procedure for doing this rather than considering every type individually.

Figure 6.2.9
An edge-marked isohedral tiling of type IH13.

6.3 ISOGONAL TILINGS

In the previous two sections we have discussed the classification of isohedral tilings. We shall now describe an analogous classification of isogonal tilings, that is, tilings whose symmetry groups act transitively on their vertices. We shall restrict attention to normal tilings, except that at the end of the section the construction of non-normal isogonal tilings will be briefly described.

Let $\mathscr{T}$ be an isogonal tiling and let V be any vertex of $\mathscr{T}$. Consider an edge E incident with V, and with the end of E near V associate a "sense of crossing". By this we mean that we associate with the end of E the clockwise or the anticlockwise sense of rotation about V. We shall say that we have a *sensed end* of E, and denote it on diagrams by a small arrow whose shaft crosses E near V (see Figure 6.3.1). This kind of assignment of arrows across the edges of the tiling must be distinguished from the more usual one along the edges of the tiles which we used in the classification of isohedral tilings.

To obtain a "vertex symbol" L of $\mathscr{T}$ we start by assigning a symbol, say a, as a label to a sensed end of an edge incident with the vertex V. Applying all the symmetries of $S(\mathscr{T})$ we obtain (because $\mathscr{T}$ is isogonal) an assignation of the symbol a to some sensed end of at least one edge at each vertex of $\mathscr{T}$. It is possible that two or more sensed ends at a given vertex are assigned the same label; it also may happen (if the sense of an end is reversed by an operation of $S(\mathscr{T})$) that the same label is assigned a second time to the same end, but with the opposite sense. In the latter case we shall consider the label as attached to an *unsensed end* and denote it on diagrams by a double-headed arrow crossing the end.

If the ends of all edges are labelled, the process is completed. Otherwise we choose an unlabelled sensed end of an edge, assign to it a new label, say b, and proceed as above. We continue in a similar manner until all ends are labelled. By Statement 4.4.1 we will have to use at most six different labels. A *vertex symbol* L of $\mathscr{T}$ is now obtained by reading off, in cyclic order, the labels attached to the ends of edges at V. If we are reading counterclockwise, then a superscript $+$ or $-$ is added to a letter to indicate counterclockwise or clockwise sensing of the end carrying that label, and no superscript is used if the end is unsensed.

An example of the derivation of a vertex symbol of an isogonal tiling is shown in Figure 6.3.1; the resulting vertex symbol is $ab^+c^+c^-b^-$.

Next we define the adjacency symbol A of $\mathscr{T}$ in the following manner. Suppose that a vertex symbol L has been chosen. Consider the edges incident with a vertex V and suppose that the edge with end labelled with the first letter appearing in the vertex symbol (say w, where w is $a, b, c, \ldots$) bears a label x at its other end. If the end w is unsensed then so is end x, and x is the first component of the adjacency symbol A. If the end w is sensed then *either* x is sensed *like* w (by which we mean that both are sensed counterclockwise, or both are sensed clockwise, about their respective vertices), *or* x is sensed *unlike* w. In the former case the first component of the adjacency symbol is x^+, in the latter x^-. The second, third, ... components of the adjacency symbol A are defined in a similar way, corresponding to the second, third, ... label appearing in the vertex symbol, until all the distinct letters in the vertex symbol are exhausted. Using this procedure it is found that the adjacency symbol of the example in Figure 6.3.1 is ac^+b^+.

Combining a vertex symbol L and the corresponding adjacency symbol A we obtain an *incidence symbol* $(L; A)$ of the isogonal tiling $\mathscr{T}$. Two isogonal tilings are said to be of the *same type* if they are topologically equivalent and their incidence symbols differ trivially. This last condition means that one of the incidence symbols can be transformed into the other by a change of labels or senses of ends of edges.

To carry out the enumeration of all isogonal types we need only to determine the topological types of isogonal tilings, and for each of these obtain a complete list of possible incidence symbols. Since each isogonal tiling is obviously homeogonal, by Statement 4.4.1 it is topologically equivalent to one of the eleven uniform tilings by regular polygons shown in Figure 2.1.5. Taking each of these eleven tilings in turn, we consider the different ways in which symmetries of the tiling can

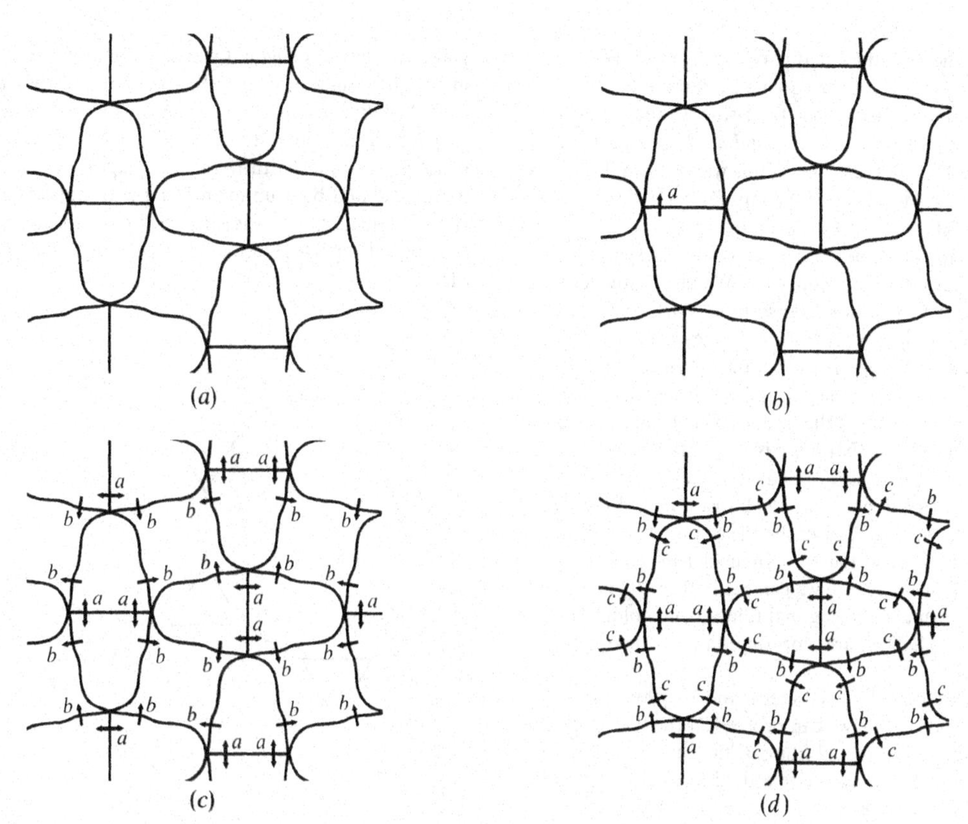

Figure 6.3.1
The labelling procedure for the ends of the edges in an isogonal tiling $\mathscr{T}'$. The tiling is shown in (*a*) and it can be verified that $S(\mathscr{T}'|V) = S(V) = d1$ in this case. Parts (*b*), (*c*) and (*d*) show successive stages in the labelling—here it is necessary to use three letters *a*, *b*, *c* to complete the process. A vertex symbol $ab^+c^+c^-b^-$ and the corresponding adjacency symbol ac^+b^+ can be read off from (*d*). This tiling is of the type IG29 in the notation of Table 6.3.1.

leave a vertex V invariant. These symmetries form a group $S(\mathscr{T}\,|\,V)$ which we shall call the *induced vertex group*, in analogy to the induced tile groups $S(\mathscr{T}\,|\,T)$ considered earlier. The possible induced vertex groups are determined by the topology of the tiling. For example, consider the tiling (3.12^2) (see Figure 6.3.2(*a*)). Since the vertex V (and therefore every vertex) has valence 3, the induced vertex group is at most *d3*, corresponding to the six permutations of the three edges meeting at V. But $S(\mathscr{T}\,|\,V)$ cannot be equal to *d3* or to *c3* since two of the edges (those separating a triangle from a 12-gon) are essentially different from the third (which separates two 12-gons). Hence the only possible groups $S(\mathscr{T}|V)$ are *c1*, which leaves every edge at V fixed, and *d1*, which interchanges the two edges that play similar roles. For each such group we can write down a corresponding vertex symbol – in the case under discussion we may choose the vertex symbols to be $a^+b^+c^+$ and ab^+b^-, as shown in Figure 6.3.2(*b*). Proceeding in this way we can determine all possible induced vertex groups for each of the eleven topological types of isogonal tilings.

The determination of possible adjacency symbols can now proceed in the obvious way. We take sequences of letters, with or without superscripts + or −, and test each whether it is a possible adjacency symbol. This procedure is analogous to that described for the isohedral tilings in Sections 6.1 and 6.2. In fact, because of the duality between the uniform tilings and the Laves tilings, it turns out that there is very little we have to do. The list of vertex symbols and adjacency symbols (and therefore of incidence symbols) for isogonal tilings is *exactly* the same as that of the tile symbols and adjacency symbols (and therefore of incidence symbols) for isohedral tilings. Hence we can read off the 93 possible incidence symbols of isogonal tilings from column (3) of Table 6.2.1. They are reproduced in column (3) of Table 6.3.1.

To make that duality precise and to justify the assertion just made, we can proceed in the following way. With each isogonal tiling $\mathcal{T}$ that has an incidence symbol $[L; A]$ we associate a dual and dually situated tiling $\mathcal{T}^*$. By the results of Chapter 4, $\mathcal{T}^*$ is homeohedral and so it is topologically equivalent to a Laves tiling $\mathcal{T}_0^*$. To each tile of $\mathcal{T}^*$ we assign a tile symbol coinciding with the vertex symbol of that vertex of $\mathcal{T}$ which it contains. This is done in a systematic manner, as indicated in Figure 6.3.3, by transferring the sense across ends of edges of $\mathcal{T}$ to directions along the edges of the tiles of $\mathcal{T}^*$ and preserving the labels. The homeomorphism between $\mathcal{T}^*$ and $\mathcal{T}_0^*$ leads to a labelling of $\mathcal{T}_0^*$ under which it is isohedral and, with the labels understood as markings on the tiles of $\mathcal{T}_0^*$, its incidence symbol is $[L; A]$. Since by Statement 6.2.1 there are 93 possible incidence symbols for $\mathcal{T}_0^*$, we conclude that there can be *at most* 93 types of isogonal tilings. In fact, it is clear that there are exactly 93 types if we agree to consider isogonal tilings with *marked edges* instead of isogonal tilings as originally defined (without any markings). The situation is exactly analogous to that in the isohedral case; there are 93 types of isogonal tilings with marked edges, some of which may not correspond to isogonal tilings with unmarked edges, but can only be represented by a uniform tiling with marked edges. Actually, it turns out that the 93 types can also be obtained and represented by isogonal tilings with marked tiles.

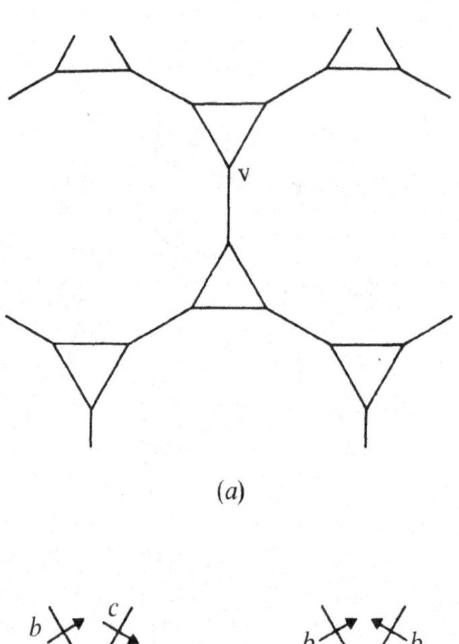

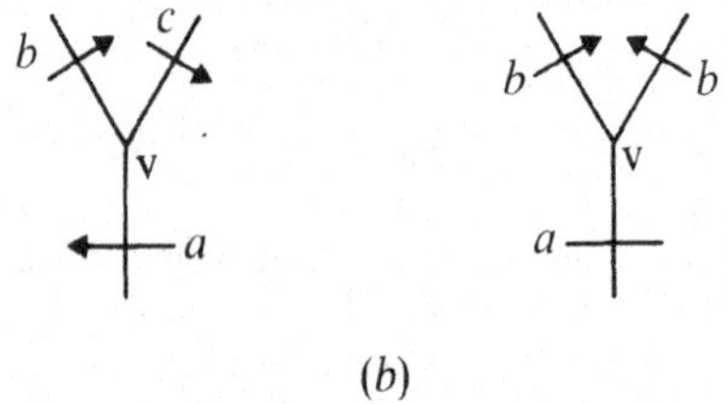

Figure 6.3.2
For the tiling of topological type (3.12^2) shown in (a), only two induced vertex groups are possible, namely *c1* and *d1*. Corresponding labellings of the ends of the edges are shown in (*b*).

To determine which of the 93 types can be represented by isogonal tilings with unmarked tiles, we proceed as follows. We examine each type in turn, and replace each labelled edge of the uniform tiling (or of any tiling topologically equivalent to it, which has all the symmetries required by the incidence symbol) by a suitable simple curve according to the following rules, which are analogous to those for isohedral tilings given in Section 6.1.

- (*a*) All mutually equivalent edges are replaced in the same manner by congruent curves.
- (*b*) If the two ends of an edge bear the same label and are sensed alike (that is, both clockwise, or both counterclockwise, about their respective vertices) then we can replace the edge by any S-curve.
- (*c*) If the two ends of an edge bear the same label and are sensed unlike each other, then the edge may be replaced by any C-curve.
- (*d*) If the two ends of an edge are sensed and have different labels, we may replace the edge by any J-curve.
- (*e*) In all other cases the edge must be taken as a straight line-segment.
- (*f*) Each curve introduced misses all the vertices of the tiling except the two at its endpoints, and it meets no other curve except at a common endpoint.

Abiding by these rules—which clearly leave the incidence symbols unchanged—it is possible to construct isogonal tilings by unmarked tiles that correspond to all types except two. These two are the types denoted IG18 and IG73 in Table 6.3.1. For the former, the incidence symbol implies that it coincides with the regular tiling by equilateral triangles and hence, when the markings are removed (or disregarded), becomes of type IG20; similarly, any tiling of type IG73 reverts, on removal of markings, to type IC76. In Figure 6.3.4 we illustrate these two types both by uniform tilings with marked edges, and also by uniform tilings with marked tiles. In Figure 6.3.5 we show examples of isogonal tilings of the other 91 types.

The above discussion can be summarized in the following comprehensive statement:

6.3.1 *There exist exactly* 93 *types of marked isogonal tilings. All can be represented by uniform tilings with marked edges, and also by uniform tilings with marked tiles. Of these,* 91 *types can be represented by* (*unmarked*) *isogonal tilings and* 63 *types can be realized by isogonal tilings with convex tiles.*

Details concerning the 93 isogonal types are presented in Table 6.3.1.

We shall now briefly consider non-normal isogonal tilings. We assume that condition N.1 is fulfilled; since all vertices have the same finite valence there are only finitely many distinct prototiles and so condition N.3 holds. Therefore, an isogonal tiling $\mathscr{T}$ can fail to be normal only by violating condition N.2, and it is easy to see that there is only one way in which this can happen—$\mathscr{T}$ must contain digons. Actually, from any normal isogonal tiling a non-normal isogonal tiling may be obtained by replacing edges with "bunches" of digons as indicated in Figure 6.3.6. This substitution must be performed subject to the requirements:

- (*a*) Each set of digons has the same symmetry group as the edge being replaced.
- (*b*) Edges of the same transitivity class are replaced by congruent sets of digons.
- (*c*) The new edges are disjoint except for common endpoints.

In Figure 6.3.7 we show two examples of non-normal isogonal tilings produced by this process. It may be applied to any isogonal tiling or marked isogonal tiling and it is not hard to establish that all non-normal tilings can be obtained in this way. Hence we have:

6.3.2 *Every isogonal tiling whose tiles are closed topological disks is either normal, or can be constructed from a normal isogonal tiling* (*with marked or unmarked edges*) *by the process described above for replacing edges in one or more transitivity classes by sets of digons.**

* See the Appendix beginning on page 653.

Table 6.3.1 THE 93 TYPES OF MARKED ISOGONAL TILINGS

The notation used and information given here is analogous to that of Table 6.2.1

Columns (7) and (8) indicate the transitivity classes of edges and tiles under symmetries of the tiling as we go cyclically round a vertex. In the case of tiles (column (8)) the notation also tells us the number of edges: $T_1, T_2, \ldots$ denote different transitivity classes of "triangles" (that is, tiles with 3 edges), $Q_1, Q_2, \ldots$ the classes of "quadrangles" (tiles with 4 edges), and so on. H stands for "hexagon", O for "octagon" and D for "dodecagon". This terminology does not, of course, imply that the tiles are polygons; it is introduced simply as a convenient way to denote the number of edges of each tile.

Column (9) indicates the number of aspects of the tiles in the notation of Table 6.2.1 except that here the various transitivity classes of tiles that appear in column (8) are listed separately.

Column (11) gives cross reference to Table 6.2.1 by indicating the isohedral type in those cases where the isogonal-type is necessarily isohedral.

Type (1)	Topological type (2)	Incidence symbol (3)	Symmetry group (4)	Induced vertex group (5)	Real-izations (6)
IG1	(3^6)	$(a^+b^+c^+d^+e^+f^+; d^+e^+f^+a^+b^+c^+)$	$p1$	$c1$	N
IG2		$(a^+b^+c^+d^+e^+f^+; b^-a^-f^+e^-d^-c^+)$	pg	$c1$	N
IG3		$(a^+b^+c^+d^+e^+f^+; c^-e^+a^-f^-b^+d^-)$	pg	$c1$	N
IG4		$(a^+b^+c^+d^+e^+f^+; a^+e^+c^+d^+b^+f^+)$	$p2$	$c1$	C, N
IG5		$(a^+b^+c^+d^+e^+f^+; a^+e^+d^-c^-b^+f^+)$	pgg	$c1$	C, N
IG6		$(a^+b^+c^+d^+e^+f^+; a^+e^-c^+f^-b^-d^-)$	pgg	$c1$	C, N
IG7		$(a^+b^+c^+d^+e^+f^+; b^+a^+d^+c^+f^+e^+)$	$p3$	$c1$	C, N
IG8		$(a^+b^+c^+a^+b^+c^+; a^+b^+c^+)$	$p2$	$c2$	C, N
IG9		$(a^+b^+c^+a^+b^+c^+; a^+c^-b^-)$	pgg	$c2$	N
IG10		$(a^+b^+a^+b^+a^+b^+; b^+a^+)$	$p3$	$c3$	N
IG11		$(a^+a^+a^+a^+a^+a^+; a^+)$	$p6$	$c6$	N
IG12		$(ab^+c^+dc^-b^-; dc^-b^-a)$	cm	$d1(s)$	N
IG13		$(ab^+c^+dc^-b^-; db^+c^+a)$	pmg	$d1(s)$	C, N
IG14		$(a^+b^+c^+c^-b^-a^-; c^-b^-a^-)$	cm	$d1(l)$	N
IG15		$(a^+b^+c^+c^-b^-a^-; a^+b^-c^+)$	pmg	$d1(l)$	C, N
IG16		$(a^+b^+c^+c^-b^-a^-; a^-c^+b^+)$	$p31m$	$d1(l)$	C, N
IG17		$(ab^+b^-ab^+b^-; ab)$	cmm	$d2$	C, N
IG18		$(ababab; ab)$	$p31m$	$d3(s)$	M
IG19		$(a^+a^-a^+a^-a^+a^-; a^-)$	$p3m1$	$d3(l)$	N
IG20		$(aaaaaa; a)$	$p6m$	$d6$	C
IG21	$(3^4.6)$	$(a^+b^+c^+d^+e^+; e^+c^+b^+d^+a^+)$	$p6$	$c1$	C, N
IG22	$(3^3.4^2)$	$(a^+b^+c^+d^+e^+; a^-e^+d^-c^-b^+)$	cm	$c1$	N
IG23		$(a^+b^+c^+d^+e^+; a^+e^+c^+d^+b^+)$	$p2$	$c1$	C, N
IG24		$(a^+b^+c^+d^+e^+; a^-e^+c^+d^+b^+)$	pmg	$c1$	C, N
IG25		$(a^+b^+c^+d^+e^+; a^+e^+d^-c^-b^+)$	pgg	$c1$	C, N
IG26		$(ab^+c^+c^-b^-; ab^-c^+)$	cmm	$d1$	C, N
IG27	$(3^2\,4.3.4)$	$(a^+b^+c^+d^+e^+; a^+d^-e^-b^-c^-)$	pgg	$c1$	C, N
IG28		$(a^+b^+c^+d^+e^+; a^+c^+b^+e^+d^+)$	$p4$	$c1$	C, N
IG29		$(ab^+c^+c^-b^-; ac^+b^+)$	$p4g$	$d1$	C, N
IG30	$(3.4.6.4)$	$(a^+b^+c^+d^+; a^-b^-d^+c^+)$	$p31m$	$c1$	C, N
IG31		$(a^+b^+c^+d^+; b^+a^+d^+c^+)$	$p6$	$c1$	C, N
IG32		$(a^+a^+b^+b^-; a^-b^-)$	$p6m$	$d1$	C, N
IG33	$(3.6.3.6)$	$(a^+b^+c^+d^+; d^+c^+b^+a^+)$	$p3$	$c1$	C, N
IG34		$(a^+b^+a^+b^-; b^-a^-)$	$p6$	$c2$	N
IG35		$(a^+b^+b^-a^-; a^-b^-)$	$p3m1$	$d1(s)$	C, N
IG36		$(a^+a^-b^+b^-; b^-a^-)$	$p31m$	$d1(l)$	C, N
IG37		$(a^+a^-a^+a^-; a^-)$	$p6m$	$d2$	C, N
IG38	(3.12^2)	$(a^+b^+c^+; a^-c^+b^+)$	$p31m$	$c1$	C, N
IG39		$(a^+b^+c^+; a^+c^+b^+)$	$p6$	$c1$	C, N
IG40		$(ab^+b^-; ab)$	$p6m$	$d1$	C, N

Transitivity classes of edges (7)	Transitivity classes of tiles (8)	Aspects (9)	Pattern type (10)	Equal tilings (11)
αβγαβγ	$T_1T_2T_1T_2T_1T_2$	T_11D; T_21D	PP1	
ααβγγβ	$T_1T_1T_1T_2T_2T_2$	T_11D1R; T_21D1R	PP2	
αβγβαγ	$T_1T_2T_1T_2T_1T_2$	T_11D1R; T_21D1R	PP2	
αβγδβε	$T_1T_1T_2T_2T_2T_1$	T_12D; T_22D	PP7	
αβγγβδ	$T_1T_1T_2T_2T_2T_1$	T_12D2R; T_22D2R	PP9	
αβγδβδ	$T_1T_1T_2T_2T_1T_2$	T_12D2R; T_22D2R	PP9	
ααββγγ	$T_1T_2T_1T_3T_1T_4$	T_13D; T_21D; T_31D; T_41D	PP21	
αβγαβγ	$TTTTTT$	$T2D$	PP8	IH84
αββαββ	$TTTTTT$	$T2D2R$	PP10	IH86
αααααα	$T_1T_2T_1T_2T_1T_2$	T_11D;T_21D	PP22	
αααααα	$TTTTTT$	$T2D$	PP45	IH90
αββαββ	$TTTTTT$	$T1D1R$	PP6	IH83
αβγαγβ	$TTTTTT$	$T2D2R$	PP13	IH85
αβααβα	$T_1T_2T_1T_2T_1T_2$	T_11; T_21	PP6	
αβγγβα	$T_1T_1T_2T_2T_2T_1$	T_12; T_22	PP13	
αββββα	$T_1T_2T_3T_2T_3T_2$	T_11; T_23; T_31	PP25	
αββαββ	$TTTTTT$	$T2$	PP20	IH91
αααααα	$TTTTTT$	$T1D1R$	(PP26)	IH89
αααααα	$T_1T_2T_1T_2T_1T_2$	T_11; T_21	PP29	
αααααα	$TTTTTT$	$T2$	PP51	IH93
αββγα	$HT_1T_2T_1T_1$	$H1D$; T_16D; T_22D	PP42	
αβγγβ	$QQTTT$	$Q1$; $T1D1R$	PP5	
αβγδβ	$QQTTT$	$Q1D$; $T2D$	PP7	
αβγδβ	$QQTTT$	$Q2$; $T2D2R$	PP11	
αβγγβ	$QQTTT$	$Q1D1R$; $T2D2R$	PP9	
αβγγβ	$QQTTT$	$Q1$; $T2$	PP19	
αβγβγ	$TTQTQ$	$Q1D1R$; $T2D2R$	PP9	
αββγγ	TTQ_1TQ_2	Q_11D; Q_21D; $T4D$	PP30	
αββββ	$TTQTQ$	$Q1D1R$; $T4$	PP35	
αβγγ	$QHQT$	$H1$; $Q3$; $T1D1R$	PP23	
ααββ	$QHQT$	$H1D$; $Q3D$; $T2D$	PP42	
ααββ	$QHQT$	$H1$; $Q3$; $T2$	PP48	
αββα	T_1HT_2H	$H1D$; T_11D; T_21D	PP21	
αααα	$THTH$	$H1D$; $T2D$	PP43	
αββα	T_1HT_2H	$H1$; T_11; T_21	PP28	
αααα	$THTH$	$H1$; $T1D1R$	PP25	
αααα	$THTH$	$H1$; $T2$	PP49	
αββ	DDT	$D1$; $T1D1R$	PP23	
αββ	DDT	$D1D$; $T2D$	PP42	
αββ	DDT	$D1$; $T2$	PP48	

Table 6.3.1 (continued)

Type (1)	Topological type (2)	Incidence symbol (3)	Symmetry group (4)	Induced vertex group (5)	Real-izations (6)
IG41	(4^4)	$(a^+b^+c^+d^+; c^+d^+a^+b^+)$	*p1*	*c1*	*N*
IG42		$(a^+b^+c^+d^+; c^+b^+a^+d^-)$	*pm*	*c1*	*N*
IG43		$(a^+b^+c^+d^+; c^-d^+a^-b^+)$	*pg*	*c1*	*N*
IG44		$(a^+b^+c^+d^+; b^-a^-d^-c^-)$	*pg*	*c1*	*N*
IG45		$(a^+b^+c^+d^+; c^-b^-a^-d^-)$	*cm*	*c1*	*N*
IG46		$(a^+b^+c^+d^+; a^+b^+c^+d^+)$	*p2*	*c1*	*C, N*
IG47		$(a^+b^+c^+d^+; c^+b^+a^+d^+)$	*p2*	*c1*	*C, N*
IG48		$(a^+b^+c^+d^+; a^-b^-c^-d^-)$	*pmm*	*c1*	*C, N*
IG49		$(a^+b^+c^+d^+; a^-b^+c^-d^+)$	*pmg*	*c1*	*C, N*
IG50		$(a^+b^+c^+d^+; c^+b^-a^+d^+)$	*pmg*	*c1*	*C, N*
IG51		$(a^+b^+c^+d^+; c^-b^+a^-d^+)$	*pgg*	*c1*	*C, N*
IG52		$(a^+b^+c^+d^+; c^-d^-a^-b^-)$	*pgg*	*c1*	*C, N*
IG53		$(a^+b^+c^+d^+; b^-a^-c^+d^+)$	*pgg*	*c1*	*C, N*
IG54		$(a^+b^+c^+d^+; a^-b^-c^-d^+)$	*cmm*	*c1*	*C, N*
IG55		$(a^+b^+c^+d^+; b^+a^+d^+c^+)$	*p4*	*c1*	*C, N*
IG56		$(a^+b^+c^+d^+; b^+a^+c^-d^-)$	*p4g*	*c1*	*C, N*
IG57		$(a^+b^+a^+b^+; a^+b^+)$	*p2*	*c2*	*C, N*
IG58		$(a^+b^+a^+b^+; a^-b^+)$	*pmg*	*c2*	*N*
IG59		$(a^+b^+a^+b^+; b^-a^-)$	*pgg*	*c2*	*N*
IG60		$(a^+b^+a^+b^+; a^-b^-)$	*cmm*	*c2*	*N*
IG61		$(a^+b^+a^+b^+; b^+a^+)$	*p4*	*c2*	*N*
IG62		$(a^+a^+a^+a^+; a^+)$	*p4*	*c4*	*N*
IG63		$(a^+a^+a^+a^+; a^-)$	*p4g*	*c4*	*N*
IG64		$(ab^+cb^-; cb^-a)$	*pm*	*d1(s)*	*N*
IG65		$(ab^+cb^-; ab^-c)$	*pmm*	*d1(s)*	*C, N*
IG66		$(ab^+cb^-; cb^+a)$	*pmg*	*d1(s)*	*C, N*
IG67		$(ab^+cb^-; ab^+c)$	*cmm*	*d1(s)*	*C, N*
IG68		$(a^+b^+b^-a^-; b^-a^-)$	*cm*	*d1(l)*	*N*
IG69		$(a^+b^+b^-a^-; a^+b^+)$	*pmg*	*d1(l)*	*C, N*
IG70		$(a^+b^+b^-a^-; a^-b^-)$	*p4m*	*d1(l)*	*C, N*
IG71		$(a^+b^+b^-a^-; b^+a^+)$	*p4g*	*d1(l)*	*C, N*
IG72		$(abab; ab)$	*pmm*	*d2(s)*	*C*
IG73		$(abab; ba)$	*p4g*	*d2(s)*	*M*
IG74		$(a^+a^-a^+a^-; a^+)$	*cmm*	*d2(l)*	*C, N*
IG75		$(a^+a^-a^+a^-; a^-)$	*p4m*	*d2(l)*	*N*
IG76		$(aaaa; a)$	*p4m*	*d4*	*C*
IG77	(4.6.12)	$(a^+b^+c^+; a^-b^-c^-)$	*p6m*	*c1*	*C, N*
IG78	(4.8^2)	$(a^+b^+c^+; a^+b^-c^-)$	*cmm*	*c1*	*C, N*
IG79		$(a^+b^+c^+; a^+c^+b^+)$	*p4*	*c1*	*C, N*
IG80		$(a^+b^+c^+; a^-b^-c^-)$	*p4m*	*c1*	*C, N*
IG81		$(a^+b^+c^+; a^-c^+b^+)$	*p4g*	*c1*	*C, N*
IG82		$(ab^+b^-; ab^-)$	*p4m*	*d1*	*C, N*
IG83	(6^3)	$(a^+b^+c^+; b^-a^-c^-)$	*cm*	*c1*	*N*
IG84		$(a^+b^+c^+; a^+b^+c^+)$	*p2*	*c1*	*C, N*
IG85		$(a^+b^+c^+; a^-b^+c^+)$	*pmg*	*c1*	*C, N*
IG86		$(a^+b^+c^+; b^-a^-c^+)$	*pgg*	*c1*	*C, N*
IG87		$(a^+b^+c^+; a^-b^-c^-)$	*p3m1*	*c1*	*C, N*
IG88		$(a^+b^+c^+; b^+a^+c^+)$	*p6*	*c1*	*C, N*
IG89		$(a^+a^+a^+; a^-)$	*p31m*	*c3*	*N*
IG90		$(a^+a^+a^+; a^+)$	*p6*	*c3*	*N*
IG91		$(ab^+b^-; ab^+)$	*cmm*	*d1*	*C, N*
IG92		$(ab^+b^-; ab^-)$	*p6m*	*d1*	*C, N*
IG93		$(aaa; a)$	*p6m*	*d3*	*C*

Transitivity classes of edges (7)	Transitivity classes of tiles (8)	Aspects (9)	Pattern type (10)	Equal tilings (11)
$\alpha\beta\alpha\beta$	$QQQQ$	$Q1D$	PP1	IH41
$\alpha\beta\alpha\gamma$	$Q_1Q_2Q_2Q_1$	$Q_11;\ Q_21$	PP3	
$\alpha\beta\alpha\beta$	$QQQQ$	$Q1D1R$	PP2	IH43
$\alpha\alpha\beta\beta$	$QQQQ$	$Q1D1R$	PP2	IH44
$\alpha\beta\alpha\gamma$	$Q_1Q_2Q_1Q_2$	$Q_11;\ Q_21$	PP5	
$\alpha\beta\gamma\delta$	$QQQQ$	$Q2D$	PP7	IH46
$\alpha\beta\alpha\gamma$	$Q_1Q_2Q_2Q_1$	$Q_11D;\ Q_21D$	PP7	
$\alpha\beta\gamma\delta$	$Q_1Q_2Q_3Q_4$	$Q_11;\ Q_21;\ Q_31;\ Q_41$	PP14	
$\alpha\beta\gamma\delta$	$Q_1Q_2Q_2Q_1$	$Q_12;\ Q_22$	PP11	
$\alpha\beta\alpha\gamma$	$Q_1Q_2Q_2Q_1$	$Q_11D1R;\ Q_22$	PP11	
$\alpha\beta\alpha\gamma$	$QQQQ$	$Q2D2R$	PP9	IH51
$\alpha\beta\alpha\beta$	$Q_1Q_2Q_1Q_2$	$Q_11D1R;\ Q_21D1R$	PP9	
$\alpha\alpha\beta\gamma$	$QQQQ$	$Q2D2R$	PP9	IH53
$\alpha\beta\gamma\delta$	$Q_1Q_2Q_3Q_1$	$Q_12;\ Q_21;\ Q_31$	PP17	
$\alpha\alpha\beta\beta$	$Q_1Q_2Q_1Q_3$	$Q_11D;\ Q_22D;\ Q_31D$	PP30	
$\alpha\alpha\beta\gamma$	$Q_1Q_2Q_1Q_3$	$Q_14;\ Q_21D1R;\ Q_32$	PP33	
$\alpha\beta\alpha\beta$	$QQQQ$	$Q1D$	PP8	IH57
$\alpha\beta\alpha\beta$	$QQQQ$	$Q2$	PP12	IH66
$\alpha\alpha\alpha\alpha$	$QQQQ$	$Q1D1R$	PP10	IH59
$\alpha\beta\alpha\beta$	$Q_1Q_2Q_1Q_2$	$Q_11;\ Q_21$	PP18	
$\alpha\alpha\alpha\alpha$	$Q_1Q_2Q_1Q_2$	$Q_11;\ Q_21$	PP31	
$\alpha\alpha\alpha\alpha$	$QQQQ$	$Q1D$	PP32	IH62
$\alpha\alpha\alpha\alpha$	$QQQQ$	$Q2$	PP34	IH73
$\alpha\beta\alpha\beta$	$QQQQ$	$Q1$	PP4	IH64
$\alpha\beta\gamma\beta$	$Q_1Q_2Q_2Q_1$	$Q_11;\ Q_21$	PP15	
$\alpha\beta\alpha\beta$	$QQQQ$	$Q1D1R$	PP13	IH58
$\alpha\beta\gamma\beta$	$QQQQ$	$Q2$	PP19	IH67
$\alpha\alpha\alpha\alpha$	$QQQQ$	$Q1$	PP6	IH62
$\alpha\beta\beta\alpha$	$QQQQ$	$Q2$	PP13	IH69
$\alpha\beta\beta\alpha$	$Q_1Q_2Q_3Q_2$	$Q_11;\ Q_22;\ Q_31$	PP39	
$\alpha\alpha\alpha\alpha$	$Q_1Q_2Q_1Q_2$	$Q_12;\ Q_21D1R$	PP35	
$\alpha\beta\alpha\beta$	$QQQQ$	$Q1$	PP16	IH72
$\alpha\alpha\alpha\alpha$	$QQQQ$	$Q1D1R$	(PP36)	IH63
$\alpha\alpha\alpha\alpha$	$QQQQ$	$Q1$	PP20	IH74
$\alpha\alpha\alpha\alpha$	$Q_1Q_2Q_1Q_2$	$Q_11;\ Q_21$	PP40	
$\alpha\alpha\alpha\alpha$	$QQQQ$	$Q1$	PP41	IH76
$\alpha\beta\gamma$	DHQ	$D1;\ H2;\ Q3$	PP46	
$\alpha\beta\gamma$	OOQ	$O1;\ Q1$	PP17	
$\alpha\beta\beta$	OOQ	$O1D;\ Q1D$	PP30	
$\alpha\beta\gamma$	O_1O_2Q	$O_11;\ O_21;\ Q2$	PP37	
$\alpha\beta\beta$	OOQ	$O2;\ Q1D1R$	PP33	
$\alpha\beta\beta$	OOQ	$O1;\ Q1$	PP38	
$\alpha\alpha\beta$	HHH	$H1$	PP5	IH12
$\alpha\beta\gamma$	HHH	$H1D$	PP7	IH8
$\alpha\beta\gamma$	HHH	$H2$	PP11	IH13
$\alpha\alpha\beta$	HHH	$H1D1R$	PP9	IH9
$\alpha\beta\gamma$	$H_1H_2H_3$	$H_11;\ H_21;\ H_31$	PP27	
$\alpha\alpha\beta$	$H_1H_2H_1$	$H_12D;\ H_21D$	PP42	
$\alpha\alpha\alpha$	HHH	$H1$	PP24	IH18
$\alpha\alpha\alpha$	HHH	$H1D$	PP44	IH11
$\alpha\beta\beta$	HHH	$H1$	PP19	IH17
$\alpha\beta\beta$	$H_1H_1H_2$	$H_12;\ H_21$	PP47	
$\alpha\alpha\alpha$	HHH	$H1$	PP50	IH20

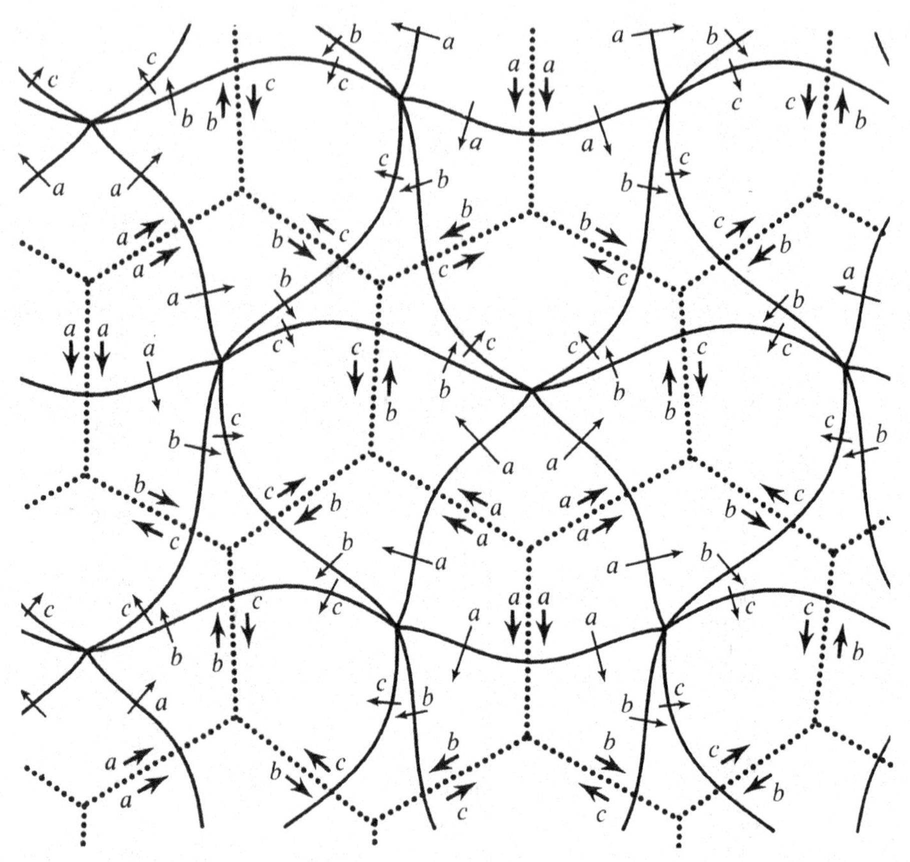

Figure 6.3.3
An illustration of how the labelling of an isogonal tiling $\mathscr{T}$ (solid lines) can be transferred to a dually situated homeohedral tiling $\mathscr{T}^*$ (dotted lines).

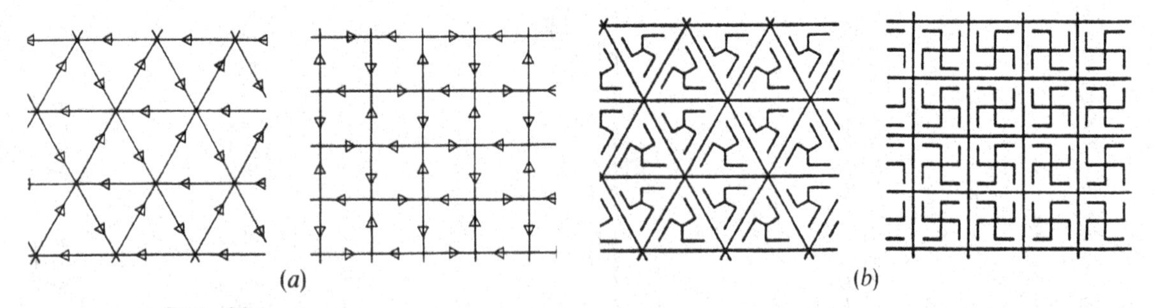

Figure 6.3.4
Tilings of types IG18 and IG73. These types cannot be represented by unmarked tilings—either the edges must be marked as in (*a*) (compare Exercise 6.2.11) or the tiles must be marked as in (*b*).

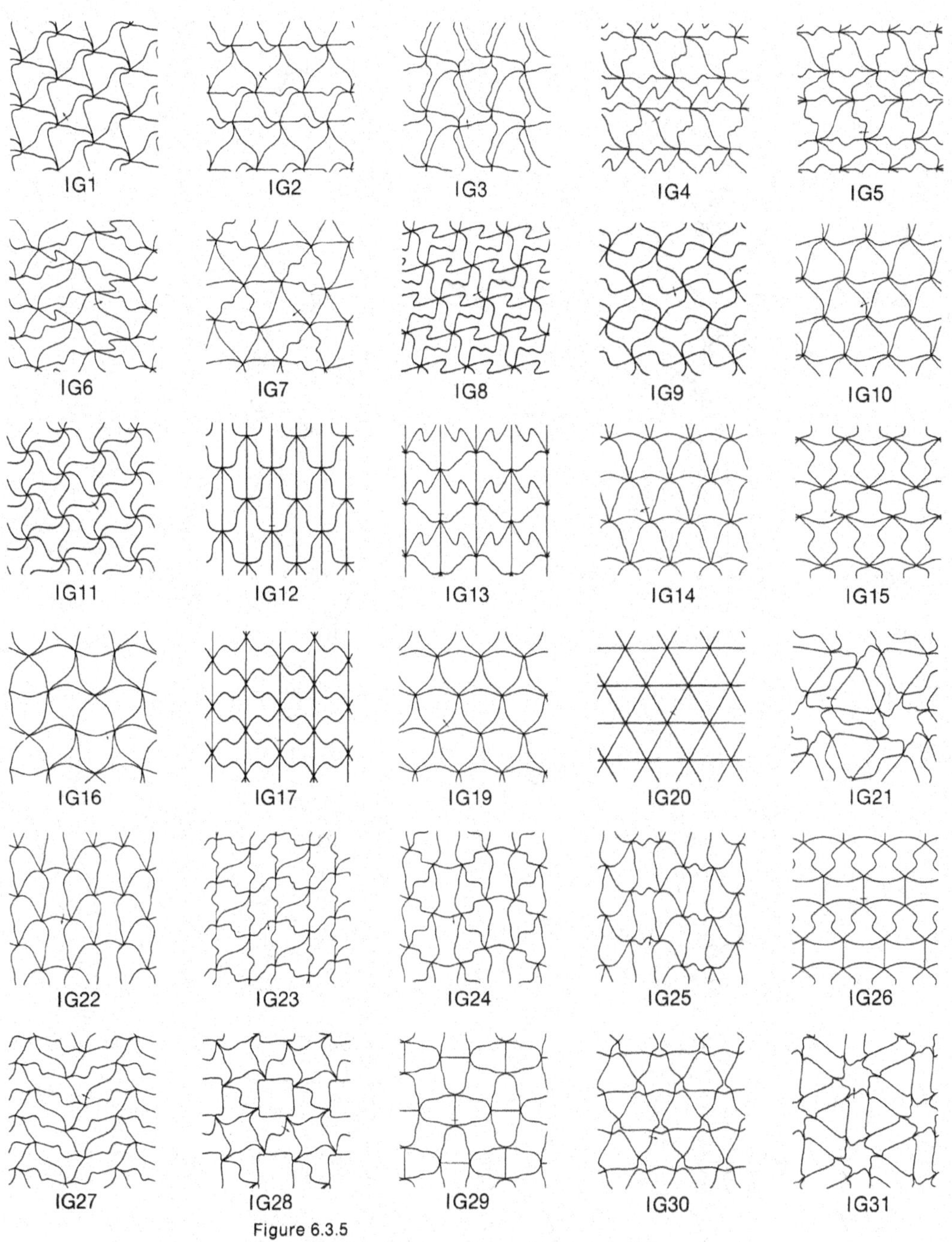

Figure 6.3.5
Examples of the 91 types of (unmarked) isogonal tilings.

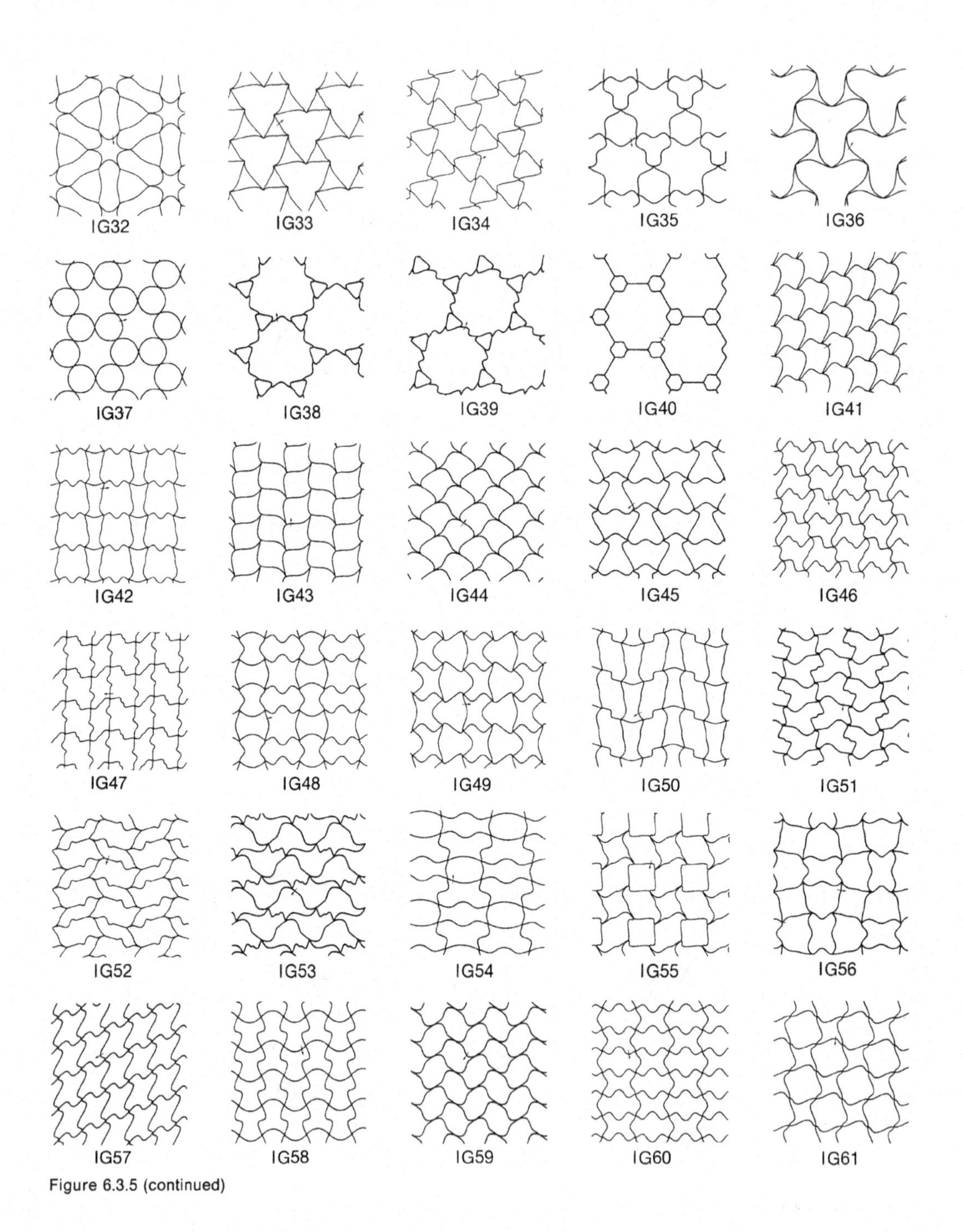

Figure 6.3.5 (continued)

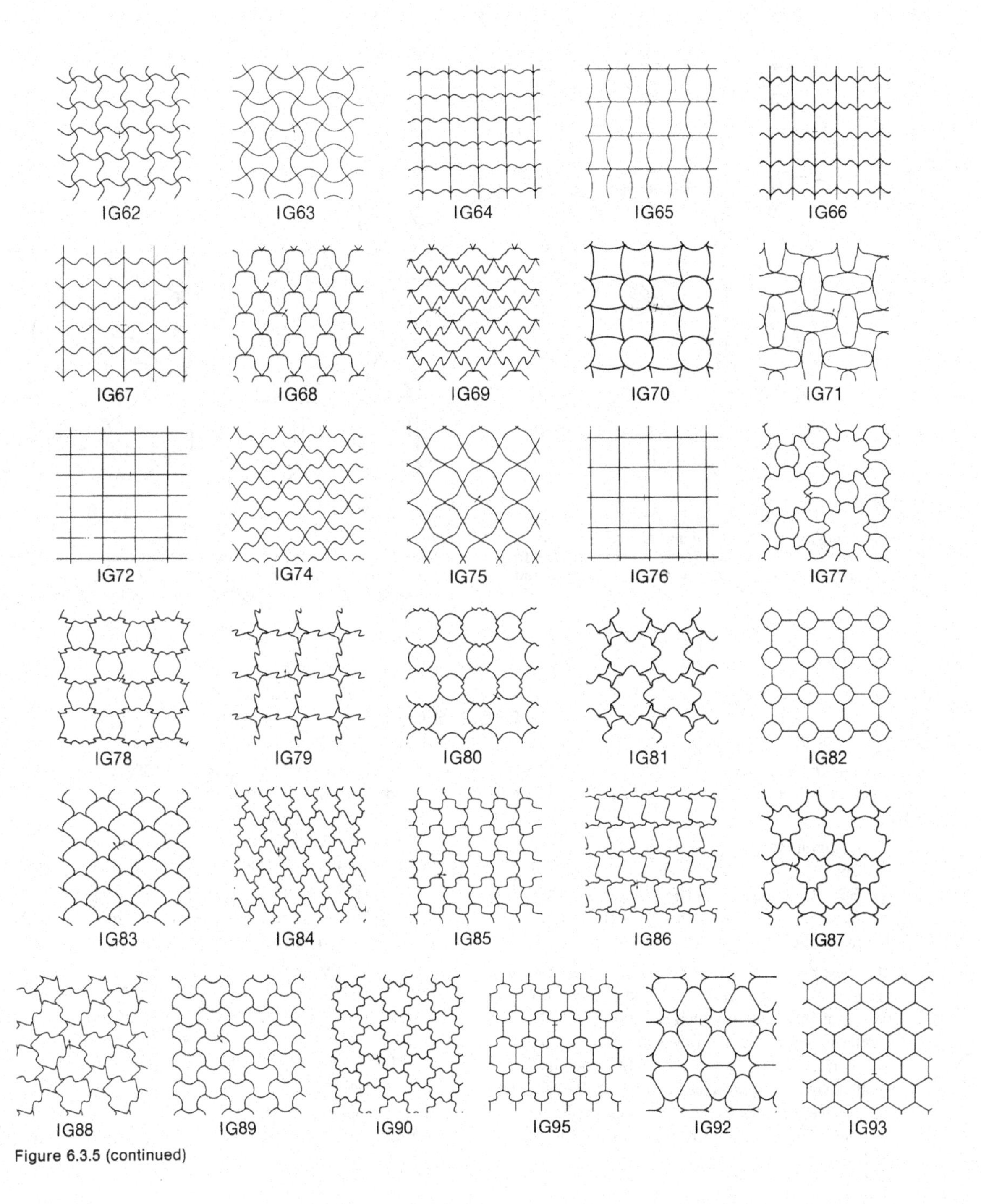

Figure 6.3.5 (continued)

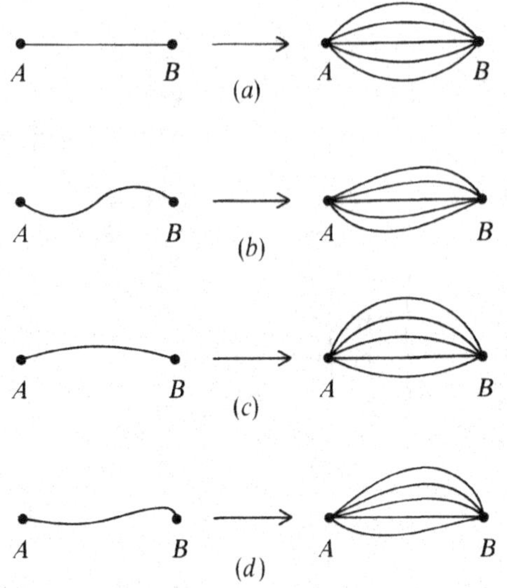

Figure 6.3.6
Replacing an edge in a normal isogonal tiling by a "bunch" of digons. Using this procedure all non-normal types of isogonal tilings can be obtained. The four kinds of edges are shown: (*a*) a line-segment, (*b*) an S-curve, (*c*) a C-curve, and (*d*) a J-curve.

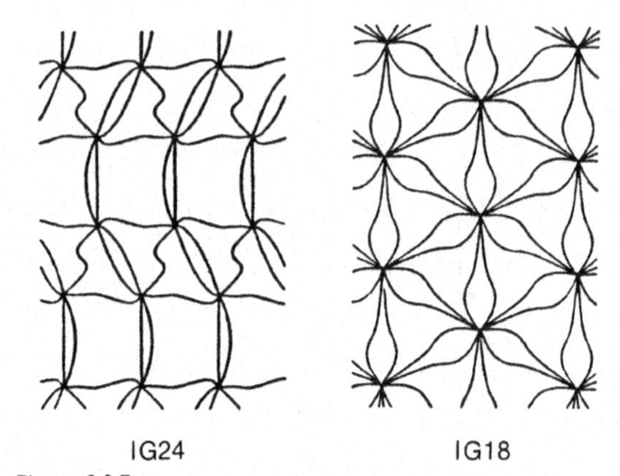

Figure 6.3.7
Two examples of non-normal isogonal tilings containing digons obtained from normal tilings by the replacement method indicated in Figure 6.3.6 and described in the text.

EXERCISES 6.3

1. Determine the incidence symbols and isogonal types for the tilings in Figures 1.3.9, 1.3.11, 1.5.4, 2.5.3 and 4.1.3.
2. Find an isogonal tiling by convex tiles for each of the types IG6, IG16, IG21, IG27, IG29, IG36, IG38, IG39, IG65, IG69, IG88 and IG92. (The types IG36, IG38 and IG39 were missed by Šubnikov [1916] and Šubnikov & Kopcik [1972].)
3. (*a*) Find an isogonal tiling of type IG1 in which each edge is a C-curve, and show that there is no tiling of this type in which each edge is a S-curve.
   (*b*) Determine whether the isogonal types IG6, IG10, IG14, IG26, IG36, IG39, IG42, IG47, IG54 and IG81 have representatives in which each edge is (i) a S-curve, or (ii) a C-curve. (A straight line-segment is, for these purposes, not counted as a S-curve or a C-curve.)
4. (*a*) Find an isogonal tiling of type IG88 in which all edges are straight line-segments but not all tiles are convex.
   (*b*) Determine all the isogonal types that admit representatives which have straight line-segments as edges but not all tiles of which are convex.
5. (*a*) Find an isogonal tiling of type IG50 in which each tile has reflective symmetry (that is, its symmetry group is *d1* or a supergroup of *d1*).
   (*b*) Show that there is no isogonal tiling of type IG51 in which each tile has reflective symmetry.
   (*c*) Find which types of isogonal tilings can have all tiles with symmetry groups (*i*) *c3*, (*ii*) *c4*, (*iii*) *c6*, or supergroups of these.
6. Explain how to construct, for each of the 91 isogonal types, a representative which is a uniform tiling with either (*a*) marked tiles, or (*b*) marked edges.

7. Refine the construction indicated in Figure 6.3.3 to prove that for each isogonal tiling $\mathcal{T}$ it is possible to find a dually situated isohedral tiling $\mathcal{T}^*$. Give examples in which $\mathcal{T}^*$ has an incidence symbol different from all possible incidence symbols of $\mathcal{T}$. Show that if we admit markings on the tiles of $\mathcal{T}^*$ then the construction may be carried out in such a way that $\mathcal{T}^*$ is a marked isohedral tiling with the same incidence symbol as $\mathcal{T}$.

   Provide examples that establish the assertion in Section 4.2 that it is impossible to deduce the types of (unmarked) isogonal tilings from those of (unmarked) isohedral tilings.
8. Find an unmarked isogonal tiling which is monohedral but not isohedral.
9. For the topological types (3^6) and (6^3) determine all the isogonal types that admit monotoxal representatives.
10. Verify that every two isogonal tilings of topological type (4.6.12) are of the same isogonal type.

6.4 ISOTOXAL TILINGS

We recall that a tiling $\mathcal{T}$ is isotoxal if the symmetry group $S(\mathcal{T})$ is transitive on the edges of $\mathcal{T}$. We shall present here a classification of isotoxal tilings which is similar in spirit but different in detail from the classifications of isohedral and isogonal tilings we have given in Sections 6.2 and 6.3. Our first objective is to define edge symbols, adjacency symbols and incidence symbols analogous to those of the previous sections. We shall again restrict attention to normal tilings, and only at the end of the section shall we briefly consider the isotoxal tilings which are not normal in that they include digons or vertices of valence 2.

Let us suppose that $\mathcal{T}$ is a normal isotoxal tiling (see, for example, the tilings of Figure 6.4.1). Let T be any tile of $\mathcal{T}$ and, as in the case of isohedral tilings, let us assign a symbol, say a, to any *directed* (oriented) edge of T. Applying the operations of the symmetry group $S(\mathcal{T})$ will yield a corresponding assignment of this symbol a to certain edges of other tiles of $\mathcal{T}$. As in the isohedral case, not only may two or more edges of the same tile be assigned the symbol a, but it may also happen that this same symbol is assigned a second time to the same edge of T but in a reversed direction. In this case we regard a as label for an *undirected* (unoriented) edge of T.

Each edge E of $\mathcal{T}$ is an edge of two tiles, and the fact that $S(\mathcal{T})$ is transitive on these edges implies that at least one of the two edges of tiles that coincide with E must be labelled a. If both these edges of tiles bear the symbol a, then necessarily all the edges of all the tiles will be labelled and there is nothing further to be done. If not, we label the other directed edge of a tile at E with the symbol b and apply the operations of $S(\mathcal{T})$ in an exactly similar manner. In this way, all the edges of all the tiles will be labelled.

To define the *edge symbol* L of $\mathcal{T}$ we first write down the symbols assigned to the edges of the tiles at E (either aa or ab) and then add superscripts + or −, or no superscript to these symbols as follows:

(*a*) If the two edges of tiles are oriented in opposite directions then the superscripts are + +.

(*b*) If the two edges of tiles are oriented in the same direction, then the superscripts are + −.

(*c*) If the two edges of tiles are unoriented, then no superscripts are used.

Hence, ignoring trivial changes of notation, there are just five possible edge symbols, namely a^+b^+, a^+a^+,

a^+a^-, ab and aa. It will be seen that, in effect, the edge symbol is obtained by reading off the symbols that occur as we proceed cyclically around an edge, and the superscripts indicate the relative orientations of the edges of tiles meeting at that edge.

In a similar manner the *adjacency symbol* of $\mathcal{T}$ is defined. Here we proceed cyclically around each of the two tiles of which an edge coincides with E, reading off the labels in order. We use superscripts $+$, $-$ or no superscript, to indicate whether the label relates to an edge of tile which is oriented coherently, or oppositely, or is unoriented. If the two tiles which abut on E have identical symbols, then we need state only one of them; otherwise both must be given in the adjacency symbol.

As examples, consider the two isotoxal tilings of Figure 6.4.1. The first has topological type $\langle 4^2; 4^2 \rangle$, edge symbol a^+b^+ and adjacency symbol $a^+b^+b^-a^-$; the second has topological type $\langle 3.6; 4^2 \rangle$, edge symbol ab and adjacency symbol aaa, $bbbbbb$.

The incidence symbol of the tiling is obtained by putting the edge symbol L and adjacency symbol A together in the form $\langle L; A \rangle$. The incidence symbols of the tilings in Figure 6.4.1 are thus $\langle a^+b^+; a^+b^+b^-a^- \rangle$ and $\langle ab; aaa, bbbbbb \rangle$, respectively.

Two isotoxal tilings are defined to be of *the same type* if they are of the same topological type and if their incidence symbols differ trivially. This means that one can be changed into the other by changing notation and possibly reversing the orientations.

The enumeration of possible types of normal isotoxal tilings follows exactly the same lines as that of isohedral and isogonal tilings described in the previous sections. We begin by listing the topological types. Since each isotoxal tiling is clearly homeotoxal, we can use the list of possible topological types of homeotoxal tilings determined in Section 4.5; these types are $\langle 3^2; 6^2 \rangle$, $\langle 3.6; 4^2 \rangle$, $\langle 4^2; 3.6 \rangle$, $\langle 4^2; 4^2 \rangle$ and $\langle 6^2; 3^2 \rangle$. Taking each in turn we attempt to assign edge and adjacency symbols in all possible ways. The number of possibilities turns out to be very limited. There are only five different edge symbols, and the edges of a triangular tile can be labelled in only two essentially different ways, namely $a^+a^+a^+$ or aaa. In

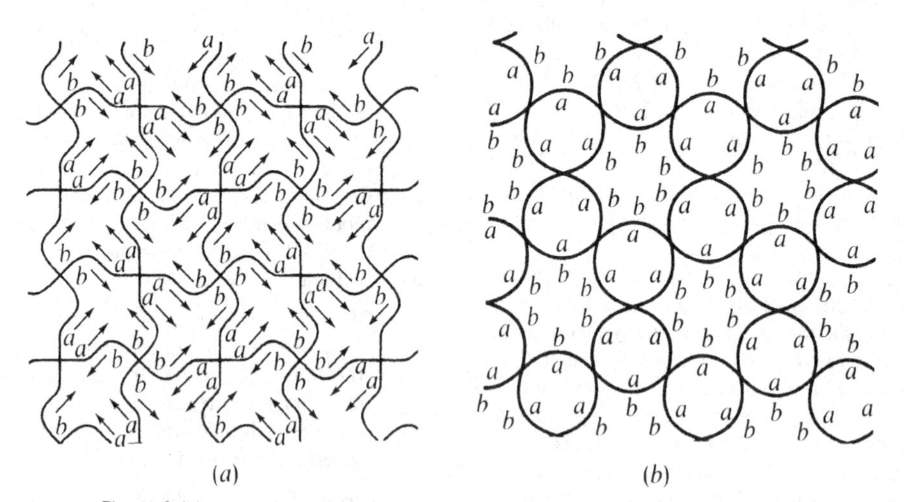

Figure 6.4.1
The labelling process for an isotoxal tiling. This labelling leads to the edge symbols and adjacency symbols indicated in the text.

the case of hexagonal tiles there are five possibilities and for quadrangular tiles there are nine. But many of these can be eliminated immediately since it is easily checked that they do not correspond to a labelling that can be carried out consistently for all the tiles of any tiling of the given topological type.

We can thus determine all the possible incidence symbols; there are 30 of these and they are listed in Table 6.4.1. They do not necessarily correspond to isotoxal tilings as originally defined, but to types of *marked* isotoxal tilings, that is, types that can be represented by tilings in which either the tiles are marked, or the edges are marked.

To convert an isotoxal tiling by marked tiles into an isotoxal tiling by unmarked tiles, we use a procedure analogous to those described previously. We start with the appropriate "standard" tiling of Figure 4.5.2, or any distortion of it that has all the symmetries required by the type in question, and then replace the edges by simple curves according to the following rules:

(*a*) If the edge symbol is a^+a^+, all edges are replaced by congruent S-curves.
(*b*) If the edge symbol is *ab*, all edges are replaced by congruent C-curves.
(*c*) If the edge symbol is a^+b^+, all edges are replaced by congruent J-curves.
(*d*) If the edge symbol is a^+a^- or *aa*, each edge is left as a straight line-segment.
(*e*) The new edges are pairwise disjoint except possibly at their endpoints.

Moreover the replacements must be done in such a way that the symmetry group $S(\mathcal{T})$ is not changed.

Carrying out this procedure for each of the 30 types we arrive at the following result:

6.4.1 *There exist precisely* 30 *types of marked normal isotoxal tilings;* 26 *of these can be represented by normal tilings with unmarked tiles.*

Examples of the 26 (unmarked) types are shown in Figure 6.4.2. The four types that cannot be represented in such a way are IT3, IT21, IT22 and IT28. Tilings of these types, with marked edges or with marked tiles, are shown in Figure 6.4.3. In each of these cases the edge symbol is a^+a^-, so that the edges must be straight and then the unmarked tiling has a larger symmetry group than that implied by the incidence symbol, so the unmarked tiling is of a different type.

Upon comparing Table 6.4.1 with the corresponding tables in the previous sections we see that 25 of the isotoxal types are also isohedral, isogonal, or both. Appropriate cross-references are given in Table 6.4.1. In particular, we draw attention to the type IT14. This is the only type of isotoxal tiling which is neither isohedral nor isogonal. On the other hand, in addition to the three regular tilings, there exist eight types (IT2, IT17, IT18, IT19, IT20, IT24, IT27, IT29) which are isohedral, isogonal and isotoxal. Two types (IT3, IT21), realizable only by marked isotoxal tilings, also have these three properties.

We now turn to consider non-normal isotoxal tilings, restricting attention to tilings in which each tile is a closed topological disk. Since at most two non-equivalent tiles meet at each edge we deduce that every isotoxal tiling is either monohedral or dihedral (actually, it must even be isohedral or 2-isohedral). It is thus uniformly bounded and the only way it can fail to be normal is by violating condition N.2. Hence it must contain digons (tiles with only two adjacents). Due to the duality between digons and pseudovertices ("vertices" of valence 2, as defined in Section 4.2) it is convenient to determine at the same time the isotoxal tilings that contain digons and those that contain pseudovertices. The arguments of Section 4.2 can be extended to enumerate the various topological types that occur. It can be shown that, in addition to the five topological types already mentioned, in this more general situation fourteen more are possible, namely:

$$\langle 2.3; 12^2\rangle, \langle 2.4; 6.12\rangle, \langle 2.4; 8^2\rangle, \langle 2.6; 4.12\rangle,$$
$$\langle 2.6; 6^2\rangle, \langle 2.8; 4.8\rangle, \langle 2.12; 4.6\rangle$$

and

$$\langle 12^2; 2.3\rangle, \langle 6.12; 2.4\rangle, \langle 8^2; 2.4\rangle, \langle 4.12; 2.6\rangle,$$
$$\langle 6^2; 2.6\rangle, \langle 4.8; 2.8\rangle, \langle 4.6; 2.12\rangle$$

Table 6.4.1 THE 30 TYPES OF MARKED ISOTOXAL TILINGS

Most of the information given here is analogous to that in Tables 6.2.1 and 6.3.1. The notation is explained in the text.

Column (9) indicates the dual of each type and an asterisk shows that the type is self-dual.

Column (11) shows whether the isotoxal type is either isohedral or isogonal and gives the corresponding list number in Tables 6.2.1 and 6.3.1.

Column (12) list all the non-normal isotoxal types that can be derived by the procedures described on pages 6.4.6 and 6.4.7.

Column (13) is explained in Exercise 6.4.6.

Type (1)	Topological type (2)	Incidence symbol (3)	Symmetry group (4)	Induced edge group (5)	Realizations (6)	Transitivity classes of vertices (7)
IT1	$\langle 3^2; 6^2\rangle$	$\langle a^+b^+; a^+a^+a^+, b^+b^+b^+\rangle$	*p3*	*c1*	*N*	$\alpha\alpha$
IT2		$\langle a^+a^+; a^+a^+a^+\rangle$	*p6*	*c2*	*N*	$\alpha\alpha$
IT3		$\langle a^+a^-; a^+a^+a^+\rangle$	*p31m*	*d1(l)*	*M*	$\alpha\alpha$
IT4		$\langle ab; ababab\rangle$	*p3m1*	*d1(p)*	*N*	$\alpha\alpha$
IT5		$\langle aa; aaaaaa\rangle$	*p6m*	*d2*	*C*	$\alpha\alpha$
IT6	$\langle 3\cdot 6; 4^2\rangle$	$\langle a^+b^+; a^+a^+a^+, b^+b^+b^+b^+b^+b^+\rangle$	*p6*	*c1*	*N*	$\alpha\alpha$
IT7		$\langle a^+b^+; a^+a^+a^+, b^+b^-b^+b^-b^+b^-\rangle$	*p31m*	*c1*	*N*	$\alpha\alpha$
IT8		$\langle ab; aaa, bbbbbb\rangle$	*p6m*	*d1(p)*	*C, N*	$\alpha\alpha$
IT9	$\langle 4^2; 3\cdot 6\rangle$	$\langle a^+b^+; a^+b^+a^+b^+\rangle$	*p6*	*c1*	*N*	$\alpha\beta$
IT10		$\langle a^+b^+; a^+b^+b^-a^-\rangle$	*p31m*	*c1*	*N*	$\alpha\beta$
IT11		$\langle a^+a^-; a^+a^-a^+a^-\rangle$	*p6m*	*d1(l)*	*C*	$\alpha\beta$
IT12	$\langle 4^2; 4^2\rangle$	$\langle a^+b^+; a^+a^+a^+a^+, b^+b^+b^+b^+\rangle$	*p4*	*c1*	*N*	$\alpha\alpha$
IT13		$\langle a^+b^+; a^+a^+a^+a^+, b^+b^-b^+b^-\rangle$	*p4g*	*c1*	*N*	$\alpha\alpha$
IT14		$\langle a^+b^+; a^+a^-a^+a^-, b^+b^-b^+b^-\rangle$	*pmm*	*c1*	*N*	$\alpha\beta$
IT15		$\langle a^+b^+; a^+b^+a^+b^+\rangle$	*p4*	*c1*	*N*	$\alpha\beta$
IT16		$\langle a^+b^+; a^+b^+b^-a^-\rangle$	*p4g*	*c1*	*N*	$\alpha\beta$
IT17		$\langle a^+b^+; a^+b^-a^+b^-\rangle$	*pgg*	*c1*	*N*	$\alpha\alpha$
IT18		$\langle a^+b^+; a^+b^-b^+a^-\rangle$	*cm*	*c1*	*N*	$\alpha\alpha$
IT19		$\langle a^+a^+; a^+a^+a^+a^+\rangle$	*p4*	*c2*	*N*	$\alpha\alpha$
IT20		$\langle a^+a^+; a^+a^-a^+a^-\rangle$	*cmm*	*c2*	*C, N*	$\alpha\alpha$
IT21		$\langle a^+a^-; a^+a^+a^+a^+\rangle$	*p4g*	*d1(l)*	*M*	$\alpha\alpha$
IT22		$\langle a^+a^-; a^+a^-a^+a^-\rangle$	*p4m*	*d1(l)*	*M*	$\alpha\beta$
IT23		$\langle ab; aaaa, bbbb\rangle$	*p4m*	*d1(p)*	*N*	$\alpha\alpha$
IT24		$\langle ab; abab\rangle$	*p4g*	*d1(p)*	*N*	$\alpha\alpha$
IT25		$\langle aa; aaaa\rangle$	*p4m*	*d2*	*C*	$\alpha\alpha$
IT26	$\langle 6^2; 3^2\rangle$	$\langle a^+b^+; a^+b^+a^+b^+a^+b^+\rangle$	*p3*	*c1*	*N*	$\alpha\beta$
IT27		$\langle a^+a^+; a^+a^+a^+a^+a^+a^+\rangle$	*p6*	*c2*	*N*	$\alpha\alpha$
IT28		$\langle a^+a^-; a^+a^-a^+a^-a^+a^-\rangle$	*p3m1*	*d1(l)*	*M*	$\alpha\beta$
IT29		$\langle ab; ababab\rangle$	*p31m*	*d1(p)*	*N*	$\alpha\alpha$
IT30		$\langle aa; aaaaaa\rangle$	*p6m*	*d2*	*C*	$\alpha\alpha$

Transitivity classes of tiles (8)	Type of dual tiling (9)	Pattern type (10)	Equal tilings (11)	Derived tilings (12)	Type of centered tiling (13)
$T_1 1D$; $T_2 1D$	IT26	PP21	IG10		IH33
$T2D$	IT27	PP43	IG11, IH90	IT2b, 2d	IH34
$T1D1R$	IT29	PP25	IG18, IH89	IT3b	IH36
$T_1 1$; $T_2 1$	IT28	PP28	IG19	IT4d	IH35
$T2$	IT30	PP49	IG20, IH93	IT5b, 5d, 5bd, 5db	IH37
$T2D$; $H1D$	IT9	PP42	IG34		IH31
$T1D1R$; $H1$	IT10	PP23	IG36		IH30
$T2$; $H1$	IT11	PP48	IG37	IT8d	IH32
$Q3D$	IT6	PP42	IH34		IH31
$Q3$	IT7	PP23	IH36		IH30
$Q3$	IT8	PP48	IH37	IT11b	IH32
$Q_1 1D$; $Q_2 1D$	IT15	PP30	IG61		IH55
$Q_1 1D$; $Q_2 2$	IT16	PP33	IG71		IH56
$Q_1 1$; $Q_2 1$	*IT14	PP14			IH48
$Q2D$	IT12	PP30	IH61		IH55
$Q4$	IT13	PP33	IH71		IH56
$Q1D1R$	*IT17	PP9	IG59, IH59		IH52
$Q1$	*IT18	PP5	IG68, IH68		IH45
$Q1D$	*IT19	PP31	IG62, IH62	IT19b, 19d	IH61
$Q1$	*IT20	PP18	IG74, IH74	IT20b, 20d	IH60
$Q1D1R$	IT24	PP35	IG73, IH63	IT21b	IH71
$Q2$	IT23	PP39	IH75	IT22b	IH70
$Q_1 1$; $Q_2 1$	IT22	PP39	IG75	IT23d	IH70
$Q2$	IT21	PP35	IG63, IH73	IT24d	IH71
$Q1$	*IT25	PP40	IG76, IH76	IT25b, 25d, 25bd, 25db	IH76
$H1D$	IT1	PP21	IH10		IH33
$H1D$	IT2	PP43	IG90, IH11	IT27b, 27d	IH34
$H1$	IT4	PP28	IH19	IT28b	IH35
$H1$	IT3	PP25	IG89, IH18	IT29d	IH36
$H1$	IT5	PP49	IG93, IH20	IT30b, 30d, 30bd, 30db	IH37

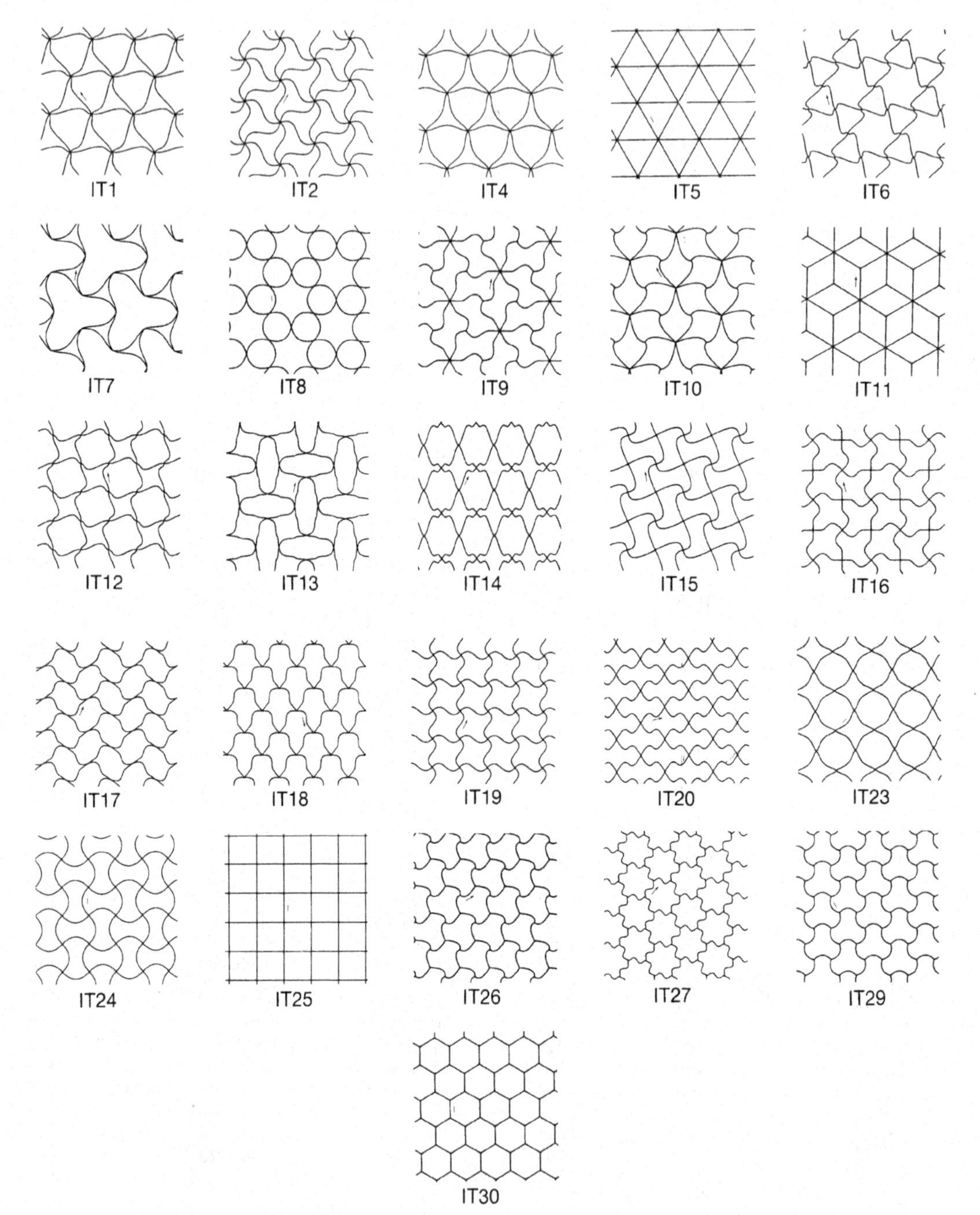

Figure 6.4.2
Examples of the 26 types of (unmarked) isotoxal tilings.

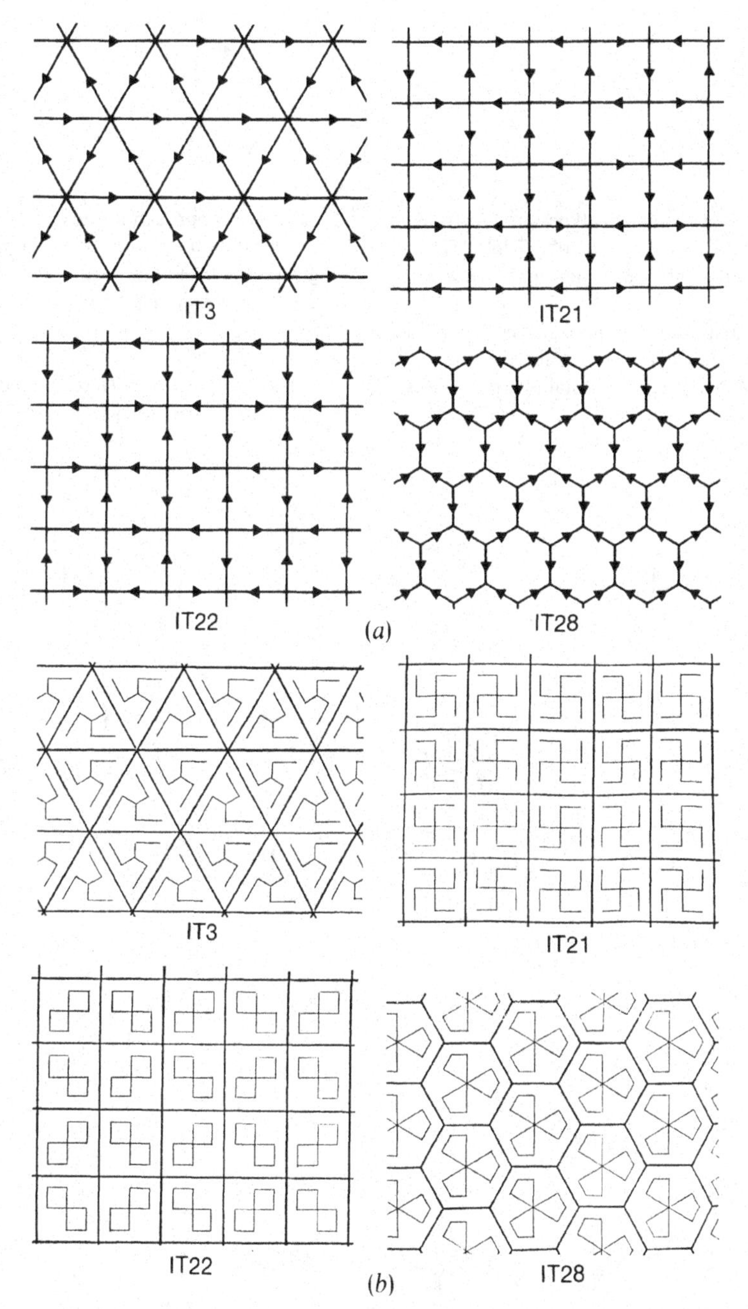

Figure 6.4.3
The four types of isotoxal tilings that cannot be represented by unmarked tilings—either the edges must be marked as in (*a*), or the tiles must be marked as in (*b*).

The first seven of these topological types are illustrated by the tilings IT5b, IT11b, IT25b, IT5db, IT30b, IT25bd, IT30db in Figure 6.4.4 and the remaining seven are their duals.

Next, the possible incidence symbols for each topological type can be determined, and so a complete enumeration of the types of marked non-normal isotoxal tilings can be accomplished.

A simpler approach to the enumeration problem is based on the observation that each of the fourteen topological types listed above can be derived from the five normal types by applying one or both of the following operations:

(*a*) *Edge bifurcation*. Here each edge is replaced by a digon. If the vertices incident with the original

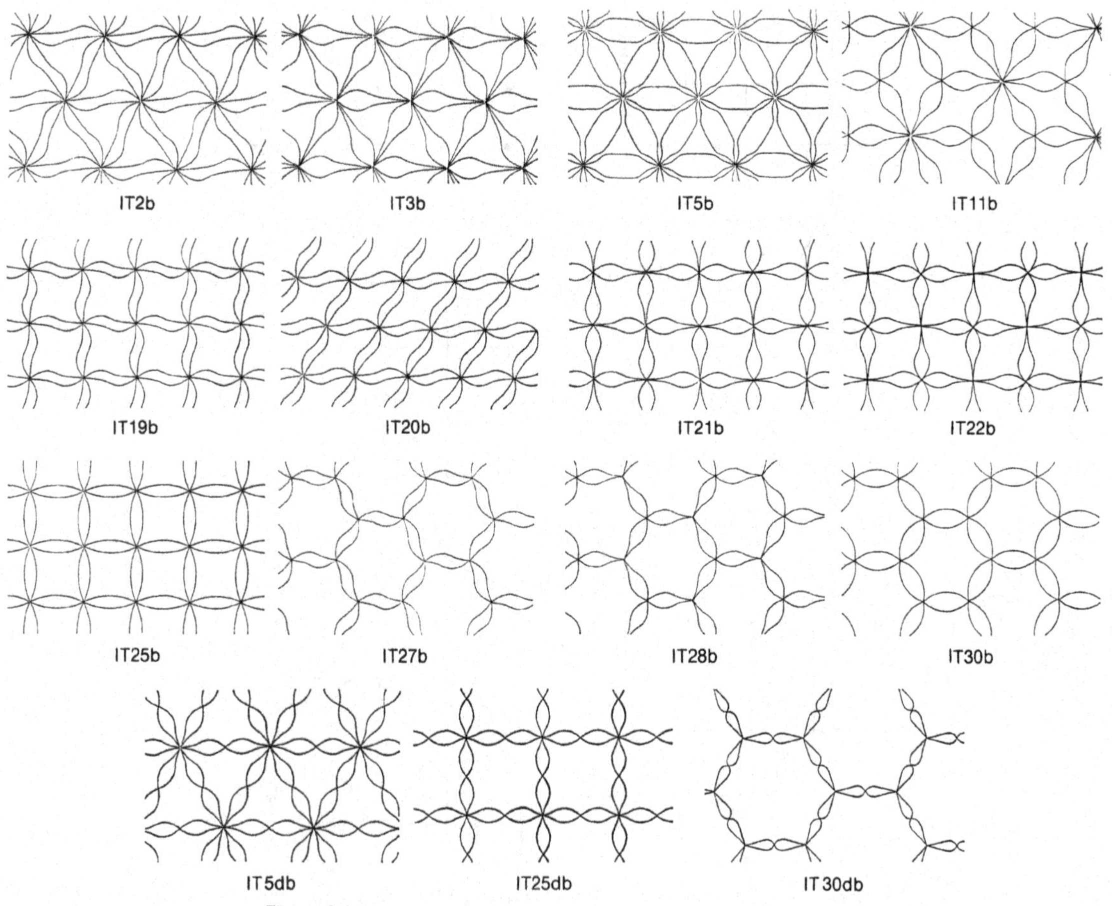

Figure 6.4.4
The fifteen types of (non-normal) isotoxal tilings that include digons.

edge are A and B then the vertices of the digon in the new tiling are A and B.

(*b*) *Edge division.* Here each edge is replaced by two edges meeting in a pseudo-vertex. If the vertices incident with the original edge are A and B, then the endpoints of the new edges are A, C and B, where C is the new pseudo-vertex.

These two operations may be applied not only to isotoxal tilings, but also—under certain conditions—to tilings with labelled edges. The method of replacement is illustrated in Figure 6.4.5. If the edge symbol is a^+a^+, a^+a^- or aa, the edge bifurcation yields a new type with edge symbol a^+b^+, a^+b^+ or ab respectively; the new adjacency symbol is obtained by adjoining b^+b^+, b^+b^- or bb to that of the original tiling. If the edge symbol is a^+a^+, ab or aa then edge division may be applied in an analogous manner, but it cannot be applied to the other three types of labelled edges.

Following this procedure we can obtain, starting for example with a tiling of type IT30, a tiling IT30b of topological type $\langle 2.6; 6^2 \rangle$ by edge bifurcation, a tiling IT30d of type $\langle 12^2; 2.3 \rangle$ by edge division, and two further tilings (IT30bd of type $\langle 4.12; 2.6 \rangle$ and IT30db of type $\langle 2.12; 4.6 \rangle$) by applying both operations consecutively. All these new tilings are said to be derived from IT30; a full list of the derived tilings is given in column (11) of Table 6.4.1. In fact, as is easily checked, all derived isotoxal types can be realized by (unmarked) tilings even if the original isotoxal types admit no such realizations.

In all we obtain fifteen additional isotoxal types with digons, of which examples are given in Figure 6.4.4, and fifteen additional types with pseudo-vertices. The latter are not illustrated since tilings representing them can be simply obtained by inserting a pseudo-vertex at the midpoint of each edge of tilings of types IT2, IT4, IT5, IT8, IT19, IT20, IT23, IT24, IT25, IT27, IT29, IT30, IT5b, IT25b and IT30b.

The completeness of the list of non-normal isotoxal types so obtained is a consequence of the following fact, which can easily be verified. Any non-normal isotoxal tiling of one of the fourteen topological types determined above, and having tiles with labelled edges, can be reduced to one of the 30 normal types of Table 6.4.1. by applying the inverses of edge bifurcation and edge division. Hence we have proved the following result:

6.4.2 *There exist* 30 *types of non-normal isotoxal tilings whose tiles are topological disks. Of these* 15 *contain digons and* 15 *contain pseudo-vertices.**

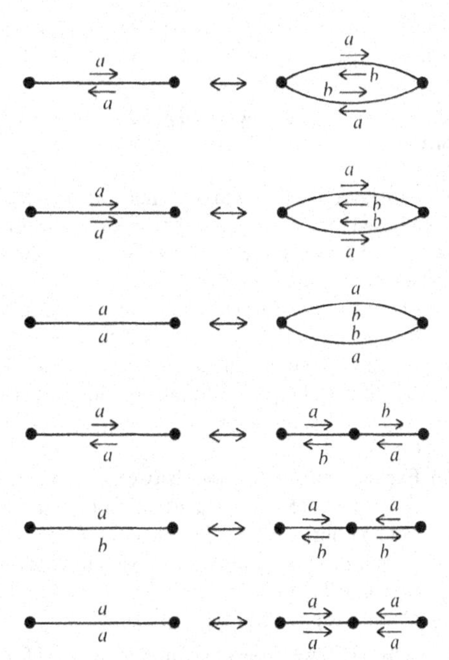

Figure 6.4.5
Edge bifurcation and edge division of labelled edges in an isotoxal tiling. These procedures enable all non-normal types to be obtained from normal isotoxal tilings.

* See the Appendix beginning on page 653.

EXERCISES 6.4

1. Determine the incidence symbols and the isotoxal types for the tilings in Figures 1.3.3, 1.3.9, 1.3.11, 2.5.3(*d*), 2.5.4(*j*), (*l*), (*n*) and 2.5.7(*b*), (*d*).

2. For each isotoxal type marked *C* in column (6) of Table 6.4.1 find a realization with convex tiles.

3. Prove that no tiling of a type marked only *N* in column (6) of Table 6.4.1 has all tiles convex.

4. Show that no normal isotoxal tiling with unmarked tiles can be of one of the types marked *M* in column (6) of Table 6.4.1.

5. Using the relations between labels indicated in Figure 6.4.6 show that each isotoxal type has a unique dual isotoxal type. Verify the assertions about dual types in column (9) of Table 6.4.1.

6. In Figure 6.4.7 we show one representative $\mathscr{T}$ for each of the five topological types of isotoxal tilings, together with a tiling $\mathscr{T}_c$ arising from $\mathscr{T}$ by *tile centering* (see Section 4.5). If a marking (or labelling) is imparted to $\mathscr{T}$ in such a way that it remains isotoxal of a certain type, these markings or labels can be considered as markings on $\mathscr{T}_c$.
   (*a*) Show that the marked tiling $\mathscr{T}_c$ is isohedral, and that the isohedral type of $\mathscr{T}_c$ is determined by the isotoxal type of the marked (or labelled) $\mathscr{T}$. (For each isotoxal type $\mathscr{T}$ the corresponding isohedral type of $\mathscr{T}_c$ is indicated in column (13) of Table 6.4.1.)
   (*b*) Explain how the relations between $\mathscr{T}$ and $\mathscr{T}_c$ can be used to derive the classification of marked isotoxal tilings from the classification of marked isohedral tilings. Use this approach to verify the enumeration given in Table 6.4.1 of the isotoxal types of topological type $\langle 4^2; 3.6\rangle$.

7. In Figure 6.4.8 we show one representative $\mathscr{T}$ for each of the five topological types of isotoxal tilings, together with a tiling $\mathscr{T}_a$ arising from $\mathscr{T}$ by a process called *edge centering*, which is analogous to the process of tile centering considered in Exercise 6.
   (*a*) Find a method of interpreting markings or labellings of $\mathscr{T}$ as markings on $\mathscr{T}_a$ in such a way that $\mathscr{T}_a$ is a marked isogonal tiling whenever the markings of $\mathscr{T}$ make it into a marked isotoxal tiling. Show that the isotoxal type of $\mathscr{T}$ determines the isogonal type of $\mathscr{T}_a$. (The numbers in column

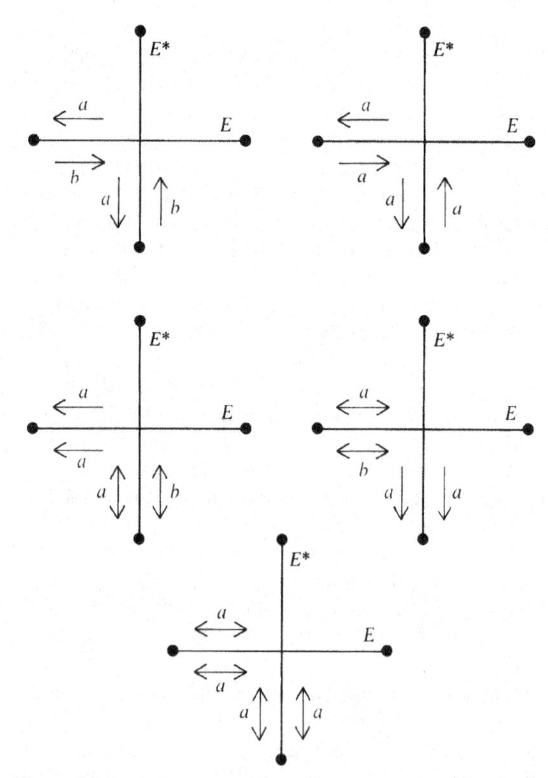

Figure 6.4.6
Relations between labelled edges of dual isotoxal tilings (see Exercise 6.4.5).

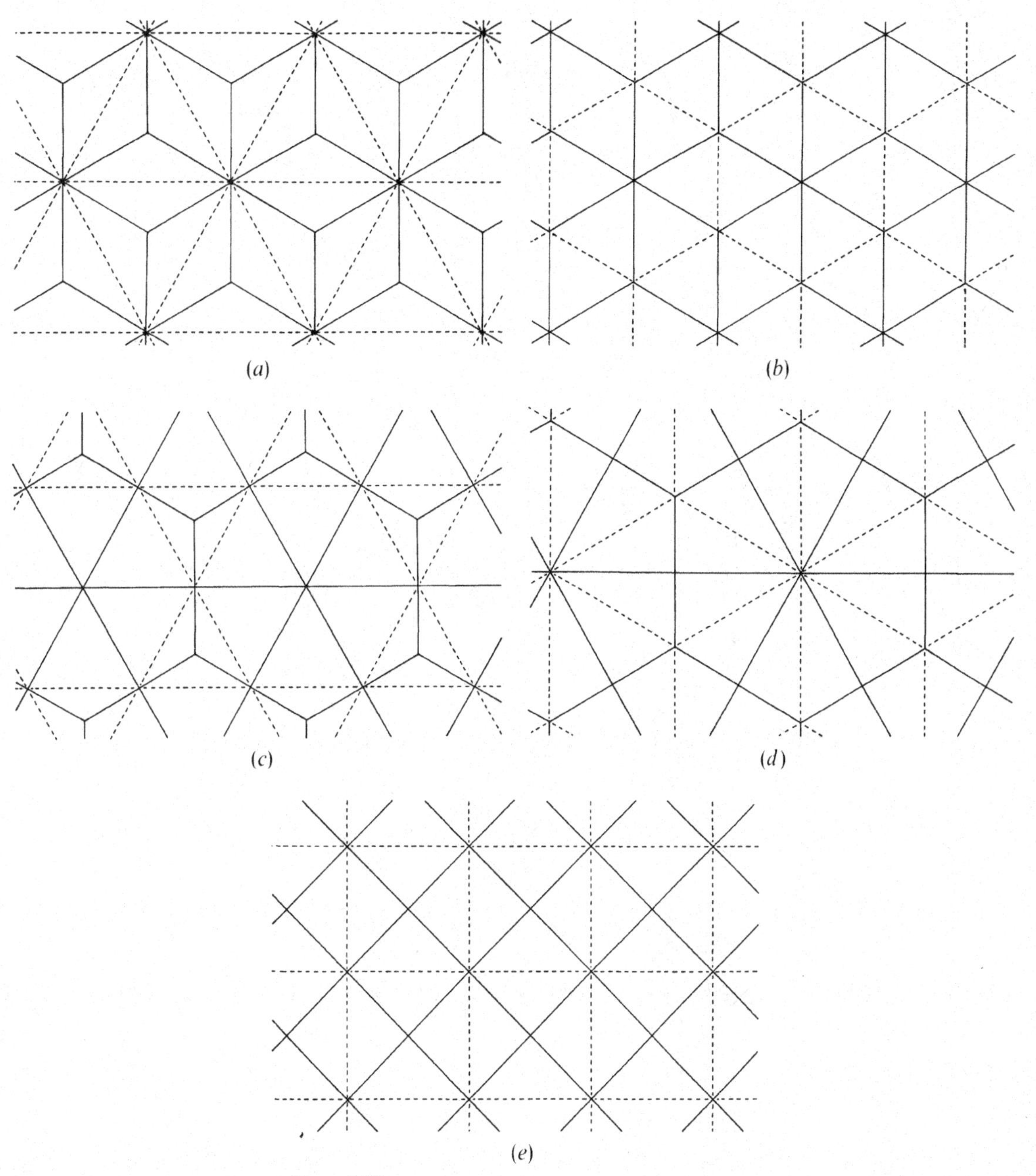

Figure 6.4.7
Tilings that are obtained from the five topological types of isotoxal tilings by the procedure known as tile centering (see Exercise 6.4.6).

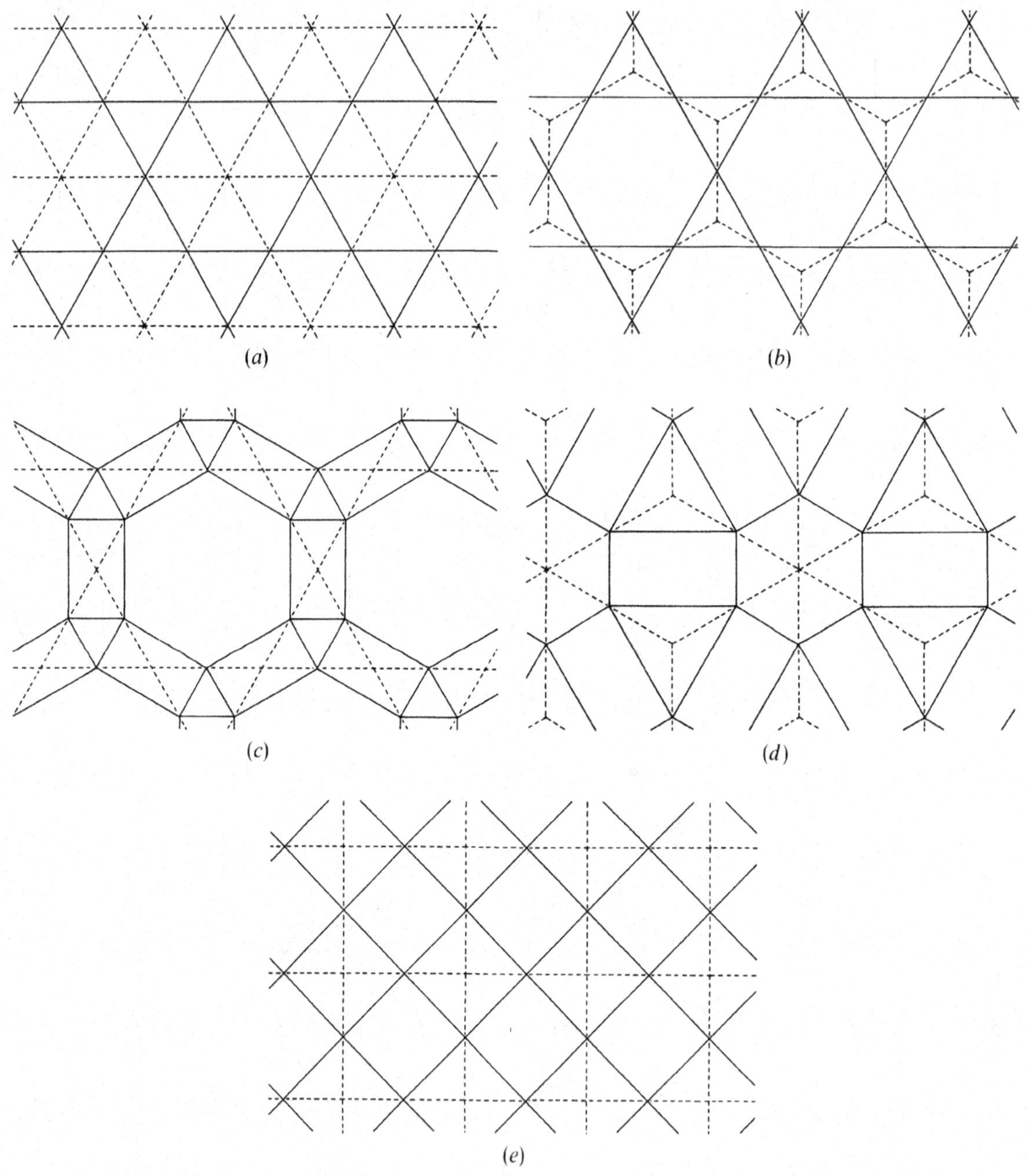

Figure 6.4.8
Tilings that are obtained from the five topological types of isotoxal tilings by the procedure known as edge centering (see Exercise 6.4.7).

(13) of Table 6.4.1 indicate the isogonal type of $\mathcal{T}_{a}$.)

(*b*) Use part (*a*) to explain how the classification of isotoxal tilings could be derived from the classification of isogonal ones, and apply this approach to find all types of isotoxal tilings of topological type $\langle 3^2; 6^2 \rangle$.

8. Obtain representatives of the fifteen non-normal types of isotoxal tilings that contain pseudo-vertices as duals of some of the tilings in Figure 6.4.6.

6.5 STRIP TILINGS AND STRIPED PATTERNS

Tilings which are not normal because they contain unbounded tiles present some interesting features. We shall not attempt to treat systematically all the possibilities, but will examine just one special kind of such tilings. Suggestions for additional investigations appear in the exercises, and a discussion of the results available in the literature is given in Section 6.6.

A *topological strip* is any closed set which is the image of a closed strip (determined by a pair of parallel straight lines) under a *bounded homeomorphism* φ of the plane onto itself; here a homeomorphism φ is called *bounded* provided there exists a constant c such that ϕ displaces no point of the plane through a distance greater than c.

A *strip tiling* is any family of topological strips with pairwise disjoint interiors, the union of which covers the plane. A *marked strip tiling* is any strip tiling in which each tile carries one copy of the motif of a pattern. Notice that here unbounded motifs are permitted (see, for example, Figure 6.5.1).

Our object in this section is the classification of isohedral strip tilings and of isohedral marked strip tilings. In doing so we shall present another illustration of the use of tile and adjacency symbols in a classification problem. It will be seen that our procedures have to be modified only slightly.

Our discussion is restricted to (marked and unmarked) strip tilings, and so the initial classification by topological type (which was the first step in the case of normal isohedral tilings in Section 6.2) does not arise here. Moreover, for marked strip tilings it is clearly enough to consider the case in which the tiles are strips bounded by parallel straight lines, since any decrease of symmetry due to the shape of the tiles can as well be produced by a suitable choice of the motif. Hence we shall assume that this choice has been made in the following discussion of marked strip tilings. As in Section 6.2, we shall study the marked tilings by introducing appropriate labels on the tiles, then obtain the tile symbols, and finally determine all adjacency symbols.

The induced tile groups are of three kinds.

(*a*) *Finite groups*. Clearly, each is a subgroup of *d2*, and so there are the five possibilities illustrated in Figure 6.5.1 (*a*).

(*b*) *Strip groups* exhibited in Section 1.4. There are seven possibilities, illustrated in Figure 6.5.1(*b*).

(*c*) *Continuous strip groups*. By this we mean groups that contain translations of type λt as λ runs through all the real numbers; see Section 5.5. There are just two possibilities, shown in Figure 6.5.1(*c*).

Allocation of tile symbols is straightforward in the finite case—as we can see from Figure 6.5.1(*a*) it follows the same general principles as with bounded tiles. In the case of strip groups some modifications have to be introduced. Suppose that, as before, we place a lettered arrow along the edge of a tile. Then application of the

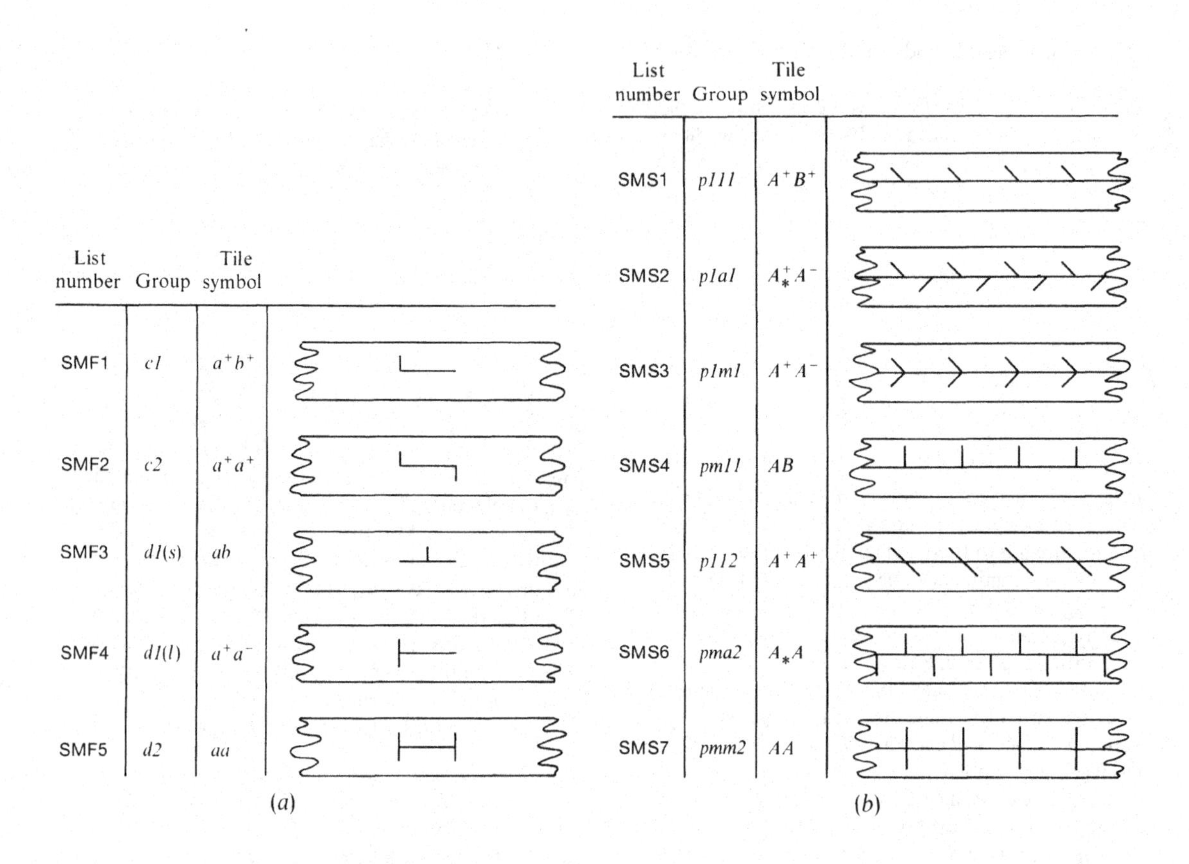

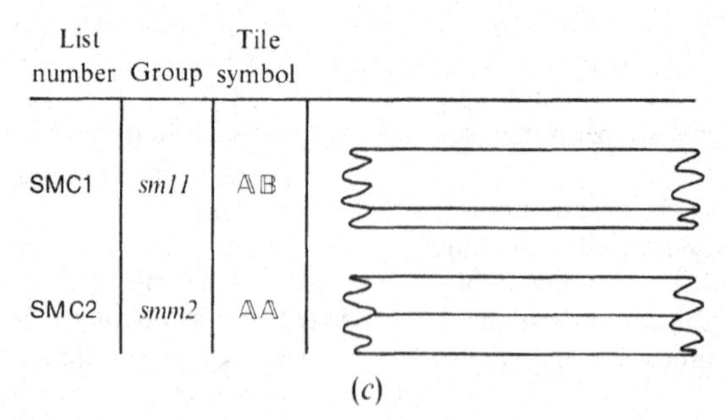

Figure 6.5.1
The groups that map a strip onto itself together with corresponding markings and tile symbols. Parts (*a*), (*b*) and (*c*) show the finite, strip and continuous strip groups, respectively.

tile group will have the effect of translating it—as well as possibly moving it to the other side of the tile or reversing it. In the latter case we use a double-headed arrow as previously. A few of the possibilities are shown in Figure 6.5.2. We must therefore consider the situation where each edge has an infinity of lettered arrows assigned to it. To avoid clumsy notation for the tile symbol, we shall write

A^+ for an edge $\ldots a^+a^+a^+a^+ \ldots$,

A^- for an edge $\ldots a^-a^-a^-a^- \ldots$,

A for an edge $\ldots aaaa \ldots$,

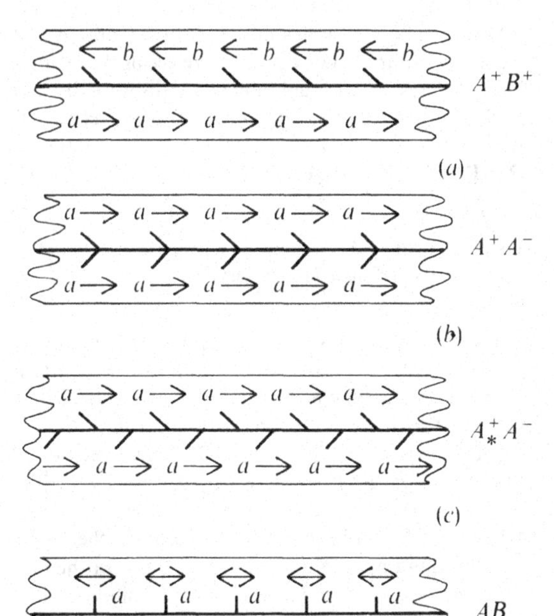

Figure 6.5.2
Examples showing how labelled arrows are allocated to the edges of marked infinite strip tiles, together with the corresponding tile symbols.

and so on. In addition, it is necessary for the tile symbol to distinguish between cases such as those shown in Figures 6.5.2(*b*) and (*c*). In the first the arrows are *opposed* (that is, reflected in the center line of the tile), whereas in the other they are *staggered* (that is, related by a glide-reflection). The latter possibility is denoted by placing a subscript asterisk ($_*$) between the symbols—so that the tiles of Figure 6.5.2 correspond to A^+B^+, A^+A^-, $A^+{}_*A^-$ and AB, respectively. Other cases are illustrated in Figure 6.5.1(*b*).

For the two continuous strip groups, the method of allocating the tile symbols should be obvious from Figure 6.5.1(*c*). To distinguish the continuous case we use a different style of type for the symbols $\mathbb{A}$ and $\mathbb{B}$.

To find all possible marked strip tilings that are isohedral we follow the procedure described in Section 6.2, allocating adjacency symbols in all possible ways. Only one point of difference arises: in the case of tiles with strip groups we must allow for the fact that the labels on two adjacent tiles may either be opposite or staggered (see Figure 6.5.3). Again, the latter is denoted by an asterisk ($_*$), in this case added as a subscript to the appropriate letter. For the continuous groups the definition of the adjacency symbols is completely straight forward.

Applying this procedure we eventually obtain:

6.5.1 *There exist precisely* 41 *types of isohedral marked strip tilings; of these,* 27 *types can be realized by strip tilings with unmarked tiles.*

These types are listed in Table 6.5.1, and a tiling of each type is shown in Figure 6.5.4. Table 6.5.1 contains additional information besides the tile and adjacency symbols, the induced tile group and symmetry group of the tiling. As in the case of normal isohedral tilings we indicate in column (7) the possible *realizations* of the tiling. The symbol M means that the type exists only as a marked strip tiling, while N means that it can also be realized by an unmarked tiling, that is, we can modify the edges and remove the markings and arrive at an isohedral tiling of the same type.

In column (8) we suggest a notation for the types of

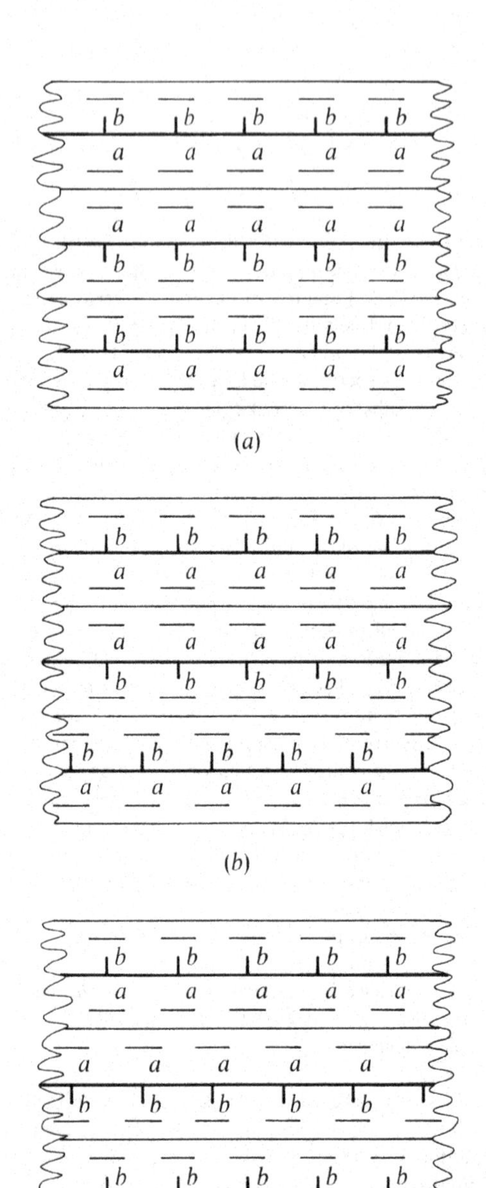

Figure 6.5.3
Examples illustrating the definition of adjacency sumbols for marked strip tilings in the case where the tile symbol is AB and the induced tile group is *pm11*. The adjacency symbols are: (*a*) AB, (*b*) AB_* and (*c*) A_*B_*.

marked tiling patterns that arise if we consider, as in Section 5.4, the interiors of the tiles (together with the markings) as copies of the motif. Although these patterns are not discrete (since their motifs are unbounded), in the case of STF1 to STF12 they correspond exactly to twelve of the strip patterns of Section 5.2, and so we have allocated to them the same reference numbers. (The justification for this is that in each case the symmetry group, the induced tile group and the set of motif-transitive groups are the same.) For types STS1 to STS26 and STC1 to STC3 the patterns are of new types, not previously considered. In column (9) we give the symbols used by Wollny [1969].

If we delete the tiles entirely and consider only the arrangement of markings in the plane, we obtain a pattern with unbounded motifs. We recall that some kinds of patterns with unbounded motifs were discussed in Section 5.5. Here we are interested in patterns that need not be filamentary, but instead satisfy conditions P.1, P.2, P.3, DP.2, DP.3, UP.2 and

UP.3 *The motif M is a closed, connected set and the symmetries of M include translations.*

For obvious reasons we shall call these *striped patterns.* With only one exception, the striped patterns are of the same type as the marked tiling patterns discussed above, and the same notation will be used for them. The exceptional case is STC1, where deletion of the tiles leaves a set of equidistant parallel lines; this results in extra symmetries, and the pattern becomes of type SPC3. For this reason we have shown the type of marked tiling pattern SPC1 in parentheses in the table.

However, there are some additional possibilities in the case of striped patterns, namely those in which only a finite number of copies of the motif is allowed. We have:

6.5.2 *There exist precisely six types of non-trivial striped patterns with a finite number of copies of the motif; each contains just two copies.*

Table 6.5.1 ISOHEDRAL MARKED STRIP TILINGS

The meanings of the first six columns are evident from their headings.

Column (7) indicates realizations; N means that the type can be realized by an unmarked strip tiling and M means that the type can only be realized by marked strips.

Column (8) lists the pattern type which is obtained by deleting the tiles (but leaving the markings as motifs) from a marked strip tiling of the given type.

Column (9) gives the symbol for the type in the notation of Wollny [1969].

Tiling type (1)	Tile symbol (2)	Adjacency symbol (3)	Motif type (4)	Symmetry group (5)	Induced tile group (6)	Realization (7)	Pattern type (8)	Wollny's symbol (9)
STF1	a^+b^+	b^+a^+	SMF1	$p1111$	$c1$	N	PS1	$\pi_1(\mathscr{F}_1^0)$
STF2	a^+b^+	b^-a^-	SMF1	$p1a1$	$c1$	N	PS2	$\pi_4(\mathscr{F}_1^3)$
STF3	a^+b^+	a^-b^-	SMF1	$pm11$	$c1$	M	PS5	$\pi_2(\mathscr{F}_1^2)$
STF4	a^+b^+	a^+b^+	SMF1	$p112$	$c1$	N	PS7	$\pi_6(\mathscr{F}_2^0)$
STF5	a^+b^+	a^-b^+	SMF1	$pma2$	$c1$	N	PS9	$\pi_6(\mathscr{F}_2^2)$
STF6	a^+a^+	a^+	SMF2	$p112$	$c2$	N	PS8	$\pi_5(\mathscr{F}_2^0)$
STF7	a^+a^+	a^-	SMF2	$pma2$	$c2$	M	PS10	$\pi_4(\mathscr{F}_2^2)$
STF8	ab	ba	SMF3	$p1m1$	$d1(s)$	N	PS4	$\pi_2(\mathscr{F}_1^1)$
STF9	ab	ab	SMF3	$pmm2$	$d1(s)$	M	PS14	$\pi_4(\mathscr{F}_2^1)$
STF10	a^+a^-	a^-	SMF4	$pm11$	$d1(l)$	M	PS6	$\pi_1(\mathscr{F}_1^2)$
STF11	a^+a^-	a^+	SMF4	$pma2$	$d1(l)$	N	PS11	$\pi_3(\mathscr{F}_2^2)$
STF12	aa	a	SMF5	$pmm2$	$d2$	M	PS15	$\pi_3(\mathscr{F}_2^2)$
STS1	A^+B^+	B^+A^+	SMS1	$p1$	$p111$	N	SPS1	$\pi_1(p1)$
STS2	A^+B^+	B^-A^-	SMS1	pg	$p111$	N	SPS2	$\pi_7(pg)$
STS3	A^+B^+	$A_*^-B_*^-$	SMS1	pg	$p111$	N	SPS3	$\pi_2(pg)$
STS4	A^+B^+	A^-B^-	SMS1	pm	$p111$	M	SPS4	$\pi_3(pm)$
STS5	A^+B^+	$A^-B_*^-$	SMS1	cm	$p111$	N	SPS5	$\pi_3(cm)$
STS6	A^+B^+	A^+B^+	SMS1	$p2$	$p111$	N	SPS6	$\pi_2(p2)$
STS7	A^+B^+	$A^+B_*^-$	SMS1	pgg	$p111$	N	SPS7	$\pi_2(pgg)$
STS8	A^+B^+	A^+B^-	SMS1	pmg	$p111$	N	SPS8	$\pi_2(pmg)$
STS9	$A_*^+A^-$	A_*^-	SMS2	pg	$p1a1$	N	SPS9	$\pi_1(pg)$
STS10	$A_*^+A^-$	A^-	SMS2	pm	$p1a1$	M	SPS10	$\pi_1(cm)$
STS11	$A_*^+A^-$	A^+	SMS2	pgg	$p1a1$	N	SPS11	$\pi_1(pgg)$
STS12	A^+A^-	A^-	SMS3	pm	$p1m1$	M	SPS12	$\pi_2(pm)$
STS13	A^+A^-	$A_*{}^-$	SMS3	cm	$p1m1$	N	SPS13	$\pi_2(cm)$
STS14	A^+A^-	A^+	SMS3	pmg	$p1m1$	N	SPS14	$\pi_1(pmg)$
STS15	AB	BA	SMS4	pm	$pm11$	N	SPS15	$\pi_3(pm)$
STS16	AB	B_*A_*	SMS4	cm	$pm11$	N	SPS16	$\pi_{10}(cm)$
STS17	AB	A_*B_*	SMS4	pmg	$pm11$	N	SPS17	$\pi_{13}(pmg)$
STS18	AB	AB	SMS4	pmm	$pm11$	M	SPS18	$\pi_2(pmm)$
STS19	AB	AB_*	SMS4	cmm	$pm11$	N	SPS19	$\pi_2(cmm)$
STS20	A^+A^+	A^+	SMS5	$p2$	$p112$	N	SPS20	$\pi_1(p2)$
STS21	A^+A^+	A_*^-	SMS5	pgg	$p112$	N	SPS21	$\pi_9(pgg)$
STS22	A^+A^+	A^-	SMS5	pmg	$p112$	M	SPS22	$\pi_{11}(pmg)$
STS23	A_*A	A_*	SMS6	pmg	$pma2$	N	SPS23	$\pi_{12}(pmg)$
STS24	A_*A	A	SMS6	cmm	$pma2$	M	SPS24	$\pi_1(cmm)$
STS25	AA	A	SMS7	pmm	$pmm2$	M	SPS25	$\pi_1(pmm)$
STS26	AA	A_*	SMS7	cmm	$pmm2$	N	SPS26	$\pi_3(cmm)$
STC1	$\mathbb{A}\mathbb{B}$	$\mathbb{B}\mathbb{A}$	SMC1	$sm1$	$sm11$	M	(SPC1)	
STC2	$\mathbb{A}\mathbb{B}$	$\mathbb{A}\mathbb{B}$	SMC1	$sm1$	$sm11$	M	SPC2	
STC3	$\mathbb{A}\mathbb{A}$	$\mathbb{A}$	SMC2	smm	$smm2$	N	SPC3	

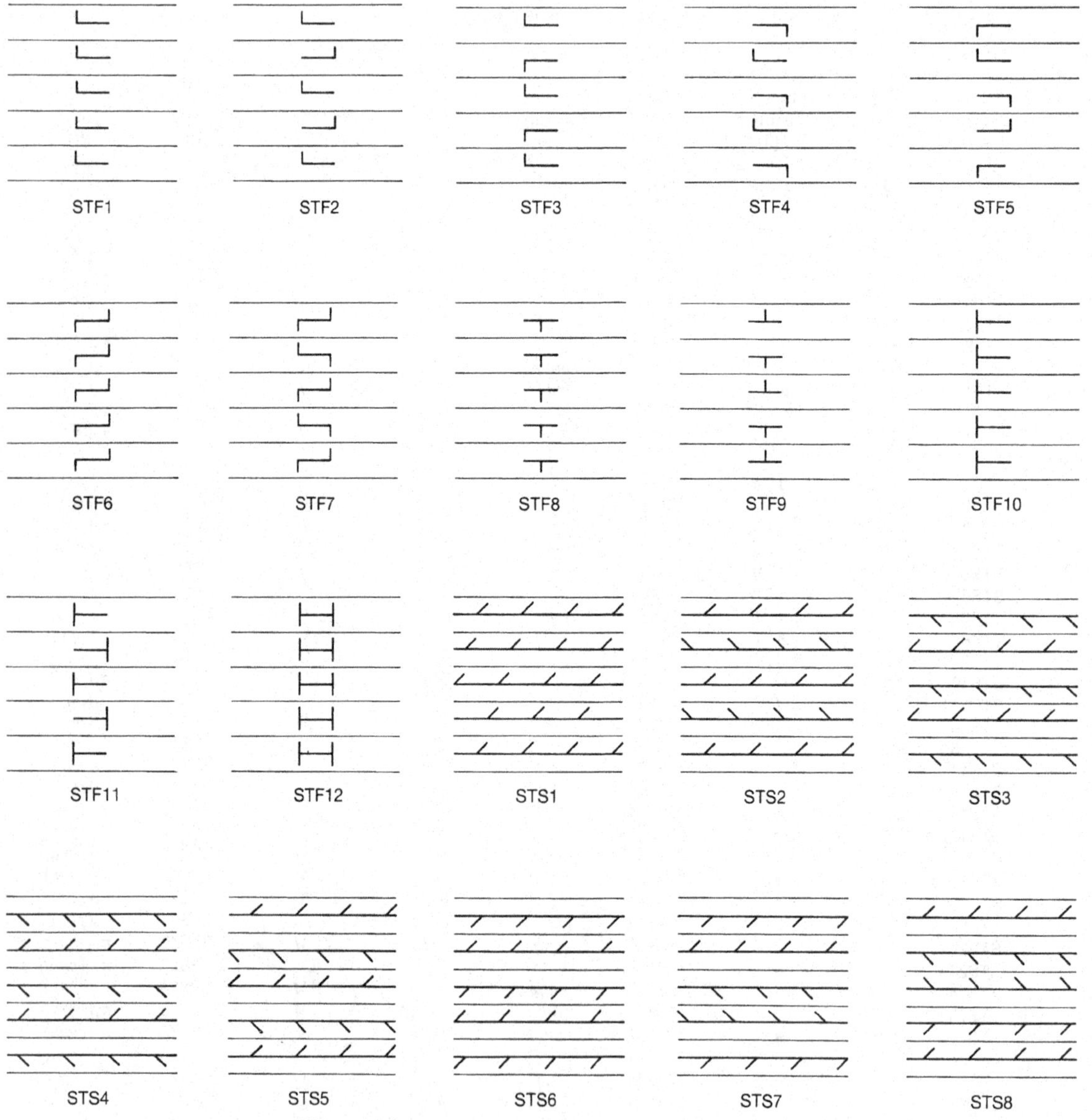

Figure 6.5.4
Examples of the 41 types of isohedral marked strip tilings.

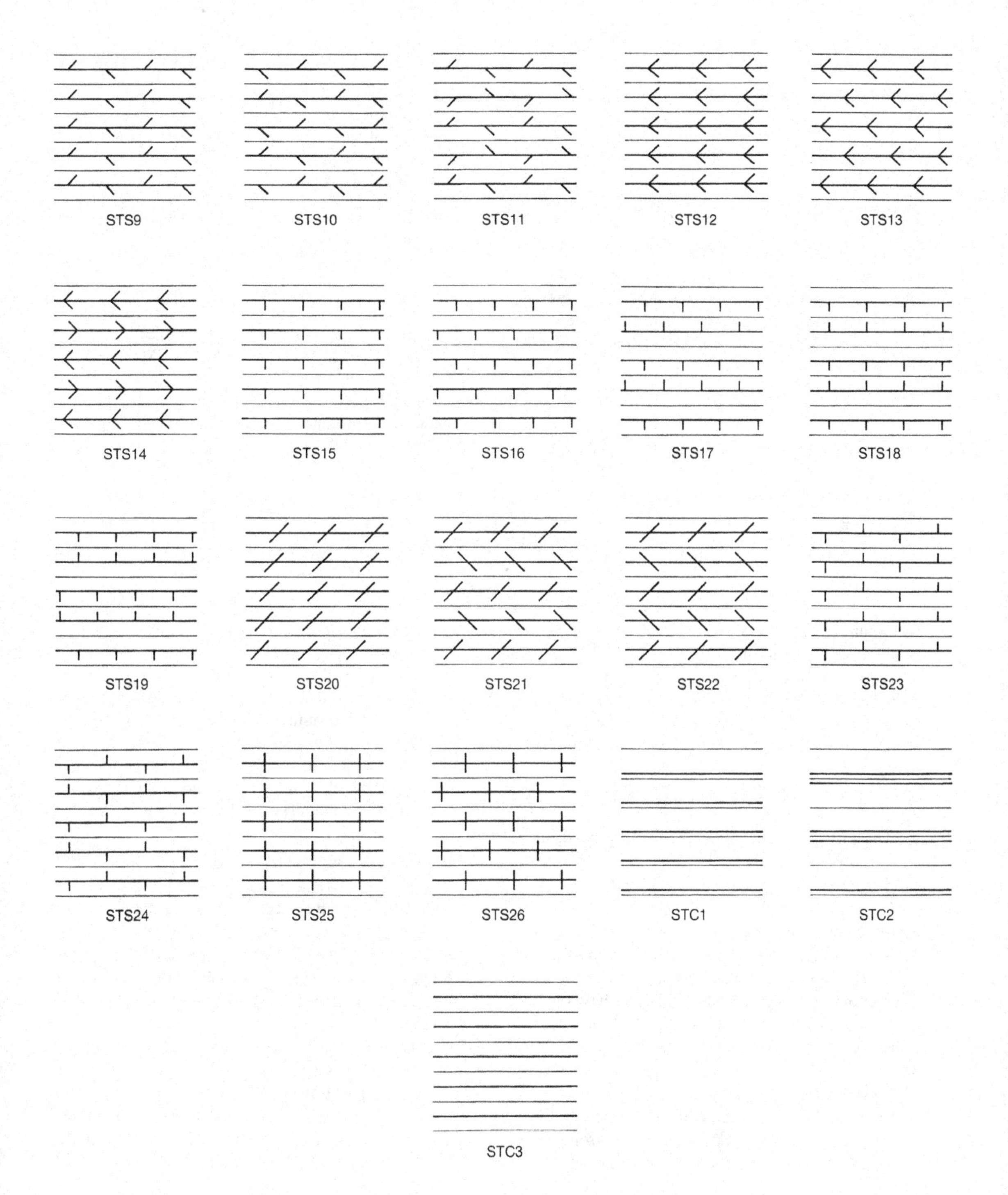
STS9
STS10
STS11
STS12
STS13
STS14
STS15
STS16
STS17
STS18
STS19
STS20
STS21
STS22
STS23
STS24
STS25
STS26
STC1
STC2
STC3

Examples of each of these six types are shown in Figure 6.5.5.

If Condition DP.3 is not assumed, that is, if patterns with a single copy of the motif are admitted, then there exist eight additional types of striped patterns. These eight types of trivial patterns are illustrated by the markings on the tiles of types SMS1 to SMS7 and SMC2 in Figure 6.5.1.

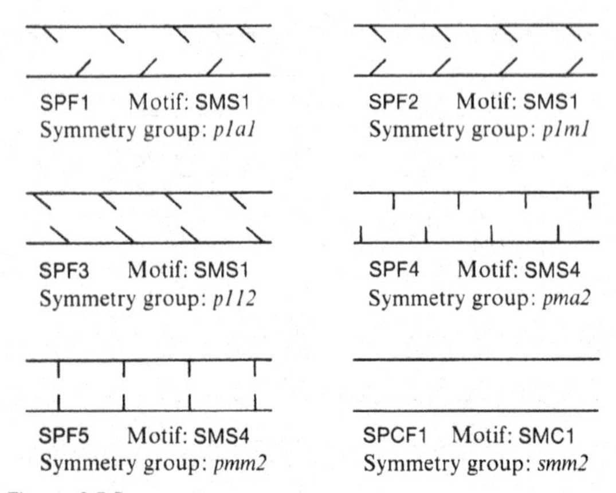

Figure 6.5.5
Examples of the six types of striped patterns involving exactly two copies of the motif.

EXERCISES 6.5

1. Determine the isohedral type of each of the isohedral strip tilings in Figure 6.5.6.
2. Construct isohedral strip tilings of the types marked *N* in column (7) of Table 6.5.1.
3. Show that there exist no unmarked isohedral strip tilings of any of the types marked *M* in column (7) of Table 6.5.1.
4. Determine the type of each of the striped patterns in Figure 6.5.7, namely:
   (*a*) The pattern that includes the white cat heads in part (*a*) of the figure.
   (*b*) The patterns indicated in black in parts (*b*) and (*c*).
5. Adapt the method of incidence symbols to the classification of marked isohedral *sector tilings*, that is, tilings whose tiles are infinite "wedges" radiating from a point (see Figure 6.5.8). Which of these types can be represented by unmarked isohedral sector tilings?
6. (*a*) In Figure 6.5.9 we show five types of isohedral tilings using a right-angled semi-infinite strip *T* as prototile. Determine the symmetry group, induced tile group, and set of tile-transitive subgroups in each case. Show also that every isohedral tiling with prototile *T* is of one of the types shown in Figure 6.5.9.
   (*b*) Determine all types of marked isohedral tilings whose prototile consists of the semi-infinite strip *T* to which a suitable marking has been added. Show how incidence symbols can be used for the identification and classification of these tilings.
   (*c*) Decide which of the types found in part (*b*) can be realized by unmarked isohedral tilings by semi-infinite strips.
7. In this section we remarked that the strip tilings of types STF1 to STF12 lead to marked tiling patterns, and hence to strip patterns by considering the motifs alone. In a similar manner, by fragmenting the motifs (see Figure 6.5.10) the types STS1 to STS26 of Figure 6.5.4, and the types SPF1 to SPF5 of Figure 6.5.5 correspond to 26 periodic patterns and 5 strip patterns, respectively.
   (*a*) Use this correspondence and the list of strip pattern types given in Section 5.2 to verify that the enumeration of SPF types is complete.
   *(*b*) Use this correspondence and the list of periodic pattern types given in Section 5.2 to verify that the enumeration of STS types is complete.

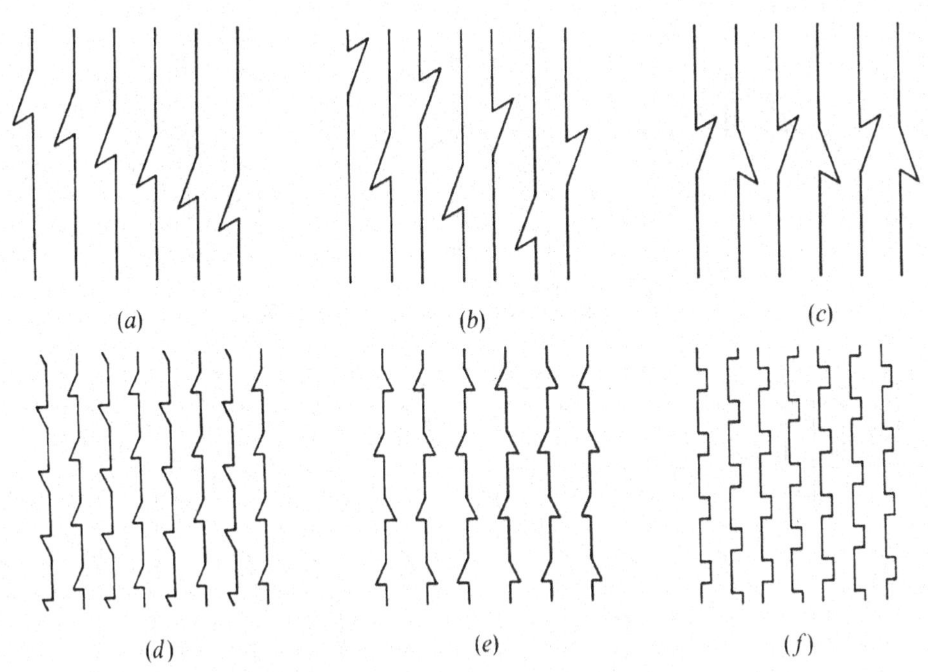

Figure 6.5.6
Examples of isohedral (unmarked) strip tilings (see Exercise 6.5.1).

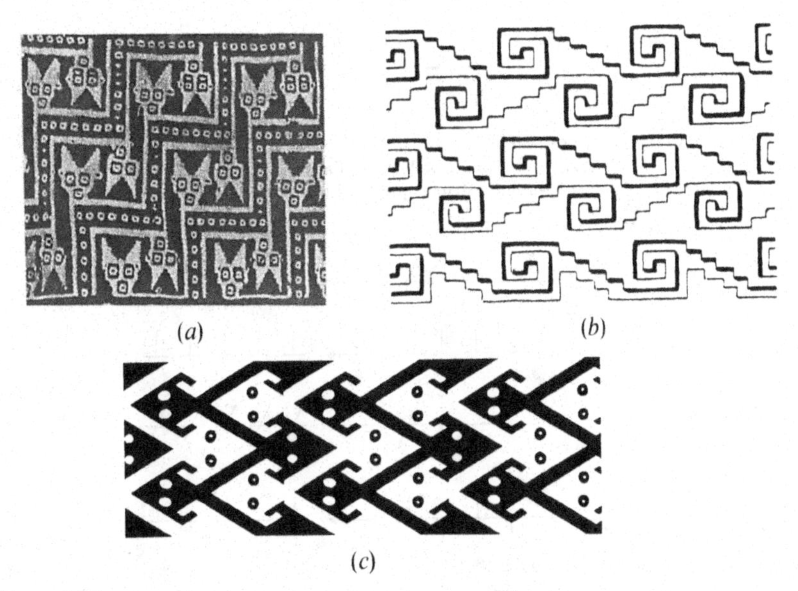

Figure 6.5.7
Examples of striped patterns from folk art; (*a*) is from Izumi [1964], (*b*) from Appleton [1950], and (*c*) from Humbert [1970] (see Exercise 6.5.4).

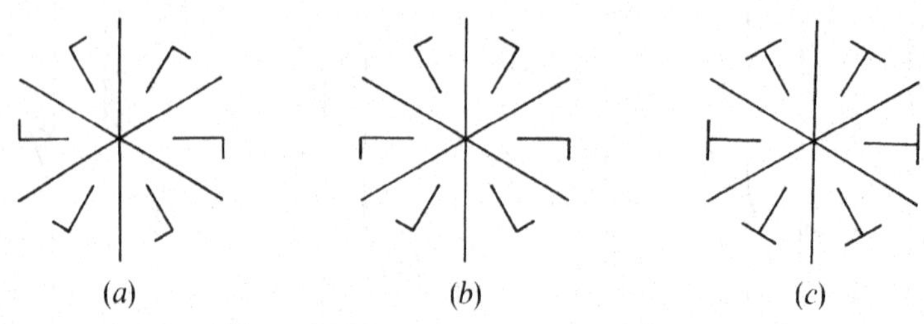

Figure 6.5.8
Examples of sector tilings (see Exercise 6.5.5).

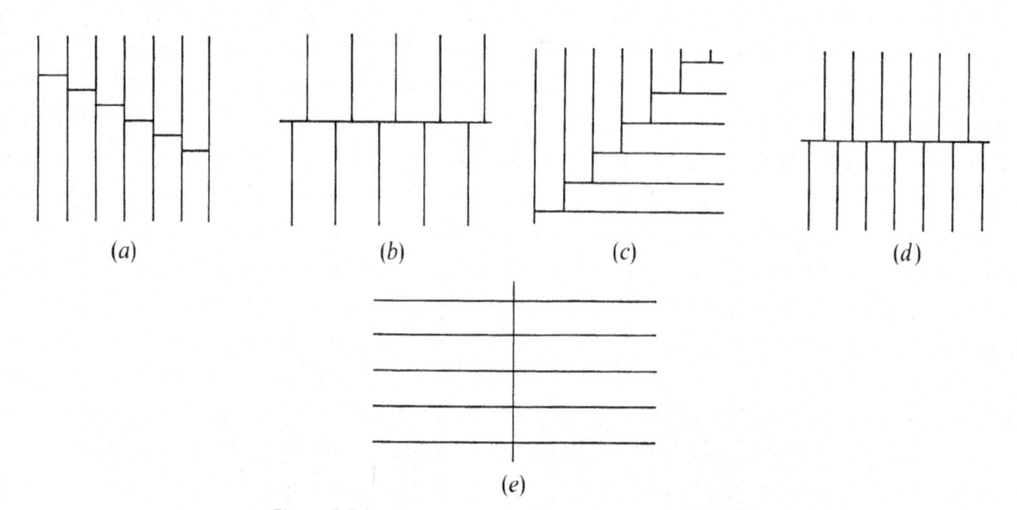

Figure 6.5.9
The five types of isohedral tilings that have a right-angled semi-infinite strip as prototile (see Exercise 6.5.6).

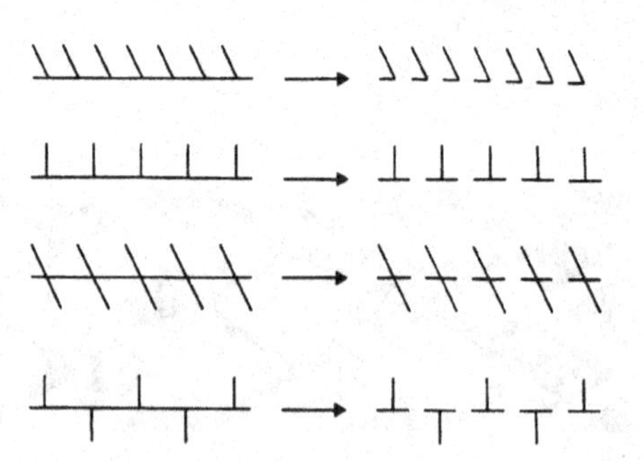

Figure 6.5.10
Unbounded motifs may be fragmented to change striped patterns into discrete patterns with the same symmetry group.

6.6 NOTES AND REFERENCES*

In view of the large role that regularity and symmetry play in mathematics, it is surprising to note that the development of the classifications of tilings with transitivity properties described in this chapter was exceedingly slow, and that much of it occurred outside the usual mathematical publications.

Isohedral tilings in which the tiles have the shapes of animals, birds, etc. have been used very effectively by several artists, of which the best known is M. C. Escher; see Escher [1958], [1971], [1982], [1983]. (For comments on Escher's designs see Coxeter [1981], Dress [1985a], Grünbaum [1985], MacGillavry [1965], Shephard [1985]. A report on the unpublished classifications Escher devised for isohedral and other tilings is given in Schattschneider [1985].) Other isohedral designs can be found in Teeters [1974], Ranucci & Teeters [1977], Chow [1979], [1980]. The papers by Chow describe how such tilings can be designed by the use of a computer.

The interest in isohedral tilings, or in special classes of such tilings, goes back at least to Fedorov [1885]. At first he was concerned with "parallelogons", that is, convex polygons which admit isohedral tilings using only translates of the prototile. (A recent treatment of this topic is given by Wollny [1974b] who, however, makes no reference to the earlier work of Fedorov or Delone on parallelogons.) In a later paper Fedorov [1899] mentioned non-convex parallelogons, extended his investigations to "planigons" (that is, prototiles of isohedral tilings) and presented a classification of the corresponding tilings. Still later (see Fedorov [1916]) he returned to the topic of classification; however, he appears to have been motivated chiefly by an argument with Šubnikov [1916] and many of his statements are ill-considered and not supported by facts. For example, Fedorov asserts that all isohedral tilings (of the plane) can be obtained as limits of "isohedral polyhedra circumscribed about spheres when the radius of the sphere becomes infinite". But there is no adequate explanation of the limiting process and it seems impossible that any such procedure can lead to valid results. (Except for bipyramids and trapezohedra, no isohedral polyhedron can have more than 120 faces, or an axis of 6-fold symmetry—properties which one would expect to be retained in any reasonable interpretation of a "limiting process"—yet these are clearly possible for isohedral tilings.) Fedorov also asserts that it is possible to derive all the types of isogonal tilings of the plane (presumably by convex polygons) from the types of isohedral tilings (by convex tiles) using duality. But as we have already pointed out in Sections 2.7 and 4.2, there is no duality theorem which justifies such a statement. Moreover, the possible existence of a result of this kind must surely be eliminated by the fact that the number of types of isogonal tilings by convex tiles is 63, while the corresponding number for isohedral tilings with convex tiles is 47. (Actually, if Figure 8 in his paper had been drawn correctly, Fedorov would have noticed that it contradicts some of his assertions regarding duality.) However, the main criticism of Fedorov's work is that he never gives precise definitions, nor states exactly the principles on which his classifications are based. This, and the corresponding vagueness of other authors, make it very difficult to compare their results with one another, or with the treatment presented here. A discussion of some discrepancies between Fedorov's assertions and the correct results appears in Delone, Galiulin & Štogrin [1979].

Šubnikov & Kopcik [1972, Chapter 2] also discuss parallelogons and planigons, and as their treatment provides a good example of the lack of precision that seems to pervade this subject, we shall discuss it in some detail. Having stated that parallelogons are either parallelograms or convex hexagons (thus missing the non-convex hexagonal parallelogons discussed by Fedorov) they continue (page 173 of the English translation): "In order to pass from parallelogons to *planigons*, i.e., to equal polygons occupying the plane without gaps or overlapping *and* lying in positions differing by virtue of rotations around axes or reflections in symmetry planes, it is sufficient to divide each parallelogon into equal parts in accordance with the proposed symmetry of the planigon." They seem to be asserting (and this interpretation is confirmed by the caption of their Figure

* See the Appendix beginning on page 653.

172) that every isohedral tiling by convex tiles can be obtained by dissecting the tiles of a suitable tiling by parallelogons—but this is manifestly false with their definition of parallelogons, as can be seen from the tilings in Figure 9.1.4. (Although inappropriate here, the idea of cutting up the tiles of an isohedral tiling into congruent tiles turns out to be useful in the discovery of anisohedral polygons, as discussed in Section 9.3.) But even if we agree to consider only those isohedral tilings which can be obtained by this method (and so exclude several of Fedorov's planigons), the treatment by Šubnikov and Kopcik is sloppy and inconsistent. For example, their Figure 172 purports to show all methods of "dividing parallelogons into equal parts (planigons)" and they claim that there are exactly 48 ways of doing this. But this claim is meaningless since no explanation is given of when two "ways of dividing" are to be considered the same or different; moreover, the partition labelled "30" is obviously inadmissible. As examples of the inconsistency we note that the two bisections of a non-square rectangle by its medians are counted as "different" while all bisections of a rectangle by slanting segments through the center are regarded as the "same"—irrespective of whether the segment meets the long or short sides of the rectangle. Further, the two diagonals of a rhomb are considered as leading to different partitions, while the three diagonals of a hexagon with symmetry group *c2* are regarded as the "same". Other such inconsistencies abound and one is forced to conclude that if any reasonable set of criteria were adopted and rigorously applied, then the number of possibilities would almost certainly differ from 48.

The efforts of Haag [1911] are even more difficult to interpret. He starts from the observation (compare Section 5.4) that the Dirichlet tiling associated with a dot pattern is an isohedral tiling, and he then attempts to describe all possible "types". However, the method by which he proceeds is not adequately explained, and some of his examples (Haag [1911, Figure 3]) are clearly not Dirichlet tilings of any dot pattern. But the greatest deficiency of Haag's treatment is the lack of any clear and precise definition of what constitutes a "type" of tiling. In the absence of such a definition, discussions are necessarily vague and the whole effort becomes useless. The same remarks apply to his later attempts at classification (Haag [1923], [1926], [1932], [1936]).

The work of the MacMahons (P. A. MacMahon [1921], [1922], P. A. MacMahon & W. P. D. MacMahon [1922], W. P. D. MacMahon [1925]) is of interest from several points of view. In particular, attention is given to non-convex and even non-polygonal tiles; tiles with colored edges, and tilings in which adjacent tiles must satisfy certain conditions regarding the colors of abutting edges are considered—these can be viewed as precursors of the Wang tiles which we shall discuss in Chapter 11; and a variant of adjacency conditions is introduced for the first time for uncolored tiles as well. Among the shortcomings of these papers, besides a lack of precision in defining the types, is the restriction to directly congruent copies of the prototiles and, in W. P. D. MacMahon [1925], to edge-to-edge tilings by polygons. For additional comments see Andrews [1985].

The enumeration of isohedral tilings by Sinogowitz [1939] is significant for the attempt to define the notion of "type" of tiling, as well as for the results obtained. Restricting attention to tilings by convex polygons, Sinogowitz describes a procedure which appears similar to the one we used in Sections 6.1 and 6.2 but relies on verbal explanations instead of any formal incidence symbols. Without giving details of the derivation, Sinogowitz produces a list of 93 types. If he had been enumerating marked isohedral tilings his list would have been correct for it coincides (except for minor discrepancies) with that given in Table 6.2.1. But since his aim—and the goal he believed he had achieved—was the enumeration of types with convex tiles, he *ought* to have arrived at a list of 47 types, namely those with an entry *C* in column (6) of Table 6.2.1.

Delone [1959] based his classification on a kind of adjacency symbol, the precursor of the incidence symbol described in this chapter. Delone's symbols are appropriate only if the induced tile group is *c1*, and he showed that there exist 46 types of such tilings. In fact he made the same mistake as Sinogowitz; he would have been correct if he had been enumerating tilings by *marked* tiles, but to achieve his stated aim he should have eliminated the types IH48, IH80 and IH87 (in our notation) for which no unmarked representatives exist. Even in the recent version of Delone's enumeration (the full account appears in Delone, Dolbilin & Štogrin [1978], and an expository survey in Delone, Galiulin & Štogrin [1979]) the adjacency symbols have not been adapted to deal with tilings whose induced tile groups are non-trivial. To classify tilings in which tiles have non-trivial stabilizers, Delone, Dolbilin and Štogrin convert them to tilings with trivial stabilizers by partitioning all tiles in a suitable manner. The enumeration yields 93 types—however, the objects classified are not tilings, but "pairs" ($\mathcal{T}$,G) consisting of a tiling $\mathcal{T}$ and a tile-transitive subgroup G of $S(\mathcal{T})$, analogous to the "restricted dot patterns" we considered in Exercise 5.3.11. This classification appears to be equivalent to the classification of marked tilings presented in Section 6.2, but to be harder to adapt to various uses. We have not verified whether (and in precisely what sense) such an equivalence holds. Comparisons are hard to make since the treatment in Delone, Dolbilin & Štogrin [1978] is rather informal; in the beginning, the tilings are assumed to have convex tiles and to be edge-to-edge—but later non-convex tiles are used, and the precise ramifications are not clear. (There is at least one error in the enumeration presented in Delone, Galiulin & Štogrin [1979, Table 2]; the type $P_{3B,5}$ should have been listed as $P_{3A,11}$, and consequently $P_{3B,6}$ should be $P_{3B,5}$.)

Probably due to the difficulties of transition from type of "pairs" to type of tilings, Delone and his co-workers do not obtain the classification of isohedral tilings into 81 types, which we have seen in Section 6.2. Other authors appear to have experienced similar difficulties in reconciling the concepts of marked and unmarked tilings, or of "pairs" and tilings. Hence the original aim—the enumeration of types of isohedral tilings—is often modified, forgotten or replaced by a classification of "pairs" or related concepts. This occurred, for example, with Heesch's [1968b] attempt at classifying all isohedral tilings, and with Wollny's [1974a] classification of isohedral tilings of topological type [4^4].

The classification of isohedral tilings by incidence symbols which was explained in Sections 6.1 and 6.2 was first carried out in Grünbaum & Shephard [1977c]. The adjacency diagrams are presented here for the first time; this is also a development of an idea first suggested in Delone [1959]. Various applications of the incidence symbols method (other than those discussed here) are given in Grünbaum & Shephard [1979a].*

Isogonal tilings which are not uniform appear first in Sohncke [1874]; to illustrate the various types of dot patterns he found, Sohncke used the dots of these patterns as vertices of isogonal tilings with convex tiles—but he was not interested in the tilings themselves.

Subsequent developments, which we shall now relate, led to an almost unbelievable number of errors. Šubnikov [1916] attempted to determine all types of isogonal edge-to-edge tilings by convex tiles—however, he *stated* that he was classifying *mono*gonal tilings with straight-line edges and did not mention the convexity or edge-to-edge conditions. In spite of these misleading formulations, his intention is clear from the context. It is remarkable that the same careless and incorrect set of definitions is used as the starting point in the corresponding discussion in Šubnikov & Kopcik [1972] published more than half a century later. (Since the other shortcomings of Šubnikov [1916] are also repeated in Šubnikov & Kopcik [1972] and the latter work is readily available in English, we shall from now on refer only to it and quote page numbers of the English translation.) As with previous authors, Šubnikov and Kopcik never define a *type* of isogonal tiling. Whatever interpretation was intended, the types IG36, IG38 and IG39 (in our notation) should have been included among those shown on pages 176 to 179. (It is interesting to note that a tiling of type IG36 is presented in a different context on page 208 of Šubnikov & Kopcik [1972].) In addition, Šubnikov and Kopcik give the following construction for isogonal edge-to-edge tilings by convex polygons, implying that it yields all such tilings (pp. 176, 179): Starting from a "system of equivalent points" (that is, a dot pattern) . . . "We join each point of the system to its nearest neighbor with a segment of straight line. We continue the process as long as no intersections of the straight lines occur except at the points under consideration. If there are several equal shortest distances, we draw all the corresponding straight

* See the Appendix beginning on page 653.

lines, provided that these do not intersect, but do not draw any which do intersect . . . We shall call the isogons just constructed "complete," in contrast to the "incomplete" ones, which may be derived from the complete isogons by eliminating certain line segments in such a way that the remaining segments divide the plane into convex polygons." However, it is clear that following these instructions it is not possible to construct the isogonal tilings shown in Figure 6.6.1; the number of such examples could easily be increased.

Next to investigate isogonal tilings was Sauer [1937]. He restricted attention to polygonal tiles and to tilings in which translations and rotations alone act transitively (that is, to tilings which are vertex-transitive under one of the groups *p1*, *p2*, *p3*, *p4*, or *p6*). Again, no definition of "type" is given, although this word is used and a completeness of the enumeration is claimed. However, the type IG7 was certainly missed by Sauer, and his treatment can at best be described as a set of instructions on how to construct certain kinds of isogonal tilings.

Recently, Fedotov [1978] attempted to determine all types of isogonal tilings, in connection with his investigations of representations of space-time by discrete sets. Mentioning none of the previous work on isogonal tilings, he gave vague verbal descriptions of "types", never stopping to explain what kinds of tilings hc is considering, or how different "types" can be distinguished. Whichever way one tries to interprete his statements, the enumeration is woefully inadequate. A detailed critique of Fedotov's paper can be found in Grünbaum & Shephard [1981b].

The classification of isogonal tilings presented in Section 6.3 is adapted from Grünbaum & Shephard [1978a].

The classification of isotoxal tilings presented in Section 6.4 is based on Grünbaum & Shephard [1978b]. The only previous investigation of such tilings is Heesch [1933a]. Due to the vagueness of his formulations it is not possible to compare Heesch's results with ours, but in any reasonable interpretation his classification is incomplete.

The only treatment of strip tilings in the literature is in the works of Wollny [1969], [1974b], [1981], [1983], [1984]. He actually investigated and classified a much wider class of tilings with unbounded tiles. For marked strip tilings our classification, although based on different considerations, coincides with Wollny's. For the other results of Wollny the reader is urged to consult the original works.

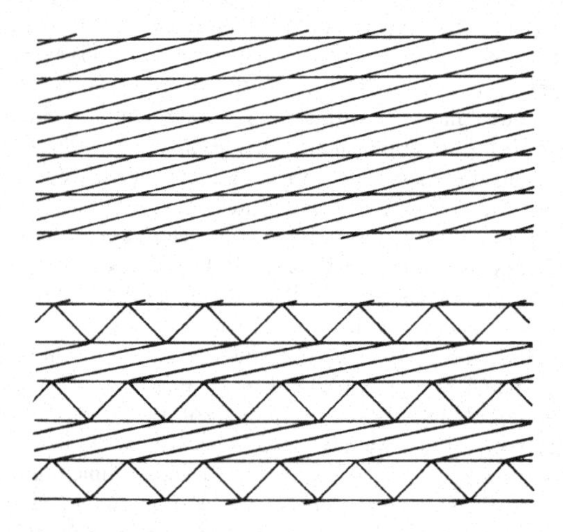

Figure 6.6.1
Two examples of isogonal tilings by convex polygons which cannot be obtained by the method of Šubnikov [1916] and Šubnikov & Kopcik [1972].

7

CLASSIFICATION WITH RESPECT TO SYMMETRIES

Figure 7.0.1
An example of a line arrangement, that is, a set $\mathscr{S}$ of lines on which the symmetry group of $\mathscr{S}$ acts transitively. Such arrangements can be classified into homeomeric types as described in this chapter. There are infinitely many such types; in Section 7.7 we shall describe how to construct them.

7

CLASSIFICATION WITH RESPECT TO SYMMETRIES

The classification of patterns described in Chapter 5 (by symmetry, motif-transitive and induced groups) and the classification of tilings described in Chapter 6 (using incidence symbols) may seem to have little in common. The purpose of this chapter is to show that they represent different aspects of one basic idea.

Because of the generality of the procedure to be described, and also because it is presented here for the first time in full detail, it seems appropriate to explain the underlying idea in intuitive terms. We begin with one of the simplest applications, namely to discrete patterns.

Let $\mathscr{P}_1$ and $\mathscr{P}_2$ be two patterns with isomorphic symmetry groups $S(\mathscr{P}_1)$ and $S(\mathscr{P}_2)$ (see Section 1.4). Denote by Ψ any mapping which maps the copies of the motif in $\mathscr{P}_1$ onto the copies of the motif in $\mathscr{P}_2$, and moreover does this in a one-to-one manner. Then the existence of such a mapping Ψ which is "compatible" with every symmetry of $\mathscr{P}_1$ and whose inverse Ψ^{-1} is "compatible" with every symmetry of $\mathscr{P}_2$ is enough to ensure that $\mathscr{P}_1$ and $\mathscr{P}_2$ are of the same pattern type as defined in Chapter 5; moreover, the converse statement is true as well (see Statement 7.2.1). A formal definition of compatibility will be given in Section 7.1; it may be thought of as meaning that if a symmetry $s \in S(\mathscr{P}_1)$ sends the copy M_i of the motif of $\mathscr{P}_1$ onto the copy M_j, then a corresponding symmetry $s' \in S(\mathscr{P}_2)$ sends the copy $\Psi(M_i)$ of the motif of $\mathscr{P}_2$ onto the copy $\Psi(M_j)$.

A simple variation of this idea turns out to have far-reaching applications: instead of one-to-one mappings Ψ between copies of the motifs, we consider homeomorphisms of the planes. We recall that in the introduction to Chapter 4 we explained how a homeomorphism h from a plane E_1 onto a plane E_2 can be described in terms of "stretching without tearing". Let $\mathscr{R}_1$ and $\mathscr{R}_2$ be geometric objects (such as patterns, tilings, or families of curves) that lie in E_1 and E_2 respectively and whose symmetry groups $S(\mathscr{R}_1)$ and $S(\mathscr{R}_2)$ are isomorphic. If there exists a homeomorphism h that maps $\mathscr{R}_1$ onto $\mathscr{R}_2$ then the objects are called *homeomorphic*; but if, in addition, h is "compatible" with every symmetry in $S(\mathscr{R}_1)$ and h^{-1} is "compatible" with every symmetry in $S(\mathscr{R}_2)$, then $\mathscr{R}_1$ and $\mathscr{R}_2$ are called *homeomeric*. We can think of compatibility of h with the symmetries of $\mathscr{R}_1$ as meaning that it maps lines of reflective symmetry of $\mathscr{R}_1$ onto lines of reflective symmetry of $\mathscr{R}_2$, centers of n-fold rotational symmetries of $\mathscr{R}_1$ onto centers of n-fold rotational symmetries of $\mathscr{R}_2$, and so on. Thus h will map a group diagram of $S(\mathscr{R}_1)$ (see Section 1.4) onto a group diagram of $S(\mathscr{R}_2)$. (Although it is helpful to think of homeomerism in terms of group diagrams, this description is not completely adequate due to the fact that the translation vectors are not uniquely determined. In the formal definition of compatibility which will be given in Section 7.1, we shall see that *every* symmetry has to be taken into account.)

The central result concerning tilings (see Statement 7.2.2) is that if two tilings are of the same isohedral, isogonal or isotoxal type, as defined by incidence symbols, then they are homeomeric, and conversely. Thus homeomerism yields exactly the same classifications as those in Chapter 6. However, we can apply it to geometric objects other than tilings. In the case of discrete patterns it yields a finer classification than that by pattern type. A little consideration will show why this is so. The use of homeomorphisms instead of one-to-one correspondences introduces topological properties of the patterns into the discussion. So, for example, if the motif of a pattern $\mathscr{P}$ is an open set, then the homeomeric type of $\mathscr{P}$ will depend upon whether the closures of the copies of the motif are disjoint or not. This is just one way in which discrete patterns of the same pattern type can turn out to be of different homeomeric types; there are also other ways in which this can happen, examples of which will be given when we discuss the classification of segment patterns in Section 7.4.

The above classifications of geometric objects by homeomerism seem to us to be completely natural. The use of homeomerism is not restricted to objects in the plane but applies to any "geometry" in the sense of the "Erlangen program" of Klein [1872]. Actually, it fits so well into that program that it is hard to understand why its formulation has been delayed by more than a century. Perhaps the explanation is that investigators were so blinded by the elegant group-theoretic aspects that they were unable to appreciate the underlying geometric phenomena which they were attempting to explain.

In spite of its many advantages, classification by homeomerism has a serious deficiency; it yields no information on *how many* different types of objects of a given kind exist. It seems as if the consequent enumeration problems must be dealt with by a separate method in each case. Sometimes no such method is known, and then we have nothing better to suggest than a simple empirical listing of types. There remain many open problems in this area.

The main purpose of this chapter is to give details of the homeomeric classification of several different types of geometric objects, such as patterns with a prescribed motif. Some of these have been previously investigated, others are new. Further applications and variants of the same basic idea will be discussed in the next two chapters. These do not, by any means, exhaust the possibilities, but should be considered as examples of the applicability of the technique to a variety of problems.

7.1 COMPATIBILITY

We begin our discussion by explaining, in detail, the meaning of the word "compatibility" as applied to homeomorphisms. Before giving a formal definition, let us consider some examples. Figure 7.1.1 shows three topologically equivalent tilings $\mathscr{T}_1$, $\mathscr{T}_2$ and $\mathscr{T}_3$. The symmetry group of each contains a reflection in a line L_i; we denote this reflection by r_i, $i = 1, 2, 3$. It is easy to see that there exists a homeomorphism ϕ of the plane which not only maps $\mathscr{T}_1$ onto $\mathscr{T}_2$, but also maps the line L_1 onto the line L_2. We describe this situation by saying that the homeomorphism ϕ is "compatible" with the reflection r_1. On the other hand, although there are infinitely many homeomorphisms ϕ mapping $\mathscr{T}_1$ onto $\mathscr{T}_3$, in no case is the line of reflection L_1 mapped onto a line of reflection in $\mathscr{T}_3$. We say that each such ϕ is *not* compatible with the reflection r_1.

Exactly similar considerations hold with symmetries of other kinds—translations, rotations and glide-reflections—and we may apply the same idea to other geometric objects such as patterns, systems of lines, curves or circles in the plane, etc.

We now give a formal definition. Let $\mathscr{R}_1$ and $\mathscr{R}_2$ be geometric objects lying in planes E_1, E_2, respectively. Let $\phi: E_1 \to E_2$ be a homeomorphism which maps $\mathscr{R}_1$ onto $\mathscr{R}_2$, and let s be any symmetry of $\mathscr{R}_1$ (that is, any isometry of E_1 that maps $\mathscr{R}_1$ onto itself). Then ϕ is said to be *compatible* with s if there exists a symmetry s' of $\mathscr{R}_2$ such that

$$s'(\phi(P)) = \phi(s(P))$$

for every point P in the plane E_1.

The condition is sometimes expressed by saying that s' exists so that the diagram

$$\begin{array}{ccc} E_1 & \xrightarrow{\phi} & E_2 \\ {\scriptstyle s}\downarrow & & \downarrow{\scriptstyle s'} \\ E_1 & \xrightarrow{\phi} & E_2 \end{array}$$

is commutative. This means that, starting from any point P of E_1 we arrive at the same point of E_2 whether we first apply s and then ϕ, or first apply ϕ and then s'. Yet another way of expressing the condition is to say that

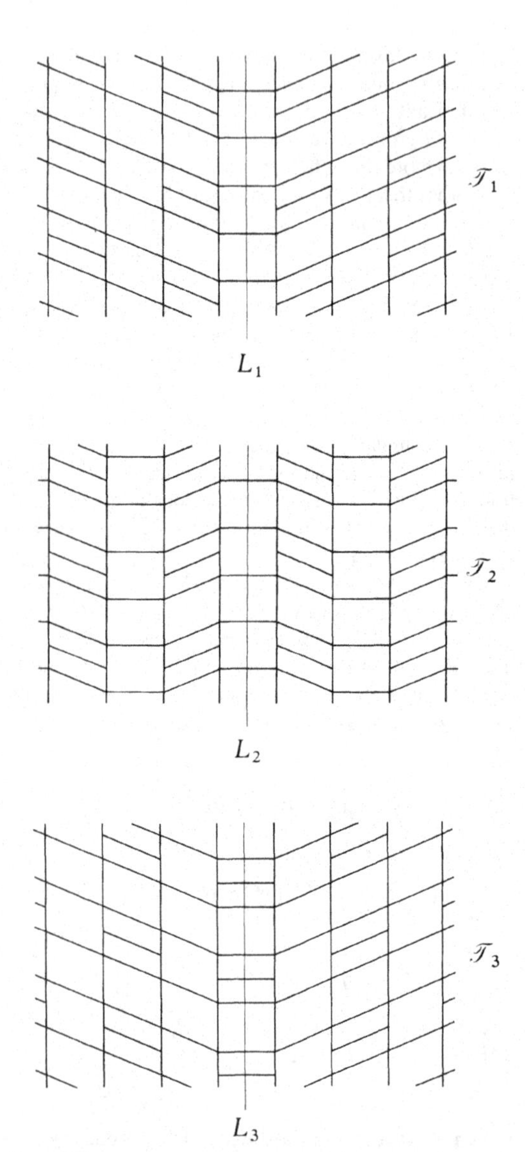

Figure 7.1.1.
There exists a homeomorphism ϕ which maps tiling $\mathcal{T}_1$ onto $\mathcal{T}_2$ and is compatible with reflection in the line L_1. There is no such homeomorphism mapping $\mathcal{T}_1$ onto $\mathcal{T}_3$.

$\phi s \phi^{-1}: E_2 \to E_2$ is a symmetry s' of $\mathcal{R}_2$. In the terminology of group theory, s and s' are *conjugate elements*. In this case it follows immediately that s' must be of the same kind as s, that is, a translation, reflection, n-fold rotation, or glide-reflection. We shall usually write $\sigma(s)$ instead of s'.

For the most part we shall be interested in the case where a homeomorphism ϕ is compatible with several, if not all, of the symmetries of $\mathcal{R}_1$. For example, if $\mathcal{R}_1$ is a periodic tiling and ϕ is compatible with all the translations in $S(\mathcal{R}_1)$, then $\mathcal{R}_2$ is necessarily also periodic, and we shall say that $\mathcal{R}_1$ is *periodically homeomorphic* to $\mathcal{R}_2$.

If ϕ is compatible with *every* symmetry of $\mathcal{R}_1$ then, for brevity, we shall say that it is *compatible with the symmetry group* $S(\mathcal{R}_1)$. In this case the set of symmetries $\sigma(s)$, for all $s \in S(\mathcal{R}_1)$, forms a subgroup of $S(\mathcal{R}_2)$ which is isomorphic to $S(\mathcal{R}_1)$ in the sense of Section 1.4, the isomorphism being represented by $\sigma: S(\mathcal{R}_1) \to S(\mathcal{R}_2)$. It is possible, of course, for $S(\mathcal{R}_1)$ to be isomorphic to a proper subgroup of $S(\mathcal{R}_2)$, that is, not the whole of $S(\mathcal{R}_2)$. This happens in the case of the tilings $\mathcal{T}_1$ and $\mathcal{T}_2$ of Figure 7.1.1.

For further examples we refer back to Section 4.1. For the tilings of Figure 4.1.3 there exists a homeomorphism from one to the other which is compatible with the symmetry group. But no such homeomorphism compatible with the symmetry group exists for the tilings of Figure 4.1.4, in spite of the fact that these are topologically equivalent.

The idea of compatibility extends in an obvious manner to combinatorial isomorphisms, as defined in Section 4.1. We explained there, in the proof of Statement 4.1.1, how a homeomorphism of a tiling induces a combinatorial isomorphism. Thus a symmetry, being a homeomorphism of a tiling onto itself, induces a combinatorial isomorphism in the same way: if s is the symmetry of $\mathcal{T}$ then the induced combinatorial isomorphism will be denoted by

$$s_E: \mathscr{E}(\mathcal{T}) \to \mathscr{E}(\mathcal{T})$$

where $\mathscr{E}(\mathcal{T})$ is the set of elements (vertices, edges and tiles) of $\mathcal{T}$. Such a mapping s_E will be called a *combina-*

torial symmetry of $\mathscr{T}$. We shall say that a combinatorial isomorphism $\Phi:\mathscr{E}(\mathscr{T}_1) \to \mathscr{E}(\mathscr{T}_2)$ is *compatible with the combinatorial symmetry* s_E of $\mathscr{T}_1$ if there exists a combinatorial symmetry s'_E of $\mathscr{T}_2$ such that

$$\begin{array}{ccc} \mathscr{E}(\mathscr{T}_1) & \xrightarrow{\Phi} & \mathscr{E}(\mathscr{T}_2) \\ {\scriptstyle s_E}\downarrow & & \downarrow{\scriptstyle s'_E} \\ \mathscr{E}(\mathscr{T}_1) & \xrightarrow{\Phi} & \mathscr{E}(\mathscr{T}_2) \end{array}$$

is commutative.

For examples we again refer to the diagrams of Section 4.1. For the tilings of Figure 4.1.3 the combinatorial isomorphism from one to the other, which is indicated by the labelling of the elements, is compatible with all the symmetries of the tiling. On the other hand there is no such combinatorial isomorphism for the tilings of Figure 4.1.4.

These statements are consequences of the following fundamental result. It represents a strengthening of Statement 4.1.1 and it can be proved in a similar manner.

7.1.1 *If $\mathscr{T}_1, \mathscr{T}_2$ are normal tilings, then there exists a homeomorphism ϕ mapping $\mathscr{T}_1$ onto $\mathscr{T}_2$ compatible with a given symmetry s of $\mathscr{T}_1$ if and only if there exists a combinatorial isomorphism $\Phi:\mathscr{E}(\mathscr{T}_1) \to \mathscr{E}(\mathscr{T}_2)$ which is compatible with the combinatorial symmetry s_E of $\mathscr{T}_1$.*

An immediate consequence of Statement 7.1.1 is that the corresponding result holds for homeomorphisms and combinatorial isomorphisms compatible with *all* symmetries in $S(\mathscr{T}_1)$ and not with just one symmetry. It is in this form that we shall use the result in the next section.

Similar considerations apply to discrete patterns. Define $M(\mathscr{P})$ as the set of copies of the motif in a discrete pattern $\mathscr{P}$. Then the analogues of the combinatorial isomorphism Φ and the combinatorial symmetry s_E in the above discussion are one-to-one mappings

$$\Psi: M(\mathscr{P}_1) \to M(\mathscr{P}_2)$$

and

$$s_M: M(\mathscr{P}_1) \to M(\mathscr{P}_1)$$

where s_M is induced by the symmetry $s \in S(\mathscr{P}_1)$. Then Ψ is *compatible* with s_M if there exists an $s'_M \in S(\mathscr{P}_2)$ such that

$$\begin{array}{ccc} M(\mathscr{P}_1) & \xrightarrow{\Psi} & M(\mathscr{P}_2) \\ {\scriptstyle s_M}\downarrow & & \downarrow{\scriptstyle s'_M} \\ M(\mathscr{P}_1) & \xrightarrow{\Psi} & M(\mathscr{P}_2) \end{array}$$

is commutative.

7.1.2 *If $\mathscr{P}_1$ and $\mathscr{P}_2$ are discrete patterns such that there exists a homeomorphism Φ mapping $\mathscr{P}_1$ onto $\mathscr{P}_2$ compatible with a given symmetry $s \in S(\mathscr{P}_1)$, then there exists a one-to-one mapping $\Psi: M(\mathscr{P}_1) \to M(\mathscr{P}_2)$ which is compatible with the mapping $s_M: M(\mathscr{P}_1) \to M(\mathscr{P}_1)$ corresponding to s.*

The proof of this statement is immediate—we need only take for Ψ the mapping between the sets of motifs induced by Φ. However, it is easy to see (in contrast to Statement 7.1.1) that the condition is here *not* both necessary and sufficient; the existence of a mapping Ψ does not imply the existence of a corresponding homeomorphism.

In the next section we shall need certain topological results that concern the existence of homeomorphisms under some additional conditions. The reader may, if he wishes, postpone consideration of them till later. We first recall a well known (but difficult to prove) fact from topology (see, for example, Newman [1951, p. 173]):

7.1.3 *If C_1 and C_2 are closed topological disks in the plane and if $\phi_0: \mathrm{bd}\, C_1 \to \mathrm{bd}\, C_2$ is a homeomorphism of the simple closed curve that bounds C_1 onto the simple closed curve that bounds C_2, then there exists a homeomorphism ϕ of the plane onto itself such that $\phi(C_1) = C_2$ and $\phi(A) = \phi_0(A)$ for each point $A \in \mathrm{bd}\, C_1$.*

This result may be formulated briefly by saying that any homeomorphism between the boundaries of two closed topological disks may be *extended* to a homeomorphism between the disks themselves.

By combining Statement 7.1.3 with standard topological methods of "crosscuts", the following related results may be obtained. Surprisingly, we were unable to locate them in the literature.

7.1.4 *Let C_1 and C_2 be closed topological disks, S any group of isometries which are symmetries of C_1, and let ϕ_0 be a homeomorphism of* bd C_1 *onto* bd C_2 *that is compatible with the symmetries in S. Then there exists a homeomorphism Φ which is compatible with the symmetries in S, extends ϕ_0, and maps C_1 onto C_2.*

7.1.5 *Let R_1 be a closed topological ring bounded by the simple closed curves Q_1^O on the outside and Q_1^I on the inside, and similarly let the ring R_2 be bounded by curves Q_2^O and Q_2^I. Let S be any group of symmetries of R_1, and let $\psi_O : Q_1^O \to Q_2^O$ and $\psi_I : Q_1^I \to Q_2^I$ be orientation-preserving homeomorphisms compatible with the symmetries in S. Then there exist homeomorphisms $\phi_I : Q_1^I \to Q_1^I$ and $\psi : R_1 \to R_2$, such that both ϕ_I and ψ are compatible with all the symmetries in S, and ψ simultaneously extends ψ_O and $\psi_I \circ \phi_I$.*

EXERCISES 7.1

1. Consider the eight finite patterns in Figure 7.1.2, each of which consists of four copies of an open topological disk.
   (*a*) Show that each of the patterns in parts (*b*), (*d*), (*e*), (*f*) and (*h*) of Figure 7.1.2 is the image of the pattern in (*a*) under a homeomorphism compatible with all the symmetries of the pattern in (*a*), but that no such homeomorphism exists for the patterns in (*c*) and (*g*).
   (*b*) Find a symmetry of the pattern in (*e*) which is not compatible with any homeomorphism mapping that pattern onto the pattern (*d*) or the pattern (*g*).
   (*c*) Find a homeomorphism ϕ, which maps the pattern in (*f*) onto that in (*e*), such that ϕ is compatible with all the symmetries of the pattern in (*f*). Show that its inverse ϕ^{-1} is compatible with all the symmetries of the pattern in (*e*).
   (*d*) The patterns in (*b*) and (*c*) are of the same pattern type $PF2_2$. Explain why it is impossible to find a homeomorphism, mapping one onto the other, which is compatible with all the symmetries.

2. For the two patterns of type PP19 shown in Figure 7.1.3, determine whether or not there exists a homeomorphism which maps the pattern in (*a*) onto that in (*b*) and is compatible with all the symmetries of the pattern in (*a*). Show that such a homeomorphism exists if the motif in both patterns is a (narrow) rectangular region but not if the motif is a straight line-segment.

3. Figure 7.1.4 shows three doubly infinite strips.
   (*a*) Show that each of the strips in parts (*b*) and (*c*) of Figure 7.1.4 is the image of the strip in (*a*) under a homeomorphism compatible with all the symmetries of the strip in (*a*).
   (*b*) Find a homeomorphism ϕ which maps the strip in (*a*) onto that in (*b*), is compatible with all symmetries of the strip in (*a*), and is such that its inverse ϕ^{-1} *is not* compatible with all the symmetries of the strip in (*b*).
   (*c*) Find a homeomorphism ϕ which maps the strip in (*a*) onto that in (*b*), is compatible with all the

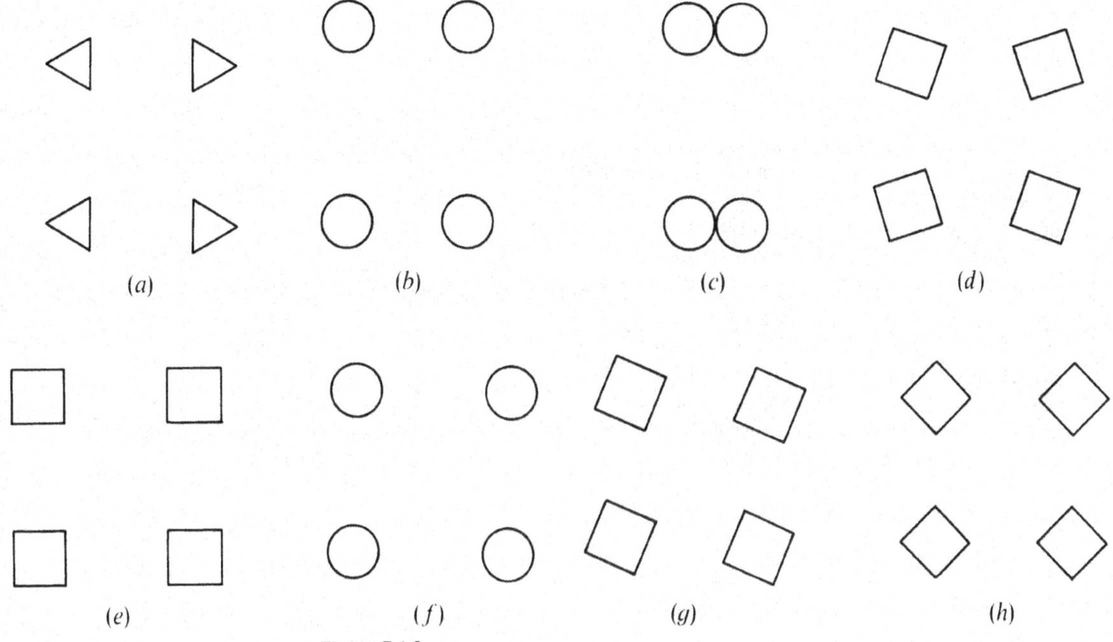

Figure 7.1.2
Examples of open disk patterns, each with four copies of the motif.

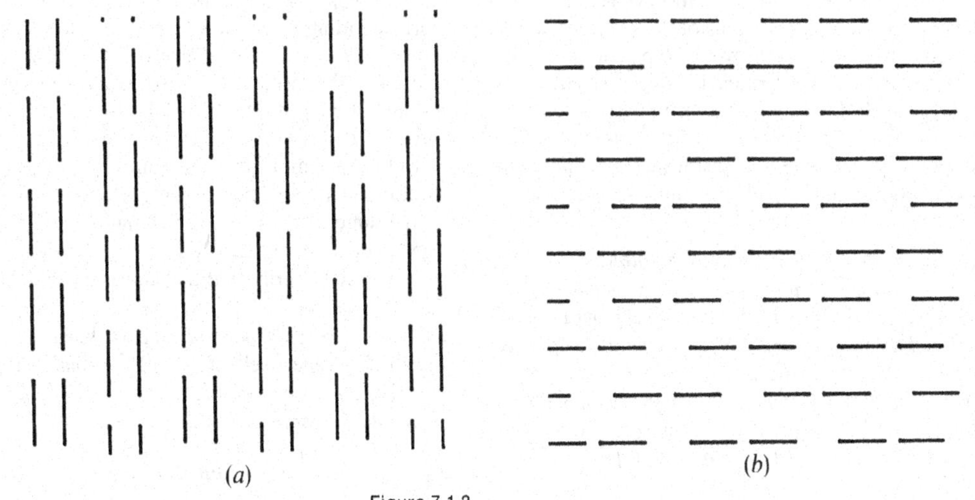

Figure 7.1.3
Two patterns of type PP19.

symmetries of the strip in (*a*), and is such that its inverse ϕ^{-1} *is* compatible with all the symmetries of the strip in (*b*).

(*d*) Show that no homeomorphism ϕ which maps the strip in (*a*) onto that in (*c*) has inverse ϕ^{-1} compatible with all the symmetries of the strip in (*c*).

4. Show that if $\mathscr{T}_1, \mathscr{T}_2, \mathscr{T}_3$ are tilings, and if $\phi_1:\mathscr{T}_1 \to \mathscr{T}_2$ and $\phi_2:\mathscr{T}_2 \to \mathscr{T}_3$ are homeomorphisms compatible with all symmetries in $S(\mathscr{T}_1)$ and $S(\mathscr{T}_2)$ respectively, then the composition $\phi = \phi_2 \circ \phi_1$ is a homeomorphism of $\mathscr{T}_1$ onto $\mathscr{T}_3$ compatible with all symmetries in $S(\mathscr{T}_1)$.

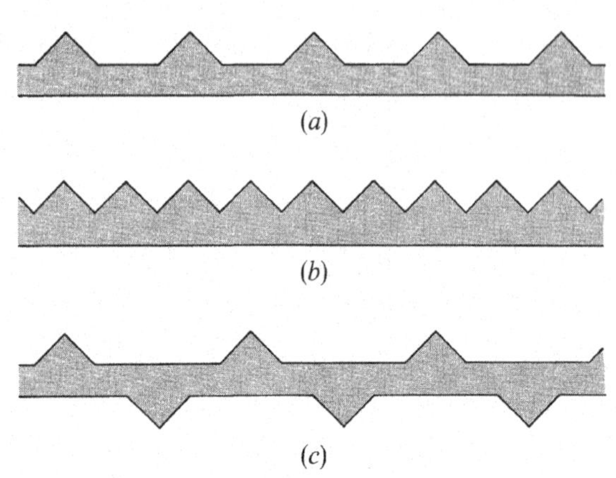

Figure 7.1.4
Three doubly infinite strips. Exercise 7.1.3 is concerned with existence of homeomorphisms between these which are compatible with their symmetries.

7.2 HOMEOMERISM

The intuitive basis of the classification of geometric objects (patterns, tilings, systems of lines, curves, circles etc.) is the assignment to the same "type" of any two objects which can be obtained from each other by mappings which do not affect the symmetries. Different choices of kinds of mappings lead to different classifications. We now formalize this idea using the concept of compatibility explained in the previous section.

7.2.1 *Let $\mathscr{P}_1$ and $\mathscr{P}_2$ be two discrete patterns, and let $\Psi: M(\mathscr{P}_1) \to M(\mathscr{P}_2)$ be a one-to-one correspondence between the set of copies of the motif of $\mathscr{P}_1$ and the set of copies of the motif of $\mathscr{P}_2$. If Ψ is compatible with every mapping $s_M: M(\mathscr{P}_1) \to M(\mathscr{P}_1)$ induced by a symmetry $s \in S(\mathscr{P}_1)$, and Ψ^{-1} is compatible with every mapping $s'_M: M(\mathscr{P}_2) \to M(\mathscr{P}_2)$ induced by a symmetry $s' \in S(\mathscr{P}_2)$, then $\mathscr{P}_1$ and $\mathscr{P}_2$ are of the same pattern type.*

From the definition of compatibility we see that each mapping s_M corresponds to a mapping $s'_M = \Psi \circ s_M \circ \Psi^{-1}$ and so each symmetry $s \in S(\mathscr{P}_1)$ corresponds to a symmetry $s' \subset S(\mathscr{P}_2)$. The converse is also true, and it is easy to check that this one-to-one correspondence is an isomorphism. Moreover, if we restrict s to be one of the symmetries of $S(\mathscr{P}_1)$ which leaves a copy M_i of the motif of $\mathscr{P}_1$ invariant, then the corresponding s' will leave the copy $\Psi(M_i)$ of the motif of $\mathscr{P}_2$ invariant, and conversely. Thus the stabilizer groups of the motifs in the two patterns are isomorphic. Finally, if we restrict s to belong to a motif-transitive subgroup of $S(\mathscr{P}_1)$ then the corresponding symmetries s' will form a motif-transitive subgroup for $\mathscr{P}_2$. These conditions are sufficient to ensure that the patterns are of the same type, as defined in Chapter 5. This completes the proof of Statement 7.2.1.

It is easy to verify that the converse of Statement 7.2.1 is also true.

Two geometric objects $\mathscr{R}_1$ and $\mathscr{R}_2$ lying in the planes E_1 and E_2 respectively are called *homeomeric* (from the Greek ὁμοιος, similar, and μερος, part) if there exists a homeomorphism $h: E_1 \to E_2$ such that:

H.1 *h is compatible with the symmetry group $S(\mathscr{R}_1)$,*

and

H.2 *h^{-1} is compatible with the symmetry group $S(\mathscr{R}_2)$.*

In particular, it will be noticed that the definition implies that the symmetry groups $S(\mathscr{R}_1)$ and $S(\mathscr{R}_2)$ must be isomorphic as defined in Section 1.4.

Homeomerism is an equivalence relation, and the corresponding equivalence classes will be called the *homeomeric types* of the objects under consideration. Analogously, patterns of the same pattern type are sometimes said to be *henomeric* (from ἑνόω-to join, to unite), or to have the same *henomeric type*.

To illustrate the definition we shall show how it applies to isohedral tilings and also to several kinds of patterns.

In the case of isohedral tilings, refer to Figure 7.2.1. All three tilings have the same symmetry group *pgg*, the same induced tile group $e = c1$, and are of the same topological type $[4^4]$. Let us see whether or not they are homeomeric.

Let P be the midpoint of the edge between T_1 and T_2 in $\mathscr{T}_1$ and let $s \in S(\mathscr{T}_1)$ be the central reflection in this point P. Then if $\mathscr{T}_1$ and $\mathscr{T}_2$ were homeomeric there would exist a homeomorphism h compatible with s, that is, such that

$$hs(T_1) = (\sigma(s))h(T_1)$$

for some central reflection $\sigma(s)$ which is a symmetry of $\mathscr{T}_2$. This condition is equivalent to $T_3 = \sigma(s)(T_4)$ where $hs(T_1) = h(T_2) = T_3$ and $h(T_1) = T_4$. But T_1 and T_2 are adjacent tiles in $\mathscr{T}_1$, so that T_3 and T_4, being images of these tiles under the homeomorphism h, must be adjacent tiles in $\mathscr{T}_2$. On the other hand, it is evident from the diagram that no two adjacent tiles in $\mathscr{T}_2$ are interchanged by any central reflection $\sigma(s)$. (In fact, in $\mathscr{T}_2$ all central reflections map tiles of $\mathscr{T}_2$ onto nonadjacent tiles.) We deduce that $T_3 \neq \sigma(s)(T_4)$ and so there is no homeomorphism compatible with s. This agrees with our previous classification since $\mathscr{T}_1$ is of type IH51 and $\mathscr{T}_2$ is of type IH52.

However, if we apply the same type of argument to tilings $\mathscr{T}_1$ and $\mathscr{T}_3$ we see that there exists a homeomorphism which is compatible with every symmetry of $\mathscr{T}_1$ and whose inverse is compatible with every symmetry of $\mathscr{T}_3$. Thus $\mathscr{T}_1$ and $\mathscr{T}_3$ are of the same homeomeric type. They are also of the same isohedral type, namely IH51. Intuitively, all we have done is to replace the edges of $\mathscr{T}_1$

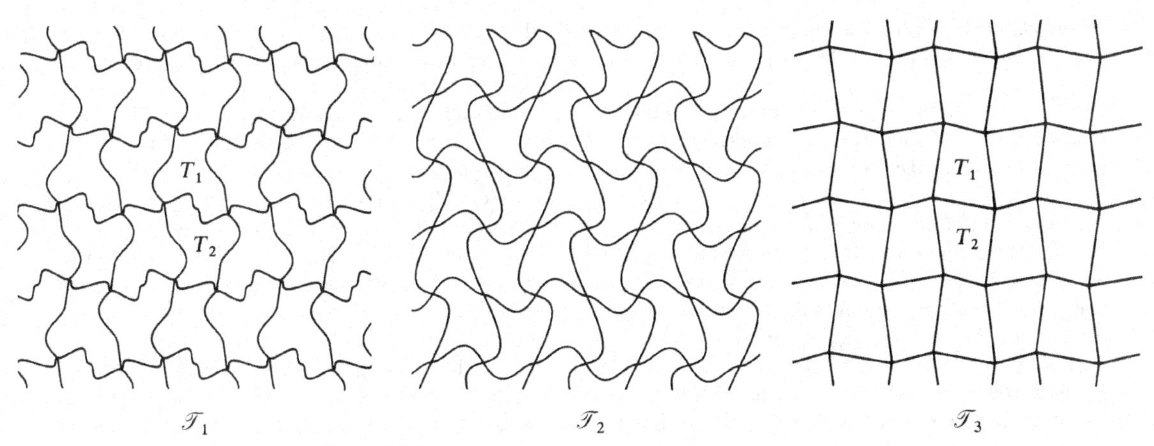

Figure 7.2.1
Three tilings with the same symmetry group (*pgg*) and the same induced tile group (*c1*) which illustrate the concept of homeomerism. Tiling $\mathscr{T}_1$ is homeomeric to $\mathscr{T}_3$, but not to $\mathscr{T}_2$.

by straight line-segments (such a replacement corresponds to the application of a homeomorphism) and in so doing we have not, in any way, affected the symmetries of the tiling.

In more general terms, if two tilings $\mathcal{T}_1$ and $\mathcal{T}_2$ are homeomeric, then those symmetries of $\mathcal{T}_1$ which map a tile $T \in \mathcal{T}_1$ onto its adjacents exactly correspond to those symmetries which map some tile $h(T) \in \mathcal{T}_2$ onto its adjacents. This clearly implies that the adjacency symbols of $\mathcal{T}_1$ and $\mathcal{T}_2$ are essentially the same and so $\mathcal{T}_1$ and $\mathcal{T}_2$ are of the same isohedral type as previously defined. On the other hand, if $\mathcal{T}_1$ and $\mathcal{T}_2$ are of the same isohedral type, then the adjacency symbols define, in an obvious manner, a combinatorial isomorphism between the tilings. Using Statement 7.1.1, this implies the existence of a homeomorphism h and of an isomorphism σ with the required properties. We deduce:

7.2.2 *The classification of isohedral tilings into types, based on the topological type and the incidence symbol as described in Section* 6.2, *coincides with the homeomeric classification of the corresponding tiling patterns. There exist* 81 *types of such tilings.*

The relation between homeomerism and pattern type is given by the following easy but important result:

7.2.3 *If $\mathcal{P}_1$ and $\mathcal{P}_2$ are discrete patterns which are homeomeric then $\mathcal{P}_1$ and $\mathcal{P}_2$ are of the same pattern type. In fact, they are of the same refined pattern type.*

This is an immediate consequence of Statement 7.2.1 and the definition of refined pattern types on page 222; the map Ψ is taken to be the one-to-one correspondence induced by the homeomorphism used in showing that the patterns are homeomeric. The converse of Statement 7.2.3 is not true, even for patterns with the same motif, as can be seen from the example in Figure 7.1.3. These patterns are of the same type but (if the motif in each is a line segment) they are not homeomeric (see Exercise 7.1.2). Hence homeomerism gives a finer classification of patterns than that introduced in Chapter 5. On the other hand, in the case of dot patterns the converse of Statement 7.2.3 is *almost* true; the homeomeric types of dot patterns coincide with the different dot pattern types, *except* for DPP48—to this dot pattern type correspond two homeomeric types (examples of which are labelled DPP48A and DPP48B in Figure 5.3.3). The exceptional case would cause tiresome complications in the formulation of some of the following statements; these are avoided by using the concept of *refined pattern types* (see page 222) which coincide with pattern types except that we count DPP48A and DPP48B as different. This distinction applies not only to dot patterns but—as mentioned in Section 5.2—to all discrete patterns.

Using this terminology, the property mentioned above can be stated succintly in the following manner.

7.2.4 *Two dot patterns have the same refined pattern type if and only if they are homeomeric.*

Since dot patterns are discrete patterns, Statement 7.2.3 implies that homeomeric dot patterns are of the same refined pattern type. To prove conversely that dot patterns $\mathcal{D}_1$ and $\mathcal{D}_2$ of the same refined pattern type are homeomeric, we need to construct a suitable homeomorphism h mapping $\mathcal{D}_1$ onto $\mathcal{D}_2$. This is a straightforward (but lengthy and tedious) process, the details of which depend on the pattern type under consideration. We shall not give them all here, but as an example illustrate the procedure in one case only. In Figure 7.2.2 we show two dot patterns $\mathcal{D}_1$ and $\mathcal{D}_2$ of type DPP19. Each pattern of this type depends on three parameters (see Table 5.3.1) and suitable distances to take as these parameters are indicated in Figure 7.2.2. Let us stretch the plane vertically in the uniform ratio c_2/c_1 and simultaneously stretch horizontally the vertical strips determined by the dots of $\mathcal{D}_1$—with the ratio of homogeneous stretching alternating between a_2/a_1 and b_2/b_1 in neighboring strips. The resulting homeomorphism h maps $\mathcal{D}_1$ onto $\mathcal{D}_2$, is compatible with $S(\mathcal{D}_1)$, and h^{-1} is compatible with $S(\mathcal{D}_2)$. The existence of this homeomorphism h is sufficient to establish that $\mathcal{D}_1$ and $\mathcal{D}_2$ are homeomeric, and so for this dot pattern type the proof of Statement 7.2.4 is completed.

With minor modifications, the same method also

yields the following result, the proof of which we omit. It clearly implies that, in the case of closed circular disk patterns, the homeomeric classification coincides with that by refined pattern type.

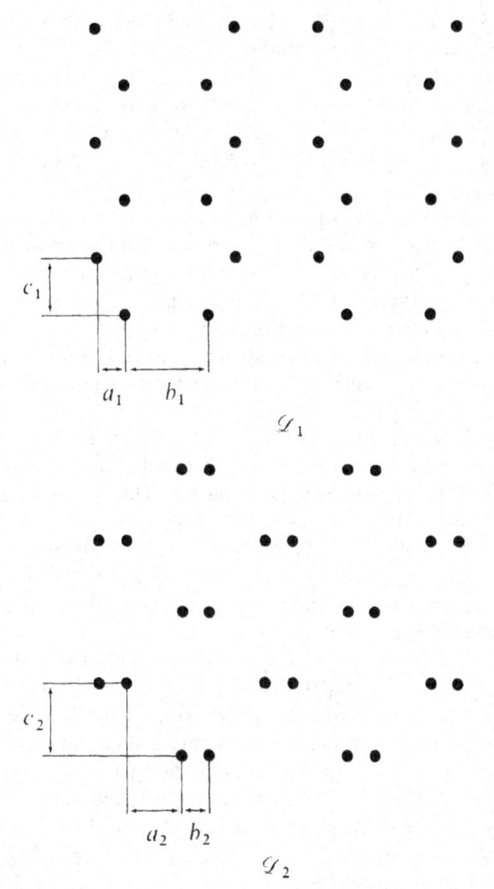

Figure 7.2.2
The construction of a homeomorphism ϕ between two dot patterns of type PP19, such that ϕ is compatible with all the symmetries of $\mathscr{D}_1$, and ϕ^{-1} is compatible with all the symmetries of $\mathscr{D}_2$.

7.2.5 *Let $\mathscr{C}_1$ and $\mathscr{C}_2$ be closed circular disk patterns of the same refined pattern type and let S be any group of symmetries which acts transitively on the motifs of $\mathscr{C}_1$. Then there exists a homeomorphism compatible with the symmetries in S which maps $\mathscr{C}_1$ onto $\mathscr{C}_2$.*

We are now ready for the central result of this section. We recall that a closed disk pattern is a discrete pattern whose motif is a closed topological disk.

7.2.6 *Two closed disk patterns $\mathscr{P}_1$ and $\mathscr{P}_2$ have the same refined pattern type if and only if they are homeomeric.*

In view of Statement 7.2.3 it is enough to establish that $\mathscr{P}_1$ and $\mathscr{P}_2$ are homeomeric if they have the same refined pattern type. The proof* of this assertion consists in the construction of a suitable homeomorphism h as required in the definition of homeomerism. Our argument depends upon the following idea. First we find closed circular disk patterns $\mathscr{C}_1$ and $\mathscr{C}_2$ which subtend $\mathscr{P}_1$ and $\mathscr{P}_2$ respectively, and then define three homeomorphisms, h_1 from $\mathscr{P}_1$ to $\mathscr{C}_1$, h_0 from $\mathscr{C}_1$ to $\mathscr{C}_2$, and h_2 from $\mathscr{P}_2$ to $\mathscr{C}_2$, each compatible with the relevant symmetries. The composition $h = h_2^{-1} \circ h_0 \circ h_1$ mapping $\mathscr{P}_1$ onto $\mathscr{P}_2$ is then the required homeomorphism. In the construction of h_1 and h_2 it is necessary to make use of additional closed disk patterns $\mathscr{Q}_1$ and $\mathscr{Q}_2$ that engulf $\mathscr{P}_1$ and $\mathscr{P}_2$, see Figure 7.2.3.

The detailed proof proceeds in several steps. The first step consists in choosing a dot pattern $\mathscr{D}_1$ associated with $\mathscr{P}_1$ in such a way that each dot D_{1i} of $\mathscr{D}_1$ is contained in the interior of the corresponding copy M_{1i} of the motif of $\mathscr{P}_1$. (This condition is automatically satisfied unless the induced motif group of $\mathscr{P}_1$ is $c1$ or $d1$; in these two cases it can be fulfilled by appropriate choice of the dots.)

The second step is the construction of a closed circular disk pattern $\mathscr{C}_1$ in which the copies C_{1i} of the motif are centered at the points D_{1i} and each C_{1i} is contained in the corresponding M_{1i}.

* This proof is quite difficult; the reader may wish to omit it and, instead, assume the truth of Statement 7.2.6.

Choosing a fixed j, we shall next apply Statement 7.1.4 to obtain a homeomorphism ϕ_j between M_{1j} and C_{1j} that is compatible with all the symmetries of $S(\mathscr{P}_1|M_{1j})$. (We take this to be the group S in Statement 7.1.4. For ϕ_0 in that statement we take any homeomorphism between the boundaries of M_{1j} and C_{1j} which is compatible with the symmetries in this group.)

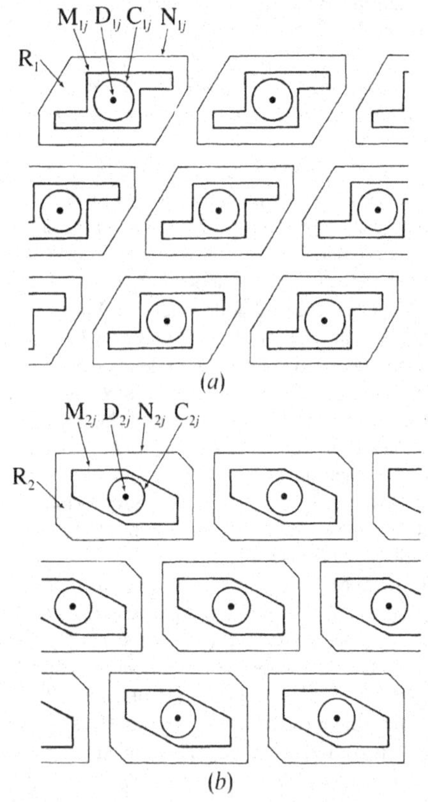

Figure 7.2.3
This illustrates the proof of Statement 7.2.5. Here M_{1j} and M_{2j} are motifs of the patterns $\mathscr{P}_1$ and $\mathscr{P}_2$, and D_{1j}, D_{2j}, C_{1j}, C_{2j}, N_{1j}, N_{2j} are motifs, respectively, of the dot patterns $\mathscr{D}_1$ and $\mathscr{D}_2$, the closed circular disk patterns $\mathscr{C}_1$ and $\mathscr{C}_2$, and the closed disk patterns $\mathscr{Q}_1$ and $\mathscr{Q}_2$, which are constructed in the proof. The topological rings R_1 and R_2 mentioned in the proof are also indicated.

Now choose a closed disk pattern $\mathscr{Q}_1$ which engulfs $\mathscr{P}_1$ and is such that the copies N_{1i} of the motif of $\mathscr{Q}_1$ contain the corresponding copies M_{1i} of the motif of $\mathscr{P}_1$ in their interiors. That this is possible is a consequence of Statement 5.1.1. Let R_1 be the topological ring with outer boundary $Q_1^O = \text{bd } N_{1j}$, and inner boundary $Q_1^I = \text{bd } M_{1j}$, and let R_2 be the topological ring with outer boundary $Q_2^O = \text{bd } N_{1j}$ and inner boundary $Q_2^I = \text{bd } C_{1j}$. Then Statement 7.1.5—with $S = S(\mathscr{P}_1|M_{1j})$, ψ_O the identity on the outer boundary and ψ_I the restriction of ϕ_j (constructed above) to Q_1^I—implies the existence of a homeomorphism ψ_j of R_1 onto R_2 compatible with all the symmetries of $S(\mathscr{P}_1|M_{1j})$.

The homeomorphism h_1 of the plane onto itself is defined as follows. If A is a point which does not belong to any N_{1i} we put $h_1(A) = A$; for $A \in M_{1j}$ we put $h_1(A) = \phi_j(A)$ and for $A \in R_1$ we put $h_1(A) = \psi_j(A)$; for points in other copies N_{1i} we define the image $h_1(A)$ by symmetry. It is then obvious that h_1 is a homeomorphism that maps $\mathscr{P}_1$ onto $\mathscr{C}_1$ and is compatible with all the symmetries in $S(\mathscr{P}_1)$.

In an exactly similar way we defined the homeomorphism h_2 which maps $\mathscr{P}_2$ onto a circular disk pattern $\mathscr{C}_2$ defined in an analogous manner. This h_2 is compatible with all the symmetries of $S(\mathscr{P}_2)$.

To define the homeomorphism h_0 we apply Statement 7.2.5 to $\mathscr{C}_1$ and $\mathscr{C}_2$ with $S = S(\mathscr{P}_1) = S(\mathscr{P}_2)$. Then h_0 maps $\mathscr{C}_1$ onto $\mathscr{C}_2$ and is compatible with all the symmetries of $S = S(\mathscr{P}_1)$.

The mapping h of the plane onto itself is defined by $h(A) = h_2^{-1}(h_0(h_1(A)))$. As a composition of homeomorphisms, it is a homeomorphism which maps $\mathscr{P}_1$ onto $\mathscr{P}_2$, and since all the homeomorphisms h_0, h_1, h_2 are compatible with all the symmetries in $S(\mathscr{P}_1)$ and in the isomorphic $S(\mathscr{P}_2)$, h is also compatible with all these symmetries. The same remarks apply to h^{-1}, so $\mathscr{P}_1$ and $\mathscr{P}_2$ are homeomeric as claimed, and the proof of Statement 7.2.6 is complete.

In general, patterns of the same refined pattern type need not be homeomeric. The patterns of Figure 7.1.3 show that Statement 7.2.6 can fail if we only require of $\mathscr{P}_1$ and $\mathscr{P}_2$ that they are compact and discrete—it is essential that their motifs are closed topological disks for the result to be true. However, as mentioned in Section 5.1, given any two compact discrete patterns $\mathscr{P}_1$ and $\mathscr{P}_2$ it is possible to engulf them by closed disk patterns $\mathscr{Q}_1$ and $\mathscr{Q}_2$ in such a way that these are subtended by $\mathscr{P}_1$ and $\mathscr{P}_2$. We may then say that $\mathscr{P}_1$ and $\mathscr{P}_2$ are of the same type if $\mathscr{Q}_1$ and $\mathscr{Q}_2$ are homeomeric, and then it follows from Statement 7.2.6 that this classification coincides with that by refined pattern type.

Arguments similar to those presented above may be used to establish the following analogous results, thereby providing "justifications" for the various classifications of tilings discussed in Chapter 6.

7.2.7 *Two isogonal tilings are of the same isogonal type if and only if they are of the same homeomeric type.*

7.2.8 *Two isotoxal tilings are of the same isotoxal type if and only if they are of the same homeomeric type.*

7.2.9 *Let $\mathscr{T}_1$ and $\mathscr{T}_2$ be two marked isohedral tilings in which the markings are closed topological disks contained in the interiors of the tiles. Then $\mathscr{T}_1$ and $\mathscr{T}_2$ are of the same isohedral type if and only if they are of the same homeomeric type.*

Analogous statements hold for marked isogonal tilings and for marked isotoxal tilings. Here two marked tilings $\mathscr{T}_1$ and $\mathscr{T}_2$ are said to be of the same homeomeric type if there exists a homeomorphism ϕ that maps $\mathscr{T}_1$ onto $\mathscr{T}_2$ in such a way that the markings of $\mathscr{T}_1$ are mapped onto the markings of $\mathscr{T}_2$, ϕ is compatible with all the symmetries of $\mathscr{T}_1$ and ϕ^{-1} is compatible with all the symmetries of $\mathscr{T}_2$.

In some instances a classification finer than the one by homeomeric type appears natural and advantageous. We shall say that two geometric objects $\mathscr{R}_1$ and $\mathscr{R}_2$ are *diffeomeric* if they are homeomeric using a homeomorphism h such that both h and its inverse h^{-1} are *differentiable maps.* (For our purposes it is convenient to call a map f from the plane E_1 to the plane E_2 differentiable if the coordinates of the point $f(P)$ in E_2 are differentiable functions of the coordinates of the point P in E_1.) For example, the patterns in Figure 7.2.4 can be shown to be diffeomeric by constructing an appropriate map h. On the other hand, the homeomeric patterns in Figure 7.2.5 are not diffeomeric since no differentiable map h can have the Z-shaped motifs of the pattern (*b*), with two corners each, as images of the corner-free segments which are the motifs of the pattern (*a*).

Although the classification by diffeomeric type is too fine for most kinds of patterns, it is appropriate in certain cases. We shall study only one example of this kind: the segment patterns in Section 7.4.

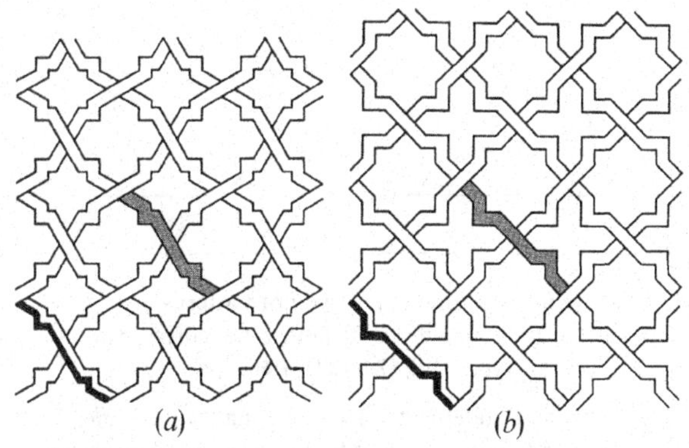

Figure 7.2.4
Examples of diffeomeric patterns. In each pattern the motif is a polygonal arc (with its endpoints deleted), one copy of which is emphasized at lower left. The construction of the required map h is not hard to explain, although a description by explicit formulas would probably be rather complicated. The shading is explained in Exercise 7.2.9.

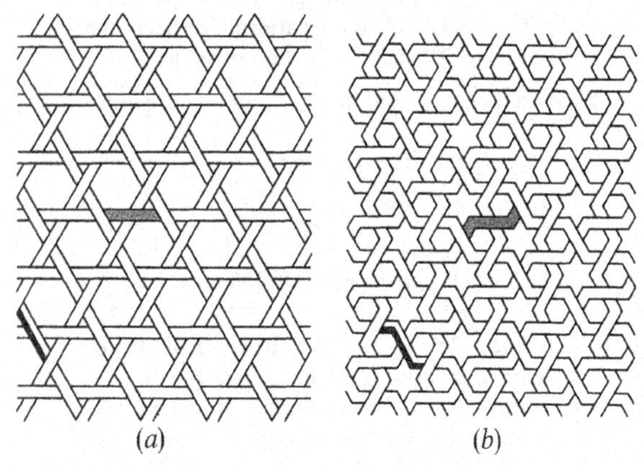

Figure 7.2.5
Examples of homeomeric but not diffeomeric patterns. Since images of smooth curves under differentiable maps are smooth, no such map can exist which carries the open segments that serve as motifs in (*a*), onto the Z-shaped motifs in (*b*). The shading is explained in Exercise 7.2.9.

EXERCISES 7.2

1. Prove in detail that every two dot patterns of type DPP39 are homeomeric. Establish the corresponding results for types DPP17 and DPP46.

2. Show that Statement 7.2.6 is no longer valid if *closed* circular disk patterns are replaced by *open* circular disk patterns.

3. (*a*) Construct an infinite family $\mathscr{F}$ of open disk patterns of type $PF3_6$ such that no two patterns in $\mathscr{F}$ are homeomeric.
   (*b*) Show that any finite, strip or periodic pattern type could be used instead of $PF3_6$ in part (*a*).

4. In this exercise we consider only periodic patterns whose motifs are closed straight line-segments.
   (*a*) For which pattern types is it possible to find two patterns which are not homeomeric?
   (*b*) Show that in no case is it possible to find three patterns of the same type, no two of which are homeomeric.

5. Show that two isohedral tilings $\mathscr{T}_1$ and $\mathscr{T}_2$ are homeomeric if and only if the tiling patterns (see Section 5.4) derived from $\mathscr{T}_1$ and $\mathscr{T}_2$ are homeomeric.

6. Show that two isotoxal tilings $\mathscr{T}_1$ and $\mathscr{T}_2$ are homeomeric if and only if the discrete periodic patterns $\mathscr{P}_t(\mathscr{T}_1)$ and $\mathscr{P}_t(\mathscr{T}_2)$ (see Section 5.4) are homeomeric.

7. Show that two isogonal tilings $\mathscr{T}_1$ and $\mathscr{T}_2$ are homeomeric if and only if the discrete periodic patterns $\mathscr{P}_g(\mathscr{T}_1)$ and $\mathscr{P}_g(\mathscr{T}_2)$ (see Section 5.4) are homeomeric.

8. (*a*) Construct a pattern which is of the same homeomeric type as the patterns in Figure 7.2.4, but has a distinct diffeomeric type.
   (*b*) Construct a pattern in which the motif is not an open segment and which is diffeomeric to the pattern in Figure 7.2.5(*a*).

9. In Figures 7.2.4 and 7.2.5 consider the shaded regions as motifs of open topological disk patterns. Are the two patterns in each figure diffeomeric?

*10. Show that Statement 7.2.4 remains valid if in its formulation "homeomeric" is replaced by "diffeomeric".

11. Show that for each type of isohedral tilings, with eight exceptions, there exists an infinity of diffeomeric types. Show that in each of the eight exceptional cases all tilings are of a single diffeomeric type.

7.3 CIRCULAR DISK PATTERNS

In this and the following two sections we shall illustrate the homeomeric classification of patterns by some specific examples. The first of these is that of patterns using a circular disk as a motif. The literature on these goes back about fifty years; details are given in the final section of this chapter.

There are two distinct cases to be considered, depending upon whether the disks are open or closed. In the case of *closed circular disk patterns* we have already shown that the homeomeric types coincide with the refined pattern types (see Statement 7.2.5) and consequently, from Statement 5.3.1, we obtain;

7.3.1 *Closed circular disks are the motifs of* 2 *families of homeomeric types of finite patterns (each family depending upon an integer n),* 7 *types of strip patterns, and* 31 *types of periodic patterns.*

Examples of these types are shown in Figures 5.3.1, 5.3.2 and 5.3.3. They correspond to the entries in Table 7.3.1 which have C in column (3).

Table 7.3.1 THE HOMEOMERIC TYPES OF OPEN CIRCULAR DISK PATTERNS

Column (1) gives a symbol for the homeomeric type; CPF, CPS and CPP stand for finite circle patterns, strip circle patterns, and periodic circle patterns, respectively. If we delete the initial C and also the hyphen and final integer we obtain the underlying pattern type in the notation of Section 5.2. Thus CPP8-1 and CPP8-2 both have underlying pattern type PP8. The fact that several homeomeric types correspond to the same pattern type vindicates our statement that in the case of open circle patterns, the homeomeric classification is finer.

Columns (2) and (3) show the superpattern type and the supermotif. The notation for the superpattern type is that of Sections 5.2 and 6.5, and the entry "Coherent" in column (2) shows that the superpattern type is trivial, that is, there is just one copy of the supermotif. The notation for the supermotifs in column (3) is that of Figures 7.3.2 and 7.3.3, except in the case of the coherent periodic patterns. In this case we have not assigned a special symbol to the motif, but used an asterisk ($*$) in column (3). All these motifs (and therefore also the corresponding periodic coherent patterns) are shown in Figure 7.3.4. Entries C in column (3) correspond to pattern types in which the closures of the disks are pairwise disjoint and so can be understood as also specifying the homeomeric types of closed circular disk patterns. These are in one-to-one correspondence with the dot pattern types of Section 5.3—the centers of the disks may be taken as the corresponding dots.

Column (4) gives the number of degrees of freedom in choosing a pattern of the type, with a fixed circular disk as motif.

Column (5) lists references to the appearance of the homeomeric type in the literature: N stands for Niggli [1927], N* for Niggli [1928], H for Haag [1929], F for Finsterwalder [1936], S for Sinogowitz [1938], and FT for Fejes Tóth [1965].

Homeomeric type (1)	Pattern type of superpattern (2)	Type of supermotif (3)	Degrees of freedom (4)		References (5)
		Finite patterns ($n \geq 2$)			
CPF2_n-1	PF2_n	C	2		
CPF2_n-2	PF3_n	CF2	1		
CPF3_n-1	PF3_n	C	1		
CPF3_n-2	Coherent	CFn	0		F-4a
		Strip patterns			
CPS7-1	PS7	C	3		
CPS7-2	PS8	CF2	2		
CPS7-3	SPF3	CS1	2		
CPS7-4	Coherent	CS4	1	N-12d,	F-4g
CPS9-1	PS9	C	3		
CPS9-2	PS10	CF2	2		
CPS9-3	PS11	CF2	2		
CPS9-4	Coherent	CS3	1	N-12c,	F-4e
CPS11-1	PS11	C	2		
CPS11-2	SPF4	CS1	1		

Table 7.3.1 (continued)

Homeomeric type (1)	Pattern type of superpattern (2)	Type of supermotif (3)	Degrees of freedom (4)	References (5)				
CPS11-3	Coherent	CS2	1	N-12b,		F-4c		
CPS11-4	Coherent	CS6	0	N-12f,		F-4d		
CPS12-1	PS12	C	3					
CPS12-2	PS13	CF2	2					
CPS12-3	PS14	CF2	2					
CPS12-4	PS15	CF4	1					
CPS13-1	PS13	C	2					
CPS13-2	PS15	CF2	1					
CPS13-3	SPF5	CS1	1					
CPS13-4	Coherent	CS5	0	N-12e,		F-4f		
CPS14-1	PS14	C	2					
CPS14-2	PS15	CF2	1					
CPS15-1	PS15	C	1					
CPS15-2	Coherent	CS1	0	N-12a,		F-4b		
			Periodic patterns					
CPP7-1	PP7	C	5				S-3	
CPP7-2	PP8	CF2	4				S-4	
CPP7-3	SPS6	CS1	4				S-5	
CPP7-4	SPS9	CS2	3				S-6	
CPP7-5	SPS20	CS4	3				S-7	
CPP7-6	SPS20	CS6	2				S-8	
CPP7-7	Coherent	*	2	N-7a,	H-1a,		S-9,	FT-23
CPP7-8	Coherent	*	2	N-9a,	H-1b,	F-9,	S-10,	FT-11
CPP7-9	Coherent	*	1	N-11f,	H-1c,		S-11,	FT-2
CPP8-1	PP8	C	3				S-1	
CPP8-2	SPS20	CS1	2				S-2	
CPP9-1	PP9	C	4				S-21	
CPP9-2	PP10	CF2	3				S-22	
CPP9-3	SPS7	CS1	3				S-23	
CPP9-4	SPS11	CS2	3				S-24	
CPP9-5	SPS21	CS4	2				S-25	
CPP9-6	SPS21	CS6	2				S-26	
CPP9-7	Coherent	*	2	N-7b,	H-3a,3b,		S-27,	FT-19
CPP9-8	Coherent	*	2	N*,		F-12,	S-28,	FT-7
CPP9-9	Coherent	*	1	N-11d,	H-3c,		S-29,	FT-4
CPP9-10	Coherent	*	1	N*,	H-3d,		S-30,	FT-3
CPP11-1	PP11	C	4				S-37	

Table 7.3.1 (continued)

Homeomeric type (1)	Pattern type of superpattern (2)	Type of supermotif (3)	Degrees of freedom (4)	References (5)
CPP11-2	PP12	CF2	3	S-39
CPP11-3	PP13	CF2	3	S-38
CPP11-4	SPS8	CS1	3	S-40
CPP11-5	SPS14	CS5	2	S-43
CPP11-6	SPS22	CS4	2	S-42
CPP11-7	SPS23	CS3	2	S-41
CPP11-8	Coherent	*	1	N-9d, H-4b, F-10, S-44, FT-8
CPP13-1	PP13	C	3	S-31
CPP13-2	SPS14	CS1	2	S-33
CPP13-3	SPS17	CS1	2	S-32
CPP13-4	SPS23	CS2	2	S-34
CPP13-5	SPS23	CS6	1	S-35
CPP13-6	Coherent	*	1	N-9b, H-4a, F-11, S-36, FT-10
CPP14-1	PP14	C	4	S-18
CPP14-2	PP15	CF2	3	S-19
CPP14-3	PP16	CF4	2	S-20
CPP15-1	PP15	C	3	S-14
CPP15-2	PP16	CF2	2	S-15
CPP15-3	SPS18	CS1	2	S-16
CPP15-4	SPS25	CS5	1	S-17
CPP16-1	PP16	C	2	S-12
CPP16-2	SPS25	CS1	1	S-13
CPP17-1	PP17	C	4	S-56
CPP17-2	PP18	CF2	3	S-58
CPP17-3	PP19	CF2	3	S-57
CPP17-4	PP20	CF4	2	S-59
CPP17-5	SPS24	CS3	2	S-60
CPP17-6	Coherent	*	1	N-7f, H-5a, F-15, S-61, FT-21
CPP19-1	PP19	C	3	S-48
CPP19-2	PP20	CF2	2	S-49
CPP19-3	SPS19	CS1	2	S-50
CPP19-4	SPS24	CS2	2	S-51
CPP19-5	SPS24	CS6	1	S-53
CPP19-6	SPS26	CS5	1	S-52
CPP19-7	Coherent	*	1	N-7c, H-2a,2b, F-26,27, S-54, FT-18
CPP19-8	Coherent	*	0	N-11e, H-2c, F-14, S-55, FT-6
CPP20-1	PP20	C	2	S-45
CPP20-2	SPS26	CS1	1	S-46
CPP20-3	Coherent	*	1	N-9c, H-5b, F-8, S-47, FT-12

Table 7.3.1 (continued)

Homeomeric type (1)	Pattern type of superpattern (2)	Type of supermotif (3)	Degrees of freedom (4)	References (5)				
CPP21-1	PP21	C	3				S-86	
CPP21-2	PP22	CF3	2				S-87	
CPP23-1	PP23	C	2				S-95	
CPP23-2	PP24	CF3	2				S-97	
CPP23-3	PP25	CF2	2				S-96	
CPP23-4	Coherent	*	1	N-8c,	H-9a,	F-31,32,	S-98,	FT-30
CPP25-1	PP25	C	2				S-92	
CPP25-2	PP26	CF3	1				S-93	
CPP25-3	Coherent	*	1	N-10b,	H-9b,9c,	F-29,	S-94,	FT-17
CPP27-1	PP27	C	3				S-90	
CPP27-2	PP28	CF2	2				S-91	
CPP28-1	PP28	C	2				S-88	
CPP28-2	PP29	CF3	1				S-89	
CPP30-1	PP30	C	3				S-62	
CPP30-2	PP31	CF2	2				S-63	
CPP30-3	PP32	CF4	2				S-64	
CPP30-4	Coherent	*	1	N-7d,	H-6a,6c,	F-16,17,18,	S-65,	FT-22
CPP33-1	PP33	C	3				S-82	
CPP33-2	PP34	CF4	2				S-84	
CPP33-3	PP35	CF2	2				S-83	
CPP33-4	Coherent	*	1	N-7e,	H-8a,		S-85,	FT-20
CPP35-1	PP35	C	2				S-78	
CPP35-2	PP36	CF2	1				S-79	
CPP35-3	Coherent	*	1	N-9e,	H-6b,	F-13,	S-80,	FT-9
CPP35-4	Coherent	*	0	N-11c,	H-8b,	F-7,	S-81,	FT-5
CPP37-1	PP37	C	3				S-74	
CPP37-2	PP38	CF2	2				S-75	
CPP37-3	PP39	CF2	2				S-76	
CPP37-4	PP41	CF8	1				S-77	
CPP38-1	PP38	C	2				S-70	
CPP38-2	PP40	CF3	1				S-71	
CPP38-3	PP41	CF4	1				S-72	
CPP38-4	Coherent	*	0	N-8a,	H-7a,	F-6,	S-73,	FT-27
CPP39-1	PP39	C	2				S-68	
CPP39-2	PP41	CF4	1				S-69	

Table 7.3.1 (continued)

Homeomeric type (1)	Pattern type of superpattern (2)	Type of supermotif (3)	Degrees of freedom (4)	References (5)				
CPP41-1	PP41	C	1				S-66	
CPP41-2	Coherent	*	0	N-9f,	H-7b,	F-5,	S-67,	FT-13
CPP42-1	PP42	C	3				S-99	
CPP42-2	PP43	CF2	2				S-100	
CPP42-3	PP44	CF3	2				S-101	
CPP42-4	PP45	CF6	2				S-102	
CPP42-5	Coherent	*	1	N-8b,	H-10a,	F-33,34,	S-103,	FT-31
CPP42-6	Coherent	*	1	N-8e,	H-10b,	F-28,	S-104,	FT-16
CPP42-7	Coherent	*	1	N-10c,	H-10c,	F-30,	S-105,	FT-15
CPP42-8	Coherent	*	0	N-11b,	H-10d,	F-25,	S-106,	FT-24
CPP46-1	PP46	C	3				S-124	
CPP46-2	PP47	CF2	2				S-126	
CPP46-3	PP48A	CF2	2				S-127	
CPP46-4	PP48B	CF2	2				S-125	
CPP46-5	PP49	CF4	1				S-128	
CPP46-6	PP50	CF6	1				S-129	
CPP46-7	PP51	CF12	1				S-130	
CPP46-8	Coherent	*	0	N-8f,	H-11a,	F-24,	S-131,	FT-28
CPP47-1	PP47	C	2				S-113	
CPP47-2	PP49	CF2	1				S-114	
CPP47-3	PP51	CF6	1				S-115	
CPP48A-1	PP48A	C	2				S-116	
CPP48A-2	PP50	CF3	1				S-117	
CPP48A-3	PP51	CF6	1				S-118	
CPP48A-4	Coherent	*	0	N-10d,	H-11d,	F-22,	S-119,	FT-14
CPP48B-1	PP48B	C	2				S-120	
CPP48B-2	PP49	CF2	1				S-121	
CPP48B-3	PP50	CF3	1				S-122	
CPP48B-4	Coherent	*	0	N-8d,	H-11b,	F-23	S-123,	FT-29
CPP49-1	PP49	C	1				S-111	
CPP49-2	Coherent	*	0	N-10a,	H-11e,	F-21	S-112,	FT-26
CPP50-1	PP50	C	1				S-109	
CPP50-2	Coherent	*	0	N-8g,	H-11c,	F-20	S-110,	FT-25
CPP51-1	PP51	C	1				S-107	
CPP51-2	Coherent	*	0	N-11a,	H-11f,	F-19	S-108,	FT-1

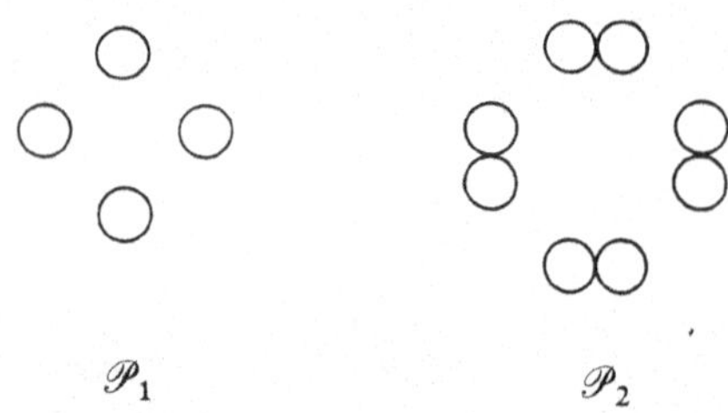

Figure 7.3.1
Two examples of open circular disk patterns. For the pattern $\mathscr{P}_1$ of type CPF3$_4$ the copies of the supermotif are the corresponding closed circular disks. For $\mathscr{P}_2$, which is of type CPF2$_4$, the supermotif consists of two touching closed circular disks. The superpattern is CPF3$_4$ in both cases.

Open circular disk patterns are more interesting since the fact that such disks can "touch" (that is, have boundary points in common) means that topological considerations enter. As a result, the homeomeric classification is finer than that by pattern type. In fact, the following is true:

7.3.2 *Open circular disks are the motifs of* 4 *families of homeomeric types of finite patterns (each depending upon a positive integer n),* 24 *types of strip patterns, and* 131 *types of periodic patterns.*

A list of all these types appears in Table 7.3.1. In order to establish Statement 7.3.2 we proceed as follows. Let the copies of the motif M of the open circular disk pattern $\mathscr{P}$ be denoted by $M_i (i \in I)$ and consider the union $\bigcup_{i \in I} \bar{M}_i$ of the closures of these. This union consists of one or more connected components $K_j (j \in J)$ each congruent to a set K which we shall call the *supermotif* of $\mathscr{P}$. (Examples of supermotifs appear in Figures 7.3.2 and 7.3.3.) The family $K_j (j \in J)$ forms a discrete pattern $\mathscr{Q}$ which we call the *superpattern* of $\mathscr{P}$. Thus, for example, in Figure 7.3.1 we show a pattern $\mathscr{P}_2$ with eight copies of the open circular disk motif; the same diagram can also be regarded as representing a superpattern $\mathscr{Q}$ with four copies of a supermotif which is the union of two closed circular disks.

The method of enumeration we shall adopt is that of determining all possible supermotifs K which can be built up from circular disks, and then determining the superpattern types that use these. The fact that the symmetry group of $\mathscr{P}$ is transitive on the disks M_i (and hence on the disks $\bar{M}_i$) places severe restrictions on the possible forms of K and also on the superpatterns $\mathscr{Q}$ for which K is the motif. Four cases need to be distinguished:

(*a*) K consists of a single circular disk. Then exactly the same patterns $\mathscr{P}$ arise as for closed disks. These correspond to the entries in Table 7.3.1 which have C in column (3).

(*b*) K consists of the union of a finite number ($n \geq 2$) of closed circular disks. The centers of the disks must lie on a circle and K is denoted by CFn, see Figure 7.3.2. In this case either the superpattern $\mathscr{Q}$ is trivial (with just one copy of K), or $\mathscr{Q}$ must be a finite, strip or periodic pattern with K as motif. In the latter cases only $n = 2, 3, 4, 6, 8$ or 12 can occur and the patterns correspond to the entries in Table 7.3.1 with CF2, CF3, CF4, CF6, CF8 or CF12 in column (3).

(*c*) K consists of the union of an infinite number of closed disks and its symmetry group is a strip group. (The six possible forms for K are shown in

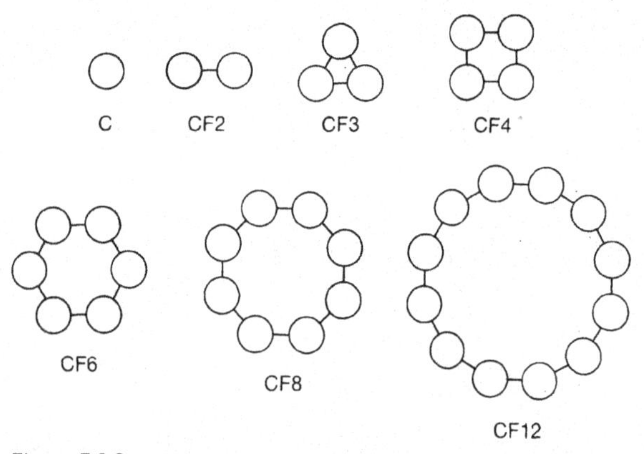

Figure 7.3.2
The motif C and the six possible finite supermotifs for open circular disk patterns. For clarity, in this and the following diagrams we have adopted a conventional representation: a pair of disk joined by a line segment is used to indicate that in the pattern there is a pair of larger disks (concentric with those shown) touching one another. Besides representing supermotifs, this diagram can also be interpreted as showing six of the infinite family of finite coherent patterns CPF3$_n$-2, each of which consists of just one copy of CFn ($n \geq 2$), see Table 7.3.1.

Figure 7.3.3.) Then $\mathscr{Q}$ can consist of just one strip, two strips, or an infinite number of strips forming one of the striped patterns described in Section 6.5. These patterns correspond to entries in Table 7.3.1 which have CS1, CS2, CS3, CS4, CS5 or CS6 in column (3).

(*d*) *K* consists of the union of an infinite number of disks and its symmetry group is one of the seventeen crystallographic groups. These patterns correspond to the entries in Table 7.3.1 which have an asterisk in column (3).

When the superpattern $\mathscr{Q}$ is trivial (which means that $\bigcup_{i \in I} \bar{M}_i$ is a connected set) the pattern $\mathscr{P}$ will be called *coherent*. As will be seen from the list of references in column (5) of Table 7.3.1, coherent patterns, especially in the periodic case, seem to have been studied more extensively than any other kind of circle patterns. All patterns in case (*d*) above are coherent.

For the non-coherent patterns of cases (*a*), (*b*) and (*c*) the various possibilities can be analysed without difficulty using a modification of the procedure described in Section 5.3. We take each supermotif *K* (shown in Figures 7.3.2 and 7.3.3) in turn and then determine which superpatterns $\mathscr{D}$ use *K* as their motif. Moreover, we must choose only those $\mathscr{Q}$ for which the induced motif group is transitive on the circular disks which make up *K*. A systematic investigation yields the list of non-coherent types given in Table 7.3.1.

For coherent patterns, such as those of case (*d*), the method of finding all the homeomeric types is, in theory at any rate, easy. We need only examine each pattern type and see when it is possible for a representative of that type to consist of mutually touching circular disks such that their closures have a connected union. Enumeration of cases can be carried out in a systematic manner. For example, let us start with a dot pattern $\mathscr{D}$ and join any two of its dots by a straight line-segment. Applying the symmetries of $S(\mathscr{D})$ we obtain a whole "network" of line segments, and we reject any such network if two of the segments intersect at a point other than an endpoint of each. The same number of segments must meet at each dot—this number is called the *valence* of the network.

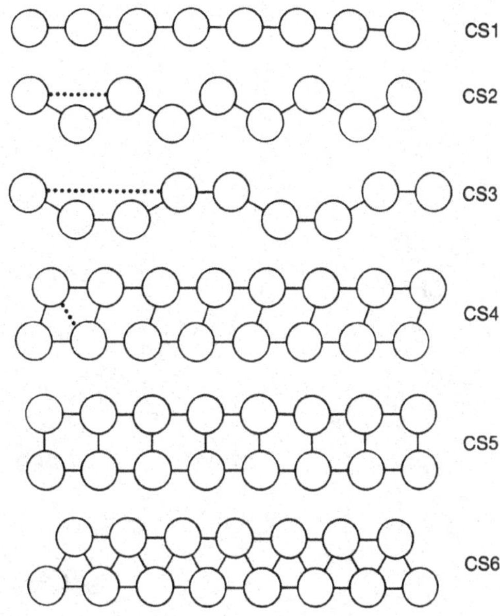

Figure 7.3.3
The six possible supermotifs made up from circular disks, whose symmetry groups are strip groups. The conventional representation explained in the caption to Figure 7.3.2 is used. Besides representing supermotifs this diagram can also be interpreted as showing the six coherent strip patterns CPS15-2, CPS11-3, CPS9-4, CPS7-4, CPS13-4 and CPS11-4 in which just one copy of the supermotif occurs, see Table 7.3.1. Dotted lines indicate distances between disks that can be varied—each such distance corresponds to a parameter, the number of which is indicated in column (4) of Table 7.3.1.

By adding line segments and applying $S(\mathscr{D})$ we can build up more and more complicated networks, though eventually the process must stop since the valence of any network cannot exceed six. In this way we obtain a complete set of connected networks.

To complete the investigation of coherent patterns, it is now necessary only to examine each connected network and determine whether it is possible to replace each dot by a circular disk in such a way that two disks touch if and only if the corresponding dots are connected by a line-segment. It turns out that for a suitable dot pattern of the chosen type and the corresponding network this can always be done; we arrive without difficulty at the 31 homeomeric types of coherent periodic circle patterns shown in Figure 7.3.4. The coherent finite patterns are those with just one "ring" of *n* touching circles CF*n* ($n \geq 2$) (see Figure 7.3.2) and the coherent strip patterns are the six types shown in Figure 7.3.3. All these are

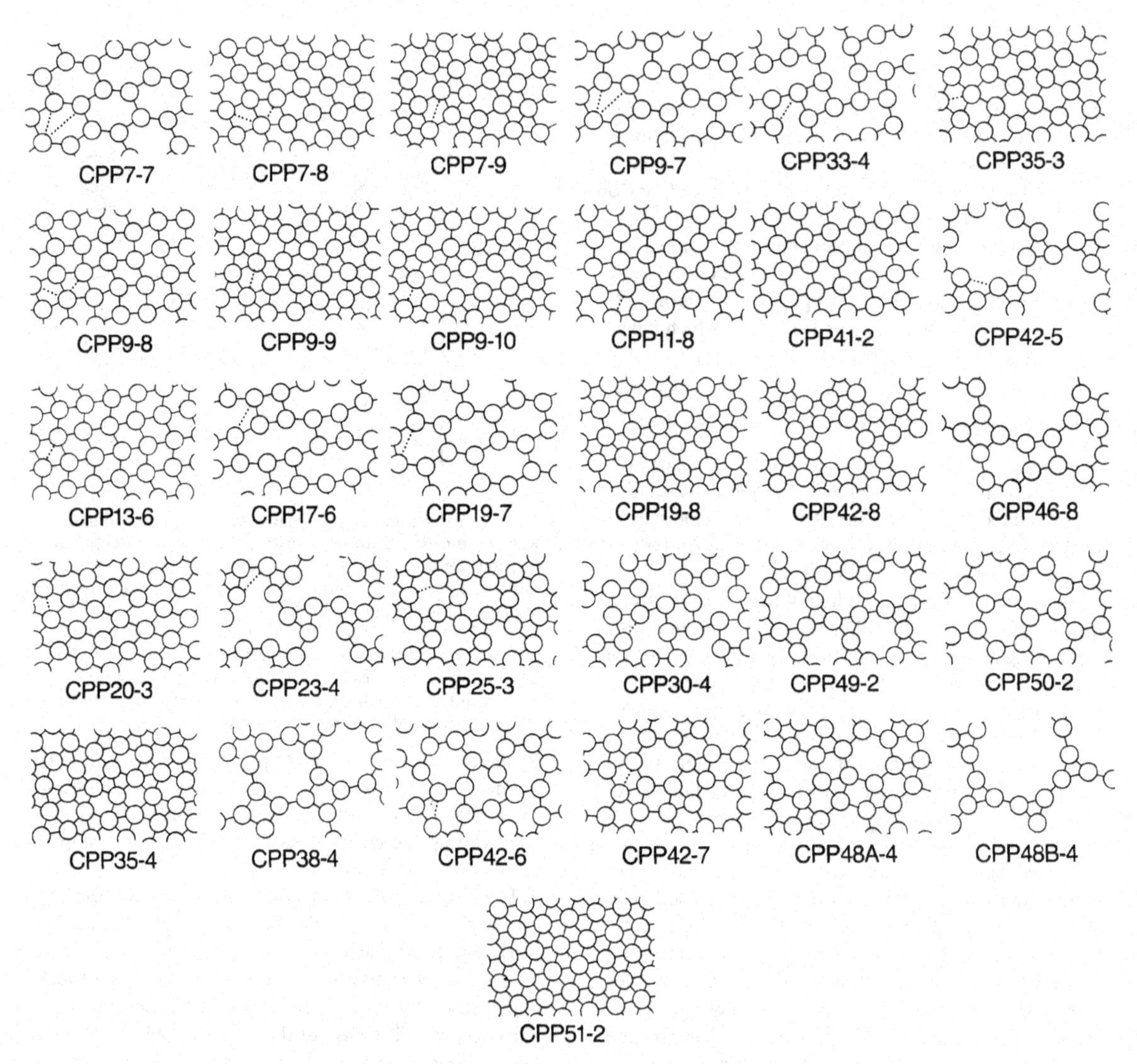

Figure 7.3.4

Examples of the 31 types of coherent periodic open circular disk patterns. The symbols used are those in column (1) of Table 7.3.1. Since each pattern consists of just one copy of the supermotif, we have not assigned separate symbols to these supermotifs (as we did in the case of the finite and strip patterns of Figures 7.3.2 and 7.3.3). In Table 7.3.1 these patterns correspond to an entry * in column (3). The conventional representation explained in the caption to Figure 7.3.2 is used in this diagram also. If the disks are deleted and the line segments extended to their centers, we obtain the corresponding network used in the enumeration of these types as explained on page 355.

indicated in Table 7.3.1 by the entry "coherent" in column (2).

The same "network" method can also be used in the non-coherent case. For these types we need only stop adding segments before a connected network is obtained. Each connected component of the network will then correspond to a copy of the supermotif. The two methods available in the non-coherent case enable us to cross-check the enumeration and so verify that the data in Table 7.3.1 is correct. This completes the proof of Statement 7.3.2.

We must stress the fact that the homeomeric classification of open circular disk patterns is finer than that by pattern type, supermotif and superpattern type. Examples establishing this are easily found. The coherent patterns CPP7-7, CPP7-8 and CPP7-9 of Figure 7.3.4 illustrate this, and four other such sets of patterns can also be found. It appears to be an "accidental" property of circle patterns that in all these cases the superpattern is trivial; in the next section examples of a corresponding phenomenon with non-coherent patterns will be given.

EXERCISES 7.3

1. Verify parts of Statement 7.3.2 by determining the homeomeric types of open circular disk patterns with one of the following properties:
   (*a*) the supermotif is CF6,
   (*b*) the supermotif is CF3,
   (*c*) the supermotif is CS3,
   (*d*) the superpattern is PP25,
   (*e*) the superpattern is PP41,
   (*f*) the superpattern is SPS23.

2. Let $\mathscr{F}$ be a family of patterns $\mathscr{P}$ depending on certain parameters. To each $\mathscr{P} \in \mathscr{F}$ assign a point $p(\mathscr{P})$ (in Euclidean or some other suitable space) in such a way that $p(\mathscr{P})$ depends continuously on the parameters of $\mathscr{P}$. Then the set $\{p(\mathscr{P}) | \mathscr{P} \in \mathscr{F}\}$ is called a *relational indicator* for the family $\mathscr{F}$. Such an indicator is useful for visualizing the relationships between the patterns in $\mathscr{F}$. For example we can speak of patterns as being close to one another, or of one pattern as being the limit of a sequence of others.
   (*a*) The finite open circular disk patterns, with a unit disk as motif, are parametrized by two numbers x, y with $x \geq y \geq 1$ which express the distances from the center of one disk to the centers of its two neighbors. To each such pattern $\mathscr{P}$ corresponds the point $p(\mathscr{P})$ in the plane with coordinates (x,y). Verify that Figure 7.3.5 is a relational indicator for these patterns.
   (*b*) Construct the analogous relational indicator for finite dot patterns and explain its relationship to the indicator of part (*a*).
   (*c*) In some cases it simplifies the relational indicator if different sets of parameters are used for the different subfamilies of $\mathscr{F}$; then subfamilies which depend on k parameters can usually be represented by flat k-dimensional subspaces of a suitable Euclidean space. In the following problems relational indicators should be understood in this sense. Verify that Figure 7.3.6 shows a relational indicator for the twelve strip patterns of types CPS7-n, CPS11-n and CPS13-n, with $n = 1, 2, 3, 4$.
   *(*d*) Extend Figure 7.3.6 to a relational indicator for

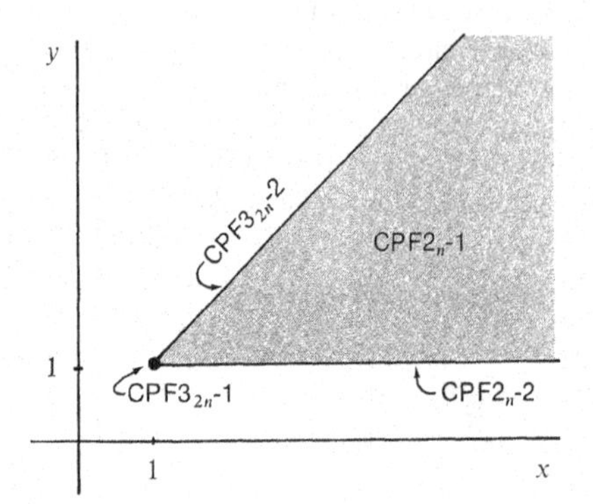

Figure 7.3.5
A relational indicator for finite open circular disk patterns. For each n, patterns of type CPF2$_n$-1 correspond to points in the gray region $\{(x,y)|1 < y < x\}$, those of type CPF2$_n$-2 and CPF3$_n$-2 to the half-lines $\{(x,y)|1 = y < x\}$ and $\{(x,y)|1 < y = x\}$ respectively, and the pattern of type CPF3$_{2n}$-1 to the point (1,1).

the family of all strip open circular disk patterns.

*(*e*) Construct a relational indicator for the family of all dot strip patterns and explain its relationship to the relational indicator of part (*d*).

*(*f*) Show that the relational indicator of the family of all periodic open circular disk patterns has precisely two connected components.

(*g*) Determine a relational indicator for the family of all periodic open circular disk patterns with

(*i*) superpattern type PP39 and supermotif CF1,

(*ii*) superpattern type PP16 and supermotif CF2,

(*iii*) superpattern type PP46 and supermotif CF1, and limits of each of these kinds of patterns.

3. (*a*) Determine the homeomeric types of finite patterns which have, as motif:

(*i*) An open circular disk from which a point, other than its center, has been deleted.

(*ii*) The union of an open circular disk with a single point on its boundary. In this case decide what differences in the enumeration arise if we insist that the pattern is discrete.

(*b*) Determine the homeomeric types of strip patterns which have as motif the union of an open circular disk with an open semicircle on its boundary. Does the enumeration change if the semicircle on the boundary is taken to be closed, or if we insist that the pattern is discrete? [For example, consider the pattern types arising from DPS12, DPS13, DPS14 and DPS15 only.]

(*c*) For each of the patterns in parts (*a*) and (*b*) determine the number of degrees of freedom.

(*d*) For cases (*i*) and (*ii*) of part (*a*) find a relational indicator.

4. Determine the homeomeric types of all (*a*) finite, (*b*) strip, and (*c*) periodic, patterns which have as motif an open square (that is, the open region which is the interior of the polygon). [For example, consider types PF2$_n$, PS11 and PP17.]

5. Determine the homeomeric types of all (*a*) finite, (*b*) strip, and (*c*) periodic, "patterns with overlapping motifs" (see Section 5.6) which have as motif a circle

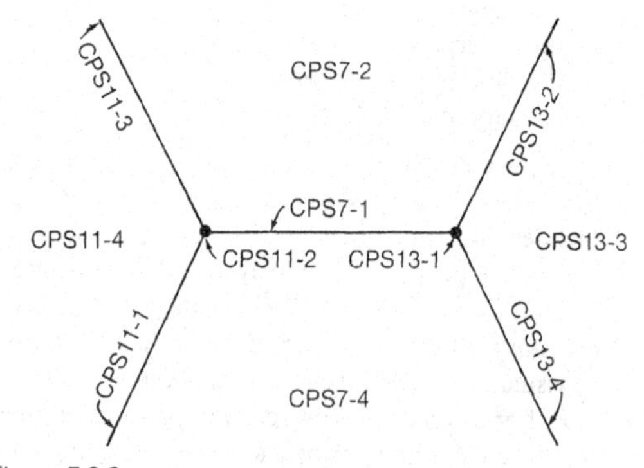

Figure 7.3.6
A relational indicator for the family of open circular disk strip patterns of types CPS7-k, CPS11-k and CPS13-k for k = 1, 2, 3 or 4. Patterns of type CPS7-3 correspond to points of the three dimensional region under the polyhedral surface shown.

(that is, the curve, *not* a disk). In each case restrict attention to such patterns in which no point is on, or inside, three or more copies of the motif. [For example, consider types PF3$_n$, PS13 and PP19.]

6. Determine the homeomeric types of all (*a*) finite, (*b*) strip, and (*c*) periodic, "layered patterns" (see Section 5.6) in which the motif is a circle (the curve, *not* a disk) and no point of the plane is on, or inside, three or more of the circles. Thus the circles in Figure 7.3.7(*a*) and (*b*) form patterns of the kind under consideration, those in (*c*) do not. Observe that the symmetries of layered patterns may include the interchange of "above" and "below" or the combination of such an interchange with an isometry of the plane. [For example, consider types PF1$_n$, PS7 and PP30.]

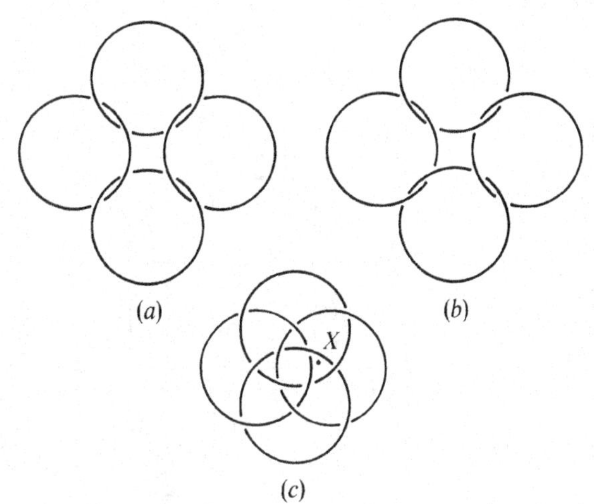

Figure 7.3.7
Examples of layered patterns with a circle as motif. The pattern in (*c*) is not to be included in the enumeration requested in Exercise 7.3.6 since some points, such as *X*, are enclosed by three or more circles.

7.4 SEGMENT PATTERNS

For our second example of homeomeric classification we consider patterns whose motifs are straight line-segments. So far as we are aware, this is the first time that a full account of these patterns has been published, which is surprising in view of the fact that many of them are very decorative (see Figure 7.4.4) and occasionally occur as so-called lattice designs (see Dye [1937], [1981]).

There are two cases to be considered, depending upon whether the motif is a line segment which is *closed* or (*relatively*) *open* (that is, has had its two endpoints removed). In the case of closed segments the patterns are discrete and the procedure of Section 5.3 can be used to determine the possible pattern types. We find 3 families of types of finite patterns (each depending upon an integer *n*), 12 types of strip patterns, and 38 types of periodic patterns. However, the homeomeric classification turns out to be finer than that by refined pattern type, and in several cases one refined pattern type corresponds to two homeomeric types. In Figure 7.4.1 we show two examples of pattern types (PP38 and PP49) for which this occurs. Examination of all the possibilities enables us to establish the following result:

7.4.1 *Closed line segments are the motifs of* 4 *families of homeomeric types of finite patterns* (*each depending on a positive integer n*), 16 *types of strip patterns, and* 50 *types of periodic patterns.*

The patterns bearing the label *AA* (whose significance will be explained shortly) in Figures 7.4.2 and 7.4.3 are representatives of each of the families of finite patterns and strip patterns whose motifs are closed line-segments. All of these types, together with the 50 types of periodic patterns, are listed in Table 7.4.1; they correspond to the entries which have *AA* in column (4). Whenever two homeomeric types are bracketed (the brackets are in column (6)) they correspond to the same pattern type and

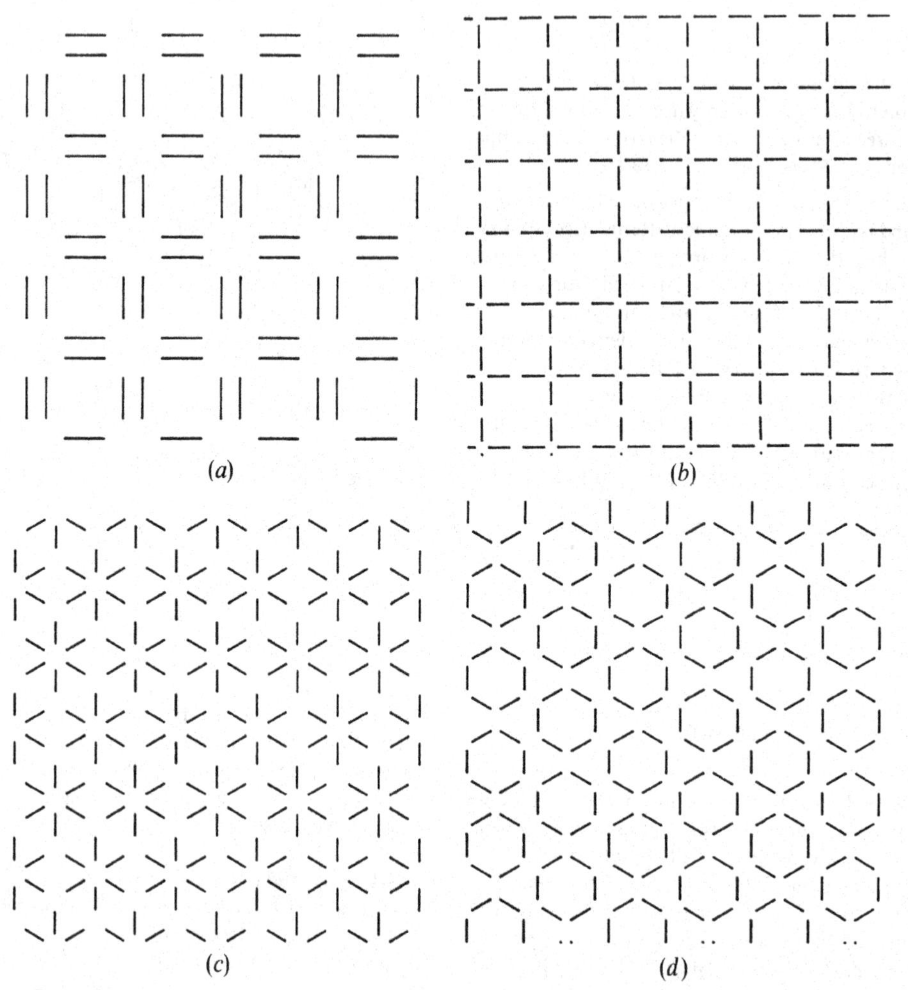

Figure 7.4.1
Examples of periodic patterns with a straight line-segment as motif which are of the same pattern type (**PP38** in (*a*) and (*b*), **PP48A** in (*c*) and (*d*)) but of different homeomeric types. Finite patterns with this same property can be found in Figure 7.4.2, and strip patterns in Figure 7.4.3.

one can be obtained from the other by rotating each copy of the motif by a right angle about its center.

For open line-segments the situation is much more complicated. As the closures of the segments may fail to be disjoint (and so lead to nondiscrete patterns) topological considerations enter and the homeomeric classification is much finer.

In order to analyse these patterns we begin by observing that the endpoint of a line-segment can have three possible positions relative to the other segments:

Table 7.4.1 THE HOMEOMERIC TYPES OF PATTERNS WITH AN OPEN STRAIGHT LINE-SEGMENT AS MOTIF

Column (1) gives a symbol for the type. LPF, LPS and LPP stand for finite segment pattern, strip segment pattern, and periodic segment pattern, respectively. If we delete the initial L and also the hyphen and final integer we obtain the pattern type in the notation of Section 5.2. Thus LPP2-1 and LPP2-2 have the same underlying pattern type PP2.

Columns (2) and (3) indicate the superpattern type and the supermotif. The notation for the superpattern type is that of Sections 5.2 and 6.5, and the entry "Coherent" means that the superpattern is trivial, that is, consists of just one copy of the supermotif. The notation for the supermotifs is that of Figures 7.4.2 and 7.4.3, except for the coherent periodic patterns. In this case we have not assigned a special symbol to the motif, but used the entry * in column (3). Some of these periodic supermotifs are shown in Figure 7.4.4.

Column (4) shows the endpoint class of each pattern type. An entry AA corresponds to a homeomeric type of (discrete) pattern that can also be realized with a closed line segment as motif.

Column (5) gives the number of degrees of freedom in choosing a pattern of the given type in which the motif is a unit line segment.

Column (6) gives information on how to construct a pattern of the desired type. A symbol of the type, or a reference to a diagram, accompanied by an asterisk (*) means that a pattern can be obtained by deleting endpoints (or vertices) and possibly also the midpoints of segments (or edges) and tiles of the pattern or tiling indicated. Similarly # means that maximal line segments are to be taken as copies of the motif, and // means that each segment is to be replaced by a pair of parallel segments. Finite patterns and strip patterns are illustrated in Figures 7.4.2 and 7.4.3, respectively. Periodic patterns for which there is no entry in column (6) can be obtained from the superpattern type and supermotif specified in columns (2) and (3), by the method of Section 5.3. If two types are bracketed in column (6) then they can be obtained from each other by rotating all the copies of the motif through a right angle, as in the examples of Figure 7.4.1.

Column (7) shows the number of distinct diffeomeric types of patterns with the given homeomeric type. An entry F indicates an infinite family (parametrized by a real-valued parameter) of diffeomeric types. The sign x means that a single diffeomeric type corresponds to each of the cases $n = 1, 2$.

Homeomeric type (1)	Pattern type of superpattern (2)	Type of supermotif (3)	Endpoint class (4)	Degrees of freedom (5)	Notes (6)	Diffeomeric types (7)
			Finite patterns			
LPF1_n-1 ($n \geq 2$)	PF1_n	L	AA	2		1
LPF1_n-2 ($n \geq 3$)	Coherent	LF1-n	AC	1		1
LPF2_n-1 ($n \geq 1$)	PF2_n	L	AA	3		1
LPF2_n-2 ($n \geq 2$)	PF3_n	LF2-1	AB	2		3
LPF2_n-3 ($n \geq 1$)	Coherent	LF2-n	AB_{2n}	1		Fx
LPF2_n-4 ($n \geq 2$)	Coherent	LF3-n	BB	1		3
LPF3_n-1 ($n \geq 2$)	PF3_n	L	AA	1	}	1
LPF3_n-2 ($n \geq 2$)	PF3_n	L	AA	1	}	1
LPF3_n-3 ($n \geq 3$)	Coherent	LF4-n	AB_n	0		1
LPF3_n-4 ($n \geq 2$)	Coherent	LF5-n	BB	0		1
			Strip Patterns			
LPS2-1	PS2	L	AA	3		1
LPS2-2	Coherent	LS1	AC	2		1
LPS3-1	PS3	L	AA	3		1
LPS3-2	PS4	LF2-1	AB	2		1
LPS5-1	PS5	L	AA	3		1

Table 7.4.1 (continued)

Homeomeric type (1)	Pattern type of superpattern (2)	Type of supermotif (3)	Endpoint class (4)	Degrees of freedom (5)	Notes (6)	Diffeomeric types (7)
LPS5-2	PS6	LF2-1	AB	2		1
LPS7-1	PS7	L	AA	4		1
LPS7-2	PS8	LF4-2	AB	2		1
LPS7-3	SPF3	LS5	BB	2		1
LPS8-1	PS8	L	AA	2		1
LPS9-1	PS9	L	AA	4		1
LPS9-2	PS10	LF2-1	AB	3		2
LPS9-3	PS11	LF4-2	AB	2		1
LPS9-4	SPF4	LS3	BB	2		1
LPS9-5	Coherent	LS2	BB	1		1
LPS10-1	PS10	L	AA	2		1
LPS10-2	Coherent	LS3	BB	1		1
LPS11-1	PS11	L	AA	2	}	1
LPS11-2	PS11	L	AA	2		1
LPS11-3	SPF4	LS5	BB	1		1
LPS12-1	PS12	L	AA	4		1
LPS12-2	PS13	LF2-1	AB	3		3
LPS12-3	PS14	LF2-1	AB	3		2
LPS12-4	PS15	LF2-2	AB_4	2		1
LPS12-5	PS15	LF3-2	BB	2		1
LPS12-6	SPF5	LS3	BB	2		1
LPS12-7	Coherent	LS4	BB_4	1		1
LPS13-1	PS13	L	AA	2	}	1
LPS13-2	PS13	L	AA	2		1
LPS13-3	PS15	LF4-2	AB	1		1
LPS13-4	SPF5	LS5	BB	1		1
LPS14-1	PS14	L	AA	2	}	1
LPS14-2	PS14	L	AA	2		1
LPS14-3	PS15	LF4-2	AB	1		1
LPS15-1	PS15	L	AA	1	}	1
LPS15-2	PS15	L	AA	1		1
LPS15-3	Coherent	LS5	BB	0		1
			Periodic Patterns			
LPP2-1	PP2	L	AA	4		1
LPP2-2	SPS9	LS1	AC	3		1
LPP3-1	PP3	L	AA	4		1
LPP3-2	PP4	LF2-1	AB	3		1
LPP5-1	PP5	L	AA	4		1
LPP5-2	PP6	LF2-1	AB	3		1

Table 7.4.1 (continued)

Homeomeric type (1)	Pattern type of superpattern (2)	Type of supermotif (3)	Endpoint class (4)	Degrees of freedom (5)	Notes (6)	Diffeomeric types (7)
LPP5-3	SPS10	LS1	*AC*	3		1
LPP5-4	Coherent	*	*BC*	2		1
LPP7-1	PP7	L	*AA*	6		1
LPP7-2	PP8	LF4-2	*AB*	4		1
LPP7-3	SPS6	LS5	*BB*	4		1
LPP8-1	PP8	L	*AA*	4		1
LPP8-2	SPS20	LS5	*BB*	2		1
LPP9-1	PP9	L	*AA*	5		1
LPP9-2	PP10	LF4-2	*AB*	3		1
LPP9-3	SPS7	LS5	*BB*	4		1
LPP9-4	SPS11	LS1	*AC*	4		1
LPP9-5	SPS11	LS3	*BB*	3		1
LPP9-6	Coherent	*	*BC*	2	LPP10-2*	1
LPP9-7	Coherent	*	*CC*	2	LPP10-3//	1
LPP10-1	PP10	L	*AA*	3		1
LPP10-2	Coherent	*	*CC*	2	Figure 7.4.4	1
LPP10-3	Coherent	*	*C*C**	1	Figure 7.4.4	1
LPP11-1	PP11	L	*AA*	5		1
LPP11-2	PP12	LF4-2	*AB*	3		1
LPP11-3	PP13	LF2-1	*AB*	4		2
LPP11-4	SPS8	LS5	*BB*	3		1
LPP11-5	SPS17	LS3	*BB*	3		1
LPP11-6	SPS23	LS2	*BB*	2		1
LPP12-1	PP12	L	*AA*	3		1
LPP12-2	SPS23	LS3	*BB*	2		1
LPP13-1	PP13	L	*AA*	3	}	1
LPP13-2	PP13	L	*AA*	3		1
LPP13-3	SPS14	LS5	*BB*	2		1
LPP13-4	SPS17	LS5	*BB*	2		1
LPP14-1	PP14	L	*AA*	5		1
LPP14-2	PP15	LF2-1	*AB*	4		2
LPP14-3	PP16	LF2-2	*AB*	3		1
LPP14-4	PP16	LF3-2	*BB*	3		1
LPP14-5	SPS18	LS3	*BB*	3		1
LPP14-6	SPS25	LS4	*BB*	2		1
LPP15-1	PP15	L	*AA*	3	}	1
LPP15-2	PP15	L	*AA*	3		1
LPP15-3	PP16	LF4-2	*AB*	2		1
LPP15-4	SPS18	LS5	*BB*	2		1
LPP16-1	PP16	L	*AA*	2		1
LPP16-2	SPS25	LS5	*BB*	1		1

Table 7.4.1 (continued)

Homeomeric type (1)	Pattern type of superpattern (2)	Type of supermotif (3)	Endpoint class (4)	Degrees of freedom (5)	Notes (6)	Diffeomeric types (7)
LPP17-1	PP17	L	AA	5		1
LPP17-2	PP18	LF4-2	AB	3		1
LPP17-3	PP19	LF2-1	AB	4		2
LPP17-4	PP20	LF2-2	AB	3		1
LPP17-5	PP20	LF3-2	BB	3		1
LPP17-6	SPS19	LS3	BB	3		1
LPP17-7	SPS24	LS2	BB	2		1
LPP17-8	SPS26	LS4	BB	2		1
LPP17-9	Coherent	*	BB_4	1	LPP18-3*	1
LPP18-1	PP18	L	AA	3		1
LPP18-2	SPS24	LS3	BB	2		1
LPP18-3	Coherent	*	B_4B_4	1	Figure 1.3.11(*d*)*	1
LPP19-1	PP19	L	AA	3	} Figure 7.1.3(*a*)	1
LPP19-2	PP19	L	AA	3	} Figure 7.1.3(*b*)	1
LPP19-3	PP20	LF4-2	AB	2		1
LPP19-4	SPS19	LS5	BB	2		1
LPP20-1	PP20	L	AA	2		1
LPP20-2	SPS26	LS5	BB	1		1
LPP21-1	PP21	L	AA	4		1
LPP21-2	PP22	LF1-3	AC	3		1
LPP21-3	PP22	LF4-3	AB_3	2		1
LPP21-4	PP22	LF5-3	BB	2		1
LPP21-5	Coherent	*	CC	2	Figure 7.4.4	1
LPP21-6	Coherent	*	CC	2	Figure 7.4.4	1
LPP21-7	Coherent	*	CC	1	Figure 7.4.4	1
LPP21-8	Coherent	*	B_3C	1	Figure 1.4.2(*j*)*	1
LPP21-9	Coherent	*	C^*C^*	1	Figure 7.4.4	1
LPP21-10	Coherent	*	C_*C_*	1	Figure 2.4.2(*h*)* for $0 < \alpha < \frac{1}{2}$	1
LPP23-1	PP23	L	AA	4		1
LPP23-2	PP24	LF1-3	AB	3		1
LPP23-3	PP24	LF4-3	AB	2		1
LPP23-4	PP24	LF5-3	BB	2		1
LPP23-5	PP25	LF2-1	AB	3		2
LPP23-6	PP26	LF2-3	AB	2		F
LPP23-7	PP26	LF3-3	BB	2		3
LPP23-8	Coherent	*	BB_3	1	Figure 1.4.2(*k*)*	1
LPP23-9	Coherent	*	B_4B_4	1	Figure 2.5.3(*d*)*	F
LPP23-10	Coherent	*	BC	1	Figure 7.4.4	3
LPP23-11	Coherent	*	B_6C	1	Figure 7.4.4	F
LPP25-1	PP25	L	AA	2	}	1
LPP25-2	PP25	L	AA	2	}	1
LPP25-3	PP26	LF4-3	AB	1		1
LPP25-4	PP26	LF5-3	BB	1		1
LPP25-5	Coherent	*	CC	1	Figure 7.4.4	1
LPP25-6	Coherent	*	C_*C_*	0	Figure 2.4.2(*h*)* for $\alpha = \frac{1}{2}$	1

Table 7.4.1 (continued)

Homeomeric type (1)	Pattern type of superpattern (2)	Type of supermotif (3)	Endpoint class (4)	Degrees of freedom (5)	Notes (6)	Diffeomeric types (7)
LPP27-1	PP27	L	AA	4		1
LPP27-2	PP28	LF2-1	AB	3		2
LPF27-3	PP29	LF2-3	AB	2		F
LPP27-4	PP29	LF3-3	BB	2		3
LPP27-5	Coherent	*	BB_6	1	Figure 7.4.4	F
LPP28-1	PP28	L	AA	2	}	1
LPP28-2	PP28	L	AA	2		1
LPP28-3	PP29	LF4-3	AB	1		1
LPP28-4	PP29	LF5-3	BB	1		1
LPP30-1	PP30	L	AA	4		1
LPP30-2	PP31	LF4-2	AB	2		1
LPP30-3	PP32	LF1-4	AC	3		1
LPP30-4	PP32	LF4-4	AB_4	2		1
LPP30-5	PP32	LF5-4	BB	2		1
LPP30-6	Coherent	*	CC	2	LPP31-2//	1
LPP30-7	Coherent	*	CC	2	Figure 7.4.4	1
LPP30-8	Coherent	*	CC	2	Figure 7.4.4	1
LPP30-9	Coherent	*	BC	1	LPP31-2*	1
LPP30-10	Coherent	*	B_4C	1	Figure 7.4.4	1
LPP30-11	Coherent	*	C^*C^*	1	Figure 7.4.4	1
LPP31-1	PP31	L	AA	2		1
LPP31-2	Coherent	*	CC	1	Figure 2.4.2(g)#	1
LPP33-1	PP33	L	AA	4		1
LPP33-2	PP34	LF1-4	AC	3		1
LPP33-3	PP34	LF4-4	AB	2		1
LPP33-4	PP34	LF5-4	BB	2		1
LPP33-5	PP35	LF2-1	AB	3		2
LPP33-6	PP36	LF2-2	AB	2		1
LPP33-7	PP36	LF3-2	BB	2		1
LPP33-8	Coherent	*	BB_4	1	Figure 7.4.4	1
LPP33-9	Coherent	*	B_4B_4	1	Figure 2.5.7(b)*	1
LPP33-10	Coherent	*	BC	2	LPP35-4*	3
LPP33-11	Coherent	*	B_4C	1	Figure 7.4.4	1
LPP35-1	PP35	L	AA	2	}	1
LPP35-2	PP35	L	AA	2		1
LPP35-3	PP36	LF4-2	AB	1		1
LPP35-4	Coherent	*	CC	1	Figure 1.4.2(n)#	1
LPP36-1	PP36	L	AA	1		1
LPP36-2	Coherent	*	C^*C^*	0	Figure 7.4.4	1
LPP37-1	PP37	L	AA	4		1
LPP37-2	PP38	LF2-1	AB	3		2
LPP37-3	PP39	LF2-1	AB	3		3
LPP37-4	PP40	LF2-2	AB	2		1
LPP37-5	PP40	LF3-2	BB	2		1
LPP37-6	PP41	LF2-4	AB	2		F

Table 7.4.1 (continued)

Homeomeric type (1)	Pattern type of superpattern (2)	Type of supermotif (3)	Endpoint class (4)	Degrees of freedom (5)	Notes (6)	Diffeomeric types (7)
LPP37-7	PP41	LF3-4	BB	2		5
LPP37-8	Coherent	*	BB_4	2	Figure 2.5.4(*l*)*	1
LPP37-9	Coherent	*	BB_8	1	Figure 7.4.4	F
LPP38-1	PP38	L	AA	2	} Figure 7.4.1(*a*)	1
LPP38-2	PP38	L	AA	2	} Figure 7.4.1(*b*)	1
LPP38-3	PP40	LF4-2	AB	1		1
LPP38-4	PP41	LF4-4	AB	1		1
LPP38-5	PP41	LF5-4	BB	1		1
LPP38-6	Coherent	*	BB_4	0	LPP40-2*	1
LPP39-1	PP39	L	AA	2	}	1
LPP39-2	PP39	L	AA	2	}	1
LPP39-3	PP41	LF4-4	AB	1		1
LPP39-4	PP41	LF5-4	BB	1		1
LPP40-1	PP40	L	AA	1		1
LPP40-2	Coherent	*	B_4B_4	0	Figure 1.2.1(*b*)*	1
LPP42-1	PP42	L	AA	4		1
LPP42-2	PP43	LF4-2	AB	2		1
LPP42-3	PP44	LF1-3	AC	3		1
LPP42-4	PP44	LF4-3	AB_3	2		1
LPP42-5	PP44	LF5-3	BB	2		1
LPP42-6	PP45	LF1-6	AC	3		1
LPP42-7	PP45	LF4-6	AB_6	2		1
LPP42-8	PP45	LF5-6	BB	2		1
LPP42-9	Coherent	*	BC	1	LPP43-2*	1
LPP42-10	Coherent	*	BC	1	LPP43-3*	1
LPP42-11	Coherent	*	B_3C	1	Figure 1.3.5#	1
LPP42-12	Coherent	*	B_6C	1	Figure 7.4.4	1
LPP42-13	Coherent	*	C^*C^*	1	Figure 7.4.4	1
LPP42-14	Coherent	*	CC	2	LPP43-4//	1
LPP42-15	Coherent	*	CC	2	Figure 7.4.4	1
LPP42-16	Coherent	*	CC	2	Figure 7.4.4	1
LPP43-1	PP43	L	AA	2		1
LPP43-2	Coherent	*	CC	1	Figure 2.4.2(*f*)#	1
LPP43-3	Coherent	*	CC	1	Figure 2.4.2(*e*)#	1
LPP43-4	Coherent	*	C^*C^*	0	Figure 7.4.4	1
LPP46-1	PP46	L	AA	4		1
LPP46-2	PP47	LF2-1	AB	3		3
LPP46-3	PP48A	LF2-1	AB	3		3
LPP46-4	PP48B	LF2-1	AB	3		3
LPP46-5	PP49	LF2-2	AB_4	2		1
LPP46-6	PP49	LF3-2	BB	2		1
LPP46-7	PP50	LF2-3	AB_6	2		F
LPP46-8	PP50	LF3-3	BB	2		5
LPP46-9	PP51	LF2-6	AB_{12}	2		F
LPP46-10	PP51	LF3-6	BB	2		5
LPP46-11	Coherent	*	BB_4	1	LPP48A-8*	3

Table 7.4.1 (continued)

Homeomeric type (1)	Pattern type of superpattern (2)	Type of supermotif (3)	Endpoint class (4)	Degrees of freedom (5)	Notes (6)	Diffeomeric types (7)
LPP46-12	Coherent	*	BB_6	1	Figure 2.5.4(*h*)*	F
LPP46-13	Coherent	*	BB_{12}	1	Figure 7.4.4	F
LPP47-1	PP47	L	AA	2	}	1
LPP47-2	PP47	L	AA	2	}	1
LPP47-3	PP49	LF4-2	AB	1		1
LPP47-4	PP51	LF4-6	AB_6	1		1
LPP47-5	PP51	LF5-6	BB	1		1
LPP47-6	Coherent	*	BB_6	0	LPP49-4*	1
LPP48A-1	PP48A	L	AA	2	}	1
LPP48A-2	PP48A	L	AA	2	}	1
LPP48A-3	PP50	LF4-3	AB_3	1		1
LPP48A-4	PP50	LF5-3	BB	1		1
LPP48A-5	PP51	LF4-6	AB_6	1		1
LPP48A-6	PP51	LF5-6	BB	1		1
LPP48A-7	Coherent	*	B_3B_6	0	Figure 4.5.2(*d*)*	1
LPP48A-8	Coherent	*	B_4B_4	0	Figure 4.5.2(*c*)*	1
LPP48B-1	PP48B	L	AA	2		1
LPP48B-2	PP48B	L	AA	2		1
LPP48B-3	PP49	LF4-2	AB	1		1
LPP48B-4	PP50	LF4-3	AB	1		1
LPP48B-5	PP50	LF5-3	BB	1		1
LPP48B-6	Coherent	*	BB_3	0	LPP49-3*	1
LPP49-1	PP49	L	AA	1	} Figure 7.4.1(*c*)	1
LPP49-2	PP49	L	AA	1	} Figure 7.4.1(*d*)	1
LPP49-3	Coherent	*	B_3B_3	0	Figure 1.2.1(*c*)*	1
LPP49-4	Coherent	*	B_6B_6	0	Figure 1.2.1(*a*)*	1

(*A*) The endpoint does not lie in the closure of any segment.

(*B*) The endpoint coincides with an endpoint of at least one other segment.

(*C*) The endpoint coincides with a relatively interior point of some other segment.

It is possible for an endpoint to have both properties (*B*) and (*C*) at the same time (see, for example, the pattern that is obtained by interpreting as copies of the motif the open sides of the large triangles in Figure 2.4.2(*h*), or the patterns LPP10-3, LPP21-9, LP30-11, LP36-2, LP42-13 and LP43-4 in Figure 7.4.4). In this case we shall regard it as being of type (*C*). In the diagrams, a conventional method of representing the three kinds of endpoints is adopted; this is explained in the caption of Figure 7.4.2. Based on these three possibilities, motifs, and therefore patterns, can be divided into six *endpoint classes AA, AB, AC, BB, BC* and *CC*. Variations of these symbols, which we shall use in the diagrams and Table 7.4.1 are:

(*a*) The addition of a subscript v (≥ 3) to a symbol B to indicate the valence of the corresponding endpoint. (If no subscript is added, then a valence of 2 is to be understood.)

(*b*) The addition of an asterisk, under certain circumstances, to a symbol C. We use C_* to mean that two segments have coinciding endpoints of type C and lie on the *same* side of a third segment, while C^* means that the corresponding segments lie on *opposite* sides of a third segment.

7.4.2 *Open line-segments are motifs of ten infinite families of homeomeric types of finite patterns. Four of these are of endpoint class AA, three of class AB, one of class AC, and two of class BB.*

Examples of all these families appear in Figure 7.4.2 and are listed in Table 7.4.1.

7.4.3 *Open line-segments are motifs of 37 homeomeric types of strip patterns. Sixteen of these are of endpoint class AA, ten of class AB, one of class AC and ten of class BB.*

Examples of these types appear in Figure 7.4.3 and are listed in Table 7.4.1.

7.4.4 *Open line-segments are motifs of 208 homeomeric types of periodic patterns. Of these, 59 are of endpoint-class AA, 50 of class AB, 8 of class AC, 64 of class BB, 12 of class BC and 24 of class CC. There are 56 types of coherent patterns.*

Some of these types are illustrated in Figure 7.4.4 and details appear in Table 7.4.1. This table, which establishes the assertions of Statement 7.4.2, 7.4.3 and 7.4.4 was prepared in the following manner.

For patterns of endpoint class AA, the homeomeric types of open segment patterns are in one-to-one correspondence with those whose motif is a closed segment. Their enumeration therefore follows from Statements 7.4.1.

Non-coherent patterns of other endpoint classes can be found by first determining the possible supermotifs (connected components of the union of the closures of the line-segments) and then the corresponding superpatterns. The procedure is analogous to that described in Section 7.3. It turns out that there are five finite families of supermotifs, of which 24 (including a single line-segment) and the five strip supermotifs, can be used in strip and periodic superpatterns.

All these supermotifs are shown in Figures 7.4.2 and 7.4.3 and symbols for them, which are used in column (3) of Table 7.4.1, are indicated. In some cases the same homeomeric type can arise from two different (homeomorphic) supermotifs used in the same superpattern. For example supermotifs LF3-2 and LF5-4 can both be used in superpattern PP16 to yield a pattern of type PP14; this occurs in Table 7.4.1 as homeomeric type LPP14-4. Again, supermotifs LF2-1 and LF4-2 can be used in superpattern PP15 to yield a pattern of type PP14, namely homeomeric type LPP14-2. In each case we have chosen the first of the two possible supermotifs to enter in column (3) of the table since for these the induced supermotif group and symmetry group of the motif are identical. This presentation also has the advantage that it prevents the pattern from changing type "accidentally". For example, using LF5-4 in a superpattern PP16 will "usually" yield LPP14-2, but may, if the horizontal and vertical spacing of the motifs happens to be equal, yield type LPP39-4.

The enumeration of the coherent patterns is more difficult. Those of endpoint class BB can be derived from the isotoxal tilings (Section 6.4) by deleting the tiles, vertices, and possibly the midpoints of the edges. In the latter case the edge must be made up of two equal and equivalent line segments placed end to end; if not, each edge must be a straight line-segment.

For coherent patterns of endpoint classes BC and CC there seems to be no easy or systematic method of obtaining them all. We have used a variety of empirical approaches, and believe that our listing is complete. However, we cannot affirm this with absolute certainty. This fact illustrates the point, mentioned in the introduction to this chapter, that homeomerism is a useful method of classification but usually yields no easy method of enumerating the possible types.

If the homeomeric classification of patterns with open line-segments as motifs is replaced by the finer diffeomeric

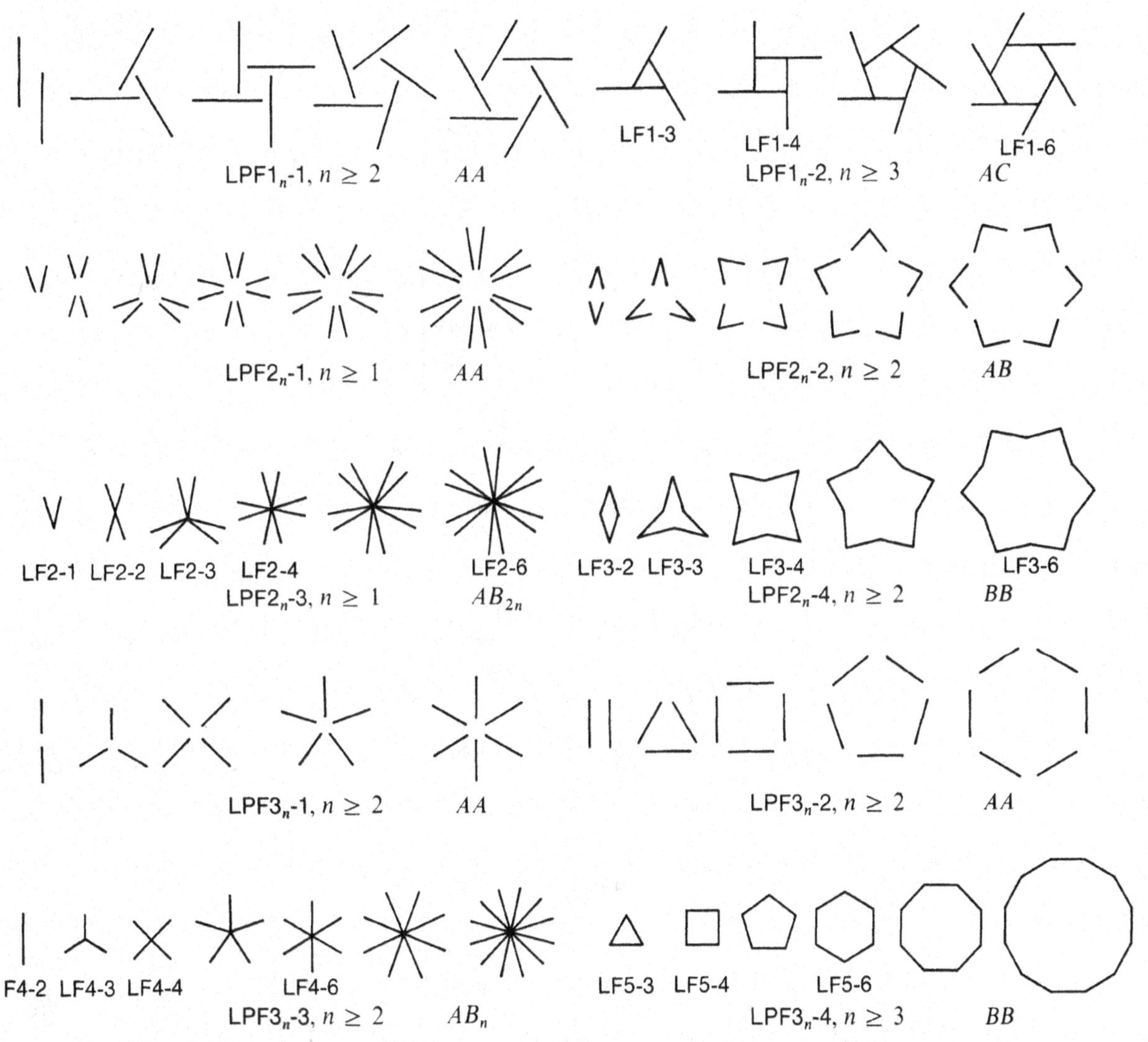

Figure 7.4.2
The homeomeric types of finite patterns with an open line segment L as motif. There are ten families of types, each illustrated by several representatives. Patterns in five of the families are coherent and some of these (indicated by a reference symbol beginning LF) can serve as supermotifs in periodic and strip patterns. The endpoint class of each pattern is also indicated. The four families in endpoint class AA can also be realized by closed line segments.

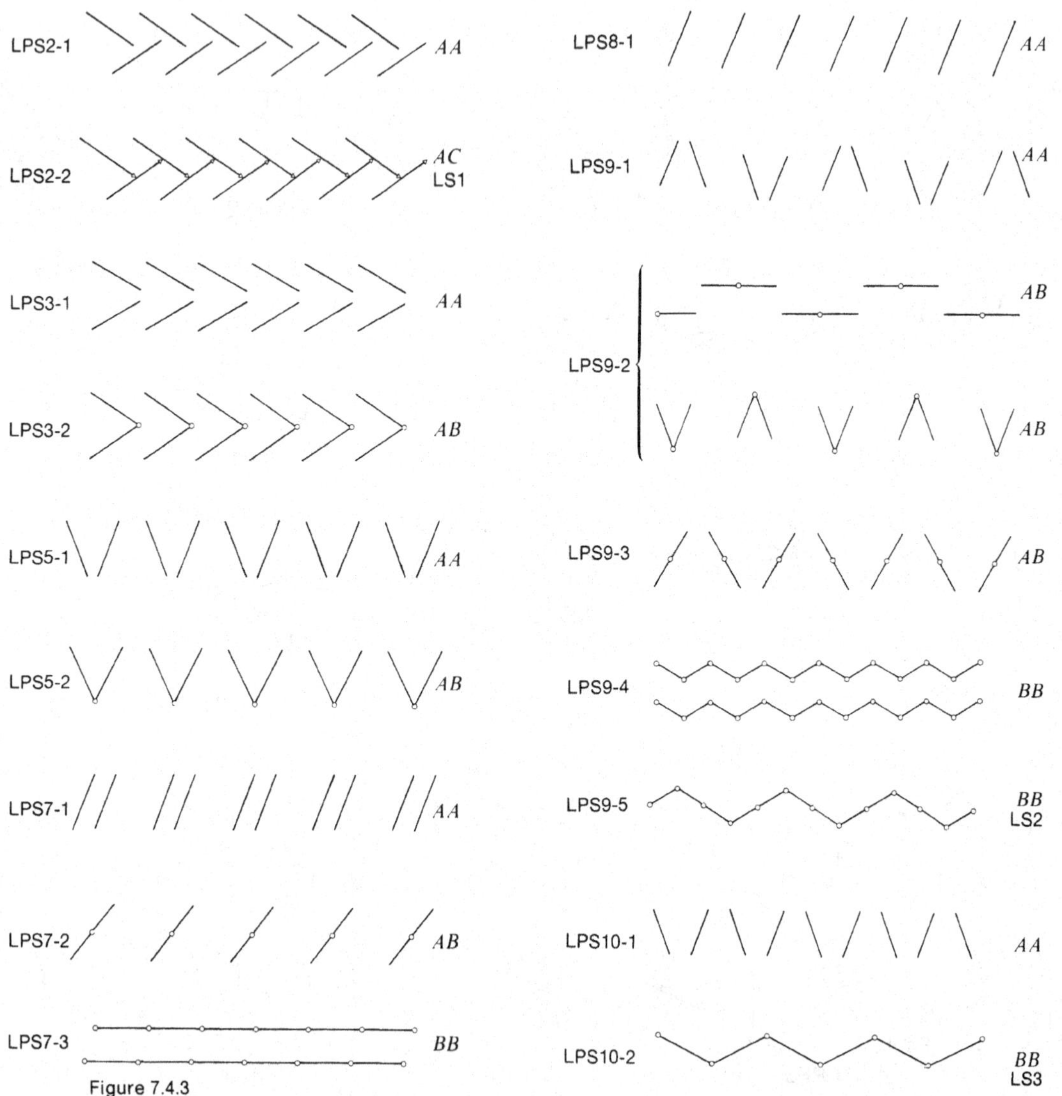

Figure 7.4.3
The homeomeric types of strip patterns with an open line segment as motif. Five of the 37 types are coherent and these are indicated by a reference symbol beginning LS. They can serve as supermotifs in other patterns. The endpoint class of each pattern is indicated as in Figure 7.4.2. The sixteen types in endpoint class *AA* correspond to types which can be realized by closed line segments. Where required for clarity, endpoints have been marked by arrowheads or dots. In cases where patterns of the same homeomeric type occur in several diffeomeric types, one representative of each diffeomeric type is shown.

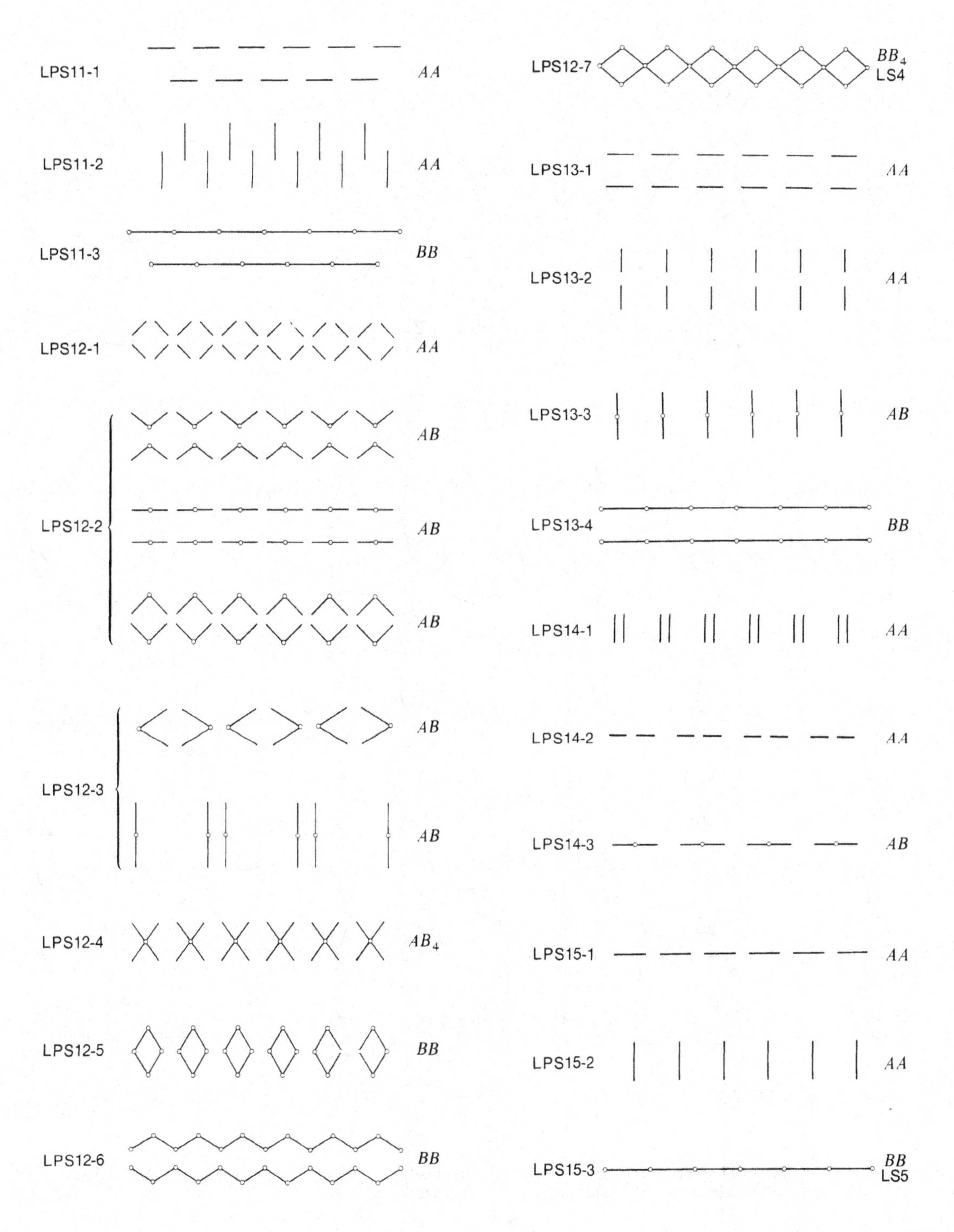
LPS11-1
AA
LPS11-2
AA
LPS11-3
BB
LPS12-1
AA
LPS12-2
AB
AB
AB
LPS12-3
AB
AB
LPS12-4
AB_4
LPS12-5
BB
LPS12-6
BB
LPS12-7
BB_4
LS4
LPS13-1
AA
LPS13-2
AA
LPS13-3
AB
LPS13-4
BB
LPS14-1
AA
LPS14-2
AA
LPS14-3
AB
LPS15-1
AA
LPS15-2
AA
LPS15-3
BB
LS5

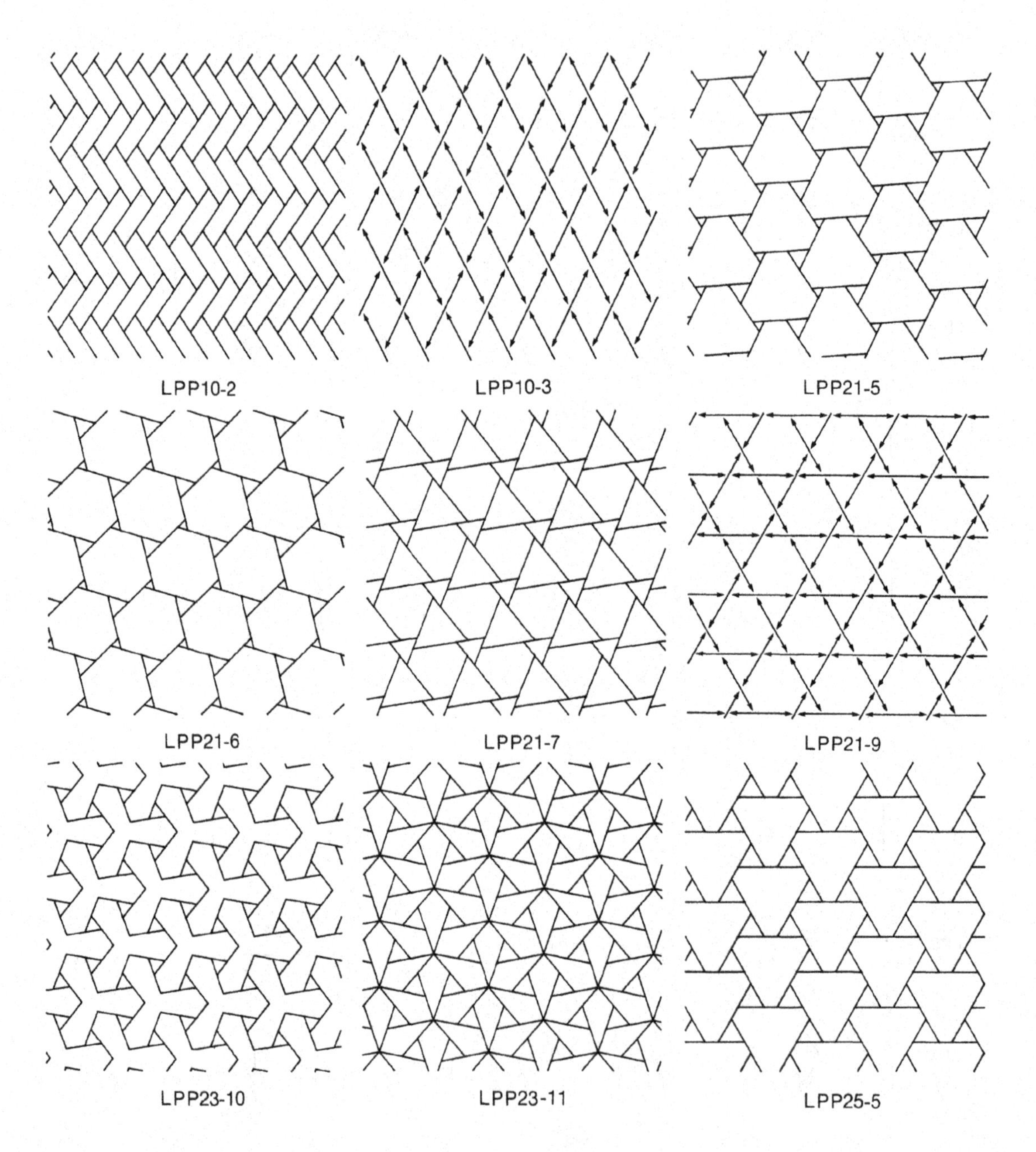

Figure 7.4.4
Examples of 24 of the 56 homeomeric types of coherent periodic patterns with an open line segment as motif. Further information about these patterns is given in Table 7.4.1.

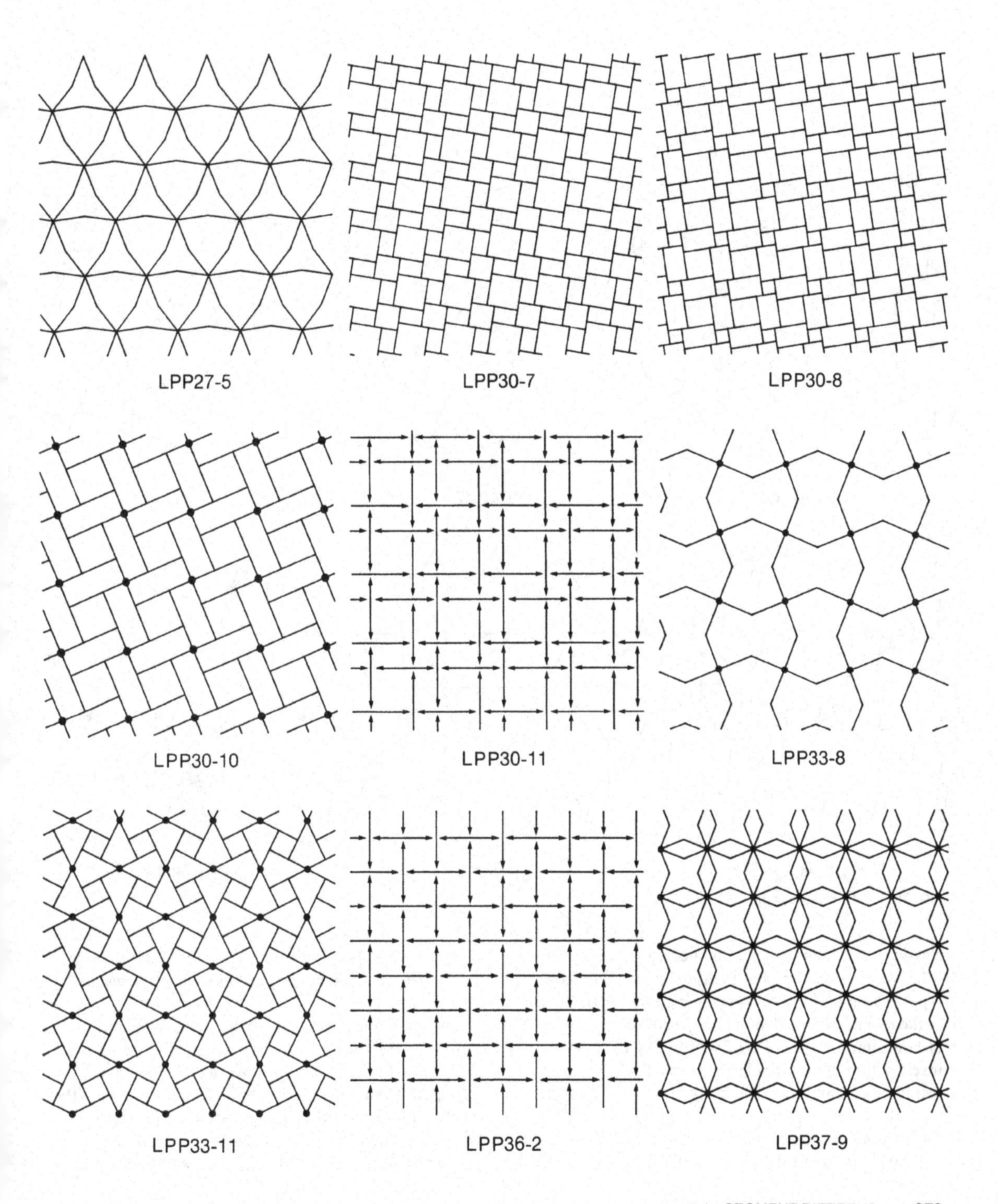
LPP27-5
LPP30-7
LPP30-8
LPP30-10
LPP30-11
LPP33-8
LPP33-11
LPP36-2
LPP37-9

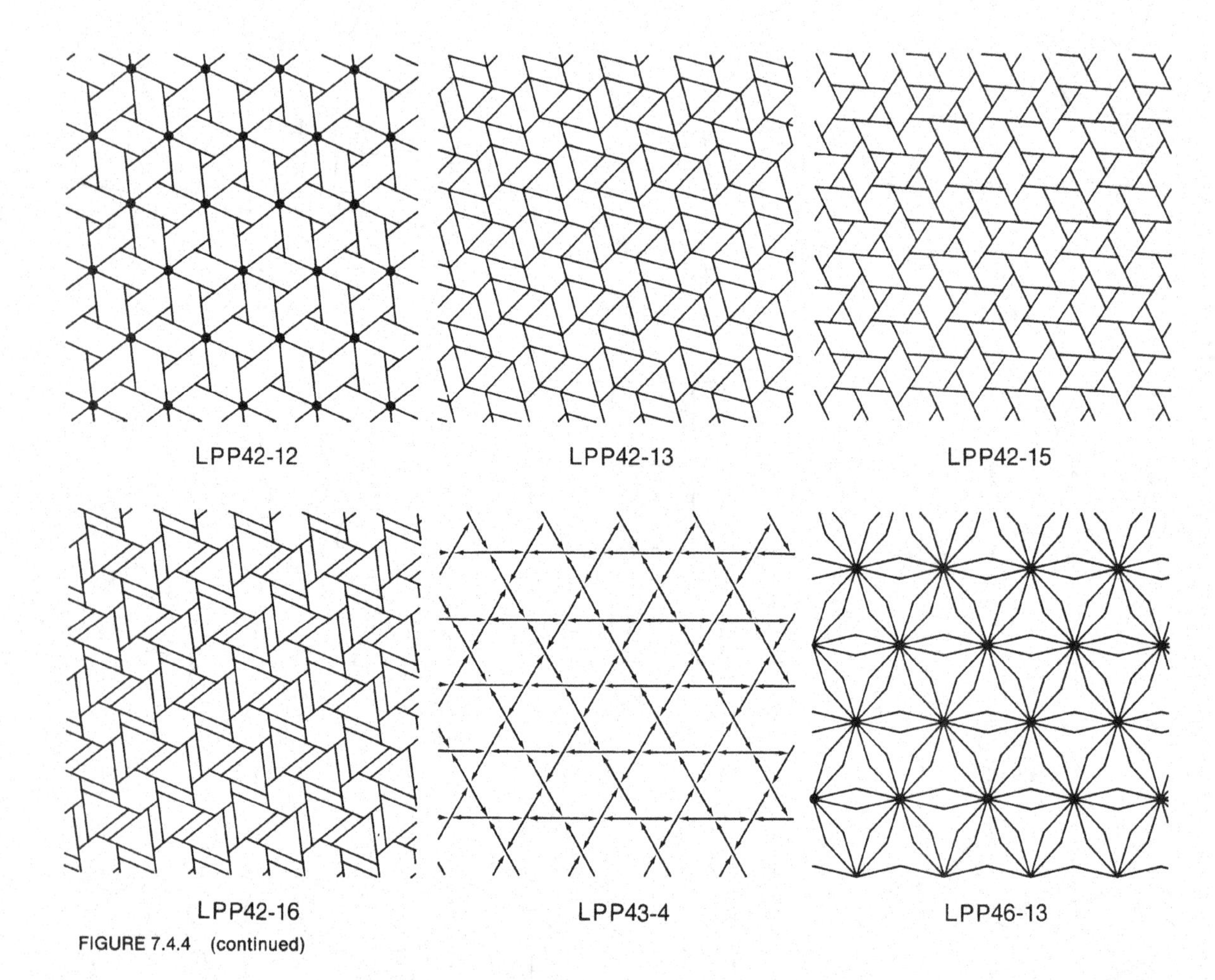

FIGURE 7.4.4 (continued)

classification, then new possibilities appear. In Figure 7.4.3 we show examples of the three possible diffeomeric types of strip patterns of homeomeric type LPS12-2, and of the two possible diffeomeric types of patterns of each of homeomeric types LPS9-2 and LPS12-3. A detailed investigation of all types leads to the following result (compare column (7) of Table 7.4.1):

7.4.5 *Open line-segments are the motifs of* 41 *diffeomeric types of strip patterns, and of* 235 *types and* 11 *families of diffeomeric types of periodic patterns; each family depends on a real-valued parameter.*

The analogous results for finite open segment patterns can be found in the first part of Table 7.4.1.

We should stress that in Statement 7.4.5 (and in the entries in Column (7) of Table 7.4.1) we assert only the existence of a certain number of diffeomeric types. We believe that our listings constitute a complete enumeration, but this has not been established so far.

EXERCISES 7.4

1. Verify the data contained in Table 7.4.1 for patterns with an open segment as motif and of pattern types PP25, PP31 and PP47.
2. Construct a relational indicator for the family of all strip patterns with an open segment as motif, which are of types
   (*a*) LPS7-2 and their limits,
   (*b*) LPS2-1 and their limits.
3. Classify the homeomeric types of all patterns which have as motif an open (or closed) line-segment, and which are continuous (that is, in some direction every translation is a symmetry of the pattern). (See Section 5.5 for the continuous strip and periodic groups.)
4. (*a*) Describe a method for determining, from the information in this section and in Chapter 5, all the homeomeric types of patterns which have as motif a half-open, half-closed line-segment (that is, a segment with just one of its endpoints included).
   (*b*) Use the method you described in part (*a*) to find all such patterns of pattern type (*i*) PP21, (*ii*) PP32.

*5. Determine the homeomeric types of all patterns which have symmetry group *p*3 or *p*4 and which have as motif:
   (*a*) an open circular disk from which two points, symmetric with respect to the center, have been removed;
   (*b*) the union of an open circular disk with two diametrically opposite points on its boundary;
   (*c*) a closed circular disk from which two diametrically opposite boundary points have been removed;
   (*d*) the union of an open square with its four corners;
   (*e*) an open semicircle.

6. Verify the information given in Table 7.4.1 about the diffeomeric types of open segment patterns of pattern type PP37. Illustrate by an example each of the sixteen types.
7. Show that with the interior of a square as motif, there exist at least three diffeomeric types of coherent strip patterns of pattern type PS2; all have the same homeomeric type.

7.5 ELLIPSE PATTERNS

Due to a combination of topological and symmetry properties, the circular disk patterns and line-segment patterns investigated in the previous two sections are rather special. For our third illustration of homeomeric classification we shall now investigate patterns with motifs of a more general character.

We begin by investigating *ellipse patterns*, that is, patterns whose motif is the region bounded by a non-circular elliptical curve. As before, there are two cases to be considered depending upon whether the motif is open or closed. In the case of *closed ellipse patterns*, we can apply the procedure described in Section 5.3 to determine the possible refined pattern types, and these correspond exactly to the homeomeric types—as was the case with closed circular disk patterns. We can prove:

7.5.1 *Closed (non-circular) ellipses are the motifs in* 3 *families (each depending on an integer n) of homeomeric types of finite patterns,* 12 *types of strip patterns and* 38 *types of periodic patterns.*

These are listed in Table 7.5.1—they correspond to the entries with the letter E in column (3). Examples of the finite and strip patterns are shown in Figures 7.5.1 and 7.5.2.

Since open ellipses can touch one another, topological

Table 7.5.1 THE HOMEOMERIC TYPES OF PATTERNS WITH AN OPEN ELLIPSE AS MOTIF

Column (1) gives a symbol for the type. EPF, EPS and EPP stand for finite ellipse pattern, strip ellipse pattern, and periodic ellipse pattern, respectively. If we delete the initial E and also the hyphen and final integer we obtain the pattern type in the notation of Section 5.2. Thus EPP43-1 and EPP43-2 both have underlying pattern type PP43.

Columns (2) and (3) indicate the superpattern type and the supermotif. The notation for the superpattern type is that of Sections 5.2 and 6.5, and the entry "Coherent" means that the superpattern is trivial, that is, consists of just one copy of the supermotif. The notation for the supermotifs is that of Figures 7.5.1 and 7.5.2 except for the coherent periodic patterns. In this case we have not assigned a special symbol to the motif but used an entry $*$ in column (3). All the periodic supermotifs are shown in Figure 7.5.3.

Column (4) gives the number of degrees of freedom in choosing a pattern of the given type in which the motif is a given (non-circular) open ellipse.

Column (5) indicates the number of ellipses which touch a given ellipse in the pattern. Those types with the entry 0 in column (5) correspond to patterns which can be realized using a closed ellipse as motif.

Column (6) gives references to the ellipse patterns published by Nowacki [1948]. The reference number to the type in Nowacki's paper is preceded by N, and examples of all these are shown in Figure 7.5.3. Four types, marked †, are also shown—these are the four types missed by Nowacki.

Homeomeric type (1)	Pattern type of superpattern (2)	Type of supermotif (3)	Degrees of freedom (4)	Contact number (5)	References (6)
		Finite patterns			
EPF1_n-1 ($n \geq 2$)	PF1_n	E	2	0	
EPF1_n-2 ($n \geq 2$)	Coherent	EF1-n	1	2	
EPF2_n-1 ($n \geq 1$)	PF2_n	E	$\begin{cases}2\ (n = 1)\\3\ (n > 1)\end{cases}$	0	
EPF2_n-2 ($n \geq 2$)	PF3_n	EF2-1	2	1	
EPF2_n-3 ($n \geq 1$)	Coherent	EF2-n	1	$\begin{cases}1\ (n = 1)\\2\ (n > 1)\end{cases}$	
EPF3_n-1 ($n \geq 2$)	PF3_n	E	1	0	
EPF3_n-2 ($n \geq 2$)	Coherent	EF3-n	0	$\begin{cases}1\ (n = 2)\\2\ (n > 2)\end{cases}$	
		Strip patterns			
EPS2-1	PS2	E	3	0	
EPS2-2	SPF1	ES7	2	2	
EPS2-3	Coherent	ES1	2	2	
EPS2-4	Coherent	ES2	1	4	
EPS3-1	PS3	E	3	0	
EPS3-2	PS4	EF2-1	2	1	
EPS3-3	SPF2	ES7	2	2	
EPS3-4	Coherent	ES3	1	3	
EPS5-1	PS5	E	3	0	
EPS5-2	PS6	EF2-1	2	1	
EPS7-1	PS7	E	4	0	
EPS7-2	PS8	EF1-2	3	1	
EPS7-3	SPF3	ES7	2	2	
EPS7-4	Coherent	ES4	2	2	

Table 7.5.1 (continued)

Homeomeric type (1)	Pattern type of superpattern (2)	Type of supermotif (3)	Degrees of freedom (4)	Contact number (5)	References (6)
EPS7-5	Coherent	ES5	2	3	
EPS7-6	Coherent	ES6	1	4	
EPS8-1	PS8	E	2	0	
EPS8-2	Coherent	ES7	1	2	
EPS9-1	PS9	E	3	0	
EPS9-2	PS10	EF1-2	2	1	
EPS9-3	SPF4	ES10	2	2	
EPS9-4	Coherent	ES8	2	2	
EPS9-5	Coherent	ES9	1	3	
EPS10-1	PS10	E	2	0	
EPS10-2	Coherent	ES10	1	2	
EPS11-1	PS11	E	2	0	
EPS11-2	SPF4	ES15	1	2	
EPS11-3	Coherent	ES11	1	2	
EPS11-4	Coherent	ES12	0	4	
EPS12-1	PS12	E	4	0	
EPS12-1	PS13	EF2-1	3	1	
EPS12-3	PS14	EF2-1	3	1	
EPS12-4	PS15	EF2-2	2	2	
EPS12-5	SPF5	ES10	2	2	
EPS12-6	Coherent	ES13	1	3	
EPS13-1	PS13	E	2	0	
EPS13-2	PS15	EF3-2	1	1	
EPS13-3	SPF5	ES15	1	2	
EPS13-4	Coherent	ES14	0	3	
EPS14-1	PS14	E	2	0	
EPS14-2	PS15	EF3-2	1	1	
EPS15-1	PS15	E	1	0	
EPS15-2	Coherent	ES15	0	2	
		Periodic patterns			
EPP2-1	PP2	E	4	0	
EPP2-2	SPS9	ES1	3	2	
EPP2-3	SPS9	ES2	2	4	
EPP3-1	PP3	E	4	0	
EPP3-2	PP4	EF2-1	3	1	
EPP3-3	SPS4	ES7	2	2	
EPP3-4	SPS12	ES3	2	3	
EPP5-1	PP5	E	4	0	
EPP5-2	PP6	EF2-1	3	1	
EPP5-3	SPS5	ES7	3	2	
EPP5-4	SPS10	ES1	3	2	

Table 7.5.1 (continued)

Homeomeric type (1)	Pattern type of superpattern (2)	Type of supermotif (3)	Degrees of freedom (4)	Contact number (5)	References (6)
EPP5-5	SPS10	ES2	2	4	
EPP5-6	SPS13	ES3	2	3	
EPP5-7	Coherent	*	2	3	N-6
EPP5-8	Coherent	*	1	5	N-8
EPP7-1	PP7	E	5	0	
EPP7-2	PP8	EF1-2	4	1	
EPP7-3	SPS6	ES7	4	2	
EPP7-4	SPS20	ES4	4	2	
EPP7-5	SPS20	ES5	4	3	
EPP7-6	SPS20	ES6	4	4	
EPP7-7	Coherent	*	3	3	N-1
EPP7-8	Coherent	*	2	4	N-2
EPP7-9	Coherent	*	1	5	N-3
EPP8-1	PP8	E	4	0	
EPP8-2	SPS20	ES7	3	2	
EPP8-3	Coherent	*	2	4	N-4
EPP8-4	Coherent	*	1	6	N-5
EPP9-1	PP9	E	4	0	
EPP9-2	PP10	EF1-2	4	1	
EPP9-3	SPS7	ES7	4	2	
EPP9-4	SPS11	ES2	3	2	
EPP9-5	SPS21	ES4	3	2	
EPP9-6	SPS21	ES5	3	3	
EPP9-7	SPS21	ES6	3	4	
EPP9-8	Coherent	*	3	3	N-10
EPP9-9	Coherent	*	3	4	N-11
EPP9-10	Coherent	*	2	5	N-12
EPP9-11	Coherent	*	2	5	N-13
EPP9-12	Coherent	*	1	6	†
EPP9-13	Coherent	*	1	6	†
EPP10-1	PP10	E	3	0	
EPP10-2	SPS21	ES7	2	2	
EPP10-3	Coherent	*	2	4	N-14
EPP10-4	Coherent	*	1	6	N-15
EPP11-1	PP11	E	5	0	
EPP11-2	PP12	EF1-2	4	1	
EPP11-3	PP13	EF2-1	4	1	
EPP11-4	SPS8	ES7	4	2	
EPP11-5	SPS14	ES3	3	3	
EPP11-6	SPS17	ES10	3	2	
EPP11-7	SPS22	ES4	3	2	
EPP11-8	SPS22	ES5	3	3	
EPP11-9	SPS22	ES6	3	4	
EPP11-10	SPS23	ES8	3	2	
EPP11-11	SPS23	ES9	2	3	
EPP11-12	Coherent	*	3	3	N-16
EPP11-13	Coherent	*	2	4	N-17
EPP11-14	Coherent	*	1	5	N-18

Table 7.5.1 (continued)

Homeomeric type (1)	Pattern type of superpattern (2)	Type of supermotif (3)	Degrees of freedom (4)	Contact number (5)	References (6)
EPP12-1	PP12	E	3	0	
EPP12-2	SPS22	ES7	2	2	
EPP12-3	SPS23	ES10	2	2	
EPP12-4	Coherent	*	1	4	N-20
EPP13-1	PP13	E	3	0	
EPP13-2	SPS14	ES15	1	2	
EPP13-3	SPS17	ES15	2	2	
EPP13-4	SPS23	ES11	2	2	
EPP13-5	SPS23	ES12	1	4	
EPP13-6	Coherent	*	1	4	N-19
EPP14-1	PP14	E	5	0	
EPP14-2	PP15	EF2-1	4	1	
EPP14-3	PP16	EF2-2	3	2	
EPP14-4	SPS18	ES10	3	2	
EPP14-5	SPS25	ES13	2	3	
EPP15-1	PP15	E	3	0	
EPP15-2	PP16	EF3-2	2	1	
EPP15-3	SPS18	ES15	2	2	
EPP15-4	SPS25	ES14	1	3	
EPP16-1	PP16	E	2	0	
EPP16-2	SPS25	ES15	1	2	
EPP16-3	Coherent	*	0	4	N-9
EPP17-1	PP17	E	5	0	
EPP17-2	PP18	EF1-2	4	1	
EPP17-3	PP19	EF2-1	4	1	
EPP17-4	PP20	EF2-2	3	2	
EPP17-5	SPS19	ES10	3	2	
EPP17-6	SPS24	ES8	3	2	
EPP17-7	SPS24	ES9	2	3	
EPP17-8	SPS26	ES13	2	3	
EPP17-9	Coherent	*	2	3	N-21
EPP17-10	Coherent	*	1	4	N-22
EPP18-1	PP18	E	3	0	
EPP18-3	SPS24	ES10	2	2	
EPP18-3	Coherent	*	1	4	N-25
EPP19-1	PP19	E	3	0	
EPP19-2	PP20	EF3-2	2	1	
EPP19-3	SPS19	ES15	2	2	
EPP19-4	SPS24	ES11	2	2	
EPP19-5	SPS24	ES12	1	4	
EPP19-6	SPS26	ES14	1	3	
EPP19-7	Coherent	*	1	3	N-23
EPP19-8	Coherent	*	0	5	N-24
EPP20-1	PP20	E	2	0	
EPP20-2	SPS26	ES15	1	2	

Table 7.5.1 (continued)

Homeomeric type (1)	Pattern type of superpattern (2)	Type of supermotif (3)	Degrees of freedom (4)	Contact number (5)	References (6)
EPP20-3	Coherent	*	1	4	N-26
EPP20-4	Coherent	*	0	6	N-27
EPP21-1	PP21	E	4	0	
EPP21-2	PP22	EF1-3	3	2	
EPP21-3	Coherent	*	2	4	N-39
EPP21-4	Coherent	*	1	6	†
EPP23-1	PP23	E	4	0	
EPP23-2	PP24	EF1-3	3	2	
EPP23-3	PP25	EF2-1	3	1	
EPP23-4	PP26	EF2-3	2	2	
EPP23-5	Coherent	*	2	3	N-41
EPP23-6	Coherent	*	1	4	N-42
EPP25-1	PP25	E	2	0	
EPP25-2	PP26	EF3-3	1	2	
EPP25-3	Coherent	*	1	4	N-43
EPP25-4	Coherent	*	0	6	N-44
EPP27-1	PP27	E	4	0	
EPP27-2	PP28	EF2-1	3	1	
EPP27-3	PP29	EF2-3	2	2	
EPP27-4	Coherent	*	1	3	N-40
EPP28-1	PP28	E	2	0	
EPP28-2	PP29	EF3-3	1	2	
EPP30-1	PP30	E	4	0	
EPP30-2	PP31	EF1-2	3	1	
EPP30-3	PP32	EF1-4	3	2	
EPP30-4	Coherent	*	2	3	N-28
EPP30-5	Coherent	*	2	4	N-29
EPP30-6	Coherent	*	1	5	†
EPP31-1	PP31	E	2	0	
EPP31-2	Coherent	*	1	4	N-30
EPP33-1	PP33	E	4	0	
EPP33-2	PP34	EF1-4	2	2	
EPP33-3	PP35	EF2-1	3	1	
EPP33-4	PP36	EF2-2	2	2	
EPP33-5	Coherent	*	2	3	N-34
EPP33-6	Coherent	*	1	4	N-35
EPP35-1	PP35	E	2	0	
EPP35-2	PP36	EF3-2	1	1	
EPP35-3	Coherent	*	1	4	N-36
EPP35-4	Coherent	*	0	5	N-37
EPP36-1	PP36	E	1	0	
EPP36-2	Coherent	*	0	4	N-33
EPP37-1	PP37	E	4	0	

Table 7.5.1 (continued)

Homeomeric type (1)	Pattern type of superpattern (2)	Type of supermotif (3)	Degrees of freedom (4)	Contact number (5)	References (6)
EPP37-2	PP38	EF2-1	3	1	
EPP37-3	PP39	EF2-1	3	1	
EPP37-4	PP40	EF2-2	2	2	
EPP37-5	PP41	EF2-4	2	2	
EPP37-6	Coherent	*	1	3	N-31
EPP38-1	PP38	E	2	0	
EPP38-2	PP40	EF3-2	1	1	
EPP38-3	PP41	EF3-4	1	2	
EPP38-4	Coherent	*	1	3	N-32
EPP39-1	PP39	E	2	0	
EPP39-2	PP41	EF3-4	1	2	
EPP40-1	PP40	E	1	0	
EPP40-2	Coherent	*	0	4	N-33
EPP42-1	PP42	E	4	0	
EPP42-2	PP43	EF1-2	3	1	
EPP42-3	PP44	EF1-3	3	2	
EPP42-4	PP45	EF1-6	3	2	
EPP42-5	Coherent	*	2	3	N-45
EPP42-6	Coherent	*	2	3	N-46
EPP42-7	Coherent	*	2	4	N-47
EPP42-8	Coherent	*	1	5	N-48
EPP43-1	PP36	E	2	0	
EPP43-2	Coherent	*	1	4	N-49
EPP46-1	PP46	E	4	0	
EPP46-2	PP47	EF2-1	3	1	
EPP46-3	PP48A	EF2-1	3	1	
EPP46-4	PP48B	EF2-1	3	1	
EPP46-5	PP49	EF2-2	2	2	
EPP46-6	PP50	EF2-3	2	2	
EPP46-7	PP51	EF2-6	2	2	
EPP46-8	Coherent	*	1	3	N-50
EPP47-1	PP47	E	2	0	
EPP47-2	PP49	EF3-2	1	1	
EPP47-3	PP51	EF3-6	1	2	
EPP47-4	Coherent	*	0	3	N-53
EPP48A-1	PP48A	E	2	0	
EPP48A-2	PP50	EF3-3	1	2	
EPP48A-3	PP51	EF3-6	1	2	
EPP48A-4	Coherent	*	0	4	N-52
EPP48B-1	PP48B	E	2	0	
EPP48B-2	PP49	EF3-2	1	1	
EPP48B-3	PP50	EF3-3	1	2	
EPP48B-4	Coherent	*	0	3	N-51
EPP49-1	PP49	E	1	0	
EPP49-2	Coherent	*	0	4	N-54

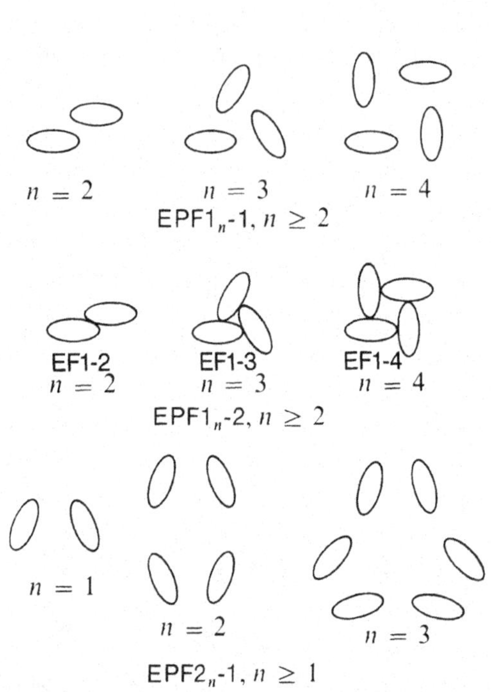

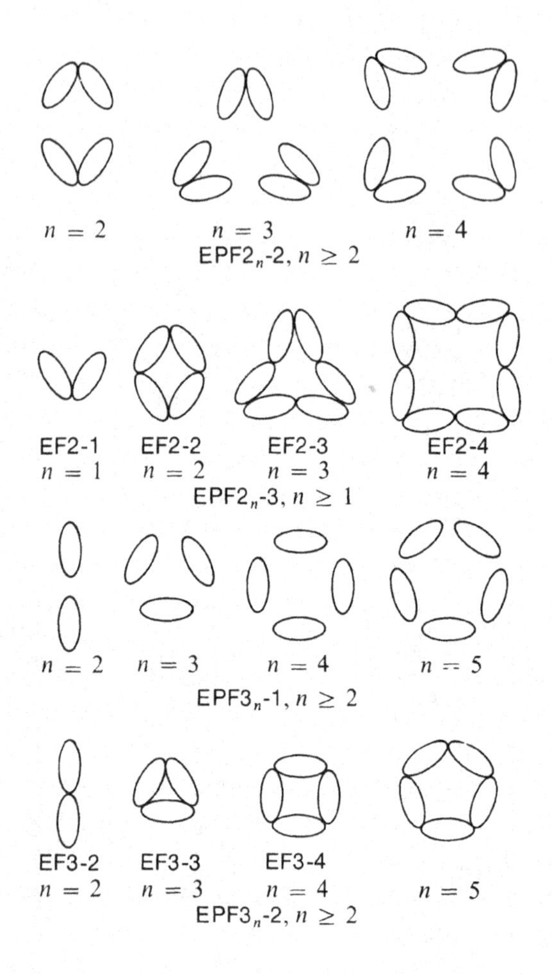

Figure 7.5.1
Representative of the seven infinite families of homeomeric types of finite patterns whose motif is an open ellipse. Near each set of patterns we give the symbol for the family and the range of values of the parameter n. In the case of coherent patterns we also indicate the symbol (beginning with EF) that will be used if the set of ellipses is used as a supermotif in the construction of superpatterns. Further information about these patterns appears in Table 7.5.1.

considerations enter and so (as we would expect) the homeomeric classification of open ellipse patterns is finer than that by refined pattern type alone. It is remarkable to report that the coherent periodic types of open ellipse patterns were determined, in essence, by Nowacki [1948]. The qualification in this statement refers to the fact that there were some small errors in Nowacki's work—he missed four types and drew one incorrectly—and also to the fact that he did not specify what constitutes a "type". However, he applied to the problem an intuitive understanding which led him to carry out, in fact if not in name, a homeomeric classification of these patterns. This illustrates our earlier remark that homeomerism is an extremely natural way to classify patterns and other geometric objects with respect to symmetry.

Our results concerning open ellipse patterns are summarized in the following three statements, and details concerning all the types appear in Table 7.5.1.

7.5.2 *Open ellipses are motifs of seven families (each depending on an integer n) of homeomeric types of finite patterns. Three of these families and one other pattern are coherent and can serve as supermotifs for superpatterns.*

Illustrations of the seven families appear in Figure 7.5.1.

7.5.3 *Open ellipses are motifs of* 43 *homeomeric types of strip patterns. Of these types,* 15 *are coherent and can serve as supermotifs in strip and periodic superpatterns.*

Illustrations of the 43 types appear in Figure 7.5.2.

7.5.4 *Open ellipses are motifs of* 192 *homeomeric types of periodic patterns, of which* 57 *types are coherent.*

In Figure 7.5.3 we show all the coherent types. The remaining 135 types can be constructed from the specification of the supermotifs and superpatterns listed in Table 7.5.1. The last column of this table gives references to the occurrences of the types in Nowacki's pioneering paper mentioned above.

To establish Statements 7.5.2, 7.5.3 and 7.5.4 it suffices to explain how Table 7.5.1 was constructed. The procedure for the non-coherent types is exactly analogous to that described in the previous two sections—we use the technique of enumerating possible finite and strip supermotifs, and then see how these can be used in the construction of superpatterns. As in the case of line-segment patterns it can happen that the same homeomeric type can arise using different supermotifs in the same superpattern type. In these cases we have entered in column (3) of Table 7.5.1 the supermotif whose symmetry group coincides with the induced motif group of the superpattern. For the coherent patterns we have had to use *ad hoc* arguments and though we believe that our enumeration is complete, this cannot be asserted with absolute certainty.

A remarkable feature of the homeomeric classification of ellipse patterns is that the eccentricity of the ellipses we use is irrelevant; exactly the same homeomeric types emerge whether we use nearly circular ellipses or ones which are very long and thin. This observation is a special case of the following result, which vindicates our earlier remark that in this section much more general patterns are being considered. We use the word *d2-oval* to mean any region M bounded by a smooth strictly convex curve such that the symmetry group $S(M)$ is the dihedral group *d2*.

7.5.5 *The homeomeric types of patterns with an open d2-oval as motif are in one-to-one correspondence with the homeomeric types of open ellipse patterns.*

In other words, the precise shape of the curve bd M is irrelevant as far as the homeomeric classification is concerned so long as it is smooth, strictly convex and has the correct symmetry group. (This situation is somewhat analogous to that we mentioned when discussing (discrete) closed disk patterns—the homeomeric types depend only on the symmetry group of the motif and not on its actual "shape".)

To prove Statement 7.5.5 we shall indicate how all the homeomeric types of open *d2*-oval patterns of a given pattern type can be constructed. Consideration of every case makes the argument long and tedious, but the underlying idea is quite simple. We shall illustrate it by one example, namely pattern type PP21 whose symmetry group is *p3*.

Let us begin by considering a *d2*-oval pattern of this type in which the copies of the motif have disjoint closures, see Figure 7.5.4(*a*). This is the first homeomeric type, and it corresponds to the ellipse pattern EPP21-1 in Table 7.5.1. We now *modify* the pattern, by which we mean that we move the copies of the motif in the plane and possibly alter their size—but we always do so in such a way that the underlying pattern type PP21 remains unchanged. If, by such a modification, two copies of the motif come into contact, then it is easy to see that further contacts necessarily occur, and the copies of the motifs will touch in threes about centers of 3-fold rotation of the symmetry group *p3* (see Figure 7.5.4(*b*)). These "triplets" are either of type EF1-3 or of type EF3-3 but even in the latter case the induced motif group must be *c3* (and *not d3*—if it were *d3* then the pattern type would have changed to PP25). The superpattern is of type PP22 and the homeomeric type corresponds to the ellipse pattern EPP21-2.

Without altering the homeomeric type, we can now modify this pattern, increasing the motifs in size and rotating the triplets about their centers until further contacts occur. There are just two possibilities: either a *d2*-oval of one triplet comes into contact with one of a

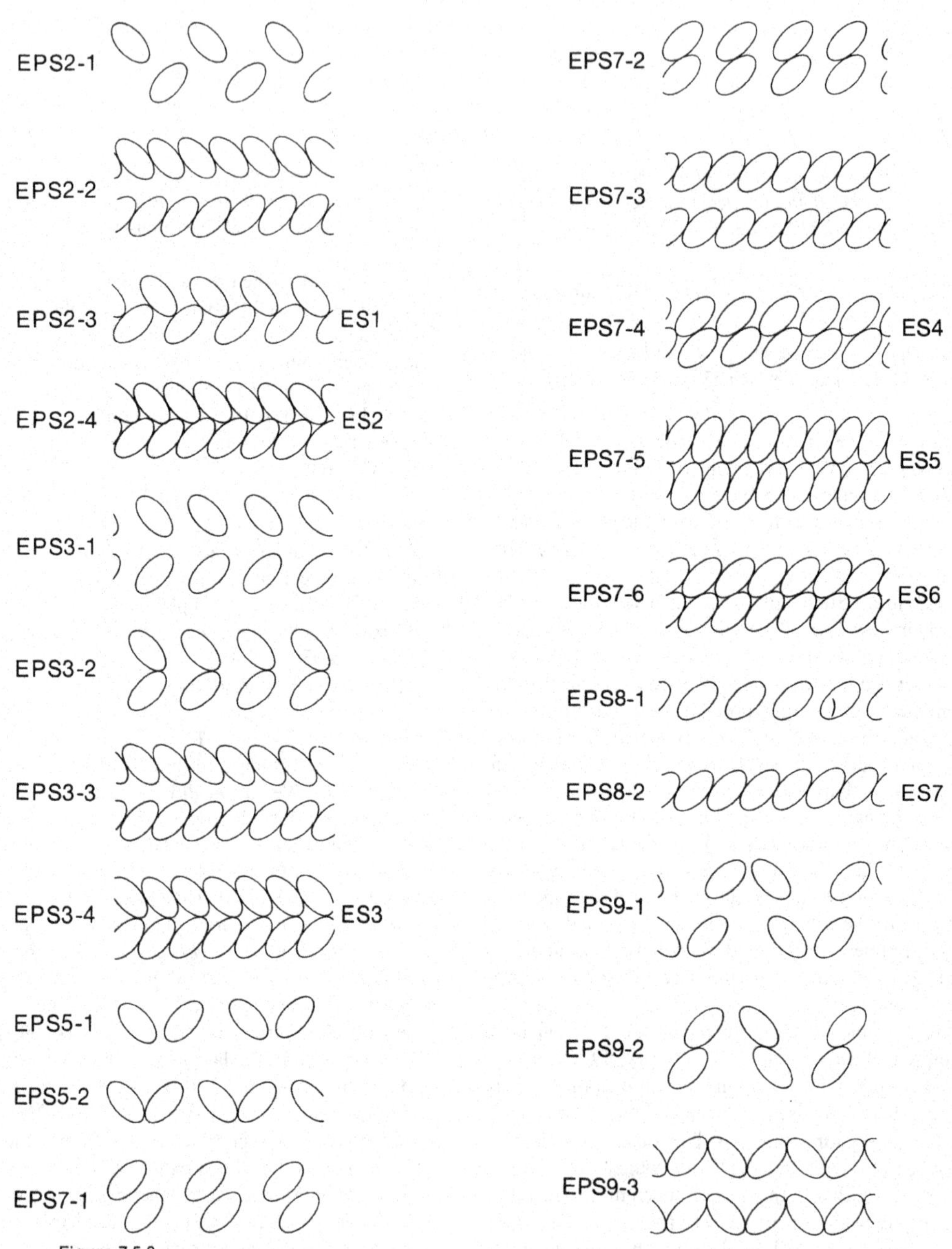

Figure 7.5.2
Representatives of the 43 homeomeric types of strip patterns whose motif is an open ellipse. The ellipse pattern types and symbols for supermotifs are given as in Figure 7.5.1. Further information about these patterns appears in Table 7.5.1.

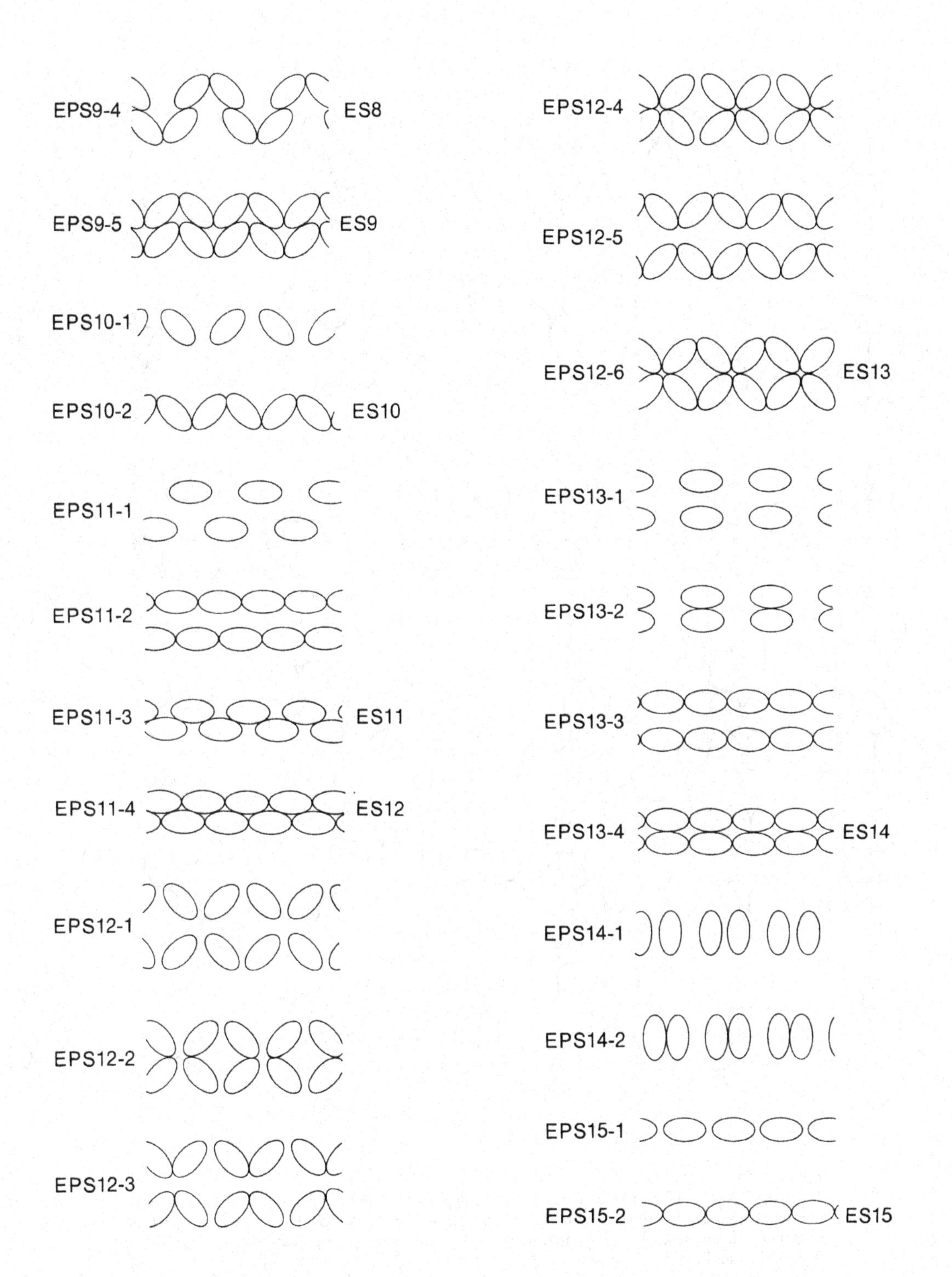
EPS9-4
ES8
EPS9-5
ES9
EPS10-1
EPS10-2
ES10
EPS11-1
EPS11-2
EPS11-3
ES11
EPS11-4
ES12
EPS12-1
EPS12-2
EPS12-3
EPS12-4
EPS12-5
EPS12-6
ES13
EPS13-1
EPS13-2
EPS13-3
EPS13-4
ES14
EPS14-1
EPS14-2
EPS15-1
EPS15-2
ES15

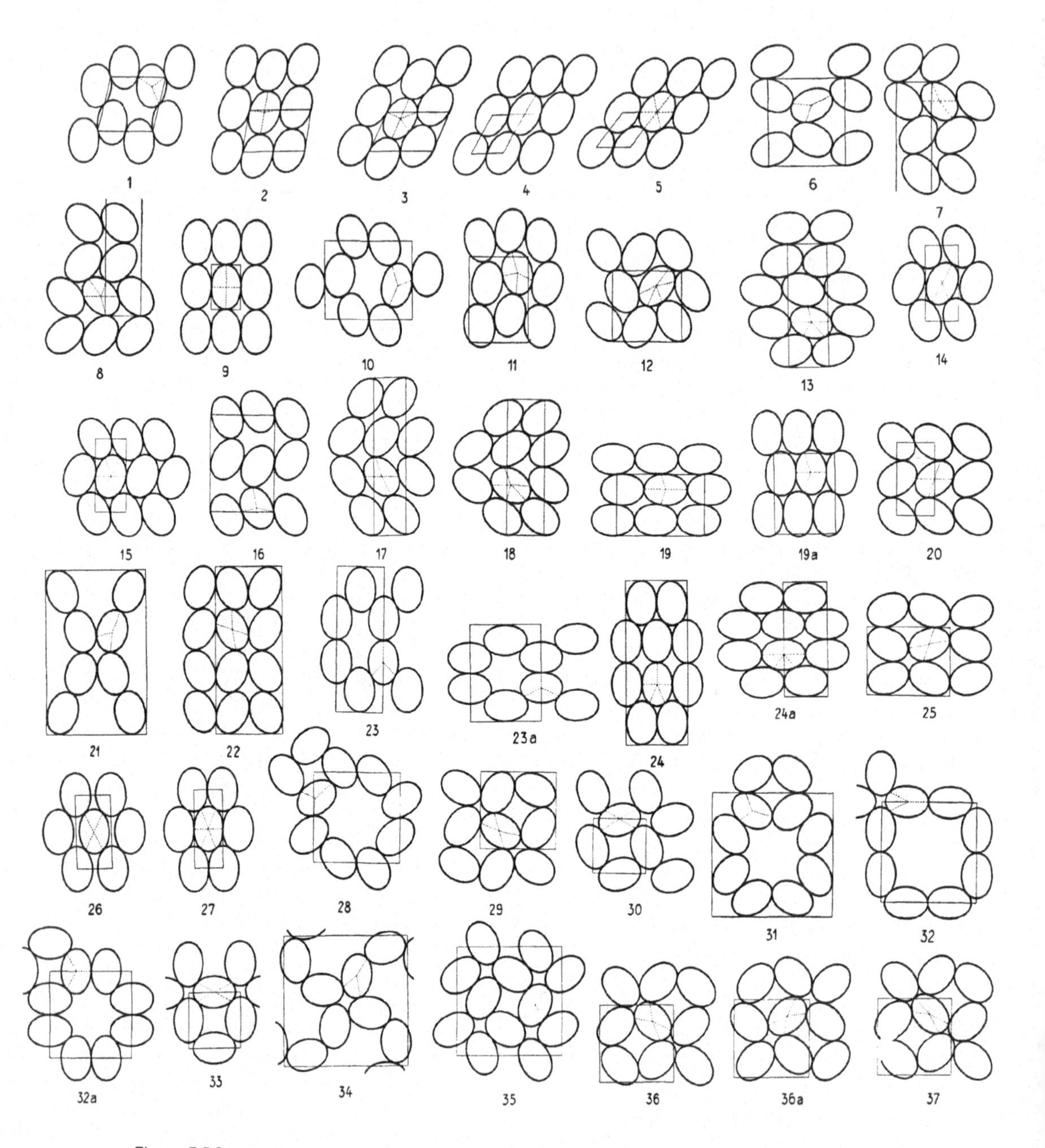

Figure 7.5.3
Representatives of the 57 homeomeric types of coherent periodic patterns whose motif is an open ellipse. Further information about these patterns appears in Table 7.5.1. The patterns in part (*a*) are reproduced from Nowacki [1948], numbered as in the original; the homeomeric type symbols can be found from Table 7.5.1. In part (*b*) we show patterns of the four types that Nowacki missed.

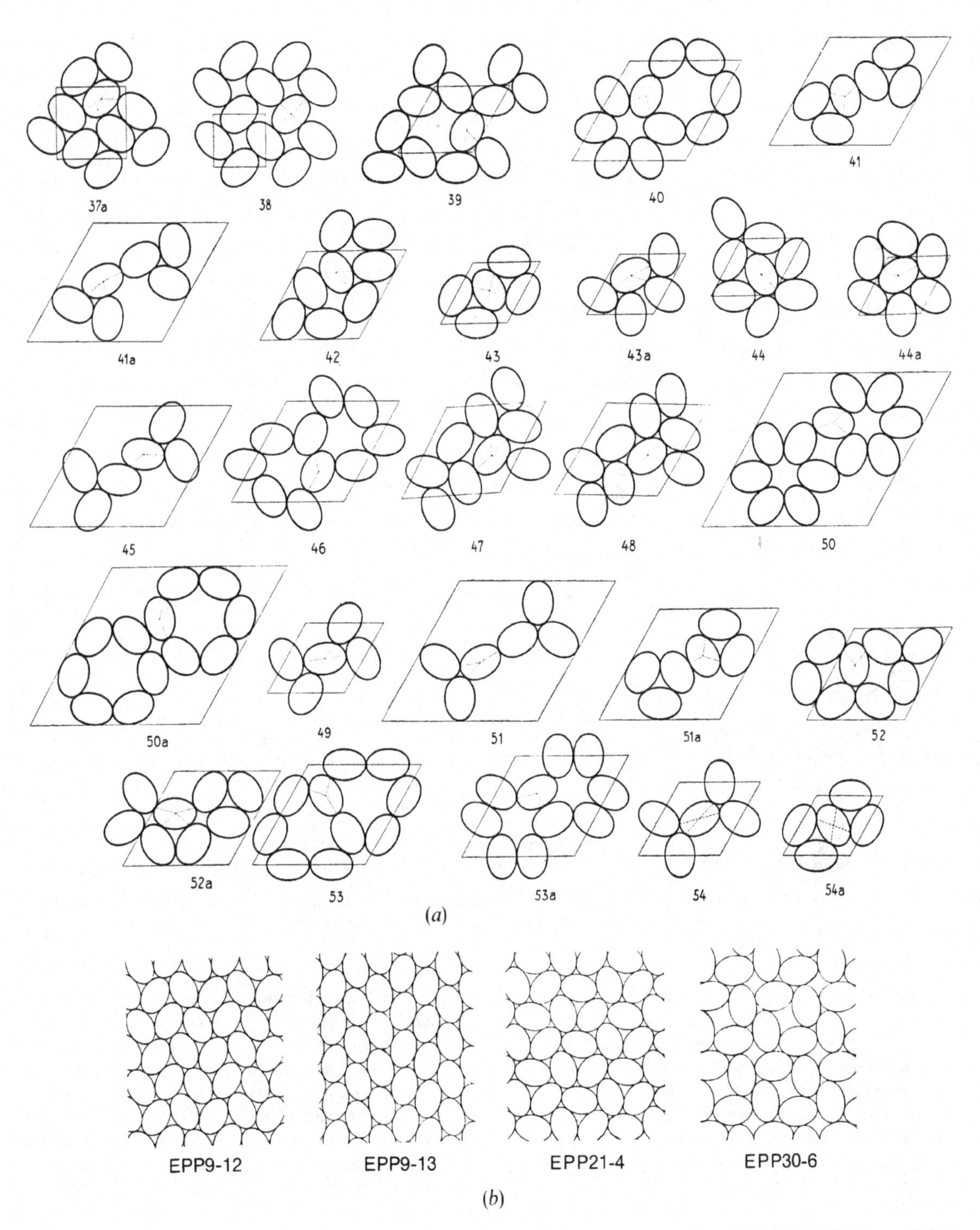
37a
38
39
40
41
41a
42
43
43a
44
44a
45
46
47
48
50
50a
49
51
51a
52
52a
53
53a
54
54a
(a)
EPP9-12
EPP9-13
EPP21-4
EPP30-6
(b)

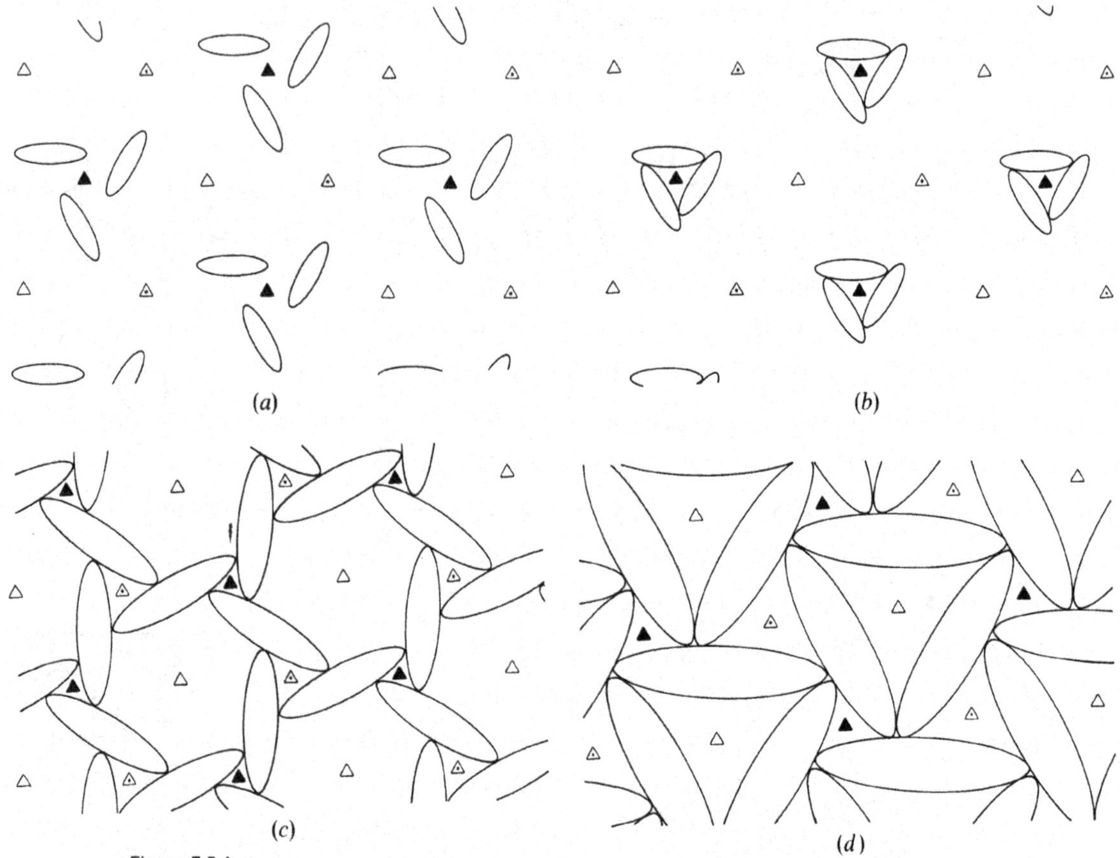

Figure 7.5.4
The construction of the four homeomeric types of pattern whose motifs are open ovals and have underlying pattern type PP21. In (*a*) the copies of the motif have disjoint closures, and in (*b*) they have been moved so as to form triplets around the centers of 3-fold rotation of the group *p3*. In (*c*) and (*d*) the triplets have been "expanded" until they touch. There are two possibilities: in (*c*) each oval has four contacts, and in (*d*) six. These are the only possibilities and we deduce that there are just four possible homeomeric types.

nearby triplet (see Figure 7.5.4(*c*)) or else it comes into contact with two such (see Figure 7.5.4(*d*)). As before, the fact that the underlying pattern is of type PP21 forces further contacts and the resulting patterns will correspond to the homeomeric types EPP21-3 and EPP21-4 respectively. No further homeomeric types can be produced by modifications and so we deduce that there are only these four possibilities which correspond exactly to the four types of ellipse patterns with underlying pattern type PP21. Using elementary continuity arguments it can be shown that each of the possibilities is actually realizable. Carrying out similar reasoning for each of the other pattern types leads to the proof of Statement 7.5.5.

Some of the assumptions in Statement 7.5.5 are necessary (see Exercises 7.5.3 and 7.5.4) while others may be relaxed (see Exercises 7.5.5 and 7.5.6). A full clarification of the details still remains to be worked out. In particular we remark that patterns corresponding to type

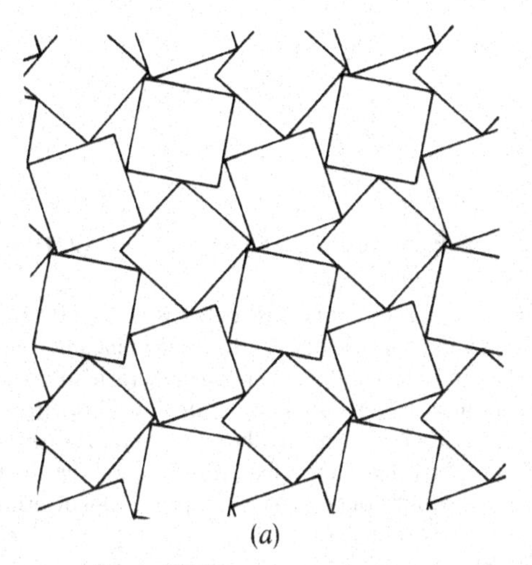

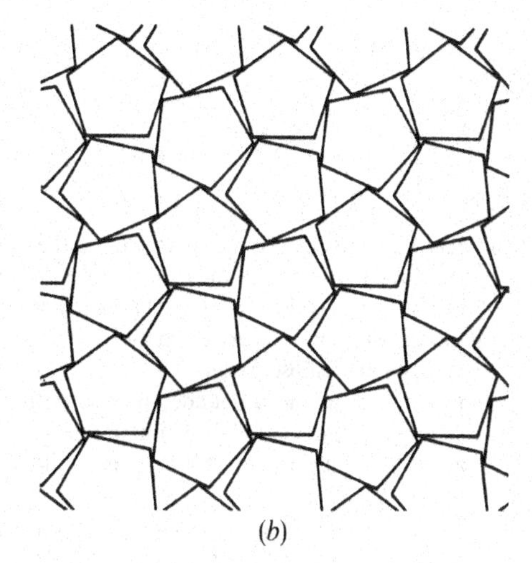

Figure 7.5.5
Some remarkable patterns by open squares and regular pentagons which are homeomeric with the type EPP21-1 of periodic ellipse patterns. It will be noticed that many of the interstices are equilateral triangles. Rotations of period three about the centers of these triangles are symmetries of the patterns.

EPP21-4 also exist with a wide variety of motifs other than $d2$-ovals. More precisely, using elementary continuity arguments we can show:

7.5.6 *If M is an open and bounded non-circular convex set, there exists a coherent pattern with motif M which has symmetry group $p3$ and in which every copy of M is contiguous to six other copies, each of which is rotated by $120°$ with respect to it.*

For example, Figure 7.5.5 shows such patterns in which the motif is either a square or a regular pentagon.

It seems possible that Statement 7.5.5 can be generalized in the following form: *if M is an open oval, then the possible homeomeric types of patterns with M as motif depend only on the symmetry group $S(M)$ of M and not on M itself.*

Here an "oval" means any region of the plane bounded by a smooth and strictly convex curve. The validity of the statement can probably be decided by methods analogous to those suggested for the proof of Statement 7.5.5.

EXERCISES 7.5

1. Prove Statement 7.5.1.
2. Verify that the entries in Table 7.5.1 corresponding to the periodic patterns PP30, PP31, PP32 (symmetry group $p4$) and PP27, PP28, PP29 (symmetry group $p3m1$) are correct. In each case describe a convenient set of parameters which specify the pattern and hence verify the entries in column (4) of the table.
3. Show that Statement 7.5.5 fails for each of the (open) motifs in Figure 7.5.6.

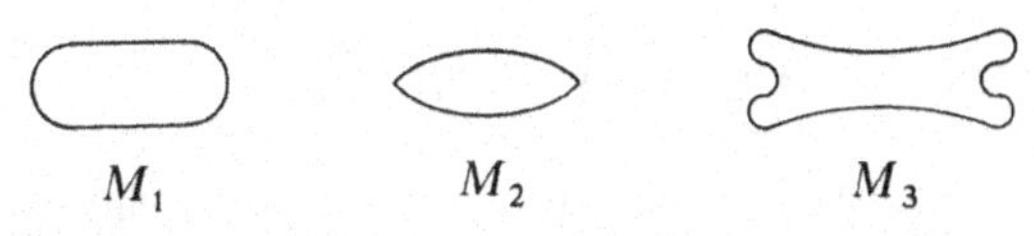

Figure 7.5.6
Examples of motifs with symmetry group $d2$ for which Statement 7.5.4 fails. Motif M_1 is smooth and convex but not strictly convex. Motif M_2 is strictly convex but not smooth. Motif M_3 is smooth and has no line segments in its boundary, but is not convex.

4. (*a*) Show that there exist motifs which are the interiors of strictly convex smooth curves with symmetry groups (*i*) *d1*, (*ii*) *c1*, (*iii*) *d4*, and for which there exist coherent periodic patterns not homeomeric to any ellipse pattern.

**(*b*) Does any open motif bounded by a smooth strictly convex curve with symmetry group *c2* admit a pattern not homeomeric to an ellipse pattern?

**5. (*a*) Show that Statement 7.5.5 remains valid if *S*(*M*) is *d10* instead of *d2*.

(*b*) Characterize all possible choices for the symmetry group *S*(*M*) which can replace *d2* in Statement 7.5.5.

6. Find a motif *M* which is an open topological disk having as boundary the union of two congruent circular arcs, such that Statement 7.5.5 is valid for *M* although *M* is not smooth.

7. (*a*) Show that if *M* is the interior of a strictly convex and smooth curve, in any pattern with motif *M* each copy of the motif is contiguous to at most six other copies.

(*b*) Find a motif *M* which is the interior of a strictly convex curve, and a periodic pattern with motif *M* in which each copy of the motif is contiguous to seven other copies.

**(*c*) What integers can replace "seven" in part (*b*)?

8. Carry out the homeomeric classification of strip patterns in which the motif *M* is the interior of a strictly convex and smooth curve and *S*(*M*) is (*a*) *d1*, (*b*) *d3*.

9. A classification of ellipse patterns which is finer than the homeomeric one results if the homeomorphisms used in the definitions of homeomerism are required to map axes of ellipses onto axes of ellipses (though possibly interchanging the minor and major axes).

(*a*) Show that in this finer classification the ellipse patterns of homeomeric type EPS9-4 determine four subtypes.

(*b*) Into how many subtypes are the patterns of homeomeric type EPS7-5 differentiated?

(*c*) How many subtypes arise from patterns of the type EPP21-2?

10. A classification of ellipse patterns that is even finer than the one considered in Exercise 9 is obtained if the homeomorphisms used in establishing homeomerism are required to map each major axis onto a major axis and each minor axis onto a minor one.

(*a*) Show that in this classification each of the types EPS14-2 and EPS15-2 splits into two subtypes.

(*b*) How many subtypes result from ellipse patterns of type EPS9-4?

(*c*) Determine all the subtypes for patterns of type EPP17-2.

**11. Does the analogue of Statement 7.5.5 hold for *d2*-ovals in the finer classification introduced in Exercise 7.5.9? The "axes" of the oval are to be interpreted as the lines of reflection in the symmetry group *d2*.

7.6 FILAMENTARY PATTERNS

For another example that exhibits the applicability of the homeomeric classification, we consider filamentary patterns. We recall from Section 5.5 that these are special curve patterns in which the motifs satisfy condition UP.2. Examples of such patterns appear in Figure 5.5.3.

In analysing these patterns we need the following concept. For a given filament *M*, let *M'* represent the set obtained by deleting a finite arc *A* from *M*, and for each such *M'* let S_A represent the intersection of all the closed two-way infinite strips that contain *M'*. Then the intersection $\bigcap_A S_A$, as *A* varies over all the finite arcs of *M*, can be shown to be a non-empty strip S_M, possibly reduced to a single line. This strip will be called the *asymptotic strip* of *M*; examples of asymptotic strips are shown in Figure 7.6.1. Here and in some other illustrations we indicate the boundary of the asymptotic strip by

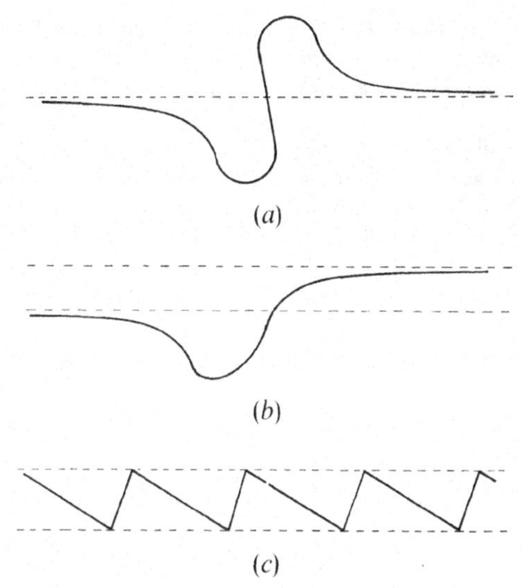

Figure 7.6.1
Examples showing the construction of asymptotic strips for various filaments. In (*a*) the strip reduces to a single line. The boundaries of the strips are, in each case, indicated by dashed lines.

dashed lines. It is clear that in any filamentary pattern the asymptotic strips must all be parallel and some, or all, of them may coincide.

For any filament M, the symmetry group $S(M)$ of M must map the asymptotic strip onto itself, and so must be either a finite group (*c1*, *c2*, *d1*, *d2*) or a strip group (*p111*, *p1a1*, *p1m1*, *pm11*, *p112*, *pma2*, *pmm2*) or, if M is a straight line, the continuous group *smm2*. No filament can have *d2*, *p1m1* or *pmm2* as its symmetry group; examples of filaments showing that the other groups are possible are given in Figure 7.6.2. It is possible for two, or infinitely many, filaments to have the same asymptotic strip; it is easily verified that all the possibilities are illustrated in Figure 7.6.2.

Guided by a feeling for "essential sameness" of filamentary patterns we find it desirable to distinguish between the types of the patterns shown in Figure 7.6.3.

These two patterns are clearly homeomeric, but differ in the behavior of the asymptotic strips: pairs of copies of the motif in the pattern of Figure 7.6.3(*b*) have the same asymptotic strip, in contrast to the situation in Figure 7.6.3(*a*). Therefore, the classification we shall describe will

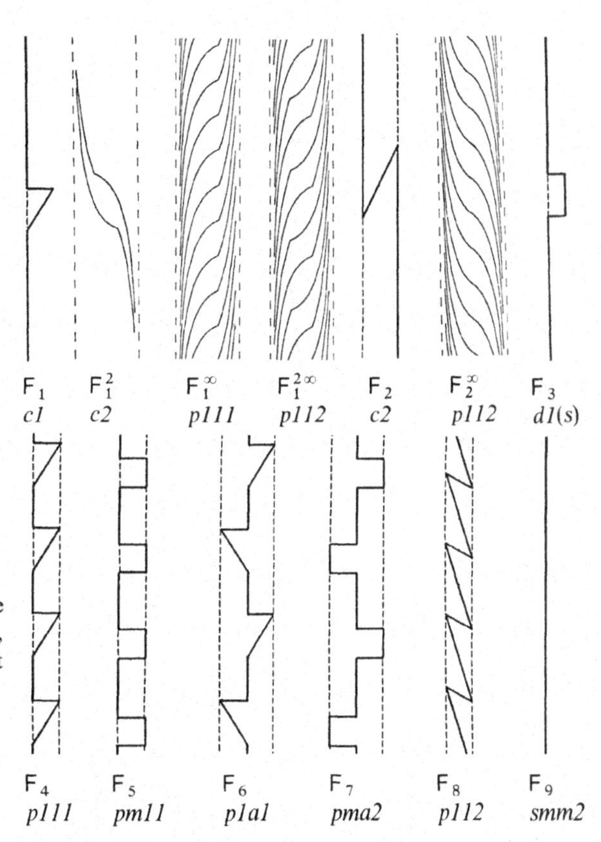

Figure 7.6.2
The different types of motifs and supermotifs used in the construction of filamentary patterns. The symbol used in Table 7.6.1, and the symmetry group, are indicated near each. The dashed lines are the boundaries of the asymptotic strips.

be a *refined homeomeric classification*, in which we require that the homeomorphisms which establish the patterns as being of the same homeomeric type must map pairs of filaments with distinct asymptotic strips onto pairs with the same property.

With this definition, we have the following result:

7.6.1 *There exist* 90 *refined homeomeric types of non-trivial filamentary patterns. Of these, ten consist of a finite number of filaments, and an additional eight are contained in a strip of finite width.*

All these types are listed in Table 7.6.1.

The method of proof of Statement 7.6.1 follows essentially the same stages as in the preceding sections. We use as supermotif the union of all filaments that have the

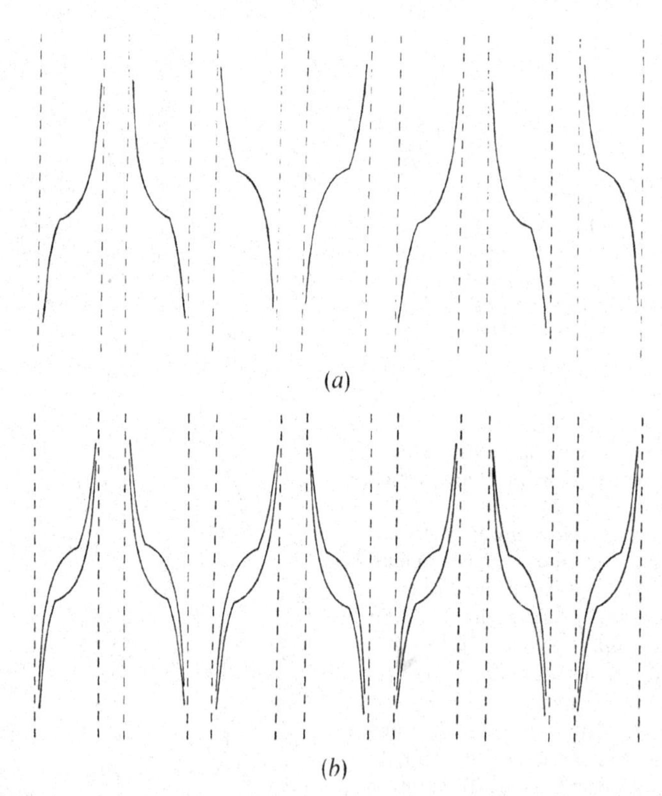

Figure 7.6.3
Two homeomeric types of filamentary patterns. In (*a*) the asymptotic strips of the filaments are disjoint; in (*b*) the asymptotic strips of pairs of filaments coincide. In order to differentiate between such patterns we refine the homeomeric classification and assign (*a*) to the type FPS9-1 and (*b*) to the type FPS9-2.

Table 7.6.1 THE REFINED HOMEOMERIC TYPES OF NON-TRIVIAL FILAMENTARY PATTERNS

Column (1) gives a symbol for the type. If we delete the initial F, the hyphen and the final integer, we obtain the pattern type of the discrete pattern obtainable by truncation or fragmentation.

Columns (2) and (3) indicate the superpattern type and the supermotif. The notation for the superpatterns is that of Sections 5.2 and 6.5; the notation for the supermotifs is that of Figure 7.6.2.

In **column (4)**, *F* means that the pattern consists of a finite number of filaments, and *S* indicates that although the number of filaments is infinite, they are all contained in a strip of finite width. The letters *a*, *b* and *c* indicate that the asymptotic strips have no boundary points in common, that they have boundary points in common on one side, or that they have boundary points in common on both sides, respectively. The notations *b*(*r*), *b*(*g*) and *b*(*t*) mean that the common boundary of two asymptotic strips is a line of reflective symmetry of the pattern, a line of glide-reflection, or that it contains centers of 2-fold rotation, respectively.

Refined homeomeric type (1)	Pattern type of superpattern (2)	Type of supermotif (3)	Comments (4)
$FPF1_2$-1	$PF1_2$	F_1	*F*
$FPF1_2$-2	Trivial	F_1^2	*F*
$FPF2_1$	$PF2_1$	F_1	*F*
$FPF3_2$	$PF3_2$	F_3	*F*
FPC2	SPCF1	F_9	*F*
FPS1-1	PS1	F_1	
FPS1-2	Trivial	F_1^∞	*S*
FPS2-1	PS2	F_1	
FPS2-2	SPF1	F_1^∞	*S*, *a*
FPS2-3	SPF1	F_1^∞	*S*, *b*
FPS2-4	SPF1	F_4	*F*
FPS3-1	SPF2	F_1^∞	*S*, *a*
FPS3-2	SPF2	F_1^∞	*S*, *b*
FPS3-3	SPF2	F_4	*F*
FPS4	PS4	F_3	
FPS5	PS5	F_1	
FPS7-1	PS7	F_1	
FPS7-2	PS8	F_1^2	
FPS7-3	SPF3	F_1^∞	*S*, *a*
FPS7-4	SPF3	F_1^∞	*S*, *b*
FPS7-5	Trivial	$F_1^{2\infty}$	*S*
FPS7-6	SPF3	F_4	*F*
FPS8-1	PS8	F_2	
FPS8-2	Trivial	F_2^∞	
FPS9-1	PS9	F_1	
FPS9-2	PS10	F_1^2	
FPS10	PS10	F_2	
FPS11	SPF4	F_5	*F*
FPS13	SPF5	F_5	*F*
FPS14	PS14	F_3	
FPP1-1	SPS1	F_1^∞	*a*
FPP1-2	SPS1	F_1^∞	*c*
FPP1-3	SPS1	F_4	
FPP2-1	SPS2	F_1^∞	*a*
FPP2-2	SPS2	F_1^∞	*c*
FPP2-3	SPS3	F_1^∞	*a*
FPP2-4	SPS3	F_1^∞	*b*
FPP2-5	SPS3	F_1^∞	*c*
FPP2-6	SPS2	F_4	
FPP2-7	SPS3	F_4	
FPP2-8	SPS9	F_6	
FPP3-1	SPS4	F_1^∞	*a*
FPP3-2	SPS4	F_1^∞	*b*
FPP3-3	SPS4	F_1^∞	*c*
FPP3-4	SPS4	F_4	
FPP4	SPS15	F_5	
FPP5-1	SPS5	F_1^∞	*a*
FPP5-2	SPS5	F_1^∞	*b*(*r*)
FPP5-3	SPS5	F_1^∞	*b*(*g*)
FPP5-4	SPS5	F_1^∞	*c*
FPP5-5	SPS5	F_4	

Table 7.6.1 (continued)

Refined homeomeric type (1)	Pattern type of superpattern (2)	Type of supermotif (3)	Comments (4)
FPP5-6	SPS10	F_6	
FPP6	SPS16	F_5	
FPP7-1	SPS6	F_1^∞	*a*
FPP7-2	SPS6	F_1^∞	*b*
FPP7-3	SPS6	F_1^∞	*c*
FPP7-4	SPS20	$F_1^{2\infty}$	*a*
FPP7-5	SPS20	$F_1^{2\infty}$	*c*
FPP7-6	SPS6	F_4	
FPP7-7	SPS11	F_6	
FPP8-1	SPS20	F_2^∞	*a*
FPP8-2	SPS20	F_2^∞	*c*
FPP8-3	SPS20	F_8	
FPP9-1	SPS7	F_1^∞	*a*
FPP9-2	SPS7	F_1^∞	*b*(*g*)
FPP9-3	SPS7	F_1^∞	*b*(*t*)
FPP9-4	SPS7	F_1^∞	*c*
FPP9-5	SPS21	$F_1^{2\infty}$	*a*
FPP9-6	SPS21	$F_1^{2\infty}$	*c*
FPP9-7	SPS7	F_4	
FPP10-1	SPS21	F_2^∞	*a*
FPP10-2	SPS21	F_2^∞	*c*
FPP10-3	SPS21	F_8	
FPP11-1	SPS8	F_1^∞	*a*
FPP11-2	SPS8	F_1^∞	*b*(*r*)
FPP11-3	SPS8	F_1^∞	*b*(*t*)
FPP11-4	SPS8	F_1^∞	*c*
FPP11-5	SPS22	$F_1^{2\infty}$	*a*
FPP11-6	SPS22	$F_1^{2\infty}$	*c*
FPP11-7	SPS8	F_4	
FPP12-1	SPS22	F_2^∞	*a*
FPP12-2	SPS22	F_2^∞	*c*
FPP12-3	SPS22	F_8	
FPP13-1	SPS17	F_5	
FPP13-2	SPS23	F_7	
FPP15	SPS18	F_5	
FPP19-1	SPS19	F_5	
FPP19-2	SPS24	F_7	
FPC3	SPC2	F_9	
FPC4	SPC3	F_9	

same asymptotic strip, and construct with these supermotifs striped (super)patterns as described in Section 6.5. The only difference is that if the supermotif is one of F_1^∞, $F_1^{2\infty}$ or F_2^∞, then the superpatterns in which the supermotifs are contiguous (that is, the asymptotic strips have common boundary points) are not homeomeric to those in which the supermotifs are not contiguous. In Figure 7.6.4 we show a typical example of such a difference.

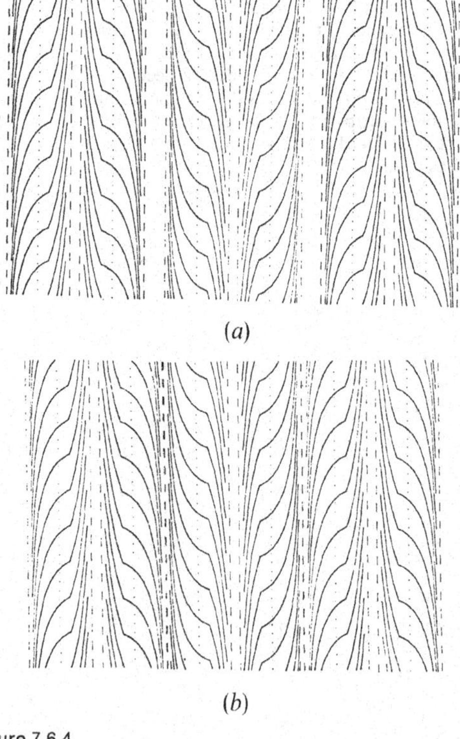

Figure 7.6.4
The homeomeric types of these two patterns are different since in (*a*) the supermotifs (of type F_1^∞) are not contiguous, and in (*b*) they are contiguous along a line that contains centers of 2-fold rotational symmetry of the pattern. In the notation of Table 7.6.1 the pattern in (*a*) is of type FPP11-1 and that in (*b*) of type FPP11-3.

To explain the notation used in Table 7.6.1 for the various types of filamentary patterns we recall from Section 5.1 that filaments of types F_1, F_2, F_3 can be *truncated*, that is, replaced by a finite subarc which has the same symmetries as the filament. Applying the same truncation to all filaments in the pattern, we obtain a discrete pattern; an example of such truncation is shown in Figure 7.6.5. Similarly, filaments of types F_4, F_5, F_6, F_7, F_8 can be *fragmented*, that is, replaced by a strip pattern which has the same symmetries as the filament and is contained in the filament. In Figure 7.6.6 we illustrate the process by which fragmentation leads to a discrete pattern.

The symbol of each type of filamentary pattern in Table 7.6.1 (except those with filaments of type F_9) consists of the pattern type of the discrete pattern obtainable from it by truncation or fragmentation, preceded by F to indicate "filament", and followed—if necessary—by distinguishing numerals.

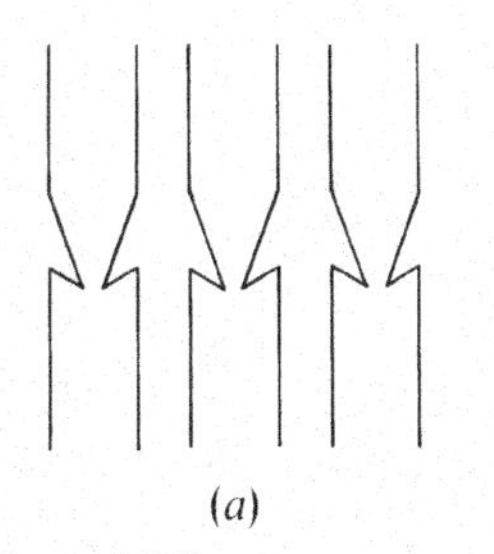

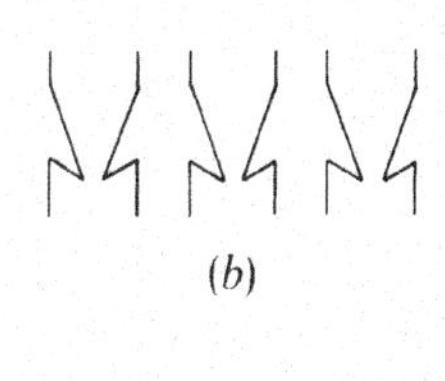

Figure 7.6.5
A filamentary pattern (of type FPS5) and a discrete pattern (of type PS5) obtained from it by truncation.

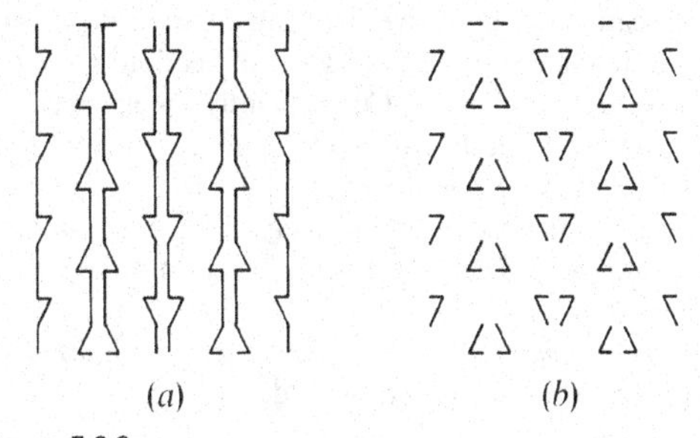

Figure 7.6.6
A filamentary pattern (of type FPP11-7) and a discrete pattern (of type PP11) obtained from it by fragmentation.

EXERCISES 7.6

1. Prove the assertion made in this section that the intersection $\bigcap_A S_A$, which we used in the definition of an asymptotic strip, is non-empty.

2. (*a*) Find a filament of type F_2 in which the asymptotic strip reduces to a single line.
   (*b*) Prove that for filaments of types F_4, F_5, F_6, F_7 and F_8, the asymptotic strip never reduces to a single line.
   (*c*) Find a filament of type F_3 in which the asymptotic strip does not reduce to a single line.

3. Show that each of the 90 types of filamentary patterns can be realized by filaments required to be unions of straight line-segments of unit length.

4. Use Table 7.6.1 to deduce the (unrefined) homeomeric classification of filamentary patterns.

5. Refine the classification of filamentary patterns given in Table 7.6.1 by requiring that the homeomorphisms used in establishing the patterns as homeomeric map asymptotic strips onto asymptotic strips. (Thus, for example, type FPS11 splits into three types according to whether the asymptotic strips of the two copies of the motif F_5 are disjoint, contiguous or overlap. Cases in which the asymptotic strips reduce, or do not reduce to a single line (see Exercise 7.6.2) must also be distinguished.)

7.7 TRANSITIVE ARRANGEMENTS OF LINES

Our final illustration of the use of homeomerism is in the classification of objects we shall call "line arrangements". These are not patterns in the sense of Section 5.1 since the lines are not disjoint. Consequently the investigation is of a rather different nature from those of the preceding four sections. Here we shall be dealing with geometric objects of which there are infinitely many homeomeric types, and our object is one of contruction rather than enumeration.

A *line arrangement* $\mathscr{L}$ is any set of distinct lines in the plane such that the following hold:

LA.1 *The symmetry group* $S(\mathscr{L})$ *of* $\mathscr{L}$ *acts transitively on the lines of* $\mathscr{L}$.

LA.2 *The complement of* $\mathscr{L}$ *in the plane is a set of polygonal regions whose closures are the tiles of a normal tiling* $\mathscr{T}(\mathscr{L})$.

Examples of line arrangements are given in Figure 7.0.1 and in Figure 7.7.1. Condition LA.2 excludes all finite sets of lines and also, for example, a set of parallel equidistant lines. It ensures, in fact, that the symmetry group $S(\mathscr{L})$ is one of the crystallographic groups (except *p1* or *p2*) discussed in Section 1.4.

We begin by describing a construction for a line arrangement. First we choose a lattice of points Λ_i of one of the three types shown in Figure 7.7.2. We also select one of the crystallographic groups G listed beneath the chosen lattice Λ_i and think of Λ_i as situated so that G acts transitively on its points. Now let L be any line which *either* contains two (and therefore infinitely many) points of Λ_i *or* is parallel to such a line. $\mathscr{L}$ is defined as the set of lines obtained by applying the isometries of G to the line L and it is easy to see that their totality satisfies our definition of a line arrangement.

Our main result is the following:

7.7.1 *Every line arrangement can be obtained by the procedure described above.*

In Figure 7.7.1 we have indicated the lattice and group used in the construction of the line arrangement shown in each diagram. For the line pattern of Figure 7.0.1 the lattice Λ_3 and the group *p6m* were used.

To prove Statement 7.7.1, consider any line L of a given line arrangement $\mathscr{L}$, and let P be an arbitrary point of L. Denote by $\mathscr{D}$ the dot pattern obtained by applying the operations of $S(\mathscr{L})$ to P. This dot pattern must be of one of the types shown in Figure 5.3.3, with the exception of DPP7 and DPP8. These two cannot arise for their symmetry groups are *p2* and condition LA.2 excludes this as a possible symmetry group of $\mathscr{L}$. Examination of patterns of the other 29 dot pattern types shows that each can be expressed in the form $\Lambda_i^1 \cup \Lambda_i^2 \cup \cdots \cup \Lambda_i^s$ where each Λ_i^j is a translate of one of the lattices Λ_i ($i = 1, 2, 3$) of Figure 7.7.2 and s is a small integer ($s \leq 12$); compare Exercise 5.3.8. Suppose the point P (on L) belongs to Λ_i^1. Then L must also contain another point P' of this same lattice Λ_i^1 for if it did not do so, application of the group $S(\mathscr{L})$ (which contains all translative symmetries of Λ) to L would yield a dense set of lines parallel to L. This would violate condition LA.2. Since L joins two points of Λ_i^1 it must either join two points of Λ_i or be parallel to such a line. This shows that $\mathscr{L}$ can be constructed by the procedure described above, and hence Statement 7.7.1 is proved.

Various choices of L and of group G can lead to the same line arrangement $\mathscr{L}$, and we know of no way by which one can tell when two different such choices necessarily lead to line arrangements of distinct homeomeric types. We do not even know whether keeping G fixed and considering L in different *parallel* positions yields only a finite number of homeomeric types. However we can assert that the total number of types is infinite, see Exercise 7.7.2.

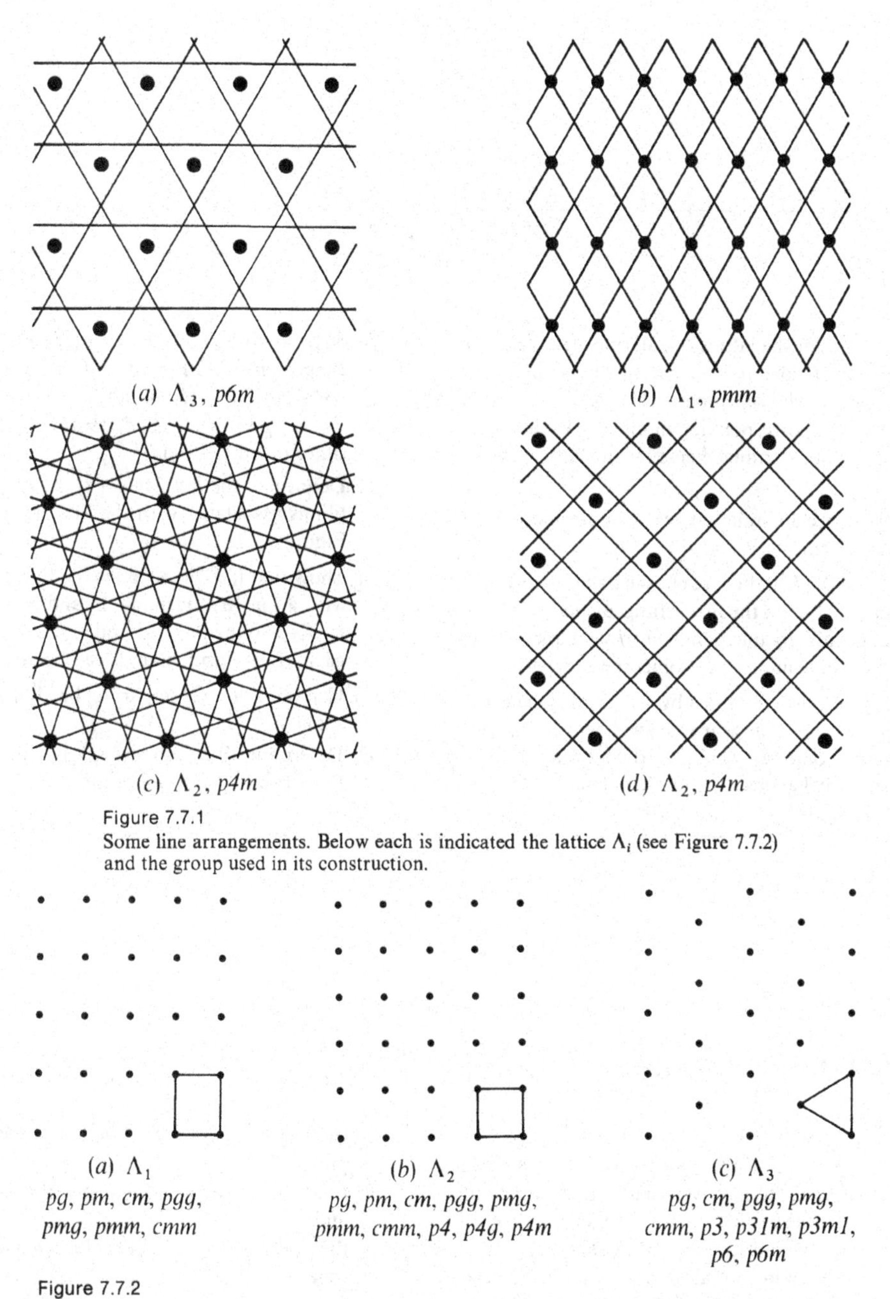

Figure 7.7.1
Some line arrangements. Below each is indicated the lattice Λ_i (see Figure 7.7.2) and the group used in its construction.

Figure 7.7.2
The lattices used to make up periodic dot patterns, and the crystallographic groups other than *p1, p2* which are transitive on them. These lattices are used to construct line arrangements as described in the text. The three lattices are based on (*a*) a rectangle, (*b*) a square, and (*c*) an equilateral triangle.

EXERCISES 7.7

1. (*a*) Show that for the lattice Λ_2, all line arrangements with lines parallel to lines of reflective symmetry of Λ_2 form two homeomeric classes.
   (*b*) What is the corresponding number of homeomeric classes of such arrangements for the lattices Λ_1 and Λ_3?
2. Show that there are infinitely many homeomeric types of line arrangements.

**3. Prove that if Λ_i, L and G are chosen as in the proof of Statement 7.7.1, then the line arrangements, obtained by starting with any line parallel to L and applying G, belong to a finite number of homeomeric types.

4. The tilings $\mathscr{T}(\mathscr{L})$ determined by line arrangements $\mathscr{L}$ (see condition LA.2) appear to be well worth deeper investigation. Since they are periodic, they are balanced, strongly balanced, prototile balanced, etc.
   (*a*) Show that the tiling $\mathscr{T} = \mathscr{T}(\mathscr{L})$ of the line arrangement $\mathscr{L}$ of Figure 7.0.1 satisfies $t_3(\mathscr{T}) = 18/34$, $t_4(\mathscr{T}) = 15/34$, $t_6(\mathscr{T}) = 1/34$, $v_4(\mathscr{T}) = 39/68$, $v_6(\mathscr{T}) = 12/68$, $v_{12}(\mathscr{T}) = 1/68$ and all other t_i and v_k are 0.
   (*b*) Determine the frequencies t_j, v_k, $t_{[Ti]}$ for the tilings associated with the line arrangements of Figure 7.7.1.
   (*c*) Show that if $\mathscr{T} = \mathscr{T}(\mathscr{L})$ is the tiling associated with a line arrangement $\mathscr{L}$ and if $t_j(\mathscr{T}) > 0$ and $v_k(\mathscr{T}) > 0$ then $j \leq 12$ and $k \leq 12$. Show that each of the upper bounds can be attained.
   **(*d*) Can both bounds in part (c) be attained for the same $\mathscr{L}$?
   **(*e*) Determine all possible values of j and k for which $t_j > 0$ and $v_k > 0$ can occur.

7.8 NOTES AND REFERENCES

The classification of patterns (either by pattern type as described in Chapter 5 or by homeomerism as described in this chapter) seems to have very few antecedents in the literature. One of the earliest attempts to formulate general principles of classification is that of Delone [1932]; these have been used in several situations and have been reformulated most explicitly by Delone, Galiulin & Štogrin [1979, pp. 240–241] in the following form:

"Let a combinatorial-topological object S be given together with a group G of combinatorially isomorphic maps of S onto itself. We call (S,G) a *pair*. Two pairs (S,G) and (S',G') are defined to be of the same type provided:

1. the objects S and S' are combinatorially isomorphic;
2. the groups G and G' are abstractly isomorphic; and
3. the groups G and G' act in the same way on the objects S and S'."

Although Delone's definition contains the germ of a useful idea, the lack of precision led (and continues to lead) to confusion and misunderstandings.

The difficulties in the use of Delone's definition have three main causes. The first is that it is a classification of pairs (*S*,*G*) rather than of the objects themselves, and the second is its dependence upon the rather vague phrase "act in the same way" in condition (3); we shall return to these shortly. The third difficulty stems from the formulation of condition (1). In the case of patterns, the "combinatorial isomorphism" of condition (1) reduces to a bijection; hence the "type" with which this definition is concerned can be understood—if *G* and *G*′ are the full symmetry groups of *S* and *S*′, and condition (3) is interpreted with a sufficient measure of goodwill—as being the pattern type (henomeric type) of Chapter 5. However, Delone and the other writers who used this definition of "type" (see, for example, Delone [1958], [1959], [1963], Delone, Dolbilin & Štogrin [1978], Delone, Galiulin & Štogrin [1979], Heesch [1968b], Niggli [1963], Sinogowitz [1939], Wollny [1974a], [1974b]. Zamorzaeva [1979]) applied it only in the context of polyhedra or tilings of various kinds; in these, combinatorial equivalence and topological equivalence (homeomorphism) coincide. In particular, if *S* and *S*′ are isohedral tilings and *G* and *G*′ their groups of symmetry, then this classification into types coincides with the classification into 81 types given in Section 6.2 and not with the classification of isohedral tilings by pattern type. On the other hand, if various tile-transitive subgroups of the full symmetry group are admitted as *G* and *G*′, the classification becomes essentially equivalent to the classification of marked isohedral tilings into 93 types. (Although not stated in the definition, it is evident from the context that the groups considered should act transitively on "parts" of the "objects".)

This approach to classification appears to be—psychologically and practically—inferior to the one we followed, even if we disregard the lack of precision. This claim is based on the fact that Delone and others, in applying the method, have all fallen into the logical trap of claiming to (or attempting to) classify *tilings* while actually classifying "pairs" (or *marked* tilings, see Section 6.6). Readers of these papers have often become victims of this confusion; see, for example, Anonymous [1960] (which is a paean to Delone, presumably by the editors of the journal "Kristallografiya"), Schwerdtfeger [1962], Horváth [1965], Burckhardt [1967a], Lampert & Čoka [1973], Hortobagyi [1975]. Moreover, we have seen that for patterns other than tiling patterns the classification by pattern type, refined pattern type, and homeomeric type are all distinct. It is therefore rather hard to evaluate many of the assertions concerning enumerations of types, which are made without precise and explicit descriptions of the classifying principles. All too often authors claim to be doing one thing while actually doing something else. In case the classification of the "pairs" (*S*,*G*) is indeed the desired aim, it seems to us that our method of achieving this goal by the artifice of introducing markings is more natural, more convenient, conceptually simpler, and more easily adaptable to various contexts (see, for example, Chapters 8 and 9). An expository and somewhat polemic discussion of classification methods for tilings and patterns can be found in Bersim & Fedotov [1980].

We believe that the homeomeric classification of patterns has not been discussed in the literature in general terms prior to Grünbaum & Shephard [1977e], [1978e], [1979a], [1980b], [1981a]. This is—in a way—only natural, since the notion of pattern has not been defined previously. On the other hand, many authors have devised classifications for specific kinds of patterns which—in fact if not in letter—amount to the homeomeric point of view. This is evident, for example, in the work of Niggli [1927], [1928] and Sinogowitz [1939] on "circle packings", which is best understood as the homeomeric classification of open circular disk patterns. Niggli considered only coherent "circle packings"; very attractive illustrations of these appear in Fejes Tóth [1965, Figure 42], but the explanation given there is incorrect. In contrast, the investigations of Haag [1929] and Finsterwalder [1936] lack a precise focus on the criteria of classification. (However, Finsterwalder has very nice drawings of many of the circle packings, not only in the plane but also on the sphere and in the hyperbolic plane.) For the homeomeric classification of circular disk patterns on the sphere, which corrects and extends Finsterwalder's work, see Grünbaum & Shephard [1985d], [1985e].

Dimotif circular disk patterns with symmetry group *p6m*, which are "stable" in the sense that no disk may be increased in radius without overlapping another disk, have been enumerated by Krötenheerdt & Reichstein [1978].

A classification of coverings of the plane by circular disks, analogous to Sinogowitz's classification of circular disk patterns, can be found in Dominyák [1967].

Various circle patterns, usually with two or more transitivity classes of motifs, have been considered by various authors in connection with the problem of densest packings of circles in the Euclidean and non-Euclidean planes, or with the related thinnest covering problem (see, for example, Fejes Tóth [1973], Fejes Tóth & Heppes [1980], Molnár [1959], [1966], [1977]).

The classification of segment patterns presented in Section 7.4 is new, except for the strip patterns which have been considered in Grünbaum & Shephard [1978e], [1981a]. (In the first of these it was erroneously stated that there exist 51 homeomeric types of periodic line-segment patterns with closed segments as motifs; the correct number is 50—see Statement 7.4.1.) A few examples of segment patterns have appeared in Islamic ornament (see, for example, El-Said & Parman [1976]) and in the design of "interlocking grids" (Gat [1978]). Schafranovsky [1961] investigated finite segment patterns, but as his criteria appear to be different from ours, the enumerations are not the same.

The topic of ellipse patterns was first introduced in Nowacki's [1948] work on coherent ellipse patterns. Matsumoto [1968] has investigated the density of ellipse patterns with symmetry group *pgg* and shown that this is always less than the maximal possible density $\pi/2\sqrt{3}$. Some of the results of Section 7.5 have been announced in Grünbaum & Shephard [1981a], but the complete classification of ellipse patterns is presented here for the first time.*

The diffeometric classification was first explicitly defined in Grünbaum & Shephard [1981a]. Unfortunately, some of the data on numbers of diffeomeric types contained in Tables 2, 3 and 4 of that paper are incorrect.

The homeomeric classifications of filamentary patterns (Section 7.6) and of transitive arrangements of lines (Section 7.7) appear not to have been considered in earlier publications. Periodic tilings generated by straight lines, in which all tiles are triangles, have been considered in Grünbaum & Shephard [1978c].

* See the Appendix beginning on page 653.

COLORED PATTERNS AND TILINGS

Figure 8.0.1
An attractive and ingenious pattern of squares from Islamic art (Bourgoin [1879, Plate 13]) perfectly colored in four colors. This pattern is of type PP48A and it can be perfectly k-colored if k has one of the values $3n$, $6n$, n^2, $2n^2$, $3n^2$ or $6n^2$ for a positive integer n (see Senechal [1979a]).*

* See the Appendix beginning on page 653.

8

COLORED PATTERNS AND TILINGS

Although the coloring of patterns and tilings is almost universal in the decorative arts, there exists only a small amount of mathematical literature devoted to the subject. Moreover, much of this is unsatisfactory because the authors tend to adopt *ad hoc* definitions and systems of classification; as a result, comparisons are often difficult to make and the scope of the achievements is not very clear.

In this chapter we shall treat the topic in a much more systematic manner—a procedure made possible by our basic approach to classifying geometric objects by symmetry, as discussed in Chapter 7. The application of these principles to colored patterns and tilings is straightforward, and the resulting classifications seem to be precisely what some earlier authors were seeking. We begin by defining the necessary concepts, which lead us to the coarsest classification—by color groups. This is refined, in the following sections, in various ways.

The subject abounds with open problems. Many of these concern the enumeration of the number of types of colorings of patterns or tilings with a given number of colors, when classified according to one of the methods. A great deal of work in this area remains to be done.

8.1 COLOR SYMMETRIES

Let $\mathscr{P}$ be a discrete pattern and suppose that to each copy of the motif in $\mathscr{P}$ we assign one of a given finite set of colors. Then we obtain what is called a *colored pattern*, which we denote by $\mathscr{P}^\wedge$. More formally, we may denote the colors by integers $1, 2, \ldots, k$ (for some k) and then define $\mathscr{P}^\wedge$ to be a pair $(\mathscr{P},\chi)$ where $\mathscr{P}$ is the (uncolored) pattern that *underlies* $\mathscr{P}^\wedge$ and χ is a function which maps each copy M_i of the motif M of $\mathscr{P}$ onto an element of the set $\{1, 2, \ldots, k\}$. Thus each copy M_i is assigned the "color" $\chi(M_i)$; the pair $M_i^\wedge = (M_i,\chi(M_i))$ is called a *colored copy* of the motif of $\mathscr{P}$ or of $\mathscr{P}^\wedge$. All the copies of the motif of $\mathscr{P}^\wedge$ assigned to a given color j (that is to say, all those M_i such that $\chi(M_i) = j$) are said to form *a color class* and it will be convenient to refer to this as the color class j. If k is minimal in the sense that each of the colors $1, 2, \ldots, k$ is assigned to at least one copy of the motif of $\mathscr{P}$, then $\mathscr{P}^\wedge$ is called a *k-colored pattern**.

All these definitions apply, with obvious modifications, to colored tilings, and we shall use symbols like $\mathscr{T}^\wedge$, or $\chi(T_i)$ and terms like "colored tile" or "k-colored tiling" without further explanation. The terminology extends that already introduced in Sections 1.3 and 2.9.

Examples of colored tilings have already been given in Figures 2.9.2 and 5.2.13. In the latter the tilings are 3-colored, and the three colors are indicated by white, black and stippling (instead of by integers 1, 2 and 3). A similar representation is employed in many of the diagrams of this chapter; various kinds of shading and hatching are used for the colors. In general, color 1 is indicated by white, and colors 2, 3, ... by increasingly dark shades of gray. Further examples of colored tilings appear in Figures 8.1.1 and 8.1.2 and many diagrams

* This definition is adopted for simplicity. For some purposes, it is more convenient to define a k-colored pattern as one in which *at most* k colors are used.

of k-colored patterns, for small values of k, are given in Sections 8.2, 8.3 and 8.4.

The definition of a symmetry of a colored tiling $\mathscr{T}^\wedge$ given in Section 1.3 extends in an obvious way to colored patterns: a *(color-preserving) symmetry of* $\mathscr{P}^\wedge$ is any isometry which maps each colored copy of a motif of $\mathscr{P}^\wedge$ into a copy of the motif of the same color. The group of all such symmetries is called the *color-preserving symmetry group* of $\mathscr{P}^\wedge$ and is denoted by $S(\mathscr{P}^\wedge)$. However, in the analysis of colored patterns, a much more useful concept is the following.

A *color symmetry* of a colored pattern $\mathscr{P}^\wedge$ is a pair $s^\wedge = (s,\theta)$ where $s \in S(\mathscr{P})$ is a symmetry of $\mathscr{P}$, and θ is a a permutation of colors which is *compatible* with s. By this we mean that $\theta(\chi(M_i)) = \chi(s(M_i))$ for each copy M_i of the motif M. In other words, whenever s maps a copy of the motif of color k onto one of color j, then it maps the whole color class k onto color class j, and this happens if and only if the permutation θ maps k onto j. We shall say that s and θ are *associated* with each other and with the color symmetry $s^\wedge = (s,\theta)$.

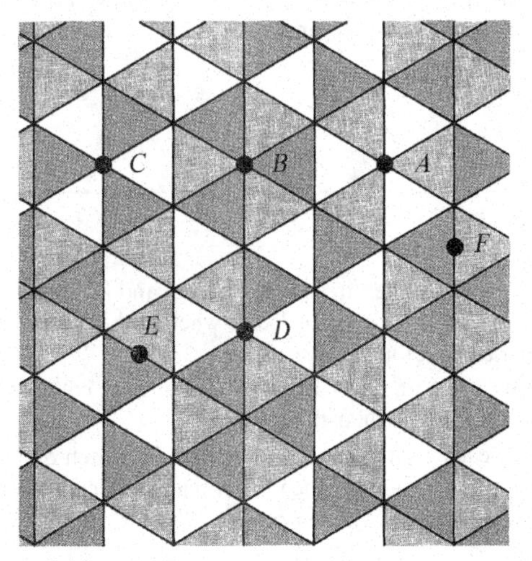

Figure 8.1.1
A 3-chromatic tiling which is not perfectly colored.

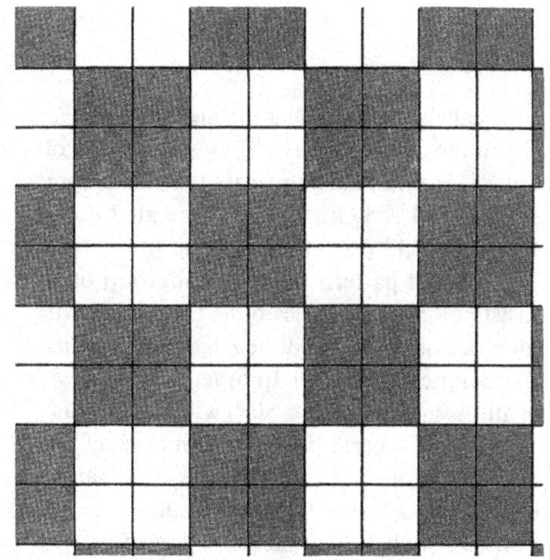
Figure 8.1.2
A 2-chromatic tiling $\mathscr{T}^\wedge$ for which the groups $S(\mathscr{T}^\wedge)$, $S_c(\mathscr{T}^\wedge)$ and $S(\mathscr{T})$ are all distinct, although each is of type *p4m*.

For example, in the 3-colored tiling $\mathscr{T}^\wedge$ of Figure 8.1.1, which has the regular tiling (3^6) as its underlying tiling $\mathscr{T}$, let a, b, c denote clockwise rotations about the points A, B, C by π, and let α, β, γ be the permutations (12), (23) and (31), respectively. Then $a^\wedge = (a,\alpha)$, $b^\wedge = (b,\beta)$ and $c^\wedge = (c,\gamma)$ are color symmetries of $\mathscr{T}^\wedge$. Other color symmetries of $\mathscr{T}^\wedge$ are $d^\wedge = (d,\delta)$ and $f^\wedge = (f,\phi)$ where d is the clockwise rotation by $\frac{1}{3}\pi$ about D, f is a rotation by π about E, $\delta = (123)$ and $\phi = (23)$. On the other hand, certain symmetries of $\mathscr{T}$—such as a half-turn about the point F—are clearly not associated with color symmetries of $\mathscr{T}^\wedge$ since there is no compatible permutation of colors.

The color-preserving symmetries of a pattern $\mathscr{P}^\wedge$ are simply those color symmetries (s,θ) for which θ is the identity permutation. In Figure 8.1.1, rotations by $2\pi/3$ about any of the points A, B or C, or reflection in the line through A, B, C are examples of such symmetries.

If $\mathscr{P}^\wedge$ is a k-colored pattern, the set $C(\mathscr{P}^\wedge)$ of all color symmetries of $\mathscr{P}^\wedge$ clearly forms a group under composition

$$(s_1,\theta_1)(s_2,\theta_2) = (s_1s_2,\theta_1\theta_2)$$

and we call $C(\mathscr{P}^\wedge)$ either the *color symmetry group of* $\mathscr{P}^\wedge$ or the *group of color symmetries of* $\mathscr{P}^\wedge$. The k-colored pattern $\mathscr{P}^\wedge$ is called *k-chromatic* if the group $C(\mathscr{P}^\wedge)$ acts transitively on the colored copies $\hat{M_i} = (M_i,\chi(M_i))$ of the motif of $\mathscr{P}^\wedge$. It is not hard to verify that the colored tiling of Figure 8.1.1 is 3-chromatic, as are all 3-colored tilings of Figure 5.2.13 except (*a*), (*f*) and (*g*).

If $\mathscr{P}^\wedge$ is a colored pattern then the subgroup of $S(\mathscr{P})$ which consists of all symmetries of $\mathscr{P}$ associated with color symmetries of $\mathscr{P}^\wedge$ is called the *underlying group* of $\mathscr{P}^\wedge$ and is denoted by $S_C(\mathscr{P}^\wedge)$. In other words, $S_C(\mathscr{P}^\wedge)$ consists of all those elements $s \in S(\mathscr{P})$ with which it is possible to associate a permutation θ such that (s,θ) is a color symmetry of $\mathscr{P}^\wedge$. Clearly $\mathscr{P}^\wedge$ is k-chromatic if and only if $S_C(\mathscr{P}^\wedge)$ is a motif-transitive subgroup of $S(\mathscr{P})$. Further, we shall say that $\mathscr{P}^\wedge$ is *perfectly colored* if $S_C(\mathscr{P}^\wedge) = S(\mathscr{P})$, that is, if *every* symmetry of $\mathscr{P}$ is associated with a color symmetry of $\mathscr{P}^\wedge$. Thus perfectly k-colored patterns and perfectly k-colored isohedral tilings are necessarily k-chromatic.

The 3-chromatic tiling of Figure 8.1.1 is not perfectly colored since, as we have already remarked, certain symmetries of $\mathscr{P}$ (such as a half-turn about the point F) are *not* associated with color symmetries. All the 3-chromatic tilings of Figure 5.2.13 (that is, all except (*a*), (*f*) and (*g*)) are perfectly colored. For further examples consider the 2-colored patterns obtained from the tilings of Figure 5.2.13 by ignoring the white tiles and taking the interiors of the black and stippled tiles as colored copies of the motif. Of these, only the patterns obtained from (*a*), (*b*), (*c*) and (*d*) are perfectly colored—the pattern obtained from (*e*), is even *not* 2-chromatic, while the remaining two, (*f*) and (*g*), are 2-chromatic but the coloring is not perfect since the rotational symmetries of the patterns are not associated with colored symmetries. Figure 8.1.2 shows a 2-chromatic tiling $\mathscr{T}^\wedge$ which is not perfectly colored; the three groups $S(\mathscr{T}^\wedge)$, $S_C(\mathscr{T}^\wedge)$ and $S(\mathscr{T})$ are distinct even though each is of type *p4m*.

Although certain patterns, such as those of type PP1, can be k-chromatically colored (and even perfectly k-colored) for all values of k, this is not generally the case. For example it is easy to see that a pattern of type PP21 has no 2-chromatic coloring and one of type PP41 has no perfect 3-coloring. The problem of finding the values of k for which perfect k-colorings are possible will be discussed in Section 8.7.

If $\mathscr{P}^\wedge$ is a k-colored pattern for which the underlying pattern type is primitive, then the only motif-transitive subgroup of $S(\mathscr{P})$ is $S(\mathscr{P})$ itself and it readily follows that $\mathscr{P}^\wedge$ is k-chromatic if and only if it is perfectly colored. Hence perfect k-colorings of primitive patterns can be used to determine and describe all the possible types of groups of color symmetries of patterns. These latter groups, known as k-color groups, form the topic of the next section.

EXERCISES 8.1

1. Determine which of the colorings of the three regular tilings shown in Figure 2.9.2 are (*a*) k-chromatic, and (*b*) perfect.
2. For each of the three regular tilings and for each integer $k \geq 2$, show that there exist at least two essentially distinct k-chromatic colorings of the tiling. (Here "essentially distinct" means that one cannot be transformed into the other by a permutation of colors.)
3. For each of the three regular tilings and for each k satisfying $2 \leq k \leq 10$, decide whether there exists a perfect k-coloring of the tiling.
4. For $k = 2$ and $k = 3$ find all k-chromatic colorings of the three regular tilings.
5. For each k satisfying $2 \leq k \leq 6$ and for each of the eight Laves tilings which is not regular, decide whether the tiling admits a k-chromatic coloring.

8.2 COLOR GROUPS

It is convenient to represent the color symmetry group $C(\mathscr{P}^{\wedge})$ of a colored pattern $\mathscr{P}^{\wedge}$ by a *color group diagram.* This is analogous to the group diagrams of symmetry groups which we introduced in Section 1.4; the symmetry elements in $S_C(\mathscr{P}^{\wedge})$ are denoted by symbols similar to those of Table 1.4.1 and near each element $s \in S_C(\mathscr{P}^{\wedge})$ is indicated the permutation θ with which it is associated in a color symmetry $s^{\wedge} = (s,\theta) \in C(\mathscr{P}^{\wedge})$. More precisely, near a symbol for the center of a *cn* or *dn* rotation we indicate the permutation associated with one counterclockwise step of rotation, near a line of reflection we indicate the permutation associated with the reflection, and near the symbol of a glide-reflection we show the permutation associated with one glide step. The identity permutation ε is indicated in the diagrams by empty parentheses ( ). Notice that on the diagram we do *not* indicate any symmetries of $\mathscr{P}$ which are not associated with color symmetries of $\mathscr{P}^{\wedge}$.

For example, in Figure 8.2.1 we show a 4-chromatic pattern $\mathscr{P}^{\wedge}$ (with underlying pattern type PP38) together with part of its color group diagram.

We shall say that the color symmetry groups $C(\mathscr{P}_1^{\wedge})$ and $C(\mathscr{P}_2^{\wedge})$ of two k-colored patterns $\mathscr{P}_1^{\wedge}$ and $\mathscr{P}_2^{\wedge}$ are *isomorphic* if their color group diagrams can be made to coincide by applying a suitable affine transformation (and also, possibly, altering one's choice of vectors corresponding to the translations—compare Section 1.4) combined with a permutation of colors in one of the patterns. The class of all mutually isomorphic color symmetry groups of k-chromatic strip or periodic discrete patterns is called a (strip or periodic) *k-color group.*

8.2.1 *For each $k \geq 2$ there exist only finitely many different strip and periodic k-color groups.*

In fact in order to determine these groups we need only choose a set of elements which generate a given strip or periodic group and then associate with each such generator, in every possible way, a permutation of $(1, 2, \ldots, k)$. Our choice of permutations is restricted by the fact that every algebraic relation between the generating symmetries must also be satisfied by the associated permutations. Since there are clearly only a finite number of possibilities for each value of k, Statement 8.2.1 is proved.

For very small values of k, the above procedure gives a practical method of enumerating the k-color groups. The only difficulty is to discover and eliminate isomorphic groups. However, this becomes rapidly more difficult as k increases and it seems necessary to use computers to carry out the enumeration. In Table 8.2.1 (which is based on results of Wieting [1981]) we list the known values for the number of k-color groups for $k \leq 12$ specified by the underlying strip or periodic group.

Using algebraic methods, the enumeration of k-color groups has been attempted by several authors. For example Senechal [1979a] gives an enumeration for various infinite classes of integers k. Additional references and historical details will be given in Section 8.8. For a complete account of Senechal's results the reader should consult the original paper; here we shall only quote an easily stated partial result:

8.2.2 *The number of periodic k-color groups, where k is a prime number greater than* 3, *equals* 16 *if* $k \equiv 1 \pmod{12}$, 15 *if* $k \equiv 7 \pmod{12}$, 14 *if* $k \equiv 5 \pmod{12}$ *and* 13 *if* $k \equiv 11 \pmod{12}$.

An analogous enumeration can easily be carried out for the strip k-color groups and we obtain the following results (communicated to us by M. Senechal); this enumeration has been carried out independently by Jarratt & Schwarzenberger [1981].

8.2.3 *For each odd integer k there are precisely* 7 *different strip k-color groups. For even values of k there are* 17 *groups if* $k \equiv 2 \pmod 4$ *and* 19 *groups if* $k \equiv 0 \pmod 4$.

In Figures 8.2.2 and 8.2.3 we illustrate the periodic 2-color and 3-color groups by perfectly colored primitive patterns. Illustrations analogous to those in Figure 8.2.2

can be found in many places (see, for example, Weber [1929], Belov & Tarhova [1956a], Šubnikov & Kopcik [1972]); the representations in Figure 8.2.3 seem to be new. In Tables 8.2.2 and 8.2.3 we present further information about the 2-color and 3-color groups.

As stated in the previous section, although the representation of the color groups by perfectly colored primitive patterns (instead of by their color group diagrams) is convenient, it can be misleading in the sense that two such representations of the same color group can "appear" very different. Examples are given in Figure 8.2.4. Another example of two "apparently" different colorings corresponding to the same color group, this time of a non-primitive pattern, will be given in Figure 8.3.1. In the following sections we shall present finer classifications of colorings of patterns that overcome, in part, this shortcoming of classification by color group alone.

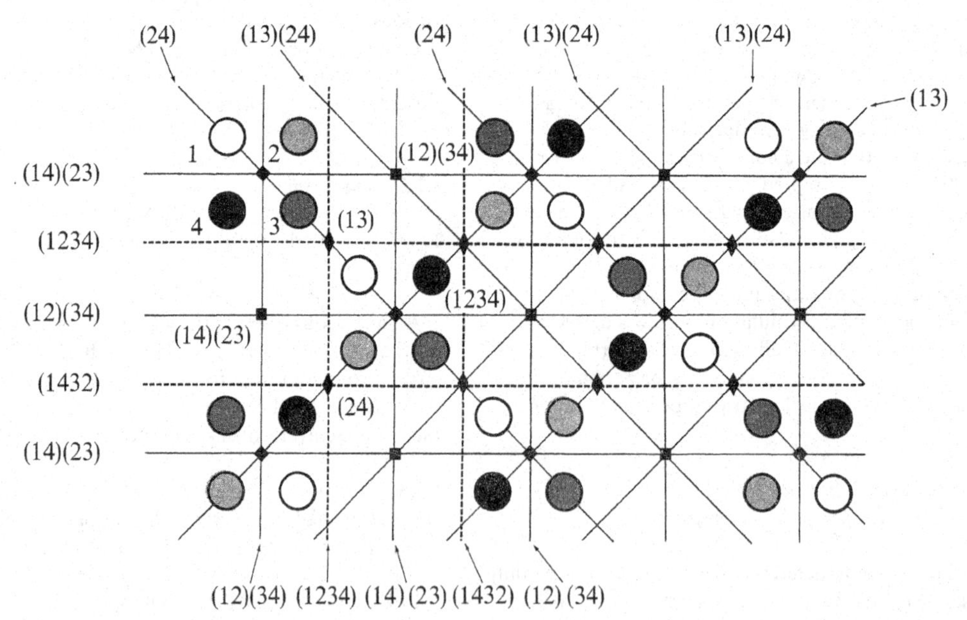

Figure 8.2.1
A 4-chromatic tiling and the corresponding color group diagram. For clarity, only a few of the symmetry elements have been indicated; in particular, no translations are shown.

Table 8.2.1 THE NUMBERS OF STRIP AND PERIODIC k-COLOR GROUPS FOR $k \leq 12$

The underlying group is indicated in the left column. The entries for the periodic groups are from Wieting [1981].

Underlying group	Number of colors										
	2	3	4	5	6	7	8	9	10	11	12
p111	1	1	1	1	1	1	1	1	1	1	1
p1a1	1	1	1	1	1	1	1	1	1	1	1
p1m1	3	1	3	1	3	1	3	1	3	1	3
pm11	2	1	2	1	2	1	2	1	2	1	2
p112	2	1	2	1	2	1	2	1	2	1	2
pma2	3	1	3	1	3	1	3	1	3	1	3
pmm2	5	1	7	1	5	1	7	1	5	1	7
Total strip groups	17	7	19	7	17	7	19	7	17	7	19
p1	1	1	2	1	1	1	2	2	1	1	2
pg	2	2	4	2	5	2	7	3	6	2	11
pm	5	2	10	2	11	2	16	3	12	2	23
cm	3	2	7	2	7	2	13	3	8	2	17
p2	2	1	3	1	2	1	4	2	2	1	3
pgg	2	1	4	1	4	1	7	2	5	1	9
pmg	5	2	11	2	11	2	19	3	12	2	26
pmm	5	1	13	1	9	1	21	2	10	1	25
cmm	5	1	11	1	8	1	21	2	9	1	22
p3	—	2	1	—	1	1	—	3	—	—	4
p31m	1	2	1	—	5	—	1	3	—	—	7
p3m1	1	2	1	—	4	—	1	3	—	—	7
p4	2	—	5	1	2	—	9	1	4	—	9
p4g	3	—	7	—	2	—	13	1	3	—	10
p4m	5	—	13	—	2	—	28	1	3	—	16
p6	1	2	1	—	5	1	1	3	—	—	8
p6m	3	2	2	—	11	—	3	3	—	—	20
Total periodic groups	46	23	96	14	90	15	166	40	75	13	219

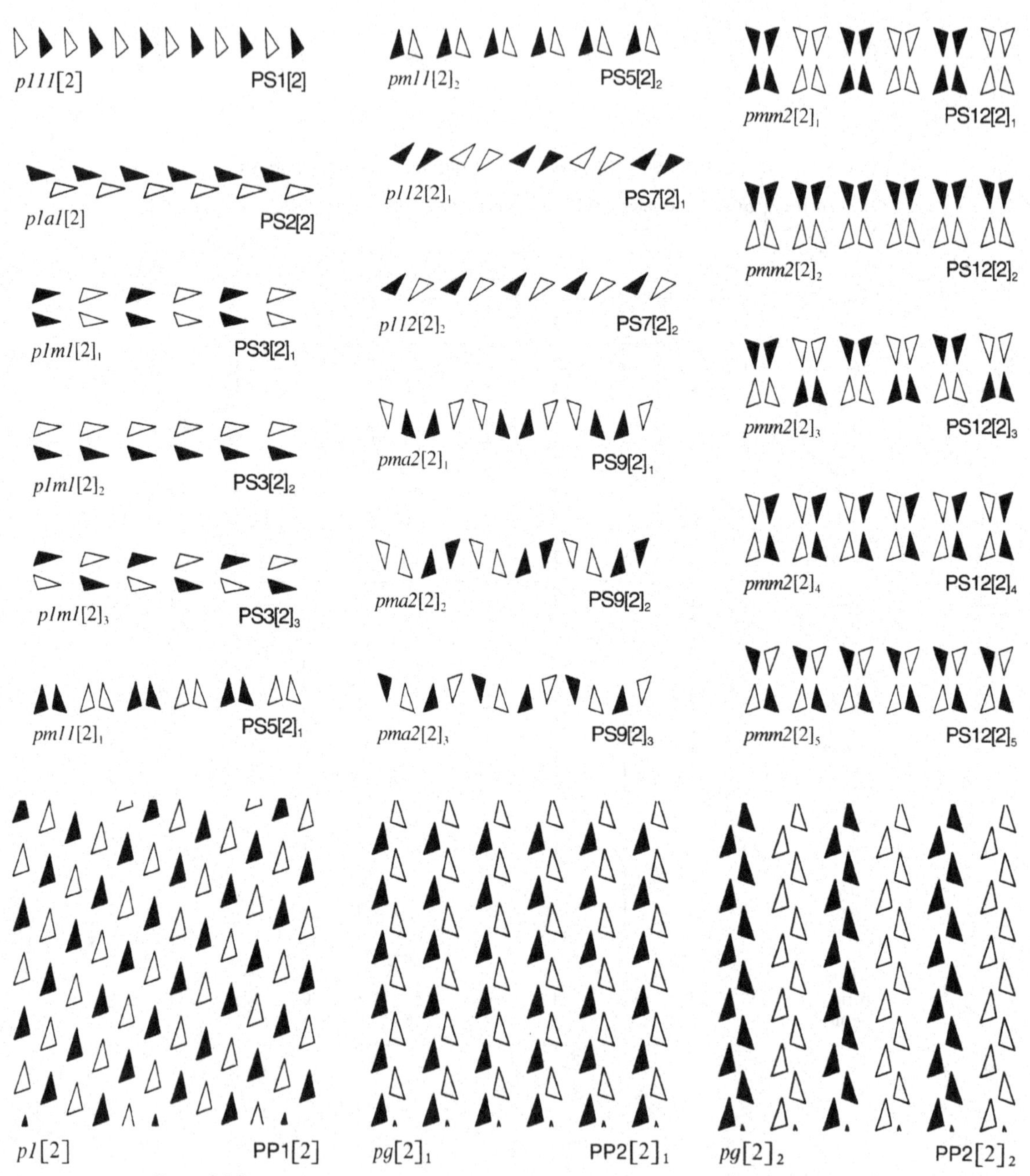

Figure 8.2.2
Perfectly 2-colored primitive patterns that illustrate the 17 strip and 46 periodic 2-color groups. Near each pattern we have indicated a symbol for its 2-color symmetry group and its color pattern type (see Section 8.3 for details).

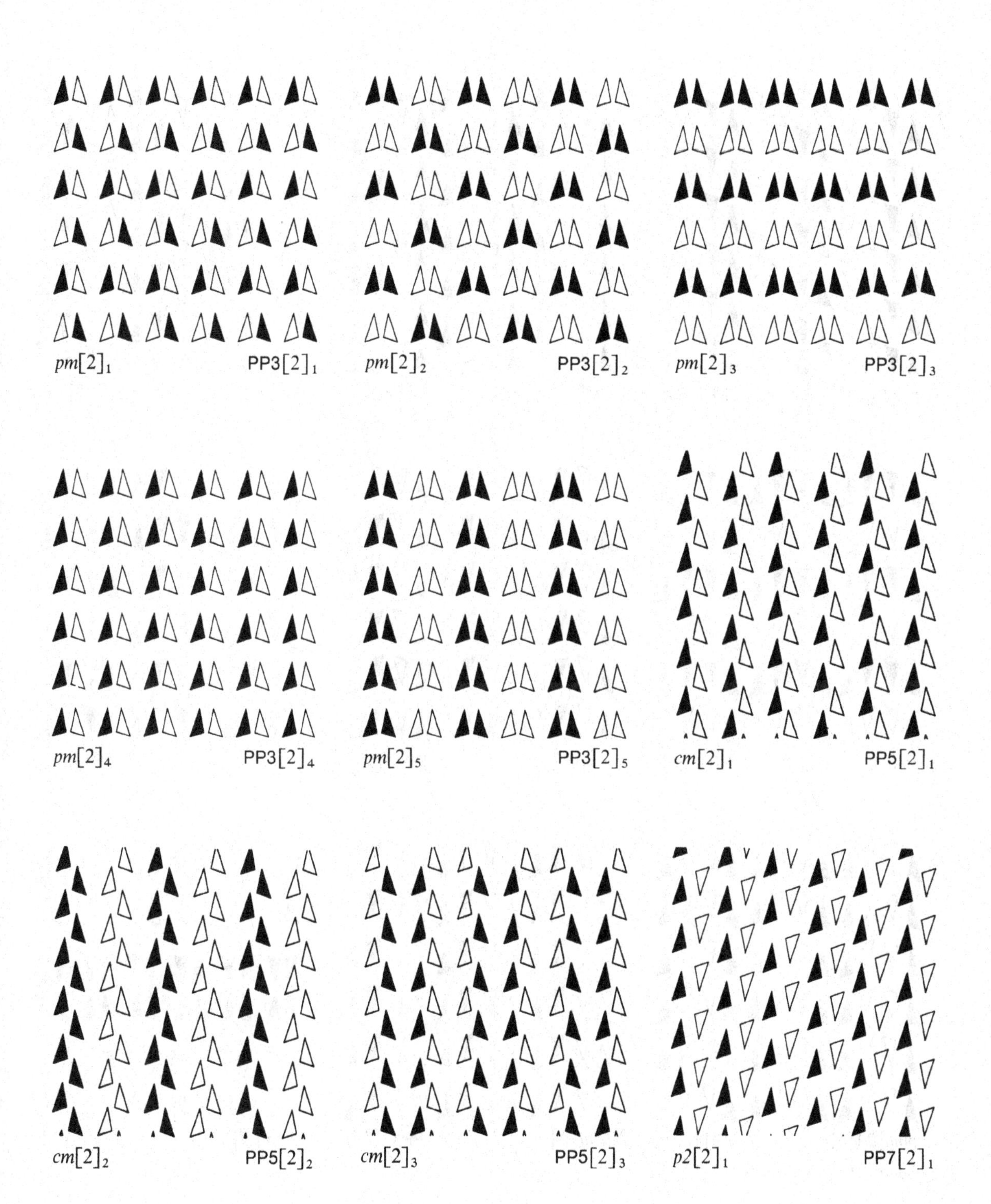

$pm[2]_1$ PP3$[2]_1$
$pm[2]_2$ PP3$[2]_2$
$pm[2]_3$ PP3$[2]_3$
$pm[2]_4$ PP3$[2]_4$
$pm[2]_5$ PP3$[2]_5$
$cm[2]_1$ PP5$[2]_1$
$cm[2]_2$ PP5$[2]_2$
$cm[2]_3$ PP5$[2]_3$
$p2[2]_1$ PP7$[2]_1$

$p2[2]_2$ PP7$[2]_2$

$pgg[2]_1$ PP9$[2]_1$

$pgg[2]_2$ PP9$[2]_2$

$pmg[2]_1$ PP11$[2]_1$

$pmg[2]_2$ PP11$[2]_2$

$pmg[2]_3$ PP11$[2]_3$

$pmg[2]_4$ PP11$[2]_4$

$pmg[2]_5$ PP11$[2]_5$

$pmm[2]_1$ PP14$[2]_1$

Figure 8.2.2 (continued)

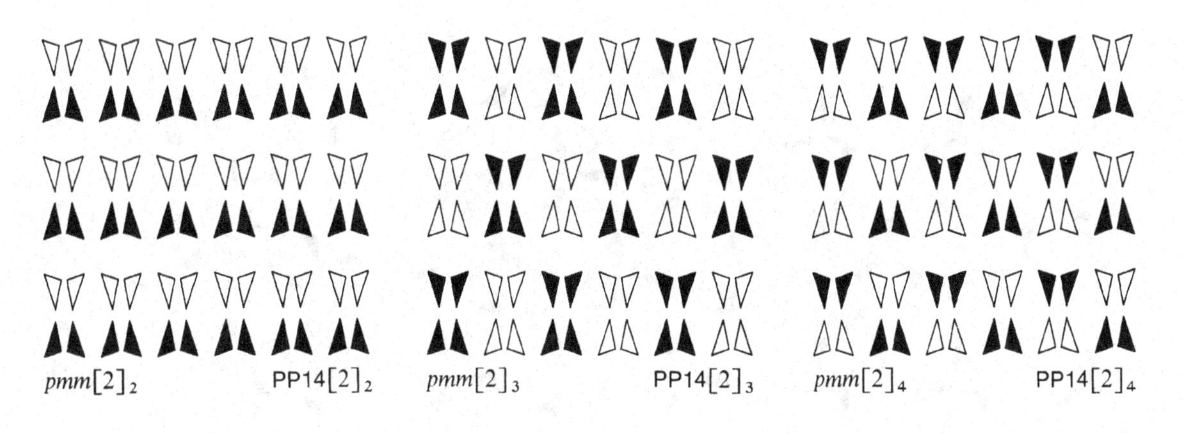

$pmm[2]_2$ PP14$[2]_2$ $pmm[2]_3$ PP14$[2]_3$ $pmm[2]_4$ PP14$[2]_4$

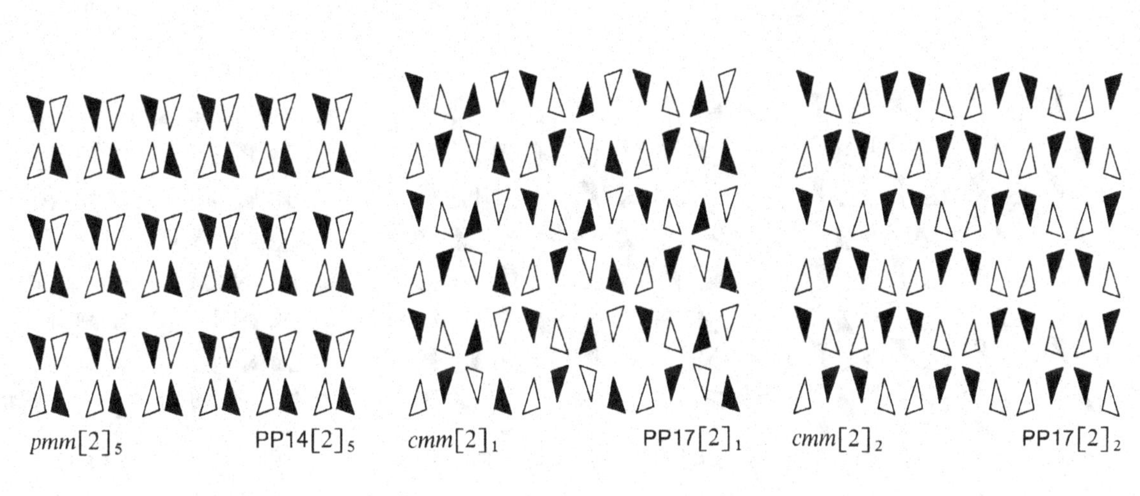

$pmm[2]_5$ PP14$[2]_5$ $cmm[2]_1$ PP17$[2]_1$ $cmm[2]_2$ PP17$[2]_2$

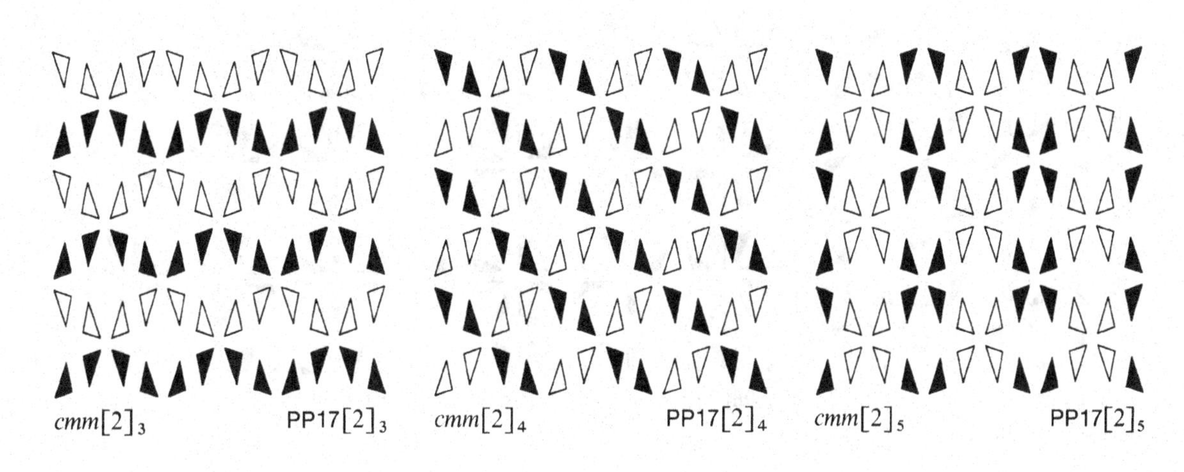

$cmm[2]_3$ PP17$[2]_3$ $cmm[2]_4$ PP17$[2]_4$ $cmm[2]_5$ PP17$[2]_5$

Figure 8.2.2 (continued)

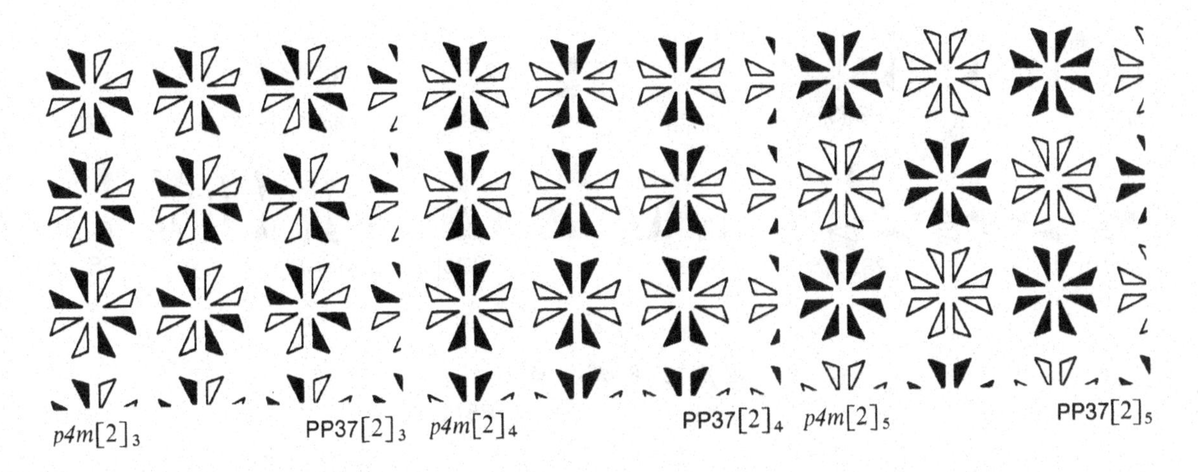
p4m[2]3
PP37[2]3
p4m[2]4
PP37[2]4
p4m[2]5
PP37[2]5

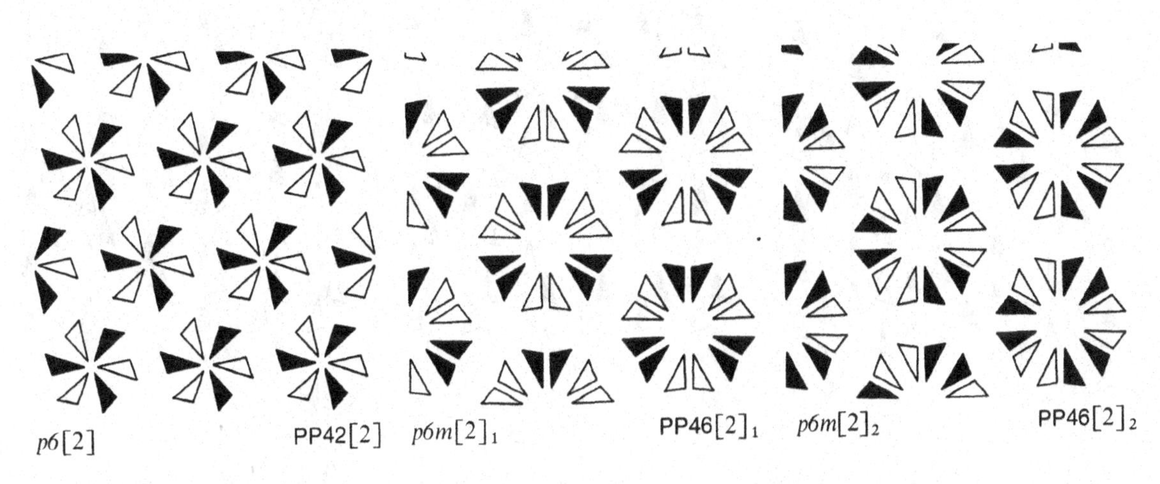
p6[2]
PP42[2]
p6m[2]1
PP46[2]1
p6m[2]2
PP46[2]2

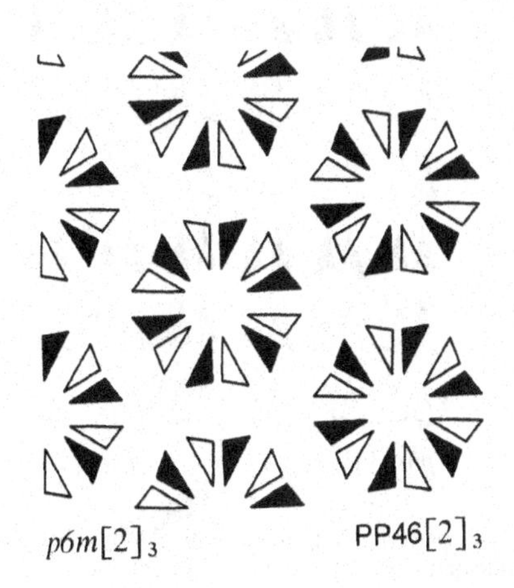
p6m[2]3
PP46[2]3

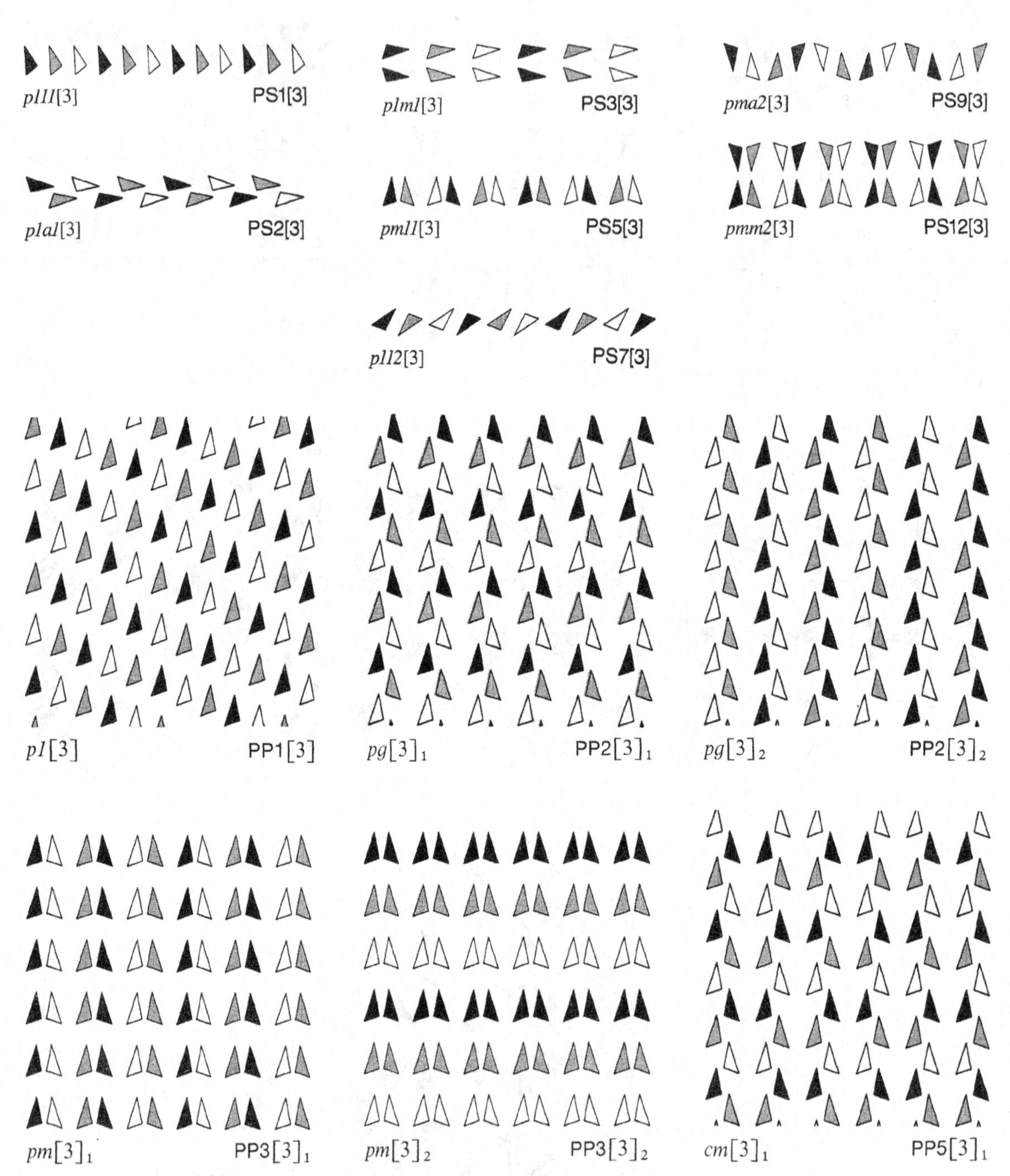

Figure 8.2.3
Perfectly 3-colored primitive patterns that illustrate the 7 strip and 23 periodic 3-color groups. Near each pattern we have indicated a symbol for its 3-color symmetry group and its color pattern type.

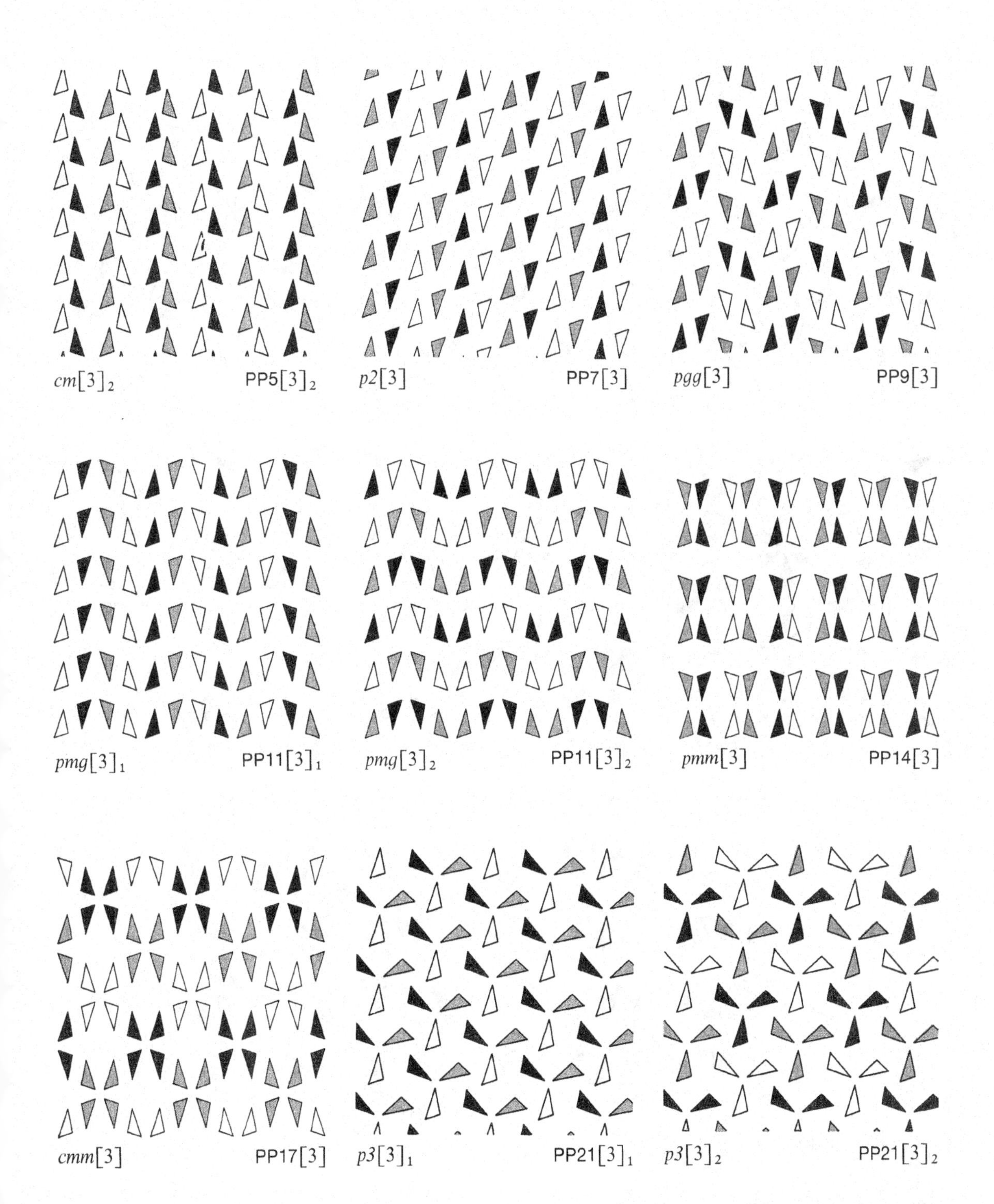
$cm[3]_2$
$PP5[3]_2$
$p2[3]$
$PP7[3]$
$pgg[3]$
$PP9[3]$
$pmg[3]_1$
$PP11[3]_1$
$pmg[3]_2$
$PP11[3]_2$
$pmm[3]$
$PP14[3]$
$cmm[3]$
$PP17[3]$
$p3[3]_1$
$PP21[3]_1$
$p3[3]_2$
$PP21[3]_2$

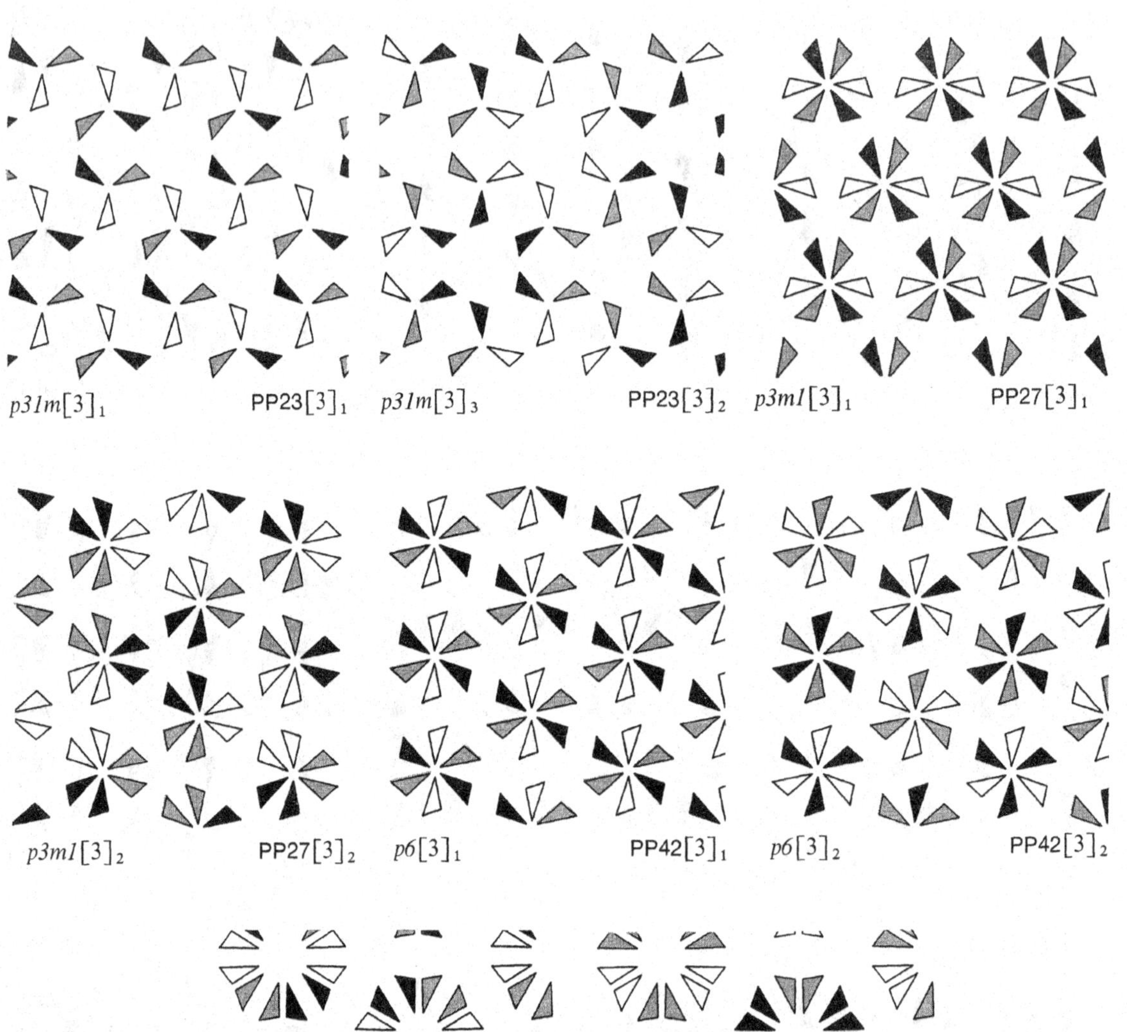

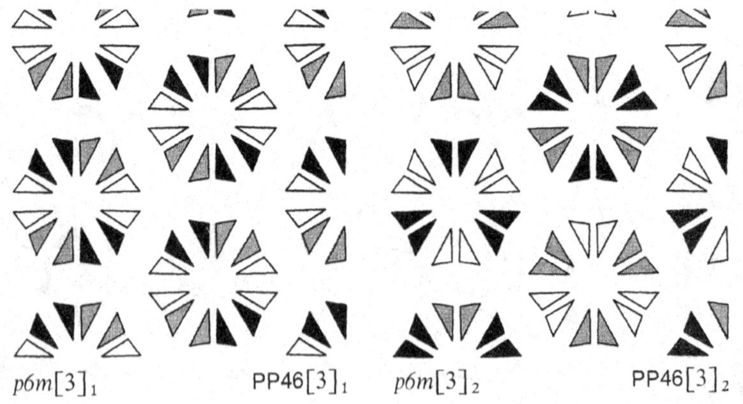

Figure 8.2.3 (continued)

Table 8.2.2 THE 46 PERIODIC 2-COLOR GROUPS

A symbol of each group is given in **column (1)**; these correspond to the symbols used in Figure 8.2.2, and were introduced in Grünbaum [1976] and Grünbaum & Shephard [1977e]. Symbols used by other authors (see, for example, Shubnikov, Belov & others [1964] or MacGillavry [1965]) are shown in **column (5)**. In **column (2)** we indicate the isohedral type of tilings which admit 2-chromatic colorings with each 2-color group. A color-tile symbol and the corresponding color-adjacency symbol of the 2-chromatic tiling are shown in **columns (3) and (4)**; these symbols will be explained in Section 8.6. The color of the starting tile is not indicated in the color-tile symbol since the color group does not depend on it. An asterisk * in **column (2)** signals that only marked 2-chromatic tilings can have this 2-color group. Another set of symbols for color groups has been proposed by Coxeter [1985]; see Crowe [1985] for a comparison of the symbols for 2-color groups used by various authors.

Symbol of the group	Type of tiling	Color-tile symbol	Color-adjacency symbol	Alternative symbol
(1)	**(2)**	**(3)**	**(4)**	**(5)**
$p1[2]$	IH1	$a^+b^+c^+d^+e^+f^+$	$d_*^+e_*^+f^+a_*^+b_*^+c^+$	p'_b1
$pg[2]_1$	IH3	$a^+b^+c^+d^+e^+f^+$	$c_*^-e^+a_*^-f_*^-b^+d_*^-$	pg'
$pg[2]_2$			$c^-e_*^+a^-f_*^-b_*^+d_*^-$	p'_b1g
$pm[2]_1$	IH42	$a^+b^+c^+d^+$	$c_*^+b_*^-a_*^+d_*^-$	p'_bg
$pm[2]_2$			$c_*^+b_*^-a_*^+d^-$	$c'm$
$pm[2]_3$			$c_*^+b^-a_*^+d^-$	p'_bm
$pm[2]_4$			$c^+b_*^-a^+d_*^-$	pm'
$pm[2]_5$			$c^+b_*^-a^+d^-$	p'_b1m
$cm[2]_1$	IH22	$a^+b^+c^+d^+e^+$	$a_*^-e^+d_*^-c_*^-b^+$	cm'
$cm[2]_2$			$a_*^-e^+d^-c^-b^+$	p'_cg
$cm[2]_3$			$a^-e^+d_*^-c_*^-b^+$	p'_cm
$p2[2]_1$	IH4	$a^+b^+c^+d^+e^+f^+$	$a_*^+e^+c_*^+d_*^+b^+f_*^+$	$p2'$
$p2[2]_2$			$a_*^+e^+c^+d^+b^+f_*^+$	p'_b2
$pgg[2]_1$	IH6	$a^+b^+c^+d^+e^+f^+$	$a_*^+e_*^-c_*^+f^-b_*^-d^-$	pgg'
$pgg[2]_2$			$a^+e_*^-c^+f_*^-b_*^-d_*^-$	$pg'g'$
$pmg[2]_1$	IH50	$a^+b^+c^+d^+$	$c_*^+b^-a_*^+d^+$	p'_bmg
$pmg[2]_2$			$c^+b_*^-a^+d_*^+$	$pm'g$
$pmg[2]_3$			$c_*^+b_*^-a_*^+d_*^+$	p'_bgg
$pmg[2]_4$			$c^+b^-a^+d_*^+$	pmg'
$pmg[2]_5$			$c^+b_*^-a^+d^+$	$pm'g'$

Table 8.2.2 (continued)

Symbol of the group	Type of tiling	Color-tile symbol	Color-adjacency symbol	Alternative symbol
(1)	(2)	(3)	(4)	(5)
$pmm[2]_1$	IH48	$a^+b^+c^+d^+$	$a^-_*b^-c^-d^-$	p'_bmm
$pmm[2]_2$	*		$a^-_*b^-c^-_*d^-$	pmm'
$pmm[2]_3$			$a^-_*b^-_*c^-d^-$	$c'mm$
$pmm[2]_4$	*		$a^-_*b^-_*c^-_*d^-$	p'_bgm
$pmm[2]_5$	*		$a^-_*b^-_*c^-_*d^-_*$	$pm'm'$
$cmm[2]_1$	IH54	$a^+b^+c^+d^+$	$a^-_*b^-_*c^-_*d^+_*$	p'_cgg
$cmm[2]_2$			$a^-b^-_*c^-d^+_*$	cmm'
$cmm[2]_3$			$a^-_*b^-c^-_*d^+$	p'_cmg
$cmm[2]_4$			$a^-_*b^-_*c^-_*d^+$	$cm'm'$
$cmm[2]_5$			$a^-b^-c^-d^+_*$	p'_cmm
$p31m[2]$	IH38	$a^+b^+c^+$	$a^-_*c^+b^+$	$p31m'$
$p3m1[2]$	IH87*	$a^+b^+c^+$	$a^-_*b^-_*c^-_*$	$p3m'$
$p4[2]_1$	IH28	$a^+b^+c^+d^+e^+$	$a^+_*c^+_*b^+_*e^+d^+$	p'_c4
$p4[2]_2$			$a^+c^+_*b^+_*e^+_*d^+_*$	$p4'$
$p4g[2]_1$	IH56	$a^+b^+c^+d^+$	$b^+a^+c^-_*d^-_*$	$p4g'm'$
$p4g[2]_2$			$b^+_*a^+_*c^-d^-$	$p4'g'm$
$p4g[2]_3$			$b^+_*a^+_*c^-_*d^-_*$	$p4'gm'$
$p4m[2]_1$	IH80	$a^+b^+c^+$	$a^-_*b^-_*c^-$	p'_c4gm
$p4m[2]_2$	*		$a^-_*b^-_*c^-_*$	$p4m'm'$
$p4m[2]_3$	*		$a^-b^-_*c^-_*$	$p4'm'm$
$p4m[2]_4$			$a^-_*b^-c^-$	$p4'mm'$
$p4m[2]_5$			$a^-b^-c^-_*$	p'_c4mm
$p6[2]$	IH21	$a^+b^+c^+d^+e^+$	$e^+_*c^+b^+d^+_*a^+_*$	$p6'$
$p6m[2]_1$	IH77	$a^+b^+c^+$	$a^-b^-_*c^-_*$	$p6'm'm$
$p6m[2]_2$			$a^-_*b^-c^-$	$p6'mm'$
$p6m[2]_3$			$a^-_*b^-_*c^-_*$	$p6m'm'$

Table 8.2.3 THE 23 PERIODIC 3-COLOR GROUPS

A symbol of each group is given in **column (1)**; these correspond to symbols used in Figure 8.2.3, and were used in Grünbaum [1976] and Grünbaum & Shephard [1977e], [1983b]. In **column (2)** we indicate the isohedral type of tilings which admit 3-chromatic colorings with each 3-color group. A color-tile symbol and the corresponding color-adjacency symbol of the 3-chromatic tiling are shown in **columns (3) and (4)**; these symbols will be explained in Section 8.6. The color of the starting tile is not indicated in the color-tile symbol since the color group does not depend on it. An asterisk * in column (2) signals that only marked 3-chromatic tilings can have this 3-color group.

Symbol of the group	Type of tiling	Color-tile symbol	Color-adjacency symbol
(1)	**(2)**	**(3)**	**(4)**
$p1[3]$	IH1	$a^+b^+c^+d^+e^+f^+$	$d^+(123)e^+(\ )f^+(321)a^+(321)b^+(\ )c^+(123)$
$pg[3]_1$ $pg[3]_2$	IH3	$a^+b^+c^+d^+e^+f^+$	$c^-(321)e^+(\ )a^-(123)f^-(123)b^+(\ )d^-(321)$ $c^-(12)e^+(321)a^-(12)f^-(13)b^+(123)d^-(13)$
$pm[3]_1$ $pm[3]_2$	IH42	$a^+b^+c^+d^+$	$c^+(\ )b^-(12)a^+(\ )d^-(13)$ $c^+(123)b^-(\ )a^+(321)d^-(\ )$
$cm[3]_1$ $cm[3]_2$	IH22	$a^+b^+c^+d^+e^+$	$a^-(\ )e^+(123)d^-(321)c^-(123)b^+(321)$ $a^-(12)e^+(\ )d^-(13)c^-(13)b^+(\ )$
$p2[2]$	IH4	$a^+b^+c^+d^+e^+f^+$	$a^+(12)e^+(321)c^+(12)d^+(13)b^+(123)f^+(13)$
$pgg[3]$	IH6	$a^+b^+c^+d^+e^+f^+$	$a^+(23)e^-(123)c^+(23)f^-(13)b^-(321)d^-(13)$
$pmg[3]_1$ $pmg[3]_2$	IH50	$a^+b^+c^+d^+$	$c^+(\ )b^-(13)a^+(\ )d^+(12)$ $c^+(123)b^-(\ )a^+(321)d^+(13)$
$pmm[3]$	IH48*	$a^+b^+c^+d^+$	$a^-(\ )b^-(12)c^-(\ )d^-(13)$
$cmm[3]$	IH54	$a^+b^+c^+d^+$	$a^-(23)b^-(\ )c^-(12)d^+(13)$
$p3[3]_1$ $p3[3]_2$	IH7	$a^+b^+c^+d^+e^+f^+$	$b^+(321)a^+(123)d^+(321)c^+(123)f^+(321)e^+(123)$ $b^+(123)a^+(321)d^+(321)c^+(123)f^+(\ )e^+(\ )$
$p31m[3]_1$ $p31m[3]_2$	IH38	$a^+b^+c^+$	$a^-(23)c^+(123)b^+(321)$ $a^-(\ )c^+(123)b^+(321)$
$p3m1[3]_1$ $p3m1[3]_2$	IH87* *	$a^+b^+c^+$	$a^-(12)b^-(23)c^-(13)$ $a^-(12)b^-(23)c^-(12)$
$p6[3]_1$ $p6[3]_2$	IH21	$a^+b^+c^+d^+e^+$	$e^+(123)c^+(123)b^+(321)d^+(\ )a^+(321)$ $e^+(12)c^+(321)b^+(123)d^+(13)a^+(12)$
$p6m[3]_1$ $p6m[3]_2$	IH77	$a^+b^+c^+$	$a^-(12)b^-(12)c^-(13)$ $a^-(\ )b^-(12)c^-(13)$

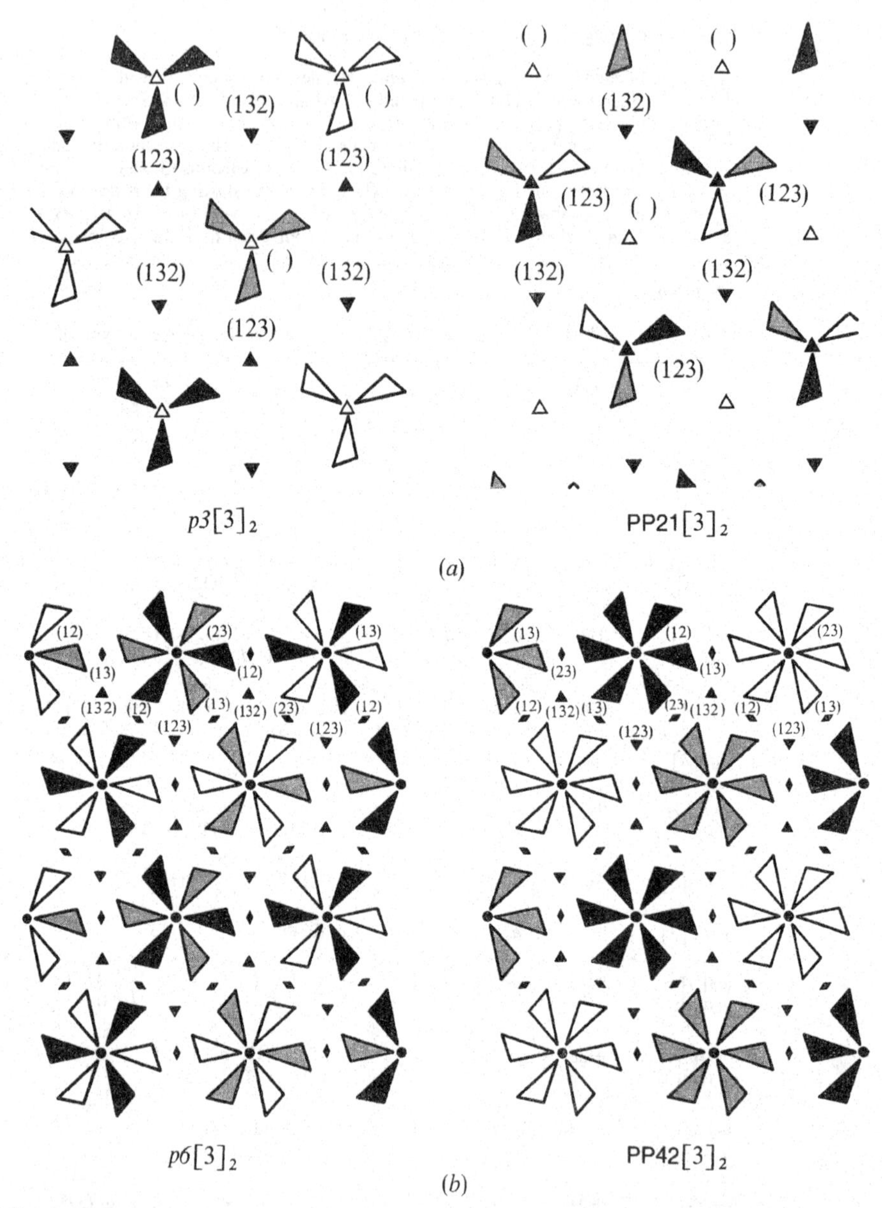

Figure 8.2.4
Examples of primitive 3-color patterns which differ in appearance and yet illustrate the same 3-color groups. In each part of the figure we show the diagram of a color group and a pattern for which this color group is the color symmetry group. In (*a*) the motifs are differently related to the three kinds of 3-fold rotational symmetries; in (*b*) the patterns differ in the choice of motif that bears color 1.

EXERCISES 8.2

1. Decide which of the k-colored patterns or tiling patterns in Figures 5.2.13 and 5.3.5 (*d*) (*e*) (*f*) are k-chromatic, $k = 2$ or 3; for those that are, draw a color-group diagram and determine its k-color group in the notation of Figures 8.2.2 and 8.2.3.
2. For each k satisfying $4 \leq k \leq 6$ find which of the seven primitive strip patterns admits a perfect k-coloring. Hence check the numbers of k-color strip groups ($k = 4,5,6$) given in Table 8.2.1.
3. (*a*) For each k and n satisfying $2 \leq k \leq 6$ and $2 \leq n \leq 12$ determine all the k-chromatic colorings of patterns of types $\mathsf{PF1}_n$, $\mathsf{PF2}_n$ and $\mathsf{PF3}_n$.

   (*b*) Determine all 2-chromatic and 3-chromatic colorings of patterns $\mathsf{PF1}_n$ for all n.
4. Determine the k-color group of each of the k-chromatic colorings of the regular tilings found in Exercise 8.1.4 (for $k = 2$ or 3).

8.3 COLOR PATTERN TYPES

In Chapter 5, in the context of uncolored patterns, we showed that the notion of pattern type can be used to refine the classification of patterns by symmetry groups alone. In the case of colored patterns the situation is completely analogous; to obtain the appropriate definitions we need only consider color symmetries instead of symmetries.

More precisely we shall say that two k-chromatic patterns $\hat{\mathscr{P}}_1$ and $\hat{\mathscr{P}}_2$ are of the same *k-color pattern type* if—possibly after an appropriate renaming (permutation) of the colors in one of the patterns—they have the same k-color symmetry groups, the same stabilizers of their colored motifs (that is, induced k-color groups), and the same set of colored-motif transitive subgroups of their color symmetry groups. An alternative, and equivalent, definition which follows the general procedure for classifying objects with respect to their symmetries (see Chapter 7) is as follows: two colored patterns $\hat{\mathscr{P}}_1$ and $\hat{\mathscr{P}}_2$ are of the same color pattern type if, possibly after an appropriate permutation of the colors in one of the patterns, there exists a one-to-one color-preserving mapping $\Psi: M(\hat{\mathscr{P}}_1) \to M(\hat{\mathscr{P}}_2)$ from the set $M(\hat{\mathscr{P}}_1)$ of colored copies of the motif of $\mathscr{P}_1$ to the corresponding set $M(\hat{\mathscr{P}}_2)$ for $\hat{\mathscr{P}}_2$, such that Ψ is compatible with the underlying group $S_C(\hat{\mathscr{P}}_1)$ and Ψ^{-1} is compatible with the underlying group $S_C(\hat{\mathscr{P}}_2)$.

To illustrate this definition consider the three 3-chromatic colorings of the pattern $\mathscr{P}$ of type PP15 shown in Figure 8.3.1. All are perfect 3-colorings, and the color symmetry group of each is the 3-color group *pmm*[3]. In Figures 8.3.1(*a*) and (*b*) the induced color group (that is, the subgroup that leaves one copy of the motif invariant) is *d1* with no changes of color. In Figure 8.3.1(*c*) it is *d1* but associated with a transposition of two of the colors as indicated on the diagram*. It follows that the 3-color pattern type of the pattern in Figure 8.3.1(*c*) is different from the 3-color pattern type of those in Figures 8.3.1(*a*) and (*b*). To show that the patterns in Figures 8.3.1(*a*) and (*b*) are of the same 3-color pattern type, we can either verify that the motif-transitive color subgroups

* Since we shall not discuss the possibilities in detail, it is not appropriate to assign special symbols for these groups.

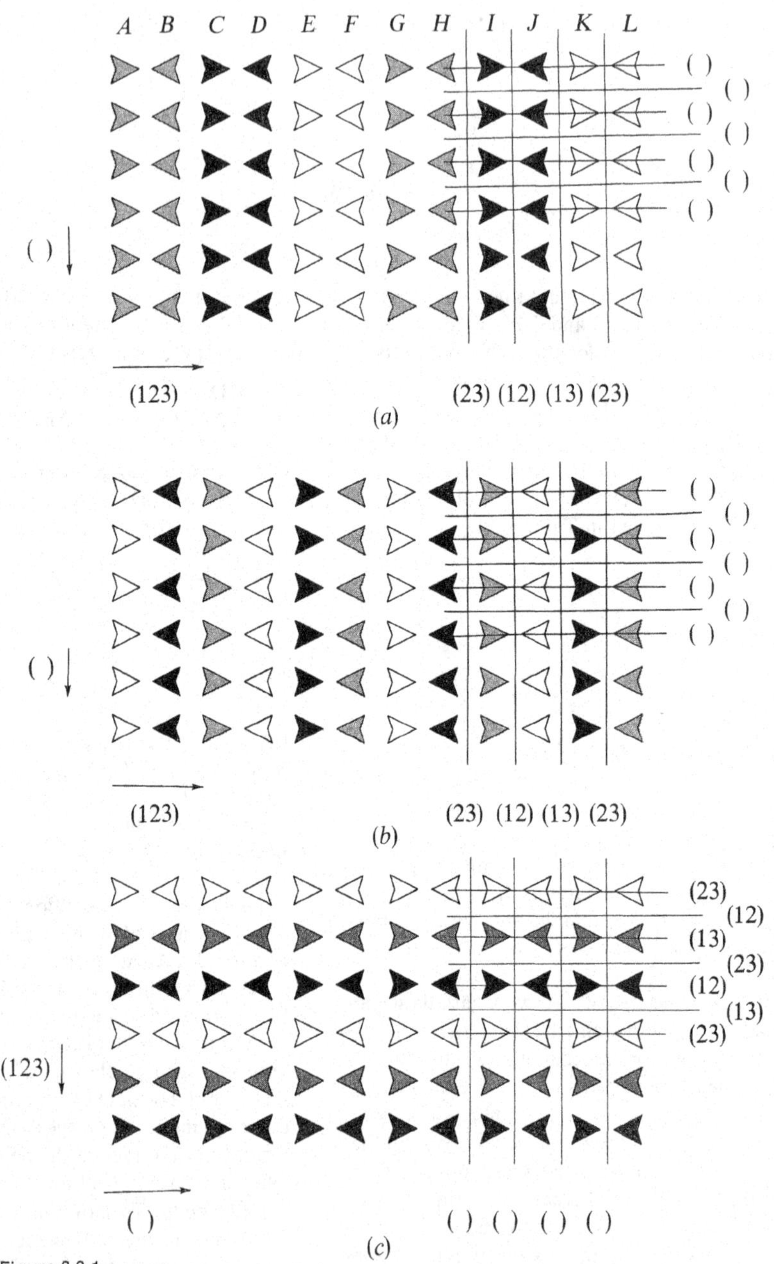

Figure 8.3.1
Examples of perfectly colored 3-chromatic patterns with color group *pmm*[3] and underlying pattern of type PP15. The patterns of (*a*) and (*b*) have the same 3-color pattern type PP15[3], while that of (*c*) has a different 3-color pattern type, namely PP15[3]*. The meaning of the letters above the patterns in parts (*a*) and (*b*) is explained in the text.

of $C(\hat{\mathscr{P}_1})$ and $C(\hat{\mathscr{P}_2})$ are the same, or else use the alternative definition given above; we define Ψ to be the one-to-one correspondence that maps the motifs in each column of Figure 8.3.1(*a*) into the motifs in the column of Figure 8.3.1(*b*) that carries the same letter $A, B, C, \ldots$ at its head. It is easy to verify that Ψ and its inverse are compatible with the groups $S_C(\hat{\mathscr{P}_1})$ and $S_C(\hat{\mathscr{P}_2})$, respectively. The fact that the colorings in Figures 8.3.1(*a*) and (*b*) "look different" is a motivation for some of the finer classifications to be introduced later.

As a second example consider the 17 strip and 46 periodic perfectly 2-colored patterns in Figure 8.2.2 and the 7 strip and 23 periodic perfectly 3-colored patterns in Figure 8.2.3. The fact that the color symmetry groups of all these are different implies that they are also of different color pattern types. We shall refer to these patterns by the symbols marked near each. It is immediate that these are the only k-color pattern types for $k = 2$ or $k = 3$ with discrete and primitive underlying patterns.

Consider now the two 2-chromatic patterns in Figure 8.3.2. They have the same underlying pattern of type PP15, their 2-color symmetry groups are both $pmm[2]_4$ and their induced groups are both $d1$ with no color

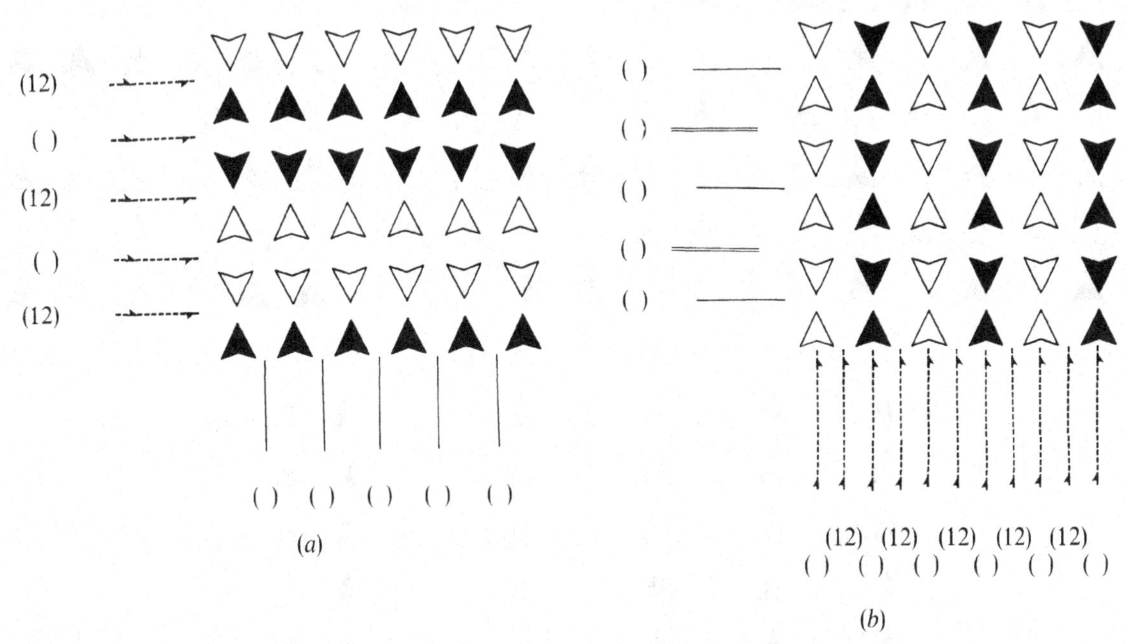

Figure 8.3.2
Two 2-chromatic colorings of a pattern of type PP15. They have the same color symmetry group $pmm[2]_1$ and the same induced group ($d1$ with no color changes), but are of different 2-color types since the sets of color subgroups that act transitively on the motifs are unequal. For example the pattern in (*a*) admits only one such subgroup of type $pmg[2]_1$ (indicated on the diagram by the lines of reflection and glide-reflection) while the pattern in (*b*) admits two such motif-transitive subgroups. (The lines of reflection of one of these are indicated by single lines and of the other by double lines; the glide-reflections are the same for both.)

changes. However they are of different 2-color pattern types since the sets of motif-transitive subgroups of their 2-color symmetry groups are different. To see this it is enough to verify that although both admit motif-transitive subgroups of type $pmg[2]_1$, the pattern in Figure 8.3.2(*a*) admits just one such group, while that in Figure 8.3.2(*b*) admits two.

With non-perfect colorings (so that $S_C(\mathscr{P}^\wedge) \neq S(\mathscr{P})$) new considerations arise. For example consider the five 2-colored patterns with underlying pattern of type PP4 shown in Figure 8.3.3. The 2-color symmetry groups of the (perfectly colored) patterns shown in Figures 8.3.3(*a*) (*c*) and (*d*) are $pm[2]_2$, $pm[2]_3$ and $pm[2]_5$, so they are of different color pattern types, namely $\mathsf{PP4}[2]_2$, $\mathsf{PP4}[2]_3$ and $\mathsf{PP4}[2]_5$, respectively. They are also different from the 2-color pattern types shown in Figure 8.2.2 since here the induced groups are non-trivial. On the other hand the 2-colored patterns in Figures 8.3.3(*b*) and (*e*) are 2-chromatic with color symmetry groups $pm[2]_2$ and $pm[2]_5$, but consideration of their color symmetries

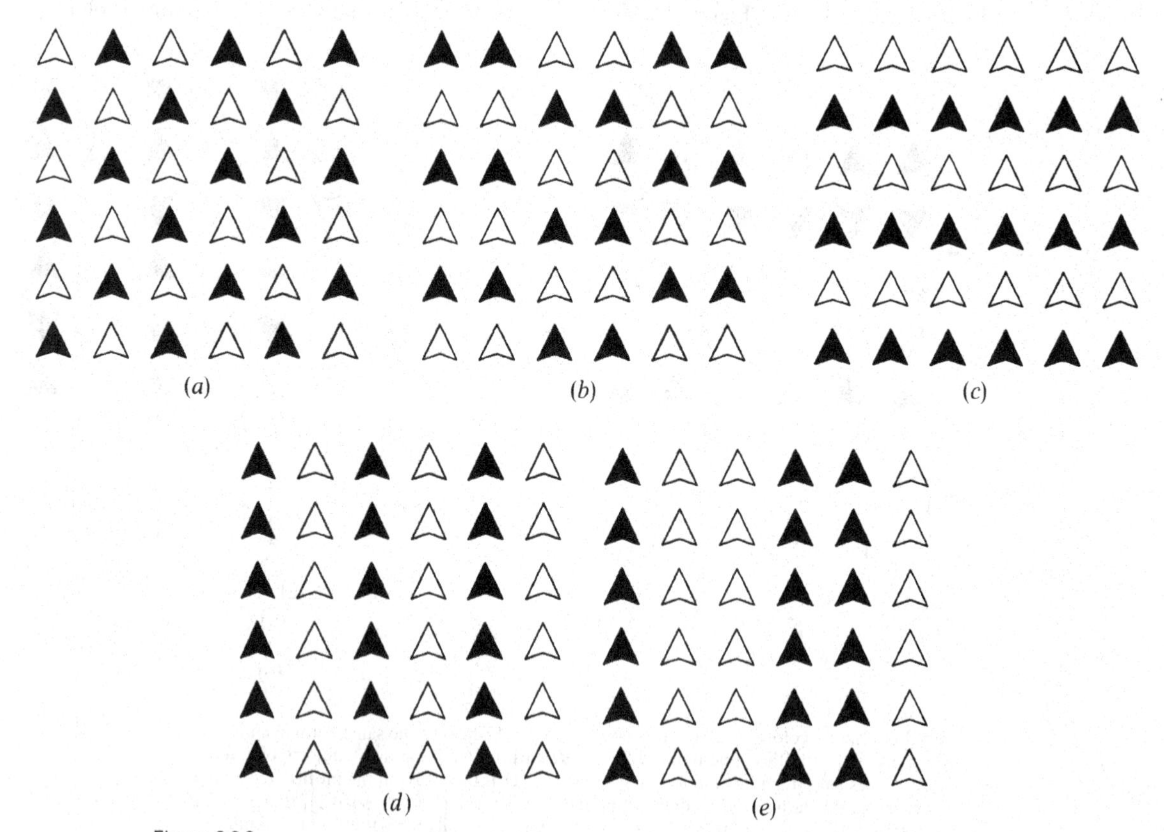

Figure 8.3.3
Examples of 2-chromatic patterns with underlying pattern of type PP4. The 2-color pattern types are (*a*) $\mathsf{PP4}[2]_2$, (*b*) $\mathsf{PP3}[2]_2$, (*c*) $\mathsf{PP4}[2]_3$, (*d*) $\mathsf{PP4}[2]_5$ and (*e*) $\mathsf{PP3}[2]_5$.

reveals that they are of 2-color pattern types PP3[2]$_2$ and PP3[2]$_5$—the types denoted by the same symbols in Figure 8.2.2. The reason for this phenomenon is that in Figures 8.3.3(*b*) and (*e*) the colorings are not perfect and in each case the underlying group $S_C(\mathscr{P}^\frown)$ induces on the copies of the motif only the trivial group. In other words, although PP4 is not primitive, it behaves as if it were primitive with respect to the color symmetry groups of these two colorings.

Another example of a similar nature is shown in Figure 8.3.4. Here we show two patterns of the same 2-color pattern type whose underlying patterns are of different pattern types, namely PP16 and PP41. This situation can only arise if at least one of the patterns is not perfectly colored, and for this reason it is convenient, when enumerating k-color pattern types, to restrict attention to perfectly colored patterns.

One way to carry out such an enumeration is analogous to that mentioned in Section 5.2. We consider each of the primitive k-colored patterns of Figures 8.2.2 (for $k = 2$) and 8.2.3 (for $k = 3$), and see whether it is possible to "move" the colored copies of the motif (in such a way that the color symmetry group is unchanged) until two or more copies "merge". The various possibilities can be deduced from those for uncolored patterns (see Figure 5.2.5), though of course here we may merge two copies of the motif only if they are of the same color. If this procedure is carried out systematically we obtain the following:

8.3.1 *There exist* 28 *strip and* 88 *periodic* 2-*color pattern types.*

8.3.2 *There exist* 15 *strip and* 59 *periodic* 3-*color pattern types.**

Examples of patterns of all these types not already illustrated in Figures 8.2.2 and 8.2.3 appear in Figures 8.3.5 and 8.3.6. It seems that the number of k-color pattern types with $k \geq 4$ is unknown.

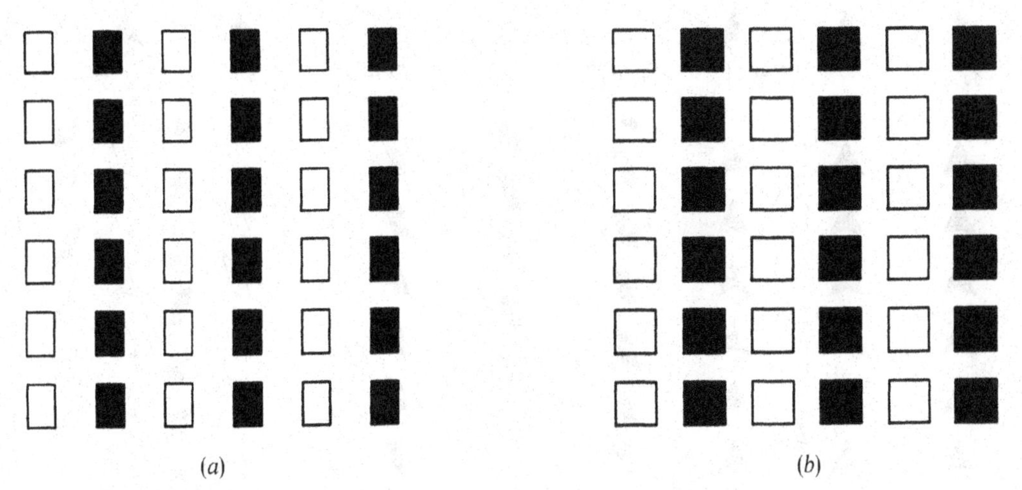

Figure 8.3.4
Two examples of 2-chromatic patterns of the same 2-color pattern type PP16[2]$_1$ with underlying discrete patterns of different types (PP16 in (*a*) and PP41 in (*b*)). The pattern in (*a*) is perfectly colored, that in (*b*) is not.

* See the Appendix beginning on page 653.

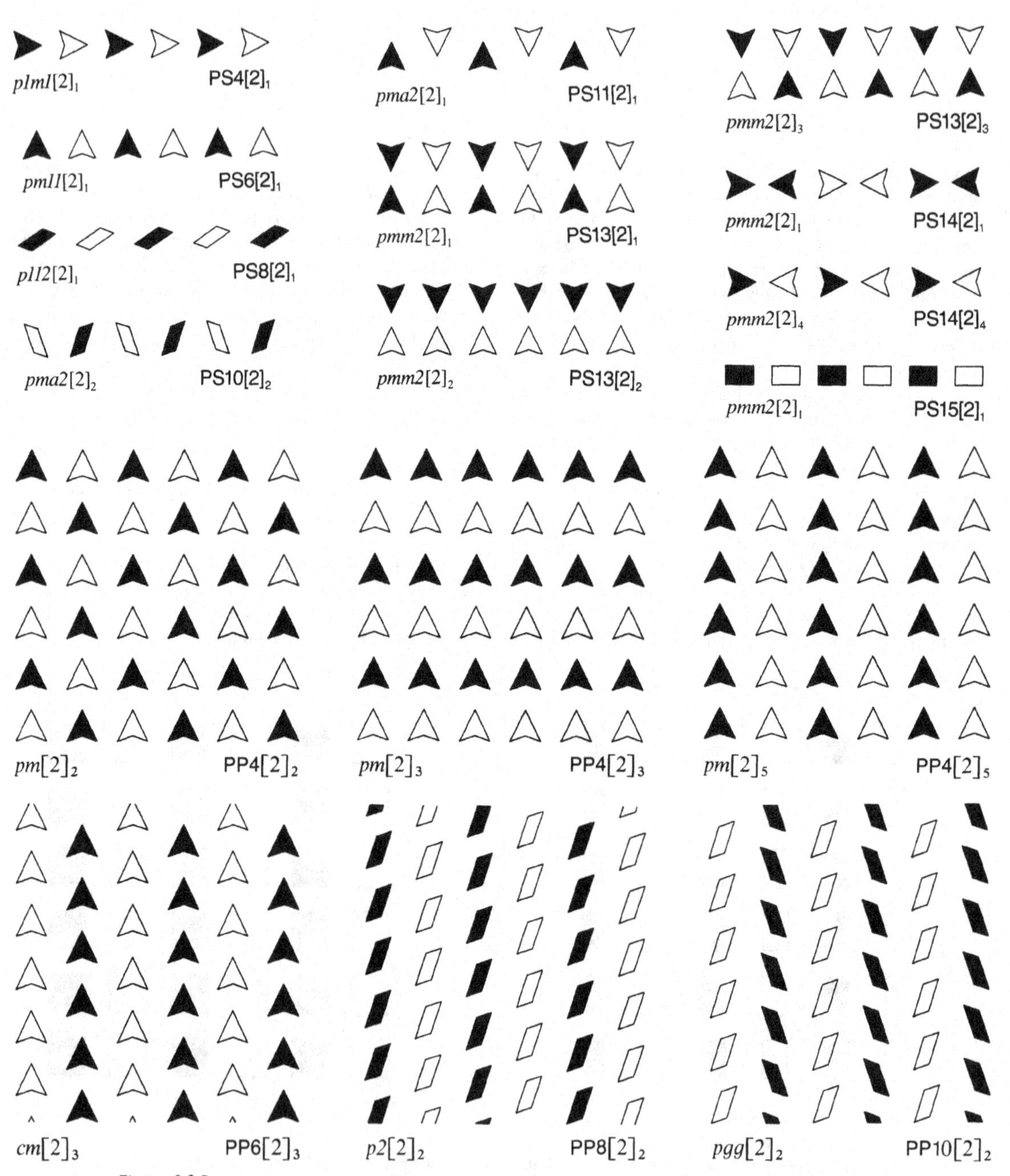

Figure 8.3.5
Examples of all the different 2-color pattern types (11 strip and 42 periodic) whose underlying patterns are non-primitive.

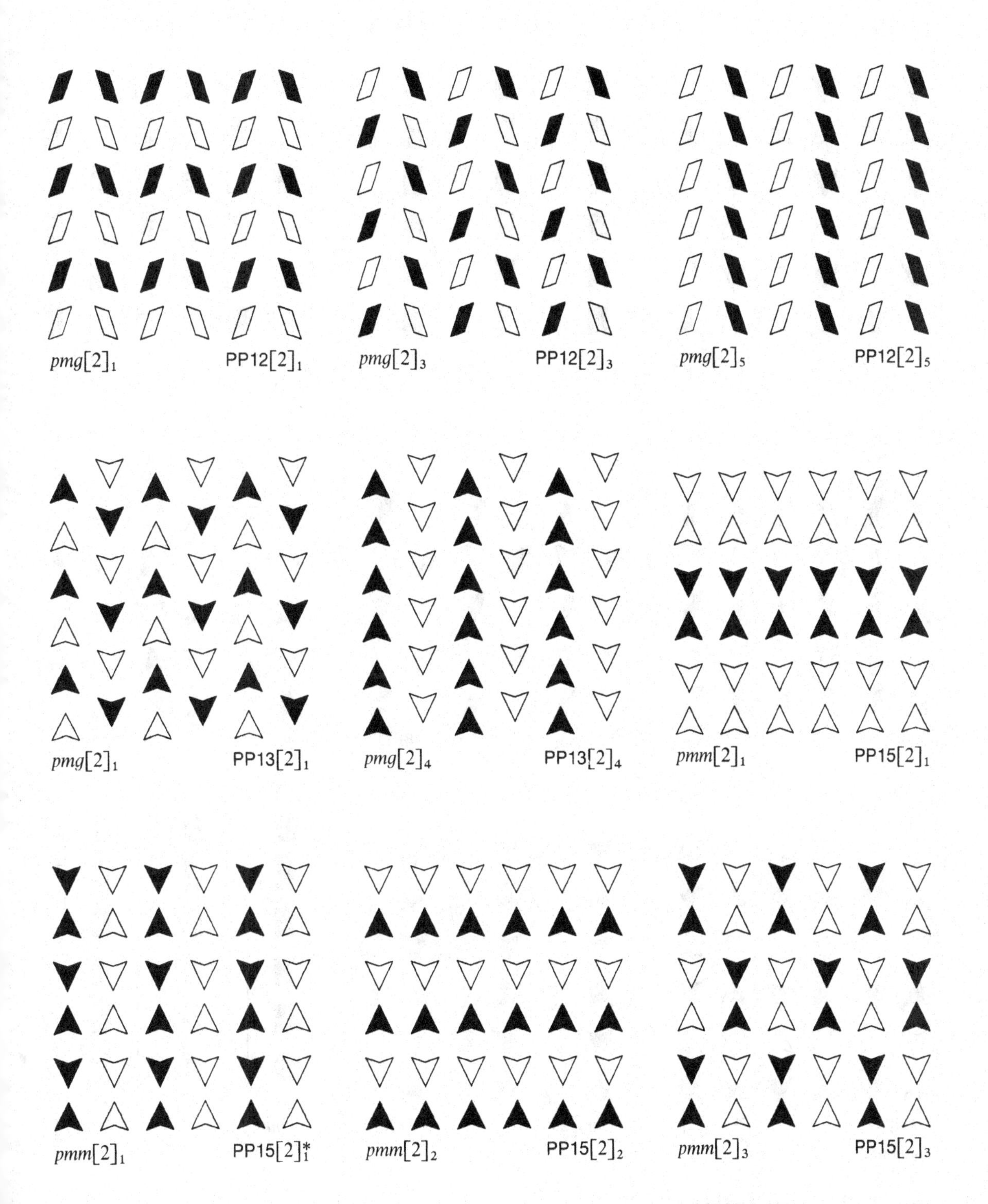

$pmg[2]_1$
PP12$[2]_1$
$pmg[2]_3$
PP12$[2]_3$
$pmg[2]_5$
PP12$[2]_5$
$pmg[2]_1$
PP13$[2]_1$
$pmg[2]_4$
PP13$[2]_4$
$pmm[2]_1$
PP15$[2]_1$
$pmm[2]_1$
PP15$[2]_1^*$
$pmm[2]_2$
PP15$[2]_2$
$pmm[2]_3$
PP15$[2]_3$

Figure 8.3.5 (continued)

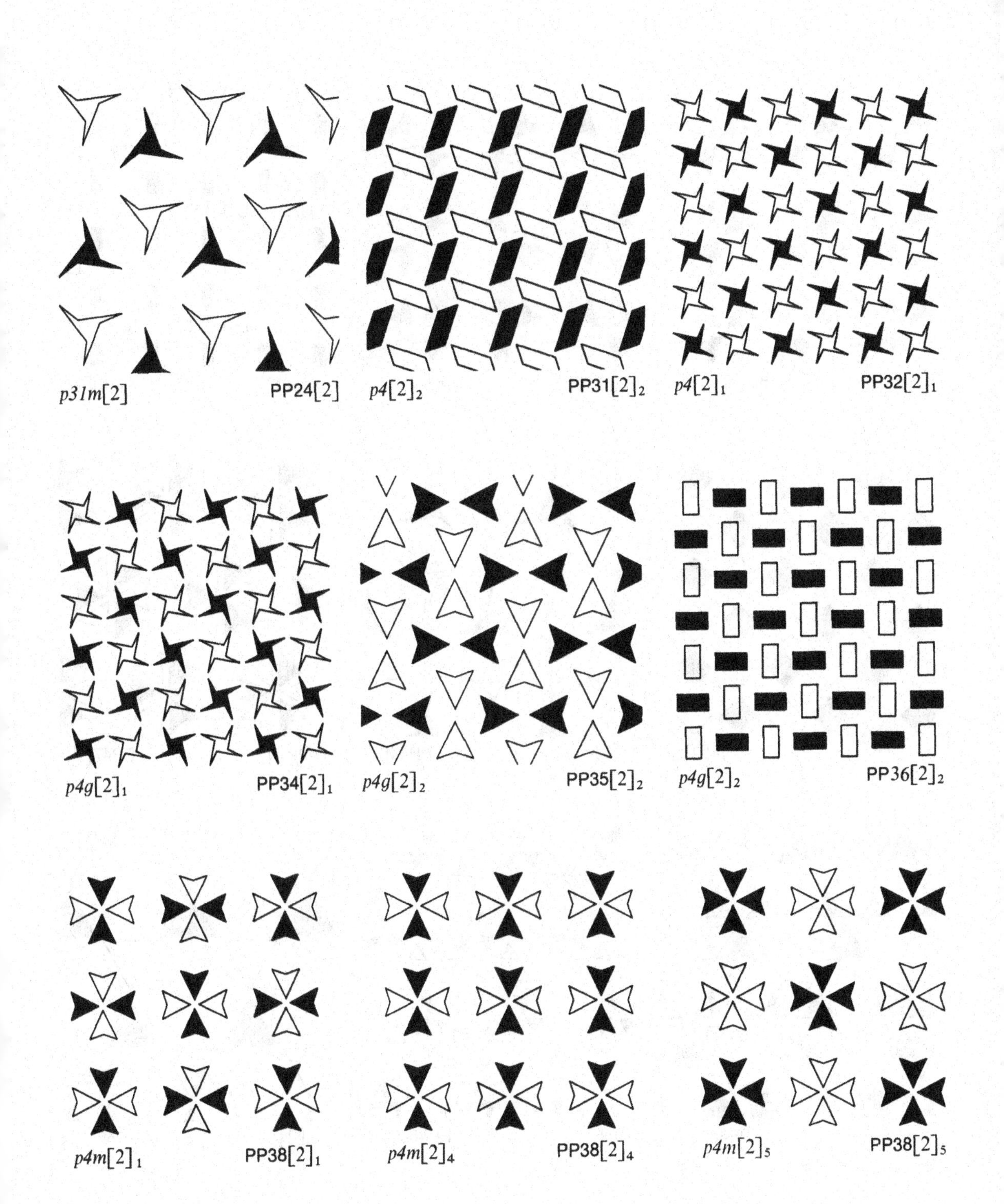
$p31m[2]$
PP24[2]
$p4[2]_2$
$\mathrm{PP31}[2]_2$
$p4[2]_1$
$\mathrm{PP32}[2]_1$
$p4g[2]_1$
$\mathrm{PP34}[2]_1$
$p4g[2]_2$
$\mathrm{PP35}[2]_2$
$p4g[2]_2$
$\mathrm{PP36}[2]_2$
$p4m[2]_1$
$\mathrm{PP38}[2]_1$
$p4m[2]_4$
$\mathrm{PP38}[2]_4$
$p4m[2]_5$
$\mathrm{PP38}[2]_5$

Figure 8.3.5 (continued)

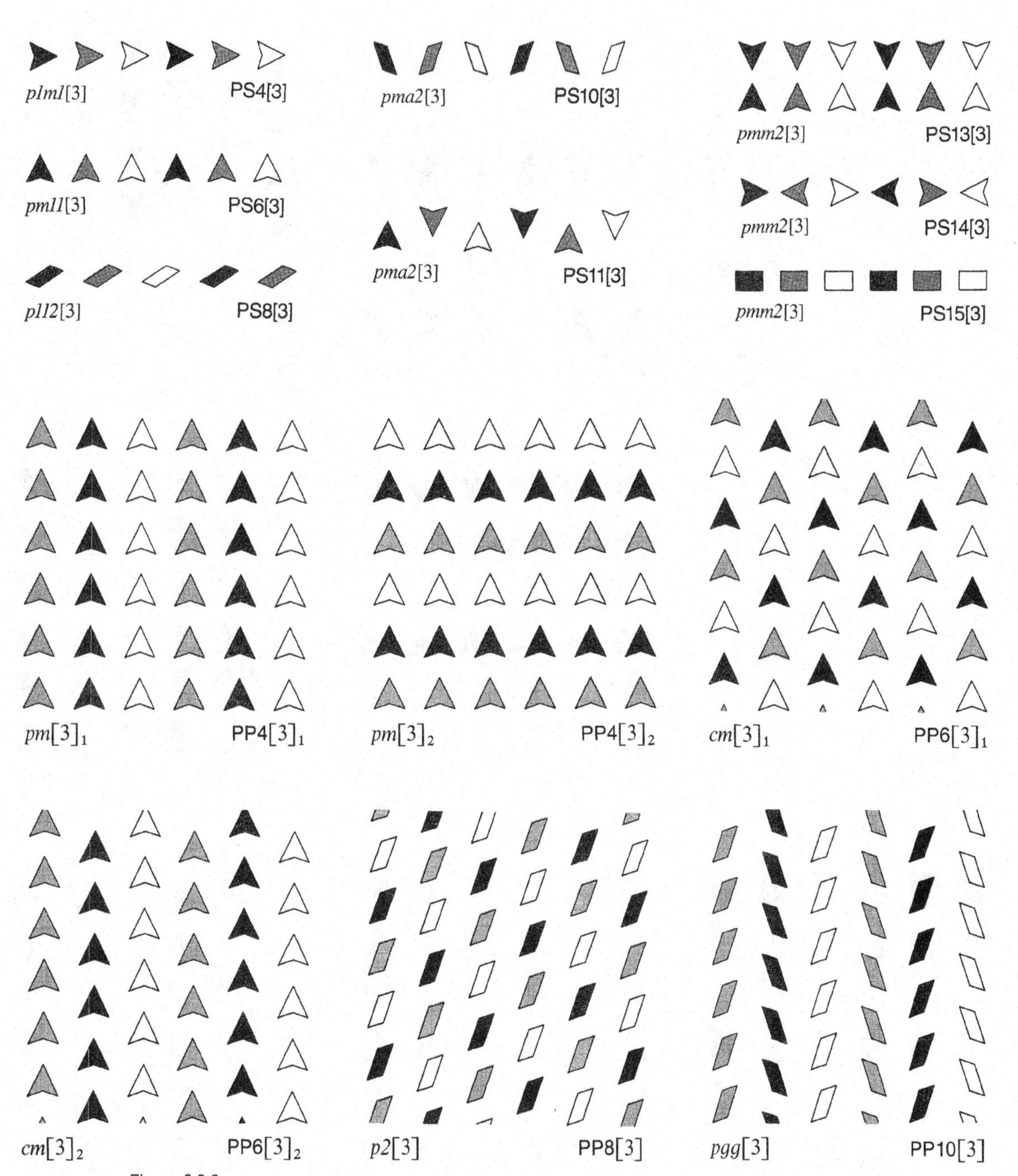

Figure 8.3.6
Examples of all the different 3-color pattern types (8 strip and 36 periodic) whose underlying patterns are non-primitive.

Figure 8.3.6 (continued)

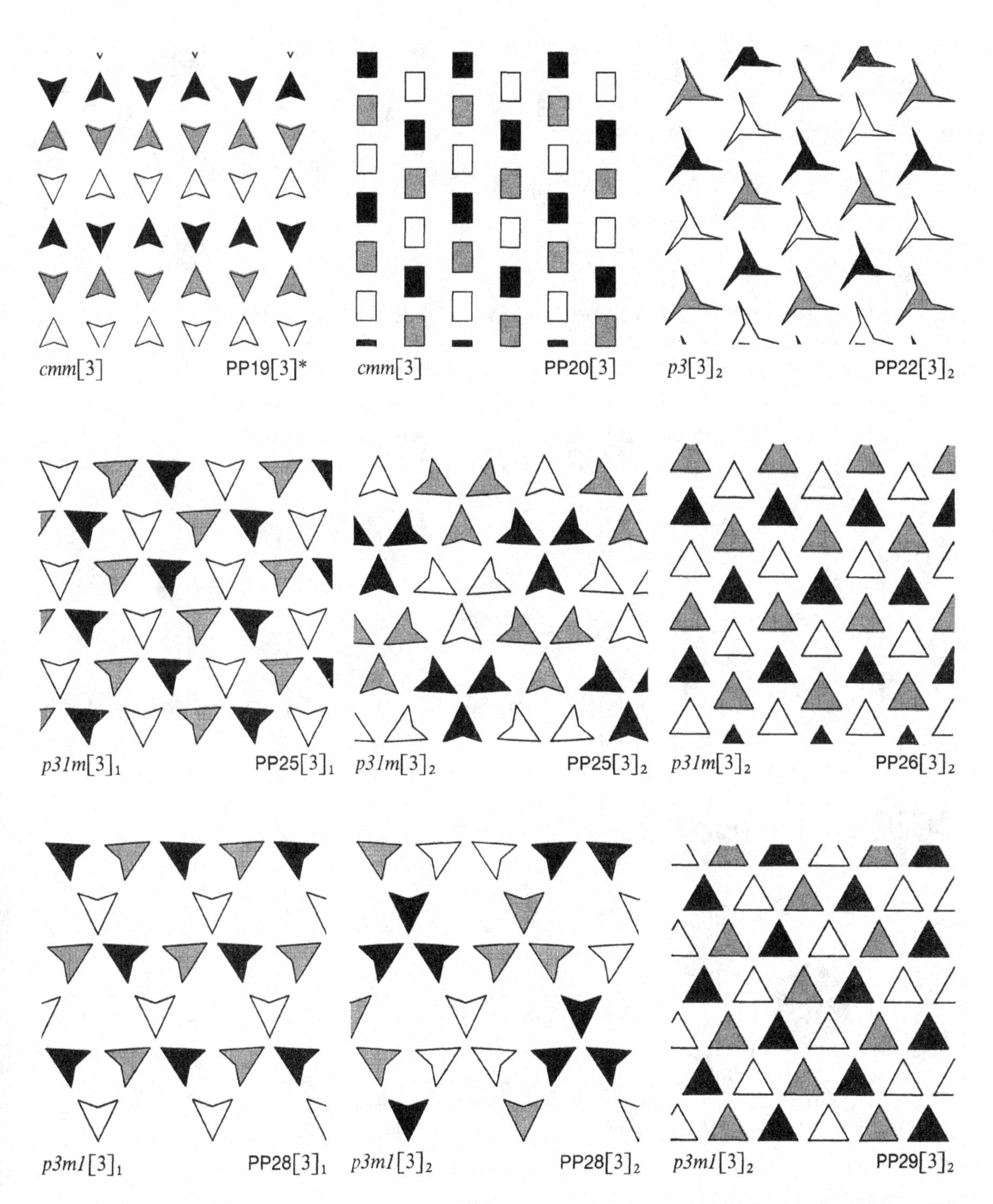
cmm[3]
PP19[3]*
cmm[3]
PP20[3]
p3[3]2
PP22[3]2
p31m[3]1
PP25[3]1
p31m[3]2
PP25[3]2
p31m[3]2
PP26[3]2
p3m1[3]1
PP28[3]1
p3m1[3]2
PP28[3]2
p3m1[3]2
PP29[3]2

Figure 8.3.6 (continued)

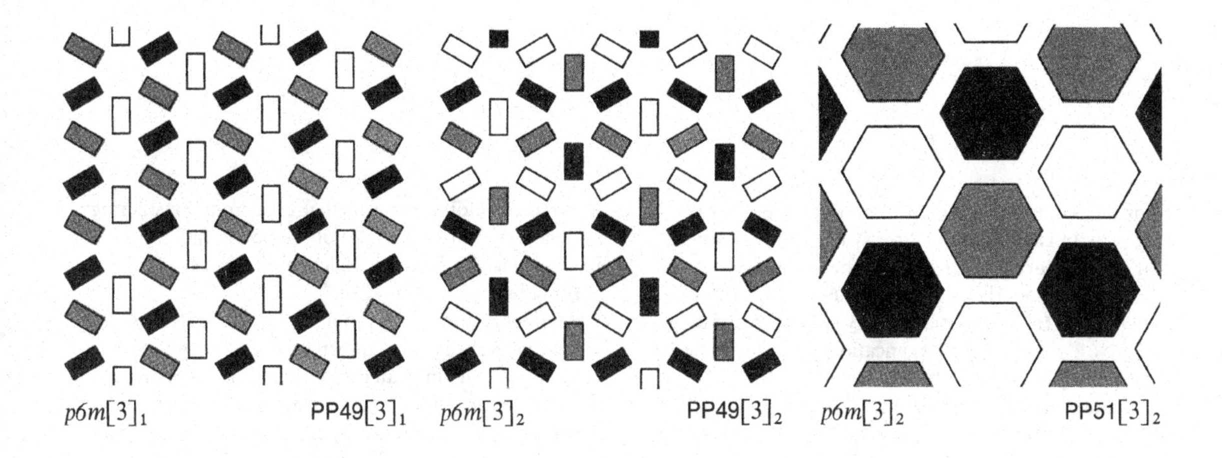

EXERCISES 8.3

1. (*a*) Determine the 3-color pattern types of those tiling patterns in Figure 5.2.13 which are 3-chromatic.
   (*b*) For the tilings of Figure 5.2.13 consider the 2-colored patterns formed by the black and stippled tiles. For those which are 2-chromatic determine the 2-color pattern types.
2. For each of the 2-chromatic patterns of types $PP38[2]_1$, $PP38[2]_4$, $PP38[2]_5$, $PP39[2]_3$ and $PP39[2]_5$ (see Figure 8.3.5) determine all the colored-motif transitive subgroups of their color symmetry groups.

*3. (*a*) Determine all the 4-color pattern types of strip patterns.
   (*b*) Determine all the *k*-color pattern types of strip patterns, where *k* is an odd integer, $k \geq 3$.

8.4 CHROMATIC COLORINGS OF GIVEN PATTERNS AND EQUICOLOR TYPES

The classification criteria we have considered in the previous sections have depended entirely on color groups—the patterns that appeared played a subordinate role. But there is both a practical and theoretical interest in taking the opposite point of view and asking the question "Given a pattern, in how many different ways can it be colored?". For such a question to be meaningful we must restrict the sort of colorings allowed and also say precisely what we mean by "different ways". The most appropriate restriction on colorings appears to be to consider only those which are *k*-chromatic, and to interpret "different ways" as meaning "of different equicolor types".

The definition of an *equicolor type* is similar to that of color pattern type given in Section 8.3 except that we ask for the color-preserving one-to-one correspondence Ψ to be compatible with *every symmetry* of $\mathscr{P}_1$ (and not just with the symmetries in the underlying group $S_C(\mathscr{P}_1^{\wedge})$) and for Ψ^{-1} to be compatible with every symmetry of $\mathscr{P}_2$. The classification by equicolor types is, of course, finer than that by color pattern types, as we can see from the examples in Figure 8.3.4. These are of different equicolor types even though they are of the same color pattern type. In fact, it follows directly from the definition that if two colored patterns are of the same equicolor type then the underlying patterns must be of the same pattern type. In the case of perfectly colored patterns, since $S_C(\mathscr{P}^{\wedge}) = S(\mathscr{P})$ the classifications by color pattern types and by equicolor types coincide.

To illustrate the above definition, consider the pattern $\mathscr{P}$ of type PP1 in Figure 8.4.1(*a*), and the 2-chromatic colorings of $\mathscr{P}$ shown in Figures 8.4.1(*b*), (*c*) and (*d*). These are equicolored as can be seen by taking for Ψ the color-preserving one-to-one correspondence indicated by the labels A, B, C on the copies of the motif. (In other words Ψ maps each copy of the motif onto one bearing the same label; it should be clear how to extend Ψ to the unlabelled copies of the motif.) The color symmetry groups of all these patterns are isomorphic; this isomorphism is defined by specifying that each vector labelled ◦ (which represents the identity so that the corresponding translation preserves colors) is mapped onto the vector with the same label, and similarly for the vectors labelled * (for which the corresponding translations interchange the colors). These pairs of vectors generate, and therefore completely define, the color symmetry groups of the patterns. Notice that differences between these three colored patterns that arise from considering the acute and obtuse angles of the polygonal motifs are irrelevant as far as the classification into equicolor types is concerned.

Another example is given in Figure 8.4.2. Here we show five 2-chromatic colorings of a rectangle pattern of type PP20. As indicated in the caption, these are of four equicolor types. Here the classification by equicolor types coincides with that by color pattern type in spite of the fact that some of the patterns are not perfectly colored.

In order to determine the different equicolor types possible in a k-chromatic coloring of a given pattern $\mathscr{P}$ we can use the information on motif-transitive subgroups of $S(\mathscr{P})$ given in Section 5.2 together with (for $k = 2$ or $k = 3$) the results on k-color pattern types given in Sections 8.2 and 8.3. To carry out a similar determination for $k \geq 4$ it would be necessary to list the 4-color pattern types first. We illustrate the procedure by a typical example.

Let $k = 2$ and let the pattern $\mathscr{P}$ be of type PP32, see Figure 8.4.3(*a*). From Table 5.2.3 we see that $S(\mathscr{P})$ has six different motif-transitive proper subgroups. Five of these are indicated by their symmetry elements in Figures 8.4.3(*b*) (*c*) (*d*) (*e*) and (*f*). With respect to these

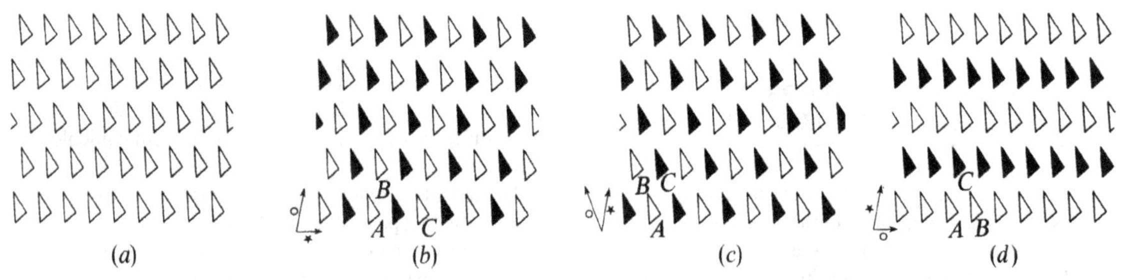

Figure 8.4.1
The pattern $\mathscr{P}$ in (*a*) is of pattern type PP1, and the three 2-chromatic colorings of $\mathscr{P}$ shown in (*b*), (*c*) and (*d*) are of 2-color pattern type PP1[2]. These three are distinguishable by the properties of their motifs such as the positions of their acute and obtuse angles relative to the translational color symmetries. The color symmetry group in each case is generated by a color-preserving translation, marked (◦), and a color-reversing translation marked ★. Consequently all three 2-color patterns are of the same equicolor type.

groups, $\mathscr{P}$ may be considered to be a pattern of types PP31, PP30, PP8, PP7 and PP7, respectively. The sixth motif-transitive proper subgroup of $S(\mathscr{P})$, namely *p1*, does not lead to any additional types since the motif of $\mathscr{P}$ has symmetry group *c4* which is a forbidden supergroup for PP1; hence it will not be considered in the remainder of the discussion.

The next step is to consider all the perfect 2-colorings of each of the pattern types PP32, PP31, PP30, PP8 and PP7 and to find all the ways in which each can be applied to the patterns of Figure 8.4.3(*a*) to (*f*). From Figure 8.4.3(*a*) we are thus led to the unique 2-coloring shown in Figure 8.4.3(*g*) which is of 2-color pattern type $\mathrm{PP32}[2]_1$. The pattern in Figure 8.4.3(*b*) can be colored as in 2-color pattern type $\mathrm{PP31}[2]_2$, but the resultant colored pattern has additional symmetries and coincides with

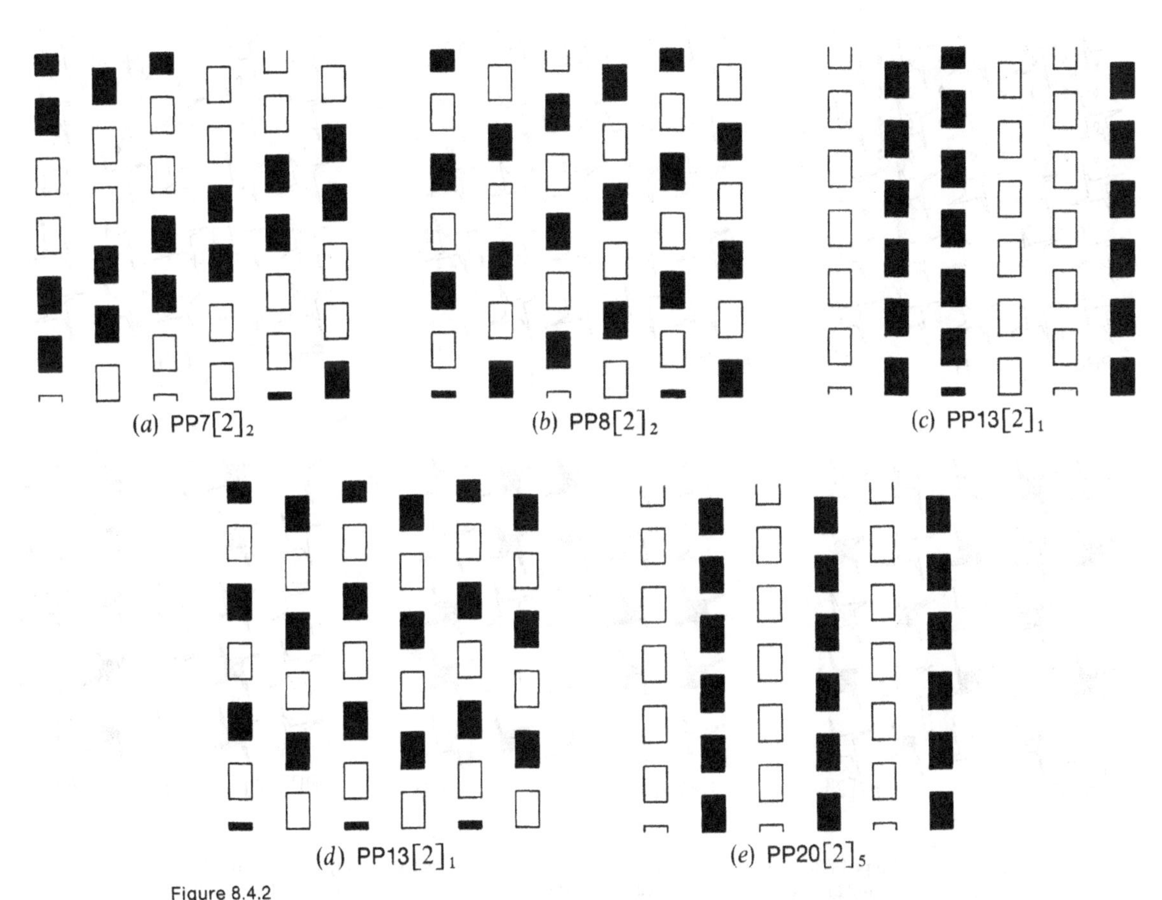

Figure 8.4.2
Five 2-chromatic colorings of a pattern of type PP20. The 2-color pattern type is indicated near each. All are of different equicolor types except for (*c*) and (*d*) which are equicolored.

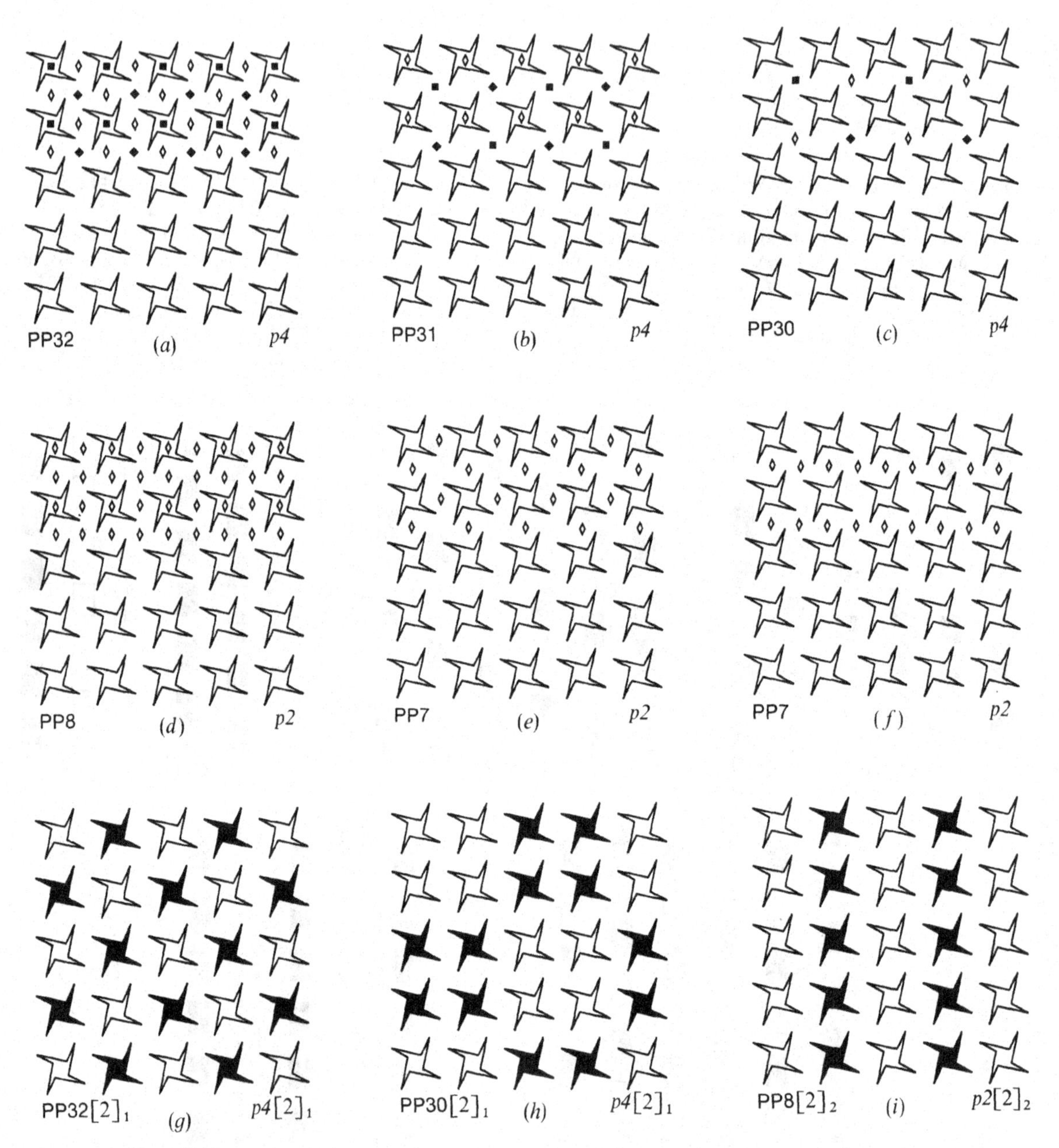

Figure 8.4.3
A pattern $\mathscr{P}$ of type PP32 is shown in (*a*) and the motif-transitive proper subgroups of $S(\mathscr{P})$ are shown in (*b*), (*c*), (*d*), (*e*) and (*f*). Near each we indicate both the subgroup and the pattern type that results when we restrict the symmetry group of the pattern to the subgroup under consideration. The last six parts of the figure (parts (*g*) to (*l*)) show the equicolor types of 2-colorings of $\mathscr{P}$. Near each we have indicated its 2-color pattern type and its 2-color group.

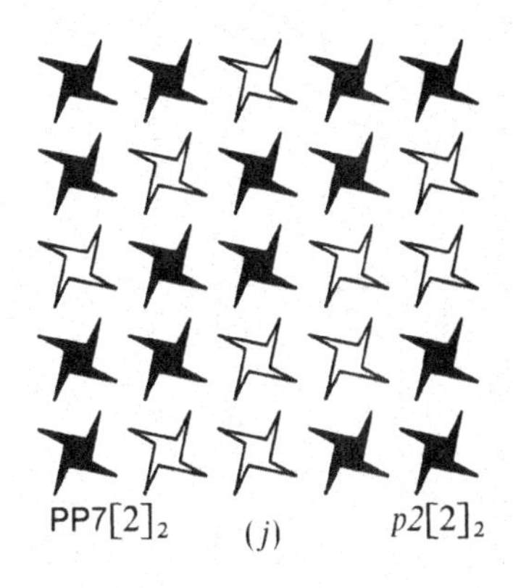

$PP7[2]_2$ (*j*) $p2[2]_2$

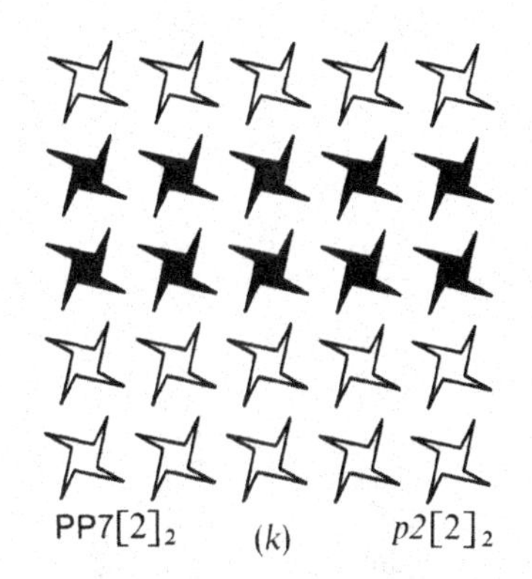

$PP7[2]_2$ (*k*) $p2[2]_2$

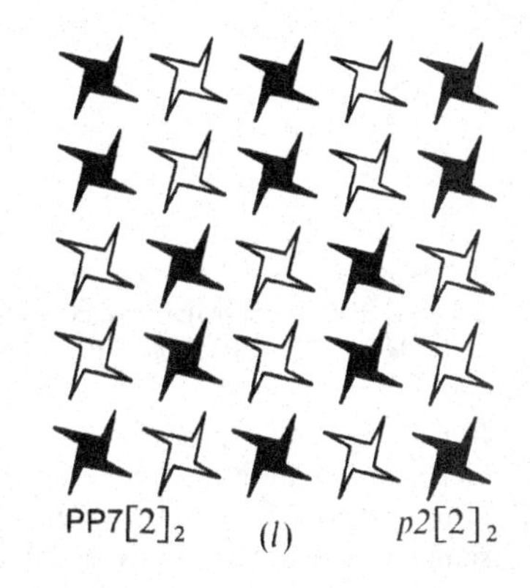

$PP7[2]_2$ (*l*) $p2[2]_2$

that of Figure 8.4.3(*g*). The pattern in Figure 8.4.3(*c*) can be colored either as $PP30[2]_1$ or $PP30[2]_2$; the first leads to the colored pattern of Figure 8.4.3(*h*), while the second again yields that of Figure 8.4.3(*g*). For the pattern in Figure 8.4.3(*d*) we can apply only the coloring of type $PP8[2]_2$—but this can be done in two ways that are inequivalent; either the centers of 2-fold rotation at A and B are color-reversing, or else those at A and C are. In the former case we obtain the pattern of Figure 8.4.3(*i*) while in the latter we again obtain that of Figure 8.4.3(*g*). For the patterns in Figures 8.4.3(*e*) and (*f*) colorings of types $PP7[2]_1$ and $PP7[2]_2$ are applicable. For Figure 8.4.3(*e*) the coloring $PP7[2]_1$ again leads to the pattern shown in Figure 8.4.3(*g*), while coloring $PP7[2]_2$ can be applied in two distinct ways. One leads to the pattern of Figure 8.4.3(*j*) and the other leads to the pattern $PP8[2]_2$ of Figure 8.4.3(*i*). Coloring the pattern of Figure 8.4.3 (*f*) by type $PP7[2]_1$ leads to the pattern $PP8[2]_2$ of Figure 8.4.3(*i*), while of the four distinct colorings by type $PP7[2]_2$ two lead again to patterns $PP32[2]_1$ and $PP8[2]_2$ of Figures 8.4.3(*g*) and (*i*) and the last two yield two new patterns shown in Figures 8.4.3(*k*) and (*l*).

Thus we obtain in all six different equicolor types of 2-chromatic colorings of patterns of type PP32. Contrast this with the fact that there is only one 2-color pattern type with underlying pattern PP32, namely $PP32[2]_1$.

Repeating a similar process for each pattern type one could obtain a complete list of equicolor types of k-chromatic colorings of discrete strip and periodic patterns. So far as we are aware, this enumeration of equicolor types, while completely straightforward, has never been carried out even for $k = 2$.

EXERCISES 8.4

1. For the 2-chromatic patterns considered in Exercise 8.3.1(*b*) determine which pairs are of the same equicolor type.
2. Determine all equicolor types of 2-chromatic and 3-chromatic colorings of dot patterns of types DPP47 and DPP48.

*3. Determine all equicolor types of 2-chromatic and 3-chromatic colorings of strip patterns.

*4. Find all equicolor types of 4-chromatic colorings of the coherent strip patterns that have an open line-segment as motif.

*5. Find all equicolor types of 4-chromatic colorings of the regular square tiling.

8.5 COLOR-HOMEOMERISM AND HOMEOCHROMATIC TYPES

In Chapter 7 we explained how the classification of (uncolored) patterns by "pattern types" can be refined using homeomerism. In the case of colored patterns similar considerations apply. In this section we shall introduce two new methods of classification—into color-homeomeric types and into homeochromatic types. Each of these is finer than the methods described in the previous sections of this chapter since it takes into account the topological structure of the patterns. Unlike the methods discussed earlier, the classifications to be described here yield an infinite number of types. For this reason they are too fine for many purposes. However, they seem to be most appropriate (in that they fit in with our intuitive ideas of when two patterns are "the same" or not) in two situations: when investigating colored patterns whose motifs are simple geometrical shapes (such as circular or elliptical disks or line segments) and in the study of colored tilings. The first of these topics will be referred to in the exercises—the number of types is large but finite and, so far as we are aware, has in no case been determined. The second topic, that of colored tilings, will be considered in some detail in the next section.

Let $\mathscr{P}_1^{\wedge}$ and $\mathscr{P}_2^{\wedge}$ be two colored patterns. If, possibly after permuting the colors in one of the patterns, we can find a color-preserving homeomorphism $h: E_1 \to E_2$ which maps $\mathscr{P}_1^{\wedge}$ into $\mathscr{P}_2^{\wedge}$, is compatible with every symmetry in $S_C(\mathscr{P}_1^{\wedge})$ and is such that h^{-1} is compatible with every symmetry in $S_C(\mathscr{P}_2^{\wedge})$, then $\mathscr{P}_1^{\wedge}$ and $\mathscr{P}_2^{\wedge}$ are said to be *color-homeomeric*.

The definition of the term *homeochromatic* is analogous: we specify that h must be compatible with every symmetry of $S(\mathscr{P}_1)$ and h^{-1} compatible with every symmetry of $S(\mathscr{P}_2)$ (that is, with the symmetry groups of the underlying patterns $\mathscr{P}_1$ and $\mathscr{P}_2$). Intuitively* we may think of two patterns as being of the same homeochromatic type if h, besides mapping $\mathscr{P}_1$ onto $\mathscr{P}_2$ also maps a group diagram for $S(\mathscr{P}_1)$ onto a group diagram for $S(\mathscr{P}_2)$ (and vice versa for h^{-1}). They are of the same color-homeomeric type if the analogous condition holds for the color-group diagrams of $C(\mathscr{P}_1^{\wedge})$ and $C(\mathscr{P}_2^{\wedge})$.

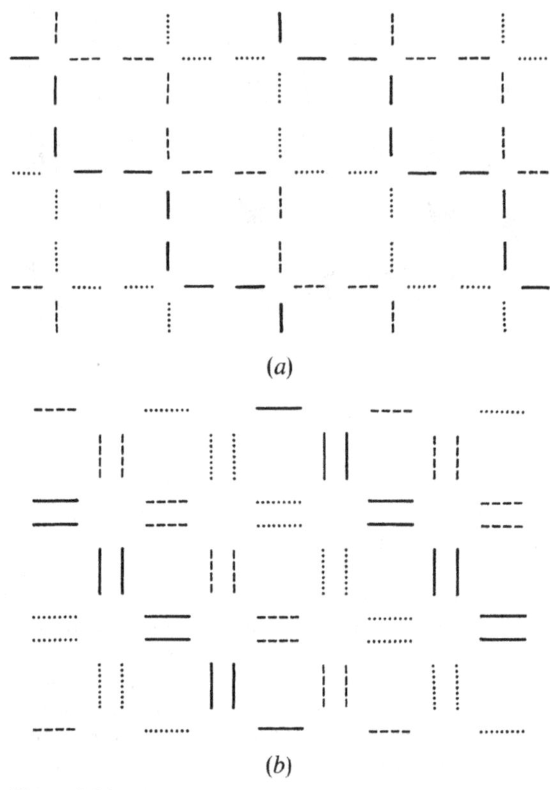

Figure 8.5.1
Two 3-chromatic colorings of line-segment patterns of type PP38. The colored patterns are of the same 3-color pattern type PP17[3] and are even color homeomeric. However the underlying patterns are not homeomeric.

These definitions have two immediate consequences. The first is that the classifications by color homeomerism and by homeochromatic type coincide when perfectly colored patterns are under consideration for then $S_C(\mathscr{P}^{\wedge}) = S(\mathscr{P})$. The second is that when two patterns are homeochromatic then the existence of a color-preserving homeomorphism, as required in the definition, implies that the underlying patterns are homeomeric. On the other hand, the corresponding statement is not true for color-homeomeric patterns. In Figure 8.5.1 we show two patterns which are of the same color-pattern type PP17[3], are equicolored and are even color-

* This interpretation is subject to the same qualifications as those expressed on page 336.

homeomeric, yet the underlying patterns are not homeomeric, being of types LPP38-1 and LPP38-2 in Table 7.4.1 (see Figures 7.4.1(*a*) and (*b*)). This kind of situation can only arise when at least one of the patterns is not perfectly colored; here $S_C(\hat{\mathscr{P}_1})$ and $S_C(\hat{\mathscr{P}_2})$ are of type *cmm* while $S(\mathscr{P}_1)$ and $S(\mathscr{P}_2)$ are of type *p4m*. Another example where color-homeomeric patterns have underlying patterns of different homeomeric types will be given in Figure 8.6.1 in the context of tiling patterns.

The above definitions are further illustrated in Figure 8.5.2. Fourteen patterns are shown, each with the interior of a quadrangle as motif, and each is perfectly colored so the classifications by color-homeomeric types and by homeochromatic types coincide. By considering the homeomeric types of the underlying (uncolored) patterns we can divide the fourteen patterns into five sets as follows:

The pattern of Figure 8.5.2(*a*).
The patterns of Figures 8.5.2(*b*), (*c*) and (*d*).
The patterns of Figures 8.5.2(*e*), (*f*), (*g*) and (*h*).
The patterns of Figures 8.5.2(*i*), (*j*) and (*k*).
The patterns of Figures 8.5.2(*l*), (*m*) and (*n*).

Clearly no pattern in one of these sets is homeochromatic to any pattern in another set.

The pattern in the first set (Figure 8.5.2(*a*)) is of color-pattern type $PP6[3]_2$ while the thirteen patterns in the other four sets are all of type $PP2[3]_2$.

In the second set it is easy to see that the patterns of Figures 8.5.2(*b*) and (*c*) are homeochromatic since a slight distortion of the motif suffices to convert one into the other and this indicates how a color-preserving homeomorphism, as required in the definition of homeochromatic type, can be constructed. It is not quite so easy to see that the patterns of Figures 8.5.2(*c*) and (*d*) are homeochromatic—here the necessary color-preserving homeomorphism h_0 can be described as one that translates the "rows" of copies of the motif alternately to the right and to the left through a distance equal to the bar at the lower left corner of the diagram. In between the rows, h_0 can be completed linearly.

In the third set, the patterns of Figures 8.5.2(*e*), (*f*) and (*h*) are of the same homeochromatic type, while that of (*g*) is of a different type. The latter assertion follows because in (*g*) each copy of the motif is contiguous to two of the same color, while in the other three patterns this does not occur. The precise definitions of the color-preserving homeomorphisms needed to prove that the patterns Figures 8.5.2(*e*), (*f*) and (*h*) are homeochromatic are slightly complicated. They can be described as similar to the homeomorphism h_0 mentioned above, shifting the "rows" of copies of the motif alternately to the right and to the left. Notice however that the homeomorphisms do not preserve the shapes of the motifs—the mappings required are indicated by the similar labelling of corresponding points of the motifs as shown to the right of the three diagrams.

The patterns of the fourth set (Figures 8.5.2(*i*), (*j*) and (*k*)) can be thought of as arising from the patterns of Figures 8.5.2(*f*), (*g*) and (*h*) by moving the "columns" nearer together until the copies of the motifs touch at their corners. This modification affects the classification in that the first two patterns (*i*) and (*j*) are not homeochromatic, while the third (*k*) is homeochromatic to (*j*) (and *not* to (*i*) as might be expected from the above discussion of the corresponding patterns of the third set). In fact it is quite easy to describe a suitable color-preserving homeomorphism that enables one to show that patterns (*k*) and (*j*) are homeochromatic.

In the fifth and final set, we notice that in each pattern (Figures 8.2.5(*l*), (*m*) and (*n*)) some copies of the motif touch others at single points while others are contiguous along line-segments. This rather more complicated topological structure is sufficient to ensure that all three patterns are of different homeochromatic types.

In the above examples, the topology of each pattern played a crucial role—different homeochromatic types were distinguished by the contiguity of copies of the motif with others of the same or different colors, and whether these contiguities were at single points or line-segments. Our final example (Figure 8.5.3) shows two pairs of patterns; the patterns of each pair are of the same equicolor type yet are not color-homeomeric in spite of the fact that the underlying patterns are of the same topological

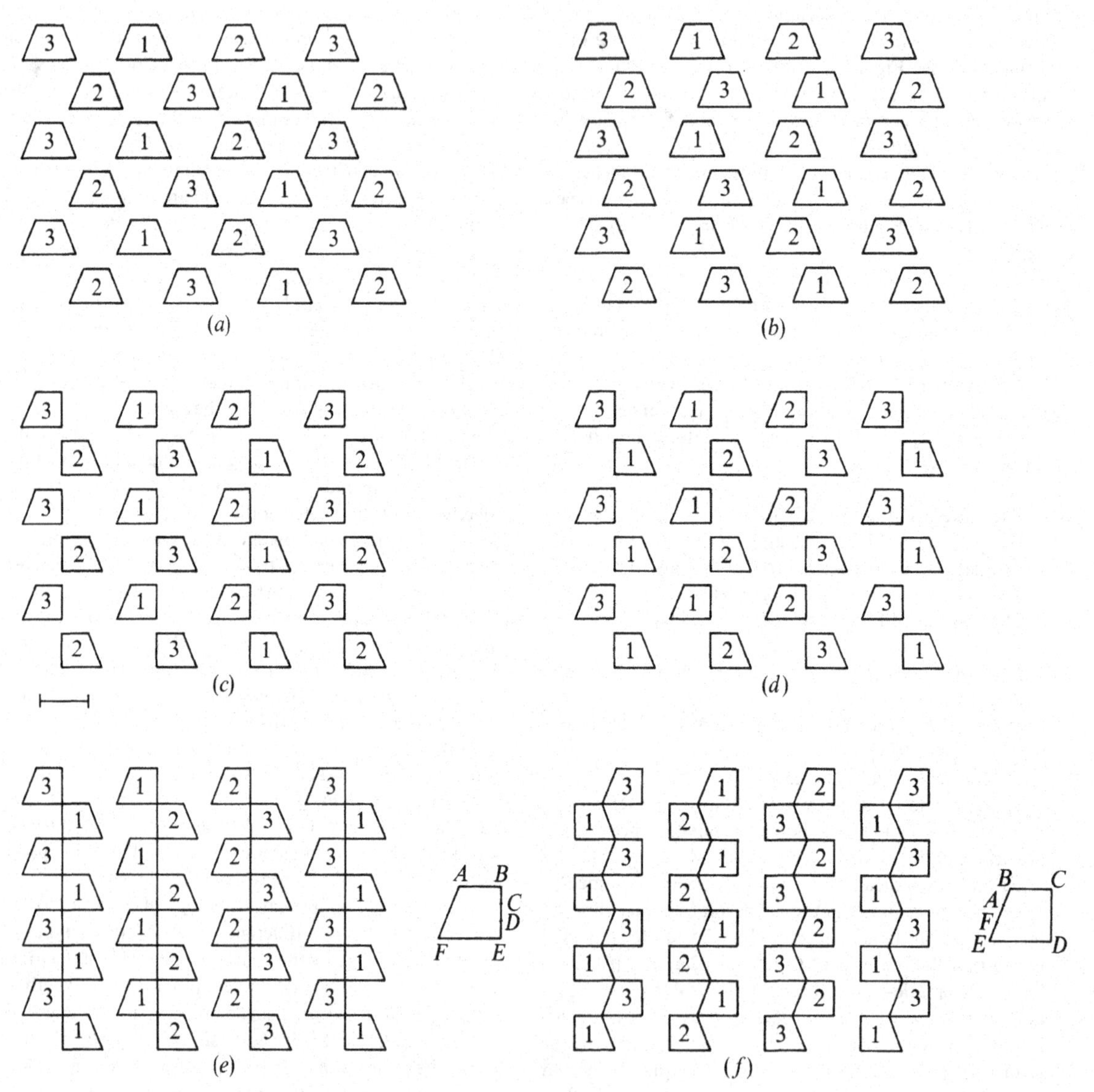

Figure 8.5.2
Fourteen perfectly colored patterns. The classification of these patterns into homeochromatic types is discussed in the text.

(g)
(h)
A B
C
D
F E
(i)
(j)
(k)
(l)
(m)
(n)

type and in Figure 8.5.3(*a*) are even homeomeric. We can explain this phenomenon by saying that in the first pattern of Figure 8.5.3(*a*) similarly colored copies of the motif are separated by a single line of reflection (of $S_C(\mathscr{P}^\wedge)$) while in the second they are not. In the remaining patterns (Figure 8.5.3(*b*)) the colored line-segments bear different relations to the lines of reflection of $S_C(\mathscr{P}^\wedge)$.

In Chapter 7 we discussed three methods of classifying (uncolored) patterns—by symmetry group, by pattern type, and by homeomeric type. In this chapter we have introduced five methods of classifying colored patterns—by color symmetry group, by color-pattern type, by equicolor type, by color-homeomeric type and by homeochromatic type. We conclude by summarizing the relationship between all these methods (see Figure 8.5.4).

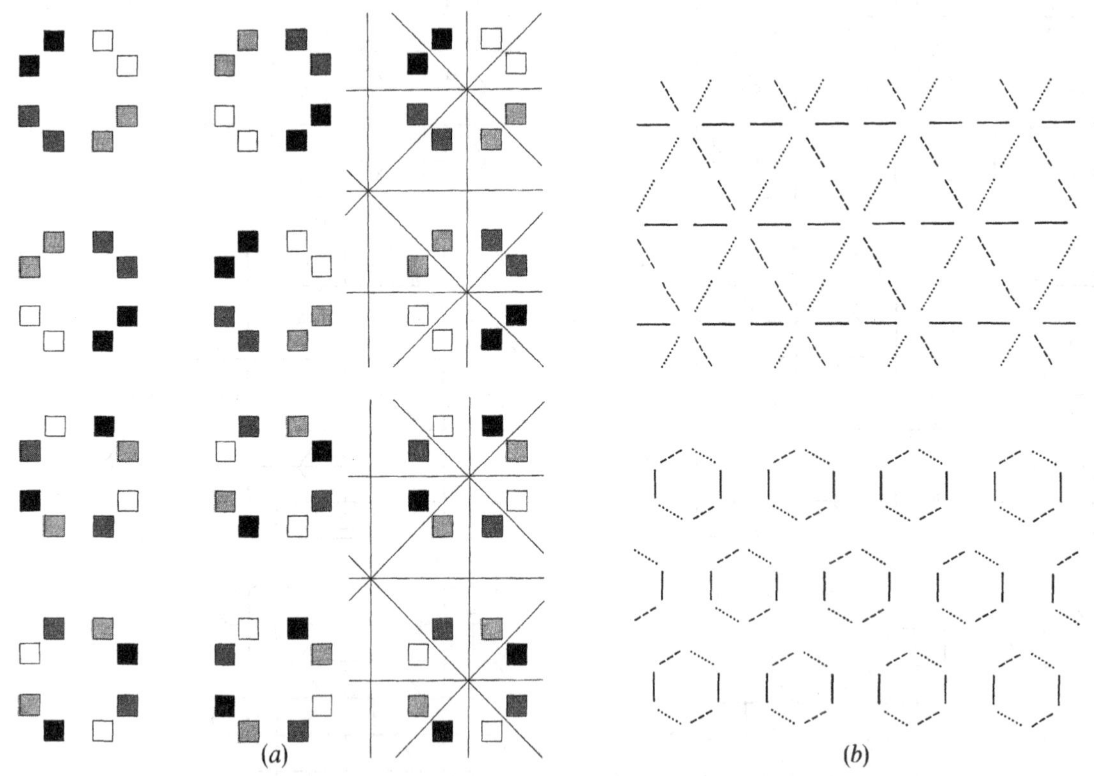

Figure 8.5.3
(*a*) Two perfect 4-colorings of a discrete pattern of type PP37; they have the same 4-color pattern type and no copy of the motif is contiguous to another, but they are not homeochromatic. Notice that corresponding lines on the right of each diagram correspond to reflective color symmetries of the same kind. (*b*) Two 3-chromatic segment patterns (with closed line-segments as motifs) of color pattern type PP47[3]$_1$. They are equicolored but not homeochromatic.

With the arrows representing implications (and so pointing from a finer classification to a coarser one) the diagram is self-explanatory. The examples given here and in the previous sections enable us to establish the following result.

8.5.1 *The relationships between the eight methods of classification are completely described by Figure 8.5.4. There are no relationships other than those indicated on the diagram.*

The finest of the classifications, that by homeochromatic type, is the most appropriate for our discussion of colored tilings which will be given in the next section.

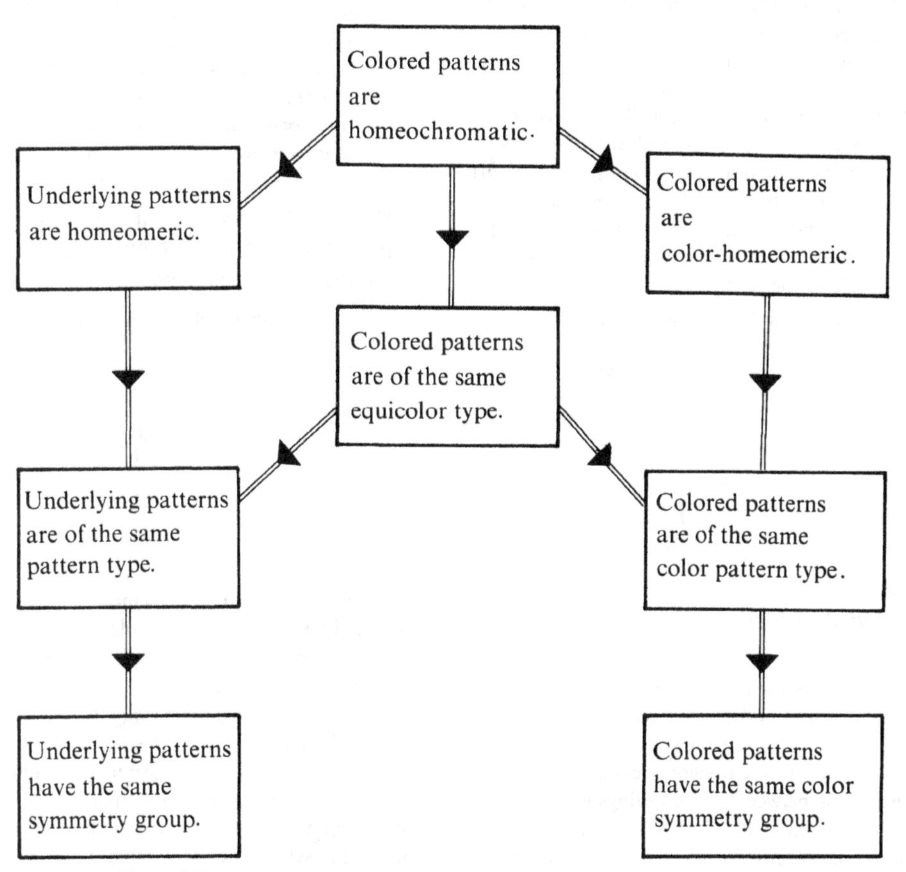

Figure 8.5.4
The relationships between the eight methods of classifying patterns and colored patterns described in this chapter and Chapter 7.

EXERCISES 8.5

1. (*a*) Show that dot patterns of type DPP41 admit 2-chromatic colorings of six different homeochromatic types.
   (*b*) Investigate which of the homeochromatic types in part (*a*) have the same (*i*) equicolor type, (*ii*) color-homeomeric type, (*iii*) color pattern type, or (*iv*) color symmetry group.
   (*c*) Consider the homeochromatic types of 2-chromatic colorings of the closed circular disk patterns of type CPP41-1 which correspond to the homeochromatic types of dot patterns considered in part (*a*). Determine which of them have the same (*i*) homeochromatic type, (ii) color-homeomeric type, or (*iii*) color pattern type, as a suitably chosen colored ellipse pattern.

*2. (*a*) Determine all the homeochromatic types of 2-chromatic colorings of each of the coherent strip patterns with an open line-segment as motif.
   (*b*) Determine which of the homeochromatic types found in part (*a*) have the same (*i*) equicolor type, (*ii*) color-homeomeric type, (*iii*) color pattern type, or (*iv*) color symmetry group.
   (*c*) Decide which of the homeochromatic types found in part (*a*) have the same (*i*) homeochromatic type, (*ii*) color-homeomeric type, or (*iii*) color pattern type, as a suitably chosen colored pattern in which the motif is an open semi-circle.

3. For each of the six coherent open circular disk strip patterns find all homeochromatic types of
   *(*a*) 3-chromatic colorings, and
   **(*b*) k-chromatic colorings for all $k \geq 4$.

8.6 COLORED TILINGS

We turn now to a detailed investigation of k-chromatic tilings. Since we wish to take into account the topology of the tilings, it is clearly appropriate to use one of the classifications described in the previous section. Also, as we wish to distinguish between colored tilings whose underlying tilings are of different isohedral types, the homeochromatic classification is the more appropriate (see Figure 8.6.1).

The procedure we shall follow is a variant of that used in Chapter 6—the incidence symbol for an isohedral tiling will be adapted so as to include information about the way in which the tiling is colored. In order to simplify the exposition we need to introduce the concept of a primitively-colored tiling.

Let $\mathscr{T}^\wedge$ be a k-chromatic tiling. In Section 8.1 we considered three groups of symmetries associated with $\mathscr{T}^\wedge$, namely $S(\mathscr{T})$ the symmetry group of the underlying tiling $\mathscr{T}$, $S_C(\mathscr{T}^\wedge)$ the group of symmetries of $\mathscr{T}$ associated with color symmetries of $\mathscr{T}^\wedge$, and $S(\mathscr{T}^\wedge)$ the group of color-preserving symmetries of $\mathscr{T}^\wedge$ (that is to say, the group of those color symmetries which map each tile onto one of the same color). From these we obtain three induced groups, as follows:

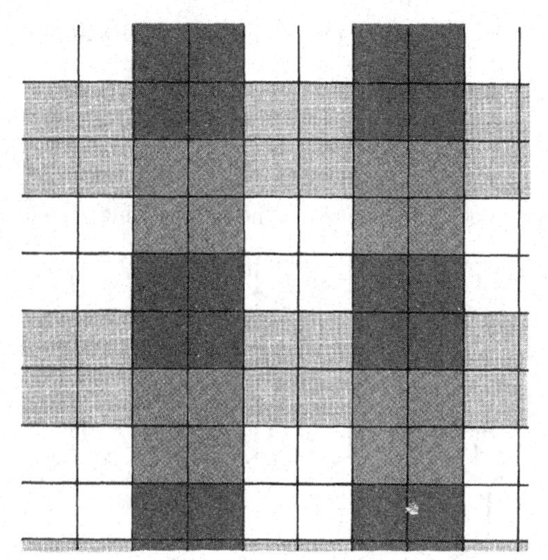

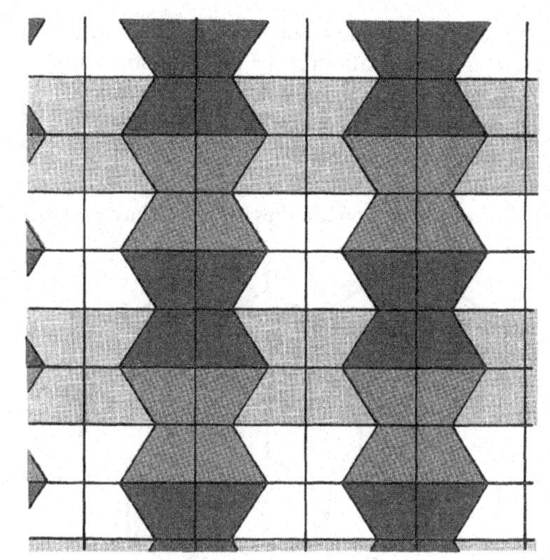

Figure 8.6.1
Two tilings which are color homeomeric but are not homeochromatic. Since the underlying tilings are not homeomeric it seems preferable to distinguish between tilings such as these. Hence the classification by homeochromatic type is the more appropriate.

(*a*) $S(\mathcal{T} \mid T)$ is the *induced tile group* of $\mathcal{T}$ consisting of those elements of $S(\mathcal{T})$ that leave a given tile T fixed;

(*b*) $S_C(\mathcal{T}^{\wedge} \mid T^{\wedge})$ is the *induced colored tile group* of $\mathcal{T}^{\wedge}$ consisting of those elements of $S_C(\mathcal{T}^{\wedge})$ that leave a given colored tile $T^{\wedge}$ fixed; and

(*c*) $S(\mathcal{T}^{\wedge} \mid T^{\wedge})$ is the *induced tile group of the colored tiling* $\mathcal{T}^{\wedge}$ consisting of those elements of $S(\mathcal{T}^{\wedge})$ that leave a given colored tile $T^{\wedge}$ fixed. Thus isometries in $S(\mathcal{T}^{\wedge} \mid T^{\wedge})$ not only map $T^{\wedge}$ onto itself, but also map every tile of $\mathcal{T}^{\wedge}$ onto one of the same color.

Each of these groups is a subgroup of the preceding one. It is possible for two of the groups, or for all three of them, to be equal (see Figures 8.6.2 and 8.6.3), but in general the three will be distinct (see Figure 8.6.4). A tiling is said to be *primitively colored* if $S_C(\mathcal{T}^{\wedge} \mid T^{\wedge}) = S(\mathcal{T}^{\wedge} \mid T^{\wedge})$, that is to say, if every color symmetry of $\mathcal{T}^{\wedge}$ that maps $T^{\wedge}$ onto itself is associated with the identity permutation of colors. An example of a primitively-colored 3-chromatic tiling is given in Figure 8.6.3(*a*). The following basic result relates the k-chromatic tilings to those which are primitively colored.

8.6.1 *Every k-chromatic tiling can be converted into a primitively colored k-chromatic tiling by adding a suitable uncolored mark to each tile.*

We begin by adding a mark M to a chosen tile T of the underlying tiling $\mathcal{T}$ of the k-chromatic tiling $\mathcal{T}^{\wedge}$ in such a way that the symmetry group of the resulting marked tile T_0 coincides with $S(\mathcal{T}^\wedge | T^\wedge)$. The tiling $\mathcal{T}$ determines a tiling pattern $\mathcal{P}_{\mathcal{H}}(\mathcal{T})$ (see Section 5.4) and the underlying group $S_C(\mathcal{T}^\wedge)$ acts transitively on the copies of the motif of $\mathcal{P}_{\mathcal{H}}(\mathcal{T})$, that is, on the interiors of the tiles of $\mathcal{T}$. Applying Statement 5.2.3 with $\mathcal{P} = \mathcal{P}_{\mathcal{H}}(\mathcal{T})$, $G = S_C(\mathcal{T}^\wedge)$, $F = S(\mathcal{T}^\wedge | T^\wedge)$ we find a subgroup H of G that acts transitively on the tiles of $\mathcal{T}$ and has F as its stabilizer. Use H to transfer copies of the mark M on T_0 onto all the other tiles of $\mathcal{T}$, and so obtain a marked tiling $\mathcal{T}_0$. If each tile of $\mathcal{T}_0$ is colored in the same way as the corresponding tile of $\mathcal{T}^\wedge$, we obtain a colored marked tiling $\mathcal{T}_0^\wedge$ which is primitively colored. Thus Statement 8.6.1 is established. It will be observed that the symmetry group $S(\mathcal{T}_0)$ of the marked tiling $\mathcal{T}_0$ is a

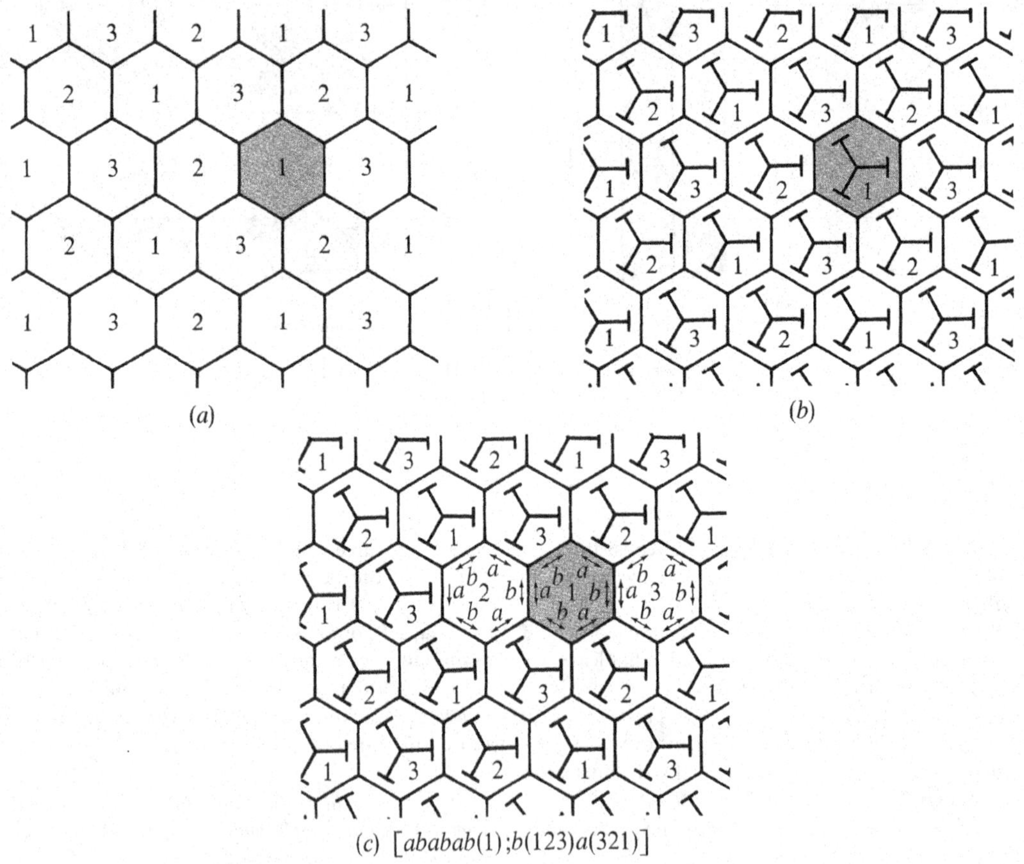

Figure 8.6.2
(*a*) An example of a 3-chromatic tiling for which $S(\mathcal{T}|T) = S_c(\mathcal{T}^\wedge|T^\wedge) = d6$ but $S(\mathcal{T}^\wedge|T^\wedge)$ is not equal to this group. (*b*) A marking of $\mathcal{T}$ that reduces $\mathcal{T}^\wedge$ to a primitively colored tiling. (*c*) The derivation of the corresponding color-incidence symbol.

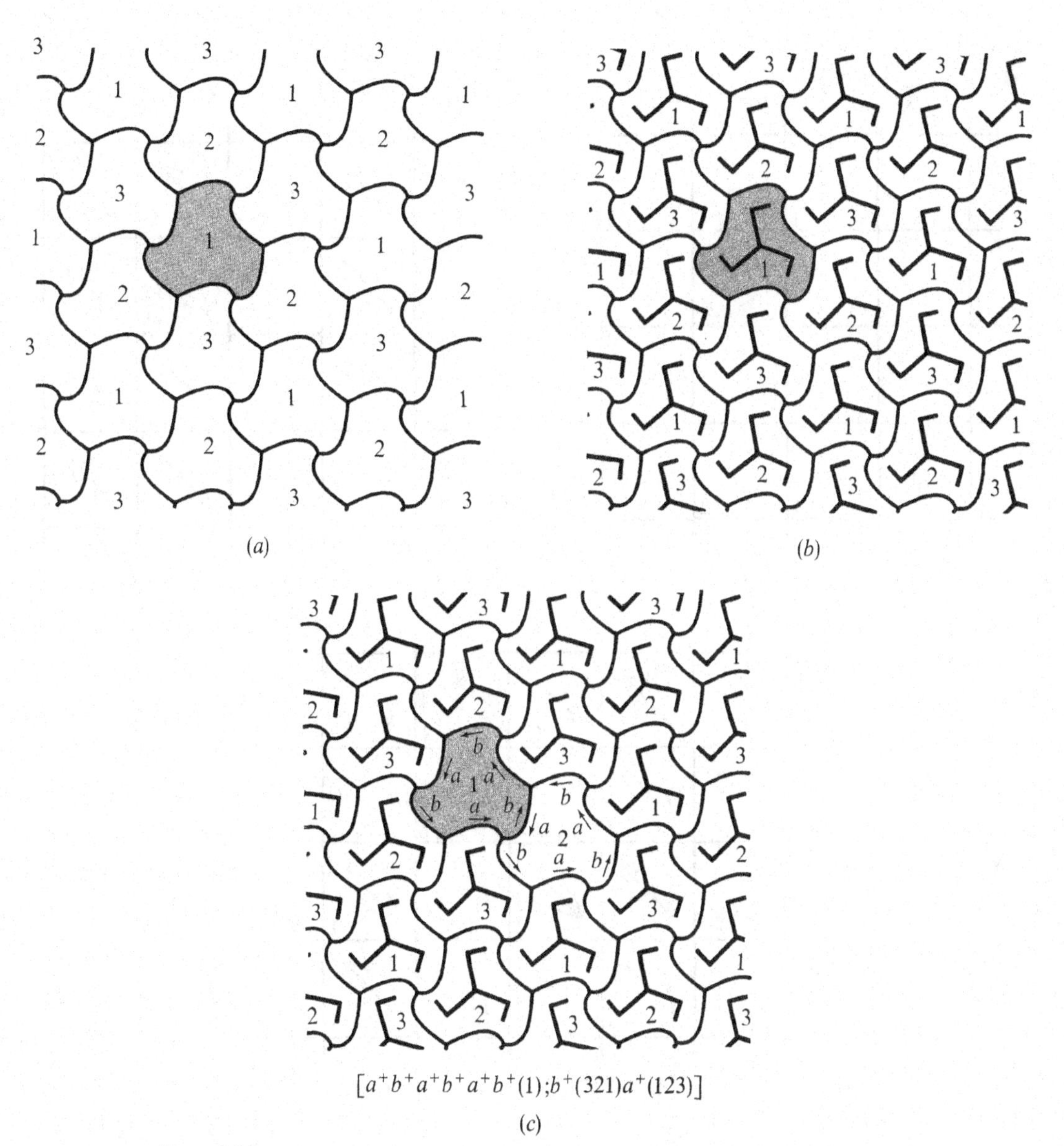

Figure 8.6.3
(*a*) An example of a 3-chromatic tiling for which $S(\mathscr{T}|T) = S_c(\mathscr{T}^\wedge|T^\wedge) = S(\mathscr{T}^\wedge|T^\wedge) = c3$.
(*b*) A marking of $\mathscr{T}$ that arises by the procedure described in the proof of Statement 7.6.1.
(*c*) The derivation of the corresponding color-incidence symbol.

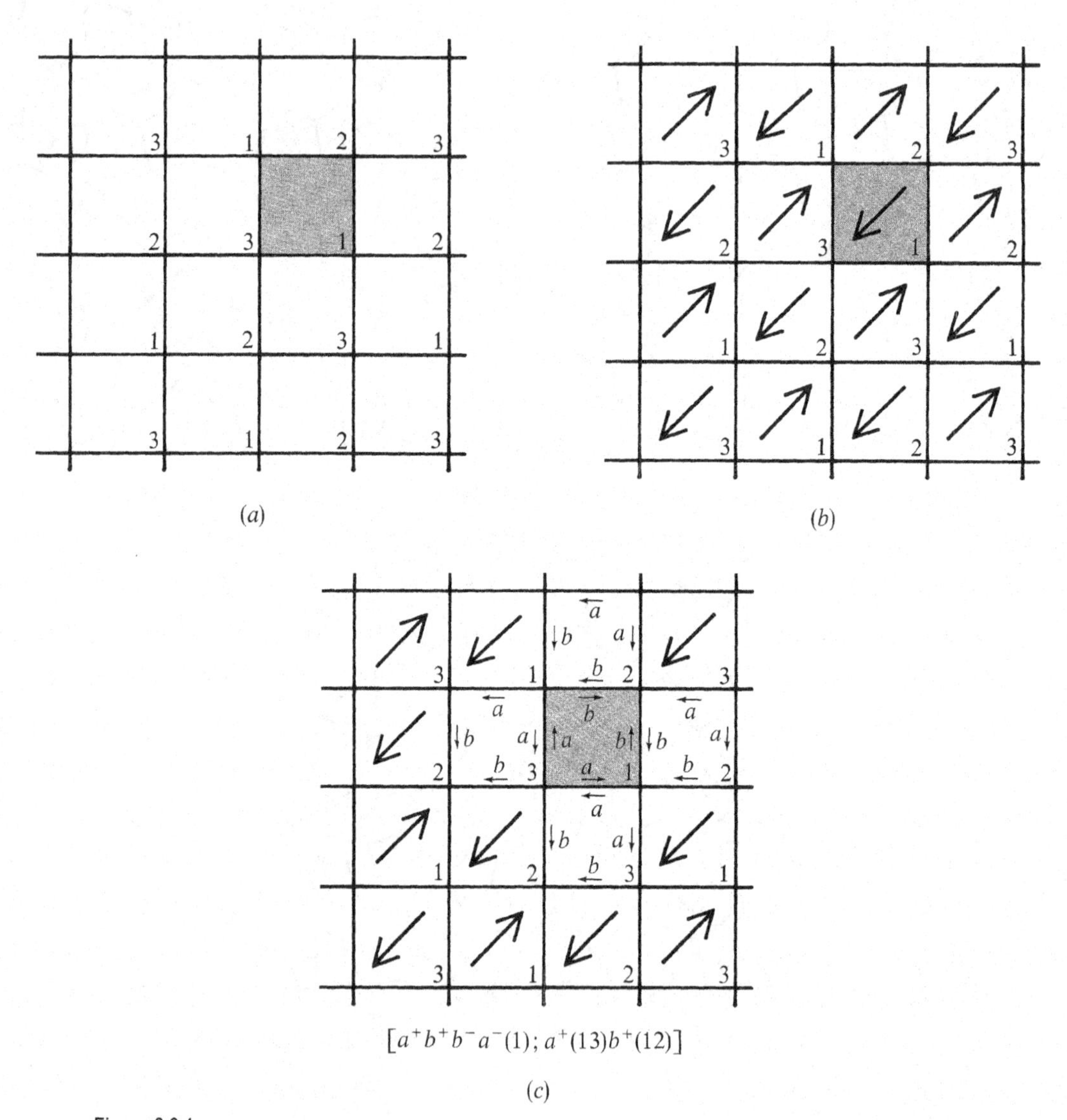

Figure 8.6.4
(*a*) A 3-chromatic tiling for which the groups $S(\mathscr{T}|T) = d4$, $S_c(\mathscr{T}^\wedge|T^\wedge) = d2$ and $S(\mathscr{T}^\wedge|T^\wedge) = d1$ are distinct. (*b*) A marking of $\mathscr{T}$ as required in Statement 8.6.1 to ensure color primitivity. (*c*) The derivation of the corresponding color-incidence symbol.

subgroup of $S_C(\mathscr{T}^\wedge)$ and moreover this group $S(\mathscr{T}_0)$ is tile-transitive.

The procedure just described is illustrated in Figures 8.6.2 to 8.6.5. In each figure part (*b*) shows a suitable marking $\mathscr{T}_0$ of the underlying tiling of the 3-chromatic tiling $\mathscr{T}^\wedge$ shown in (*a*). The colored tiling $\mathscr{T}_0^\wedge$ obtained from $\mathscr{T}_0$ by assigning the colors as in (*a*) is primitively colored, whereas (except in Figure 8.6.3) the tiling shown in (*a*) is not. Thus, in an obvious sense, the markings of Figure 8.6.3(*b*) are redundant, though that need not concern us here.

Even after the mark M on the first tile T has been chosen, in some cases $\mathscr{T}_0$ is not uniquely determined; there may be some choice in the marking of the other tiles. In Figure 8.6.5(*b*), (*d*) we show two markings of the tiling underlying that shown in (*a*), each of which corresponds to a primitive coloring. The reason for this phenomenon is that the choice of the subgroup H, whose existence is asserted by Statement 5.2.3, is not unique—in this case H can be either of type *p1* or of type *pg*.

Our next task is to explain the construction of the "color-incidence symbol" of a primitively colored k-chromatic tiling $\mathscr{T}_0^\wedge$. A *color-tile symbol* L of $\mathscr{T}_0^\wedge$ begins with the tile symbol of the underlying tiling $\mathscr{T}_0$. In other words we label the sides of each tile in $\mathscr{T}_0$ as described in Section 6.2 and then read off the labels in cyclic order as we go round a chosen tile T, adding to each label an appropriate superscript if required. The symbol L is completed by adding to the tile symbol of $\mathscr{T}_0$ the color of the tile T in parentheses.

The corresponding *color-adjacency symbol* of $\mathscr{T}_0$ is defined to be the same as an adjacency symbol for $\mathscr{T}_0$ except that after each label we insert the corresponding permutation θ of colors. Thus the letter in the adjacency symbol of $\mathscr{T}_0$ which corresponds to a tile T_i adjacent to T is followed by the permutation θ_i associated with the color symmetry which maps T onto T_i. We illustrate the procedure by reference to Figure 8.6.4(*c*). A tile symbol of $\mathscr{T}_0$ is $a^+b^+b^-a^-$ and so, if we choose T to be of color 1, the color-tile symbol of $\mathscr{T}_0^\wedge$ is $a^+b^+b^-a^-(1)$. The corresponding adjacency symbol of $\mathscr{T}_0$ is a^+b^+. The tile T_1 adjacent to T along a side a of the latter bears color 3, and it is easily checked that the color symmetry mapping T onto T_1 interchanges colors 1 and 3 and leaves color 2 unchanged. Hence the first part of the color-adjacency symbol is $a^+(13)$. The second part is determined similarly; it is $b^+(12)$.

It will be recalled from Chapter 6 that the tile and adjacency symbols are not sufficient to determine the isohedral type of a tiling uniquely—we must also know the topological type. Exactly the same is true for colored tilings and we remark that further ambiguities would arise if the color of the tile T (or *defining tile* as we shall call it) were omitted from the tile symbol. The situation is analogous to that for patterns shown in Figure 8.2.4(*b*). Here, not only does a different choice of color of the defining tile T affect the "appearance" of the tiling, it can also change its homeochromatic type.

The *color-incidence symbol* of a primitively colored k-chromatic tiling $\mathscr{T}_0^\wedge$ is defined to be an expression of the form $[L;A]$ where L is a color-tile symbol and A is a color-adjacency symbol of $\mathscr{T}_0^\wedge$. Examples of color-incidence symbols are given in Figures 8.6.2 to 8.6.5.

For the sake of uniformity we shall usually take the color of the defining tile T to be $i = 1$. For a different defining tile T', of color j, the corresponding incidence symbol is easily determined. Let (s,π) be a color symmetry which maps tile T onto T' (so that $\pi(i) = j$). If the incidence symbol corresponding to T is

$$[x_1x_2 \ldots x_n(i);\ y_1(\theta_1)y_2(\theta_2) \ldots y_m(\theta_m)],$$

then that corresponding to T' is

$$[x_1x_2 \ldots x_n(j);\ y_1(\theta'_1)y_2(\theta'_2) \ldots y_m(\theta'_m)]$$

with $\theta'_r = \pi \circ \theta_r \circ \pi^{-1}$ for $r = 1, 2, \ldots, m$. This holds also when $i = j$. Thus the incidence symbol depends essentially on the choice of defining tile and not only on its color.

Corresponding to Statement 7.2.1 we have the following result, which is proved in a similar manner.

8.6.2 *Two primitively colored k-chromatic tilings are homeochromatic if and only if they are of the same topological type and their color-incidence symbols differ trivially.*

The phrase "differ trivially" means that one can be obtained from the other by a change of labels, colors or defining tile, possibly permuting the letters in the tile symbol cyclically or reversing them, and making the corresponding modifications to the adjacency symbol. Examples will be given near the end of this section.

The use of color-incidence symbols makes possible, for each k, an enumeration of the possible homeochromatic types of primitively colored k-chromatic tilings. We take each (marked) isohedral tiling $\mathcal{T}$ in turn and consider a tile T of $\mathcal{T}$ together with all its adjacents. The incidence symbol of $\mathcal{T}$ is known so we need only list possible candidates for the permutations θ_i $(i = 1, 2, \ldots, m)$ which are to follow the various letters

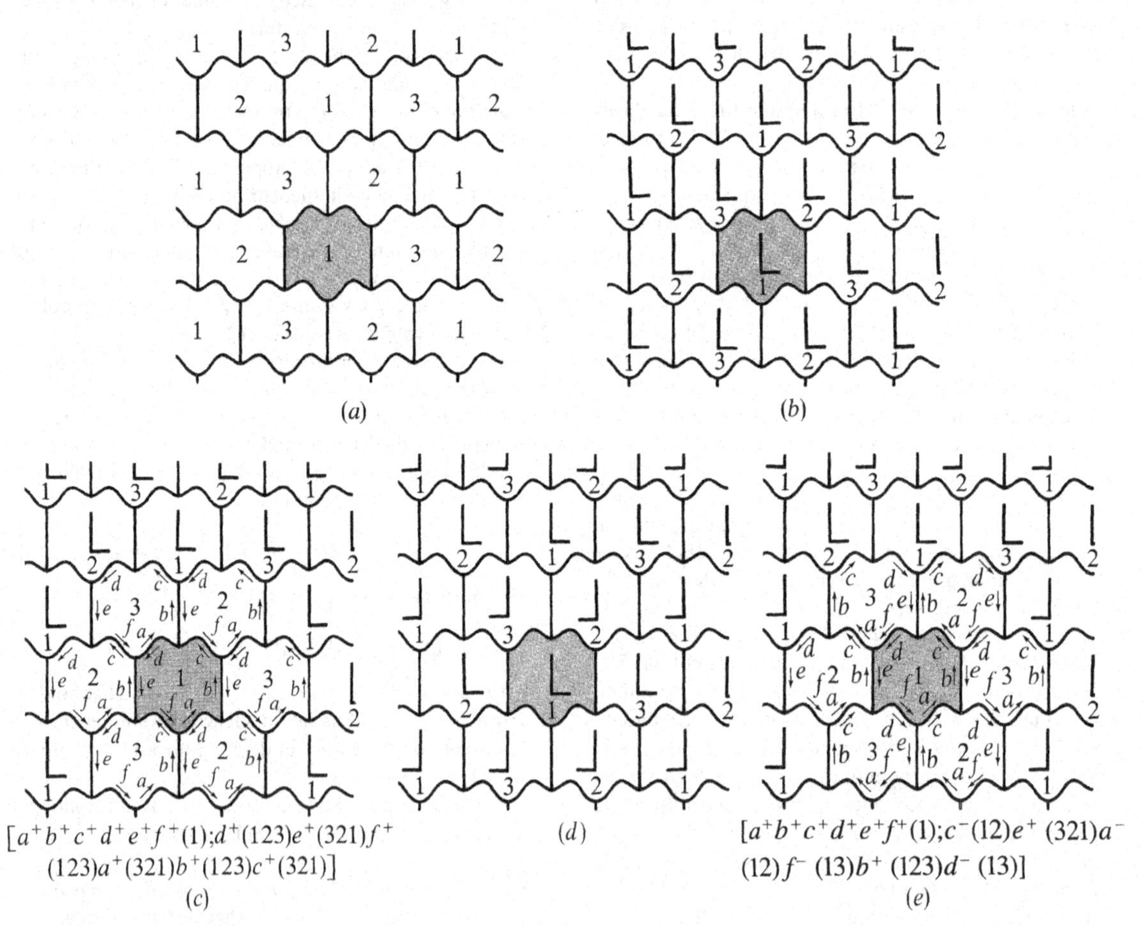

Figure 8.6.5

Parts (*b*) and (*d*) show two different markings of the tiling shown in (*a*) that reduce it to a primitively colored tiling. The derivation of the corresponding color-incidence symbols is shown in (*c*) and (*e*).

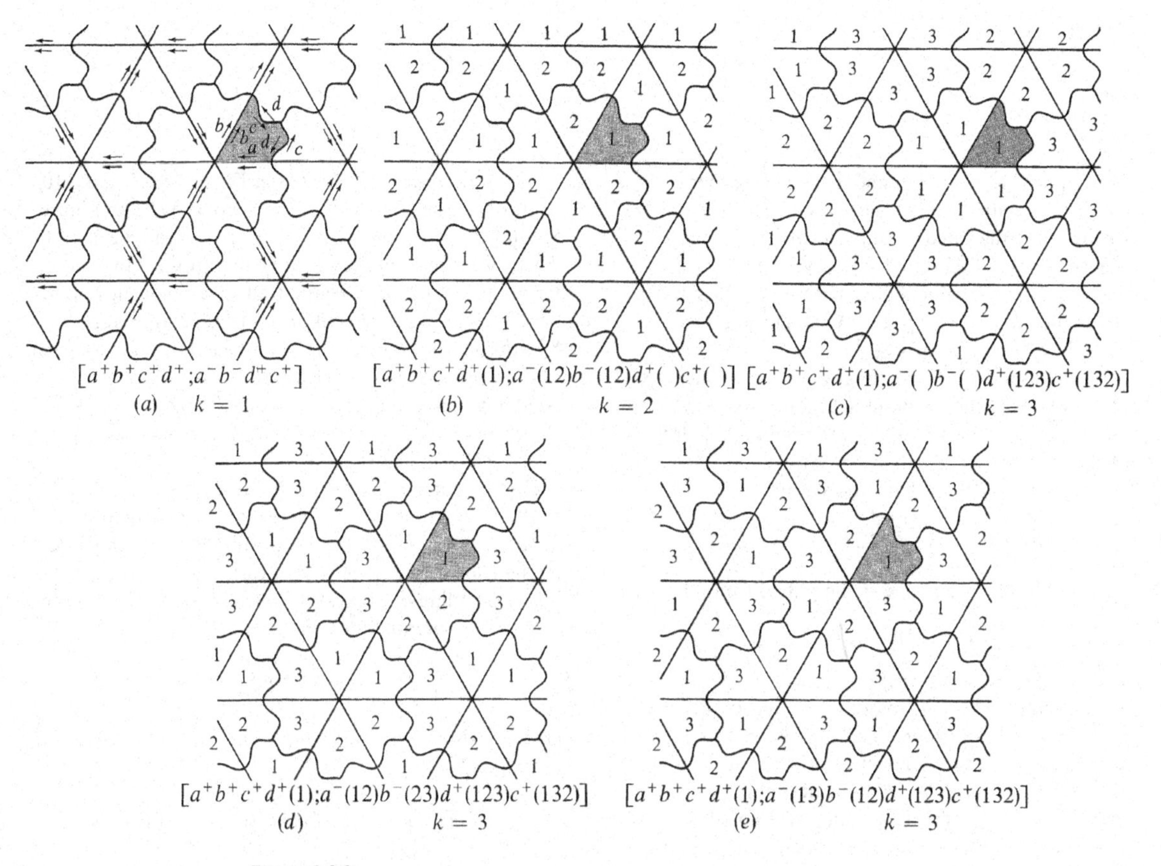

Figure 8.6.6
The primitive pattern IH30 and its color-primitive k-chromatic colorings for $k = 2$ and 3. In each case the corresponding color-incidence symbol is shown.

in the adjacency part of the color-incidence symbol. We pick out those sets of permutations which are consistent in the sense that they lead to a unique coloring of the tiles in $\mathscr{T}$, each by one of k colors. From each such set of permutations we obtain a coloring, and after eliminating duplicates, a complete list of homeochromatic types is obtained.

As an example, consider the tiling of type IH30 shown in Figure 8.6.6(a). The incidence symbol is $[a^+b^+c^+d^+; a^-b^-d^+c^+]$ so the color-incidence symbols of the colorings are of the form

$$[a^+b^+c^+d^+(1); a^-(\theta_1)b^-(\theta_2)d^+(\theta_3)c^+(\theta_4)]$$

where the θ_i are permutations of $1, 2, \ldots, k$.

For a given k, the number of cases to be considered can be reduced considerably by various means—not *every* possible set of permutations has to be considered. For example, when $k = 2$, the 3-fold rotation about a vertex where the edges labelled c and d meet can only correspond to the identity, so $\theta_3 = \theta_4 = (\ )$. Clearly then $\theta_1 = \theta_2 = (12)$ since the only other possibility, when either corresponds to the identity, cannot lead to a 2-colored tiling (see Figure 8.6.6(b)).

For $k = 3$, the 3-fold rotation mentioned above can only correspond to a permutation of colors (123) or (132). (The identity permutation is easily seen to be inadmissible.) Thus $\theta_3 = \theta_4^{-1} = (123)$ or (132). From reflections in the edges labelled a or b we see that $\theta_1^2 = \theta_2^2 = (\ )$, the identity, and so each of θ_1 and θ_2 must be either a 2-cycle (ij) or the identity. The possibilities are easily tested and we are led to the three primitively colored 3-chromatic tilings shown in Figures 8.6.6(*c*), (*d*) and (*e*). It is clear that these are of distinct homeochromatic types and so the enumeration is completed. A similar investigation could be carried out for $k \geq 4$.

In the case of 2-colored tilings the color-incidence symbol can be simplified without losing any essential information; we may omit the color of the defining tile from the tile symbol, and in the adjacency symbol we may omit ( ) and replace (12) by $_*$. Thus the symbol

$$[a^+b^+c^+d^+(1); a^-(12)b^-(12)d^+(\ )c^+(\ )]$$

of Figure 8.6.6(*b*) becomes, in simplified form,

$$[a^+b^+c^+d^+; a_*^-b_*^-d^+c^+].$$

This notation is used in Table 8.2.2 which lists primitively colored isohedral tilings whose symmetry groups are the 46 periodic 2-color groups. Table 8.2.3 contains the corresponding information for three colors. The fact that each color group can be represented in this way as the color symmetry group of a primitively colored isohedral tiling leads to an easy and systematic, if laborious, way of enumerating all the *k*-color groups for each *k*.

We can now explain how to determine all the homeochromatic types of *k*-chromatic colorings of a given isohedral tiling $\mathcal{T}$ (and not, as previously, only the primitively-colored ones). We begin by determining the pattern type of the tiling pattern $\mathcal{P}_h(\mathcal{T})$ (see column (10) of Table 6.2.1), and then all the motif-transitive subgroups of this pattern can be read off from Table 5.2.3. These groups are tile-transitive for $\mathcal{T}$. For each such group H we add marks to the tiles of $\mathcal{T}$ and so obtain a marked tiling $\mathcal{T}_0$ whose symmetry group coincides with H. Using the method of color-incidence symbols explained above, we determine all the primitive *k*-chromatic colorings of the tilings $\mathcal{T}_0$. Finally we delete the marks on the tiles and so, by Statement 8.6.1, we are left with all possible homeochromatic types of colorings of $\mathcal{T}$; repetitions may occur but these are readily eliminated.

The procedure is illustrated in Figures 8.6.7 to 8.6.10. In this example we consider a tiling of type IH90 (Figure 8.6.7(*a*)) of which the corresponding pattern type is PP44. There are three motif-transitive subgroups for this pattern type and hence three tile-transitive subgroups of $\mathcal{T}$, namely two of type *p*6 and one of type *p*2. In Figures 8.6.7(*b*), (*c*) and (*d*) some of the centers of rotation for each of these groups are indicated, and we also show corresponding marked tilings $\mathcal{T}_0$ whose symmetry groups are these three tile-transitive subgroups. The incidence symbols are also indicated.

For the tiling of Figure 8.6.7(*b*) it is easy to see that for $k = 2$ the only possible color-incidence symbol is $[a^+a^+a^+(1); a^+(12)]$, and no primitive *k*-colorings are possible for $k > 2$ (see Figure 8.6.8).

For the tiling of Figure 8.6.7(*c*) more possibilities arise. For $k = 2$, potential incidence symbols like

$$[a^+b^+c^+(1); c^+(\ )b^+(12)a^+(\ )]$$

are inconsistent and the only possible primitive 2-coloring corresponds to

$$[a^+b^+c^+(1); c^+(12)b^+(12)a^+(12)].$$

For $k = 3$, we obtain four possibilities and for $k = 4$ there are three possibilities as shown, along with the corresponding color-incidence symbols, in Figure 8.6.9.

For the tiling of Figure 8.6.7(*d*) we find three tilings with $k = 2$, four with $k = 3$ and eight with $k = 4$. These are shown in Figure 8.6.10.

At every stage we must be careful to avoid repetitions of types. Thus for the tiling of Figure 8.6.7(*c*) the symbols

$$[a^+b^+c^+(1); c^+(23)b^+(12)a^+(23)]$$

and $$[a^+b^+c^+(1); c^+(23)b^+(13)a^+(23)]$$

differ trivially since one can be obtained from the other by interchanging colors 2 and 3, leaving color 1 fixed. Figure 8.6.9(*h*) shows the corresponding colored tiling. In this particular case, however, we can say more; the two symbols lead to *identical* colored tilings since one

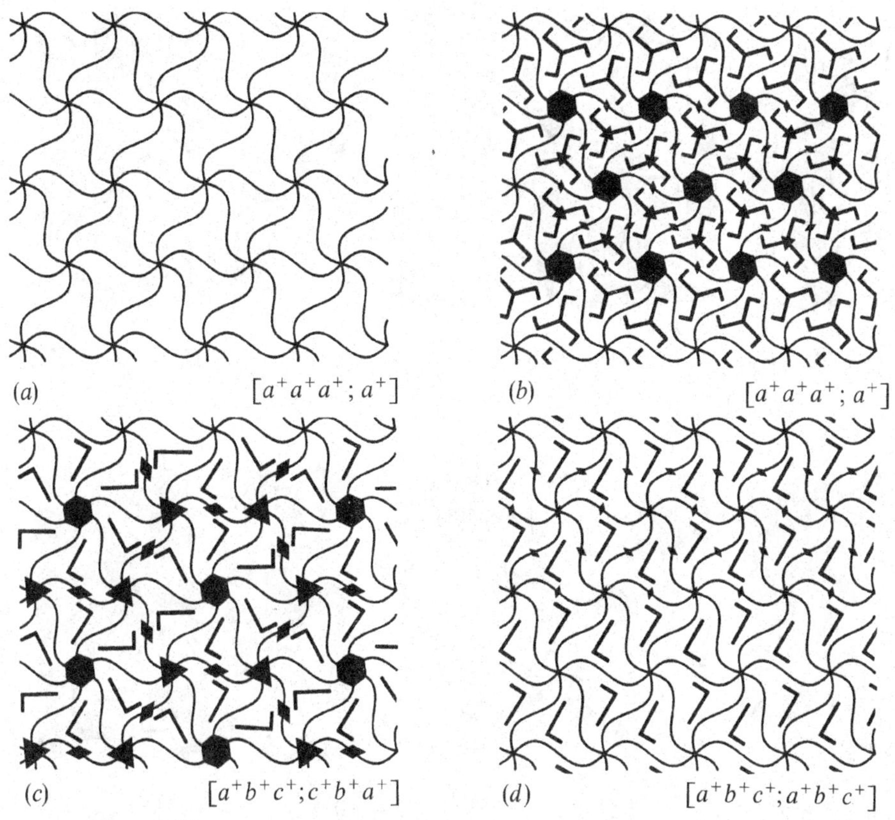

Figure 8.6.7
A tiling of isohedral type IH90 and the corresponding marked tilings that correspond to the tile-transitive subgroups of its symmetry group.

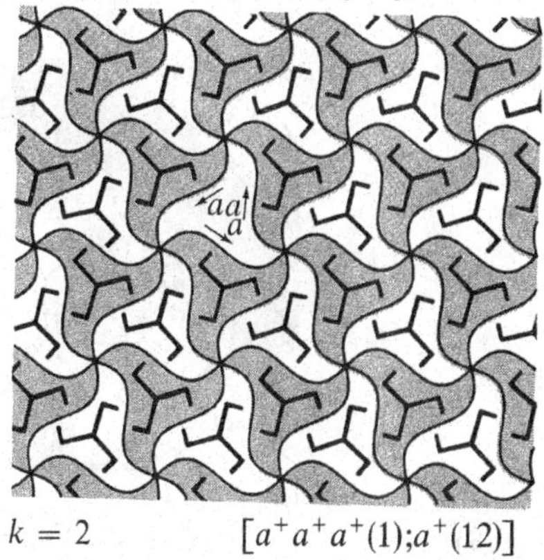

Figure 8.6.8
The only primitive coloring of the marked tiling shown in Figure 8.6.7(b).
No other k-colorings are possible with $k \geq 2$.

(a) $k = 2$
$[a^+b^+c^+(1);c^+(12)b^+(12)a^+(12)]$

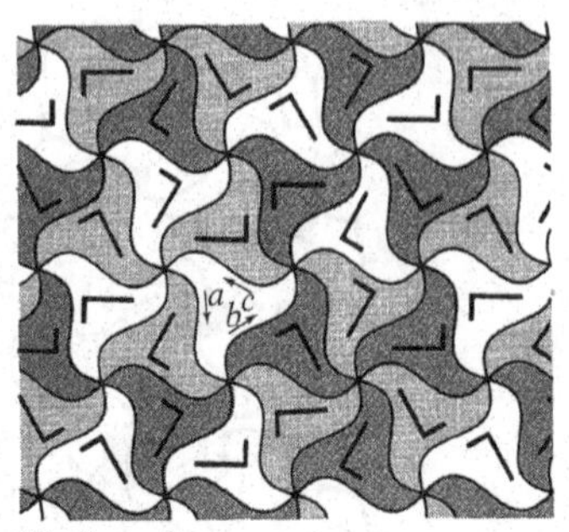

(b) $k = 3$
$[a^+b^+c^+(1);c^+(12)b^+(13)a^+(12)]$

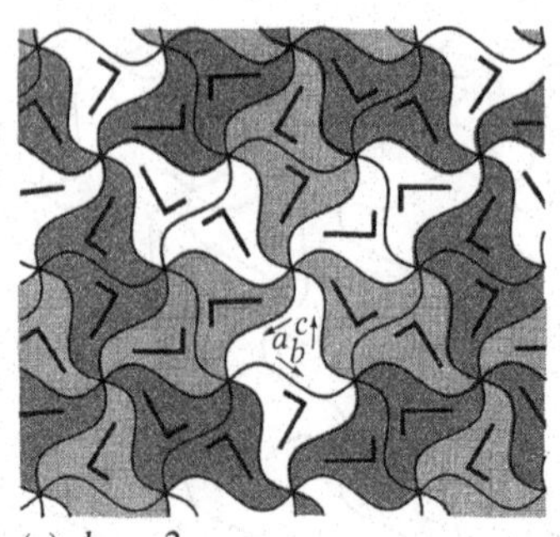

(c) $k = 3$
$[a^+b^+c^+(1);c^+(12)b^+(23)a^+(12)]$

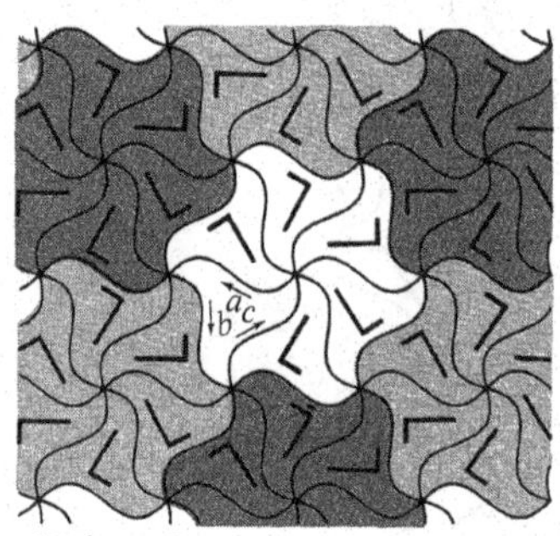

(d) $k = 3$
$[a^+b^+c^+(1);c^+(23)b^+(12)a^+(23)]$

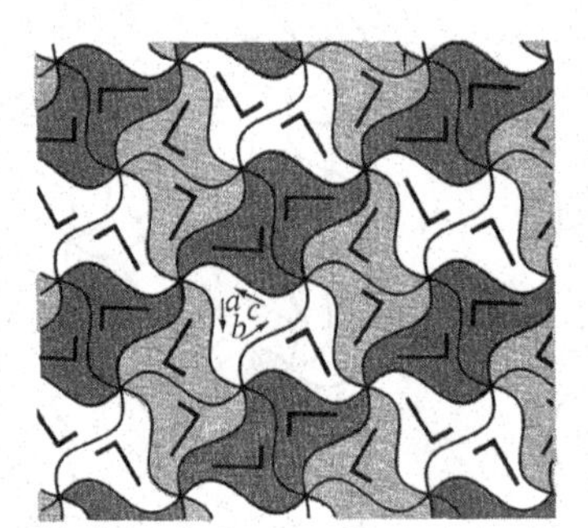

(e) $k = 3$
$[a^+b^+c^+(1);c^+(123)b^+(\ )a^+(321)]$

(f) $k = 4$
$[a^+b^+c^+(1);c^+(123)b^+(14)(23)a^+(132)]$

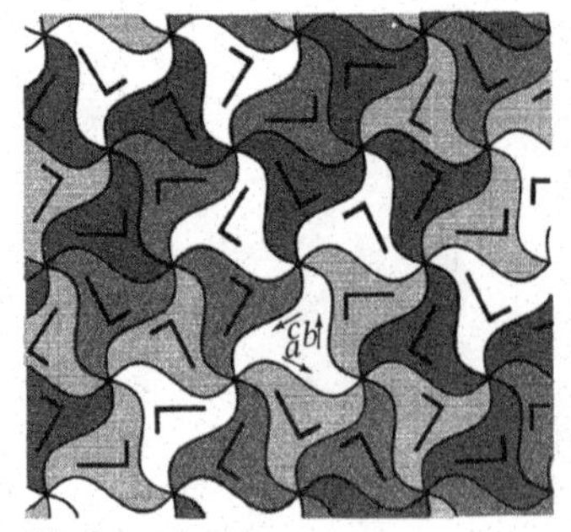

(g) $k = 4$
$[a^+b^+c^+(1);c^+(123)b^+(12)(34)a^+(132)]$

(h) $k = 4$
$[a^+b^+c^+(1);c^+(432)b^+(12)(34)a^+(234)]$

Figure 8.6.9
The primitive 2-colorings, 3-colorings and 4-colorings of the marked tiling shown in Figure 8.6.7(c).

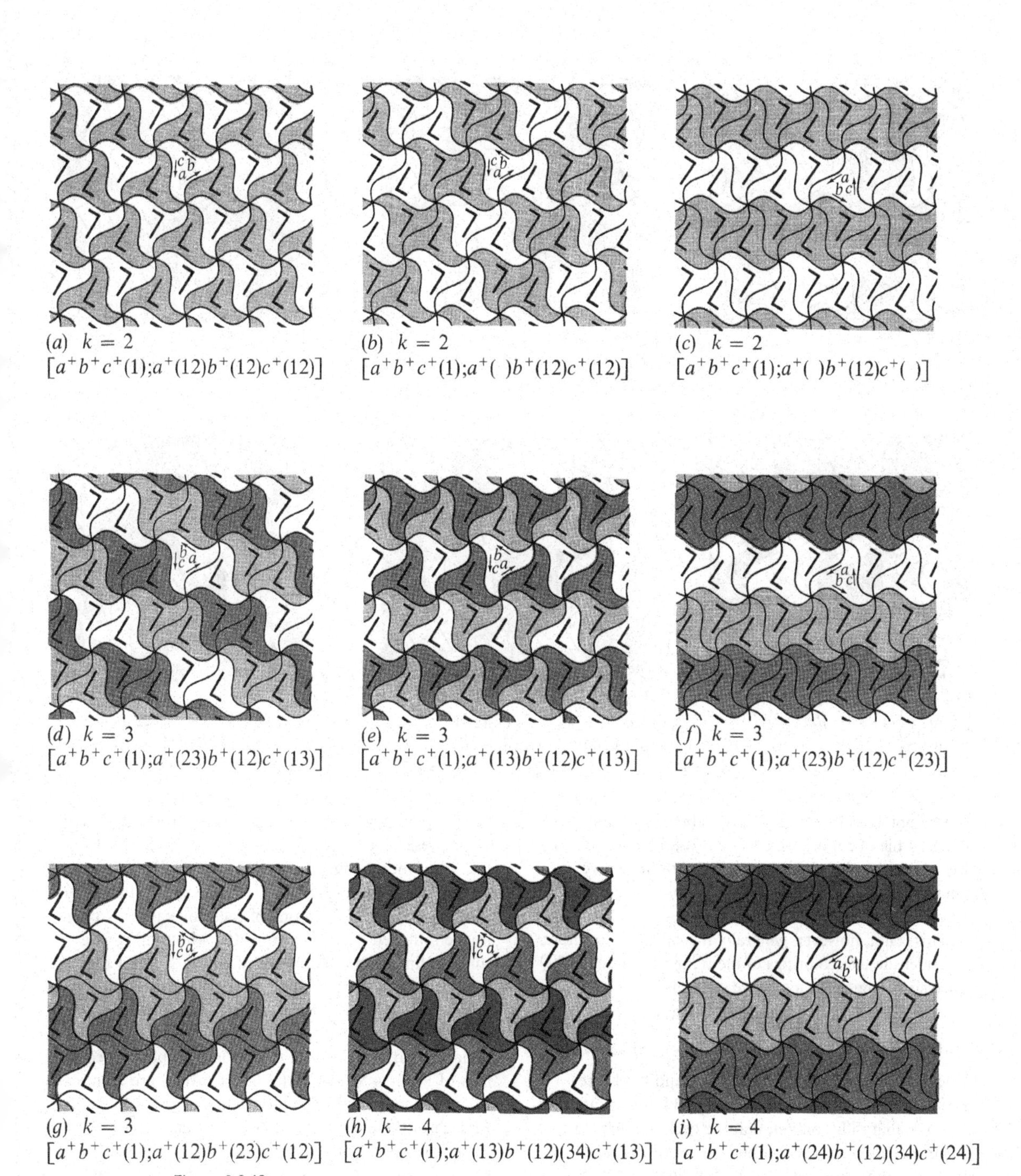

(*a*) $k = 2$ $[a^+b^+c^+(1);a^+(12)b^+(12)c^+(12)]$

(*b*) $k = 2$ $[a^+b^+c^+(1);a^+(\ )b^+(12)c^+(12)]$

(*c*) $k = 2$ $[a^+b^+c^+(1);a^+(\ )b^+(12)c^+(\ )]$

(*d*) $k = 3$ $[a^+b^+c^+(1);a^+(23)b^+(12)c^+(13)]$

(*e*) $k = 3$ $[a^+b^+c^+(1);a^+(13)b^+(12)c^+(13)]$

(*f*) $k = 3$ $[a^+b^+c^+(1);a^+(23)b^+(12)c^+(23)]$

(*g*) $k = 3$ $[a^+b^+c^+(1);a^+(12)b^+(23)c^+(12)]$

(*h*) $k = 4$ $[a^+b^+c^+(1);a^+(13)b^+(12)(34)c^+(13)]$

(*i*) $k = 4$ $[a^+b^+c^+(1);a^+(24)b^+(12)(34)c^+(24)]$

Figure 8.6.10
The primitive 2-colorings, 3-colorings and 4-colorings of the marked tiling shown in Figure 8.6.7(*d*).

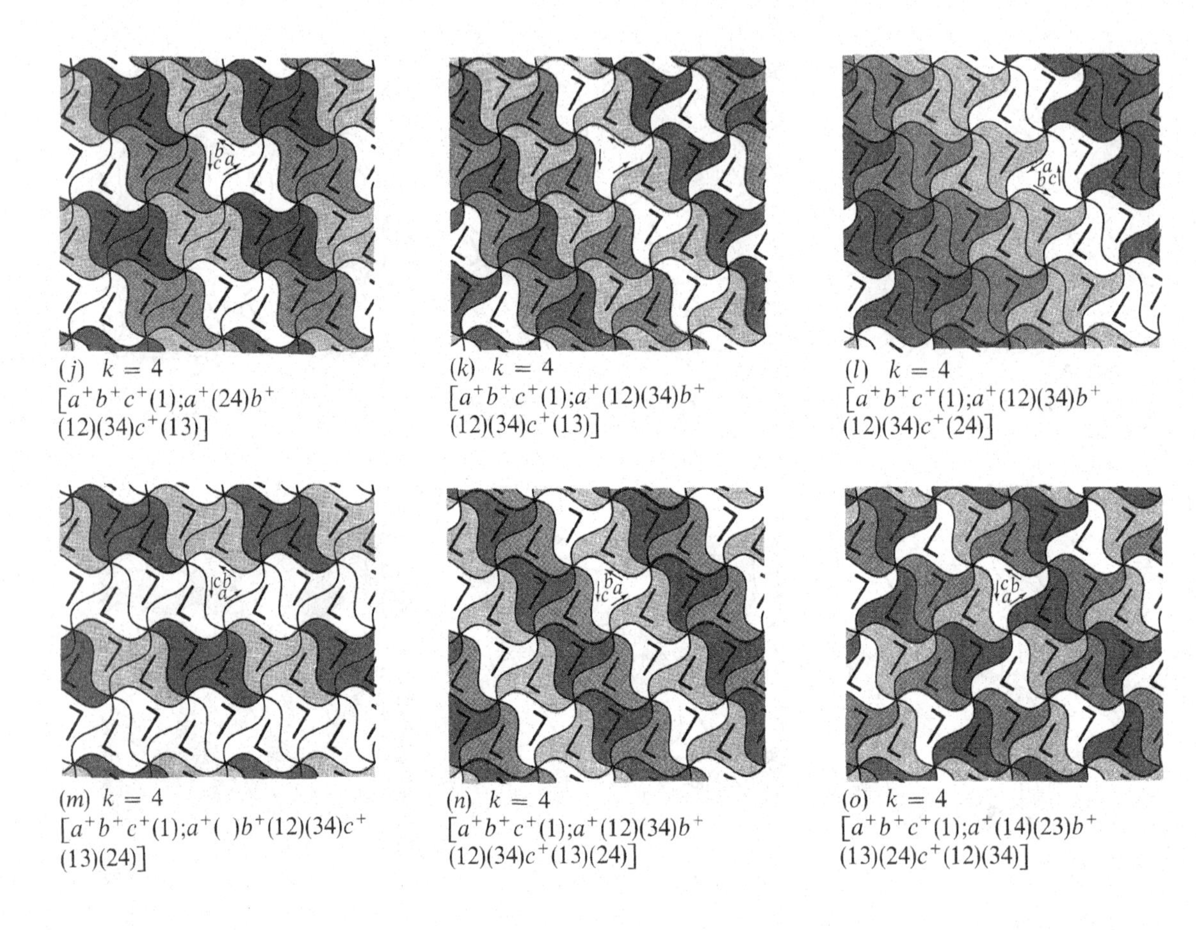

(*j*) $k = 4$
$[a^+b^+c^+(1); a^+(24)b^+(12)(34)c^+(13)]$

(*k*) $k = 4$
$[a^+b^+c^+(1); a^+(12)(34)b^+(12)(34)c^+(13)]$

(*l*) $k = 4$
$[a^+b^+c^+(1); a^+(12)(34)b^+(12)(34)c^+(24)]$

(*m*) $k = 4$
$[a^+b^+c^+(1); a^+(\ )b^+(12)(34)c^+(13)(24)]$

(*n*) $k = 4$
$[a^+b^+c^+(1); a^+(12)(34)b^+(12)(34)c^+(13)(24)]$

(*o*) $k = 4$
$[a^+b^+c^+(1); a^+(14)(23)b^+(13)(24)c^+(12)(34)]$

can be obtained from the other by choosing a different defining tile of color 1. In other words, if two copies of the tiling are colored in accordance with the above incidence symbols, then one is the image of the other under a color-preserving isometry.

This is not, however, generally the case. For example, consider the two colorings of the tiling of Figure 8.6.7(*c*) corresponding to the incidence symbols

$$[a^+b^+c^+(1);\ c^+(123)b^+(12)(34)a^+(132)]$$

and $$[a^+b^+c^+(1);\ c^+(123)b^+(13)(24)a^+(132)].$$

These are shown in Figures 8.6.9(*g*) and 8.6.11, respectively. Interchanging colors 2 and 3 leads to incidence symbols that differ only in that the cycles attached to c^+ and a^+ are replaced by their inverses. It is easy to see that the colorings cannot be brought into coincidence by applying any permutation of colors or isometry of the tiling; nevertheless they are homeochromatic. The color-preserving homeomorphism required in the definition of homeochromatic type is more difficult to describe; it corresponds to a transformation that alters the shape of each tile in one of the tilings into its mirror image followed by reflection of the whole tiling in a suitable line. The interchange of colors 2 and 3 then leads to the required color-preserving homeomorphism.

Finally, we remove the markings from the tilings shown in Figures 8.6.8, 8.6.9 and 8.6.10. When this is done several repetitions occur. For example, the colorings of Figures 8.6.8, 8.6.9(*a*) and 8.6.10(*a*) are identical, as are the colorings of Figures 8.6.9(*f*) and 8.6.10(*o*). In each of these cases the coloring of the unmarked tiling is perfect, though in the latter it is not primitive since the tiling admits, as a color symmetry, a 3-fold rotation about the center of each tile. This is associated with a cyclic permutation of three of the colors so that $S_C(\mathcal{T}^\wedge|T^\wedge) \neq S(\mathcal{T}^\wedge|T^\wedge)$.

Eliminating repetitions leaves us with a total of three types of 2-colorings, eight types of 3-colorings and ten types of 4-colorings. This is a complete enumeration of the homeochromatic types of k-chromatic colorings of a tiling of type IH90 with two, three or four colors.

The procedure we have just described for determining the homeochromatic types of colorings of isohedral tilings is completely straightforward. It is therefore surprising to report that for no value of k has the enumeration been carried out. We can, however, assert the following.

8.6.3 *For any given value of $k \geq 2$ and any isohedral tiling $\mathcal{T}$, the number of homeochromatic types of k-chromatic colorings of $\mathcal{T}$ is finite.*

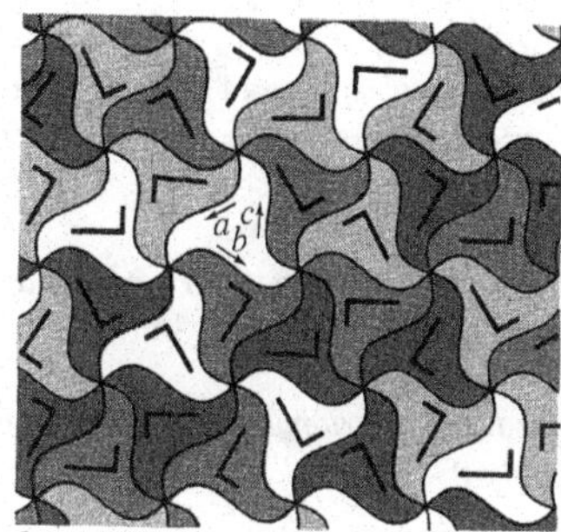

$k = 4$
$[a^+b^+c^+(1);c^+(123)b^+(13)(24)a^+(132)]$

Figure 8.6.11
This 4-coloring of the marked tiling shown in Figure 8.6.7(*c*) is of the same homeochromatic type as that in Figure 8.6.9(*g*) in spite of the fact that it is not possible to make the colorings coincide by applying a permutation of the colors in either tiling.

EXERCISES 8.6

1. Find examples, other than that shown in Figure 8.6.1, in which the classification by color-homeomeric type and by homeochromatic type differ.

*2. Find color-incidence symbols for all the 2-chromatic and 3-chromatic colorings of the regular tilings.

3. (*a*) Using the methods of this section determine all the homeochromatic types of k-chromatic colorings ($2 \leq k \leq 4$) of the Laves tiling $[3.6.3.6]$.
   *(*b*) Which of the homeochromatic types found in part (*a*) have the same (*i*) equicolor type, (*ii*) color-homeomeric type, (*iii*) color pattern type, or (*iv*) color symmetry group?

**4. List the homeochromatic types of 2-chromatic colorings of all 93 types of marked isohedral tilings.

8.7 PERFECT COLORINGS OF TILINGS

In Section 8.1 we pointed out that in general a given pattern or tiling cannot be perfectly k-colored for every value of k. In Grünbaum & Shephard [1977d] it was shown that every isohedral tiling and every pattern can be perfectly k-colored for infinitely many values of k; the permissible values of k were determined explicitly by Senechal [1979a], and the reader is referred to this paper for further details.*

Here we consider only perfect colorings of the three regular tilings; we shall prove the following results.

* See the Appendix beginning on page 653.

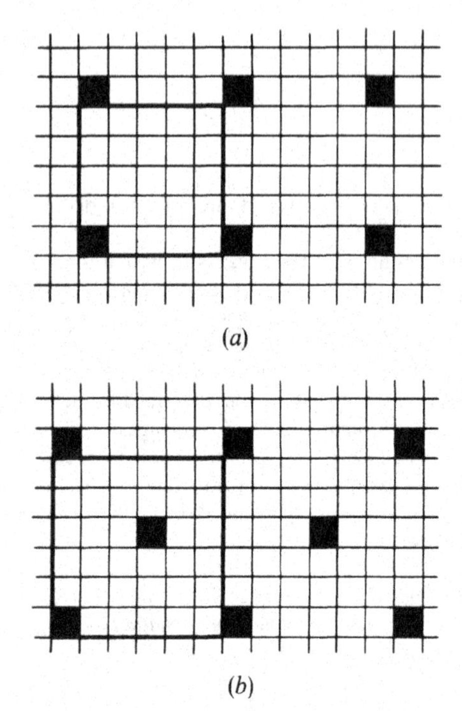

Figure 8.7.1
The possible arrangements of tiles of color class 1 in perfect colorings of the regular tiling (4^4). The cases $m = 5$ and $m = 6$ are illustrated.

8.7.1 *The regular tiling by squares* (4^4) *admits a perfect k-coloring if and only if* $k = n^2$ *or* $k = 2n^2$ *for some positive integer n.*

8.7.2 *The regular tiling by hexagons* (3^6) *admits a perfect k-coloring if and only if* $k = n^2$ *or* $k = 3n^2$.

8.7.3 *The regular tiling by triangles* (6^3) *admits a perfect k-coloring if and only if* $k = 2n^2$, $k = 6n^2$, $k = (3n - 2)^2$ *or* $k = (3n - 1)^2$.

In each case we shall construct the perfect colorings explicitly and each coloring will be unique within a permutation of the colors or a symmetry of the tiling.

We begin by considering the tiles of color class 1 in a perfect coloring of (4^4). Let m be the smallest positive integer such that a horizontal translation by m tiles brings a tile of color 1 into coincidence with another tile of color 1. The fact that the color classes are congruent enables us to deduce that *every* tile obtainable from a tile of color j by a horizontal translation of m tiles also has color j. Further, considering rotation about the center of a tile through 90° we see that exactly the same assertion follows for vertical translations also. Thus, for each j, color j is assigned to, at least, all tiles that form a square lattice Λ_j of "mesh" m.

Considering the color class 1, two possibilities arise (see Figure 8.7.1 where color class 1 is indicated by black squares; $m = 5$ in (*a*) and $m = 6$ in (*b*)):

(*i*) either all the tiles of color 1 belong to the lattice Λ_1, or
(*ii*) some other tile has color 1.

In case (*i*) we see that all the m^2 tiles in a mesh of Λ_1 must have different colors, and that these same colors are repeated in the same way in every mesh (see Figure 8.7.2(*a*) for the case $m = 5$). Hence $k = m^2$ colors occur.

In case (*ii*), consider reflections in the horizontal and vertical lines that pass through the centers of the tiles of color 1. Due to the minimality of m the only possible positions for the extra tiles of color 1 are at the centers of the meshes, and so this situation can arise only if $m = 2n$ is even. Hence each mesh of $(2n)^2 = 4n^2$ tiles contains two tiles of color 1, the number of colors is $k = 2n^2$ and we obtain an arrangement of colors like that shown in Figure 8.7.2(*b*) for the case $m = 6$.

It is easily verified that each of the colorings described above and illustrated in Figure 8.7.2 is perfect, and hence $k = m^2$ or $k = 2n^2$. The proof of Statement 8.7.1 is thus completed.

For the hexagonal tiling (3^6) the procedure is similar. Orienting the tiling as in Figure 8.7.3, so that one third of the edges are vertical, we again let m be the smallest positive integer such that a horizontal translation by m tiles brings a hexagonal tile of color 1 into coincidence with another tile of color 1. Then we can deduce, as before, that the tiles of color 1 form at least a lattice with triangular mesh of side m, see Figure 8.7.3. A fundamental region R for the corresponding group of translations is rhomb-like (indicated in Figure 8.7.3 by thickened lines) and hence, if there are no other tiles of color 1, we have $k = m^2$, this being the number of tiles in R (see Figures

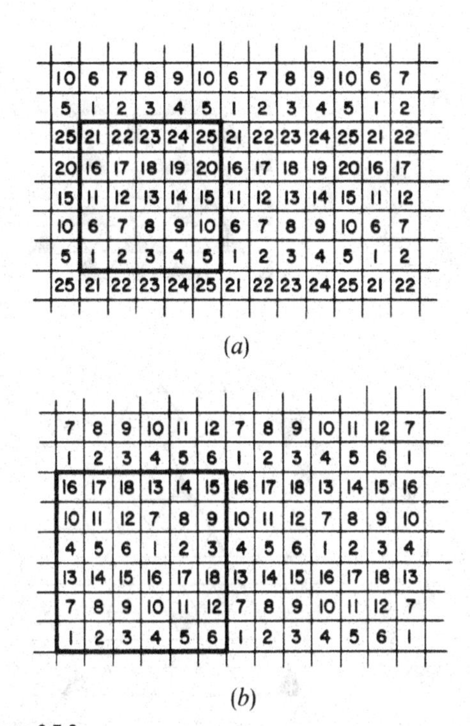

Figure 8.7.2
The perfect colorings of (4^4) corresponding to the arrangement of tiles of color class 1 shown in Figure 8.7.1. In (*a*) $m^2 = k = 25$ colors are used and in (*b*) $\frac{1}{2}m^2 = k = 18$ colors are used.

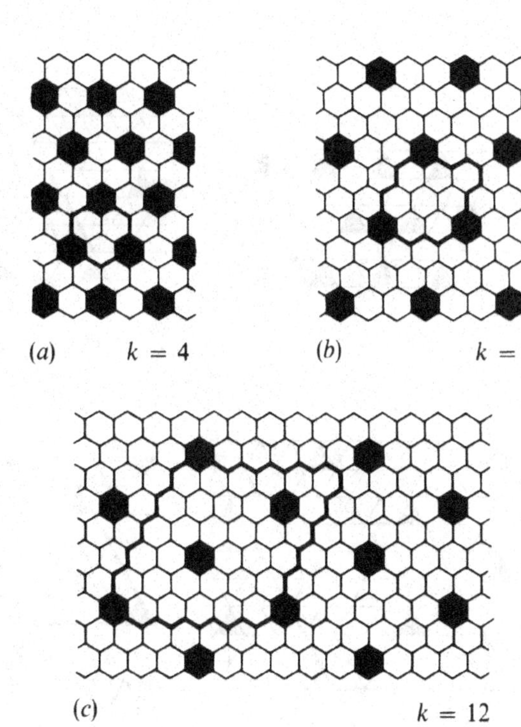

Figure 8.7.3
The possible arrangements of tiles of color class 1 in perfect colorings of the regular tiling (3^6). Cases $m = 2$, 3 and 6 are shown and the corresponding values of k are indicated.

8.7.3(*a*) and (*b*) for the cases $m = 2$ and 3). On the other hand, if another tile has color 1, then considering reflections in the altitudes of the triangles forming the meshes, we see that the only position it can occupy is the center of one (and therefore every) "triangle" in the mesh, see Figure 8.7.3(*c*). This case can only arise when $m = 3n$ for some integer n—in Figure 8.7.3(*c*) we have $n = 2$. The fact that each of the other color classes is a translation of this makes it evident how the coloring is to be completed. In all, $3n^2$ colors will be used. As each of the colorings we have described is perfect, the proof of Statement 8.7.2 is completed.

For the triangular tiling (6^3) the situation is slightly more complicated, due to the fact that the tiles occur in two different "orientations" or aspects. Considering the tiles of only one aspect, we again define m to be the smallest positive integer such that a horizontal translation through a distance mb (where b is the length of the side of a tile) brings a triangular tile of color 1 into coincidence with another tile of color 1. Again we deduce that all the tiles of one aspect of color 1 form, at least, a lattice with triangular mesh of side m. A fundamental region R for the corresponding group of translations is a rhomb (indicated in Figure 8.7.4 by thickened lines) and hence if there are no further tiles of color 1 we have $k = 2m^2$ (see Figures 8.7.4(*a*) and (*b*) for the cases $m = 2$ and $m = 3$). If there is another tile of color 1 in R then the only position it can occupy is the center of one of the triangles of the mesh. Three cases arise:

(*i*) If $m = 3n$ then the triangle in the left half of the rhomb R has a central tile of the same aspect as the tiles forming the mesh. Hence the number of colors is $k = 6n^2$ (see Figures 8.7.4(*c*) and (*d*) for the cases $n = 1$ and $n = 2$).

(*ii*) If $m = 3n - 2$ then the triangle in the left half of the rhomb R has a central tile of the opposite aspect to the tiles forming the mesh. Hence the

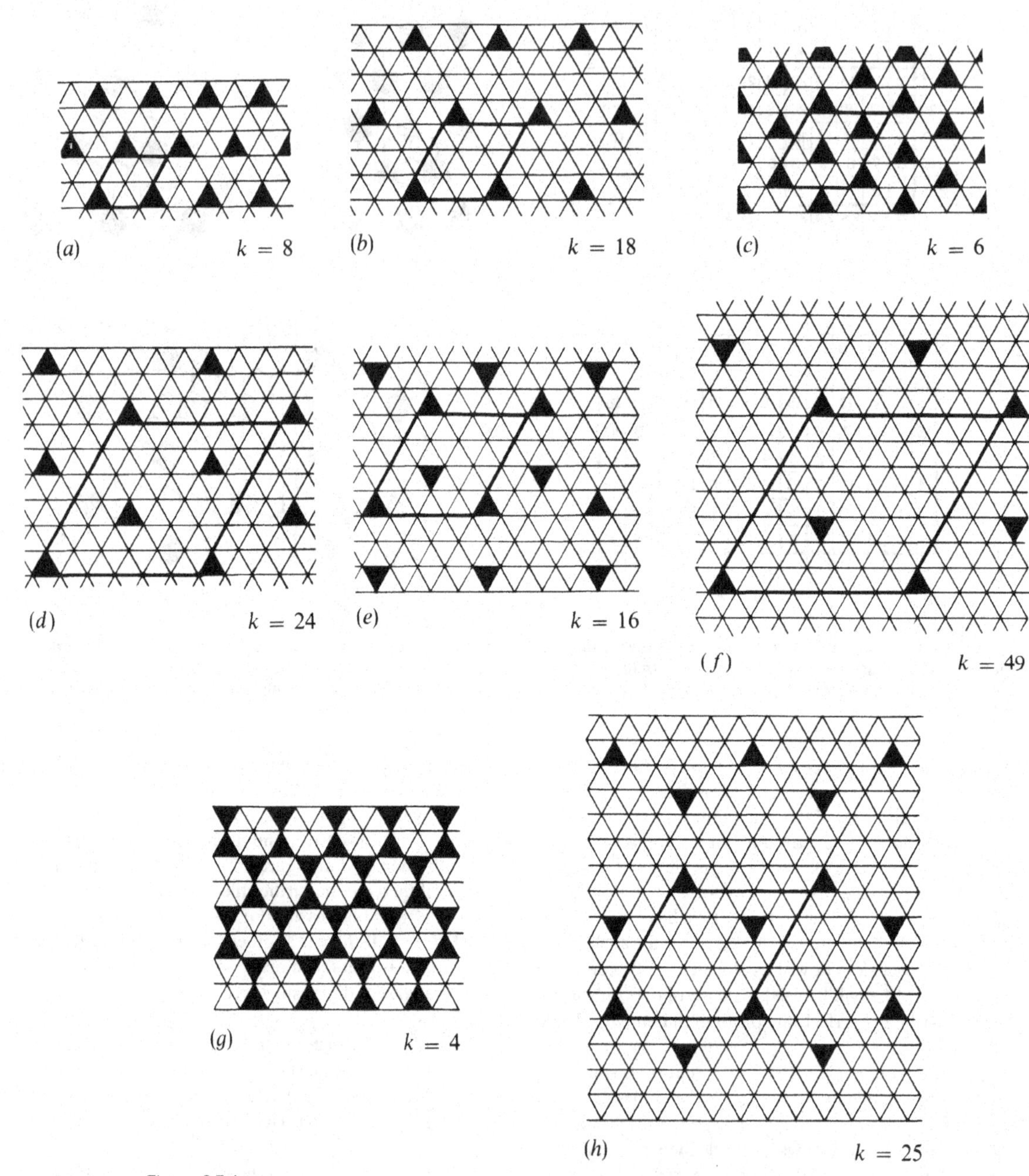

Figure 8.7.4
The possible arrangements of tiles of color class 1 in perfect colorings of the regular tiling (6^3). The corresponding values of m are: (a) and (g) $m = 2$, (b) and (c) $m = 3$, (e) $m = 4$, (h) $m = 5$, (d) $m = 6$ and (f) $m = 7$.

number of colors is $k = (3n - 2)^2$ (see Figures 8.7.4(*e*) and (*f*) for the cases $n = 2$ and $n = 3$).

(*iii*) Finally, if $m = 3n - 1$ then each of the triangles in the right half of the rhomb R has a central tile of the opposite aspect to the tiles forming the mesh. Hence the number of colors is $k = (3n - 1)^2$ (see Figure 8.7.4(*g*) and (*h*) for the cases $n = 1$ and $n = 2$).

In Figure 8.7.4 we have indicated only the color class 1, but it is evident how, as before, the coloring of the tiling can be completed in a perfect manner by k colors where k is of the stated form. Thus Statement 8.7.3 is proved, and all perfect colorings of the regular tilings have been described.

We conclude with a few remarks about perfect colorings of other isohedral tilings. If $\mathscr{T}_1$ and $\mathscr{T}_2$ are two such tilings of the same topological type, and every symmetry of $\mathscr{T}_1$ is a symmetry of $\mathscr{T}_2$, then it follows from the definitions that to every perfect coloring of $\mathscr{T}_2$ will correspond a perfect coloring of $\mathscr{T}_1$. Roughly speaking, the smaller the symmetry group, the more perfect colorings will exist. For example, a tiling of type IH62 (see Figure 6.2.4) is of the same topological type as the regular tiling (4^4), its symmetry group consists of the *direct* isometries (translations and rotations) in the symmetry group of (4^4). We deduce from Statement 8.7.1 that each tiling of type IH62 admits perfect k-colorings when $k = n^2$ or $2n^2$. However other values of k are possible; it can be shown that there also exist perfect k-colorings when $k = n^2 + p^2$ where n and p are integers such that $n > 0$ and $n \geq p \geq 0$.

This same number $k = n^2 + p^2$ also arises as a permissible value of k for perfect k-colorings of isohedral tilings of type IH28. A perfect 5-coloring of a very decorative tiling of this type is shown in Grünbaum & Shephard [1977d, Figure 10]. This tiling was designed by M. C. Escher who, however, did not seem to be aware of the possibility of using five colors—a perfect 4-coloring of this tiling by Escher is shown in MacGillavry [1965].

EXERCISES 8.7

*1. Give a detailed proof of the assertion that every isohedral tiling admits perfect k-colorings for infinitely many values of k.

*2. Determine all the types of isohedral tilings which admit perfect k-colorings for every $k \geq 2$.

*3. Determine all the values of k for which there exists a perfect k-coloring of isohedral tilings of type (*a*) IH37 (*b*) IH72; (*c*) IH73.

8.8 NOTES AND REFERENCES

In every civilization and culture, colored tilings and patterns appear among the earliest decorations. Splendid examples are shown in Jones [1856], d'Avennes [1877], Bossert [1928] and in other accounts of early art. In particular, 2-color patterns arose—early and frequently—through a device known as "counterchange" (see, for example, Christie [1929, Chapter 10]). An early paper with remarkable counterchange designs formed by diagonally divided squares—one-half black, one-half white—was published by Truchet [1704]. For a more recent treatment, with many illustrations, see Cullinane

[1976]. However, all these were more or less "accidental" occurrences, independently reinvented many times, and passed from generation to generation by artists and artisans. The only artist who deliberately and consistently tried to investigate colored patterns (more specifically, colored tilings) was M. C. Escher (see Escher [1958], [1960], [1971], [1982], [1983], MacGillavry [1965], Ernst [1976]). The idea of devising isohedral tilings in which the tiles have shapes of animals or various objects, and are colored by two or more colors, has in recent years become a popular tool in mathematical education (see, for example, Ranucci & Teeters [1977], Maletsky [1974], Teeters [1974]. Haak [1976], Brisse [1981]). (A different method of drawing families of mutually related tilings of the Escher type is investigated by Bøggild [1980a], [1980b].) In view of this, it is rather surprising to note that, except in Escher's work, k-chromatic patterns with $k \geq 3$ occur very rarely in art and decoration. Moreover, most of those that do occur are of the very simplest kinds (typically, a strip pattern in which the colors are cyclically permuted, or a periodic pattern generated by two translations, one of which leaves the colors unchanged while the other permutes them cyclically). Very few examples have the complexity of the 3-chromatic tilings shown in Figure 5.2.13. (Much of the startling and attractive art of Hans Hinterreiter is based on the idea of homeomorphically transforming colored tilings or patterns with a crystallographic group of symmetries; see Hinterreiter [1977] and Makovicky [1979].)*

In spite of the widespread use of colored patterns and tilings in art, theoretical investigations of the various kinds of colorings have been largely ignored by mathematicians and crystallographers; there are, however, some exceptions to this statement. For example, in a series of papers, Woods [1935a], [1935b], [1935c], [1936] gives the first complete, explicit and deliberate enumeration of strip and periodic 2-color groups. (See Crowe [1985] for an account of Woods' work.) Unfortunately, these nicely written papers seem to have had equally little influence on mathematics as on design. Then there is the work of Escher, already mentioned. Its influence on mathematicians and, even more, on crystallographers and designers seems to have been considerable, but there appears to be no account available of the systematic results which Escher obtained in his investigations of colored tilings. (It is to be hoped that the efforts of Doris Schattschneider in this respect will be published soon; see Schattschneider [1985].) In a different direction leads the work of Washburn [1977], who developed a theory of 2-chromatic patterns motivated by designs on the pottery of Southwest American Indian tribes. Washburn was apparently not aware of the existence of other works on 2-color groups and patterns. Although her classification of 2-chromatic patterns may be appropriate for its intended use, it is not quite consistent. For example, Washburn's classification of 2-chromatic strip patterns into 21 types falls between the classifications by the 17 2-color strip groups and by the 28 2-color pattern types.

For an account of an interesting variety of 2-colored strip patterns, see Bérczi [1985].

In a somewhat different sense, the history of 2-color groups started in the late 1920's, even before the work of Woods mentioned above. In order to describe this we shall begin by explaining some of the situations in which 2-color groups are useful. Instead of referring (as we shall do) to black-and-white patterns (or to patterns of some other two colors), it is clear that we may consider patterns in which each copy of the motif has one of two mutually exclusive properties, and we can interpret 2-color groups as expressing (generalized) symmetries of such patterns. In Figure 8.8.1 we illustrate several instances of this kind, which show the wide applicability of the concept of 2-color groups.

One further modification is of considerable geometric and historic interest. It is related to the last example in Figure 8.8.1 and deals with patterns in a plane which is assumed to be situated in 3-dimensional space and *to have two sides.* Thus a copy of the motif can be situated on either of the two sides, and symmetries of the pattern may involve *reflection in the plane itself.* It is obviously possible to relate 2-chromatic patterns in the usual sense to such (uncolored) patterns on the two-sided plane by placing motifs of one color (say white) on one "side" of the plane, those of the other color (say black) on the

* See the Appendix beginning on page 653.

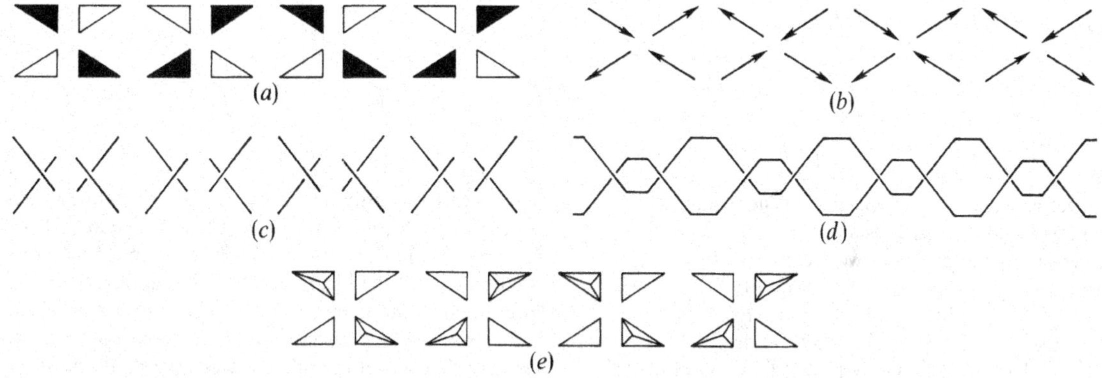

Figure 8.8.1
Strip patterns in which each copy of the motif is endowed with one of two mutually exclusive properties. The pattern in (*a*) is 2-chromatic. In (*b*) we show a pattern consisting of magnetic rods (taken as segments) in which the north pole is indicated by an arrowhead; each symmetry either preserves or else reverses all magnetic orientations. Parts (*c*) and (*d*) represent a layered pattern (see Section 5.7) of segments and a "layered filamentary pattern"; here a symmetry of the pattern may be accompanied by the reversal of the over-under relationship. The pattern in (*e*) consists of tetrahedra (three-sided pyramids) in 3-dimensional space, such that their bases form a strip pattern in the plane of the paper; some of the symmetries of the pattern of tetrahedra involve a reflection in the plane of the paper, while others do not. It is clear that the symmetry groups of these patterns are isomorphic and can be identified with the 2-color group of the 2-chromatic pattern in (*a*).

other "side"; but the converse correspondence has two shortcomings. A pattern on the two-sided plane may happen to be situated in its entirety on one side and thus correspond to a monochromatic (that is, uncolored) pattern in the usual sense. On the other hand, each copy of the motif of a pattern on the two-sided plane may be accompanied by a copy obtained by reflection in the plane itself—so that in trying to give a 2-chromatic representation in the ordinary plane we would have to assign to each copy of the motif *both* colors, white and black. Following Šubnikov (see below) this quandary is frequently resolved by saying that such patterns of the two-sided plane correspond to *gray* patterns of the ordinary plane. From this discussion it is clear that the number of strip or periodic groups of patterns on the two-sided plane equals the number of 2-color strip or periodic groups, plus twice the number of (uncolored) strip or periodic groups. Hence there are $31 = 17 + 2 \cdot 7$ strip groups, and $80 = 46 + 2 \cdot 17$ periodic ones. Similar relations hold for the numbers of strip or periodic pattern types. Moreover, it is not hard to verify that these 31 or 80 groups can be obtained from 3-dimensional crystallographic groups by restricting attention to those groups of the 3-dimensional space which map a given strip or plane onto itself. (For additional comments on this topic, see Grünbaum & Shephard [1983b].)

The historic interest of these considerations lies in the fact that 2-color groups first appeared in the literature in the guise of representations of symmetry groups of strip and periodic patterns on the two-sided plane, with no concern for or connection with 2-chromatic patterns. Speiser [1927] was the first to propose the investigation of groups of symmetries of the two-sided plane, and he showed that there are 31 strip groups. Partly as a consequence of Speiser's suggestion, several crystallographers determined the 80 periodic groups of the two-sided plane independently and simultaneously in 1929 (see Alexander & Herrmann [1929], Heesch [1929], Hermann [1929], Weber [1929]). It is interesting to note that Weber used 46 2-chromatic patterns (composed of black and white triangles) to represent by diagrams those groups of the two-sided plane which correspond neither to monochromatic nor to gray patterns. But Weber, as well as the other authors, continued stressing the 80 periodic symmetry groups of the 2-sided plane even when discussing black-and-white patterns. They clearly were not considering colored patterns and their groups as objects of independent interest. Similar attitudes were exhibited by many later authors (see, for example, Šubnikov [1930], [1940], [1946], [1951], Müller [1944], Burckhardt [1945], Dornberger-Schiff [1956], [1959], Holser [1958], [1961], Neronova & Belov [1961a], Wood [1964], Šubnikov & Kopcik [1972], and many of the references mentioned below). Brief surveys of the symmetry groups of finite, strip and periodic 2-color patterns, together with numerous references to earlier papers, can be found in Le Corre [1958] and Nowacki [1960]. Šubnikov introduced the term "antisymmetry" to denote any operation which involves interchanging the sides of the plane (in other words, which corresponds to a color-reversing symmetry in the 2-chromatic representation), and the expression soon found wide currency. To us this seems unfortunate, since this terminology may well have contributed to the delay in understanding (and even in considering) the symmetries of colored patterns. (The treatment of color groups in Jaswon & Rose [1983] is confused and misleading.)

Belov [1956b] noted the symmetry and color-symmetry groups of some Moorish patterns. He and his collaborators also used tilings colored with three or more colors (see Belov & Tarhova [1956a], [1956b], Belov [1956a], [1959], Belov & Belova [1957]), but their understanding of the concept of color symmetry seems to have been incomplete. For example, although the last of these papers utilizes 2-colored tilings to represent the 46 periodic 2-color groups, there is no mention of isohedral 2-colored tilings; moreover, in several cases in which such tilings could have been used, more complicated 2-colored tilings are shown without any explanation.*

Another discovery made, apparently, by Belov and his co-workers led to some k-chromatic tilings with $k > 2$ but also caused an unfortunate narrowing of the scope of investigation and so contributed to delays in the development of color-symmetry groups. The discovery in question amounts to the observation that each 2-chromatic pattern in the plane can be interpreted as an uncolored periodic pattern in 3-dimensional space. To do this it is only necessary to place the white copies of the motif, for example, in the plane $z = 0$ and also, by translation along the z-axis, in the planes $z = n$ for each integer n; the black copies of the motif are translated to the planes $z = n + \frac{1}{2}$ in a similar manner. Conversely, to any given periodic pattern in 3-space, in which all copies of the motif are contained in planes of the form $z = n$ and $z = n + \frac{1}{2}$ (copies in planes differing by an integral value of z being translates of each other parallel to the z-axis) there corresponds a 2-chromatic planar pattern (or a monochromatic one, or a gray one). This led Belov and co-workers to the idea of representing certain 3-dimensional patterns by k-colored tilings of the plane. The 3-dimensional patterns they considered are those in which all copies of the motif are situated in k different equidistant planes, and translation along the z-axis through k times the distance between neighboring planes is a symmetry of the pattern. In such a situation the motifs in each of k consecutive planes are assigned one of k colors, and everything is collapsed (by orthogonal projection) to a k-colored pattern in the (x,y)-plane. It was found that there are fifteen such representations of crystallographic groups in 3-space by multicolor patterns in the plane. These are shown in several publications (see Belov & Tarhova [1956a], [1956b], Belov & Belova [1957], Belov, Belova & Tarhova [1958], [1964], Šubnikov & Kopcik [1972]; there are slight differences between the various publications, and the last two should be considered as definitive). Observing that among these fifteen multicolor patterns there are four pairs which involve "mere reversal of the colour sequence", Lockwood & Macmillan [1978, Chapter 11] assert that "in fact there are only eleven polychromatic types that allow rotations". In any case, even a very hasty glance at the colored tilings in these papers makes it clear that the groups of color symmetries of chromatic planar patterns were not in their authors' minds. Indeed, the specific colors were used only as a code for the sequence of parallel planes, and so, for example, the two 4-chromatic tiling patterns in Figure 8.8.2 are shown as representations of the same 3-dimensional crystallographic group, although these 4-chromatic patterns clearly have different 4-color

* See the Appendix beginning on page 653.

(*a*)

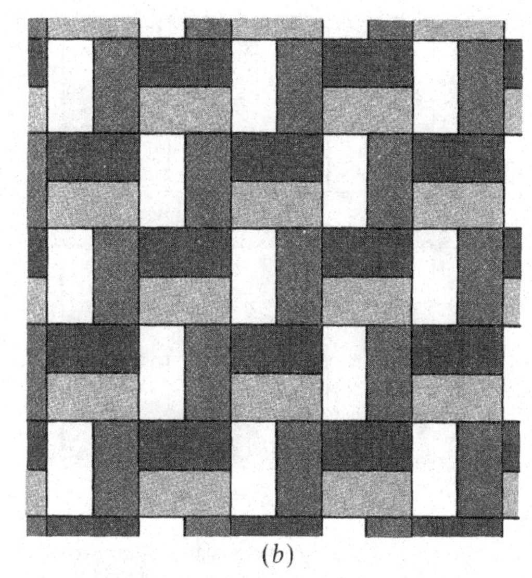
(*b*)

Figure 8.8.2
Two 4-chromatic patterns, used by Belov and co-workers to represent the same 3-dimensional crystallographic group (see Belov, Belova & Tarhova [1964, Plate I], Šubnikov & Kopcik [1972, Figure 230]). The 4-color groups of these patterns are clearly different, since the pattern in (*a*) has *p4* as its underlying group $S_C(\mathscr{P})$, and the one in (*b*) has *p4g*.

groups. Extensions of the investigations of Belov and co-workers can be found in Zamorzaev [1969], Palistrant & Zamorzaev [1971], Kopcik & Kužukeev [1972], Palistrant [1972], Zamorzaev, Galyarskii & Palistrant [1978]. It seems that the unwillingness of many crystallographers to consider any symmetries except those arising from isometries of the 3-dimensional space, hampered their understanding of color symmetries in the plane. This, in turn, led to delays in understanding the 3-dimensional analogues.

The change in attitude among crystallographers—to an acceptance of color groups as "symmetries" in which each geometric isometry is paired with a systematic change in the labels attached to the objects under consideration—seems to have occurred mainly in the 1960's, although some explicit indications occur earlier, for example in Cochran [1952], Mackay [1957], Niggli [1959]. This attitude, which is essentially the one adopted in this chapter, is fully present, for example, in van der Waerden & Burckhardt [1961], Bohm [1963], MacGillavry [1965] and Burckhardt [1966]. However, other attitudes continue to be advocated and since the same words are often used in various meanings, some confusion is inevitable.

We shall now give an account of those works which are directly concerned with colored patterns in the plane. There is an abundance of literature on various aspects of color symmetries in the 3-dimensional space, but no references are given as this is outside the scope of the book.

Loeb [1971a] gives an original, interesting and satisfactory account of the 2-color groups, using the word "configuration" instead of "group". Unfortunately, when discussing multicolor patterns, Loeb restricts the admissible color changes so severely that he obtains a total of only 54 periodic k-color "configurations" with $k \geq 3$, and many of the 2-chromatic patterns classified by Loeb earlier in the book do not qualify as 2-color "configurations". Moreover, his classification does not apply to a colored pattern (or tiling) unless a set of symmetry elements (centers of rotation, lines of reflection, etc.) compatible with the restricted types of color symmetries admitted, has been designated. As shown by the example in Figure 8.8.3, the same 4-chromatic coloring of the square tiling belongs to different "color configurations" depending on the choice of symmetry elements. Although the impression is given that color groups are being con-

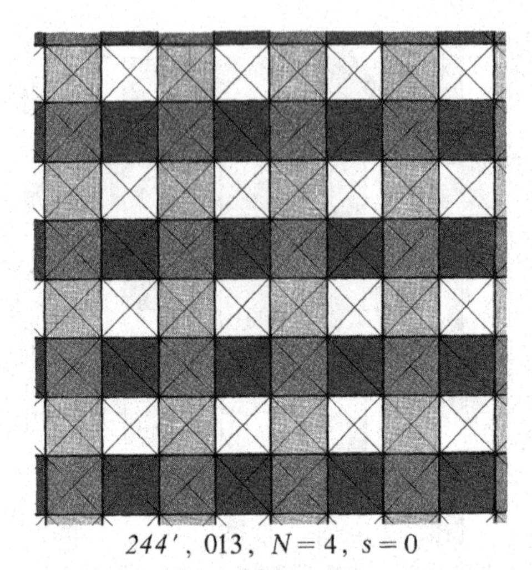

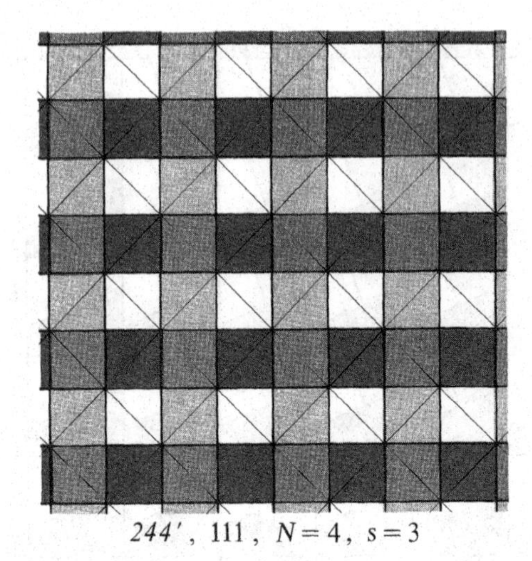

Figure 8.8.3
The same 4-chromatic coloring of the square tiling can be interpreted as representing two of Loeb's "color configurations"; Loeb's symbols are indicated near each. His "color configurations" depend not only on the 4-chromatic tiling but also on the choice of symmetry elements (centers of rotation, lines of reflection, etc); hence the classification does not lead to k-color groups.

sidered, this is not what actually happens. (This impression is reinforced by the remarks of W. T. Holser in the Foreword to Loeb [1971a, page x]. He says: "The enumeration of plane color groups [in this book] concludes with a detailed analysis of some of M. C. Escher's drawings".) In particular, colored tilings which have not only the same color groups but are even of the same color pattern type are assigned by Loeb to different "color configurations" (see Figure 8.8.4).

Macdonald & Street [1977], [1978a], [1978b] expand Loeb's investigations by removing his restrictions on the color symmetries and discuss colorings of "friezes" (strip patterns) and "crystals" (periodic patterns).

In complete analogy to the way in which isohedral or homeohedral tilings can be generalized to 2-isohedral or 2-homeohedral tilings, or monomotif patterns to dimotif patterns, it is possible to generalize k-chromatic patterns to *(k_1,k_2)-chromatic patterns.* By this we mean a monomotif or dimotif pattern $\mathscr{P}$ and a coloring $\mathscr{P}^\wedge$ of $\mathscr{P}$ such that the colored copies of the motif form two transitivity classes under the color symmetries of $\mathscr{P}^\wedge$. As examples we show in Figure 8.8.5 several (k_1,k_2)-colorings of the uniform tilings (3^6) and (3.6.3.6). In Section 4.8 we mentioned (1,1)-chromatic colorings of the Laves tilings in connection with the enumeration of 2-homeohedral tilings. Other examples of (k_1,k_2)-chromatic patterns are given in Macdonald [1979] and Roth [1982], [1983], [1984], where also more general considerations are presented. In view of its wide applicability and the challenging open problems it presents, this topic may attract considerable attention in the future.

As generalizations of 2-color symmetries (understood as "antisymmetries" in the sense of Šubnikov), several authors have investigated the "symmetry" groups of "patterns" in which the copies of the motif have not just one, but $n \geq 2$ distinct properties each of which can be present or absent ("antisymmetries of several kinds"). In principle, these are clearly special cases of the general k-colorings and could be found from them by selection. In practice, since $k = 2^n$, in most instances it is not practical to proceed in this manner; also, most of the work on "multiple antisymmetries" occurred before the general concept of color groups was well understood. Mackay [1957] dealt with periodic groups for $n = 2$ antisymmetry operations which are interchangeable to some extent. Roman [1959], Pabst [1962], [1963], Šubnikov [1962a], [1962b], [1963a], Belov, Kuncevič & Neronova [1962] determined the number of strip groups for $n = 2$ (that

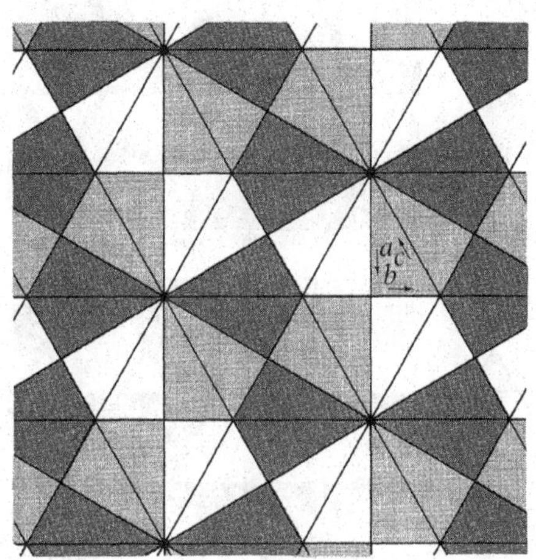

PP46[3]$_1$ $[a^+b^+c^+\ (2); a^-\ (12)b^-\ (12)c^-\ (13)]$

Loeb: $\underline{2}\ \underline{3}\ \underline{6}$, 014, $N = 3, s = 0$

PP46[3]$_1$ $[a^+b^+c^+\ (1); a^-\ (12)b^-\ (12)c^-(13)]$

Loeb: $\underline{2}\ \underline{3}\ \underline{6}$, 014, $N = 3, s = 1$

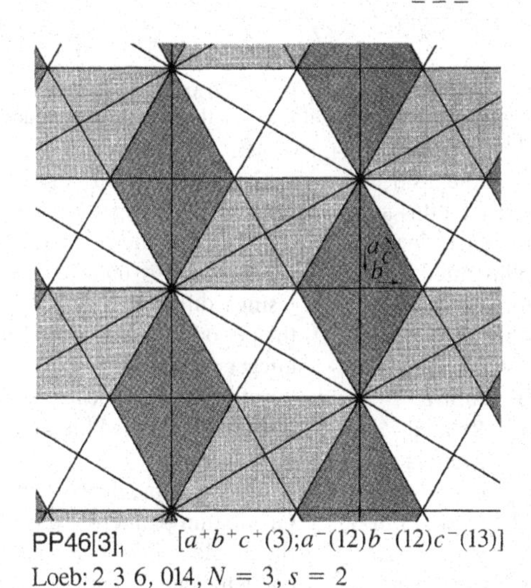

PP46[3]$_1$ $[a^+b^+c^+(3); a^-(12)b^-(12)c^-(13)]$

Loeb: $\underline{2}\ \underline{3}\ \underline{6}$, 014, $N = 3, s = 2$

Figure 8.8.4
Three 3-color "configurations" of Loeb [1971*a*, Figure 88], redrawn to use the same three colors. In Loeb's interpretation they represent distinct "3-color configurations", although they clearly have the same 3-color groups ($p6m[3]_1$ in our notation) and are of the same 3-color pattern types PP46$[3]_1$. Near each tiling we noted Loeb's symbols, as well as its color-incidence symbol in the notation of Section 8.6. The tilings belong to different homeochromatic types.

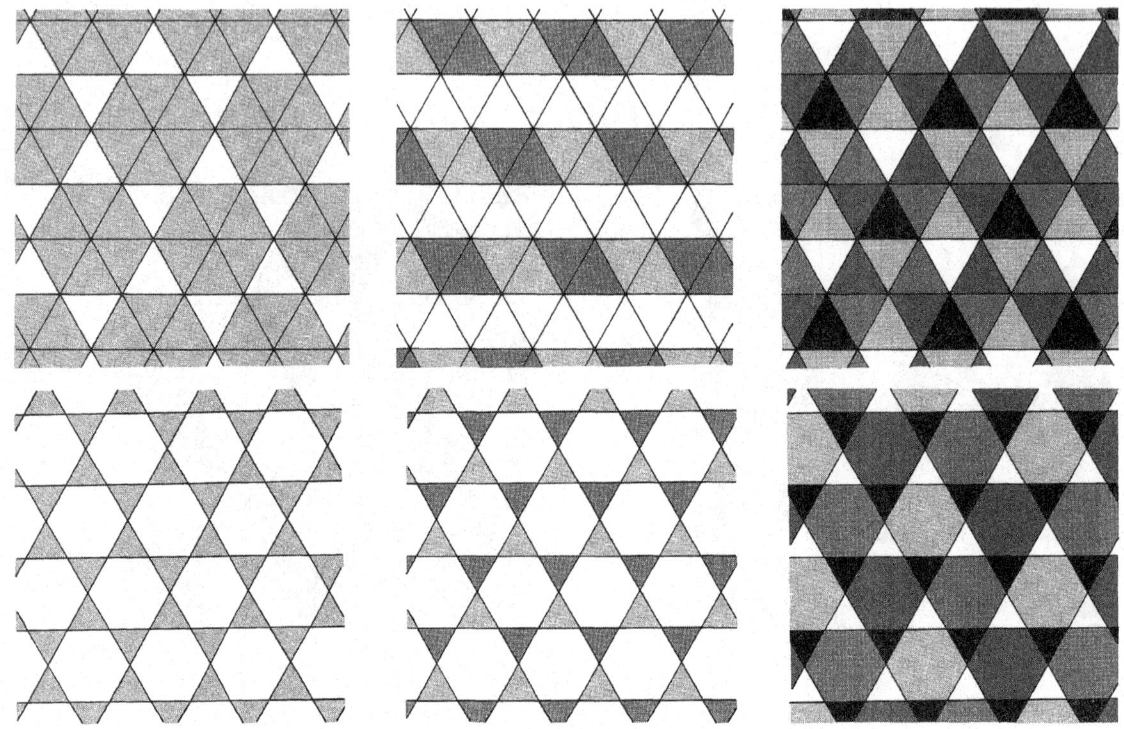

Figure 8.8.5
Examples of (k_1,k_2)-chromatic tilings with $(k_1,k_2) = (1,1)$, $(1,2)$ and $(2,3)$. In the illustrations we use white and black for the first set of colors, shades of gray for the second. The underlying tilings are the uniform tilings (3^6) and $(3.6.3.6)$.

is, the number of black-and-white groups of the 2-sided strip); Palistrant & Zamorzaev [1964] extended this to all $n \geq 2$. Zamorzaev & Palistrant [1960] determined the periodic groups for all $n \geq 2$, and illustrated those corresponding to $n = 2$ by 4-colored tilings (in Zamorzaev & Palistrant [1961]); other determinations of the periodic groups with $n = 2$ kinds of "antisymmetry" were carried out by Palistrant & Zamorzaev [1963] and Palistrant [1963]. Related material is presented in Pawley [1961], Neronova & Belov [1961a], [1961b], Šubnikov [1963a], Roman [1963], [1970], [1971], Palistrant [1965], [1966], [1967], [1968], [1972], Neronova [1966], [1967], Zamorzaev [1967]. Detailed surveys of these and related investigations are given by Donnay [1967], Kopcik [1967], [1975], Zamorzaev & Palistrant [1980], and Zamorzaev [1976]; Zamorzaev's book presents many of the enumerations in extensive tables.

The 23 periodic 3-color groups were first presented in Grünbaum [1976] and in Grünbaum & Shephard [1977e]. An independent enumeration of these groups, together with the enumeration of all periodic k-color groups for $k \leq 60$, was performed by Wieting [1981]. Using a different approach, Senechal [1979a] enumerates the periodic k-color groups for many infinite families of integers k (but not for all integers). In the same very interesting paper, for each periodic pattern type all values of k are determined for which the pattern admits a perfect k-coloring.

The color-incidence symbols first appeared in Grünbaum & Shephard [1979a] and were used in listing the 2-color and 3-color groups. The description of the color-incidence symbols given in this paper is valid only for primitively colored tilings. The 3-color pattern types were first illustrated in Grünbaum & Shephard [1983b]. A recent exposition of color symmetries and color patterns, and their history, can be found in Schwarzenberger [1984].

TILINGS BY POLYGONS

Figure 9.0.1
A tiling designed by Roger Penrose. The prototile is an 18-iamond sometimes known as "Penrose's wheelbarrow"; it does not admit isohedral tilings, and occurs in 6 direct and 6 reflected aspects in the unique monohedral tiling it admits. For other tilings by n-iamonds, see Table 9.4.1 and Figures 9.4.2 and 9.4.5.

TILINGS BY POLYGONS

In this chapter we consider those properties of tilings which depend essentially on the fact that their tiles are polygons. In contrast to Chapter 2, here we do not restrict attention to regular polygons.

One of the oldest and most challenging problems in this area is the determination of all convex polygons that can serve as prototiles of monohedral tilings. A survey of the present knowledge is given in Section 9.3, and a brief history of attempts to solve the problem appears in Section 9.6.

We recall from Section 1.1 that we use the words "corners" and "sides" for parts of a polygon, reserving "vertices" and "edges" for elements of a tiling. Thus an n-gon is a polygon with n corners and n sides, each of which is a straight line-segment joining two corners, see Figure 1.1.2. A normal tiling by polygons is called *proper* if the intersection of any two tiles is contained in a side of each tile. The tilings in Figures 9.0.1, 1.3.8(*b*), (*c*), 1.3.9 and 1.3.10, as well as many of the tilings by star polygons shown in Section 2.5, are not proper; in fact, any tiling in which a corner of a tile is not a vertex of the tiling is necessarily not proper. Every edge-to-edge tiling is proper, but there are proper tilings that are not edge-to-edge, see for example Figure 1.3.8(*a*) and the tilings illustrated in Section 2.4. Clearly every tiling by convex tiles is polygonal; it is also easy to see that it is necessarily proper.

In the first three sections of this chapter we are mainly concerned with proper tilings. However, such a restriction is not appropriate in the discussion of tilings by polyominoes, polyiamonds and polyhexes in Section 9.4 or of spiral tilings in Section 9.5.

9.1 ISOHEDRAL TILINGS BY POLYGONS

Many of the published treatments of isohedral tilings by polygons are concerned with the discovery of possible shapes of polygonal prototiles that admit such tilings, and exhibit just one tiling by each kind of tile (see, for example, Heesch & Kienzle [1963]). It seems to us that it is of greater interest to find *all* possible types of proper isohedral tilings by polygons. Here we shall show how this can be done in the case of convex tiles; in the next section we shall deal with the same problem in the non-convex case. The discussion follows the presentation in Grünbaum & Shephard [1978d] where additional details can be found.

In order to make our exposition precise we must first explain exactly what we mean when we say that two isohedral tilings by polygons are of the same type. Of course, we already have the definition of "isohedral type" given in Section 6.2, which coincides with that of "homeomeric type" discussed in Section 7.2. However, some refinement of this classification is desirable as we can see, for example, from the tilings denoted by P_3-4, P_3-5 and P_3-6 in Figure 9.1.1. These three are all of the same isohedral type IH53 but their appearance differs sufficiently for us to attempt to distinguish them in some

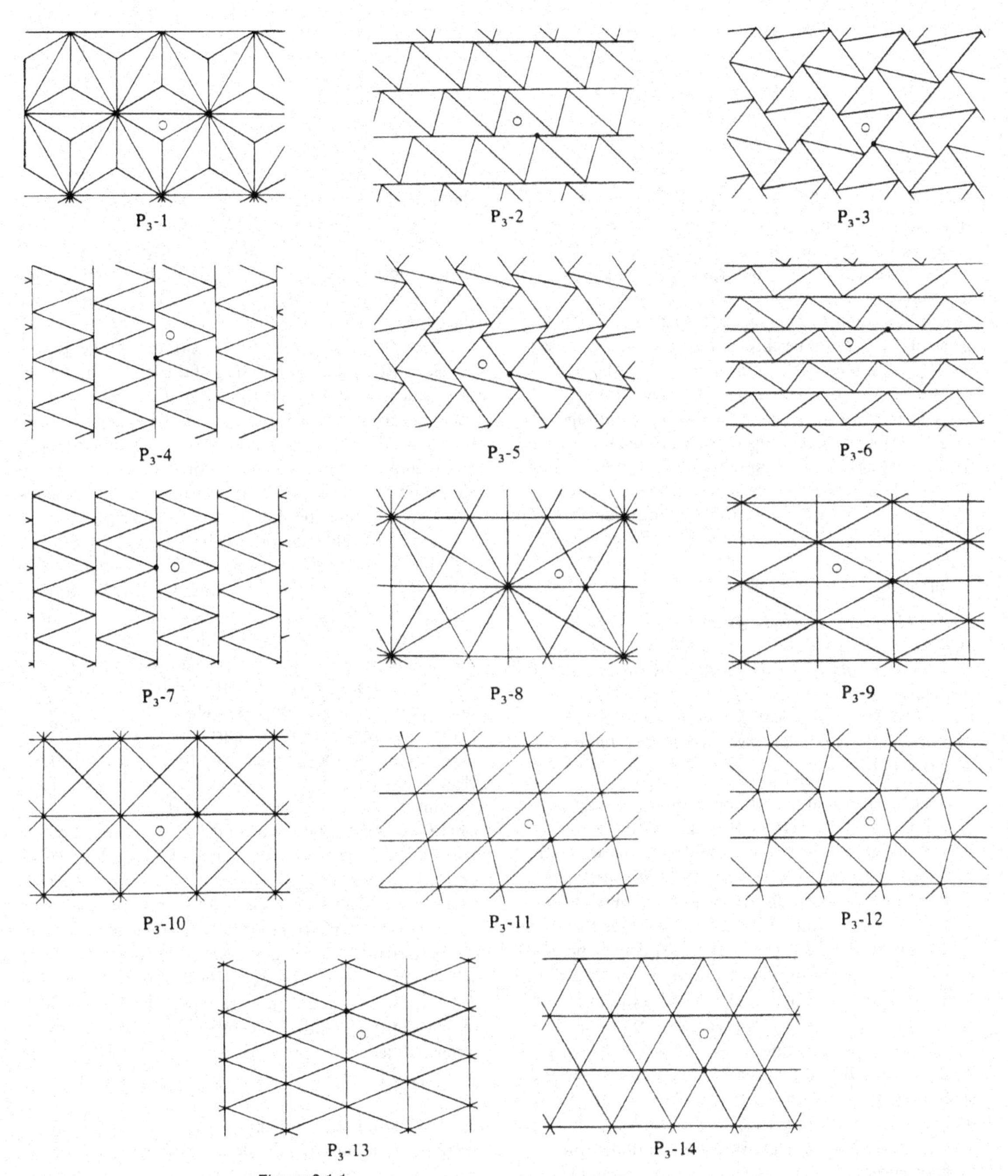

Figure 9.1.1
The fourteen polygonal isohedral types of proper tilings by triangles.

way. It turns out that this is possible, and we do it by introducing the concept of "polygonal isohedral type".

The definition of homeomeric tilings $\mathcal{T}_1$, $\mathcal{T}_2$ depended on the existence of a homeomorphism ϕ compatible with the symmetry groups of $\mathcal{T}_1$ and $\mathcal{T}_2$ (see Section 7.2). Such a homeomorphism maps the vertices of $\mathcal{T}_1$ onto those of $\mathcal{T}_2$. If ϕ can be chosen in such a way that a vertex V of $\mathcal{T}_1$ is a corner of a tile T in $\mathcal{T}_1$ if and only if $\phi(V)$ is a corner of a tile $\phi(T)$ in $\mathcal{T}_2$, then $\mathcal{T}_1$ and $\mathcal{T}_2$ are said to be of the same *polygonal isohedral type.* This simply means that every tile T of $\mathcal{T}_1$ has the same number of sides as the corresponding tile $\phi(T)$ of $\mathcal{T}_2$, and if any side E of T is made up of two or more edges E_1, $E_2, \ldots, E_n$ of $\mathcal{T}_1$, then the corresponding side $\phi(E)$ of $\phi(T)$ is made up of the same number of edges $\phi(E_1)$, $\phi(E_2), \ldots, \phi(E_n)$ of $\mathcal{T}_2$.

Our main result is as follows.

9.1.1 *There exist precisely* 107 *polygonal isohedral types of proper tilings by convex polygons; of these,* 14 *types have triangular tiles,* 56 *types have quadrangular tiles,* 24 *types have pentagonal tiles and* 13 *types have hexagonal tiles. There are no other polygonal isohedral tilings by convex tiles; in particular, there are no such tilings by n-gons with* $n \geq 7$.*

The last part of this assertion is a direct consequence of Statement 3.2.3. Illustrations of the 107 types are shown in Figures 9.1.1 to 9.1.4, and further information is given in the tables. Part of the proof of the first half of Statement 9.1.1 follows directly from the results of Section 6.2. From column (6) of Table 6.2.1 we see that 47 of the 81 isohedral types can be realized by convex polygonal tiles arranged edge-to-edge, and in the tables these types are marked by an asterisk near the symbol of the type. All the non-edge-to-edge types can be obtained from these 47 by examining each in turn and deciding whether it can also be realized by polygons with fewer sides. This actually amounts to checking whether one or more of the interior angles can be made to have the value π, so that the corresponding vertex of the tiling is no longer a corner of the tile. The process of examining each case is straightforward, though the total effort necessary is considerable. For most of the types, a polygon can be the prototile of a tiling of that polyhedral isohedral type only if its angles and sides fulfill certain conditions. The conditions involving angles are listed explicitly in the tables; those involving edge-lengths of the sides arise as a consequence of the edge-transitivities. For example, since the edge-transitivities of the type P_4-33 are $\alpha(\beta\gamma)\gamma\beta$ it follows that the length of the second side ($\beta\gamma$) of the quadrangle must be equal to the sum of the lengths of the third side (γ) and the fourth side (β). Of course, there are other restrictions on edge-lengths, in the form of inequalities that arise as consequences of convexity and of the constraints on angles and sides. For example, in type P_3-5 the triangle inequality implies that the edge in transitivity class γ must be shorter than that in class β.

We conclude this section with a description of the Dirichlet tilings associated with dot patterns (compare Sections 5.3 and 5.4). Since these are the only kinds of tilings we shall discuss in the remaining part of this section, for brevity we shall call them *D*-tilings. Thus a tiling $\mathcal{T}$ is a *D*-tiling provided there exists a dot pattern D such that each tile of $\mathcal{T}$ is the Dirichlet region of a dot in D, that is, each tile of $\mathcal{T}$ consists of all the points of the plane whose distance from a particular dot in D is less than or equal to the distance from every other dot. It is clear that each *D*-tiling is an isohedral edge-to-edge tiling by convex polygons. Therefore, from Tables 9.1.1 to 9.1.4, each *D*-tiling must be of one of the 47 types indicated by asterisks in the first column of each of these tables. However not every tiling of these types is a *D*-tiling; information about the possibilities is given in column (2) of Table 9.1.5. In fact, the relationship between the pattern type of dot pattern and the polygonal isohedral type of the corresponding *D*-tiling is rather complicated. It is possible for dot patterns of the same pattern type to lead to *D*-tilings of different types, and, on the other hand, for dot patterns of different pattern types to lead to *D*-tilings of the same type. These relationships are summarized in Table 9.1.5.

For a more precise statement it would be necessary to describe quantitatively the dependence between the parameters of the dot patterns and the resulting isohedral types of the *D*-tilings. Examples are given in Exercises 6, 7 and 8 at the end of this section.

* See the Appendix beginning on page 653.

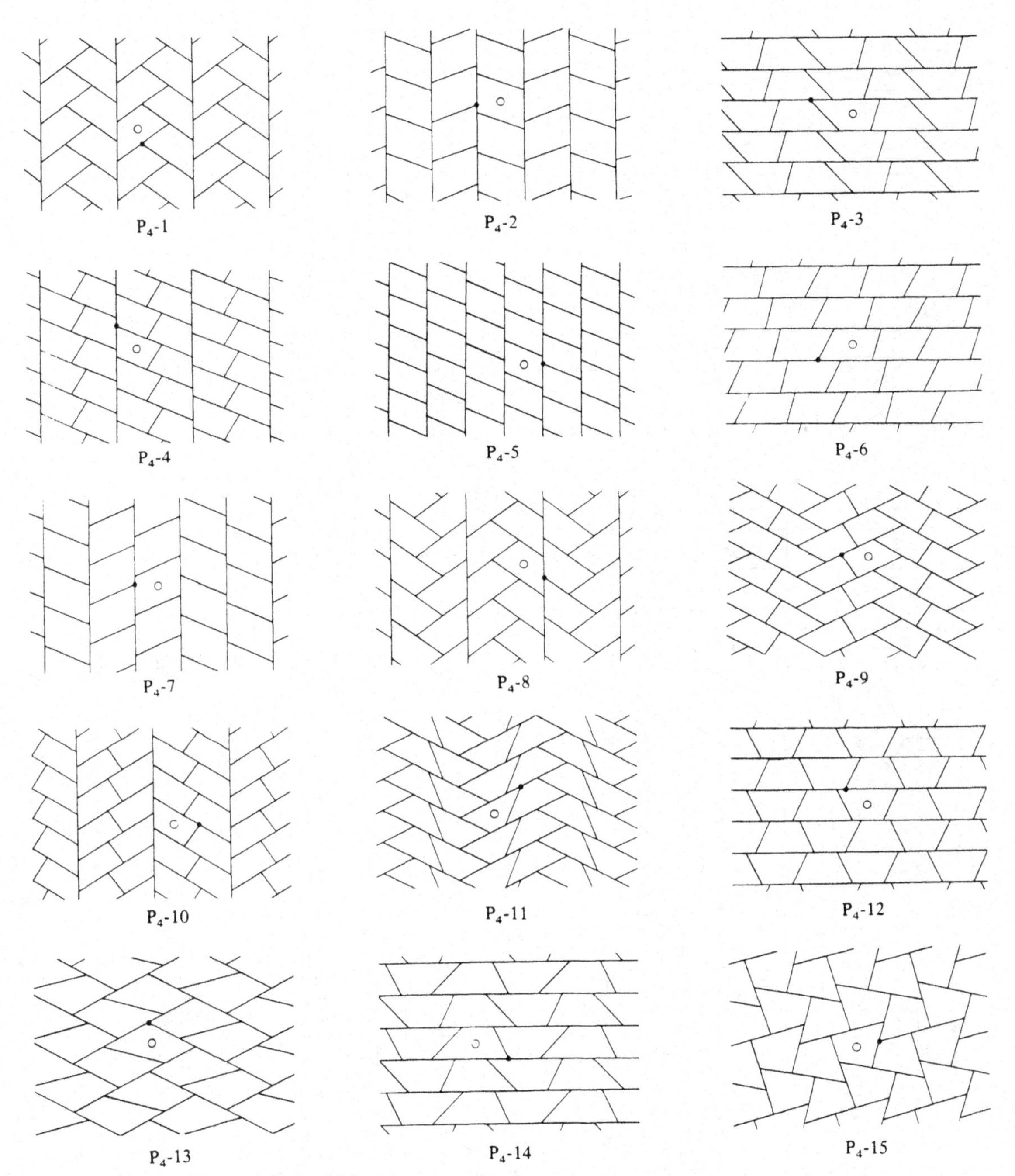

Figure 9.1.2
The 56 polygonal isohedral types of proper tilings by quadrangles.

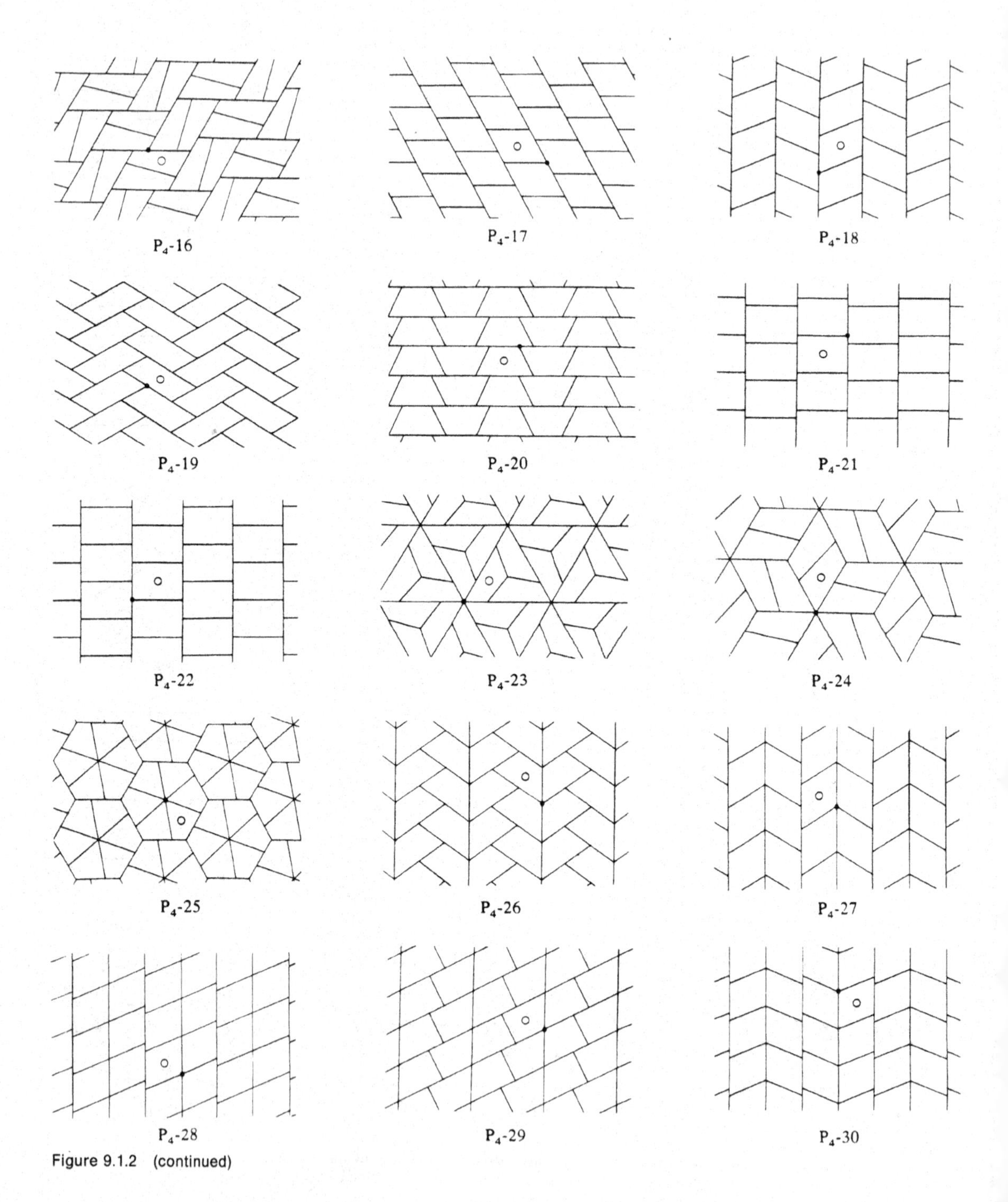

Figure 9.1.2 (continued)

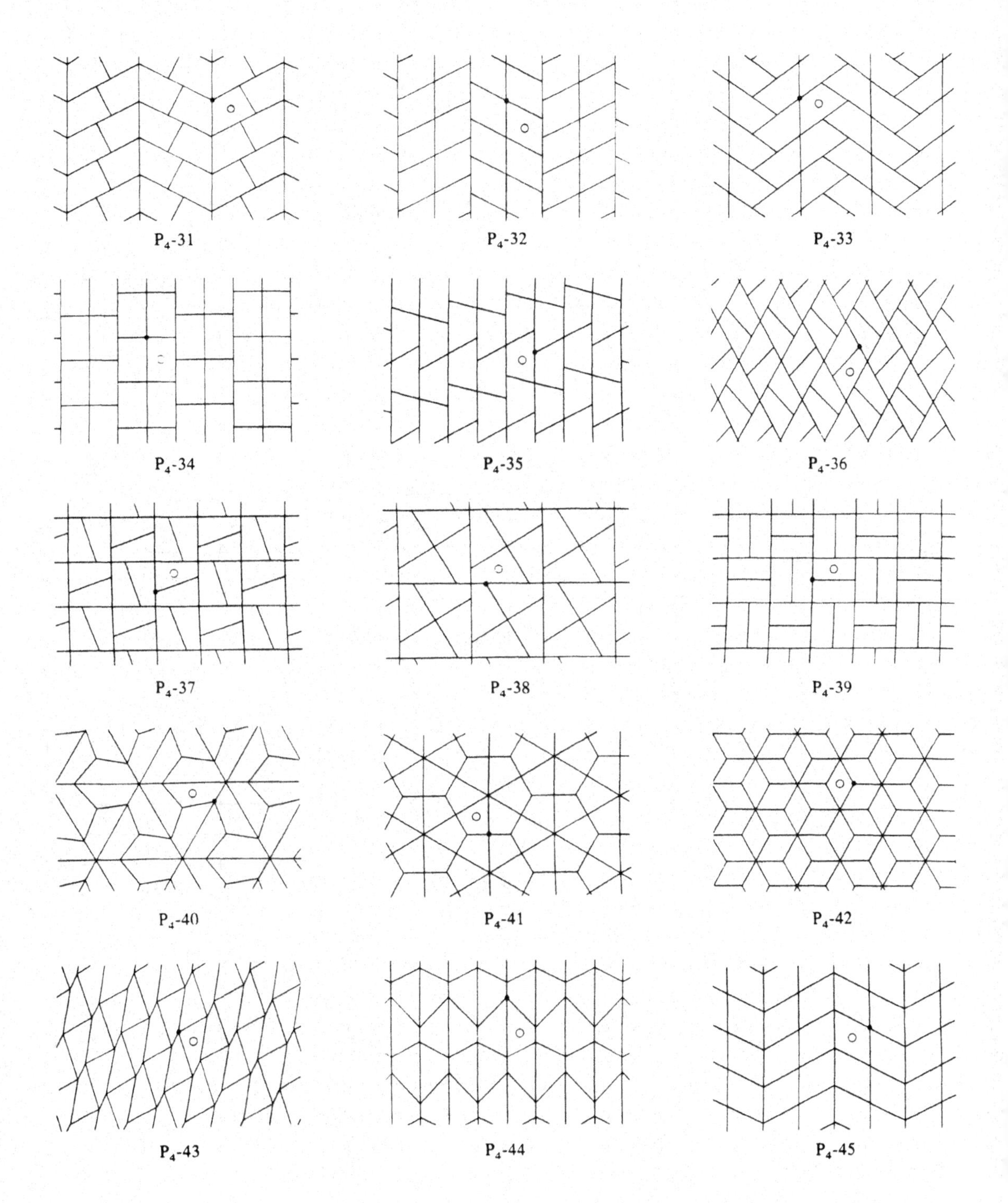
P_4-31 P_4-32 P_4-33
P_4-34 P_4-35 P_4-36
P_4-37 P_4-38 P_4-39
P_4-40 P_4-41 P_4-42
P_4-43 P_4-44 P_4-45

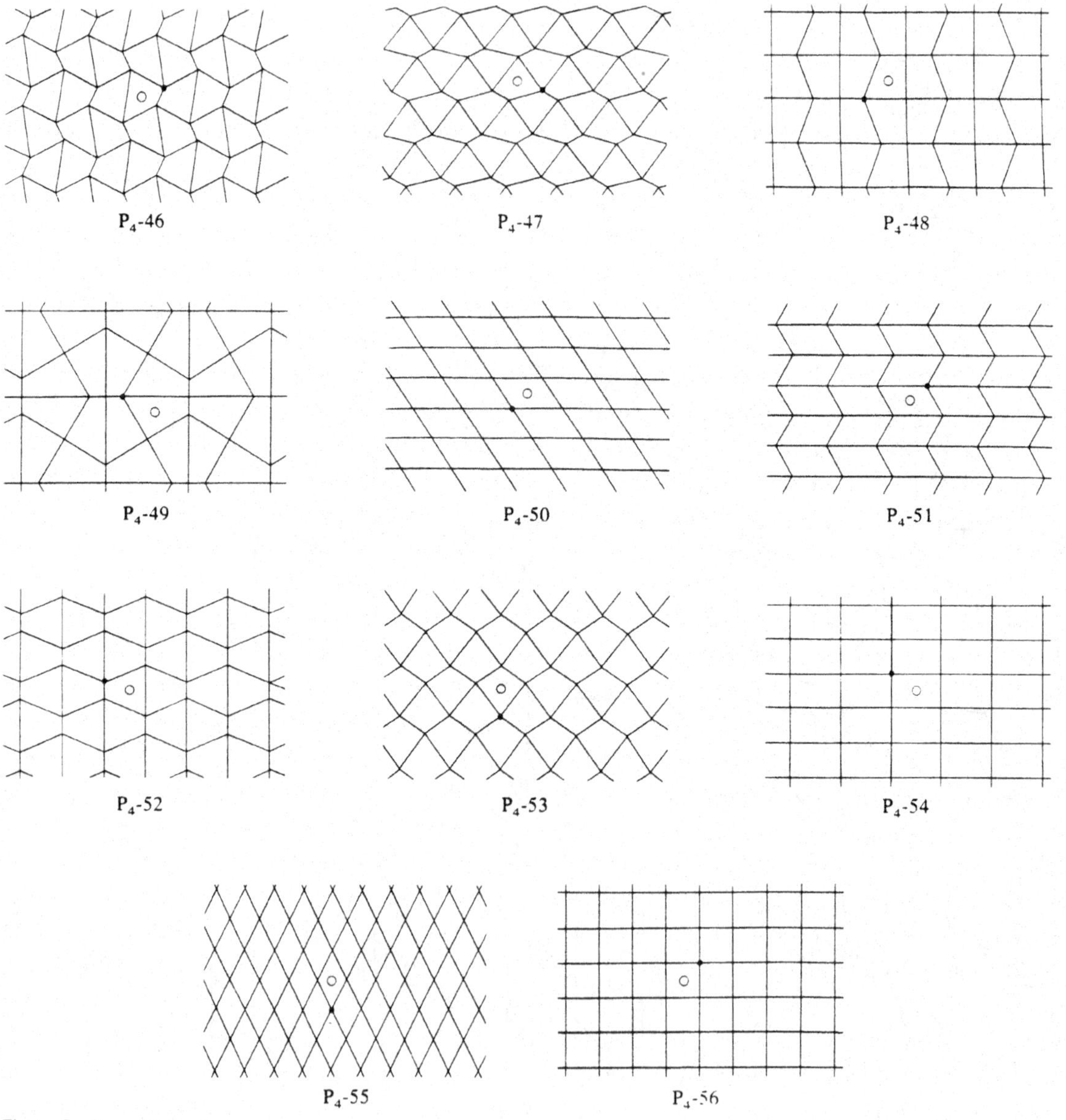

Figure 9.1.2 (continued)

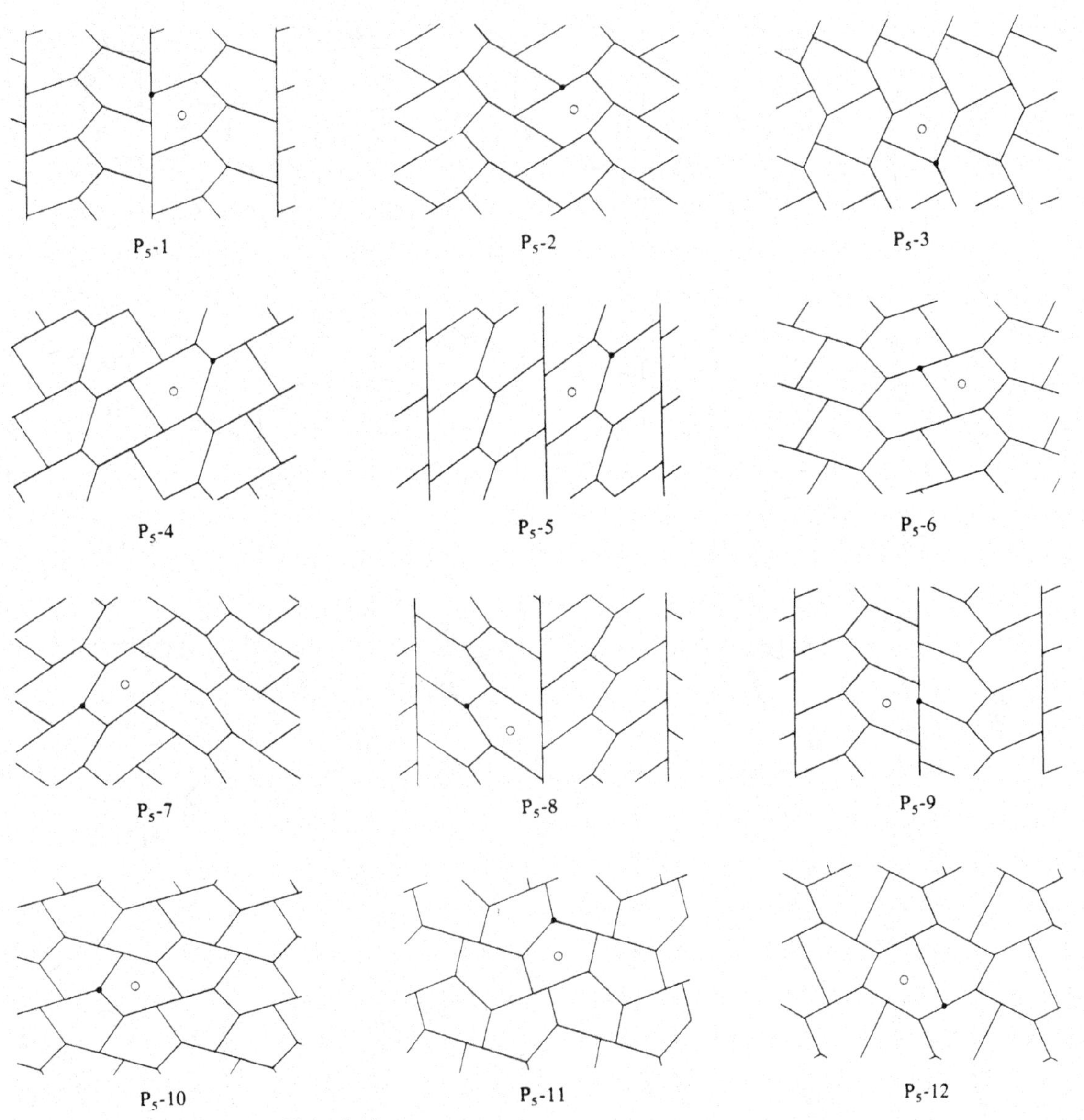

Figure 9.1.3
The 24 polygonal isohedral types of proper tilings by pentagons.

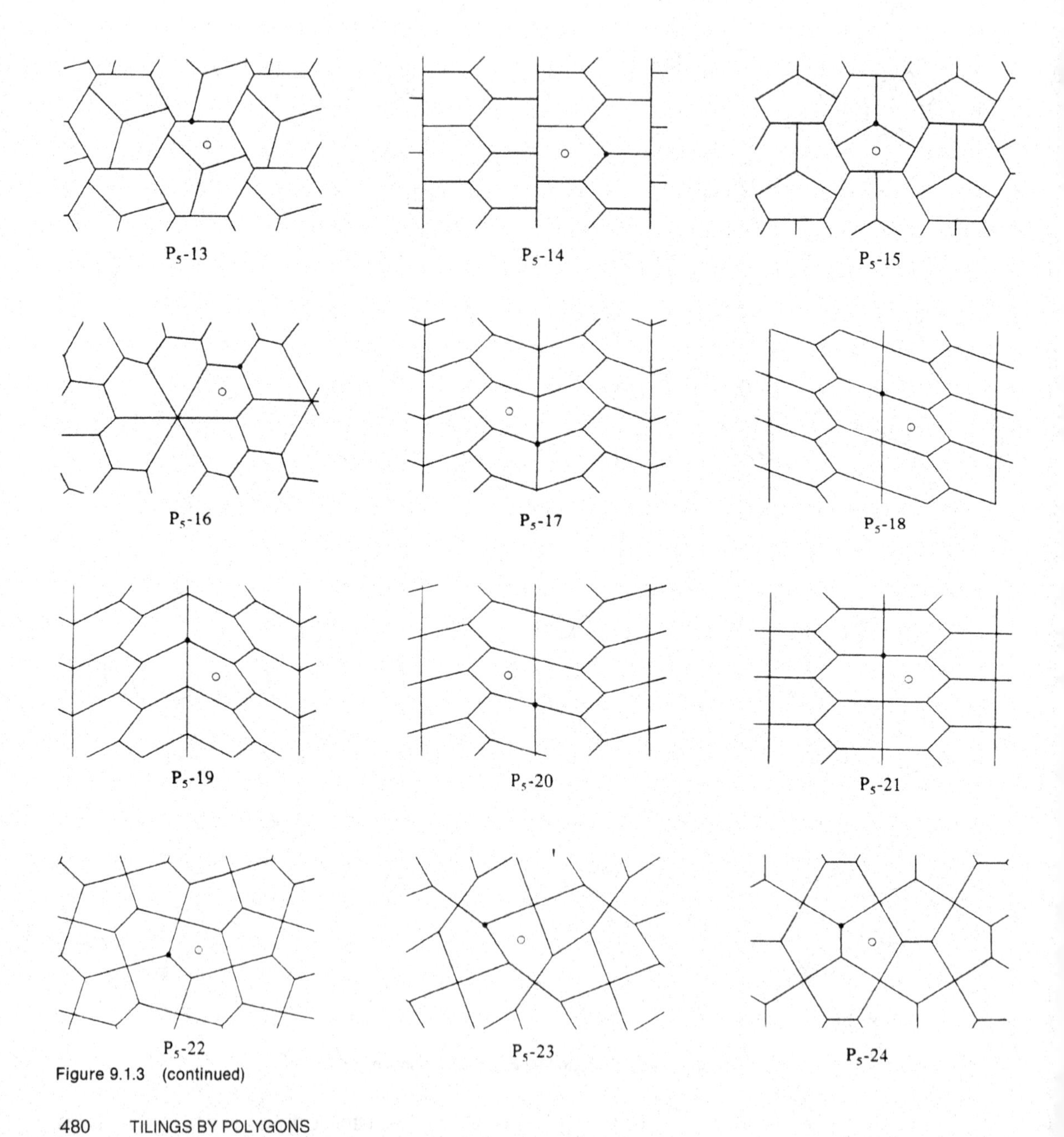

Figure 9.1.3 (continued)

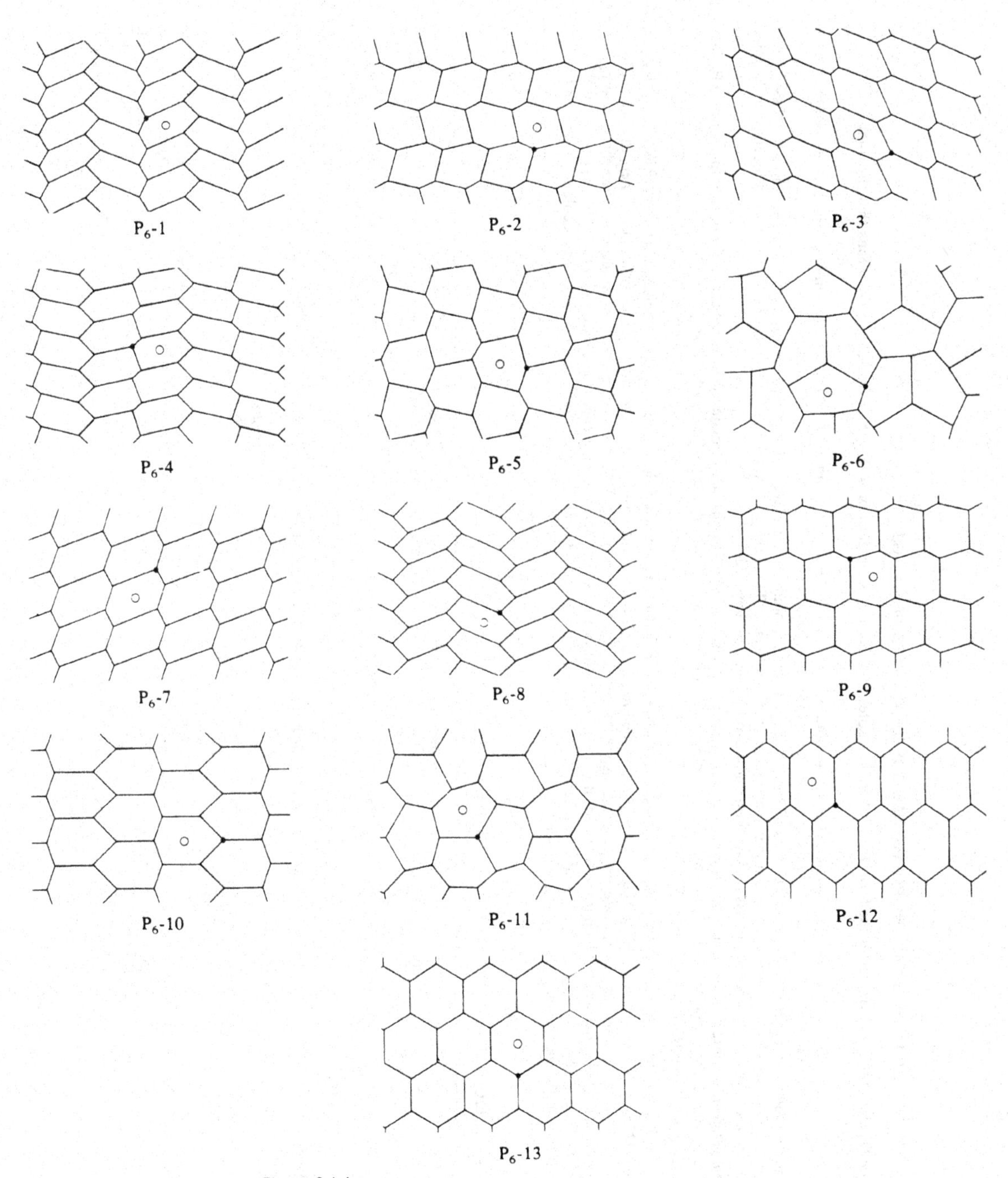

Figure 9.1.4
The thirteen polygonal isohedral types of proper tilings by hexagons.

Tables 9.1.1, 9.1.2, 9.1.3 and 9.1.4 THE POLYGONAL ISOHEDRAL TYPES OF TILINGS

Column (1) shows the list number in the form P_n-m where n denotes that the tiles are n-gons (n = 3, 4, 5 or 6). An asterisk means that the tiling is edge-to-edge.

Columns (2) to (5) give the same information as the corresponding columns in Table 6.2.1. In column (5) we also indicate, in parentheses, the symmetry group of the tile in those cases where this necessarily differs from the induced tile group.

Columns (6) and (7) indicate the transitivity classes of vertices and edges as in columns (7) and (8) of Table 6.2.1, except that, in column (6), parentheses enclose a letter that corresponds to a vertex of the tiling which is not a corner of the tile. This occurs when the two edges meeting at the vertex are collinear. In column (7), parentheses are used to enclose the symbols corresponding to two such collinear edges.

Column (8) gives the angle relations for the polygonal tile. Equalities between the edge-lengths can be deduced from column (7). For example, in type P_4-24 with edge transitivity $(\alpha\beta)\beta\gamma\alpha$, the first side $(\alpha\beta)$ of the tile is equal in length to the sum of the third side (β) and the fifth (α).

Column (9) corresponds to column (9) of Table 6.2.1.

Column (10) gives the corresponding type in the notation of Table 6.2.1. HK refers to Heesch & Kienzle [1963], S to Schattschneider [1978b].

Column (11) lists, for each type of tiling, the vertices at which the interior angles can be increased to π, thus reducing the number of corners of a tile. After each such vertex, or pair of vertices, we indicate the polygonal isohedral type of the resultant tiling. This reference is given in parentheses when the new tiling is of the same isohedral type as that from which it was derived; if it is of a different isohedral type then the reference is enclosed in [square] brackets. The latter possibility occurs when making an angle equal to π increases the symmetry group and so alters the isohedral type. The data in this column is used in the classification of tilings by non-convex polygons (see Section 9.2).

Table 9.1.1 ISOHEDRAL TYPES OF TILINGS BY TRIANGLES

List number (1)	Topological type (2)	Incidence symbol (3)	Symmetry group (4)	Tile group (5)	Vertex transitivity (6)	Edge transitivity (7)	Angle relations (8)	Aspects (9)	References (10)
*P_3-1	$[3.12^2]$	$[ab^+b^-; ab^-]$	$p6m$	$d1$	$\alpha\alpha\beta$	$\alpha\beta\beta$	$A = B = \pi/6$	6	IH40
P_3-2	$[4^4]$	$[a^+b^+c^+d^+; a^+b^+c^+d^+]$	$p2$	e	$(\alpha)\alpha\alpha\alpha$	$\alpha)\beta\gamma(\delta$	$A = \pi$	$2D$	IH46
P_3-3		$[a^+b^+c^+d^+; c^-b^+a^-d^+]$	pgg	e	$(\alpha)\alpha\alpha\alpha$	$\alpha)\beta\alpha(\gamma$	$A = \pi$	$2D, 2R$	IH51
P_3-4		$[a^+b^+c^+d^+; b^-a^-c^+d^+]$	pgg	$e\ (d1)$	$\alpha\alpha\alpha(\alpha)$	$\alpha\alpha(\beta\gamma)$	$A = C, D = \pi$	2	IH53
P_3-5			pgg	e	$(\alpha)\alpha\alpha\alpha$	$\alpha)\alpha\beta(\gamma$	$A = \pi$	$2D, 2R$	IH53
P_3-6			pgg	e	$\alpha(\alpha)\alpha\alpha$	$(\alpha\alpha)\beta\gamma$	$B = \pi$	$2D, 2R$	IH53
P_3-7		$[a^+b^+b^-a^-; a^+b^+]$	pmg	$d1(1)$	$(\alpha)\alpha\alpha\alpha$	$\alpha)\beta\beta(\alpha$	$A = \pi, B = D$	2	IH69
*P_3-8	$[4.6.12]$	$[a^+b^+c^+; a^-b^-c^-]$	$p6m$	e	$\alpha\beta\gamma$	$\alpha\beta\gamma$	$A = \pi/3, B = \pi/2$	$6D, 6R$	IH77
*P_3-9	$[4.8^2]$	$[a^+b^+c^+; a^+b^-c^-]$	cmm	e	$\alpha\alpha\beta$	$\alpha\beta\gamma$	$C = \pi/2$	$2D, 2R$	IH78
*P_3-10		$[ab^+b^-; ab^-]$	$p4m$	$d1$	$\alpha\alpha\beta$	$\alpha\beta\beta$	$A = B = \pi/4$	4	IH82
*P_3-11	$[6^3]$	$[a^+b^+c^+; a^+b^+c^+]$	$p2$	e	$\alpha\alpha\alpha$	$\alpha\beta\gamma$	—	$2D$	IH84
*P_3-12		$[a^+b^+c^+; a^-b^+c^+]$	pmg	e	$\alpha\alpha\alpha$	$\alpha\beta\gamma$	—	$2D, 2R$	IH85
*P_3-13		$[ab^+b^-; ab^+]$	cmm	$d1$	$\alpha\alpha\alpha$	$\alpha\beta\beta$	$A = B$	2	IH91
*P_3-14		$[aaa; a]$	$p6m$	$d3$	$\alpha\alpha\alpha$	$\alpha\alpha\alpha$	$A = B = C = \pi/3$	2	IH93

Table 9.1.2 ISOHEDRAL TYPES OF TILINGS BY QUADRANGLES

List number (1)	Topological type (2)	Incidence symbol (3)	Symmetry group (4)	Tile group (5)	Vertex transitivity (6)	Edge transitivity (7)	Angle relations (8)	Aspects (9)	References (10)	Related tilings (11)
P_4-1	$[3^6]$	$[a^+b^+c^+d^+e^+f^+; b^-a^-f^+e^-d^-c^+]$	pg	e	$(\alpha)\alpha\alpha\beta(\beta)\beta$	$\alpha)\alpha\beta(\gamma\gamma)(\beta$	$A = E = B + C = D + F = \pi$	$1D, 1R$	IH2	—
P_4-2		$[a^+b^+c^+d^+e^+f^+; c^-e^+a^-f^-b^+d^-]$	pg	$e\ (c2)$	$(\alpha)\beta\alpha(\beta)\alpha\beta$	$\alpha)\beta(\alpha\gamma)\beta(\gamma$	$B = E, C = F, A = D = \pi$	$1D, 1R$	IH3	—
P_4-3		$[a^+b^+c^+d^+e^+f^+; a^+e^+c^+d^+b^+f^+]$	$p2$	e	$\alpha\alpha\beta\beta(\beta)(\alpha)$	$\alpha\beta\gamma(\delta\beta\varepsilon)$	$A + B = C + D = E = F = \pi$	$2D$	IH4	—
P_4-4			$p2$	e	$(\alpha)\alpha\beta\beta(\beta)\alpha$	$\alpha)\beta\gamma(\delta\beta)(\varepsilon$	$A = E = C + D = B + F = \pi$	$2D$	IH4	—
P_4-5			$p2$	$e\ (c2)$	$(\alpha)\alpha\beta(\beta)\beta\alpha$	$\alpha)\beta(\gamma\delta)\beta(\varepsilon$	$A = D = \pi, B = E, C = F$	$1D$	IH4	—
P_4-6			$p2$	e	$\alpha(\alpha)\beta\beta(\beta)\alpha$	$(\alpha\beta)\gamma(\delta\beta)\varepsilon$	$B = E = C + D = A + F = \pi$	$2D$	IH4	—
P_4-7		$[a^+b^+c^+d^+e^+f^+; a^+e^+d^-c^-d^+f^+]$	pgg	$e\ (c2)$	$(\alpha)\alpha\beta(\beta)\beta\alpha$	$\alpha)\beta(\gamma\gamma)\beta(\delta$	$A = D = \pi, B = E, C = F$	$1D, 1R$	IH5	—
P_4-8			pgg	e	$(\alpha)\alpha\beta\beta(\beta)\alpha$	$\alpha)\beta\gamma(\gamma\beta)(\delta$	$A = E = C + D = B + F = \pi$	$2D, 2R$	IH5	—
P_4-9			pgg	e	$\alpha\alpha(\beta)\beta\beta(\alpha)$	$\alpha(\beta\gamma)\gamma(\beta\delta)$	$C = F = A + B = D + E = \pi$	$2D, 2R$	IH5	—
P_4-10			pgg	e	$\alpha(\alpha)\beta(\beta)\beta\alpha$	$(\alpha\beta)(\gamma\gamma)\beta\delta$	$B = D = C + E = A + F = \pi$	$2D, 2R$	IH5	—
P_4-11			pgg	e	$\alpha(\alpha)(\beta)\beta\beta\alpha$	$(\alpha\beta\gamma)\gamma\beta\delta$	$B = C = D + E = A + F = \pi$	$2D, 2R$	IH5	—
P_4-12		$[a^+b^+c^+d^+e^+f^+; a^+e^-c^+f^-b^-d^-]$	pgg	$e\ (d1)$	$(\alpha)\alpha\beta(\beta)\alpha\beta$	$\alpha)\beta(\gamma\delta)\beta(\delta$	$A = D = B + C = E + F = \pi$	2	IH6	—
P_4-13			pgg	e	$\alpha(\alpha)\beta\beta\alpha(\beta)$	$(\alpha\beta)\gamma\delta(\beta\delta)$	$B = F = A + E = C + D = \pi$	$2D, 2R$	IH6	—
P_4-14			pgg	e	$\alpha\alpha\beta\beta(\alpha)(\beta)$	$\alpha\beta\gamma(\delta\beta\delta)$	$A + B = C + D = E = F = \pi$	$2D, 2R$	IH6	—
P_4-15			pgg	e	$(\alpha)\alpha(\beta)\beta\alpha\beta$	$\alpha)(\beta\gamma)\delta\beta(\delta$	$A = C = B + E = D + F = \pi$	$2D, 2R$	IH6	—
P_4-16			pgg	e	$(\alpha)\alpha\beta\beta\alpha(\beta)$	$\alpha)\beta\gamma\delta(\beta\delta$	$A = F = B + E = C + D = \pi$	$2D, 2R$	IH6	—
P_4-17		$[a^+b^+c^+a^+b^+c^+; a^+b^+c^+]$	$p2$	$c2$	$\alpha(\alpha)\alpha\alpha(\alpha)\alpha$	$(\alpha\beta)\gamma(\alpha\beta)\gamma$	$B = E = \pi, A = D, C = F$	$1D$	IH8	—
P_4-18		$[a^+b^+c^+a^+b^+c^+; a^+c^-b^-]$	pgg	$c2$	$\alpha\alpha(\alpha)\alpha\alpha(\alpha)$	$\alpha(\beta\beta)\alpha(\beta\beta)$	$c = F = \pi, A = D, B = E$	$1D, 1R$	IH9	—
P_4-19			pgg	$c2$	$(\alpha)\alpha\alpha(\alpha)\alpha\alpha$	$\alpha)\beta(\beta\alpha)\beta(\beta$	$A = D = \pi, B = E, C = F$	$1D, 1R$	IH9	—
P_4-20		$[ab^+c^+dc^-b^-; db^+c^+a]$	pmg	$d1(s)$	$\alpha\alpha\alpha(\alpha)(\alpha)\alpha$	$\alpha\beta(\gamma\alpha\gamma)\beta$	$A = B, C = F, D = E = \pi$	2	IH13	—
P_4-21			pmg	$d1(s)\ (d2)$	$\alpha\alpha(\alpha)\alpha\alpha(\alpha)$	$\alpha(\beta\gamma)\alpha(\gamma\beta)$	$A = B = D = E = \frac{\pi}{2}$, $C = F = \pi$	1	IH13	
P_4-22		$[ab^+b^-ab^+b^-; ab^+]$	cmm	$d2$	$\alpha\alpha(\alpha)\alpha\alpha(\alpha)$	$\alpha(\beta\beta)\alpha(\beta\beta)$	$A = B = D = E = \frac{\pi}{2}$, $C = F = \pi$	1	IH17	—
P_4-23	$[3^4.6]$	$[a^+b^+c^+d^+e^+; e^+c^+b^+d^+a^+]$	$p6$	e	$\alpha\beta\gamma\beta(\beta)$	$\alpha\beta\beta(\gamma\alpha)$	$A = \frac{\pi}{3}, C = \frac{2\pi}{3}, E = \pi$	$6D$	IH21	—
P_4-24			$p6$	e	$\alpha(\beta)\gamma\beta\beta$	$(\alpha\beta)\beta\gamma\alpha$	$A = \frac{\pi}{3}, C = \frac{2\pi}{3}, B = \pi$	$6D$	IH21	—
P_4-25			$p6$	e	$\alpha\beta\gamma(\beta)\beta$	$\alpha\beta(\beta\gamma)\alpha$	$A = \frac{\pi}{3}, C = \frac{2\pi}{3}, D = \pi$	$6D$	IH21	—
P_4-26	$[3^3.4^2]$	$[a^+b^+c^+d^+e^+; a^-e^+d^-c^-b^+]$	cm	e	$\alpha\alpha\beta\beta(\beta)$	$\alpha\beta\gamma(\gamma\beta)$	$A + B = C + D = E = \pi$	$1D, 1R$	IH22	—
P_4-27			cm	$e\ (c2)$	$\alpha\alpha\beta(\beta)\beta$	$\alpha\beta(\gamma\gamma)\beta$	$A + B = C + E = D = \pi$	$1D, 1R$	IH22	—
P_4-28		$[a^+b^+c^+d^+e^+; a^+e^+c^+d^+b^+]$	$p2$	$e\ (c2)$	$\alpha\alpha\beta(\beta)\beta$	$\alpha\beta(\gamma\delta)\beta$	$A + B = C + E = D = \pi$	$1D$	IH23	—
P_4-29			$p2$	e	$\alpha\alpha(\beta)\beta\beta$	$\alpha(\beta\gamma)\delta\beta$	$A + B = D + E = C = \pi$	$2D$	IH23	—
P_4-30		$[a^+b^+c^+d^+e^+; a^-e^+c^+d^+b^+]$	pmg	$e\ (c2)$	$\alpha\alpha\beta(\beta)\beta$	$\alpha\beta(\gamma\delta)\beta$	$A + B = C + E = D = \pi$	$1D, 1R$	IH24	—
P_4-31			pmg	e	$\alpha\alpha(\beta)\beta\beta$	$\alpha(\beta\gamma)\delta\beta$	$A + B = D + E = C = \pi$	$2D, 2R$	IH24	—
P_4-32		$[a^+b^+c^+d^+e^+; a^+e^+d^-c^-b^+]$	pgg	$e\ (c2)$	$\alpha\alpha\beta(\beta)\beta$	$\alpha\beta(\gamma\gamma)\beta$	$A + B = C + E = D = \pi$	$1D, 1R$	IH25	—
P_4-33			pgg	e	$\alpha\alpha(\beta)\beta\beta$	$\alpha(\beta\gamma)\gamma\beta$	$A + B = D + E = C = \pi$	$2D, 2R$	IH25	—
P_4-34		$[ab^+c^+c^-b^-; ab^-c^+]$	cmm	$d1\ (d2)$	$\alpha\alpha\beta(\beta)\beta$	$\alpha\beta(\gamma\gamma)\beta$	$A = B = C = E = \frac{\pi}{2}, D = \pi$	1	IH26	—
P_4-35	$[3^2.4.3.4]$	$[a^+b^+c^+d^+e^+; a^+d^-e^-b^-c^-]$	pgg	$e\ (d1)$	$(\alpha)\alpha\beta\alpha\beta$	$\alpha)\beta\gamma\beta(\gamma$	$A = \pi, B = E, C = D$	2	IH27	—
P_4-36			pgg	e	$\alpha\alpha\beta(\alpha)\beta$	$\alpha\beta(\gamma\beta)\gamma$	$A + B = D = \pi$	$2D, 2R$	IH27	—

Table 9.1.2 (continued)

List number (1)	Topological type (2)	Incidence symbol (3)	Symmetry group (4)	Tile group (5)	Vertex transitivity (6)	Edge transitivity (7)	Angle relations (8)	Aspects (9)	References (10)	Related tilings (11)
P_4-37		$[a^+b^+c^+d^+e^+; a^+c^+b^+e^+d^+]$	$p4$	e	$\alpha\alpha\beta(\alpha)\gamma$	$\alpha\beta(\beta\gamma)\gamma$	$A + B = D = \pi, C = E = \frac{\pi}{2}$	$4D$	IH28	—
P_4-38			$p4$	e	$(\alpha)\alpha\beta\alpha\gamma$	$\alpha)\beta\beta\gamma(\gamma$	$A = B + D = \pi, C = E = \frac{\pi}{2}$	$4D$	IH28	—
P_4-39		$[ab^+c^+c^-b^-; ac^+b^+]$	$p4g$	$d1$ $(d2)$	$\alpha\alpha\beta(\alpha)\beta$	$\alpha\beta(\beta\beta)\beta$	$A = B = C = E = \frac{\pi}{2}, D = \pi$	2	IH29	—
*P_4-40	[3.4.6.4]	$[a^+b^+c^+d^+; a^-b^-d^+c^+]$	$p31m$	e	$\alpha\beta\alpha\gamma$	$\alpha\beta\gamma\gamma$	$B = \frac{\pi}{6}, D = \frac{\pi}{3}$	$3D, 3R$	IH30	—
*P_4-41		$[a^+a^-b^+b^-; a^-b^-]$	$p6m$	$d1$	$\alpha\beta\alpha\gamma$	$\alpha\alpha\beta\beta$	$A = C = \frac{\pi}{2}, B = \frac{\pi}{3}, D = \frac{2\pi}{3}$	6	IH32	—
*P_4-42	[3.6.3.6]	$[a^+b^-b^+b^-; a^-b^-]$	$p6m$	$d2$	$\alpha\beta\alpha\beta$	$\alpha\alpha\alpha\alpha$	$A = C = \frac{2\pi}{3}, B = D = \frac{\pi}{3}$	3	IH37	—
*P_4-43	$[4^4]$	$[a^+b^+c^+d^+; a^+b^+c^+d^+]$	$p2$	e	$\alpha\alpha\alpha\alpha$	$\alpha\beta\gamma\delta$	—	$2D$	IH46	$A(P_3\text{-}2)$
*P_4-44		$[a^+b^+c^+d^+; a^-b^+c^-d^+]$	pmg	e	$\alpha\beta\beta\alpha$	$\alpha\beta\gamma\delta$	$A + D = B + C = \pi$	$2D, 2R$	IH49	—
P_4-45		$[a^+b^+c^+d^+; c^+b^-a^+d^+]$	pmg	e $(c2)$	$\alpha\beta\beta\alpha$	$\alpha\beta\alpha\gamma$	$A = C, B = D$	$1D, 1R$	IH50	—
*P_4-46		$[a^+b^+c^+d^+; c^-b^+a^-d^+]$	pgg	e	$\alpha\alpha\alpha\alpha$	$\alpha\beta\alpha\gamma$	—	$2D, 2R$	IH51	$A(P_3\text{-}3)$
*P_4-47		$[a^+b^+c^+d^+; b^-a^-c^+d^+]$	pgg	e	$\alpha\alpha\alpha\alpha$	$\alpha\alpha\beta\gamma$	—	$2D, 2R$	IH53	$A(P_3\text{-}5)$, $B(P_3\text{-}6)$, $D(P_3\text{-}4)$
*P_4-48		$[a^+b^+c^+d^+; a^-b^-c^-d^+]$	cmm	e	$\alpha\beta\gamma\alpha$	$\alpha\beta\gamma\delta$	$B = C = \frac{\pi}{2}, A + D = \pi$	$2D, 2R$	IH54	—
*P_4-49		$[a^+b^+c^+d^+; b^+a^+c^-d^-]$	$p4g$	e	$\alpha\beta\alpha\gamma$	$\alpha\alpha\beta\gamma$	$A + C = \pi, B = D = \frac{\pi}{2}$	$4D, 4R$	IH56	—
*P_4-50		$[a^+b^+a^+b^+; a^+b^+]$	$p2$	$c2$	$\alpha\alpha\alpha\alpha$	$\alpha\beta\alpha\beta$	$A = C, B = D$	$1D$	IH57	—
*P_4-51		$[a^+b^+a^+b^+; a^-b^+]$	pmg	$c2$	$\alpha\alpha\alpha\alpha$	$\alpha\beta\alpha\beta$	$A = C, B = D$	$1D, 1R$	IH58	—
*P_4-52		$[ab^+cb^-; ab^+c]$	cmm	$d1(s)$	$\alpha\alpha\alpha\alpha$	$\alpha\beta\gamma\beta$	$A = B, C = D$	2	IH67	—
*P_4-53		$[a^+b^+b^-a^-; a^+b^+]$	pmg	$d1(l)$	$\alpha\alpha\alpha\alpha$	$\alpha\beta\beta\alpha$	$B = D$	2	IH69	$A(P_3\text{-}7)$
*P_4-54		$[abab; ab]$	pmm	$d2(s)$	$\alpha\alpha\alpha\alpha$	$\alpha\beta\alpha\beta$	$A = B = C = D = \frac{\pi}{2}$	1	IH72	—
*P_4-55		$[a^+b^-a^+a^-; a^+]$	cmm	$d2(l)$	$\alpha\alpha\alpha\alpha$	$\alpha\alpha\alpha\alpha$	$A = C, B = D$	1	IH74	—
*P_4-56		$[aaaa; a]$	$p4m$	$d4$	$\alpha\alpha\alpha\alpha$	$\alpha\alpha\alpha\alpha$	$A = B = C = D = \frac{\pi}{2}$	1	IH76	—

Table 9.1.3 ISOHEDRAL TYPES OF TILINGS BY PENTAGONS

List number (1)	Topological type (2)	Incidence symbol (3)	Symmetry group (4)	Tile group (5)	Vertex transitivity (6)	Edge transitivity (7)	Angle relations (8)	Aspects (9)	References (10)	Related tilings (11)
P_5-1	$[3^6]$	$[a^+b^+c^+d^+e^+f^+;$ $b^-a^-f^+e^-d^-c^+]$	pg	e	$\alpha(\alpha)\alpha\beta\beta\beta$	$(\alpha\alpha)\beta\gamma\gamma\beta$	$B = A + C = \pi$	$1D, 1R$	IH2, S-15	$D(P_4$-1), $E[P_4$-18]
P_5-2			pg	e	$(\alpha)\alpha\alpha\beta\beta\beta$	$\alpha)\alpha\beta\gamma\gamma(\beta$	$A = B + C = \pi$	$1D, 1R$	IH2, S-17	$E(P_4$-1), $D[P_4$-19]
P_5-3		$[a^+b^+c^+d^+e^+f^+;$ $c^-e^+a^-f^-b^+d^-]$	pg	e	$\alpha(\beta)\alpha\beta\alpha\beta$	$(\alpha\beta)\alpha\gamma\beta\gamma$	$B = A + F = \pi$	$1D, 1R$	IH3, S-19	$C[P_4$-20], $E[P_4$-19]
P_5-4		$[a^+b^+c^+d^+e^+f^+;$ $a^+e^+c^+d^+b^+f^+]$	$p2$	e	$\alpha\alpha(\beta)\beta\beta\alpha$	$\alpha(\beta\gamma)\delta\beta\varepsilon$	$C = D + E = \pi$	$2D$	IH4, HK-P5-1, S-24	$B(P_4$-3), $A(P_4$-4), $F(P_4$-6)
P_5-5			$p2$	e	$\alpha\alpha\beta(\beta)\beta\alpha$	$\alpha\beta(\gamma\delta)\beta\varepsilon$	$D = C + E = \pi$	$2D$	IH4, S-10	$A(P_4$-5), $B(P_4$-4)
P_5-6		$[a^+b^+c^+d^+e^+f^+;$ $a^+e^+d^-c^-b^+f^+]$	pgg	e	$\alpha\alpha\beta\beta\beta(\alpha)$	$\alpha\beta\gamma\gamma(\beta\delta)$	$F = A + B = \pi$	$2D, 2R$	IH5, S-12	$C(P_4$-9), $D(P_4$-10), $E(P_4$-11)
P_5-7			pgg	e	$\alpha\alpha(\beta)\beta\beta\alpha$	$\alpha(\beta\gamma)\gamma\beta\delta$	$C = D + E = \pi$	$2D, 2R$	IH5, S-13	$A(P_4$-8), $B(P_4$-11), $F(P_4$-9)
P_5-8			pgg	e	$\alpha\alpha\beta(\beta)\beta\alpha$	$\alpha\beta(\gamma\gamma)\beta\delta$	$D = C + E = \pi$	$2D, 2R$	IH5, S-11	$A(P_4$-7), $B(P_4$-10)
P_5-9			pgg	e	$(\alpha)\alpha\beta\beta\beta\alpha$	$\alpha)\beta\gamma\gamma\beta(\delta$	$A = B + F = \pi$	$2D, 2R$	IH5, S-16	$C(P_4$-8), $D(P_4$-7)
P_5-10		$[a^+b^+c^+d^+e^+f^+;$ $a^+e^-c^+f^-b^-d^-]$	pgg	e	$\alpha\alpha(\beta)\beta\alpha\beta$	$\alpha(\beta\gamma)\delta\beta\delta$	$C = D + F = \pi$	$2D, 2R$	IH6, HK-P5-3, S-20	$A(P_4$-15), $B[P_4$-20], $E(P_4$-13)
P_5-11			pgg	e	$\alpha\alpha\beta(\beta)\alpha\beta$	$\alpha\beta(\gamma\delta)\beta\delta$	$D = C + F = \pi$	$2D, 2R$	IH6, S-21	$A(P_4$-12), $B(P_4$-15), $E(P_4$-16)
P_5-12			pgg	e	$\alpha\alpha\beta\beta(\alpha)\beta$	$\alpha\beta\gamma(\delta\beta)\delta$	$E = A + B = \pi$	$2D, 2R$	IH6, S-14	$C(P_4$-13), $D(P_4$-16), $F(P_4$-14)
P_5-13		$[a^+b^+c^+d^+e^+f^+;$ $b^+a^+d^+c^+f^+e^+]$	$p3$	e	$(\alpha)\beta\alpha\gamma\alpha\delta$	$\alpha)\alpha\beta\beta\gamma(\gamma$	$A = \pi,\ B = D = F = \frac{2\pi}{3}$	$3D$	IH7, HK-P5-10 S-22	—
P_5-14		$[a^+b^+c^+c^-b^-a^-;$ $a^+b^-c^+]$	pmg	$d1$	$(\alpha)\alpha\beta\beta\beta\alpha$	$\alpha)\beta\gamma\gamma\beta(\alpha$	$A = \pi,\ B = F = \frac{\pi}{2},\ C = E$	2	IH15, S-18	$D[P_4$-22]
P_5-15		$[a^+b^+c^+c^-b^-a^-;$ $a^-c^+b^+]$	$p31m$	$d1$	$\alpha\beta\gamma(\beta)\gamma\beta$	$\alpha\beta(\beta\beta)\beta\alpha$	$A = C = E = \frac{2\pi}{3},$ $B = F = \frac{\pi}{2},\ D = \pi$	3	IH16, HK-P5-8, S-23	—
*P_5-16	$[3^4.6]$	$[a^+b^+c^+d^+e^+;$ $e^+c^+b^+d^+a^+]$	$p6$	e	$\alpha\beta\gamma\beta\beta$	$\alpha\beta\beta\gamma\alpha$	$A = \frac{2\pi}{3},\ C = \frac{\pi}{3}$	$6D$	IH21, S-2	$B(P_4$-24), $D(P_4$-25), $E(P_4$-23)
*P_5-17	$[3^3.4^2]$	$[a^+b^+c^+d^+e^+;$ $a^-e^+d^-c^-b^+]$	cm	e	$\alpha\alpha\beta\beta\beta$	$\alpha\beta\gamma\gamma\beta$	$A + B = \pi$	$1D, 1R$	IH22, S-6	$C(P_4$-26), $D(P_4$-27)
*P_5-18		$[a^+b^+c^+d^+e^+;$ $a^+e^+c^+d^+b^+]$	$p2$	e	$\alpha\alpha\beta\beta\beta$	$\alpha\beta\gamma\delta\beta$	$A + B = \pi$	$2D$	IH23, S-4	$C(P_4$-29), $D(P_4$-28)
*P_5-19		$a^+b^+c^+d^+e^+;$ $a^-e^+c^+d^+b^+]$	pmg	e	$\alpha\alpha\beta\beta\beta$	$\alpha\beta\gamma\delta\beta$	$A + B = \pi$	$2D, 2R$	IH24, S-5	$C(P_4$-31), $D(P_4$-30)
*P_5-20		$[a^+b^+c^+d^+e^+;$ $a^+e^+d^-c^-b^+]$	pgg	e	$\alpha\alpha\beta\beta\beta$	$\alpha\beta\gamma\gamma\beta$	$A + B = \pi$	$2D, 2R$	IH25, S-7	$C(P_4$-33), $D(P_4$-32)
*P_5-21		$[ab^+c^+c^-b^-;$ $ab^-c^+]$	cmm	$d1$	$\alpha\alpha\beta\beta\beta$	$\alpha\beta\gamma\gamma\beta$	$A = B = \frac{\pi}{2},$ $C = E$	2	IH26, S-1	$D(P_4$-34)
*P_5-22	$[3^2.4.3.4]$	$[a^+b^+c^+d^+e^+;$ $a^+d^-e^-b^-c^-]$	pgg	e	$\alpha\alpha\beta\alpha\beta$	$\alpha\beta\gamma\beta\gamma$	$C + E = \pi$	$2D, 2R$	IH27, S-8	$A(P_4$-35), $D(P_4$-36)
*P_5-23		$[a^+b^+c^+d^+e^+;$ $a^+c^+b^+e^+d^+]$	$p4$	e	$\alpha\alpha\beta\alpha\gamma$	$\alpha\beta\beta\gamma\gamma$	$C = E = \frac{\pi}{2}$	$4D$	IH28, HK-P5-9, S-9	$A(P_4$-38), $D(P_4$-37)
*P_5-24		$[ab^+c^+c^-b^-;$ $ac^+b^+]$	$p4g$	$d1$	$\alpha\alpha\beta\alpha\beta$	$\alpha\beta\beta\beta\beta$	$A = B, C = E = \frac{\pi}{2}$	4	IH29, S-3	$D(P_4$-39)

Table 9.1.4 ISOHEDRAL TYPES OF TILINGS BY HEXAGONS

List number (1)	Topological type (2)	Incidence symbol (3)	Symmetry group (4)	Tile group (5)	Vertex transitivity (6)	Edge transitivity (7)	Angle relations (8)	Aspect (9)	References (10)	Related tilings (11)
$^*P_6$-1	$[3^6]$	$[a^+b^+c^+d^+e^+f^+; b^-a^-f^+e^-d^-c^+]$	pg	e	$\alpha\alpha\alpha\beta\beta\beta$	$\alpha\alpha\beta\gamma\gamma\beta$	$A + B + C = D + E + F = 2\pi$	$1D, 1R$	IH2	$A(P_5$-2), $B(P_5$-1), $AD[P_4$-19], $AE(P_4$-1), $BE[P_4$-18]
$^*P_6$-2		$a^+b^+c^+d^+e^+f^+; c^-e^+a^-f^-b^+d^-]$	pg	e	$\alpha\beta\alpha\beta\alpha\beta$	$\alpha\beta\alpha\gamma\beta\gamma$	$A + B + F = C + D + E = 2\pi$	$1D, 1R$	IH3	$B(P_5$-3), $AD(P_4$-2), $BC[P_4$-20], $BE[P_4$-19]
$^*P_6$-3		$[a^+b^+c^+d^+e^+f^+; a^+e^+c^+d^+b^+f^+]$	$p2$	e	$\alpha\alpha\beta\beta\beta\alpha$	$\alpha\beta\gamma\delta\beta\varepsilon$	$A + B + F = C + D + E = 2\pi$	$2D$	IH4, HK-P6-4	$A(P_5$-5), $B(P_5$-4), $AD(P_4$-5), $AE(P_4$-4), $BC(P_4$-3), $BE(P_4$-6)
$^*P_6$-4		$[a^+b^+c^+d^+e^+f^+; a^+e^+d^-c^-b^+f^+]$	pgg	e	$\alpha\alpha\beta\beta\beta\alpha$	$\alpha\beta\gamma\gamma\beta\delta$	$A + B + F = C + D + E = 2\pi$	$2D, 2R$	IH5	$A(P_5$-9), $B(P_5$-6), $C(P_5$-7), $D(P_5$-9), $AC(P_4$-8), $AD(P_4$-7), $BC(P_4$-11), $BD(P_4$-10), $BE(P_4$-9)
$^*P_6$-5		$[a^+b^+c^+d^+e^+f^+; a^+e^-c^+f^-b^-d^-]$	pgg	e	$\alpha\alpha\beta\beta\alpha\beta$	$\alpha\beta\gamma\delta\beta\delta$	$A + B + E = C + D + F = 2\pi$	$2D, 2R$	IH6, HK-P6-10	$A(P_5$-11), $B(P_5$-10), $E(P_5$-12), $AC(P_4$-15), $AD(P_4$-12), $AF(P_4$-16), $BC[P_4$-20], $BF(P_4$-13)
$^*P_6$-6		$[a^+b^+c^+d^+e^+f^+; b^+a^+d^+c^+f^+e^+]$	$p3$	e	$\alpha\beta\alpha\gamma\alpha\delta$	$\alpha\alpha\beta\beta\gamma\gamma$	$B = D = F = \frac{2\pi}{3}$	$3D$	IH7, HK-P6-20	$A(P_5$-13)
$^*P_6$-7		$[a^+b^+c^+a^+b^+c^+; a^+b^+c^+]$	$p2$	$c2$	$\alpha\alpha\alpha\alpha\alpha\alpha$	$\alpha\beta\gamma\alpha\beta\gamma$	$A = D, B = E, C = F$	$1D$	IH8	$BE(P_4$-17)
$^*P_6$-8		$[a^+b^+c^+a^+b^+c^+; a^+c^-b^-]$	pgg	$c2$	$\alpha\alpha\alpha\alpha\alpha\alpha$	$\alpha\beta\beta\alpha\beta\beta$	$A = D, B = E, C = F$	$1D, 1R$	IH9	$CF(P_4$-18), $AD(P_4$-19)
$^*P_6$-9		$[ab^+c^+dc^-b^-; db^+c^+a]$	pmg	$d1(s)$	$\alpha\alpha\alpha\alpha\alpha\alpha$	$\alpha\beta\gamma\alpha\gamma\beta$	$A = B, C = F, D = E$	2	IH13	$CF(P_4$-21), $DE(P_4$-20)
$^*P_6$-10		$[a^+b^+c^+c^-b^-a^-; a^+b^-c^+]$	pmg	$d1(l)$	$\alpha\alpha\beta\beta\beta\alpha$	$\alpha\beta\gamma\gamma\beta\alpha$	$B = F, C = E, F + A + B = 2\pi$	2	IH15	$A(P_5$-14), $AD[P_4$-22]
$^*P_6$-11		$[a^+b^+c^+c^-b^-a^-; a^-c^+b^+]$	$p31m$	$d1(l)$	$\alpha\beta\gamma\beta\gamma\beta$	$\alpha\beta\beta\beta\beta\alpha$	$A = C = E = \frac{2\pi}{3}, B = F$	3	IH16	$D(P_5$-15)
$^*P_6$-12		$[ab^+b^-ab^+b^-; ab^+]$	cmm	$d2$	$\alpha\alpha\alpha\alpha\alpha\alpha$	$\alpha\beta\beta\alpha\beta\beta$	$A = B = D = E, C = F$	1	IH17	$CF(P_4$-22)
$^*P_6$-13		$[aaaaaa; a]$	$p6m$	$d6$	$\alpha\alpha\alpha\alpha\alpha\alpha$	$\alpha\alpha\alpha\alpha\alpha\alpha$	$A = B = C = D = E = F = \frac{2\pi}{3}$	1	IH20	

Table 9.1.5 THE DIRICHLET TILINGS OF DOT PATTERNS

For each of the 47 types of edge-to-edge isohedral tilings by convex polygons listed in **column (1)** we give the following information.

Column (2) indicates the number of independent parameters needed to specify a tiling of this type, together with a letter with the following meaning:

- *a* No tiling of this type is a *D*-tiling.
- *b* Some, but not all, tilings of this type are *D*-tilings.
- *c* Every tiling of this type is a *D*-tiling.

Columns (3) and (4) give the pattern type (or types) of the dot patterns that yield *D*-tilings of the given type and the number of parameters needed to specify these dot patterns. Letters in column (3) have the following meanings:

- *d* Dot patterns of this type yield the same *D*-tiling as the previous row.
- *e* For some values of the parameters, dot patterns of this type yield *D*-tilings of a different polygonal isohedral type but with tiles having the same number of sides.
- *f* For some values of the parameters, dot patterns of this type yield *D*-tilings by polygons with a larger number of sides.
- *g* For some values of the parameters, dot patterns of this type yield *D*-tilings by polygons with a smaller number of sides.

Columns (5) and (6) show the number of parameters needed to specify the prototile of a *D*-tiling of the type under consideration, and the number of degrees of freedom in choosing the dot of a suitable dot pattern inside each tile to yield the required *D*-tiling.

Column (7) lists the symbol for the *D*-tiling used by Delone, Galiulin & Štogrin [1979]. For *D*-tilings in which the stabilizer of each tile is trivial, the same notation was used by Delone [1959].

(1)	(2)	(3)	(4)	(5)	(6)	(7)
P_3-1	1*c*	48B	2	1	1	3*C*3
P_3-8	1*c*	46	3	1	2	3*D*
P_3-9	2*c*	17 *f*	4	2	1	3*B*1
P_3-10	1*c*	38	2	1	1	3*B*5
		37 *d*	3	1	2	
P_3-11	3*b*	7 *f*	5	3	0	3*A*1
P_3-12	3*b*	11 *f*	4	3	0	3*A*4
P_3-13	2*b*	19 *f*	3	2	0	3*A*11
P_3-14	1*c*	50	1	1	0	3*A*10
		47 *d*	2	1	1	
		27 *d*	3	1	2	
P_4-40	2*c*	23	3	2	1	4*B*2
P_4-41	1*c*	48A	2	1	1	4*B*3
P_4-42	1*c*	49	1	1	0	4*C*5
		28 *d*	2	1	1	
P_4-43	5*b*	7 *fg*	5	3	0	4*A*1
P_4-44	4*b*	11 *fg*	4	3	0	4*A*6
P_4-45	3*a*					
P_4-46	4*a*					
P_4-47	4*b*	9 *f*	4	3	0	4*A*2
P_4-48	3*c*	17 *g*	4	3	1	4*A*5
P_4-49	2*c*	33	3	2	1	4*A*5
P_4-50	3*a*					
P_4-51	3*a*					
P_4-52	3*b*	19 *fg*	3	3	0	4*A*19
P_4-53	3*b*	13 *f*	3	2	0	4*A*20
P_4-54	2*c*	16	2	2	0	4*A*28
		15 *d*	3	2	1	
		14 *d*	4	2	2	
P_4-55	2*a*					
P_4-56	1*c*	41	1	1	0	4*A*36
		39 *d*	2	1	1	
P_5-16	3*c*	42	3	3	0	5*C*
P_5-17	6*a*					
P_5-18	7*b*	7 *fg*	5	4	0	5*A*1
P_5-19	7*b*	11 *g*	4	4		5*A*3
P_5-20	6*a*					
P_5-21	3*b*	19 *g*	3	3	0	5*A*5
P_5-22	6*b*	9 *fg*	4	2	0	5*B*2
P_5-23	3*b*	30	3	2	0	5*B*1
P_5-24	2*c*	35	2	2	0	5*B*3
P_6-1	5*a*					
P_6-2	5*a*					
P_6-3	7*b*	7 *g*	5	5	0	6, 4
P_6-4	6*b*	9 *eg*	4	4	0	6, 5
P_6-5	6*b*	9 *eg*	4	4	0	6, 2
P_6-6	3*b*	21	3	3	0	6, 1
P_6-7	5*b*	8	3	3	0	6, 8
P_6-8	3*a*					
P_6-9	4*b*	13 *eg*	3	3	0	6, 11
P_6-10	4*b*	13 *eg*	3	3	0	6, 10
P_6-11	2*b*	25	2	2	0	6, 15
P_6-12	3*b*	20	2	2	0	6, 14
P_6-13	1*c*	51	1	1	0	6, 20

EXERCISES 9.1

1. Let T denote the equilateral triangle, T' a triangle with angles 30°, 60°, 90°, T'' a triangle with angles 45°, 45°, 90°, and T''' a triangle with angles 30°, 30°, 120°. For each of the fourteen types in Table 9.1.1 draw a tiling of that type with prototile T, T', T'' and T''' or show that such a tiling does not exist.
2. Which of the types in Table 9.1.2 can be realized by tilings in which the prototile is (*a*) a square, (*b*) a non-square rectangle, or (*c*) a non-square rhomb?
3. (*a*) Show that for each type in Table 9.1.2 there exists a quadrangle Q which is not the prototile of any tiling of that type.
   (*b*) Determine as small a set of types from Table 9.1.2 as possible with the property that every quadrangle Q admits a tiling of at least one of these types.
4. For (*a*) pentagons, (*b*) hexagons, all sides of which have length 1, find necessary and sufficient conditions on the angles of these polygons to make them prototiles of proper isohedral tilings.
5. Verify parts of Statement 9.1.1 by showing:
   (*a*) A proper isohedral tiling of type IH27 is of one of the types P_4-35, P_4-36 or P_5-22.
   (*b*) A proper isohedral tiling of type IP53 is of one of the types P_3-4, P_3-5, P_3-6 or P_4-47.
6. (*a*) Let the prototile of a tiling of type P_3-11 have sides of lengths x, y, 1, where $y < x < 1$ and $x + y > 1$. Show that the relational indicator of all such tilings of type P_3-11 is a triangular region, and that the *D*-tilings among them correspond precisely to that part of the relational indicator in which $x^2 + y^2 > 1$.
   (*b*) Construct the analogous relational indicator for the tilings and *D*-tilings of type P_3-12.
   (*c*) Show that apart from isohedral tilings of types P_3-11 and P_3-12, all polygonal edge-to-edge isohedral tilings by triangles are *D*-tilings.
7. (*a*) Let the parametrization of the dot patterns of type DPP13 be standardized by taking as unit the distance between adjacent dots along a line of reflective symmetry. Let the other two parameters be the distances x and y from two such points to the nearest point not on that line, so that $x > y > 0$ and $x + y > 1$. Construct the relational indicator for the family of all dot patterns of type DPP13 and show how the circular arc $x^2 + y^2 = 1$ partitions this relational indicator into parts that correspond to *D*-tilings of types P_4-53, P_6-9 and P_6-10.
   (*b*) Construct the analogous relational indicator for dot patterns of type DPP19 and partition it so as to correspond to *D*-tilings of types P_3-13, P_4-52 and P_5-21.
8. Explain why dot patterns of types DPP39 and DPP41 yield the same *D*-tilings.

9.2 PROPER ISOHEDRAL TILINGS BY NON-CONVEX POLYGONS

In a proper tiling by polygons, each edge is a straight line-segment joining two vertices of the tiling. Hence it may be possible to convert an arbitrary tiling $\mathcal{T}$ into a proper polygonal tiling by replacing each edge of $\mathcal{T}$ by a line-segment joining the corresponding vertices. If a new tiling $\mathcal{T}_s$ results from these replacements, then $\mathcal{T}_s$ is said to be obtained from $\mathcal{T}$ by *straightening*. Two possibilities arise:

(*a*) $\mathcal{T}_s$ is of the same isohedral type as $\mathcal{T}$; in this case we say that the type can be realized by polygons; or

(*b*) whatever tiling of the isohedral type of $\mathcal{T}$ is chosen, the straightened tiling is always of a different type. This situation occurs if straightening necessarily introduces new symmetries, so that the symmetry group $S(\mathcal{T}_s)$ is larger than that of $\mathcal{T}$. A typical example is isohedral type IH62; here straightening necessarily produces the regular tiling by squares, which is of type IH76.

Examination of each of the 81 isohedral types shows that the distinction between cases (*a*) and (*b*) *does not depend upon whether the resulting polygonal tiles are convex or not*. We cannot reproduce here the full details of the proof of this assertion, but it is made plausible by the following argument. Let us, as in Section 6.2, label the oriented or unoriented edges of each tile of $\mathcal{T}$ and then assign the same labels to the corresponding edges of $\mathcal{T}_s$; labelled in this manner $\mathcal{T}_s$ is clearly of the same isohedral type as $\mathcal{T}$. We are concerned with the question whether the type is changed by removing the labels. But this obviously depends *only* on the symmetries of a tile T and its relationship to its adjacents, and *not* on the convexity character of T.

Now it is easy to check, from the diagrams given in Section 6.2, that $\mathcal{T}$ can always be chosen so that the tiles of $\mathcal{T}_s$ are convex polygons. We deduce the following:

9.2.1 *Any isohedral type of tiling that can be represented by a proper tiling with polygonal tiles can necessarily be represented by one with convex tiles. Hence any polygonal tiling of one of the* 34 *isohedral types marked N in column* (9) *of Table 6.2.1 is necessarily not proper.*

To find all proper isohedral tilings by non-convex polygons it therefore suffices to restrict attention to the 47 types marked by asterisks in the diagrams and tables of the previous section. In fact, using the notion of "polygonal isohedral type" defined in Section 9.1, we see now that these figures give all the information about the non-convex case as well. But it is more appropriate to refine our definition of "type" still further, as follows.

Two tilings $\mathcal{T}_1$ and $\mathcal{T}_2$ by (convex or non-convex) polygons are of the same *refined polygonal isohedral type* if they are of the same polygonal isohedral type and, in addition, satisfy the following condition:

> For each vertex V of $\mathcal{T}_1$ that lies on the boundary of a tile T of $\mathcal{T}_1$, the interior angle of T at V is less than, equal to, or greater than π according as the interior angle of $\phi(T)$ at $\phi(V)$ is less than, equal to, or greater than π. Here ϕ is, as before, the homeomorphism used in showing that $\mathcal{T}_1$ and $\mathcal{T}_2$ are of the same polygonal isohedral type.

Informally, this condition means that the convexity character of each tile at a vertex of the tiling is unchanged by the mapping ϕ.

To determine the refined types of tilings we must therefore decide which angles of the polygonal tiles can exceed π. It turns out that this is easy to do. We already know when it is possible to increase an angle at a vertex to the value π—the possibilities have been indicated in the tables of the previous section—and it is in precisely the same cases that it is possible to increase the angle to a value greater than π, leading to a non-convex tile.

To illustrate the definitions and the process just described, we consider the tilings in Figures 9.2.1 and 9.2.2. From the tiling by hexagons of type P_6-10 shown in Figure 9.2.1(*a*) we obtain a tiling by pentagons (Figure 9.2.1(*b*)) and one by quadrangles (Figure 9.2.1(*c*)) by increasing angles to π. Increasing these angles still further leads to the hexagonal tilings of Figures 9.2.1(*d*) and (*e*); these are also of type P_6-10 according to the definition of the previous section—but adopting the refined definition the tilings in Figures 9.2.1(*d*) and (*e*) are of different types. Similar is the example involving pentagons shown in Figure 9.2.2. Here three isohedral tilings by non-convex pentagons can be derived, and these are distinct from the original P_5-4 and from each other according to the refined definition.

Using the refined notion of type and examining the various possibilities that occur, we arrive at the following result:

9.2.2 *There are* 96 *refined polygonal isohedral types of proper tilings by non-convex polygons* (6 *by quadrangles,* 48 *by pentagons and* 42 *by hexagons*). *Each of these types is specified by the corners of the tiles which are non-convex; these are listed in the tables of Section 9.1 as corners at which the angles can be made equal to* π.

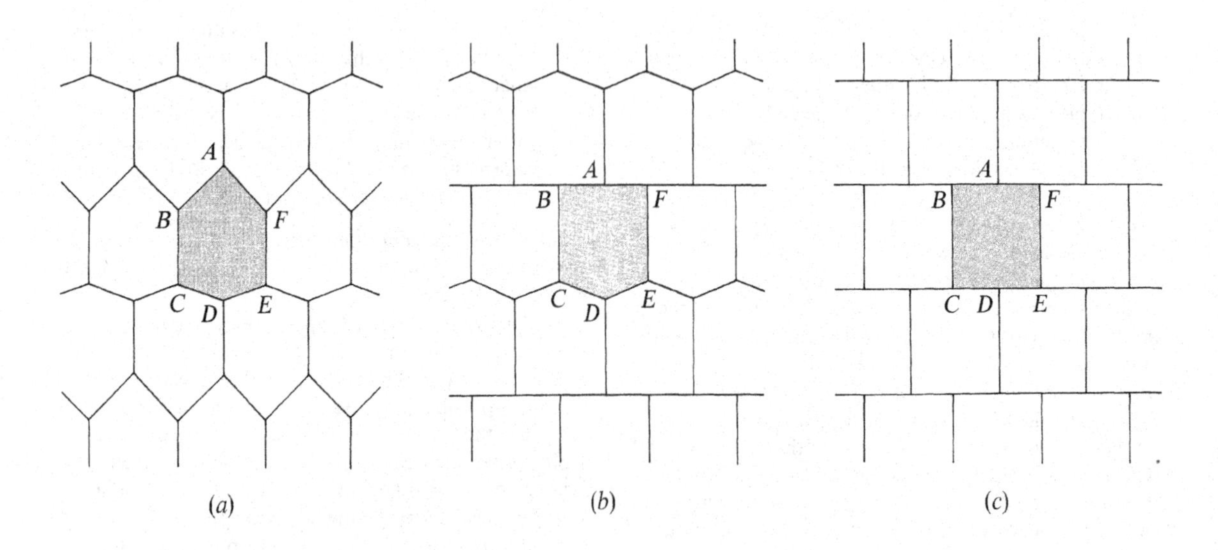

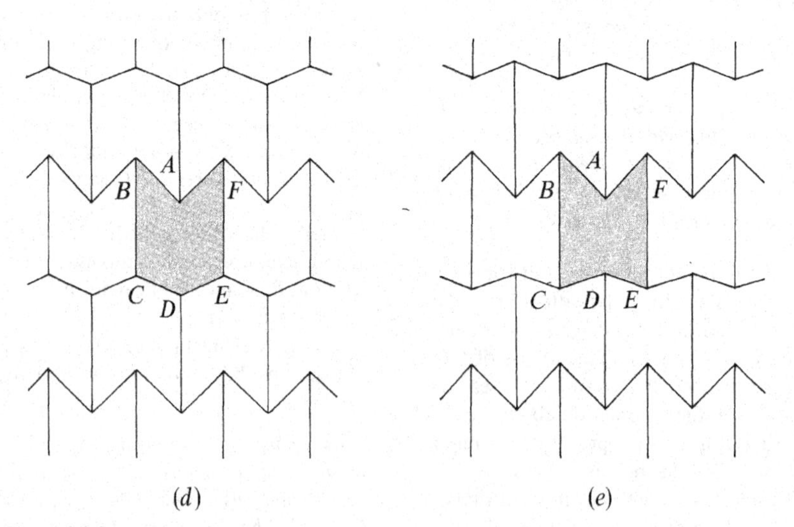

Figure 9.2.1
(*a*) The isohedral tiling by convex hexagons of type P_6-10. (*b*) (*c*) Tilings by pentagons and quadrangles obtained by increasing the angles of the hexagonal prototile to π. (*d*) (*e*) Tilings by non-convex hexagons obtained by making the same angles greater than π.

Figure 9.2.2
The tiling P_5-4 and the three tilings by non-convex pentagons that can be obtained from it.

EXERCISES 9.2

1. Determine the restricted polygonal isohedral types of tilings which have as prototile (*a*) a quadrangle with symmetry group *d1* and with angles 15°, 60°, 15° and 270°, and (*b*) a pentagon with angles 60°, 120°, 45°, 270°, 45° and with two equal sides adjacent to the 270° angle.

2. (*a*) Prove that no pentagonal prototile of a proper isohedral tiling can have two adjacent angles of more than 180° each.

   (*b*) Show that the assertion of part (*a*) remains valid if the word "proper" is deleted.

   (*c*) Show that the assertion of part (*a*) remains valid if the word "monohedral" is substituted for "isohedral".

3. Determine the restricted polygonal isohedral types with non-convex tiles that can be derived from the types (*a*) P_5-2, and (*b*) P_6-1.

9.3 OTHER MONOHEDRAL TILINGS BY CONVEX POLYGONS

The results of Section 9.1 enable us to compile a list of convex polygons which admit isohedral tilings. It comprises all triangles and quadrangles, five types of pentagons, and three types of hexagons. Details are given in Table 9.3.1, and it is a straightforward matter to check that this list is complete. It must be pointed out, however, that many types of isohedral tilings are possible only if the prototile satisfies certain additional conditions. For example, every triangular tile admits a tiling of type P_3-11, but it admits a tiling of type P_3-1 only if its angles are $\pi/6$, $\pi/6$ and $2\pi/3$. Again, every pentagon with two parallel sides (type 1 in Table 9.3.1) admits a tiling of type P_5-4, but it admits a tiling of type P_5-5 only if the parallel sides are of equal length.

Many of these prototiles also admit monohedral tilings which are not isohedral. If the tiling contains parallel edge-lines (as with types P_3-1, P_4-1, P_5-1, for example), then it is possible to translate or reflect the "strips" of tiles between a pair of edge-lines in such a way as to destroy isohedrality. However, there are also non-trivial ways in which non-isohedral tilings can be constructed;

Table 9.3.1 CONVEX POLYGONS WHICH ADMIT MONOHEDRAL TILINGS

Column (1) gives a list of the convex polygons which are, at present, known to admit monohedral tilings; it is not known whether or not this list is complete. For pentagons and hexagons a drawing of the polygon is given together with the relationships that must hold between its angles and between its sides.

Columns (2) and (3) give *one* example of a monohedral tiling using the polygon as prototile, and the symmetry group of this tiling. Even when the polygon is in its most general form, the tiling is usually not unique. Whenever an isohedral tiling is possible, this has been chosen as representative tiling and its type is indicated by the reference symbol in Figures 9.1.1 to 9.1.4. If no isohedral tiling is possible, then a reference is given to the part of Figure 9.3.2 in which an example of the tiling is shown.

Polygon (1)		Representative tiling and its symmetry group (2)	(3)
Triangle		P_3-11	*p2*
Quadrangle		P_4-43	*p2*
Pentagon			
Type 1	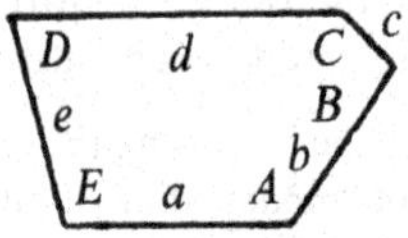 $D + E = \pi$	P_5-4	*p2*
Type 2	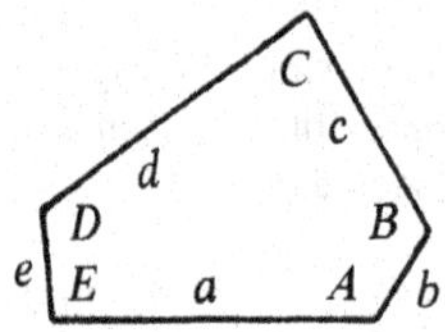 $C + E = \pi$ $a = d$	P_5-10	*pgg*

Table 9.3.1 (continued)

Polygon (1)		Representative tiling (2)	and its symmetry group (3)
Type 3	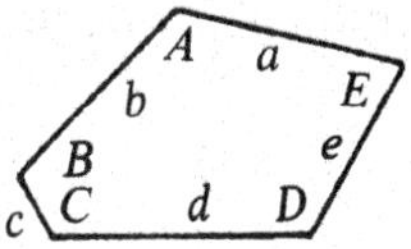 $A = C = D = \frac{2}{3}\pi$ $a = b$ $d = c + e$	P_5-13	$p3$
Type 4	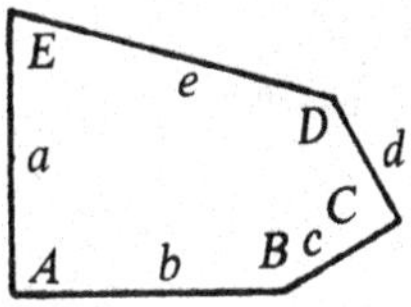 $A = C = \frac{1}{2}\pi$ $a = b$ $c = d$	P_5-23	$p4$
Type 5	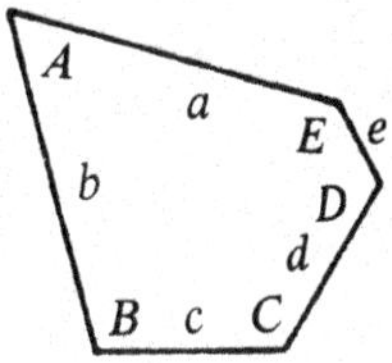 $C = 2A = \frac{2}{3}\pi$ $a = b$ $c = d$	P_5-16	$p6$
Type 6	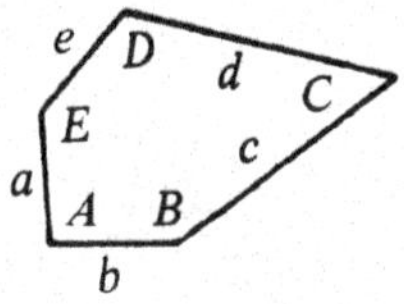 $C + E = \pi$ $A = 2C$ $a = b = e, c = d$	Figure 9.3.2(*a*)	$p2$
Type 7	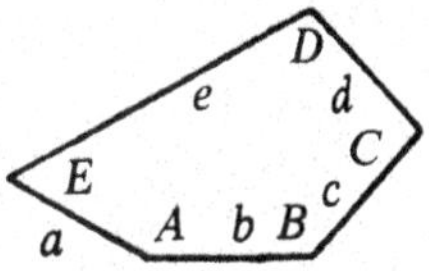 $2B + C = 2\pi$ $2D + A = 2\pi$ $a = b = c = d$	Figure 9.3.2(*b*)	pgg
Type 8	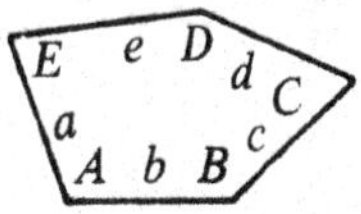 $2A + B = 2\pi$ $2D + C = 2\pi$ $a = b = c = d$	Figure 9.3.2(*c*)	pgg
Type 9	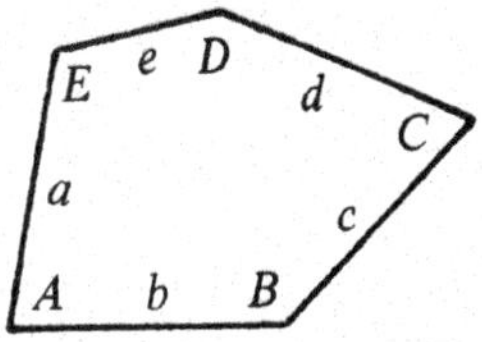 $2E + B = 2\pi$ $2D + C = 2\pi$ $a = b = c = d$	Figure 9.3.2(*d*)	pgg

Table 9.3.1 (continued)

Polygon (1)		Representative tiling and its symmetry group (2)	(3)
Type 10	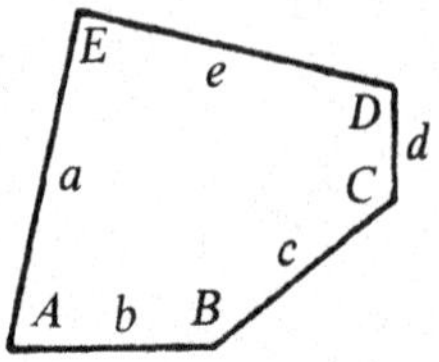 $E = \frac{1}{2}\pi$, $A + D = \pi$ $2B - D = \pi$, $2C + D = 2\pi$ $a = e = b + d$	Figure 9.3.2(*e*) and (*i*)	*p*2, *cmm*
Type 11	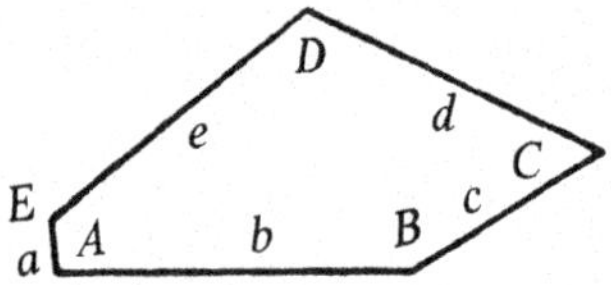 $A = \frac{1}{2}\pi$, $C + E = \pi$ $2B + C = 2\pi$ $d = e = 2a + c$	Figure 9.3.2(*f*)	*pgg*
Type 12	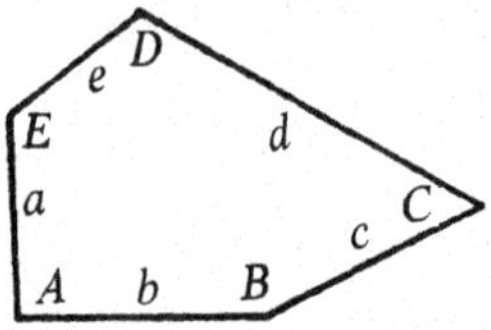 $A = \frac{1}{2}\pi$, $C + E = \pi$ $2B + C = 2\pi$ $2a = c + e = d$	Figure 9.3.2(*g*)	*pgg*
Type 13	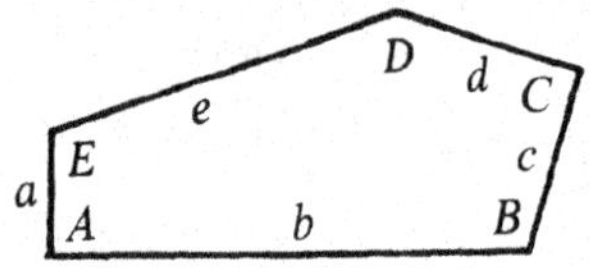 $A = C = \frac{1}{2}\pi$ $B = E = \pi - \frac{1}{2}D$ $c = d$, $2c = e$	Figure 9.3.2(*h*)	*pgg*
Hexagon			
Type 1	 $A + B + C =$ $D + E + F = 2\pi$ $a = d$	P_6-3	*p*2
Type 2	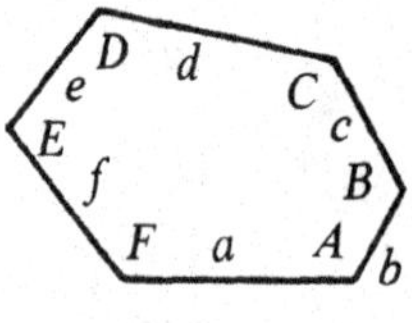 $A + B + D =$ $C + E + F = 2\pi$ $a = d$ $c = e$	P_6-5	*pgg*
Type 3	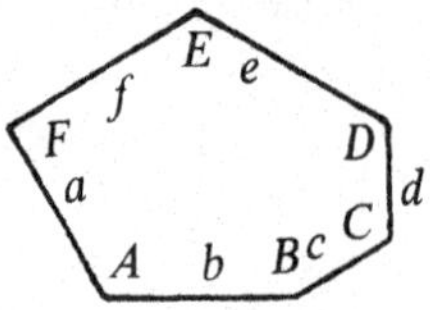 $A = C = E = \frac{2}{3}\pi$ $a = b$ $c = d$ $e = f$	P_6-6	*p*3

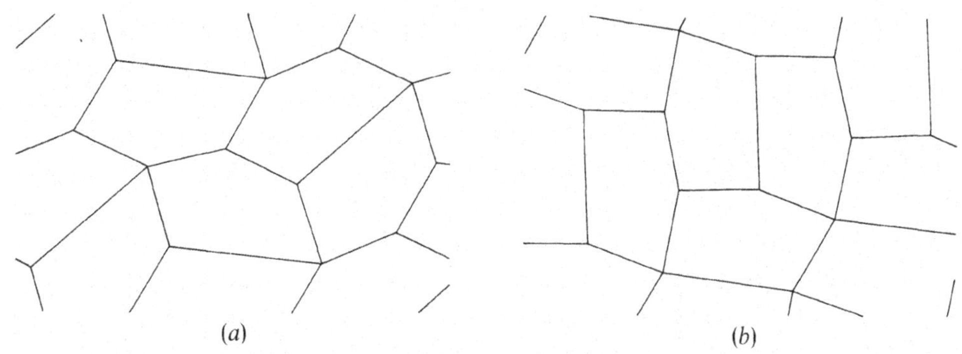

Figure 9.3.1
Two non-isohedral tilings discovered by Marjorie Rice. Tiling (*a*) is by pentagons of type 1 and tiling (*b*) is by pentagons of type 2 in Table 9.3.1. Each of these pentagons also admits an isohedral tiling.

in Figure 9.3.1 we show two such tilings, recently discovered by Marjorie Rice. These use pentagons of types 1 and 2.

Our main aim here is to describe the convex *anisohedral* polygons, that is, polygons which admit a monohedral tiling but do not admit any isohedral tiling. From Table 9.3.1 it follows immediately that the only polygons that could be anisohedral are pentagons and hexagons. However, it was established by Reinhardt [1918] and confirmed in whole or in part by such later authors as Heesch & Kienzle [1963] and Bollobás [1963] that no convex hexagon is anisohedral. It follows that in searching for convex anisohedral polygons it is only necessary to consider pentagons. The first three examples of such pentagons were described by Kershner [1968] and appear as types 6, 7, 8 in Table 9.3.1. Five further examples, types 9 to 13, have since been discovered by Marjorie Rice and Richard James. Tilings of all eight types are shown in Figure 9.3.2; in each case tiles marked by an asterisk form a patch $\mathscr{A}$. If we represent by R the union of the tiles in $\mathscr{A}$ then R admits an isohedral tiling of the plane. Such a patch $\mathscr{A}$ will be called an *isohedral repeating unit* (or a *translative repeating unit* if only translations of R are required). This remark suggests a method by which one can attempt to find other anisohedral tiles. We start from any one of the 81 types of isohedral tilings described in Section 6.2 and then investigate whether the shape of the tile can be modified in such a manner that it can be expressed as a union of congruent non-overlapping pentagons. It seems as though this approach, although laborious, might be systematically applied to find other examples. An alternative method, used successfully by both Kershner and Rice, is to postulate various linear relations between the angles of a pentagon and then, by trial and error, see if these enable the corners of the tiles to fit together in such a way that a tiling is possible. This approach was pioneered as long ago as 1925 by W. P. D. MacMahon who, if he had not restricted his attention to isohedral tilings, might have made a more substantial contribution to the subject.

At present the situation is rather unsatisfactory, in that we do not know whether the list in Table 9.3.1 is complete. It seems as if the number of possible arrangements of tiles is so great that every attempt at systematic enumeration has, up to the present, missed several possibilities, and it may well be that a number of other types of pentagons remain to be discovered. For a fuller description of this topic see Gardner [1975a], [1975c] and Schattschneider [1978b], [1981]; Table 9.3.1 is based on information given in Schattschneider's papers.

Hirschhorn & Hunt [1981] made a computer search for equilateral convex pentagons which are prototiles of monohedral tilings. They claim that their enumeration is complete, but it has not been published so far and ought to be independently verified. Particular cases of monohedral tilings by convex pentagons have been considered by Dunn and others (see Dunn [1971]).

Throughout this section we have restricted attention to convex polygons. About the non-convex case we have very little information (except concerning isohedral tilings). Some examples were given by Palmer [1979] and Rice & Schattschneider [1980]; other examples can be extracted from information given in the next section.

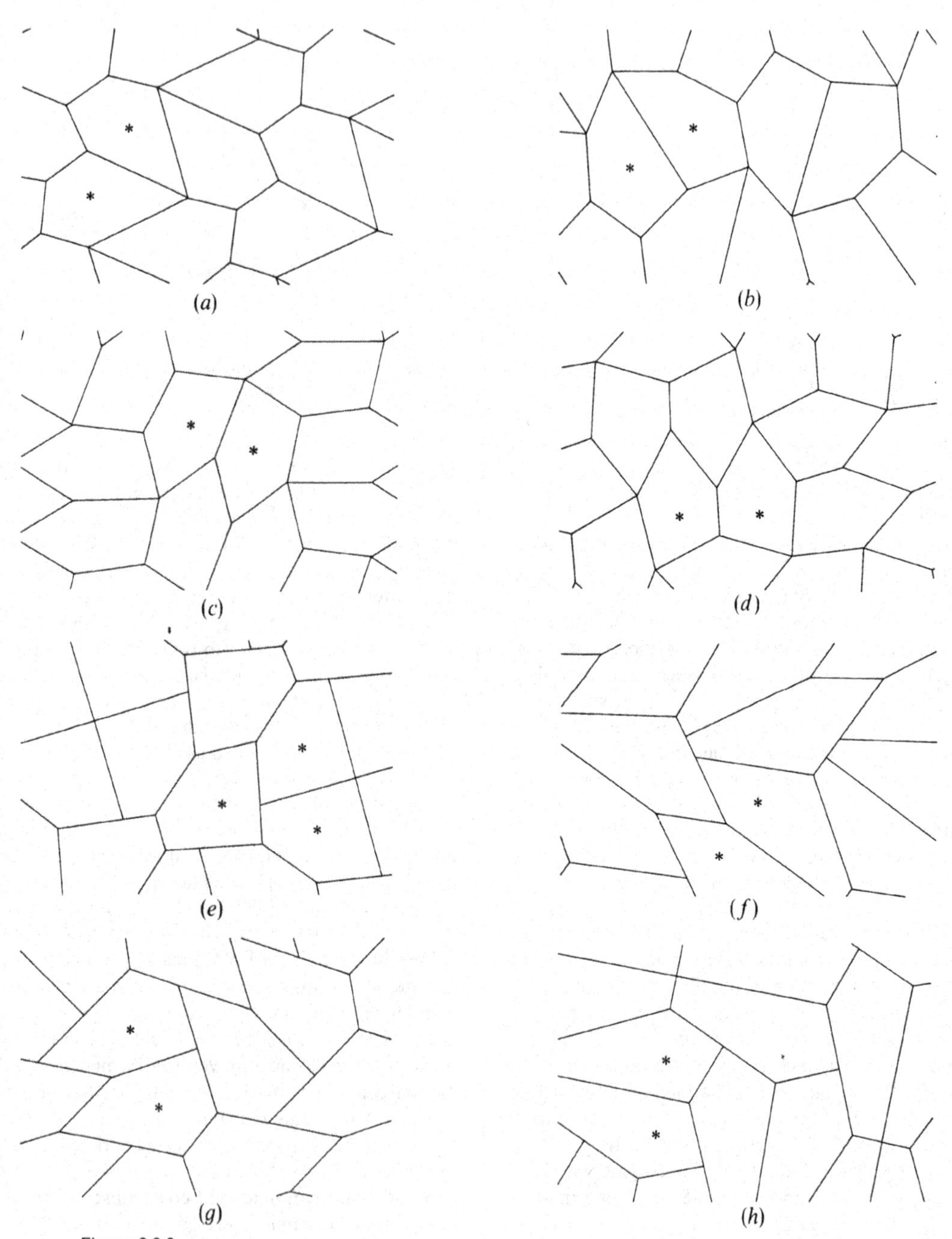

Figure 9.3.2
Examples of monohedral tilings by anisohedral convex pentagons. These pentagons are of types 6 to 13 in Table 9.3.1.

Schattschneider [1982] aims to enumerate all the equilateral pentagons, both convex and non-convex, which tile the plane. However, her list depends upon an extension of the computer search by Hirschhorn and Hunt mentioned above. At the present time this search appears not to have been carried out, and so the completeness of these results remains in doubt. For a problem concerning one of the tilings by non-convex pentagons, see Cundy [1983]. So far as we are aware, the existence of non-convex anisohedral hexagons has never been ruled out.

A listing of both convex and non-convex hexagons which tile isohedrally appears in Heesch & Kienzle [1963]; the completeness of this enumeration seems not to have been independently verified.

There are many open problems concerning monohedral tilings by polygons. One such question, unresolved despite its apparent simplicity, is whether every monohedral tiling can be approximated arbitrarily closely by monohedral tilings with polygonal prototiles.

EXERCISES 9.3

1. (*a*) Show that each of the types 4, 7 and 8 of pentagons in Table 9.3.1 has precisely one equilateral representative, and that each of the types 1 and 2 has infinitely many equilateral representatives.

   (*b*) Show that there exists no equilateral pentagon of any of the eight types in Table 9.3.1 which are not mentioned in part (*a*).

**2. To each tile T that admits a monohedral tiling of the plane corresponds an integer n defined as the smallest number of copies of T in an isohedral repeating unit. Hilbert believed that $n = 1$ for every tile which admits a monohedral tiling, while Heesch found a tile for which $n = 2$, see Figure 1.3.8. For a tile of type 10 in Table 9.3.1, the value of n is 3. Can a tile be found to correspond to *any* given value of n? If not, what are the possible values of n?

**3. Do there exist pentagons or hexagons which tile isohedrally only in improper tilings although they admit proper monohedral tilings?

9.4 POLYIAMONDS, POLYOMINOES AND POLYHEXES*

Let a region U of the plane be the union of n equilateral triangles selected from the regular tiling (3^6). Then if U is a topological disk, we call it an *n-iamond*, or a *polyiamond* if we do not wish to mention n explicitly. Corresponding terms for the unions of tiles from the regular tiling (4^4) are *n-omino* and *polyomino*, and from (6^3) are *n-hexe* and *polyhexe*. The name polyomino was invented by S. W. Golomb in 1953, although traces of the concept occur in Aubry [1907]. The name polyiamond was proposed by T. H. O'Beirne in 1961 and the word polyhexe seems to have been coined by analogy.

It should be noted that our definitions differ slightly from those adopted by some authors. For example, we do not consider a tile such as that of Figure 9.4.1 to be a polyomino because it is not a topological disk, whereas some authors would describe it as a 7-omino. It is interesting to note that although the problem was raised long ago (for polyominoes by Quijano [1908]) no formulae are known which give the numbers of different

* See the Appendix beginning on page 653.

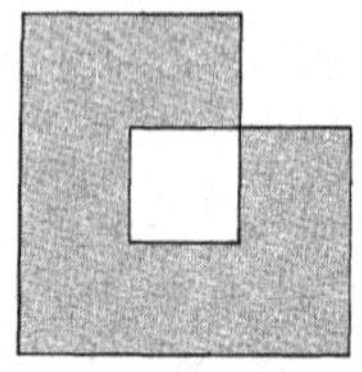

Figure 9.4.1
A union of seven squares that is not considered to be a polyomino since it is not a topological disk.

n-ominoes, n-iamonds or n-hexes as functions of n. This is an unsolved combinatorial problem. When we state the numbers of different tiles of a given kind, these numbers have been determined empirically (see, for example, Lunnon [1971], [1972], Ellard [1982]).

Diagrams of all 6-ominoes, 7-ominoes, and 8-ominoes appear in March & Matela [1974]. Bounds for the number of convex n-ominoes have been given by Klarner [1965] and Klarner & Rivest [1974]. Sampson & Trofanenko [1979] give an algorithm for generating the distinct n-hexes for a given value of n.

The first problem we shall consider is that of determining which polyiamonds, polyominoes and polyhexes are prototiles of isohedral or monohedral tilings. The results available are displayed in Tables 9.4.1, 9.4.2 and 9.4.3. (Golomb [1966] established that every n-omino with $n \leq 6$ admits a monohedral tiling of the plane as well as a number of related results.) In each case, following the value of n, we give drawings of the tiles (except when their number is too large!) and state the smallest possible number of aspects that can occur in a tiling with that prototile. It is clear that in the case of polyiamonds and polyhexes the maximum possible number of aspects needed is $6D$, $6R$ (that is, six direct and six reflected aspects) and in the case of polyominoes it is $4D$, $4R$. There exist 18-iamonds for which this maximum is attained (see Figure 9.0.1 for one of these) but we know of no polyhexe attaining the bound. (Regarding polyominoes see Exercise 9.4.5.) Whenever possible, we give in the tables reference to tilings that avoid reflections. In Figures 9.4.2, 9.4.3 and 9.4.4 we show examples of some of the tilings. It is of interest that the "smallest" polyomino for which a tiling is possible only if we use reflections is a certain 7-omino; this is noted, along with certain other information, in the footnotes to the tables. The problem of tiling a rectangle by congruent polyominoes, which is clearly relevant to that of tiling the whole plane, has been considered by Klarner [1965], [1969]. A method of finding some polyominoes that tile the plane has been discussed by Gilbert [1983]. The literature of recreational mathematics contains much material on tiling various shapes by polyominoes, polyhexes, polyiamonds and related kinds of tiles. Besides books by Golomb [1965] and Gardner [1966], [1978], the interested reader should consult the *Journal of Recreational Mathematics*, which publishes many relevant articles. Related questions concerning packings of polyominoes under various conditions have also been considered; see, for example, Bantegnie [1977], Klamkin & Liu [1980b].

To state our second problem (or, rather, group of problems) it is convenient to use the following definitions; they are formulated for polyominoes, though similar definitions clearly hold for polyiamonds and polyhexes. A *basic set* of n-ominoes is a set that contains one n-omino of each shape; a *reflective set* of n-ominoes contains one n-omino of each shape that has reflective symmetry, and one direct and one reflected copy of every other n-omino; a *full set* of n-ominoes contains one copy of each aspect of every n-omino. When working with basic sets we can rotate, reflect and translate the tiles; with reflective sets, we may rotate and translate but not reflect, and with full sets we may only translate. The question to be considered is whether there is a tiling in which the union of the n-ominoes of a basic set (or a reflective or a full set) form either a translative repeating unit, or else an isohedral repeating unit. No general results seem to be known; the information available to us is presented in Figure 9.4.5.

A related problem is whether it is possible, for each n, to have a tiling which contains copies of all n-ominoes and which is equitransitive. If it were true that each basic set admits a translative repeating unit, the answer to this problem would obviously be affirmative. But it may be so even if, as seems likely to be the case, some basic sets do not admit translative repeating units.*

The question whether a given set of polyominoes admits a tiling of the plane was considered by Golomb [1970]; he established that its algorithmic decidability is equivalent to the decidability of the tiling problem for Wang tiles which we will discuss in detail in Chapter 11. Using sophisticated algebraic tools, Barnes [1982] gives

* See the Appendix beginning on page 653.

Table 9.4.1 MONOHEDRAL TILINGS BY *n*-IAMONDS

For each value of *n* in **column (1)**, the number of different *n*-iamonds is indicated in **column (4)** and drawings of the tiles (for $n \leq 7$) appear in **column (2)**. In **column (3)** we give the minimum number of aspects that occur in an isohedral tiling by the given *n*-iamond, reflections of the prototile being avoided where possible.

Notes for **column (2)**:

(*a*) This tile also admits a non-isohedral tiling with aspects 4*D* (see Figure 9.4.2(*a*)).
(*b*) For a proof that this 7-iamond will not tile, see Gardner [1971, p. 250].
(*c*) These results are claimed by David Bird, Gregory J. Bishop, Andrew L. Clarke, John W. Harris and Wade Philpott, see Gardner [1975b].
(*d*) It is not known whether all these admit isohedral tilings.
(*e*) Determined by David Bird (see Gardner [1975b]) but not independently confirmed.

n **(1)**	Tiles **(2)**	Aspects **(3)**	Number of tiles **(4)**
1		2	1
2		1	1
3		2	1
4		1*D*	3
		1	
		2	
5	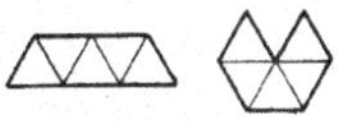	2	4
		2*D*	
6	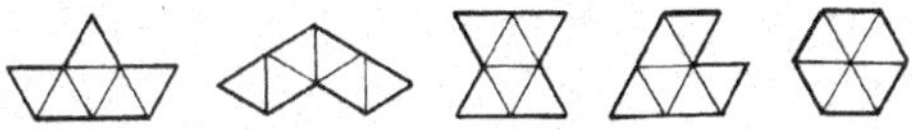	1	12
	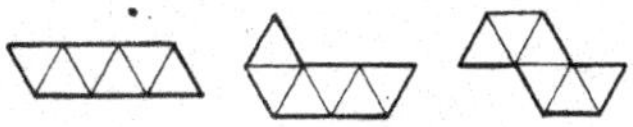	1*D*	
	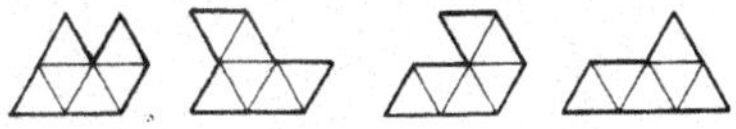	2*D*	

Table 9.4.1 (continued)

n (1)	Tiles (2)		Aspects (3)	Number of tiles (4)
7	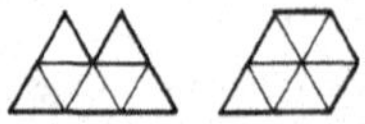		2	24
	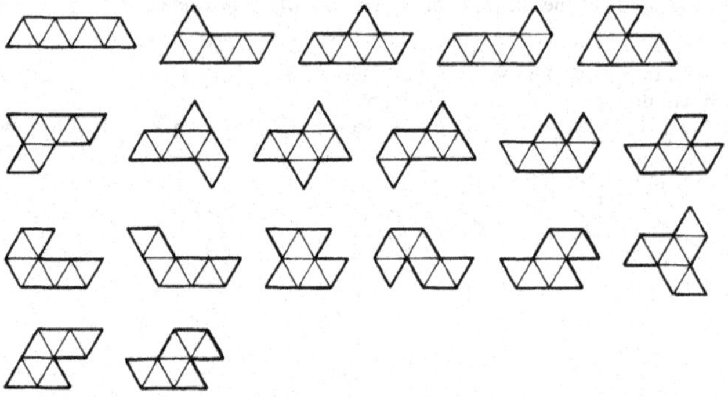		2D	
	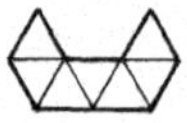		6	
	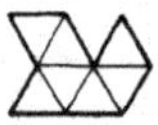	(*a*)	6D	
		(*b*)	No tiling possible	
8	All are prototiles of monohedral tilings	(*c*)(*d*)		66
9	All are prototiles of monohedral tilings except the following 20: 	(*d*)(*e*)		159

Table 9.4.2 MONOHEDRAL TILINGS BY *n*-OMINOES

The information given in this table corresponds to that in Table 9.4.1.

Notes for **column (2)**:

(*a*) See Gardner [1975b].
(*b*) This tile also admits a monohedral (but not isohedral) tiling with two aspects (see Figure 9.4.3(*a*)).
(*c*) This tile also admits a monohedral (but not isohedral) tiling with aspects 1*D* 1*R* (see Figure 9.4.3(*b*)).
(*d*) This is the smallest polyomino for which reflections are necessary in any monohedral tiling (see Figure 9.4.3(*d*)).
(*e*) Determined by David Bird (see Gardner [1975b]) but not independently confirmed.

n (1)	Tiles (2)	Aspects (3)	Number of tiles (4)
1		1	1
2	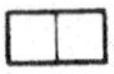	1	1
3	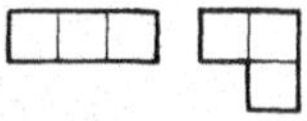	1	2
4	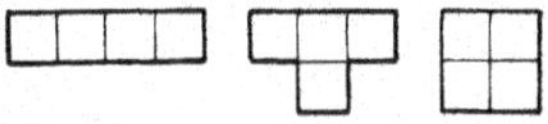	1	5
	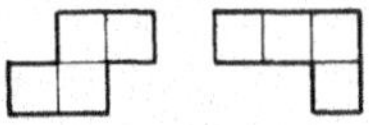	1*D*	
5	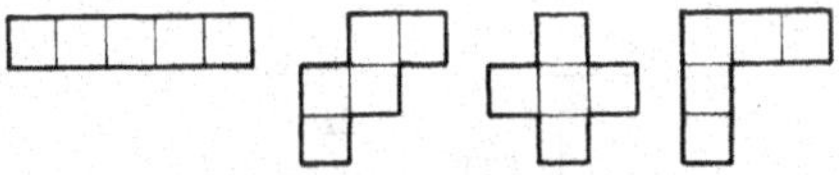	1	12
	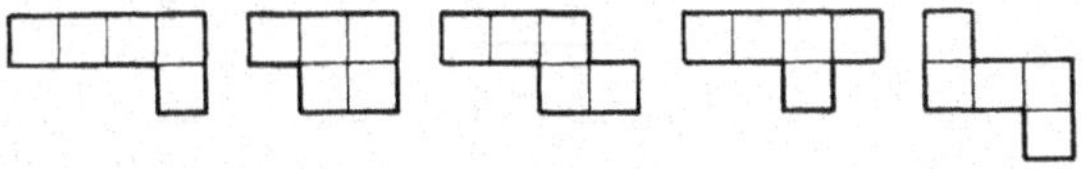	1*D*	
	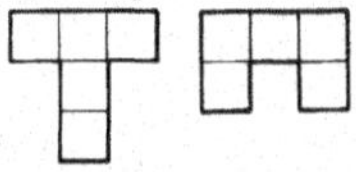	2	
	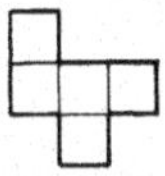	2*D*	
6	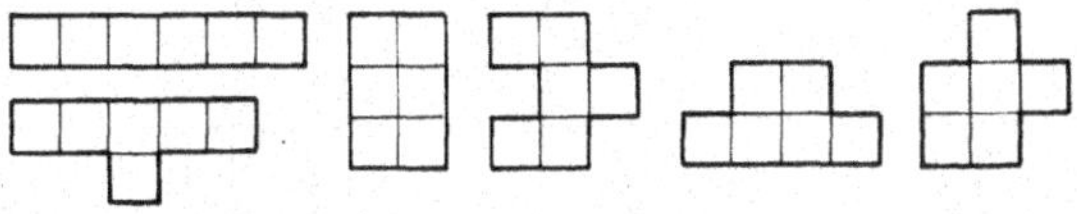	1	35

Table 9.4.2 (continued)

n (1)	Tiles (2)	Aspects (3)	Number of tiles (4)
	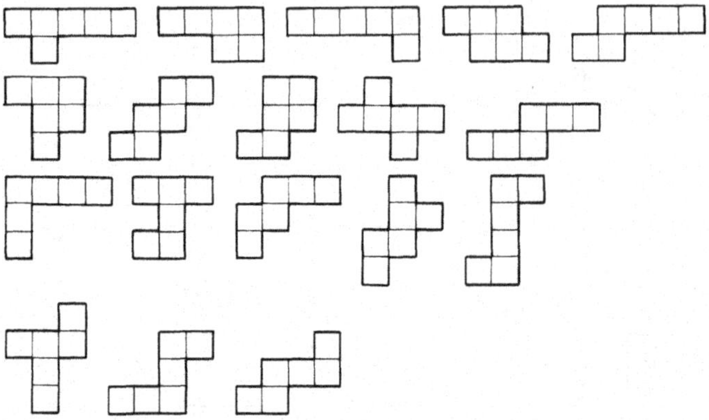	1*D*	
	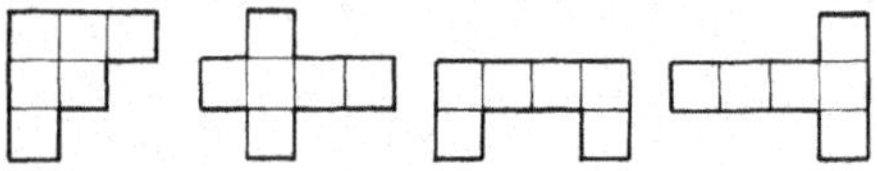	2	
	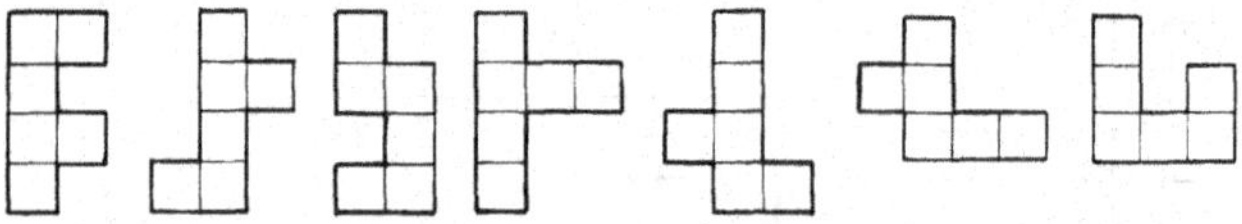	2*D*	
7	101 types (*a*)	1, 1*D*, 2 or 2*D*	107
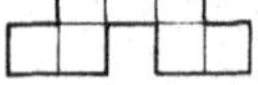	(*b*)	4	
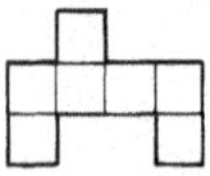	(*c*)	1*D*, 1*R* or 4*D*	
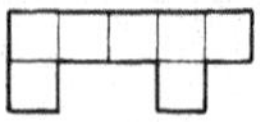	(*d*)	2*D*, 2*R*	
	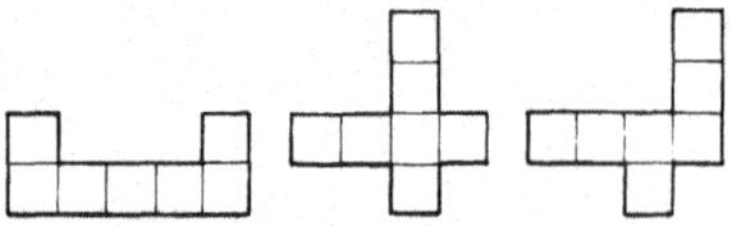	No tilings possible	

Table 9.4.2 (continued)

n (1)	Tiles (2)	Aspects (3)	Number of tiles (4)
8	All are prototiles of monohedral tilings except the following twenty:  (*e*)		363

Table 9.4.3 MONOHEDRAL TILINGS BY n-HEXES

The information in this table corresponds to that in Table 9.4.1.

Notes for **column (2)**:

(*a*) This tiling also admits a monohedral but not isohedral tiling with aspects 1*D* 1*R*.
(*b*) This tile also admits a monohedral but not isohedral tiling with two aspects that differ by a rotation of $\frac{1}{3}\pi$.
(*c*) Determined by David Bird (see Gardner [1975*b*]) but not independently confirmed.

n (1)	Tiles (2)	Aspects (3)	Number of tiles (4)
1		1	1
2	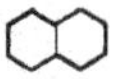	1	1
3	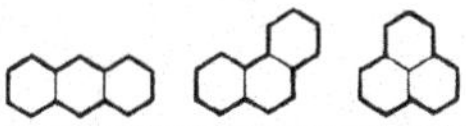	1	1

Table 9.4.3 (continued)

n (1)	Tiles (2)	Aspects (3)	Number of tiles (4)
4	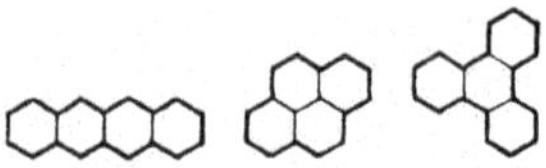	1	7
	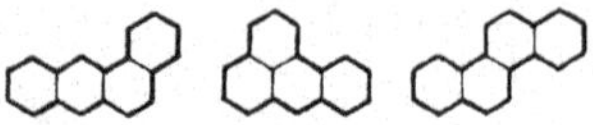	1*D*	
	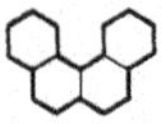	2	
5	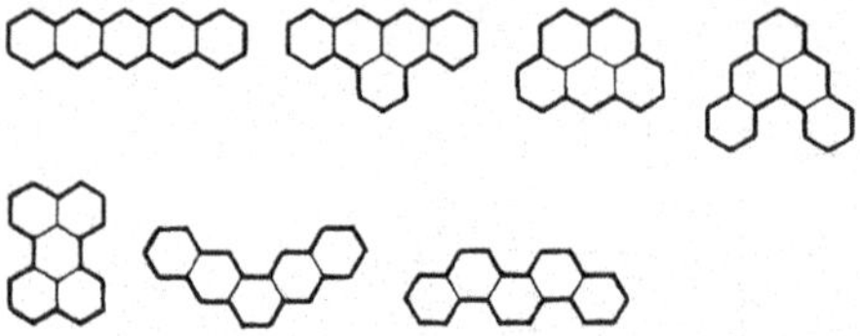	1	22
	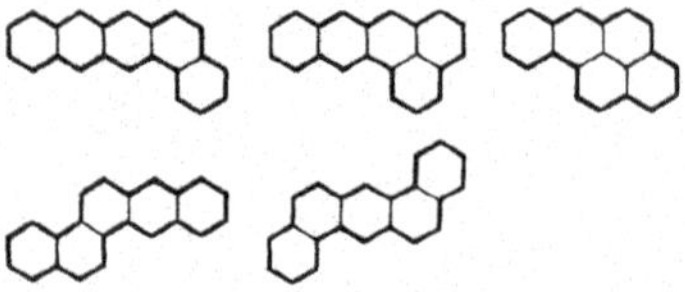	1*D*	
	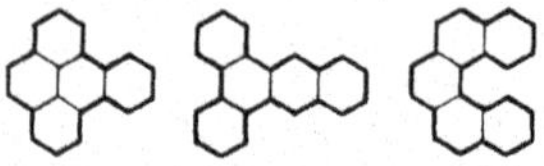	2	
	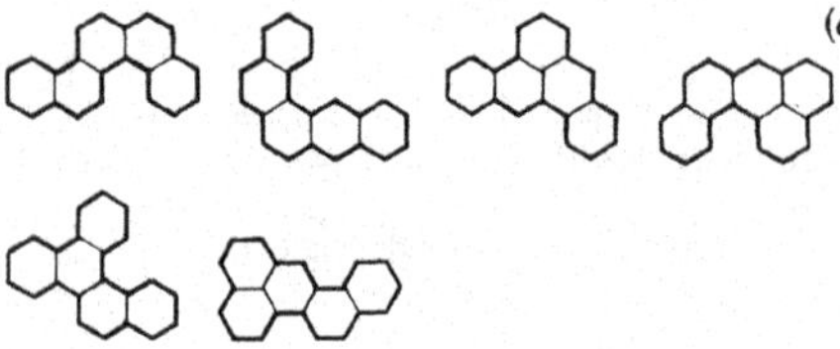 (*a*)	2*D*	
	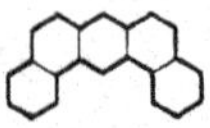 (*b*)	4	

Table 9.4.3 (continued)

n (1)	Tiles (2)	Aspects (3)	Number of tiles (4)
6	All are prototiles of monohedral tilings except the following four:	(c)	84

a powerful method of deciding whether a given set of polyominoes will, in a generalized sense, tile the plane.*

A natural generalization of polyominoes, polyiamonds and polyhexes can be based on any isohedral tiling. If $\mathcal{T}$ is an isohedral tiling and n a positive integer, we call an *n*-$\mathcal{T}$-tile any closed topological disk which is the union of n tiles of $\mathcal{T}$; if the value of n is not relevant we shall speak of a *poly-*$\mathcal{T}$*-tile*. Clearly polyiamonds, polyominoes and polyhexes are poly-$\mathcal{T}$-tiles that correspond to the choice of $\mathcal{T}$ as one of the regular tilings. (In a different generalization, Krishnamurti & Roe [1979] investigate "poly-$\mathcal{T}$-tiles"

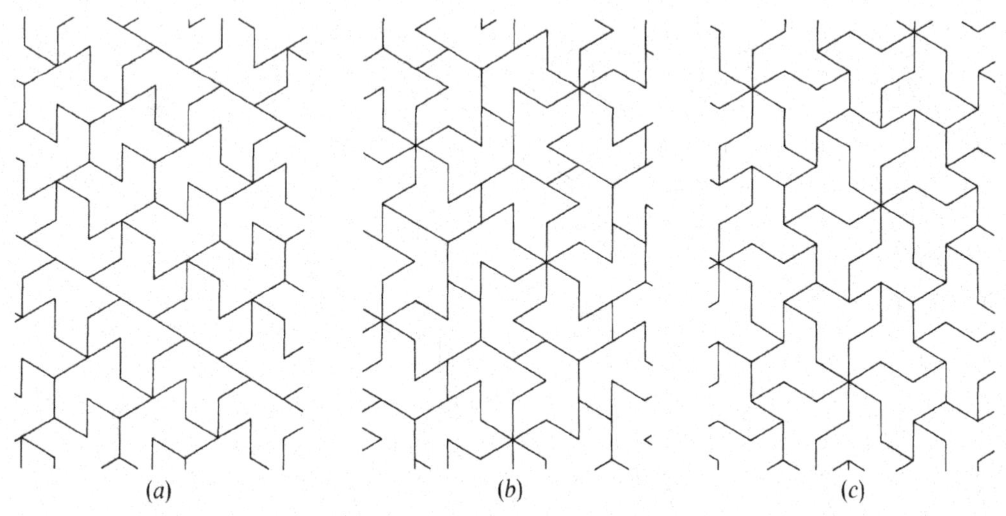

Figure 9.4.2
Examples of monohedral tilings by polyiamonds. The prototile in (*a*) is the same 7-iamond as in (*b*). In (*a*) four direct aspects are used; the tiling is not isohedral. In the isohedral tiling (*b*) six direct aspects are used. Part (*c*) shows an isohedral tiling using two direct aspects of another 7-iamond.*

* See the Appendix beginning on page 653.

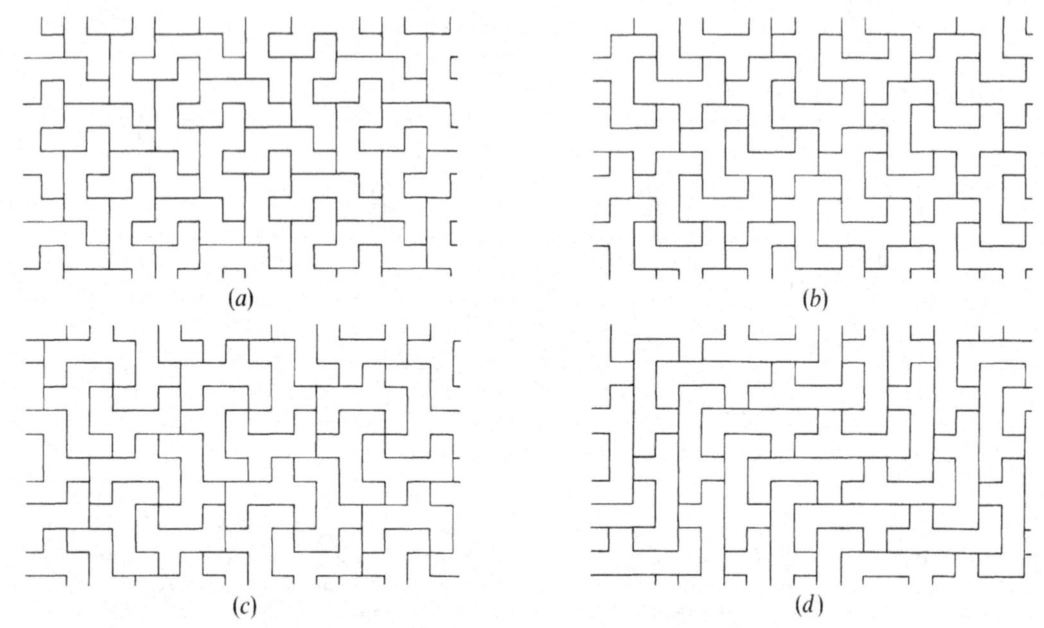

Figure 9.4.3

Examples of tilings by 7-ominoes. The tiling (*a*) requires two aspects of the prototile, one of which is a rotation of the other through $\frac{1}{2}\pi$. Tilings (*b*) and (*c*) use the same 7-omino as prototile—in (*b*) with aspects $1D$, $1R$ and in (*c*) with aspects $4D$. In (*d*) we show the smallest n-omino for which reflected aspects necessarily occur in any tiling. The tilings in (*c*) and (*d*) are isohedral, those in (*a*) and (*b*) are not.*

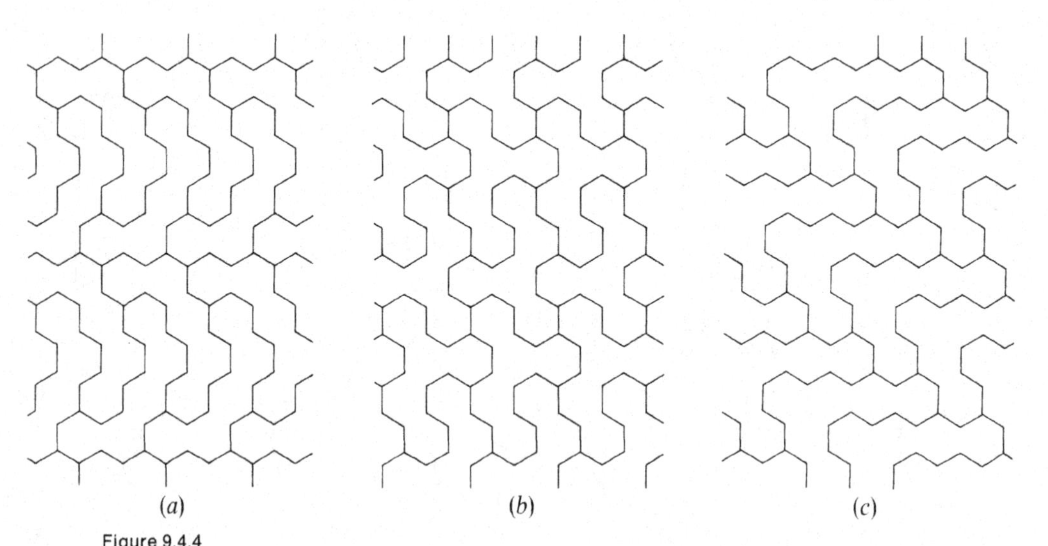

Figure 9.4.4

Some tilings by polyhexes. In the monohedral tiling (*a*) the 5-hexe prototile occurs in two aspects related by a $\frac{1}{3}\pi$ rotation. In the isohedral tiling (*b*) the 5-hexe prototile occurs in two direct aspects related by a half-turn. In the isohedral tiling (*c*) the 6-hexe prototile occurs in only one direct aspect.

* See the Appendix beginning on page 653.

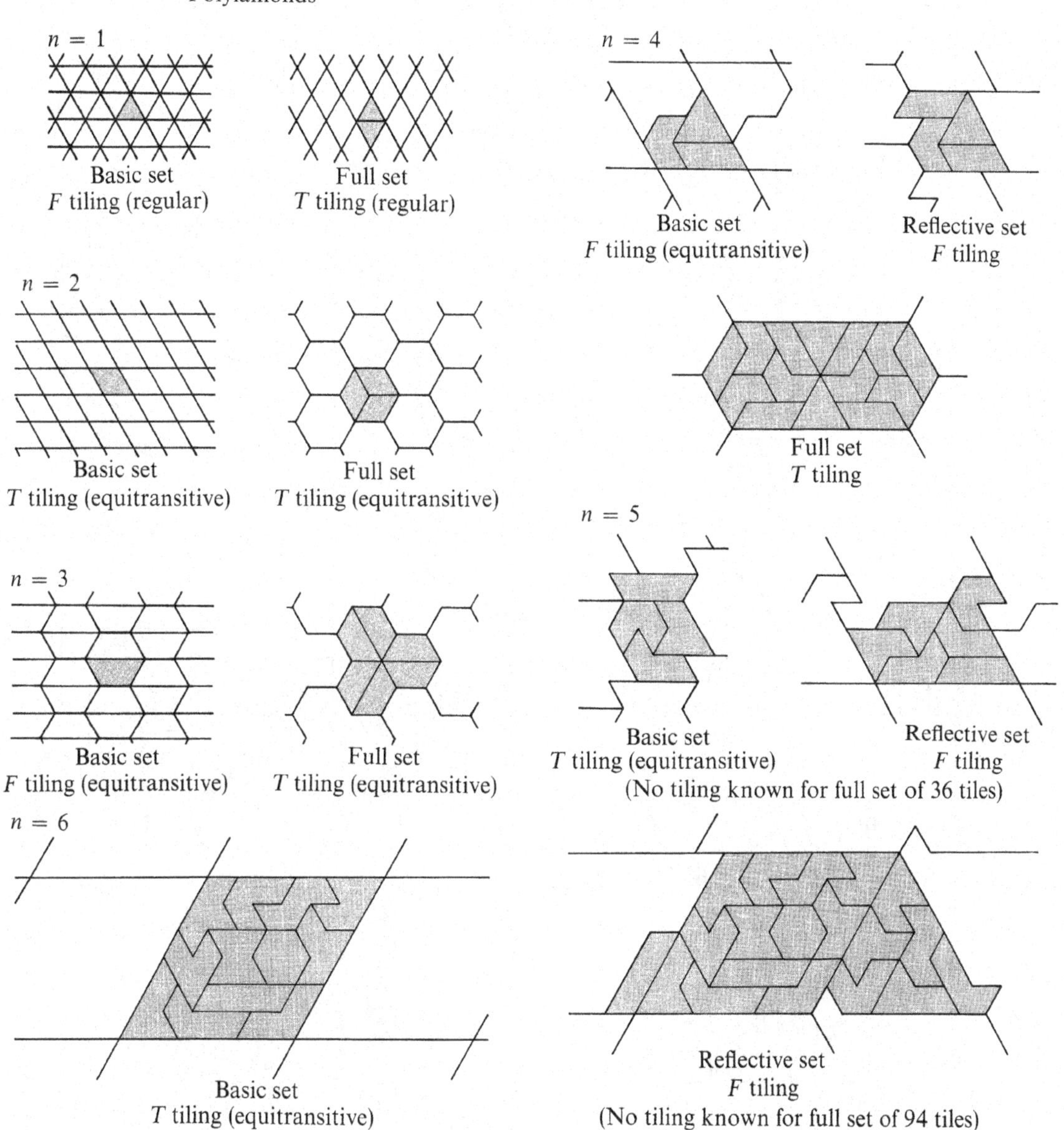

Figure 9.4.5

The known tilings by basic, reflective or full sets of polyiamonds, polyominoes and polyhexes. T indicates that the set forms a translative repeating unit, and F that the set forms a fundamental region for the symmetry group of the tiling.

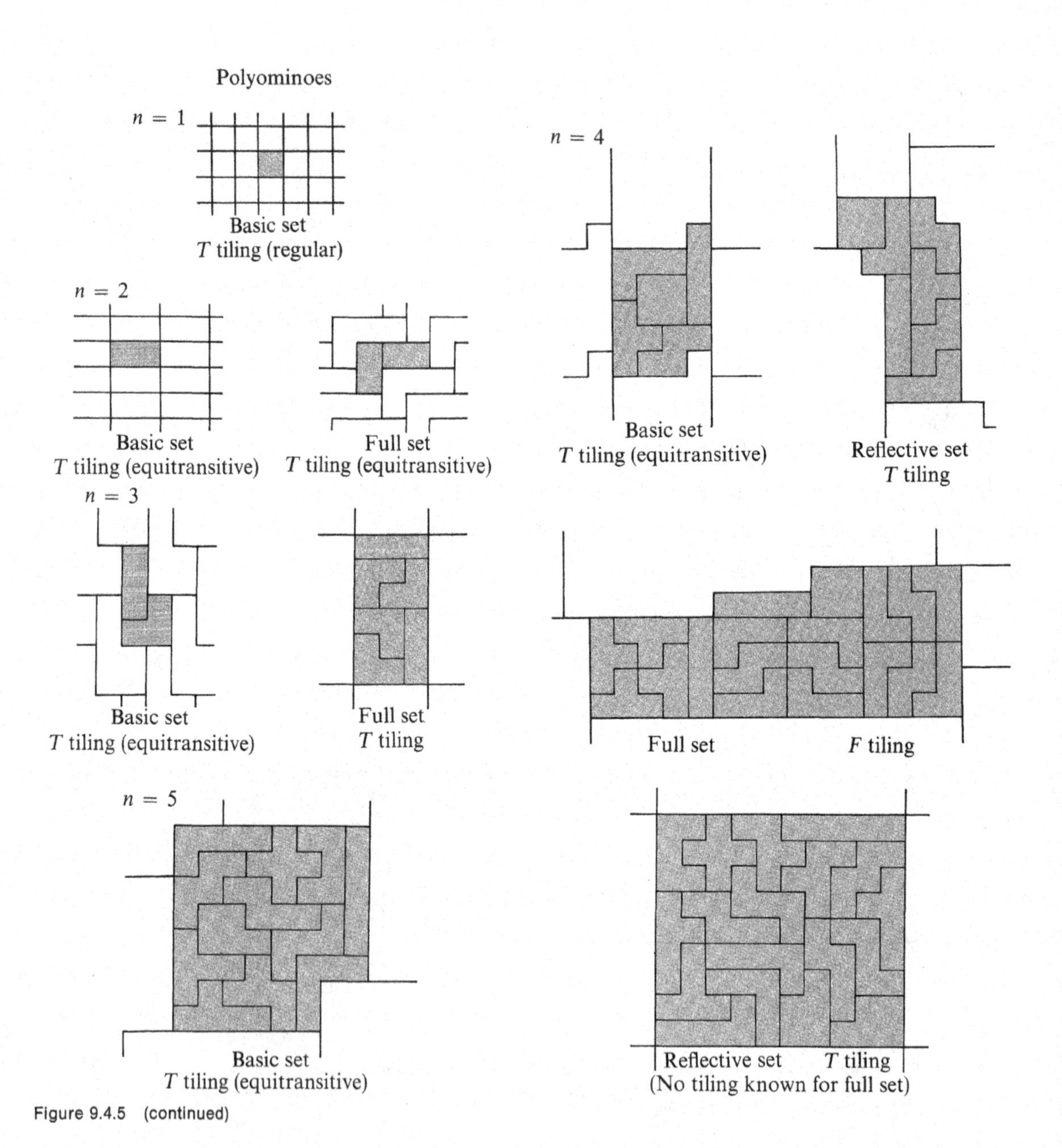

Figure 9.4.5 (continued)

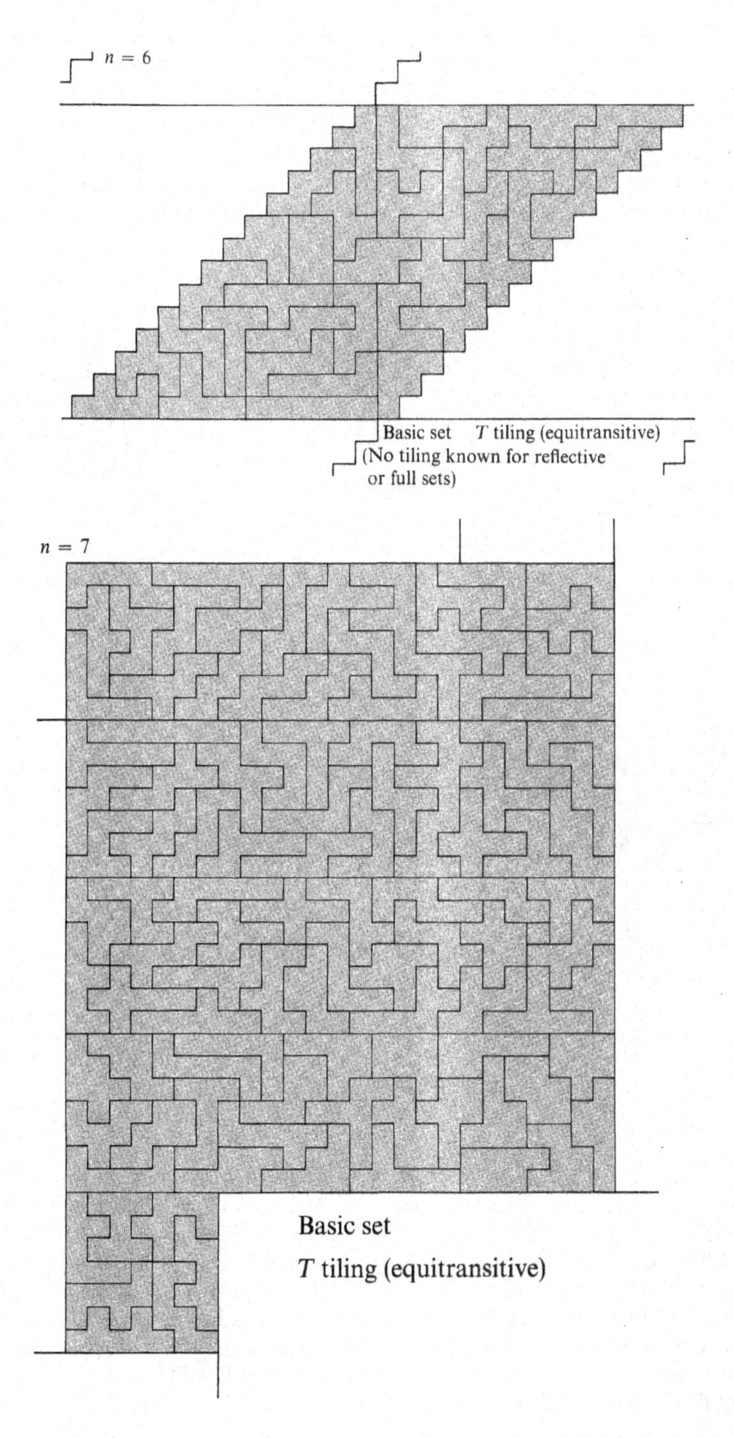

Basic set

T tiling (equitransitive)

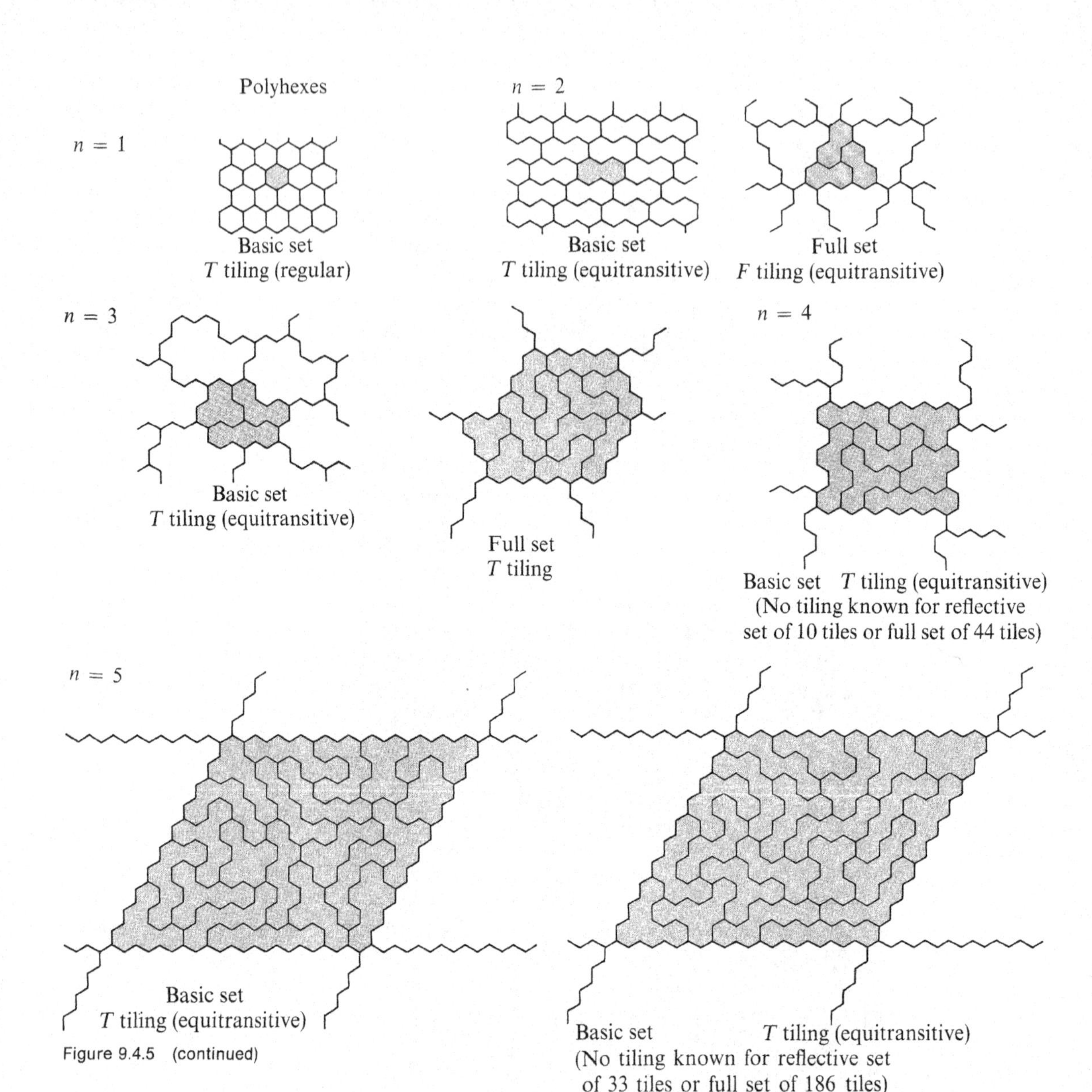

Figure 9.4.5 (continued)

for the case in which $\mathscr{T}$ is not isohedral but is one of the uniform tilings, and these authors present several algorithms for enumerating them. See Bidwell [1975] for a related concept.) Tiles which are unions of congruent triangles with angles 45°, 45° and 90° (not necessarily taken from the Laves tiling [4.8.8]) have been called "polyaboloes"; tilings using them can be derived from the results of Tominago [1981].

The idea of amalgamating tiles of a given tiling to form bigger tiles has attracted the attention of several authors. See, for example, Holladay [1981], Holroyd [1983]. It has also been used (see Bell, Diaz, Holroyd & Jackson [1983], Gibson & Lucas [1984], Lucas & Gibson [1984a], [1984b], Bell, Diaz & Holroyd [1985]) in a hierarchical addressing system for surfaces in computer applications. Related applications can be found also in Burt [1980].*

* See the Appendix beginning on page 653.

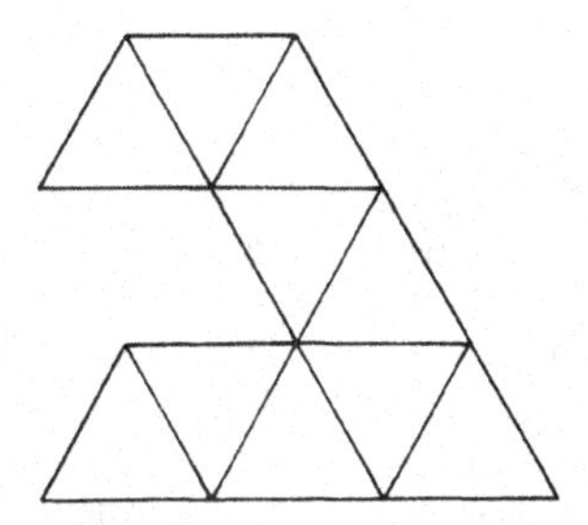

Figure 9.4.6
A 10-iamond that admits a monohedral tiling, but does not admit any isohedral tiling.

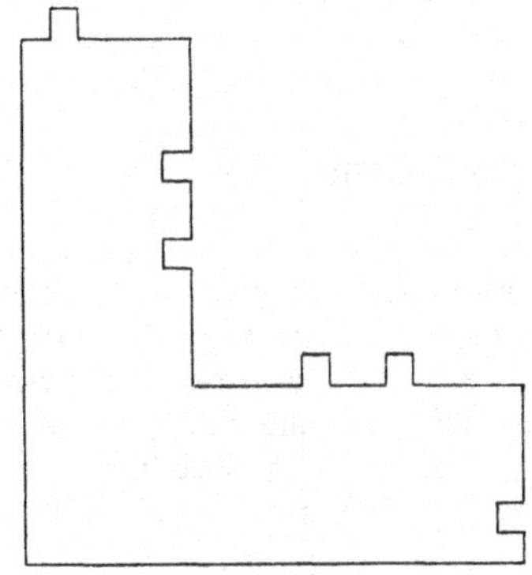

Figure 9.4.7

1. Let T denote the 7-omino prototile of Figure 9.4.3(a).
   (a) Show that T admits no isohedral tiling with at most 3 aspects.
   (b) Determine all isohedral tilings with prototile T.
   (c) Show that there are uncountably many monohedral tilings with prototile T and just two aspects.
2. Let T be the 5-hexe prototile of the tiling in Figure 9.4.4(a).
   (a) Show that T admits no isohedral tiling with at most 3 aspects.
   (b) Determine all isohedral tilings with prototile T.
3. Consider the 7-iamond prototile T of Figure 9.4.2(a).
   (a) Show that T does not admit any isohedral tiling with at most 5 aspects.
   (b) Find a monohedral tiling with 12 aspects of prototile T.
   (c) Find all isohedral tilings with prototile T.
4. In this problem we are concerned with tiles that are prototiles of monohedral tilings but admit no isohedral tilings.
   (a) Show that the 10-iamond of Figure 9.4.6 (brought to our attention by Ludwig Danzer) has this property.
   **(b) Does any 10-iamond other than the one in Figure 9.4.6 have this property?
   *(c) Do any n-iamonds with $n \leq 9$ have this property?
   (d) Find a polyomino and a polyhexe with this property.
   **(e) Determine the least n and k such that there exist an n-omino and a k-hexe with this property.
5. (a) Show that the 180-omino in Figure 9.4.7 is monomorphic and that the only tiling with this prototile has $4D$, $4R$ aspects.
   (b) Find an n-omino with $n < 180$ which tiles only with $4D$, $4R$ aspects.
6. Among the many possible problems on poly-$\mathcal{T}$-tiles we mention the following as examples.
   (a) Show that, for each $\mathcal{T}$, the number of 2-$\mathcal{T}$-tiles is at most 5 and find a tiling $\mathcal{T}$ for which that bound is attained.
   (b) Find a tiling $\mathcal{T}$ that is not regular but for which all 2-$\mathcal{T}$-tiles are congruent.
   *(c) Find the maximum possible number of non-congruent 3-$\mathcal{T}$-tiles.
   (d) Find a 2-$\mathcal{T}$-tile, in a suitably chosen isohedral tiling $\mathcal{T}$ which is not the prototile of any monohedral tiling.
   (e) Does each basic set of 2-$\mathcal{T}$-tiles admit a translative repeating unit (or an isohedral repeating unit)? Answer the same question for reflective and full sets of 2-$\mathcal{T}$-tiles. [For example, consider types IH7, IH22 and IH56.]

9.5 SPIRAL TILINGS

In Chapter 3 we mentioned Voderberg's discovery of a tile of which two copies can completely surround a third, see Figure 3.2.4. In his paper Voderberg [1936] remarked that this tile admits a monohedral "spiral tiling" (see Figure 9.5.1), and added in a later note (Voderberg [1937]) that other spirals are possible.

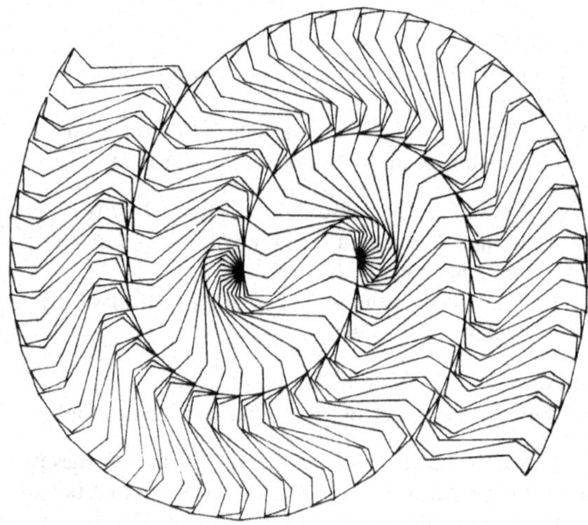

Figure 9.5.1
The spiral tiling discovered by Voderberg [1936]. Its prototile has some other remarkable properties; see Figure 3.2.4.

Such spirals are visually sufficiently striking to merit further investigation, and we give here a brief account of some of the possibilities. The main difficulty in a mathematical treatment of this topic is to define exactly what we mean by a "spiral"—to some extent, at least, the "spiral effect" is psychological. This is illustrated by the contrast between Figures 9.5.2 and 9.5.4. In Figure 9.5.2(*a*) we show a tiling of the plane by isosceles triangles; the two equal edges are taken to be of unit length, and the vertical angle is π/n, where n is any positive integer (in the diagram $n = 12$). If we split the tiling into two half-planes and slide one of these half-planes relative to the other through 1, 2 or 3 units of distance we get the tilings of Figures 9.5.2(*b*), (*c*) and (*d*) and similarly for larger distances. A displacement through a distance $m \geq 1$ yields a tiling which may, with some justification, be called a "spiral" with $2m$ arms. However, these "spirals" are not very convincing or interesting. A much more satisfactory result is obtained if we replace the unit edges by suitable curves, as in Figure 9.5.3. This is precisely the way Voderberg's tiling arises—it can be constructed by replacing each unit edge of the 2-armed spiral in Figure 9.5.2(*b*) by an S-curve which consists of 5 straight line-segments (see Figure 9.5.3(*a*)).

Using the technique just described it is easy to construct any number of (very decorative) spirals with an even number of "arms". From the replacement curves in Figures 9.5.3(*b*) and (*c*) we obtain, for example, the spiral tilings of Figure 9.5.4.

It is much more challenging to find spirals with an odd number of arms. One example has already been displayed in Figure 1.2.3(*b*), and it was shown there how the same prototile (see Figure 9.5.3(*d*) or Figure 1.2.3(*a*)) also admits several other remarkable monohedral tilings (see Figures 1.0.1 and 1.2.3(*c*)). In Figure 9.5.5 we show a 3-armed spiral with the same prototile. (For a generalization of this construction see Exercise 9.5.1.)

The tile of Figure 9.5.3(*e*), which may be called a reflexed $(2n + 1)$-gon (in the figure we show the case $n = 3$) also admits many remarkable tilings (see Simonds [1977], Hatch [1978] and Fielker [1978]). In Figure 9.5.6 we show a 1-armed spiral using this prototile. In contrast to the tilings mentioned in the previous paragraph, this prototile has symmetry group *d1* and so only *directly* congruent copies of the tile are used in any tiling. A number of tilings using reflexed 8-gons, some of a spiral character are shown in Simonds [1978] and it seems as if there are many possibilities using these reflexed polygons still to be explored. However, at present, we know of no prototile which admits a monohedral r-armed spiral tiling for any *odd* value of $r \geq 5$.*

* See the Appendix beginning on page 653.

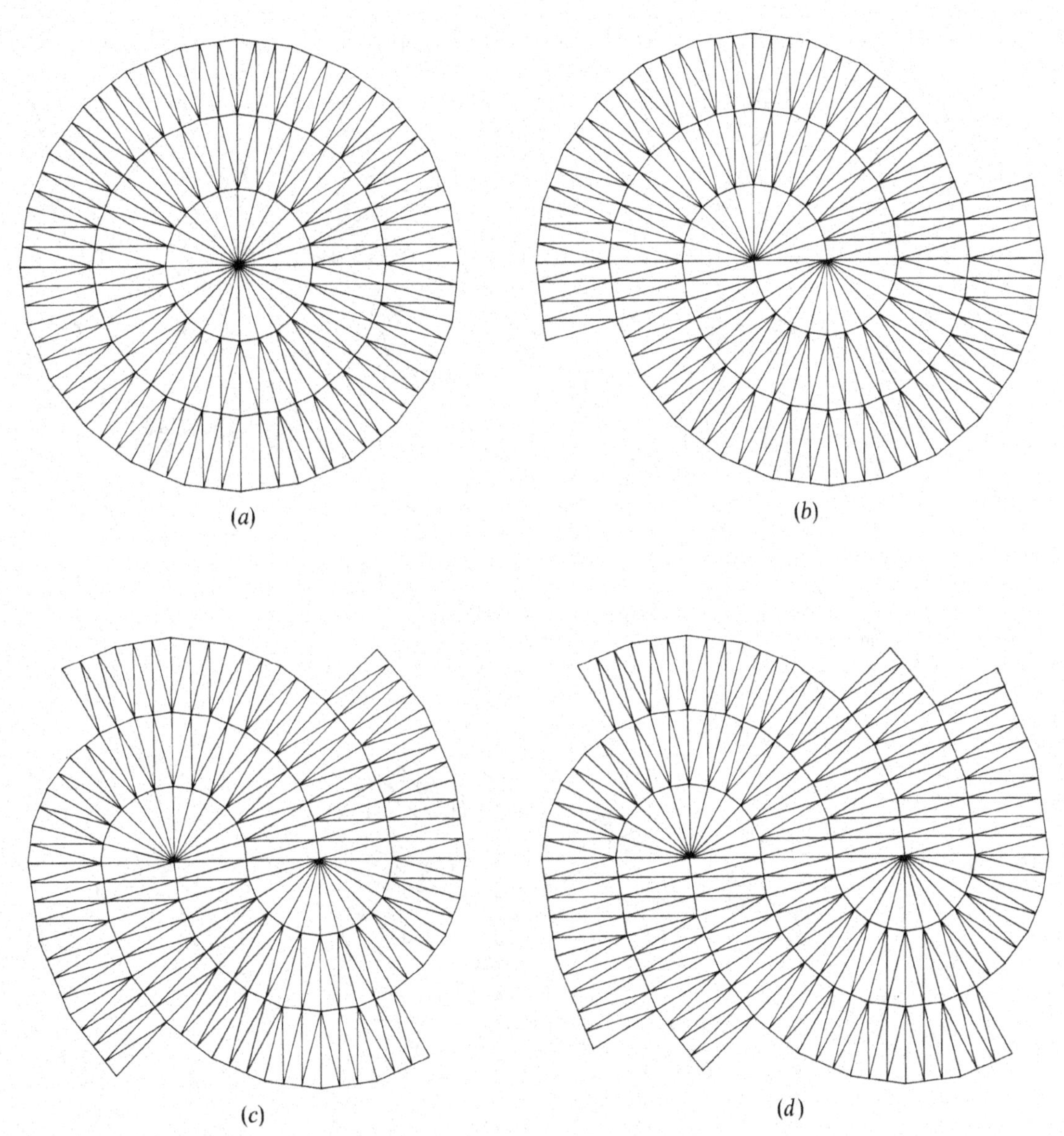

Figure 9.5.2
The construction of "spiral" tilings from a monohedral tiling by isosceles triangles.

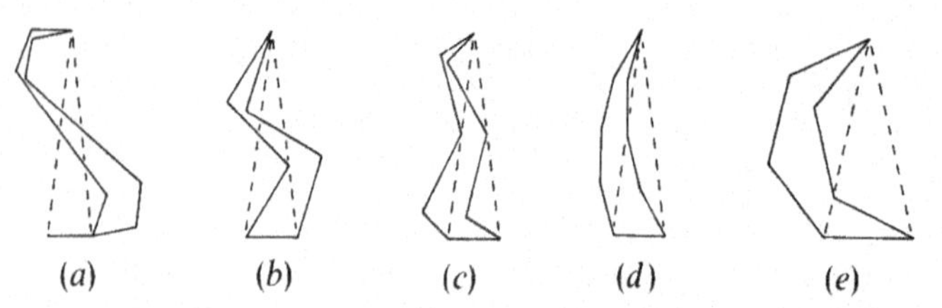

Figure 9.5.3
Some tiles that can replace the isosceles triangles in Figure 9.5.2 to make more convincing "spirals".

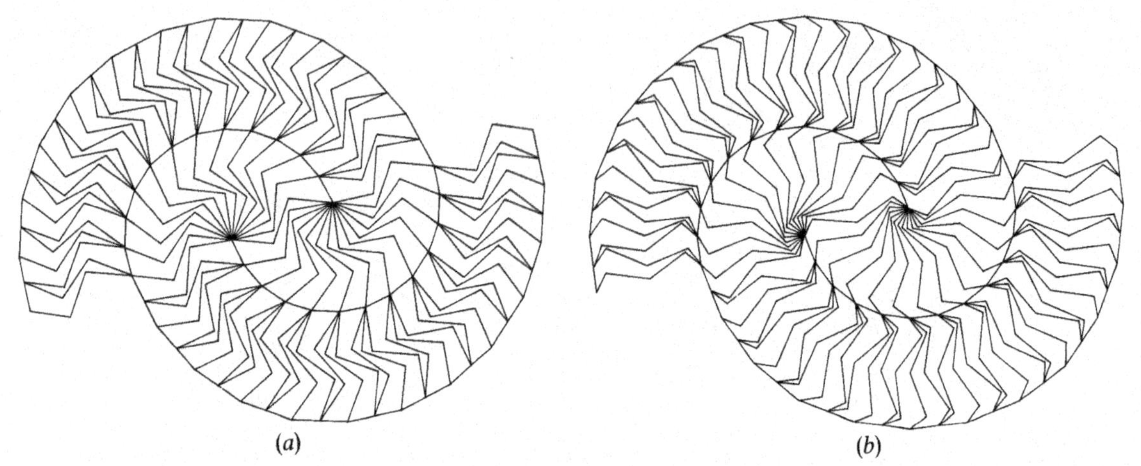

Figure 9.5.4
Two examples of spiral tilings obtained from those in Figure 9.5.2 by replacing the tiles by those in Figure 9.5.3.

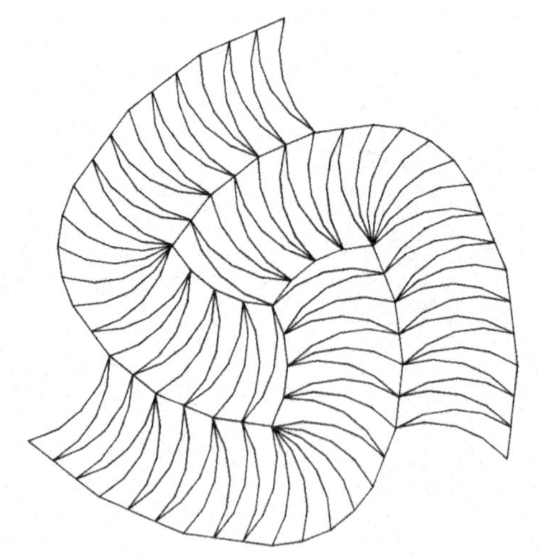

Figure 9.5.5
A 3-armed spiral with the prototile shown in Figure 9.5.3(*d*).

Figure 9.5.6
A one-armed spiral whose prototile is a reflexed 7-gon.

EXERCISES 9.5

1. Let r be a positive integer, $r \geq 2$, and let T be an isosceles triangle whose vertical angle is $\pi/3r$ and whose sides are 1, 1 and $2 \sin \pi/6r$. We replace each unit edge by r consecutive sides of a $6r$-gon of unit circumradius to obtain a tile T_r (see Figure 9.5.3(d) for the case $r = 4$). Show that for each r the tile T_r admits a 1-armed spiral, a 3-armed spiral, and many other spiral tilings.
2. With the prototile of Figure 9.5.3(d) construct a 6-armed spiral tiling with symmetry group $c6$.
3. One possible, mathematically precise, definition of "spiral tiling" is the following. A monohedral tiling $\mathcal{T}$ with prototile T is a *spiral* if it is possible to *mark* T by one or several arcs so that the union of the corresponding arcs on *all* the tiles of $\mathcal{T}$ consists of a finite number of unbounded simple curves. Although this definition has several drawbacks (among them the disregard for the psychological aspect—see parts (f) and (g) below), we shall make use of it in the following problems.
   (a) Verify that in the spiral tilings of Figures 1.0.1, 1.2.3(b), 9.5.4, 9.5.5 and 9.5.6 it is possible to find an edge of the prototile and use it as the "marked arc" in the above definition. (Note that two marked edges of tiles may coincide at an edge of the tiling.)
   (b) Consider the tiling $\mathcal{T}$ of Figure 9.5.7(b) with the prototile T shown in Figure 9.5.7(a). Show that $\mathcal{T}$ is a spiral by choosing as the marked arc on T the union of several edges. Show also that its spiral character is made even more pronounced if the marking consists of a single arc that crosses T.
   (c) Construct a 2-armed spiral in which each tile is crossed three or more times by spiralling curves formed by suitable markings.
   (d) Use the prototile T_2 of Exercise 1, marked with a suitable arc, to construct a 4-armed spiral tiling with symmetry group $c4$.
   (e) Using the prototile T_4 (see Figure 9.5.3(d)) marked with a suitable arc, construct a 2-armed spiral in which the two arcs form two curves, each unbounded in one direction only.
   (f) By marking with suitable arcs, show that the reg-

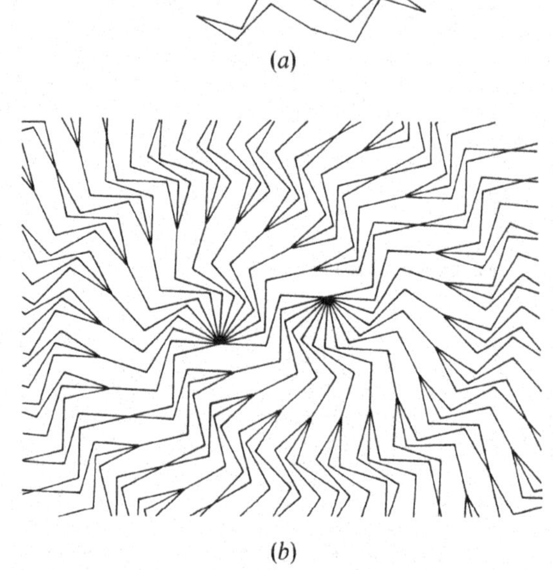

Figure 9.5.7
The prototile (*a*) of a 2-armed spiral (*b*), both with symmetry group *c2*. The prototile is the union of two copies of the tile of Figure 9.5.3(*b*).

ular tiling (4^4) by squares can be interpreted as a spiral with 1, 2, 3, 4, 8, . . . arms.

** (*g*) For which values of *n* is it possible to mark the tiles of (4^4) so that an *n*-armed spiral results?

(*h*) Using suitably marked arcs show that the regular tiling (3^6) can be interpreted as a spiral with 1, 2 or 3 arms.

(*i*) Using a suitable arc as marking show that the regular tiling (6^3) can be interpreted as a 1-armed spiral.

**(*j*) For which values of *n* is it possible to mark a regular hexagon by (one or several) arcs so as to make it the prototile of an *n*-armed spiral?

(*k*) Consider the attractive patch in Figure 9.5.8 (from Lord & Wilson [1984, p. 167]). (1) Decide whether it is possible to extend the patch to a monohedral tiling of the whole plane. (2) Is it possible to carry out the extension in such a way that, with a suitable marking of the prototile, it becomes a spiral tiling?

4. For certain integers *n* it is easy to find spiral tilings in which the prototile occurs in precisely *n* aspects.
   (*a*) Verify that $n = 30$ in Voderberg's spiral (Figure 9.5.1) and that $n = 24$, 24 and 48 in the spirals of Figure 9.5.4.
   (*b*) Show that there exist spiral tilings (in the sense of Exercise 3) with just four aspects of the prototile.
   (*c*) Show that there is no upper bound on the number of aspects possible in spirals.
   ** (*d*) Show that there is no spiral tiling with just three aspects of the marked prototile.
   ** (*e*) Show that if two spiral tilings use the same prototile then it occurs in the same number of aspects in both.
   ** (*f*) Show that in each spiral the prototile occurs in just a finite number of aspects.

Figure 9.5.8
A spiral tiling from Lord & Wilson [1984].

9.6 NOTES AND REFERENCES

The first attempt to determine all polygons which will tile the plane was made in the dissertation of Reinhardt [1918]. He observed that all triangles and quadrangles, five families of pentagons and three families of hexagons have this property. Thus he discovered every type which tiles isohedrally (compare Table 9.3.1). Many later authors investigated this problem or parts of it. The MacMahons (P. A. MacMahon [1921], [1922], P. A. MacMahon & W. P. D. MacMahon [1922], W. P. D. MacMahon [1925]) restricted themselves to edge-to-edge tilings in which only directly congruent copies of the prototile are allowed. The book of Heesch & Kienzle [1963], which reports on work done by Heesch mostly in the nineteen-thirties, gives also examples of non-convex polygons with at most six sides which admit isohedral tilings, without requiring that the tilings are proper. Additional references can be found in Grünbaum & Shephard [1978d] and in the excellent survey Schattschneider [1978b]; Sections 9.1 and 9.2 are mainly based on the former, Section 9.3 relies heavily on the latter.

Until 1935 it seems to have been generally believed that every polygon which admits a monohedral tiling also admits an isohedral tiling. The eighteenth of Hilbert's famous problems (Hilbert [1900]) asks whether there exists a tile that admits a monohedral—but no isohedral—tiling of 3-dimensional space. From the context it appears that Hilbert assumed that the corresponding planar problem has a negative solution. A similar (though rather vaguely expressed) assumption had been made earlier by Fedorov [1885, Section 64]. The same opinion was shared by Reinhardt, who was Hilbert's assistant during some of the years of World War I, and whose dissertation we mentioned at the start of this section; see also Reinhardt [1929]. A positive answer to Hilbert's three-dimensional problem was found by Reinhardt [1928a]; in that paper he asserted that no such tiles exist in the plane, and that his proof of this will appear soon. However, Heesch [1935] described a counterexample—a tile that admits a periodic tiling of the plane, but no isohedral tiling. Heesch's tile (see Figure 1.3.8) is a non-convex polygon, and admits only improper tilings. Variants of that tile were given in Heesch, Heesch & Loef [1944], Heesch [1968b] and Milnor [1976]. The latter is an account of developments related to Hilbert's eighteenth problem; a section is devoted to monohedral tilings but goes no further than Heesch's contribution made in 1935.

One of the aims of Reinhardt's thesis was the determination of *all* convex tiles which admit monohedral tilings of the plane; he expected to establish that each such tile admits an isohedral tiling. He did not succeed, and he acknowledged that his list of pentagons might not be complete. However, he asserted that if any pentagons were missed, their discovery "could be done by the above method; but carrying out such a discussion is highly cumbersome, very laborious, and offers little satisfaction. Moreover there is a certain probability that no other types of pentagons [besides the five families] will be discovered." (translated from Reinhardt [1918, p. 85]). Actually, although he was at that stage considering only edge-to-edge tilings by convex pentagons, Reinhardt missed some tiles which are prototiles of homogeneous (see Section 4.3) monohedral tilings (such as those in Figures 9.3.2(*a*), (*c*) and (*g*). Nevertheless, there persisted for a long time the widespread impression that Reinhardt had proved that there exist no anisohedral tiles. Possibly this misunderstanding was caused by the relative inaccessibility of his thesis (Reinhardt [1918]), and by the unnecessarily complicated formulations of results in it.

The next advance was due to Kershner [1968], [1969]; he restricted attention to convex polygons and attempted to find all tiles which admit monohedral tilings. Kershner produced examples of convex pentagons which admit periodic monohedral tilings but no isohedral ones (see Figure 9.3.2) and thus improved on Heesch's example mentioned above. Moreover, the tiling shown in Figure 9.3.2(*a*) uses (unlike Heesch's example) only directly congruent tiles. The three families of pentagons discovered by Kershner satisfy all the requirements assumed in Reinhardt's thesis, and should have been included in his list.

Kershner believed and asserted that his list of monohedral convex prototiles is complete, and this was generally accepted as fact. An expository article by Gardner was written under this assumption (Gardner [1975a]). However, several readers (in particular Richard James and Marjorie Rice, whose tilings we mentioned in Section 9.3) pointed out that this assumption is incorrect (see Gardner [1975c] and Schattschneider [1978b]). This is one of the few examples in recent times in which a popular article in a non-professional journal has led to the refutation of presumed facts that had been accepted by the mathematical community. We must stress that the problem is still open.* Even the following (rather extreme) possibility has not been disproved: there exists a convex pentagon which tiles using only directly congruent copies and is such that no monohedral tiling it admits is periodic. For a discussion of related problems, and of analogues for monohedral tilings of space, see Grünbaum & Shephard [1980c]. Other results on tiling 3-dimensional space can be found in Danzer, Grünbaum & Shephard [1983], Grünbaum, Mani-Levitska & Shephard [1984], Schulte [1985].

Tilings by polygons have been used as a tool in the determination of densities of packings of the plane by arbitrary convex sets. See, for example, Kuperberg [1982] and the references given there.

The material about polyominoes, polyiamonds and polyhexes is scattered throughout the literature of recreational mathematics. We can do no more than mention the principal references: Golomb [1965] and Gardner [1975b], as well as many other publications of these authors. For much of the material in Section 9.4 we are indebted to Gardner. The results themselves, often unpublished by their originators, are due to R. Bantegnie [1977], A. W. Bell, David Bird, G. J. Bishop, L. Clarke, J. H. Conway, J. W. Harris, T. H. O'Beirne, R. Penrose and others.

Voderberg's original spiral tiling (Figure 9.5.1) has been reproduced several times (see Goldberg [1955], Fejes Tóth [1965, Figure 45], Gardner [1977]). The explanation of its construction—by modifying the edges of one of the tilings by isosceles triangles in Figure 9.5.2—is due to Goldberg. Much of the material in Section 9.5 originated in 1976 in an attempt to find other kinds of spirals, and in particular those with an odd number of arms. In spite of the fact that spiral tilings are very decorative, there is extremely little literature on the subject. It seems to have been ignored by artists and designers, and the only mathematical treatments of which we know, apart from the accounts in Grünbaum & Shephard [1979b], [1981e], are the short papers and notes by Simonds, Hatch and Fielker mentioned in the text.

Two recent developments should be mentioned. Hirschhorn & Hunt [1985] give a detailed account of their enumeration of all equilateral convex pentagons which tile monohedrally; no new types were discovered. In contrast to that—vindicating the cautionary statement made above—a new convex pentagon which tiles monohedrally was found by Rolf Stein; see "A new pentagon tiler", *Mathematics Magazine* 58(1985), 308. In the notation used in Table 9.3.1, Stein's pentagon can be described by $A = \pi/2$, $C + E = \pi$, $2B + C = 2D + E = 2\pi$, $2a = a + c = d = e$; it is uniquely determined up to similarity, and $E = 110.67667°$.*

* See the Appendix beginning on page 653.

10

APERIODIC TILINGS*

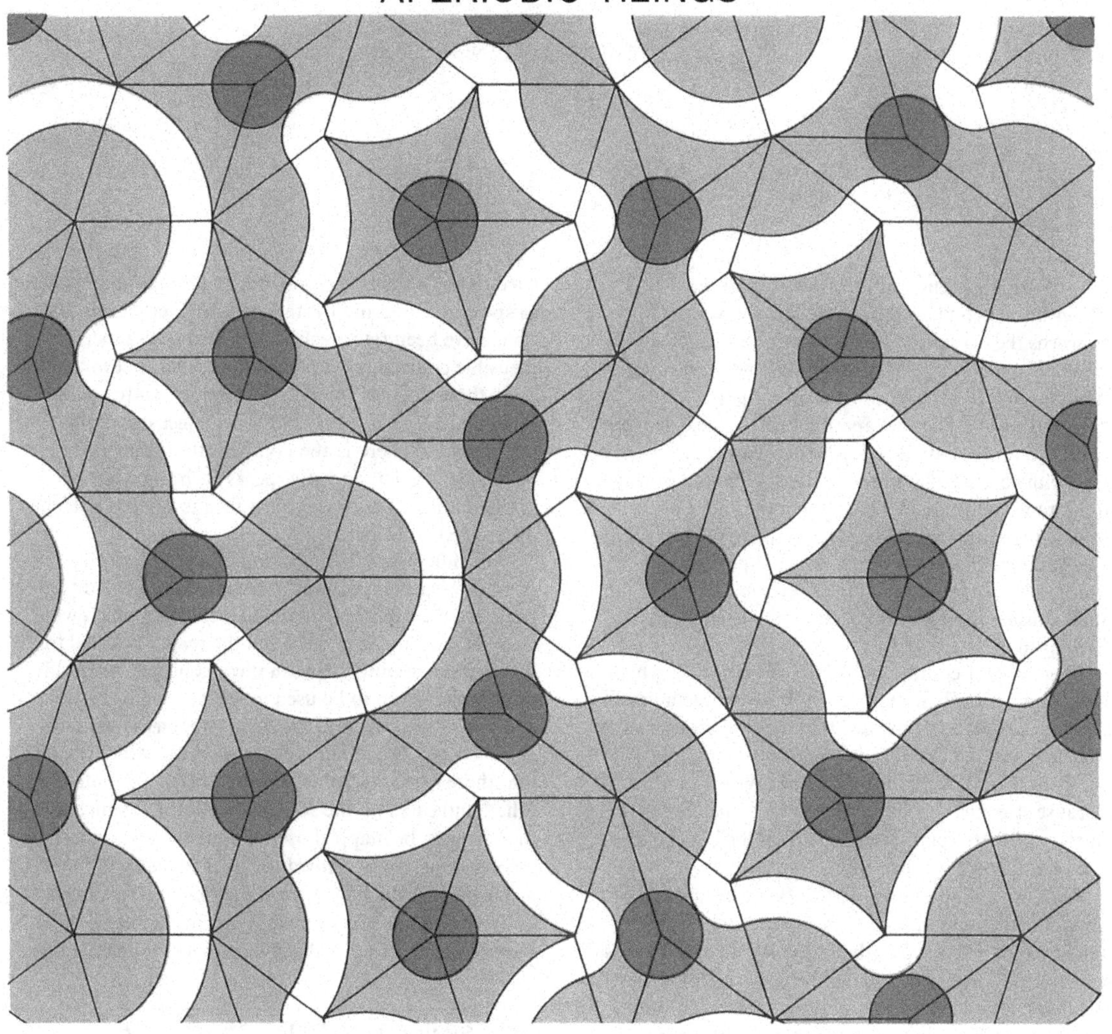

Figure 10.0.1
A tiling by Penrose rhombs. Each of the two quadrangular tiles is marked as in Figure 10.3.25 and they must be placed so that the gray marks match, that is, pass over the edges of the tiling from each tile to its adjacent tiles. The dark gray marks form small circular disks, as shown, while the light gray marks form shapes which, if they close up to form "rings", necessarily have 5-fold symmetry (see Statement 10.3.4). This is an example of an aperiodic tiling; although the two tiles admit an uncountable infinity of different tilings, no such tiling has a translation as a symmetry.

* See the Appendix beginning on page 653.

10

APERIODIC TILINGS

One of the most remarkable discoveries in the theory of tilings has taken place during the last few years—it concerns the existence of sets of prototiles which admit infinitely many tilings of the plane, yet no such tiling is periodic. Sets of prototiles with this property will be called *aperiodic*. The first aperiodic sets to be discovered consisted of what are now known as Wang tiles—square tiles with colored edges which must match and of which only translations are allowed. Wang tiles have many interesting properties and it will be convenient to discuss these all together in the next chapter. Aperiodic sets of Wang tiles will form the topic of Section 11.1. Here we shall consider tilings only in the original sense, that is, as defined in Section 1.1.

There are, of course, many sets of prototiles which admit non-periodic tilings. Even a 2-by-1 rectangle has that property. However, the essential feature of an *aperiodic* set of prototiles is that *every* tiling admitted by them is *necessarily* non-periodic. We stress this fact because it seems that there has been some confusion in the past between the terms "aperiodic" in the sense used here, and "non-periodic".

There is very little published work on the subject of aperiodic tiles and many results are presented here for the first time. We are greatly indebted to Roger Penrose and John Conway for much of the material, to Robert Ammann for allowing us to publish an account of his remarkable tilings, and to many others who have read this chapter in manuscript form and made helpful comments and suggestions.

Unlike the material in the earlier sections of the book many of our assertions here are *not* supported by published proofs. In some cases they depend on information passed to us by word of mouth and although we have tried to be as precise as possible, to fill in as many details as space permits, and to state clearly whether our assertions have been rigorously established or are merely plausible conjectures, we warn the reader that in a few cases there is some doubt as to the exact status of our statements. This is inevitable in a subject still in its infancy—but there is the compensation that open problems abound and it appears to be an area of mathematics where many new results are likely to be discovered soon.

One difficulty is that of terminology. The words "composition", "decomposition", "recomposition" and "inflation" are used with very specific meanings, which will be explained in due course, and the reader is urged to distinguish carefully between these concepts, especially as the terms seem to be used in various senses by workers in this field. In particular we must emphasise the difference between two tilings being *congruent*, which means that there exists an isometry which maps one onto the other, and *equal* (in the sense of Section 1.1) which means that one can be mapped onto the other by a similarity. Here a similarity is defined to be an isometry followed by a change of scale, that is, an *expansion* or *contraction.*

By way of introduction we begin, in Section 10.1, with a short account of the tilings known as similarity tilings.

10.1 SIMILARITY TILINGS

The name "similarity tiling" can be applied to two different concepts. The first, which is the source of many designs (see Exercise 5.1.14), is that of a tiling which admits similarities as symmetries. Diagrams of some of these tilings appear in Figure 10.1.1 and further examples can be

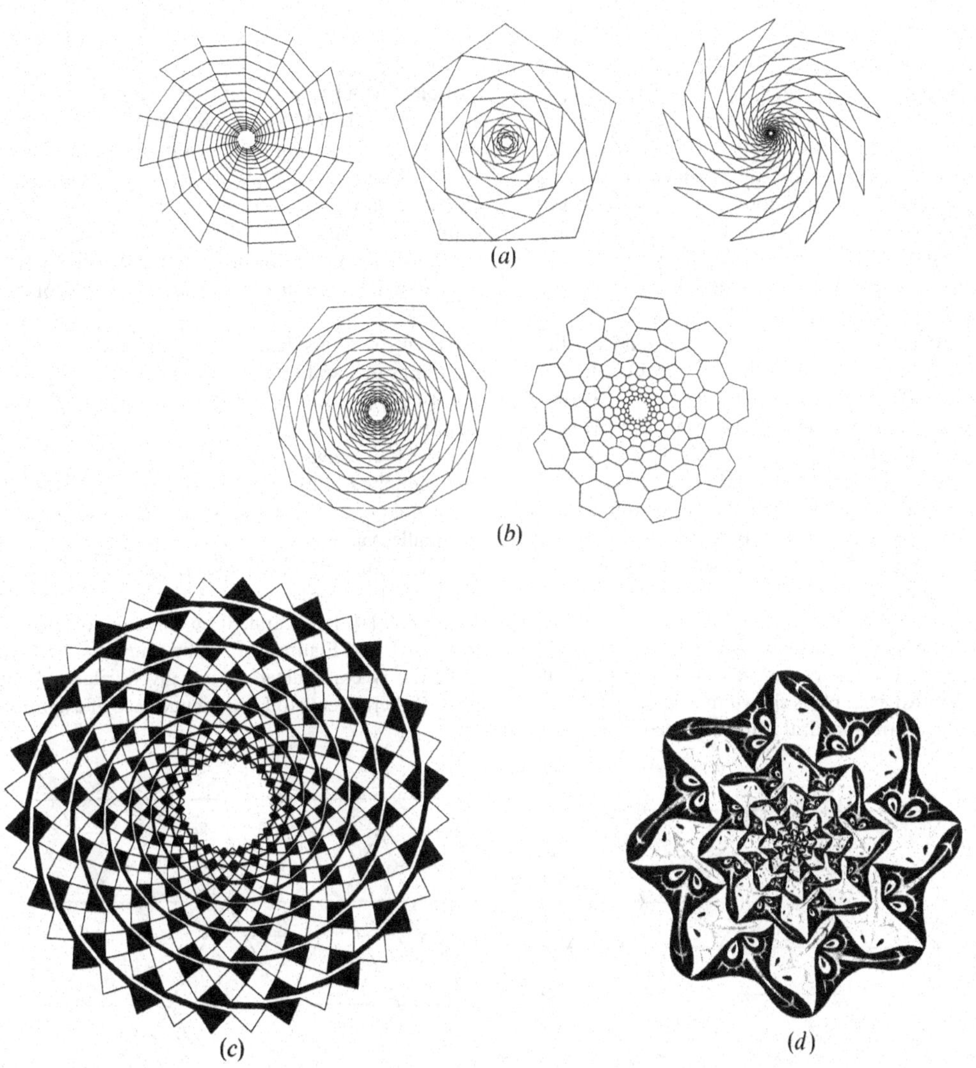

Figure 10.1.1
Examples of tilings which admit similarities as symmetries. The tilings in (*a*) and (*b*) have one transitivity class of tiles (and one transitivity class of vertices) under similarity symmetries; the tilings in (*b*) admit reflective symmetries, those in (*a*) do not. Disregarding the coloring, the tiling in (*c*) has two transitivity classes of tiles (and two of vertices) under similarity symmetries; the colored tiling, with the "squares" colored gray, is the well-known "Fraser spiral", invented by Fraser [1908] in his study of optical illusions. The tiling in (*d*) is the central part of the woodcut "Path of Life I" by Escher [1971]; it also has two transitivity classes of tiles and two of vertices.

found in Šubnikov [1960], Beard [1973], Escher [1971], Galyarskii [1974a], Lockwood & Macmillan [1978, Chapter 10] and Zamorzaeva [1979]. Zamorzaeva [1981] investigates similarity tilings which arise as Dirichlet tilings of suitable "dot patterns" with groups of similarity symmetries. See also Roman [1972] for related material. None of these tilings is normal for they possess singular points and the tiles are not uniformly bounded (see Exercise 10.1.1).

Here, however, we shall use the term "similarity tiling" in a different sense. To explain this we introduce the idea of "composition". We shall say that a tiling $\mathcal{T}_1$ is obtained by *composition* from a tiling $\mathcal{T}_2$ if each tile of $\mathcal{T}_1$ is a union of tiles of $\mathcal{T}_2$. Thus each edge of $\mathcal{T}_1$ will be a union of edges of $\mathcal{T}_2$, and each vertex of $\mathcal{T}_1$ will be a vertex of $\mathcal{T}_2$. The idea of composition has already been implicit in several parts of this book—for example, a tiling by polyominoes (Section 9.4) is obtained by composition from the regular tiling (4^4) by squares.

It is possible for two unequal monohedral tilings to be equal (though not, of course, congruent) to compositions of each other. An example is shown in Figure 10.1.2. In Figure 10.1.2(*a*) an isohedral tiling of type [$3^4.6$] is exhibited as a composition of the regular tiling (3^6), and in Figure 10.1.2(*b*) this relationship is reversed. Further examples of such compositions may be found in Grossman [1948]. Thus composition does not induce a partial ordering on the set of all tilings. If every tile of $\mathcal{T}_1$ is a union of k tiles of $\mathcal{T}_2$, then $\mathcal{T}_1$ will be called a *k-composition* of $\mathcal{T}_2$. In particular a tiling will be called a *similarity tiling* if it is equal to a (proper) composition of itself.

To illustrate this definition we mention that the regular tilings (4^4) and (3^6) and the Laves tilings [4.6.12] and [4.8^2] (see Figure 2.7.1) are similarity tilings. Other examples appear in Figures 10.1.3 to 10.1.6. The first two of these four tilings are periodic.

A monohedral tiling $\mathcal{T}$ is called a *k-similarity* tiling if $\mathcal{T}$ is equal to a k-composition of itself, and, moreover, no smaller value of $k > 1$ has this property. For example, the tilings [4^4], [6^3], [4.6.12] and [4.8^2] correspond to the values $k = 4$, 4, 3 and 2. The tilings of Figures 10.1.4, 10.1.5 and 10.1.6 correspond to the values 9, 4 and 4, and Figure 10.1.3 shows how a k-similarity tiling can be constructed for any given value of k. In general there seems to be an abundance of examples of

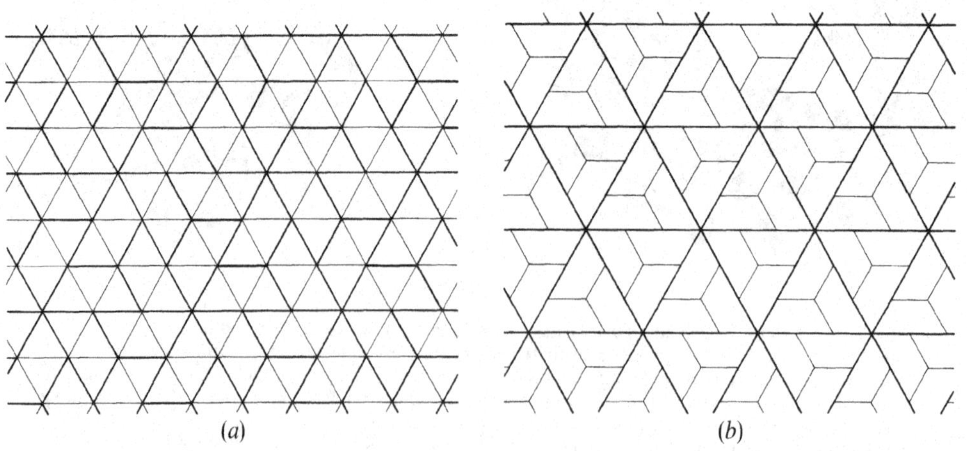

Figure 10.1.2
Unequal monohedral tilings, each of which can be obtained from the other by composition.

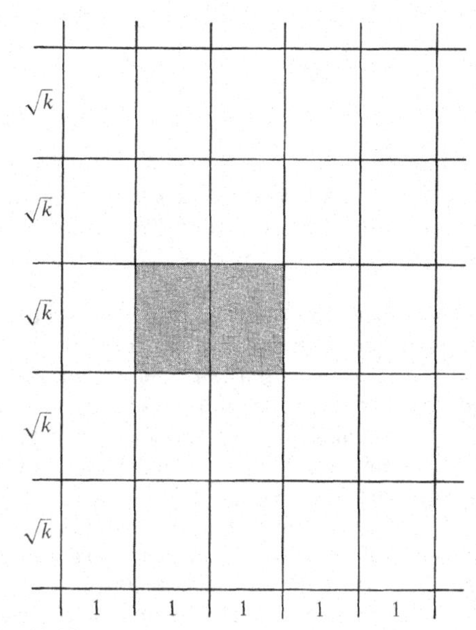

Figure 10.1.3
A k-similarity tiling. In this diagram $k = 2$; the union of the two small gray tiles is a tile of the composed tiling; it is similar in shape to each of the original tiles. This construction works for any positive integer value of k.

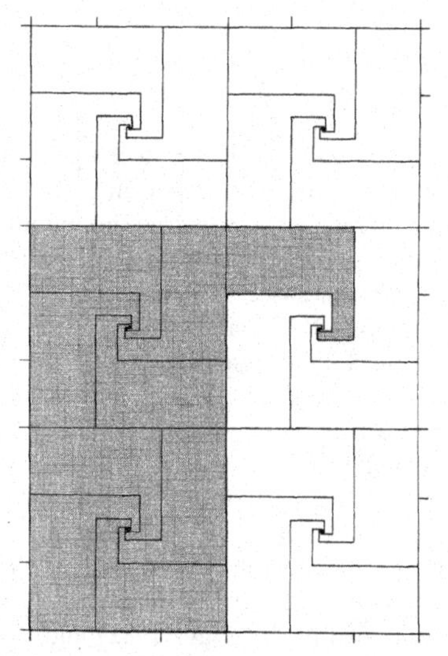

Figure 10.1.4
A periodic 9-similarity tiling. The union of the nine shaded tiles is a tile of the composed tiling; it is similar in shape to each of the original tiles.

Figure 10.1.5
A 4-similarity tiling. The tiling obtained by rotating through an angle of $\frac{1}{2}\pi$ about the marked point and then expanding in the ratio 2:1 is a 4-composition of the original tiling. The union of the four gray tiles is a tile of the composed tiling. This tiling is non-periodic.

Figure 10.1.6
Another non-periodic 4-similarity tiling. The similarity consists of rotation about the marked point through an angle of $\frac{2}{3}\pi$, followed by expansion in the ratio 2:1. The union of the four gray tiles is a tile of the composed tiling.

k-similarity tilings when k is a perfect square, but they are comparatively rare when k is prime.

A systematic method of constructing monohedral k-similarity tilings for a given k can be based on results concerning k-rep tiles (see Langford [1940], Sibson [1940], Golomb [1964], Davies [1966], Beck & Bleicher [1971], Valette & Zamfirescu [1974], Giles [1979a], [1979b], [1979c]). A *k-rep tile* is defined as any tile T that can be dissected into k congruent parts T' each of which is similar to T. For such a tile we find a similarity which maps one of the parts T' onto the tile T; repeated applications of this operation to T yield a tiling which can be modified in an obvious way so as to form a k-similarity tiling. Additional results on k-rep tiles were obtained by Dekking [1982] and Doyen & Landuyt [1983]. For related matters see also Giles [1980].*

For some k-similarity tilings, such as those of Figures 10.1.3 and 10.1.4, there are many different ways in which unions of sets of k tiles of $\mathcal{T}$ can be taken as the tiles of a tiling similar to $\mathcal{T}$. In other cases, such as those of Figures 10.1.5 and 10.1.6, this composition process is unique. (This would not be so, of course, if k were not uniquely defined by minimality.) The following result may appear surprising at first sight.

10.1.1 *If the k-composition process described above for a monohedral k-similarity tiling is unique, then* $\mathcal{T}$ *is not periodic.*

In fact we shall show that $\mathcal{T}$ cannot possess a translation as a symmetry. For suppose that there were such a translation t through a distance d. Then uniqueness of composition implies that t must also be a symmetry of the k-composed tiling. Applying this argument repeatedly to further k-compositions we obtain tilings with arbitrarily large tiles. But it is impossible for t to be a symmetry of any tiling in which every tile contains a circular disk of diameter larger than d, and hence we arrive at a contradiction which proves Statement 10.1.1.

Figures 10.1.5 and 10.1.6 illustrate the above. The corresponding values of k are 4 and in each case the process of 4-composition is unique.

The proof of Statement 10.1.1 is typical of a procedure which we shall use many times in this chapter to show that a tiling is non-periodic. In fact, if one could find a prototile which only admits k-similarity tilings with uniqueness of k-composition we would have discovered a single aperiodic tile. Unfortunately no such tile is known! But in the following sections we shall show how nearly we can achieve this goal.

EXERCISES 10.1

1. In this exercise the term "similarity tiling" will have the meaning ascribed to it in the first paragraph of the section—that is, a tiling which is mapped onto itself by a similarity that is not an isometry.
   (*a*) Construct a similarity tiling $\mathcal{T}$ in which each tile is a closed topological disk and every two tiles are equivalent under some similarity that maps $\mathcal{T}$ onto itself.
   *(*b*) Prove that every tiling $\mathcal{T}$ of the kind described in part (*a*) has infinitely many singular points, one and only one of which belongs to all tiles of $\mathcal{T}$.
   **(*c*) Describe a homeomeric classification of tilings of the kind described in part (*a*). [A possible approach to this problem is to show that the group $S(\mathcal{T})$ of symmetries of $\mathcal{T}$ is cn for a suitable n, and that this n characterizes the homeomeric type of $\mathcal{T}$.]
2. (*a*) Find an isohedral tiling $\mathcal{T}_1$ of type [3.4.6.4]

* See the Appendix beginning on page 653.

which is a composition of the regular tiling $\mathcal{T}_2$ of type (3^6) and is such that $\mathcal{T}_2$ is also a composition of $\mathcal{T}_1$.

(*b*) Show that the regular tiling (3^6) is not the composition of any tiling of type $[4^4]$.

**(*c*) Determine all pairs of different topological types of isohedral tilings $\mathcal{T}_1$, $\mathcal{T}_2$ such that $\mathcal{T}_1$ is a composition of $\mathcal{T}_2$ and $\mathcal{T}_2$ is a composition of $\mathcal{T}_1$.

3. (*a*) Use the decomposition of the larger trapezium into four smaller ones indicated in Figure 10.1.7(*a*) to generate a monohedral non-periodic 4-similarity tiling. [Hint: Use rotations about a suitable point and expansion in the ratio 2:1. Find the exact position of the point that makes this construction possible.]

(*b*) Use the decomposition indicated in Figure 10.1.7(*b*) to generate a monohedral non-periodic 3-similarity tiling. About which point should the rotation be performed?

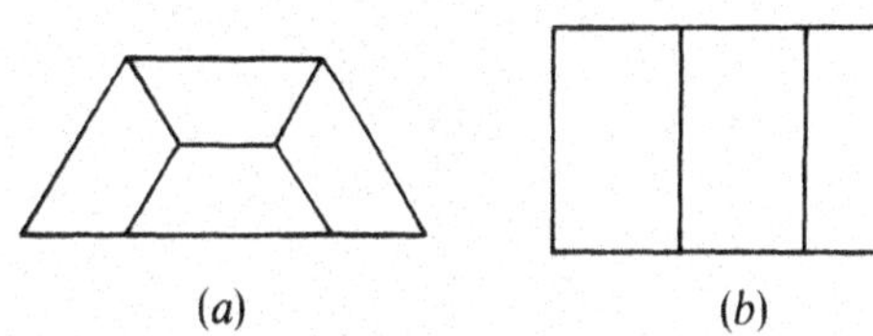

Figure 10.1.7
Examples of k-rep tiles. In (*a*) $k = 4$ and in (*b*) $k = 3$.

(*c*) Find a 2-rep tile which is not a parallelogram and use it to generate a monohedral non-periodic 2-similarity tiling.

**4. Can a closed topological disk be both a 2-rep tile and also a 3-rep tile?

5. Justify the assertion made in this section that the composition process (with minimal k) is not unique for the tilings in Figures 10.1.3 and 10.1.4.

10.2 ROBINSON'S APERIODIC TILES

This section is principally devoted to describing the aperiodic set of six prototiles discovered by Raphael M. Robinson in 1971 and denoted here by R1. It is of a somewhat different nature from the aperiodic sets to be described later. In Robinson's original paper [1971] he discusses a number of variants but here we shall restrict ourselves to the one set. At the end of the section we shall also mention a similar aperiodic set of six tiles recently discovered by Robert Ammann.

The six Robinson tiles are shown in Figure 10.2.1. Basically the tiles are squares with modifications to their corners and sides. If, for the moment, we ignore the modifications of the sides and just look at the corners, we see that the tiles are of two types, exemplified by Figure

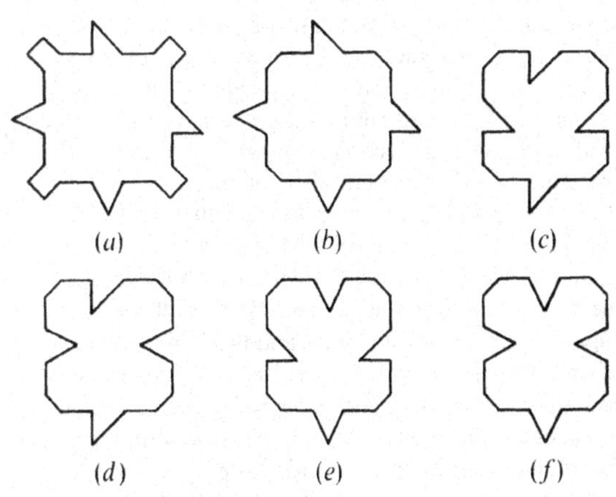

Figure 10.2.1
Robinson's aperiodic set R1 of six tiles.

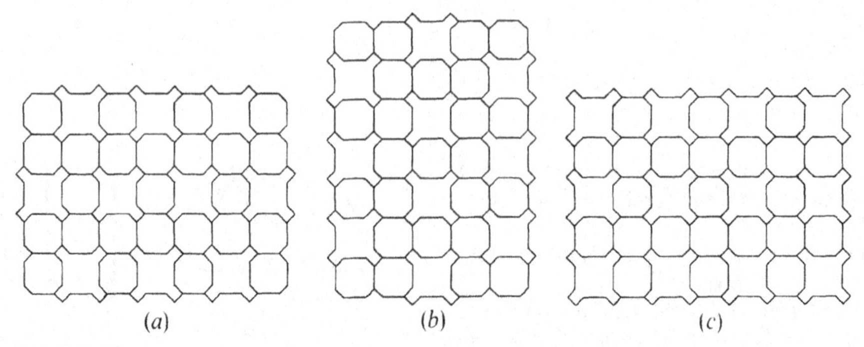

Figure 10.2.2
Three examples of tilings by cornered and cornerless tiles. In a tiling by Robinson's aperiodic set R1 the arrangements of cornered tiles shown in (*a*) and (*b*) cannot occur.

10.2.1(*a*) on the one hand, and Figures 10.2.1(*b*)–(*f*) on the other. The first type we shall call *cornered*, and the second type *cornerless*. Then it is evident that in any tiling the cornered tiles must only occur *either* as alternate tiles in every second row, *or* as alternate tiles in every second column, *or* both, see Figure 10.2.2. In fact, as we shall see shortly, it turns out that the modifications to the sides force the cornered tiles to lie in alternate rows and alternate columns (as in Figure 10.2.2(c)) in each of two half-planes.

Now consider the sides. These have, as will be seen, projections and indentations of two types, the first of which is symmetric and the second is asymmetric. For ease of representation and description we shall indicate these modifications to the sides by arrows on the tiles, and mark the cornered tiles by quarter-circles as in Figure 10.2.3. A single arrowhead corresponds to one of the symmetric projections on the sides, a double arrow-head to an asymmetric projection as shown. The arrow-tails correspond to the indentations in an analogous way. With this representation, precisely one quarter-circle must occur at each vertex of the tiling, and each arrowhead on a side must match an arrowtail on the adjacent tile. Following Robinson, the tiles of types (*a*) and (*b*) in Figure 10.2.3 are called *crosses*, and those of types (*c*)–(*f*) *arms*. If oriented as in Figure 10.2.3, the crosses will be said to *face upwards* and *to the right*, these being the directions of the double arrows. By a *back* of the cross we will mean the direction of either single arrow.

Since only crosses are cornered, these must occupy alternate positions in alternate rows (say), and clearly the only tiles that will fit between them are arms. Moreover, the crosses must always face each other, or be back to back. If they face each other then the other double arrows on the cross must point in the same direction, either up or down. Hence somewhere we must have an arrangement of three tiles like *ABC* of Figure 10.2.4. Tile *B* is drawn as of type (*e*) but clearly it could also be of type (*c*), or its reflection.

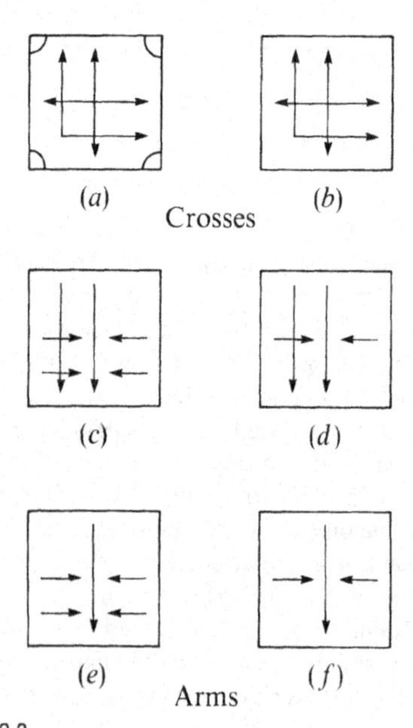

Figure 10.2.3
Marked square tiles corresponding to the tiles of Robinson's aperiodic tiles shown in Figure 10.2.1.

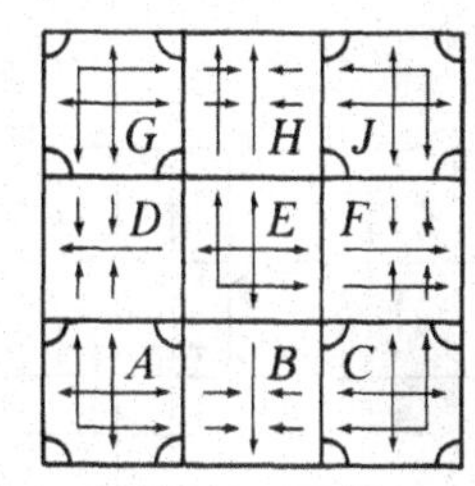

Figure 10.2.4
A 3×3 block of Robinson's tiles.

Now tile D must be an arm (for a cross has only arrowheads on its sides) and to fit against A it must be of type (c), (d) or (e). In any case its right side will carry the tail of one or two arrows, and the same will clearly be true of the left side of tile F. We deduce that E must be a cross, and we shall suppose that it faces upwards and to the right as shown. This choice for E forces B and D to be of type (e) and F to be of type (c). Of course the cross E is necessarily cornerless.

It is now evident that H cannot be a cross and so, because of the corner shapes, G and J must be cornered tiles and therefore crosses. Thus H must be an arm of type (c). We arrive at a patch of nine tiles as shown, the only possible variation being in the orientation of the central cross and the four arms adjacent to it.

Having constructed a 3×3 block of tiles we now show how to extend this to a 7×7 block. Let us assume our given 3×3 block is as in Figure 10.2.4 with its central cross facing upwards and to the right (see Figure 10.2.5). Then the tiles along its upper edge (P, Q and R) must be arms because no cross has arrowtails on its sides, and the only possibility is that P, Q, R are of the types (f), (e), (f) (as drawn) or of the types (d), (c), (d) (in which case each horizontal arrow would be replaced by a double one running from right to left). Similar possibilities arise for the tiles S, T, U, and from this we deduce that the central tile V must be a cornerless cross. Arguing in this way, to complete the 7×7 block we must have a cross at the center with four radiating arms of three tiles each, and a 3×3 block of tiles in each quadrant determined by these radiating arms. The only possible variation is in the orientation of the central cross and the arrows in the arms radiating from it—we have shown the cross facing downwards and to the right, but the other three orientations are possible as well.

In a similar manner we may extend the 7×7 block to a 15×15 block, and so on. At each stage the direction in which the square block is extended depends upon the chosen orientation of the central cross. For example if the central cross of a 7×7 block faces downwards and to the right, then further 7×7 blocks must be placed below and to the right of the given 7×7 block to complete the required 15×15 block. Hence we see that if we choose all the central crosses with the same orientation then we arrive at a tiling of a quarter plane. If all the crosses face to the right, say, and infinitely many face up and infinitely many face down, then we arrive at a tiling of the right half-plane. Similarly, if infinitely many

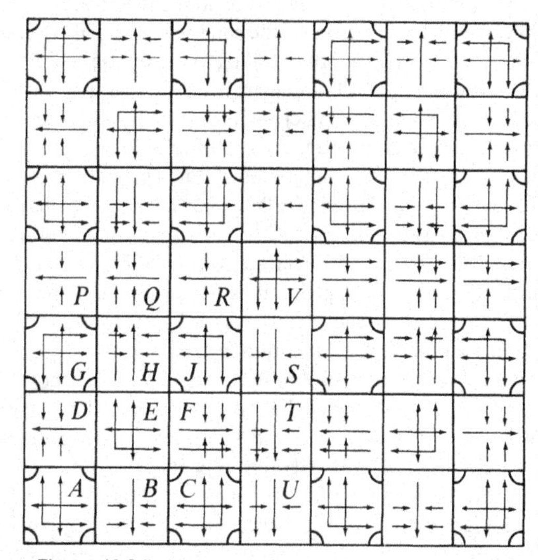

Figure 10.2.5
A 7×7 block of Robinson's tiles that extends the 3×3 block shown in Figure 10.2.4.

central crosses face in each of the four directions, we arrive at a tiling of the whole plane.

Except in the last case it is necessary to assemble four quarter-planes, two quarter-planes and one half-plane, or two half-planes to complete the tiling of the whole plane. This is done by using "corridors" or "faults" consisting of infinite rows or columns of tiles of types (d) or (f) as shown, for example, in Figure 10.2.6. The word "fault" seems appropriate for it is only across such rows or columns of tiles that the orderly square grid arrangement of cornered tiles can be broken.

The above construction clearly yields all possible tilings by the Robinson prototiles, and it is easy to deduce that there exist an uncountable infinity of such tilings (see Exercise 10.2.1). Moreover we can now establish the fundamental result:

10.2.1 *The set* R1 *of six Robinson prototiles is aperiodic.*

To prove this we first remark that if the tiles of a tiling are basically square tiles with modified edges (like the Robinson tiles) and the tiling is periodic, then translations parallel to the sides of the squares necessarily occur among the symmetries of the tiling (see Exercise 10.2.2). However, the method of construction implies that for each n we can find facing crosses which are at distance 2^n apart (namely the central crosses of $(2^n-1)\times(2^n-1)$ blocks) in both rows and columns. But clearly this implies that no translation parallel to the sides of the squares through a distance less than 2^n is a symmetry of the tiling. Since 2^n can be arbitrarily large, no translative symmetry of the tiling exists. Thus the tiling is non-periodic, the tiles form an aperiodic set, and Statement 10.2.1 is proved.

On the other hand, as Robinson points out, each such tiling is "almost periodic" in the sense that (except possibly across fault lines) the 3×3 blocks repeat with period 8, the 7×7 blocks with period 16, and so on. In general, a block of $(2^n-1)\times(2^n-1)$ tiles repeats horizontally and vertically with period 2^{n+1}. Thus by choosing a period large enough we can ensure that an arbitrarily small fraction of the tiles fail to repeat. Moreover if we choose any patch of tiles, which does not have two perpendicular fault lines, from a Robinson tiling $\mathscr{T}$, it will be repeated

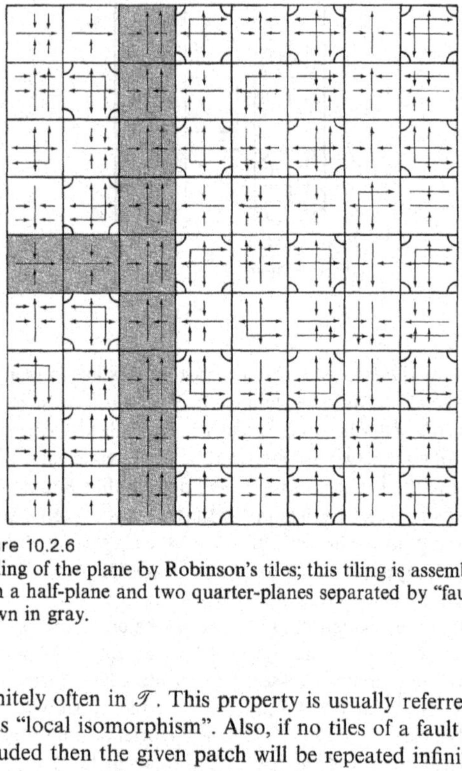

Figure 10.2.6
A tiling of the plane by Robinson's tiles; this tiling is assembled from a half-plane and two quarter-planes separated by "faults" shown in gray.

infinitely often in $\mathscr{T}$. This property is usually referred to as "local isomorphism". Also, if no tiles of a fault are included then the given patch will be repeated infinitely often in every tiling by Robinson tiles.

Another property of the Robinson tiles is that, in any tiling, the fraction

$$\frac{1}{2^2}+\frac{1}{2^4}+\frac{1}{2^6}+\cdots=\frac{1}{3}$$

of the tiles are crosses and the remaining $\frac{2}{3}$ are arms. Of the crosses, $\frac{3}{4}$ are cornered and the rest are cornerless.

It is of some interest to consider the pattern formed by the lines of the double arrows in the tiling $\mathscr{T}$. These form bigger and bigger squares, the center of each square being a corner of one of double the size (see Figure 10.2.7) and arranged in such a way that they do not cross the fault lines. If we delete the smallest squares then we are left with an analogous pattern of squares of exactly twice the size. This corresponds to the tiling that would be

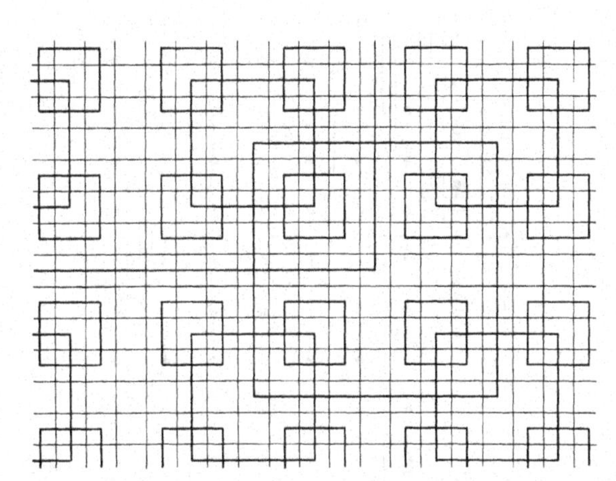

Figure 10.2.7
The overlapping squares formed by the lines of double arrows in a tiling by Robinson's tiles.

obtained by omitting all rows and columns that contain cornered crosses, and then making the central crosses of the original 3×3 blocks into cornered tiles. This property is analogous to the similarity property of tilings described in Section 10.1.

In 1977 Ammann described another set of tiles (set A1) based on squares, a set which is of a similar nature to the Robinson set R1 described above. It has however features which make it of independent interest. This set also contains six tiles, and is shown in a slightly modified form in Figure 10.2.8 (see Exercise 10.2.4). One of the tiles is cornered and the other five are cornerless. The edges have projections and indentations of three different shapes. In Figure 10.2.9 we show an equivalent set of marked square tiles, the various projections being represented by single, double and triple arrows. The matching rule is similar to that of the Robinson tiles. We must ensure that exactly one quadrant of a circle (corresponding to a cornered tile) lies at each vertex, and the three sorts of arrows must match head against tail. Since all the double arrows run vertically, say, and the triple arrows horizontally, the five tiles (*a*) to (*e*) cannot be turned

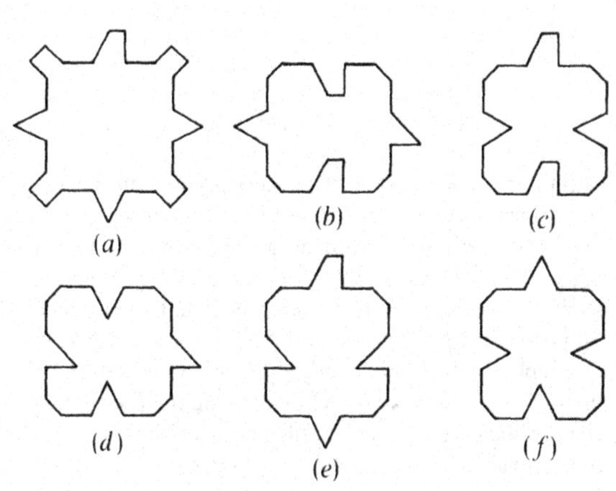

Figure 10.2.8
Ammann's aperiodic set A1 of six tiles.

through $\frac{1}{2}\pi$. Only reflections in horizontal and vertical lines are allowed. Tile (*f*) on the other hand can occur in any orientation. It can be shown that:

10.2.2 *The set* A1 *of six Ammann prototiles is aperiodic.*

We shall not give a complete proof of this statement as we did in the case of the Robinson tiles; we leave the detailed analysis to the reader (see Exercise 10.2.3). Here we confine ourselves to a few remarks. The tiling can contain a "fault" consisting of an infinite row or column

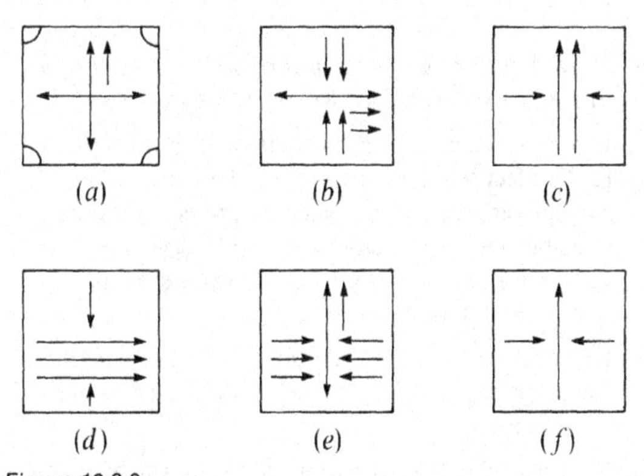

Figure 10.2.9
Marked square tiles corresponding to the Ammann tiles of Figure 10.2.8.

of tiles of types (*c*), (*d*) or (*f*). Except across fault lines the cornered squares must occur in every alternate row and every alternate column (across fault lines the rows of squares can be "staggered"). An example of the tiling is shown in symbolic form in Figure 10.2.10. Here we have indicated the marked tiles, omitting all the single arrows and replacing the double and triple arrows by single lines. This simplified method of marking the tiles is sufficient to indicate each sort of tile uniquely, though it does not yield an equivalent matching condition. The tiles bearing the mark ● are cornered. As will be seen, we have an arrangement of bigger and bigger H's (in contrast to the pattern of squares for the Robinson tiles shown in Figure 10.2.7). If we delete the smallest H's then we are left with a similar pattern of larger H's. This corresponds to the tiling obtained by deleting the rows and columns containing the cornered tiles (*a*) of Figure 10.2.8 and then replacing the tiles (*e*) which occurred at the centers of the smallest H's by tiles of type (*a*).

The artistic potential of constructions similar to those underlying Figures 10.2.7 and 10.2.10 is investigated by Christensen [1982].

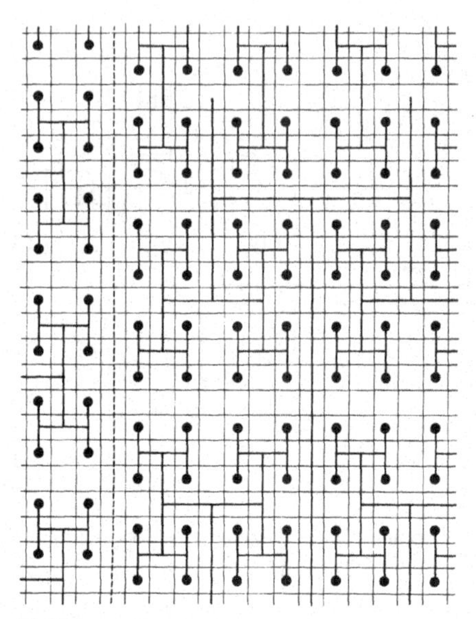

Figure 10.2.10
A schematic representation of a tiling by the Ammann aperiodic set A1. The notation is explained in the text. Rows of cornered tiles (marked ●) can be staggered across a fault line.

EXERCISES 10.2

1. Show that there exists an uncountable infinity of distinct tilings using the six Robinson tiles.
2. Prove the assertion made just after Statement 10.2.1 to the effect that if the prototiles of a periodic tiling are squares, possibly modified by projections and indentations on their edges, then there exists a translation parallel to the edges of the squares which is a symmetry of the tiling.
*3. Explain in detail how to construct a tiling using the Ammann tiles, and hence verify that all such tilings are non-periodic.
*4. Show that the Ammann tiles still force non-periodicity even if the projections on the edges shown in Figure 10.2.8 are each replaced by a projection which is symmetrical, that is, unchanged by a reflection which interchanges the two ends of the edge. (This is the original form of the tiles as described by Ammann—in the text we have introduced asymmetrical projections in order to simplify the construction of the tilings.)
5. Prove the assertion on page 528 concerning the frequencies of the various kinds of tiles in a Robinson tiling.
6. Determine the frequencies of the various kinds of tiles in a tiling of the set A1 of Ammann tiles.

10.3 THE PENROSE APERIODIC TILINGS

In 1973 and 1974 Roger Penrose discovered three sets of aperiodic prototiles which we shall now describe.

The first set, denoted here by P1, consists of six prototiles, see Figure 10.3.1. Unlike the tilings of the previous section, these are based on rhombs, regular pentagons, pentacles and "half-pentacles", with edges modified by projections and indentations. (We have slightly altered the shapes originally proposed by Penrose [1974] in order to make the presentation simpler.) There are three different kinds of projections and corresponding indentations. In Figure 10.3.2 we show the same tiles with these modifications omitted and replaced by labels 0, 1, 2, $\bar{0}$, $\bar{1}$, $\bar{2}$. The matching condition for these marked tiles is that 0 must fit against $\bar{0}$, 1 against $\bar{1}$, and 2 against $\bar{2}$. Of course, without the matching condition for the edges, the tiles of the Penrose set P1 shown in Figure 10.3.2 admit periodic tilings. Some examples (omitting piece (*b*)) can be found in Caris [1980], along with other tilings using various convex or starshaped tiles.

In Figure 10.3.3 a patch of the tiling is shown and in view of this it may seem "obvious" that the whole plane can be tiled using the set P1. In fact it is not quite so easy

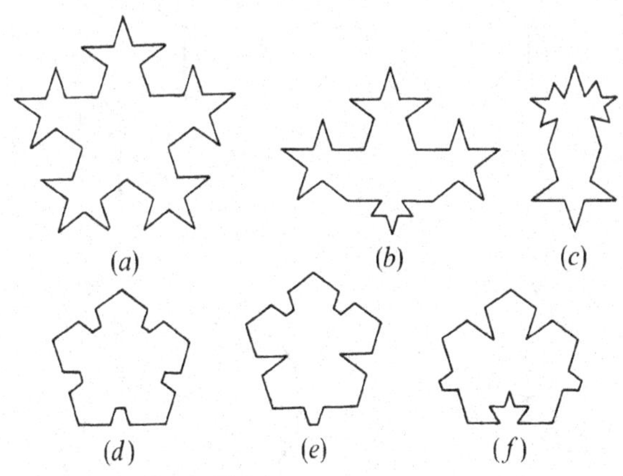

Figure 10.3.1
Penrose's aperiodic set P1 of six tiles.

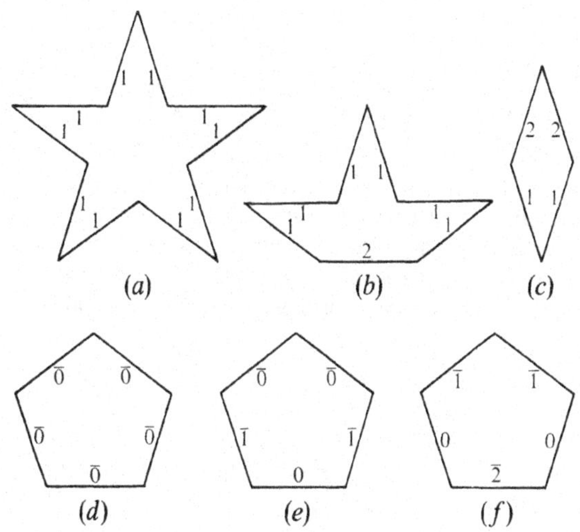

Figure 10.3.2
Six edge-labelled tiles that correspond to the Penrose tiles of Figure 10.3.1. On adjacent tiles the numbers 0, 1 and 2 must fit against $\bar{0}$, $\bar{1}$ and $\bar{2}$, respectively.

as it seems, especially if one cuts the tiles out of cardboard and tries to fit them empirically! Our first task will be to prove that these prototiles admit a tiling of the plane and we shall then proceed to show that any such tiling is non-periodic. Both proofs will depend on a process analogous to "composition" as described in Sections 10.1 and 10.2.

Consider the six patches of tiles shown in Figure 10.3.4. Each such patch can be transformed into a new tile by deleting the edges in its interior (that is to say, taking the union of the smaller tiles) and replacing certain quadruples of labelled edges on the boundary, between the marked vertices, according to the scheme indicated in Figure 10.3.5. These larger tiles are similar to the original ones and have corresponding edge markings. Edges labelled 1 occur in two enantiomorphic forms which explains why two different sets of four edges on the patches correspond to the same labelled edge on the larger tile—if the edges were oriented the correspondence would be one-to-one. The larger tiles are $\tau^2 = 2.618034\ldots$ times as large as the original ones. (Here $\tau = (1+\sqrt{5})/2 = 1.618034\ldots$ is the positive root of the quadratic equation $\tau^2 - \tau - 1 = 0$, and is known as the *golden number*.)

It is convenient to generalize the meaning of the word "composition" to cover the procedure just described. As explained in Section 1.5, we can often think of each tile in a tiling as being of a basic shape with edges modified or labelled in such a way as to specify a matching condition. Then *composition* will mean the process of taking unions of tiles so as to build up larger tiles which are basically the same shapes as those from which we started and whose edge modifications and labels (induced by those of the original tiles) specify a matching condition equivalent to the original one. It may happen (as in the case of the Penrose set P1 just described) that the edge

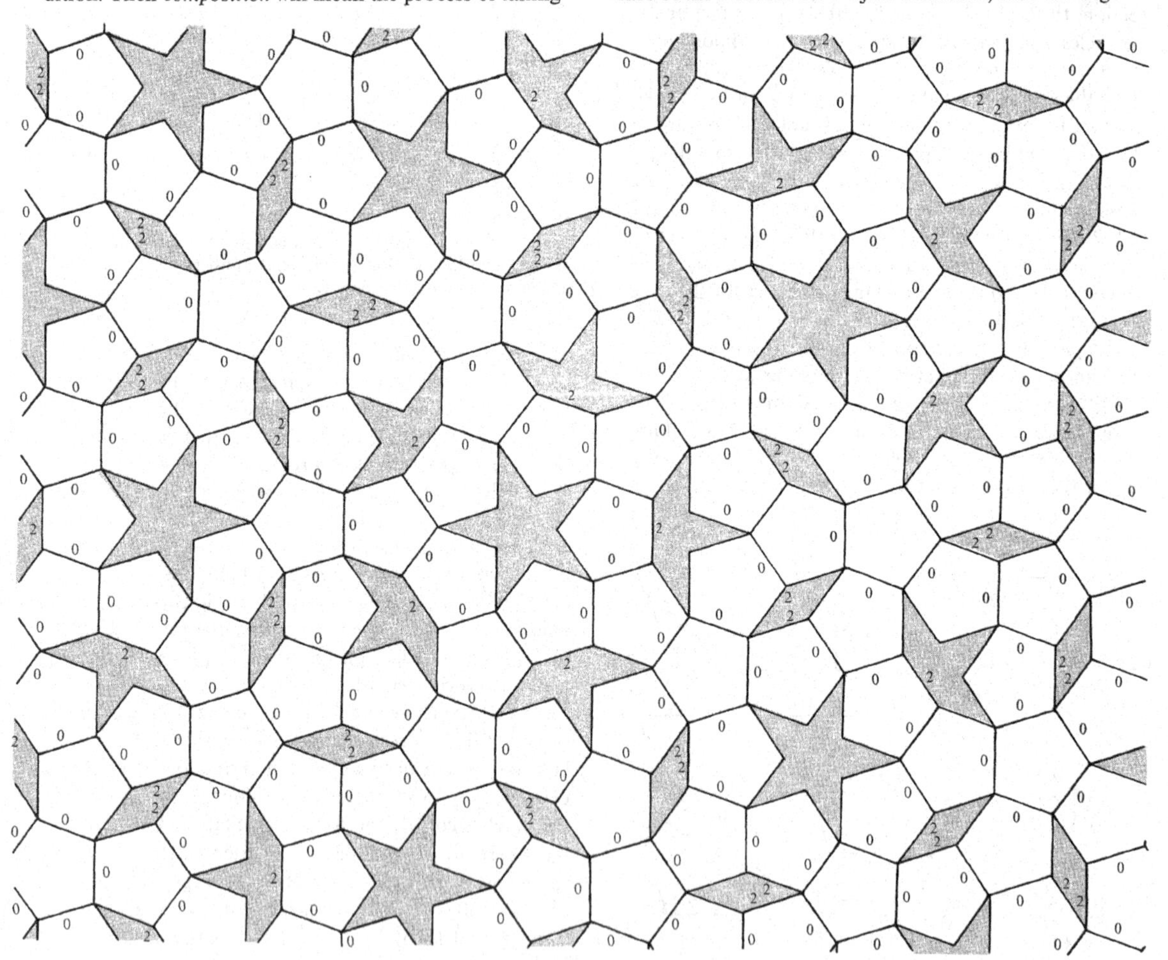

Figure 10.3.3
A tiling by the set P1 of Penrose prototiles. For clarity, only labels 0 and 2 are indicated—labels 1 and those with bars are omitted.

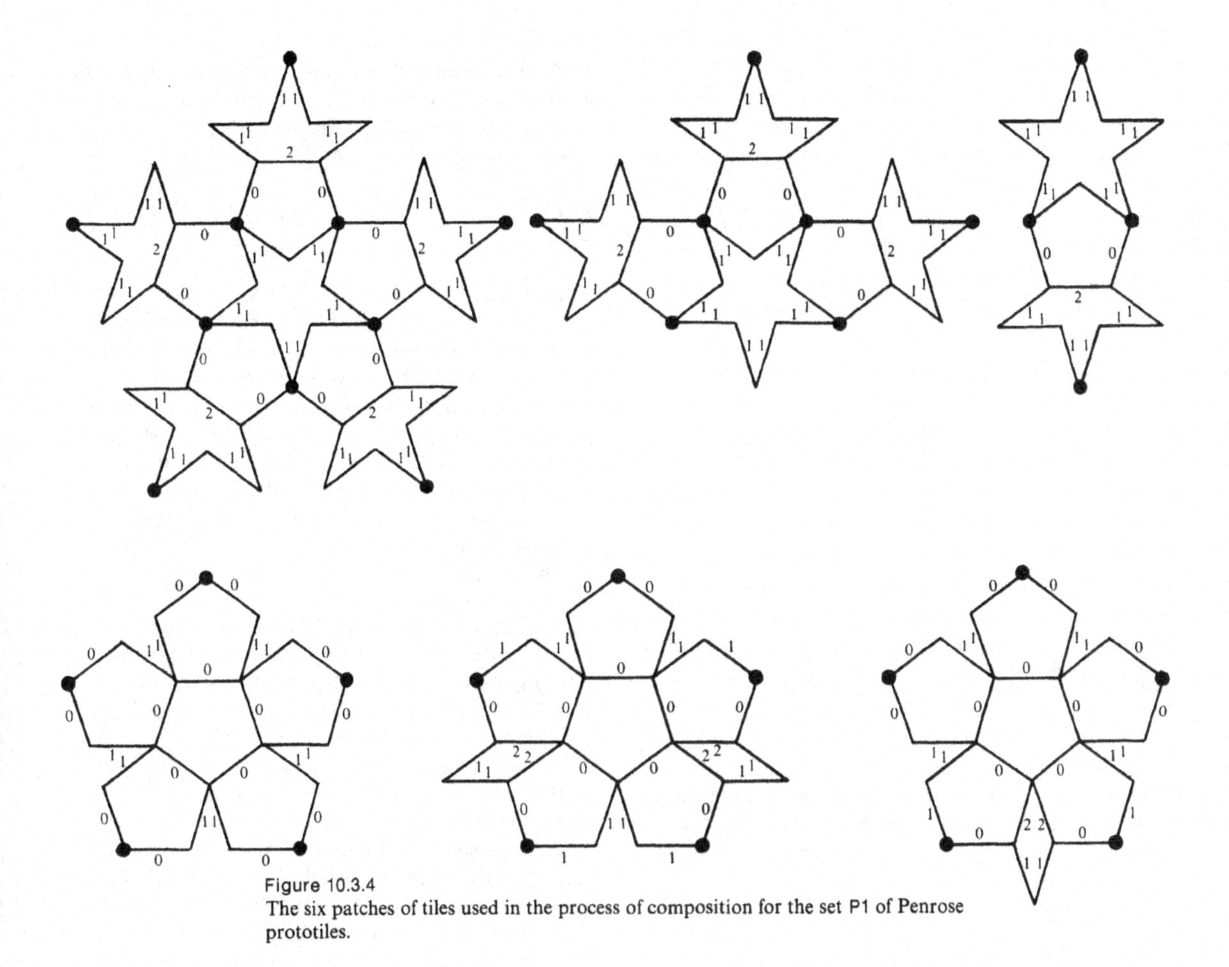

Figure 10.3.4
The six patches of tiles used in the process of composition for the set P1 of Penrose prototiles.

modifications and labels on the larger tiles are very different from those on the original tiles—as far as the definition is concerned, this is of no consequence so long as the matching conditions are equivalent. We shall also generalize the word *decomposition*. It will mean the inverse of the process of composition—the cutting up of large tiles into smaller ones.

Decomposition enables us to prove that the set P1 of Penrose prototiles admits a tiling of the plane. Start from any patch $\mathscr{A}$ of tiles (even a single tile) and enlarge it in the ratio $\tau^2:1$. Then decomposition of the tiles in the enlarged patch $\mathscr{A}$ yields a patch $\mathscr{A}'$ made up of tiles of the original size. By repeating this process we obtain arbitrarily large patches of tiles which implies, using the Ex-

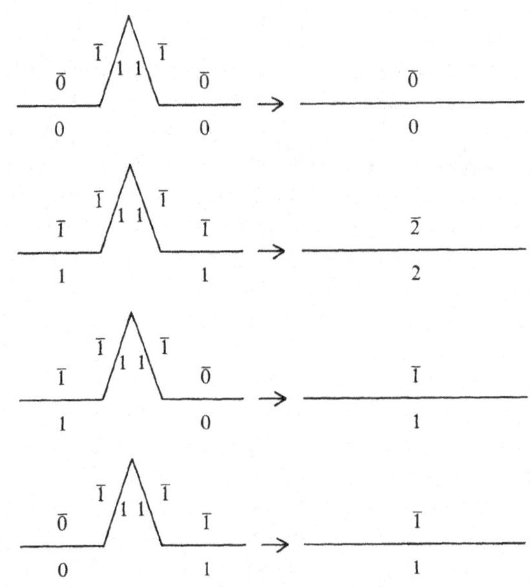

Figure 10.3.5
This diagram shows how the sides of the patches in Figure 10.3.4 are replaced in the process of composition.

tension Theorem (Statement 3.8.1), that the whole plane can be tiled. This process of increasing the size of the tiles by expansion and then decomposing them into tiles of the original size, will be called *inflation*.

To show that any tiling using the set P1 is non-periodic we use composition in the same way as in the proof of Statement 10.1.1. We start with a tiling $\mathscr{T}$ and cut it up into patches of the kinds shown in Figure 10.3.4. It can be shown that it is possible to do this in a unique way. (See Exercise 10.3.1—the details are similar, but more complicated, for the set P1 than the proof of the analogous result for *A*- tiles and *B*-tiles which is given on page 542.) Consequently, composition applied to $\mathscr{T}$ yields a unique tiling which will be denoted by $\tau^2\mathscr{T}$. In Figure 10.3.6 we show the tiling $\tau^2\mathscr{T}$ superimposed on $\mathscr{T}$. By repeating the process we obtain a tiling $\tau^{2n}\mathscr{T}$ with arbitrarily large tiles. A translation through a distance d cannot be a symmetry of $\tau^{2n}\mathscr{T}$ if n is chosen sufficiently large (so that each tile contains a circular disk of diameter greater than d). We deduce that no translation is a symmetry of $\mathscr{T}$ and so $\mathscr{T}$ is non-periodic. We have proved:

10.3.1 *The set* P1 *of six Penrose prototiles is an aperiodic set.*

An alternative method of describing decomposition has recently been suggested by Ammann. In any tiling by the set P1 we mark each tile by the lines and labels indicated in Figure 10.3.7. We then delete the original tiles and regard the interior lines, appropriately labelled, as the edges and vertices of a new tiling. This uses the same set P1 of prototiles except that each is reduced in size in the ratio τ^{-1}:1, see Figure 10.3.8. This procedure can be called *semi-decomposition* since if it is applied twice we obtain decomposition as previously defined. On the other hand we remark that an inverse process of "semi-composition" is not easy to describe in an analogous way; a tiling cannot be converted into one by larger tiles by any unique system of marking. As will be seen from Figure 10.3.8, the pentagons (*d*) of Figure 10.3.2 must be marked in two different ways according to the disposition of the surrounding tiles.

An interesting and attractive variant of the aperiodic set P1 is shown in Figure 10.3.9. It was designed by Penrose, and is obtained by replacing every edge by a circular arc of some fixed radius. If the arcs are chosen suitably (so that different lengths of arc correspond to the edges marked 0, 1, 2 in Figure 10.3.2) then it seems likely that a non-periodic tiling will necessarily result, though, so far as we are aware, all the possibilities have not yet been completely analysed. Certainly a much greater variety of tilings can occur than are admitted by the original set P1. For example any circular patch (of which there are an infinite number in any tiling) can be rotated through an arbitrary angle, and other modifications are also possible.

Recently Penrose [1978] has remarked that the set P1 can be reduced in size to an aperiodic set of *five* tiles. To do this we replace tiles (*b*), (*c*) and (*f*) in Figure 10.3.2 by two tiles, one of which is the union of a half-pentacle

(b) and a pentagon (f) and the other the union of a rhomb (c) and two pentagons (f); the new tiles are such that the edges marked 2 and $\bar{2}$ are eliminated.

The second set of aperiodic tiles, denoted here by P2, was discovered by Penrose in 1974 and is more remarkable than the first in that it contains only two tiles. The first mention of these tiles in the literature appears to be that of Guy [1976]; the material presented here is based on Robinson's notes [1975], an article by Gardner [1977] which fully explores some of the recreational aspects of these tiles, and private communications from Penrose, Ammann, and Conway. To the latter we are especially

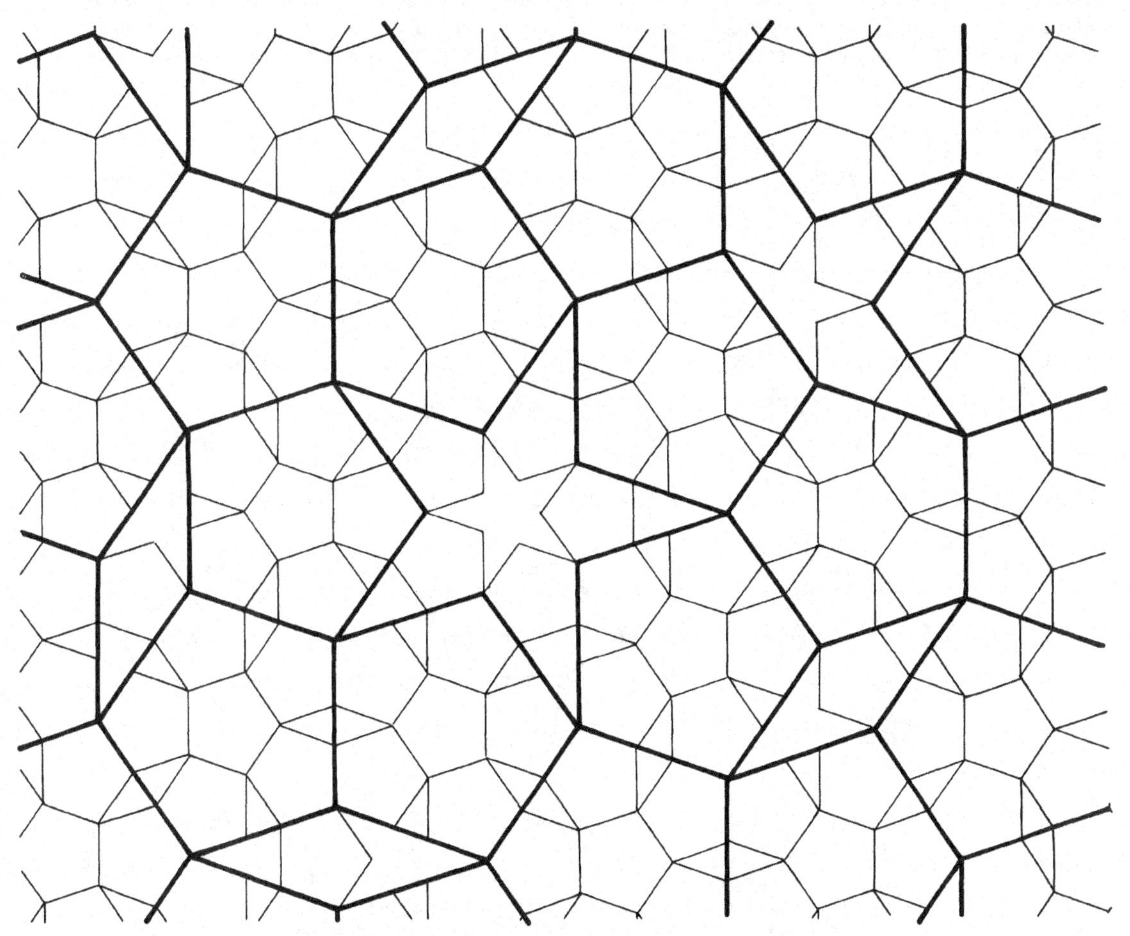

Figure 10.3.6
An example of composition in a tiling by the set P1 of Penrose prototiles.

indebted for much of the material in this section and in Sections 10.5 and 10.6. For an interesting account of how Penrose discovered the sets P1 and P2 see Penrose [1978]. These tiles are, we believe, the only ones described in this book which are covered by patents in various countries.

The tiles, which are usually called *Penrose kites and darts* are shown in Figure 10.3.10. Each is a quadrangle with two sides of length 1 and two sides of length τ (the golden number). The angles are as marked and the corners are colored with two colors, say red and green, as shown. These colors will usually be indicated in our diagrams by small solid and open circles. The matching condition is that we must put equal edges together and also match the colors at the vertices. An example of a tiling by these tiles is shown in Figure 10.3.11.

If we do not wish to use colors and matching conditions involving them then it is possible to achieve an equivalent condition using the shapes of the tiles only. For this we first note that as each tile has an axis of

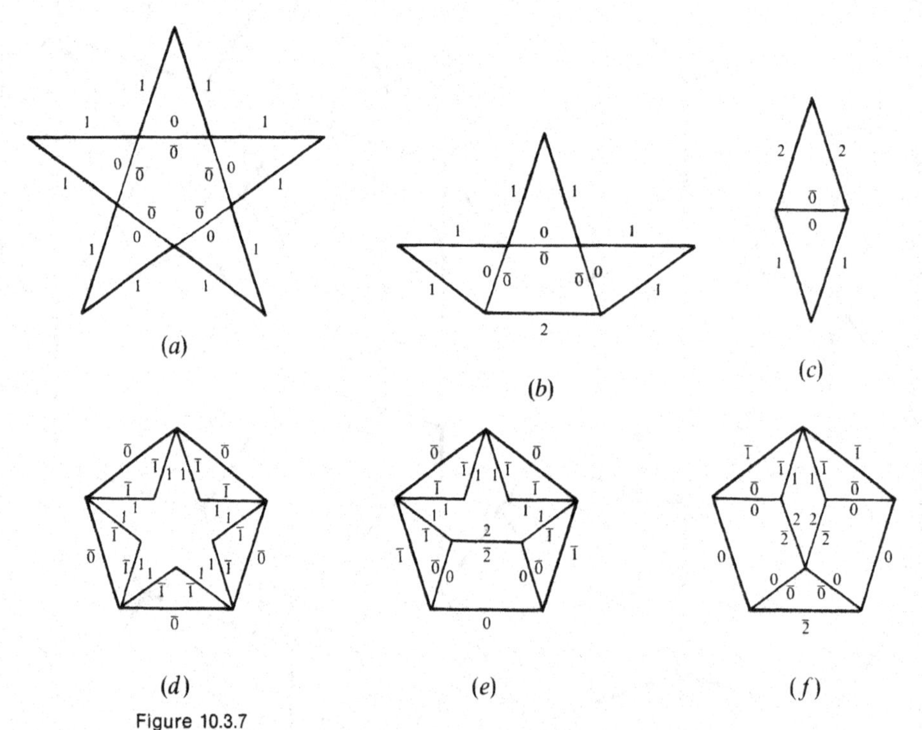

Figure 10.3.7
If, in any tiling by the set P1 of Penrose prototiles we mark each tile as shown here, and then delete the edges of the tiles leaving only the markings, we arrive at another tiling by the set P1 but reduced in size in the ratio $\tau^{-1}:1$. This is the process of semi-decomposition. For clarity, the markings on the original tiles are shown outside the corresponding edges.

symmetry it is never necessary to use reflections of the prototiles—direct congruences are sufficient. Hence we need only replace the edges by suitable J-curves (of two different kinds) to obtain the required result, see Figure 10.3.12. Penrose has suggested an amusing variant of his tiling (in the style of Escher) in which the J-curves are chosen in such a way that the tiles resemble two sorts of birds, see Figure 10.3.13! The relationship between this tiling and that by kites and darts is indicated on the diagram.

A simple way to investigate the properties of these Penrose tiles is due to Robinson [1975]. (For some

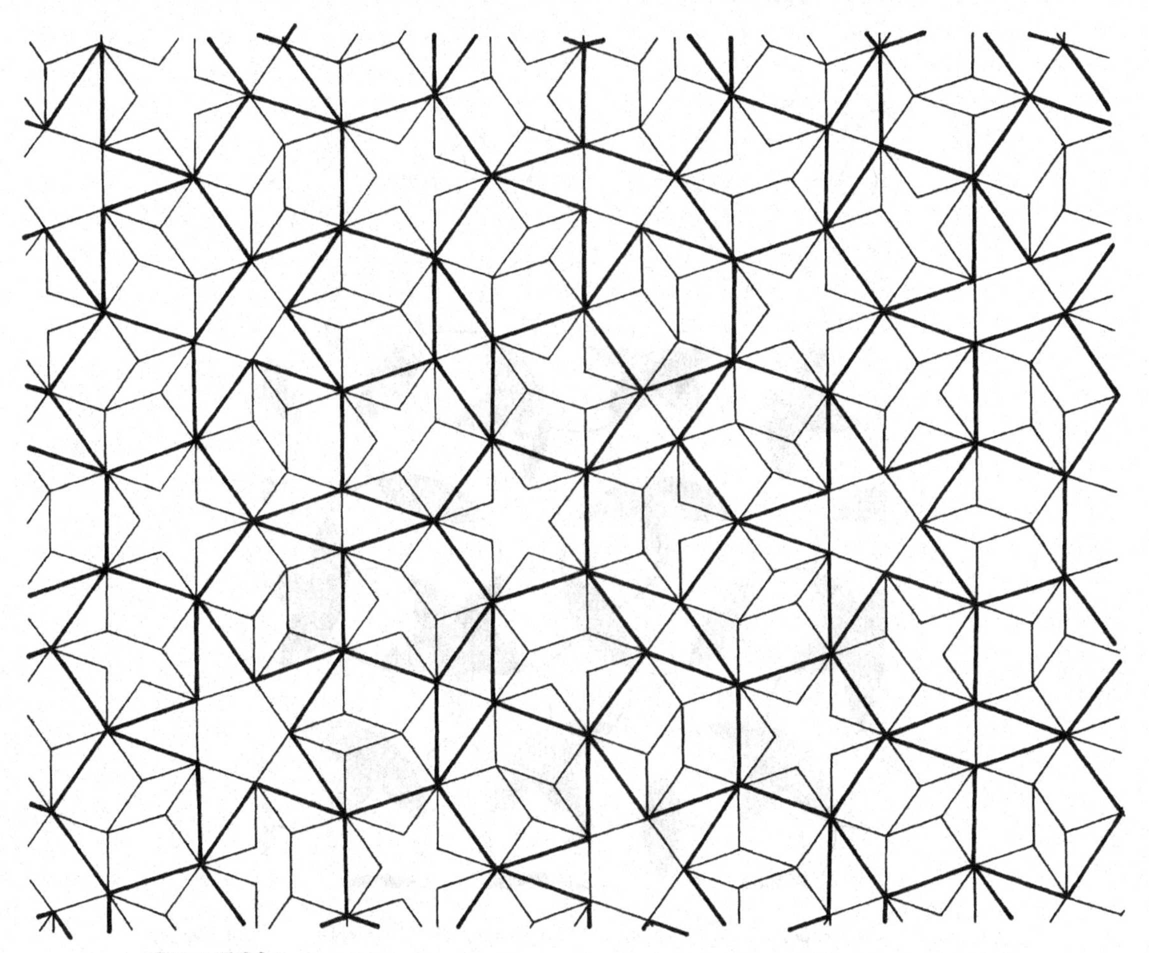

Figure 10.3.8
Superimposed tilings illustrating semi-decomposition for the set P1 of Penrose prototiles. For clarity, the markings on the larger tiles have been omitted.

additional remarks on these triangular tiles see Dekking [1983].) He proposed cutting each kite and dart along its axis of symmetry into two triangles, as indicated by the dashed lines in Figure 10.3.10. In this way we obtain the triangular tiles of Figure 10.3.14(*a*). These will be referred to as *A*-tiles, the larger and smaller being denoted by L_A and S_A, respectively. The matching condition for these tiles is a little more complicated than for the set P2; each monochromatic edge (that is, one which joins two vertices of the same color) must be oriented. This orientation is conveniently represented on the diagrams by an arrow, as shown in Figure

Figure 10.3.9
A modification by Penrose of his set P1 of aperiodic prototiles. Each edge is replaced by a circular arc whose center is the "point" of a pentacle or half-pentacle. A small portion of the original P1 tiling is reproduced at the top of the diagram to show the relationship between the tilings. Three of the prototiles have been colored black and three are white. It is conjectured that these "curvilinear" tiles are also aperiodic.

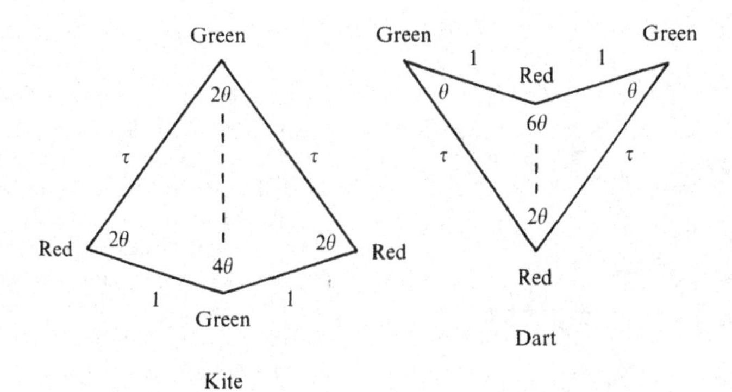

Figure 10.3.10
The Penrose aperiodic set P2 (kites and darts). The sides are of two lengths in the ratio τ:1 and the angle $\theta = \frac{1}{5}\pi$.

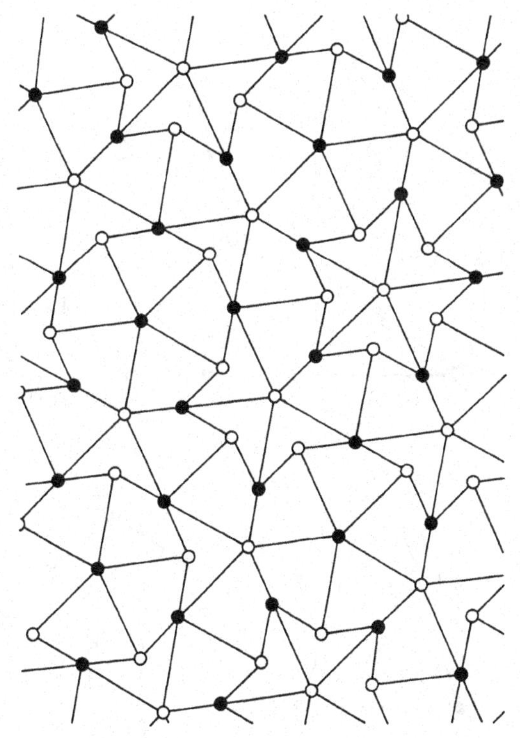

Figure 10.3.11
A tiling by Penrose kites and darts. Red and green vertices are denoted by open and solid circles.

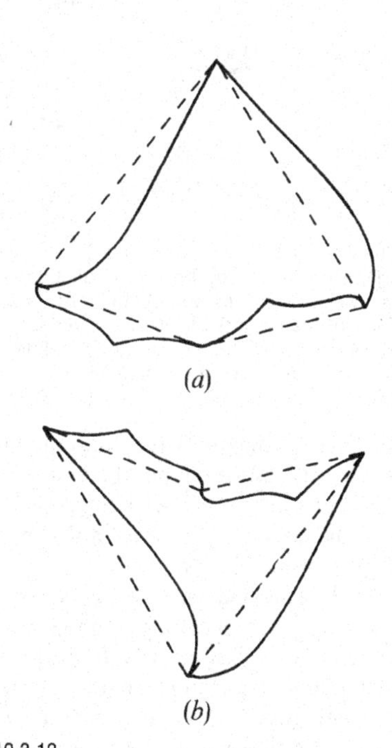

Figure 10.3.12
Modifications of the Penrose kite and dart which enable one to avoid the necessity of coloring the vertices. The long and short sides are replaced by two J-curves as shown.

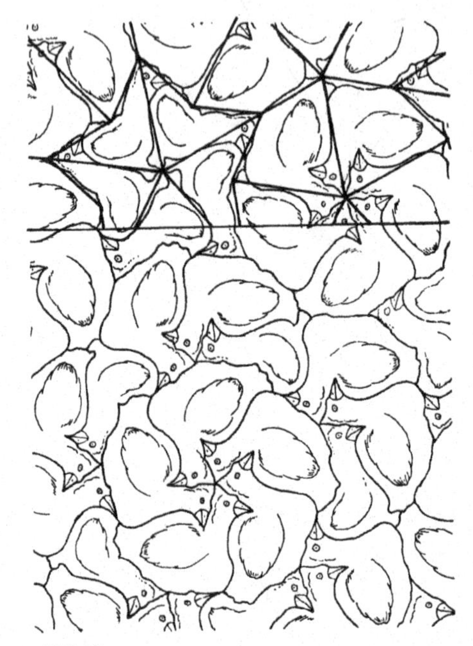

Figure 10.3.13
An amusing modification by Penrose of his tiling by kites and darts. The two sorts of birds are obtained by replacing the edges of the kites and darts by suitably chosen J-curves as in Figure 10.3.12. Some of the original tiling by kites and darts is shown in the upper part of the diagram.

10.3.14(*a*). In any tiling by A-tiles not only must the colors at the vertices match, but the monochromatic edges of adjacent tiles must be oriented in the same direction. (The latter is referred to in Robinson's paper as the "balance condition".)

Any tiling by kites and darts can be converted into an A-tiling by the process of cutting each tile into two pieces as described above, and conversely, from any A-tiling we can derive a tiling using the tiles P2 by removing the monochromatic edges—this is equivalent to amalgamating the A-tiles in mirror-image pairs (see Figures 10.3.15(*a*) and (*b*)).

In Figure 10.3.14(*b*) we show how the A-tiles can be composed to form two new tiles, called B-tiles, for which the same matching condition applies. The larger and smaller B-tiles are denoted by L_B and S_B, and we note that $L_A = S_B$. Composition applied to these B-tiles produces the $\tau A'$-tiles shown in Figure 10.3.14(*c*). Here τ means that each of the new tiles is similar in shape to one of the original A-tiles but enlarged in the ratio τ:1 and the prime (′) means that the colors red and green are interchanged. We orient the monochromatic edges as shown. It is clear how we can proceed. By composition the $\tau A'$-tiles lead to $\tau B'$-tiles, the $\tau B'$-tiles to $\tau^2 A$-tiles, and so on.

The composition process just described can be applied to tilings as well as to the individual tiles. By this we

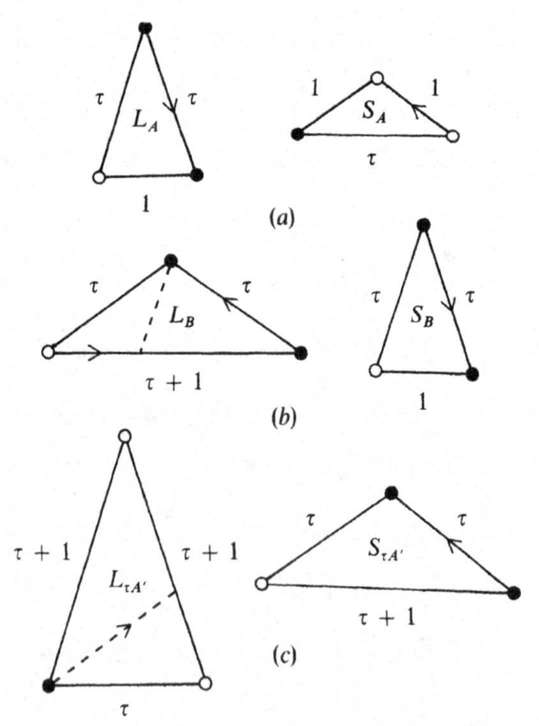

Figure 10.3.14
The triangular tiles used in Robinson's analysis of the tilings by Penrose kites and darts.

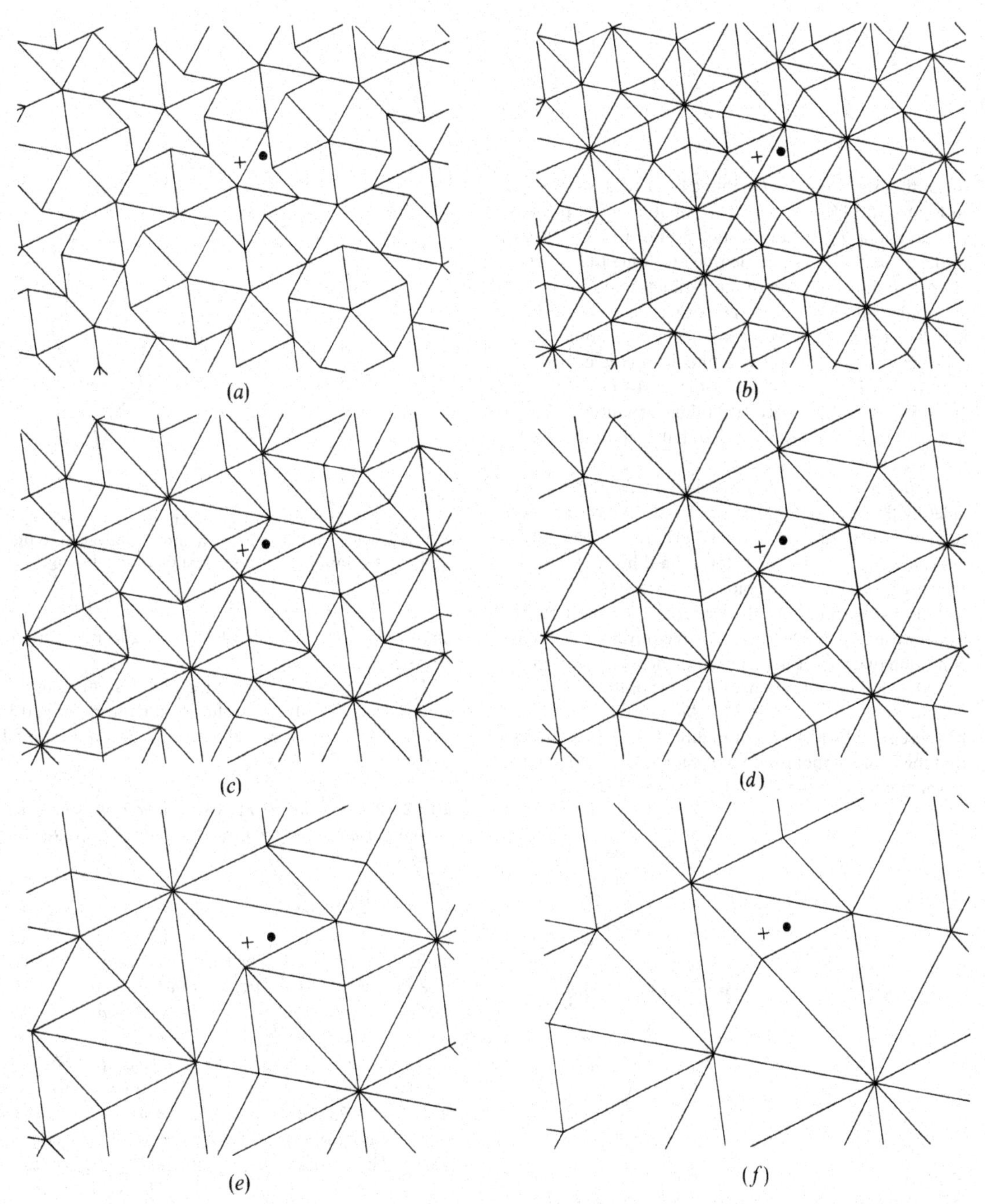

Figure 10.3.15
A Penrose tiling and the corresponding A-tiling, B-tiling, $\tau A'$-tiling, $\tau B'$-tiling and $\tau^2 A$-tiling.

mean that starting from *any* tiling by A-tiles (such as that in Figure 10.3.15(b)) it is always possible to compose the tiles so that a B-tiling results (see Figure 10.3.15(c)), and to do so uniquely. In fact the rule for doing this is very simple—on the diagram of the A-tiling we delete every short edge that joins a red vertex to a green vertex. From any B-tiling we may, in a similar manner, obtain a $\tau A'$-tiling by deleting those monochromatic edges that separate an L_B-tile from an S_B-tile, and so on. In Figure 10.3.15 we show the successive tilings produced from that in part (b) and it is easily verified that each of the compositions is unique.

The B-tilings seem to be especially interesting for, by deleting the longest edge of each L_B tile and the shortest edge of each S_B tile, we obtain two rhombs. These, which are shown in Figure 10.3.16, form the third set P3 of Penrose prototiles. Here colors of vertices as well as lengths and orientations of edges are to match, though as before, it is possible to obtain an equivalent matching condition (without using colored vertices or oriented edges) by modifying the edges of the rhombs in a suitable manner, see Figure 10.3.17. An example of a tiling by these rhombs is shown in Figure 10.3.18. It is strange that the Penrose rhombs do not seem to have attracted as much attention as the kites and darts—having such simple shapes with edges of equal length they seem to us to be extremely elegant.

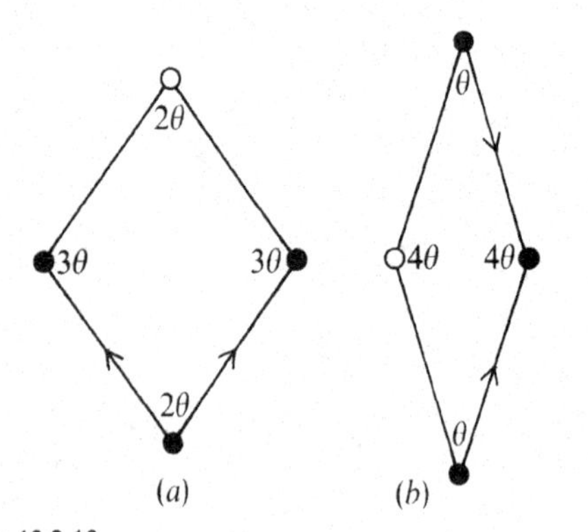

Figure 10.3.16
The Penrose aperiodic set P3 (the Penrose rhombs). The angle θ is $\frac{1}{5}\pi$, and red and green vertices are denoted by open and solid circles.

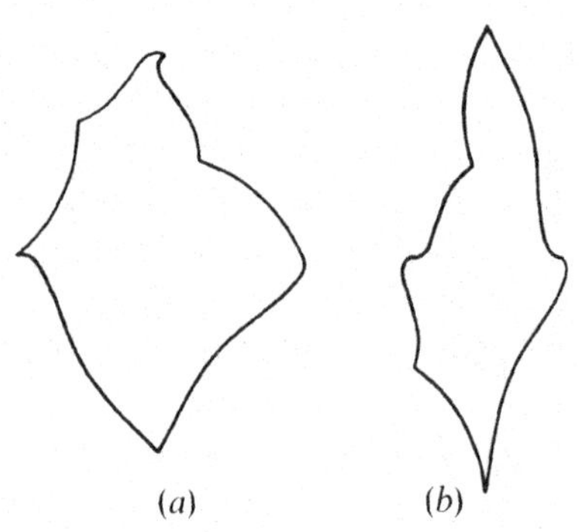

Figure 10.3.17
Modifications to the edges of the Penrose rhombs of Figure 10.3.16 which yield a matching condition equivalent to that implied by coloring the vertices and orienting the edges.

The processes of composition and decomposition for tilings described above (and illustrated in Figure 10.3.15) can be used exactly as in the proof of Statement 10.3.1 to establish the following.

10.3.2 *With the given matching conditions, the two A-tiles form an aperiodic set of prototiles. So do the two B-tiles.*

Immediately we deduce:

10.3.3 *With the given matching conditions, the set* P2 *(the Penrose kite and dart) forms an aperiodic set of prototiles. So does the set* P3 *of Penrose rhombs.*

The processes of composition and decomposition defined for the A-tiles and B-tiles have analogues for both the sets P2 and P3. For example, in Figure 10.3.19, we show how decomposition can be applied to kites and darts of set P2. The decomposition of the underlying A-tiles yields B-tiles which lead to the patches of Penrose rhombs as shown. As some of the B-tiles are half-rhombs these may be completed to form whole rhombs with the

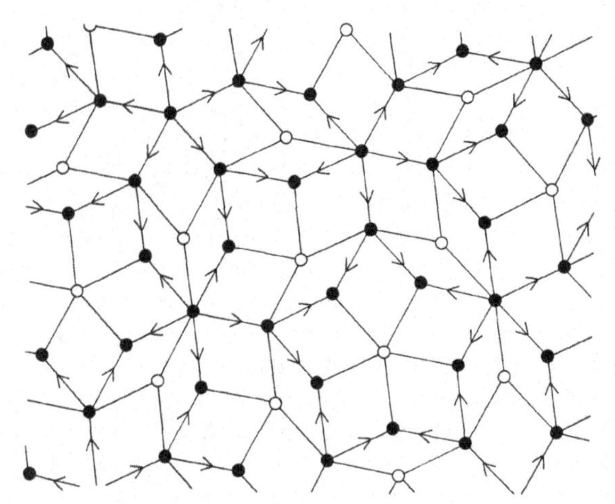

Figure 10.3.18

A tiling by Penrose rhombs. This arises by amalgamating pairs of triangles in a *B*-tiling. Red and green vertices are denoted by open and solid circles.

"star" consisting of five darts at its center. Similarly there are just two tilings by the Penrose rhombs which have symmetry group *d5*.

We conclude this section with some remarks on variations on and interrelations between the sets P1, P2 and P3 of aperiodic prototiles; the reader will find additional details in Penrose [1978]. Already the connection between tilings by sets P2 and P3 will be clear from their relationships with the *A*-tilings and *B*-tilings. It can be shown that the following procedure applied to a tiling using the set P1 yields one by kites and darts. We mark each of the P1 tiles as indicated in Figure 10.3.21(*a*). If these markings are regarded as indicating the edges and vertices of a new tiling—the edges and vertices of the original tiles being deleted—we obtain a tiling by set P2. Conversely, if we start from any tiling by the set P2 and mark the tiles as in Figure 10.3.21(*b*), then these markings indicate a new tiling by

effect that the resultant patch is larger than one might expect. A second decomposition, also shown in Figure 10.3.19, brings us back to kites and darts, and again the completion of half-tiles makes the patch grow even larger. Notice that the double decomposition interchanges the colors red and green on the vertices. In Figure 10.3.20 we show successive decompositions (augmented by completion of the half-tiles) applied to a patch of tiles with symmetry group *d5*.

Although they are aperiodic, the sets of Penrose tiles P1, P2 and P3 admit tilings with non-trivial symmetry groups (*d1* or *d5*). The construction of a tiling with symmetry group *d5* by sets P2 or P3 is illustrated by Figure 10.3.20. It will be seen how repeated inflation can lead to larger and larger patches of tiles, each with symmetry group *d5*, and an obvious modification of the Extension Theorem (Statement 3.8.3) enables us to deduce the existence of tilings of the whole plane with this same symmetry group. In fact there are precisely two such tilings by kites and darts, one of which has a "sun" consisting of five kites at its center, and the other has a

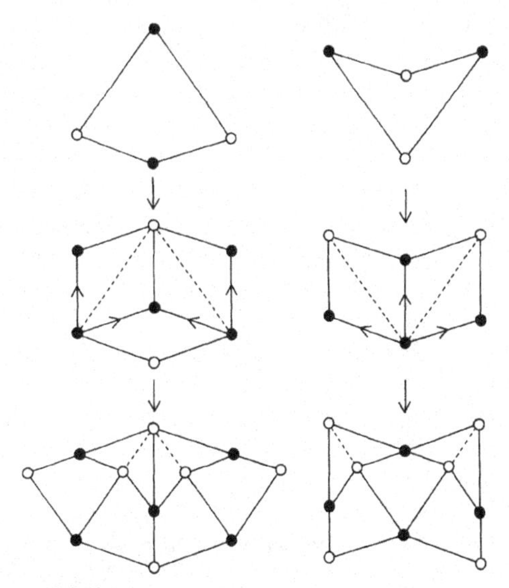

Figure 10.3.19

Decomposition of the Penrose kite and dart. At each stage the half-tiles are completed to form whole tiles.

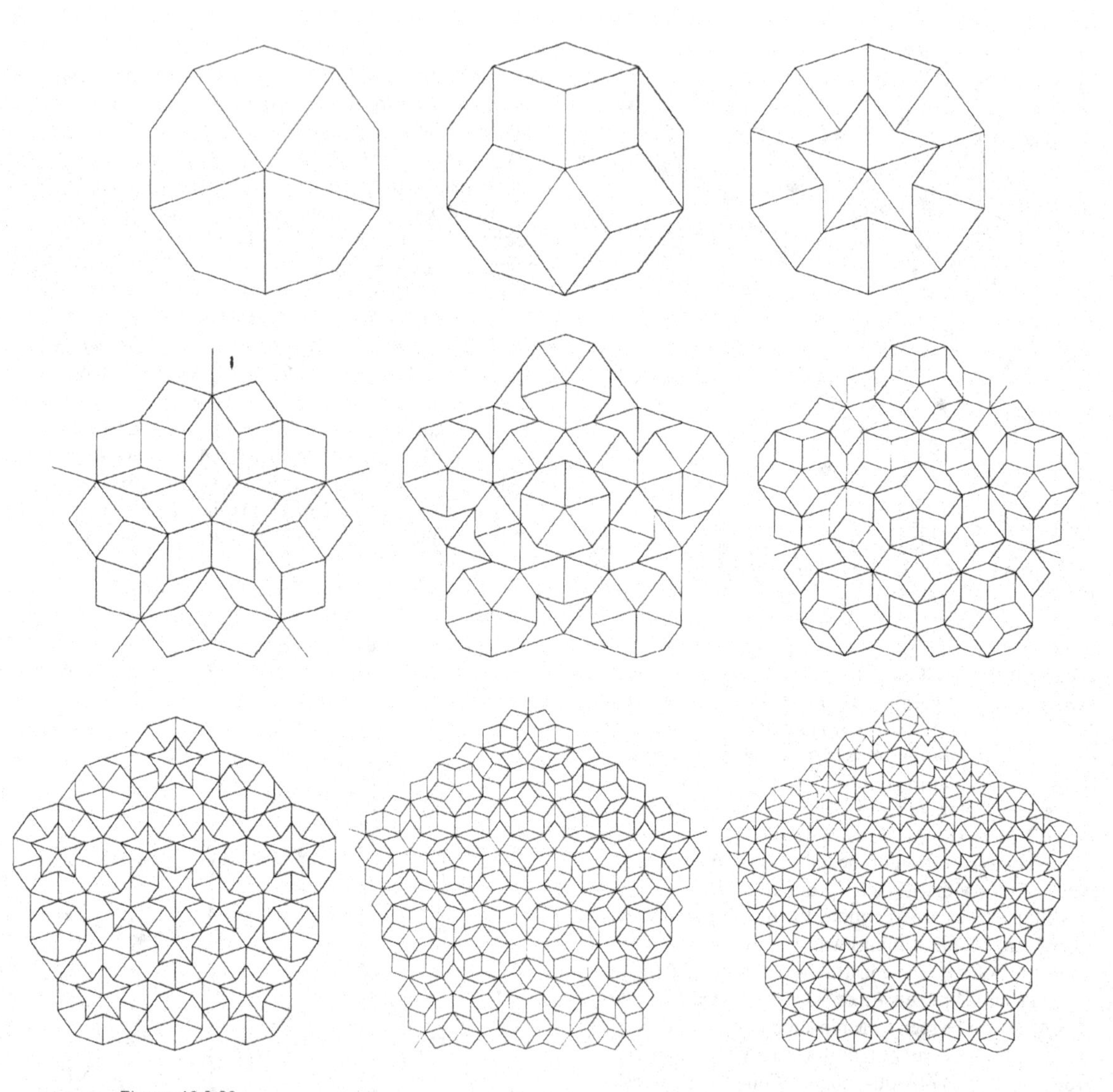

Figure 10.3.20
Successive decompositions of patches of tiles of the Penrose sets P2 and P3. These correspond to the decompositions of the underlying A-tiles and B-tiles shown in Figure 10.3.14. Expanding these patches appropriately and using the Extension Theorem (Statement 3.8.1) enables us to construct tilings with symmetry group $d5$. We are indebted to Fred Lunnon for supplying these diagrams drawn by the computer at University College, Cardiff, Wales. Another computer program for drawing a tiling by Penrose tiles was described by Regener [1978].*

* See the Appendix beginning on page 653.

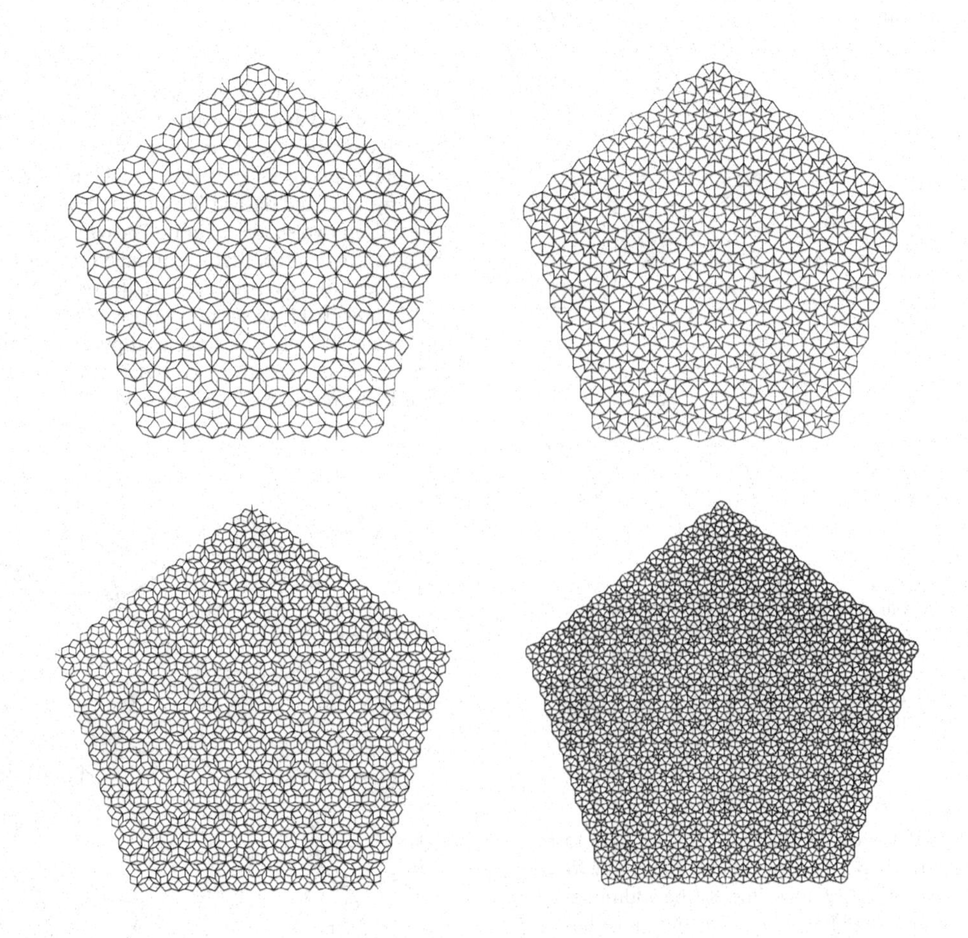

tiles of the set P1. The relationship is illustrated in Figure 10.3.22 where the two tilings are shown superimposed. Of course the relationships between the three tilings can be used to prove the aperiodicity of any two of them once the aperiodicity of the third has been established.

The relationship between two tilings described in the previous paragraph seems to be of considerable interest since it generalizes the idea of decomposition. Given any two sets of prototiles $\mathscr{S}_1$ and $\mathscr{S}_2$, if it is possible to mark the tiles of $\mathscr{S}_1$ in such a way that on *any* tiling $\mathscr{T}_1$ admitted by $\mathscr{S}_1$, the corresponding marks are the edges and vertices of *some* tiling $\mathscr{T}_2$ admitted by $\mathscr{S}_2$, then we shall say that $\mathscr{T}_2$ is obtained from $\mathscr{T}_1$ (or $\mathscr{S}_2$ from $\mathscr{S}_1$) by *recomposition.* We have already met several instances of recomposition. The earliest example arises from duality. If a set $\mathscr{S}_2$ of prototiles admits a unique tiling (as with the tiles indicated by dashed lines in Figure

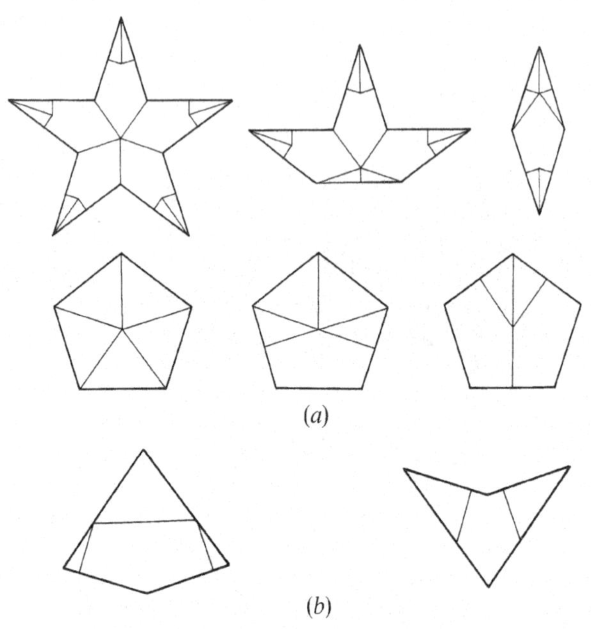

Figure 10.3.21
The markings that must be added to Penrose's aperiodic set P1 and to the kites and darts (set P2) to show that each is a recomposition of the other. For clarity the labels on the edges of the tiles have been omitted—they can be deduced from the fact that the tiles are in the same relative positions in the diagram as in Figures 10.3.1 and 10.3.10. The exact positions of the markings can be deduced from Figure 10.3.22.

2.7.3(*a*)) then the dual (indicated by solid lines in the figure) is a recomposition of it. The same is true for the tiling of Figure 2.7.3(*b*) if we impose the additional condition that the tiling is edge-to-edge, but in neither case is the opposite relation true. Another example occurred on page 534 where semi-decomposition was defined as a recomposition of the original tiling. In this terminology, as we see from Figures 10.3.21 and 10.3.22, each of the sets P1 and P2 (or the tilings admitted by them) is a recomposition of the other. We shall find the idea of recomposition convenient in the future to describe the relationship between various sets of aperiodic tiles.

Several suggestions have been made for replacing the colored vertices and oriented edges of the sets P2 and P3 by marks which lead to an equivalent matching condition. Two particularly attractive variants are shown in Figures 10.3.23 and 10.3.25, the first of which is due to Conway and the second adapted from one suggested by Penrose. In each case the tiles are marked with curves or bands of two different colors—in Figure 10.3.23 these are indicated by solid and dashed lines and in Figure 10.3.25 by bands in two shades of gray. The curves or bands on the tiles must match, that is, continue from each tile to its adjacent tiles over the edges of the tiling. In Figures 10.3.24 and 10.0.1 we show examples of tilings marked in these ways. They have a somewhat psychedelic appearance! Apart from aesthetic considerations the following remarkable result can be established:

Figure 10.3.22
Tilings by the Penrose aperiodic sets P1 and P2 superimposed to illustrate recomposition (see Figure 10.3.21). This diagram is taken from Penrose [1978].

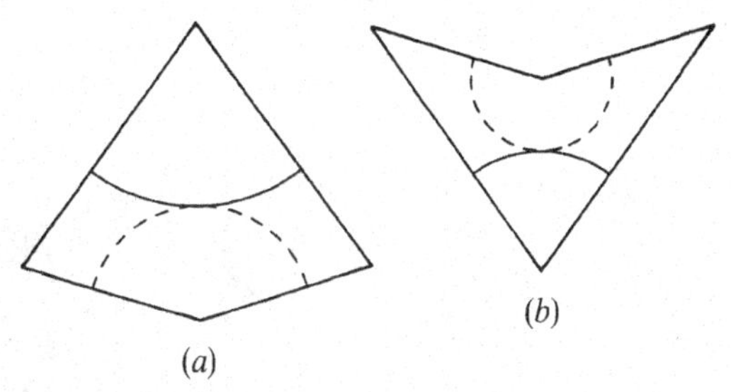

Figure 10.3.23
Markings on the Penrose kite and dart that yield a matching condition equivalent to that of Figure 10.3.10.

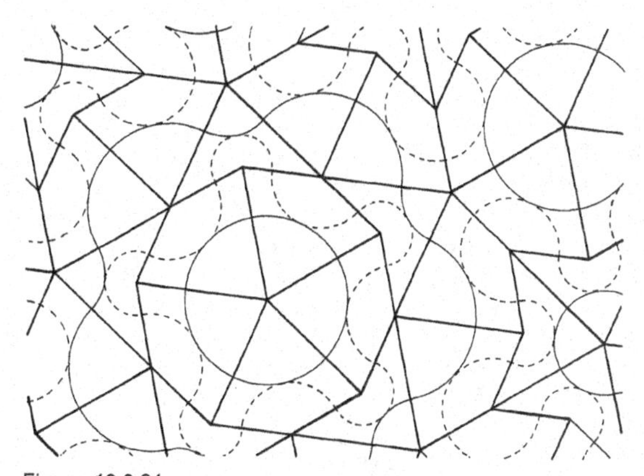

Figure 10.3.24
An example of a tiling by kites and darts when they have been marked as in Figure 10.3.23.

10.3.4 *With the tiles marked in the manner just described, every closed curve or band has symmetry group d5.*

In fact Conway has shown that in either tiling at most two curves or bands of each color are not closed, and that the patch of tiles inside any closed curve or band also has symmetry group *d5*. We shall make use of this fact in our discussion of the cartwheel tiling in Section 10.5.

The proof of Statement 10.3.4, of which details are requested in the exercises, depends upon repeated application of composition. In this way each closed curve is "shrunk" (relative to the size of the tiles) until it is so small that it *must* have the required symmetry group.

Yet another way of marking the kites and darts (or, indeed, of either of the two other sets of Penrose tiles) which yields an equivalent matching condition is by means of straight line segments which must continue as straight lines across the edges of the tiling. These lines are known as Ammann bars after their originator, and they are of great importance in further investigations (see Section 10.6). In Figure 10.3.26 we show how Ammann bars may be defined for the kites and darts, and in Figure 10.3.27 we give an example of a tiling on which the bars have been marked.

Although the Penrose rhombs (set P3) are convex, it is not possible to modify them so that they remain convex and at the same time so that the matching condition depends only on the shape of the tiles and not on any coloring or orientation of the vertices or edges. (This fact is suggested by the edge-replacements shown in Figure 10.3.17, but a rigorous proof is far from obvious since there is no necessity for the vertices in the modified tiling to lie in the same positions in the plane as those of the original tiling. An example where this phenomenon occurs is shown in Figure 10.3.13 where it will be observed that the vertices are not in the same positions as in the original kite and dart tiling.) This raises the interesting problem of finding the smallest sets of aperiodic prototiles, each of which is an unmarked convex polygon. Ammann has produced a remarkable example of such a set by recomposition from the set P3. It contains only three tiles—one hexagon and two pentagons, see Figure 10.3.28.

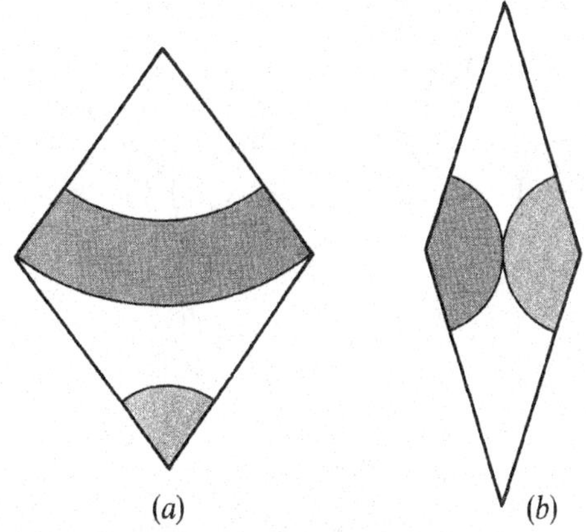

Figure 10.3.25
Markings on the Penrose rhombs which yield a matching condition equivalent to that of Figure 10.3.16. An example of a tiling marked in this way is shown in Figure 10.0.1.

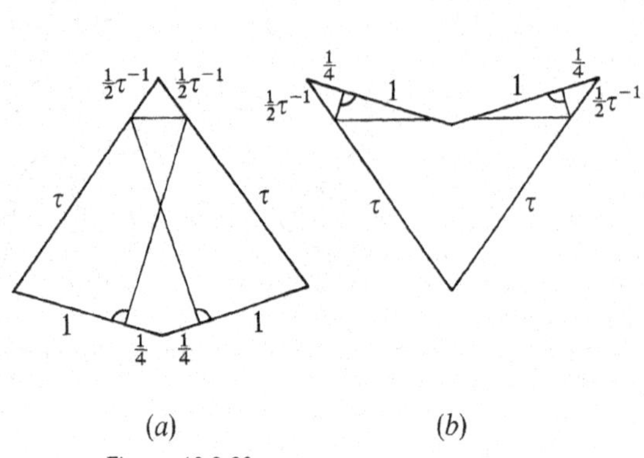

Figure 10.3.26
Ammann bars on the Penrose kite and dart.*

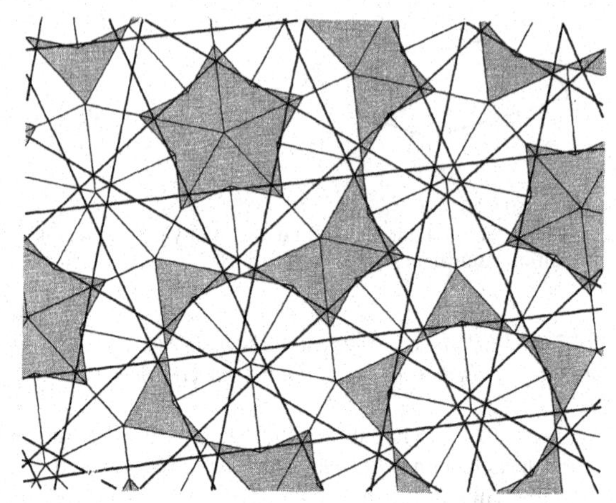

Figure 10.3.27
A tiling by Penrose kites and darts with the tiles marked by Ammann bars as in Figure 10.3.26.

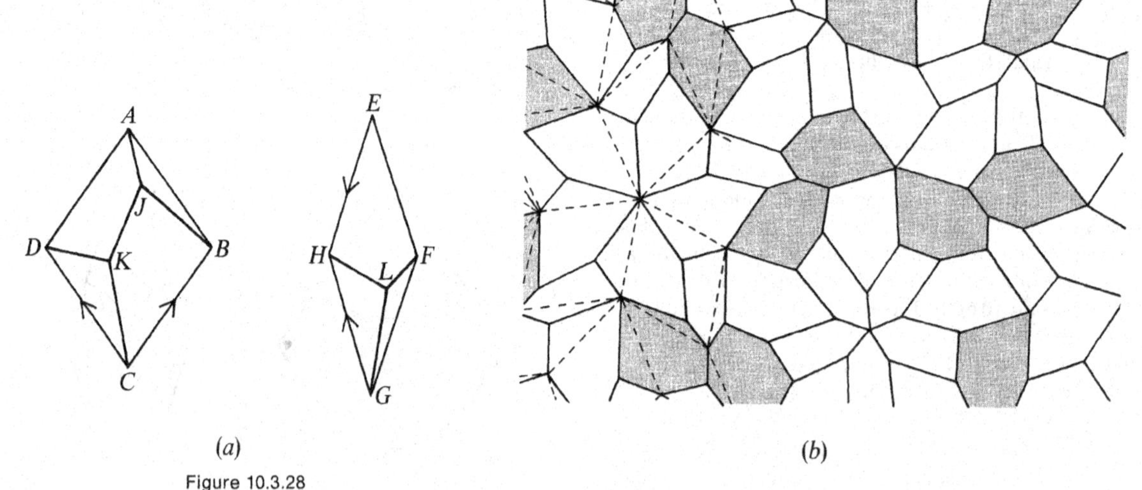

Figure 10.3.28
An aperiodic tiling due to Ammann which uses three unmarked convex polygons as prototiles—one hexagon and two pentagons. This tiling is a recomposition of one by Penrose rhombs (set P3) and is obtained from it using the markings shown in (*a*). These markings are completely determined by the choice of the point L, for we must have $AJ = LF$, $JB = LG = KC$, and $DK = HL$. For aperiodicity L must be chosen so that FL, GL, HL and JK are of different lengths. In (*b*) the relationship between a tiling by these convex polygons and a tiling by the set P3 is indicated by dashed lines on the left.

* See the Appendix beginning on page 653.

EXERCISES 10.3

*1. Verify that the composition process indicated in Figures 10.3.4, 10.3.5 and 10.3.6 can be carried out in any tiling by the set P1. [Hint: Every pentacle tile (*a*) is surrounded by five pentagonal tiles (*f*) fitting between its arms, and against the edges labelled 2 of these fit either rhombs (*c*) or half-pentacles (*b*). Each of these half-pentacles belongs to the same patch as the original star but none of the rhombs does so. In this way patches of types (*a*), (*b*) or (*c*) of Figure 10.3.4 are determined. Pentagonal tiles (*d*) lie at the centers of pentagonal patches (*d*), (*e*) and (*f*).]

2. (*a*) Find tilings of the plane with the modified Penrose tiles P1 shown in Figure 10.3.9 different from those obtained from a tiling using the P1 tiles by rotating non-overlapping circular patches.

**(*b*) Prove that this set of tiles is aperiodic without insisting that the tiles are placed edge-to-edge.

3. Show how Ammann bars can be defined for the aperiodic sets P1 and P3.

4. (*a*) Show that the recomposition of the set P2 shown in Figure 10.3.29(*a*) leads to a set of nine convex pentagons, one of which is regular and which, if the parameters are chosen correctly, forms an aperiodic set. (This construction is due to L. Danzer.)

(*b*) Show that the recomposition of the set P3 shown in Figure 10.3.29(*b*) leads to a set of six convex pentagons which, if the parameters are chosen correctly, forms an aperiodic set.

*5. Give a complete proof of Statement 10.3.4.

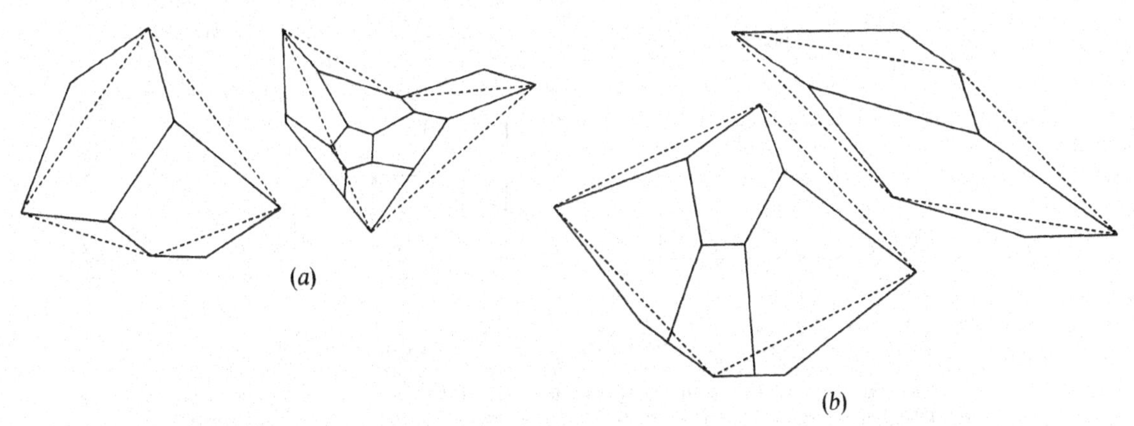

Figure 10.3.29
(*a*) An aperiodic set of nine convex pentagons (one of which is regular) obtained by recomposition of the Penrose kite and dart. This construction is due to L. Danzer. (*b*) An aperiodic set of six convex pentagons obtained by recomposition of the Penrose rhombs. The lines of subdivision are chosen so that no two are parallel and no angles but the original ones can be fitted around a point.

10.4 THE AMMANN APERIODIC TILES*

In 1977 Robert Ammann discovered several new sets of aperiodic prototiles. Descriptions of these do not appear in the literature, and we are very grateful to Ammann for allowing us to publish them here for the first time.

Like the Penrose tilings, the properties of the first two sets of Ammann prototiles (sets A2 and A3) are related to the golden number $\tau = \frac{1}{2}(\sqrt{5} + 1)$. The first set (A2) is shown in four equivalent variants in Figure 10.4.1; these variants arise from the various ways in which the matching conditions are specified. In Figure 10.4.1(*a*) lettered arrows are used, as explained in the caption. In Figure 10.4.1(*b*) the two tiles are marked with parts of small black ellipses, and we specify that the tiles must be fitted together so that *complete* ellipses are formed. (See Figure 10.4.2 for an example of the tiling.) It will be noticed that each tile occurs only in four orientations—that shown in Figure 10.4.1, reflection in a horizontal line, reflection in a vertical line, and reflection in both (that is, central reflection or rotation through 180°). Because of these reflections it is not possible to replace the markings by a matching condition that depends *only* on the shapes of the tiles unless a third tile is introduced as shown in Figure 10.4.1(*c*). The small elliptical tile, whose sole purpose is to ensure that the other two tiles fit together in the correct manner, will be called a *key* tile. Its shape is,

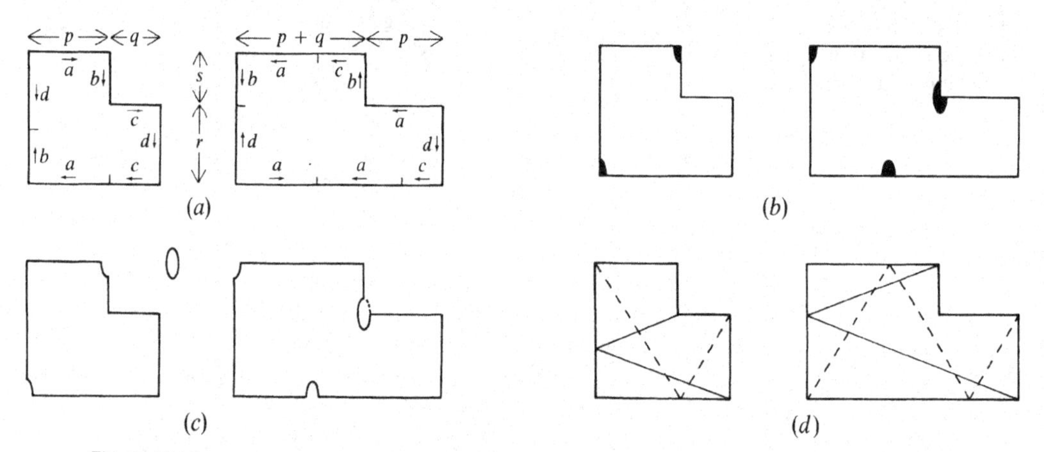

Figure 10.4.1
Four variants of Ammann's set A2 of aperiodic tiles. In (*a*) we show the dimensions of the tiles; *p*, *q*, *r* and *s* can take any positive values. The variants depend upon how the matching condition is specified. Each arrowed edge must fit against an edge with the same label and with an arrow pointing in the same direction. Using an obvious extension of the terminology of Chapter 6, the tile symbols are $a^+b^+c^+d^+c^+a^+b^+d^-$ and $a^-c^-b^-a^-d^+c^+a^+a^-d^+b^-$ and the adjacency symbol is $a^-b^-c^-d^-$. In (*b*) the black markings must be matched in such a way that complete ellipses are formed, see Figure 10.4.2. In (*c*) the tiles are unmarked but three tiles are necessary—the small elliptical tile is a key tile whose sole purpose is to ensure that the two larger tiles fit in the correct way. In (*d*) the matching condition is specified by Ammann bars on the tiles which must continue straight across the edges of the tiling, see Figure 10.4.4. This variant is possible only if the tiles are "golden" which means that $p/q = r/s = \tau$. In every tiling by any variant of this set A2, each prototile occurs in only four aspects; the matching conditions prevent rotations through $\frac{1}{2}\pi$.

* See the Appendix beginning on page 653.

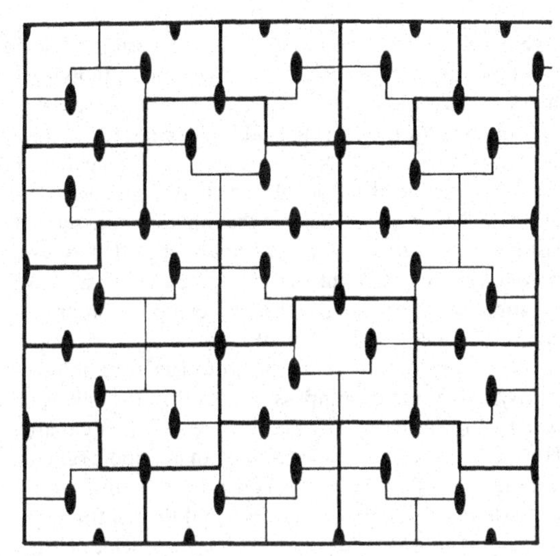

Figure 10.4.2
An example of a tiling by Ammann's aperiodic set A2. Heavier lines indicate the tiling obtained by composition.

to some extent, arbitrary, so long as its symmetry group is $d2$, the indentations in the other tiles are modified in an appropriate way and it will not tile the plane on its own. The dotted curve on the larger tile in Figure 10.4.1(c) shows a possible modification which clearly leads to an equivalent matching condition. Figure 10.4.2 can also be interpreted as showing a tiling by these three tiles if we think of the black ellipses as copies of the key tile, instead of markings on the larger tiles.

In Figure 10.4.1(a) we indicate the sizes of the tiles in terms of the four numbers p, q, r and s. Whatever positive values are chosen for these, the tiles can be composed as in Figure 10.4.3—in Figure 10.4.2 the composed tiling is indicated by heavier lines.

If we choose the values of p, q, r and s appropriately, several additional properties arise and new variants become possible. To begin with, as we see from Figure 10.4.3, the dimensions of the composed tiles are given in terms of the original ones by replacing p, q, r and s by $p + q$, p, $r + s$ and r respectively. It follows that the composed tiles will be similar to the original ones if and only if $p/q = (p + q)/p$ and $r/s = (r + s)/r$—that is, if $p/q = r/s = \tau$, the golden number. If this condition holds then we shall say that the tiles are *golden*. Note that even with golden tiles the ratio p/r can take any positive value—this is a consequence of the fact that no rotations (except through 180°) of the tiles are required in constructing a tiling (nor even permitted by the matching condition).

In the case of golden tiles there is another way of marking the tiles that ensures that they match correctly and lead to an aperiodic tiling. This is shown in Figure 10.4.1(d) and uses Ammann bars, that is, line segments

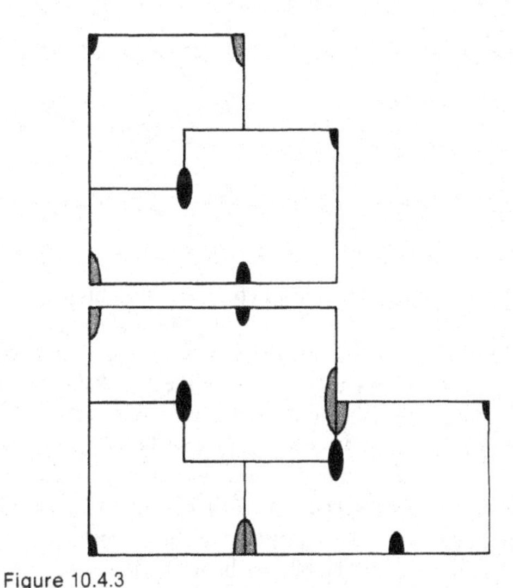

Figure 10.4.3
Composition of the prototiles of Ammann's aperiodic set A2. The composed tiles are similar in shape to the original ones if they are "golden", that is, their dimensions are such that $p/q = r/s = \tau$. In this diagram we have chosen the second variant of the set A2 shown in Figure 10.4.1(b). The gray markings are those that arise simply by enlarging the tiles—the black markings are those that arise from building up the tiles from the original ones. It will be seen that both sets of markings lead to an equivalent matching condition and hence composition (in the generalized sense of page 532) is defined.

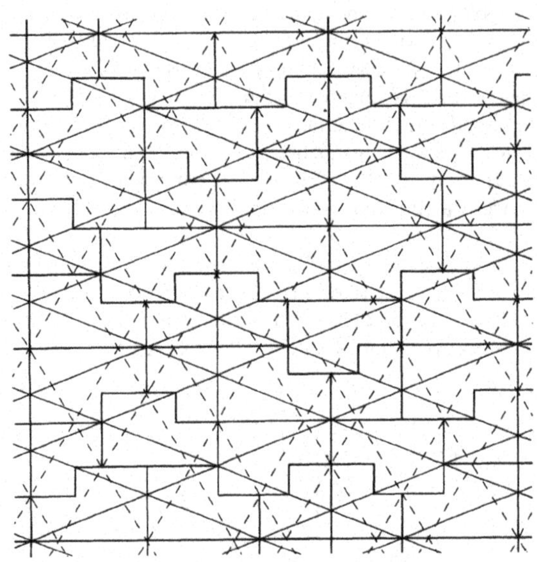

Figure 10.4.4
An example of a tiling by Ammann's aperiodic set A2 on which the Ammann bars have been marked.

marked on the tiles which must continue straight across the edges of any tiling. In Figure 10.4.4 we show an example of a tiling in which the bars have been marked.

Figure 10.4.2 shows how a tiling by the set A2 can be composed and there seems to be strong evidence that every tiling by the set A2 can be composed in a similar manner and moreover the composition process is unique. No proof of this fact has been published so far, but a proof has been obtained and will appear soon (Grünbaum & Shephard [1985c]). Hence we have the following result.

10.4.1 *Each of the four variants of the Ammann prototiles (set* A2*) shown in Figure* 10.4.1 *is an aperiodic set.*

If the tiles are golden and we put $r/p = \tau^{\frac{1}{2}}$, then it will be seen that the two tiles are similar in shape; here $r{:}p{:}s{:}q = \tau^{\frac{3}{2}}{:}\tau{:}\tau^{\frac{1}{2}}{:}1$. (One tile is obtained from the other by enlarging it in the ratio $\tau^{\frac{1}{2}}{:}1$ and rotation through a right angle.) However it does not seem possible to devise a simple matching condition for these tiles involving markings which are such that the similarity between the tiles extends to the markings also. A suggested method is shown in Figure 10.4.5(*a*) but then a complicated and seemingly artificial set of rules for matching has to be adopted. Nevertheless, the fact that an aperiodic set consisting of two similar tiles exists can be regarded as a step forward in the long-term aim of finding a single aperiodic prototile. For these similar tiles, the composition process can be simplified, see Figure 10.4.5(*b*).

Even if we do not require the prototiles to be golden, it is easy to see that it is impossible to adjust the values of p, q, r and s in such a way that the tiles are of equal area. However, as Ammann pointed out, an aperiodic set containing two prototiles of equal area can be constructed by taking $p = 3$, $q = s = 1$, $r = 2$ and then cutting up the tiles as shown in Figure 10.4.6(*a*). This yields an aperiodic set in which both tiles have area 5; an example of a tiling by these is shown in Figure 10.4.6(*b*).

Yet another variant of these tiles, in which $p = q = r = s = 1$ and the matching condition is specified by coloring the edges, will be described in Section 11.1.

The second set of aperiodic tiles discovered by Ammann (set A3) is shown in Figure 10.4.7. Again several variants are possible and are illustrated in the figure. In Figure 10.4.7(*a*) the matching condition is specified by coloring some of the vertices. We specify that similarly colored vertices must fit together, and also that the prototiles must not be reflected—only direct congruences are allowed. The composition process for this variant is shown in Figure 10.4.8, and in Figure 10.4.9 we show an example of the tiling. In Figure 10.4.7(*b*) we give a second variant in which some parts of the edges have been replaced by J-curves as shown. Here the correct matching is ensured entirely by the shape of the tiles, and clearly these shapes also forbid reflections. The third and fourth variants make use of Ammann bars, and are shown in Figures 10.4.7(*c*) and (*d*). In Figure 10.4.10 we show an example of the tiling in which both kinds of bars have been marked.

We have the following result (see Grünbaum & Shephard [1985c]).

10.4.2 *Each of the four variants of the Ammann prototiles (set A3) shown in Figure 10.4.7 is an aperiodic set.*

We remarked at the beginning of this section that the properties of the aperiodic sets A2 and A3 seem to be related to the golden number τ. More precisely, if we compose the golden A2 tiles, or the A3 tiles, we obtain tiles which are similar to the original ones but whose size is increased by a factor τ. In the remaining two sets of Ammann prototiles to be described (A4 and A5), composition leads to a similarity in which size is increased by a factor $\sqrt{2}$. There seems no doubt that the irrationality of both τ and $\sqrt{2}$ has some connection with the aperiodicity of the tiles, but it is an open question what other irrational numbers can arise in this way.

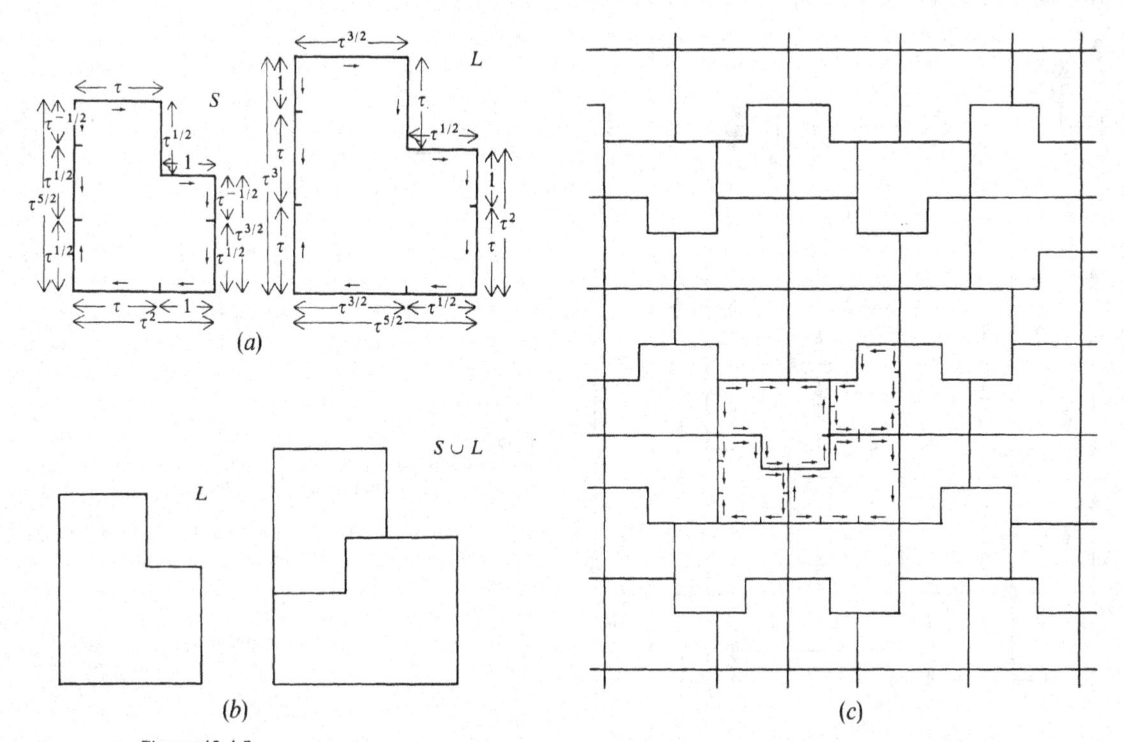

Figure 10.4.5
A modification of Ammann's aperiodic set A2 which consists of two similar tiles. In (*a*) we show the dimensions of the tiles, and in (*b*) a simplification of the composition process that is possible in this case. In (*c*) an example of the tiling is shown. If the tiles are to be similarly marked then the matching condition is necessarily more complicated—one suggestion is indicated in (*a*). Here the sides of the tiles are divided into "edges" by inserting new "vertices" as shown. In the tiling *either* edges of equal length must abut and the arrows point in the same direction *or* one edge of one tile must fit against two collinear edges of its adjacent and the arrows on one tile must point in the opposite direction to that on the other. This matching condition is illustrated by marking some of the tiles in (*c*).

The third set of Ammann tiles (A4) is shown in Figure 10.4.11 in two variants. The matching condition for the first requires completing the black "arrows", and the second uses unmarked tiles of which one is a key tile in the shape of an arrow. The dotted lines in Figure 10.4.11(*b*) suggest another equivalent variant. In Figure 10.4.12 we show the process of composition for these tiles, and notice that although the process of composition can be carried out whatever the value of p/q, the composed tiles are

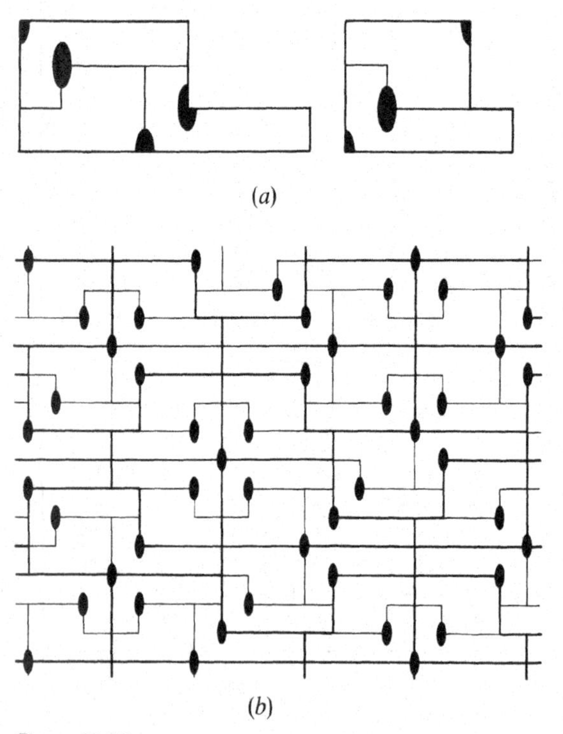

Figure 10.4.6
Although the Ammann set A2 remains aperiodic whatever the values of p, q, r and s it is impossible to adjust the values of these four quantities so as to produce two tiles of equal area. However Ammann has shown how to construct a set of aperiodic prototiles of equal area which is a recomposition of the set A2 by marking the tiles as shown in (*a*). Here $p = 3$, $q = s = 1$ and $r = 2$. Part (*b*) shows an example of a tiling by these tiles.

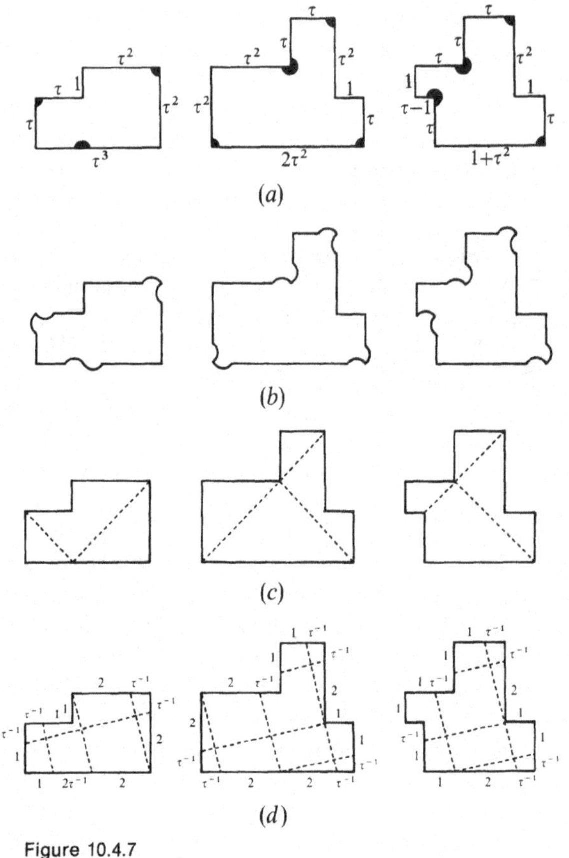

Figure 10.4.7
Four variants of Ammann's aperiodic set A3. In (*a*) the tiles must be fitted so that complete black circles are formed, see Figure 10.4.9, and in (*c*) and (*d*) the matching condition is specified by Ammann bars, see Figure 10.4.10. In (*c*) the position of the bars is clear from the diagram, and in (*d*) their positions are as marked.

similar to the original ones if and only if $p/q = \sqrt{2}$. In Figure 10.4.13 we show an example of a tiling by either variant. As shown in Grünbaum & Shephard [1985c], we have:

10.4.3 *Each of the variants of the Ammann prototiles (set* A4) *shown in Figure* 10.4.11 *is an aperiodic set.*

The final set of prototiles (A5) is shown in Figure 10.4.14 in two variants. For the first the matching condition is that arrows and semicircles must be completed, and for the second we have four unmarked tiles of which two

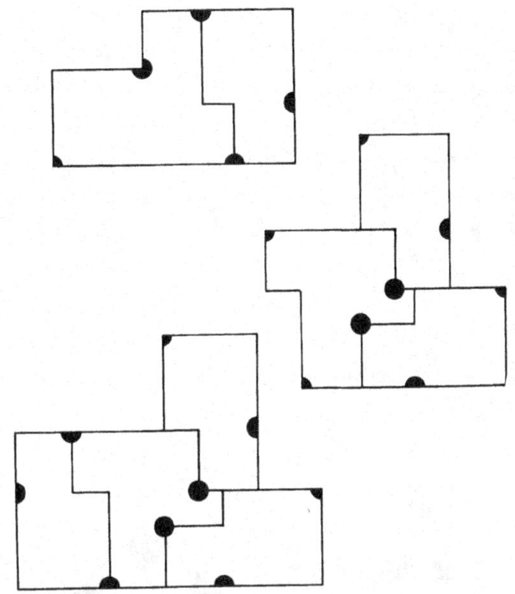

Figure 10.4.8
Composition of the tiles in Ammann's aperiodic set A3. The first variant (Figure 10.4.7 (*a*)) of the tiles is shown, and composition is in the generalized sense explained on page 532.

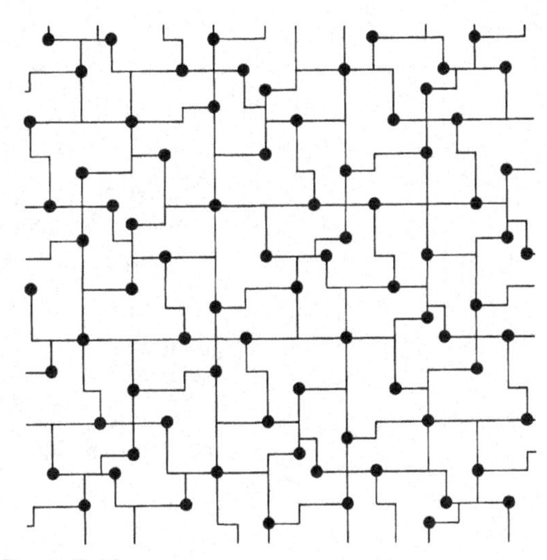

Figure 10.4.9
An example of a tiling by Ammann's aperiodic set A3. The matching condition uses the marks shown in Figure 10.4.7(*a*).

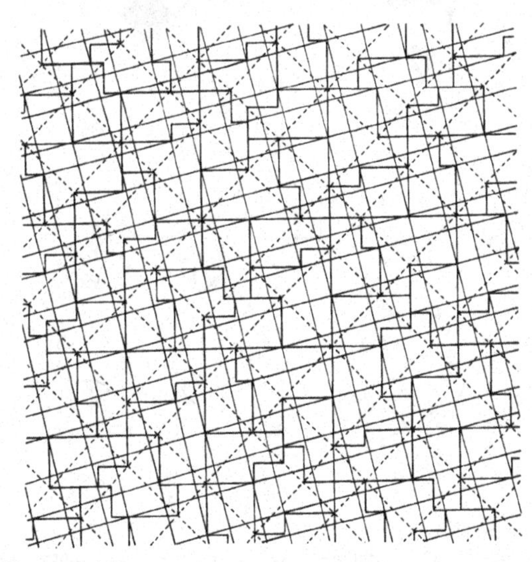

Figure 10.4.10
An example of a tiling by Ammann's aperiodic set A3. The matching condition is by Ammann bars as shown in Figures 10.4.7(*c*) and 10.4.7(*d*).

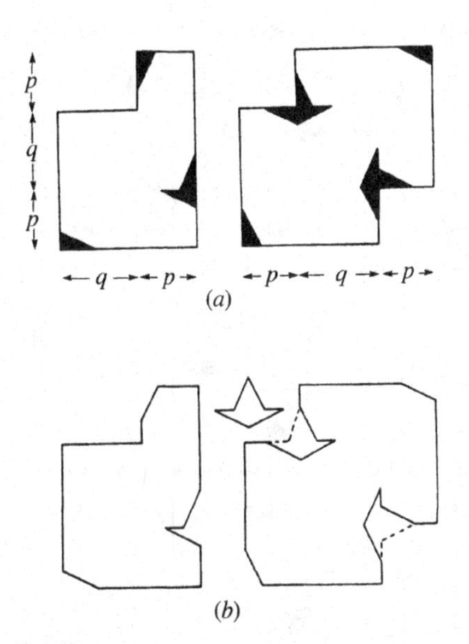

Figure 10.4.11
Two variants of Ammann's aperiodic set A4. In (*b*) one of the tiles, shaped like a small arrow, acts as a key tile. For an example of the tiling see Figure 10.4.13. The dimensions of the tiles are indicated in (*a*); the numbers p and q can take any positive value.

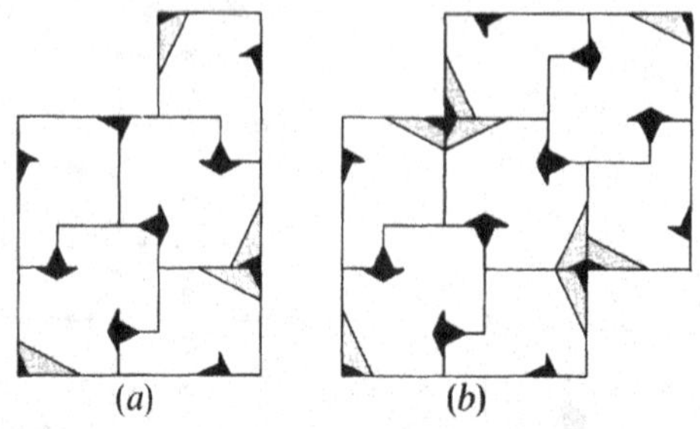

Figure 10.4.12
Composition of the tiles of Ammann's aperiodic set A4. Composition is possible whatever the values of p and q (see Figure 10.4.11(a)) but the composed tiles are similar in shape to the original ones if and only if $p/q = \sqrt{2}$.

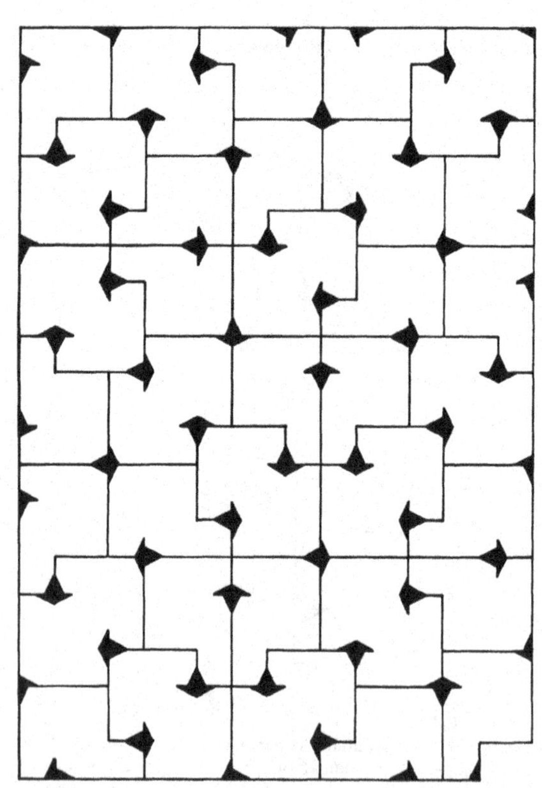

Figure 10.4.13
A tiling by Ammann's aperiodic set A4. The tiles are marked as in Figure 10.4.11(a).

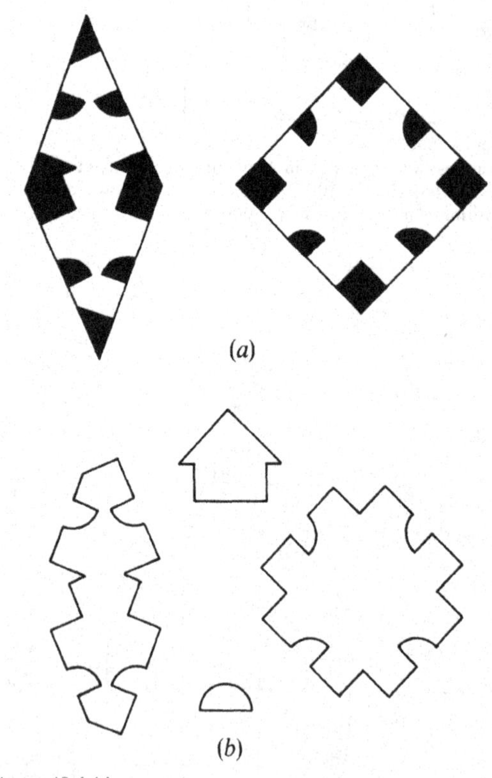

Figure 10.4.14
Two variants of Ammann's aperiodic set A5. In (a) there are two marked tiles and in (b) the tiles are unmarked but two additional key tiles are necessary to ensure that they fit together in the correct manner. An example of a tiling by this set is shown in Figure 10.4.16.

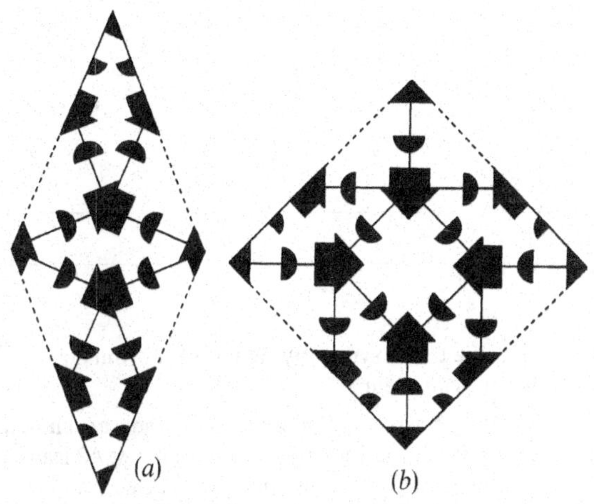

Figure 10.4.15
Composition of the prototiles of Ammann's aperiodic set A5.

Figure 10.4.16
An example of a tiling by Ammann's aperiodic set A5.

Figure 10.4.17
Tilings using the Ammann aperiodic sets A4 and A5 superimposed to show the relationship between them.

are key tiles. It will be observed that one tile is basically a square and the other is a rhomb with angles 45° and 135°. In this case the composition process is a little more complicated in that it involves half-tiles (so that decomposition could be more accurately described as recomposition); see Figure 10.4.15. An example of a tiling by either variant is shown in Figure 10.4.16.

Using composition, we arrive at the following.

10.4.4 *Each of the variants of the Ammann prototiles (set* A5) *shown in Figure* 10.4.14 *is an aperiodic set.*

A tiling by the set A4 can be obtained from one by the set A5 using recomposition. This is illustrated by the superposition of the tilings shown in Figure 10.4.17.*

* See the Appendix beginning on page 653.

EXERCISES 10.4

1. Use tile and adjacency symbols (as in Figure 10.4.1(*a*)) to specify the matching conditions for the sets A3, A4 and A5.

**2. Devise other methods of marking the tiles of Figure 10.4.5(*a*), so that the similarity of the tiles extends to the markings also, yet gives an equivalent matching condition.

*3. On the assumption that tilings by sets A2, A3, A4 and A5 are strongly metrically prototile balanced, calculate in each case the ratios between the numbers of tiles of each kind used in tiling the plane.

*4. Prove that the set of prototiles shown in Figure 10.4.6 is aperiodic.

*5. Describe the construction of a tiling using the set A2 that has symmetry group *d2*, and also one that has only the trivial symmetry. What other symmetry groups can occur?

6. Show that, whatever values of *p*, *q*, *r* and *s* are chosen, repeated composition of the tiles of the set A2 leads to a sequence of tiles which approach closer and closer to golden tiles in shape.

*7. Construct a tiling with symmetry group *c4* using the set A3. What other symmetry groups are possible?

*8. Assuming that set A4 is aperiodic, use the recomposition process shown in Figure 10.4.17 to prove that the set A5 is aperiodic.

*9. Verify that it is not possible to tile the plane with the first two tiles of Figure 10.4.7(*a*). (In fact, quite a large patch can be tiled using only these two tiles.)

10.5 LOCAL ISOMORPHISM AND THE CARTWHEEL TILING

We now resume our discussion of tilings by the Penrose kites and darts (set P2) defined in Section 10.3. Many of our results, such as those relating to local isomorphism and indicator sequences, extend in an obvious way to the other aperiodic tilings discussed in the previous two sections. Here we confine attention to the set P2 for the simple reason that this set has attracted more attention than the others, and consequently more is known (even if not yet published!) about it. The fundamental property underlying all our results is the uniqueness of composition and decomposition. As will be seen, there are many open problems in this area, and proofs of some of these are requested as double-starred exercises at the end of the section.

We shall begin by describing a very special tiling by Penrose kites and darts known as the *cartwheel tiling*, see Figure 10.5.1(*a*). It is constructed in the following way. We begin with an *ace* consisting of a dart and two kites, see Figure 10.5.2(*a*). If we inflate this we obtain the patch of Figure 10.5.2(*b*) and subsequent inflations yield the patches of Figures 10.5.2(*c*), (*d*) and (*e*) and so on. The patches will be denoted by $\mathscr{C}_0, \mathscr{C}_1, \mathscr{C}_2, \ldots$, as shown in the figure. In this context we are using "inflation" to mean decomposition as shown in Figure 10.3.19 (which converts a tiling by kites and darts into another by kites and darts) followed by an appropriate enlargement which sends the tiles to their original size. Each patch so pro-

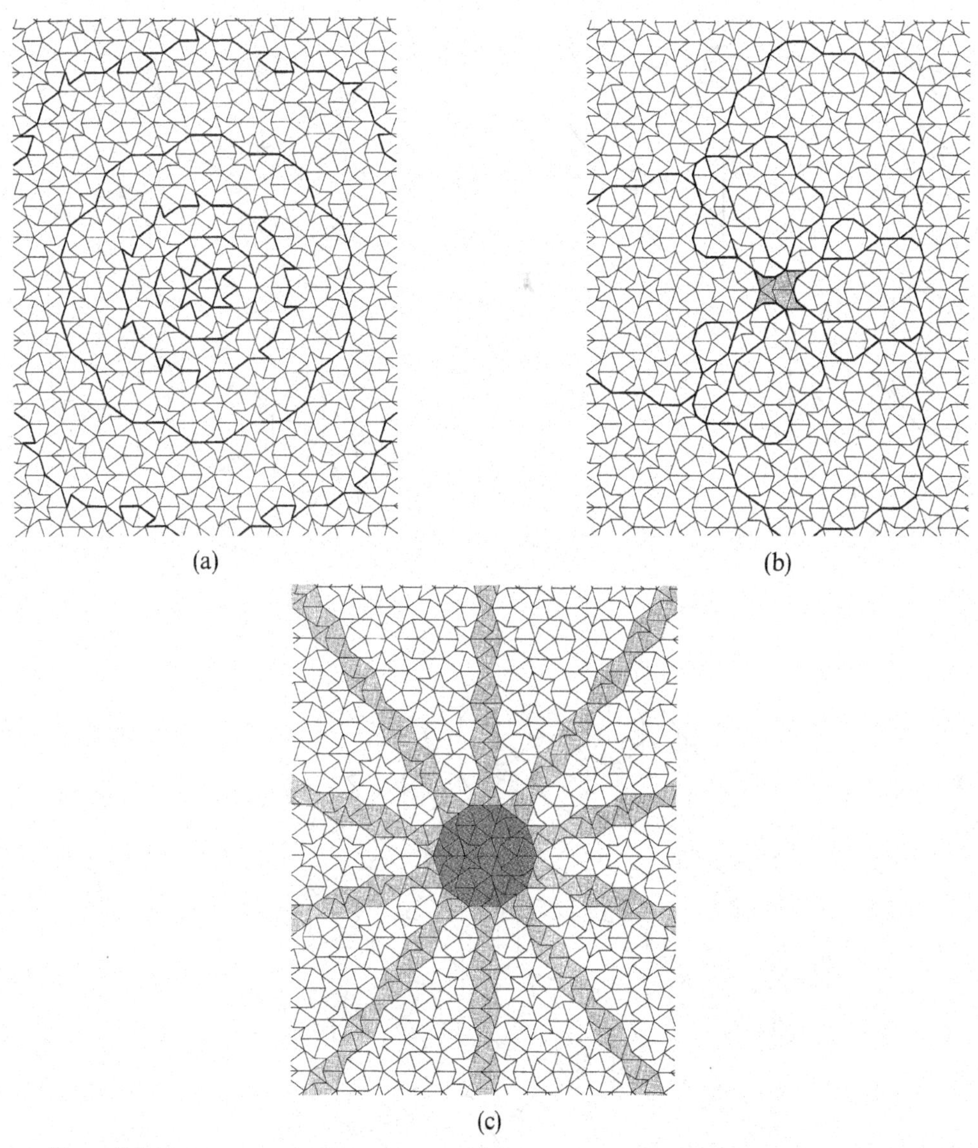

Figure 10.5.1
The cartwheel tiling. In (*a*) we show the patches $\mathscr{C}_0, \mathscr{C}_1, \mathscr{C}_2, \mathscr{C}_3, \mathscr{C}_4, \ldots$, each of which is obtained from the previous one by inflation followed by rotation through 180°. $\mathscr{C}_0$ is the ace, $\mathscr{C}_2$ is the first-order cartwheel, $\mathscr{C}_4$ is the second-order cartwheel, and larger cartwheels occur in the tiling, all with the same center. Each patch $\mathscr{C}_i$ has symmetry group *d1* but the union of the tiles in each cartwheel is a polygon with symmetry group *d5*. In (*b*) we show patches with symmetry group *d5*. It can be shown that every tile in this tiling, apart from the seven shaded tiles, belongs to a patch of this kind. Moreover these seven tiles are the only tiles in *any* tiling by kites and darts which do not belong to a patch with symmetry group *d5*. In (*c*) we have marked by shading the first-order cartwheel and the ten semi-infinite Conway worms that radiate from its sides.

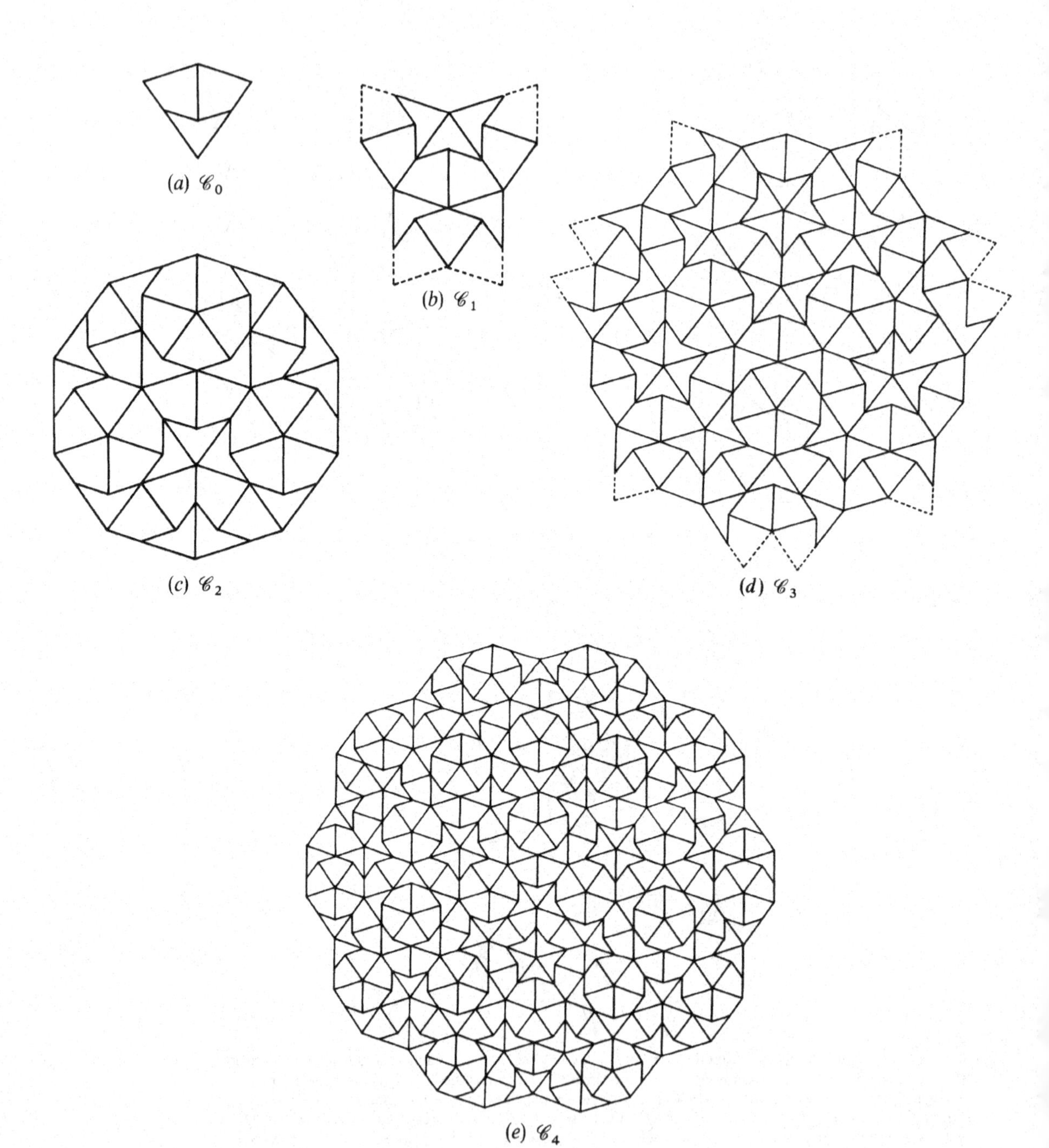

Figure 10.5.2
The first four patches that arise from successive inflations of the ace shown in (*a*). The patches $\mathscr{C}_2$ and $\mathscr{C}_4$ shown in (*c*) and (*e*) are the first- and second-order cartwheels.

duced has symmetry group $d1$, with just one reflection as symmetry, but it is convenient to trim a few tiles from the boundaries of the odd-numbered patches (as indicated by the dashed lines in Figures 10.5.2(*b*) and (*d*)) so that the resulting even-numbered patches $\mathscr{C}_2, \mathscr{C}_4, \mathscr{C}_6, \ldots$ have the property that the union of the tiles in each patch is a polygon with symmetry group $d5$ (see Figures 10.5.2(*c*) and (*e*)). These (even-numbered) patches are called *cartwheels*—in particular $\mathscr{C}_2$ is a first-order cartwheel, $\mathscr{C}_4$ is a second-order cartwheel, and in general $\mathscr{C}_{2n}$ is called a *cartwheel of order n*.

It will be apparent from Figure 10.5.2, and easily proved using inflation, that for each $n > 1$ the cartwheel $\mathscr{C}_{2n}$ contains its predecessor $\mathscr{C}_{2(n-1)}$ and, moreover, does so concentrically. We thus have an increasing sequence of patches which, by the Extension Theorem (Statement 3.8.1), enables us to deduce the existence of a tiling which contains a copy of every cartwheel $\mathscr{C}_{2n}$ ($n \geq 1$) centered at some fixed point of the plane; it is the inclusion limit of the cartwheels. This is the cartwheel tiling shown in Figure 10.5.1(*a*). The cartwheels are indicated in the diagram by thickened lines, and we have also indicated in the same way the intermediate patches $\mathscr{C}_m$ (m odd). It will be observed that in the tiling the patch $\mathscr{C}_n$ does not arise directly by inflation of $\mathscr{C}_{n-1}$—each occurs in an inverted position, that is, rotated about its center by 180°.

10.5.1 *Except for the seven tiles shown gray in Figure 10.5.1(b), every tile in the cartwheel tiling lies in a patch of tiles whose symmetry group is d5.*

For brevity we shall refer to a patch of tiles with symmetry group $d5$ as a *d5-patch**. In Figure 10.5.1(*b*) we have indicated several $d5$-patches, and it will be seen that these include all the tiles inside $\mathscr{C}_3$ except for the seven gray tiles. In particular they include all the tiles in the "ring" between $\mathscr{C}_1$ and $\mathscr{C}_3$. Now inflate the tiling: $d5$-patches go into larger $d5$-patches, and the ring between $\mathscr{C}_1$ and $\mathscr{C}_3$ goes into the ring between $\mathscr{C}_2$ and $\mathscr{C}_4$. Successive inflations show that, for each n, every tile of the ring between $\mathscr{C}_n$ and $\mathscr{C}_{n+2}$ belongs to a $d5$-patch, and hence Statement 10.5.1 is proved by induction.

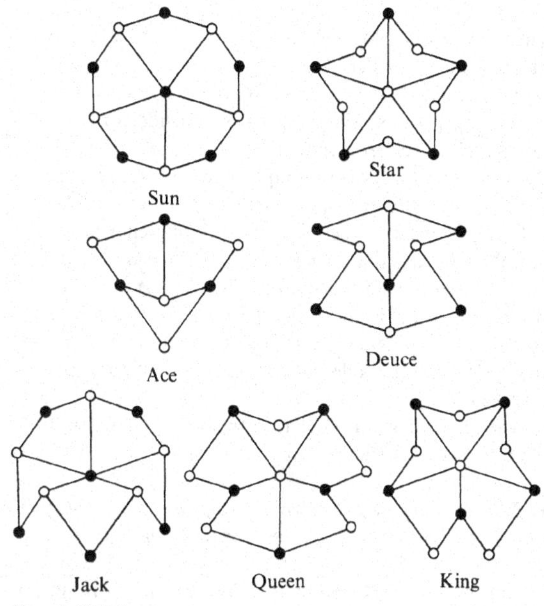

Figure 10.5.3
The seven kinds of vertex neighborhoods that occur in every tiling by Penrose kites and darts.

Let O be the center of all the cartwheels. It is easily verified that inflation sends every $d5$-patch $\mathscr{A}$ (center A) into a $d5$-patch $\mathscr{A}'$ (center A') such that A, O, A' are collinear and $\tau|OA| = |OA'|$. The converse is also true. By the use of "deflation" (by which we mean the opposite of inflation) it is clear that to every $d5$-patch $\mathscr{A}'$ corresponds a smaller $d5$-patch $\mathscr{A}$ nearer to and on the opposite side of O, unless $\mathscr{A}'$ is the "sun patch" consisting of five kites (see Figure 10.5.3). In the latter case we may think of the deflated patch as being so small that it disappears completely!

The above remark enables us to show that the gray tiles of Figure 10.5.1(*b*) do not lie in any $d5$-patch. For suppose one of these tiles T belonged to a $d5$-patch $\mathscr{A}_1$.

* Notice that the cartwheels themselves are not $d5$-patches for each has symmetry group $d1$ and *not* $d5$.

Then repeated double deflations would lead to a sequence of smaller and smaller *d5*-patches $\mathscr{A}_2, \mathscr{A}_3, \ldots$, all containing T. Eventually we would arrive at a *d5*-patch $\mathscr{A}_n$ containing T and so small that $\mathscr{A}_n$ was inside, for example, $\mathscr{C}_4$. However, from a diagram of the cartwheel tiling it is clear that this does not happen and we deduce that T does not belong to *any d5*-patch.

In this connection there is a curious difference of opinion. Penrose says that in *every* tiling by kites and darts *every* tile lies in a *d5*-patch unless the tiling is the cartwheel *and* the tile is one of the seven gray tiles in Figure 10.5.1(*b*). On the other hand, Conway says that this is *not true*! We hope the difference of opinion will soon be resolved.

One reason for introducing cartwheels is on account of the following remarkable fact:

10.5.2 *In every tiling by Penrose kites and darts, every tile T lies in a cartwheel $\mathscr{C}_{2n}$ of every order $n \geq 1$.*

To prove this we begin by noticing that in any tiling $\mathscr{T}$ by Penrose kites and darts every point P of the plane lies inside an ace (Figure 10.5.2(*a*)). To see this, suppose first that P belongs to a dart. Then necessarily two kites fit against the two shorter edges of the dart, and then these three tiles together form an ace. On the other hand, if P lies inside a kite, then it is easy to show that this kite must share at least one of its shorter edges with a dart and then the other shorter edge of the dart will necessarily fit against a second kite. Again an ace is formed.

Now let n be any positive integer, T be the tile of $\mathscr{T}$ that contains P, and $\mathscr{T}^{(2n)}$ be the tiling obtained from $\mathscr{T}$ by composing $2n$ times. Since, by the above argument, P belongs to an ace in $\mathscr{T}^{(2n)}$, if we decompose $2n$ times, P must belong to an nth order cartwheel $\mathscr{C}_{2n}$ in $\mathscr{T}$. This cartwheel contains T and so Statement 10.5.2 is proved.

By a *vertex neighborhood* $\mathscr{N}(V)$ of a vertex V in a tiling $\mathscr{T}$ we mean a minimal patch of tiles that completely surrounds V. It is straightforward to check that there are exactly seven different kinds of vertex neighborhoods as shown in Figure 10.5.3. To these Conway has given the descriptive, if somewhat fanciful, names indicated in the diagram.

10.5.3 *In every tiling by Penrose kites and darts, every one of the seven kinds of vertex neighborhoods necessarily occurs and does so infinitely often.*

To prove this we observe from Figure 10.5.2(*e*) that the second order cartwheel $\mathscr{C}_4$ contains all seven kinds of vertex neighborhood. (In fact the first-order cartwheel contains six kinds—only the star is missing.) Since, by Statement 10.5.2, every tile belongs to a second-order cartwheel, Statement 10.5.3 follows immediately.

Statement 10.5.3 enables us to prove a property of the tilings known as *local isomorphism*. This is, perhaps, one of the most unexpected features of these aperiodic tilings; we already mentioned it briefly in Section 10.2 in connection with the set R1 of Robinson tiles.

10.5.4 *Every patch $\mathscr{A}$ of tiles in a tiling $\mathscr{T}$ by Penrose kites and darts is congruent to infinitely many patches in every tiling by the same prototiles.*

To prove this, as before we denote by $\mathscr{T}^{(n)}$ the tiling obtained from $\mathscr{T}$ by composing it n times. By choosing n large enough we can ensure tnat the minimal distance τ^n between two vertices of $\mathscr{T}^{(n)}$ exceeds the diameter $d(\mathscr{A})$ of the patch $\mathscr{A}$. Then clearly $\mathscr{A}$ will contain at most one vertex of $\mathscr{T}^{(n)}$, and, suitably extending $\mathscr{A}$ if necessary, we may assume without loss of generality that it contains exactly one vertex V of $\mathscr{T}^{(n)}$. Let $\mathscr{N}(V)$ be the corresponding vertex neighborhood of $\mathscr{T}^{(n)}$. Then the union of the tiles in $\mathscr{N}(V)$ will contain the whole of the patch $\mathscr{A}$.

For any other tiling $\mathscr{T}_1$ by Penrose kites and darts, denote by $\mathscr{T}_1^{(n)}$ the tiling obtained by composing n times and let V_1 be any vertex of $\mathscr{T}_1^{(n)}$ such that the vertex neighborhood $\mathscr{N}(V_1)$ is congruent to $\mathscr{N}(V)$. Then decomposing $\mathscr{N}(V_1)$ n times will yield a patch of tiles congruent to that obtained by decomposing $\mathscr{N}(V)$ n times, and hence containing a patch congruent to $\mathscr{A}$. Since V_1 may be chosen from $\mathscr{T}_1^{(n)}$ in infinitely many ways, we deduce that $\mathscr{T}_1$ contains infinitely many patches congruent to $\mathscr{A}$, and so Statement 10.5.4 is proved.

Suppose that the minimal distance between two copies of a given patch $\mathscr{A}$ in $\mathscr{T}$ is denoted by $p(\mathscr{A})$. Then the above proof leads to a (crude) estimate for the maximum possible value of $p(\mathscr{A})$ in terms of the diameter $d(\mathscr{A})$ of $\mathscr{A}$. In any tiling $\mathscr{T}$ the maximum distance apart of two vertices of the same kind (that is, with the same kind of vertex neighborhood) is $3\tau + 2 = \tau^4$. This can easily be checked from the diagram of $\mathscr{C}_4$ (Figure 10.5.2(*e*)) remembering, from Statement 10.5.2, that every tile in every tiling lies inside a copy of $\mathscr{C}_4$. The maximum actually occurs when the chosen vertices are centers of stars. Consequently, in $\mathscr{T}^{(n)}$ nearby copies of patch $\mathscr{A}$ must contain points at a distance at most τ^{n+4} apart, where n is chosen as small as possible consistent with the inequality $\tau^n \geq d$. Then $p(\mathscr{A}) \leq \tau^{n+4} \leq \tau^5 d(\mathscr{A})$. The value of τ^5 is approximately 11.09.

Empirical evidence shows that the constant τ^5 in the above inequality is much too large. Penrose and Ammann independently claim that its value should be $\frac{1}{2}+\tau =$ 2.118034 . . . and that this is the best possible.

Statement 10.5.4 has many surprising implications. For example it shows that it is impossible to decide whether two tilings are identical or not by examining a finite portion of each. So a tiling is not uniquely determined by any patch of tiles, however large. This situation contrasts sharply with that described in Section 1.5 where examples were given of tilings which were determined by very small patches of tiles.

It also means that every diagram in this chapter of a tiling by the set P2 can be taken to represent *any* tiling by kites and darts in that it portrays a patch of it. We were justified in describing the tiling shown in Figure 10.5.1 as *the* cartwheel tiling only because we could describe, using the cartwheels of Figure 10.5.2, how it could be extended indefinitely as far as we like in a precisely determined manner.

10.5.5 *Every tiling by Penrose kites and darts is metrically prototile balanced.*

We recall from Section 3.6 that this is equivalent to the assertion that it is meaningful to speak about the *average* number of kites and the *average* number of darts per unit area of the plane. Moreover, our proof will show that the ratio of these two averages is τ, the golden section ratio.

Let us denote by kL_A and kS_A the results of enlarging the large and small A-tiles (L_A and S_A) by a factor k. Then from Figure 10.3.14 we see that we may write, symbolically,

$$\tau L_A = 2L_A + S_A,$$
$$\tau S_A = L_A + S_A,$$

to mean that each τL_A tile is composed of two L_A tiles and one S_A tile, and τS_A is composed of one L_A tile and one S_A tile. We deduce

$$\tau^2 L_A = 5L_A + 3S_A,$$
$$\tau^2 S_A = 3L_A + 2S_A,$$

and, by induction for all n,

$$\tau^n L_A = f_{2n+1}L_A + f_{2n}S_A,$$
$$\tau^n S_A = f_{2n}L_A + f_{2n-1}S_A, \qquad (10.5.6)$$

where f_n is the nth term in the sequence

1, 1, 2, 3, 5, 8, 13, 21, ...

This sequence is defined recursively by

$$f_{n+2} = f_{n+1} + f_n$$

and f_n is called the *nth Fibonacci number*. We note that the continued fraction expansion of $1/\tau$ is

$$\frac{1}{1+}\frac{1}{1+}\frac{1}{1+}\frac{1}{1+}\cdots$$

and that the nth covergent of this is f_n/f_{n+1}. Hence f_n/f_{n+1} tends to $1/\tau$ as $n \to \infty$ (in fact the convergence is very rapid).

Let S be any region of the plane that is covered *exactly* by copies of the tiles $\tau^n L_A$ and $\tau^n S_A$ from a $\tau^n A$-tiling. Then if $\rho(S)$ is the ratio of number of large A-tiles L_A to the number of small A-tiles S_A in such a region, equations 10.5.6 lead to

$$\frac{f_{2n}}{f_{2n-1}} \leq \rho(S) \leq \frac{f_{2n+1}}{f_{2n}} \qquad (10.5.7)$$

which shows that $\rho(S)$ is closely approximated by τ when n is large.

Now consider *any* convex region R of the plane, with area A and perimeter P. Since the longest side of $\tau^n L_A$ or $\tau^n S_A$ is τ^{n+1} we deduce that there exists a subregion S of R, of area at least $A - \tau^{n+1}P$ which is exactly covered by copies of the tiles $\tau^n L_A$ and $\tau^n S_A$ (see Figure 10.5.4). If P/A is small, inequalities 10.5.7 imply that the ratio of L_A tiles to S_A tiles (and therefore the ratio $\rho(R)$ of kites to darts) in R is also closely approximated by τ. Hence taking any sequence of regions R, increasing in size (so that $P/A \to 0$), we see that the ratio of kites to darts in such regions tends to τ. This is clearly enough to prove that every tiling by kites and darts is metrically prototile balanced and so Statement 10.5.5 is proved.

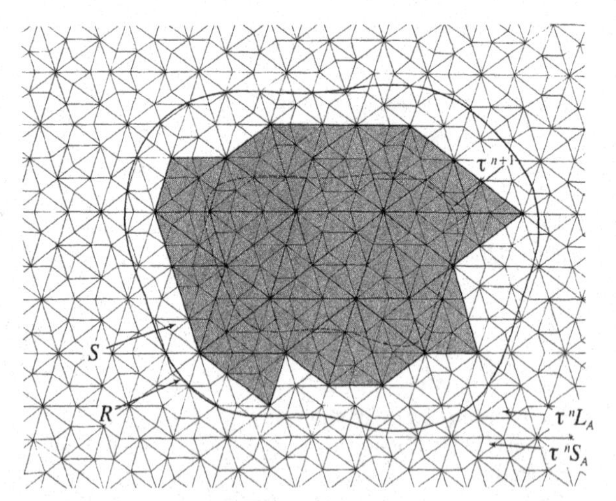

Figure 10.5.4
This diagram illustrates the proof that every tiling by Penrose kites and darts is metrically prototile balanced (Statement 10.5.5).

Penrose pointed out that the irrationality of τ, and therefore of the limit of $\rho(R)$ as R grows larger, provides an alternative proof that all tilings by kites and darts are non-periodic. This follows from the observation that in *any* periodic tiling (which is, of course, necessarily metrically prototile balanced) the ratio of the frequencies of the copies of the prototiles is necessarily rational.

We now return to a discussion of the cartwheel tiling. In Figure 10.5.5(a) we show two small patches of tiles known as *short* and *long bow ties*. A sequence of bow ties placed end to end is called a *Conway worm*. Examination of Figure 10.5.1(c) shows that in the cartwheel tiling Conway worms radiate from each of the ten sides of the first order cartwheel $\mathscr{C}_2$. In fact, each of these worms is of infinite length—a consequence of the fact that inflation of a worm leads to a set of tiles which contains a longer worm (that is, one containing more bow ties) running down its center. This will be clear from Figure 10.5.5(b) where we show the result of inflating the two sorts of bow ties—it will be seen how the new bow ties, or parts of bow ties, arise and how they fit together to form a longer worm. Of the ten worms that radiate from the sides of the cartwheel, six terminate and the other four join up in pairs across the cartwheel so as to form two endless worms which cross one another. Further remarks on the worms will be given shortly, but first we shall prove the following.

10.5.8 *Every tiling by Penrose kites and darts contains arbitrarily long finite Conway worms.*

This is an immediate consequence of the local isomorphism property (Statement 10.5.4). For we can find patches of tiles in the cartwheel tiling which contain arbitrarily long pieces of one of the endless worms that exist in that tiling—and local isomorphism then implies that replicas of such patches must occur in every other tiling.

It is easy to see that if a tiling contains two endless worms that cross then the tiling must be the cartwheel tiling. For this we first observe that whenever two worms cross they must have an ace in common, see Figure 10.5.6(a). Repeated inflation leads to new worms, as de-

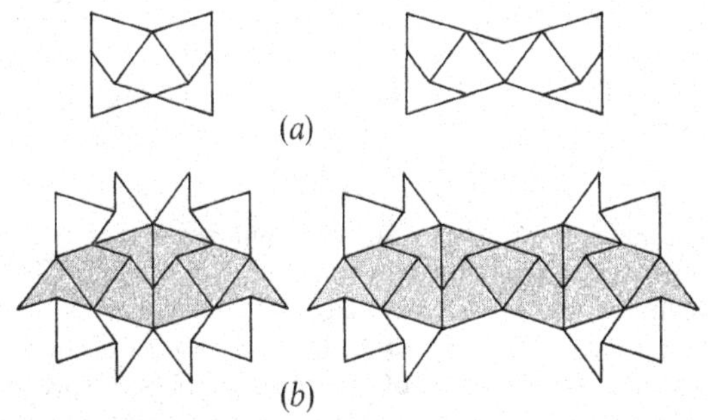

Figure 10.5.5
(*a*) Long and short bow ties. (*b*) Patches obtained by inflating the bow ties.

scribed above and also to an infinite sequence of concentric cartwheels of all orders. This proves that the tiling is the cartwheel tiling.

The configuration of the worms in a tiling repays careful study. Whenever two worms meet at an angle of 72° there are two possibilities—either they cross as in Figure 10.5.6(*a*), or one "interrupts" the other as in Figure 10.5.6(*b*). If the acute angle between the worms is 36° then one necessarily interrupts the other—it is impossible for them to cross. Since two endless worms cannot be parallel, this implies, of course, that no tiling can have more than two endless worms in it, so the cartwheel tiling is maximal in this respect. On the other hand, the local isomorphism property implies the existence of infinitely many finite worms, some of which are marked in Figure 10.5.7. At every point where two worms cross or one interrupts the other, they determine a first-order cartwheel centered at their point of intersection. In particular, the only way in which a worm W_1 can end is by its being interrupted by a worm or set of worms $\mathscr{W}$, and then another worm W_2 (a "continuation" of W_1) begins on the other side of $\mathscr{W}$, see Figures 10.5.6(*b*) and 10.5.7. We can visualize this situation by thinking of the tiling as being built up of rows of bow ties which "weave under and over each other"—the worms in the tiling are then the "visible" portions of these rows. In fact some of the rows are so fragmented by interruptions ("passing under" others) that the resulting worms are scarcely recognizable as such.

Although a worm is not laterally symmetric, it will be seen from the diagrams that the two zigzag sides are reflections of each other in the central line of the worm—this applies not only to the shape but also to the coloring of the vertices. Consequently every endless worm can be *turned* (that is, reflected in its center line) without violating the matching conditions. In the case of a finite worm the situation is more complicated for turning it moves the vertices on its ends and this necessitates certain modifications in the tiling if the matching conditions are to remain satisfied. Such modifications can be achieved by turning the worms that interrupt the given worm in a suitable manner. A full analysis of the situation is difficult, and, we believe, has never been carried out, though the theory of Ammann bars gives much information on

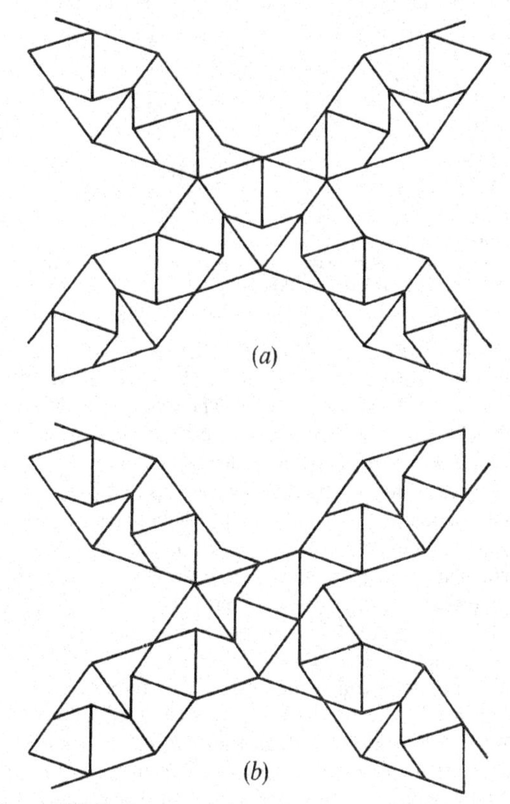

Figure 10.5.6
When two worms meet at 72° they either cross as in (*a*) and have an ace in common, or one interrupts the other as in (*b*).

Figure 10.5.7
Some worms in a tiling by Penrose kites and darts. All these worms can be turned simultaneously.

the subject (see Section 10.6). To illustrate some of the possibilities we show, in Figure 10.5.7, a tiling in which some worms have been marked. These have been chosen in such a way that they can be turned simultaneously even though many of them are finite.

The cartwheel tiling naturally leads us to the consideration of holes in a tiling. Let B be any bounded set, for example a circular disk, and let us denote by $\mathscr{U}$ any arrangement of Penrose kites and darts which tiles the plane except for part of the interior of B. The infinite patch will be called, for obvious reasons, a *tiling with a hole*, the subset of B not covered by tiles being the *hole*. It may, of course, be posssible to adjoin tiles to $\mathscr{U}$ in such a way as to fill up the hole completely, but in general this will not be possible. In the latter case we say that the hole is *essential* (see Figure 10.5.8).

There exists an equivalence relation between tilings with holes. We shall say that $\mathscr{U}_1$ and $\mathscr{U}_2$ are equivalent if one can be transformed into the other by removing, adjoining or rearranging any finite number of tiles, perhaps also reflecting the tiling in a line. Notice that in the definition we have not asserted that the hole must be connected—it may consist of any finite number of components—the definition of equivalence applies equally to holes of this kind. A fundamental problem is to enumerate the equivalence classes of holes. Conway has recently proved that there are exactly 61 such classes and we shall now describe these.

Figure 10.5.8
An example of an "essential" hole in a tiling. Notice that the Ammann bars, shown as heavy lines in the diagram, are out of alignment on the two sides of the hole.

Consider the cartwheel tiling of Figure 10.5.1(c), and remove all the tiles inside the first-order cartwheel $\mathscr{C}_2$. Then each of the semi-infinite worms radiating from the sides of the decagonal hole may be turned to either of two positions. This leads to $2^{10} = 1024$ tilings each with a 10-sided hole. Two such tilings will be equivalent if one can be obtained from the other by reflection or rotation, and it is easy to check that there are just 62 equivalence classes. Exactly one of these corresponds to the original cartwheel tiling and so to a hole that can be filled up. The other 61 possibilities are representative of all the other equivalence classes of tilings with essential holes.

We indicate briefly, without giving full details, how this result can be proved. In Section 10.3 the matching condition for the kites and darts was expressed in terms of Ammann bars (see Figures 10.3.26 and 10.3.27 and Section 10.6). It is easy to see that a hole is essential if and only if one or more bars on each side of the hole are out of alignment with the corresponding bars on the other side of it (see Figure 10.5.8). To complete the proof it is necessary to show that every essential hole is equivalent to one in which all these non-alignments occur within a first-order cartwheel. (If there are one or two non-alignments then it is easy to see that the hole is equivalent to one in which these non-alignments occur within a first-order cartwheel. If there are more non-alignments then the fact that these occur within the same cartwheel is a consequence of the fact, noted in Section 10.6, that the bars in two directions *almost* determine the bars in the other three directions uniquely.) From this we see that the holes described above are representative of the 61 equivalence classes.

The 61 types of essential holes can be reduced in size. This is done by adding an ace inside each of the 10 edges of the 10-sided hole. In this way the hole is reduced to the shape of a *decapod*—that is, the union of ten L_B tiles with one vertex in common, see Figure 10.5.9(a).

Of the 62 decapods, the one that can be filled up is shown in Figure 10.5.9(b) and is called "Batman". Examples of some of the other 61 are shown in Figure 10.5.9(c) together with names that have been suggested for them. If we think of each decapod as a large "foreign" tile introduced into a tiling, then 60 of them force a unique tiling

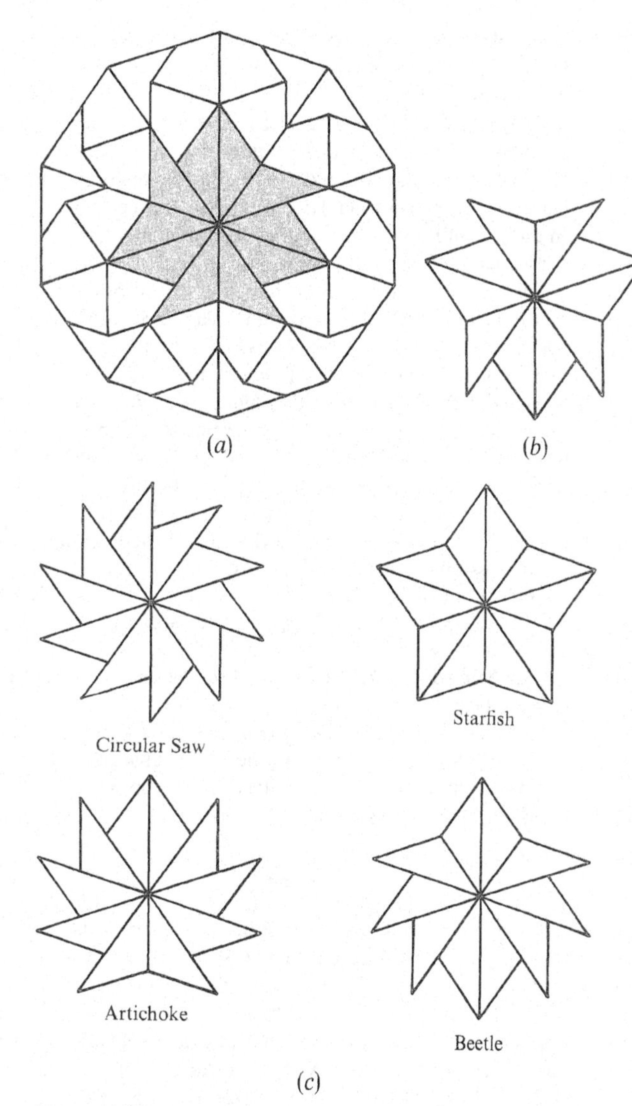

Figure 10.5.9
Examples of decapods. In (a) we show how a decagonal hole can be reduced to a decapod by introducing ten kites. The decapods of (a) and (b) are called Asterix and Batman, respectively. In all there are 62 different decapods.

of the plane—that is, there is only one way in which the rest of the plane can be covered by Penrose kites and darts. This is true even though no matching conditions are imposed on the boundary of the decapod by, for example, coloring its vertices. Apart from these 60 the remaining two (Batman and Asterix, both named after famous cartoon characters) do not force a unique tiling. In the case of Batman this is, of course, an immediate consequence of the local isomorphism property.

The final topic that we shall discuss in this section is that of index sequences. These can be defined for any of the aperiodic tilings we have discussed so far, though here we shall consider in detail only tilings by Penrose kites and darts. The importance of index sequences lies in the fact that they enable the geometric and combinatorial properties of the tilings to be expressed in numerical form, so that algebraic and number-theoretic methods become available. Here we can only give definitions and indicate some of the ideas—there remain many possibilities for investigations. Our treatment is adapted from Robinson [1975].

We begin with a tiling $\mathscr{T}$ by Penrose kites and darts and write $\mathscr{T}_A$ for the corresponding A-tiling (see Section 10.3 and Figures 10.3.15(*a*) and (*b*)). Successive compositions yield a B-tiling $\mathscr{T}_B$, a $\tau A'$-tiling $\mathscr{T}_{\tau A'}$, a $\tau B'$-tiling $\mathscr{T}_{\tau B'}$, and so on, as shown in Figure 10.3.15. For any point P which does not lie on an edge of $\mathscr{T}_A$ the index sequence $i(\mathscr{T},P)$ is defined as an infinite sequence of integers

$$(x_0, x_1, x_2, \ldots)$$

where $x_0 = 1$ or 0 according as P lies in a small or large tile of $\mathscr{T}_A$; $x_1 = 1$ or 0 according as P lies in a small or large tile of $\mathscr{T}_B$; $x_2 = 1$ or 0 according P lies in a small or large tile of $\mathscr{T}_{\tau A'}$; and so on. Each composition yields a new term in the sequence which is thus uniquely defined. Since a small A-tile does not form part of a small B-tile, and a small B-tile does not form part of a small $\tau A'$-tile (see Figure 10.3.14), we deduce that in the index sequence no two consecutive terms can be equal to 1. There are no other restrictions on the way that the terms are arranged. In fact we have the following general result.

10.5.9 *Every infinite sequence whose terms are* 0 *or* 1 *and in which no two consecutive terms are equal to* 1, *is the index sequence* $i(\mathscr{T},P)$ *of some point* P *in some tiling* $\mathscr{T}$ *by Penrose kites and darts.*

The proof of Statement 10.5.9 follows from the fact that the index sequence gives explicit instructions as to how the tiling $\mathscr{T}_A$ (and therefore $\mathscr{T}$) is to be constructed. For example, if the sequence begins $i(\mathscr{T},P) = (0, 0, 1, 0, \ldots)$ then $x_0 = 0$ shows that P must lie in a large A-tile, $x_1 = 0$ shows that this A-tile must lie in a large B-tile, $x_2 = 1$ shows that this B-tile must lie in (in fact be equal to) a small $\tau A'$-tile, and so on. In this way we obtain a nested sequence of tiles which, in an obvious manner, determine either a tiling of the whole plane or of an unbounded patch of tiles containing P. In the latter case, the union of the tiles in the patch is necessarily either a half-plane or a sector with angle $\pi/5$ and the tiling can be completed by reflecting in the lines forming the boundaries of the patch. We deduce the following.

10.5.10 *If* σ *is a symmetry of a tiling* $\mathscr{T}$, *then the index sequences* $i(\mathscr{T},P)$ *and* $i(\mathscr{T},\sigma P)$ *are identical. Conversely, if index sequences* $i(\mathscr{T},P)$ *and* $i(\mathscr{T}',P')$ *are identical then* $\mathscr{T} = \mathscr{T}'$ *and* P *either lies in the same* A*-tile as* P' *or is an image of such a point under a symmetry of* $\mathscr{T}$.

From the definition of an index sequence further properties can be derived, such as the following.

10.5.11 *The index sequences* $i(\mathscr{T},P_1)$ *and* $i(\mathscr{T},P_2)$ *of any points* P_1, P_2 *in the same tiling* $\mathscr{T}$ *differ only in a finite number of terms. Conversely, if two index sequences differ in a finite number of terms, then they correspond to two points in the same tiling.*

From Statements 10.5.9 and 10.5.11 we see that there is a one-to-one correspondence between tilings $\mathscr{T}$ by Penrose kites and darts and classes of index sequences—two such sequences belong to the same class if they eventually agree term by term. This and the fact that there exists an uncountable infinity of sequences using the digits 0 and 1, with no consecutive terms equal to 1, enables us to deduce:

10.5.12 *There exists an uncountable infinity of distinct tilings by Penrose kites and darts.*

On the other hand if two index sequences $i(\mathcal{T},P)$ and $i(\mathcal{T},P')$ agree term by term for the first n terms, then the method of construction explained in the proof of Statement 10.5.6 tells us that some patch of tiles containing P in $\mathcal{T}$ must be identical with some patch of tiles containing P' in $\mathcal{T}'$. The size of these patches depends on the value of n. This property is clearly related to that of local isomorphism.

As illustrations of the above statements we refer to Figure 10.3.15. Two points are marked, one by x and the other by •. The index sequence of the former is (0, 1, 0, 0, 1, . . .) and of the latter is (1, 0, 1, 0, 1, . . .). It is clear that these sequences differ only in their first three terms, in accordance with Statement 10.5.11.

Examining the index sequences of other tilings by kites and darts enables us to identify some tilings which are of interest.

10.5.13 *A tiling by Penrose kites and darts has symmetry group d5 with a sun at its center (Figure* 10.3.20) *if and only if the index sequence of any point ultimately agrees term by term with*

(0, 0, 1, 0, 0, 0, 1, 0, 0, 0, 1, 0, 0, 0, 1, 0, . . .).

If, on the other hand, it has a star at its center, then the index sequence of any point must ultimately agree term by term with

(1, 0, 0, 0, 1, 0, 0, 0, 1, 0, 0, 0, 1, 0, . . .).

10.5.14 *A tiling by Penrose kites and darts is the cartwheel tiling if and only if the index sequence of any point has only a finite number of non-zero terms—that is, if it ultimately agrees term by term with*

(0, 0, 0, 0, 0, . . .).

Statement 10.5.14 emphasises the very special nature of the cartwheel tiling and characterizes it in a very elegant and unexpected way. It is also the starting point of a theory of indexing the tiles recently developed by Fred Lunnon in connection with the drawing of tilings by computer. Some information on this is given in the Exercise 10.5.4.

All that we have said about index sequences can be applied, with very little modification, to tilings by the Penrose set P3 also. Here, of course, the first term in the index sequence signifies whether the given point belongs to a large or a small B-tile. We leave details of this case to the reader.

EXERCISES 10.5

*1. Characterize the index sequences of tilings by kites and darts which have symmetry group $d1$.

*2. Define an index sequence for a point P in a tiling by the aperiodic set P1. (Here there will be six different terms in the sequence corresponding to the six prototiles, but there will be restrictions on the order in which these can occur.) Find a necessary and sufficient condition on the sequence for the tiling to have symmetry group $d5$.

**3. Define index sequences for each of the Ammann tilings defined in the previous section and investigate their properties.

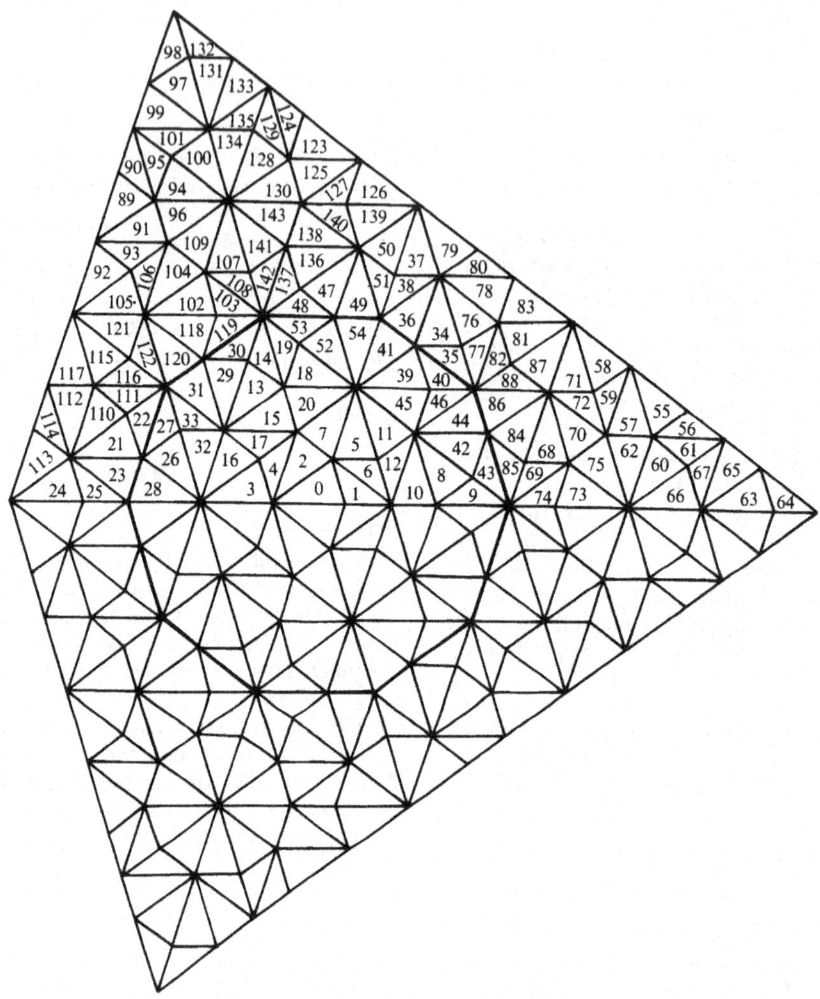

Figure 10.5.10
An *A*-tiling corresponding to part of the cartwheel tiling. The numbers are the indexes of the tiles defined as in Exercise 10.5.4.

**4. If P is any point in a tile T of a cartwheel tiling $\mathscr{T}$, and $i(\mathscr{T},P) = (x_0,x_1,x_2,\ldots)$ then we define the *index* $i(T)$ of the tile T to be

$$i(T) = \sum_{i=0}^{\infty} x_i f_{i+2},$$

where f_i is the ith Fibonacci number ($f_1 = 1, f_2 = 1, f_3 = 2, f_4 = 3, f_5 = 5$ and, generally, $f_{i+2} = f_{i+1} + f_i$).

In Figure 10.5.10 we show some of the indexes for tiles of $\mathscr{T}$. (This approach is due to Fred Lunnon.)

(*a*) Show that two distinct tiles have the same index if and only if one is the image of the other under reflection in the line of symmetry of $\mathscr{T}$.

(*b*) Show that every non-negative integer occurs as the index of some tile (and therefore of exactly two tiles) in $\mathscr{T}$.

(*c*) Let $\mathscr{T}_{\tau A'}$ be the tiling obtained by composing the

A-tiles of $\mathscr{T}_A$ twice to form $\tau A'$-tiles, and let $i'(T')$ be the index of a tile T' in $\mathscr{T}_{\tau A'}$. Prove that

$$i(T') = [\tau^2 \min_{T \subset T'} i(T)]$$

where τ is the golden number and the square brackets signify the integer part.

**5. Complete the proof of the statement that there are exactly 61 equivalence classes of holes, as given in outline in this section.

*6. Let $v_1(R), v_2(R), \ldots, v_r(R)$ be the numbers of vertices of each of the seven kinds, of a given tiling $\mathscr{T}$, that lie inside a circular disk of radius R. Show that as $R \to \infty$ the ratios of these seven numbers tend to definite limits (independent of the choice of $\mathscr{T}$) and determine the values of these ratios.

7. Using repeated composition, prove the assertion (made on page 565) that two endless worms cannot be parallel.

10.6 AMMANN BARS, MUSICAL SEQUENCES AND FORCED TILES

In Sections 10.3 and 10.4 we defined Ammann bars and showed how they can be used to specify the matching conditions for the aperiodic sets P2, A2 and A3. They can also be used in a similar manner for the sets P1 and P3, see Exercise 10.3.3. The results of this section have analogues for each of these five sets though, as in the previous section, we choose to formulate them in terms of the Penrose kites and darts.

Let $\mathscr{A}$ be a given patch of tiles; then it is easy to see that in many cases some adjacent tiles are *forced* by $\mathscr{A}$. In other words, however $\mathscr{A}$ is extended these tiles necessarily lie in certain predetermined positions. The simplest example has already been mentioned: given any dart, two kites are forced to fit against its shorter edges. A more elaborate example is shown in Figure 10.6.1. Here $\mathscr{A}$ is taken to be the vertex neighborhood called a jack, and this forces the gray tiles to lie in the positions shown. Thus the original five tiles force eleven others.

It is not so obvious, however, that a given patch can also force tiles to lie in non-adjacent positions. The existence of such forced tiles was discovered independently by Roger Penrose and Clive Bach. Following Conway's terminology, the set of tiles (both adjacent and non-adjacent) forced by a given patch $\mathscr{A}$ is called its *empire*.

The problem of finding the empires associated with certain small patches of tiles (the vertex neighborhoods) is the main topic of this section. In Figure 10.6.6 we show parts of the empires associated with the star, deuce, jack, queen and king; on the other hand, the ace and sun force no tiles at all. The diagrams are based on informa-

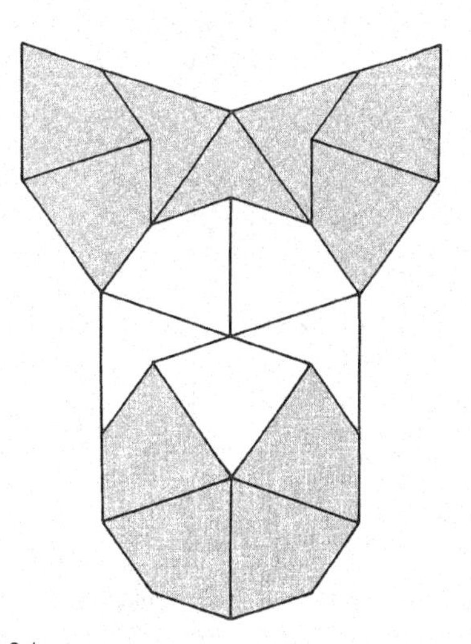

Figure 10.6.1
The vertex neighborhood known as a jack and the eleven adjacent forced tiles (shown gray).

tion supplied by Robert Ammann, and we are grateful to him for allowing us to reproduce them here. We must also express our gratitude to John H. Conway for his invaluable help and advice in presenting the material of this section.

As an approach to the problem we consider the configuration of Ammann bars in some detail. In Figure 10.3.27 we showed the Ammann bars determined by a certain tiling; from our present point of view it is the converse which is important, namely that the bars determine the tiling uniquely. This becomes apparent if we check the possible configurations of tiles and bars that can arise in a tiling. For example in Figure 10.6.2 we show a set of bars and note that whenever a pentagon like P occurs it must have arisen from a star of five darts, and whenever a pentagram like Q occurs it must have arisen from a sun of five kites. More generally a kite appears in the tiling when and only when three bars form a small triangle like one of the five in the pentagram Q. In fact such a triangle always occurs whenever two bars meet at an angle of 36°. Further, when four bars form a small rhomb such as that

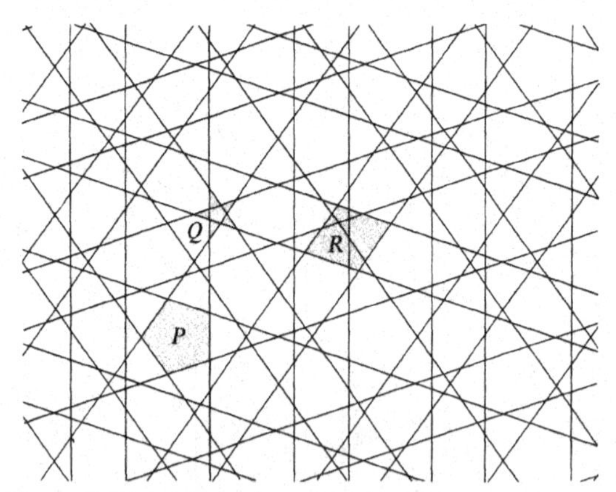

Figure 10.6.2
A set of Ammann bars. From these the original tiling by Penrose kites and darts can be reconstructed. A pentagon P corresponds to a star of five darts and a pentagram Q to a sun of five kites. Whenever the bars form a rhomb like R, two other bars must cross in the center as shown and a third bar must cut off one of the obtuse angles.

shown at R, two other bars must cross at its center, and a third bar must cut off one of its obtuse angles. We see from these considerations that bars in just two directions place strong constraints on those in the other three directions. In fact, Conway has shown that they *almost* determine them uniquely; at most one bar in each of the other three directions is not determined—this "indeterminate" bar runs along a worm that can be turned.

If we consider just one set of parallel bars then the distances between two adjacents bars take only two different values. Thus a perpendicular transversal is cut by the bars into long and short intervals denoted by L and S, the former being τ (the golden ratio) times as long as the latter, see Figure 10.6.3. At first sight the intervals may appear to be mixed randomly but on closer inspection it will be seen that this is not so—two S intervals can never be adjacent, nor can three L intervals. Any such sequence of S and L intervals that is obtained from the Ammann bars of a tiling by Penrose kites and darts is known as a *musical sequence*. We shall now investigate the properties of such sequences.

To begin with, suppose that we compose a tiling as in Figure 10.6.3, and denote the intervals that arise from the new Ammann bars by L' and S'. Then symbolically we can write

$$L' = (\tfrac{1}{2}L,S,\tfrac{1}{2}L),\qquad S' = (\tfrac{1}{2}L,\tfrac{1}{2}L) \tag{10.6.1}$$

to indicate how the sequence of L' and S' intervals is related to the L and S intervals. The two terms $\frac{1}{2}L$ which arise from adjacent intervals in the composed tiling are to be combined to form a single term L, as will be apparent from Figure 10.6.3. We shall say that the sequence of L' and S' intervals is obtained by *composing* the sequence of L and S intervals since it is analogous to, and obtained from, the composition of the underlying tiling. Decomposition of the sequences is defined in a similar way.

10.6.2 *If we compose or decompose a musical sequence we obtain a musical sequence.*

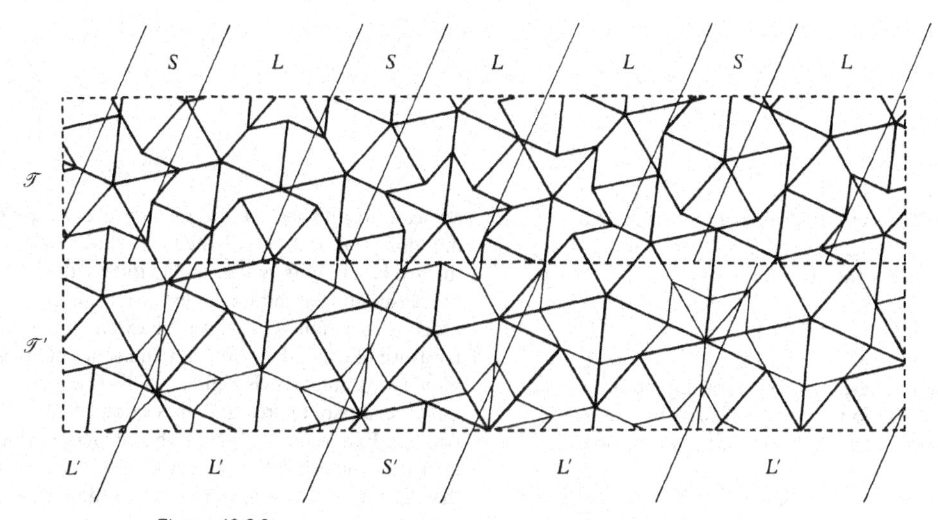

Figure 10.6.3
This diagram shows the relationship between the Ammann bars of a tiling $\mathscr{T}$ and those of the composed tiling $\mathscr{T}'$. It will be seen that the latter bisect the longer intervals between the bars of $\mathscr{T}$.

In the earlier sections of this chapter we used composition and decomposition to establish many properties of aperiodic tilings. Composition and decomposition of musical sequences enable us to establish one-dimensional analogues of these. For example, we have:

10.6.3 *Every musical sequence is non-periodic—that is, it cannot be written as an infinity of repetitions of a finite block of terms.*

10.6.4 *Musical sequences exhibit the property of local isomorphism. By this we mean that any finite block of consecutive terms in any musical sequence occurs infinitely often in every other musical sequence.*

Not only can these statements be proved by methods completely analogous to those used for tilings (details of which are requested in the exercises) but the interrelationship between musical sequences and tilings leads to other results. In particular we mention the following:

10.6.5 *Given any two sets of parallel Ammann bars (each corresponding to a musical sequence of L and S intervals) making an angle of either 36° or 72° with each other, there exists a corresponding tiling by Penrose kites and darts.*

Let $\mathscr{T}$ be any tiling with Ammann bars parallel to the given ones. Select any finite block of intervals from one of the given musical sequences. Then by Statement 10.6.4 this will be repeated (infinitely often) as a block of intervals between the parallel bars of $\mathscr{T}$. Thus if we select two such blocks, one from each of the given sequences, it will be possible to realize them as the intervals between the bars (in the given directions) determined by a finite patch of tiles of $\mathscr{T}$. There will, moreover, be only a finite number of possibilities for such a patch.

Now let us increase the sizes of the blocks of intervals chosen from the musical sequences, and so obtain larger and larger patches of tiles in $\mathscr{T}$. Finiteness implies that we can apply König's Lemma as in the proof of the Extension Theorem for Wang tiles given in the next chapter (Statement 11.2.1), and so use these patches to construct a tiling with the required properties.

Musical sequences were defined in terms of Ammann bars of Penrose tilings, and it is natural to ask whether they can be characterized independently. For an *infinite* sequence this question will be effectively answered by the analytical representation given in Statement 10.6.9 below. But already it is possible to give a very simple rule for deciding whether a given *finite* block of terms (*L*'s and *S*'s) belongs to a musical sequence or does not. We make use of the fact that composition of a musical sequence leads to a sequence which is also musical. For example, consider the two blocks

$L\,S\,L\,L\,S\,L\,S\,L\,L\,S$ and $L\,S\,L\,L\,S\,L\,S\,L\,S\,L.$

Composition corresponds to decreasing the size of the blocks according to the rules: replace S by L, $L\,L$ by S and omit single terms L. In the two cases we obtain

$L\,S\,L\,L\,S\,L$ and $L\,S\,L\,L\,L$

$L\,S\,L$ ××××××

L

Since the first block becomes, after three reductions, a single term (in this case L), we deduce that the given block is part of a musical sequence. But in the second case this is not so since after one reduction we obtain a sequence of five terms involving three consecutive L's. This is not part of a musical sequence and so the given block is itself not part of a musical sequence.

In the last section we saw the importance of the cartwheel tiling in the study of the Penrose kites and darts—here we introduce the analogous musical sequence. We recall that the cartwheel tiling arose by repeatedly inflating an ace. In the case of sequences we repeatedly inflate the block $L\,S$ using the substitutions of equations 10.6.1 to obtain the following:

$$
\begin{array}{c}
L\,.\,S \\
L\,S\,.\,L\,L \\
L\,S\,L\,L\,.\,S\,L\,S\,L \\
L\,S\,L\,L\,S\,L\,S\,.\,L\,L\,S\,L\,L\,S\,L \\
L\,S\,L\,L\,S\,L\,S\,L\,L\,S\,L\,L\,.\,S\,L\,S\,L\,L\,S\,L\,S\,L\,L\,S\,L \\
\cdots\cdots\cdots\cdots\cdots\cdots
\end{array}
\qquad (10.6.6)
$$

The dot, or decimal point, marks the "center" of each block of intervals, and each block is symmetric about this point except for the two intervals ($L\,.\,S$ or $S\,.\,L$) adjacent to it. These two terms are reversed by each inflation. For each $n \geq 1$ the $(n + 2)$nd block contains the nth block in its center, so the third, fifth, seventh, . . . blocks are longer and longer extensions of the original block $L\,.\,S$. It follows that we can extend the sequence as far as we like in either direction and so a doubly infinite sequence of intervals is defined. (This is an inclusion limit—compare the remarks in Section 3.8 in connection with the Extension Theorem.) This sequence, which is invariant under inflation except for the interchange of the central intervals, has been called the *middle-C sequence* by Conway on account of the musical analogy. In Figure 10.6.4 we show the Ammann bars determined by the first-order cartwheel, and it will be seen that in each of the five directions the bars define a short and a long interval. From this it follows that the sequence determined by any one of the five sets of bars in a cartwheel tiling is a middle-C sequence.

It is useful to have a method of generating a musical sequence analytically. For this we can proceed as follows.

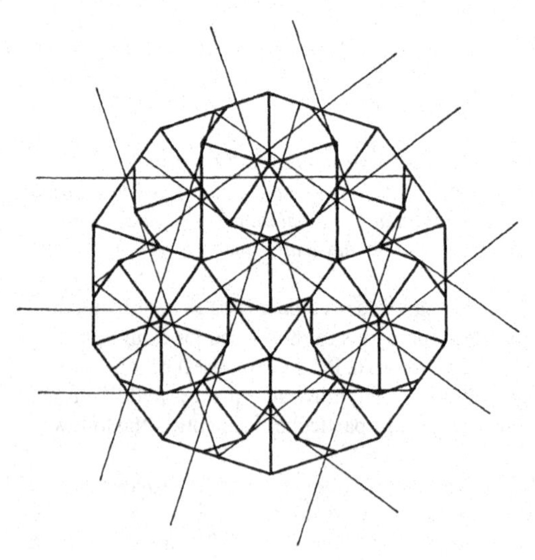

Figure 10.6.4
The Ammann bars on a first-order cartwheel.

Consider Table 10.6.1. In the first row of the table we have the sequence of integers, and in the second row τn correct to 2 decimal places. In the third row $[\tau n]$ represents the integer part of τn, and in the fourth row these terms are differenced. Finally, in the fifth row we substitute S and L for 1 and 2, respectively. Extending the table it will be seen that the last row consists of exactly the same terms as the middle-C sequence defined above. This is necessarily so for we can show that it is invariant under inflation (except for the interchange of the two central terms). Here "inflation" corresponds to multiplying by τ, which yields the sequence of letters shown in Table 10.6.2. The first five rows correspond to those of Table 10.6.1 and the last three rows correspond to the substitutions

$$3 = (+1, 1, 1+),$$

$$2 = (+1, 1+).$$

Combining the terms with the + signs as shown, it will be seen that this is the numerical counterpart of the substitutions defined in equations (10.6.1.) The fact that $\tau^2 = \tau + 1$ enables us to identify the inflated sequence with the original one and so establish the above assertions. We have proved:

Table 10.6.1 THE ANALYTIC GENERATION OF A MUSICAL SEQUENCE

$n =$	-3		-2		-1		0		1		2		3	
$\tau n =$	-4.86		-3.24		-1.62		0		1.62		3.24		4.86	
$[\tau n] =$	-5		-4		-2		0		1		3		4	
Difference		1		2		2	•	1		2	•	1		
Substitution		S		L		L		S		L		S		

Table 10.6.2 INFLATION OF A MUSICAL SEQUENCE GENERATED ANALYTICALLY

$n =$	-3		-2		-1		0		1		2		3
$\tau^2 n =$	-7.86		-5.24		-2.62		0		2.62		5.24		7.86
$[\tau^2 n] =$	-8		-6		-3		0		2		5		7
Difference		2		3		3	•	2		3		2	
Substitution		S'		L'		L'		S'		L'		S'	
	+ 1		1 + 1	1	1 + 1	1	1 + 1		1 + 1	1	1 + 1		1 +
	2		2	1	2	1	2		2	1	2		2
	L		L	S	L	S	L		L	S	L		L

10.6.7 *The analytical procedure described above and illustrated in Table 10.6.1 yields the middle-C sequence.*

From this analytical representation of a musical sequence many facts follow immediately. In most cases the proofs are completely straightforward and details are left to the reader.

10.6.8 *Suppose that there are* $x(k)$ *terms L and* $k-x(k)$ *terms S in a block of k consecutive terms in a musical sequence. Then*

$$\lim_{k\to\infty} \frac{x(k)}{k - x(k)}$$

exists and has the value τ *(the golden number). Alternatively, and equivalently,*

$$\lim_{k\to\infty} \frac{x(k)}{k} = \frac{1}{\tau}.$$

10.6.9 *Any musical sequence can be defined by differencing the sequence* $[\tau n+c]$ *as n runs through the integers, and then substituting S and L for 1 and 2 respectively. Here c is any real number; without loss of generality it can be chosen to satisfy* $0 \leq c < 1$.

To prove this we observe from Statement 10.6.4 and Table 10.6.1 that any *finite* block of intervals from a musical sequence can be written in the form $[\tau n+c_1]$ for a suitable value of c_1 satisfying $0 \leq c_1 < 1$. Taking larger and larger blocks from the given sequence leads to a sequence of values of c_1 which must, by a simple application of compactness, contain a subsequence which converges to c. This is the number required in the expression $[\tau n + c]$.

10.6.10 *For any value of k, the number* $x(k)$ *of L intervals defined in Statement 10.6.9 can take only two values, namely x and* $x + 1$ *where*

$$\frac{x}{k} < \frac{1}{\tau} < \frac{x+1}{k}. \qquad (10.6.11)$$

Thus the possible numbers of L intervals in a block of k consecutive intervals in a musical sequence arise as the numerators of those fractions with denominator k that approximate $1/\tau$ most closely from above and below. It follows that the *total length* of a block of k consecutive intervals (that is, the sum of the lengths of those intervals) can take only two different values, the smaller and larger of which we denote by $s(k)$ and $l(k)$ respectively.

For any given value of k, the analytical representation and the local isomorphism property enable us to write down all possible blocks of k consecutive intervals that occur in any musical sequence. For example, when $k = 4$, there are five such blocks, namely

L S L L .

S L L . S

L L . S L

L . S L S

. S L S L

and these are found as the blocks containing, or adjacent to, the decimal point in the middle-C sequence. Here it will be noticed that $k = 4$ and $x(k) = 2$ or 3 so that, in accordance with inequalities (10.6.11) we have

$$2/4 < 1/\tau < 3/4.$$

For every k there are just $k+1$ different blocks obtained from the middle-C sequence in the above manner, and it will be noticed that the first and last of these (the blocks adjacent to the decimal point) correspond to the lengths $s(k)$ and $l(k)$ or vice versa.

In the previous section we mentioned that $1/\tau$ is closely approximated by fractions f_n/f_{n+1} where f_n is the nth Fibonacci number. It is therefore hardly surprising that blocks of length k, where k is a Fibonacci number, have rather special properties. The following blocks, of the indicated lengths

$l(1)$: *L*

$s(1)$: *S*

$l(2)$: $L\ L$

$s(3)$: $S\ L\ S$

$l(5)$: $L\ L\ S\ L\ L$

$s(8)$: $S\ L\ S\ L\ L\ S\ L\ S$

. . . .

occur alternately to the left and right of the decimal point in the middle-C sequence. Empirical evidence suggests that all such blocks are not only centrally symmetric, but they are also unique in the sense that no other blocks (with the same number of intervals) have the same lengths.

The study of these musical sequences is of considerable interest in its own right, but here we are concerned with their application to finding forced tiles in a tiling by Penrose kites and darts. For this purpose we note that whenever two or more parallel bars are given, then other bars are fixed by them. To see how this can occur, see Figure 10.6.5. We suppose that we are given two bars b_1 and b_2 at distance L apart. The next interval can be L or S so the bar b_3 can lie in one of two positions, but the next bar b_4 is determined since it must lie at distance $l(3)$ from the first bar. This follows because 3 is a Fibonacci number so there is only one block of length $s(3)$, namely SLS, and this is excluded because we know that the first interval is L. Similarly b_5 and b_6 can each lie in one of two positions, but b_7 is determined since it must lie at distance $l(6)$ from b_1. This is a consequence of the fact that the only blocks of length 6 which begin with an L contain four L intervals and so are of length $l(6)$. Continuing in this way we see that b_8 is not determined, but b_9 is, and so on.

It follows that whenever we are given a patch of tiles which determines at least two bars in a given direction then other bars are necessarily determined, and hence some tiles may be forced. As a simple example, consider the deuce in Figure 10.6.6(*b*). Then the bars intersecting the dark gray region are those immediately determined by the tiles in the deuce and those adjacent tiles that are

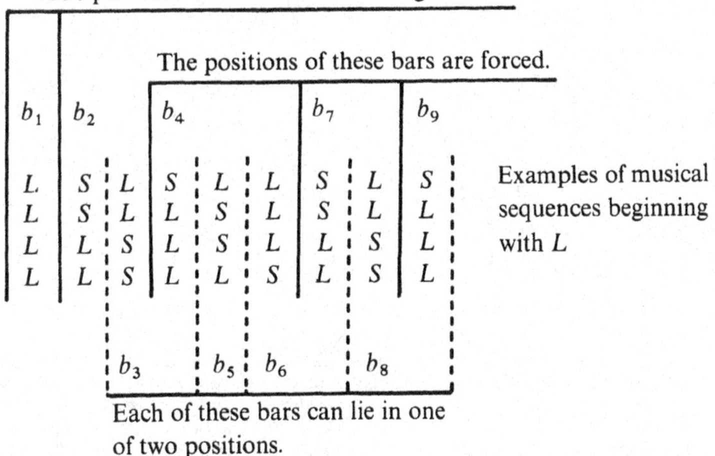

Figure 10.6.5
The Ammann bars that are determined by two given bars with the interval L between them.

forced by them. These determine, as above, systems of bars (in two directions) shown by solid lines. The small triangles that result force kites (colored light gray) to lie in the positions shown. No other tiles are forced by these bars and so the deuce's empire consists of a collinear set of kites (some of which appear in the figure) stretching to infinity in both directions.

For the other vertex neighborhoods the situation is more complicated since bars in more than two directions are determined by the given tiles and by the adjacents that are immediately forced by them. In these cases it seems better to consider those bars which are *not* forced. (In Figure 10.6.6 we follow the convention that solid bars are forced and those shown dashed are not.) Each of the bars which is not forced can occupy just one of two positions, and moving it from one position to the other corresponds to turning a row of bow ties (visible as worms) as described in the previous section. The tiles belonging to these worms are not forced and the remaining tiles, shown shaded in the diagrams, form the required empire. In the figure only a small part of the empire can be shown (namely that lying close to the vertex neighborhood which determines it). In each case the empire extends to infinity. It is surprising how many tiles are forced by each of the vertex neighborhoods and the king's empire is truly remarkable in this respect.

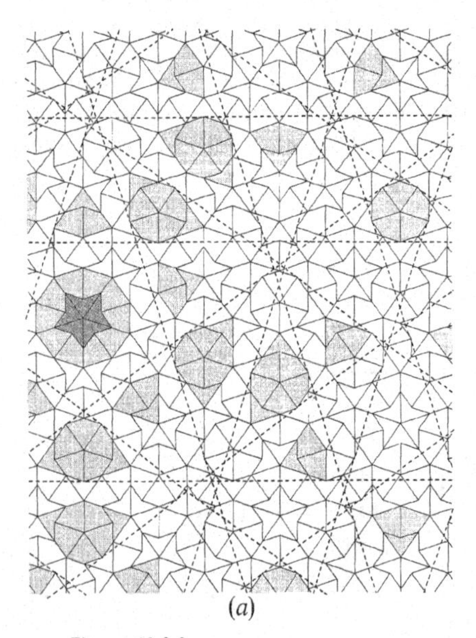

(*a*)

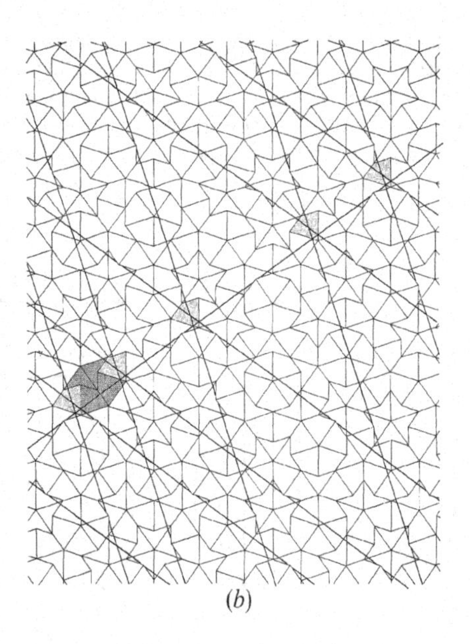

(*b*)

Figure 10.6.6
The empires (forced tiles) associated with vertex neighborhoods. In (*b*) we show the empire of the deuce. It will be seen to consist of a collinear set of kites. The solid lines are the Ammann bars which are fixed in position by the original patch. A kite is forced wherever three bars meet in a small triangle. In the other parts of the figure—(*a*) star, (*c*) jack, (*d*) queen, (*e*) king—the bars which are *not* fixed are shown by dashed lines (all others are fixed). Each dashed line therefore corresponds to a worm that can be turned. The forced tiles are those that do not lie on any of these worms. The other two vertex neighborhoods, the ace and the sun, do not force any tiles. This diagram is based on information supplied by Robert Ammann.

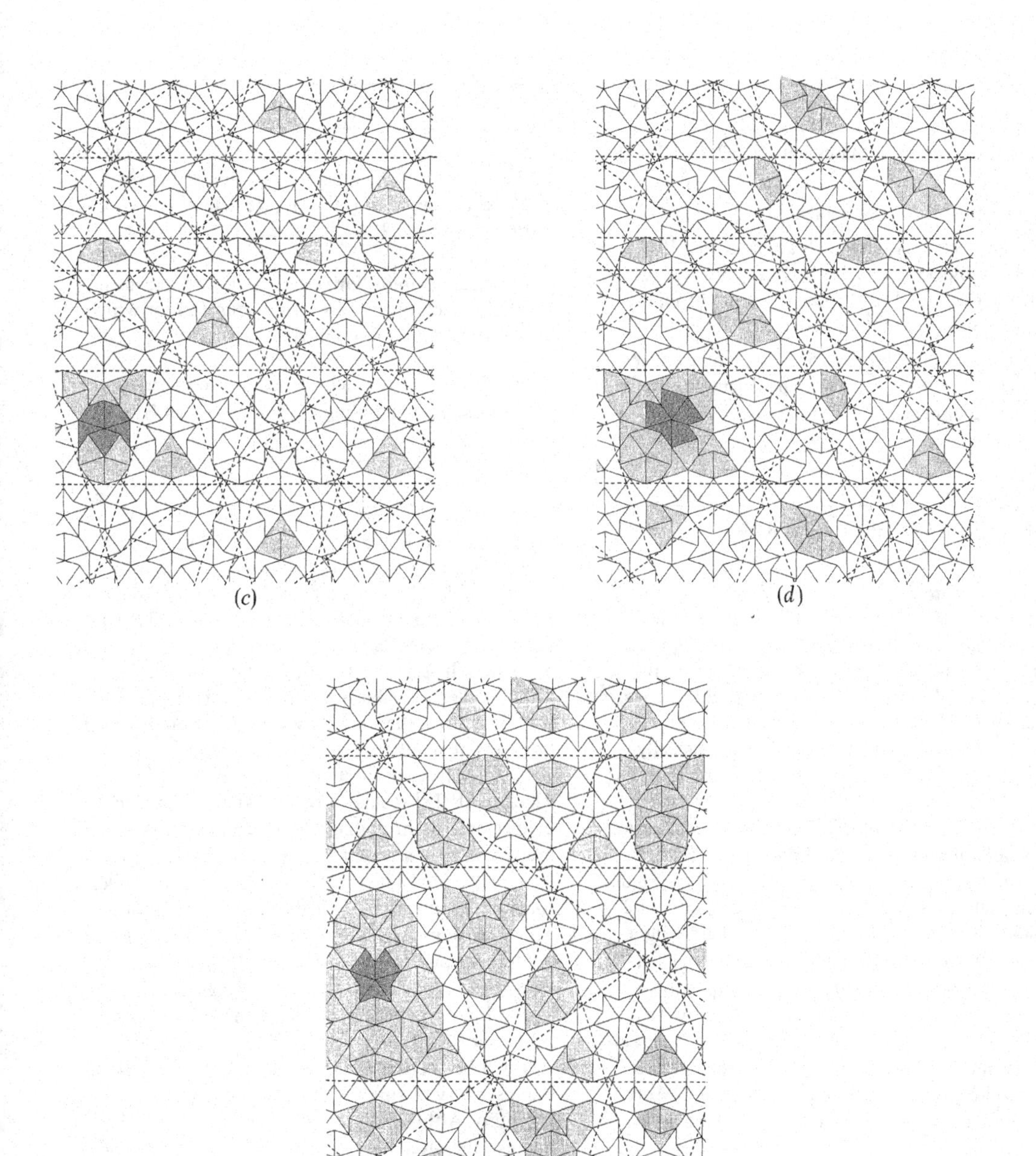
(c)
(d)
(e)

EXERCISES 10.6

*1. Give detailed proofs of Statements 10.6.2, 10.6.3 and 10.6.4.

*2. Complete the proof of Statement 10.6.5.

**3. Verify the assertion following Statement 10.6.10 that all possible blocks of k terms from any musical sequence can be found containing, or adjacent to, the decimal point in a middle-C sequence.

**4. Prove the assertions on page 577 concerning blocks of terms in a musical sequence whose lengths are the Fibonacci numbers.

**5. Construct the empires of some patches of tiles larger than the vertex neighborhoods discussed in the text. For example, what is the empire determined by a first-order cartwheel?

**6. Construct empires for vertex neighborhoods using tiles of each of the sets (other than P2) for which Ammann bars are defined.

*7. What is the empire determined by a cornered cross (just one tile) in a tiling by Robinson's set R1?

10.7 NOTES AND REFERENCES

The first attempts to discover aperiodic sets of prototiles were concerned entirely with Wang tiles (square tiles with colored edges). In the next chapter we shall give an account of these and explain how it is possible, by replacing the colored edges of the tiles by suitable curves, to convert an aperiodic set of n Wang tiles into an aperiodic set of n *unmarked* tiles, each of which is basically square. However, these sets are of little interest since they contain unnecessarily large numbers of tiles; with some ingenuity it is sometimes possible to be more economical and so obtain much smaller aperiodic sets.

For example, the Läuchli set of forty Wang tiles (Figure 11.1.3) can yield an aperiodic set of eight tiles in the following way. First reverse the colors (indicated by light and heavy lines) on the right and upper edges of each tile. The matching condition has to be altered appropriately—abutting edges on adjacent tiles must now bear unlike colors. After this modification it turns out that all forty tiles can be obtained from just eight square tiles, suitably marked, using rotations and reflections. Moreover it is easy to convert these into eight unmarked basically square tiles by suitable edge-replacements so as to obtain the required aperiodic set.

As a second example, consider the ten tiles of Figure 11.1.18. These lead directly to ten unmarked, basically square, tiles. However Robinson noticed that a slightly weaker matching condition (sufficient to ensure aperiodicity) could be obtained by deleting some of the arrows and using cornered and cornerless tiles as explained in Section 10.2. This led him to the discovery of his set R1 of six tiles. It seems to us very probable that this set R1 is the smallest possible aperiodic set of basically square tiles, or at least, the smallest such set that can be obtained from sets of Wang tiles by modifying their edges.

The construction of a "hierarchical" tiling by Mackay [1975] seems to be an independent discovery of aperiodicity through composition and decomposition, motivated by a search for new mathematical tools for crystallography. The tiling appears to use prototiles and some (not explicitly specified) matching conditions which are somewhat similar to those of the Penrose P1 set. In later publications Mackay [1976], [1979], [1981], [1982] considers other aperiodic tilings (in the plane and in higher dimensions); the planar case is equivalent to the set P3.

It will be apparent from Sections 10.3 and 10.4 that the discovery of small aperiodic sets requires great ingenuity. Although every new aperiodic sets is clearly of interest, the ultimate goal must be to discover a *single* aperiodic tile, that is, one that only admits monohedral non-periodic tilings. Though the existence of such a tile may appear unlikely, one must remember that only a few years ago, the existence of aperiodic sets containing just two tiles seemed essentially impossible. Yet, as we have seen, several such sets are now known (P2, P3, and A2, A4, A5, if marked tiles are allowed). In this connection the discovery of an aperiodic set containing two *similar* tiles (Figure 10.4.5) is significant and may point the way to future progress. In particular, we suggest that if a single aperiodic tile exists then it seems likely that a relatively complicated matching condition (such as that described in Section 12.5) may be required.

In the case of unmarked *convex* tiles, there is good reason to suppose that the smallest possible number in an aperiodic set is three, and this number is achieved by the polygons of Figure 10.3.28. On the other hand, there is no proof that a *single* convex polygon cannot be an aperiodic tile, and the possibility that such exists cannot be ruled out entirely.

All the properties of aperiodic sets discovered so far, and described in this chapter, depend either directly or indirectly on the processes of composition and decomposition. A natural question to ask is whether this is an essential property of aperiodic tiles, or whether aperiodic sets depending on some new and different principle await discovery.

The most significant advance in the subject since the original discoveries has undoubtedly been a realization of the importance of Ammann bars. It seems that these were independently discovered by several people, but were named after Ammann since he was the first to realize their utility in connection with the problem of specifying the "empires" associated with small patches of tiles (see Section 10.6). It could be argued that our treatment here should be reversed—instead of thinking of the tilings as giving rise to systems of bars, it is the systems of bars which are fundamental and the only function of the tiles is to give a practical realization to them. From this point of view, the correct approach to the discovery of new aperiodic sets must be to find tiles which *force* systems of bars with the required properties.

Moreover, this approach suggests a much wider field for investigation; the bars described so far lie in (at most) five different directions, and are related to the properties of the golden number τ (and so to $\sin \pi/5$). Is it possible that similar systems of bars (in $n > 5$ directions) exist which are related to the properties of the irrational number $\sin \pi/n$? There is evidence that this is the case, and Ammann has recently claimed that he can construct aperiodic sets of tiles (analogous to the Penrose rhombs P3) corresponding to bars related, as just described, to any value of $n \geq 5$. Obviously there are many open problems.

We briefly mention two ways in which the topic of this chapter can be extended; to give details would lead us far from the subject of this book. The first is that of aperiodic tilings in $n \geq 3$ dimensions. (In this connection we report Ammann's recent discovery of an aperiodic set of two tiles in three-dimensional space.) The second is that of aperiodic tilings in the hyperbolic plane; here the whole concept of aperiodicity must be clarified, and the interested reader should consult the pioneering paper of Robinson [1978] and the note at the end of Penrose [1978].

Aperiodic sequences of symbols, defined recursively by substitution, seem to have attracted some attention (see, for example, the bibliography of Dejean [1972]). However little of this seems relevant to the special properties of "musical sequences" into which a deeper investigation than that described in Section 10.6 appears desirable. Possibly this would lead us to a better understanding of the concept of aperiodicity.

A new approach to the analysis of the Penrose tilings has recently been suggested by N. G. de Bruijn [1981]. This approach uses "pentagrids", which bear a superficial resemblance to Ammann bars, but have certain advantages from a computational point of view.

A set of equidistant parallel lines is called a *grid*. Suppose we are given five such grids, each obtained from a given grid Γ by rotating it through multiples of $2\pi/5$ and applying translations (perpendicular to the lines of the grid) through distances $\gamma_0, \gamma_1, \gamma_2, \gamma_3, \gamma_4$, where the γ_i are real numbers. Then if

$$\gamma_0 + \gamma_1 + \gamma_2 + \gamma_3 + \gamma_4 = 0$$

the superposition of the five grids is a *pentagrid*, and a

pentagrid is called *regular* if no three lines are concurrent. The closures of the polygonal regions into which the lines of a regular pentagrid partition the plane clearly form a tiling with 4-valent vertices, and we shall refer to such a tiling as a *pentagrid tiling*.

The relevance of pentagrids is a consequence of the following remarkable result.

Every tiling by Penrose rhombs (set P3) *is dual to a pentagrid tiling, and conversely.*

Moreover, the Penrose tiling and pentagrid tiling can be relatively situated in such a way that corresponding edges are perpendicular. (It is not, however, possible to do this so that the tilings are dually situated.)

Two interesting geometric interpretations of the Penrose tiling are also mentioned in the de Bruijn paper. The first, which is due to R. M. A. Weiringa, makes use of a special method of indexing the vertices of a tiling by Penrose rhombs. Each index is an integer 1, 2, 3 or 4 and any two vertices of the tiling joined by an edge have indices which differ by 1. There is not room to give precise details of how the indexing is carried out; they can be found in de Bruijn [1981, Section 5]. Weiringa remarks that if we think of the tiling as lying in a horizontal plane in 3-dimensional space, and then we imagine each vertex raised vertically through distance $i/2$, where i is the index of the vertex, we obtain an edge-to-edge arrangement of *congruent* rhombs in 3-dimensional space. The angles of the rhombs satisfy $\tan \alpha = \pm 2$. Such an arrangement of rhombs has been called a *Weiringa roof* (in the hope that some enterprising architect will use it for the ceiling of a large room). Thus every tiling by Penrose rhombs can be described as the result of projecting a Weiringa roof orthogonally onto a suitable plane.

The second interpretation involves consideration of a dissection C of 5-dimensional space into unit cubes with vertices at all points with integer coordinates. It is possible to find a (2-dimensional) plane Π such that, if we project orthogonally onto Π the centers of the cubes of C which have non-empty intersection with Π, then the set of points we obtain is the set of vertices of a tiling by Penrose rhombs. Moreover, the vertices of all such tilings can be obtained in this way by suitable choice of Π.

F. P. M. Beenker [1982] has tried to extend de Bruijn's theory of pentagrids to tilings by squares and rhombs (with angles 45°, 135°, 45°, 135°) and so was led to the consideration of tetragrids. He remarks, however, that it is unfortunately impossible to find matching conditions for the square and rhomb which would force non-periodicity; it seems that he was unaware of the existence of Ammann's set A5.

More recently, in connection with the discovery of certain aluminum-manganese alloys which solidify with a high degree of icosahedral symmetry (see Shechtman, Blech, Gratias & Cahn [1984]), de Bruijn's approach has been applied to projections of the 6-dimensional tiling by hypercubes onto a 3-dimensional subspace. This yields a non-periodic tiling of the 3-dimensional space (which seems to be in fact aperiodic) and which appears to provide an explanation of the icosahedral alloys (see Kramer & Neri [1984], as well as A. L. Robinson [1985] and the references given there). It would appear that a very challenging area of research is opening here, made even more interesting by its relations to the experimental sciences. For some results in a related direction, see, Levine & Steinhardt [1984]; for more theoretical developments, see, Lunnon & Pleasants [1985].

It should be noted that earlier attempts have already been made to generalize aperiodic tilings to three dimensions. Various sets of polyhedra have been exhibited which certainly admit non-periodic tilings. However, as far as we are aware, in no case has a proof been given that the prototiles considered are, in fact, aperiodic. See, for example, Mackay [1981], [1982], Kramer [1982], Mosseri & Sadoc [1983].*

We conclude with a remark that seems necessary to clear up certain misconceptions. A recent paper of Harborth [1977b] claims the discovery of several new aperiodic sets of tiles. However these are only "aperiodic" if we impose the additional condition that we restrict our attention to tilings in which copies of every prototile *must* appear; without this condition periodic tilings are possible. In fact, several sets of tiles which are "aperiodic" in this weaker sense have already been given (see, for example, the tilings of Figures 1.5.9 and 1.5.10) and the discovery of such sets is very easy.

* See the Appendix beginning on page 653.

11

WANG TILES*

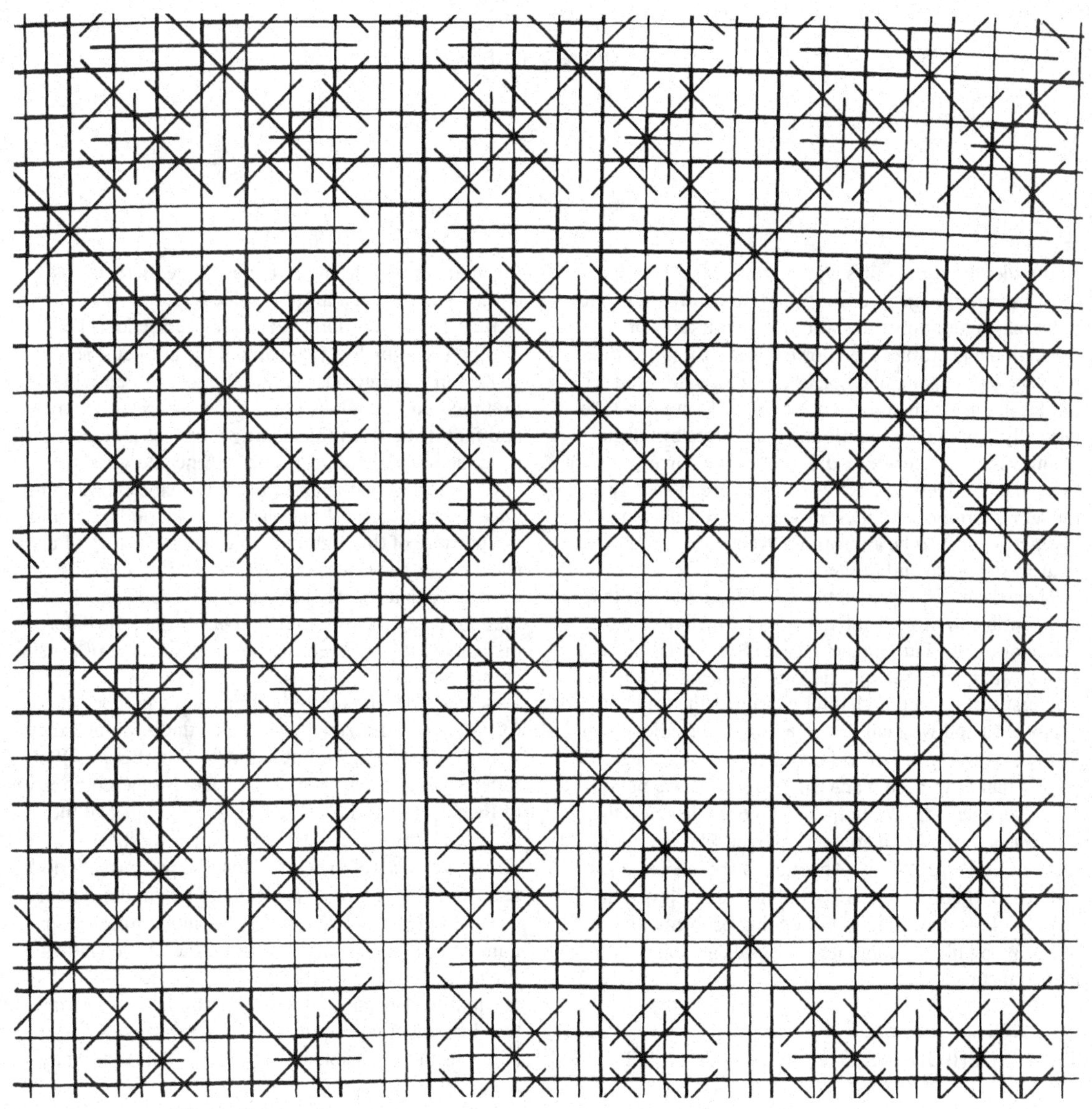

Figure 11.0.1
One of the tilings admitted by Läuchli's aperiodic set of Wang tiles shown in Figure 11.1.3. The construction of this tiling is explained in Section 11.1.

* See the Appendix beginning on page 653.

11

WANG TILES

Wang tiles are square tiles with colored edges which must be placed edge-to-edge; colors on contiguous edges must match and only translations (not rotations or reflections) of the prototiles are allowed. It is easy to see that to every set of Wang tiles corresponds an unmarked set of tiles in our original sense of Section 1.1—we can take basically square tiles and modify their edges by substituting suitable J-curves, a different curve corresponding to each color. Nevertheless, Wang tiles are of such theoretical importance and their properties are sufficiently distinctive for it to be appropriate to devote this chapter entirely to an account of them.

Their theoretical importance stems from the fact that it is possible to find sets of Wang tiles which mimic the behavior of any Turing machine (see Section 11.4) and so they are relevant to questions of mathematical logic. The first section of this chapter continues the topic of Chapter 10 and we shall describe some aperiodic sets of Wang tiles. We give a brief historical sketch of the attempts that have been made (since Berger's original discovery of aperiodic Wang tiles in 1966) to decrease the number of tiles in such an aperiodic set. The remaining sections deal with Wang's Theorem (a special case of the Extension Theorem which applies to Wang tiles), a discussion of the decidability of the tiling problem, and the use of Wang tiles to compute various elementary arithmetic functions.

11.1 APERIODIC SETS OF WANG TILES

In 1966 R. Berger discovered the first aperiodic set of tiles. It contains 20,426 Wang tiles, and although to us this number now seems extremely large, Berger's discovery was truly remarkable since it refuted Wang's conjecture that no aperiodic sets exist (see Section 11.3). It also preceded by five years any of the aperiodic sets described in the previous chapter. Berger himself remarked that the number of tiles in his set is unnecessarily large and that he believed that some proper subset would also have the property of aperiodicity. Since Berger's discovery, several mathematicians have made efforts to reduce the number of Wang tiles in an aperiodic set, and the purpose of this section is to trace the history of their efforts over the last twenty years. At the time of writing we believe that the smallest known set contains just 16 tiles. Quite apart from the numbers of tiles, the different methods that have been used seem to us to be of considerable interest.

Berger himself managed to reduce the number of tiles to 104 and he described these in his thesis, though they were omitted from the published version (Berger [1966]). Later Knuth [1968, p. 384] gave details of a modification of Berger's aperiodic set, reducing the number of tiles to 92. The method of construction, which we shall now describe, is typical of the early efforts to produce aperiodic sets, being based on the principle of "expanding squares".

The fundamental idea is that of "superimposing" a number of related tilings. Consider the 21 tiles shown in Figure 11.1.2(a). Only the first four of these (labelled α, β, γ and δ) bear the colors denoted by 1, 2, 3 or 4, and the way in which these colors are arranged implies that they admit just *one* possible tiling, namely that denoted by $\mathcal{T}_1$ in Figure 11.1.1.

Let us consider the structure of the tilings admitted by the remaining 17 tiles together with a "blank" tile which has no color labels. The color on the right side of an a-tile implies that it must be followed by a row of R-tiles, and that such a row must be terminated by a K-tile. (Exceptionally the row of R-tiles may be empty and then the K-tile will abut against the a-tile.) After the K-tile will follow a row (possibly empty) of L-tiles, then a b-tile, a row (possibly empty) of P-tiles, and then another a-tile. Similar considerations apply to the columns, and it is easy to see that the only possible tilings that use the a-, b-, c- and d-tiles are similar to those labelled $\mathscr{T}_2$, $\mathscr{T}_3$ and $\mathscr{T}_4$ in Figure 11.1.1. In fact *all* such tilings can be obtained from any one of these three by altering the

α	β	α	β	α	β	α	β	α	β	α	β	α	β	α	β	α	β	α	β	α	β	α	β
γ	δ	γ	δ	γ	δ	γ	δ	γ	δ	γ	δ	γ	δ	γ	δ	γ	δ	γ	δ	γ	δ	γ	δ
α	β	α	β	α	β	α	β	α	β	α	β	α	β	α	β	α	β	α	β	α	β	α	β
γ	δ	γ	δ	γ	δ	γ	δ	γ	δ	γ	δ	γ	δ	γ	δ	γ	δ	γ	δ	γ	δ	γ	δ
α	β	α	β	α	β	α	β	α	β	α	β	α	β	α	β	α	β	α	β	α	β	α	β
γ	δ	γ	δ	γ	δ	γ	δ	γ	δ	γ	δ	γ	δ	γ	δ	γ	δ	γ	δ	γ	δ	γ	δ
α	β	α	β	α	β	α	β	α	β	α	β	α	β	α	β	α	β	α	β	α	β	α	β
γ	δ	γ	δ	γ	δ	γ	δ	γ	δ	γ	δ	γ	δ	γ	δ	γ	δ	γ	δ	γ	δ	γ	δ
α	β	α	β	α	β	α	β	α	β	α	β	α	β	α	β	α	β	α	β	α	β	α	β
γ	δ	γ	δ	γ	δ	γ	δ	γ	δ	γ	δ	γ	δ	γ	δ	γ	δ	γ	δ	γ	δ	γ	δ
α	β	α	β	α	β	α	β	α	β	α	β	α	β	α	β	α	β	α	β	α	β	α	β
γ	δ	γ	δ	γ	δ	γ	δ	γ	δ	γ	δ	γ	δ	γ	δ	γ	δ	γ	δ	γ	δ	γ	δ
α	β	α	β	α	β	α	β	α	β	α	β	α	β	α	β	α	β	α	β	α	β	α	β
γ	δ	γ	δ	γ	δ	γ	δ	γ	δ	γ	δ	γ	δ	γ	δ	γ	δ	γ	δ	γ	δ	γ	δ
α	β	α	β	α	β	α	β	α	β	α	β	α	β	α	β	α	β	α	β	α	β	α	β
γ	δ	γ	δ	γ	δ	γ	δ	γ	δ	γ	δ	γ	δ	γ	δ	γ	δ	γ	δ	γ	δ	γ	δ
α	β	α	β	α	β	α	β	α	β	α	β	α	β	α	β	α	β	α	β	α	β	α	β
γ	δ	γ	δ	γ	δ	γ	δ	γ	δ	γ	δ	γ	δ	γ	δ	γ	δ	γ	δ	γ	δ	γ	δ
α	β	α	β	α	β	α	β	α	β	α	β	α	β	α	β	α	β	α	β	α	β	α	β
γ	δ	γ	δ	γ	δ	γ	δ	γ	δ	γ	δ	γ	δ	γ	δ	γ	δ	γ	δ	γ	δ	γ	δ
α	β	α	β	α	β	α	β	α	β	α	β	α	β	α	β	α	β	α	β	α	β	α	β
γ	δ	γ	δ	γ	δ	γ	δ	γ	δ	γ	δ	γ	δ	γ	δ	γ	δ	γ	δ	γ	δ	γ	δ
α	β	α	β	α	β	α	β	α	β	α	β	α	β	α	β	α	β	α	β	α	β	α	β
γ	δ	γ	δ	γ	δ	γ	δ	γ	δ	γ	δ	γ	δ	γ	δ	γ	δ	γ	δ	γ	δ	γ	δ

(*a*) $\mathscr{T}_1$

Figure 11.1.1
The first four tilings of an infinite sequence which are "superimposed" in defining an aperiodic set containing 92 Wang tiles.

lengths of the rows of *R*-, *L*-, *P*-, *S*-, *T*- and *X*-tiles and of the columns of *Y*-, *U*-, *D*- and *Q*-tiles. It will be convenient to refer to these tilings as of the *second kind.* Notice that $\mathscr{T}_1$ and all the tilings of the second kind are (or can be) periodic; we now show that superimposing them subject to suitable restrictions will yield a non-periodic tiling, and hence lead to an aperiodic set of Wang tiles.

Suppose that we superimpose any tiling of the second kind on the tiling $\mathscr{T}_1$. By this we simply mean that each tile will carry a label α, β, γ or δ from $\mathscr{T}_1$ and possibly also a label *a*, *b*, *c*, *d*, *N*, *J*, *K*, *R*, *L*, *P*, *S*, *T*, *X*, *Y*, *U*, *D* or *Q* from the second tiling. (There will be no second label if the corresponding tile in the second tiling is a "blank".) We impose the condition that the tiles labelled *a*, *b*, *c* or *d* (and *only* these four kinds of tile) can be superimposed on an α-tile from $\mathscr{T}_1$, and this immediately reduces the number of possibilities for the second tiling It must be equal to the tiling $\mathscr{T}_2$ shown in Figure

a	K	b	P	a	K	b	P	a	K	b	P	a	K	b	P	a	K	b	P	a	K	b	P
J	N			J	N			J	N			J	N			J	N			J	N		
c	S	d	T	c	S	d	T	c	S	d	T	c	S	d	T	c	S	d	T	c	S	d	T
Q				Q				Q				Q				Q				Q			
a	K	b	P	a	K	b	P	a	K	b	P	a	K	b	P	a	K	b	P	a	K	b	P
J	N			J	N			J	N			J	N			J	N			J	N		
c	S	d	T	c	S	d	T	c	S	d	T	c	S	d	T	c	S	d	T	c	S	d	T
Q				Q				Q				Q				Q				Q			
a	K	b	P	a	K	b	P	a	K	b	P	a	K	b	P	a	K	b	P	a	K	b	P
J	N			J	N			J	N			J	N			J	N			J	N		
c	S	d	T	c	S	d	T	c	S	d	T	c	S	d	T	c	S	d	T	c	S	d	T
Q				Q				Q				Q				Q				Q			
a	K	b	P	a	K	b	P	a	K	b	P	a	K	b	P	a	K	b	P	a	K	b	P
J	N			J	N			J	N			J	N			J	N			J	N		
c	S	d	T	c	S	d	T	c	S	d	T	c	S	d	T	c	S	d	T	c	S	d	T
Q				Q				Q				Q				Q				Q			
a	K	b	P	a	K	b	P	a	K	b	P	a	K	b	P	a	K	b	P	a	K	b	P
J	N			J	N			J	N			J	N			J	N			J	N		
c	S	d	T	c	S	d	T	c	S	d	T	c	S	d	T	c	S	d	T	c	S	d	T
Q				Q				Q				Q				Q				Q			
a	K	b	P	a	K	b	P	a	K	b	P	a	K	b	P	a	K	b	P	a	K	b	P
J	N			J	N			J	N			J	N			J	N			J	N		
c	S	d	T	c	S	d	T	c	S	d	T	c	S	d	T	c	S	d	T	c	S	d	T
Q				Q				Q				Q				Q				Q			

Figure 11.1.1 *(continued)* (*b*) $\mathscr{T}_2$

11.1.1(*b*), and must lie in one of four positions relative to $\mathscr{T}_1$ (either as shown, or obtained from this by vertical or horizontal displacements through any even number of edge-lengths).

Now superimpose a third tiling, also of the second kind, on the tiling already constructed (so that each tile may now carry up to three labels). Here we impose the condition that the tiles labelled *a, b, c* or *d* (and *only* these four kinds of tile) of the third tiling can coincide with an *N*-tile from $\mathscr{T}_2$. Again there are just four possibilities for the third tiling, namely the tiling $\mathscr{T}_3$ of Figure 11.1.1(*c*), and the tilings obtained by vertical or horizontal displacements through multiples of four edge-lengths.

A fourth tiling of the second kind is superimposed on the first three subject to the condition that *a, b, c,* or *d* coincides with an *N*-tile from $\mathscr{T}_3$ and so we obtain a tiling $\mathscr{T}_4$, and so on. Proceeding in this way we obtain an infinite sequence of tilings of the second kind $\mathscr{T}_2$, $\mathscr{T}_3$,

	Q								Q								Q						
P	a	R	K	L	b	P	P	P	a	R	K	L	b	P	P	P	a	R	K	L	b	P	P
	D		Y						D		Y						D		Y				
	J	X	N						J	X	N						J	X	N				
	U								U								U						
T	c	S	S	S	d	T	T	T	c	S	S	S	d	T	T	T	c	S	S	S	d	T	T
	Q								Q								Q						
	Q								Q								Q						
	Q								Q								Q						
P	a	R	K	L	b	P	P	P	a	R	K	L	b	P	P	P	a	R	K	L	b	P	P
	D		Y						D		Y						D		Y				
	J	X	N						J	X	N						J	X	N				
	U								U								U						
T	c	S	S	S	d	T	T	T	c	S	S	S	d	T	T	T	c	S	S	S	d	T	T
	Q								Q								Q						
	Q								Q								Q						
	Q								Q								Q						
P	a	R	K	L	b	P	P	P	a	R	K	L	b	P	P	P	a	R	K	L	b	P	P
	D		Y						D		Y						D		Y				
	J	X	N						J	X	N						J	X	N				
	U								U								U						
T	c	S	S	S	d	T	T	T	c	S	S	S	d	T	T	T	c	S	S	S	d	T	T
	Q								Q								Q						
	Q								Q								Q						

(c) $\mathscr{T}_3$

$\mathscr{T}_4, \ldots$. It is easy to verify that every tile will bear two, three or four labels (and no more, in spite of the fact that an infinite number of tilings are involved). For example, the tile marked by heavy lines in Figure 11.1.1 will carry the labels δNd—δ from $\mathscr{T}_1$, N from $\mathscr{T}_2$ and d from $\mathscr{T}_3$. In $\mathscr{T}_n$ $(n \geq 4)$ the corresponding tile is blank and so does not contribute any labels.

In Table 11.1.1 we give a list of all possible labels that can arise. The constraints on the relative positions of consecutive tilings in the sequence result from the fact that labels a, b, c and d only occur in combination with α or δ, and in the case of δ the label N must also be present.

In order to interpret the tiles with several labels as Wang tiles we proceed as follows; an example showing the tile δNd appears in Figure 11.1.2(b). If we "add" the tiles together as in the diagram we obtain one whose four edges are labelled $4y$, $3t$, 3 and $4xs$. We must interpret

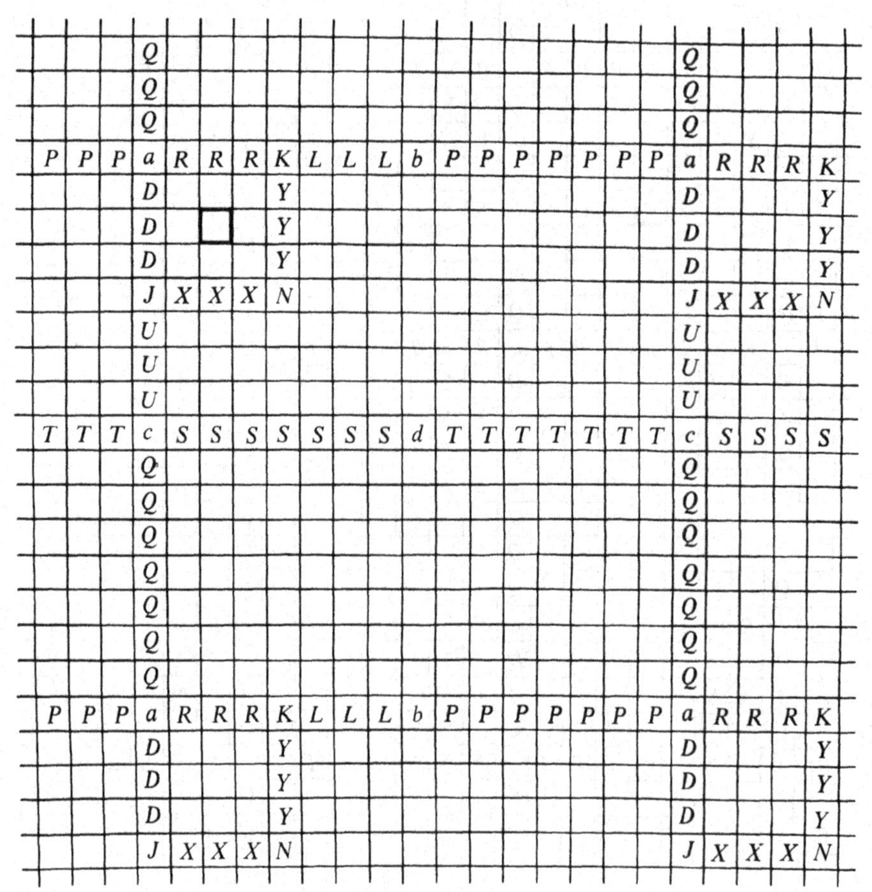

Figure 11.1.1 (*continued*) (d) $\mathscr{T}_4$

each of these (and other combinations of letters and numbers that arise) as different colors. Thus the upper edge of the tile δND labelled $4y$ must fit against the lower edge of some tile carrying the same "color" $4y$ on its lower edge.

The 92 symbols listed in Table 11.1.1 therefore lead to 92 Wang tiles. Any tiling $\mathscr{T}$ admitted by these can be interpreted as resulting from the superposition of an infinite sequence of tilings, and any symmetry of $\mathscr{T}$ must necessarily be a symmetry of each of the component tilings. However the latter involve tiles with labels a, b, c, d at the vertices of lattices of arbitrarily large squares. Thus the symmetry group of $\mathscr{T}$ contains no translations, and therefore is non-periodic; hence the 92 Wang tiles form an aperiodic set.

In 1966 Hans Läuchli managed to reduce the number of Wang tiles in an aperiodic set to 40, and although these were known to workers in this field, details were

Table 11.1.1 NINETY-TWO WANG TILES WHICH FORM AN APERIODIC SET

The notation is explained by the following example:

$$\beta Y \begin{Bmatrix} - \\ U \\ Q \end{Bmatrix} \begin{Bmatrix} P \\ T \end{Bmatrix} \text{ stands for the six tiles}$$

$$\beta YP,\ \beta YT,\ \beta YUP,\ \beta YUT,\ \beta YQP,\ \beta YQT.$$

The number of tiles is indicated in parentheses after each symbol. The interpretation of the symbols in terms of coloring of the tiles is indicated in Figure 11.1.2(a) and is explained in the text. (From Knuth [1968].)

$\alpha \begin{Bmatrix} a \\ b \\ c \\ d \end{Bmatrix}$	(4)	$\beta Y \begin{Bmatrix} - \\ U \\ Q \end{Bmatrix} \begin{Bmatrix} P \\ T \end{Bmatrix}$	(6)	$\beta \begin{Bmatrix} - \\ U \\ D \\ Q \end{Bmatrix} \begin{Bmatrix} P \\ S \\ T \end{Bmatrix}$	(12)
$\beta K \begin{Bmatrix} - \\ U \\ Q \end{Bmatrix}$	(3)	$\gamma \begin{Bmatrix} - \\ X \end{Bmatrix} \begin{Bmatrix} L \\ P \\ S \\ T \end{Bmatrix} \begin{Bmatrix} - \\ Q \end{Bmatrix}$	(16)	$\gamma R \begin{Bmatrix} - \\ Q \end{Bmatrix}$	(2)
$\gamma J \begin{Bmatrix} L \\ P \\ S \\ T \end{Bmatrix}$	(4)	$\delta X \begin{Bmatrix} L \\ P \\ S \\ T \end{Bmatrix} \begin{Bmatrix} - \\ Q \end{Bmatrix}$	(8)	$\delta Y \begin{Bmatrix} - \\ U \\ Q \end{Bmatrix} \begin{Bmatrix} P \\ T \end{Bmatrix}$	(6)
$\delta N \begin{Bmatrix} a \\ b \\ c \\ d \end{Bmatrix}$	(4)	$\delta J \begin{Bmatrix} L \\ P \\ S \\ T \end{Bmatrix}$	(4)	$\delta K \begin{Bmatrix} - \\ U \\ Q \end{Bmatrix}$	(3)
$\delta \begin{Bmatrix} R \\ L \\ P \\ S \\ T \end{Bmatrix} \begin{Bmatrix} - \\ U \\ D \\ Q \end{Bmatrix}$	(20)				

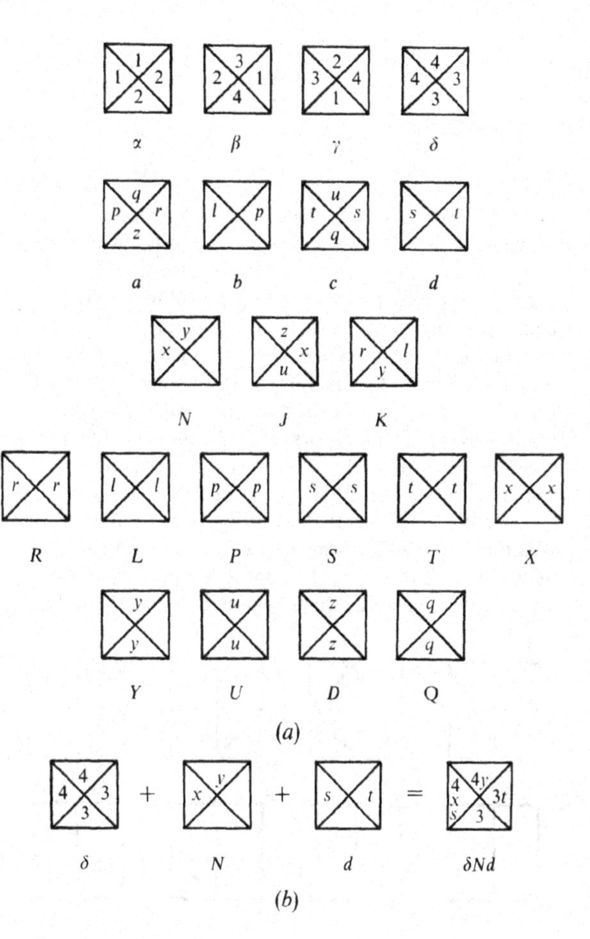

Figure 11.1.2
In (b) we show how one of the 92 aperiodic Wang (δNd) tiles is defined in terms of three "component" tiles. All 21 "component" tiles are shown in (a). Each string of symbols (such as $4y$, $4xs$, 3 or $3t$ on the tile δNd) is to be interpreted as a different color, and on adjacent tiles colors must match.

not published until 1975 when Wang [1975] reproduced drawings of the tiles. Unfortunately there was an error in Wang's diagram, so we give here, in Figure 11.1.3, a corrected version. Instead of listing the colors on the edges, it is convenient to specify the matching condition by coloring the edges in only two colors (shown by light and heavy lines in the diagram) and using line segments on the tiles which must continue across edges in a similar manner to the Ammann bars. These lines enable one to "see" the structure of the tiling, of which an example is shown in Figure 11.0.1.

The proof of the fact that any tiling by these 40 tiles is non-periodic follows similar lines to the argument (given in detail in Section 10.2) for Robinson's set R1 of six tiles. Here we only give an outline of the method. By a "cross" we mean any one of the first four tiles in the top row of Figure 11.1.3. It is easy to verify that any cross, together with its eight neighbors, must form a 3×3 block like one of those shown in Figure 11.1.4(a) or 11.1.4(b). (In the diagram we have chosen the first of the four crosses—the one that "points" vertically upwards—but the blocks corresponding to the other three crosses are similar in appearance.) Hence we can distinguish two sorts of crosses according to the arrangement of their neighboring tiles. Moreover, by investigating the possible arrangement of tiles adjacent to these 3×3 blocks, one can verify that a cross of the first sort forces a cross of the second sort to lie in one of the two positions marked X (see Figure 11.1.4(a)) and a cross of the second sort forces four crosses of the first sort to lie in

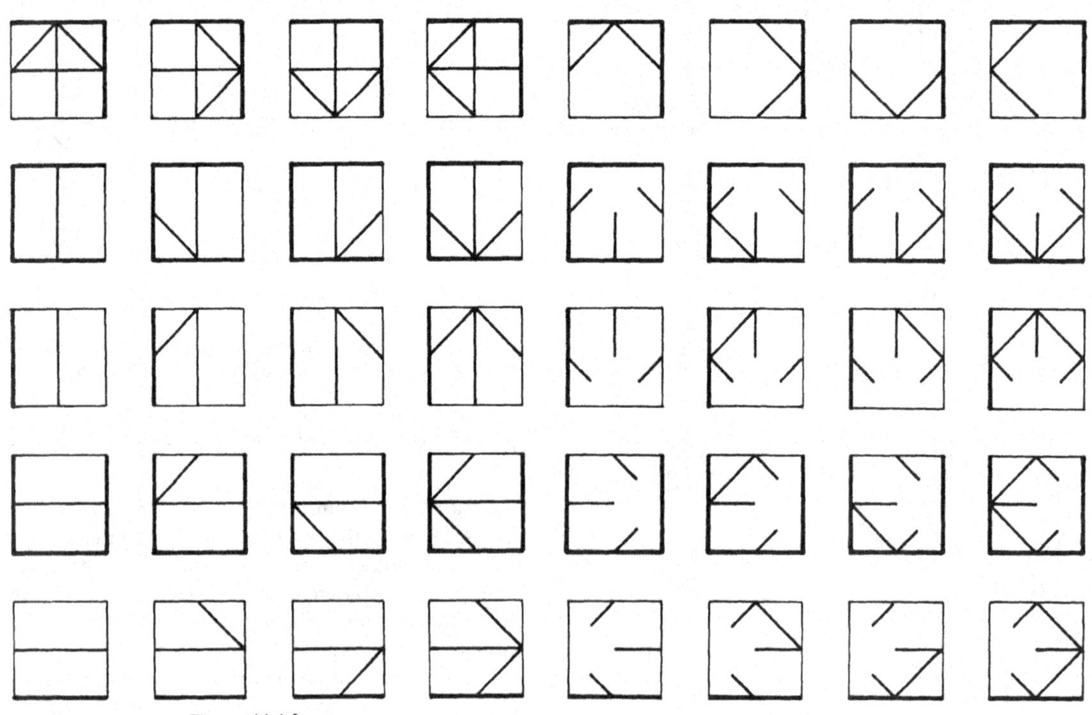

Figure 11.1.3
The 40 aperiodic Wang tiles used in Läuchli's aperiodic tiling shown in Figure 11.0.1.

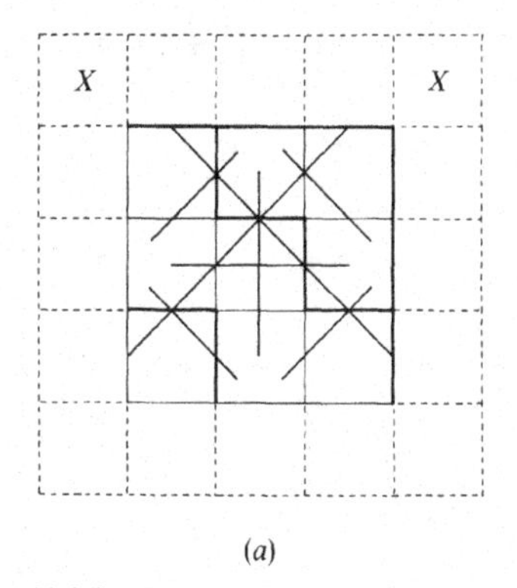

(a)

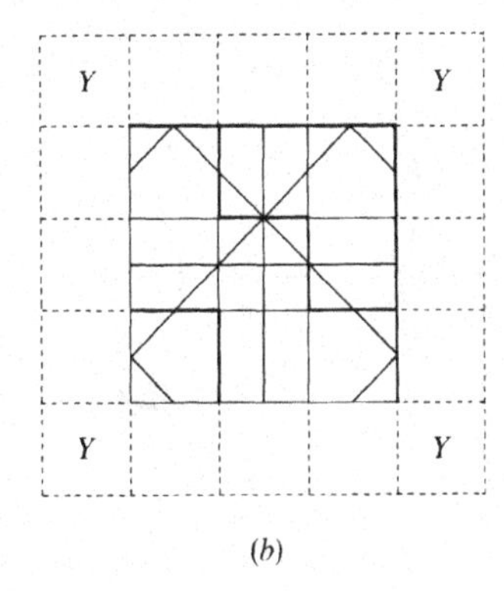

(b)

Figure 11.1.4
The 3×3 blocks that can be built up from Läuchli's tiles (Figure 11.1.3). The construction of the Läuchli tiling is similar to the Robinson tiling described in Section 10.2; these 3×3 blocks are used to construct 7×7 blocks, 15×15 blocks and so on.

the four positions marked Y(Figure 11.1.4(b)). These facts enable us to deduce the structure of the tiling: there exist 3×3 blocks (with a cross of the first sort at their centers) of which four can be assembled to form a 7×7 block (with a cross of the second sort at its center) by inserting rows and columns (each of 3 tiles) between the 3×3 blocks. Four of these 7×7 blocks can be assembled in an analogous manner to form a 15×15 block (with a cross of the second sort at its center) and so on. The tiling thus contains $(2^n - 1) \times (2^n - 1)$ blocks for each integer n. This implies that in some of the rows, and in some of the columns, crosses are at a distance $2^n - 1$ apart and as n is arbitrarily large, the tiling is necessarily non-periodic. Thus the 40 tiles of Figure 11.1.3 form an aperiodic set.

In 1967 Raphael M. Robinson (who was unaware of Läuchli's results) stated that an aperiodic set of 52 Wang tiles existed (Robinson [1967]), and in his paper of 1971 he reduced this number to 35. No details of these sets were published, but he gave a proof that 56 tiles were sufficient (see Exercise 11.1.1).

It seems likely that these numbers cannot be substantially reduced for aperiodic sets of Wang tiles constructed on the same basic principle—hence the discovery of smaller sets of aperiodic Wang tiles did not occur until a new approach was discovered. This new idea was due to Roger Penrose who showed how his set P2 of kites and darts could be used to produce an aperiodic set of 34 tiles. He described his construction in a letter to Robinson who managed to modify it in such a way as to reduce the number of tiles to 32. The method of constructing this set will now be described.

Robinson observed that any tiling by kites and darts $\mathcal{T}$ can be "cut up" as in Figure 11.1.5 into small patches (see Exercise 11.1.4). These patches are of eight different shapes, shown in Figure 11.1.6, and each occurs in exactly four aspects, namely those that arise by reflection in a horizontal line, in a vertical line, or in both these lines (central reflection). Each patch is made up of Robinson's A-tiles—that is, of kites, darts, half-kites and half-darts. Regarding each patch as a tile in a composed tiling $\mathcal{T}'$, we see that $\mathcal{T}'$ is homeohedral (it is homeomorphic to the regular tiling $[4^4]$).

We shall now explain how $\mathcal{T}'$ can be converted into a tiling by Wang tiles. Since, for Wang tiles, only translations are allowed, we must take each of the eight prototiles of $\mathcal{T}'$, in each of its four aspects. Then we distort these tiles into squares and replace the matching condi-

Figure 11.1.5
Cutting up a tiling by Penrose kites and darts is the first stage in the construction of an aperiodic set containing 32 Wang tiles, see Exercise 11.1.2.

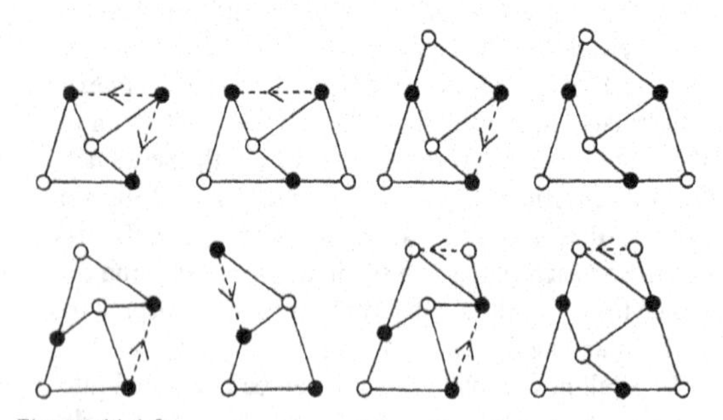

Figure 11.1.6
The eight patches of tiles that arise when a tiling by Penrose kites and darts is cut up as illustrated in Figure 11.1.5. Each patch occurs in exactly four different aspects arising from those shown by reflection in horizontal and vertical lines.

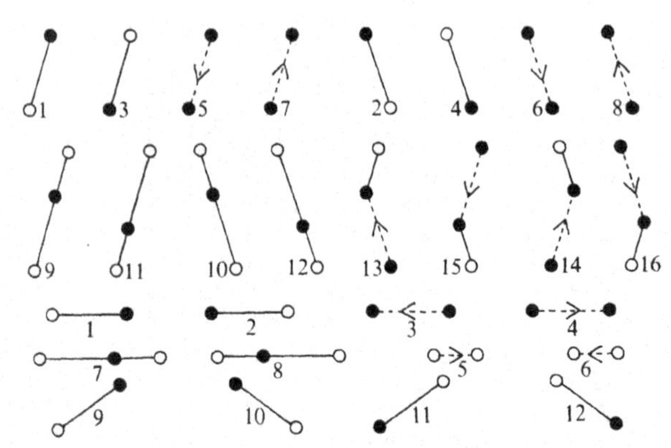

Figure 11.1.7
The patches of Figure 11.1.6, each in its four orientations, have 28 different kinds of edges—16 "vertical edges" as shown in the top two rows of this diagram, and 12 "horizontal edges" as shown in the last three rows. Each of these edges is replaced by a color, as shown, in order to convert the patches into Wang tiles.

tions inherited from the kites and darts by coloring the edges. One method of doing this is shown in Figure 11.1.7—for the "vertical edges" 16 different colors are required and for the "horizontal edges" we need 12 colors. With the suggested numbering we arrive at the 32 Wang tiles shown in Figure 11.1.8.

The fact that these 32 tiles form an aperiodic set is clear. Any tiling using these prototiles can be distorted into one using the tiles of Figure 11.1.6 each in four aspects, and this, in turn, leads to a unique tiling by kites and darts. The aperiodicity of the latter ensures that the 32 tiles form an aperiodic set.

Surprisingly both Penrose and Robinson failed to notice that if they had used Penrose rhombs (set P3) instead of the kites and darts (set P2) in the construction described above, not only would the method of cutting up the tiles have been easier to describe, but they would have been able to reduce the number of tiles in an aperiodic set to 24. In Exercise 11.1.2 we give sufficient information for the reader to be able to construct such a set for himself.

A different construction for an aperiodic set of 24 tiles was, however, discovered by Robinson in 1977. He ob-

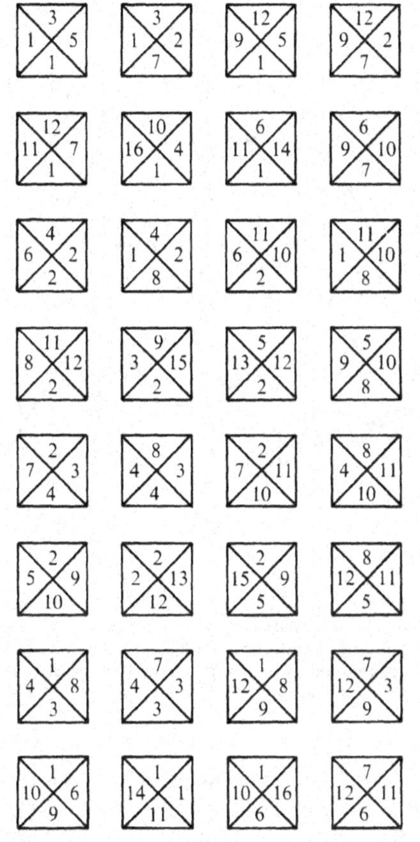

Figure 11.1.8
The 32 aperiodic Wang tiles obtained from the patches of Figure 11.1.6 when they are distorted into squares and their edges colored as in Figure 11.1.7.

served that the aperiodicity of the Ammann set A2 does not depend on the values of p, q, r and s in Figure 10.4.1 so he put them all equal to 1. Each tile is used in four aspects, so we obtain a set of eight tiles of which only translations are required. Moreover the required matching conditions can be translated into colorings of the edges of these tiles. To convert them into Wang tiles we cut the larger Ammann tile into five squares, and the smaller into three squares, introducing new colors on the internal edges so that they necessarily fit together in the required manner. At first it would appear that $(5+3)\times 4=32$ distinct Wang tiles are required, but it is possible to be more economical—we can arrange for two of the Wang tiles used in the building up of each smaller Ammann tile to be used also in the larger. In this way the number of tiles is reduced to 24. These are shown in Figure 11.1.9; eight colors 1–4 and 11–14 are used to give the correct matching conditions on the edges of the Ammann tiles, and sixteen colors 20–23, 30–33, 40–43 and 50–53 are used for the internal edges.

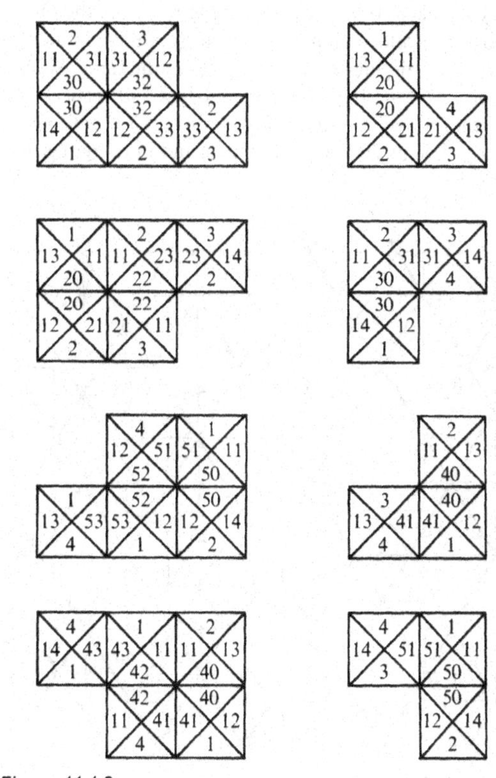

Figure 11.1.9
The conversion of the Ammann prototiles (set A2) into 24 Wang tiles that form an aperiodic set.

It seems unlikely that the procedure just described can lead to smaller sets of tiles. But this question is unimportant in view of a new idea, due to Ammann, which enables an aperiodic set of 16 Wang tiles to be defined. This set is mentioned in Robinson [1978]. The new approach uses Ammann bars, already discussed at length in Section 10.6. We are grateful to Ammann for allowing us to reproduce details here.

To illustrate the method we show, in Figure 11.1.10, a tiling $\mathcal{T}$ using the Ammann set A2. The four sets of parallel Ammann bars on $\mathcal{T}$ are indicated, two by solid and two by dashed lines. As mentioned previously, these bars may be taken to be markings on the tiles, and the required matching condition is given by specifying that the bars must continue across the edges of the tiling. We now recompose the tiling, deleting the edges and vertices of $\mathcal{T}$, interpreting the solid bars as edges of a new tiling $\mathcal{T}'$ by rhombs and parallelograms, and interpreting the other (dashed) bars as markings on the tiles of $\mathcal{T}'$. The matching condition is given, as before, by specifying that the dashed bars continue straight across the edges of the tiling. Since $\mathcal{T}$ can be reconstructed from the arrangement of bars, and $\mathcal{T}$ is aperiodic, we deduce that $\mathcal{T}'$ is also aperiodic. If only translations of the prototiles are allowed then it is easy to see from Figure 11.1.11 that there are exactly 16 prototiles as shown in Figure 11.1.12.

The final task is to convert these 16 prototiles into Wang tiles. To do this we suppose that a suitable transformation (consisting of a shear followed by appropriate expansions and contractions) is applied so that we have 16 square tiles, and then it is only necessary to replace the matching condition specified by the dashed bars by one using colored edges. In Figure 11.1.13 we show one way in which this can be done, each tile in Figure 11.1.13

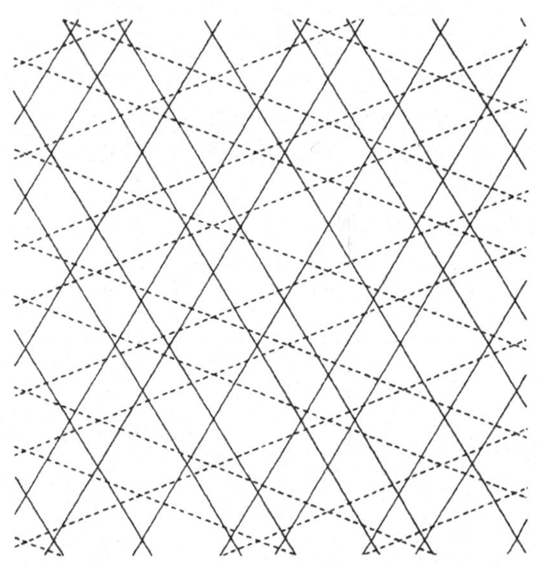

Figure 11.1.11
The Ammann bars of Figure 11.1.10 after the tiles have been detected. The solid bars are to be regarded as the edges of a new tiling by rhombs and parallelograms, the dashed bars are to be regarded as markings on the tiles specifying the matching condition.

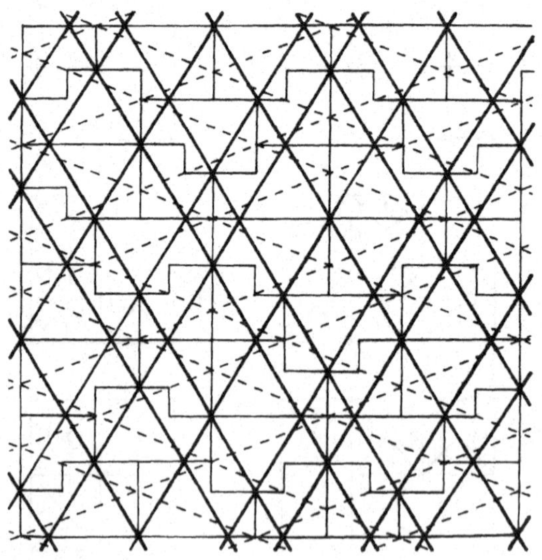

Figure 11.1.10
A tiling by the set A2 of Ammann prototiles with the four families of Ammann bars indicated, two by solid and two by dashed lines.

Figure 11.1.12
The 16 tiles that arise as indicated in Figure 11.1.11.

being in a corresponding position to that in Figure 11.1.12 from which it was derived.

Several variations of this procedure are possible. Ammann originally suggested using another set of bars joining the elliptical markings on the tiles as shown in Figure 11.1.14. The analogous procedure then yields the same set of Wang tiles shown in Figure 11.1.13.

The application of the method to the Ammann bars on a tiling by Penrose kites and darts is shown in Figure 11.1.15. Here we consider two sets of parallel bars as the edges of the new tiles and three sets of bars as markings. As will be seen this yields a set of 24 tiles. However, as Conway remarked this number can be reduced by omitting one set of bars—two alone are sufficient for the matching condition since the third is (essentially) uniquely determined by them. When this is done (the dashed bars being deleted) then the number of tiles is again 16. In fact we obtain exactly the same tiles as before as shown in Figures 11.1.12 and 11.1.13.

At the time of writing, this number 16 is the least number of Wang tiles in any known aperiodic set. There remains the possibility, of course, that further investigations will enable this number to be reduced further.

For tilings constructed from the Ammann bars, the existence of decomposition for the underlying tilings yields an analogous process for the Wang tiles. This is illustrated in Figure 11.1.16. Here we show 16 blocks of one, two or four tiles. If each of the 16 tiles of Figure 11.1.13 is replaced by the block that occupies the same relative position in Figure 11.1.16, we arrive at another tiling using the same 16 prototiles. An example is shown in Figure 11.1.17 where the blocks are indicated by heavy lines. The converse process can also be carried out. Any tiling using the 16 Wang prototiles can be cut up into the blocks of Figure 11.1.16 in a unique way, and so transformed into another tiling by replacing each block by a

Figure 11.1.13
The 16 Wang tiles that correspond to the tiles of Figure 11.1.12. These form the smallest known aperiodic set.

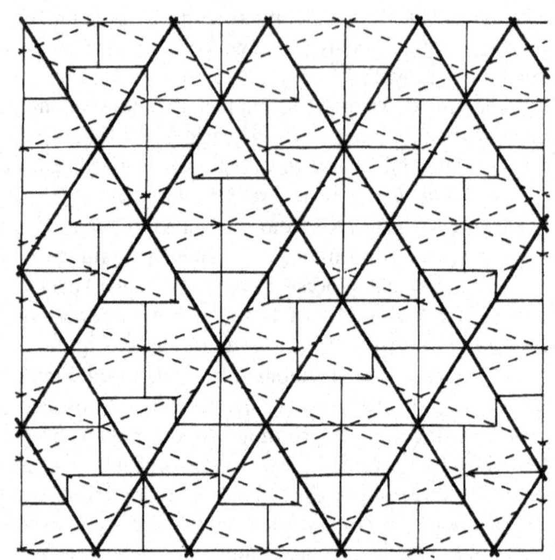

Figure 11.1.14
Another set of bars which can also be used to construct the aperiodic set of Wang tiles shown in Figure 11.1.13.

tile. This corresponds to composition of the underlying tiling.

The reduction in the number of Wang tiles in an aperiodic set from over 20,000 to 16 has been a notable achievement. Perhaps the minimum possible number has now been reached. If, however, further reductions are possible then it seems certain that new ideas and methods will be required. The discovery of such remains one of the outstanding challenges in this field of mathematics. One can, of course, look at the problem from the opposite point of view. Is it possible to prove that, for example, 15 tiles are not enough? It is difficult to see how any such proof could be constructed, and the only result we know in this direction is an unpublished theorem of Robinson that no aperiodic set of *four* Wang tiles can exist.

A related question is this. For what integers n is it possible to find an aperiodic set $\mathscr{S}$ of n prototiles such that no proper subset of $\mathscr{S}$ admits a tiling of the plane? In this section we have described or reported on sets with $n = 104, 92, 56, 52, 40, 35, 34, 32, 24$ and 16, and it is not difficult to show that integer multiples of these numbers are also possible. But other values of n exist. For example Robinson (to whom we are grateful for information and helpful comments concerning the material of this section) has recently shown that $n = 17$ is also possible.

For additional information on the relationship between aperiodic sets of Wang tiles and the aperiodic sets discussed in the previous chapter, see Section 10.7.

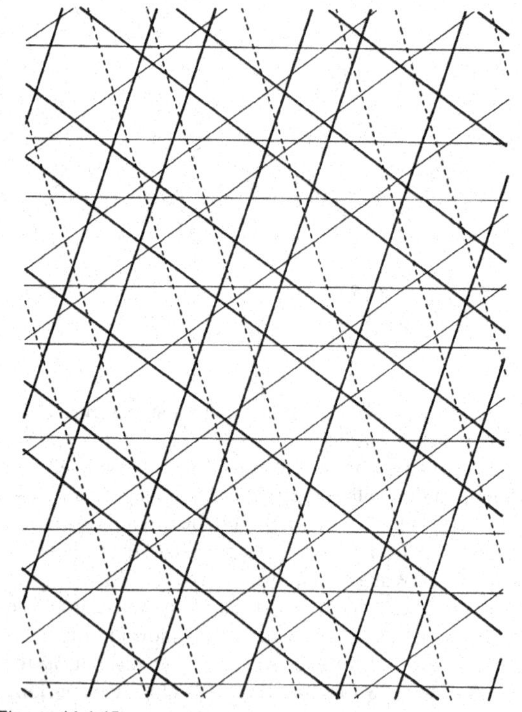

Figure 11.1.15
A set of Ammann bars obtained from a tiling by Penrose kites and darts (set P2). If the dashed bars are omitted and the heavy lines regarded as edges of a new tiling, then we obtain again the aperiodic set of Wang tiles shown in Figure 11.1.13.

Figure 11.1.16
This diagram shows how the Wang tiles of Figure 11.1.13 can be "decomposed".

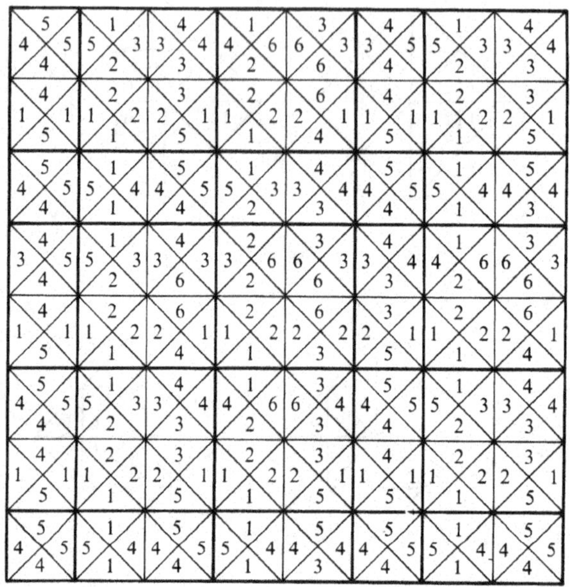

Figure 11.1.17
A tiling illustrating "composition" and "decomposition" of the 16 Wang tiles of Figure 11.1.13. The heavy lines divide the tiling into patches of the kinds shown in Figure 11.1.16.

EXERCISES 11.1

1. Figure 11.1.18 shows ten prototiles; the matching condition is that the head of each arrow must fit against the tail of an arrow on an adjacent tile. Show that these prototiles form an aperiodic set. (Rotations and reflections are allowed. It will be observed that the ten tiles can be obtained from those in Figure 10.2.3 by substituting arrows for the corner markings. These additional arrows impose a stronger condition than the corner markings in the Robinson set.)

   By considering all rotations and reflections, show that these ten tiles lead to 56 prototiles which will tile the plane using translations only, and will do so only non-periodically, and hence obtain an aperiodic set of 56 Wang tiles. (This construction is described in Robinson [1971].)

2. Figure 11.1.19 shows how a tiling by Penrose rhombs (set P3) can be cut up into patches consisting of rhombs, half-rhombs and quarter-rhombs. Use a method similar to that given in the text on pages 591–592 to convert these patches into an aperiodic set of 24 Wang tiles.

*3. We know (see page 572) that a tiling by Penrose kites and darts is almost specified uniquely by the Ammann bars in three directions. At first sight it would appear that in Figure 11.1.15 we could omit two of the parallel sets of bars (and not just the one shown dashed). Explain why this is not so, and why the method does not yield aperiodic sets with less than 16 tiles.

Figure 11.1.19
Cutting up a tiling by Penrose rhombs (set P3) into eight kinds of patches. This procedure can be used to construct an aperiodic set of 24 Wang tiles (see Exercise 11.1.2).

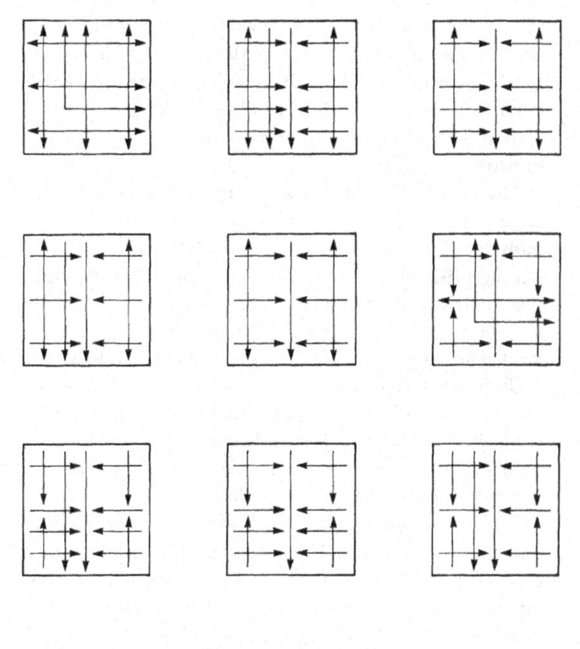

Figure 11.1.18
Marked tiles used in the construction of an aperiodic set containing 56 Wang tiles (see Exercise 11.1.1).

*4. Explain exactly how a tiling by Penrose kites and darts can be cut up (as in Figure 11.1.5) into patches of the shapes shown in Figure 11.1.6. (A possible approach to this problem is to show how each of the ten possible orientations of the first-order cartwheel can be cut into patches of the required shapes, and then show how each of the cuts can be "reconciled" wherever the cartwheels meet or overlap. A certain amount of ingenuity is required! The result will then follow from Statement 10.5.2.)

11.2 WANG'S THEOREM

Wang's Theorem can be simply described as the restriction of the Extension Theorem (Statement 3.8.1) to sets of Wang tiles. Although it can be deduced from the Extension Theorem as a special case, we give here an independent proof. The justification for doing this is not merely historical—Wang's Theorem was the first form of the Extension Theorem to be stated and proved (see Knuth [1968] and Wang [1975])—but it also introduces ideas useful in other contexts. In particular it makes use of an important graph-theoretic lemma known as König's Infinity Lemma which has many applications. One of these, together with some implications of Wang's Theorem, is explained later in this section.

We begin by stating the theorem.

11.2.1 Wang's Theorem. *Let $\mathscr{S}$ be a given finite set of Wang prototiles. If it is possible, for arbitrarily large values of n, to assemble $n \times n$ blocks of tiles satisfying the color-matching conditions, then $\mathscr{S}$ admits a tiling of the plane.*

We must emphasize that only translations of the prototiles are permitted. If we also allowed reflections or rotations through an angle π, the theorem would become trivial for then every set of prototiles would admit a tiling of the plane (see Exercise 11.2.3).

The fact that $\mathscr{S}$ is finite is essential to the proof of the theorem, as we can see from a simple modification of the example given in Section 3.8 (see Figure 3.8.3). We take the set Q of square tiles from the regular tiling (4^4) which lie in a quadrant of the plane, and assign a different color to every edge e of Q; the side of each tile which abuts on e is taken to bear that color. Then Q clearly contains arbitrarily large $n \times n$ blocks of tiles satisfying the color-matching condition—yet no tiling of the plane is possible using the tiles of Q as prototiles.

For the proof of Statement 11.2.1 we need to construct a graph G (consisting of *nodes*, some of which are joined by *arcs*). The nodes are assigned to different "levels" denoted by non-negative integers 0, 1, 2, ... On level 0 there is just one node $n^{(0)}$. On level 1, the nodes are denoted by $n_1^{(1)}, \ldots, n_r^{(1)}$ and these correspond in a one-to-one manner with the r prototiles of $\mathscr{S}$. Generally, for $k \geq 1$, the nodes on the kth level are $n_1^{(k)}, \ldots, n_{r_k}^{(k)}$, and these correspond in a one-to-one manner to all the distinct $(2k-1) \times (2k-1)$ blocks of tiles that can be constructed from $\mathscr{S}$ subject to the matching condition. To complete our description of G we need to specify which pairs of nodes are to be joined by arcs. To begin with we join $n^{(0)}$ to each of the r nodes on level 1. Then a node $n_s^{(k)}$ on the kth level is to be joined to a node $n_t^{(k+1)}$ on the $(k+1)$st level if and only if the $(2k-1) \times (2k-1)$ block corresponding to $n_s^{(k)}$ can be surrounded by tiles from $\mathscr{S}$ to form the $(2k+1) \times (2k+1)$ block that corresponds to $n_t^{(k+1)}$. No other pairs of nodes are to be joined.

The graph G has the property that it is connected, possesses a finite number of nodes on each level, and has infinitely many nodes in all. König's Infinity Lemma (König [1927], Knuth [1968, p. 381]), asserts that, under these conditions, there exists an infinite path starting at $n^{(0)}$ and passing through exactly one node of G at each level. The proof of this fact is easy; we can select the nodes recursively. Having constructed part of the path we may extend it by choosing a node on the next level connected to it and such that this node has an infinite number of successors (that is, nodes on higher levels which are connected to it by paths).

The nodes of this infinite path correspond to larger and larger square blocks of tiles, each of which is an extension on all sides of the previous block. Clearly this gives a method of tiling the plane using the tiles of $\mathscr{S}$, yielding a tiling we call the inclusion limit of the blocks; hence Statement 11.2.1 is proved.

The proof fails if $\mathscr{S}$ is infinite since the Infinity Lemma only applies if the number of nodes on each level is finite. In particular, it is apparent that the construction described in the proof cannot be carried out in this case.

If a set $\mathscr{S}$ of prototiles admits a tiling of the plane, but does not do so if any one tile is removed from $\mathscr{S}$, then $\mathscr{S}$ is called *minimal*.

11.2.2 *Suppose $\mathscr{S}$ is a finite minimal set of Wang prototiles that admits a tiling of the plane. Then there exists an* *integer n such that a copy of every prototile in $\mathscr{S}$ occurs in* every $n \times n$ *block of tiles in* every *tiling admitted by $\mathscr{S}$.*

To prove this, let us suppose that the statement is false. Then for some prototile $T \in \mathscr{S}$ there will exist arbitrarily large square blocks which do not contain copies of T. However, Wang's Theorem then implies that $\mathscr{S}$ admits a tiling from which tiles congruent to T are absent. This contradicts the assumption that $\mathscr{S}$ is minimal, and so proves Statement 11.2.2.

In Section 11.1 we gave examples of Wang tiles which admit a tiling of the plane but do not admit a *periodic* tiling. The following two statements show that even for such tilings, blocks of tiles repeat infinitely often and so the tilings are, in an intuitive sense, *almost* periodic.

11.2.3 *Given a finite set $\mathscr{S}$ of Wang prototiles that admits a tiling of the plane, and any integer n, there exists a tiling $\mathscr{T}$ using $\mathscr{S}$ in which every $n \times n$ block of tiles that occurs in $\mathscr{T}$ does so infinitely often.*

Let $\mathscr{T}_0$ be any tiling using the prototiles $\mathscr{S}$. Since the number of tiles in $\mathscr{S}$ is finite, $\mathscr{T}_0$ must contain infinitely many copies of at least one prototile; let $\mathscr{M}_1 \subset \mathscr{S}$ be the (non-empty) set of all such prototiles. Next, let $\mathscr{M}_2$ be the set of all 2×2 blocks of tiles which occur infinitely often in $\mathscr{T}_0$; clearly each such block contains only tiles of $\mathscr{M}_1$.

It is obvious that $\mathscr{M}_2$ is not empty since, on the one hand, only a finite number of 2×2 blocks of tiles can be built up with the tiles of $\mathscr{S}$, and on the other hand, copies of those tiles of $\mathscr{S}$ which do not belong to $\mathscr{M}_1$ appear only in a finite number of 2×2 blocks. We now proceed recursively, defining $\mathscr{M}_n$ for all $n \geq 1$ to be the set of all $n \times n$ blocks of tiles that occur infinitely often in $\mathscr{T}_0$. As before, we can see that $\mathscr{M}_n$ is, for each n, non-empty, and that all its "sub-blocks" belong to $\mathscr{M}_1, \mathscr{M}_2, \ldots, \mathscr{M}_{n-1}$.

We now proceed as in the proof of Statement 11.2.1, letting the nodes of the graph G correspond to the blocks in $\mathscr{M}_1, \mathscr{M}_3, \mathscr{M}_5$, etc. An infinite path in G corresponds to a new tiling $\mathscr{T}_1$. From the construction it will be seen that, for all n, every $n \times n$ block that occurs in $\mathscr{T}_1$ necessarily occurs infinitely often in $\mathscr{T}_0$. In effect, we have "eliminated" from $\mathscr{T}_0$ those blocks that occur only finitely often. However, $\mathscr{T}_1$ may still contain some square blocks that occur only finitely often in $\mathscr{T}_1$; therefore we repeat the construction and so obtain a tiling $\mathscr{T}_2$. Continuing in this way we arrive at a sequence of tilings $\mathscr{T}_0, \mathscr{T}_1, \mathscr{T}_2, \ldots$. Since there are only a finite number of possible $n \times n$ blocks of tiles that can be constructed from $\mathscr{S}$, it is clear that the process must terminate; the final tiling $\mathscr{T}$ in the sequence has the property that every $n \times n$ block of tiles which it contains occurs infinitely often. This is the tiling required to establish Statement 11.2.3.

EXERCISES 11.2

1. A finite set $\mathscr{S}$ of Wang prototiles has the property that it is possible to lay copies of the tiles in an infinite horizontal row, with contiguous edges matching, and such that, for all m, any row of m colors occurring consecutively along the top edges of the tiles also occurs somewhere along the bottom edges. Prove that $\mathscr{S}$ admits a tiling of the plane.

2. (*a*) Instead of using a set of n Wang prototiles in the theorems of this section, it is possible to use a set of $2n+1$ hexagons with colored sides as prototiles, reflections and rotations being allowed. Show that this can be achieved by the following construction, due to Wang [1975]. As usual, letters and numbers correspond to the various colors.

One of the hexagons is colored as in Figure 11.2.1(a), where 1, 2, 3, 4, 5, 6 represent new colors; every Wang prototile T colored as in Figure 11.2.1(b) is replaced by a pair of hexagons, colored as in Figure 11.2.1(c), where x is a new color, different for each prototile T. (It will be noted that this construction introduces $6+n$ new colors.)

*(b) Conversely, show that every tiling which uses as prototiles a set of m regular hexagonal tiles with colored sides is equivalent to a tiling using $36m$ Wang tiles as prototiles. [HINT: Show that each of the 12 possible orientations of a hexagon can be replaced by a suitable set of three Wang tiles.]

3. Verify the assertion at the beginning of this section that if we were allowed either to reflect Wang tiles, or rotate them through an angle of 180°, then every set of Wang prototiles would admit a tiling of the plane, and so Statement 11.2.1 would become trivial.

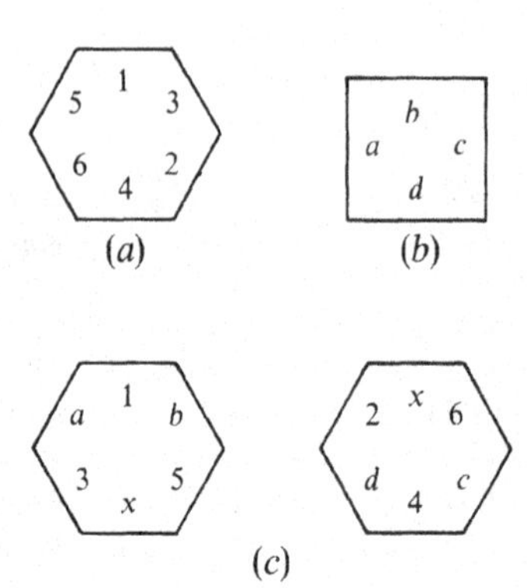

Figure 11.2.1
The construction of a set of $2n+1$ regular hexagons with colored sides that have similar properties to a set of n Wang tiles (see Exercise 11.2.2).

**4. Decide the validity of the following assertion: Given a finite set $\mathcal{S}$ of Wang prototiles that admits a tiling of the plane, there exists a tiling $\mathcal{T}$ using $\mathcal{S}$ in which every square block that occurs in $\mathcal{T}$ does so infinitely often.

11.3 THE DECIDABILITY OF THE TILING PROBLEM

In Section 1.1 we remarked on the difficulty of deciding whether or not a given prototile admits a monohedral tiling of the plane. This is a particular instance of the *Tiling Problem*: given a set $\mathcal{S}$ of prototiles, does there exist an algorithm or standard procedure for deciding whether $\mathcal{S}$ admits a tiling? The Tiling Problem is said to be *decidable* if there exists an algorithm which will yield a solution of it for any given set $\mathcal{S}$ of prototiles in a finite number of steps or trials.

Questions of decidability have long interested mathematical logicians, and about twenty years ago Hao Wang began an investigation into the decidability of the Tiling Problem. The following is a simplified version of his approach. He observed that if a set $\mathcal{S}$ of prototiles admits a tiling, then one of the following three possibilities must hold:

(a) $\mathcal{S}$ admits only periodic tilings. The simplest example of this occurs when $\mathcal{S}$ consists of just one regular hexagon for this only admits the regular (periodic) tiling (6^3).

(b) $\mathcal{S}$ admits both periodic and non-periodic tilings. This occurs, for example, if $\mathcal{S}$ consists of a square tile.

(c) $\mathcal{S}$ admits only non-periodic tilings, in other words, is an aperiodic set.

Wang then showed that the Tiling Problem is decidable if we only consider sets $\mathcal{S}$ which satisfy (a) and (b). He went on to conjecture (in 1961) that possibility (c) could

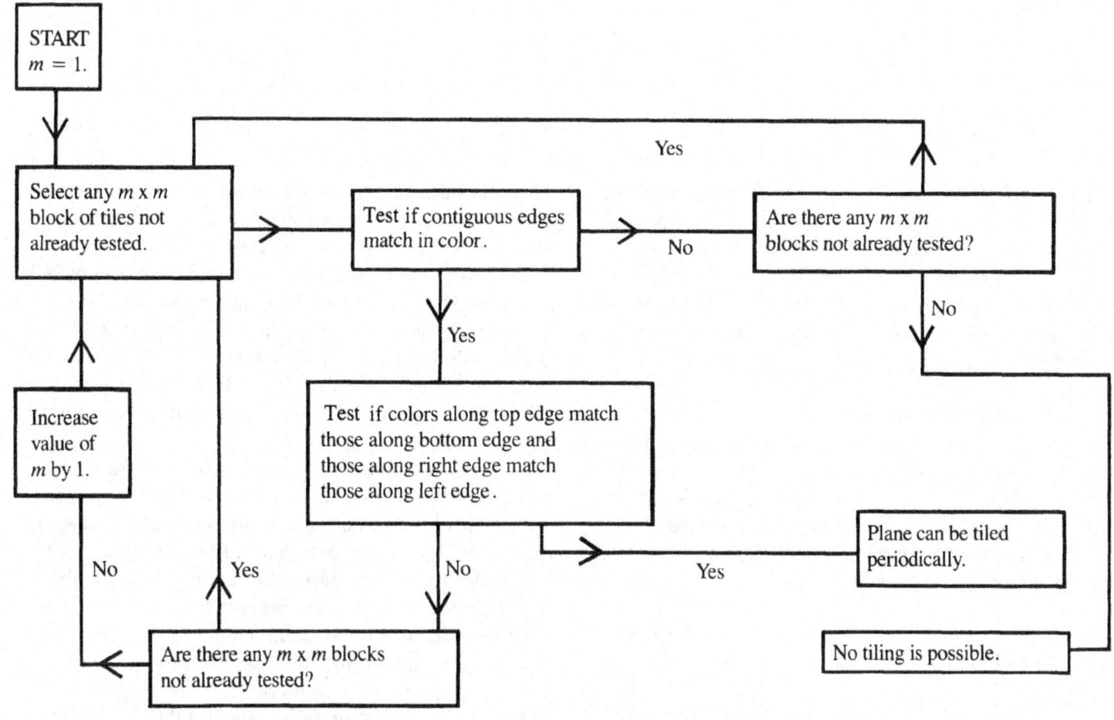

Figure 11.3.1
The flow diagram of a suggested algorithm to decide whether a given set of Wang tiles admits a tiling of the plane.

not occur. Although, with the advantage of hindsight, we now know this conjecture to be false, at the time it was made it seemed quite natural—not only were no aperiodic sets known, but no one had any idea how such a set could be constructed.

Berger's discovery of an aperiodic set in 1966, described in Section 11.1, upsets Wang's argument, and in fact it is now known that the Tiling Problem is undecidable.

In the rest of this Section we shall explain the relevance of aperiodicity to the question of decidability by giving a simple algorithm for trying to decide whether a given set $\mathcal{S}$ admits a tiling. For simplicity, as in the original investigations we shall assume that $\mathcal{S}$ is a set of Wang tiles. We make a sequence of "trials", each trial to consist of assembling on $m \times m$ block of tiles for $m = 1, 2, 3, \ldots$ and, for each m, to do so in all possible ways. We then *test* each arrangement as follows (see Figure 11.3.1):

(*a*) If contiguous edges of the tiles do not match in color then we reject the arrangement under consideration and proceed to the next.

(*b*) If all arrangements, for some value of m, are rejected because of (*a*) then we know no tiling of the plane is possible.

(*c*) For each arrangement which is admitted by (*a*) we see whether the colors on the left side of the block match those on the right, and the colors along the top match those along the bottom. If this is so, then clearly a periodic tiling can be constructed using translations of the given block. If this condition is not fulfilled, then we reject the given block and proceed to the next.

If Wang's conjecture were true then the process would

always terminate after a finite number of steps. For there are only a finite number of arrangements of tiles for each value of m, so *either* (because of Wang's Theorem, Statement 11.2.1) we would eventually arrive at the situation described in (*a*) and no tiling of the plane would be possible, *or* we would arrive at the situation described in (*c*) and a periodic tiling would exist. However, if we apply the algorithm to an aperiodic set of Wang tiles then clearly it will not terminate.

It must be emphasized that this argument shows that the existence of aperiodic sets implies that the suggested algorithm fails. It does not, by itself, prove that the Tiling Problem is undecidable. The fact that it is undecidable has been proved by Berger [1966] and Robinson [1971] and the reader should consult these papers for details.

11.4 COMPUTING BY TILES

A Turing machine is a primitive form of computer. It can be described as a device which is able to read symbols from an infinite "memory tape" and to act on these symbols in accordance with a specified "program", either by writing new symbols on the tape or by moving the tape backwards or forwards. The theory of Turing machines plays a significant role in mathematical logic for it can be shown that such a machine can compute all *recursive* functions, that is, functions whose values can be calculated in a finite number of steps. These include all the familiar functions of arithmetic such as addition, multiplication, exponentiation and so on. In fact, in spite of the simplicity and apparent inefficiency, a Turing machine is just as versatile as any digital computer that has been, or ever can be, constructed. There is an extensive literature on the subject, references to which can be found in Kleene [1967].

Wang [1975] pointed out that it is possible to simulate the operation of any Turing machine by means of a suitable tiling—in effect a set of tiles can be found which will effectively carry out any computation that is possible by a Turing machine. The basic idea is to use rows of tiles to simulate the tape in the machine, successive rows corresponding to consecutive "states" of the machine.

Instead of describing this simulation in detail, it seems more interesting to present the material in the form of a number of examples which illustrate the principles involved. We use Wang tiles and the tiling will occupy only one-quarter of the plane—there will be an upper boundary and a left boundary as indicated in Figures 11.4.1(*a*) and 11.4.2(*a*). We shall assume that these boundaries are colored with the color denoted by 0, so that each tile which abuts against a boundary must carry the color 0 on the appropriate edge.

Our first example is very simple—we show how tiles can be used to perform the addition of positive integers a, b, where, for convenience, we assume that $2 \leq a < b$. In Figure 11.4.1(*b*) we show a set of 16 prototiles which will perform this operation. The "input" of the numbers a and b into the tiling is carried out by placing two of the tiles marked A in positions a and b in the top row. (See Figure 11.4.1(*a*) for the case $a=5$, $b=9$. To simplify the diagram we have marked the colors on the edges of the tiling rather than on the edges of the tiles, and we have omitted the color 1. Every unlabelled edge is assumed to be of color 1.) If no other copies of the tile A are used, then a unique tiling is forced, and this will contain a single copy of the tile S in the top row. If tile S lies in position s, then $s=a+b$, and so indicates the sum of the given integers. (See Exercise 11.4.7.)

The principle on which the tiling is constructed may be described briefly as follows. The tile $*$, which is forced to fit in the top left corner, initiates a "signal" passing diagonally downwards to the right using the tiles in the first column (*i*) of Figure 11.4.1(*b*). This signal is indicated by the colors 3 and 4. The tiles A send signals (using color 2) vertically downwards. When the signal from the left tile A meets the diagonal signal it is turned horizontally using the lowest tile in column (*ii*) and is then propagated using color 5. When this horizontal signal meets the vertical signal from the right tile A it is turned diagonally upwards (using the lowest tile in column (*iii*)) and then propagated diagonally upwards (using the tiles in column (*iv*)) until it meets the upper boundary where it is absorbed by tile S. Thus S is forced to lie in the correct position as shown. The remaining three tiles, in column (*v*), can be regarded as "neutral" since they only serve

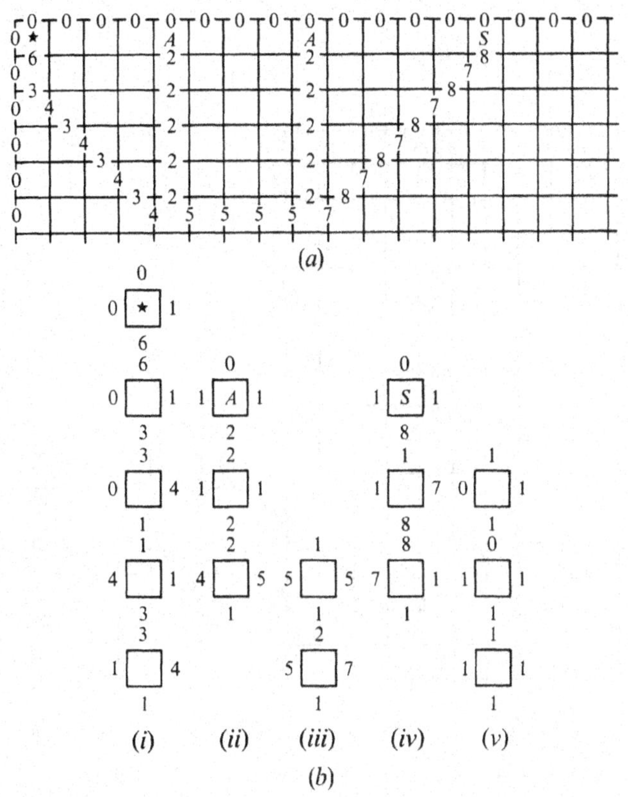

Figure 11.4.1
A set of tiles which perform the operation of addition of unequal positive integers.

to fill up the spaces between the tiles conveying the signals.

In this example we have not attempted to reduce either the number of tiles, or the number of colors, to a minimum. Our main purpose was to show how the use of "signals" in the tiling can perform the necessary operation.

Various modifications and elaborations of this example immediately suggest themselves. In particular, it is simple to find a set of tiles which force a tiling "illustrating" any sequence of positive integers defined by means of addition or repeated addition. We illustrate this remark by our second example (Figure 11.4.2(*a*)) in which the positions of the tiles *F* in the top row are

$$1, 2, 3, 5, 8, 13, 21, 34, 55, \ldots.$$

This is the well-known Fibonacci sequence. (See page 563 for the recursive definition of this sequence and also its relation to continued fractions. Here we have, for obvious reasons, deleted the initial 1.) The 17 prototiles shown in Figure 11.4.2(*b*) force the unique tiling of Figure 11.4.2(*a*).

In a case like this, it is necessary to specify the first few terms of the sequence for the rest to be determined by the recursion. For the Fibonacci sequence the first two terms are forced by tiles F', F'' which will only fit in the positions shown. After that, the tiles *F* are determined by a modification of the procedure for addition described above. The various signals that are used will be apparent from an inspection of the diagram.

Our third, and final, example is due to E. F. Moore and M. Fieldhouse; it was communicated to us by H. Wang. Here 30 prototiles (Figure 11.4.3(*b*)) force a unique tiling of the quarter plane in which tiles *P* and *C* occur at the prime and composite positions in the top row. The unit 1 is indicated by a special tile * in the first position.

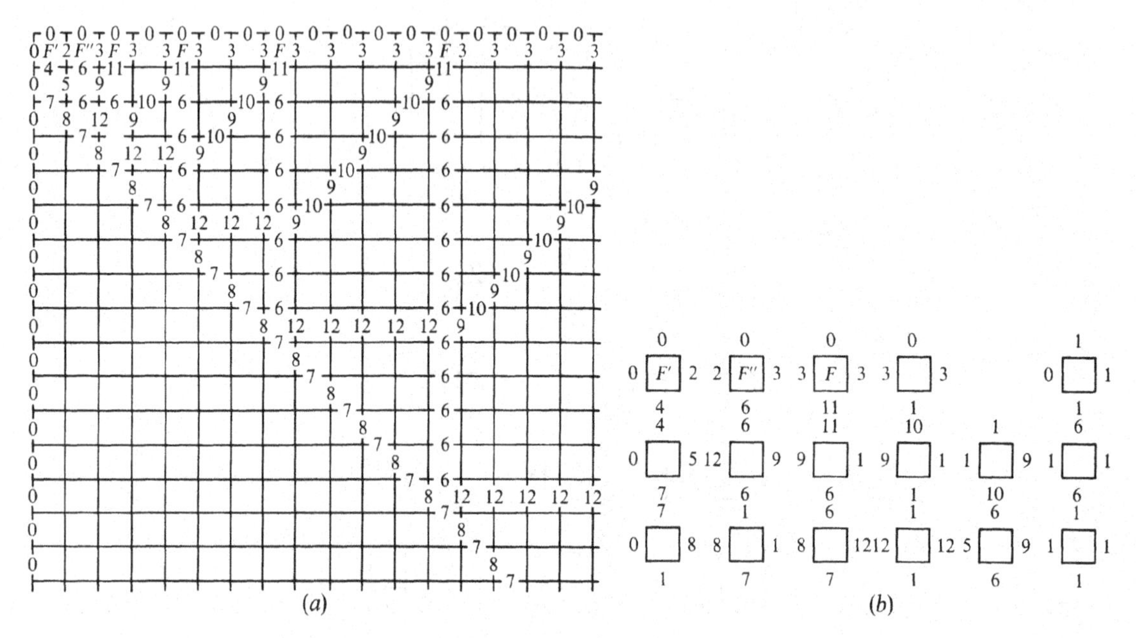

Figure 11.4.2
A tiling which indicates the Fibonacci sequence 1, 2, 3, 5, 8, 11, ...

The tiling is shown in Figure 11.4.3(*a*). Here the various tiles are specified by letters (instead of denoting the colors of their edges by numbers) and it will be seen how they build up rows of square blocks which increase in size as we move downwards. These blocks are indicated by thicker lines. The tiles marked with an asterisk are used to generate the blocks down the left boundary, and the tiles with superscript C transmit a signal to the effect that the number is composite. These "composite" signals are generated by the tiles B^* (which is only used once) and D_4 (which occurs at the top right corner of each square block that does not abut on the left boundary of the tiling). They are absorbed by the tiles C on the top row, and D_6 on the bottom right corner of certain blocks. When the signal is absorbed in this way it is regenerated higher up by a tile D_4.

The increasing blocks of tiles signal to the top row when a number has a proper divisor greater than 2. Divisibility by 2 is taken care of by the alternating A- and B-tiles in the second row. Either can transmit the composite signal. The special tile B^* is introduced simply to ensure that a prime tile P occurs in position 2.

The design of this and similar tilings clearly requires a great deal of ingenuity, especially if one attempts to be "economical" and use a small number of prototiles to achieve the required result. In the exercises, the discovery of several such tilings is requested.

For additional information relating to the topics of this and the previous section, the reader should consult the papers of Wang [1961], [1963], [1975]. A more popular account, including an explanation of the connection between tilings and Turing Machines, is given in Wang [1965]. Related material can be found in Hanf [1974], Myers [1974] and Heidler [1976].

Somewhat related to the material of this section is the mathematically and aesthetically appealing work of Miller [1968], [1970], [1980] and ApSimon [1970] on "forests of stunted trees".

★	P	P	C	P	C	P	C	C	C	P	C	P	C	C	C	P
A^*	B^*	A	B	A	B^C	A	B^C	A^C	B^C	A	B^C	A	B^C	A^C	B^C	A
D_1^*	D_3^*	D_4^*	D_1	D_2	D_4	D_1	D_2^C	D_4	D_1^C	D_2	D_4	D_1	D_2^C	D_4	D_1^C	D_2
E^*	F	D_5^*	E	F	D_5	E	F^C	D_5	E^C	F	D_5	E	F^C	D_5	E^C	F
G^*	E	D_6^*	G	E	D_6	G	E^C	D_6	G^C	E	D_6^C	G	E^C	D_6^C	G^C	E
D_1^*	D_2	D_3^*	D_4^*	D_1	D_2	D_2	D_4	D_1	D_2^C	D_2	D_4	D_1	D_2^C	D_2^C	D_4	D_1
E^*	F	G	D_5^*	E	F	G	D_5	E	F^C	G	D_5	E	F^C	G^C	D_5	E
G^*	E	F	D_5^*	G	E	F	D_5	G	E^C	F	D_5	G	E^C	F^C	D_5	G
G^*	G	E	D_6^*	G	G	E	D_6	G	G^C	E	D_6^C	G	G^C	E^C	D_6	G
D_1^*	D_2	D_2	D_3^*	D_4^*	D_1	D_2	D_2	D_2	D_4	D_1	D_2^C	D_2	D_2^C	D_4	D_1	D_2
E^*	F	G	G	D_5^*	E	F	G	G	D_5	E	F^C	G	G^C	D_5	E	F
G^*	E	F	G	D_5^*	G	E	F	G	D_5	G	E^C	F	G^C	D_5	G	E
G^*	G	E	F	D_5^*	G	G	E	F	D_5	G	G^C	E	F^C	D_5	G	G
G^*	G	G	E	D_6^*	G	G	G	E	D_6	G	G^C	G	E^C	D_6	G	G
D_1^*	D_2	D_2	D_2	D_3^*	D_4^*	D_1	D_2	D_2	D_2	D_2	D_4	D_1	D_2^C	D_2	D_2	D_2
E^*	F	G	G	G	D_5^*	E	F	G	G	G	D_5	E	F^C	G	G	G
G^*	E	F	G	G	D_5^*	G	E	F	G	G	D_5	G	E^C	F	G	G
G^*	G	E	F	G	D_5^*	G	G	E	F	G	D_5	G	G^C	E	F	G
G^*	G	G	E	F	D_5^*	G	G	G	E	F	D_5	G	G^C	G	E	F
G^*	G	G	G	E	D_6^*	G	G	G	G	E	D_6	G	G^C	G	G	E

(*a*)

Tile	top	left	right	bottom
★	0	0	2	9
A^*	9	0	9	1
B^*	2	9	1	10
E^*	4	0	4	1
G^*	1	0	1	1
P	0	2	2	2
A	2	1	2	1
B	5	2	1	1
E	4	1	4	1
F	1	4	1	4
G	1	1	1	1
C	0	2	2	5
A^C	5	1	2	6
B^C	5	2	1	6
E^C	7	1	4	6
F^C	6	4	1	7
G^C	6	1	1	6
D_1^*	1	0	3	4
D_3^*	10	3	11	1
D_4^*	1	11	8	11
D_5^*	11	1	1	11
D_6^*	11	4	1	10
D_1	1	8	3	4
D_2	1	3	3	1
D_4	6	3	8	3
D_5	3	1	1	3
D_6	3	4	1	1
D_1^C	6	8	3	7
D_2^C	6	3	3	6
D_6^C	3	1	1	6

(*b*)

Figure 11.4.3
A tiling in which the tiles labelled P and C occur at the prime and composite positions in the top row.

EXERCISES 11.4

1. Design tilings of the quarter-plane by Wang tiles which calculate
   (*a*) the sum $a+b+c$,
   (*b*) the product ab, and
   **(*c*) (a,b), the greatest common divisor of a and b.
   Here, a, b, c are unequal positive integers (each greater than 1) which are indicated by the positions of certain designated tiles in the top row of the tiling.
2. Modify the previous constructions and design tilings of the quarter-plane by Wang tiles which calculate
   (*a*) the sum $a + b$,
   (*b*) the product ab
   of positive integers a, b, equal or not. Indicate a by suitably positioning a designated tile in the top row of the tiling, and b by placing another designated tile in the appropriate place in the first column.
3. Construct sets of Wang tiles which force a designated tile to lie at positions
   (*a*) 1, 4, 9, 16, 25, . . . (squares),
   (*b*) 1, 3, 6, 10, 15, 21, . . . (triangular numbers),

(c) 1, 2, 4, 8, 16, 32, . . .   (powers of 2),
in the top row of a tiling of the quarter plane.

4. Show that the set of 32 prototiles of Figure 11.4.4 forces a tiling of the quarter plane in which * occurs in position 1, and the remaining tiles in the top row are P and C lying at the prime and composite positions respectively. (This example, also due to Moore and Fieldhouse and communicated to us by H. Wang, is, in some ways, more economical than that described in the text. Only blocks of tiles of odd size are used, and both prime and composite numbers are signalled vertically using the tiles with superscripts P and C.)

*5. Design tilings of the quarter plane by Wang tiles which will calculate *either* $a+b$ *or* ab according to the choice of an "operation tile" S (=sum) or P (=product) placed at the top left corner of the tiling. Here a and b are any unequal positive integers greater than 1.

6. Design tilings of the lower half-plane by Wang tiles which calculate
(a) the sum $a+b$,
(b) the product ab
of distinct arbitrary (positive or negative) integers a, b. (Note that it will be necessary to designate one tile as indicating the zero position in the top row.)

7. Explain what happens if more than two A-tiles are introduced into the top row of Figure 11.4.1. Also show how the tiles can be modified so that they still calculate the sum of two unequal positive integers indicated by A-tiles, yet no tiling exists with more than two A-tiles.

Tile	Top	Left	Right	Bottom
★	0	0	2	12
A^*	12	0	12	1
B^*	5	12	1	1
E^*	4	0	4	1
G^*	1	0	1	1
P	0	2	2	5
A^P	5	1	2	6
E^P	7	1	4	6
F^P	6	4	1	7
G^P	6	1	1	6
C	0	2	2	8
A^C	8	1	2	9
E^C	10	1	4	9
F^C	9	4	1	10
G^C	9	1	1	9
B	8	2	1	1
E	4	1	4	1
F	1	4	1	4
G	1	1	1	1
D_1^*	1	0	13	4
D_2^*	1	13	13	1
D_4^*	6	13	11	3
D_1^P	6	11	3	7
D_2^P	6	3	3	6
D_1^C	9	11	3	10
D_2^C	9	3	3	9
D_4^C	9	3	11	3
D_1	1	11	3	4
D_2	1	3	3	1
D_4	1	3	11	3
D_5	3	1	1	3
D_6	3	4	1	1

Figure 11.4.4
Another set of prototiles which determine prime and composite integers (see Exercise 11.4.4).

12

TILINGS WITH UNUSUAL KINDS OF TILES

Figure 12.0.1
One of Badoureau's uniform tilings using "hollow tiles", see Section 12.3.

12

TILINGS WITH UNUSUAL KINDS OF TILES

In this chapter we present a number of results on various interesting, but rarely considered, kinds of tilings. In the first three sections we explore some of the new possibilities that arise when we relax the condition that the tiles are topological disks, in the fourth section we consider multiple coverings of the plane by tiles, and in the last section we discuss various adjacency conditions. In each case there seems to be very little available literature, and the subjects abound with open problems.

12.1 TILES WITH CUTPOINTS

A tile, or plane set, T is said to *have cutpoints* if

(*a*) T is the union of a finite number c of closed topological disks $S_1, \ldots, S_c$, and
(*b*) there exists a finite subset P of T such that $T \backslash P$ is a non-connected set with c components whose closures are the sets $S_1, \ldots, S_c$.

Thus, in Figure 1.1.1, the tiles shown in (*c*) and (*d*) have cutpoints but the others do not. Further possibilities are illustrated in Figure 12.1.1. If P is minimal in the sense that no proper subset of P has the properties stated above, then P is called the *set of cutpoints of* T, the number of such points is denoted by p, and c is called the number of *components* of T.

In Figures 12.1.1(*a*), (*b*) and (*c*) we show tiles with one cutpoint and $c = 2$, 3 and 5 components. The tiles of Figures 12.1.1(*d*) and (*e*) each have two cutpoints, and those of Figures 12.1.1(*f*) and (*g*) have three and four, respectively. On the other hand, the tiles of Figure 12.1.2 do not satisfy the definition and are not tiles with cutpoints

For any prescribed values of $p \geq 3$ and $c \geq 3$ there are evidently many different topological types of tiles with cutpoints, and the number of different kinds of tiling

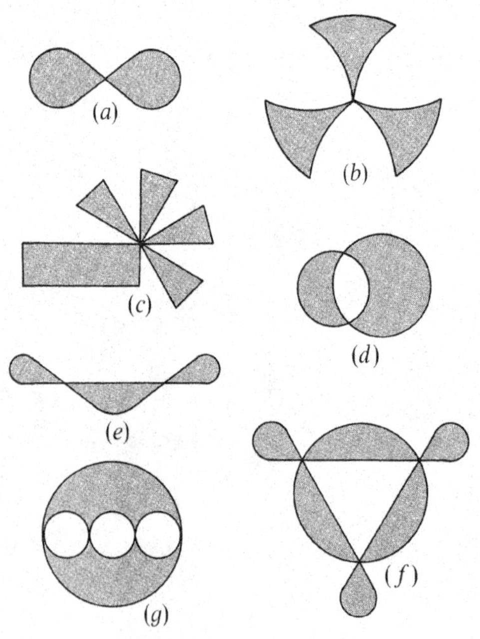

Figure 12.1.1
Examples of tiles with cutpoints.

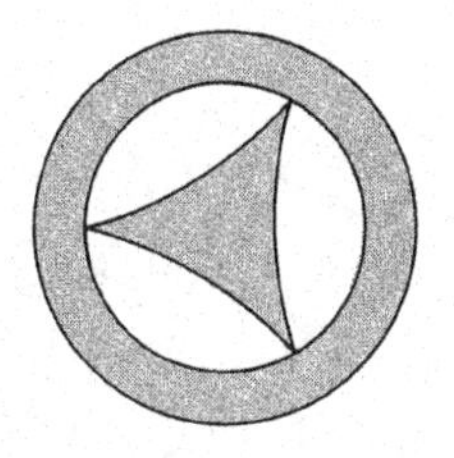
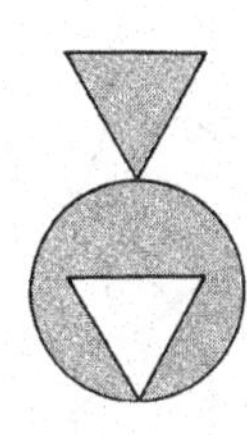

Figure 12.1.2
Two examples of sets which do not satisfy the definition of "tiles with cutpoints".

using such tiles is unmanageably large. Consequently, in the following we shall restrict attention to isohedral tilings or to homeohedral ones; even these occur in great profusion.

To begin with, consider the following construction. Start with any isohedral tiling $\mathcal{T}_0$, with prototile T_0, for which the induced tile group $S(\mathcal{T}_0|T_0)$ is non-trivial. Then it is possible to decompose T_0 into two or more sets, each of which is congruent to a set T with one cutpoint, in such a manner that the corresponding decomposition of every tile of $\mathcal{T}_0$ leads to an isohedral tiling $\mathcal{T}$ with prototile T. Examples of this construction appear in Figure 12.1.3. Here, and in later diagrams, in order to avoid the possibility of misinterpretation, the different parts of one tile have been shaded, colored or labelled in some distinctive manner. However, tilings constructed using the method just described are of little interest since, in each case, uniting all the tiles with a common cutpoint enables one to recover the original isohedral tiling $\mathcal{T}_0$, and the properties of $\mathcal{T}$ are trivially related to those of $\mathcal{T}_0$.

To avoid such possibilities we define a *restricted* tiling by tiles with cutpoints to be one in which no two tiles have a cutpoint in common. An example of a restricted isohedral tiling is given in Figure 12.1.4(*b*). Here the prototile, shown in Figure 12.1.4(*a*), has one cutpoint and two components. In Figure 12.1.5 we give examples where the tiles have two, or three, cutpoints, and other restricted isohedral tilings appear in the later diagrams. In each, the cutpoints are indicated by small black disks.

In order to investigate and classify restricted tilings, the following procedure is useful. Let $\mathcal{T}$ be a given tiling, and with each cutpoint associate a small closed topological disk containing the cutpoint in its interior. Then to each tile T of $\mathcal{T}$ *adjoin* the disks associated with its cutpoints, and *delete* from T the interiors of those disks which are associated with the cutpoints of the neighboring tiles. If the topological disks are chosen suitably then each tile T_0 which results from this construction will be a topological disk; moreover, if the original tiling $\mathcal{T}$ is isohedral then the new tiling $\mathcal{T}_0$, with prototile T_0, will also be isohedral, see Figure 12.1.4(*c*).

This construction can be reversed. We start with an isohedral tiling $\mathcal{L}$ whose prototile B is a topological disk, and a group G of isometries which acts transitively on the tiles of $\mathcal{L}$. Choose any integers $p \geq 1$ and $c \geq p+1$ and mark B by selecting $m = c+p-1$ points on its boundary. This selection of points must be made in such a way that when each tile of $\mathcal{L}$ is marked in the corresponding manner, the following two conditions are satisfied:

(*a*) the marked tiling has group G, and
(*b*) all the chosen points are distinct.

Now partition the m points on the boundary of B into p subsets $S_1, \ldots, S_p$ and join each pair of points in the same subset by an arc lying in the interior of B. This must be done so that the following additional conditions are fulfilled:

(*c*) each point must be joined to at least one other point (and therefore each subset S_i must contain at least two points),
(*d*) the arcs joining pairs of points of distinct subsets must be disjoint, and
(*e*) every two points of the same subset S_i are separated (on the boundary of B) by at least one vertex of the tiling. Thus if $\mathcal{L}$ is a tiling by n-gons then no subset S_i can contain more than n points; if S_i contains j vertices of $\mathcal{L}$ then S_i contains at most $n-j$ points.

Now we identify, in the usual topological sense, the

Figure 12.1.3
Isohedral tilings by tiles with cutpoints constructed using the method described in the text. These are not restricted tilings since two or more tiles may have cutpoints in common.

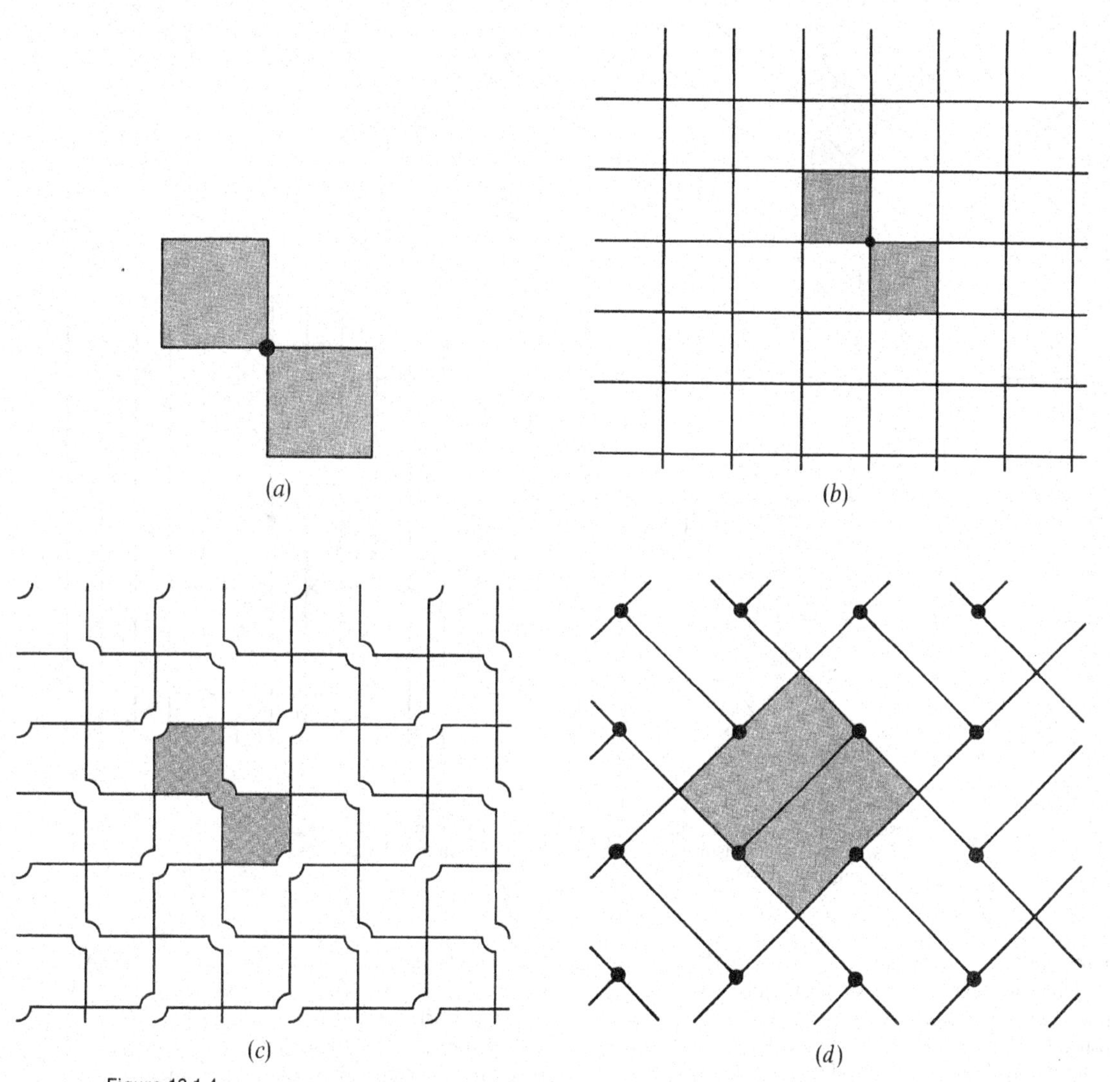

Figure 12.1.4
An illustration of the procedure for converting a restricted isohedral tiling by tiles with cutpoints (*b*) into one in which the tiles do not have cutpoints (*c*). The process can also be reversed, using identification, as indicated in (*d*).

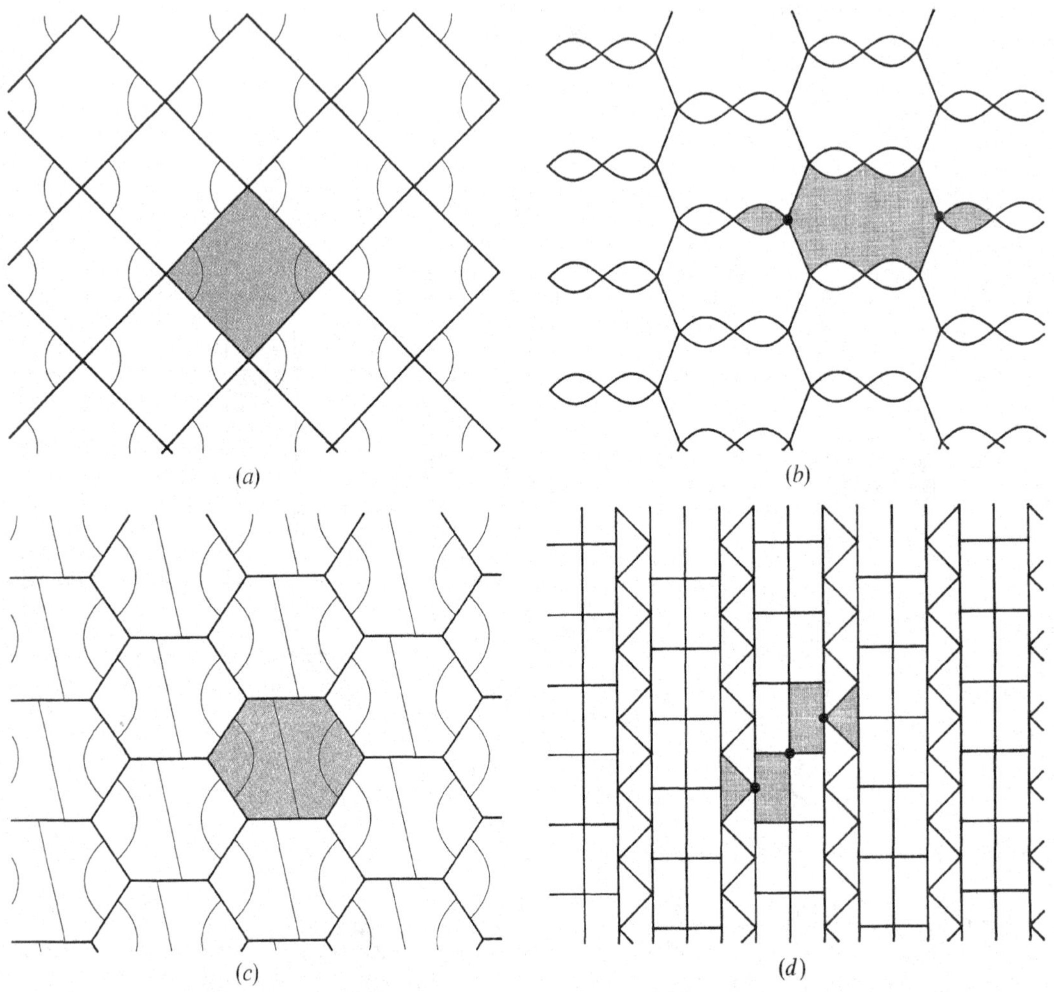

Figure 12.1.5
Examples of the identification procedure. Applied to the two Laves tilings on the left we obtain the isohedral tilings on the right, in which the tiles have two and three cutpoints, respectively.

points of each set S_i and its associated arcs. The resulting tiling is easily seen to be topologically tile-transitive and, if the shapes of the tiles are chosen suitably, it will be isohedral. Each tile will have p cutpoints and c components. A simple illustration of this procedure is given in Figure 12.1.4. Here $p = 1$ and $c = 2$; the marking of the tiles of the regular tiling $[4^4]$ is shown in Figure 12.1.4(*d*), and corresponding identification leads back to the restricted isohedral tiling of Figure 12.1.4(*b*). Another example, with $p = 1$ and $c = 3$, is given in Figure 12.1.6. Here we start with the Laves tiling $[3^2.4.3.4]$ and we note that one of the chosen points on the boundary of each tile coincides with a vertex—a possibility that is allowed by the conditions stated above.

The necessity of conditions (*a*) to (*e*) is clear; we remark that if condition (*e*) is violated, then a tile will be incident "more than once" with a cutpoint of an adjacent, and so will either fail to be a "tile with cutpoints" as we have defined it, or the tiling will not be restricted. These two possibilities are illustrated in Figure 12.1.7. Further it is not hard to see that the conditions are sufficient and, in fact, all topological types of restricted isohedral tilings can be obtained using the above construction.

It should be observed, however, that if one starts from a Laves tiling $\mathscr{L}$ and applies the above method to produce a restricted isohedral tiling $\mathscr{T}$, then the reverse procedure (that of replacing cutpoints by small topological disks) may lead to a Laves tiling $\mathscr{L}'$ different from $\mathscr{L}$. This is illustrated in Figure 12.1.8, and investigated in more detail in Exercise 12.1.6.

The following deductions are immediate:

12.1.1 *If the tiles of a restricted isohedral tiling have p cutpoints and c components then* $5p \geq c - 1 \geq p$.

These inequalities follow from $p \leq m/2$, $p \geq m/n$ (consequences of conditions (*c*) and (*e*) above), $m = c + p - 1$, and $n \leq 6$. In particular, in the case of a restricted tiling every tile must be simply connected (examples of non-simply connected tiles appear in Figures 12.1.1(*d*), (*f*) and (*g*) and 12.1.7(*b*)). This fact provides an alternative proof of the right inequality of Statement 12.1.1.

Since there are only a finite number of possible arrangements of $m = c + p - 1$ points on the boundary of a tile, and only a finite number of ways in which these can be partitioned in accordance with conditions (*a*) to (*e*) we deduce the following.

12.1.2 *There exists only a finite number of distinct topological types of restricted isohedral tilings in which each tile has a given number* $p \geq 1$ *of cutpoints. The same is true if we prescribe the number of components instead of the number of cutpoints. All these types can be obtained*

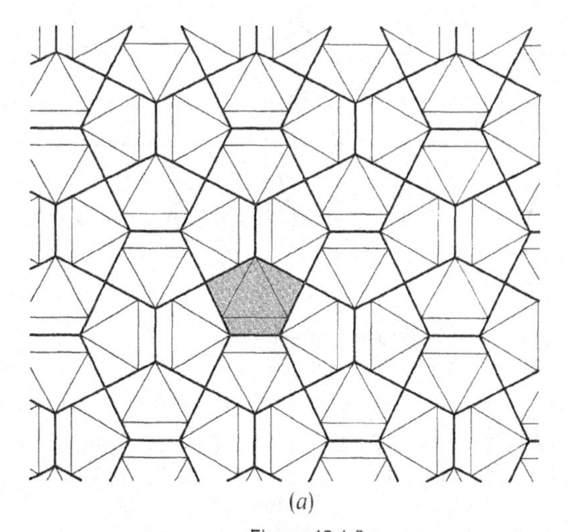

(*a*)

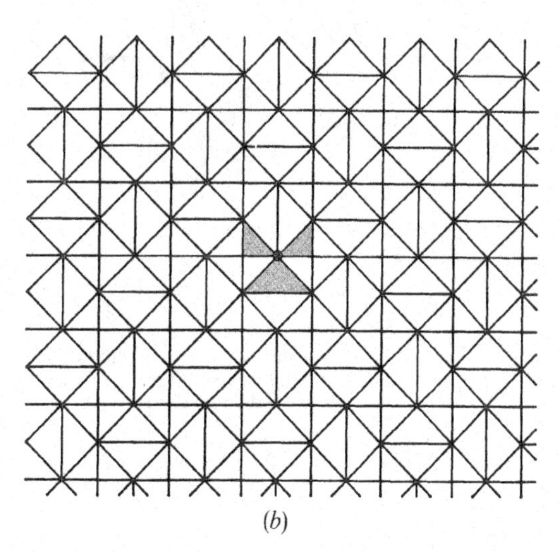

(*b*)

Figure 12.1.6
Another example of the identification procedure. Notice that here one of the chosen points is a vertex of the Laves tiling.

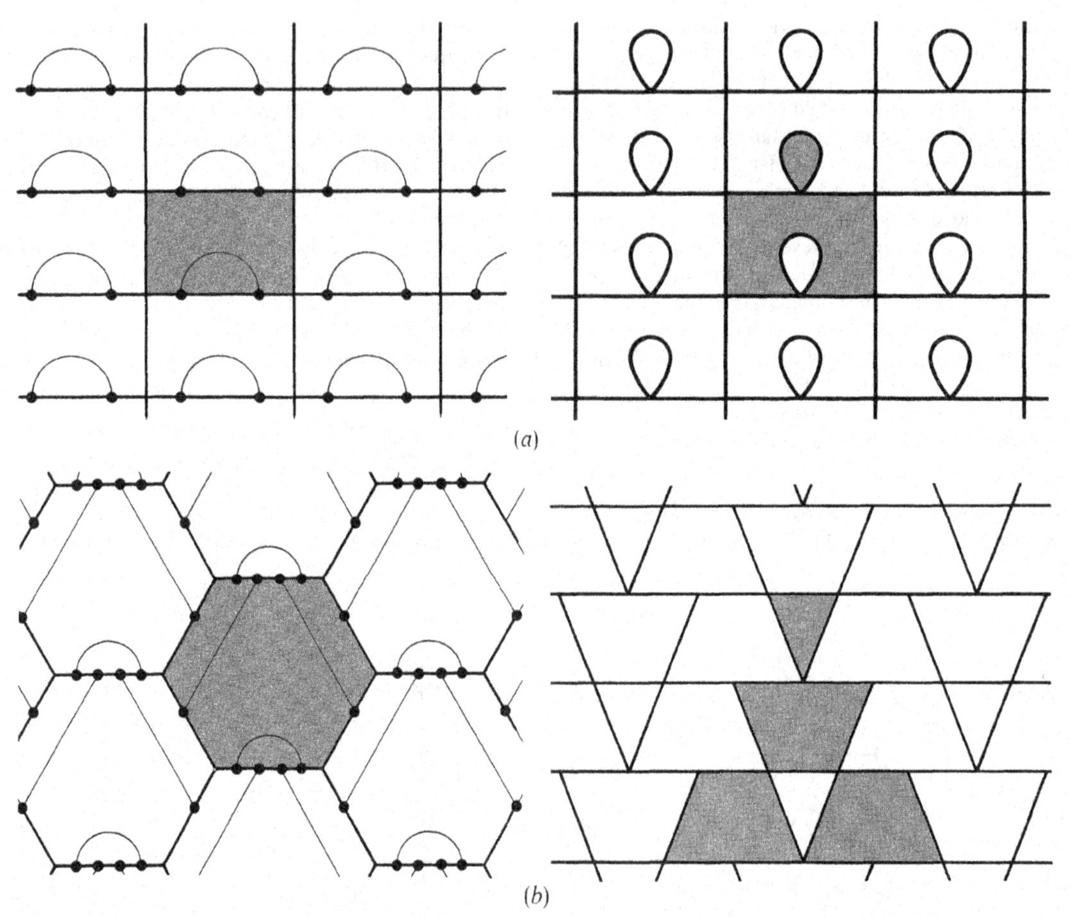

Figure 12.1.7
These diagrams show the necessity of condition (*e*) in the construction of isohedral tilings by the identification procedure. In (*a*) the tile is *not* a tile with cutpoint, and in (*b*) the tiling is *not* restricted.

from the eleven Laves tilings using the construction described above.

We have not attempted an enumeration of possible types, but content ourselves with displaying, in Figure 12.1.9, the six types of tiling derivable from the Laves tiling $[3.12^2]$ whose tiles have *one* cutpoint. Here $m=2$ or 3. Two of the many tilings derivable from $[3.12^2]$ with $p=2$ are shown in Figure 12.1.10; in each case $m=4$. A systematic investigation is not hard in principle, but in many cases the number of possibilities is very large. In contrast, we remark that *no* isohedral

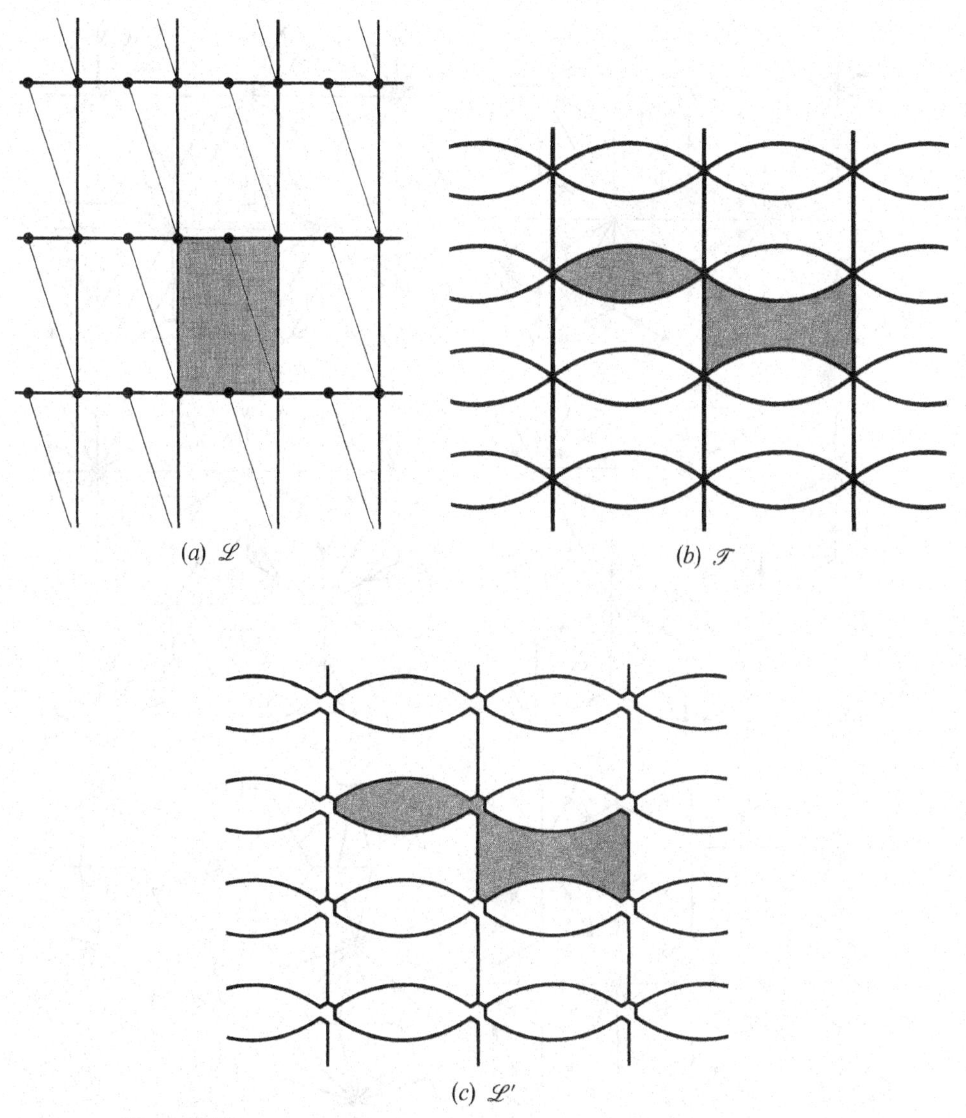

Figure 12.1.8
From the Laves tiling $\mathscr{L}$ (type $[4^4]$) of (*a*) the restricted tiling $\mathscr{T}$ of (*b*) is obtained by identification. If $\mathscr{T}$ is converted back to a Laves tiling (*c*) by replacing each cutpoint by a small disk, the resultant tiling $\mathscr{L}'$ is distinct from $\mathscr{L}$; it is of type $[3^6]$.

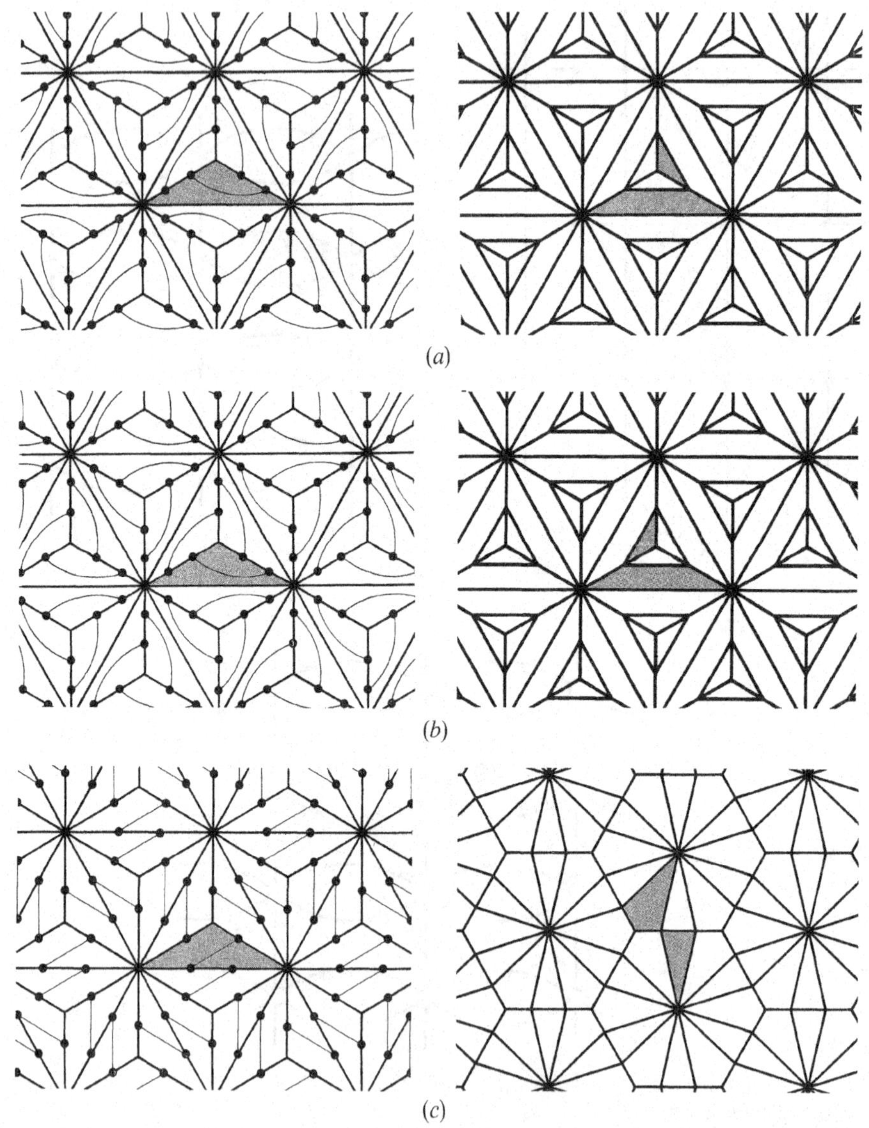

Figure 12.1.9
The six types of isohedral tilings whose tiles have one cutpoint, which can be derived from the Laves tiling [3.12^2] by identification.

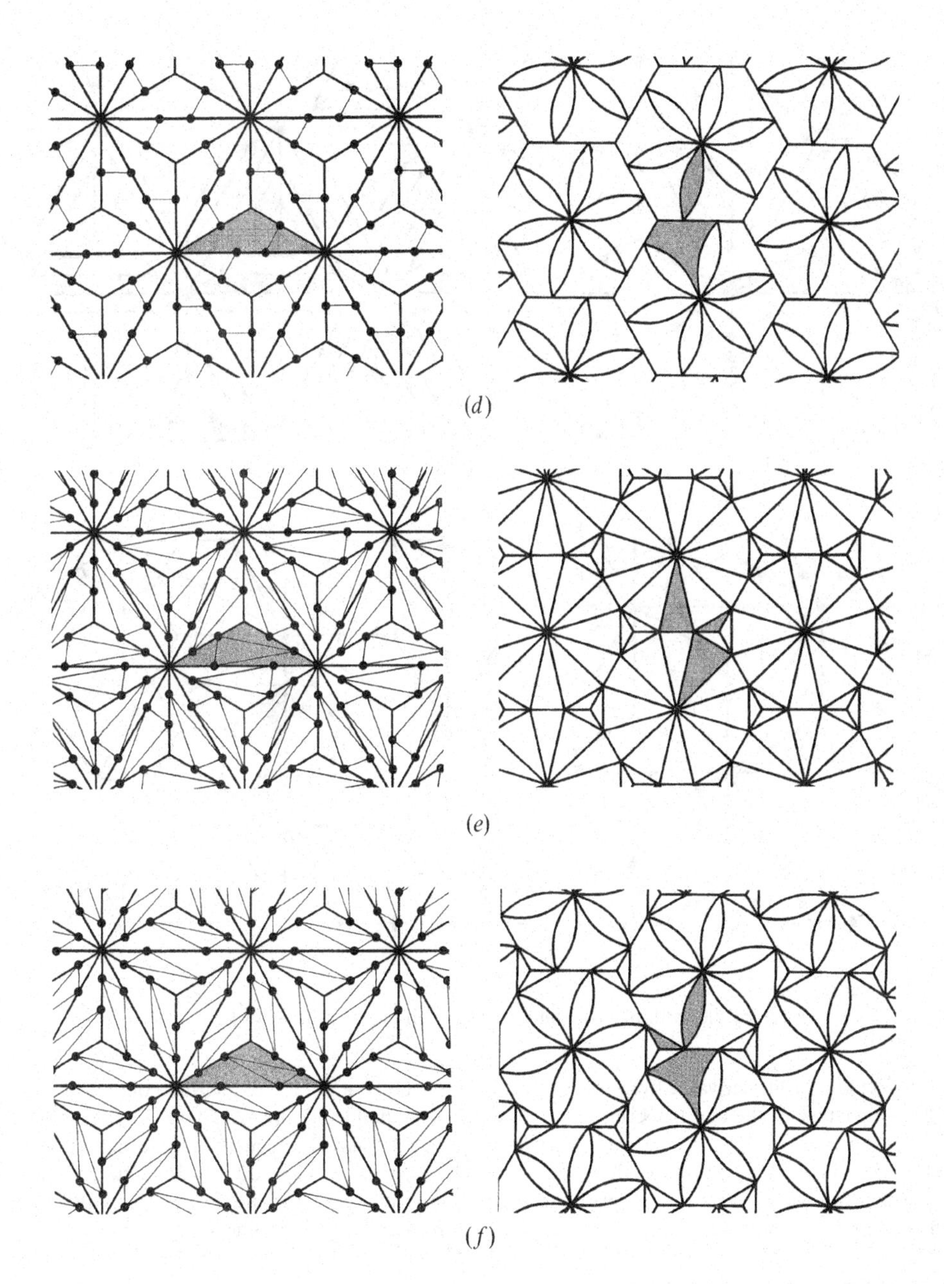
(d)

(e)

(f)

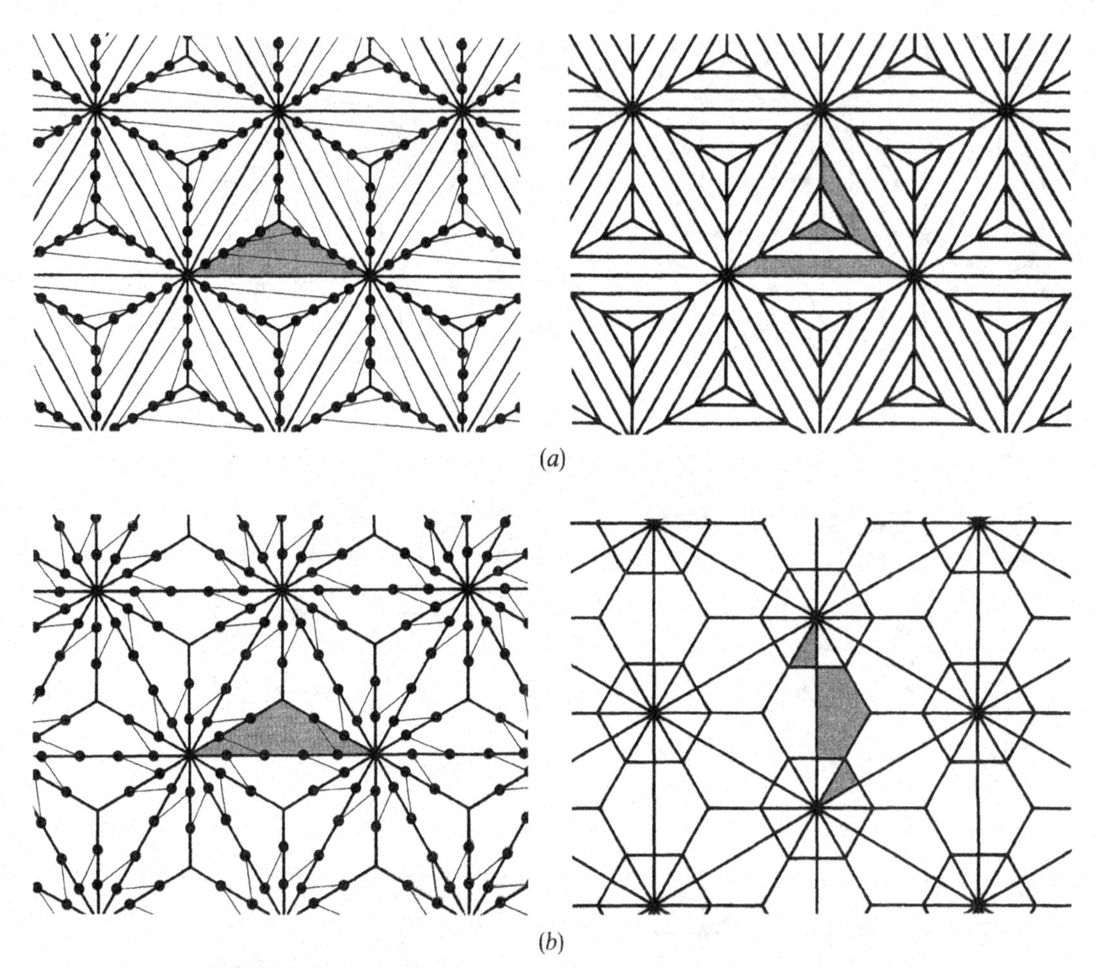

Figure 12.1.10
Two tilings, each by tiles with two cutpoints, derived from the Laves tiling [3.12^2] by identification.

tiling of any kind (restricted or not) is derivable from the Laves tiling [4.6.12] using either of the constructions described above.

Our treatment of tilings using tiles with cutpoints has necessarily been somewhat superficial. It is easy to construct tilings which are *partly-restricted* in the sense that tiles share some (but not all) of their cutpoints with other tiles (see, for example, Figure 12.1.7(*b*)). Isotoxal, isogonal, 2-homeohedral and similarity tilings using tiles with cutpoints are possible. The number of directions for investigation seems very large and the great majority of these seems to be completely unexplored.

EXERCISES 12.1

*1. Determine all the topological types of restricted isohedral tilings in which the prototile has *one* cutpoint and which are derivable, by the method described in this section, from the Laves tilings (*a*) $[3.8^2]$; (*b*) $[3.6.3.6]$.

*2. Give examples of restricted tilings by tiles with one cutpoint which are (*a*) isotoxal; (*b*) isogonal; (*c*) 2-homeohedral; (*d*) similarity tilings (in the sense used in Figure 10.1.1).

*3. Give a modification of the construction described early in this section (just prior to the definition of restricted tilings) which yields tilings with tiles that have $p = 2, 3, \ldots$ cutpoints. Is such a construction possible for each p whenever the induced tile group is non-trivial?

*4. Give an example of
   (*a*) a restricted isohedral tiling in which each tile has three cutpoints, and
   (*b*) a partly restricted isohedral tiling in which each tile has three cutpoints.

*5. Generalize the construction of restricted isohedral tilings described in connection with Figure 12.1.6 by allowing arcs as well as points to be chosen on the boundary of the tile, and later identified (shrunk) to a point. Of particular interest is the case where an arc contains two or more vertices of the tile.
   (*a*) Determine the tiling which results from the regular tiling $[3^6]$ by identifying to a single point an edge and the midpoints of two other edges (as indicated in Figure 12.1.11) of each hexagon.

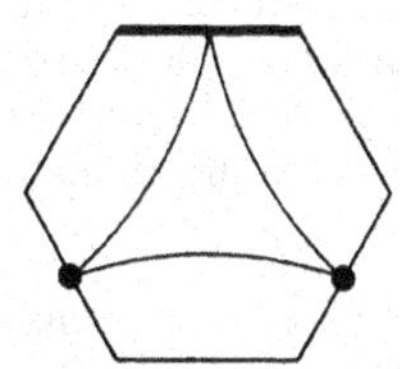

Figure 12.1.11
Identifications used in Exercise 12.1.5(*a*).

Show that the construction can be formulated in such a way that this leads to a unique tiling.
   (*b*) Describe how to obtain the same tiling by the construction in the original form, in which only points are chosen on the boundary of each tile of a Laves tiling.
   (*c*) Describe the additional tilings obtainable if the points taken in Figure 12.1.11 as midpoints of edges are allowed to move to other positions on the edges that carry them.
   (*d*) Perform the investigations indicated in parts (*a*), (*b*), and (*c*) if the set to be shrunk consists of the edge and points indicated in Figure 12.1.12.

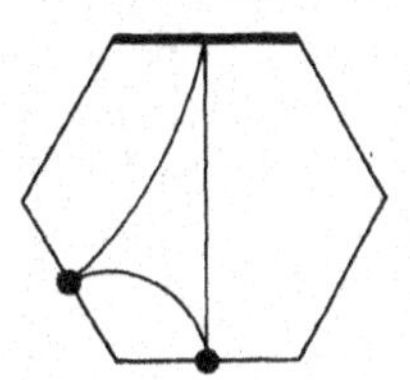

Figure 12.1.12
Identifications used in Exercise 12.1.5(*d*).

   (*e*) Prove that all restricted isohedral tilings can be obtained by the original construction, not using arcs on the boundary of a tile.

6. An isohedral tiling $\mathscr{T}$ by tiles with cutpoints is constructed by identification from a Laves tiling $\mathscr{L}$ as described above. One of the m points is chosen on the boundary of T so as to coincide with a vertex of T whose valence exceeds 3. Show that the tiling which results by replacing the cutpoints of the tiles of $\mathscr{T}$ by small topological disks, as described above, leads to a tiling $\mathscr{L}'$ whose topological type differs from that of $\mathscr{L}$.

7. (*a*) Using the prototile of Figure 12.1.13, construct an isohedral (not restricted) tiling which can reasonably be described as being of type IH63. We recall

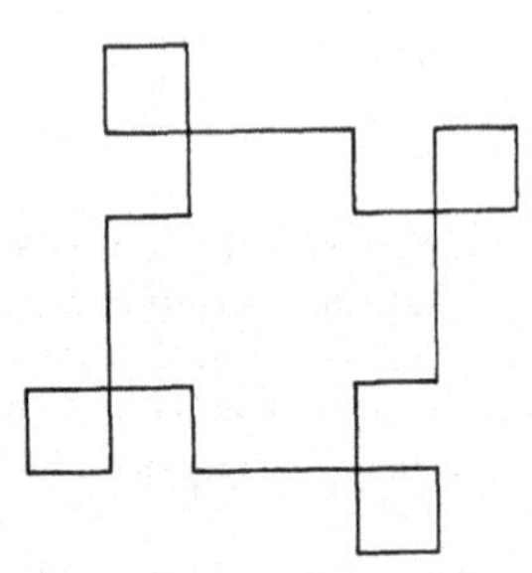

Figure 12.1.13
This tile admits an isohedral tiling which may be reasonably described as of type IH63 (see Exercise 12.1.7).

that there exist no unmarked tilings of this type in which the prototile is a topological disk.

*(*b*) Give a formal definition of "incidence symbols" for (at least some) tilings by tiles with cutpoints, which justifies the description in part (*a*).

*(*c*) Show that each of the 12 types of marked isohedral tilings, which we found in Section 6.2 not to be realizable by unmarked tilings in which the prototile is a topological disk, can be realized by (unmarked) tilings in which the prototiles have cutpoints. (For an analogous representation of the crystallographic groups in 3-dimensional space see Baškirev [1959].)

8. An isohedral tiling by tiles with cutpoints is called *idiomeric* if the stabilizer of each tile T acts transitively on the components of T (see Section 12.2 for an explanation of the name). For example, the tiling shown in Figure 12.1.4(*b*) is idiomeric, the one in Figure 12.1.6(*b*) is not.
   (*a*) Show that there exist precisely four topologically distinct types of idiomeric tilings by tiles with cutpoints, which can be derived from the regular tiling $[4^4]$ by the method explained in the text.
   (*b*) Show that there exists only a finite number of distinct topological types of idiomeric tilings by tiles with cutpoints.
   **(*c*) Determine all the types mentioned in part (*b*).

12.2 TILINGS WITH DISCONNECTED TILES

The possibility of using disconnected tiles in a tiling has already been mentioned briefly in the first chapter (see Figure 1.2.4); here we consider this topic in more detail. Because of the great abundance of tilings, in order to make the subject manageable, we shall begin by restricting attention to tilings which are isohedral and for which the prototile is the union of a finite number of disjoint topological disks.

If T has k components, then any monohedral tiling with prototile T will be called a *k-component tiling*, and we shall write $T = \{C_1, C_2, \ldots, C_k\}$. Two components C_i and C_j of T are *equivalent* if there exists an element of the induced tile group $S(\mathcal{T}|T)$ which maps C_i onto C_j. This relation partitions the set $\{C_1, \ldots, C_k\}$ into equivalence classes, and if the number of such classes is t, then $\mathcal{T}$ will be called a *t-class* tiling. Clearly $1 \leq t \leq k$ and in Figure 12.2.1 we give examples of k-component t-class tilings for various small values of k and t. In this and later diagrams we mark the various components of a tile with the same numeral, and color one of the tiles gray so that the shape of the prototile can be seen more easily. The following statement follows immediately from the definitions.

12.2.1 *Let $\mathcal{T}$ be a k-component t-class isohedral tiling and let $\mathcal{T}_1$ be the tiling obtained by regarding each component of each tile of $\mathcal{T}$ as a tile of $\mathcal{T}_1$. Then $\mathcal{T}_1$ is a p-isohedral tiling for some p satisfying $1 \leq p \leq t$.*

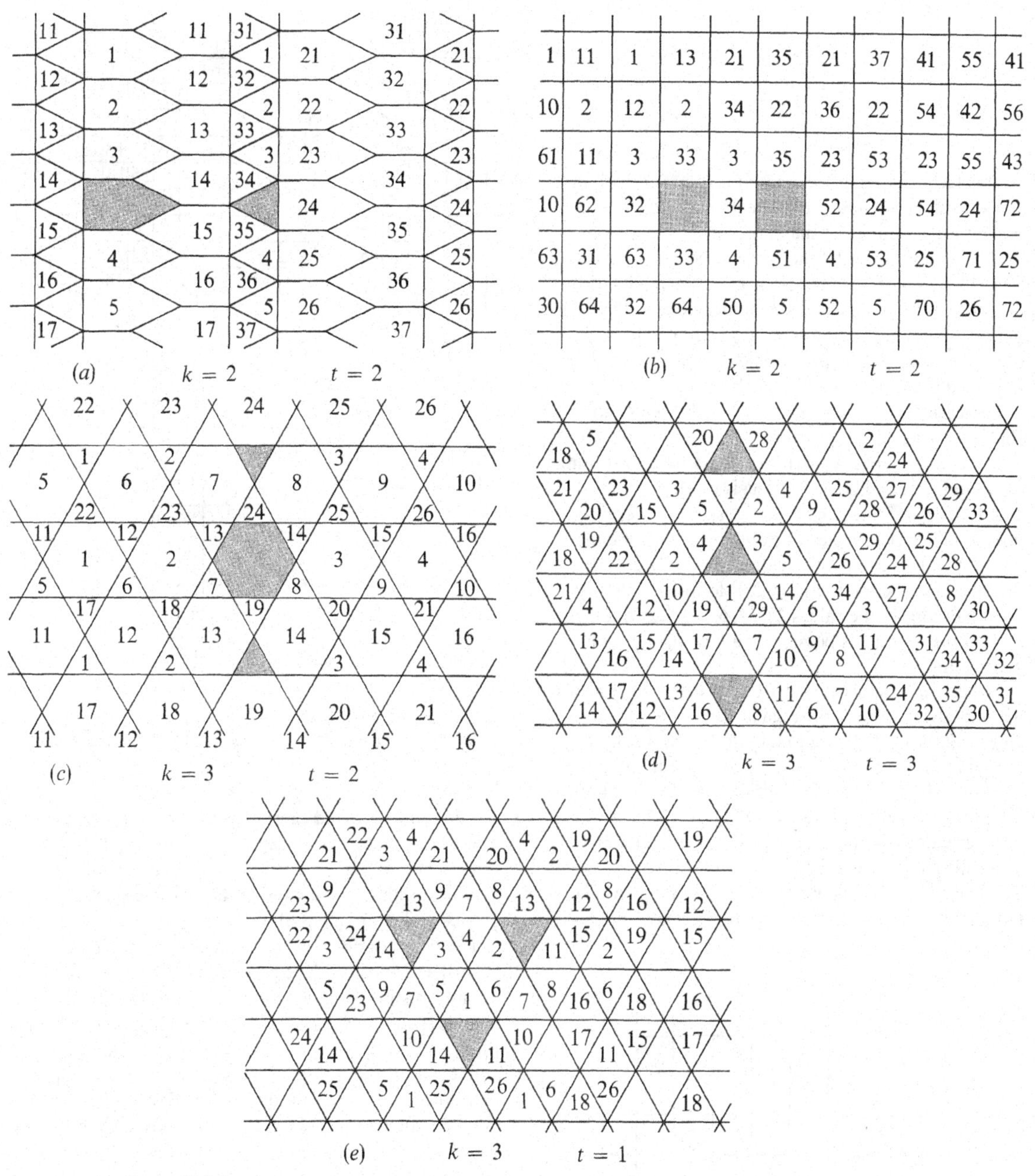

Figure 12.2.1
Examples of isohedral tilings with disconnected tiles. The various components of each tile are labelled with the same numeral and one tile is colored gray to show the shape of the prototile. The number of components k and the number of transitivity classes of components t is indicated below each tiling.

This fact helps us in the search for tilings $\mathscr{T}$ with prescribed values of t and k. For example, to find k-component 2-class tilings we may take for $\mathscr{T}_1$ any isohedral or 2-isohedral tiling and then determine which unions of sets of k tiles of $\mathscr{T}_1$ can be taken as tiles of $\mathscr{T}$. Unfortunately we know no general way to decide exactly which unions of tiles may be chosen. This lack of a definite procedure leads to many open problems, some of which are indicated in the exercises.

The case $t=1$ seems to be of special interest and we shall use word *idiomeric* (from the Greek words ιδιος = same and μερος = part) for 1-class isohedral tilings. An example of an idiomeric tiling is given in Figure 12.2.1(*e*). From Statement 12.2.1 we see that every k-component idiomeric tiling can be obtained by taking the unions of k tiles from an isohedral tiling of one of the 81 types described in Section 6.2. It follows easily that k must have one of the values 2, 3, 4, 6, 8 or 12, and that each of these values is possible in a k-component idiomeric tiling.

For example, let us consider the special case in which the underlying tiling is the regular tiling (4^4) by squares and $k=2$. We use $T(x,y)$ to denote the prototile shown in Figure 12.2.2. Although $T(x,y)$ admits a tiling for *all* relevant values of x and y, it admits an idiomeric tiling for certain special values only. In Figure 12.2.3 we show idiomeric tilings using $T(2, 1)$, $T(3, 0)$ and $T(3, 3)$ and it is not difficult to show that such tilings do not exist with $T(2, 0)$ or $T(2, 2)$. (The prototile of the tiling shown in Figure 12.2.1(b) is $T(0, 2)$ but this tiling is not idiomeric.)

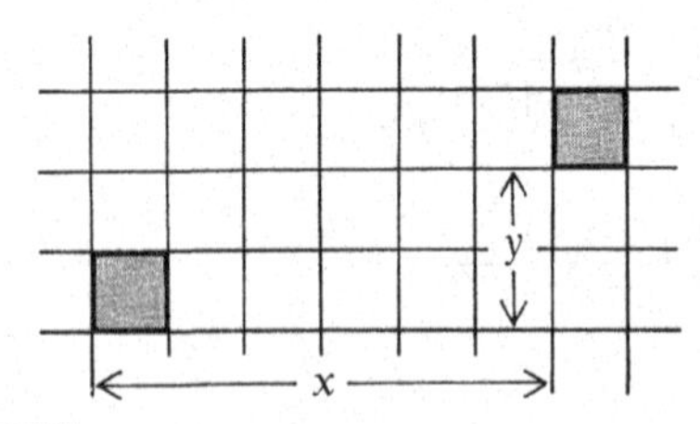

Figure 12.2.2
The tile denoted by $T(x,y)$ in the text. This is a 2-block since it is the union of two squares from the regular tiling (4^4).

(*a*) (*b*) (*c*)

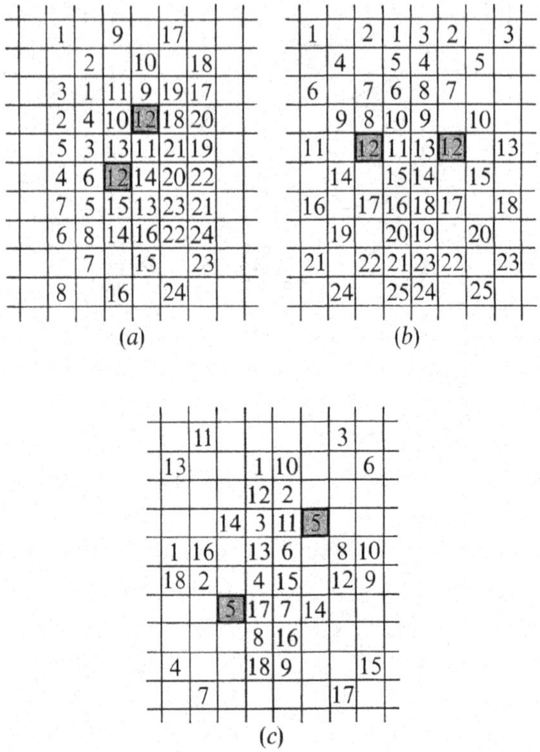

Figure 12.2.3
Idiomeric tilings with prototiles $T(2,1)$, $T(3,0)$ and $T(3,3)$.

We do not know, and cannot even make a conjecture, how to characterize those tiles $T(x,y)$ for which idiomeric tilings exist, though it seems possible that this question may be decidable by analysing the possible symmetry groups of 2-component tilings.

The tiles described in the previous paragraph are known as 2-blocks. More generally, a *k-block* is the union of any k distinct squares from the regular tiling (4^4). If the k squares in a k-block T are disjoint, then any isohedral tiling with prototile T is a k-component tiling as defined above, and if the union of the squares in T is a topological disk then T is a k-omino (see Section 9.4). However there are other possibilities since k-blocks may have cutpoints these arise if two squares of T have in common a single point which is a corner of each. An example of a 4-block with 2 components, one of which has a cutpoint, is shown in Figure 12.2.4(*a*), and three examples of monohedral tilings with this prototile are also shown in Figure 12.2.4. Two of these are isohedral and one is 2-isohedral. Other examples of tilings by 4-blocks are shown in Figures 12.2.8 to 12.2.12.

Monohedral tilings by k-blocks lead to many interesting questions and we shall now examine some of these. In order to describe a k-block it is convenient to use the following notation. Let us suppose that the regular tiling (4^4) is placed in such a position in the plane that the centers of the squares form a unit lattice Λ* (the points with integer coordinates). Then we may refer to any square by the coordinates (x,y) of its center, and refer to a k-block as

$$\{(x_1,y_1), (x_2,y_2), \ldots, (x_k,y_k)\}$$

where (x_i,y_i) is the center of the ith constituent square. For example, the tile $T(x,y)$ of Figure 12.2.2 may be denoted by $\{(0,0), (x,y)\}$ in this notation. We shall use

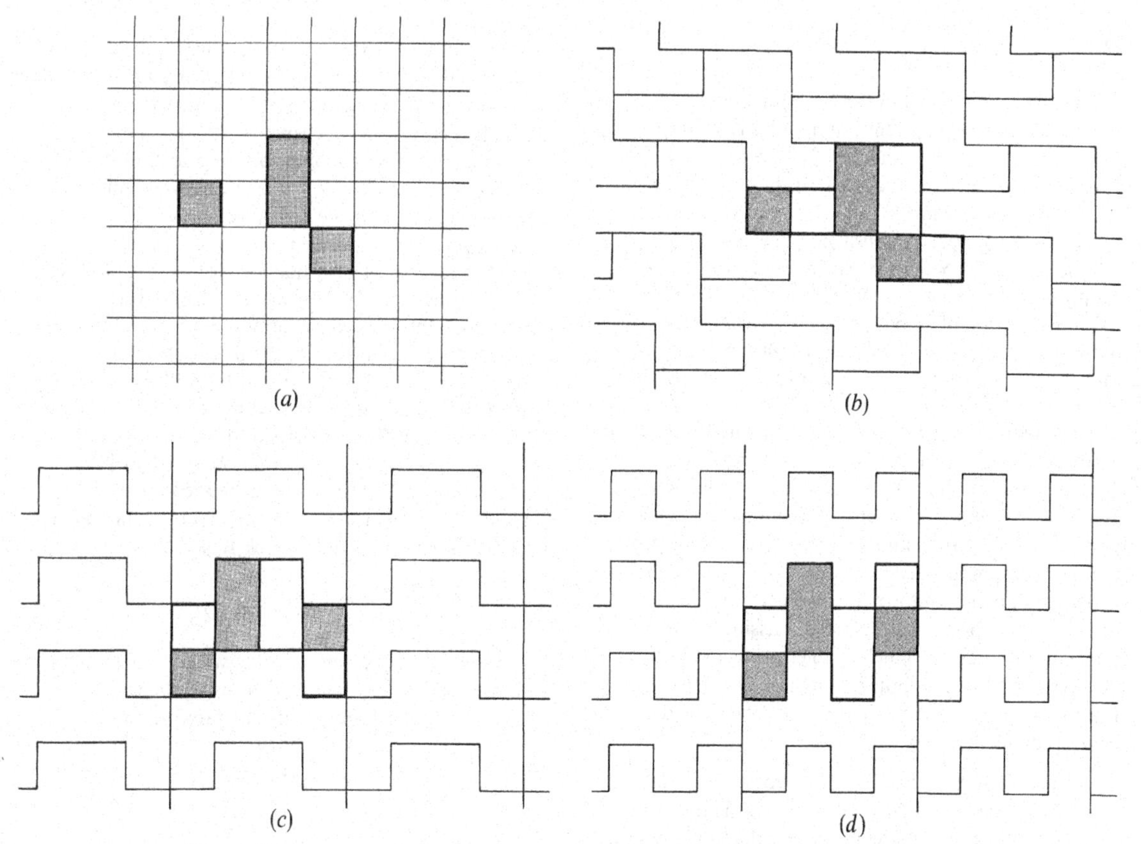

Figure 12.2.4
Examples of monohedral tilings using the 4-block shown in (a). The tilings shown in (c) and (d) are isohedral; that in (b) is 2-isohedral.

* See the Appendix beginning on page 653.

phrases like "collinear squares" or "a line joining two squares" with the obvious interpretation in terms of the points of Λ. Also, when describing a prototile it will usually be convenient to assume that one of the squares lies at the origin (0,0).

A very difficult open question, brought to our attention by David Klarner, is the following. Let K be the largest value of k such that *every* k-block admits a monohedral tiling. What is the value of K? The only information we have is the following.

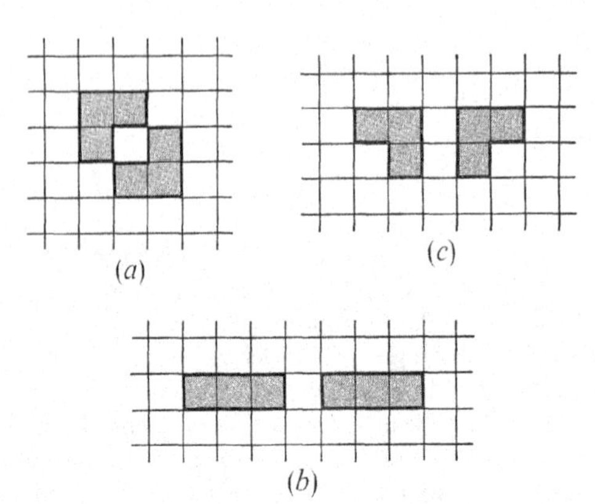

Figure 12.2.5
Three 6-blocks that do not admit a tiling of the plane.

12.2.2 *The value of K is either* 4 *or* 5.

The fact that $K < 6$ follows because there exist 6-blocks which do not admit tilings. Three examples are shown in Figure 12.2.5.* The tile of 12.2.5(*a*) obviously does not admit a tiling; the fact that the tile of Figure 12.2.5(*b*) does not is much more difficult to prove, but this has recently been established independently by Klarner and K. P. Villanger. Their proofs, which are simple in principle but very lengthy in that they involve the consideration of all possible arrangements of a large number of tiles, have never been published. A proof that the tile of Figure 12.2.5(*c*) does not admit a tiling is asked for in Exercise 12.2.6.

To show that $K \geq 4$ is more difficult since, in order to establish this inequality it is necessary to show that every 4-block admits a tiling. This was established a few years ago by Don Coppersmith, but his proof has never been published and we only know of it from an outline of the argument communicated to us privately. We are indebted to Don Coppersmith for permission to include an account of his proof, but before discussing it we shall consider the very much simpler problem of showing that every 3-block B admits a tiling (see Gordon [1980]). For this we must distinguish two cases, the second of which is divided into two subcases.

1. The squares in the 3-block B are not collinear.
2. The squares in B are collinear and the line of collinearity is (*i*) parallel to, or (*ii*) not parallel to, the sides of the squares in B.

For case 1 we observe that the translates of B and of $-B$ (the reflection of B in a point, which is the same as rotating B through 180°) can be placed in such a way that the centers of their squares form a lattice, see Figure 12.2.6. It is clear how one can *replicate* this lattice, by which we mean take disjoint translates of it, in such a way as to cover the whole plane. Hence a monohedral tiling using B as prototile can be constructed.

For case 2(*i*) the existence of a tiling by B follows immediately from the fact that any collinear 3-block can be used to tile a "strip" of squares; see Figure 12.2.7 where a possible procedure for doing this is indicated. In effect this is a 1-dimensional tiling problem and we remark that it is one of the few such problems which is non-trivial. It has been discussed by Koutsky & Sekanina [1958], Sands & Swierczkowski [1960] and Honsberger [1976, pp. 84–87]. See also Adler & Holroyd [1981] and Gordon [1980] for this and related results.*

Case 2(*ii*) can be reduced to case 2(*i*) by a suitable linear transformation. If the 3-block B is

$$\{(0,0), (ap,aq), (bp,bq)\}$$

with p and q relatively prime, then we first tile the plane with the 3-block $\{(0,0), (a,0), (b,0)\}$ using the method of case 2(*i*) and then apply the linear transformation

$$x' = px - qy$$

$$y' = qx + py.$$

This represents a rotation followed by an expansion in the ratio $(p^2+q^2)^{\frac{1}{2}}$ and so leads to a set of tiles whose

* See the Appendix beginning on page 653.

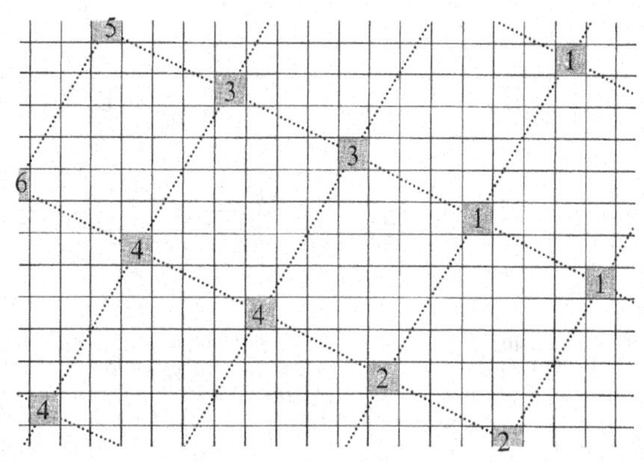

Figure 12.2.6
If B is a 3-block whose squares are not collinear, then translations of B and of $-B$ (the reflection of B in a point) can be used to cover the squares of a lattice as shown. Replication of the lattice enables the plane to be covered and hence leads to a tiling with prototile B. In this diagram B is the 3-block $\{(0,0), (3,5), (4,-2)\}$.

centers form a (square) lattice. The plane can be completely covered without overlaps by replicating this lattice (using p^2+q^2 copies), and so the required tiling by B is obtained. This completes the proof that every 3-block admits a tiling of the plane, and so demonstrates that $K \geq 3$.

To prove the corresponding result for 4-blocks we use similar arguments, but the details are much more complicated and lengthy. Here we reproduce parts of Coppersmith's ingenious proof and ask for other parts in the exercises. Again two cases need to be considered, but here each is divided into two subcases.

1. The 4-block B is such that the line l joining two suitably chosen squares of B is not parallel to the line l' that joins the other two. Also either (*i*) one of the lines l or l' or (*ii*) neither of the lines l or l', is parallel to a side of a square in B.
2. The tile B consists of four collinear squares, the line of collinearity being either (*i*) parallel to, or (*ii*) not parallel to, a side of a square in B.

In each case subcase (*ii*) can be reduced to subcase (*i*) by using a linear transformation in a similar manner to that described above for case 2 of the 3-block argument (see Exercise 12.2.8). Here we describe the proof for case 1(*i*) only. (The difficulty in proving case 2(*i*) stems from the fact that certain collinear 4-blocks do not admit a tiling of a strip of squares and hence the proof for 3-blocks indicated in Figure 12.2.7 does not generalize. In Figures 12.2.10 and 12.2.11 we give examples of tilings by collinear 4-blocks constructed by Coppersmith's method.)

Let us take B to be the 4-block

$$\{(0,0), (0,b), (c,e), (d,f)\}$$

with $e \neq f$. We use the notation $2^q!b$ to mean that 2^q is the highest power of 2 that is a divisor of b (that is, b is not divisible by 2^{q+1}). Also, of two integers n and m we say that n is more even than m if $2^q!n$, $2^r!m$ and $q > r$. The first stage in tiling the plane by B is to replicate the tile 2^q times with horizontal step 1 (where $2^q!b$), that is, we take B and then $2^q - 1$ further copies of B each obtained from the previous one by a translation of 1 unit (the length of the side of a square) in the direction of the x-axis. We then replicate this set of tiles infinitely often using a horizontal step of 2^{q+1} units. In this way we obtain a set of tiles $\mathscr{B}$ which fills the row $y = 0$ and half-fills the rows $y = e$ and $y = f$. Two possibilities must now be distinguished:

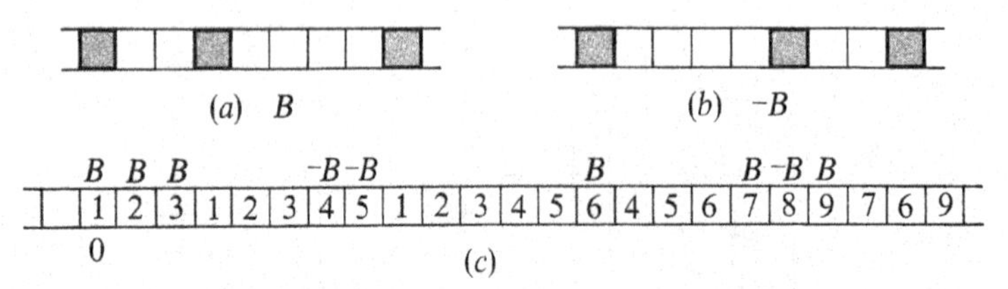

Figure 12.2.7
How to tile a single strip of tiles using any collinear 3-block B. Here, as an example, we have taken B to be the 3-block $\{(0,0), (3,0), (8,0)\}$ shown in (*a*), and $-B$ means the reflection of B in a point, see (*b*). Starting from any point O on the strip (see (*c*)) we work to the right filling each empty square that we come across, if possible, by the left component of B. If this is not possible, then we use the left component of $-B$. In the diagram the letters B and $-B$ above the squares indicate the left component of each tile that is used. The proof that the above procedure is valid depends upon showing that it never terminates, that is, it is *always* possible to use either B or $-B$ in covering the first empty square in the half-strip, which therefore can be completely tiled. In practice it has been found that a *finite* strip can always be covered by this procedure. It is not known how the length $f(a,b)$ of the shortest strip which can be tiled (or which can be tiled by the method described above) by the 3-block $B = \{(0,0), (a,0), (b,0)\}$ is related to the positions of the three squares in the prototile B (that is, to a and b). However it is known (Adler & Holroyd [1981]) that each collinear 3-block tiles a suitable finite strip in some way. Gordon [1980] established that $f(a,b) \leq 3 \,.\, 2^{a+b-2}$, even if only the algorithm just described is used.

I. The integer $(f-e)$ is more even than either e or f. We first replicate $\mathscr{B}$ with a vertical step of 2^{q+1} units. If $2^m!(f-e)$ then we replicate again, infinitely often, using a vertical step of 2^m units and incorporating horizontal shifts to ensure that each half-filled row becomes completely filled. In this way we obtain a set of tiles which fills all rows $y \equiv 0$ or $e \pmod{2^m}$ and all the other rows are completely empty. If $2^p!e$ (where $p < m$ by assumption) then we replicate this set 2^p times with a vertical step of unit length, and then replicate vertically again 2^{m-p-1} times with a step of 2^{p+1}. This yields a tiling of the plane by $\mathscr{B}$ (see Figure 12.2.8).

II. Either e or f is more even than $(f-e)$. (Notice that cases I and II exhaust all possibilities.) Assume, without loss of generality that e is more even than $(f-e)$, that $2^r!(f-e)$ and $2^s!e$, and so $s > r$. First rotate the set of tiles $\mathscr{B}$ through 180° and translate it in such a way that the half-filled rows become completely filled. We now have a set of tiles which completely fills rows $y = 0$, $y = e$, $y = f$ and $y = e+f$ and all the other rows are completely empty. Replicate this set 2^{s-r-1} times by vertical steps of 2^{r+1} units, then replicate infinitely often with vertical steps of 2^{s+1} units. Now all rows with $y \equiv 0 \pmod{2^r}$ are filled and elsewhere is emptiness. Finally we replicate 2^r times with vertical step 1 to complete the tiling of the plane, see Figure 12.2.9. This completes the proof for 4-blocks in the case 1(*i*).

In Figure 12.2.10 and 12.2.11 we give examples of tilings by collinear 4-blocks that were obtained by Coppersmith's methods. Of course, in many cases other tilings are also possible, see Figure 12.2.12.

An interesting feature of Coppersmith's proof is that in no case is it necessary to use reflections of the given 4-block B—translations and rotations through angles of 90°, 180° or 270° always suffice.

The question naturally arises as to whether Statement 12.2.2 can be refined to specify the exact value of K. To settle this it would be necessary either to show that every 5-block admits a tiling, or to find a 5-block for which no tiling is possible. Either task appears formidable!

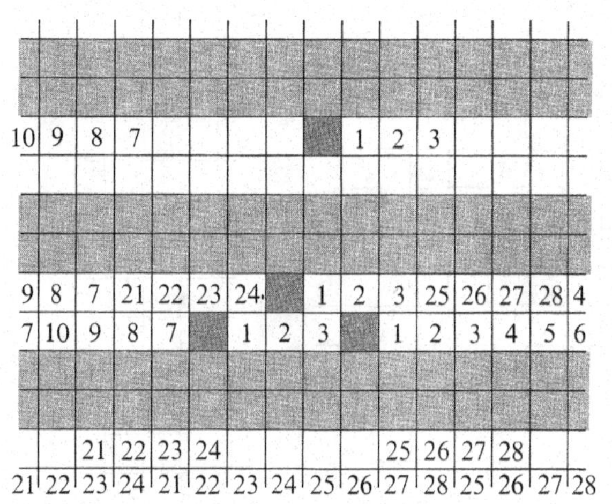

Figure 12.2.8
A tiling using the 4-block {(0,0), (4,0), (2,1) (3,5)} that is obtained by Coppersmith's method in the proof of case 1(i)I (see the text). First the horizontal white strips are tiled as shown, and then replication using two copies with a vertical step of 2 enables the whole plane to be covered. The corresponding values of the parameters, as used in the text, are $q = 2$, $m = 2$, $p = 0$.

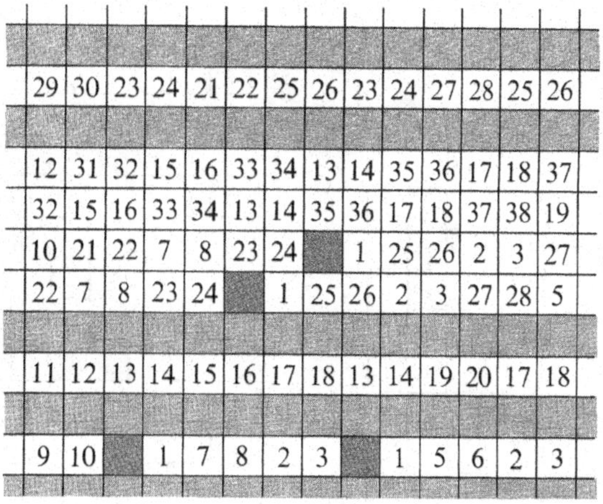

Figure 12.2.9
A tiling using the 4-block {(0,0), (6,0), (3,4), (5,5)} that is obtained using Coppersmith's method in the proof of case 1(i)II (see the text). First the horizontal white strips are tiled as shown, and replication using infinitely many copies with a vertical step of 8 enables the whole plane to be covered. The corresponding values of the parameters used in the text are $q = 1$, $r = 0$, $s = 2$.

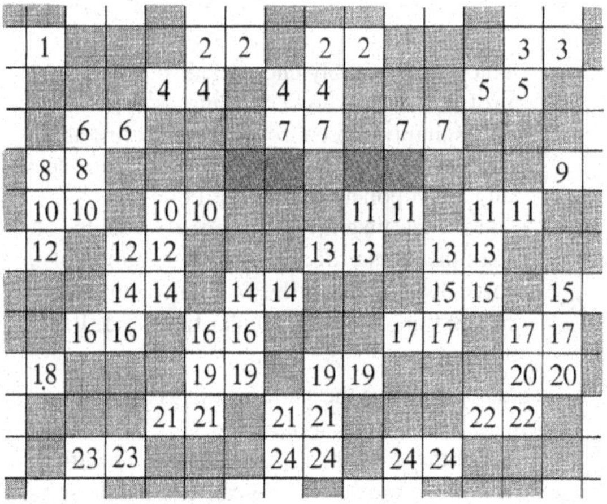

Figure 12.2.10
A tiling using the collinear 4-block $B = \{(0,0), (1,0), (3,0), (4,0)\}$. First the white regions are tiled using horizontal copies of B as shown, and then the gray areas are tiled, in an exactly similar manner, using vertical copies of B. One copy of B is indicated in dark gray. This is one of the tilings that arise in Coppersmith's proof that every collinear 4-block admits a tiling of the plane (see Exercise 12.2.9).

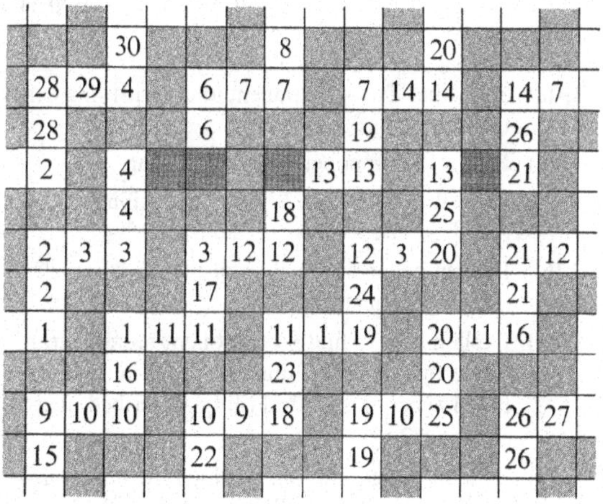

Figure 12.2.11
A tiling using the collinear 4-block $B = \{(0,0), (1,0), (3,0), (8,0)\}$. First the white zigzags are tiled using horizontal copies of B in alternate rows and the vertical copies of B in alternate columns. Replication using two copies with a diagonal step then enables the whole plane to be tiled. One copy of B is indicated in dark gray. This is another tiling that arises in Coppersmith's proof that every collinear 4-block admits a tiling of the plane (see Exercise 12.2.10).

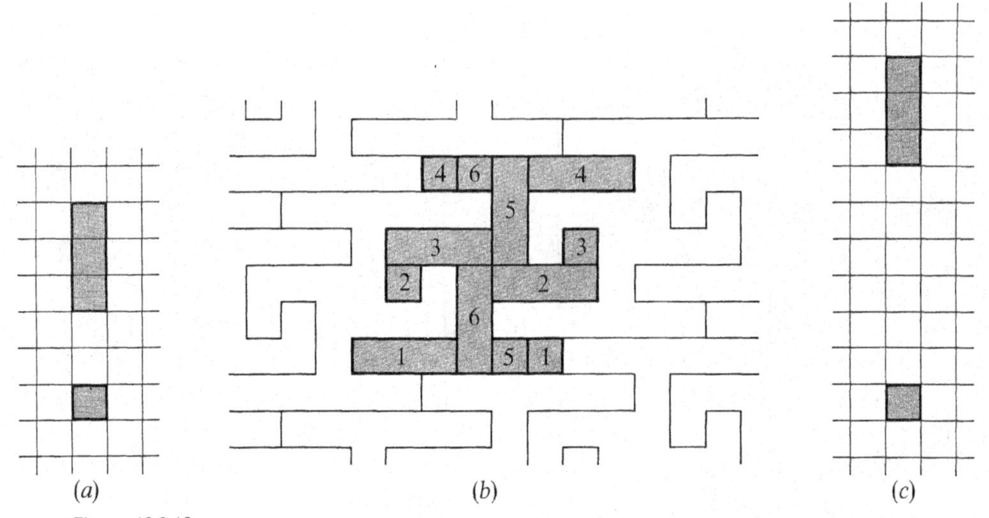

Figure 12.2.12
Another tiling using collinear 4-blocks. In (*b*) we show a region (gray) that can be tiled using the 4-block $B = \{(0,0), (3,0), (4,0), (5,0)\}$ shown in (*a*). Replication leads to a tiling of the whole plane. The same diagram (*b*) can also be interpreted as a tiling by the 4-block $\{(0,0), (7,0), (8,0), (9,0)\}$ shown in (*c*).

EXERCISES 12.2

1. (*a*) Show that if a k-component idiomeric tiling exists then $k \leq 12$.
   (*b*) Find a 12-component idiomeric tiling.
   (*c*) Show that if a t-class k-component tiling exists then $k \leq 12t$.
   **(*d*) Determine for what values of t there exist t-class $(12t)$-component tilings.
2. (*a*) Explain how to find k-component monohedral tilings by taking the union of k tiles in each of the 81 types of isohedral tilings described in Section 6.2.
   (*b*) Find an isohedral tiling in which the union of two suitably chosen disjoint tiles is not the prototile of any 2-component isohedral tiling.
   **(*c*) Does there exist an isohedral tiling such that the union of *any* two disjoint tiles is the prototile of an isohedral tiling?
   (*d*) Find which of the twenty types of isohedral tilings with topological type $[3^6]$ (that is IH1 to IH20 in Table 6.2.1) have the property that the union of a suitably chosen pair of tile is the prototile of an idiomeric tiling.
   (*e*) In a suitably chosen isohedral tiling, find three disjoint tiles whose union is not the prototile of any 3-component monohedral tiling.
   (*f*) Find at least eight types of isohedral tilings other than the regular ones for which the union of any three disjoint tiles is the prototile of a monohedral tiling. (For the regular tilings, see Exercise 12.)
3. Let k, t and p have the same meanings as in Statement 12.2.1, so that $k \geq t \geq p \geq 1$.
   (*a*) Show that all triplets (k,t,p) compatible with this condition and $k = 2$ or 3 can actually occur.
   **(*b*) Determine which triplets (k,t,p) can occur with $k > 3$. Does the answer change if we restrict attention to isohedral tilings?
4. A k-component monohedral tiling will be called *pseudo-idiomeric* if for each tile T the induced group $S(\mathcal{T} | T)$ is transitive on the k components of T, yet $\mathcal{T}$ is not isohedral.

(*a*) Find a 2-component monohedral pseudo-idiomeric tiling which is not isohedral.

**(*b*) Does there exist a 2-component prototile which admits a monohedral pseudo-idiomeric tiling, but does not admit an (isohedral) idiomeric tiling?

5. (*a*) Show that there exist, for all $n > 0$, idiomeric tilings by 2-blocks $T(1,n)$, $T(0, 2n + 1)$, $T(2n + 1, 2n + 1)$ and $T(n, n + 1)$, where $T(x, y)$ is the prototile shown in Figure 12.2.2.
   (*b*) Find some examples of tiles $T(x,y)$ other than those listed in (*a*), which admit idiomeric tilings.
   (*c*) Find some examples of tiles $T(x,y)$, other than $T(0,2)$ and $T(2,2)$ which do not admit idiomeric tilings.
6. (*a*) Show that the 6-block of Figure 12.2.5(*c*) does not admit a tiling of the plane.
   (*b*) Find some 6-blocks, other than those in Figure 12.2.5 which do not admit a tiling of the plane.
   (*c*) Find a 6-block with six components that does not admit a tiling of the plane.
7. Develop a system of incidence symbols for isohedral tilings by 2-blocks and show that these can be used to classify tilings in an analogous manner to that described in Chapter 6. Use incidence symbols to show that there exist exactly six types of idiomeric tilings with the prototile $T(1,2)$.

[The next three exercises are concerned with additional constructions used by Coppersmith in his proof that every 4-block admits a tiling of the plane.]

8. Establish that every 4-block tiles the plane in the case 1(*ii*) on page 627. [If the line joining two of the squares has slope p/q, where p and q are relatively prime integers, and if m and n are integers such that $mp+nq = 1$, then the transformation $(x',y') = (nx+my, -px+qy)$ transforms B into a block of the type considered in case 1(*i*).]

*9. Let B be the collinear 4-block $\{(0,a), (0,b), (0,c), (0,d)\}$.
   (*a*) If a, b, c, d are congruent to 0, 1, 3, 4 modulo 8, or can be made so by a suitable rotation and translation, show that B admits a tiling of the plane in a manner similar to that indicated in Figure 12.2.10.
   (*b*) If a, b, c, d are congruent to 0, 1, 4, 7 modulo 8, or can be made so by a suitable rotation and translation, show that B admits a tiling of the plane in a manner similar to that indicated in Figure 12.2.10 except that now the gray areas are to be tiled using horizontal copies of B and the white areas using vertical copies.
   (*c*) Show that if no two of a, b, c, d are congruent modulo 8, and the tile is not one of the types mentioned in parts (*a*) and (*b*) of this exercise, then it is possible to tile a single strip of tiles using B. This leads, by replication, to a tiling of the plane.
10. (*a*) Let B be the collinear 4-block $\{(0,a), (0,b), (0,c), (0,d)\}$ and suppose that the most even difference between any two of a, b, c, d is divisible by 2^q (but not by 2^{q+1}), and the next most even difference is divisible by 2^m (but not 2^{m+1}). Also suppose that $q \gtrsim 3$ and $q > m$. Show that the plane can be tiled using copies of B in a manner similar to that indicated in Figure 12.2.11 (namely by tiling zigzags with horizontal and vertical blocks in alternate rows and columns).
   (*b*) Show that in all other cases where a collinear 4-block B has no two of a, b, c, d differing by a multiple of 8, the block B admits a tiling of a single strip of tiles. Hence, by replication construct a tiling of the plane.
11. Show that, besides the five 4-ominoes (Table 9.4.2) there exist 17 4-blocks with one component. Construct a monohedral tiling with each of these.
12. (*a*) Show that every union of three tiles in the regular tiling (6^3) by hexagons is the prototile of a monohedral tiling.
   (*b*) Find a simply connected set of six tiles from (6^3) whose union does not admit a tiling of the plane.
   (*c*) Establish corresponding results for the regular tiling (3^6) by triangles.

12.3 HOLLOW TILES

In Section 2.5 we mentioned Kepler's two interpretations of a star polygon. In the first of these, a polygon is thought of as a "circuit" consisting of vertices and edges rather than as a boundary of a "piece" of the plane. It was Badoureau [1878], [1881] who first conceived the possibility of using such polygons to tile the plane. The fact that such tiles cannot, in any sense, "cover" the plane means that it is necessary to redefine many of the concepts, and in particular to explain exactly what a tiling is in this new sense.

Following Poinsot [1810], who was anticipated to some extent by Girard [1625], Meister [1769] and others, we redefine an n-gon P to be a sequence of $n \geq 3$ distinct points $V_1, \ldots, V_n = V_0$ (the *vertices* or *corners* of P) of which no three are collinear, and n line segments $[V_i, V_{i+1}]$ for $i = 0, \ldots, n-1$ (the *edges* or *sides* of P). Such an n-gon will be denoted by $P = [V_1, \ldots, V_n]$ and we shall call it either a *polygonal circuit* or a *hollow polygon* to distinguish it from a polygon in the more usual sense. The boundary of a polygon is a polygonal circuit, but the converse is not generally true—see for example the pentagram $\{5/2\}$ of Figure 2.5.1(a).

There are two important respects in which polygonal circuits differ from polygons: their edges may intersect at points which are not vertices, and they may not define a topological disk or, for that matter, a piece of the plane in any useful sense. Although many of our definitions apply more generally, for simplicity we shall (like Badoureau) be concerned mainly with *regular* polygonal circuits, that is, with the n-gons $\{n/d\}$ defined in Section 2.5. Here the word "regular" means that the symmetry group of the n-gon is transitive on the set of $2n$ flags consisting of a vertex and an incident edge (see Section 1.3).

Before we can explain the new meaning of the word "tiling" it is necessary to introduce some additional terminology. A family $\mathscr{P}$ of distinct polygonal circuits is called *edge-sharing* if every edge of each polygonal circuit of $\mathscr{P}$ is an edge of precisely one other polygonal circuit of $\mathscr{P}$. Edge-sharing families are the analogues of, and generalize, the concept of an edge-to-edge tiling defined in Section 1.1. When considering edge-sharing families it is unnecessary to distinguish between corners and sides of the polygonal circuits, and vertices and edges of the family. It will always be convenient to use the latter terms.

An edge-sharing family $\mathscr{P}$ is called *connected* if, for every two polygonal circuits P, P' in $\mathscr{P}$, there exists a finite sequence

$$P = P_0, P_1, \ldots, P_k = P'$$

of polygonal circuits in $\mathscr{P}$ such that each consecutive pair have an edge in common. If $\mathscr{P}$ is an edge-sharing family and V is any vertex of $\mathscr{P}$ then we define the *vertex figure* $V(\mathscr{P})$ of $\mathscr{P}$ at V as follows. A point X is a vertex of $V(\mathscr{P})$ if and only if $[V,X]$ is an edge of $\mathscr{P}$, and two vertices X, Y of $V(\mathscr{P})$ define an edge $[X,Y]$ of $V(\mathscr{P})$ if and only if $[V,X]$ and $[V,Y]$ are distinct edges of the same polygonal circuit P in $\mathscr{P}$. The edge $[X, Y]$ of $V(\mathscr{P})$ is said to *correspond* to P. Sometimes we shall find it helpful in the diagrams to label the edges of $V(\mathscr{P})$ by the symbols of the corresponding polygonal circuits. If $V(\mathscr{P})$ consists of just one polygonal circuit for each vertex V, then $\mathscr{P}$ will be called *unicursal*.

These definitions are illustrated in Figure 12.3.1, which shows two finite edge-sharing unicursal families of regular polygonal circuits. In each case, the vertex figures of two vertices are shown. In Figure 12.3.1(b) it will be noticed that edges overlap in pairs—a possibility that is not excluded by the definition.

A *tiling* by polygonal circuits is defined to be any connected, unicursal, locally finite, edge-sharing family. As usual, "locally finite" means that any disk in the plane meets only a finite number of polygons.

This definition of a tiling differs from that of Section 1.1 in a very important respect—it does not require that the polygons "spread out" over the whole plane. In fact we get a situation somewhat analogous to that which we encountered while studying patterns in Chapter 5; there are finite tilings (see, for example, Figures 12.3.1, 12.3.2 and 12.3.3), strip tilings (Figure 12.3.4) periodic tilings (Figure 12.3.5) and, in fact, many other possibilities as well (Exercise 12.3.1).

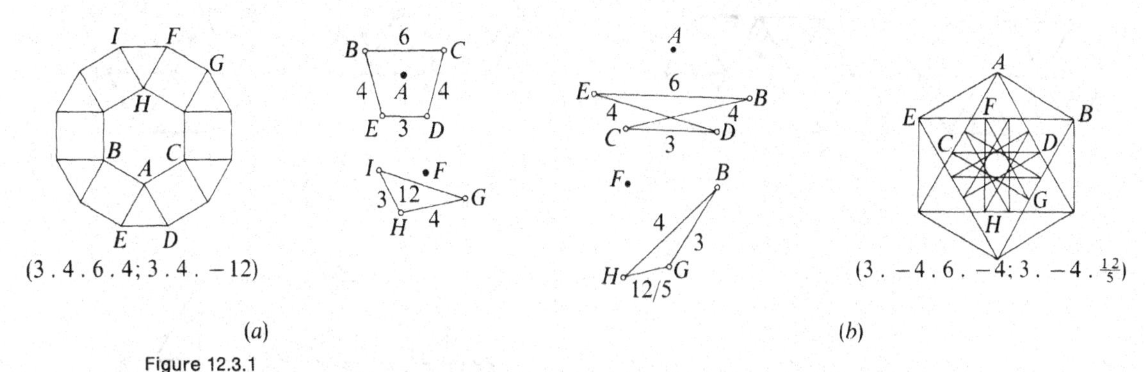

Figure 12.3.1
Two finite edge-sharing families of regular polygonal circuits. The vertex figures of two vertices in each family are shown. In (*a*) the polygonal circuits used are {3}, {4}, {6} and {12}, and in (*b*) they are {3}, {4}, {6} and {12/5}. The symbols written below each family are explained later in this section; they are analogous to those introduced in Section 2.2 for 2-uniform tilings.

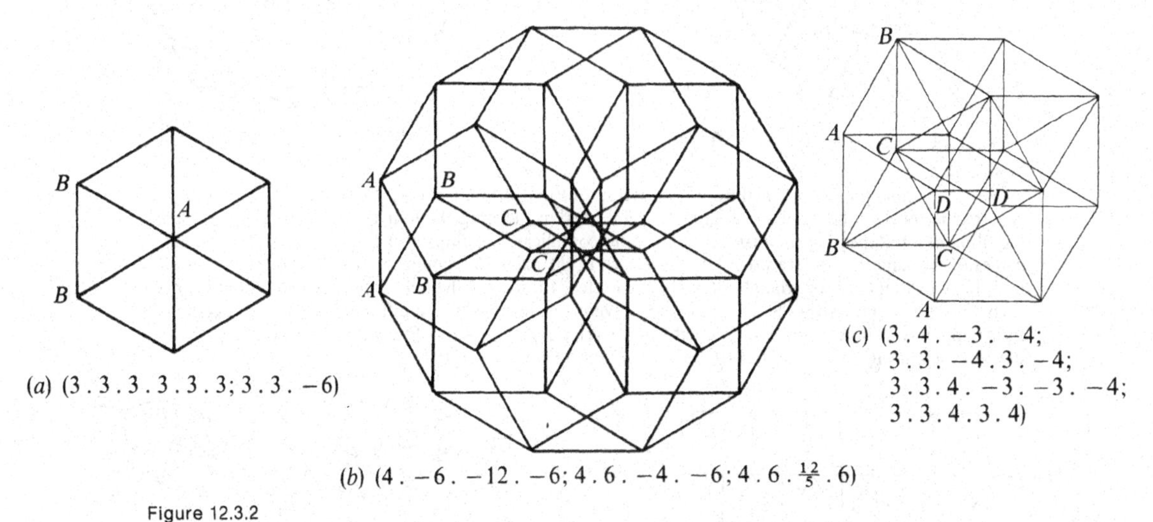

Figure 12.3.2
Three more edge-sharing families of regular polygonal circuits. These are (*a*) 2-uniform, (*b*) 3-uniform and (*c*) 4-uniform. In each case the transitivity classes of the vertices are indicated by the lettering and the corresponding symbol is also given. Each of these families is a tiling which we call a rosette. It is conjectured that all finite tilings can be obtained from the five rosettes shown here and in Figure 12.3.1 by the process of "addition" explained in the text and illustrated in Figure 12.3.3.

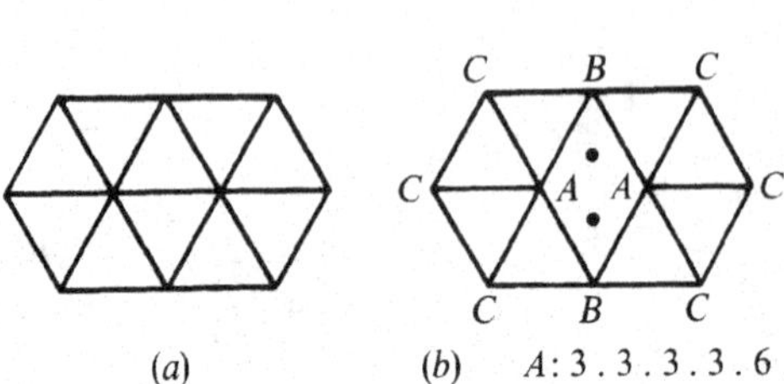

(*a*)

(*b*)

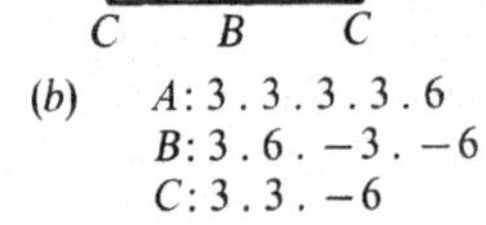

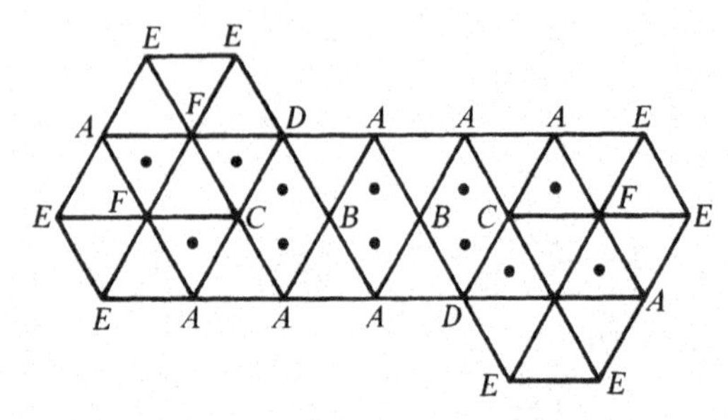

(*c*)

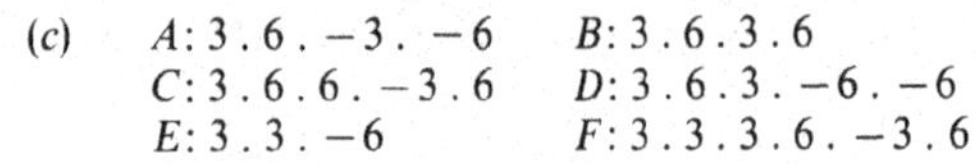

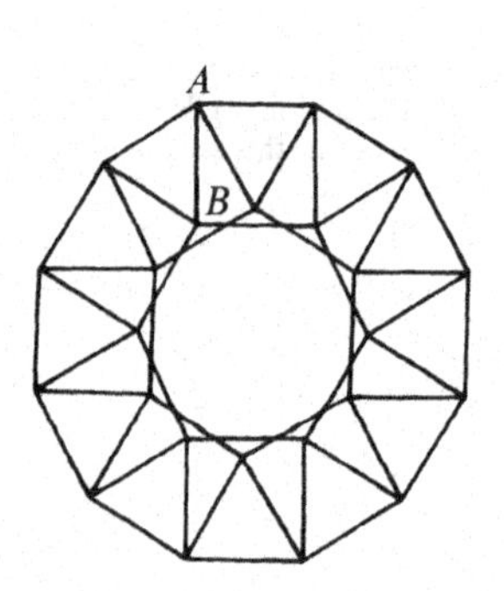

(*d*) *A*: 3 . 4 . −3 . −4 *B*: 3 . 4 . 6 . 4

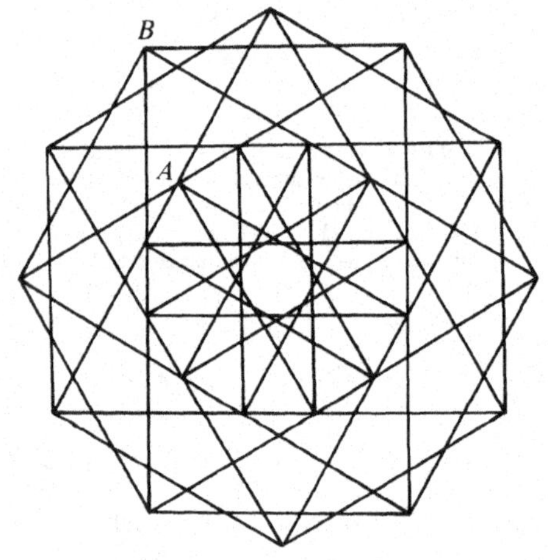

(*e*) *A*: 3 . 4 . −3 . −4 *B*: 3 . −4 . 6 . −4

Figure 12.3.3
Examples of finite families formed from the rosettes of Figures 12.3.1 and 12.3.2 by addition. In (*a*) two rosettes of the kind shown in Figure 12.3.2(*a*) overlap; deleting the four triangles in this area yields the tiling of (*b*). Repeating the process yields more complicated examples such as that of (*c*) where eight copies of the same rosette are used. Here, and also in (*b*), each dot signifies that after the rosettes have been overlapped two triangles have to be deleted. In order to help with the interpretation, near each diagram we give symbols indicating the polygons incident with the different kinds of vertices. In (*d*) and (*e*) we show examples obtained from overlapping rosettes of Figures 12.3.1(*a*) or 12.3.1(*b*) by deleting two hexagons or two star-dodecagons, respectively.

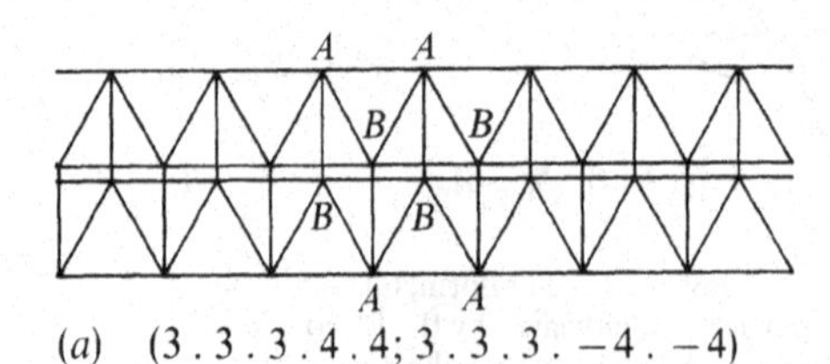

(*a*) (3 . 3 . 3 . 4 . 4; 3 . 3 . 3 . −4 . −4)

(*b*) (3 . 4 . −3 . −4; 3 . 3 . 4 . −3 . −3 . −4; 3 . 3 . 4 . 3 . 4)

Figure 12.3.4
Two tilings by triangles and squares. These are *k*-uniform with $k = 2$ in (*a*) and $k = 3$ in (*b*). Symbols for the tilings are indicated below each.

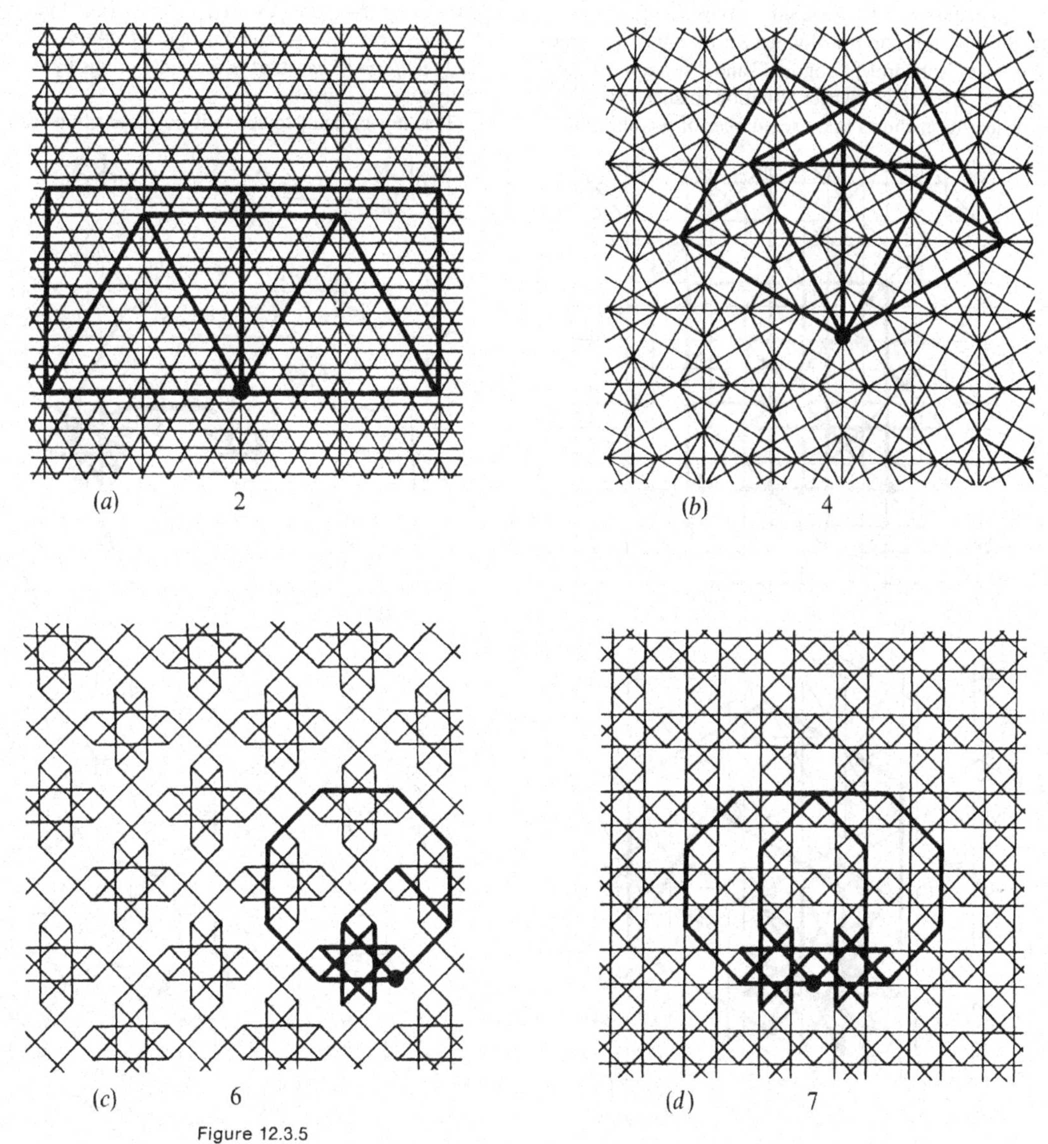

Figure 12.3.5
The 25 uniform tilings listed in Table 12.3.1 are shown here and in Figure 2.1.5.

The definition of a symmetry group extends to tilings in this new sense in the obvious manner and hence such terms as "isohedral", "isogonal" and "isotoxal" are also defined. Even if we restrict attention to such types, a little consideration shows that the number of possibilities is unmanageably large. Consequently we (like Badoureau in the papers cited above) propose to discuss in detail only *uniform tilings*, that is, isogonal tilings by regular polygons.

It is not difficult to see that no finite tilings can be

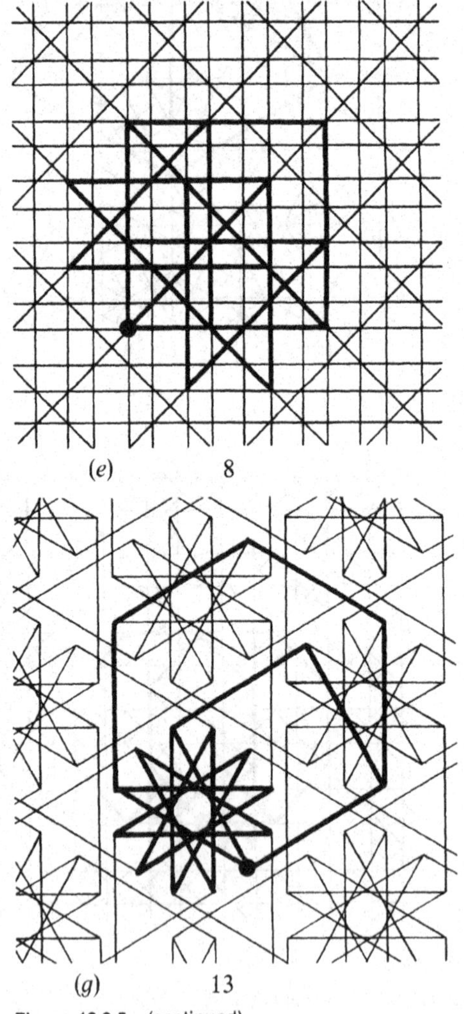

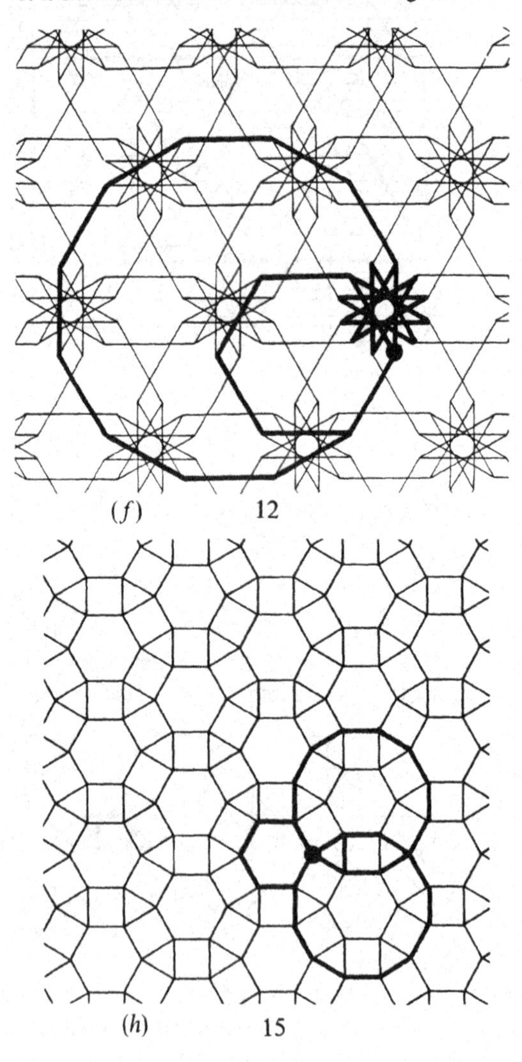

Figure 12.3.5 (continued)

uniform—their vertices will belong to $k \geq 2$ transitivity classes and so, with the obvious adaptation of the terminology of Section 2.2, they may be called *k-uniform*. Figures 12.3.1 and 12.3.2 show five such tilings with $k = 2$, 3 or 4. These seem to us to be especially attractive and we propose calling them *rosettes*. Moreover we conjecture that *all* finite k-uniform tilings can be obtained from these five rosettes by a process called *addition*. This is illustrated in Figure 12.3.3. Addition, in this sense, is carried out by placing two

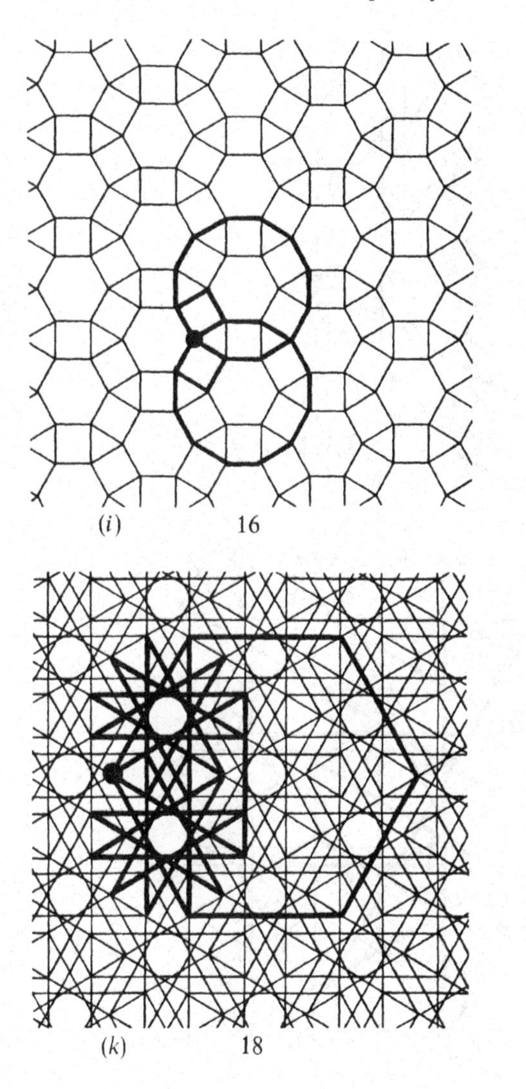

(*i*) 16

(*k*) 18

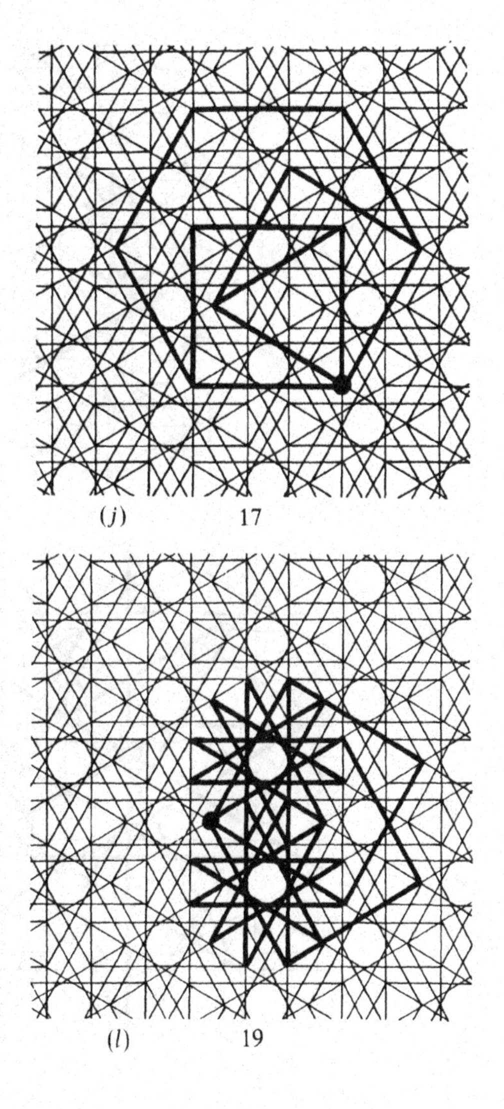

(*j*) 17

(*l*) 19

finite tilings in such a position that one, or several, polygons coincide, and then deleting the duplicated polygons. This process will not always lead to a new tiling since the resulting family may fail to be unicursal and connected but in many cases it will yield a tiling.

The idea of addition can also be extended to construct strip tilings from finite tilings, and periodic tilings from either strip or finite tilings, though certainly not all k-uniform strip and periodic tilings can be obtained in this way.

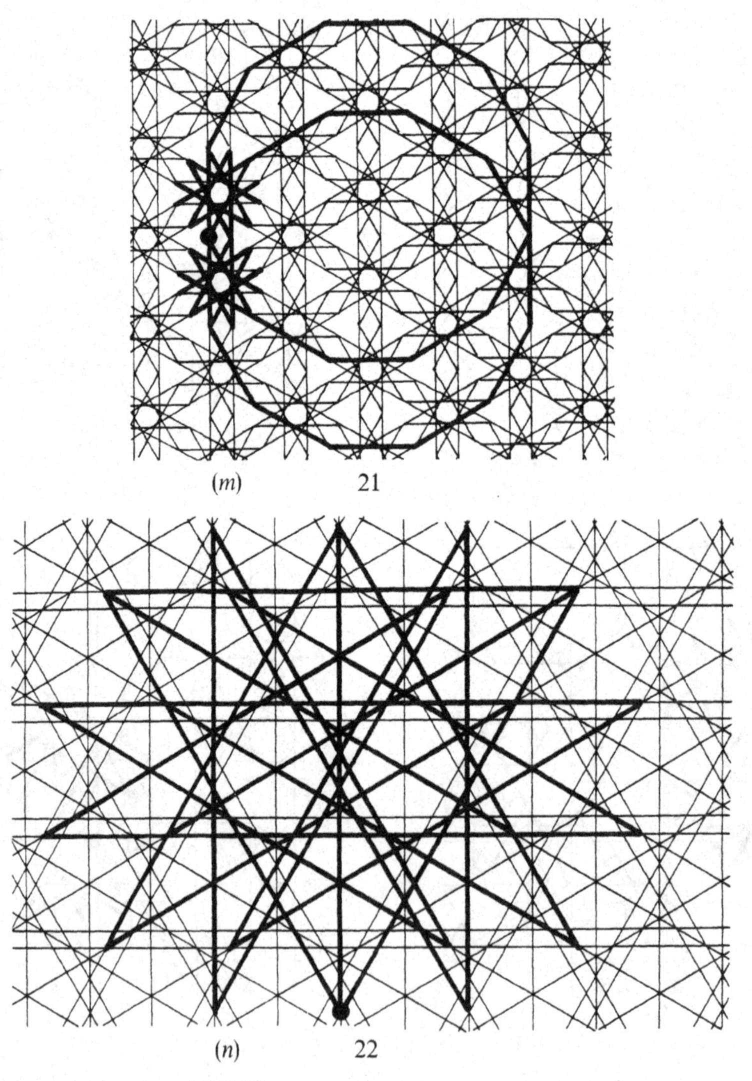

(*m*) 21

(*n*) 22

It is easy to see that there exist no uniform strip tilings, and so we obtain the following result:

12.3.1 *All uniform tilings by hollow tiles are periodic. They are listed in Table* 12.3.1 *and diagrams of the tilings appear in Figure* 12.3.5.

In order to help with the interpretation of the diagrams we have emphasized the polygons incident with the marked vertex. Note that the tiling in Figure 12.3.5(*h*) can be interpreted in three different ways as a uniform tiling by hollow tiles, as can that in Figure 12.3.5(*j*). Eleven tilings in the table correspond exactly to the uniform tilings of Section 2.1, the other fourteen are new; of these, nine tilings involve star polygons. Table 12.3.1 is based on the results of Badoureau [1881] augmented by those of Miller [1933] and Coxeter, Longuet-Higgins & Miller [1954] who produced a table of possibilities but without adequate explanation; it is taken from Grünbaum, Miller & Shephard [1982]. For a proof of Statement 12.3.1 we follow essentially the same method as that used for the proof of Statement 2.1.3. We know the size of the angles that can occur at the vertices of hollow regular polygons, and hence we can determine possible vertex figures. We must then check which of these vertex figures corresponds to a uniform tiling. The details are intricate and the checking is tedious, but we believe that the enumeration presented here is complete.

It is helpful to have a notation for the uniform and k-uniform tilings by hollow tiles analogous to that used in Chapter 2. For any vertex V of a tiling by regular polygonal circuits a symbol consisting of a row of rational numbers, separated by dots, is constructed as follows. Let $\{m_1\}, \{m_2\}, \ldots, \{m_k\}$ be the regular polygonal circuits which contain V as a vertex. We assume that they are named in such an order that two have an edge in common if and only if they are adjacent in this list. (As usual $\{m_k\}$ is considered to be adjacent to $\{m_1\}$.) With V we associate the symbol $\varepsilon_1 m_1 . \varepsilon_2 m_2 \ldots \varepsilon_k m_k$ where each ε_j is $+$ or $-$. The sign ε_1 may be chosen arbitrarily and then the other ε_j are determined consecutively by the rule that ε_j and ε_{j+1} are the same or differ according as the centers of the corresponding m_j-gon and m_{j+1}-gon lie on opposite sides of their common edge or on the same side of it, respectively. If the tiling is uniform then the symbols corresponding to all the vertices are the same. This symbol is put in parentheses and forms the symbol for the tiling. If the tiling is k-uniform then the k symbols corresponding to the k different kinds of vertices are separated by semicolons and then put in parentheses. The reader is invited to check that the symbols assigned to the tilings and vertices in the diagrams of this section are correct. As usual, if convenient we choose from among the many possibilities that symbol which is lexicographically first, and use superscripts to abbreviate.

Table 12.3.1 contains further information about the tilings. A polygonal circuit can be assigned one of two orientations, and a tiling $\mathscr{T}$ is said to be *orientable* if all the polygonal circuits of $\mathscr{T}$ can be oriented coherently, that is, so that every two circuits that share an edge induce opposite orientations on it. As will be seen from column (4) of the table, only three of the uniform tilings are not orientable.

If P is any polygonal circuit in $\mathscr{T}$, and A is any point of the plane not on an edge of P, then we can define the P-density of A as follows. Let L be any ray (half-line) with end-point A, which contains no vertex of P. To each edge e of P we assign an *index* with value 0 if e does not meet L, $+1$ if e crosses L in a counterclockwise direction about A, and -1 if e crosses L in a clockwise direction about A (see Figure 12.3.6). The *P-density* of A is defined as the sum of all these indices. The validity of the definition depends upon the elementary (but not trivial) fact that the sum of the indices is independent of the choice of L; it depends only on A, the polygonal circuit P, and the orientation of P. Intuitively we may think of the P-density as the number of times P winds round A in a counterclockwise direction. It can be shown further that any two points in the same connected component of the complement of P in the plane necessarily have the same P-density which we refer to as the P-density of the component. In Figure 12.3.6 the P-densities of the various components are indicated.

Now let $\mathscr{T}$ be any tiling by (oriented) polygonal circuits. If A is any point not on any edge of $\mathscr{T}$, the *$\mathscr{T}$-density* of A is the sum of the P-densities of A for all

Table 12.3.1 UNIFORM TILINGS BY HOLLOW TILES

Information is given about the 25 kinds of uniform tilings by hollow (finite) regular polygons.

Column (1) gives the list number; the eleven tilings with list numbers 1, 3, 5, 9, 10, 11, 14, 20, 23, 24 and 25 are shown in Figure 2.1.5 and the remaining 14 in Figure 12.3.5.

Column (2) gives the dot pattern type of the set of vertices in the notation of Section 5.3, and **column (3)** the symbol for the tiling.

Column (4) shows the symmetry group of the tiling and in **column (5)** we indicate by *O* or *N* whether the tiling is orientable or not.

Column (6) shows the density and the remaining columns give references to the works of Badoureau [1881], Coxeter, Longuet-Higgins & Miller [1954], Miller [1933] and Kepler [1619]. A star in **column (8)** means that the tiling is mentioned without a symbol being assigned, and a star in **column (10)** means the diagram contains an error. In **column (3)** an entry (*E*) indicates the tiling occurs in two enantiomorphic forms and the symbol (*R*) denotes a regular tiling.

List number (1)	Dot pattern type DPP (2)	Tiling symbol (3)	Symmetry group (4)	Orient-ability (5)	Density (6)	References: Badoureau (7)	Coxeter et al. (8)	Miller (9)	Kepler (10)
1	19	(3 . 3 . 3 . 4 . 4)	*cmm*	*O*	1	—	*	105	*M**
2	19	(3 . 3 . 3 . −4 . −4)	*cmm*	*O*	1	—	*	106	—
3	35	(3 . 3 . 4 . 3 . 4)	*p4g*	*O*	1	66	$\vert 244 = s\left\{{4 \atop 4}\right\}$	102	*N*
4	35	(3 . 3 . −4 . 3 . −4)	*p4g*	*O*	1	—	$\vert 2\frac{4}{3}\frac{4}{3} = s'\left\{{4 \atop 4}\right\}$	103	—
5	38	(4 . 8 . 8)	*p4m*	*O*	1	65	$24\vert 4 = t\{4,4\}$	79	*V*
6	38	(4 . −8 . 8/3)	*p4m*	*O*	1	152	$2\frac{4}{3}4\vert = t'\left\{{4 \atop 4}\right\}$	84	—
7	38	(8 . 8/3 . −8 . −8/3)	*p4m*	*O*	0	151	$\frac{4}{3}4\frac{2}{\infty}\vert$	83	—
8	38	(−4 . 8/3 . 8/3)	*p4m*	*O*	1	—	$24\vert\frac{4}{3} = t'\{4,4\}$	80	—
9	41	(4 . 4 . 4 . 4) (*R*)	*p4m*	*O*	1	—	$4\vert 24 = \{4,4\}$	71	*E*
10	42	(3 . 3 . 3 . 3 . 6) (*E*)	*p6*	*O*	1	—	$\vert 236 = s\left\{{3 \atop 6}\right\}$	104	*L*
11	46	(4 . 6 . 12)	*p6m*	*O*	1	63	$236\vert = t\left\{{3 \atop 6}\right\}$	96	*Mm*
12	46	(6 . −12 . 12/5)	*p6m*	*O*	2	158	$3\frac{6}{5}6\vert$	91	—
13	46	(4 . −6 . 12/5)	*p6m*	*O*	1	155	$23\frac{6}{5} = t'\left\{{3 \atop 6}\right\}$	100	—
14	48A	(3 . 4 . 6 . 4)	*p6m*	*O*	1	64	$36\vert 2 = r\left\{{3 \atop 6}\right\}$	93	*Ii*

15	48A	$(-3\,.\,12\,.\,6\,.\,12)$	*p6m*	*O*	2	64	$\frac{3}{2}6\,\|\,6$	94	*Ii*
16	48A	$(4\,.\,12\,.\,-4\,.\,-12)$	*p6m*	*N*	—	64	$26^{3/2}_{3}\,\|$	95	*Ii*
17	48A	$(3\,.\,-4\,.\,6\,.\,-4)$	*p6m*	*O*	1	156	$\frac{3}{2}6\,\|\,2 = r'\{{3 \atop 6}\}$	97	—
18	48A	$(3\,.\,12/5\,.\,-6\,.\,12/5)$	*p6m*	*O*	2	156	$36\frac{6}{5}$	98	—
19	48A	$(4\,.\,12/5\,.\,-4\,.\,-12/5)$	*p6m*	*N*	—	156	$2\frac{6}{5}\,{}^{3/2}_{3}\,\|$	99	—
20	48B	$(3\,.\,12\,.\,12)$	*p6m*	*O*	1	61	$23\,\|\,6 = t\{6,3\}$	86	*S*
21	48B	$(12\,.\,12/5\,.\,-12\,.\,-12/5)$	*p6m*	*N*	—	154	$\frac{6}{5}6^{3}_{\infty}\,\|$	90	—
22	48B	$(-3\,.\,12/5\,.\,12/5)$	*p6m*	*O*	1	—	$23\,\|\,\frac{6}{5} = t'\{6,3\}$	87	—
23	49	$(3\,.\,6\,.\,3\,.\,6)$	*p6m*	*O*	1	62	$2\,\|\,36 = \{{3 \atop 6}\}$	76	*P*
24	50	$(6\,.\,6\,.\,6)$ (*R*)	*p6m*	*O*	1	—	$\{6,3\}$	73	*F*
25	51	$(3\,.\,3\,.\,3\,.\,3\,.\,3\,.\,3)$ (*R*)	*p6m*	*O*	1	—	$6\,\|\,23 = \{3,6\}$	72	*D*

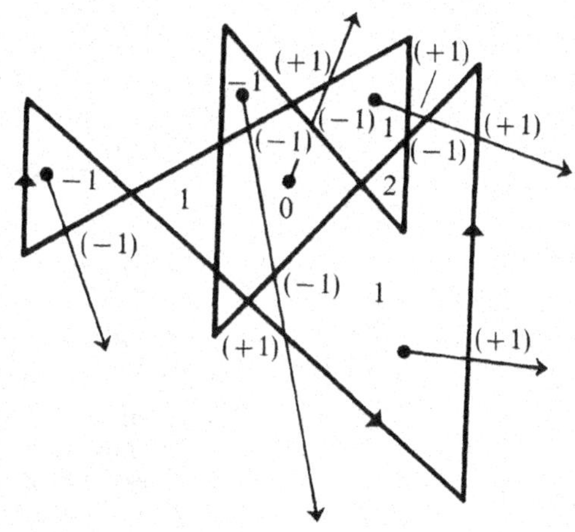

Figure 12.3.6
The *P*-densities of the various components of the complement of a polygonal circuit *P*. Various points *A* and half-lines *L* are indicated. The numbers in parentheses are the contributions of the crossings at the intersections of *L* with the polygonal circuit.

polygonal circuits in $\mathscr{T}$, provided this sum exists. In particular we shall say that $\mathscr{T}$ has density d if for every point A which does not lie on an edge of $\mathscr{T}$, the $\mathscr{T}$-density is d. It is clear that $\mathscr{T}$ can have a density in this sense only if it is orientable, and we shall conventionally choose the orientations to be such that $d \geq 0$. For example the tilings of Figures 12.3.1 and 12.3.2 each have density 0; every tiling in the original sense of Section 1.1 has density 1 when interpreted as a tiling by hollow tiles, and the uniform tilings of Table 12.3.1 have densities of 0, 1, 2, or in some cases densities do not exist. These densities are indicated in column (6) of the table.

In order to simplify our treatment we have used only "finite" regular polygonal circuits. But we could also consider tilings using apeirogons and zigzags (see Figure 12.3.7) which satisfy the condition of flag-transitivity and so may be considered to be regular. These introduce new possibilities, some of which are indicated in the exercises, and a few examples are shown in Figure 12.3.8. Both Badoureau [1881] and Coxeter, Longuet-Higgins & Miller [1954] considered tilings that include apeirogons; a detailed discussion of all the possibilities appears in Grünbaum, Miller & Shephard [1982].

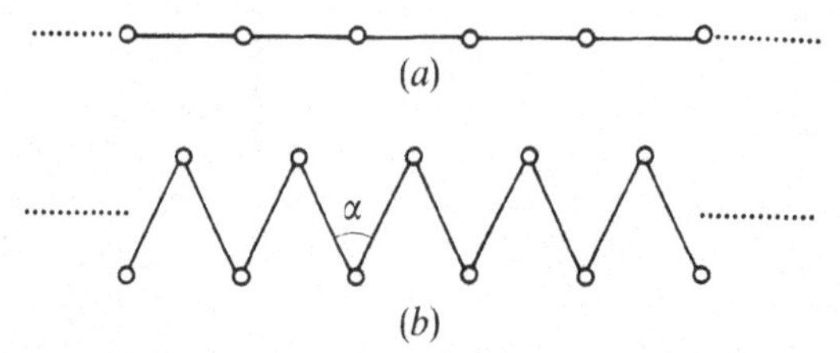

Figure 12.3.7
The two kinds of "infinite" regular polygons that can be used in tilings by hollow tiles, namely (a) an apeirogon, and (b) a zigzag. Examples of tilings using these polygons appear in Figure 12.3.8.

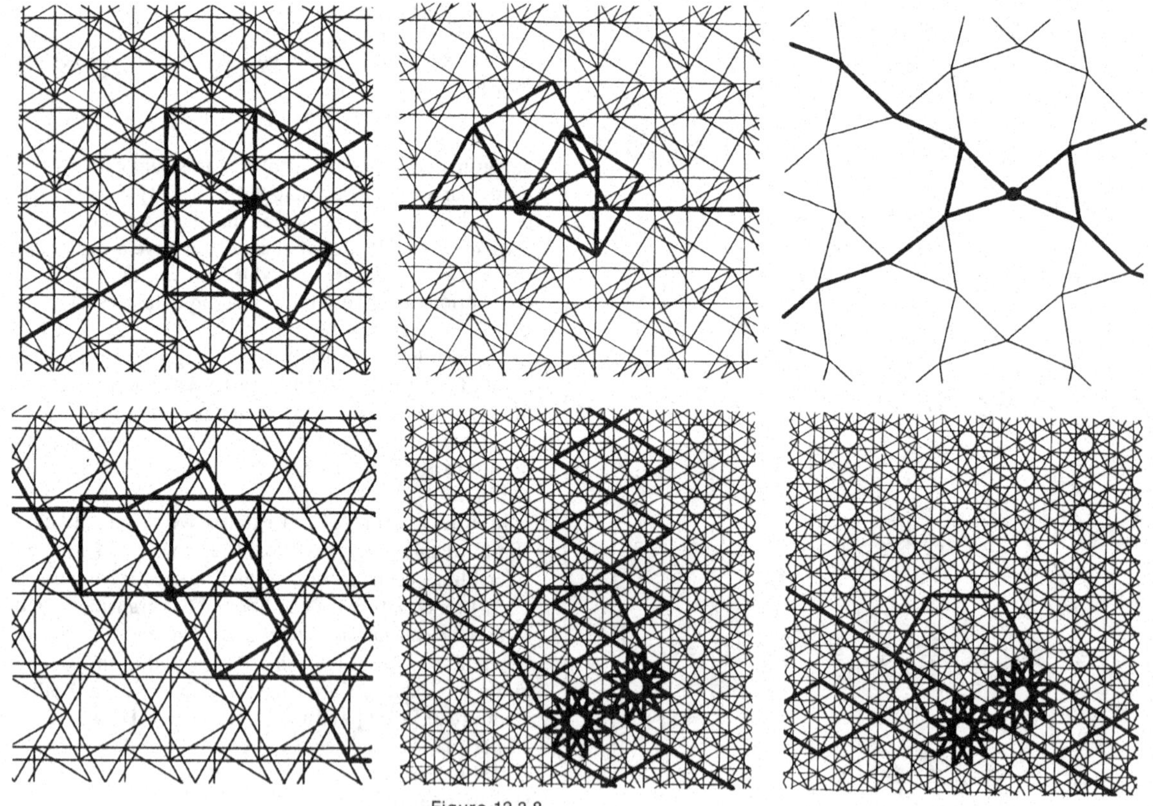

Figure 12.3.8
Tilings using apeirogons and zigzags.

EXERCISES 12.3

1. Construct tilings by regular polygons in which
   (*a*) the smallest convex subset of the plane which contains the tiling is a half-plane;
   (*b*) the convex hull of the tiling contains no half-plane and is contained in no strip.

2. Give examples of periodic tilings formed by "addition" from the rosettes of Figures 12.3.1 and 12.3.2. Are any such tilings uniform?

3. Give examples of families of polygons obtained by "addition" from the five rosettes which are (*a*) connected but not unicursal: (*b*) not connected; (*c*) connected but not simply connected.

**4. Prove the conjecture that every finite tiling by hollow regular polygons can be obtained from the five rosettes by the process of "addition".

*5. In Chapter 9 we showed that if P is any quadrangle, any pentagon with two parallel sides, or any hexagon with two parallel and equal sides, then there exists a periodic monohedral tiling with P as prototile. Is this still true if P is a hollow tile so that edges are allowed to intersect at points other than vertices? If the answer is in the negative, find necessary and sufficient conditions in each case for a polygon of the given type to admit a periodic tiling.

6. Construct at least one uniform tiling not mentioned in the text which contains:
   (*a*) finite polygons and zigzags,
   (*b*) zigzags and apeirogons,
   (*c*) finite polygons, zigzags and apeirogons,
   (*d*) zigzags only,
   (*e*) regular star polygons, zigzags and possibly other polygons.

7. Show that if one considers apeirogons as hollow regular polygons, then at least the following uniform tilings exist in addition to those given in Table 12.3.1 and in Figure 12.3.8(*a*): $(3^3.\infty)$, $(3.4.3{-}4.3.\infty)$, $(\frac{8}{3}.4.\frac{8}{3}.\infty)$, $(\frac{8}{3}.8.\infty)$, $(\frac{12}{5}.6.\frac{12}{5}.\infty)$, $(\frac{12}{5}.12.\infty)$, $(4^2.\infty)$, $(4.-8.\infty.-8)$, $(3.\infty.-3.\infty)$, $(3.\infty.3.\infty.3.\infty)$, where ∞ represents an apeirogon.

12.4 MULTIPLE TILINGS

Applying the terminology of the previous section to the case where the tiles are topological disks, we shall say that a tiling $\mathscr{T} = \{T_1, T_2, \ldots\}$ is of *density d* if every point of the plane which does not lie on the boundary of a tile lies in the interiors of exactly d tiles. If $d = 1$ we obtain, of course, the definition of Section 1.1, and if $d \geq 2$ we shall refer to the tiling as a *multiple tiling*. The idea of a multiple tiling is suggested by that of a Riemann surface—in fact a multiple tiling can be interpreted as a tiling of such a surface.

A multiple tiling is called *edge-disjoint* if no edge or part of an edge (except possibly isolated points) lies on the boundary of more than two tiles. Intuitively this means that the edges do not "overlap" in the manner shown in Figure 12.3.1(*b*) for hollow tiles. Thus in an edge-disjoint tiling any point of the plane that lies on the boundaries of three or more tiles must either be a vertex of the tiling, or a point where two or more edges cross each other.

A multiple tiling is called *vertex-disjoint* if the tiles incident with any vertex V of the tiling form a single circuit round that vertex. In other words, extending the terminology of Section 12.3 in the obvious way, the vertex figure is a single polygonal circuit or, equivalently, the tiling is unicursal. Naturally the polygonal circuit may wind around the vertex several times; the property of being vertex-disjoint implies that no two vertices of such a circuit coincide. In order to avoid uninteresting examples, in the following discussion we shall usually, but not always, restrict attention to edge-disjoint, vertex-disjoint and edge-to-edge tilings.

Petersen [1888] considered the possibility of constructing monohedral multiple tilings by regular polygons. He found that the only way in which this could be done is by superimposing several copies of the regular tilings (3^6), (4^4) or (6^3). However it seems to be an open question just how many of these tilings can be superimposed so that the resulting multiple tiling is regular in the sense that its symmetry group is transitive on the flags (see Section 1.3). If (unlike Petersen) we restrict attention to edge- and vertex-disjoint tilings, there are just three possibilities (see Figure 12.4.1 and Exercise 12.4.5(*a*)), but otherwise the problem is more difficult (see Exercise 12.4.6). Coxeter [1948] discusses superimposed regular tilings forming what he calls "compound tessellations", but these are not necessarily regular in the sense in which we use the term. Nevertheless his results yield useful information about multiple tilings. Similar questions for uniform tilings do not seem to have ever been considered; some open problems concerning these are suggested in the exercises.

For a list of regular compounds in the Euclidean or hyperbolic plane and on the 2-sphere, see Coxeter [1964]; for applications of these compound tilings in crystallography, see Takeda & Donnay [1965].

More interesting are multiple tilings which are *irreducible*, that is, cannot be obtained by superimposing several tilings of lower density. Examples were given by Marley [1974] and in Figure 12.4.2 we show irreducible tilings of densities 2, 3, 9 and 19 by convex polygons. Each of these is isohedral and edge-disjoint and, except for that of Figure 12.4.2(*a*), vertex-disjoint. The caption to this figure explains the construction of the tilings by what we call the "punctured planes" method. This method can be adapted to prove the following general result:

12.4.1 *For every $d \geq 3$ there exists a multiple tiling of density d by (not necessarily convex) hexagons, which is irreducible, edge-disjoint, vertex-disjoint and isohedral.*

From the example shown in Figure 12.4.2(*b*) we know that this result is true for $d = 3$. For larger values of d we construct the punctured plane with square holes shown in Figure 12.4.3. A method of splitting this plane

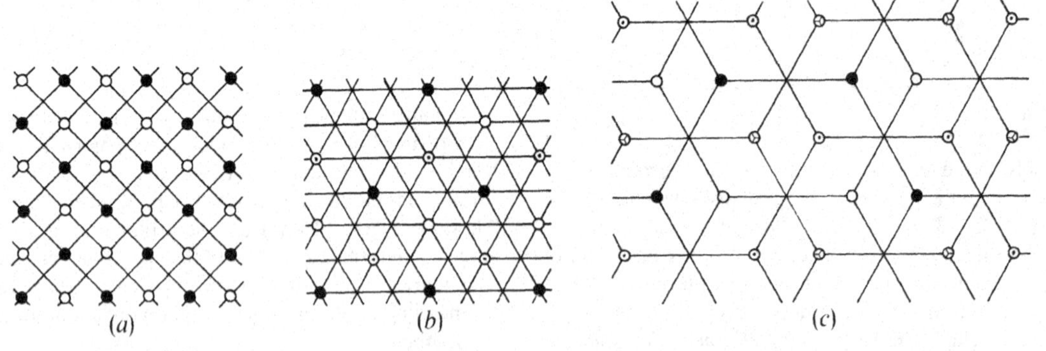

Figure 12.4.1
The three regular edge-disjoint and vertex-disjoint multiple tilings. Each of these is reducible and consists of (*a*) two layers of the tiling (4^4), (*b*) three layers of (3^6), and (*c*) four layers of (6^3). The vertices of each layer are indicated by small circles marked differently for the various layers.

into non-convex hexagons is indicated—and this can be extended in an obvious way to any value of $d \geq 4$. The tiling results from superimposing $d + 1$ of these punctured planes as explained in the caption to Figure 12.4.2.

For certain values of d it is possible to construct multiple tilings with additional properties. For example, an easy generalization of the example shown in Figure 12.4.2(*c*) yields tilings by either convex pentagons, convex hexagons or convex octagons which have as density any odd $d \geq 7$. If $d = p^2 + q^2 - 1$, where p and q are either

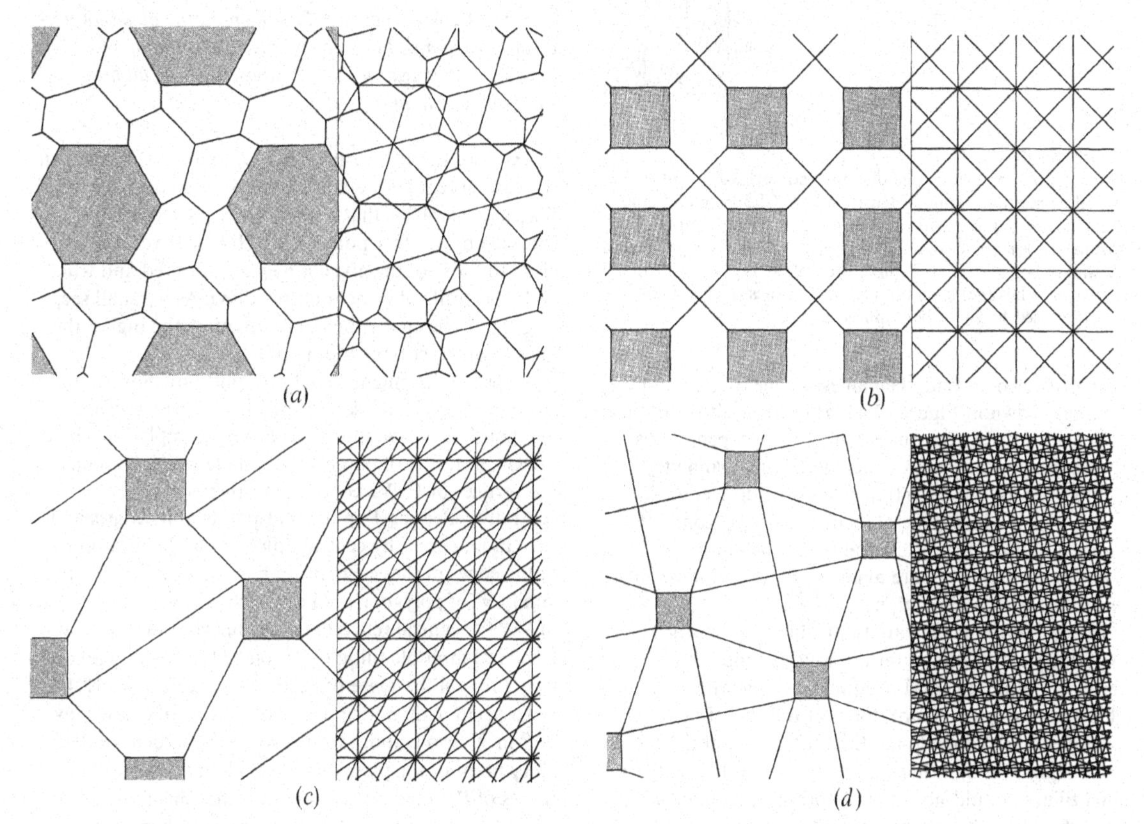

Figure 12.4.2
Examples of multiple tilings constructed using the "punctured plane" method. One begins with a plane punctured by a lattice arrangement of holes (shown in gray) which are either hexagonal as in (*a*) or square as in (*b*), (*c*) and (*d*). Each of these planes is split into polygons as shown on the left and then $d + 1$ translates of these "layers" are superimposed in such a way that every point of the plane, which is not on the edge of a tile, lies in exactly one hole, and therefore in d polygons. The appearance of the tiling is shown on the right of each part of the figure. The tilings shown are (*a*) by convex hexagons of density 2, (*b*) by convex centrally symmetric hexagons of density 3, (*c*) by convex pentagons of density 9, and (*d*) by convex quadrangles of density 19. All these tilings are edge-disjoint and, except for that in (*a*), they are also vertex-disjoint.

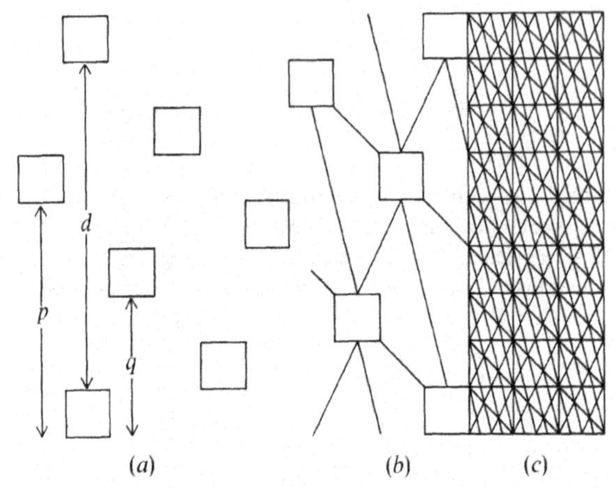

Figure 12.4.3
The construction of an edge-disjoint and vertex-disjoint isohedral multiple tiling of density $d \geq 4$. Choose p and q so that $2 \leq q < p$ and $p + q = d + 1$. In (*a*) we show how the parameters p, q are used in the construction of a plane punctured by square holes. In (*b*) this punctured plane is cut up into non-convex hexagons, and in (*c*) $d + 1$ copies of this plane are superimposed to yield the required tiling.

both even or both odd, then an easy generalization of the example shown in Figure 12.4.2(*d*) yields multiple tilings of density d by convex quadrangles, convex pentagons, convex hexagons or convex octagons, with symmetry group *c4* or *d4*. All these tilings are isohedral, edge- and vertex-disjoint, and some of them are also isogonal. If we do not insist upon isohedrality or edge- and vertex-disjointness then there are many more possibilities, most of which await investigation.

In spite of the great variety of tilings which these methods yield, we know of no multiple edge- and vertex-disjoint isohedral irreducible tiling with convex tiles and of density 2, nor, for that matter, of any monohedral irreducible multiple tiling by parallelograms or triangles, though there seems no *a priori* reason why such tilings should not exist. Another open question is that of finding the minimum possible density of irreducible edge- and vertex-disjoint multiple isohedral tilings by convex n-gons for a given value of n. For example, we believe that for $n = 6$, 7 and 8, the minimum densities are 3, 5 and 7, respectively. But we have no proof of this, and do not even know whether this minimum density increases monotonically with n.

Problems of a different character arise if we ask what shapes of convex polygonal tiles admit monohedral multiple tilings (of any density) rather than trying to find the minimum density for a given kind of tile. The following is the most general result that we have been able to prove:

12.4.2 *A convex polygon P is the prototile of a (not necessarily edge-disjoint) isohedral multiple tiling if there exists an affine transform of P whose vertices all have rational coordinates.*

Later we shall show that this condition is necessary in the case of regular n-gons. An interesting consequence of Statement 12.4.2 is that *every* convex polygon can be replaced by another polygon which has its vertices as close to those of the given polygon as desired and which is the prototile of some multiple tiling. As we shall see, in general, the closer the approximation, the higher the density of the corresponding tiling.

To prove Statement 12.4.2 we shall show how to construct a tiling $\mathscr{T}$ whose prototile is the given convex polygon P. The conditions on P clearly imply that there exists a lattice Λ of points in the plane which contains all the vertices of P. We construct further copies of P by centrally reflecting P in the mid-points of its edges, and then more copies by central reflection in the mid-points of the edges of these. Proceeding in this way we obtain a family $\mathscr{T}$ of polygons each of which is a translate of P or of $-P$ (its central reflection) and we claim that $\mathscr{T}$ is an isohedral multiple tiling of the plane (or possibly an ordinary tiling in the sense of Section 1.1), see Figure 12.4.4.

To begin with, notice that every vertex of every copy of P in $\mathscr{T}$ also belongs to Λ. From this we deduce that every point of the plane belongs to a finite number of copies of P (in fact, it is easily seen that this number cannot exceed twice the number of lattice points in P). By construction every edge of every tile in $\mathscr{T}$ is an edge of just one other tile lying on the other side of the edge, and so we deduce that at each point of the plane not on an edge, the density of $\mathscr{T}$ equals a constant d, that is, each point belongs to the interiors of exactly d tiles. This completes the proof of Statement 12.4.2. The tiling obtained in this way will be vertex-disjoint and, if the sides of P do not contain any points of L other than at their vertices and no two sides are parallel, then it will be edge-disjoint also.

It is well known that except when $n = 3$, 4 or 6, a regular n-gon cannot satisfy the condition of Statement 12.4.2; no affine transform has vertices with rational co-ordinates (see, for example, Post [1978]). If the condition in Statement 12.4.2 were necessary as well as sufficient, we could deduce from this fact that Petersen's result, quoted earlier, is true. However it is very easy to give a simple direct argument in this case. Suppose that a multiple tiling $\mathscr{T}$ has, as prototile, a regular n-gon P. Define S as the set of points in the plane which are the centers of the tiles of $\mathscr{T}$. It is clear that $\mathscr{T}$ is necessarily isohedral (which implies that S is a dot pattern in the sense of Section 5.3) and that $S(\mathscr{T}|P)$ must be the dihedral group dn ($n \geq 3$) leaving the center of P fixed. However, from our knowledge of dot patterns it follows that the only admissible values of n are 3, 4 or 6, leading to the desired result.

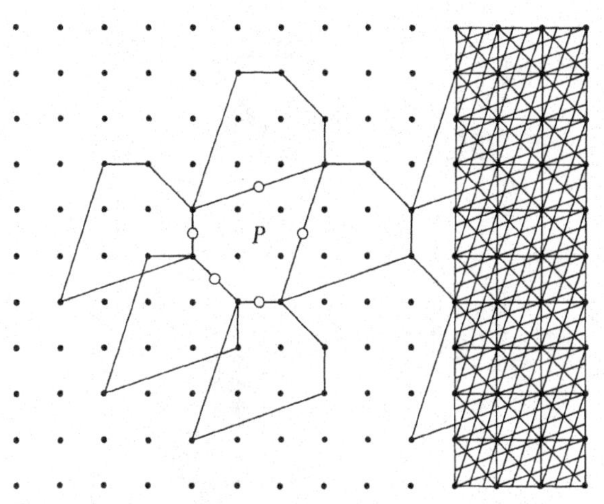

Figure 12.4.4
The construction of a multiple tiling using the convex pentagon P as prototile. The tiles are obtained by repeated central reflections in the mid-points of the edges. On the left we show P, the mid-points of its edges and the images of P under central reflections in these. Various other copies of the prototile are shown. On the right is a drawing of the tiling with all its tiles shown. The tiling is both edge-disjoint and vertex-disjoint. This method of construction is used in the proof of Statement 12.4.2.

EXERCISES 12.4

*1. Find some multiple tilings that cannot be constructed using either the punctured planes method (see Figure 12.4.2), or the method indicated in the proof of Statement 12.4.2.

**2. Find some values of $d(n)$—the minimal density of an irreducible edge- and vertex-disjoint monohedral multiple tiling by convex n-gons. Does $d(n)$ increase monotonically with n?

**3. Do there exist monohedral edge- and vertex-disjoint edge-to-edge multiple tilings by triangles or by parallelograms?

4. Prove that it is possible to construct multiple tilings by "almost regular" n-gons P that satisfy the following conditions.

*(*a*) If n is even then all the edges of P are the same length and the angles take only two different values which may be made arbitrarily close to each other.

*(*b*) If n is odd, all but one of the sides of P are equal and all but one of the angles are equal. The exceptional side and angle can be made arbitrarily close in size to the others.

(*c*) If $n = 8$ or 12 then it is possible for all the angles of P to be equal and the edges to be of two different lengths which may be made arbitrarily close to each other. (An example of a tiling by such 8-gons is shown in Figure 12.4.5.)

5. (*a*) Show that the only edge- and vertex-disjoint regular multiple tilings are those shown in Figure 12.4.1.

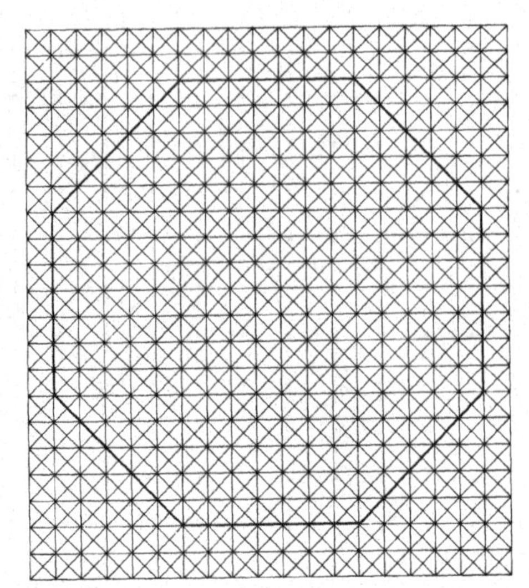

Figure 12.4.5
A multiple tiling (of density 239) by an "almost regular" octagon P. The angles of P are $\frac{3}{4}\pi$ and its sides differ in length by just over 1%.

(*b*) Show that if we do not require vertex-disjointness then there are two more such tilings.

**(*c*) Show that if a regular multiple tiling of density d uses square tiles (and is not necessarily edge- or vertex-disjoint) then $d=k^2$ or $2k^2$ for some integer k.

**(*d*) Show that if there exists a regular multiple tiling of density d by equilateral triangles, then $d=k^2$ or $3k^2$.

**(*e*) Show that if there exists a regular multiple tiling of density d by regular hexagons, then $d=k^2$ or $3k^2$.

(*f*) Show that for every $d \geq 2$ there is an isohedral multiple tiling by squares of density d which is edge- and vertex-disjoint.

(*g*) Determine for which numbers d there exist isohedral edge- and vertex-disjoint tilings of density d by regular hexagons, or by equilateral triangles.

**(*h*) For which isohedral types and which values of $d \geq 2$ do there exist multiple tilings of density d obtained from superposition of d copies of a single tiling if one insists that the multiple tiling be edge- and vertex-disjoint as well as isohedral.

6. The irreducible tiling by octagons shown in Figure 12.4.5 has the property that all its tiles are translates of each other; we call such tilings *translational.* (For additional results on translational tilings see Groemer [1978].)
   (*a*) Show that there exists no irreducible translational tiling of any density $d>1$ with a hexagonal prototile.
   (*b*) Construct an irreducible translational tiling of some density $d>1$ in which the prototile is a suitable decagon.
   **(*c*) What is the least density $d>1$ in any irreducible translational tiling?

12.5 GENERAL ADJACENCY CONDITIONS

At the beginning of the book we defined a tiling $\mathcal{T}$ as a family of sets that covers the plane without gaps or overlaps. We can think of $\mathcal{T}$ as an (unbounded) jigsaw puzzle in which the pieces (tiles) are copies of the prototiles and the way in which they can be fitted together depends solely on their shapes. As we developed the subject it was found convenient to place further restrictions on the pairs of tiles that could abut at an edge of the tiling; these took the form of markings or colorings that had to satisfy certain specified conditions which we shall call *adjacency conditions* (or matching conditions).

The first example occurred in Chapter 6. We introduced a tile symbol which told us how the sides of a polygonal tile were to be labelled with lettered arrows, and an adjacency symbol which specified the relationship between the sides of the two tiles that abutted at an edge. This procedure can be generalized to tilings in which there is more than one prototile—one needs to introduce a tile symbol for each of the prototiles and to include in the adjacency symbol all the different letters that occur. A condition of this kind will be called a *symbolic adjacency condition.*

In Chapter 11 we considered square tiles with colored sides (Wang tiles) that had to match in color, and in Chapter 10 we considered several other adjacency conditions. These included colored vertices that had to match (Figure 10.3.10), straight lines that had to continue across the edges of the tilings (the Ammann bars of Figures 10.3.26, 10.4.1(*d*), 10.4.7(*c*), (*d*)) and more complicated conditions such as adding markings to the sides of the tiles and specifying that these had to be put together to form simple geometrical shapes such as circular disks (Figure 10.4.7(*a*)), elliptical disks (Figure 10.4.1(*b*)) or arrows (Figures 10.4.11(*a*) and 10.4.14(*a*)). It is not hard to see, as has already been mentioned in some cases, that many of these conditions can be expressed as symbolic adjacency conditions or even in terms of the shapes of (suitably modified) tiles.

The purpose of this section is two-fold. Firstly we shall consider the use of symbolic adjacency conditions for monohedral tilings and secondly we shall describe some more complicated adjacency conditions that appear to be useful in tiling problems but which so far have not been described in the literature.

The problem of fitting together a given set of square or triangular tiles with colored edges so that adjacent edges match in color has been considered by several writers in recreational mathematics. The classical work is MacMahon [1921]; more recent references can be found in McCormack [1978].

Consider the tilings of Figures 12.5.1, 12.5.2 and 12.5.3.

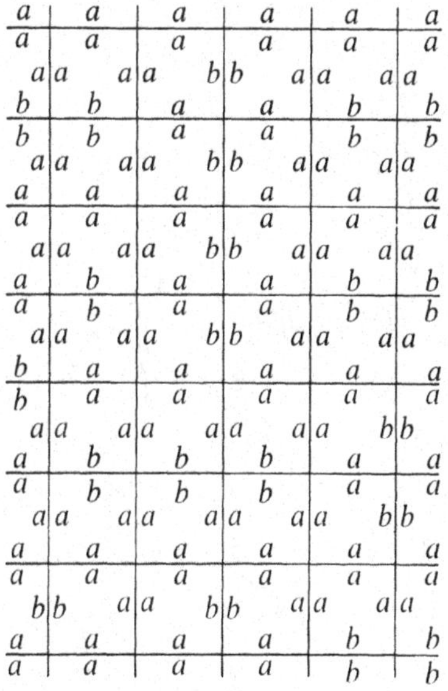

(*a*) 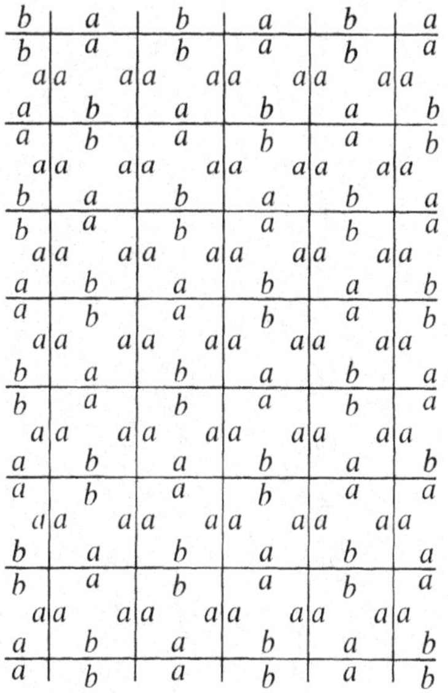 (*b*)

Figure 12.5.1
Two monohedral tilings whose prototile is a square with sides labelled *a* and *b*. The tile symbol is *aaab* and the adjacency symbol is *ab*. The tiling shown in (*b*) is isohedral, that in (*a*) is not.

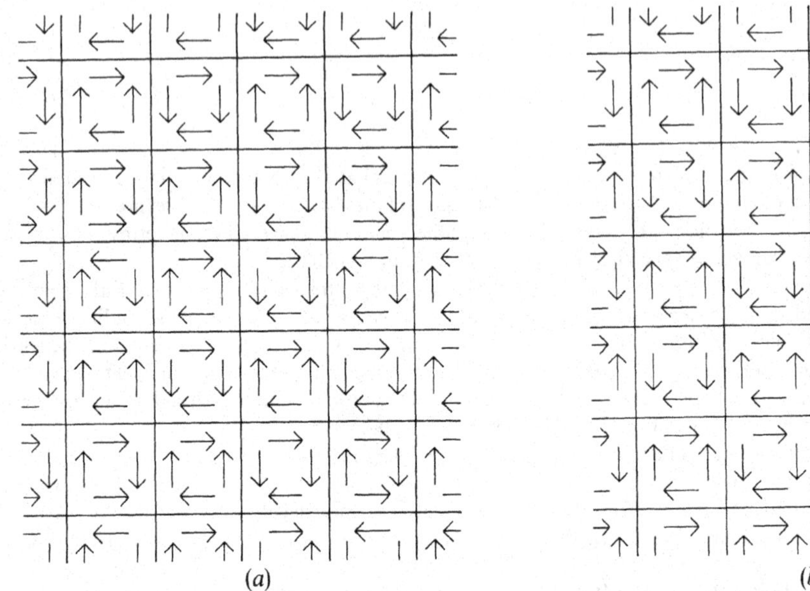

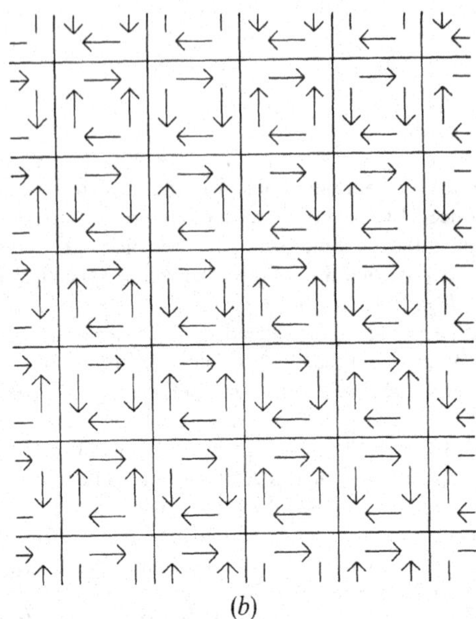

Figure 12.5.2
Two monohedral tilings with oriented edges, with arrows indicating oriented edges labelled a. The tile symbol is $a^+a^+a^+a^-$ and the adjacency symbol is a^+. The tiling shown in (*b*) is isohedral, that in (*a*) is not.

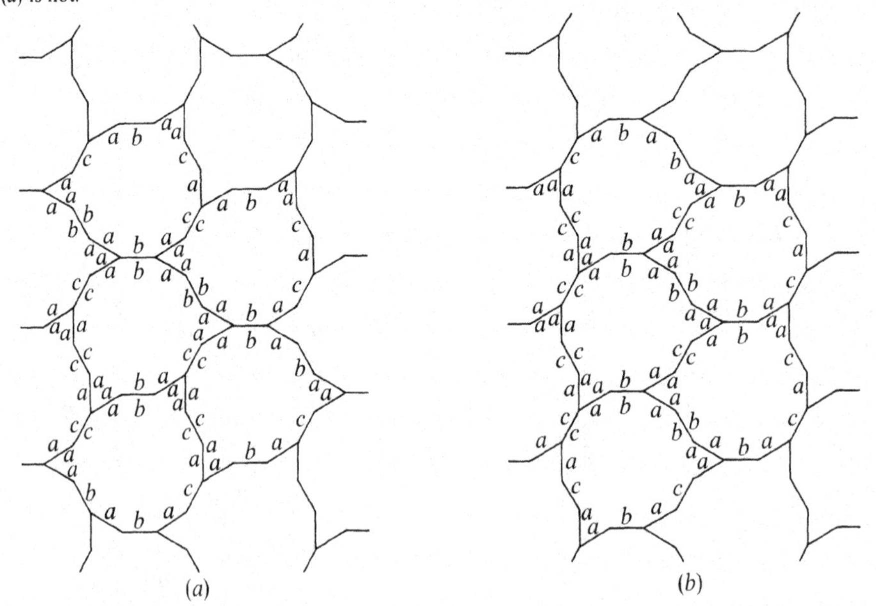

Figure 12.5.3
Two tilings by an irregularly shaped tile (the union of a regular 12-gon and two equilateral triangles) with tile symbol *aabacaabababacac* and adjacency symbol *abc*. The tiling in (*b*) is isohedral, that in (*a*) is not.

For each of these a symbolic adjacency condition is given in the caption of the diagram. Unlike the situation described in Chapter 6, however, these symbolic conditions do not specify a unique tiling nor even determine whether the tilings are isohedral or not. In each case, examples of an isohedral tiling and of a monohedral (non-isohedral) tiling are given. It is of some interest to find a criterion for the symbolic adjacency condition to correspond to a unique tiling, and to this end we shall introduce the concept of a primitively labelled tile.

Let T be any prototile whose sides are labelled by lettered arrows and let $S(T)$ be the symmetry group of this labelled tile, that is, the set of isometries that map T *and its labels* onto themselves. Then the labelling is said to be *primitive* if, whenever two sides of T bear the same label, one can be mapped onto the other by a symmetry in $S(T)$, and whenever a label corresponds to an unoriented side of T, then this side can be reversed by a symmetry of $S(T)$. In Figure 12.5.4 we show some examples of primitively and imprimitively labelled tiles.

12.5.1 *If a prototile T is primitively labelled and admits a tiling of the plane in accordance with a given adjacency symbol, then this tiling is unique and, moreover, it is isohedral.*

Thus symbolic adjacency symbols can lead to non-isohedral (monohedral) tilings only if the prototiles are imprimitively labelled; see, for example, Figures 12.5.1, 12.5.2 and 12.5.3.

The proof of Statement 12.5.1 depends upon the observation that the relative positions of two adjacent tiles are uniquely determined by the adjacency symbol if and only if the labelling is primitive (see Figure 12.5.5). This, and the simple connectedness of the plane, imply that a primitive labelling leads to a unique tiling $\mathcal{T}$. Further, if one maps a copy T_1 of the prototile onto a copy T_2 then this same mapping will map $\mathcal{T}$ onto itself. Thus $\mathcal{T}$ is isohedral.

Statement 12.5.1 cannot be strengthened. There exist imprimitively labelled tiles which only lead to isohedral

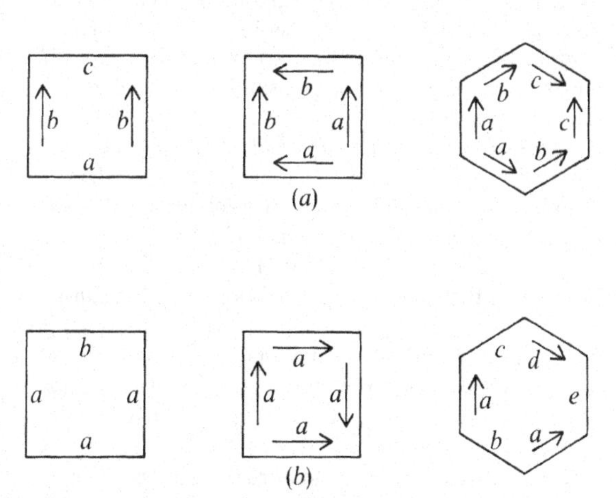

Figure 12.5.4
Tiles with symbols (*a*) ab^+cb^-, $a^+b^+b^-a^-$, $a^+b^+c^+c^-b^-a^-$, and (*b*) *aaab*, $a^+a^+a^+a^-$, $a^+cd^+ea^-b$. The labellings in (*a*) are primitive. Monohedral tilings which are not isohedral are possible only when the labelling is not primitive, as is the case with the tiles shown in (*b*).

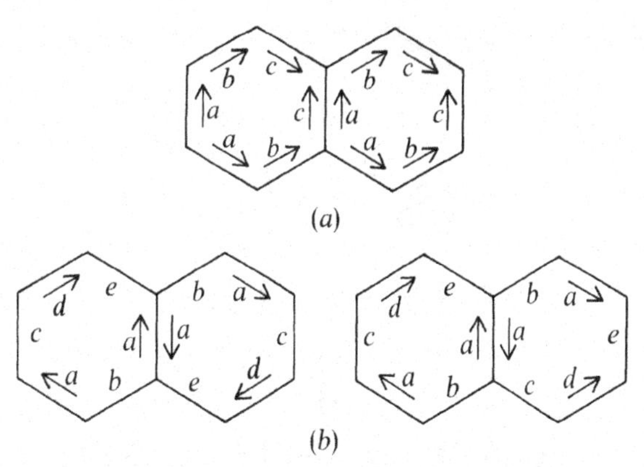

Figure 12.5.5
When the edges of the tiles are primitively labelled, as in (*a*), the relative positions of adjacent tiles are uniquely determined by an adjacency symbol. If the labelling is not primitive, as in (*b*), then there may be more than one possibility. This is the basis of the proof that monohedral tilings with primitively labelled tiles and a given adjacency symbol are unique and necessarily isohedral (Statement 12.5.1).

tilings and others that admit monohedral, but no isohedral tilings. For an example of the former see Figure 12.5.6, and for the latter, label the sides of any of the pentagonal tiles of types 6 to 13 in Table 9.3.1 with the same letter. Then, as we have seen, monohedral (but not isohedral) tilings are possible.

There are, of course, many kinds of adjacency conditions in addition to those we have considered here. A simple example of such arises if we color the sides of the tiles and then specify which colors must *not* abut. For example, if three colors are used, we might consider the adjacency condition which says that at each edge of the tiling two *different* colors must occur. So far as we are aware, tilings with such an adjacency condition have never been studied. A more complicated condition related to this has already been mentioned in Chapter 10 (see Figure 10.4.5). Here the sides of the tiles are marked with arrows of various lengths and the adjacency condition tells us which arrows are permitted on sides of adjacent tiles. The use of this condition enabled us to find an aperiodic set containing two similar and similarly labelled tiles.

Our final example shows how one can devise adjacency conditions which permit the solution of a specific problem. By a *cumulative* adjacency condition we mean one that relates tiles which are not adjacent. An example of such appears in Figure 12.5.7. Here the tiles are marked by arrows; the tail of an arrow is allowed to fit against either the head or tail of another arrow, but the heads of two arrows are allowed to abut only if both are *backed by equal numbers of arrows pointing in the same direction.* Thus in Figure 12.5.7, each of a pair of "facing" arrows is backed by a row of three arrows.

Cumulative matching conditions seem to present many unexplored possibilities. The example in Figure 12.5.7 shows a single prototile which admits, with the given adjacency condition, a *countable* infinity of tilings, thus providing a (partial) answer to the question raised on page 47.

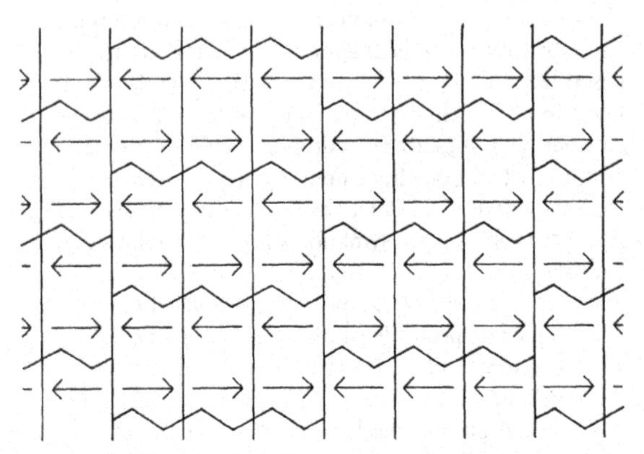

Figure 12.5.7
A tiling with a cumulative adjacency condition. With this tile, and the adjacency condition described in the text, a countable infinity of different tilings is possible.

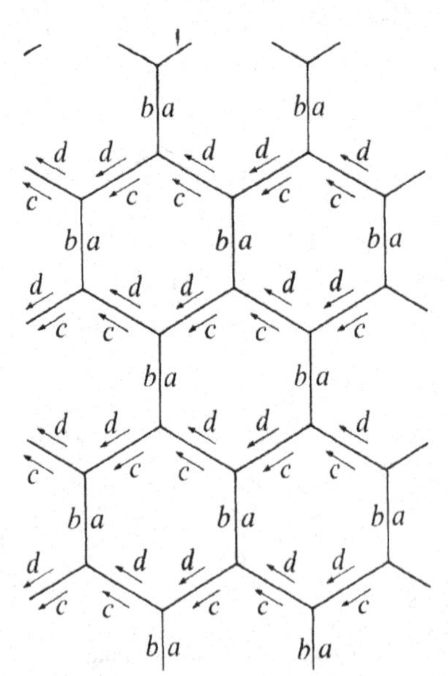

Figure 12.5.6
A tiling whose labelling is not primitive. The tile symbol is $ac^-c^-bd^+d^+$ which, with the adjacency symbol bd^-ac^-, leads to an isohedral tiling.

APPENDIX

The pages that follow are meant to help the reader find out some of the development in the understanding and use of tilings and patterns that occurred since the original publication. We are not under illusion that this is a complete accounting of the progress, but hope that it will be of some assistance in locating additional sources.

For each item we have supplied an indication for the place in the text where this information would fit appropriately; we also gave precise reference to each of the publications.

Page 45, right column. Before the first complete paragraph insert:

The wallpaper groups have been explained and presented in too many venues to be listed here. However, a novel and systematic notation for the various kinds of symmetry groups is clearly presented and beautifully illustrated in the first part of *The Symmetries of Things.* (John H. Conway, Heidi Burgiel and Chaim Goodman-Strauss. A.K. Peters, Wellesley, MA, 2008). Regrettably, this book lacks adequate references to other publications.

Page 56, line 10 of right column:

For a recent view of the question whether all symmetry groups appear in the Alhambra see B. Grünbaum (What symmetry groups are present in the Alhambra, Notices of the Amer. Math. Society 53(2006), 670–673), where additional references can be found.

Page 56, end of Chapter 1.

A different approach to symmetries occurring in the fabrics of preColumbian Peru is explained in B. Grünbaum (Periodic ornamentation of the fabric plane: Lessons from Peruvian fabrics. Symmetry 1(1990), 45–68; reprinted in: *Symmetry Comes of Age. The Role of Pattern in Culture*, D. K. Washburn and D. W. Crowe, eds. Univ. of Washington Press, Seattle 2004. Pp. 18–64. Also highly relevant is *Embedded Symmetries,* D. K. Washburn, ed., Univ. of New Mexico Press, Albuquerque 2004.

Page 64, end of text insert:

A method of describing and characterizing those periodic tilings in which all vertices are of species number 3 has been devised by C. B. Jones (Periodic tilings with vertices of species number 3. *Structural Topology* 20(1993), 49–54).

Page 68, end of text.

D. Chavey (Periodic tilings and tilings by regular polygons. I. Bounds on the number of orbits of vertices, edges and tiles. Mitteilungen Mathem. Seminar Giessen 164(1984), 37–50; Tilings by regular polygons. II. A catalog of tilings. Computers and Mathem. Applic. 17(1989), 147–165) investigates the possible numbers of orbits of the vertices, edges and tiles, and provides examples for certain values of these parameters.

Page 70, Section 2.3.

k-isohedral tilings have been considered (under various additional restrictions) by J. E. S. Sokolar (The hexagonal parquet tiling: *k*-isohedral monotiles with arbitrarily large *k*. *Math. Intelligencer* vol. 29, no. 2 (2007), 33–38.

The definition of equitransitive tilings can be generalized by omitting the requirement that the tiles be regular polygons. Concerning this concept it was proved in L. Danzer, B. Grünbaum and G. C. Shephard (Equitransitive tilings, or how to discover new mathematics. *Mathematics Magazine* 60(1987), 67–89) that in any equitransitive normal tiling of the plane, the maximal number of sides in a tile is 66, this bound being attained for six different tilings with symmetry group p6m. Several related results and problems are also mentioned. The maximal possible numbers of sides of tiles in equitransitive tilings with any of the other 16 symmetry groups have been determined by J. E. Georges and A. M. Matthews (Maximal polygons for equitransitive periodic tilings. *Pi Mu Epsilon Journal* 8(1988), 557–571).

Page 76.

The algebraic identity $1^3 + 2^3 + ... + n^3 = (n(n+1)/2)^2$, leads to the question whether it is possible to tile the square of side n(n+1)/2 by a collection of squares which contains k squares of side k for k = 1,2,...,n. The answer is negative for $n \leq 6$, affirmative for $11 \leq n \leq 33$, and open for all other n. For details and references to previous work on the topic S. T. Ahearn and C. H. Jepsen (On tiling an $m \times m$ square with m squares. *Crux Math.* 19(1993), 189–191).

Page 78, Figure 2.4.8.

The uniqueness of the Duijvestijn's squared square was proved (together with other results) by I. Gambini (A method for cutting squares into distinct squares, *Discrete Applied Mathematics* 98(1999), 65–80).

Page 79, first two paragraphs of the left column.

A survey of related topics was given by P. Schmitt (Problems in discrete and combinatorial geometry, *Handbook of Convex Geometry* Vol. A, pp. 449–483. North-Holland, Amsterdam 1993. A new presentation, and some new results, are given by F. V. Henle and J. M. Henle (Squaring the plane. Amer. Math. Monthly 115(2008), 3–12).

Page 81, Problem 2.4.6b.

The conjecture that there are precisely five types of unilateral and equitransitive tilings by squares of three sizes $x < y < x + y$, was disproved by H. Martini, E. Makai and V. Soltan (Unilateral tilings of the plane with squares of three sizes. *Beiträge Algebra Geom.* 39(1998), 481–495). The fact that there are precisely eight such tilings was independently established by D. Schattschneider (Unilateral and equitransitive tilings by squares. *Discrete Comput. Geometry* 24(2000), 519–525) and A. Bölcskei (Classification of unilateral and equitransitive tilings by squares of three sizes. *Beiträge Algebra Geom.* 41(2000), 267–277).

Page 84 of text, Caption of Figure 2.5.4:

David Lovler has found two additional tilings of this type.

Page 85, Figure 2.5.4.

David Lovler (private communication) has found an additional uniform tiling that includes a regular star polygon; see attached drawing. It is remarkable by being the only known chiral tiling of this kind.

Page 91.

A superb exposition of results on dissections is G. N. Frederickson (*Dissections: Plain and Fancy.* Cambridge Univ. Press 1997); his other books (*Hinged Dissections.* Cambridge Univ. Press 2002; *Piano-hinged Dissections: Time to Fold.* A. K. Peters, Wellesley, MA, 2006) present various new ideas on dissections.

Page 92, Figure 2.6.1.

It was brought to our attention by David Eppstein that H. Lindgren (*Recreational Problems in Geometric Dissections and How to Solve Them.* Dover, New York 1972) has, in Figure 16.6, a 4-dissection tiling of the regular 9-gon; this improves the result indicated in Figure 2.6.1(d).

Page 93, Figure 2.6.2.

David Eppstein has sent us several 5-dissection tilings for |12/5|, thus improving the result in Figure 2.6.2(h). With his permission, the most elegant of these 5-dissection tilings is indicated below.

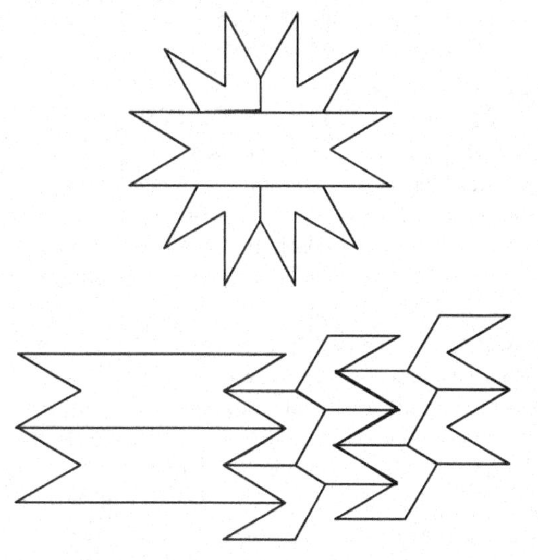

Page 102, Section 2.9.

Questions of colorings of Archimedean tilings related to this section but distinct from the ones considered here are the topic of the paper by D. W. Crowe (Precise perfect colorings of Archimedean tessellations. *Visual Mathematics* (Art and Science Electronic Journal of ISIS-Symmetry) Math. 1(1999), no. 1, 9 HTML documents; approx. 10 pp.; per Math. Reviews 2000g:05049).

Page 104 of text, Figure 2.9.2:

In the last part, the central small triangle of each group of four white triangles should have color 4

Page 125, Statement 3.2.3.

A more streamlined proof is given by D. Kazanci and A. Vince (A property of normal tilings, *Amer. Math. Monthly* vol. 111, no. 9(2004), 813–816).

Page 147 of the text, Figure 3.7.1:

The length of the arrow should equal twice the width of the strips.

Page 155.

Examples of prototiles showing that r = 2 is possible for Heesch's problem have been found by Ludwig Danzer and by Anne Fontaine (see A. Fontaine, "An infinite number of plane figures with Heesch number two", *J. Combinat. Theory, Series A* 57 (1991) 151–156). The late Robert Ammann (private communication) found the prototile shown below, which establishes that "Heesch number" r = 3 is possible; additionally, with this prototile it is possible to surround a single point four times, without it being possible to surround a point five times. At the time, it was tempting to conjecture that Ammann's example is best possible in both respects. However, there are examples with Heesch number 4 (E. Friedman, Heesch tiles with surround numbers 3 and 4, *Geombinatorics* 8(1999), 101–103, and also an example with Heesch number 5 (see http://math.uttyler.edu/cmann/math/heesch/heesch.htm, http://www.ics.uci.edu/~eppstein/junkyard/heesch/). Casey Mann rediscovered Ammann's tile, as well as a pentahex tile with Heesch number 3. Mann also found a heptiamond with Heesch number 3. Both new tiles are still unpublished. Mann can be contacted at cmann@uttyler.edu, and (as of February 2009) some of his results are available at http://www.math.uttyler.edu/polyformDB. A *modified Heesch number* which allows gaps in the surrounding patches was considered by some of these authors. Not surprisingly, a few of the tiles have a modified Heesch number larger than the usual Heesch number, but this seems to be of lesser interest than the original question.

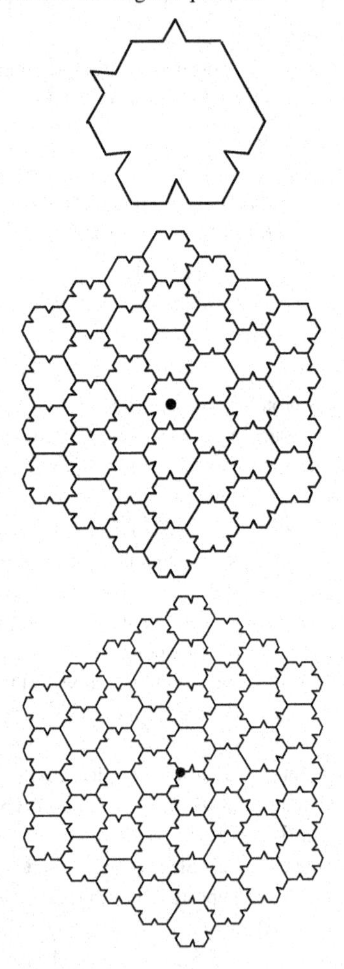

Page 162, left column, line -18.

A condition weaker than normality has been used by P. M. Soardi (Recurrence and transience of the edge graph of a tiling of the euclidean plane. *Mathematische Annalen* 287(1990), 613–626) in the study of random walks on edge-graphs of tilings.

Page 165, Chapter 4.

A somewhat different approach to topological aspects of tilings in presented by D. Renault (The uniform locally finite tilings of the plane, *J. Combinatorial Theory, Series B,* 98(2008), 651–671).

Page 183, Section 4.6.

For each $n \geq 1$, there exist only finitely many topological types of n-homeohedral tilings, and each type has an n-isohedral representative. This, as well as some related facts, was established by A. W. M. Dress and D. Huson (On tilings of the plane. *Geometriae Dedicata* 24(1987), 295–310) and by P. Schmitt (n-homeohedral types of tilings. *Geometriae Dedicata* 32(1989), 319–327). In Exercise 4.6.4(b) this was posed as an open problem.

Page 187 of the text:

In the leftmost column of tilings, the second from top is isomorphic to the second from bottom.

Page 190, end of text.

B. Grünbaum, H.-D. Löckenhoff, G. C. Shephard and A. H. Temesvári (The enumeration of normal 2-homeohedral tilings, *Geometriae Dedicata* 19(1985), 109–174) found 508 distinct normal 2-homeohedral tilings. A. W. Dress and D. Huson (On tiling the plane, *Geometriae Dedicata* 24(1987), 295–310) proved that there are precisely 508 such tilings.

A classification of special types of 2-homeohedral tilings was carried out by A. W. Dress (The 37 combinatorial types of "Heaven and Hell" patterns in the Euclidean plane, M. C. Escher: Art and science (Rome 1985), 35–45. North-Holland, Amsterdam 1986), by A. W. Dress and D. H. Huson (Heaven and hell tilings, *Structural Topology* 17(1991), 25–42), and by A. W. M. Dress and R. Scharlau (The 37 combinatorial types of minimal, non-transitive, equivariant tilings of the Euclidean plane. *Discrete Math.* 60(1986), 121–138).

Page 197, end of Section 4.7.

Results related to Statements 4.7.11 and 4.7.12 (and originating in B. Grünbaum and G. C. Shephard, Incidence symbols and their applications, in Proc. Symposia Pure Math. 34(1979), 199–244) may be found in J. Siagiova and M. E. Watkins (Covalence sequences of planar vertex-homogeneous maps, *Discrete Math.* 307(2007), 599–614), where the terminology is different.

Non-normal 2-homogeneous tilings, and graph-theoretic "isoperimetric inequalities" for them were investigated by A. Calogero (Strong isoperimetric inequality for the edge graph of a tilings of the plane. *Archiv der Mathematik* 61(1993), 584–595).

Page 200, before the last full paragraph insert:

The completeness of the enumeration by Grünbaum, Löckenhoff, Shephard and Temesvári of the 508 normal 2-homeohedral tilings was established by A. W. M. Dress and D. Huson (On tilings of the plane. *Geometriae Dedicata* 24(1987), 295–310).

Page 202.

A different view of tiling questions in the hyperbolic plane is investigated in J. W. Di Paola, Tiling in the hyperbolic plane. *Congressus Numerant.* 32(1981), 267–278. On the other hand, a computational approach to tilings in various settings in considered by O. Delgado Friedrichs, A. Dress and D. Huson (Tilings and symbols: a report on the uses of symbolic calculation in tiling theory. *Sém. Lothar. Combin.* 34(1995), Art. B34a, 14 pp. (electronic); MR1399748 = 97d:52033) and by N. P. Dolbilin, A. W. M. Dress and D. H. Huson (Two finiteness theorems for periodic tilings od d-dimensional Euclidean space. *Discrete and Computation Geometry* 20(1998), 143–153).

Page 204, Section 5.1.

A different approach to the basic definitions concerning patterns is presented by T. W. Knight (Infinite patterns and their symmetries, *Leonardo* vol. 31 no.4(1998), 305–312). Since the proposed characterization of a motif is not restricted (as in our presentation), in most cases the same pattern can be interpreted in several ways as consisting of different motifs, which may lead to different symmetry groups.

Page 215, Exercise 5.1.10.

A classification of dimotif patterns, analogous to the one presented in Section 5.2 for monomotif patterns, is given by R. L. Roth (The classification of bipatterns in the plane. *Journal of Geometry* 42(1992), 132–147). He finds that there are 112 types of periodic dimotif patterns, and 29 types of dimotif strip patterns.

Page 242, Figure 5.3.3.

The illustration of DPP23 is wrong, as was pointed out to us by Paul Dhooghe. A correct pattern is shown below.

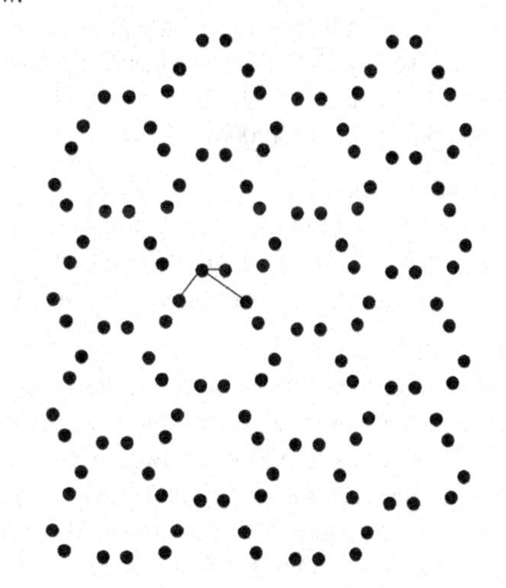

Page 252.

K. Bongartz and R. Scholle (Any tile-transitive Dirichlet tiling comes from an orbit of its symmetry group. Bergische Universität, Wuppertal. Preprint 1990; apparently unpublished) proved that if a Dirichlet tiling associated with a discrete family of points is isohedral, then it can also be obtained as the Dirichlet tiling of a transitive dot pattern.

Page 265, line 2.

Added information about such "interlace patterns", in particular with reference to Islamic ornaments, is contained in the paper by B. Grünbaum and G. C. Shephard (Interlace patterns in Islamic and Moorish art, *Leonardo* 25(1992), 331–339 + color plate; Reprinted in *The Visual Mind: Ary and Mathematics*, M. Emmer, ed., The MIT Press, Cambridge, MA 1993, pp. 147–155 + color plate).

Page 291, end of text.

A different method of enumeration of the 93 types of marked isohedral tilings is presented by A. W. M. Dress and R. Scharlau (Zur Klassifikation äquivarianter Pflasterungen, *Mitteilungen Mathem. Seminar Giessen* 164(1984), 83–136).

An algebraic description of the 46 marked isohedral tilings with trivial stabilizer (induced tile group of symmetries) was given by G. H. Greco (Algebraic properties of basic isohedral marked tilings, *Math. Proc. Cambridge Philos. Soc.* 140(2006), 417–423.

Page 299, end of text.

D. Renault (The uniform locally finite tilings of the plane, *J. Combinat. Theory* Series B, 98(2008), 651–671) describes a more general approach to isogonal tilings.

Page 317, end of text.

Additional information about isotoxal tilings and more general edge-transitive planar graphs can be found in B. Grünbaum and G. C. Shephard (Edge-transitive planar graphs, *J. Graph Theory* 11(1987), 141–155) and S. Graves, T. Pisanski and M. E. Watkins (SIAM *J. Discrete Mathematics* 23(2008), 1–18).

Page 331.

D. Girault-Beauquier and M. Nivat (Tiling the plane with one tile. In *Topology and category theory in computer science*. Oxford Univ. Press, New York, 1991. Pp. 291–333; for polyominoes also: On translating one polyomino to tile the plane, *Discrete and Computational Geometry* 6(1991, 575–592) have established that if T is a prototile of a monohedral tiling using translates only, then T admits an isohedral tiling using translates only; moreover, every tiling by translates of T has translational symmetries in at least one direction.

Page 333, before the first complete paragraph in right column.

For a different approach to classification of tilings which leads, among other results, to an independent enumeration of the 93 types of marked isohedral tilings, see Z. Lucic' and E. Molnár (Combinatorial classification of fundamental domains of finite area for planar discontinuous isometry groups. *Archiv der Mathematik* 54(1990), 511–520).

Page 400, first complete paragraph in the right column

Detailed estimates for the density of certain ellipse patterns were given by M. Tanemura and T. Matsumoto (On the density of the $p31m$ packing of ellipses. *Zeitschrift für Kristalographie* 198(1992), 89–99).

Page 401 of text, Figure 8.0.1:

The pattern is not perfectly colored. The underlying group (see page 404) consists of all rotational symmetries of the pattern.

Page 425, Statement 8.3.2.

An apparently related paper, which I was unable to obtain, is by M. A. Hann and B. G. Thomas (3-color interchange patterns, *J. Textile Institute* 98(2007), 539–547).

Page 459.

K. Bongartz and D. Mertens (On the enumeration of homeochromatic classes of k-chromatic tilings. *Discrete Math.* 85(1990), 17–42) list all homeochromatic classes of k-chromatic tilings for $k \leq 12$ colors.

Page 464, left column, new paragraph following the first complete paragraph:

An interesting analysis of the use of black and white frieze types in practical ornaments in a community is given by A. V. James, D. A. James and L. N. Kalisperis (A unique art form: The friezes of Pirgí, *Leonardo*, vol. 37, no. 3(2004), 234–242).

For a different approach to strip and plane color groups, with a very convenient notation, see H. S. M. Coxeter (The seventeen black and white frieze types. *C. R. Math. Rep. Acad. Sci. Canada*, 7(1985), 327–331;

and: Coloured symmetry. In: *M. C. Escher – Art and Science. Proc. Internat. Congress on M. C. Escher, Rome 1985.* H. S. M. Coxeter, M. Emmer, R. Penrose and M. L. Teuber, eds. North-Holland, Amsterdam 1986, pp. 15–33).

Page 466, after the first complete paragraph:

An extension of the classification of pattern types to the two-sided plane has been carried out by P. R. Cromwell (The henomeric types of 2-sided patterns in the plane. *Geometriae Dedicata* 137(2008), 63–84). This classification applies to 2-layered patterns or those with interlaced motifs. Among 2-sided patterns Cromwell finds 17 infinite families of finite pattern types, 68 types of strip patterns, and 264 types of periodic ones.

Page 474, left column, Statement 9.1.1.

As was observed by M. I. Shtogrin (Action centers of planigons. [In Russian]. *Matematicheskie Zametki* 44(1988), 262–278. English translation: *Mathematical Notes* 44(1988), 627–635) there is one additional polygonal isohedral type of proper tilings by convex quadrangles. For this type the entry in Table 9.1.2 (on page 484) would consist of the following line:

P_4–57 [3^6] [$a^+b^+c^+d^+e^+f^+$; $a^+e^-c^+f^-b^-d^-$] *pgg e* $\alpha(\alpha)\beta(\beta)\alpha\beta$ $(\alpha\beta\chi)\delta\beta\varepsilon$ $A+E=D+F=B=C=\pi$, $A=D$ 1*D*, 1*R* IH6 –

A tiling of the type P4-57 is shown below, with the same conventions as in Figure 9.1.2.

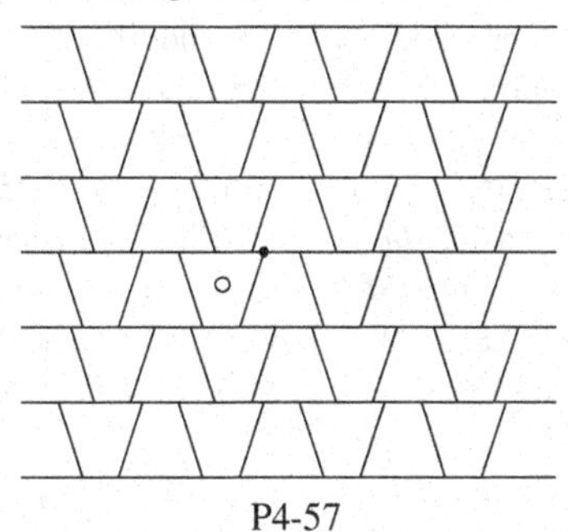

P4-57

Page 497, Section 9.4.

Detailed expositions of various aspects of polyominoes are contained in books by G. E. Martin (*Polyominoes: A Guide to Puzzles and Problems in Tiling*. Math. Assoc. of America, Washington, DC, 1991) and S. Golomb (*Polyominoes: Puzzles, Patterns, Problems, and Packings*. Revised ed. Princeton Univ. Press, 1994).

Page 498, last paragraph.

Concerning Golomb's question, the algorithmic decidability in the case of a single polyomino is considered by D. Beauquier and N. Nivat (On translating on polyomino to tile the plane, *Discrete & Computational Geometry* 6(1991), 575–592), and by I. Gambini and L Vuillon (An algorithm for deciding if a polyomino tiles the plane, *Theoretical Informatics and Applications,* 41(2007), 147–155).

Page 505, left column, line 2.

Investigations of the possibility of tilings with rotational symmetry by certain polyominoes and polyiamonds can be found in a paper by H. Fukuda, N. Mutoh, G. Nakamura and D. Schattschneider (A method to generate polyominoes and polyiamonds for tilings with rotational symmetry. *Graphs and Combinatorics* 23(2007), Supplement 1, pp. 259–267).

Page 505 of text, Figure 9.4.2:

One edge on the left margin of the middle panel is misplaced.

Page 506 of text, Figure 9.4.3:

The tiling in part (c) is not isohedral, although there exist isohedral tilings with this tile.

Page 510, end of Section 9.4:

N. MacKinnon (Some thoughts on polyomino tiles. *Mathematical Gazette* 74(1990), 31–33) introduced the concept of "idiot-proof tiles" (IPTs). A polyomino is "idiot proof" if any placement of two copies of that polyomino on the square tiling can be extended to a monohedral tiling of the plane provided the two original ones neither overlap nor enclose an area. Examples of IPTs are the L-triomino and the W pentomino. Among naturally arising questions which appear to be open are: Does there exist an infinity of IPTs? In particular, are all zigzag polyominoes IPTs? Are there any tiles for which any placement of three tiles (with the obvious restrictions) can be completed to a monohedral tiling? Affirmative answers to these questions are given in B. Grünbaum and G. C. Shephard (Idiot-proof tiles. *Mathematical Gazette* 75(1991), 143–147), together with additional results and other questions.

Sallows *et al.* (L. Sallows, M. Gardner, R. Guy and D. Knuth, Serial isogons of 90 degrees. *Math. Magazine* 64(1991), 315–324) deal with polyominoes the sides of which have successive integers as lengths. They show that each such polyomino must have a number of sides which is a multiple of 8, and find that the smallest (the only one with eight sides) is the prototile of an isohedral tiling; the four polyominoes of this kind with 16 sides do not seem to tile. For an attractive presentation of these and many related questions see A. K. Dewdney (An odd journey along even roads leads to home in Golygon City. *Scientific American*, July 1990, pp. 118–121).

W. P. Thurston (Conway's Tiling groups, *Amer. Math. Monthly* 97(1990), 757–773) gives an account of a technique, discovered by John Horton Conway, that uses finitely presented groups to decide in some cases whether a planar region can be tiled by copies of a prototile.

Many of the topics treated in this section have been supplemented by the results of G. C. Rhoads (Planar tilings by polyominoes, polyhexes, and polyiamonds, *J. of Computational and Applied Mathematics* 174(2005), 329–353).

J. E. S. Socolar (The hexagonal parquet tiling: *k*-isohedral monotiles with arbitrarily large *k*, *Mathematical Intelligencer* 29 No. 2 (2007), 33–38) discusses problems of monohedrally tiling the plane with copies of tiles obtained as unions of congruent copies of smaller tiles. It should be noted that the meaning of the terms "tiling", "k-isohedral" and others does not match the meaning attributed to them in this book and other publications.

Page 512, end of text.

The last problem was affirmatively solved by P. Gailiunas (Spiral tilings, Visual Math. (electronic), vol. 2 no. 3(2000), #2).

Page 518, left column, line 12.

This problem was explicitly stated in S. Berczi and D. Nagy (Periodicity of extremal geometric arrangements (densest packings, thinnest coverings, tessellations). *Acta Geologica Acad. Sci. Hungar.* 23(1980), 173–200).

Page 518, last paragraph.

A simpler proof of the result of Hirschhorn and Hunt was given by O. Bagina (Tiling the plane with congruent equilateral convex pentagons. *J. Combinatorial Theory* Ser. A 105(2004), 221–232).

Page 519, Chapter 10.

After the publication of "Tilings and Patters", the study of aperiodic tilings took off at an ever-accelerating pace. This was due to the publication of a short paper (D. Schechtman, I. Blech, D. Gratias and J. W. Cahn, Metalic phase with long-range orientational order and no translational symmetry. *Phys. Review Letters* 53(1984), 1951–1953) that revolutionized crystallography. The paper showed that aperiodic (or, as much of the literature calls it, "quasiperiodic") order is possible in solid-state materials; in particular, this meant that the "crystallographic restriction" dogma needs modification. The hundreds of papers published on this topic, many in journals devoted to the physical sciences, present a body of material that is impossible to survey or summarize in this brief account. Fortunately, there are several books that can provide the interested reader with guidance to the vast literature. As examples of such sources we mention: M. Senechal, *Quasicrystals and Geometry* (Cambridge University Press 1995, xv + 286 pp.); C. Radin, *Miles of Tiles*. (Student Mathematical Library, vol. 1. Amer. Math. Soc., Providence, RI, 1999. xii + 120 pp.); L. Sadun (*Topology of Tiling Spaces*. University L'ecture Series Vol. 46. Amer. Math. Soc., Providence, RI., 2008, x + 118 pp.)

A list of references to a number of earlier papers, and a new set of two tiles that tile only aperiodically (based on the tiling in Figure 10.1.5) is given by C. Goodman-Strauss (A small aperiodic set of planar tiles, *European J. of Combinatorics,* 20(1999), 375–384).

Page 524, left column, new paragraph following the first complete paragraph

By relaxing the condition of simple connectivity of tiles, various reptiles with holes are constructed by F. Jordan and S.-M. Ngai (Reptiles with holes, *Proc. Edinburgh Math. Soc.* vol. 48, no. 3(2005), 651–671).

Page 544, last line.

For another approach to the computer generation of Penrose tilings and related patterns see J. Rangel-Mondragon and S. J. Abas (Computer generation of Penrose tilings. Computer Graphics Forum 7(1988), 29–37).

Page 548 of text, Figure 10.3.26:

In part (a), a short line is missing near the left corner.

Page 550, Section 10.4.

Detailed proofs about the various aperiodic sets of Ammann tiles described here can be found in the paper by R. Ammann, B. Grünbaum and G. C. Shephard (Aperiodic tiles, *Discrete Comput. Geometry* 8(1992), 1–25).

The only information available about the person and life of Robert Ammann is contained in the paper by M. Senechal (The mysterious Mr. Ammann, *Math. Intelligencer* vol. 26, no 4(2004), 10–21).

Page 557 of the text:

The tilings in Figures 10.4.16 and 10.4.17 are incorrectly drawn.

Page 582, after the penultimate paragraph add

Early descriptions of aperiodic sets of prototiles for 3-dimensional tilings have been given by A. Katz (Theory of matching rules for the 3-dimensional Penrose tilings. *Commun. Math. Phys.* 118(1988), 263–288) and L. Danzer (Three-dimensional analogs of the planar Penrose tilings and quasicrystals. Discrete Math. 76(1989), 1–7).

Page 583, Chapter 11.

Various results on topics related to Wang tiles are presented by E. Goles and I. Rapaport (Tiling allowing rotations only, *Theoretical Computer Science* vol. 218(1999), 285–295, and in other articles coauthored by E. Goles.

Page 625L, -1.

The lattice Λ leads to another set of interesting tiling problems as well: What can be said about **tilings of** Λ, where a tiling $\mathscr{T}$ of Λ is a family of subsets of Λ such that their convex hulls are disjoint and their union covers Λ. Various unexpected results on monohedral translational tilings of Λ are established by L. Danzer, G. Murphy and J. Reay (Translational prototiles on a lattice. *Math. Magazine* 64(1991), 3–12).

Page 626, left column, line -14.

D. Coppersmith (Each four-celled animal tiles the plane. *J. Comb. Theory* A40(1985), 444–449) gives a proof of the result in the title.

Page 626, right column, line -11.

D. J. Newman (Tesselation of integers. *J. Number Theory* 9(1977), 107–111) finds necessary and sufficient conditions for solving the problem of tiling the 1-dimen-sional strip by translates of a k-block, where k is a power of a prime. The condition is especially simple if k = 3: a 3-block (0,0), (p,0), (q,0), with p relatively prime to q, tiles by translates if and only if p and q are congruent, in some order, to 1 and 2 (mod 3).

Publications listed as "to appear" in the References, that have been published in the meantime.

G. E. Andrews [1985]

Percy Alexander MacMahon, *Collected Papers.* Vol. II. Edited and with a preface by G. A. Andrews, and an introduction by G.-C. Rota. MIT Press, Cambridge, MA 1986.

P. J. Campbell [1983]

P. J. Campbell, The geometry of decoration on prehistoric Pueblo pottery from Starkweather Ruin. *Computers and Mathematics with Applications* 17(1989), 731–749.

D. W. Crowe [1985]
D. W. Crowe, The mosaic patterns of H. J. Woods. Computers and Mathematics with Applications 12B(1986), 407–411.

D. J. Crowe and D. K. Washburn [1983]
D. J. Crowe and D. K. Washburn, Groups and geometry in the ceramic art of San Ildefonso. Algebras, Groups and Geometries 2(1985), 263–277.

A. W. M. Dress and R. Scharlau [1985]
A. W. M. Dress and R Scharlau, The 37 combinatorial types of minimal, non-transitive, equivariant tilings of the Euclidean plane. *Discrete Math.* 60(1986), 121–138.

B. Grünbaum, Z. Grünbaum and G. C. Shephard [1985]
B. Grünbaum, Z. Grünbaum and G. C. Shephard, Symmetry in Moorish and other ornaments. Computers and Mathematics with Applications 12B(1986), 641–653

B. Grünbaum and G. C. Shephard [1985e]
B. Grünbaum and G. C. Shephard, Circular disk patterns on a sphere. Studia Sci. Math. Hungarica 21(1986), 303–327

B. Grünbaum and G. C. Shephard [1985f]
B. Grünbaum and G. C. Shephard, Isonemal fabrics. Amer. Math. Monthly, 95(1988), 5–30.

REFERENCES

The last line of each reference contains an indication (in square brackets) of the place (or places) in the book where this item is mentioned. The Introduction is referred to by the letter I, the letter R means that the item is mentioned in some other of the references, and the remaining indications are to the Section(s) of the book.

L. A. ABBOTT AND A. LINDENMAYER

1981 Models for growth of clones in hexagonal cell arrangements: Applications in Drosophila wing disc epithelia and plant epidermal tissues. *J. Theor. Biology* 90(1981), 495–514. **[3.10]**

A. ADLER AND F. C. HOLROYD

1981 Some results on one-dimensional tilings. *Geometriae Dedicata* 10(1981), 49–58. **[12.2]**

W. AHRENS

1901 *Mathematische Unterhaltungen und Spiele.* Teubner, Leipzig, 1901. **[2.10]**
Second ed., 1910.

1921 Third ed., 1921. **[2.10]**

K. ALBARN, J. M. SMITH, S. STEELE AND D. WALKER

1974 *The Language of Pattern.* Harper & Row, New York, 1974. **[I, 5.6]**

E. ALEXANDER

1929 Systematik der eindimensionalen Raumgruppen. *Zeitschrift für Kristallographie* 70(1929), 367–382. **[1.6]**

E. ALEXANDER AND K. HERRMANN

1929 Die 80 zweidimensionalen Raumgruppen. *Zeitschrift für Kristallographie* 70(1929), 328–345. **[8.8]**

G. L. ALEXANDERSON AND K. SEYDEL

1978 Kürschak's tile. *Math. Gazette* 62(1978), 192–196. **[2.6]**

P. S. ALEXANDROV (EDITOR)

1971 *Die Hilbertschen Probleme.* Geest & Portig, Leipzig, 1971. **[R]**

A. ANDREINI

1902 Richerche intorno ai poliedri ed alle reti autocorrelative. *Atti reale Istituto Veneto di Scienze, Lettere ed Arti,* 62(1902/03), 147–173 and 729–764. **[3.10]**

1907 Sulle reti di poliedri regolari e semiregolari e sulle corrispondenti reti correlative. *Mem. Soc. Ital. delle Scienze,* Ser. 3, 14(1907), 75–129. **[2.10, 3.10]**

G. E. ANDREWS

1985 *Collected Papers of P. A. MacMahon,* Vol. 2. (To appear). **[6.6]**

ANONYMOUS

1960 *Boris Nikolaevič Delone* (On the occasion of his 70th birthday) [In Russian]. *Kristallografiya* 5(1960), 339–340.
English translation: *Soviet Physics—Crystallography* 5(1960), 321–322. **[7.8]**

H. G. ApSimon
1970 Periodic forests whose largest clearings are of size 3. *Philos. Trans. Royal Soc. London*, Series A, 266(1970), 113–121. **[11.4]**

L. H. Appleton
1950 *Indian Art of the Americas*. Scribner, New York, 1950. **[6.5]**

P. F. Ash and E. D. Bolker
1985 Recognizing Dirichlet tessellations. *Geometriae Dedicata* 19(1985), 175–206. **[5.6]**

O. Aslanapa
1965 *Türkische Fliesen und Keramik in Anatolien*. Baha Matbaasi, Istanbul, 1965.

A. Aubry
[I]
1907 Problème 3224. *Intermèd. Math.* 14(1907), 122–124. **[9.3]**

W. Audsley and G. Audsley
1882 *Outlines of Ornament in the Leading Styles*. Scribner and Welford, London, 1882.
New edition: *Designs and Patterns from Historic Ornament*. Dover, New York, 1968. **[I, 5.1]**

P. d'Avennes
1877 *L'art arabe d'après les monuments du Kaire depuis le VIIe siècle jusqu'à la fin du XVIIIe*. Morel, Paris, 1877.
Selection of plates republished as *Arabic Art in Color*. Dover, New York, 1978. **[8.8]**

D. Avis and B. K. Bhattacharya
1982 Algorithms for computing d-dimensional Voronoi diagrams and their duals. *Tech. Report* SOCS-82-5, McGill University, Montreal, 1982. **[5.6]**

Z. Babakhanov
1962 *Islamic Historical Monuments in the USSR* [In Russian, English, and French]. N.p., 1962. **[2.5]**

A. Baddeley
1977 A fourth note on recent research in geometrical probability. *Advan. Appl. Prob.* 9(1977), 824–860. **[5.6]**

A. Badoureau
1878 Sur les figures isocèles. *C. R. Acad. Sci. Paris* 87(1878), 823–825. **[2.5, 2.10, 12.3]**
1881 Mèmoire sur les figures isoscèles, *J. École Polytechn.* 30(1881), 47–172. **[2.5, 2.10, 12.3]**

B. Bager
1966 *Nature as Designer. A Botanical Art Study.* Reinhold, New York, 1966.
Paperback reprint: Van Nostrand, New York, 1976. **[I]**

G. Bain
1951 *Celtic Art. The Methods of Construction.* W. MacLellan, Glasgow, 1951.
Reprint: Dover, New York, 1973. **[I, 5.1, 5.2]**

I. A. Baltag and V. P. Garit
1974 Complete derivation of Fedorov groups of the pseudoeuclidean plane [In Russian], in *Investigations in Discrete Geometry*, A. M. Zamorzaev, ed. Štiinca, Kišinev, 1974, pp. 91–107. **[4.8]**

R. Bantegnie
1977 Étalements cristallographiques. *Acta Mathematica Academiae Scientiarum Hungaricae* 30(1977), 283–302. **[6.2, 9.4, 9.6]**
1978 Parties fondamentales des groupes cristallographiques plans. Mimeographed notes, Besançon, 1978. **[6.2]**

W. Barlow
1894 Über die geometrischen Eigenschaften starrer Strukturen und ihre Anwendung auf Kristalle. *Zeitschrift für Kristallographie* 23(1894), 1–63. **[1.6]**

F. W. BARNES

1982 Algebraic theory of brick packing, I, II. *Discrete Math.* 42(1982), 7–26 and 129–144. **[9.4]**

D. W. BARNETTE

1970 The graphs of polytopes with involutory automorphisms. *Israel J. Math.* 9(1970), 290–298. **[4.8]**

D. W. BARNETTE AND B. GRÜNBAUM

1969 On Steinitz's theorem concerning convex 3-polytopes and on some properties of planar graphs, in *The Many Facets of Graph Theory*, G. Chartrand and S. F. Kapoor, eds. *Lecture Notes in Mathematics*, Vol. 110(1969), 27–40. Springer, Berlin, Heidelberg, and New York, 1969. **[4.8]**

W. BARRETT

1939 A note on some networks of polygons. *Math. Notes Edinburgh Math. Soc.* 31(1939), 23–24. **[3.10]**

T. BASKAN AND A. M. MACBEATH

1982 Centralizers of reflections in crystallographic groups. *Math. Proceedings of the Cambridge Philosophical Society* 92(1982), 79–91. **[4.8]**

N. M. BAŠKIREV [N. M. BASHKIREV]

1959 A generalization of Fedorov's stereohedra method [In Russian]. *Kristallografiya* 4(1959), 466–472.
English translation: *Soviet Physics—Crystallography*, 4(1959), 442–447. **[1.4, 12.1]**

R. S. BEARD

1973 *Patterns in Space.* Creative Publications, Palo Alto, CA, 1973. **[2.10, 3.10, 5.1, 10.1]**

A. BECK AND M. N. BLEICHER

1971 Einlagerung konvexer Mengen in eine ähnliche Menge. *Selecta Mathematica* 3(1971), 56–89. **[10.1]**

F. P. M. BEENKER

1982 Algebraic theory of non-periodic tilings of the plane by two simple blocks: A square and a rhombus. Report 82-WSK-04, Eindhoven University of Technology, 1982. **[10.7]**

S. B. M. BELL, B. M. DIAZ AND F. C. HOLROYD

1985 Tesseral addressing and tabular tesseral arithmetic of quadtrees. (To appear). **[9.4]**

S. B. M. BELL, B. M. DIAZ, F. C. HOLROYD AND M. J. JACKSON

1983 Spatially referenced methods of processing raster and vector data. *Image and Vision Computing* 1(4)(1983), 211–220. **[9.4]**

N. V. BELOV

1956a The one-dimensional infinite crystallographic groups [In Russian]. *Kristallografiya* 1(1956), 474–476.
English translation: *Soviet Physics—Crystallography* 1(1956), 372–374.
Also in: Shubnikov, Belov et al. [1964], 222–227. **[8.8]**

1956b Moorish patterns of the Middle Ages and the symmetry groups [In Russian]. *Kristallografiya* 1(1956), 610–613.
English translation: *Soviet Physics—Crystallography* 1(1956), 482–483. **[8.8]**

1959 On the nomenclature of the 80 plane groups in three dimensions [In Russian]. *Kristallografiya* 4(1959), 775–733.
English translation: *Soviet Physics—Crystallography* 4(1959), 730–733. **[8.8]**

N. V. BELOV AND E. N. BELOVA

1957 Mosaics for 46 plane (Shubnikov) antisymmetry groups and for 15 (Fedorov) color groups [In Russian]. *Kristallografiya* 2(1957), 21–22.
English translation: *Soviet Physics—Crystallography* 2(1957), 16–18. **[8.8]**

N. V. Belov, E. N. Belova and T. N. Tarhova [Tarkhova]

1958 More about the color symmetry groups [In Russian]. *Kristallografiya* 3(1958), 618–620.
English translation: *Soviet Physics—Crystallography* 3(1958), 625–626. **[8.8]**

1964 Polychromatic plane groups. In Shubnikov, Belov et al. [1964], 228–237. **[8.8]**

N. V. Belov, T. S. Kuncevič [Kuntsevich] and N. N. Neronova

1962 Shubnikov (antisymmetry) groups for infinite double-sided ribbons [In Russian]. *Kristallografiya* 7(1962), 805–808.
English translation: *Soviet Physics—Crystallography* 7(1963), 651–654. **[8.8]**

N. V. Belov and T. N. Tarhova [Tarkhova]

1956a Color symmetry groups [In Russian]. *Kristallografiya* 1(1956), 4–13.
English translation: *Soviet Physics—Crystallography* 1(1956), 5–11.
In part also in: Shubnikov, Belov et al. [1964], 211–219. **[8.2, 8.8]**

1956b Color symmetry groups [In Russian]. *Kristallografiya* 1(1956), 619–621.
English translation: *Soviet Physics—Crystallography* 1(1956), 487–489. **[8.8]**

S. Bérczi

1985 Escherian and non-Escherian developments of new frieze-groups in the Hanti and the old Hungarian communal art. M. C. Escher Congress, Rome, 1985. **[8.8]**

A. Berendsen

1967 *Tiles. A General History.* Viking, New York, 1967. **[I]**

R. Berger

1966 The undecidability of the domino problem. *Memoirs Amer. Math. Soc.* No. 66(1966), 72 pp. **[11.1, 11.3]**

N. N. Bersim and V. P. Fedotov

1980 Classification problems for tilings and patterns [In Russian]. *Abstracts of Lectures, All-Union Symposium on Symmetry Theory and Its Generalizations*, A. M. Zamorzaev et al., eds. Kišinev, 1980, 9–10. **[7.8]**

K. Bezdek

1977 Mosaics with each face having the same number of neighbors [In Hungarian]. *Matematikai Lapok* 28(1977/80), 297–303. **[3.2]**

S. Bezuszka, M. Kenney and L. Silvey

1977 *Tessellations: The Geometry of Patterns.* Creative Publications, Palo Alto, CA, 1977. **[2.10]**

K. K. Bhandary and K. Girgis

1977 Coordination polyhedra and structure of alloys: Binary alloys of vanadium with Group IIIB and IVB elements. *Acta Crystallographica* A33(1977), 903–913. **[2.1]**

J. K. Bidwell

1975 Extensions of polyominoes. *Mathematics Teaching* No. 71(1975), 16–17. **[9.4]**

S. Bilinski

1948 Homogene mreže ravnine (Homogeneous nets in the plane) [In Serbo-Croat]. *Rad Jugoslav. Akad. Znan. Umjet. Zagreb* 271(1948), 1–119. **[2.10, 4.8]**

1949 Homogene Netze der Ebene. *Bull. Internat. Acad. Yougoslave. Cl. Sci. Math. Phys. Tech.* (N.S.) 2(1949), 63–111. **[4.8]**

Y. Billiet

1980 The subgroups of finite index of the space groups: Determination via conventional coordinate systems. *MATCH, Communications in Mathematical Chemistry* 9(1980), 177–190. **[1.4]**

E. Bindel

1964 *Harmonien im Reiche der Geometrie in Anlehnung an Keplers "Weltharmonik".* Verlag Freies Geistesleben, Stuttgart, 1964. **[2.5, 2.10]**

G. D. Birkhoff

1933 *Aesthetic Measure.* Harvard University Press, Cambridge, MA, 1933, **[1.6]**

P. Blum and F. Bertaut

1954 Contribution à l'étude des borures à teneur élevée en bore. *Acta Crystallographica* 7(1954), 81–86. **[1.2]**

T. Bodrogi

1959 *Oceanian Art.* Corvina, Budapest, 1959. **[I]**

J. D. Bøgglid

1980a Breeding tessellations. *Mathematics Teaching* 90(1980), 32–36. **[8.8]**

1980b *Puzzles and Tesselations.* Privately published pamphlet, Copenhagen, 1980. **[8.8]**

S. A. Bogomolov

1932 *The derivation of the regular systems by Fedorov's method.* Parts 1 and 2 [In Russian]. ONTI, Leningrad, 1932–1934. **[1.6]**

J. Bohm

1963 Zur Systematik der kristallographischen Symmetriegruppenarten. *Neues Jahrbuch für Mineralogie* 100(1963), 113–124. **[8.8]**

B. Bollobás

1963 Filling the plane with congruent convex hexagons without overlapping. *Ann. Univ. Sci. Budapest. Sect. Math.* 6(1963), 117–123. **[9.3]**

V. G. Boltyanskii

1956 *Equivalent and Equidecomposable Figures* [In Russian]. Moscow, 1956.
English translation: *Equivalent and Equidecomposable Figures.* Heath, Boston, 1963. **[2.10]**

J. Borrego

1968 *Space Grid Structures.* MIT Press, Cambridge, MA, 1968. **[2.10]**

H. T. Bossert

1928 *Farbige Dekorationen. Beispiele dekorativer Wandmalerei von Altertum bis zur Mitte des 19. Jahrunderts.* Wasmuth, Berlin, 1928.
English translation: *An Encyclopaedia of Colour Decoration.* Gollancz, London, 1928.
Reprint: Wasmuth, Tübingen, 1955.
English translation: *Folk Art of Primitive Peoples.* Praeger, New York, 1960. **[I, 8.8]**

H. G. Bothe

1964 Parkettierungen der Ebene durch Kongruente Bogen. Ph.D. Thesis, Berlin, 1964. **[1.6]**

J. Bourgoin

1873 *Théories de l'ornement.* A Lévy, Paris, 1873. **[I]**

1883 Second edition: Ducher, Paris, 1883. **[I]**

1879 *Les éléments de l'art arabe: le trait des entrelacs.* Firmin-Didot, Paris, 1879.
Paperback reprint of plates: *Arabic Geometrical Pattern and Design.* Dover, New York, 1973. **[I, 2.5, 2.10, 5.6, 8.0]**

1880 *Grammaire élémentaire de l'ornement, pour servir à l'histoire, à la théorie et à la pratique des arts et à l'enseignement.* Delagrave, Paris, 1880.
New edition: *L'ornement.* Editions d'aujourd'hui, Paris, 1978. **[I]**

1901 *Études architectoniques et graphiques.* Tome 2. *Leçons de graphique élémentaire.* Charles Schmid, Paris, 1901. **[I]**

D. W. Boyd

1971 The disk-packing constant. *Aequat. Math.* 7(1971), 182–193. **[3.10]**

1973 The osculatory packing of a three

dimensional sphere. *Canad. J. Math.* 25(1973), 303–322. **[3.10]**

R. Boyd
1949 In a laboratory of design. *Scripta Math.* 15(1949), 183–192. **[5.1]**

A. D. Bradley
1933 *The Geometry of Repeating Design and Geometry of Design for High Schools.* Columbia University, Teachers College Press, New York, 1933.
Reprint: AMS Press, New York 1972. **[1.6, 2.10, 5.6]**

K. D. Brassel and D. Rief
1979 A procedure to generate Thiessen polygons. *Geographical Analysis* 11(1979), 289–303. **[5.6]**

A. Bravais
1850 Mémoire sur les systèmes formés par des points distribués regulièrement sur un plan ou dans l'espace. *J. de l'École Polytechn.* 19(1850), 1–128.
English translation: *On the systems formed by points regularly distributed on a plane or in space.* Monograph No. 4, American Crystallographic Association, 1949. **[5.6]**

M. Breen
1983a A characterization theorem for tilings having countably many singular points. *J. of Geometry* 21(1983), 131–137. **[3.10]**
1983b Some tilings of the plane whose singular points form a perfect set. *Proc. Amer. Math. Soc.* 89(1983), 477–479. **[3.2, 3.10]**
1985 Tilings whose members have finitely many neighbors. *Israel J. Math.* 52(1985), 140–146. **[3.10]**

F. Brisse
1981 La symétrie bidimensionnelle et le Canada. *Canadian Mineralogist* 19(1981), 217–224. **[8.8]**

H. Brown, R. Bülow, J. Neubüser, H. Wondratschek, and H. Zassenhaus
1978 *Crystallographic Groups of Four-Dimensional Space.* Wiley, New York, 1978. **[1.6]**

F. E. Browder (editor)
1976 Mathematical Developments Arising from Hilbert Problems. *Proc. Sympos. Pure Math.* vol. 28. Amer. Math. Soc., Providence, RI, 1976. **[R]**

M. Brückner
1900 *Vielecke und Vielflache.* Teubner, Leipzig, 1900. **[2.10]**

N. G. de Bruijn
1981 Algebraic theory of Penrose's non-periodic tilings. *Nederl. Akad. Wetensch. Proc.*, Ser. A 84(1981), 39–66. **[10.7]**

E. Buchman
1981 The impossibility of tiling a convex region with unequal equilateral triangles. *Amer. Math. Monthly* 88(1981), 748–753. **[2.4]**

L. Budde
1969 *Antike Mosaiken in Kilikien.* Bonzers, Recklinghausen, 1969. **[I]**

F. J. Budden
1972 *The Fascination of Groups.* Cambridge University Press, London, 1972. **[1.4, 1.6, 5.6]**

M. J. Buerger
1956 *Elementary Crystallography.* Wiley, New York, 1956. **[1.6, 5.6]**
1971 *Introduction to Crystal Geometry.* McGraw-Hill, New York, 1971. **[5.6]**

M. J. Buerger and J. S. Lukesh
1937 Wallpaper and atoms. How a study of Nature's crystal patterns aids scientist and artist. *Technology Review* 39(1937), 338–342 and 370. **[1.4]**

J. J. Burckhardt
1945 Die Bewegungsgruppen der doppelt zählenden Ebene. *Festschrift Andreas*

Speiser, O. Füssli Verlag, Zürich, 1945, pp. 153–159. **[8.8]**
1966 *Die Bewegungsgruppen der Kristallographie*. Second ed. Birkhäuser, Basel and Stuttgart, 1966. **[1.4, 1.6, 5.6, 8.8]**
1967a Review of Horváth [1965]. *Zentralblatt für Math.* 136(1967), 155.
Reprinted in *Math. Reviews* 41(1971), #4389. **[1.6, 7.8]**
1967b Zur Geschichte der Entdeckug der 230 Raumgruppen. *Archive for History of Exact Sciences* 4(1967/68), 235–246. **[1.6]**

M. Burt
1982 The wandering vertex method. *Structural Topology* 6(1982), 5–12. **[4.1]**

P. J. Burt
1980 Tree and pyramid structures for coding hexagonally sampled binary images. *Computer Graphics and Image Processing* 14(1980), 271–280. **[9.4]**

H. Burzlaff, W. Fischer and E. Hellner
1968 Die Gitterkomplexe der Ebenengruppen. *Acta Crystallographica* A24(1968), 57–67. **[5.3, 5.6]**

J. H. Cadwell
1966 *Topics in Recreational Mathematics.* Cambridge University Press, London, 1966. **[1.6, 5.6]**

A. F. Calvert
1904 *The Alhambra.* George Philip & Son, London, 1904. **[I]**

P. J. Campbell
1983 The geometry of decoration on prehistoric Pueblo pottery from Starkweather ruin. (To appear.) **[1.6]**

G. Caris
1980 *De Gulden Snede en het Werk van Caris.* Exposition catalog, Technische Hogeschool, Eindhoven, 1980. **[10.3]**

M. Caspar
1939 *Welt-Harmonik.* Oldenbourg, Munich, 1939. **[2.5]**

D. L. Caspar and A. Klug
1962 Physical principles in the construction of regular viruses. In *Basic Mechanisms in Animal Virus Biology, Cold Spring Harbor Symposia on Quantitative Biology* 27(1962), 1–24. **[1.6]**

D. P. Chavey
1984a Periodic tilings and tilings by regular polygons. I: Bounds on the number of orbits of vertices, edges, and tiles. *Mitteilungen Mathem. Seminar Giessen* 164(1984), 37–50. **[1.3, 2.3, 2.10]**
1984b *Periodic tilings and tilings by regular polygons.* Ph.D. thesis, University of Wisconsin, Madison, 1984. **[1.3, 2.3, 2.10]**

W. W. Chow
1979 Automatic generation of interlocking shapes. *Computer Graphics and Image Processing* 9(1979), 333–353. **[6.6]**
1980 Interlocking shapes in art and engineering. *Computer-Aided Design* 12(1980), 29–34. **[6.6]**

A. H. J. Christensen
1982 Recursive patterns or the garden of forking paths. *Leonardo* 15(1982), 177–182. **[10.2]**

A. H. Christie
1929 *Pattern Design. An Introduction to the Study of Formal Ornament.* Clarendon Press, Oxford, 1929.
Reprint: Dover, New York, 1969. **[I, 1.3, 2.10, 5.1, 5.2, 5.6, 8.8]**

H. Clouzot
1931 *Tissus nègres.* Calavas, Paris, 1931. **[I]**

W. Cochran
1952 The symmetry of real periodic two-dimensional functions. *Acta*

Crystallographica 6(1952), 630–633. **[8.8]**

L. COLLATZ

1975 Einige Beziehungen zwischen Graphen, Geometrie und Kombinatorik, in *Numerische Methoden bei graphentheoretischen und kombinatorischen Problemen*, L. Collatz, G. Meinardus, and H. Werner, eds. Birkhaüser, Basel, pp. 27–56. **[4.8]**

1977 Graphen bei Ornamenten und Verzweigungsdiagrammen, in *Numerische Methoden bei Optimierungsaufgaben*, Band 3, L. Collatz, G. Meinardus, and W. Wetterling, eds., Internat. Series of Numerical Mathematics vol. 36, pp. 23–46. Birkhaüser, Basel, 1977. **[4.8]**

J. P. CONLAN

1976 Derived tilings. *J. Combinatorial Theory* A20 (1976), 34–40. **[1.6]**

J. H. CONWAY

1965 Problem 5328. Triangle tesselations of the plane. *Amer. Math. Monthly* 72(1965), 915.
Solution by D. C. Kay, *ibid.* 73(1966), 903–904. **[2.4]**

J. H. CONWAY AND H. T. CROFT

1964 Covering a sphere with congruent great-circle arcs. *Proc. Cambridge Phil. Soc.* 60(1964), 787–800. **[1.6]**

S. COTLEUR

1977 *Stained Glass Tessellation Posters.* Creative Publications, Palo Alto, CA, 1977. **[2.10]**

H. S. M. COXETER

1947 *Regular Polytopes.* Methuen, London, 1947.
Second ed., Macmillan, New York, 1963.
Third ed., Dover, New York, 1973. **[2.10]**

1948 Configurations and maps. *Reports of a Mathematical Colloquium* (2) 8(1948), 18–38. **[2.10, 12.4]**

1961 *Introduction to Geometry.* Wiley, New York, 1961.
Second ed., Wiley, New York, 1969. **[1.3, 1.6, 3.7]**

1964 Regular compound tessellations of the hyperbolic plane. *Proc. Roy. Soc.*, Ser. A 278(1964), 147–167. **[4.7, 12.4]**

1979 The non-Euclidean symmetry of Escher's picture "Circle Limit III". *Leonardo* 12(1979), 19–25. **[4.8]**

1981 Angels and devils. In Klarner [1981], pp. 197–209. **[6.6]**

1985 Coloured symmetry. M. C. Escher Congress, Rome. 1985. **[8.2]**

H. S. M. COXETER, M. S. LONGUET-HIGGINS AND J. C. P. MILLER

1954 Uniform polyhedra. *Philos. Trans. Royal Soc. London* (A) 246(1953/54), 401–450. **[12.3]**

H. S. M. COXETER AND W. O. J. MOSER

1972 *Generators and Relations for Discrete Groups.* Third ed. Springer, Berlin, 1972. **[1.4, 1.6, 4.7, 4.8]**

1980 Fourth ed., Springer, Berlin, 1980 **[1.4]**

I. K. CRAIN

1972 Monte-Carlo simulation of the random Voronoi polygons: Preliminary results. *Search* 3(1972), 220–221. **[5.6]**

K. CRITCHLOW

1970 *Order in Space.* Viking Press, New York, 1970. **[2.10]**

1976 *Islamic Patterns. An Analytical and Cosmological Approach.* Schocken Books, New York, 1976. **[2.10, 5.6]**

D. W. CROWE

1971 The geometry of African art. I. Bakuba art. *J. of Geometry* 1 (1971), 169–182. **[1.6]**

1975 The geometry of African art. II. A catalog of Benin patterns. *Historia Math.* 2(1975), 253–271. **[1.6]**

1981 The geometry of African art. III. The smoking pipes of Begho, in *The Geometric Vein: The Coxeter Festschrift,* C. Davis et al., eds. Springer-Verlag, New York, 1981, pp. 177–189. **[1.6]**

1982 Symmetry in African art. *Ba Shiru. Journal of African Languages and Literature* 11(1982), 57–71. **[1.6]**

1985 The mosaic patterns of H. J. Woods. Computers and Mathematics with Applications, 12B(1986), 407–411. **[8.2, 8.8]**

D. W. Crowe and D. K. Washburn

1983 The geometry of decoration on historic San Ildefonso pueblo pottery, Algebra, Groups and Geometries 2(1985), 263–277. **[1.6]**

1984 Flowcharts as an aid to the symmetry classification of patterned design. (To appear.) **[1.6]**

S. H. Cullinane

1976 Diamond Theory. Preprint, 1976. **[8.8]**

H. M. Cundy [M. Cundy]

1979 p3m1 or p31m? *Math. Gazette* 63(1979), 192–193. **[1.4]**

1983 Problem 67.D. *Math. Gazette* 67(1983), 139–140. **[9.3]**

H. M. Cundy and A. P. Rollett

1951 *Mathematical Models.* Clarendon Press, Oxford, 1951.
Second ed., 1961. **[2.10]**

L. Danzer, B. Grünbaum and G. C. Shephard

1982 Can all tiles of a tiling have five-fold symmetry? *Amer. Math. Monthly* 89(1982), 568–570 and 583–585. **[2.5]**

1983 Does every type of polyhedron tile three-space? *Structural Topology* 8(1983), 3–14. **[9.6]**

R. O. Davies

1966 Replicating boots. *Math. Gazette* 50(1966), 175. **[10.1]**

L. F. Day

1903 *Pattern Design.* Batsford, London, 1903.
New edition. Taplinger, New York, 1979. **[I]**

I. Debroey and F. Landuyt

1979 Homogeneous tilings by regular convex polygons. *Math. Magazine* 52(1979), 272. **[2.3, 2.10]**

1981 Equitransitive edge-to-edge tilings by regular convex polygons. *Geometriae Dedicata* 11(1981), 47–60. **[2.3, 2.10]**

F. Dejean

1972 Sur un théoréme de Thue. *J. Combinatorial Theory* 13(1972), 90–99. **[10.7]**

F. M. Dekking

1982 Replicating superfigures and endomorphisms of free groups. *J. Combinatorial Theory* A32(1982), 315–320. **[10.1]**

1983 Pentagon tilings. *Nieuw Archief voor Wiskunde* (4) 1(1983), 63–69. **[10.3]**

A. Delandtsheer and P. Vanden Cruyce

1980 Orbits of edge-to-edge tilings by equilateral triangles, squares and regular hexagons. *J. of Geometry* 15(1980), 119–139. **[2.3]**

B. N. Delone [Delaunay]

1932 Neue Darstellung der geometrischen Kristallographie. *Zeitschrift für Kristallographie* 84(1932), 109–149. **[7.8]**

1958 The theory of stereohedra [In Russian]. *Uspehi Math. Nauk* 13, No. 4(1958), 221–223. **[7.8]**

1959 The theory of planigons [In Russian]. *Izv. Akad. Nauk SSSR Ser. Mat.* 23(1959), 365–386. **[2.10, 4.8, 5.2, 6.1, 6.2, 6.6, 7.8, 9.1]**

1963 On regular partitions of spaces [In Russian]. *Priroda* 1963, No. 2, pp. 60–63. **[7.8]**

1971 Zum achtzehnten Hilbertschen Problem, in Alexandrov [1971], 254–258. **[1.6]**

B. N. Delone, N. P. Dolbilin and M. I. Štogrin
1978 Combinatorial and metric theory of planigons [In Russian]. *Trudy Matematičeskogo Instituta Steklov. Akad. Nauk SSSR* 148(1978), 109–140.
English translation: *Proc. of the Steklov Institute of Mathematics* 1980, issue 4, 111–141. **[6.6, 7.8]**

B. N. Delone, R. V. Galiulin and M. I. Štogrin
1979 The contemporary theory of regular decompositions of Euclidean space [In Russian], in Fedorov [1979], 235–260. **[6.6, 7.8, 9.1]**

C. W. Dodge
1972 *Euclidean Geometry and Transformations.* Addison-Wesley, Reading, MA, 1972. **[1.3]**

I. Dominyák
1967 Regular systems of circles [In Hungarian]. *Magyar Tudományos Akadémia. Matematikai és Fizikai Tudományok Osziályának Közleményei* 17(1967), 331–375. **[7.8]**

J. D. H. Donnay
1967 Generalized symmetry and magnetic space groups. *Trans. Amer. Crystallogr. Assoc.* 3(1967), 74–95. **[8.8]**

J. D. H. Donnay and G. Donnay
1985 Symmetry and antisymmetry in Maori rafter designs. *Empirical Studies of the Arts* 3(1985), 23–45. **[1.6]**

K. Dornberger-Schiff
1956 On order-disorder structures (OD-structures). *Acta Crystallographica* 9(1956), 593–601. **[8.8]**
1959 On the nomenclature of the 80 plane groups in three dimensions. *Acta Crystallographica* 12(1959), 173. **[8.8]**

M. Dostal and R. Tindell
1978 The Jordan curve theorem revisited. *Jahresber. d. Deutsch. Math.-Verein.* 80(1978), 111–128. **[3.1]**

J. Doyen and M. Landuyt
1983 Dissections of polygons. *Annals of Discrete Math.* 18(1983), 315–318. **[10.1]**

A. W. M. Dress
1985a A mathematician's comment on M. C. Escher's essay "Regelmatige Vlakverdeling". M. C. Escher Congress, Rome, 1985. **[6.6]**
1985b On the classification of local disorder in globally regular and spatial patterns. (To appear). **[3.10]**

A. W. M. Dress and R. Scharlau
1984 Zur Klassifikation äquivarianter Pflasterungen. *Mitt. Math. Sem. Giessen* 164(1984), 83–136. **[4.8]**
1985 The 37 combinatorial types of minimal, non-transitive, equivariant tilings of the Euclidean plane. (To appear). **[4.8]**

C. Dresser
1862 *The Art of Decorative Design.* Day and Son, London, 1862. **[I]**

A. J. W. Duijvestijn, P. J. Federico and P. Leeuw
1982 Compound perfect squares. *Amer. Math. Monthly* 89(1982), 15–32. **[2.4]**

D. Dunham, J. Lindgren, and D. Witte
1981 Creating repeating hyperbolic patterns. *Computer Graphics* 15(1981), 215–223. **[4.8]**

J. A. Dunn
1971 Tessellations with pentagons. *Math. Gazette* 55(1971), 366–369.
Additional correspondence, *ibid.* 56(1972), 332–335. **[9.3]**

A. Dürer
1625 *Unterweisung der Messung mit dem Zirkel und Richtscheit.*
English translation: *The Painter's Manual*, Abaris, New York, 1977. **[1.5]**

C. J. Du Ry
1970 *Art of Islam.* Abrams, New York, 1970. **[I]**

D. S. Dye
1937 *A Grammar of Chinese Lattice.* Harvard University Press, Boston, 1937.
New edition: *Chinese Lattice Designs.* Dover, New York, 1974. **[I, 2.4, 5.6, 7.4]**
1981 *The New Book of Chinese Lattice Designs.* Dover, New York, 1981. **[I, 2.4, 7.4]**

V. Eberhard
1891 *Zur Morphologie der Polyeder.* Teubner, Leipzig, 1891. **[3.9]**

S. Eberhart
1975a New and old problems. *Mathematical-Physical Correspondence* 12(1975), 4–5 **[2.5]**
1975b Solutions to old problems. *Mathematical-Physical Correspondence* 13(1975), 22–24. **[2.5]**

A. L. Edmonds, J. H. Ewing and R. S. Kulkarni
1982 Torsion free subgroups of Fuchsian groups and tessellations of surfaces. *Inventiones Mathematicae* 69(1982), 331–346. **[4.8]**

M. J. Edmonds
1976 *Geometric Designs in Needlepoint.* Van Nostrand Reinhold, New York, 1976. **[I]**

E. B. Edwards
1932 *Dynamarhythmic Design.* Century, New York, 1932.
Reprint: *Pattern and Design with Dynamic Symmetry.* Dover, New York, 1967. **[I]**

H. G. Eggleston
1953 On closest packing by equilateral triangles. *Proc. Cambridge Philos. Soc.* 49(1953), 26–30. **[3.10]**
1957 *Problems in Euclidean Space: Application of Convexity.* Pergamon Press, New York, 1957. **[3.10]**

R. B. Eggleton
1974 Tiling the plane with triangles. *Discrete Math.* 7(1974), 53–65. **[2.4]**
1975 Where do all the triangles go? *Amer. Math. Monthly* 82(1975), 499–501. **[2.4]**

D. Ellard
1982 Poly-iamond enumeration. *Math. Gazette* 66(1982), 310–314. **[9.4]**

I. El-Said and A. Parman
1976 *Geometric Concepts in Islamic Art.* World of Islam Festival Publ. Co., London, 1976. **[5.6, 7.8]**

J. Enciso
1947 *Sellos del Antiguo México.* México, 1947.
English translation: *Design Motifs of Ancient Mexico.* Dover, New York, 1953. **[I]**

G. Erdtman
1952 *Pollen Morphology and Plant Taxonomy.* Almquist & Wiksell, Stockholm, 1952. **[I]**

J. W. Erickson, P. Tollin, J. F. Richardson, S. K. Burley and J. B. Bancroft
1982 The structure of an unusual ordered aggregate of papaya mosaic virus protein. *Virology* 118(1982), 241–245. **[1.6]**

R. O. Erickson
1973 Tubular packing of spheres in biological fine structure. *Science* 181(1973), 705–716. **[1.6]**

B. Ernst
1976 *The Magic Mirror of M. C. Escher.* Random House, New York, 1976. **[I, 8.8]**

M. C. Escher
1958 *Regelmatige vlakverdeling.* Stichting De Roos, Utrecht, 1958. **[6.6, 8.8]**
1960 Antisymmetrical arrangements in the plane, and regular three-dimensional bodies as sources of inspiration to an artist. Abstract 14.1, *Acta Crystallographica* 13(1960), 1083. **[8.8]**

1971 *The World of M. C. Escher*, J. L. Locher, ed. Abrams, New York, 1971. **[I, 3.10, 4.1, 6.0, 6.6, 8.8, 10.1]**

1982 *M. C. Escher: His life and Complete Graphic Work*. J. L. Locher, ed. Abrams, New York, 1982. **[I, 6.6, 8.8]**

1983 *M. C. Escher's Universe of Mind Play*. Odakyu Department Store, Tokyo, 1983. **[I, 4.8, 6.6, 8.8]**

G. EWALD

1971 *Geometry: An Introduction.* Wadsworth, Belmont, CA, 1971. **[1.3]**

H. FEDERER

1969 *Geometric Measure Theory*. Springer, New York, 1969. **[3.10]**

P. J. FEDERICO

1979 Squaring rectangles and squares. A historical review with annotated bibliography, in *Graph Theory and Related Topics*, A. Bondy and U. S. R. Murty, eds. Academic Press, New York, 1979, 173–196. **[2.4]**

E. S. FEDOROV [E. VON FEDOROW]

1885 *Elements of the Theory of Figures* [In Russian]. Imp. Acad. Sci., St. Petersburg, 1885.
Annotated new edition: Akad. Nauk SSSR, 1953. **[1.6, 3.10, 6.6, 9.6]**

1891 Symmetry in the plane [In Russian]. *Zapiski Rus. Mineralog. Obščestva*, Ser. 2, 28(1891), 345–390 + 2 plates. **[1.6, 5.6]**

1899 Reguläre Plan- und Raumtheilung. *Abh. K. Bayer. Akad. Wiss. (II Klasse)* 20(1899), 465–588. **[6.6]**
Russian translation: Fedorov [1979]. **[6.6]**

1916 Systems of planigons as typical isohedra in the plane [In Russian]. *Bull. Acad. Imp. Sci.*, Ser. 6, Vol. 10(1916), 1523–1534. **[3.10, 6.6]**

1979 *Regular Partition of Plane and Space* [In Russian; translation of Fedorov [1899]]. B. N. Delone et al., eds. Nauka, Leningrad, 1979. **[6.6]**

V. P. FEDOTOV

1978 Some remarks on discrete chronogeometry [In Russian]. *Sibirskii Matematicheskii Zhurnal* 19(1978), 186–192.
English translation: *Siberian Math. Journal* 19(1978), 132–136. **[2.10, 6.6]**

G. FEJES TÓTH

1971 Über Parkettierungen konstanter Nachbarzahl. *Studia Scientiarum Math. Hungarica* 6(1971), 133–135. **[3.2]**

1983 New results in the theory of packing and covering. In *Convexity and Its Applications*, P. M. Gruber and J. M. Wills, eds. Birkhäuser, Basel, 1983, 318–359. **[I]**

L. FEJES TÓTH

1953 *Lagerungen in der Ebene, auf der Kugel und im Raum.* Springer, Berlin, 1953.
Second ed., 1972. **[I, 2.1, 2.10]**

1960 On shortest nets with meshes of equal area. *Acta Math. Acad. Scientiarum Hungaricae* 11(1960), 363–370. **[2.1]**

1963 Isoperimetric problems concerning tessellations. *Acta Math. Acad. Scientiarum Hungaricae* 14(1963), 343–351. **[2.1]**

1965 *Reguläre Figuren.* Akadémiai Kiadó, Budapest, 1965.
English translation: *Regular Figures.* Pergamon, New York, 1964.
[1.6, 2.10, 4.7, 4.8, 5.6, 7.3, 7.8, 9.6]

1969a Remarks on a theorem of R. M. Robinson. *Studia Scientiarum Math. Hungarica* 4(1969), 441–445. **[3.2]**

1969b Scheibenpackungen konstanter Nachbarnzahl. *Acta Math. Acad. Sci. Hungar.* 20(1969), 375–381. **[3.2, 3.10]**

1973 Five-neighbour packing of convex discs. *Periodica Mathematica Hungarica* 1(1973), 221–229. **[7.8]**

1975 Tessellation of the plane with convex polygons having a constant number of neighbors. *Amer. Math. Monthly* 82(1975), 273–276. **[3.2]**

L. Fejes Tóth and A. Heppes

1980 Multi-saturated packings of circles. *Studia Scientiarum Math. Hungarica* 15(1980), 303–307. **[7.8]**

R. Fenn

1983 What is the geometry of a surface? *Amer. Math. Monthly* 90(1983), 87–98. **[4.8]**

A. G. Fesenko

1981 Densest lattice packings of congruent polygonal objects [In Ukrainian]. *Dopovidi Akad. Nauk Ukrain. RSR*, Ser. 1, 1981. No. 1, pp. 77–80. **[I]**

J. V. Field

1979 Kepler's star polyhedra. *Vistas in Astronomy* 23(1979), 109–141. **[2.10]**

D. Fielker

1978 Untitled note. *Mathematics Teaching* 84(1978), 34. **[9.5]**

S. Finsterwalder

1936 Regelmässige Anordnungen gleicher sich berührender Kreise in der Ebene, auf der Kugel und auf der Pseudosphäre. *Abh. Bayer. Akad. Wiss., Math.-naturwiss. Abteilung*, Neue Folge, 38(1936), 42 pp. **[7.3, 7.8]**

P. A. Firby and C. F. Gardiner

1982 *Surface Topology*. Wiley, New York, 1982. **[4.2]**

P. Fisher

1971 *Mosaic. History and Technique.* McGraw-Hill, New York, 1971. **[I]**

A. Florian and H. Florian

1975 Reguläre hyperbolische Mosaike und Newtonsche Zahlen, III. *Sitzungsber. Österreich. Akad. Wiss. Math.-naturw. Klasse*, 184(1975), 29–40. **[4.8]**

B. Foerster

1961 *Pattern and Structure*. Allied Masonry Council, Washington, DC, 1961. **[I]**

A. Fontaine and G. E. Martin

1983a Tetramorphic and pentamorphic prototiles. *J. Combinatorial Theory* A 34(1983), 115–118. **[1.5]**

1983b Polymorphic prototiles. *J. Combinatorial Theory* A 34(1983), 119–121. **[1.5]**

1984a An enneamorphic prototile. *J. Combinat. Theory* A 37(1984), 95–96. **[1.5]**

1984b Polymorphic polyominoes. *Math. Magazine* 57(1984), 275–283. **[1.5]**

P. Fořtova-Šámalová and M. Vilímková

1963 *Egyptian Ornament*. Allan Wingate, London, 1963. **[I]**

E. Fourrey

1907 *Curiosités géométriques*. Vuibert, Paris, 1907.
Second ed., 1920.
Third ed., 1938. **[1.6, 2.10]**

F. C. Frank and J. S. Kasper

1958 Complex alloy structures regarded as sphere packings. I. Definitions and basic principles. *Acta Crystallographica* 11(1958), 184–190. **[2.10]**

1959 Complex alloy structures regarded as sphere packings. II: Analysis and classification of representative structures. *Acta Crystallographica* 12(1959), 483–499. **[2.10]**

J. Fraser

1908 A new visual illusion of direction. *British J. of Psychology* 2(1908), 307–320 + 9 plates. **[10.1]**

R. Fricke and F. Klein

1897 *Vorlesungen über die Theorie der automorphen Funktionen*. Vol. 1. Teubner, Leipzig, 1897. **[1.6, 4.8]**

P. Gács

1972 Packing of convex sets in the plane with a great number of neighbours. *Acta Math. Acad. Scientiarum Hungaricae* 23(1972), 383–388. **[3.2]**

E. I. GALYARSKII

1967 Conical similarity groups of symmetry and antisymmetry of a different kind [In Russian]. *Kristallografiya* 12(1967), 202–207.
English translation: *Soviet Physics—Crystallography* 12(1967), 169–174. **[5.6]**

1974a Mosaics for two-dimensional similarity symmetry and antisymmetry groups [In Russian]. *Investigations in Discrete Geometry*, A. M. Zamorzaev, ed. Štiinca, Kišinev, 1974, pp. 49–63. **[5.6, 10.1]**

1974b Two dimensional groups of similarity color symmetry and antisymmetry of another kind [In Russian]. *Investigations in Discrete Geometry*, A. M. Zamorzaev, ed. Štiinca, Kišinev, 1974, pp. 63–77. **[5.6]**

E. I. GALYARSKII AND A. M. ZAMORZAEV

1963 Similarity symmetry and antisymmetry groups [In Russian]. *Kristallografiya* 8(1963), 691–698.
English translation: *Soviet Physics—Crystallography* 8(1964), 553–558. **[5.6]**

D. GANS

1969 *Transformations and Geometries*. Appleton-Century-Croft, New York, 1969. **[1.3]**

M. GARDNER

1966 *New Mathematical Diversions from "Scientific American"*. Simon and Schuster, New York, 1966. **[9.4]**

1971 *Sixth Book of Mathematical Games from "Scientific American"*. Scribner, New York, 1971. **[1.2, 3.10, 9.4]**

1975a On tessellating the plane with convex polygon tiles. *Scientific American*, July 1975, pp. 112–117. **[1.2, 9.3, 9.4]**

1975b More about tiling the plane: The possibilities of polyominoes, polyiamonds, and polyhexes. *Scientific American*, August 1975, pp. 112–115. **[9.4, 9.6]**

1975c A random assortment of puzzles. *Scientific American*, December 1975, pp. 116–119. **[9.3, 9.6]**

1977 Extraordinary nonperiodic tiling that enriches the theory of tiles. *Scientific American*, January 1977, pp. 110–121. **[9.6, 10.3]**

1978 *Mathematical Magic Show*. Vintage Books, New York, 1978. **[9.4]**

1979 Some packing problems that cannot be solved by sitting on the suitcase. *Scientific American*, October 1979, pp. 18–26. **[2.4]**

J. GARRIDO

1952 Les groupes de symétrie des ornements employés par les anciennes civilisations du Mexique. *Comptes Rendus Acad. Sci. Paris* 235(1952), 1184–1186. **[1.6]**

P. C. GASSON

1983 *Geometry of Spatial Forms*. Wiley, Chichester, 1983. **[5.6]**

D. GAT

1978 Selfsupporting interlocking grids for multiple applications (Sigma). *Architectural Science Review*, December 1978, 105–109. **[7.8]**

J. D. GERGONNE

1818 Recherches sur les polyèdres, renfermant en particulier un commencement de solution du problème proposé à la page 256 du VIIe volume des Annales, par un abonné. *Annales de Mathématiques Pures et Appliquées* 9(1818/1819), 321–344. **[2.10]**

S. GERMAIN

1969 *Les mosaiques de Timgad*. Centre Nat. Rech. Sci., Paris, 1969. **[I]**

M. GERSPACH

1890 *Les tapiseries coptes*. Maison Quantin, Paris, 1890.

1975 New edition: *Coptic Textile Design*. Dover, New York, 1975. **[I]**

G. Gerster
1968 *L'art éthiopien. Églises rupestres.* Editions Zodiaque, La Pierre-qui-Vire, France, 1968. **[I]**

M. Ghyka
1946 *The Geometry of Art and Life.* Sheed and Ward, New York, 1946.
Reprint: Dover, New York, 1977. **[2.10]**

L. Gibson and D. Lucas
1984 Vectorization of raster images using hierarchical methods. (To appear.) **[9.4]**

E. N. Gilbert
1962 Random subdivisions of space into crystals. *Annals of Mathematical Statistics* 33(1962), 958–972. **[5.6]**

W. J. Gilbert
1983 An easy way to construct spacefillings. *Structural Topology* 8(1983), 25–32. **[9.4]**

J. Giles, Jr,
1979a Infinite-level replicating dissections of plane figures. *J. Combinatorial Theory* A26(1979), 319–327. **[10.1]**
1979b Construction of replicating superfigures. *J. Combinatorial Theory* A26(1979), 328–334. **[10.1]**
1979c Superfigures replicating with polar symmetry. *J. Combinatorial Theory* A26(1979), 335–337. **[10.1]**
1980 The Gypsy method of superfigure construction. *J. Recreational Math.* 13(1980/1981), 29–33. **[10.1]**

A. Girard
1626 *Table des sines, tangentes & secantes, selon le raid de 100000 parties. Avec un traicté succint de la trigonometrie tant des triangles plans, que spheriques. Où sont plusieurs operations nouvelles, non auparavant mises en lumière, très-utiles & necessaires, non seulement aux apprentifs; mais aussi aux plus doctes practiciens des mathematiques.* Elzevier, La Haye, 1626. **[12.3]**

F. Göbel
1979 Geometrical packing and covering problems. "Packing and Covering in Combinatorics", A. Schrijver, ed. *Mathematical Centre Tracts* 106(1979), 179–199. **[1.2]**

A. Godard
1962 *L'art de l'Iran.* Arthaud, Paris, 1962.
English translation: *The Art of Iran.* Praeger, New York, 1965. **[I]**

M. Goldberg
1955 Central tesselations. *Scripta Math.* 21(1955), 253–260. **[1.3, 9.6]**

M. Goldberg and B. M. Stewart
1964 A dissection problem for sets of polygons. *Amer. Math. Monthly* 71(1964), 1077–1095. **[2.6]**

S. W. Golomb
1964 Replicating figures in the plane. *Math. Gazette* 48(1964), 403–412. **[10.1]**
1965 *Polyominoes.* Scribner, New York, 1965. **[9.4, 9.6]**
1966 Tiling with polyominoes. *J. Combinatorial Theory* 1(1966), 280–296. **[9.4]**
1970 Tiling with sets of polyominoes. *J. Combinatorial Theory* 9(1970), 60–71. **[9.4]**

B. Gordon
1980 Tilings of lattice points in Euclidean n-space. *Discrete Math.* 29(1980), 169–174. **[12.2]**

W. C. Graustein
1931 On the average number of sides of polygons of a net. *Annals of Math.* (2) 32(1931), 149–153. **[3.10]**

P. J. Green and R. Sibson
1978 Computing Dirichlet tessellations in the plane. *The Computer Journal* 21(1978), 168–173. **[5.4]**

H. Groemer
1978 On multiple space subdivisions by zonotopes. *Monatsh. Math.* 86(1978/79), 183–188. **[12.4]**

H. D. GROSSMAN

1948 Fun with lattice-points. *Scripta Mathematica* 14(1948), 157–159. **[10.1]**

B. GRÜNBAUM

1967 *Convex Polytopes.* Interscience, London, 1967. **[3.5, 3.9, 4.8]**

1975 Polytopal graphs. *Studies in Graph Theory*, D. R. Fulkerson, ed. Studies in Mathematics, vol. 12. Math. Assoc. of America 1975, 201–224. **[3.2, 3.5]**

1976 Color symmetries and colored patterns. Mimeographed notes, University of Washington, January 1976. **[8.2, 8.8]**

1984 The emperor's new clothes: full regalia, G-string, or nothing? *Math. Intelligencer* 6(1984), 47–53. **[1.6]**

1985 Mathematical challenges in Escher's geometry. M. C. Escher Congress, Rome, 1985. **[6.6]**

B. GRÜNBAUM, Z. GRÜNBAUM AND G. C. SHEPHARD

1985 Symmetry in Moorish and other ornaments. *Computers and Mathematics with Applications.* (To appear). **[1.6]**

B. GRÜNBAUM, H. D. LÖCKENHOFF, G. C. SHEPHARD AND A. H. TEMESVÁRI

1985 The enumeration of normal 2-homeohedral tilings. *Geometriae Dedicata* 19(1985), 109–174. **[4.8]**

B. GRÜNBAUM, P. MANI-LEVITSKA AND G. C. SHEPHARD

1984 Tiling three-dimensional space with polyhedral tiles of a given isomorphism type. *J. London Math. Soc.* (2) 29(1984), 181–191. **[9.6]**

B. GRÜNBAUM, J. C. P. MILLER AND G. C. SHEPHARD

1982 Uniform tilings with hollow tiles, in *The Geometric Vein—The Coxeter Festschrift.* C. Davis et al., eds. Springer-Verlag, New York, 1982, 17–64. **[2.5, 2.10, 12.3]**

B. GRÜNBAUM AND G. C. SHEPHARD

1977a Tilings by regular polygons. *Math. Magazine* 50(1977), 227–247 and 51(1978), 205–206.
Italian translation: *Archimede* 32(1980), 15–45. **[2.3, 2.5, 2.10]**

1977b Patch-determined tilings. *Math Gazette* 61(1977), 31–38. **[1.6]**

1977c The eighty-one types of isohedral tilings in the plane. *Math. Proc. Cambridge Phil. Soc.* 82(1977), 177–196. **[6.6]**

1977d Perfect colorings of transitive tilings and patterns in the plane. *Discrete Math.* 20(1977), 235–247. **[8.7]**

1977e Classification of plane patterns. Mimeographed notes distributed at the special session on "Tilings, Patterns, and Symmetries," Summer meeting of the American Math. Soc., Seattle, August 1977, 66 pp. **[5.6, 7.8, 8.2, 8.8]**

1978a The ninety-one types of isogonal tilings in the plane. *Trans. Amer. Math. Soc.* 242(1978), 335–353 and 249(1979), 446. **[6.6]**

1978b Isotoxal tilings. *Pacific J. Math.* 76(1978), 407–430. **[6.6]**

1978c Do maximal line-generated triangulations of the plane exist? *Amer. Math. Monthly* 85(1978), 37–41. **[7.8]**

1978d Isohedral tilings of the plane by polygons. *Comment. Math. Helvet.* 53(1978), 542–571. **[9.1, 9.6]**

1978e The homeomeric classification of tilings. *C. R. Math. Reports Acad. Sci. Canada* 1(1978), 57–60. **[7.8]**

1979a Incidence symbols and their applications. *Proc. Symp. Pure Math.* 34(1979), 199–244. **[4.7, 5.6, 6.6, 7.8, 8.8]**

1979b Spiral tilings. *Math Teaching* 88(1979), 50–51. **[9.6]**

1980a Satins and twills: An introduction to the geometry of fabrics. *Math. Magazine* 53(1980), 139–161 and 313. **[5.6]**

1980b Some remarks on mathematical taxonomy. *2. Kolloquium über diskrete*

Geometrie, Salzburg, 1980, pp. 179–188. **[7.8]**

1980c Tilings with congruent tiles. *Bull. Amer. Math. Soc. N. S.* 3(1980), 951–973. **[9.6]**

1981a A hierarchy of classification methods for patterns. *Zeitschrift für Kristallographie* 154(1981), 163–187. **[7.8]**

1981b Some remarks on Fedotov's paper on discrete chronogeometry [In Russian]. *Sibirsk. Mat. Zh.* 22(1981), 220–226. English version: *Siberian Math. J.* 22(1981), 164–169. **[6.6]**

1981c Patterns on the 2-sphere. *Mathematika* 28(1981), 1–35. **[4.8]**

1981d The geometry of planar graphs, in *Combinatorics*, H. N. V. Temperley, ed. *London Math. Soc. Lecture Note Series* 52(1981), 124–150. **[4.8]**

1981e Some problems on plane tilings. In Klarner [1981], pp. 167–196. **[9.6]**

1982a The theorems of Euler and Eberhard for tilings of the plane. *Resultate der Mathematik* 5(1982), 19–44. **[3.9]**

1982b Spherical tilings with transitivity properties, in *The Geometric Vein—The Coxeter Festschrift*. C. Davis et al., eds. Springer-Verlag, New York, 1982, 65–98. **[4.8]**

1982c Analogues for tilings of Kotzig's theorem on minimal weights of edges. *Ann. of Discrete Math.* 12(1982), 129–140. **[4.8]**

1983a The 2-homeotoxal tilings of the plane and the 2-sphere. *J. Combinatorial Theory* B 34(1983), 113–150. **[4.8]**

1983b Tilings, patterns, fabrics and related topics in discrete geometry. *Jahresber. Deutsch. Math. Verein.* 85(1983), 1–32. **[5.6, 8.2, 8.6]**

1985a A catalogue of isonemal fabrics. *Annals of the New York Acad. Sciences* 440(1985), 279–298. **[5.6]**

1985b An extension to the catalogue of isonemal fabrics. *Discrete Math.* (to appear). **[5.6]**

1985c The aperiodic tilings of R. Ammann. (In preparation). **[10.4]**

1985d Patterns of circular disks on a 2-sphere *3. Kolloquium über diskrete Geometrie*, Salzburg, 1985, pp. 243–251. **[7.8]**

1985e Circular disk patterns on a sphere. *Studia Scient. Math Hungarica* (to appear). **[7.8]**

1985f Isonemal fabrics. (To appear). **[7.8]**

H. W. Guggenheimer

1967 *Plane Geometry and Its Groups.* Holden-Day, San Francisco, 1967. **[1.3, 1.4, 1.6]**

J. Guiart

1963 *The Arts of the South Pacific.* Golden Press, New York, 1963. **[I, 5.2]**

A. M. Günzburg

1929 Die Grundzüge der Lehre von der Symmetrie auf Linien und in Ebenen. *Zeitschrift für Kristallographie* 71(1929), 81–94. **[1.6]**

R. K. Guy

1976 The Penrose pieces. *Bull. London Math. Soc.* 8(1976), 9–10. **[10.3]**

F. Haag

1911 Die regelmässigen Planteilungen. *Zeitschrift für Kristallographie* 49(1911), 360–369. **[6.6]**

1923 Die regelmässigen Planteilungen und Punktsysteme. *Zeitschrift für Kristallographie* 58(1923), 478–489. **[6.6]**

1926 Die Planigone von Fedorow. *Zeitschrift für Kristallographie* 63(1926), 179–186. **[6.6]**

1929 Die Kreispackungen von Niggli. *Zeitschrift für Kristallographie* 70(1929), 353–366. **[7.3, 7.8]**

1932 Strukturformeln für Ebenenteilungen. *Zeitschrift für Kristallographie* 83(1932), 301–307. **[3.10, 6.6]**

1933 Die Grundgleichung für Ebenen- und Raumteilung. *Zeitschrift für Kristallographie* 86(1933), 153–155. **[3.10]**

1936 Die Polygone der Ebenenteilungen. *Zeitschrift für Kristallographie* 96(1936), 78–80. **[6.6]**

S. HAAK

1976 Transformation geometry and the artwork of M. C. Escher. *Math. Teacher* 69(1976), 647–652. **[8.8]**

G. D. HALSEY AND E. HEWITT

1978 Eine gruppentheoretische Methode in der Musiktheorie. *Jahresber. Deutsch. Math.-Verein.* 80(1978), 151–207. **[1.6]**

W. HANF

1974 Nonrecursive tilings of the plane, I. *J. Symbolic Logic* 39(1974), 283–285. **[11.4]**

E. H. HANKIN

1925a *The Drawing of Geometric Patterns in Saracenic Art.* Memoirs of the Archaeological Survey of India, No. 15, Calcutta, 1925. **[5.6]**

1925b Examples of methods of drawing geometrical arabesque patterns. *Math. Gazette* 12(1925), 370–373. **[5.6]**

1934 Some difficult Saracenic designs. *Math. Gazette* 18(1934), 165–168 and 20(1936), 318–319. **[5.6]**

F. A. HANSON

1985 From symmetry to anthropophagy: The cultural context of Maori art. *Empirical Studies of the Arts* 3(1985), 47–61. **[1.6]**

H. HARBORTH

1977a Prescribed numbers of tiles and tilings. *Math. Gazette* 61(1977), pp. 296–299. **[1.5, 1.6]**

1977b Nichtperiodische Parkettierungen der Ebene. *Math. Naturwiss. Unterricht* 30(1977), 453–456. **[1.5, 1.6, 10.7]**

R. D'HARCOURT

1975 *Textiles of Ancient Peru and Their Techniques.* University of Washington Press, Seattle, 1975. **[I]**

G. HATCH

1978 Tessellations with equilateral reflex polygons. *Math. Teaching* 84(1978), 32. **[9.5]**

E. W. HAURY

1937 Pottery types at Snaketown, in *Excavations of Snaketown. Material Culture*, by H. H. Gladwin, E. W. Haury, E. B. Sayles, and N. Gladwin. *Medalion Papers*, No. 25(1937), pp. 169–229. **[I]**

E. HEESCH, H. HEESCH AND J. LOEF

1944 *System einer Flächenteilung und seine Anwendung zum Werkstoff- und Arbeitsparen.* Hauptausschuss Maschinen und Apparate, Reichsminister für Rüstung und Kriegsproduktion, [Berlin], 1944. **[9.6]**

H. HEESCH

1929 Zur Strukturtheorie der ebenen Symmetriegruppen. *Zeitschrift für Kristallographie* 71(1929), 95–102. **[8.8]**

1930 Zur systematischen Strukturtheorie. IV: Über die Symmetrien zweiter Art in Kontinuen und Semidiskontinuen. *Zeitschrift für Kristallographie* 73(1930), 346–356. **[5.6]**

1932 Über topologisch reguläre Teilungen geschlossener Flächen. *Nachr. Ges. Wiss. Göttingen, Math.-Phys. Kl.* 1932, 268–273. **[4.8]**

1933a Über topologisch gleichwertige Kristallbindungen. *Zeitschrift für Kristallographie* 84(1933), 399–407. **[3.10, 4.8, 6.6]**

1933b Ueber Kugelteilung. *Comment. Math. Helvet.* 6(1933/34), 443–453. **[4.8]**

1935 Aufbau der Ebene aus kongruenten Bereichen. *Nachr. Ges. Wiss. Göttingen*, New Ser. 1(1935), 115–117. **[1.3, 4.8, 9.6]**

1968a Parkettierungsprobleme. *Der Mathematikunterricht*, 1968, Issue 4, 5–45. **[2.9, 2.10]**

1968b *Reguläres Parkettierungsproblem.* Westdeutscher Verlag, Cologne and Opladen, 1968. **[1.2, 1.3, 1.6, 2.9, 2.10, 3.8, 4.8, 6.6, 7.8, 9.6]**

1968c Eine Betrachtung der 11 homogenen Ebenenteilungen. *Der Mathematikunterricht*, 1968, Issue 4, 66–78. **[2.9, 2.10]**

H. HEESCH AND O. KIENZLE

1963 *Flächenschluss. System der Formen lückenlos aneinanderschliessender Flachteile.* Springer, Berlin, 1963. **[I, 9.1, 9.3, 9.6]**

K. HEIDLER

1976 Die Unentscheidbarkeit des Eck-Dominoproblems. *Math. Phys. Semesterber.* 23(1976), 237–250. **[11.4]**

N. F. M. HENRY AND K. LONSDALE, EDITORS.

1965 *International Tables for X-Ray Crystallography.* Volume I, *Symmetry Groups.* Kynoch Press, Birmingham, 1965. **[5.6]**

D. HENSLEY

1978 Fibonacci tiling and hyperbolas. *The Fibonacci Quarterly* 16(1978), 37–40. **[2.4]**

H. HERDA

1981 Tiling the plane with incongruent regular polygons. *The Fibonacci Quarterly* 19(1981), 437–439.
Also in *Convexity and Related Combinatorial Geometry*, D. C. Kay and M. Breen, eds. Dekker, New York and Basel, 1982, 225–228. **[2.4]**

C. HERMANN

1929 Zur systematischen Strukturtheorie. III: Ketten- und Netzgruppen. *Zeitschrift für Kristallographie* 69(1929), 250–270. **[8.8]**

1935 *Internationale Tabellen zur Bestimmung von Kristallstrukturen.* Vol. 1: *Gruppentheoretische Tafeln.* Borntraeger, Berlin, 1935. **[5.6]**

D. HILBERT

1900 Mathematische Probleme. *Göttinger Nachrichten* 1900, pp. 253–297;
Also *Arch. Math. Phys.* Ser. 3, Vol. 1(1901), 44–63 and 213–237.
Reprinted in Alexandrov [1971].
English translation: Mathematical problems. *Bull. Amer. Math. Soc.* 8(1902), 437–479.
Reprinted in Browder [1976], pp. 1–34. **[1.2, 1.6, 9.6]**

D. HILBERT AND S. COHN-VOSSEN

1932 *Anschauliche Geometrie.* Springer, Berlin, 1932.
Reprint: Dover, New York, 1944.
English translation: *Geometry and the Imagination.* Chelsea, New York, 1952. **[1.6]**

H. HILTON

1903 *Mathematical Crystallography and the Theory of Groups of Movements.* Clarendon Press, Oxford, 1903.
Reprint: Dover, New York, 1963. **[1.6]**

H. HINTERREITER

1977 Exhibition Catalogue 1977. Galerie & Edition Schlégl, Zürich, 1977 **[8.8]**

M. D. HIRSCHHORN AND D. C. HUNT

1981 Equilateral convex pentagons which tile the plane. Mimeographed note, Kensington, N.S.W., Australia, 1981. **[9.3]**

1985 Equilateral convex polygons which tile the plane. *J. Combinat. Theory* A 39(1985), 1–18. **[2.7, 9.6]**

J. D. Hoag
1977 *Islamic Architecture.* Abrams, New York, 1977. **[I]**

D. R. Hofstadter
1983 Parquet deformations: Patterns of tiles that shift gradually in one dimension. *Scientific American*, July 1983, pp. 14–20. **[4.1]**

H. Holden
1975 Fibonacci tiles. *The Fibonacci Quarterly* 13(1975), 45–49. **[2.4]**

E. Holiday
1970 *Altair Design.* Pantheon, London, 1970. **[2.1]**
1978 *Altair Design. 4.* Random House, New York, 1978. **[1.1]**

K. Holladay
1981 Growth rules for a class of polygonal animals. *Congressus Numerantium* 33(1981), 109–128. **[9.4]**
1983 2-isohedral triangulations. *Geom. Dedicata* 15(1983), 155–170. **[4.4]**

F. Holroyd
1983 The geometry of tiling hierarchies. *Ars Combinatoria* 16B(1983), 211–244. **[9.4]**

W. T. Holser
1958 Point groups and plane groups in a two-sided plane and their subgroups. *Zeitschrift für Kristallographie* 110(1958), 266–281. **[8.8]**
1961 Classification of symmetry groups. *Acta Crystallographica* 14(1961), 1236–1242. **[8.8]**

R. Honsberger
1970 *Ingenuity in Mathematics.* Random House, New York, 1970. **[2.4]**
1976 *Mathematical Gems II.* Mathematical Association of America, 1976. **[12.2]**

S. Horemis
1970 *Optical and Geometrical Patterns and Designs.* Dover, New York, 1970. **[5.1, 5.6]**

C. P. Hornung
1932 *Handbook of Designs & Devices.* Harper and Brothers, New York, 1932. Revised edition: Dover, New York, 1946. **[I, 5.1]**
1975 *Allover Patterns for Designers and Craftsmen.* Dover, New York, 1975. **[I]**

I. Hortobagyi
1975 Review of Lampert and Čoka [1973]. *Zentralblatt für Math.* 287(1975), 311. **[7.8]**

J. Horváth
1964 Über die regulären Mosaiken der hyperbolischen Ebene. *Ann. Univ. Sci. Budapest. Sec. Math.* 7(1964), 49–53. **[4.7, 4.8]**
1965 Bemerkungen zur Theorie der Planigone. *Ann. Univ. Sci. Budapest. Sect. Math.* 8(1965), 147–153. **[7.8]**

W. S. Huff
1983 The parquet deformations from the basic design studio of William S. Huff. Mimeographed notes, SUNY at Buffalo, 1983. **[4.1]**

C. Humbert
1970 *Ornamental Design.* Viking Press, New York, 1970. **[6.5]**

P. Huneke
1977 Smooth tesselations of a sphere (a la Betty Collings) and other surfaces. Abstract 747-57-1, *Notices Amer. Math. Soc.* 24(1977), p. A-50. **[4.8]**

W. Hurewicz and H. Wallman
1941 *Dimension Theory.* Princeton University Press, Princeton, 1941. **[3.10]**

M. Ickis
1959 *The Standard Book of Quilt Making and Collecting.* Dover, New York, 1959. **[I]**

M. S. IPŞIROGLU
1971 *Das Bild im Islam.* A. Scholl, Vienna and Munich, 1971. **[I, 2.10]**

G. VAN ITERSON
1970 *Mathematische und Mikroskopisch-Anatomische Studien über Blattstellungen.* Fischer, Jena, 1970. **[1.6]**

S. IZUMI
1964 *The Costume and Textile Ornament of the Pre-Inca Cultures.* San-ichi Shobo, Tokyo, 1964. **[I, 6.5]**

J. D. JARRATT AND R. L. E. SCHWARZENBERGER
1981 Coloured frieze groups. *Utilitas Mathematica* 19(1981), 295–303. **[8.2]**

M. A. JASWON AND M. A. ROSE
1983 *Crystal Symmetry. Theory of Colour Crystallography.* Ellis Horwood, Chichester, 1983. **[8.8]**

M. JEGER
1966 *Transformation Geometry.* Wiley, New York, 1966. **[1.3]**

S. JENDROL AND E. JUCOVIČ
1972 On a conjecture by B. Grünbaum. *Discrete Math.* 2(1972), 35–49. **[3.9]**

O. JONES
1856 *The Grammar of Ornament.* Quartich, London, 1856.
Reprinted: 1910, 1928. **[I, 8.8]**

E. JUCOVIČ
1973 Analogues of Eberhard's theorem for 4-valent 3-polytopes with involutory automorphisms. *Discrete Math.* 6(1973), 249–254. **[3.9]**

W. JUSTEMA
1968 *The Pleasures of Pattern.* Reinhold, New York, 1968. **[5.2]**

F. KÁRTESZI
1957 Eine Bemerkung über das Dreiecksnetz der hyperbolischen Ebene. *Publicationes Math. Debrecen* 5(1957), 142–146. **[4.7, 4.8]**

N. D. KAZARINOFF AND R. WEITZENKAMP
1973 Squaring rectangles and squares. *Amer. Math. Monthly* 80(1973), 877–888. **[2.4]**

O. H. KELLER
1930 Über die lückenlose Einfüllung des Raumes mit Würfeln. *J. reine angew. Math.* 163(1930), 231–248. **[2.10]**

C. F. KELLEY AND W. L. MOWLL
1912 *A Test-Book of Design.* Houghton Mifflin, Boston, 1912. **[I]**

G. KEPES (EDITOR)
1965 *Structure in Art and in Science.* Brazillier, New York, 1965. **[I]**
1966 *Module, Proportion, Symmetry, Rhythm.* Brazillier, New York, 1966. **[I]**

I. KEPPLER [J. KEPLER]
1619 *Harmonice Mundi.* Lincii, 1619.
German translation: M. Caspar [1939].
Also: *Johannes Kepler Gesammelte Werke*, M. Caspar, ed. Band VI. Beck, Munich, 1940.
Reprint: Culture et Civilisation, Brussels, 1968. **[2.0, 2.5, 2.10, 12.3]**

B. v. KERÉKJÁRTÓ
1923 *Vorlesungen über Topologie.* Springer, Berlin, 1923. **[1.6]**

R. B. KERSHNER
1968 On paving the plane. *Amer. Math. Monthly* 75(1968), 839–844. **[9.3, 9.6]**
1969 On paving the plane. *APL Technical Digest* 8(1969), No. 6, pp. 4–10. **[9.6]**

R. KIKUCHI
1956 Shape distribution of two-dimensional soap froths. *J. of Chemical Physics* 24(1956), 861–867. **[3.10]**

S. KIM
1976 Non-sharing dissections. Unpublished manuscript. **[2.4]**

J. M. KINGSTON
1957 Mosaics by reflection. *Math. Teacher* 50(1957), 280–286. **[2.10]**

A. KISS
1973 *Roman Mosaics in Hungary*. Akadémiai Kiadó, Budapest, 1973. **[I]**

A. I. KITAIGORODSKII
1955 Organic crystal chemistry [In Russian]. *Akad. Nauk SSSR*, Moscow 1955.
English translation: *Organic Chemical Crystallography*. Consultants Bureau, New York, 1961. **[5.1]**

M. S. KLAMKIN AND A. LIU
1980a Note on a result of Niven on impossible tessellations, *Amer. Math. Monthly* 87(1980), 651–653. **[3.10]**
1980b Polyominoes on the infinite checkerboard. *J. Combinatorial Theory* A28(1980), 7–16. **[9.4]**

D. A. KLARNER
1965 Some results concerning polyominoes. *The Fibonacci Quarterly* 3(1965), 9–20. **[9.4]**
1969 Packing a rectangle with congruent *n*-ominoes. *J. Combinatorial Theory* 7(1969), 107–115. **[9.4]**
1981 *The Mathematical Gardner*, D. A. Klarner, ed. Prindle, Weber & Schmidt, Boston, 1981. **[R]**

D. A. KLARNER AND R. L. RIVEST
1974 Asymptotic bounds for the number of convex *n*-ominoes. *Discrete Math.* 8(1974), 31–40. **[9.4]**

S. C. KLEENE
1967 *Introduction to Metamathematics.* North-Holland, Amsterdam, 1967. **[11.4]**

F. KLEIN
1872 *Vergleichende Betrachtungen über neuere geometrische Forschungen. Programm zum Eintritt in die philosophische Fakultät und den Senat der k. Friedrich-Alexanders-Universität zu Erlangen.* A. Deichert, Erlangen, 1872.
Annotated reprint: *Mathematische Annalen* 43(1893), 63–100; also in: *Gesammelte Mathematische Abhandlungen*, Vol. 1, pp. 460–497 (Berlin 1921); *Geometrie*, K. Strubecker, ed., Wissenschafliche Buchgesellschaft, Darmstadt, 1972, pp. 118–155; *Das Erlanger Programm*, Ostwalds Klassiker der exakten Wissenschaften, vol. 253 (Geest & Portig, Leipzig 1974); and *Mathematical Intelligencer*, vol. 0(1977), 23–30.
English translation: A comparative review of recent researches in geometry, *Bull. New York Math. Soc.* 2(1893), 215–249. **[7.1]**

B. KLOTZEK
1983 Some discontinuous groups in non-Euclidean geometries. *Blätter zu den Potsdamer Forschungen* 1/83 (preprint), Potsdam, 1983. **[4.8]**

G. KNIGHT
1984a The geometry of Maori art—Rafter patterns. *New Zealand Math. Magazine* 21(1984), 36–40. **[1.6]**
1984b The geometry of Maori art—Weaving patterns. *New Zealand Math. Magazine* 21(1984), 80–86. **[1.6]**

D. E. KNUTH
1968 *The Art of Computer Programming*, Vol. 1. Addison-Wesley, Reading, MA, 1968. **[11.1, 11.2]**
1973 *The Art of Computer Programming*, Vol. 3. Addison-Wesley, Reading, MA, 1973. **[5.6]**

E. KOCH AND W. FISCHER
1978 Complexes for crystallographic point groups, rod groups and layer groups. *Zeitschrift für Kristallographie* 147(1978), 21–38. **[1.6]**

D. KÖNIG
1927 Über eine Schlussweise aus dem Endlichen ins Unendliche. *Acta Litt. Sci. Szeged* 3(1927), 121–130. **[11.2]**

V. A. Kopcik [Koptsik]
1967 A general sketch of the development of the theory of symmetry and its applications in physical crystallography over the last 50 years [In Russian]. *Kristallografiya* 12(1967), 755–774.
English translation: *Soviet Physics—Crystallography* 12(1968), 667–683. **[8.8]**
1975 Advances in theoretical crystallography. Color symmetry of defect crystals. *Kristall und Technik* 10(1975), 231–245. **[8.8]**

V. A. Kopcik [Koptsik] and Ž. M. Kužukeev [Zh.-N. M. Kuzhukeev]
1972 Derivation of the three-, four-, and six-color Belov space groups from tables of irreducible representations [In Russian]. *Kristallografiya* 17(1972), 705–711.
English translation: *Soviet Physics–Crystallography* 17(1973), 622–627. **[8.8]**

K. Koutský and M. Sekanina
1958 On the decomposition of the straight line in the congruent three-point sets [in Czech]. *Časopis pěst. matem.* 83(1958), 317–326. **[12.2]**

M. Kraitchik
1942 *Mathematical Recreations.* W. W. Norton, New York, 1942.
Second ed., Dover, New York, 1953. **[2.10]**

P. Kramer
1982 Non-periodic central space filling with icosahedral symmetry using copies of seven elementary cells. *Acta Crystallographica* A38(1982), 257–264. **[10.7]**

P. Kramer and R. Neri
1984 On periodic and non-periodic space fillings of E^m obtained by projection. *Acta Crystallographica* A40(1984), 580–587. **[10.7]**

S. Kravitz
1964 Tessellations and polyhedrons. *Product Engineering*, Vol. 35, No. 15(1964), 119–122. **[2.10]**

R. Krishnamurti and P. H. O'N. Roe
1979 On the generation and enumeration of tessellation designs. *Environment and Planning* B6(1979), 191–260. **[9.4]**

O. Krötenheerdt
1969 1970a 1970b Die homogenen Mosaike n-ter Ordnung in der euklidischen Ebene. I, II, III. *Wiss. Z. Martin-Luther-Univ. Halle-Wittenberg, Math.-Natur. Reihe* 18(1969), 273–290, 19(1970), 19–38 and 97–122. **[2.2]**

O. Krötenheerdt and J. Reichstein
1978 Über stabile reguläre Kreispackungen n-ter Ordnung mit speziellen Untersuchungen für $n = 2$. *Wissenschafliche Zeitschrift Martin-Luther-Universität Halle-Wittenberg* 27(1978), 53–73. **[7.8]**

W. Kuperberg
1982 Packing convex bodies in the plane with density greater than 3/4. *Geometriae Dedicata* 13(1982), 149–155. **[9.6]**

C. Kuratowski
1924 Sur les coupures irréductibles du plan. *Fund. Math.* 6(1924), 130–145. **[1.6]**

J. Kürschak
1898 Über das regelmässige Zwölfeck. *Math. Naturw. Ber. Ung.* 15(1898), 196–197. **[2.6]**

H. Lalvani
1977 Transpolyhedra. Dual transformations by explosion-implosion. *Papers in Theoretical Morphology*, 1. H. Lalvani, New York, 1977. **[4.1]**
1981 *Multi-dimensional periodic arrangements of transforming space structures.* Ph.D. thesis, University of Pennsylvania, 1981. **[4.1]**

1982a Structures on hyper-structures. *Structural Topology* 6(1982), 13–20. **[4.1]**

1982b Structures on hyper-structures. Multidimensional periodic arrangements of transforming space structures. *Papers in Theoretical Morphology*, 3. H. Lalvani, New York, 1982. **[4.1]**

D. Lampert and G. Čoka

1973 The density of a regular system of disks. *Ann. Univ. Sci. Budapest. Sect. Math.* 16(1973), 69–85. **[7.8]**

C. D. Langford

1940 Uses of a geometric puzzle. *Math. Gazette* 34(1940), 209–211. **[10.1]**

C. Lantuejoul

1978 Computation of the histograms of the number of edges and neighbours of cells in a tesselation, in *Geometrical Probability and Biological Structures: Buffon's 200th Anniversary*, R. E. Miles and J. Serra, eds. Lecture Notes in Biomathematics. Springer, Berlin, 1978, pp. 323–329. **[3.10, 5.6]**

S. Lantz

1976 *Trianglepoint*. Viking Press, New York, 1976. **[I]**

J. L. Larsen and C. W. Gull

1977 *The Patchwork Quilt Design & Coloring Book*. Butterick, New York, 1977. **[I]**

F. Laves

1930 Die Bau-Zusammenhänge innerhalb der Kristallstrukturen, I. *Zeitschrift für Kristallographie* 73(1930), 202–265. **[5.3]**

1931a Ebenenteilung in Wirkungsbereiche. *Zeitschrift für Kristallographie* 76(1931), 277–284. **[3.10, 5.6]**

1931b Ebenenteilung und Koordinationszahl. *Zeitschrift für Kristallographie* 78(1931), 208–241. **[3.9, 4.8]**

B. Lavrič

1980 Parketiranje ravnine s konveksnimi mnogokotniki [In Slovenian; English summary]. (Tiling the plane with convex polygons) *Obzornik mat. fiz.* 27(1980), 97–101. **[3.10]**

J. E. Lawrenson and H. Wondratschek

1976 The extraordinary orbits of the 17 plane groups. *Zeitschrift für Kristallographie* 143(1976), 471–484. **[5.3]**

C. L. Lawson

1977 Software for C^1 surface interpolation. In *Mathematical Software*, III. Academic Press, New York, 1977, pp. 161–193. **[5.6]**

Y. Le Corre

1958 Les groupes de symétrie bicolores et leurs applications. *Bulletin de la Société Française de Mineralogie et de Cristallographie* 81(1958), 120–125. **[8.8]**

D. T. Lee

1980 Two-dimensional Voronoi diagrams in the L_p-metric. *J. Association for Computing Machinery* 27(1980), 604–618. **[5.6]**

D. T. Lee and R. L. Drysdale, III

1981 Generalization of Voronoi diagrams in the plane. *SIAM J. Comput.* 10(1981), 73–87. **[5.6]**

L. P. Lee

1973 The conformal tetrahedric projection with some practical applications. *Cartographic Journal* 10 (1973), 22–28. **[2.10]**

J. Lehner

1964 *Discontinuous Groups and Automorphic Functions*. Mathematical Surveys No. 8, Amer. Math. Society, Providence, RI, 1964. **[4.8]**

L. Lévy

1891 Sur les pavages à l'aide de polygones réguliers. *Bull. de la Société Philomatique de Paris*, (8) 3(1891), 46–50. **[2.8, 2.10]**

1894 Question 262. *Interméd. Math.* 1(1894), 147, and 7(1900), 153. **[2.8, 2.10]**

D. LEVINE AND P. J. STEINHARDT

1984 Quasi-crystals: A new class of ordered structures. *Physical Review Letters* 53(1984), 2477–2480. **[10.7]**

A. B. LEWIS

1925 *Decorative Art of New Guinea: Incised Designs.* Anthropology Design Series, No. 4. Field Museum of Natural History, Chicago, 1925.
Reprinted in: *Decorative Art of New Guinea.* Dover, New York, 1973. **[I]**

F. T. LEWIS

1931 A comparison between the mosaic of polygons in a film of artificial emulsion and the pattern of simple epithelium in surface view (cucumber epidermis and human amnion). *Anatomical Record* 50(1931), 235–265. **[3.10]**

H. LINDGREN

1972 *Recreational Problems in Geometric Dissections and How to Solve Them.* Dover, New York, 1972. **[2.6, 2.10]**

J. LINHART

1977 Scheibenpackungen mit nach unten beschränkter Nachbarnzahl. *Studia Scientiarum Math. Hungarica* 12(1977), 281–293. **[3.2]**

D. V. LITTLE

1974 A third note on recent research in geometrical probability. *Advan. Appl. Prob.* 6(1974), 103–130. **[5.6]**

C. L. LIU

1968 *Introduction to Combinatorial Mathematics.* McGraw-Hill, New York, 1968. **[4.7]**

H.-D. LÖCKENHOFF

1968 *Über die Zerlegung der Ebene in zwei Arten topologisch verschiedener Flächen.* Ph.D. thesis, Philipps-Universität Marburg, Marburg, 1968. **[4.8]**

E. H. LOCKWOOD AND R. H. MACMILLAN

1978 *Geometric Symmetry.* Cambridge University Press, London, 1978 **[8.8, 10.1]**

A. L. LOEB

1971a *Color and Symmetry.* Wiley, New York, 1971. **[1.6, 5.6, 8.8]**

1971b Structure and patterns in science and art. *Leonardo* 4(1971), 339–346. **[5.6]**

1976 *Space Structures. Their Harmony and Counterpoint.* Addison-Wesley, Reading, MA, 1976. **[3.10, 4.8, 5.4, 5.6]**

1978 Algorithms, structures and models, in *Hypergraphics. Visualizing Complex Relationships in Art, Science and Technology*, D. W. Brisson, ed. AAAS Selected Symposium 24, pp. 49–64. **[5.6]**

E. A. LORD AND C. B. WILSON

1984 *The Mathematical Description of Shape and Form.* Wiley, New York, 1984. **[9.6]**

D. LUCAS AND L. GIBSON

1984a Introduction to GBT. (To appear.) **[9.4]**

1984b A system for hierarchical addressing in Euclidean space (To appear.) **[9.4]**

J. LUIS AND M. AMORÓS

1968 *Molecular Crystals: Their Transforms and Diffuse Scattering.* Wiley, New York, 1968. **[5.1]**

W. F. LUNNON

1971 Counting polyominoes, in *Computers in Number Theory*, A.O.L. Atkin and B. J. Birch, eds. Academic Press, London and New York, 1971, 347–372. **[9.4]**

1972 Counting hexagonal and triangular polyominoes, in *Graph Theory and Computing*, R. C. Read, ed. Academic Press, London and New York, 1972, 87–100. **[9.4]**

W. F. LUNNON AND P. A. B. PLEASANTS

1985 Quasicrystallographic tilings. (To appear). **[10.7]**

A. M. MACBEATH

1967 The classification of non-Euclidean plane crystallographic groups. *Canadian J. Math.* 19(1967), 1192–1205. **[4.8]**

S. O. MACDONALD

1979 Combinatorics—A branch of group theory? in *Combinatorial Mathematics* VI, A. F. Horadam and W. D. Wallis, eds. Armidale, Australia, 1978. Lecture Notes in Math., vol. 748. Springer-Verlag, Berlin, Heidelberg, and New York, 1979, pp. 11–20. **[8.8]**

S. O. MACDONALD AND A. P. STREET

1977 On crystallographic colour groups, in *Combinatorial Mathematics* IV, L. R. A. Casse and W. D. Wallis, eds. Adelaide, 1975. Lecture Notes in Mathematics, vol. 560. Springer-Verlag, Berlin, Heidelberg, and New York, 1977, pp. 149–157. **[8.8]**

1978a The analysis of colour symmetry, in *Combinatorial Mathematics* V. Canberra, 1977. Lecture Notes in Mathematics, vol. 686. Springer-Verlag, Berlin, Heidelberg, and New York, 1978, pp. 210–222. **[8.8]**

1978b The seven friezes and how to colour them, *Utilitas Math.* 13(1978), 271–292. **[8.8]**

C. H. MACGILLAVRY

1965 *Symmetry Aspects of M. C. Escher's Periodic Drawings.* Oosthoek, Utrecht, 1965.
Reprinted as: *Fantasy & Symmetry. The Periodic Drawings of M. C. Escher.* Abrams, New York, 1976. **[1.6, 6.6, 8.2, 8.7, 8.8]**

A. L. MACKAY

1957 Extensions of space-group theory. *Acta Crystallographica* 10(1957), 543–548. **[8.8]**

1969 The structure of structure. Some problems in solid state chemistry. *Chimia* 23(1969), 433–437. **[1.4]**

1975 Generalized crystallography. *Izvještaj Jugoslavenskog Centra za Kristalografiju* 10(1975), 15–36. **[10.7]**

1976 Crystal symmetry, *Physics Bulletin,* November 1976, 495–497. **[10.7]**

1979 What is a mock-lattice? *Collected Abstracts, Symposium über Mathematische Kristallographie,* Riederalp (Wallis, Switzerland), August 1979, pp. 16–17. **[10.7]**

1981 De nive quinquangula—On the pentagonal snowflake [In Russian]. *Kristallografiya* 26(1981), 909–918.
English version: *Soviet Physics—Crystallography* 26(1981), 517–522. **[10.7]**

1982 Crystallography and the Penrose pattern. *Physica* 114A(1982), 609–613. **[10.7]**

P. A. MACMAHON

1921 *New Mathematical Pastimes.* Cambridge University Press, London, 1921. **[6.6, 9.6, 12.5]**

1922 The design of repeating patterns for decorative work. *J. Roy. Soc. Arts* 70(1922), 567–578.
Related discussion, *ibid.,* pp. 578–582. **[6.6, 9.6]**

P. A. MACMAHON AND W. P. D. MACMAHON

1922 The design of repeating patterns. *Proc. Roy. Soc. London* 101(1922), 80–94. **[6.6, 9.6]**

W. P. D. MACMAHON

1925 The theory of closed repeating polygons in Euclidean space of two dimensions.

Proc. London Math. Soc. (2) 23(1925), 75–93. **[6.6, 9.6]**

R. H. MACMILLAN
1979 Pyramids and pavements: Some thoughts from Cairo. *Math. Gazette* 63(1979), 251–255. **[I]**

J. S. MADACHY
1966 *Mathematics on Vacation.* Scribner, New York, 1966. **[2.10]**

W. MAGNUS
1974 *Noneuclidean Tesselations and Their Groups.* Academic Press, New York and London, 1974. **[2.6, 4.8]**

P. MAHLO
1908 *Topologische Untersuchungen über Zerlegung in ebene und sphärische Polygone.* (Ph.D. thesis, Friedrichs-Universität, Halle-Wittenberg.) Kaemmerer, Halle, 1908, 99 pp. + 7 plates. **[2.6]**

E. MAKOVICKY
1979 The crystallographic art of Hans Hinterreiter. *Zeitschrift für Kristallographie* 150(1979), 13–21. **[8.8]**

E. MAKOVICKY AND M. MAKOVICKY
1977 Arabic geometrical patterns—a treasury for crystallographic teaching. *Jahrbuch für Mineralogie Monatshefte,* 1977, No. 2, pp. 58–68. **[5.6]**

E. M. MALETSKY
1974 Designs with tessellations. *Math. Teacher* 67(1974), 335–338. **[8.8]**

B. B. MANDELBROT
1977 *Fractals. Form, Chance and Dimension.* W. H. Freeman, San Francisco, 1977.
New edition: *The Fractal Geometry of Nature.* W. H. Freeman, San Francisco, 1982. **[3.10]**

P. MANI
1971 Automorphismen von polyedrischen Graphen. *Math. Ann.* 192(1971), 279–303. **[4.8]**

P. MANI-LEVITSKA, B. GUIGAS AND W. E. KLEE
1979 Rectifiable n-periodic maps. *Geometriae Dedicata* 8(1979), 127–137. **[4.8]**

L. MARCH AND R. MATELA
1974 The animals of architecture: some census results on N-omino populations for $N = 6, 7, 8$. *Environment and Planning* B1(1974), 193–216. **[1.6, 9.4]**

L. MARCH AND P. STEADMAN
1971 *The Geometry of Environment.* RIBA Publication, London, 1971.
Paperback reprint: MIT Press, Cambridge, MA, 1974. **[I]**

G. C. MARLEY
1974 Multiple subdivisions of the plane. *Mathematics Magazine* 47(1974), 202–206. **[12.4]**

G. C. MARS (EDITOR)
1925 *Brickwork in Italy.* American Face Brick Association, Chicago, 1925. **[I]**

G. E. MARTIN
1982 *Transformation Geometry. An Introduction to Symmetry.* Springer-Verlag, New York, Heidelberg, and Berlin, 1982. **[1.3, 1.4]**

T. MATSUMOTO
1968 Proof that the *pgg* packing of ellipses has never the maximum density. *Zeitschrift für Kristallographie* 126(1968), 170–174. **[7.8]**
1979 Orbit types in the plane, in *Collected Abstracts, Symposium on Mathematical Crystallography,* Riederalp (Wallis, Switzerland), 7–9 August 1979, pp. 10–11. **[5.6]**

J. MCCORMACK
1978 The 24 coloured squares problem. Query 135, *Notices Amer. Math. Society* 25(1978), 67.
Response, *ibid.,* 145. **[12.5]**

W. H. M'CREA
1933 On nets of polygons occurring in nature. *Math. Notes Edinburgh Math. Soc.* 28(1933), 8–12. **[3.4, 3.10]**

L. P. MEISSNER
1982 Fibonacci residue cycles. Preprint, Lawrence Berkeley Laboratory. **[5.6]**

A. L. F. MEISTER
1769 Generalia de genesi figurarum planarum et inde pendentibus earum affectionibus. *Novi Commentarii Societatis Regiae Scientiarum Gottingensis* 1(1769/1770), 144–180 and plates III to XI. **[12.3]**

T. MENTEN
1975 *Japanese Border Designs.* Dover, New York, 1975. **[I]**

W. MERETZ AND B. L. VAN DER WAERDEN
1966 Statistische Theorie der aequalen Zellteilung. *Naturwissenschaften* 53(1966), 8–11. **[3.10]**

H. MESCHKOWSKI
1960 *Ungelöste und unlösbare Probleme der Geometrie.* Vieweg und Sohn, Braunschweig, 1960.
1966 English translation: *Unsolved and Unsolvable Problems in Geometry.* Ungar, New York, 1966. **[2.4]**

F. S. MEYER
1888 *Handbuch der Ornamentik.* Leipzig, 1888. English translation: *Handbook of Ornament.* Dover, New York, 1957. **[I]**

P. T. MIELKE
1983 A tiling of the plane with triangles. *Two-Year College Math. J.* 14(1983), 377–381. **[2.4]**

J. C. P. MILLER
1933 *On stellar constitution, on statistical geophysics, and on uniform polyhedra.* Ph.D. thesis, Cambridge, 1933. **[12.3]**
1968 Periodic forests of stunted trees, in *Computers in Mathematical Research,* R. F. Churchhouse and J.-C. Herz, eds. North Holland, Amsterdam, 1968, pp. 149–167. **[11.4]**
1970 Periodic forests of stunted trees. *Philos, Trans. Royal Soc. London,* Series A, 266(1970), 63–111. **[11.4]**
1980 On rotational tessellations and copses. *Philos. Trans. Royal Soc. London,* Series A, 293(1980), 599–641. **[11.4]**

J. MILNOR
1976 Hilbert's Problem 18: On crystallographic groups, fundamental domains, and on sphere packings, in Browder [1976], 491–506. **[1.6, 9.6]**

H. MINKOWSKI
1907 *Diophantische Approximationen.* Teubner, Leipzig, 1907. **[2.10]**

J. MOLNAR
1959 Unterdeckung und Überdeckung der Ebene durch Kreise. *Ann. Univ. Sci. Budapest. Sect. Math.* 2(1959), 33–40. **[7.8]**
1966 Collocazioni di cerchi con esigenza di spazio. *Ann. Univ. Sci. Budapest. Sect. Math.* 9(1966), 71–86. **[7.8]**
1977 On the ρ-system of unit circles. *Ann. Univ. Sci. Budapest. Sect. Math.* 20(1977), 195–203. **[7.8]**

P. A. P. MORAN
1966 A note on recent research in geometrical probability. *J. Appl. Prob.* 3(1966), 453–463. **[5.6]**
1969 A second note on recent research in geometrical probability. *Advan. Appl. Prob.* 1(1969), 73–89. **[5.6]**

R. MOSSERI AND J. F. SADOC
1983 Two and three dimensional non-periodic networks obtained from self-similar tiling. (To appear.) **[10.7]**

E. MÜLLER [MUELLER]
1944 *Gruppentheoretische und Strukturanalytische Untersuchungen der Maurischen Ornamente aus der Alhambra in Granada.* (Ph.D. Thesis, Univ. Zürich.) Baublatt, Rüschlikon, 1944. **[1.6, 8.8]**
1946 El estudio de ornamentos como applicacion de la teoria de los grupos

de orden finito. *Euclides (Madrid)* 6(1946), 42–52. **[1.6]**

D. MYERS

1974 Nonrecursive tilings of the plane, II. *J. Symbolic Logic* 39(1974), 286–294. **[11.4]**

S. B. NADLER, JR.

1978 *Hyperspaces of Sets.* Marcel Dekker, New York, 1978. **[3.8]**

N. N. NERONOVA

1966 Classification principles for symmetry groups and groups of a different kind of antisymmetry. I: Scheme of the crystallographic symmetry groups and groups of a different kind of antisymmetry [In Russian]. *Kristallografiya* 11(1966), 495–504.
English translation: *Soviet Physics—Crystallography* 11(1967), 445–452. **[8.8]**

1967 Classification principles for the symmetry and antisymmetry groups. II: Special spatial elements as the basis of classification for symmetry groups [In Russian]. *Kristallografiya* 12(1967), 3–10.
English translation: *Soviet Physics—Crystallography* 12(1967), 3–8. **[8.8]**

N. N. NERONOVA AND N. V. BELOV

1961a A single scheme for the classical and black-and-white crystallographic symmetry groups [In Russian]. *Kristallografiya* 6(1961), 3–12.
English translation: *Soviet Physics—Crystallography* 6(1961), 1–9. **[8.8]**

1961b Color antisymmetry mosaics [In Russian]. *Kristallografiya* 6(1961), 831–839.
English translation: *Soviet Physics—Crystallography* 6(1962), 672–678. **[8.8]**

M. H. A. NEWMAN

1951 Elements of the topology of plane sets of points. Second ed. Cambridge University Press, London, 1951. **[3.1, 4.1, 7.1]**

A. NIGGLI

1959 Zur Systematik und gruppentheoretischen Ableitung der Symmetrie-, Antisymmetrie-und Entartungssymmetriegruppen. *Zeitschrift für Kristallographie* 111(1959), 288–300. **[8.8]**

1963 Zur Topologie, Metrik und Symmetrie der einfachen Kristallformen. *Schweiz. Mineral. und Petrograph. Mitt.* 43(1963), 49–58. **[7.8]**

P. NIGGLI

1919 *Geometrische Kristallographie des Diskontinuums.* Borntraeger, Leipzig, 1919. **[1.6, 5.6]**

1924 Die Flachensymmetrien homogener Diskontinuen. *Zeitschrift für Kristallographie* 60(1924), 283–298. **[1.6]**

1926 Die regelmässige Punktverteilung längs einer Geraden in einer Ebene. (Symmetrie von Bordürmuster.) *Zeitschrift für Kristallographie* 63(1926), 255–274. **[1.6]**

1927 Die topologische Strukturanalyse, I. *Zeitschrift für Kristallographie* 65(1927), 391–415. **[5.6, 7.3, 7.8]**

1928 Die topologische Strukturanalyse, II. *Zeitschrift für Kristallographie* 68(1928), 404–466. **[7.3, 7.8]**

J. NIMAN AND J. NORMAN

1978 Mathematics and Islamic art. *Amer. Math. Monthly* 85(1978), 489–490. **[5.6]**

I. NIVEN

1978 Convex polygons which cannot tile the plane. *Amer. Math. Monthly* 85(1978), 785–792. **[3.10]**

J. Norman and S. Stahl
1979 *The Mathematics of Islamic Art. A Packet for Teachers of Mathematics, Social Studies, and Art*. Metropolitan Museum of Art, New York, 1979. **[5.6]**

W. Nowacki
1933 Der Begriff "Voronoischer Bereich". *Zeitschrift für Kristallographie* 85(1933), 331–332. **[5.6]**
1935 *Homogene Raumteilung und Kristallstruktur*. (Dissertation, ETH Zurich 1935. 91 pp.) Leemann, Zurich, 1935. **[4.8, 5.6]**
1948 Symmetrie und physikalisch-chemische Eigenschaften kristallisierter Verbindungen. V. Über Ellipsen-packungen in der Kristallebene. *Schweiz. Mineralog. Petrog. Mitt.* 28(1948), 502–508. **[5.6, 7.5, 7.8]**
1960 Zur Symmetrielehre. 1: Überblick über "zweifarbige" Symmetriegruppen. *Fortschritte der Mineralogie* 38(1960), 96–107. **[8.8]**
1972 Bemerkungen zur Geschichte der Raumgruppen-Symbole von Fedorow, Schoenflies and Hermann-Mauguin. *Zeitschrift für Kristallographie* 135(1972), 145–158. **[1.4]**

P. G. O'Daffer and S. R. Clemens
1976 *Geometry: An Investigative Approach*. Addison-Wesley, Menlo Park, CA, 1976. **[1.6, 2.10, 5.6]**

M. O'Keeffe and B. G. Hyde
1980 Plane nets in crystal chemistry. *Philos. Trans. Royal Soc. London*. Series A, 295 (1980), 553–618. **[3.10, 4.8]**

O. Ore
1967 *The Four-Color Problem*. Academic Press, New York, 1967. **[3.1]**

W. Ostwald
1922a *Die Harmonie der Formen*. Unesma, Leipzig, 1922. **[I]**
1922b *Die Welt der Formen*. Unesma, Leipzig, 1922–1925. **[I]**

H. Ōuchi
1977 *Japanese Optical and Geometrical Art*. Dover, New York, 1977. **[I]**

T. Öz
1957 *Turkish Ceramics*. The Turkish Press, Broadcasting and Tourist Department, Ankara, 1954. **[I]**

A. Pabst
1962 The 179 two-sided, two-colored, band groups and their relations. *Zeitschrift für Kristallographie* 117(1962), 128–134. **[8.8]**
1963 Derivation of the two-sided, two-colored, band groups with their associated point groups and a numerical correction. *Zeitschrift für Kristallographie* 119(1963), 148–150. **[8.8]**

R. Padwick and T. Walker
1977 *Pattern: Its Structure and Geometry*. Ceolfrith Press, Sunderland (England), 1977. **[I]**

A. F. Palistrant
1963 Groups of layer symmetry and antisymmetry of different types [In Russian]. *Kristallografiya* 8(1963), 783–785.
English translation: *Soviet Physics—Crystallography* 8(1964), 624–626. **[8.8]**
1965 Planar point groups of symmetry and different types of antisymmetry [In Russian]. *Kristallografiya* 10(1965), 3–9.
English translation: *Soviet Physics—Crystallography* 10(1965), 1–5. **[8.8]**
1966 Two-dimensional groups of color symmetry and color antisymmetry

of different kinds [In Russian]. *Kristallografiya* 11(1966), 707–713.
English translation: *Soviet Physics—Crystallography* 11(1967), 609–613. **[8.8]**

1967 Groups of colored symmetry and of different kinds of antisymmetry of layers [In Russian]. *Kristallografiya* 12(1967), 194–201.
English translation: *Soviet Physics—Crystallography* 12(1967), 162–168. **[8.8]**

1968 Planar point groups with color symmetry and different classes of antisymmetry [In Russian]. *Kristallografiya* 13(1968), 955–959.
English translation: *Soviet Physics—Crystallography* 13(1969), 833–836. **[8.8]**

1972 Color symmetries and various antisymmetries of edgings and strips [In Russian]. *Kristallografiya* 17(1972), 1096–1102.
English translation: *Soviet Physics—Crystallography* 17(1973), 977–981. **[8.8]**

A. F. PALISTRANT AND A. M. ZAMORZAEV

1963 The symmetry and antisymmetry groups of layers [In Russian]. *Kristallografiya* 8(1963), 166–173.
English translation: *Soviet Physics—Crystallography* 8(1963), 120–126. **[8.8]**

1964 Groups of symmetry and of different types of antisymmetry of borders and ribbons [In Russian]. *Kristallografiya* 9(1964), 155–161.
English translation: *Soviet Physics—Crystallography* 9(1964), 123–128. **[8.8]**

1971 Complete derivation of multicolor two-dimensional and layer groups [In Russian]. *Kristallografiya* 16(1971), 681–689.
English translation: *Soviet Physics—Crystallography* 16(1972), 594–601. **[8.8]**

M. PALMER

1979 Tessellations of non-convex pentagons. *Math. Teaching*, No. 86 (March 1979), 8–11. **[9.3]**

K. PARLASCA

1959 *Die Römischen Mosaiken in Deutschland. Römisch-Germanische Forschungen*, Band 23. W. de Gruyter, Berlin, 1959. **[I]**

G. S. PAWLEY

1961 Mosaics for color antisymmetry groups [In Russian]. *Kristallografiya* 6(1961), 109–110.
English translation: *Soviet Physics—Crystallography* 6(1961), 87–89. **[8.8]**

P. PEARCE

1978 *Structure in Nature Is a Strategy for Design*. MIT Press, Cambridge, MA, 1978. **[2.10]**

P. PEARCE AND S. PEARCE

1978 *Polyhedra Primer*. Van Nostrand Reinhold, New York, 1978. **[4.8]**

D. PEDOE

1976 *Geometry and the Visual Arts*. Dover, New York, 1976. **[1.5]**

R. PENROSE

1974 The role of aesthetics in pure and applied mathematical research. *Bull. Inst. Math. Appl.* 10(1974), 266–271. **[10.3]**

1978 Pentaplexity. *Eureka* 39(1978), 16–22.
Reprint: *Mathematical Intelligencer* 2(1979), 32–37, and *Geometrical Combinatorics*, F. C. Holroyd and R. J. Wilson, eds. Pitman, London, 1984, pp. 55–65. **[10.3, 10.7]**

L. S. PENROSE AND R. PENROSE

1958 Puzzles for Christmas. *New Scientist*, December 1958. **[4.8]**

A. W. Petersen
1888 Om Planers bedækning ved Hjælp af et System af regulære *n*-Kanter. *Tidsskrift for Mathematik* (Copenhagen) Ser. 5, vol. 6(1888), 182–184. **[12.4]**

R. Pfister
1938 *Les toiles imprimées de Fostat et l'Hindoustan.* Les Éditions d'Art et d'Histoire, Paris, 1938. **[I]**

H. C. Plummer
1950 *Brick and Tile Engineering. Handbook of Design.* Structural Clay Products Institute, Washington DC, 1950. **[I]**

L. Poinsot
1810 Mémoire sur les polygones et les polyèdres. *J. École Polytechn.* 10(1810), 16–48. **[12.3]**

G. Pólya
1924 Über die Analogie der Kristallsymmetrie in der Ebene. *Zeitschrift für Kristallographie* 60(1924), 278–282. **[1.6, 5.6]**

C. Pomerance
1977 On a tiling problem of R. B. Eggleton. *Discrete Math.* 18(1977), 63–70. **[2.4]**

K. A. Post
1978 Regular polygons with rational vertices. *Math Gazette* 62(1978), 205–206. **[12.4]**

O. Prochnow
1934 *Formenkunst der Natur.* Wasmuth, Berlin, 1934. **[I]**

R. M. Proctor
1969 *The Principles of Pattern for Craftsmen and Designers.* Van Nostrand, New York, 1969. **[I, 5.1, 5.2, 5.6]**

G. Quijano
1908 Problème 3430. *Interméd. Math.* 15(1908), 195. **[9.4]**

E. R. Ranucci
1971 Space-filling in two dimensions. *Math. Teacher* 64(1971), 587–593. **[4.8]**

E. R. Ranucci and J. L. Teeters
1977 *Creating Escher-Type Drawings.* Creative Publication, Palo Alto, CA, 1977. **[6.6, 8.8]**

E. Regener
1978 Programming the Penrose pattern. Proc. 9th Southeastern Conference on Combinatorics, Graph Theory, and Computing, Boca Raton, 1978. *Congressus Numerantium* 21(1978), 593–606. **[10.3]**

K. Reinhardt
1918 *Über die Zerlegung der Ebene in Polygone.* (Inaugural-Dissertation, Univ. Frankfurt a.M.) R. Noske, Borna and Leipzig, 1918. **[3.1, 3.9, 3.10, 9.3, 9.6]**
1927 Zwei Beweise für einen Satz über die Zerlegung der Ebene. *Tôhoku Math. J.* 28(1927), 221–225. **[3.10]**
1928a Zur Zerlegung der euklidischen Räume in kongruente Polytope. *S.-Ber. Preuss. Akad. Wiss. Berlin,* 1928, 150–155. **[9.6]**
1928b Über die Zerlegung der hyperbolischen Ebene in konvexe Polygone. *Jahresberichte der Deutschen Mathematiker-Vereinigung* 37(1928), 330–332. **[3.10]**
1929 Über der Zerlegung der euklidischen Ebene in Kongruente Bereiche. *Jahresber. Deutsch. Math.-Verein.* 38(1929), 12(italics). **[9.6]**
1934 Aufgabe 170. *Jahresber. Deutsch. Math.-Verein.* 44(1934), 41(italics). **[3.2]**

J. Revault
1973 *Designs & Patterns from North African Carpets & Textiles.* Dover, New York, 1973. **[I]**

D. RHYNSBURGER

1973 Analytic delineation of Thiessen polygons. *Geographical Analysis* 5(1973), 133–144. **[5.6]**

M. RICE AND D. SCHATTSCHNEIDER

1980 The incredible pentagonal versatile. *Mathematics Teaching* 93(1980), 52–53. **[9.3]**

C. A. RICHMOND

1937 Repeating designs in surfaces of negative curvature. *Amer. Math. Monthly* 44(1937), 33–35. **[4.8]**

B. RINČEN

1969 Decorative folk ornaments of Mongols [In Russian]. No place, no date. **[I]**

J. RIORDAN

1958 *An Introduction to Combinatorial Analysis.* Wiley, New York, 1958. **[4.7]**

P. ROBIN

1887 Carrelage illimité en polygones réguliers. *La Nature* 15(1887), 95–96. **[2.1, 2.10]**

A. L. ROBINSON

1985 Doing crystallography in six dimensions. *Science* 228(1985), 314–315. **[10.7]**

R. M. ROBINSON

1967 Seven polygons which admit only nonperiodic tilings of the plane (Abstract). *Notice Amer. Math. Soc.* 14(1967), 835. **[11.1]**

1971 Undecidability and nonperiodicity of tilings of the plane. *Inventiones Math.* 12(1971), 177–909. **[10.2, 11.1, 11.3]**

1975 Comments on the Penrose tiles. Mimeographed notes, September 1975. 8 pages. **[10.3, 10.5]**

1978 Undecidable tiling problems in the hyperbolic plane. *Inventiones Math.* 44(1978), 259–264. **[10.7, 11.1]**

1979 Multiple tilings of *n*-dimensional space by unit cubes. *Mathematische Zeitschrift* 166(1979), 225–264. **[2.10]**

C. A. ROGERS

1964 *Packing and Covering.* Cambridge Mathematical Tracts 54, Cambridge University Press, London, 1964. **[5.6]**

1970 *Hausdorff measure.* Cambridge University Press, London, 1970. **[3.10]**

K. ROGERS

1976 *Advanced Calculus.* Merrill, Columbus, OH, 1976. **[3.8]**

T. ROMAN

1959 Symmetry of 4-dimensional border ornaments [In Russian]. *Dokl. Akad. Nauk. SSSR* 128(1959), 1122–1124. **[8.8]**

1963 Les groupes de symétrie des ornements bicolores en relief. *Bull. Math. de la Soc. Sci. Math. Phys. de la R.P.R.* 7(55)(1963), 231–273. **[8.8]**

1970 Les colonnes cylindriques bicolores. *Zeitschrift für Kristallographie* 132(1970), 372–384. **[8.8]**

1971 Les colonnes cylindriques transparentes unicolores et bicolores. *Acta Crystallographica* A27(1971), 323–331. **[1.6, 8.8]**

1972 Die ein und zweidimensionalen Gruppen von verallgemeinerten Symmetrien. *Mathematica* 14(37)(1972), 167–182. **[10.1]**

B. I. ROSE AND R. D. STAFFORD

1981 An elementary course in mathematical symmetry. *Amer. Math. Monthly* 88(1981), 59–64. **[1.4]**

J. ROSEN

1972 *Symmetry discovered.* Cambridge University Press, London, 1975. **[5.6]**

R. L. ROTH

1982 Color symmetry and group theory. *Discrete Math.* 38(1982). 273–296. **[8.8]**

1983 Compound color symmetry and strong equivalence. *Communications in Algebra* 11(1983), 2029–2049. **[8.8]**

1984 Local color symmetry. *Geometriae Dedicata* 17(1984), 99–108. **[8.8]**

C. L. SAFFORD AND R. BISHOP
1972 *America's Quilts and Coverlets*. Dutton, New York, 1972. **[I, 2.10]**

H. SALADIN
1926 *L'Alhambra de Grenade*. A. Morance, Paris, 1926. **[I]**

L. F. SALAZAR, A. M. HUTCHESON, P. TILLON, AND H. R. WILSON
1978 Optical diffraction studies of particles of potato virus T. *J. General Virology* 39(1978), 333–342. **[1.6]**

J. R. SAMPSON AND S. C. TROFANENKO
1979 Aspects of shape. *Canad. J. of Operations Research and Information Processing* 17(1979), 138–150. **[9.4]**

A. D. SANDS AND S. ŠWIERCZKOWSKI
1960 Decomposition of the line in isometric three-point sets. *Fundamenta Math.* 48(1960), 361–362. **[12.2]**

L. A. SANTALÓ
1976 *Integral Geometry and Geometric Probability. Encyclopedia of Mathematics and Its Applications*, vol. 1. Addison-Wesley, Reading, MA, 1976. **[1.6]**
1980 Random lines and tessellations in a plane. *Stochastica* 4(1980), 3–13. **[5.6]**

A. SAYARI, Y. BILLIET AND H. ZARROUK
1978 Complete list of subgroups and changes of standard setting of two-dimensional space groups. *Acta Crystallographica* A34(1978), 553–555. **[1.4]**

R. SAUER
1937 Ebene gleicheckige Polygongitter. *Jahresber. Deutsch. Math.-Verein.* 47(1937), 115–124. **[6.6]**

I. I. SCHAFRANOVSKY
1961 Erweiterung der Kristallformenlehre. *Fortschritte der Mineralogie* 39(1961), 187–195. **[7.8]**

D. SCHATTSCHNEIDER
1978a The plane symmetry groups. Their recognition and notation. *Amer. Math. Monthly* 85(1978), 439–450. **[1.4, 1.6]**
1978b Tiling the plane with congruent pentagons. *Math. Magazine* 51(1978), 29–44. **[2.7, 9.1, 9.3, 9.6]**
1980 Will it tile? Try the Conway Criterion! *Math. Magazine* 53(1980), 224–233. **[1.2]**
1981 In praise of amateurs. In Klarner [1981], pp. 140–166. **[9.3]**
1982 A (complete) catalogue of equilateral pentagons that tile. Mimeographed note, July 1982. **[9.3]**
1985 M. C. Escher's classification system for his colored periodic drawings. M. C. Escher Congress, Rome, 1985. **[6.6, 8.8]**

F. L. SCHAUERMANN
1892 *Theory and Analysis of Ornament Applied to the Work of Elementary and Technical Schools*. Sampson Low, Marston, London, 1892. **[I]**

K. SCHERER
1983 The impossibility of a tesselation of the plane into equilateral triangles whose sidelengths are mutually different, one of them being minimal. *Elemente der Math.* 38(1983), 1–4. **[2.4]**

V. SCHLEGEL
1883 Theorie der homogen zusammengesetzten Raumgebilde. *Nova Acta Leop.-Carol. Deutsch. Akad. Naturforscher* 44(1883), 339–459 + 9 plates. **[4.8]**

P. SCHMITT
1985a Pairs of tiles admitting finitely or countably infinitely many tilings. *Geometriae Dedicata* 20(1986), 133–142. **[1.5]**
1985b Sets of tiles with a prescribed number to tilings. (To appear). **[1.5]**

1985c σ-morphic sets of prototiles. (To appear). **[1.5]**

A. SCHOENFLIES

1891 *Kristallsysteme und Kristallstruktur.* Teubner, Leipzig, 1891. **[1.6]**

1923 *Theorie der Kristallstruktur.* Borntraeger, Leipzig 1923. **[5.6]**

A. SCHUBNIKOW (A. V. ŠUBNIKOV)

1929 Über die Symmetrie des Kontinuums. *Zeitschrift für Kristallographie* 72(1929), 272–290. **[5.6]**

E. SCHULTE

1985 The existence on non-tiles and non-facets in three dimensions. *J. Combinatorial Theory* A38(1985), 75–81. **[9.6]**

R. L. E. SCHWARZENBERGER

1974 The 17 plane symmetry groups. *Math. Gazette* 58(1974), 123–131. **[1.6]**

1984 Colour symmetry. *Bull London Math. Society* 16(1984), 209–240. **[8.8]**

H. SCHWERDTFEGER

1962 Review of Delone [1959]. *Math. Reviews* 23(1962), #A3503. **[7.8]**

A. W. SEABY

1948 *Pattern without Pain.* Batsford, London, 1948. **[I]**

M. SENECHAL

1979a Color groups. *Discrete Appl. Math.* 1(1979), 51–73. **[8.0, 8.2, 8.7, 8.8]**

1979b Themes and variations. *Criss-Cross Art Communications*, Nos. 7–8 (1979), 40–45. **[2.6]**

1980 The genesis of growth twins [In Russian]. *Kristallografiya* 25(1980), 908–915.
English version: *Soviet Physics—Crystallography*, 25(1980), 520–524. **[1.5]**

1984 Morphisms of crystallographic groups: kernels and images. (To appear). **[1.4]**

1985a Tiling the torus. (To appear). **[4.8]**

1985b Escher designs on surfaces. M. C. Escher Congress, Rome, 1985. **[4.8]**

M. SENECHAL AND G. FLECK (EDITORS)

1977 *Patterns of Symmetry.* University of Massachusetts Press, Amherst, 1977. **[I]**

F. W. SHAFFER

1979 *Indian Designs from Ancient Ecuador.* Dover, New York, 1979. **[I]**

D. SHECHTMAN, I. BLECH, D. GRATIAS AND J. W. CAHN

1984 Metallic phase with long-range orientational order and no translational symmetry. *Physical Review Letters* 53(1984), 1951–1953. **[10.7]**

A. O. SHEPARD

1948 *The Symmetry of Abstract Design with Special Reference to Ceramic Decoration.* Publication 574, Carnegie Institution of Washington, Washington, DC. 1948. **[I, 1.6]**

1956 *Ceramics for the Archaeologist.* Publication 609, Carnegie Institution of Washington, Washington, DC, 1956. **[I, 1.6]**

G. C. SHEPHARD

1985 What Escher might have done. M. C. Escher Congress, Rome, 1985. **[6.6]**

A. V. SHUBNIKOV [ŠUBNIKOV], N. V. BELOV, AND OTHERS

1964 *Colored Symmetry.* Pergamon Press, Oxford, 1964. **[8.2]**

R. SIBSON

1940 Comments on Note 1464. *Math. Gazette* 24(1940), 343, **[10.1]**

1978 Locally equiangular triangulations. *The Computer Journal* 21(1978), 243–245. **[5.6]**

1980 The Dirichlet tessellation as an aid in data analysis. *Scandinavian J. Statistics* 7(1980), 14–20. **[5.6]**

M. SIJELMASSI

1974 *Les arts traditionnels au Maroc.* Flammarion, Pairs, 1974. **[I]**

J. M. Dos Santos Simoes

1969 *Azulejaria em Portugal nos Séculos XV e XVI.* Fundação C. Gulbenkian, Lisbon, 1969. **[I]**

D. R. Simonds

1977 Central tessellations with an equilateral pentagon. *Mathematics Teaching* 81(1977), 36–37. **[9.5]**

1978 Untitled note. *Mathematics Teaching* 84(1978), 33. **[9.5]**

I. M. Singer and J. A. Thorpe

1967 *Lecture Notes on Elementary Topology and Geometry.* Scott, Foresman, Glenview, IL, 1967. **[3.7]**

U. Sinogowitz

1939 Die Kreislagen und Packungen kongruenter Kreise in der Ebene. *Zeitschrift für Kristallographie* 100(1939), 461–508. **[5.6, 6.6, 7.3, 7.8]**

V. Smart

1971 *Roopa Samhita. Design through the Ages.* Kala-Ravi Trust, Ahmedabad, 1971. **[I, 5.2]**

N. L. Smirnova and L. I. Potešnova

1966 Regular systems of points of three-dimensional groups as combinations of planar sets [In Russian]. *Vestnik Moskov. Univ. Ser. Geolog.* No. 6(1966), 41–45. **[5.6]**

C. S. Smith

1954 The shape of things. *Scientific American*, January 1954, 58–64. **[3.10]**

1965 Structure, substructure, superstructure. Pages 29–41 in Kepes [1965]. **[I]**

1978 Structural hierarchy in science, art, and history, in *On Aesthetics in Science*, J. Wechsler, ed. MIT Press, Cambridge, MA, 1978, pp. 9–53. **[3.10]**

G. S. Smith, Q. Johnson and P. C. Nordine

1965 The crystal structure of ScB_2C_2. *Acta Crystallographica* 19(1965), 668–673. **[1.2]**

L. Sohncke

1874 Die regelmässigen ebenen Punktsysteme von unbegrenzter Ausdehnung. *J. reine und angew. Math.* 77(1874), 47–101. **[5.3, 5.6, 6.6]**

D. M. Y. Sommerville

1905 Semi-regular networks of the plane in absolute geometry. *Trans. Roy. Soc. Edinburgh* 41(1905), 725–747 + 12 plates. **[2.8, 2.10]**

A. Speiser

1927 *Die Theorie der Gruppen von endlicher Ordnung.* Second ed. Springer, Berlin, 1927.

1956 Fourth ed., Birkhauser, Basel, 1956. **[1.4, 1.6, 5.6, 8.8]**

A. Speltz

1921 *Das farbige Ornament aller historischen Stile.* Wiesike, Brandenburg, 1921. **[I]**

A. Spilhaus

1983 World ocean maps: the proper places to interrupt. *Proc. Amer. Philos. Soc.* 127(1983), 50–60. **[2.10]**

1984 Plate tectonics in geoforms and jigsaws. *Proc. Amer. Philos. Soc.* 128(1984), 257–269. **[2.10]**

1985 *The Puzzle of the Plates. Spilhaus Repeating World Map.* American Geophysical Union, Washington, DC, 1985. **[2.10]**

R. Sprague

1939 Beispiel einer Zerlegung des Quadrats in lauter verschiedene Quadrate. *Math. Zeitschrift.* 45(1939), 607–608. **[2.4]**

S. K. Stein

1974 Algebraic tiling. *Amer. Math. Monthly* 81(1974), 445–462. **[1.6, 2.10]**

H. Steinhaus

1950 *Mathematical Snapshots.* Oxford University Press, New York, 1950.

1960 Second ed., 1960. **[2.10]**

E. Steinitz

1903 Review of Andreini [1902] [In German].

Fortschritte der Math. 34(1903), 538–540. **[3.10]**

1922 Polyeder und Raumeinteilungen. *Enzykl. Math. Wiss.* Vol. 3(Geometrie), Part 3AB12, pp. 1–139 (1922). **[4.8]**

E. STEINITZ AND H. RADEMACHER

1934 *Vorlesungen über die Theorie der Polyeder.* Springer, Berlin, 1934. **[4.8]**

P. S. STEVENS

1974 *Patterns in Nature.* Little, Brown, Boston, 1974. **[2.10]**

1980 *Handbook of Regular Patterns.* MIT Press, Cambridge, MA, 1980. **[1.6]**

A. H. STONE

1985 Closed tilings of Euclidean spaces. *Suppl. Rendic. Circ. Mat. Palermo.* (To appear). **[3.10]**

A. B. STOTT

1910 Geometrical deduction of semiregular from regular polytopes and space fillings. *Verhandelingen der Koninklijke Akademie van Wetenschappen te Amsterdam (Eerste Sectie)* 11(1910), No. 1, 24 pp. + 3 plates. **[4.1]**

YU. G. STOYAN

1975 *The Packing of Geometric Objects* [In Russian]. Naukova Duma, Kiev, 1975. **[I]**

A. V. ŠUBNIKOV [SHUBNIKOV, SCHUBNIKOW]

1916 On the question of the structure in crystals. [In Russian] *Bull. Acad. Imp. Sci.*, Ser. 6, 10(1916), 755–799. **[2.10, 3.10, 4.8, 6.3, 6.6]**

1930 Über die Symmetrie des Semikontinuums. *Zeitschrift für Kristallographie* 73(1930), 430–433. **[8.8]**

1940 *Symmetry. The Laws of Symmetry and Their Application in Science, Technology and Applied Art* [In Russian]. Akad. Nauk SSSR, Moscow, 1940. **[8.8]**

1946 *Atlas of Crystallographic Groups of Symmetry* [In Russian]. Akad. Nauk SSSR, Moscow—Leningrad, 1946. **[8.8]**

1951 *Symmetry and Antisymmetry of Finite Figures* [In Russian]. Akad. Nauk SSSR, Moscow—Leningrad, 1951.
English translation in: Shubnikov, Belov et al. [1964], pp. 1–172. **[5.6, 8.8]**

1960 Symmetry of similarity [In Russian]. *Kristallografiya* 5(1960), 489–496.
English translation: *Soviet Physics—Crystallography* 5(1961), 469–476. **[5.6, 10.1]**

1962a Symmetry and antisymmetry groups (classes) of finite strips [In Russian]. *Kristallografiya* 7(1962), 3–6.
English translation: *Soviet Physics—Crystallography* 7(1962), 1–4. **[8.8]**

1962b Black-white groups of infinite ribbons [In Russian]. *Kristallografiya* 7(1962), 186–191.
English translation: *Soviet Physics—Crystallography* 7(1962), 145–149. **[8.8]**

1963a Incompleteness of "A unified scheme of crystallographic groups" [In Russian]. *Kristallografiya* 8(1963), 131–132.
English translation: *Soviet Physics—Crystallography* 8(1963), 101–102. **[8.8]**

1963b The information contained in a regular system of points [In Russian]. *Kristallografiya* 8(1963), 943–944.
English translation: *Soviet Physics—Crystallography* 8(1964), 760–761. **[5.6]**

A. V. ŠUBNIKOV [SHUBNIKOV] AND V. A. KOPCIK [KOPTSIK]

1972 *Symmetry in Science and Art* [In Russian]. Nauka Press, Moscow, 1972.
English translation: *Symmetry in Science and Art.* Plenum Press, New York, 1974. **[I, 1.6, 4.8, 5.6, 6.3, 6.6, 8.2, 8.8]**

G. TACHELLA
1931 Sulle divisione semiregolari del piano. *Atti Soc. Ligust. Sci.* No 10(1931), 18–25. **[2.10]**

H. TAKEDA AND J. D. H. DONNAY
1965 Compound tessellations in crystal structures. *Acta Crystallographica* 19(1965), 474–476. **[2.10, 12.4]**

J. R. TARRANT
1973 Comments on the Lösch central place system. *Geographical Analysis* 5(1973), 113–121. **[2.10]**

J. L. TEETERS
1974 How to draw tessellations of the Escher type. *Math. Teacher* 67(1974), 307–310. **[6.6, 8.8]**

A. H. TEMESVARI
1979 Über doppelhomogene Zerlegungen. *Ann. Univ. Sci. Budapest. Sect. Math.* 22/23 (1979/1980), 163–196. **[4.8]**
1980 Über die doppelhomogene Zerlegungen in der Euklidischen Ebene, in *2. Kolloquium über Diskrete Geometrie*, A. Florian, ed. Universität Salzburg, Salzburg, 1980, pp. 216–224. **[4.8]**

C. THOMASSEN
1980 Planarity and duality of finite and infinite graphs. *J. Combinatorial Theory* B29(1980), 244–271. **[4.8]**
1984 Plane representations of graphs. Instructional lectures at the Silver Jubilee Conference on Combinatorics, Waterloo (Ontario), 1982 (to appear). **[4.8]**

D. W. THOMPSON
1925 On the thirteen semi-regular solids of Archimedes, and on their development by the transformation of certain plane configurations. *Proc. Royal Soc. London*, Series A, 107(1925), 181–188. **[2.10]**

W. P. THURSTON
1982 Three dimensional manifolds, Kleinian groups and hyperbolic geometry. *Bull. Amer. Math. Society* (N.S.) 6(1982), 357–381. **[3.1]**

M. TOMINAGO
1981 Tiling hexabolo with pentabolos [In Japanese]. *Sugaku Seminar*, Vol. 20, No. 4(1981), 36–38. **[9.4]**

M. TROWELL
1960 *African Design.* Praeger, New York, 1960. **[I, 5.2]**

S. TRUCHET
1704 Mémoire sur les combinaisons. *Histoire de l'Academie Royale des Sciences*, Année 1704, 363–372 and plates 12–20. **[8.8]**

J. TURNER
1968 A graph-theoretic model for periodic discrete structures. *J. Franklin Institute* 285(1968), 52–58. **[2.10, 3.10]**

W. T. TUTTE
1948 The dissection of equilateral triangles into equilateral triangles. *Proc. Cambridge Philos. Soc.* 44(1948), 463–482. **[2.4]**
1965 The quest of the perfect square. *Amer. Math. Monthly* 72(1965), No. 2, Part II, pp. 29–35. **[2.4]**
1966 Squared rectangles. *Proc. IBM Sci. Comput. Sympos. Combinatorial Problems* (Yorktown Heights, NY, 1964). IBM Data Process. Div., White Plains, NY, 1966, 3–9. **[2.4]**
1981 Dissections into equilateral triangles, in Klarner [1981], 127–139. **[2.4]**

A. VALETTE
1980 On simplicial tilings of the plane. *Bull. Soc. Math. Belg.* Ser. A.32(1980), 11–19. **[3.10]**
1981 Tilings of the plane by topological disks. *Geometriae Dedicata* 11(1981), 447–454. **[3.2, 3.10]**

G. VALETTE AND T. ZAMFIRESCU
1974 Les partages d'un polygone convexe en 4 polygones semblables au premier. *J.*

Combinatorial Theory 16(1974), 1–16. **[10.1]**

I. VERMES

1971 Über die Parkettierungsmöglichkeit der hyperbolischen Ebene durch nicht-total asymptotische Vielecke. *Beiträge zur Algebra und Geometrie* 1(1971), 9–13. **[4.8]**

1974 Über ebene hyperbolische Mosaike. *Ann. Univ. Sci. Budapest. Sect. Math.* 17(1974), 131–137. **[4.8]**

1979 Ausfüllung der hyperbolischen Ebene durch kongruente Hyperzykelbereiche. *Periodica Math. Hungar* 10(1979), 217–229. **[4.8]**

H. VODERBERG

1936 Zur Zerlegung der Umgebung eines ebenen Bereiches in kongruente. *Jahresber. Deutsch. Math.-Verein.* 46(1936), 229–231. **[1.0, 3.2, 9.5]**

1937 Zur Zerlegung der Ebene in kongruente Bereiche in Form einer Spirale. *Jahresber. Deutsch. Math.-Verein.* 47(1937), 159–160. **[3.2, 9.5]**

D. WADE

1976 *Pattern in Islamic Art.* Overlook Press, Woodstock. NY, 1976. **[5.6]**

B. L. VAN DER WAERDEN AND J. J. BURCKHARDT

1961 Farbgruppen. *Zeitschrift für Kristallographie* 115(1961), 231–234. **[8.8]**

T. R. S. WALSH

1972 Characterizing the vertex neighbourhoods of semi-regular polyhedra. *Geometriae Dedicata* 1(1971), 117–123. **[3.10, 4.8]**

H. WANG

1961 Proving theorems by pattern recognition. II. *Bell System Tech. J.* 40(1961), 1–42. **[11.4]**

1963 Dominoes and the AEA case of the decision problem, in *Proc. Symp. Math. Theory of Automata, New York, 1962.* Polytechnic Press, New York, 1963, pp. 23–56. **[11.4]**

1965 Games, logic and computers. *Scientific American*, November 1965, 98–106. **[11.4]**

1975 Notes on a class of tiling problems. *Fundam. Math.* 82(1975), 295–305. **[11.1, 11.2, 11.4]**

D. K. WASHBURN

1977 A symmetry analysis of Upper Gila Area ceramic design. *Papers of the Peabody Museum of Archaeology and Ethnology*, Vol. 68(1977). **[I, 1.6, 8.8]**

1983 Symmetry analysis of ceramic design: Two tests of the method on Neolithic material from Greece and the Aegean, in *Structure and Cognition in Art*, D. K. Washburn, ed. Cambridge University Press, London, 1983. **[1.6]**

R. WATSON

1973 Semi-regular tessellations. *Math. Gazette* 57(1973), 186–188. **[2.10]**

L. WEBER

1929 Die Symmetrie homogener ebener Punktsysteme. *Zeitschrift für Kristallographie* 70(1929), 309–327. **[8.2, 8.8]**

G. WEGNER

1971 Bewegungsstabile Packungen konstanter Nachbarnzahl. *Studia Scientiarum Math. Hungarica* 6(1971), 431–438. **[3.2]**

A. F. WELLS

1956 *The Third Dimension in Chemistry.* Clarendon Press, Oxford, 1956. **[5.6]**

1970 *Models in Structural Inorganic Chemistry.* Oxford University Press, New York, 1970. **[5.6]**

1977 *Three-Dimensional Nets and Polyhedra.* Wiley, New York, 1977. **[4.8]**

W. VON WERSIN

1953 *Das elementare Ornament und seine Gesetzlichkeit.* Maier, Ravensburg, 1953. **[I]**

R. L. WHEEDEN AND A. ZYGMUND

1977 *Measure and Integral.* Marcel Dekker, New York, 1977. **[3.1]**

G. E. WHEELER

1958 Cell face correlations and geometrical necessity. *Amer. J. of Botany* 45(1958), 439–449. **[3.10]**

W. A. WHITWORTH

1884 Problem 6904. *Mathematical questions with their solutions, from the Educational Times,* 40(1884), 59. **[3.10]**

T. W. WIETING

1981 *The Mathematical Theory of Plane Chromatic Ornaments.* Marcel Dekker, New York, 1981. **[8.2, 8.8]**

J. A. WIGHT

1975 *Mediaeval Floor Tiles.* John Baker, London, 1975. **[I]**

J. B. WILKER

1967 Open disk packings of a disk. *Canad. Math. Bull.* 10(1967), 395–415. **[3.10]**

1981 Topologically equivalent two-dimensional isometries. *Topology and Its Applications* 12(1981), 105–114. **[4.8]**

H. C. WILKIE

1966 On non-Euclidean crystallographic groups. *Math. Zeitschrift* 91(1966), 87–102. **[4.8]**

T. H. WILLCOCKS

1951 A note on some perfect squared squares. *Canad. J. Math.* 3(1951), 304–308. **[2.4]**

R. WILLIAMS

1972 *Natural Structure.* Eudaemon Press, Moorpark, CA, 1972.
Reprinted as *The Geometrical Foundation of Natural Structure.* Dover, New York, 1979. **[2.10, 4.8]**

1977 *Art Form Designs to Color.* 2. Grosset and Dunlap, New York, 1977. **[1.1]**

W. WOLLNY

1969 *Reguläre Parkettierung der euklidischen Ebene durch unbeschränkte Bereiche.* Bibliographisches Institut, Mannheim, 1969. **[3.10, 5.3, 5.6, 6.5, 6.6]**

1974a Die regulären Parketts mit dem Quadratnetz. *Geometriae Dedicata* 3(1974), 41–60. **[3.10, 6.6, 7.8]**

1974b Die Fundamentalbereichstypen der Translationsgruppe *p1*. *Geometriae Dedicata* 3(1974), 385–397. **[3.10, 6.6, 7.8]**

1981 Dualität von Cayley-Diagramm und Fundamentalbereichsparkett bei den diskreten 1-dimensionalen Bewegungsgruppen. *Geometriae Dedicata* 11(1981), 31–46. **[6.6]**

1983 Merkwürdige Fundamentalbereichstypen der 17 diskreten Bewegungsgruppen. *Geometriae Dedicata* 11(1983), 13–19. **[3.10, 6.6]**

1984 Contributions to Hilbert's eighteenth problem. *Pacific J. Math.* 112(1984), 451–495. **[3.10, 4.8, 6.6]**

H. WONDRATSCHEK

1976 Extraordinary orbits of space groups. Theoretical considerations. *Zeitschrift für Kristallographie* 143(1976), 460–470. **[5.3]**

1979 Orbits, lattice complexes, and orbit classes: Theory and applications. *Collected Abstracts, Symposium of Mathematical Crystallography,* Riederalp (Wallis, Switzerland), 7–9 August 1979, p. 9. **[5.6]**

1980 Crystallographic orbits, lattice complexes, and orbit types. *MATCH, Communications in Mathematical Chemistry* 9(1980), 121–125. **[5.6]**

E A. WOOD

1964 The 80 diperiodic groups in three dimensions. *Bell Telephone System Technical Publications.* Monograph 4680(1964). **[8.8]**

J. H. WOODS

1935a The geometrical basis of pattern design.

I: Point and line symmetry in simple figures and borders. *Journal of the Textile Institute* (*Manchester*) 26(1935), T197–T210. **[8.8]**

1935b The geometrical basis of pattern design. II: Nets and sateens. *Journal of the Textile Institute* (*Manchester*) 26(1935), T293–T308. **[8.8]**

1935c The geometrical basis of pattern design. III: Geometrical symmetry in plane patterns. *Journal of the Textile Institute* (*Manchester*) 26(1935), T341–T357. **[8.8]**

1936 The geometrical basis of pattern design. IV: Counterchange symmetry in plane patterns. *Journal of the Textile Institute* (*Manchester*) 27(1936), T305–T320. **[8.8]**

P. B. Yale

1968 *Geometry and Symmetry*. Holden-Day, San Francisco, 1968. **[1.3]**

K. Yoneyama

1917 Theory of continuous set of point. *Tôhoku Math. J.* 12(1917), 43–158. **[1.6]**

M. Zacharias

1931 Review of Tachella [1931]. *Jahrbuch über die Fortschritte der Mathematik* 57(1931), 1528. **[2.10]**

A. M. Zamorzaev

1967 Quasisymmetry *P*-symmetry) groups [In Russian]. *Kristallografiya* 12(1967), 819–825.
English translation: *Soviet Physics—Crystallography* 12(1968), 717–722. **[8.8]**

1969 Color-symmetry space groups [In Russian]. *Kristallografiya* 14(1969), 195–200.
English translation: *Soviet Physics—Crystallography* 14(1969), 155–159. **[8.8]**

1976 *Theory of simple and multiple anti-symmetry* [In Russian]. Štiinca, Kišinev, 1978. **[8.8]**

A. M. Zamorzaev, E. I. Galyarskii and A. F. Palistrant

1978 *Color symmetry, its generalizations and applications* [In Russian]. Štiinca, Kišinev, 1978. **[8.8]**

A. M. Zamorzaev and A. F. Palistrant

1960 The two-dimensional Shubnikov groups [In Russian]. *Kristallografiya* 5(1960), 517–524.
English translation: *Soviet Physics—Crystallography* 5(1960), 497–503. **[8.8]**

1961 Mosaics for 167 two-dimensional Shubnikov groups (three lowest orders) [In Russian]. *Kristallografiya* 6(1961), 163–176.
English translation: *Soviet Physics—Crystallography* 6(1961), 127–140. **[8.8]**

1980 Antisymmetry, its generalizations and geometrical applications. *Zeitschrift für Kristallographie* 151(1980), 231–248. **[8.8]**

E. A. Zamorzaeva

1979 Classification of regular divisions of the plane for similarity symmetry groups [In Russian]. *Dokladi Akademii Nauk SSSR* 247(1979), 276–279.
English translation: *Soviet Math. Dokl.* 20(1979), 689–692. **[7.8, 10.1]**

1981 Regular Dirichlet partitions for two-dimensional groups of similarity symmetries [In Russian]. *Dokladi Akad. Nauk SSSR* 260(1981), 343–345. **[10.1]**

C. Zaslavsky

1973 *Africa Counts*. Prindle, Weber & Schmidt, Boston, 1973. **[1.6]**

B. Zaslow

1977 A guide to analyzing prehistoric ceramic decorations by symmetry and pattern mathematics. Anthropological Research Papers No. 2, Arizona State University, Tempe 1977. **[1.6]**

1980 Mirror orientation in Hohokam designs and the chronology of early Hohokam phases. *The Kiva* 45(1980), 211–225. **[1.6]**

1981 Pattern dissemination in the prehistoric Southwest and Mesoamerica. Anthropological Research Papers No. 25, Arizona State University, Tempe 1981. **[1.6]**

B. ZASLOW AND A. E. DITTERT, JR.

1977a The pattern technology of the Hohokam. Anthropological Research Papers No. 2, Arizona State University, Tempe 1977. **[1.6]**

1977b Pattern theory used as an archaeological tool: A preliminary statement. *Southwestern Lore* 43(1977), 18–24. **[1.6]**

H. ZEITLER

1967 Über Netze aus regulären Polygonen in der hyperbolischen Geometrie. *Elem. Math.* 22(1967), 56–62. **[4.7]**

H. ZIESCHANG, E. VOGT AND H.-D. COLDEWAY

1970 *Flächen und ebene diskontinuierliche Gruppen.* Lecture Notes in Math. vol. 122. Springer, Berlin, 1970.
Expanded English translation: *Surfaces and Planar Discontinuous Groups.* Lecture Notes in Math., vol. 835. Springer-Verlag, New York, 1980. **[4.8]**

INDEX

Page numbers in heavy type refer to definitions, statements of theorems, lemmas, etc. Page numbers in italic type refer to illustrations. The letter *n* after a page number indicates a footnote. Abbreviations are included in the main index. References beginning with a symbol that represents an integer (such as *n*-Archimedean, *n*-uniform, *n*-hedral, etc.) appear after the alphabetical section.